AF598669

ALLE ZEIT WACH
1842

Handbook of Geochemistry

Vol. II/4

Elements Kr (36) to Ba (56)

Executive Editor: K. H. Wedepohl

Springer-Verlag Berlin Heidelberg New York

ISBN 3-540-06879-1 Springer-Verlag Berlin · Heidelberg · New York
ISBN 0-387-06879-1 Springer-Verlag New York · Heidelberg · Berlin

Library of Congress Catalog Card Number: 78-85402

Typesetting, printing, and binding: Universitätsdruckerei H. Stürtz AG, Würzburg

2121/3140-5 4 3 2 1 0

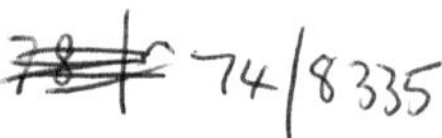

LIST OF CHAPTERS

Chemical Elements Occurring in Nature

Atomic numbers (in heavy print) are identical with chapter numbers in Volume II of this handbook. Elements with very close relations, such as the noble gases, the lanthanides and the platinum group elements, are combined and covered under one element of the respective group. Elements printed in italics will not be included in this book.

STANDARD SECTIONS OF CHAPTERS

(Chapters have the atomic numbers of the elements)

A. Crystal chemistry
B. Isotopes in nature
C. Abundance in cosmos, meteorites, lunar materials and tektites
D. Abundance in rock-forming minerals; phase equilibria; minerals of the respective element
E. Abundance in common igneous rock types
F. Behavior in magmatogenic processes (pegmatites, gas transport etc.) and ore deposition
G. Behavior during weathering and alteration of rocks; abundance in soils
H. Solubilities of compounds which control concentrations of the element in natural waters; adsorption processes; valence states in natural environments
I. Abundance in natural waters and in the atmosphere
K. Abundance in common sediments and sedimentary rock types
L. Biogeochemistry; abundance in coal, crude oil ect.
M. Abundance in common metamorphic rock types
N. Behavior in metamorphic processes
O. Relations to other elements, economic importance etc.

References (Common abbreviations: SB = Strukturbericht; SR = Structure Reports)

LIST OF SYMBOLS DESIGNATING ANALYTICAL METHODS

S	Optical emission sprectrography or spectrometry: arc, spark etc.
A	Atomic absorption spectrometry
E	Electrolysis
F	Optical emission spectrometry: flame photometry
X	X-ray spectrometry and X-ray fluorescence spectrometry
W	Chemical (wet conventional and rapid) methods etc.
C	Colorimetric, spectrophotometric methods
P	Polarographic methods
N/R	Neutron activation, radiometric measurements
I	Isotope dilution, mass spectrometric measurements
M	Microprobe analysis
L	Fluorimetric measurements
Mass	Spark source mass spectrometry etc.
	Or combinations of these symbols

Krypton 36

and the other Noble Gases 10, 18, 54, 86

see Helium 2

Rubidium 37

A	G. Cocco, L. Fanfani and P. F. Zanazzi	(Istituto di Mineralogia dell'Università di Perugia, Perugia, Italy)
B—C*, D—E, G, I—N	K. S. Heier	(Mineralogisk-Geologisk Museum, Oslo, Norway)
	and	
	G. K. Billings	(Syn An Inc. P.O. Box 1735 Socorro, New Mexico, U.S.A.)
***C—IV**	E. Steinnes,	(Institut for Atomenergi, Kjeller, Norway)
	K. S. Heier and G. K. Billings	

37-A. Crystal Chemistry

Rubidium, the fourth member of the alkali metal family, has the electronic configuration $1s^2\,2s^2\,2p^6\,3s^2\,3p^6\,3d^{10}\,4s^2\,4p^6\,5s^1$. After the first ionization, the rubidium ion attains the configuration of the noble gas krypton. This explains the chemical behavior of Rb and its tendency to achieve the oxidation state +1. Rb^+ has an ionic radius very similar to that of K^+, the preceding alkali metal in the Periodic Table; its usually accepted value (c. f., this handbook, Vol. I, Table 12-8, p. 389) is 1.49 Å, only 11% larger than that of K^+ (1.33 Å), and therefore the crystal chemistry of these two elements is very similar. In nature Rb does not form minerals of its own, but it is dispersed especially in K-minerals, where it is concentrated in the later crystallizates. The highest rubidium contents are observed in amazonites, microclines, muscovites, lepidolites, zinnwaldites and biotites of special occurrences. The similarity of ionic radii associates Rb^+ with Tl^+ (ionic radius 1.49 Å), for example in muscovite and microcline. In pollucite and rhodizite, rubidium seems to replace cesium (ionic radius 1.65 Å) to some extent; however, the crystallochemical properties of this element differ considerably from rubidium.

Rb^+ exhibits an ionic radius similar to that of the complex cation H_3O^+; this substitution has been observed in synthetic compounds belonging to the metatorbernite group (Schulte, 1965).

As for the other alkali metals, the present crystallochemical review on Rb considers only halogenides and oxygen-containing compounds. Reliable information on Rb-halogen and Rb-oxygen bond lengths are listed in Table 37-A-1.

I. Halogen Compounds

The simple Rb halides have the "NaCl" structure type. Rb-halogen distances are: Rb—F, 2.815 Å; Rb—Cl, 3.290 Å; Rb—Br, 3.444 Å; Rb—I, 3.671 Å (after Sysiö, 1969). $RbHF_2$ as the analogous K compound, is known in two forms; the form stable at room temperature is tetragonal and becomes cubic above 180° C; both structures are of a fluorite-like type (Kruh *et al.*, SR 1956, 215). In $RbLiF_2$, Rb^+ ions are surrounded by eight fluorine atoms with distances ranging from 2.78 to 3.16 Å (average value, 2.95 Å; Burns and Busing, 1965). Several double fluorides have been obtained: $RbCaF_3$ and $RbMnF_3$ are cubic, $RbZnF_3$ is tetragonal, $RbMgF_3$ is monoclinic, but all the structures belong to the various modifications of the perowskite type (Ludekens and Welch, SR 1952, 166; Hoppe *et al.*, SR 1961, 308). $RbCoF_3$ has the same arrangement and is cubic above and tetragonal below 101° K (Nouet *et al.*, 1969). $RbNiF_3$ is isotypic with the hexagonal modification of $BaTiO_3$: the mean $Rb^{[12]}$—F value is 2.93 Å (Babel, 1969). Rb_3AlF_6 has tetragonal symmetry at room temperature. At 375° C it undergoes a transition in a cubic phase with a behavior closely analogous to that of Cs_3AlF_6 (Holm, 1965). $RbBF_4$ has a baryte-like atomic arrangement (Hoard and Blair, SB 1935, 419); each Rb^+ is

Table 37-A-1

		Range Å	Mean Å	Reference
RbF	$Rb^{[6]}$—F	2.815	2.82	SYSIÖ (1969)
$RbLiF_2$	$Rb^{[8]}$—F	2.78—3.16	2.95	BURNS and BUSING (1965)
$RbNiF_3$	$Rb_{I}^{[12]}$—F	2.922—3.043	2.98	BABEL (1969)
	$Rb_{II}^{[12]}$—F	2.796—2.942	2.90	
$RbBF_4$	$Rb^{[12]}$—F	2.92—3.37	3.07	CLARK and LYNTON (1969)
α-$RbFeF_4$	$Rb^{[6]}$—F	2.61—3.14	2.78	TRESSAUD *et al.* (1969)
β-$RbFeF_4$	$Rb^{[8]}$—F	2.97	2.97	TRESSAUD *et al.* (1969)
γ-Rb_2BeF_4	$Rb_{I}^{[6]}$—F	2.87—3.07	2.97	MUSTAFAEV *et al.* (1966)
	$Rb_{II}^{[8]}$—F	2.61—2.96	2.85	
$RbBe_2F_5$	$Rb^{[6]}$—F	2.82—3.02	2.91	ILJUKHIN and BELOV (SR 1961, 303)
$RbPaF_6$	$Rb^{[10]}$—F	2.81—3.17	2.96	BURNS *et al.* (1968)
RbCl	$Rb^{[6]}$—Cl	3.290	3.29	SYSIÖ (1969)
$RbNiCl_3$	$Rb^{[12]}$—Cl	3.480—3.640	3.56	ASMUSSEN *et al.* (1969)
α-$RbMnCl_3 \cdot 2H_2O$	$Rb^{[8]}$—Cl	3.377—3.573	3.48	JENSEN (1967)
β-$RbMnCl_3 \cdot 2H_2O$	$Rb^{[8]}$—Cl	3.313—3.622	3.43	JENSEN (1967)
RbBr	$Rb^{[6]}$—Br	3.444	3.44	SYSIÖ (1969)
RbI	$Rb^{[6]}$—I	3.671	3.67	SYSIÖ (1969)
$RbAg_4I_5$	$Rb^{[6]}$—I	3.62	3.62	BRADLEY and GREENE (1967)
Rb_2O	$Rb^{[4]}$—O	2.92	2.92	HELMS and KLEMM (SB 1939, 85)
Rb_2O_2	$Rb^{[6]}$—O	2.85	2.85	FÖPPL (SR 1957, 234)
$Rb_2Ti_6O_{13}$	$Rb^{[8]}$—O	2.77—3.14	2.92	ANDERSSON and WADSLEY (1962)
$Rb_xMn_xTi_{2-x}O_4$	$Rb^{[8]}$—O	2.93—3.20	3.06	REID *et al.* (1968)
$RbPO_3$	$Rb^{[7]}$—O	2.90—3.20	2.99	CORBRIDGE (SR 1956, 300)
β-$Rb_2[SO_4]$	$Rb_{I}^{[10]}$—O	2.90—3.23	3.10	WYCKOFF (1965)
	$Rb_{II}^{[9]}$—O	2.82—3.22	2.98	
$RbAl[SO_4]_2 \cdot 12H_2O$	$Rb^{[6]}$—O	3.07	3.07	LARSON and CROMER (1967)
$Rb(AmO_2)[CO_3]$	$Rb^{[12]}$—O	3.03—3.10	3.06	ELLINGER and ZACHARIASEN (SR 1954, 510)
$Rb(UO_2)[NO_3]_3$ by neutron diffraction	$Rb^{[14]}$—O	2.94—3.29	3.19	BARCLAY *et al.* (1965)

surrounded by twelve F atoms with distances in the range 2.92—3.37 Å (calculated from CLARK and LYNTON, 1969).

$RbFeF_4$ exhibits a reversible polymorphic transformation at 650° C; in the orthorhombic α-form, each Rb^+ has six F neighbors with distances from 2.61 to 3.14 Å (the average value is 2.78 Å). In the tetragonal β-form, the coordination polyhedron around Rb^+ is a square prism with Rb—F distances of 2.97 Å (TRESSAUD *et al.*, 1969).

Rb_2BeF_4 shows three polymorphic modifications. In the γ-form Rb is in six- and eight-fold coordination with Rb—F distances from 2.61 to 3.05 Å (MUSTAFAEV *et al.*, 1966). In $RbBe_2F_5$ there are brucite-type layers of Rb octahedra, the $Rb^{[6]}$—F mean distance is 2.91 Å (ILJUKHIN and BELOV, SR 1961, 303). In contrast to $RbNbF_6$ and $RbTaF_6$ (COX, SR 1956, 224) which have the $KOsF_6$ structure, in $RbPaF_6$ the atomic arrangement resembles that of K_2ZrF_6 (BURNS *et al.*, 1968); Rb^+ has ten nearest fluorine atoms in the range 2.81—3.17 Å.

The hexagonal atomic arrangements of $RbNiCl_3$ and $RbNiBr_3$ are related to that of $BaNiO_3$; the average $Rb^{[12]}$—Cl distance is 3.56 Å (ASMUSSEN *et al.*, 1969). About the same bond lengths are found in isostructural $RbCoCl_3$ (ENGBERG and SOLING, 1967). The α- and β-forms of $RbMnCl_3 \cdot 2H_2O$ are orthorhombic and triclinic respectively. In both structures the Rb^+ ions have eight Cl^- ions as nearest neighbors with average bond lengths of 3.48 and 3.43 Å (JENSEN, 1967). In $RbAg_4I_5$, Rb ions are surrounded by six I^- ions; the distance Rb—I is 3.62 Å (BRADLEY and GREENE, 1967).

II. Oxygen Compounds

$Rb_2^{[4]}O^{[8]}$ has the anti-fluorite arrangement with Rb—O distances of 2.92 Å (HELMS and KLEMM, SB 1939, 85). Rb_2O_2 is orthorhombic; the alkali-cation is irregulary coordinated by six oxygen atoms with Rb—O distances of 2.85 Å (FÖPPL, SR 1957, 234). $RbPaO_3$ has the perovskite structure (KELLER, 1965). $Rb_2Ti_6O_{13}$ is isomorphous with $Na_2Ti_6O_{13}$ and $K_2Ti_6O_{13}$; Rb^+ is in eightfold coordination with a $Rb^{[8]}$—O mean distance of 2.92 Å (ANDERSSON and WADSLEY, 1962). In the non-stoichiometric compound $Rb_xMn_xTi_{2-x}O_4$ with $0.60 < x < 0.80$, Rb^+ is surrounded by eight oxygen atoms at distances up to 3.20 Å. Two additional oxygen atoms at 3.56 Å are probably unbonded (REID *et al.*, 1968). In $^1_\infty$ $Rb[PO_3]$, Rb^+ ions are coordinated to seven oxygen atoms in a range from 2.90 to 3.20 Å (mean value of $Rb^{[7]}$—O is 2.99 Å; CORBRIDGE, SR 1956, 300). At room temperature Rb_2SO_4 is orthorhombic with the β-K_2SO_4 arrangement. Coordination around the two non-equivalent Rb^+ ions is ten- and nine-fold with average distances of 3.10 and 2.98 Å respectively (calculated from WYCKOFF, 1965). A hexagonal form occurs at temperatures above 657° C. This modification is isotypic with the high-temperature form of cesium and potassium sulphates (TABRIZI *et al.*, 1968). Rb_2CrO_4 is isotypic with the orthorhombic modification of Rb_2SO_4; average values of Rb—O distances are 3.19 and 3.09 Å (SMITH and COLBY, SR 1940, 158). Many double rubidium sulphates with the chemical formula $Rb_2M_2^{2+}[SO_4]_3$, where M^{2+} can be Ca, Cd, Co, Fe^{2+}, Mg, Mn^{2+} or Ni, have the langbeinite atomic arrangement (GATTOW and ZEMANN, 1958).

According to LEDSHAM and STEEPLE (1968), $RbAl[SO_4]_2 \cdot 12H_2O$ and $RbCr[SO_4]_2 \cdot 12H_2O$ belong to the same class of α-alums as Tl- and K-alums, in contrast to Cs alums which have been classified as β-alums. In the aluminum alum Rb^+ is surrounded by six water molecules at 3.07 Å (LARSON and CROMER, 1967). Rb_2MnO_4 is orthorhombic and isotypic with the analogous compounds of potassium and cesium (DUQUENOY, 1969). $Rb_2Cr_2O_7$, which is known at room temperature in triclinic and monoclinic modifications isotypic with those of $K_2Cr_2O_7$, undergoes a phase change on heating at 330° C (ERDEY *et al.*, 1965).

The crystal structure of $RbAmO_2[CO_3]$ is isotypic with that of $KPuO_2[CO_3]$; the mean value of $Rb^{[12]}$—O distances is 3.06 Å (ELLINGER and ZACHARIASEN, SR 1954, 510). In $RbUO_2[NO_3]_3$ (BARCLAY *et al.*, 1965), the Rb^+ ion has a rather large coordination number; 14 oxygen atoms surround the alkali cation with distances from 2.94 to 3.29 Å.

Revised manuscript received: June 1970

References: Section 37-A

ANDERSON, S., WADSLEY, A. D.: The structures of $Na_2Ti_6O_{13}$ and $Rb_2Ti_6O_{13}$ and the alkali metal titanates. Acta Cryst. **15**, 194 (1962).

ASMUSSEN, R. W., LARSEN, T. K., SOLING, H.: The crystal structure of $RbNiCl_3$ and $RbNiBr_3$. Acta Chem. Scand. **23**, 2055 (1969).

BABEL, D.: Die Kristallstrukturen der hexagonalen Fluorperowskite. Z. Anorg. Allgem. Chem. **369**, 117 (1969).

BARCLAY, G. A., SABINE, T. M., TAYLOR, J. C.: The crystal structure of rubidium uranyl nitrate: A neutron-diffraction study. Acta Cryst. **19**, 205 (1965).

BRADLEY, J. N., GREENE, P. D.: Relationship of structure and ionic mobility in solid MAg_4I_5. Trans. Faraday Soc. **63**, 2516 (1967).

BURNS, J. H., BUSING, W. R.: Crystal structures of rubidium lithium fluoride, $RbLiF_2$, and cesium lithium fluoride, $CsLiF_2$. Inorg. Chem. **4**, 1510 (1965).

— LEVY, H. A., KELLER, O. L. JR.: The crystal structure of rubidium hexafluoroprotoactinate (V), $RbPaF_6$. Acta Cryst. B **24**, 1675 (1968).

CLARK, M. J. R., LYNTON, H.: Crystal structures of potassium, ammonium, rubidium and cesium tetrafluoborates. Can. J. Chem. **47**, 2579 (1969).

DUQUENOY, G.: Synthèses entre solides à partir d'un supraoxyde alcalin-manganates de potassium, rubidium ou cesium. Compt. Rend. C **268**, 828 (1969).

ENGBERG, A., SOLING, H.: On the crystal structures of $RbCoCl_3$ and Rb_3CoCl_5. Acta Chem. Scand. **21**, 168 (1967).

ERDEY, L., LIPTAY, G., BIDLO, G.: Polymorphous transformation of rubidium dichromate. J. Inorg. Nucl. Chem. **27**, 2451 (1965).

GATTOW, G., ZEMANN, J.: Double sulfates of the langbeinite type, $A_2^{I}B_2^{II}(SO_4)_3$. Z. Anorg. Allgem. Chem. **293**, 233 (1958).

HOLM, J. L.: Phase transitions and structure of the high-temperature phases of some compounds of the cryolite family. Acta Chem. Scand. **19**, 261 (1965).

JENSEN, S. J.: The crystal structures of α- and of β-$RbMnCl_3 \cdot 2H_2O$. Acta Chem. Scand. **21**, 889 (1967).

KELLER, C.: Ternary or quaternary protactinium oxides with perovskite structure. J. Inorg. Nucl. Chem. **27**, 321 (1965).

LARSON, A. C., CROMER, D. T.: Refinement of the alum structures. III. X-ray study of the α-alums, K, Rb and $NH_4Al(SO_4)_2 \cdot 12H_2O$. Acta Cryst. **22**, 793 (1967).

LEDSHAM, A. H. C., STEEPLE, H.: The classification of the chromium alums. Acta Cryst. B **25**, 398 (1969).

MUSTAFAEV, N. M., ILYUKHIN, V. V., BELOV, N. V.: The crystal structures of K and Rb orthofluoroberyllates and their relationship to compounds of general formula M_2BX_4. Soviet Phys.-Cryst. **10**, 676 (1966).

NOUET, J., KLEINBERGER, R., DE KOUCHKOVSKY, R.: Etude radiocristallographique à basse température de la perowskite fluorée $RbCoF_3$. Compt. Rend. B **269**, 986 (1969).

REID, A. F., MUMME, W. G., WADSLEY, A. D.: A new class of compound $M_x^{+}A_x^{3+}Ti_{2-x}O_4$ $(0.60 < x < 0.80)$ typified by $Rb_xMn_xTi_{2-x}O_4$. Acta Cryst. B **24**, 1228 (1968).

SCHULTE, E.: Zur Kenntnis der Uranglimmer. Neues Jahrb. Mineral. Monatsh. 242 (1965).

SYSIÖ, P. A.: On the additivity of crystal radii in alkali halides. Acta Cryst. B **25**, 2374 (1969).

TABRIZI, D., GAULTIER, M., PANNETIER, G.: Analyse radiocristallographique des formes "basse" (β) et "haute" (α) temperature de sulfate de césium Cs_2SO_4. Bull. Soc. Chim. France 935 (1968).

TRESSAUD, A., GALY, J., PORTIER, J.: Structure cristalline des variétés basse et haute température du fluoferrite de rubidium $RbFeF_4$. Bull. Soc. Franc. Minéral. Crist. **92**, 335 (1969).

WYCKOFF, R. W. G.: Crystal Structures (second ed.) vol. 3. New York and London: Interscience Publisher 1965.

Revised manuscript received: June 1970

37-B. Isotopes in Nature

Natural rubidium consists of two isotopes, ^{85}Rb (72.15%) and ^{87}Rb (27.85%) which do not show any natural fractionation, (Shields *et al.*, 1963). ^{87}Rb is radioactive.

The radioactive decay of $^{87}Rb \xrightarrow{\beta^-} {}^{87}Sr$ is of great importance in geochronology. Because of the long half-life of ^{87}Rb (5×10^{10} yrs., see p. 385 of Volume I of this handbook), the method is particularly useful in the dating of old rocks. With the improvement of chemical preparation methods and mass spectrometric techniques and instrumentation, the method is applied to increasingly younger systems, and systems with unfavorable, or low, Rb/Sr ratios. An important development in the Rb/Sr dating has been "the whole rock method". It was found that Rb-minerals, especially K-feldspar and biotite, frequently gave discordant Rb/Sr ages presumably caused by movement of radiogenic Sr and cation exchange between minerals in a rock volume. However, the rocks themselves could frequently be regarded as a closed system with respect to Rb and Sr, indicating that radiogenic Sr released from Rb-minerals either equilibrated with common Sr in neighboring minerals, e.g. plagioclase, or remained on the grain boundaries.

The large fraction of low energy β particles in the spectrum of ^{87}Rb makes it difficult to determine the absolute beta activity. Estimates have varied within wide limits, and there is some uncertainty in their determination. Recent summaries of the problem are given by Leutz *et al.* (1962), Heier and Adams (1964) and Wetherill (1966). Half-lives of 4.7×10^{10} yrs. (decay constant of 1.47×10^{-11} yr.$^{-1}$) and 5.0×10^{10} yrs. (decay constant of 1.390×10^{-11} yr.$^{-1}$) are both in use. Wetherill (1966) mentions that the 4.7×10^{10} yrs. half-life is more likely the correct one.

(For additional information on radiogenic Sr see Sect. 38-B.)

Revised manuscript received: January 1970

37-C. Abundance in Cosmos, Meteorites, Tektites and Lunar Materials

I. Cosmic Abundance

Estimates of the cosmic abundances of the alkali elements are compiled byHEIER and ADAMS (1964) and CAMERON (1966). They are given here for Rb in Table 37-C-1.

The solar K/Rb ratio (by weight) is 777.

Table 37-C-1. *Cosmic atomic abundance estimates of rubidium (silicon 1 × 10⁶).* (*Compilation from* HEIER *and* ADAMS, 1964; CAMERON, 1966)

	Rb
GOLDSCHMIDT (1937)	6.8
BROWN (1949)	7.1
UREY (1952)	15
SUESS and UREY (1956)	6.5
ALLEN (1961 b)	7.1
CAMERON (1966)	5.0
CLAYTON *et al.* (1961)	4.24
Sun (CAMERON, 1966)	9.5

II. Rb in Meteorites

Data on rubidium in different types of meteorites is given in Table 37-C-2. A thorough discussion of rubidium (and cesium) in stone meteorites is given by SMALES *et al.* (1964).

Table 37-C-2. *Rubidium contents and K/Rb ratios in chondrites etc.*

Type	ppm Rb	K/Rb	Reference
Ordinary chondrites			
Richardton	2.96	276	1
Forest City	2.75; 2.9; 3.12	298; 289	1, 2, 6
ANHM 2406	2.9; 3.5; 3.9		2, 3, 4
Beardsley	4.90 (10.67)	186	1, 6
Nininger 1349	4.83		1
Holbrook	2.22; 2.79	396	1, 6
ANHM 1162	2.3		1
Modoc	3.45; 3.15	258; 263	1, 5
Chandakapur	3.5		2
Ochausk	2.1	357	2
Ochausk, A	1.64		5
Ochausk, B	1.62		5
Bjurböle	2.9	296	2
Limerick	2.3; 2.22		2, 5

Table 37-C-2 (Continued)

Type	ppm Rb	K/Rb	Reference
Chateau Renard	2.9		2
Gilgoin	3.02; 3.06		5
Allegan	2.17—2.24		5
Crumlin	3.92—3.78		5
Mangwendi	2.38—2.36		5
Beaver Ck.	1.28—1.35		5
Marion	2.54—2.92		5
Long Island	1.89—1.80; 1.93		5
Bluff A	1.04—1.06		5
Bluff B	0.84—0.82		5
Bremervörde	3.62		5
Olivenza	2.10		5
Merva	2.67		5
Khor Teniki	1.74		5
Elm Creek	0.96		5
Dhurmsala	2.36		5
Hessle	1.60		5
Futtepuhr	2.00		5
Hendersonville	1.86		5
Eli Elwah	2.54		5
Mt. Browne	1.56		5
Atoka	3.01		5
Esterville	1.16; 0.17		5, 6
Homestead	2.81; 3.07	320	5, 6
Bath	2.40		6
Farmington	2.0		6
Nakhla	2.95		6
Estacado	2.45		6
New Concord	3.54		6
Knyahinya	1.38—1.49		5
Achondrites			
Pasamonte	0.21; 0.51	2,000	1, 4
Nuevo Laredo	0.38; 0.36	990	1
Sioux County	0.18	1,250	1
Nininger	0.29	1,196	1
Moore County	0.16; 0.13	1,200; 4,046	1, 7
Johnstown	0.139		5
Bishopville	1.78	741	7
Carbonaceous chondrites			
Ivuna	2.3	250	5
Mighei	1.72—1.66; 3	266	5, 8
Felix	1.36	308	5
Orgueil IA, B, C	1.78; 1.76; 1.74	195	9
Orgueil II	2.08	237	9
Murray	1.57	266	9
Mokia, A, B	1.20; 1.21	242—277	9
Lance	1.42	273	9

Table 37-C-2 (Continued)

Type	ppm Rb	K/Rb	Reference
Pallasites			
Imilac	0.033		5
Bolivia	1.15		5

References: 1. GAST (1960a); 2. WEBSTER *et al.* (1958); 3. HERZOG and PINSON (1956); 4. SCHUMACHER (1956); 5. SMALES *et al.* (1964); 6. PINSON *et al.* (1965); 7. COMPSTON *et al.* (1965); 8. PINSON (1954); 9. MURTHY and COMPSTON (1966).

Analytical methods: 1. I; 2. N; 3. I; 4. N; 5. N, I; 6. I; 7. I; 8. S; 9. I.

The ordinary chondrites form a fairly uniform group with respect to Rb concentration. The average of the 44 entries in Table 37-C-2 is 2.4 ppm with a standard error of the mean $(s/\sqrt{n}) = 0.14$ ppm [excluding the 10.67 ppm value for Beardsley given by PINSON *et al.* (1965)]. The carbonaceous chondrites are similar to the ordinary chondrites in Rb content as well as K/Rb ratios but the achondrites form a distinctive group characterized by low Rb contents and high K/Rb ratios.

III. Tektites

Existing data on alkali elements, including Rb, in tektites was reviewed by HEIER and ADAMS (1964). Rubidium, and particularly the K/Rb ratio, is considered important for theories of tektite origin (TAYLOR and AHRENS, 1959; PINSON, PHILPOTTS, and SCHNETZLER, 1965). Some recent data on Rb content and K/Rb ratios of the different important tektite types are given in Table 37-C-3. It would appear that the different tektite groups can be separated on the basis of Rb-content. The philippinites and indochinites form one group. They grade upwards with only slight overlap into

Table 37-C-3. *Rubidium content and K/Rb ratios in tektites*

No. of anal.	Type	ppm Rb	std. error of mean	K/Rb	std. error of mean	Reference
23	Australites	94	3.5	250	14	TAYLOR and SACHS (1964) (19 anal.); PINSON *et al.* (1965) (4 anal.)
24	Bediasites	64	2	275	9	CHAO (1963) (21 anal.); PINSON *et al.* (1965) (3 anal.)
30	Moldavites	138	2	213	2	PINSON *et al.* (1965)
7	Philippinites	117	2	172	2	PINSON *et al.* (1965)
8	Indochinites	119	3	171	2	PINSON *et al.* (1965)
1	Javaite	98		162		PINSON *et al.* (1965)
1	Ivory Coast	65		209		PINSON *et al.* (1965)

the moldavites which have the maximum Rb-content of any tektite group. The australites are transitional between the philippinite-indochinites and the bediasites but do not overlap with the bediasites. The bediasites with the lowest Rb content have the highest K/Rb ratios but there is a continuous gradation between bediasites, australites and moldavites which form one group clearly separate from the philippinites and indochinites.

The bulk of the australite and bediasite analyses was by optical spectrography which may account for their greater spread in K/Rb ratios.

Revised manuscript received: January 1970

IV. Lunar Materials

Data on Rb in lunar fines and common lunar rock types are given in Table37-C-4. Like K, Rb is depleted in lunar rocks. The highest concentrations are found in KREEP-type rocks (about 15 ppm). Mare basalts, except Apollo 11 type A, have concentrations close to 1 ppm. Even lower concentrations are found in lunar anorthosites. K/Rb ratios of lunar basalts are similar to those of terrestrial basalts. In lunar anorthositic rocks the K/Rb ratio varies within limits. This might be an analytical effect caused by uncertainties in the determination of potassium at such low levels.

A major contribution to the analytical data for Rb in lunar material comes from isotope dilution mass spectrometry (in connection with age determinations). Other techniques used are X-ray fluorescence and neutron activation.

Table 37-C-4. *Rubidium content of lunar rocks and fines* (in ppm)[a]

Rock type	$\bar{x}$	s	Range	n	K/Rb
Mare basalts					
Apollo 11 A	5.6	0.33	5.1 –5.9	6	400–530
Apollo 11 B	0.89	0.19	0.74–1.2	7	580–810
Apollo 12	0.99	0.18	0.64–1.27	17	290–710
Apollo 15	0.87	0.16	0.68–1.2	22	420–690
Apollo 17	0.8	0.5	0.3 –1.9	8	[b]
KREEP *type rocks*					
Apollo 12 KREEP	16.7		15.3–18.0	2	310–330
Apollo 12 sample 12013	38			1	480
Apollo 14 breccias	16.1	7.2	5.5–33	14	200–460
Apollo 14 basalts	13.1		12.7–13.5	2	280–330
Apollo 15 KREEP	14.7	1.3	13.2–16.1	5	310–330
Apollo 16 KREEP	9.4		7.0–11.5	4	240–320
Apollo 17 noritic breccias	6.2	1.4	3.9–8.1	7	280–490
Highland rocks					
Apollo 15 anorthositic rocks	0.28		0.16–0.46	3	180–730
Apollo 16 anorthositic rocks, >31% Al_2O_3	0.5		0.02–1.3	5	[b]
Apollo 16, 25–31% Al_2O_3	2.0	1.7	0.2–7.1	18	140–1660
Apollo 16, 21–24% Al_2O_3	4.6	1.5	2.5–6.5	8	130–370
Apollo 17, anorthositic gabbros	1.3		0.4–2.1	4	[b]

Table 37-C-4 (continued)

Rock type	$\bar{x}$	s	Range	n	K/Rb
Fines					
Apollo 11, sample 10084	2.93			1	380
Apollo 12	6.2	0.93	4.9–7.7	6	220–360
Apollo 12, high K	9.7		8.5–10.8	2	350–370
Apollo 14	14.6	1.50	13.0–18.3	10	270–360
Apollo 15	4.4	1.23	1.7–6.2	23	280–370
Apollo 16	2.6	0.61	1.3–3.8	41	230–490
Apollo 17	2.4	0.98	0.9–4.3	35	290–720
Luna 16	1.9	0.24	1.5–2.4	11	430–880
Luna 20	1.6		1.5–1.6	3	350–390

[a] $\bar{x}$ is derived by first averaging all reliable existing data on one sample; the different samples are then averaged (each sample is given equal weight). Standard deviations (s) are calculated from (n) which is the number of samples for which data exist.
[b] Not calculated because uncertain data.

For references from which concentration data are obtained see footnote of Table 11-C-8.

No significant differences between lunar and terrestrial isotopic abundances (^{85}Rb/^{87}Rb) have been found by Wanless, R. K., Loweridge, W. D., Stevens, R. D.: Proceedings of the Apollo 11 Lunar Science Conference **2**, 1729 (1970). Barnes, I. L., Garner, E. L., Gramlich, J. W., Macklan, L. A., Moody, J. R., Moore, L. J., Murphy, T. J., Shields, W. R.: Proceedings of the fourth Lunar Science Conference, **2**, 1197 (1973).

Manuscript received: September 1974

37-D. Rubidium Abundance in Rock-Forming Minerals

The similarity of K and Rb ions is illustrated by the fact that Rb is always incorporated in K-minerals and forms no minerals of its own. However, Rb in lepidolite as in amazonite and late stage hydrothermal microclines from pegmatites may exceed several percent (HEIER and ADAMS, 1964).

It follows that the most important Rb containing minerals are the micas and K-feldspars, and most of the Rb in the crust is contained in these minerals. The Rb-content of these minerals increases with increasing "differentiation" of the host rock and is at a maximum in the late stage pegmatite K-minerals. The Rb^+ ion is larger than K^+ and is preferentially incorporated in the "12 coordinated" K-sites in the micas. This is demonstrated by the K/Rb ratios of co-existing K-feldspar and mica, which is always lowest in the latter (HEIER and ADAMS, 1964; LANGE *et al.*, 1966). The breakdown of micas under granulite facies metamorphism may contribute to the high K/Rb ratios of these rocks but is not the entire explanation as their K-feldspar phase is also depleted in Rb. When biotite and muscovite co-exist in rocks, Rb seems preferentially to enter the biotite but maximum Rb concentrations are found in muscovites of pegmatites which grade into lepidolites.

RHODES (1969) studied Rb concentrations in K-feldspars separated from granites. The granites were divided into three groups which were labelled sub-volcanic granites, contact aureole granites and regional aureole granites. It was concluded that there were no significant statistical differences in Rb contents of feldspar from the three granite groups, and the structural state of the feldspar, orthoclase or microline, was not significant to the Rb content. The principal factor influencing the Rb content of the K-feldspar was the bulk composition of the host rock. The Rb content increases and the K/Rb ratio decreases from granodiorite to leucocratic granite.

IIYAMA (1968) studied the distribution of Rb between potassium feldspar and plagioclase crystallized in equilibrium at 600° C and 1,000 bars. Rb was preferentially fixed in the K-feldspar phase. The partition coefficient D = ppm Rb in K-f./ppm Rb in plagioclase varied with increasing anorthite content from 9.6 to 40.

Histograms showing the distribution of Rb in K-feldspar, biotite and muscovite are shown in Fig. 37-D-1a, b, c. These Rb determinations are by quantitative X-ray spectrography or isotope dilution.

Table 37-D-1 gives ranges of Rb concentrations and K/Rb ratios in some common rock forming minerals (see also SHAW, 1968).

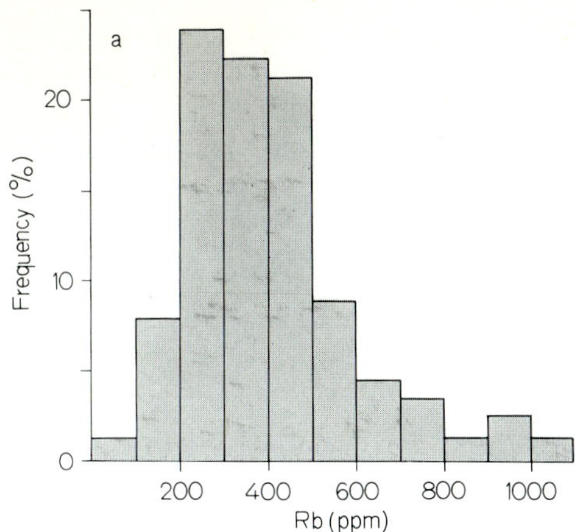

Fig. 37-D-1a. Histogram showing the distribution of Rb in 203 K-feldspars

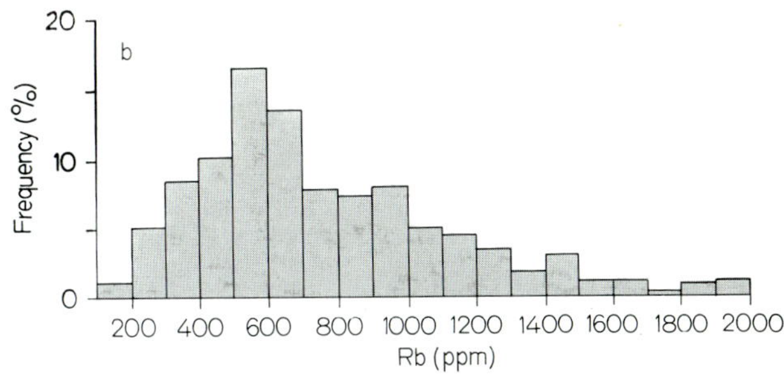

Fig. 37-D-1b. Histogram showing the distribution of Rb in 261 biotites (10 values above 2,000 ppm, max. 4,145 ppm)

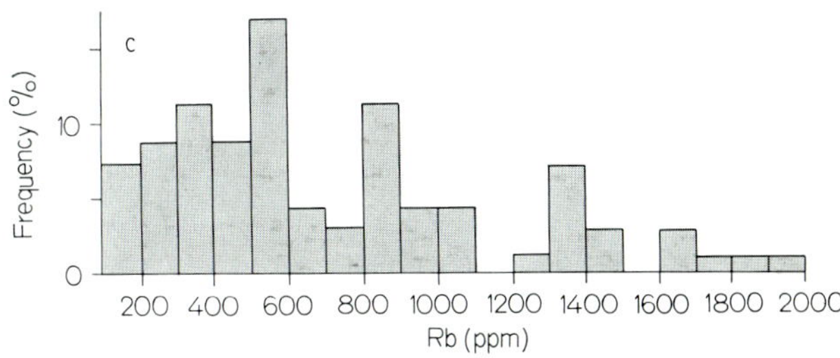

Fig. 37-D-1c. Histogram showing the distribution of Rb in 70 muscovites (13 values above 2,000 ppm, max. 22,370 ppm)

Table 37-D-1. *Range of Rb (in ppm) and K/Rb in some non-pegmatitic rock-forming minerals. ( ) number of samples*

Ref.	Biotite		Muscovite		Amphibole		Olivine		Pyroxene	
	Rb	K/Rb	Rb	K/Rb	Rb	K/Rb	Rb	K/Rb	Rb	K/Rb
1	(10) 1,000 to 1,750	(6) 40 to 76			(2) 10	(2) 1,200 to 1,400				
2	(16) 570 to 2,525	(16) 29 to 115	(4) 125 to 1,000	(4) 70 to 424						
3		(8) 91 to 275				(8) 735 to 2,320				
4	(14) 122 to 1,809				(50) 0.2 to 167	(50) 100 to 5,000			(2) 0.1 to 3	(2) 243 to 670
5							(3) 0.02 to 0.09	(3) 113 to 361	(9) 0.06 to 1.0	(9) 67 to 714
6	(9) 593 to 1,192	(9) 65 to 129								
7	(16) 280 to 1,474	(16) 35 to 219								
8	(139) 290 to 580	(16) 95 to 246								
11					(3) 1.5 to 14	(3) 171 to 1,630	(4) 0.004 to 0.066	(4) 199 to 221	(14) 0.033 to 3.6	(14) 105 to 980
12							(4) 0.009 to 0.55	(4) 318 to 594	(2) 0.05 to 1.34	(2) 346 to 474
	K-feldspar		Plagioclase		Nepheline		Analcite		Garnet	
	Rb	K/Rb	Rb	K/Rb	Rb	K/Rb	Rb	K/Rb	Rb	K/Rb
1	(5) 400 to 600	(4) 147 to 255								
2	(18) 255 to 865	(7) 148 to 415	(4) 5 to 30	(2) 382 to 847						
4			(2) 1 to 28	(2) 280 to 342						

6	(16) 191 to 500	(16) 245 to 554								
7	(16) 103 to 609	(16) 108 to 620								
9	(5) 117 to 168	(5) 560 to 784	(6) 4 to 35	(6) 780 to 2,100	(8) 50 to 120	(8) 370 to 918				
10							(5) 15 to 25	(2) 492 to 576		
11									(10) 0.019 to 3.9	(10) 106 to 318
12			(7) 0.14 to 7.66	(7) 568 to 2,151						
13	(70) 177 to 1,650	(70) 65 to 589								

References: 1. Sen *et al.* (1959); 2. Zartman (1964); 3. Chiang (1965); 4. Hart and Aldrich (1967); 5. Stueber (1969); 6. White (1966); 7. Lange *et al.* (1966); 8. Mishchenko and Orsa (1965); 9. Heier (1966); 10. Wilkinson (1959); 11. Griffin and Murthy (1969); 12. Murthy and Griffin (1970); 13. Rhodes (1969).

Revised manuscript received: January 1970

37-E. Abundance in Common Magmatic Rock Types; Terrestrial Abundance

I. Ultramafic Rocks

Data on Rb in ultramafic rocks up to 1962 was summarized by HEIER and ADAMS (1964). TUREKIAN and WEDEPOHL (1961) estimated an average concentration of 0.2 ppm Rb in ultramafic rocks. Their estimate was based on an average of 40 ppm K and a K/Rb ratio of 200. Some recent Rb determinations and K/Rb ratios are given in Table 37-E-1. It will appear from this table that the Rb concentration in ultramafic rocks is non-uniform.

Table 37-E-1. *Rubidium concentrations and K/Rb ratios in ultramafic rocks (by isotope dilution)*. (From STUEBER and MURTHY, 1966; see also ALLSOPP *et al.*, 1969)

	Rb (ppm)	K/Rb
Alpine-type ultramafic rocks:		
Dun Mtn., New Zealand, dunite	0.111	215
Papua, dunite	0.302	203
Shikoku, Japan, dunite	0.099	244
Kalgoorli, Australia, serpentinite	0.284	419
Tulameen, British Columbia, dunite	0.130	204
Cantwell, Alaska, dunite	0.072	268
Mt. Albert, Quebec, peridotite	0.158	375
Addie-Webster, North Carolina, dunite	0.077	195
Bett's Cove, Newfoundland, serpentinite	1.036	228
Siopurssuit, Greenland, dunite	2.42	266
Tinaquillo, Venezuela, peridotite	0.093	277
Konya, Turkey, dunite	0.140	263
Almklovdalen, Norway, dunite	0.131	337
Ultramafic inclusions:		
Kerguelen Islands, peridotite inclusion	0.420	452
Ross Island, Antarctica, dunite inclusion	1.15	188
Galapagos Islands, peridotite inclusion	0.352	349
Kapfenstein Austria, peridotite inclusion	0.271	115
Chihuahua, Mexico, peridotite inclusion	0.413	368
Ludlow, California, peridotite inclusion	0.398	196
Monduli, Tanganyika, peridotite inclusion	3.67	198
Wesselton Pipe, South Africa, peridotite inclusion	4.48	143
Bultfontein Pipe, South Africa, peridotite inclusion	2.03	230
Kakanui, New Zealand, garnet peridotite	1.73	542
Ultramafic zones in stratiform sheets:		
Muskox, NWT Canada, serpentinite	7.75	172
Muskox, NWT Canada, pyroxenite	3.47	317
Kola Penninsula, Russia, pyroxenite	1.12	353
Stillwater, ultramafic zone	0.382	348
Highwood Mtns., Montana, peridotite	142	102

II. Eclogites

Eclogites are often considered as plutonic or high pressure equivalents of basalts. On the other hand they trend through garnet peridotites into ultramafic rocks. Table 37-E-2 lists Rb concentrations in some "eclogitic" rocks.

Table 37-E-2. *Rubidium concentrations and K/Rb ratios in eclogites.* (From HEIER and COMPSTON, 1966; see also ALLSOPP *et al.*, 1969)

	ppm Rb	K/Rb	Method
Metamorphic environment:			
Volda, Norway	3.6	333	X
Volda, Norway	4.40	270	I
Almklovdalen, Norway	1.5	667	X
Eiksundsdal, Norway	0.3	300	X
Eiksundsdal, Norway	0.153	401	I
Wüstuben, Germany	9.1	275	X
Mittenbachgraben, Austria	2.1	138	X
Mittenbachgraben, Austria	1.81	163	I
Volcanic environments:			
Roberts Victors mine, S.A.	57.3	176	X
Roberts Victors mine, S.A.	9.1	341	X
Roberts Victors mine, S.A.	9.9	333	X
Roberts Victors mine, S.A.	82.8	214	X
Roberts Victors mine, S.A.	12.9	349	X
Roberts Victors mine, S.A.	22.8	294	X
Roberts Victors mine, S.A.	10.7	280	X
Jägersfontein S.A.	3.8	263	X
Kimberley S.A.	3.5	200	X
Kimberley S.A.	57	193	X
Bultfontein S.A.	3.0	433	X
Bultfontein S.A.	3.45	318	X
Dodoma	6.2	210	X
Salt Lake Crater, Oahu, Hawaii	0.54	370	X
Salt Lake Crater, Oahu, Hawaii	0.591	282	I
Visser Pipe Tanganyika (from STUEBER and MURTHY, 1966)	7.52	95	I

III. Anorthosites

Though anorthosites are often grouped with ultramafic rocks they are chemically and probably also genetically unrelated. The ultramafic rocks listed in Table 37-E-1 are derived from the Earth's mantle and brought into the crust as inclusions in volcanic rocks, or tectonically emplaced in orogenic belts, or formed by crystal fractionation in cooling gabbroic intrusions ultimately derived from the mantle. Though the site of anorthosite generation is uncertain when considering massive anorthosite bodies rather than stratiform sheets clearly associated with gabbro intrusions, much evidence indicates a deep crustal origin (GREEN, 1966).

Some recent information on Australian and Norwegian anorthosites are listed in Table 37-E-3.

Table 37-E-3. *Rubidium concentrations and K/Rb ratios in anorthosites (by X-ray fluorescence spectrometry)*

	ppm Rb	K/Rb	Reference
Gabbroic anorthosite, Musgrave Ranges, Central Australia	3	1,733	Lambert and Heier (1968)
Anorthosite, Langöy, North Norway	3.6	1,778	Heier (unpubl.)
	2.2	2,727	Heier (unpubl.)
	1.6	3,125	Heier (unpubl.)
	3.2	2,125	Heier (unpubl.)
	1.7	2,647	Heier (unpubl.)
	3.3	2,182	Heier (unpubl.)
Anorthosite, Flakstadöy, North Norway	2.5	2,160	Heier (unpubl.)
	2.7	2,370	Heier (unpubl.)
	2.3	1,870	Heier (unpubl.)

The K/Rb ratios in anorthosites are much higher than those of ultramafic rocks and eclogites listed in Tables 37-E-1 and 37-E-2 respectively. This could entirely reflect the mineralogical control exerted by plagioclase. Similar high K/Rb ratios are, however, approached by medium and high pressure granulite facies rocks of various bulk chemistry and high contents of K-feldspar in addition to intermediate plagioclase (Lambert and Heier, 1968; Heier, unpublished). Such areas constitute the characteristic environment of anorthosites in nature.

IV. Other Igneous Rocks

Variation diagrams of Rb in a number of igneous rock series were given by Nockolds and Allen (1953, 1954, 1956).

Table 37-E-4. *Published estimates of Rb concentrations (in ppm) in igneous rocks*

	1	2	3	4	5	6	7
Low Ca-granite						170	170
Granite	170	600	830	550	455—910		
High Ca-granite						110	99
Syenite	110				550	110	124
Monzonite							136
Neph. syenite			439	91			197
Qtz. diorite							76
Diorite	110						77
Gabbro	30	20		18	(27)		28
Basalt						30	47
Andesite							73
Dacite							97
Phonolite		330					
Trachyte							238
Rhyolite							217

1. Horstman (1957); 2. Wager and Mitchell (1943, German gabbro); 3. Goldschmidt *et al.* (1934); 4. Goldschmidt *et al.* (1933); 5. Sahama (1945); 6. Turekian and Wedepohl (1961); 7. Heier and Adams (1964).

For additional reference, on granites see also: Kolbe and Taylor (1966) and Rooke (1964), and on rhyolites Ewart *et al.* (1968).

Previous estimates of Rb concentrations in igneous rocks are given in Tables 37-E-4 and 37-E-5. Recent emphasis on Rb—Sr dating and the definition of initial $^{87}Sr/^{86}Sr$ ratios has increased the amount of data on Rb concentration in igneous rocks. Most of the Rb determinations after 1962 are by X-ray fluorescence spectrography and isotope dilution.

Table 37-E-5. *Rubidium concentrations in igneous rocks after* HURLEY *et al.* (1962) *(column 1), and* HEDGE (1966) *(column 2) (analyses by a number of different methods)*

	ppm Rb (1)	No. of analyses (column 1)	ppm Rb (2)
Granite	196	(290)	190
Granodiorite	122	(9)	110
Syenite	136	(14)	
Diorite and andesite	88	(21)	40
Gabbro and basalt	32	(331)	
Olivine basalt	18	(11)	
Alkali basalt			30
Oceanic tholeiitic basalt			3
Eclogite	6.9	(1)	
Anorthosite	0.6	(1)	
Kimberlite, S. Africa	171	(1)	
Mica-augite peridotite	495	(1)	
Pyroxenite	0.5	(1)	

a) Volcanic Rocks

Recent studies of alkali element concentration (and $^{87}Sr/^{86}Sr$ ratios) have resulted in a division of basaltic rocks according to environment:

1. tholeiitic basalts of the ocean floors,
2. tholeiitic and alkali basalts of oceanic islands,
3. circum-oceanic basalts which trend into andesites,
4. continental tholeiitic basalts,
5. continental alkali basalts.

The data on Rb concentrations in volcanic rocks is reviewed in Table 37-E-6. Additional data on rhyolites is by EWART *et al.* (1968) and EWART and TAYLOR (1969).

Added in proof: HART and NALWALK report Rb contents between 1.3 and 3.4 ppm, and K/Rb ratios of 430 to 2,030 in submarine basalts from the Puerto Rico trench (Geochim. Cosmochim. Acta **34**, 145—155 (1970).

Table 37-E-6. *Rubidium concentrations in volcanic rocks*

	ppm Rb	N=() $s/\sqrt{n}$	K/Rb	N=() $s/\sqrt{n}$	References
Basalts, Atlantic (dredged)	2.57	(20) 0.59	1,468	(14) 249	1, 2, 3, 4
Basalts, Pacific (dredged)	2.10	(12) 0.5	1,270	(3) 286	2, 3, 4

Table 37-E-6 (Continued)

	ppm Rb	N = () $s/\sqrt{n}$	K/Rb	N = () $s/\sqrt{n}$	References
Basalts, Atlantic Islands	25	(17) 4	428	(4) 92	1, 4, 5, 6, 7
Basalts, Pacific Islands (excl. Hawaii)	28	(21) 3	418	(9) 55	1, 6, 8
Basalts, Indian Ocean	17	(4) 3.5	407	(4) 25	9
Alkaline basalts, Indian Ocean	36	(2) 1	320	(2) 14	9
Andesites, Indian Ocean	50	(3) 4	332	(3) 7	9
Tholeiitic basalt, Hawaii	8	(24) 0.6	543	(17) 24	1, 4, 5 10, 11, 12
Alkaline basalt, Hawaii	15	(6) 1.4	419	(6) 33	4, 10, 11, 12
Nepheline basalt, basanite, etc., Hawaii	26	(12) 2.5	417	(12) 12	4, 10, 11, 12
Hawaiite, Hawaii	33	(9) 3.7	465	(9) 29	4, 10, 11, 12
Mugearite, excl. of one value of 10.5 ppm Rb, Hawaii	66	(4) 3.2	383	(4) 10	4, 10, 11
Trachyte, Hawaii	121	(7) 1.5	307	(7) 11	4, 10, 11, 12
Circum-oceanic:					
Tholeiitic basalt, Japan	5.9				6
Basalt, Japan	12.0				6
Olivine basalt, Japan	41				6
Andesite, Japan	20	(8) 3.0	419	(7) 24	5, 13
Basalt, New Zealand	20	(25) 1.6	493	(25) 22	14, 15
Andesite, New Zealand	45	(31) 4	287	(29) 15	13, 14, 15, 16
Spilitic rocks, Puerto Rico	40	(36) 5	650	(31) 21	17
Andesitic rocks, Puerto Rico	30	(26) 35	686	(26) 44	17
Continental rocks:					
Basalts	26	(36) 3	296	(3) 29	16, 6, 5, 18, 19
Basalts, Antarctica and Tasmania	31	(54) 7	238	(45) 33	19, 21, 22, 24

Table 37-E-6 (Continued)

	ppm Rb	N=() $s/\sqrt{n}$	K/Rb	N=() $s/\sqrt{n}$	References
Basalts, Karroo	11	(16) 2	465	(16) 20	20, 23
Andesites	46	(4) 7.5	363		6, 19
Dacites	55	(4) 9	270	(3) 7.3	14, 6
Rhyolites, New Zealand	107	(73) 2.1	250	(73) 3	14
Rhyolites, North America	191	(32) 8.5			5, 6, 25, 26
Acid volcanic glasses	139	(38) 5.6	237	(35) 12	5, 27, 28
Potash-rich volcanics	381	(22) 26.5	160	(10) 13	29

References to Table 37-E-6

	Analytical method
1. Faure and Hurley (1963)	I
2. Gast (1965)	I
3. Tatsumuto *et al.* (1965)	I
4. Bence (1966)	I
5. Hedge and Walthall (1963)	I
6. Hedge (1966)	I
7. Heier *et al.* (1967)	X
8. Engel *et al.* (1965), not original data	I
9. McDougall and Compston (1965)	I
10. Lessing *et al.* (1963)	X
11. Lessing and Catanzaro (1964)	X
12. Hamilton (1965)	I
13. Taylor and White (1966)	X
14. Ewart and Stipp (1967a)	X
15. Ewart and Stipp (1967b)	X
16. Gunn (1965)	X
17. Lidiak *et al.* (unpubl.)	A
18. Pidgeon and Compston (1965)	I
19. Heier *et al.* (1965)	X
20. Erlank and Hofmeyr (1966)	X
21. Compston *et al.* (1968)	X
22. Gunn (1965)	X
23. Alsopp (1965)	I
24. Gunn (1966)	X
25. Hurley *et al.* (1962)	I
26. Muelhburger *et al.* (1966)	I
27. Carmichael and McDonald (1962)	X
28. Butler and Smith (1962)	X
29. Hurley *et al.* (1966)	I

b) Plutonic Rocks

Average Rb concentrations in plutonic rocks are listed in Tables 37-E-4 and 37-E-5. Rb-concentrations in different parts of the layered Skaergaard intrusion were given by HAMILTON (1963). MOORBATH and BELL (1965) gave Rb concentrations in Tertiary volcanic and plutonic rocks at Skye. DAWSON (1962) gave Rb concentrations in 14 kimberlites from Africa. A survey of 214 post-1961 analyses (references given in the index) gave an average of 276 ppm Rb in granites. Because much recent data on Rb in granites have been made for the purpose of whole rock Rb/Sr dating, one could expect a sampling bias towards high Rb-concentration. Of the 214 analyses referred to above, 116 were done for the purpose of Rb—Sr dating. Their average Rb concentration was 203 ppm. Thus in this case there is no bias towards high Rb concentrations in granite samples collected for dating purpose.

c) Nepheline Syenites

GERASIMOVSKII (1966) compared the geochemistry of agpaitic and miascitic nepheline syenites. According to this date, Rb, together with Tl and Cs, is less in the agpaitic nepheline syenites than in the miascitic ones; the average Rb content is 250 ppm *vs.* 430—950 ppm, and it was also shown that this may be related to the K content which is also lower in the agpaitic rocks. The data for the agpaitic rocks of the Lovozero Massif is given in Table 37-E-7 together with data for the Ilimaussaq intrusion, S. W. Greenland. They are compared with the miascitic nepheline syenites of Stjernøy (North Norway) and Blue Mountain (Ontario, Canada) (HEIER, 1964, 1965). The miascitic rocks analysed by HEIER have low Rb concentrations and high K/Rb ratios. In comparing the agpaitic and miascitic nepheline syenites, GERASIMOVSKII (1966) concluded that Rb is lower in the agpaitic types. This is not supported by the data on the Stjernøy and Blue Mountain nepheline syenites. GERASIMOVSKII referred to early Rb determinations in miascitic rocks reported by GOLDSCHMIDT (1954), and it is possible that they are in error.

Table 37-E-7. *Rubidium concentrations (in ppm) and K/Rb ratios of the agpaitic Lovozero Massif (Kola peninsula)*, (GERASIMOVSKII, 1966), *Ilimaussaq intrusion (S. W. Greenland)*, (HAMILTON, 1964), *and the miascitic nepheline syenites of Stjernøy, North Norway, and Blue Mountain, Ontario*, (HEIER, 1965)

	Lovozero Massif phase			Ilimaussaq intrusion			Stjernøy neph. syenite	Ontario, Canada neph. syenite		
	1	2	3	augite syenite	granite	agpaite		BM. 1	BM. 2	BM. 3
Rb	130	230	320	200	411	333	115	44	70	70
K/Rb	287	183	116	212	78	90	579	795	560	737

Phase 1 = nepheline syenites.

Phase 2 = average of urtites, foyaites, lujaurites, amphibole lujaurites.

Phase 3 = average of eudialyte-lujaurites, porphyritic lujaurites, porphyritic lujaurites with lovozerite; poikilitic sodalite-syenites, towites.

Stjernøy = average of 16 analyses.

V. K/Rb Ratios

Here follows a general discussion of K/Rb ratios.

As a trace element with ionic and atomic properties very similar to that of K, Rb should be camouflaged by K in K-minerals and the two elements should be strongly positively correlated. Mainly because of larger ionic size, Rb is enriched in low temperature K-minerals and in fractional crystallization it will be concentrated in the residual liquid. When a rock system undergoes some sort of differentiation, Rb is expected to be concentrated relative to K in the "felsic" fractions and the K/Rb ratio will *decrease* in a sequence from mafic to acid rocks. Thus, even though Rb is completely contained in K-minerals and forms no mineral on its own, it is not perfectly camouflaged by K in GOLDSCHMIDT's original sense of this term.

The K/Rb relationship has proved most valuable in evaluating the geochemistry of Rb. It was first emphasized by AHRENS *et al.* (1952) who demonstrated that only rarely does the K/Rb ratio for common terrestrial rock types fall outside the limits of 160—300. (The original Rb-values corrected by a factor of 0.393, TAYLOR *et al.*, 1956.) Subsequent research has proven AHRENS' statement to be generally correct and most rocks of the continents have K/Rb ratios between the limits of 160 and 300, with an average of about 230 (HEIER and ADAMS, 1964). It was long realized that the mafic rocks trended towards and beyond the upper limit (300) of this ratio (NOCKOLDS and ALLEN, 1953; HORSTMAN, 1957; HEIER, 1960; LIEBENBERG, 1960; GAST, 1960; HEIER and ADAMS, 1964), and that Rb enrichment resulting in K/Rb ratios less than 160 occurred in "differentiated" granites and pegmatites (e.g. TAYLOR *et al.*, 1956; HEIER and TAYLOR, 1959). However, the general similarity between the "average" K/Rb ratio of igneous rocks and that of ordinary chondrites (Table 37-C-2: average K/average Rb = 350) did lend support to the concept of near constancy of this ratio during a range of geologic processes.

In the range from mafic to felsic rocks the Rb concentration increases two orders of magnitude while the K increase is only about one order of magnitude. A constancy of the K/Rb ratio is therefore not possible. LESSING *et al.* (1963) pointed out that the volcanic rocks in Hawaii were characterized by high K/Rb ratios (average about 500) and subsequent research has demonstrated that high K/Rb ratios are characteristic of basaltic rocks with the exception of continental tholeiites (Table 37-E-6). The maximum K/Rb ratios are observed in basalts dredged from the deep ocean floors, and these are also very low in K. Since much of this material is glassy, it cannot be explained as accumulates and it is considered a primary undifferentiated basaltic magma[1].

It is not understood why many continental tholeiitic basalts have low K/Rb ratios. It may reflect conditions in the upper mantle or a selective contamination process affecting the magma as it ascends through the Rb rich environment in the continental crust (HEIER *et al.*, 1965; COMPSTON *et al.*, 1968). It is relevant to these considerations that continental alkali basalts have relatively higher K/Rb ratios than continental tholeiitic basalts, (426 ± 48, ABBOT, 1967).

---

[1] Added in proof: HART (Earth and Planetary Science Letters **6**, 295—303, 1969) shows that high K/Rb ratios appear to be a primary feature of submarine tholeiitic magmas, though the trend of decreasing K/Rb with increasing K content is shown to be a probable result of later alteration, both during low grade metamorphism and during exchange with sea water.

The high K/Rb ratios in basalts have led to speculation about the value of this ratio in the upper mantle. LESSING *et al.* (1963) proposed that the ratio for Hawaiian tholeiitic basalts (512) is the K/Rb ratio of the upper oceanic mantle. TAUBENECK (1965) argued that their data indicated an even higher K/Rb ratio of the upper mantle, and GAST (1965) suggested that the K/Rb ratio of the upper mantle exceeds 1,500. Though not arguing against this general idea, HEIER and COMPSTON (1966) pointed out that K/Rb ratio in ultramafic rocks tend to be less than 300 (Table 37-E-1) indicating, perhaps, that the K/Rb ratios in basalts have little bearing on this ratio in the mantle but rather reflect the different mantle mineralogies with depth that give rise to basalt magmas through partial melting.

If a high K/Rb ratio in the mantle can be verified, it follows that the geochemical chondrite model is not valid for the overall terrestrial abundance of the elements. For this reason GAST (1965) advocated an achondritic model. K/Rb ratios in meteorites are given in Table 37-C-2.

Recent work has demonstrated that K and Rb are significantly fractionated by processes normally associated with regional metamorphism (see Section 37-N).

VI. Terrestrial abundance

Various estimates of terrestrial Rb concentrations are given in Table 37-E-8. It is now considered that the estimates of HEIER and ADAMS (1964) are too high primarily for two reasons.

1. They were based on an overall chondritic composition of the earth containing 2.9 ppm Rb. As discussed in Chapter 19 this assumption is now discredited. HURLEY (1968) estimates that the earth is depleted in Rb relative to chondrites by a factor of 0.14.

2. The estimate of Rb concentration in different spheres of the earth was based on the assumed K-distribution and accepting a probably too low K/Rb ratio (230) for material from the deeper part of the crust.

Restrictions on the Rb concentration in the upper mantle may be based on consideration of $^{87}Sr/^{86}Sr$ ratios of Recent oceanic basalts. These are in the region of 0.702 to 0.704 indicating that the average Rb/Sr ratio in the source area cannot have exceeded 0.03 through geologic time. If we assume the Sr content of the earth to be that of chondritic meteorites (10 ppm), and all Sr to be concentrated in the upper mantle (about 10% of the earth), the maximum Rb content of this region is 3 ppm. Since Sr is concentrated in the crust relative to the mantle, 3 ppm in the upper mantle is therefore an upper limit. It would appear that the true distribution in the earth is somewhere between columns 4 and 5, Table 37-E-8. The considerably lower than previously accepted Rb concentrations in crustal rocks used in columns 4 and 5 results from the recognition of basalts with very low Rb contents on the deep ocean floor, and for the continents from the recognition of significant Rb depletion in high pressure granulite facies rocks (LAMBERT and HEIER, 1968; HEIER, unpublished data). Similar results were derived by HURLEY (1968) who estimated 100 ppm Rb in the upper continental crust and 15 ppm Rb in the lower crust assumed to approximate granulite facies composition. This is similar to the estimates made in Table 37-E-8, columns 4 and 5. HURLEY (1968) estimated 0.38 ppm Rb for the whole earth and a concentration in the crust of 45%.

Table 37-E-8. *Rubidium distribution in the earth (in ppm)* [a]

	Approximate mass fraction	1	2	3	4	5
Core	31.5					
Mantle	68.1	3.6				
lower mantle	57.1					
upper mantle	11.0				4.2	0.7
Crust	0.4	91	74	75	37	37
oceanic crust	0.1	35		30	2.5	2.5
continental crust	0.3	113	87	90	48	48
Earth	100	2.9			0.61	0.22
% of terrestrial Rb in continental crust		11			24	65

[a] See text for estimate by HURLEY (1968).

1. HEIER and ADAMS (1964). 2. FAURE and HURLEY (1963). 3. TAYLOR (1964). 4. Pyrolite composition of upper mantle (K = 0.1079%, K/Rb as in ultrabasic rocks, Table 37-E-2); $^1/_3$ upper sialic crust with 2.6% K, K/Rb = 230; $^2/_3$ lower crust of granulite facies rocks, gabbro and anorthosite averaging 1.5% K; 15 ppm Rb, K/Rb = 1,000. 5. As in column 4 but K/Rb = 1,500 in upper mantle.

Revised manuscript received: January 1970

37-G. Behavior during Weathering and Abundance in Soils

The general characteristics of processes controlling Rb distribution in soils is similar to that of the other alkali elements, see Chapter 3.

In weathering Rb is closely linked to K. Adsorption may play an important role in the concentration of Rb relative to K in the late stages of weathering. GOLDSCHMIDT (1954) found that Rb was held in adsorption positions more firmly than K. BUTLER (1957) stated that Rb concentration was related to K only when illite formed from weathered micas and feldspars.

HARRIS and ADAMS (1969) found a continuous decrease in the K/Rb ratio during weathering, probably by a combination of ion exchange and adsorption (see Chapter 3). WEBBER and JELLEMA (1965) reported a Rb soil/Rb rock ratio of 1.07 and a mean Rb soil content of 100 for 300 samples of Quebec soils.

Table 37-G-1. *Change in Rb content (ppm) and K/Rb ratio with weathering.* (From HORSTMAN, 1957)

		Fresh rock	Weathered rock					
Rhyolite[a]	Rb	200	100	160				
	K/Rb	200	160	180				
Granite[a]	Rb	170	40	200	60	350	110	
	K/Rb	210	100	40	70	100	40	
Granite[a]	Rb	220	250	90				
	K/Rb	200	140	100				
Rhyolite[a]	Rb	250	210	40				
	K/Rb	140	150	70				
Andesite[a]	Rb	130	170					
	K/Rb	230	180					
Granodiorite[a]	Rb	230	90	60				
	K/Rb	130	140	80				
Lamprophyre[a]	Rb	120	160	80				
	K/Rb	250	230	110				
Granite Gneiss[b]	Rb	130	200	140	150	60	10	10
	K/Rb	250	220	210	310	320	410	110
Diabase[b]	Rb	20	50					
	K/Rb	470	250					
Amphibolite[b]	Rb	10	20					
	K/Rb	210	140					
Siltstone[c]	Rb	110	100					
	K/Rb	240	200					
Siltstone[c]	Rb	150	150					
	K/Rb	250	220					
Sandstone[c]	Rb	180	150					
	K/Rb	200	160					

[a] BROCK (1943).
[b] GOLDICH (1938).
[c] VAN HOUTEN (1953); (for references a—c see HORSTMAN, 1957).

The data of Table 37-G-1 show that the K/Rb ratio decreases with weathering. The trace element geochemistry of soils was summarized by VINOGRADOV (1959). His averages for Rb in soils from different countries are listed in Table 37-G-2. He derived an average of 150 ppm Rb for all soils investigated (500 determinations). The mean Rb content of soil samples (this chapter) is 140 ± 27 ppm Rb ($s/\sqrt{n}$): BUTLER (1953), BUTLER (1954), MCLAUGHLIN (1955), VINOGRADOV (1950), BURRIDGE and AHN (1965). The range is 1.5—1,800 ppm and the median is 35 ppm.

Table 37-G-2. *Rubidium and cesium contents of soils of various countries.* (From VINOGRADOV, 1959)

	Range (ppm Rb)	Average ppm Rb	Average ppm Cs
U.S.S.R.:			
Various soils (42)	10—89	60	X.O
Various soils (40)	10—150	100	5
Europe:			
Various soils (22)	20—190	33	5
U.S.A.			
Various soils (26)	10—100	20	
Scotland:			
Various soils (6)	30—600	260	
Various soils (161)	20—1,000	270	
Japan:			
Various soils (28)			1
Various soils (15)			1
Great Britain:			
Various soils (21)	140—350	200	

Revised manuscript received: January 1970

37-I. Abundance in Natural Waters

I. Continental Waters

Durum and Haffty (1963) in their study of Rb in some of the large rivers of North America, found that the median Rb content was 1.5 ppb and the range was 0—8 ppb. Kharkar *et al.* (1968) found an average of 1.1 ppb from twelve rivers, using neutron activation. Sreekumaran *et al.* (1968) found an average of 1.8 ppb and a range of 0.8—2.9 ppb in 9 samples of North American rivers and a value of 1.1 ppb for Lake Mead.

De Villiers (1962) found a Rb value of 8 to 150 ppb in the Orange River, see also Galitsyn and Slavyanova (1965).

The mean content of 56 samples of river water is 1.3 ± 0.02 ppb Rb ($s/\sqrt{n}$) (Durum *et al.*, 1960; Miller, 1961; Livingstone, 1963). The range is 0.2—7.4 ppb and the median is 1.1 ppb. Pantcheve (1965) reported Rb values up to 0.8 ppb in 19 samples of groundwater. Galitsyn and Slavyanova (1965) analysed 300 ground and river waters. They did not detect Rb (detection limit = 8 ppb).

II. Hydrothermal Waters

Ellis and Wilson (1960) found the Wairakei (New Zealand) thermal waters to be enriched in Rb with respect to K. Golding and Speer (1961) analyzed thermal waters in New Zealand and found Rb values of >0.1 ppm. They found high Rb contents in rhyolitic areas and low Rb contents in basaltic and andesitic areas. Ellis and Mahon (1964) also studied the thermal waters of New Zealand. They found a range in Rb content from less than 0.1 ppm to 3.1 ppm (K/Rb = 60—280). Ellis and Mahon (1967) studied the leaching of several elements including Rb with water held at different temperatures for 14 days in contact with different rock types (see also Mahon, 1967).

Uzumasa (1963) analyzed 80 hot spring waters from Japan and found that 25% contained <0.1 ppm Rb, 63% contained 0.1—1 ppm Rb, and 12% contained >1 ppm Rb. He also noted that the hot spring waters contained more Rb than seawater and river water. Ristic *et al.* (1956) found Rb concentrations of 0.1—1 ppm in a study of mineral waters.

The Salton Sea geothermal waters contain unusually high Rb contents: 75 and 60 ppm (Helgeson, 1968) and 169 ppm (White, 1965).

Excluding these values, the mean content of Rb in 180 samples of thermal waters is 0.82 ± 0.08 ppm Rb ($s/\sqrt{n}$). The range is 0.01—7.7 ppm and the median is 0.32. The data are from Ellis and Wilson (1960), Golding and Speer (1965), Ellis and Mahon (1964), Glover (1967), and Mahon (1967).

III. Seawater

The Rb concentration of seawater is higher than the Rb concentration of river and groundwaters, but is still very much less than the Rb concentration in sediments.

The Rb concentration in the oceans is fairly constant with respect to various locations (BOLTER *et al.*, 1964), and with depth (FABRICAND *et al.*, 1966). The Rb/chlorinity ratio appears to be constant with depth. SMITH *et al.* (1965) reported an average of 123 ppb Rb for 16 samples of various oceans and depths converted to $35^0/_{00}$ salinity. The range was 112—134 ppb.

Residence time in seawater for Rb (GOLDBERG, 1965) has been calculated as 2.7×10^5 years based on a 0.12 ppm concentration of Rb in seawater and using the data of DURUM and HAFFTY (1963). This is the same value GOLDBERG and ARRHENIUS (1958) calculated for Rb on the basis of other data.

The presently accepted Rb concentration in seawater is 0.12 ppm (GOLDBERG, 1965).

IV Interstitial Waters

SAPPO (1960) found Rb concentrations in oilfield waters of 13.2—23.2 ppb. BILLINGS *et al.* (1969) reported a mean Rb content of 1.18 ppm for Canadian oilfield brines and calculated a weighted mean of 0.88 ppm Rb for all subsurface waters of the Western Canada sedimentary basin.

The mean Rb content of 139 samples of formation waters (WHITE, 1965; BILLINGS, unpubl. data) is 3.59 ± 0.41 ppm Rb ($s/\sqrt{n}$). The range is 0.01—18.8 ppm and the median is 1.6 ppm.

Table 37-I-1. *Population statistics for Rb in natural waters (in ppm, excluding rivers)*

Subdivision	$\bar{X}$	s	$s/\sqrt{n}$	Range	Median	n
Rivers (ppb)	1.3	1.0	0.1	0.2 — 7.4	1.1	56
Hydrothermal waters	0.82	1.11	0.08	0.01— 7.7	0.32	180
Formation waters	3.59	4.86	0.41	0.01—18.8	1.6	139
Sea water	0.12					

Revised manuscript received: January 1970

37-K. Abundance in Common Sediments and Sedimentary Rock Types

Population statistics for Rb in the sedimentary rocks are given in Table 37-K-1.

Table 37-K-1. *Population statistics for Rb in sedimentary rocks and soils (in ppm)*

Subdivision	$\bar{X}$	s	$s/\sqrt{n}$	Range	Median	n
Soils	140	285	27	1.5—1,800	35	116
Dolomites [a]	45	42	4	6 — 370	33	120
Silty-limestones	75	41	3	6 — 167	79	130
Sands and sandstones	46	23	4	9 — 100	40	33
Shales	164	98	6	20 — 663	143	253
Argillaceous sediments	128	21	2	28 — 220	125	116

[a] See text for precautionary comment.

I. Carbonate Sediments and Rocks

Landergren (1948) reported 5 ppm Rb in marine siderite ores. Horstman (1957) did not detect Rb (determination limit 5 ppm) in limestones and dolomites containing from 1 to 5 per cent silicates.

Billings and Ragland (1968) found the Rb content of the acid soluble portion of Recent carbonate sediments to range from 0.3 to 3.4 ppm Rb. Most of the Rb in carbonates resides in the insoluble fraction.

Degens *et al.* (1958) found a range of 100—1,100 ppm, average 480 ppm Rb (10 values), and 80—650 ppm, average 360 ppm Rb (8 values), in the less than 2μ size fraction of insoluble residues of marine Vanport and Putnam Hill limestones and fresh water Freeport and Upper Kittaning limestones respectively.

These values seem high but if the 2μ fraction is only a small per cent of the rock, they compare reasonably well with the limestone data of Table 37-K-1.

Graf (1960) compiled data on Rb in Scottish sedimentary carbonate rocks published by the Geological Survey of Great Britain. Seventy-five per cent of the 183 analyses contained Rb below the given sensitivity limit of 30 ppm. The range was from less than 30 ppm to 800 ppm Rb with an average of 60 ± 11 ppm Rb.

The mean Rb content of 120 dolomites is 45 ± 4 ppm ($s/\sqrt{n}$). The range is 6 to 370 ppm. The median is 33 ppm. The data are from Weber (1964). An additional 162 of Weber's samples were below the 6 ppm detection limit.

Havard (1967) determined Rb in 130 Devonian silty-limestones. The mean was 75 ± 3 ppm Rb ($s/\sqrt{n}$).

II. Argillaceous Sediments and Shales

Rb is concentrated relative to K in shales (see section 37-G). Heier and Adams (1964) fount the average K/Rb ratio in shales to be 150. In a study of the Gulf of

Mexico sediments WELBY (1958) found a suggestion of an increase in Rb content with distance from the shore (his Rb data should be corrected for systematic error, see TAYLOR, 1960).

SPENCER (1966) found an association of Rb with K in Silurian shales. Diagenetic recrystallization of illite to orthoclase results in exclusion of Rb, HORSTMAN (1957). DEGENS *et al.* (1957) found marine shales to have a higher Rb content than fresh water shales. They found a significant correlation between Rb and the illite/kaolinite ratio. CAMPBELL and WILLIAMS (1965) suggested that K/Rb ratios of 250—300 represented non-marine to brackish-water shales and ratios of 150—200 represented marine shales (Canadian Cretaceous samples).

FENNER and HAGNER (1967) reported a range of 16—220 ppm Rb in 31 samples of shales with a mean of 110 ppm Rb.

LEASK (1967) found the K/Rb ratio of Precambrian argillites to be higher than that of younger shales (approximate mean 270 *vs.* 150). He concluded that this resulted from continued recycling of younger sediments (see discussion of K/Rb in soils). The mean Rb content of the Precambrian argillites was 142 ppm Rb.

HAVARD (1967) reported a mean Rb content of Devonian black shales of 130 ppm, average K/Rb ratio of 346. The average K_2O content of these shales was 4.5% (maximum of 7.9%) which is considerably higher than that of normal shales (see Chapter 19). HAVARD suggested that the high K_2O content and K/Rb ratio resulted from authigenic growth of K-minerals. The Chattanooga black shale is also high in K_2O content (6%—12%), LANDIS (1962).

GORHAM and SWAINE (1965) found the following average Rb concentrations of lake sediments: reduced muds — 55 ppm; oxidized muds — 45 ppm; oxidate crusts — 70 ppm. Three samples of glacial clays ranged from 30 to 40 ppm Rb (K from 1.45 to 1.61 per cent).

The mean content of 253 shales is 164 ± 6 ppm Rb ($s/\sqrt{n}$): DEGENS *et al.* (1957), HORSTMAN (1957), NICHOLLS and LORING (1962), HIRST and DUNHAM (1963), AUDLEY-CHARLES (1965), SPENCER (1966), HAVARD (1967), LEASK (1961), and WILLIAMS (1967). The range is 20—663 ppm Rb and the median is 143 ppm.

The mean Rb content of 116 Recent argillaceous sediments is 128 ± 2 ppm Rb ($s/\sqrt{n}$): HORSTMAN (1957), WELBY (1958), and HIRST (1962). The range is 28—220 ppm Rb and the median is 125 ppm.

III. Psammitic Sedimentary Rocks

The bulk of Rb in these rocks resides in the feldspar and sheet-mineral fraction and its concentration showes an inverse relationship to the SiO_2 content.

Arkose and feldspathic sandstones and sandstones with argillitic matrix will therefore contain more Rb than pure quartz sandstones. HORSTMAN (1957) found a range from 20 to 100 ppm Rb in four sandstones (average 60 ppm). The highest value was from a glauconitic sand. AUDLEY-CHARLES (1965) reported eleven analyses of cherts and radiolarites which averaged 56 ± 7 ppm Rb ($s/\sqrt{n}$) with 9 additional analyses below 40 ppm (detection limit).

The mean Rb content of 33 sandstones and Recent sands is 46 ± 4 ppm Rb ($s/\sqrt{n}$): HORSTMAN (1957), WELBY (1958), HIRST (1962). The range is 9—100 ppm Rb and the median is 40 ppm.

IV. Evaporites

STEWART (1963) reviewed the geochemistry of evaporites. Rb was only reported in a blue halite from Wintershall, Germany (5 ppm).

GALITSYN and SLAVAGANOVA (1965) found Rb values up to 1.6 ppm in lakes where halite precipitation had begun. The K/Rb ratio is extremely high in sylvites (BILLINGS; unpubl. data, Rb less than 10 ppm).

BRAITSCH (1966) reported that the partition (wt. per cent Rb in carnallite/wt. per cent Rb in solution) appears to have a negative temperature coefficient. Experimental research in the KCl—RbCl—H_2O system showed a decrease in Rb uptake by sylvite with increasing temperature (McIntire, 1963). The Rb content of sylvite in some German deposits ranged from 47—252 ppm and from 990—1,670 ppm in carnallite (BRAITSCH, 1966). KÜHN (1963) found a range of 80—1,820 ppm Rb in carnallite from Germany and Canada as average for different beds.

V. Deep Sea Sediments

GOLDSCHMIDT *et al.* (1934) reported 391 ppm Rb in pelagic clay. HORSTMAN (1957) gave 160 and 60 ppm Rb in two composites of four pelagic clays and 10 ppm in deep sea carbonates.

HORSTMAN (1957) reported 10 ppm Rb in a composite of 6 Globigerina oozes. WELBY (1958) analyzed 19 Globigerina oozes. The mean was 30 ± 3 ppm Rb ($s/\sqrt{n}$); range 10—50 ppm Rb, median 27 ppm (corrected data, see TAYLOR, 1960).

YOUNG (1954) and WEDEPOHL (1960) stated that the Rb contents of both near-shore and deep sea clays are similar.

Revised manuscript received: January 1970

37-L. Biogeochemistry

Rb appears not to be an essential component of living matter but rather a toxic agent and/or partial substitute for K. K/Rb ratios in plants are of the order of 1,000 (FLORKON and MASON, 1962). MALLETTE *et al.* (1960) concluded that Rb competes with K in adsorption processes.

LESTER (1958) found that some species of bacteria could use Rb as a substitute for K in their life processes. BUROVINA and NESTEROV (1962), in their studies on marine animals, found Rb concentrated in organisms with respect to sea water. BLACK and MITCHELL (1952) reported a range of 25 to 250 ppm Rb in the dry ash of algae.

COLLANDER (1941), EPSTEIN and HAGAN (1952), MENZEL and HEALD (1955) and FRIED and NOGGLE (1958) found that the K/Rb ratio in plants bears a constant ratio to the K/Rb ratio in the substrate solution. FRIED *et al.* (1959) studied the K/Rb ratios in the soil plant system.

AIDINYAN (1959) found that the Rb content of plants correlates with the K content. Rb tends to concentrate in the delicate vegative and reproductive organs of plants as does K. Intensive leaching by plants causes Rb to concentrate in the soil, especially the humus zone.

Data for Rb in coal and seaweed are given in Table 37-L-1. BOWEN (1966) reviewed the biochemistry of Rb.

Table 37-L-1. *Rubidium in coal and seaweed* (TUPPER and LORING, 1961)

	Rb ppm
Coal Sample:	
Vitrain SRI	33
Vitrain IV 6	46
Vitrain SR I	35
Vitrain IV 25	30
Vitrain V 25 II	55
Vitrain IV 6	65
Vitrain IV 3, 8	100
Vitrain HBno 26	250
(SMALES and SALMON, 1955)	
Seaweed:	Rb ppm (wet plant weight)
Ascophyllum nodasum	1.4
Focus gervatus	1.8
Focus vesiculosus	2.4
Laminaria digitate	2.2
Laminaria saccharina	1.4
Porphyra umbilicalis	0.61
Rhody menia palamata	1.08
Average	1.58
(SMALES and SALMON, 1955)	

Table 37-L-1 (Continued)

	Rb ppm
Coal Ash:	
Fencehouses boiler coal	92
Hetton, D.C. boiler coal	141
Mordon boiler coal	175
Malton cooking coal	79
Average	122

Revised manuscript received: January 1970

37-M. Abundance in Common Metamorphic Rocks

There are no indications of systematic variations in the Rb concentrations during regional metamorphism until the level of granulite facies, but more research is needed. Recent determinations of Rb in schists are by EVANS (1964), PIDGEON and COMPSTON (1965) and WHITE (1966). Considerable movement of Rb is observed in granite aureoles and xenoliths in granites. BOWLER (1959) studied contact metamorphic aureoles and xenoliths in granites from south-west England. All the aureoles showed culmination of Rb towards the granite contact, rising far above the levels in shales which were not thermally metamorphosed. These trends were accentuated in the small xenoliths (see also HEIER and ADAMS, 1964; EVANS, 1964; MOORBATH and SHACKLETON, 1966).

Rb concentrations of gneisses are reported by HURLEY *et al.* (1962), HEDGE and WALTHALL (1963), LAMBERTH (1964), GUNN (1965), ZARTMAN (1965), PIDGEON and COMPSTON (1965), GOLDICH *et al.* (1966), WHITE (1966), FERRARA and GRAVELLE (1966) and MOORBATH and SHACKLETON (1966). The average of 78 gneisses is 186 ppm Rb with a standard deviation of 90 ppm ($s/\sqrt{n}=10.2$). Because of the overlap in values for granites (including granodiorites) and gneisses, it is not possible to distinguish between them on the basis of Rb-content.

Data on Rb concentrations in granite and gneisses of the Precambrian in Australia are given by LAMBERT and HEIER (1968). They divide the gneisses into four groups: "acid", and "sub-acid", intermediate, "basic", and compare areas of different metamorphic grade. They were able to show that no systematic loss of Rb took place before conditions of medium to high pressure granulite facies. Only the granulites of the Musgrave and Fraser ranges are in this category, and the sub-acid and intermediate rocks in these areas are significantly depleted in Rb relative to other areas. The low pressure granulites which are iepresented at Cape Naturaliste, Eyre Peninsula and East Kimberley are not depleted in Rb. Relative depletion of Rb in medium to high pressure granulite facies rocks was first noted by HEIER (1960) in rocks from Norway (see also HEIER and ADAMS, 1964; HEIER, 1964, 1965). The Rb depletion in medium to high pressure granulite facies iocks may be more significant than that shown by the Australian data (HEIER, unpubl. data).

Revised manuscript received: January 1970

37-N. Behavior in Metamorphic Reactions

Recent work has demonstrated that K and Rb are significantly fractionated by processes normally associated with regional metamorphism. Heier (1960) pointed out that granulite facies rocks had significantly higher K/Rb ratios than associated amphibolite facies rocks. This was verified by Lambert and Heier (1968) using more sensitive analytical methods. They found that sub-acid and intermediate rocks of medium to high pressure granulite facies had significantly higher K/Rb ratios (300—600) than corresponding rocks in amphibolite facies (150—250). Investigations in Norway, including and extending the area discussed by Heier (1960), verifies the Rb depletion in medium to high pressure granulite facies rocks (K/Rb of 500 to 1,000, Heier, unpubl. data). If rocks of this nature are typical of the lower continental crust very high K/Rb ratios can be expected in these regions ($\sim$1000). Anorthosites, which typically are associated with these rocks have similar high K/Rb ratios (Table 37-E-3). This is taken into consideration in the estimates of the terrestrial abundance of Rb (Table 37-E-8).

Revised manuscript received: January 1970

References: Sections 37-B to 37-E, 37-G, 37-I to 37-N

ABBOT, M. J.: Petrology of the Nandewar Volcano. Ph. D. dissertation, Dept. of Geology, Australia National University, Canberra (1965).

— K and Rb in a continental igneous rock suite. Geochim. Cosmochim. Acta **31**, 1035—1041 (1967).

AHRENS, L. H.: The significance of the chemical bond for controlling the geochemical distribution of the elements. Part I. Phys. Chem. Earth **5**, 1—54 (1964).

—, W. H. PINSON, and M. M. KEARNS: Association of rubidium and potassium and their abundance in igneous rocks and meteorites. Geochim. Cosmochim. Acta **2**, 229—242 (1952).

AIDINYAN, R. KH.: Distribution of rare alkalies in colloids of soils and the participation of vegetation in this process. Geokhimiya 346—357 (1959).

ALDRICH, L. T., G. W. WETHERILL, C. R. TILTON, and G. L. DAVIS: Half life of ^{87}Rb. Phys. Rev. **103**, 1045—1047 (1956).

ALLER, L. H.: The abundance of the elements. New York: Interscience Publishers 1961.

ALLSOPP, H. L.: Rb—Sr and K—Ar measurements on the Great Dyke of Southern Rhodesia. J. Geophys. Rex. **70**, 977—984 (1965).

—, and P. KOLBE: Isotopic age determinations on the Cape Granite and intruded Malmesbury sediments, Cape Peninsula, South-Africa. Geochim. Cosmochim. Acta **29**, 1115—1130 (1965).

—, L. O. NICOLAYSEN, and P. HAHN-WEINHEIMER: Rb/K ratios and Sr-isotopic compositions of minerals in eclogitic and peridotitic rocks. Earth and Planetary Sci. Letters **5**, 231—244 (1960).

AUDLEY-CHARLES, M. G.: A geochemical study of Cretaceous ferromanganiferrous sedimentary rocks from Timor. Geochim. Cosmochim. Acta **29**, 1153—1174 (1965).

— Some aspects of the chemistry of Cretaceous siliceous sedimentary rocks from Timor. Geochim. Cosmochim. Acta **29**, 1175—1192 (1965a).

BÄHNISH, I. G., u. E. HUSTER: Neubestimmung der Halbwertszeit des ^{87}Rb. Naturwissenschaften **41**, 495—496 (1954).

— — u. W. WALCHER: Zu Halbwertszeit und Zerfallsschema des ^{87}Rb. Naturwissenschaften **39**, 379—380 (1952).

BEARD, G. B., and W. H. KELLY: Nucl. Phys. **28**, 570 (1961).

BEUCE, A. E.: The differentiation history of the earth by rubidium—strontium isotopic relationships M.I.T.-1381 — 14. Fourteenth Annual Progress Report for 1966, 35—78 (1966).

BILLINGS, G. K., B. HITCHON, and D. R. SHAW: Geochemistry and origin of formation waters in the western Canada sedimentary basin. Part 2. Alkali metals. In: The Geochemistry of Subsurface Brines (E. E. ANGINO and G. K. BILLINGS, eds.). Chem. Geol. **4**, 211—223 (1969).

—, and P. C. RAGLAND: Geochemistry and mineralogy of the Recent reef and lagoonal sediments south of Belize, Br. Honduras. Chem. Geol. **3**, 135—153 (1968).

BLACK, W. A. P., and R. L. MITCHELL: Trace elements in the common brown algae and in sea water. J. Marine Biol. Assoc. **30**, 575—584 (1952).

BOLTER, E., K. K. TUREKIAN, and D. F. SCHUTZ: The distribution of Rb, Cs, Ba in the oceans. Geochim. Cosmochim. Acta **28**, 1459—1466 (1964).

BORSI, S., G. FERRARA, and E. TONGIORGI: Rb/Sr and K/Ar ages of intrusive rocks of Adamello and M. Sabion (Trentino, Italy). Earth and Planetary Sci. Letters **1**, 55—57 (1966).

BOWEN, H. J. M.: Trace elements in biochemistry. 241 pp. New York: Acad. Press 1966.

BOWLER, C. M. L.: The distribution of the five alkali elements and fluorine in some granites and associated aureoles from the southwest of England. Ph.D. dissertation, University of Bristol (1959).

BRAITSCH, O.: Bromine and rubidium as indicators of environment during sylvite and carnallite deposition of the Upper Rhine valley evaporites. In: J. Z. RAU, ed., Second Symposium on Salt, Northern Ohio Geol. Soc. Inc. Cleveland, 293—301 (1966).

BROWN, H.: A table of relative abundances of nuclear species. Rev. Mod. Phys. **21**, 625—634 (1949).

BUROVINA, I. V., and V. P. NESTEROV: Quantitative determination of Rb in biological objects by the isotope dilution method. Biofizika **7**, 233—235 (1962).

BURRIDGE, J. C., and P. M. AHN: A spectrographic survey of representative Ghana forest soils. J. Soil Sci. **16**, 296—309 (1965).

BUTLER, J. R.: The geochemistry and mineralogy of rock weathering. (1) The Lizard area, Cornwall. Geochim. Cosmochim. Acta **4**, 157—178 (1953).

— The geochemistry and mineralogy of rock weathering. (2) The Nordmarka Area, Oslo. Geochim. Cosmochim. Acta **6**, 268—281 (1954).

—, P. BOWDEN, and A. Z. SMITH: K/Rb ratios in the evolution of the younger granites of Northern Nigeria. Geochim. Cosmochim. Acta **26**, 89—100 (1952).

—, and A. Z. SMITH: Zirconium, niobium and certain other trace elements in some alkali igneous rocks. Geochim. Cosmochim. Acta **26**, 945—953 (1962).

CABELL, M. J., and A. A. SMALES: The determination of rubidium and caesium in rocks, minerals and meteorites by neutron-activation analysis. Analyst. **82**, 390—406 (1957).

CAMERON, A. G. W.: Abundances of the elements. Handbook of Physical Constants — Revised edition (S. P. CLARK, JR., ed.). Geol. Soc. Am. Mem. **97**, 7—10 (1966).

CAMPBELL, F. A., and G. D. WILLIAMS: Chemical composition of shales of the Manville group (Lower Cretaceous) of Central Alberta, Canada. Bull. AAPG. **49**, 81 (1965).

CARMICHAEL, I., and A. MCDONALD: The geochemistry of some natural acid glasses from the North Atlantic Tertiary volcanic province. Geochim. Cosmochim. Acta **25**, 189—222 (1961).

CHAO, E. C. T.: The petrographic and chemical characteristics of tektites. Tektites (J. A. O'KEEFE, ed.), p. 51—94, Chicago Univ. Press 1963.

CHAUDHURY, P. K., SEN: Radioactivity of rubidium. Proc. Nat. Inst. Sci. India **8**, 45 (1942).

CHIANG, M. C.: Element partition between hornblende and biotite in the rocks from Loon Lake aureole, Chandos township, Ontario. M. Sc. Thesis, McMaster University (1965) (ref. SHAW, 1968).

CLAYTON, FOWLER, HULL, and ZIMMERMANN: Ann. Phys. (N.Y.) **12**, 331 (1961).

CLIFFORD, T. N., D. C. REX, and N. J. SNELLING: Radiometric age data for the Urungwe and Miami granites of Rhodesia. Earth and Planetary Sci. Letters **2**, 5—12 (1967).

COLLANDER, R.: Acta Botan. Fenn. **29**, 1—12 (1941).

COMPSTON, W., J. F. LOVERING, and M. J. VERNON: The rubidium-strontium age of the Bishopville aubrite and its component enstatite and feldspar. Geochim. Cosmochim. Acta **29**, 1085—1099 (1965).

—, I. MCDOUGALL, and K. S. HEIER: Geochemical comparison of the Mesozoic basaltic rocks of Antarctica, South Africa, South America and Tasmania. Geochim. Cosmochim. Acta **32**, 129—149 (1968).

CURRAN, S. C., D. DIXON, and H. W. WILSON: The natural radioactivity of rubidium. Phys. Rev. **84**, 151—152 (1951).

DAWSON, J. B.: Basutoland kimberlites. Bull. Geol. Soc. Am. **73**, 545—560 (1962).

DEGENS, E. T., E. G. WILLIAMS, and M. L. KEITH: Environmental studies of Carboniferous sediments. I. Geochemical criteria for differentiating marine from fresh water shales. Bull. Am. Assoc. Petrol. Geologists **41**, 2427—2455 (1957).

— — — Environmental Studies of Carboniferous sediments. Part II. Application of geochemical criteria. Bull. Am. Assoc. Petrol. Geologists **42**, 981—987 (1958).

DEVILLIERS, P. R.: The chemical composition of the water of the Orange river at Vioolsdrit, Cape Providence. Rep. Suid-Afrika, Dept. Myhwese. Ann. Geol. Ophame **1**, 197—208 (1962).

DODSON, M. H., and L. E. LONG: Age of Lundy granite, Bristol Channel. Nature **195**, 975—976 (1962).

DURUM, W. H., and J. HAFFTY: Implications of minor element content of some major streams of the world. Geochim. Cosmochim. Acta **27**, 1—11 (1963).

DURUM, W. H., S. G. HEIDEL, and L. J. TISON: World-wide runoff of dissolved solids, Publ. No. 51 of Int. Assoc. Sci. Hydrologists, 618—628 (1960).

EGELKRAUT, K., and H. LOUTZ: β-spectrum and half-life of ^{87}Rb. Z. Physik **161**, 13—19 (1961).

EKLUND, S.: Studies in nuclear physics. Excitation by means of X-rays. Activity of ^{87}Rb. Arkiv Mat., Astron., Fysik **33 A**, No. 14 (1946).

ELLIS, A. J., and W. A. J. MAHON: Natural and hydrothermal systems of experimental hot water/rock interactions. Geochim. Cosmochim. Acta **28**, 1323—1357 (1964).

— — Natural hydrothermal systems and experimental hot water/rock interactions. (Part II.) Geochim. Cosmochim. Acta **31**, 519—538 (1967).

—, and S. H. WILSON: The geochemistry of alkali metal ions in the Wairakei hydrothermal system, N. Zealand. J. Geol. Geophys. **3**, 539—617 (1960).

— — The geochemistry of alkali metal ions in the Wairakei hydrothermal system, New Zealand. J. Sci. Technol. **4**, 415—430 (1960).

ENGEL, A. E. J., C. G. ENGEL, and R. G. HAVENS: Chemical characteristics of oceanic basalts and the upper mantle. Bull. Geol. Soc. Am. **76**, 719—734 (1965).

EPSTEIN, E., and C. E. HAGAN: A kinetic study of the absorption of alkali cations by barley roots. Plant Physiol. **27**, 457—474 (1952).

ERLANK, A. J., and P. K. HOFMEYR: K/Rb and K/Cs ratios in Karroo dolerites from South Africa. J. Geophys. Res. **71**, 5439—5445 (1966).

EVANS, B. W.: Fractionation of elements in the pelitic hornfelses of the Cashel-Lough Wheelaun intrusion, Connemara, Eire. Geochim. Cosmochim. Acta **28**, 127—156 (1964).

EWART, A., and J. STIPP: Unpublished data (1968).

—, and S. R. TAYLOR: Trace element geochemistry of the rhyolitic volcanic rocks, Central North Island, New Zealand, Phenocryst data. Contr. Mineral. and Petrol. **22**, 127—146 (1969).

— —, and A. C. CAPP: Trace and minor element geochemistry of the rhyolitic volcanic rocks, Central North Island, New Zealand. Contr. Mineral. and Petrol. **18**, 76—104 (1968).

FABRICAND, B. P., E. S. IMBIMBO, M. E. BREY, and J. A. WESTON: Atomic absorption analyses for Li, Mg, K, Rb, and Sr in ocean waters. J. Geophys. Res. **71**, 3917—3921 (1966).

FAIRBAIRN, H. W., G. FAURE, W. H. PINSON, P. M. HURLEY, and J. L. POWELL: Initial ratio of strontium 87 to strontium 86, whole-rock age, and discordant biotite in the Monteregian igneous province, Quebec. J. Geophys. Res. **68**, 6515—6522 (1963).

—, P. M. HURLEY, and W. H. PINSON: The relation of discordant Rb—Sr mineral and whole rock ages in an igneous rock to its time of crystallization and to the time of subsequent Sr^{87}/Sr^{86} metamorphism. Geochim. Cosmochim. Acta **23**, 135—144 (1961).

— — — Preliminary age study and initial $^{87}Sr/^{86}Sr$ of Nova Scotia granitic rocks by the Rb—Sr whole-rock method. Bull. Geol. Soc. Am. **75**, 253—258 (1964).

FAURE, G., and P. M. HURLEY: The isotopic composition of strontium in oceanic and continental basalts. Application to the origin of igneous rocks. J. Petrol. **4**, 31—50 (1963).

FENNER, P., and A. F. HAGNER: Correlation of variations in trace elements and mineralogy of the Esopus-Formation, Kingston, New York. Geochim. Cosmochim. Acta **31**, 237—261 (1967).

FERRANA, G., and M. GRAVELLE: Radiometric ages from Western Ahaggar (Sahara) suggesting an eastern limit for the West African Craton. Earth and Planetary Sci. Letters **1**, 319—324 (1966).

FLEISCHER, M.: Summary of new data on rock samples G-1 and W-1, 1962—1965. Geochim. Cosmochim. Acta **29**, 1263—1284 (1965).

FLINTA, J., and S. EKLUND: On the radioactivity of ^{87}Rb. Arkiv Fysik **7**, 401—412 (1954).

FLORKIM, M., and H. S. MASON (eds.): Comparative Biochemistry, vol. IV. New York: Academic Press 1962.

FLYNN, K. F., and L. E. GLENDENIN: Written communication to Aldrich and Wetherill (1958).

— — Half-life and beta spectrum of ^{87}Rb. Phys. Rev. **116**, 744—748 (1959).

FRIED, M., G. HAWKES, and W. Z. MACKIE: Rb to K relations in the soil-plant system. Soil Sci. Soc. Am. Proc. **23**, 360—362 (1959).
—, and J. C. NAGGLE: Multiple site uptake of individual cations by roots as affected by hydrogen ion. Plant Physiol. **33**, 139—144 (1958).
FRITZE, K., and F. STRESSMANN: Z. Naturforsch. **11** a, 277 (1956).
GALITSYN, M. S., and L. V. SLAVYANOVA: Rb in subsurface and surface waters of the Caspian depression. Dokl. Akad. Nauk. SSSR **165**, 678—681 (1965).
GAST, P. W.: Alkali metals in stone meteorites. Geochim. Cosmochim. Acta **19**, 1—4 (1960a).
— Limitations on the composition of the upper mantle. J. Geophys. Res. **65**, 1287—1297 (1960b).
— Terrestrial ratio of potassium to rubidium and the composition of the earth's mantle. Science **147**, 853—860 (1965).
GERASIMOVSKII, V. I.: Geochemical features of agpaitic nepheline-syenites. Chemistry of the Earth's Crust, vol. 1 (A. P. VINOGRADOV, ed.), p. 104—118.: Oldbourne Press 1966.
GLENDENIN, L. E.: Present status of the decay constants. Ann. N.Y. Acad. Sci. **91**, 166—176 (1961).
GLOVER, R. B.: The chemistry of thermal waters at Rotorua. New Zealand J. Sci. **10**, 70—96 (1967).
GOLDBERG, E. D.: Minor elements in sea water. Chem. Oceanography, vol. 1, chap. 5. Academic Press 1965.
—, and G. O. S. ARRHENIUS: Chemistry of Pacific pelagic sediments. Geochim. Cosmochim. Acta **13**, 153—212 (1958).
GOLDICH, S. S., E. G. LIDIAK, C. E. HEDGE, and F. G. WALTHALL: Geochronology of the midcontinent region, United States. J. Geophys. Res. **71**, 5389—5408 (1966).
GOLDING, R. M., and M. G. SPEER: Alkali ion analysis of thermal waters in New Zealand. New Zealand J. Sci. **4**, 203—213 (1965).
GORHAM, E., and D. SWAINE: The influence of oxidizing and reducing conditions upon the distribution of some elements in lake sediments. Limnol. Oceanog. **10**, 268—279 (1965).
GOLDSCHMIDT, V. M.: Geochemische Verteilungsgesetze der Elemente. IX. Die Mengenverhältnisse der Elemente und der Atomarten. Skrifter Norske Videnskaps-Akad. Oslo, I: Mat-Naturv. Kl. No. 4 (1937).
— Geochemistry. 730 pp. Oxford University Press 1954.
—, H. BAUER u. H. WITTE: Zur Geochemie der Alkalimetalle II. Nachr. Ges. Wiss. Goettingen, Math.-Physik. Kl. IV, N.F. **1**, No. 4, 39—55 (1934).
—, H. BERMAN, H. HAUPTMANN u. C. PETERS: Zur Geschichte der Alkalimetalle. Nachr. Ges. Wiss. Goettingen, Math.-Physik. Kl. III. **34**; IV **35**, 235—244 (1933).
GRAF, D. L.: Geochemistry of carbonate sediments and sedimentary carbonate rocks. Pt. III, Minor element distribution. Illinois State Geol. Surv., Circ. **301**, 71 pp. (1960).
GREEN, T. H.: High pressure experiments on the genesis of anorthosites: Petrology of the upper mantle. Department of Geophysics and Geochemistry, Australian National University, Publ. No. 444 (preprint), p. 206—233 (1966).
GRIFFIN, W. L., and V. RAMA MURTHY: Distribution of K, Rb, Sr and Ba in some minerals relevant to basalt genesis. Geochim. Cosmochim. Acta **33**, 1389—1414 (1969).
GUNN, B. M.: K/Rb and K/Ba ratios in Antarctic and New Zealand tholeiites and alkali basalts. J. Geophys. Res. **70**, 6241—6247 (1965).
— Modal and element variation in Antarctic tholeiites. Geochim. Cosmochim. Acta **30**, 881—920 (1966).
HAHN, O., u. M. ROTENBACH: Über die Radioaktivität des Rubidiums. Phys. Z. **20**, 194—202 (1919).
HAMILTON, E. I.: The isotopic composition of strontium in the Skaergaard intrusion, East Greenland. J. Petrol. **4**, 383—391 (1963).
— The geochemistry of the northern part of the Ilimaussaq intrusion. Medd. Grönland, Köbenhavn, 104 pp. (1964).
— Distribution of some trace elements and the isotopic composition of strontium in Hawaiian lavas. Nature **206**, 251—253 (1965).

HARRISS, R. C., and J. A. S. ADAMS: K—Rb fractionation in the sedimentary cycle. Bull. Geol. Soc. Am. (in press) (1969).
HART, S. R., and L. T. ALDRICH: Fractionation of potassium/rubidium by amphiboles: Implications regarding mantle composition. Science **155**, 325—327 (1967).
HAVARD, K. R.: Mineralogy and geochemistry: Epshaw Formation, southern Alberta, Unpubl. M. S. thesis, Univ. of Calgary (1967).
HAXEL, O., F. G. HOUTERMANS, and H. KEMMERICH: On the half-life of ^{87}Rb. Phys. Rev. **74**, 1886—1887 (1948).
HEDGE, C. E.: Variations in radiogenic strontium found in volcanic rocks. J. Geophys. Res. **71**, 6119—6126 (1966).
—, and F. G. WALTHALL: Radiogenic strontium-87 as an index of geologic processes. Science **140**, 1214—1217 (1963).
HEIER, K. S.: Petrology and geochemistry of high-grade metamorphic and igneous rocks on Langöy, Northern Norway. Norg. Geol. Undersokelse **207**, 246 pp. (1960).
— Geochemistry of the nepheline syenite on Stjernöy, North Norway. Norsk Geol. Tidsskr. **44**, 205—215 (1964).
— Rubidium/strontium and strontium 87/strontium 86 ratios in deep crustal material. Nature **202**, 477—478 (1964).
— A geochemical comparison of the Blue Mountain (Ontario, Canada) and Stjernöy (Finnmark, North Norway) nepheline syenites. Norsk Geol. Tidsskr. **45**, 41—52 (1965).
— Metamorphism and the chemical differentiation of the crust. Geol. Foren. Stockholm Forh. **87**, 249—256 (1965).
— Some crystallochemical relations of nephelines and feldspars on Stjernöy, North Norway. J. Petrol. **1**, 95—113 (1966).
—, and J. A. S. ADAMS: The geochemistry of the alkali metals. Phys. Chem. Earth. **5**, 253—381 (1964).
—, and C. BROOKS: Geochemistry and the genesis of the Heemskirk granite, West Tasmania. Geochim. Cosmochim. Acta **30**, 633—643 (1966).
—, B. W. CHAPPELL, P. A. ARRIENS, and J. W. MORGAN: The geochemistry of four Icelandic basalts. Norsk Geol. Tidsskr. **46**, 427—437 (1967).
—, and W. COMPSTON: K/Rb ratios of eclogites. Earth and Planetary Sci. Letters **1**, 293—294 (1966).
— —, and I. MCDOUGALL: Thorium and uranium concentrations, and the isotopic composition of strontium in the differentiated Tasmanian dolerites. Geochim. Cosmochim. Acta **29**, 643—659 (1965).
—, and S. R. TAYLOR: Distribution of Li, Na, K, Rb, Cs, Pb and Tl in southern Norwegian pre-Cambrian alkali feldspars. Geochim. Cosmochim. Acta **15**, 284—304 (1959).
— — A note on the geochemistry of alkaline rocks. Norsk Geol. Tidsskr. **44**, 197—203 (1964).
HELGESON, H. C.: Geologic and thermodynamic characteristics of the Salton Sea geothermal system. Am. J. Sci. **266**, 129—166 (1968).
HENDERSON, C. M. B.: Minor element chemistry of leucite and pseudoleucite. Mineral. Mag. **35**, 506—603 (1965).
HERZ, N., and C. V. DUTRA: Trace elements in alkali feldspars, Quadrilatero Ferrifero, Minas Gerais, Brazil. Amer. Mineral. **51**, 1593—1607 (1966).
HERZOG, L. F., and W. H. PINSON: Rubidium/strontium age, elemental and isotopic abundance studies of stony meteorites. Am. J. Sci. **254**, 555—566 (1956).
HIRST, D. M.: The geochemistry of modern sediments from the Gulf of Paria — II. The location and distribution of trace elements. Geochim. Cosmochim. Acta **26**, 1147—1188 (1962).
—, and K. C. DUNHAM: Chemistry and petrography of the Marl Slate of S. E. Durham, England. Econ. Geol. **58**, 912—940 (1963).
HORSTMAN, E. L.: The distribution of lithium, rubidium, and caesium in igneous and sedimentary rocks. Geochim. Cosmochim. Acta **12**, 1—28 (1957).
HURLEY, P. M.: Absolute abundance and distribution of Rb, K and Sr in the Earth. Geochim. Cosmochim. Acta **32**, 273—283 (1968).

Hurley, P. M., H. W. Fairbairn, W. H. Pinson, Jr.: Rb—Sr isotopic evidence in the origin of potash-rich lavas of Western Italy. Earth and Planetary Sci. Letters **5**, 301—306 (1966).

—, H. Hughes, G. Faure, H. W. Fairbairn, and W. H. Pinson: Radiogenic strontium-87 model of continent formation. J. Geophys. Res. **67**, 5315—5334 (1962).

Huster, E.: A redetermination of the half-life of Rb-87. Nuclear processes in geologic settings. Natl. Acad. Sci-Natl. Res. Council Publ. **400**, 195—202 (1956).

— Written communication to H. Hinterberger (1959).

Iiyama, J. T.: Etude experimentale de la distribution d'elements en traces entre deux feldspaths. Bull. Soc. Franc. Mineral. Crist. **91**, 130—140 (1968).

Kharkor, D. P., K. K. Turekian, and K. K. Bertine: Stream supply of dissolved silver, molybdenum, antimony, selenium, chromium, cobalt, rubidium, and cesium to the oceans. Geochim. Cosmochim. Acta **32**, 285—298 (1968).

Kemmerich, M.: Die Halbwertszeit des Rubidiums^{87}. Z. Physik **126**, 399—409 (1948).

Kolbe, P., and S. R. Taylor: Geochemical investigation of the granitic rocks of the Snowy Mountains area, New South Wales. J. Geol. Soc. Australia **13**, 1—25 (1968).

Kühn, R.: Rubidium als geochemisches Leitelement bei der lagerstättenkundlichen Charakterisierung von Carnalliten und natürlichen Salzlösungen. Neues Jahrb. Mineral. **5**, 107—115 (1963).

Lambert, I. B., and K. S. Heier: Geochemical investigations of high grade regional metamorphic and associated rocks in the Australian Shield. Lithos **1**, 30—53 (1968).

Lambert, R. St. J.: Isotopic age determinations on gneisses from the Tauernfenster, Austria. Geolog. Bundesanstalt, Verhandl. **1**, 16—27 (1964).

Landergren, S.: On the geochemistry of Swedish iron ores and associated rocks. A study of iron formation. Sveriges Geol. Undersokn. Ser. C., No. 496, **42**, No. 5 (1948).

Landis, E. R.: Uranium and other trace elements in Devonian and Mississippian black shales in the central midcontinent area. U.S. Geol. Surv. Bull. **1107-E**, 289—336 (1962).

Lange, I. M., R. C. Reynolds, and J. B. Lyons: K/Rb ratios in coexisting K-feldspars and biotites from some New England granites and metasediments. Chem. Geol. **1**, 317—328 (1966).

Leask, D. M.: The geochemistry of Precambrian argillites: Purcell system of Southern Alberta and British Columbia M. Sc. thesis, Univ. of Calgary (1967).

Lessing, P., and E. J. Catanzaro: ^{87}Sr/^{86}Sr ratios in Hawaiian lavas. J. Geophys. Res. **69**, 1599—1601 (1964).

—, R. W. Decker, and R. C. Reynolds, Jr.: Potassium and rubidium distribution in Hawaiian lavas. J. Geophys. Res. **68**, 5851—5855 (1963).

Lester, G.: Requirement for K by bacteria. J. Bacteriol. **75**, 426—428 (1958).

Leutz, H., H. Wenninger u. K. Ziegler: Die Halbwertszeit des ^{87}Rb. Z. Physik **169**, 409—416 (1962).

Lewis, G. M.: The natural radioactivity of rubidium. Phil. Mag. [7] **43**, 1070—1074 (1952).

Libby, W. F.: Simple absolute measurement technique for beta radioactivity. Application to naturally radioactive rubidium. Anal. Chem. **29**, 1566—1570 (1957).

Lidiak, E. G., G. K. Billings, and R. C. Harris: Potassium: rubidium ratios in North-Central Puerto Rican flow basalts. (unpubl.) (1965).

Little, E. J., and M. M. Jones: A complete table of electronegativities. J. Chem. Educ. **37**, 231 (1960).

Livingstone, D. A.: Chemical composition of rivers and lakes. U.S. Geol. Surv. Profess. Papers **440**-G (1963).

Long, L. E.: Preliminary Rb—Sr investigation of Tertiary granite and granophyre from Skye. Geochim. Cosmochim. Acta **28**, 1870—1873 (1964).

— Rb—Sr chronology of the Carn Chuinneag intrusion, Ross-Shire, Scotland. J. Geophys. Res. **69**, 1589—1597 (1964).

MacGregor, M. H., and M. L. Weidenbeck: Decay of Rb-87. Phys. Rev. **86**, 420 (1952).

— — The third forbidden beta-spectrum of Rb-87. Phys. Rev. **94**, 138 (1954).

Mahon, W. A. J.: Natural hydrothermal systems and the reaction of hot water with sedimentary rocks. New Zealand J. Sci. **10**, 206—221 (1967).

Mallette, M. F., P. M. Althouse, and C. O. Clagett: Biochemistry of plants and animals. New York: J. Wiley & Sons 1960.

Mason, B.: Meteorites. New York: John Wiley & Sons 1962.

— The carbonaceous chondrites. Space Sci. Reviews **1**, 621—646 (1962).

McDougall, I., and W. Compston: Strontium isotope composition and potassium—rubidium ratios in some rocks from Reunion and Rodriguez, Indian Ocean. Nature **207**, 252—253 (1965).

McIntire, W. C.: Trace element partition coefficients — a review of theory and applications to geology. Geochim. Cosmochim. Acta **27**, 1209—1264 (1963).

McLaughlin, R. J. W.: Geochemical changes due to weathering under varying climatic conditions. Geochim. Cosmochim. Acta **8**, 109—130 (1955).

McNair, A., and H. W. Wilson: Phil. Mag. **64**, 563 (1961).

Menzel, R. G., and W. R. Heald: Distribution of K, Rb, Cs, Ca and Sr within plants grown in nutrient soils. Soil Sci. **80**, 287—293 (1955).

Miller, J. P.: Solutes in small streams draining single rock types, Sangre de Cristo Range, New Mexico. U.S. Geol. Surv. Water Supply Papers **1535-F** (1961).

Mishchenko, V. S., and V. I. Orsa: On the distribution of Li, Rb, and K in the granitoids of the middle Dnjeper region. Geochem. Internat. **2**, 293—300 (1965).

Moorbath, S., and J. D. Bell: Strontium isotope abundance studies and rubidium-strontium age determinations on Tertiary igneous rocks from the Isle of Skye, North-West Scotland. Petr. **6**, 37—66 (1965).

—, and R. M. Shackleton: Isotopic ages from the Precambrian Mona complex of Anglesey, North Wales (Great Britain). Earth and Planetary Sci. Letters **1**, 113—117 (1966).

Morgan, J. W., and K. S. Heier: Uranium, thorium and potassium in six U.S.G.S. standard rocks. Earth and Planetary Sci. Letters **1**, 158—160 (1966).

Muehlberger, W. R., C. E. Hedge, R. E. Denison, and R. F. Marvin: Geochronology of the midcontinent region. United States. J. Geophys. Res. **71**, 5409—5426 (1966).

Mühlhoff, W.: Aktivität von Kalium und Rubidium gemessen mit dem Elektronenzählrohr. Ann. Physik **1**, 205—224 (1930).

Murthy, V. Rama, and W. L. Griffin: K-Rb fractionation by plagioclase feldspars. Chem. Geol., in press (1970).

Nicholls, G. D., and D. H. Loring: The geochemistry of some British carboniferous sediments. Geochim. Cosmochim. Acta **26**, 181—223 (1962).

Nockolds, S. R., and R. Allen: The geochemistry of some igneous rock series. Geochim. Cosmochim. Acta **4**, 105—142 (1953).

— — The geochemistry of some igneous rock series. Pt. II. Geochim. Cosmochim. Acta **5**, 245—285 (1954).

— — The geochemistry of some igneous rock series — III. Geochim. Cosmochim. Acta **9**, 34—77 (1956).

Orban, G.: Untersuchungen über die Radioaktivität der Alkalimetalle mit der Nebelstrahlmethode. Sitzber. Akad. Wiss. Wien, Math.-Naturw. Kl. Abt. IIa **140**, 121 (1931).

Ovchinnikova, G. V.: Geochimija **5**, 392 (1960).

Pauling, L.: The nature of the chemical bond. 450 pp. 2nd ed.: Cornell University Press 1948.

Pentcheva, E.: The distribution of rare and dispersed elements in Bulgarian saline underground waters. Compt. Rend. Acad. Bulgare Sci. **18** (2), 149—151 (1965).

Pidgeon, R. T., and W. Compston: The age and origin of the Cooma granite and its associated metamorphic zones, New South Wales. J. Petr. **6**, 193—222 (1965).

Pinson, W. H.: Trans. Am. Geophys. Union **35**, 360 (1954).

Pinson, W. H., Jr., J. A. Philpotts, and C. C. Schnetzler: K/Rb ratios in tektites. J. Geophys. Res. **70**, 2889—2894 (1965).

—, L. L. Schnetzler, E. Beiser, H. W. Fairbairn, and P. M. Hurley: Rb—Sr age of stony meteorites. Geochim. Cosmochim. Acta **29**, 455—466 (1965).

Powell, R. M., W. H. Pinson, Jr., H. W. Fairbairn, and R. F. Cormier: Test of the half-life of ^{87}Rb (Abstract). Bull. Geol. Soc. Am. **68**, 1782 (1957).

Rausch, W., u. W. Schmidt: Vorträge auf Gauvereinstagg. der Dtsch. Physik. Ges. in Bad Pyrmont im April 1960. Hinterberger 1960.

Rhodes, J. M.: On the chemistry of potassium feldspars in granitic rocks. Chem. Geol. **4**, 373—392 (1969).

RISTIC, S., S. ARSENIJEVIC, and V. MILUTINOVIC: Spectrochemical determination of Rb and Cs in the mineral waters of Niska Banja. Glassnik Hem. Drustva Beograd **21**, 283—290 (1956).

ROOKE, J. M.: Element distribution in some acid igneous rocks of Africa. Geochim. Cosmochim. Acta **28**, 1187—1198 (1964).

SAPPO, P. V.: The Rb and Cs contents of the strata water of a series of rocks in the Surakhany and Balakhany — Sabunchi-Romany fields. Trudy Azorbaidzhan. Nauch-Issled. Inst. po Dobyche Nef T., No. 10, 206—211 (1960).

SAHAMA, TH. G.: Spurenelemente der Gesteine im südlichen Finnisch-Lappland. Bull. Comm. Geol. Finlande **136**, 15—67 (1945).

SCHUMACHER, E.: Helv. Chim. Acta **36**, 538 (1956).

— Isolierung von K, Rb, Sr, Ba und seltenen Erden aus Steinmeteoriten. Helv. Chim. Acta **39**, 531—537 (1956).

SEN, N., S. R. NOCKOLDS, and R. ALLEN: Trace elements in minerals from rocks of the Southern California batholith. Geochim. Cosmochim. Acta **16**, 58—78 (1959).

SHAW, D. M.: A review of K—Rb fractionation trends by covariance analysis. Geochim. Cosmochim. Acta **32**, 573—601 (1968).

SHIELDS, W. R., E. L. GARNER, C. E. HEDGE, and S. S. GOLDICH: Survey of ^{85}Rb/^{87}Rb ratios in minerals. J. Geophys. Res. **68**, 2331—2334 (1963).

SMALES, A. A., T. C. HUGHES, D. MAPPER, C. A. J. MCINNES, and R. K. WEBSTER: The determination of rubidium and caesium in stony meteorites. Geochim. Cosmochim. Acta **28**, 209—233 (1964).

—, and L. SALMON: Determination by radioactivation of small amounts of Rb and Cs in sea water and related materials of geochemical interest. Analyst **80**, 37—47 (1955).

SMITH, K. C., K. C. PILLAI, T. F. CHOW, and T. R. FOLSOM: Determination of rubidium in seawater. Limnol. Oceanog. **10**, 226—232 (1965).

SPENCER, D.: Factors affecting element distributions in a Silurian graptolite band. Chem. Geol. **1**, 211—249 (1966).

SREEKUMARAN, C., K. C. PILLAI, and T. R. FOLSOM: The concentrations of lithium, potassium, rubidium and cesium in some western American rivers and marine sediments. Geochim. Cosmochim. Acta **32**, 1229—1234 (1968).

STEWART, F. H.: Marine evaporites. U.S. Geol. Surv. Profess. Papers **440**-Y (1963).

STRASSMANN, F., u. E. WALLING: Abscheidung des reinen Strontium Isotops 87. Ber. Deut. Chem. Ges., Abt. B **71**, 1—9 (1938).

STUEBER, A. M.: Abundances of K, Rb, and Sr isotopes in ultramafic rocks and minerals from western North Carolina. Geochim. Cosmochim. Acta **33**, 543—554 (1969).

—, and V. RAMA MURTHY: Strontium isotope and alkali element abundances in ultramafic rocks. Geochim. Cosmochim. Acta **30**, 1243—1259 (1966).

SUESS, H. E., and H. C. UREY: Abundances of the elements. Rev. Mod. Phys. **28**, 53—74 (1956).

TATSUMUTO, M., C. E. HEDGE, and A. E. J. ENGEL: Potassium, rubidium, strontium, thorium, uranium and the ratio of ^{87}Sr to ^{86}Sr in oceanic tholeiitic basalt. Science **150**, 886—888 (1956).

TAUBENECK, W. H.: An appraisal of some potassium—rubidium ratios in igneous rocks. J. Geophys. Res. **70**, 475—478 (1965).

TAYLOR, S. R.: Occurrence of alkali metals in some Gulf of Mexico sediments; ammended Rb values and K/Rb ratios. J. Sediment. Petrol. **30**, 217—320 (1960).

— Trace element abundances and the chondritic Earth model. Geochim. Cosmochim. Acta **28**, 1989—1998 (1964).

—, and L. H. AHRENS: The significance of K/Rb ratio for theories of tektite origin. Geochim. Cosmochim. Acta **15**, 370—372 (1959).

—, C. H. EMELEUS, and C. S. EXLEY: Some anomalous K/Rb ratios in igneous rocks and their petrological significance. Geochim. Cosmochim. Acta **70**, 224—229 (1956).

TUPPER, W. M., and D. H. LORING: A short note on the relative abundance of Rb and Sr in Vitrain ashes from coals in Nova Scotia. Geochim. Cosmochim. Acta **24**, 314—316 (1961).

TUREKIAN, K. K., and K. H. WEDEPOHL: Distribution of the elements in some major units of the earth's crust. Bull. Geol. Soc. Am. **72**, 175—192 (1961).

UREY, H. C.: The abundances of the elements. Phys. Rev. **88**, 248—252 (1952).

UZUMASA, Y.: Chemical investigations of hot springs in Japan; Rb and Cs in hot springs. Nippon Kagaku Zasshi **84**, 726—731 (1963).

VINOGRADOV, A. P.: The geochemistry of rare and dispersed chemical elements in soils (English translation). New York: Consultants Bureau 1959.

WAGER, L. R., and R. L. MITCHELL: Preliminary observations on the distribution of trace elements in the rocks of the Skaergaard intrusion, Greenland. Mineral. Mag. **26**, 283—296 (1943).

WEBBER, G. R., and J. N. JELLEMA: Comparison of chemical composition of soils and bedrock of Mont St. Hilaire, Quebec, Can. J. Earth Sci. **2**, 44—58 (1965).

WEBER, J. N.: Trace element composition of dolostones and dolomites and its bearing on the dolomite problem. Geochim. Cosmochim. Acta **28**, 1817—1868 (1964).

WEBSTER, R. K., J. W. MORGAN, and A. A. SMALES: Some recent Harwell analytical work on geochronology. Trans. Am. Geophys. Union **38**, 543—546 (1957).

— — — Caesium in chondrites. Geochim. Cosmochim. Acta **15**, 150—152 (1958).

WEDEPOHL, K. H.: Spurenanalytische Untersuchungen an Tiefseetonen aus dem Atlantik. Ein Beitrag zur Deutung der geochemischen Sonderstellung von pelagischen Tonen. Geochim. Cosmochim. Acta **18**, 200—231 (1959).

WELBY, C. W.: Occurrence of alkali metals in some Gulf of Mexico sediments. J. Sediment Petrol. **28**, 431—452 (1958).

WETHERILL, G. W.: Radioactive decay constants and energies. Handbook of Physical Constants — Revised edit. (S. P. CLARK). Geol. Soc. Am. Mem. **97**, 587 pp. (1966).

WHITE, A. J. R.: Genesis of migmatites from the Palmer region of South Australia. Chem. Geol. **1**, 165—200 (1966).

WHITE, D. E.: Saline waters of sedimentary rocks. Am. Ass. Petr. Geol., Mem. 4 (YOUNG and GALLEY, eds.) (1965).

WILKINSON, J. F. G.: The geochemistry of a differentiated teschenite sill near Gunnedah, N.S.W. Geochim. Cosmochim. Acta **16**, 123—150 (1959).

WILLIAMS, H. H.: Some aspects of ion exchange in shales, unpubl. M.S. thesis, Univ. of Calgary (1967).

YOUNG, E. J.: Bull. Geol. Soc. Am. **65**, 1329 (1954).

ZARTMAN, R. E.: A geochronologic study of the Lone Grove pluton from the Llano uplift, Texas. J. Petrol. **5**, 359—408 (1964).

— Rubidium—strontium age of some metamorphic rocks from the Llano uplift, Texas. J. Petrol. **6**, 28—36 (1965).

Revised manuscript received: January 1970

Strontium 38

A	K. FISCHER	(Lehrstuhl für Kristallographie, Universität des Saarlandes, Saarbrücken, Germany)
B	Z. E. PETERMAN and C. HEDGE	(United States Geological Survey, Denver, Col. 80225, U.S.A.)
C	B. H. MASON	(National Museum of Natural History, Washington, D. C. 20560, U.S.A.)
D	A. M. STUEBER	(Dept. of Geology, Miami University, Oxford, Ohio 45056, U.S.A.)
E, F	G. FAURE	(Dept. of Geology and Mineralogy, The Ohio State University, Columbus, Ohio 43210, U.S.A.)
G	K. H. WEDEPOHL	(Geochemisches Institut der Universität, Göttingen, Germany)
H, I	J. M. GIESKES	(Scripps Institution of Oceanography, La Jolla, Calif. 92093, U.S.A.)
K, L	J. VEIZER	(Department of Geology, University of Ottawa, Ottawa, Ontario K1N 6N5, Canada)
M, N	G. P. SIGHINOLFI	(Istituto di Mineralogia e Petrologia, Università di Modena, Italy)

38-A. Crystal Chemistry

I. General

At present only about 25 strontium minerals are known. In most of them, Sr is bound to oxygen (or OH, H_2O) exclusively and its valency is always two. The ionic radius of Sr^{2+} is between those of Ca^{2+} and Ba^{2+} and is somewhat smaller than the Pb^{2+} radius. Therefore, in a number of minerals and artificial compounds, Sr is replaced by these atoms and also replaces them; for example, celestite $Sr[SO_4]$ is isotypic with barite $Ba[SO_4]$, anglesite $Pb[SO_4]$ and baryto-celestite, $(Sr, Ba)[SO_4]$. Strontianite $Sr[CO_3]$ is isotypic with witherite $Ba[CO_3]$, cerussite $Pb[CO_3]$, and aragonite $Ca[CO_3]$ (but not with calcite which forms an isomorphous series with carbonates of smaller cations). Belovite, $Sr_5(OH)[PO_4]_3$, belongs to the apatite series, and in natural brewsterite, $(Sr, Ba, Ca)[Al_2Si_6O_{16}] \cdot 5H_2O$, Sr is partially replaced by Ba and Ca. Synthetic $SrCu[Si_4O_{10}]$ and $Sr_2[Al^{[4]}SiO_7]$ are members of the gillespite and melilite groups, respectively. In addition, $Eu(OH)_2 \cdot H_2O$ is isostructural with both $Sr(OH)_2 \cdot H_2O$ and $Ba(OH)_2 \cdot H_2O$ (Bärnighausen, 1966).

In Table 38-A-1, a few minerals and selected artificial compounds with fairly or well known crystal structures are listed. The coordination numbers of Sr^{2+} vary from 6 to 12 (or possibly even more, cf. $Sr[B_4O_7]$). A coordination number of 9 appears to be rather stable in a coordination polyhedron which can be described as a trigonal prism with 6 atoms at the corners and 3 additional atoms above the centers of its prismatic faces (cf. also computations of Dunitz and Orgel, 1960). Besides a cubic coordination polyhedron in SrF_2, other polyhedrons with a coordination number of 8 are listed in Table 38-A-1. The polyhedrons for higher coordination numbers are in most cases irregular. They appear to be dominated by the anions of the crystal structures which provide the stronger binding forces. The reported Sr—(O, OH, H_2O) distances range from 2.5 to 3.25 Å.

II. Crystal Chemistry of Some Minerals and Selected Artificial Compounds

In $Sr(OH)_2 \cdot H_2O$, the eight-fold coordination of Sr by oxygen atoms is dominated by the Sr—(OH, H_2O) bonds (Bärnighausen and Weidlein, 1967). The coordination polyhedron can be described in two different ways: as a distorted trigonal prism of (OH) plus two H_2O molecules with slightly longer distances (see Table 38-A-1) or as a distorted quadratic antiprism (as also reported for $Sr(OH)_2 \cdot 8H_2O$ and $Sr[Cr_2O_7]$). A well-defined coordination number of 9 has been found for two independent Sr positions in $Sr[B(OH)_4]_2$, and for a single Sr in $SrH[AsO_4] \cdot H_2O$ (Sr-haidingerite) and in $Sr[VO_3]_2 \cdot 4H_2O$. The coordination polyhedron is very similar in all the three compounds and no systematic difference of Sr—O distances between the atoms at the corners of the trigonal prism and those

Table 38-A-1. *Coordination of Sr in some crystal structures*

Mineral or artificial compound	Ligands and their distances to Sr in Å	Coordination polyhedron	References
$Sr^{[6]}O^{[6]}$	6 O 2.58	regular octahedron	1
$Sr^{[8]}(OH)_2 \cdot H_2O$	6 OH 2.59—2.69 2 H_2O 2.70; 2.74	distorted trigonal prism above top of two side faces of this prism	2
$Sr^{[8]}(OH)_2 \cdot 8H_2O$	8 H_2O 2.60	quadratic antiprism	3
$Sr^{[12]}Ti^{[6]}O_3$	8 O 2.76	cube-octahedron	4
$Sr^{[8]}F_2{}^{[4]}$	8 F 2.60	cube	5
$Sr^{[9]}[B^{[4]}(OH)_4]_2$	Sr_1: 9 OH 2.53—2.97 Sr_2: 9 OH 2.52—2.84	distorted trigonal prism with 3 OH on top of prismatic faces	6
$Sr^{[9]}[B_4{}^{[4]}O_7]$	9 O 2.52—2.84 (6 O 3.04—3.20)	see text	7
$Sr^{[10]}[B_3{}^{[3]}B_3{}^{[4]}O_9.(OH)_2] \cdot 3H_2O$ tunellite	6 O 2.64—2.90 4 H_2O 2.61—2.98	irregular, see text	8
$Sr^{[12]}[SO_4]$ celestite	10 O 2.48—2.99 2 O 3.25	see text	9
$Sr^{[8]}[Cr_2O_7]$	8 O 2.53—2.66	distorted quadratic antiprism	10
$Sr^{[12]}Sr_2{}^{[10]}[PO_4]_2$	6 + 6 O 2.63; 3.10 10 O 2.48—2.72	distorted icosahedron, see text	11
$Sr^{[9]}H[AsO_4] \cdot H_2O$ (Sr-haidingerite)	7 O 2.57—3.12 2 H_2O 2.68; 2.69	distorted trigonal prism with "centered" side faces	12
$Sr^{[9]}[V^{[4]}O_3]_2 \cdot 4H_2O$	6 H_2O 2.62; 2.73 3 O 2.54—2.74	distorted trigonal prism on top of the prismatic faces	13

References: 1. GERLACH, SB 1913—1928, 118. 2. BÄRNIGHAUSEN and WEIDLEIN (1967). 3. SMITH (SR 1954, 520), PREISINGER (*ibid.*). 4. DONNAY *et al.* (1963). 5. VAN ARKEL (SB 1913—1928, 186). 6. KRAVCHENKO (1965), cf. KUTSCHABSKY (1965). 7. PERLOFF and BLOCK (1966); cf. WITZMANN and BEULICH (1965), KROGH-MOE (1964). 8. CLARK (1964). 9. GARSKE and PEACOR (1965). 10. WILHELMI (1966). 11. ZACHARIASEN (SR 1947—1948, 388). 12. BINAS (1966). 13. SEDLACEK and DORNBERGER-SCHIFF (1965).

on top of the prismatic faces could be detected (cf., however, $Sr(OH)_2 \cdot H_2O$). The coordination number in $Sr[B_4O_7]$ is not well defined; in addition to the 9 oxygen atoms at a distance of 2.84 Å or less (forming an irregular polyhedron, point symmetry m), three oxygen atoms occur at distances of 3.04 and 3.05 Å, two oxygen atoms at 3.15 Å and one oxygen atom at 3.20 Å. A coordination number of 10 has been reported for tunellite, $Sr[B_6O_9(OH)_2] \cdot 3H_2O$. The coordination polyhedron, which fills "holes" in the borate sheets, has no symmetry and exhibits no simple geometrical form. For information on the two different coordination polyhedrons of $Sr_3[PO_4]_2$, which is isostructural with $Ba_3[PO_4]_2$, see subsection 56-A-II. The (idealized) 12-fold coordination in celestite consists of a five-membered ring on one side and a parallel six-membered ring on the other side, the latter having one oxygen on top of its center. For additional information on the 10- or 12-fold coordination, see subsection 56-A-II (barite). In the zeolite brewsterite, $(Sr, Ba, Ca)^{[9]}[Al_2Si_6O_{16}] \cdot 5H_2O$, the coordination polyhedron must not be typical for Sr because in its usual cation position substantial proportions of Ba and Ca occur (PERROTTA and SMITH, 1964).

Revised manuscript received: July 1970

References: Section 38-A

BÄRNINGHAUSEN, H.: Gitterkonstanten und Raumgruppe der isotypen Verbindungen $Eu(OH)_2 \cdot H_2O$, $Sr(OH)_2 \cdot H_2O$ und $Ba(OH)_2 \cdot H_2O$. Z. Anorg. Allgem. Chem. **342**, 233 (1966).

— WEIDLEIN, J.: Die Kristallstruktur von Strontiumhydroxid-Monohydrat. Acta Cryst. **22**, 252 (1967).

BINAS, H.: Die Struktur des Sr-Haidingerits, $SrHAsO_4 \cdot H_2O$. Z. Anorg. Allgem. Chem. **347**, 140 (1966).

CLARK, J. R.: The crystal structure of tunellite, $SrB_6O_9(OH)_2 \cdot 3H_2O$. Am. Mineralogist **49**, 1549 (1964).

DONNAY, J. D. H., DONNAY, G., COX, E. G., KENNARD, O., KING, M. V. (eds.): Crystal Data Determinative Tables, 2nd ed. Am. Cryst. Assoc., Monograph No. 5 (1963).

DUNITZ, J. D., ORGEL, L. E.: Stereochemistry of ionic solids. Advan. Inorg. Chem. Radiochem. **2**, 1 (1960)

GARSKE, D., PEACOR, D. R.: Refinement of the structure of celestite $SrSO_4$. Z. Krist. **121**, 204 (1965).

KRAVCHENKO, V. B.: The crystal structure of the monoclinic modification of $SrB_2O_4 \cdot 4H_2O = Sr[B(OH)_4]_2$. Zh. Strukt. Khim. **6**, 835 (1965) (translated).

KROGH-MOE, J.: The crystal structure of strontium diborate, $SrO \cdot 2B_2O_3$. Acta Chem. Scand. **18**, 2055 (1964).

KUTSCHABSKY, L.: Zur Kristallstruktur des $Sr[B(OH)_4]_2$. Z. Chem. **5**, 110 (1965).

PERLOFF, A., BLOCK, S.: The crystal structure of the strontium and lead tetraborates, $SrO \cdot 2B_2O_3$ and $PbO \cdot 2B_2O_3$. Acta Cryst. **20**, 274 (1966).

PERROTTA, A. J., SMITH, J. V.: The crystal structure of brewsterite. Acta Cryst. **17**, 857 (1964).

SEDLACEK, P., DORNBERGER-SCHIFF, K.: Das Strukturprinzip des Strontiummetavandat, $Sr(VO_3)_2 \cdot 4H_2O$. Acta Cryst. **18**, 407 (1965).

WILHELMI, K.-A.: The crystal structure of strontium dichromate, $SrCr_2O_7$. Arkiv Kemi **26**, 149 (1966).

WITZMANN, H., BEULICH, W.: Beitrag zur Struktur wasserfreier Strontiumborate. Naturwissenschaften **52**, 157 (1965).

Revised manuscript received: July 1970

38-B. Isotopes in Nature*

I. Introduction

There are four naturally occurring isotopes of strontium with the following approximate abundances: $^{84}Sr = 0.55\%$, $^{86}Sr = 9.75\%$, $^{87}Sr = 6.96\%$, $^{88}Sr = 82.74\%$.

The isotopic abundances of ^{84}Sr, ^{86}Sr, and ^{88}Sr are constant in nature. These species are neither radioactive nor the decay products of any naturally radioactive isotope. Geological processes do not produce any fractionation of strontium isotopes that can be detected with present analytical techniques. There are small natural variations in the abundance of ^{87}Sr, however, due to variable increments of ^{87}Sr produced by the radioactive beta decay of ^{87}Rb which constitutes about 28 percent of natural rubidium (see Section 37-B). The variations of ^{87}Sr are small because rubidium occurs in low abundance in most natural materials and the half-life of ^{87}Rb is long (50×10^9 years). The ^{87}Rb—^{87}Sr chronometer has been extremely useful in determining geological and cosmological ages and also the isotopic composition of strontium has been useful as a tracer of various geological processes.

a) Analytical Methods

Strontium isotopic compositions are determined by solid source mass spectrometry, and strontium from natural samples is first purified and concentrated, usually through the use of ion exchange resins.

Variations in ^{87}Sr are currently expressed as variations in the $^{87}Sr/^{86}Sr$ ratio (as ^{86}Sr is constant in nature). Although natural terrestrial processes do not fractionate strontium isotopes, this is observed in mass spectrometric analyses. The amount of instrument-induced fractionation will vary with the mode of ionization used and various other analytical parameters, but fractionation in the $^{87}Sr/^{86}Sr$ ratio is usually removed by "normalizing" this ratio until it corresponds to a $^{86}Sr/^{88}Sr$ ratio of 0.1194.

Analytical uncertainties in $^{87}Sr/^{86}Sr$ have been markedly reduced during the last decade as instruments and analytical techniques have improved. Most laboratories now quote precision values in the range of 0.03 to 0.07 percent, and analytical uncertainties in some analyses of lunar samples are reported as being less than 0.01 percent.

There are three standards which are commonly used in strontium isotope work: ocean water, which has a $^{87}Sr/^{86}Sr$ of 0.7091 (Wasserburg *et al.*, 1969; Hildreth and Henderson, 1971); Eimer and Amend $SrCO_3$, Lot no. 492327, which has a $^{87}Sr/^{86}Sr$ of 0.7080 (Hildreth and Henderson, 1971); and the newer National Bureau of Standards $SrCO_3$, no. 987, which has a $^{87}Sr/^{86}Sr$ of 0.7101 (National Bureau of Standards, 1971).

---

* Publication authorized by the Director, U.S. Geological Survey.

b) Principles of the Use of Strontium as an Isotope Tracer

The use of strontium isotopes as natural tracers in geological processes was an outgrowth of Rb—Sr geochronology. GAST (1960) used the initial $^{87}Sr/^{86}Sr$ ratio of Rb-poor meteorites as the primordial value for the earth and then calculated Rb/Sr of the earth's mantle from the difference between this primordial value and that of young basalts which presumably were derived from the mantle. The pertinent equation for such calculations is:

$$(^{87}Sr/^{86}Sr)_t = (^{87}Sr/^{86}Sr)_i + (^{87}Rb/^{86}Sr)_t\, e^{\lambda t}\text{-}1,$$

where the subscript i indicates the initial ratio and t the ratio after the elapse of a given amount of time. Lambda (λ) has a value of 1.39×10^{-11}/year. The above equation can be approximated by:

$$(^{87}Sr/^{86}Sr)_t = (^{87}Sr/^{86}Sr)_i + 2.9\ (Rb/Sr)\ \lambda t,$$

because ^{87}Rb is a known portion of total Rb, ^{86}Sr is nearly a constant portion of total Sr, and because the half-life of ^{87}Rb is so long that the exponential in the decay can often be ignored.

The data that were available to GAST (1960) were limited in quantity and quality, but he was able to conclude that the Rb/Sr of the mantle was in the range of 0.014 to 0.060. GAST further concluded, from available geochemical data, that the Rb/Sr of average crustal rocks was an order of magnitude higher than the value he had calculated for the earth's mantle. In other words, $^{87}Sr/^{86}Sr$ in the crust should increase at a rate about 10 times that of the rate of increase in the mantle. This observation led to the use of strontium isotopes in numerous studies which have attempted to determine whether a particular igneous rock was derived from the mantle or by efusion of pre-existing crust.

HURLEY *et al.* (1962) applied these concepts to igneous rocks of diverse ages and oncluded that the sialic crust of the North American continent has grown thro ugh

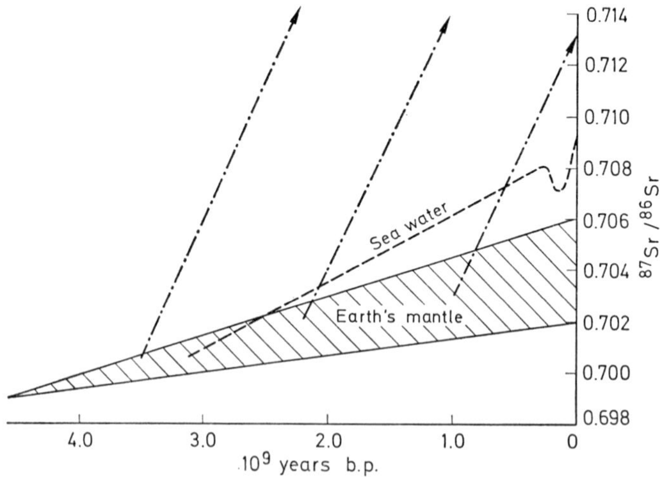

Fig. 38-B-1. A generalized diagram of the evolution of $^{87}Sr/^{86}Sr$ with time, in the earth's mantle, sea water, and three segments of continental crust (dot-dash lines — each having a Rb/Sr ratio of 0.20)

geological time by additions of new material from the mantle, and that most reworked crustal rocks must be relatively young.

The precision in $^{87}Sr/^{86}Sr$ values obtained prior to 1962 was not sufficient, however, to measure the progressive increase in initial $^{87}Sr/^{86}Sr$ ratios through geological time. Such a plot was first made by HEDGE and WALTHALL (1963) and is presented in generalized form in Fig. 38-B-1.

Initial $^{87}Sr/^{86}Sr$ ratios may be measured directly in young rocks or in older rocks that have low Rb/Sr ratios. For other rocks, the initial $^{87}Sr/^{86}Sr$ ratio may be calculated if the age, Rb/Sr, and present $^{87}Sr/^{86}Sr$ ratio are known.

A more lengthy development of the principles and applications of strontium isotopes was presented recently by FAURE and POWELL (1972).

II. Primordial $^{87}Sr/^{86}Sr$ Ratios

The types of stone meteorites known as basaltic achondrites have very low Rb/Sr ratios and hence these can be used to estimate the strontium isotopic composition at the time of formation of planetary objects in the solar system. The precision of these measurements has improved over the years and has gradually tightened about an initial $^{87}Sr/^{86}Sr$ value near 0.699 (SCHUMACHER, 1956; HERZOG and PINSON, 1956; WEBSTER *et al.*, 1957; GAST, 1960; GAST, 1962; HEDGE and WALTHALL, 1963). Recently PAPANASTASSIOU and WASSERBURG (1969) have made very precise initial $^{87}Sr/^{86}Sr$ determinations on six basaltic achondrites and found that they all had the same initial ratio of 0.69898 ± 0.00006. Most meteorites, other than basaltic achondrites, have high enough Rb/Sr ratios to have produced a significant enrichment in $^{87}Sr/^{86}Sr$ and thus will yield a Rb—Sr age. With only a few exceptions, all of the Rb—Sr meteorite ages are essentially 4.6×10^9 years (see PAPANASTASSIOU and WASSERBURG, 1969, for summary and discussion).

Initial $^{87}Sr/^{86}Sr$ ratios have been determined for numerous lunar basalts, whose ages range from 3.1 to 4.0 b.y. (c.f. WASSERBURG and PAPANASTASSIOU, 1971). Many of these lunar initial ratios are very primitive and approach closely the initial ratio of achondritic meteorites. The low Rb/Sr ratio of most lunar basalts and their low initial $^{87}Sr/^{86}Sr$ ratios both indicate that the Rb/Sr ratio of the source material of lunar basalts is quite low.

III. The Mantle

Probably the best insight into the present-day strontium isotopic composition of the earth's mantle has come from the measurements of $^{87}Sr/^{86}Sr$ of young, oceanic basaltic rocks. These rocks are rich in strontium and are therefore not readily affected by any natural contamination caused by assimilation of strontium from sources in the earth's crust. Also the young lavas which erupted on oceanic islands presumably have issued only through earlier oceanic basalt; assimilation of wallrocks thus should not significantly alter the $^{87}Sr/^{86}Sr$ values.

a) Oceanic Basalts

The available $^{87}Sr/^{86}Sr$ ratios of oceanic basalts are summarized in Fig. 38-B-2. The values range from a low of 0.7012 for a tholeiite from the Juan de Fuca Rise (HEDGE and PETERMAN, 1970) to a high of 0.7066 for an alkaline basalt from Samoa

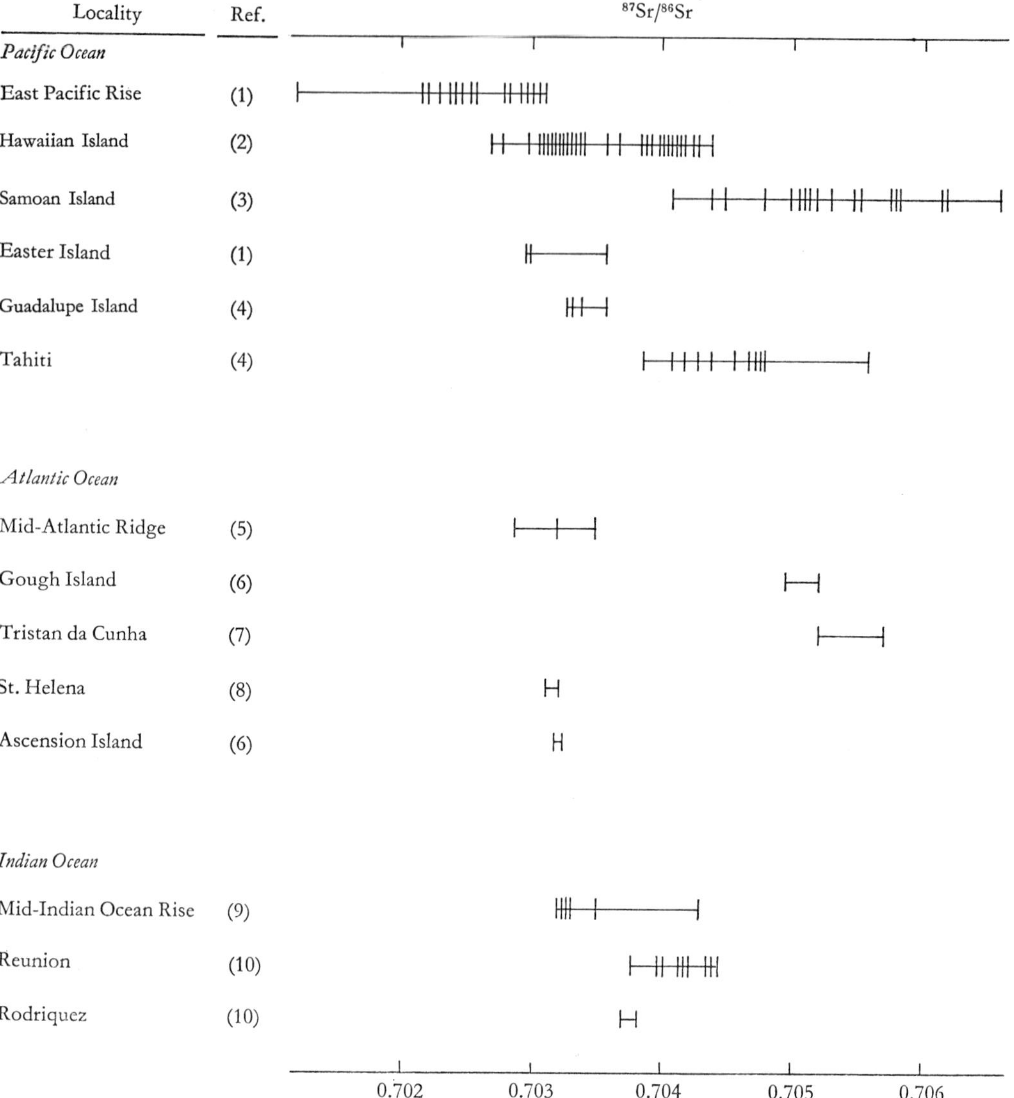

Fig. 38-B-2. $^{87}Sr/^{86}Sr$ values in oceanic basalts. Data sources are: 1, Hedge and Peterman (1970); 2, Powell *et al.* (1965), Hedge (1966), Powell and DeLong (1966); 3, Hubbard (1971), Hedge *et al.* (1972); 4, Peterman and Hedge (1971); 5, Tatsumoto *et al.* (1965); 6, Gast *et al.* (1964); 7, Bence and Hurley (1967); 8, Hedge (1966); 9, Subbarao and Hedge (1973); 10, McDougall and Compston (1965)

(Hedge *et al.*, 1972). Within each ocean basin, the low-potassium ridge tholeiites tend to have the lowest $^{87}Sr/^{86}Sr$ ratios. There are also some data which indicate that tholeiites from various oceanic spreading centers may have differing $^{87}Sr/^{86}Sr$ ratios. Hart (1971) found a small difference in $^{87}Sr/^{86}Sr$ ratio between composite samples from the Mid-Atlantic Ridge and the East Pacific Rise, with the Mid-

Atlantic Ridge composite having the higher ratio. There is certainly a suggestion of a difference in the means of the two sets of data shown in Fig. 38-B-2. More distinctive, however, is the difference between the theoliites of the East Pacific Rise (including the Gordo and Juan de Fuca Rises) and those of the Mid-Indian Ocean Rise.

Among the island basalts, the more potassic basalts such as those from Samoa, Gough and Tristan da Cunha, tend to have the higher $^{87}Sr/^{86}Sr$ ratios (Hedge and Peterman, 1970; Peterman and Hedge, 1971). There are exceptions, however, as for example, the more alkaline rocks from the Hawaiian Islands which have lower $^{87}Sr/^{86}Sr$ ratios than many of the tholeiites (Powell *et al.*, 1965; Powell and DeLong, 1966).

The Samoan data are remarkable in that there is a rather orderly progression of increasing $^{87}Sr/^{86}Sr$ ratio along the length of the island chain (Hedge *et al.*, 1972). Simple patterns of $^{87}Sr/^{86}Sr$ varying in space or time have not yet been reported in other island groups.

b) Evolution of the Mantle $^{87}Sr/^{86}Sr$ Ratio

$^{87}Sr/^{86}Sr$ measurements of oceanic basalts indicate that the earth's mantle has a $^{87}Sr/^{86}Sr$ ratio that today ranges from 0.702 to 0.706. The primordial value, as indicated by studies of meteoritic and lunar material, was lower (0.699) and probably had very little, if any, variation. The evolution of the mantle ratio has been of course directly tied to the evolution of the earth as a whole and Sr isotopes are an important constraint that must be satisfied in earth models.

The simplest model of mantle-strontium evolution assumes that the mantle was originally homogeneous in $^{87}Sr/^{86}Sr$ (at a value of 0.699), but that Rb/Sr varied from about 0.016 to 0.038 (these are the ratios necessary to produce the present ratios of 0.702 to 0.706 over a 4.6 b.y. span of time). Volcanism would then tap various parts of the mantle producing magmas with varying $^{87}Sr/^{86}Sr$ ratios. This is, in effect, an "infinite reservoir" model in that only previously untouched parts of the mantle would be forming magmas (Fig. 38-B-3a).

Such a simple model was dealt a severe blow when it was discovered (Tatsumoto *et al.*, 1965) that tholeiitic basalts from oceanic rises did not have sufficient

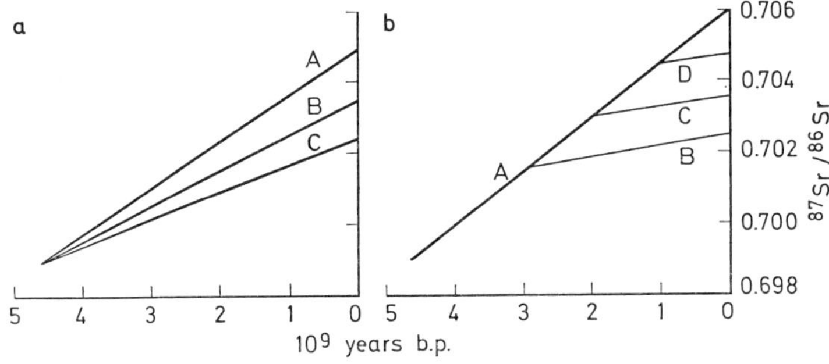

Fig. 38-B-3. Evolution of $^{87}Sr/^{86}Sr$ in the earth's mantle according to the two models. Model (a) is the "infinite reservoir" model in which different parts of the mantle (A, B, and C) each have unique Rb/Sr ratios. Model (b) is a "depletion" model in which the undepleted mantle is represented by line A; B, C, and D are mantle segments which have been depleted at various times in the past

Rb/Sr ratios to have generated even their low $^{87}Sr/^{86}Sr$ ratios over 4.6 b.y. This observation has been confirmed by later studies (HEDGE and PETERMAN, 1970; HART, 1971) and was an important factor leading to various "depletion" models for the mantle. GAST (1968) proposed a depletion model in which ocean-ridge tholeiites are derived from a mantle source that has been depleted in Rb (and other large cations) by previous magmatism. Such depletion models have gained considerable acceptance in accounting for chemical and isotopic differences within basalts (c.f. HEDGE and NOBLE, 1971; PETERMAN *et al.*, 1971; HART *et al.*, 1972). According to this model (Fig. 38-B-3b), basalts with relatively high $^{87}Sr/^{86}Sr$ ratios can be explained as being derived from a mantle that has been either less depleted or depleted more recently than the source of low $^{87}Sr/^{86}Sr$ basalts (PETERMAN and HEDGE, 1971).

ARMSTRONG (1968) has proposed a fundamentally different model in which crustal material is recycled through the mantle. According to this model the mantle material itself has very low Rb/Sr and $^{87}Sr/^{86}Sr$ ratios and higher values of these ratios are found by tapping portions of the mantle which include recycled crustal material.

c) Ultramafic Rocks

Strontium isotopic analyses of ultramafic rocks are another means of understanding something about the Rb—Sr system in the mantle. This approach is complicated, however, by the very low strontium contents of many ultramafic rocks, which makes them susceptible to natural or laboratory contamination. Almost any natural or laboratory contaminant will have a higher $^{87}Sr/^{86}Sr$ ratio than mantle strontium and therefore contamination will tend to raise the value of ultramafic samples. This point is stressed because many of the ultramafic samples analyzed have $^{87}Sr/^{86}Sr$ ratios that are higher than any that have been found in oceanic basalts.

This discussion of strontium isotopes in ultramafic rocks may be divided into two groups: 1. ultramafic inclusions in basalts and kimberlites; and 2, alpine-type ultramafics.

The $^{87}Sr/^{86}Sr$ ratio of ultramafic inclusions ranges from 0.7025 to 0.7095. An appreciable number of these inclusions have $^{87}Sr/^{86}Sr$ higher than those that have been found in oceanic basalts; however, the ratios of continental basalts do cover this entire range. More important, the $^{87}Sr/^{86}Sr$ ratio of the inclusions is, in many instances, higher than that of the basalt in which they are included (LEGGO and HUTCHINSON, 1968; LAUGHLIN *et al.*, 1971; HUTCHINSON and DAWSON, 1970; KUDO *et al.*, 1972; PAUL, 1971). In some cases the ratio of the inclusions is like that of the basalts (e.g. STUEBER and MURTHY, 1966; O'NEIL *et al.*, 1970), but inclusions having $^{87}Sr/^{86}Sr$ lower than the enclosing basalts are rare (DASCH and GREEN, 1972). Kimberlites tend to have very high ratios (MITCHELL and CROCKETT, 1971) — in many instances higher than the contained xenoliths (MANTON and TATSUMOTO, 1971). However, BERG and ALLSOPP (1972), working with samples carefully selected for freshness, have recently found lower $^{87}Sr/^{86}Sr$ ratios in kimberlites than have previously been reported.

Another intriguing aspect about the $^{87}Sr/^{86}Sr$ ratios of ultramafic inclusions is that the various mineral phases commonly have different strontium isotopic compositions (ALLSOPP *et al.*, 1968; PETERMAN *et al.*, 1970a; COMPSTON and LOVERING, 1969; PAUL, 1971; KUDO *et al.*, 1972; DASCH and GREEN, 1972).

When different xenoliths from one suite, or the various minerals from one xenolith, have varying $^{87}Sr/^{86}Sr$ ratios, there is usually some positive correlation with Rb/Sr which suggests an "age" of the order of 10^9 years. The usual interpretation is that such xenoliths are accidental fragments of shallow mantle material which were picked up by the ascending magma.

Many of the alpine-type ultramafic rocks also have higher $^{87}Sr/^{86}Sr$ ratios than oceanic basalts (Stueber and Murthy, 1966; Stueber, 1969; Lanphere, 1968; Bonatti *et al.*, 1970). The Rb and Sr concentrations in these rocks tend to be lower than those of ultramafic inclusions, and their Rb/Sr ratios show little correlation with their $^{87}Sr/^{86}Sr$ ratios. The interpretation most often given to strontium isotopic data from alpine ultramafic rocks is that they represent depleted, residual mantle which has yielded magma at some time well in the geological past. A long time interval between depletion and emplacement of the ultramafic bodies into the crust is required to produce the $^{87}Sr/^{86}Sr$ differences between the alpine ultramafic rocks and undepleted mantle (as inferred from oceanic basalts).

IV. The Continental Crust

The applicability of strontium isotopes as geological tracers in studying processes of crustal formation and evolution is based largely on the partitioning of Rb and Sr between the crust and mantle. Therefore, it is important to know as accurately as possible the abundances of Rb and Sr, which, along with geological age, dictate the $^{87}Sr/^{86}Sr$ ratios in the continental crust, and in water and weathering products containing strontium derived therefrom.

a) Crustal Abundances

Average $(^{87}Sr/^{86}Sr)_t$ values for particular segments of crustal terrains can be determined from their mean age and mean Rb/Sr ratio. Estimates of mean Rb/Sr ratios have been made by direct sampling of large areas (e.g. Shaw *et al.*, 1967), by estimates of the proportions of major rock types within the crust (e.g. Gast, 1960; Hurley *et al.*, 1962), and by analyzing waters and sediments from closed basins Hart and Tilton, 1966). Most of these studies have dealt directly with upper crustal rocks; little is known about the isotopic and trace element character of the lower crust but some inferences have been made on the basis of data available for high-grade metamorphic rocks.

Many estimates of the average Rb/Sr ratio of the upper continental crust have been published and a few are listed below:

Table 38-B-1

	Average Rb/Sr	$^{87}Sr/^{86}Sr$
Gast (1960) — "silicic" crust	0.33	0.728[a]
Hurley *et al.* (1962) —"silicic" crust	0.25	0.720—0.730
Shaw *et al.* (1967) — Canadian Shield	0.35	0.730[a]
Hart and Tilton (1966) — Superior Province	0.25	0.730

[a] Calculated assuming an average age of 2×10^9 years and an initial $^{87}Sr/^{86}Sr$ ratio of 0.701.

The estimate by HART and TILTON (1966) is based on $^{87}Sr/^{86}Sr$ analyses of water and sediment from Lake Superior. The lake-water has a value of 0.718 whereas the sediments are higher with 0.739. The differences are due to differential weathering of primary minerals, and this observation allows additional inferences to be drawn concerning the $^{87}Sr/^{86}Sr$ values of sea water. If the average age of the exposed crust is about 2×10^9 years, the present-day $^{87}Sr/^{86}Sr$ ratio of sea water sets a minimum Rb/Sr ratio of 0.09 for the crust. Allowance for strontium contained in pelagic clays, which from the Atlantic Ocean have a mean $^{87}Sr/^{86}Sr$ ratio of about 0.72 (DASCH, 1969b), would increase the estimated Rb/Sr ratio to within the range shown above.

Geophysical data have formerly been interpreted as indicating a lower crust of approximate basaltic composition. Recent studies, however, have shown that intermediate rocks of granulite grade may satisfy the geophysical features of the lower crust (see LAMBERT, 1971, for a review). HURLEY (1968), using limited available data for granulite-grade rocks, estimated 15 ppm Rb and 500 ppm Sr in the lower crust, or a Rb/Sr ratio of 0.03. There is some indication that during granulite facies metamorphism, Rb/Sr ratios are lowered, probably by Rb loss. In southwest Australia, Precambrian rocks of granulite grade have a mean Rb/Sr ratio of 0.21 compared with 0.35 for amphibolite-grade rocks (LAMBERT, 1971). Similarly, granulites of northern Norway have a Rb/Sr ratio of 0.12 compared with 0.46 for comparable lower grade rocks (HEIER and THORESEN, 1971). Thus, it appears likely that in a given segment of the continental crust, both Rb/Sr and $^{87}Sr/^{86}Sr$ ratios decrease with increasing depth.

b) Sea Water

Strontium, because of its long residence time, is isotopically well-mixed in ocean water. Average $^{87}Sr/^{86}Sr$ ratios based on reliable isotopic measurements agree within analytical errors: 0.7089 (FAURE *et al.*, 1965; HAMILTON, 1966); 0.7091 (HILDRETH and HENDERSON, 1971; MURTHY and BEISER, 1968; PETERMAN *et al.*, 1970; WASSERBURG *et al.*, 1969), and 0.7092 (KAUSHAL and WETHERILL, 1969). These values represent samples from all of the oceans, and no variations in $^{87}Sr/^{86}Sr$ have been reported. The best average value for sea water, based on a composite sample from the Atlantic, Pacific and Arctic Oceans, is 0.70906 (HILDRETH and HENDERSON, 1971). The $^{87}Sr/^{86}Sr$ ratio of sea water is significantly lower than the average value for the large reservoir of strontium contained in upper crustal continental rocks. However, the carbonate-free fractions of pelagic sediments probably contain the necessary strontium with higher $^{87}Sr/^{86}Sr$ ratios to balance the isotopic budget (see Subsection IV/c/1).

$^{87}Sr/^{86}Sr$ ratios of sea water in the geological past can be determined by analyses of marine limestones, fossil shell material, or other minerals that formed in equilibrium with sea water. Although only a few analyses are available for Precambrian time (e.g. HEDGE and WALTHALL, 1963; M.I.T. 1965; PERRY *et al.*, 1971), $^{87}Sr/^{86}Sr$ evolution curves for sea water and the mantle apparently converge at approximately 3×10^9 years ago (Fig. 38-B-1). Significant $^{87}Sr/^{86}Sr$ variations in sea water during Phanerozoic time (Fig. 38-B-4) have been documented by PETERMAN *et al.* (1970) and DASCH and BISCAYE (1971) from data on carbonates. The most notable feature of this curve is the excursion to low $^{87}Sr/^{86}Sr$ ratios during the Mesozoic Era which PETERMAN *et al.* (1970) suggested was due to an increased contribution of volcano-

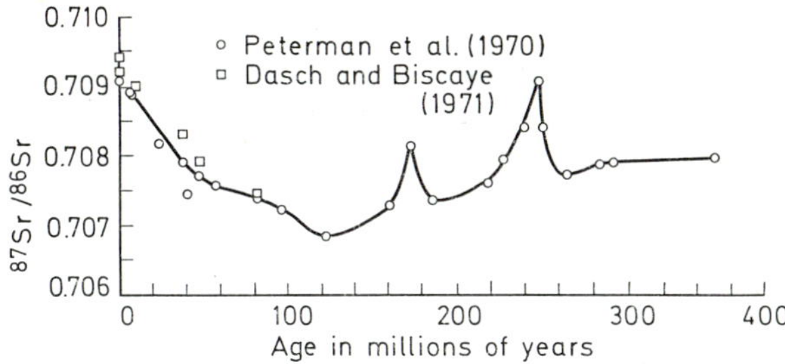

Fig. 38-B-4. $^{87}Sr/^{86}Sr$ values in sea water through Phanerozoic time, from data on carbonates of fossil shells according to PETERMAN *et al.* (1970) and DASCH and BISCAYE (1971)

genic strontium. Periods of higher $^{87}Sr/^{86}Sr$ ratios may be related to exposure and erosion of older crustal rocks during times of regional continental emergence. ARMSTRONG (1971) suggested that continental glaciation may also provide a means of increasing the $^{87}Sr/^{86}Sr$ ratio in sea water by accentuated erosion of old crustal rocks from the shield areas.

c) Sediments and Sedimentary Rocks

1. Pelagic Sediments

Strontium in pelagic sediments is contained mainly in the carbonate fractions although significant concentrations may occur in the detrital and authigenic phases. $^{87}Sr/^{86}Sr$ values of the carbonate are the same as of the sea water in which it formed and because of the high strontium content, a small amount of carbonate will mask the isotopic composition of the silicate fraction. Strontium isotopic data for the silicate fractions (mainly clays) show that these have not equilibrated their strontium with that of sea water. Silicate fractions of sediments from the Atlantic Ocean have $^{87}Sr/^{86}Sr$ ratios in the range of 0.704 to 0.743 (BISCAYE and DASCH, 1971; DASCH, 1969b; MURTHY and BEISER, 1968). Regional variations of $^{87}Sr/^{86}Sr$ ratios in these sediments reflect the age and composition of the source areas as modified by chemical weathering and sorting in the depositional environment. Sediments in the western North Atlantic are characterized by $^{87}Sr/^{86}Sr$ ratios which average 0.72 (DASCH, 1969b), whereas those in the Argentine Basin in the South Atlantic have lower $^{87}Sr/^{86}Sr$ ratios: 0.71 (BISCAYE and DASCH, 1971; DASCH, 1969b).

Only limited data are available for sediments from other oceans. DASCH (1969b) reports $^{87}Sr/^{86}Sr$ values of 0.7046 and 0.7089 for two samples from the Indian Ocean and 0.7044 and 0.7072 for two samples from the southeastern Pacific near South America. CHURCH (1971) analyzed six bulk samples (including the carbonate phase) from the northeastern Pacific and obtained $^{87}Sr/^{86}Sr$ values ranging from 0.7081 to 0.7165.

Some non-carbonate phases of pelagic sediments which form in equilibrium with sea water, such as phillipsite, contain strontium of the same isotopic composition as sea water (PUSHKAR and PETERSON, 1967). Iron-rich sediments from the East Pacific Rise are dominated by strontium of sea-water isotopic composition as a result of absorption and of incorporation of biogenic materials (DASCH *et al.*, 1971).

2. Greywackes (Trench Sediments)

Strontium isotope data for greywackes are useful, in conjunction with petrological and chemical data, in providing information on the age and composition of source terrains. $^{87}Sr/^{86}Sr$ values for greywackes from the western United States and New Zealand, which contain significant detritus from young volcanic and/or plutonic rocks, are summarized in Fig. 38-B-5, and most of these rocks have ratios in the range of 0.704 to 0.708. Sediments in reservoirs along the Columbia River are greywacke-like sands (WHETTEN, 1966), and data for these illustrate the relation between age and composition of the source rocks and the strontium isotopic composition (PETERMAN and WHETTEN, 1972).

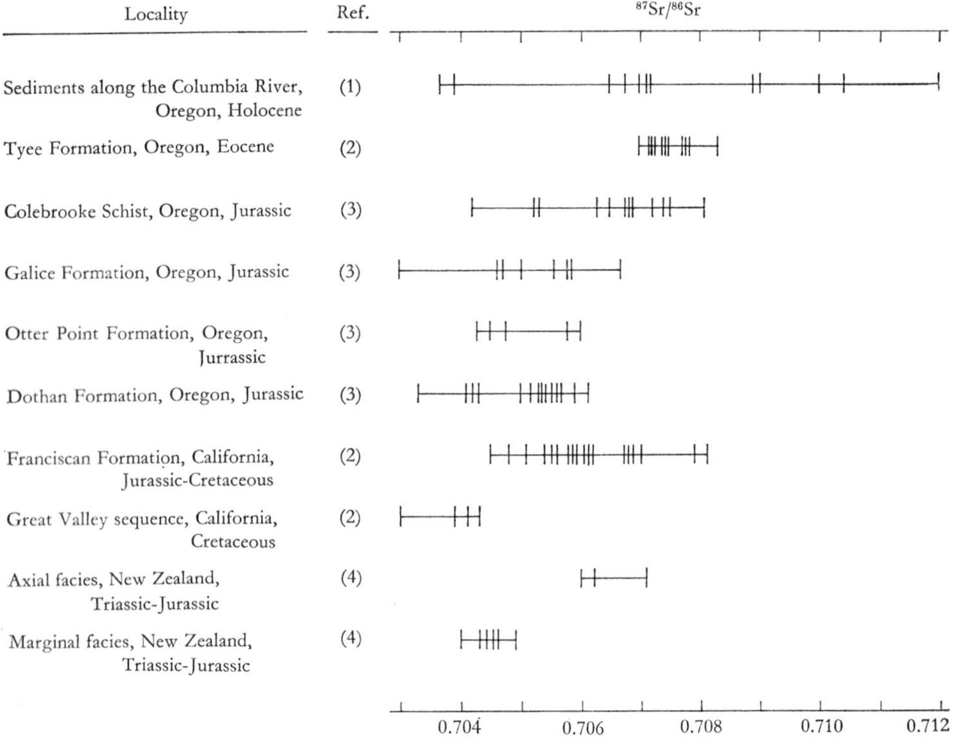

Fig. 38-B-5. $^{87}Sr/^{86}Sr$ values of greywackes from the western United States and New Zealand. Data sources are: (1) PETERMAN and WHETTEN, 1972; (2) PETERMAN *et al.*, 1967; (3) COLEMAN, 1972; (4) EWART and STIPP, 1968

3. Shales and Other Sediments

Much strontium isotope information exists for sedimentary rocks other than those discussed in the previous subsections, especially for pelitic rocks. Shales have received considerable attention from the standpoint of Rb—Sr dating as first proposed by COMPSTON and PIDGEON (1962). DASCH (1969b) reviewed much of the literature on attempts at dating shales and discussed the problems in the method in light of his studies on oceanic sediments. Rb—Sr isochrons determined on shales may provide useful information on $(^{87}Sr/^{86}Sr)_0$ ratios which reflect provenance, although the assumption of uniform initial ratios at the time of deposition may not always be correct. Analyses of shales with variable carbonate contents may produce erroneous isochrons which are mixing lines (CHAUDHURI and BROOKINS, 1969).

Other strontium isotope studies dealing with sediments are directed toward specific problems such as the age of detritus (MUFFLER and DOE, 1968), characterization of the isotopic nature of old crustal domains (HART and TILTON, 1966), and origin of sediments and brines in the Red Sea rift (FAURE and JONES, 1969).

V. Applications to Petrogenesis

Probably the most interesting aspect of strontium isotopes is their application to questions about the sources of magmas. As crustal rocks contain relatively more ^{87}Sr than do mantle rocks, it should therefore be possible to use strontium isotopes to determine if granites, andesites, and other rocks of debatable genesis, have originated from a mantle source, a crustal source, or some combination thereof. The results of these studies are often disappointingly inconclusive when used by themselves, but when combined with other types of geological, geochemical, and isotopic data, they can usually provide at least some constraints as to acceptable petrogenic models.

a) Continental Volcanic Rocks

If continental volcanic rocks originate within a mantle that is like the suboceanic mantle, and if there is no assimilation during the transit of the magmas upwards through the continental crust, then the $^{87}Sr/^{86}Sr$ ratio of continental volcanic rocks would be the same as that of oceanic volcanic rocks. In many continental basalts the $^{87}Sr/^{86}Sr$ ratios are within the range of those of oceanic basalts. However, many continental basalts are enriched in radiogenic ^{87}Sr relative to ocean basalts. About the highest $^{87}Sr/^{86}Sr$ ratios that have been reported for basaltic rocks are those of the Jurassic dolerites of Tasmania (HEIER *et al.*, 1965) and similar rocks from Antarctica (COMPSTON *et al.*, 1968; FAURE and ELLIOT, 1971), all of which have a ratio of about 0.711. Not as high, but still unusual for large volumes of basaltic lava, are the values of 0.707 found by LEEMAN and MANTON (1971) for the bulk of the lavas from the Snake River Plain of Idaho. HEIER *et al.* (1965) suggested that the mantle beneath Tasmania might have a considerably higher Rb/Sr ratio than the suboceanic mantle, and DOE (1968) hypothesized that the subcontinental basalt source might, in general, have different Rb/Sr and U/Pb ratios than the suboceanic mantle. Such differences were also favored by LEEMAN and MANTON (1971) and DAVIES *et al.* (1970) to account for the large volumes of continental basalts with high $^{87}Sr/^{86}Sr$ ratios encountered in their respective studies. If the mantle beneath the continents has a higher Rb/Sr ratio than that under the oceans, then the mantle ^{87}Sr evolution models shown in Fig. 38-B-3 are only appropriate for the oceanic regions, and different evolution trends are required for the subcontinental mantle. Some possible subcontinental evolutionary trends have been drawn by STUEBER and MURTHY (1966), DAVIES *et al.* (1970), and LEEMAN and MANTON (1971).

An alternative interpretation of elevated $^{87}Sr/^{86}Sr$ ratios in some continental basalts is that they were derived from a mantle having $^{87}Sr/^{86}Sr$ like that of the suboceanic mantle, but that radiogenic ^{87}Sr has been incorporated from the crust. Such an interpretation is supported by the observation that silicic lavas, associated with continental basalts, almost invariably have higher $^{87}Sr/^{86}Sr$ ratios than do the basalts (HEDGE, 1966; HOEFS and WEDEPOHL, 1968; BELL and POWELL, 1969; DICKINSON *et al.*, 1969; BELL and POWELL, 1970; PETERMAN *et al.*, 1970). Similar trends have been found in intrusive series (HAMILTON, 1963; MOORBATH and BELL, 1965; PANKHURST, 1969). Even more direct evidence for magmatic assimilation of crustal Sr has been afforded by studies in which early-formed phenocrysts are not in strontium-isotopic equilibrium with the enclosing lava (NOBLE and HEDGE, 1969; DASCH, 1969a), and by the $^{87}Sr/^{86}Sr$ ratio of different samples of a single lava flow having distinct $^{87}Sr/^{86}Sr$ ratios (LAUGHLIN *et al.*, 1972).

Anatexis of deeper parts of the continental crust frequently fits the strontium-isotopic data, particularly of the more silicic lavas (c.f. HURLEY *et al.*, 1966), but this is probably not a generally suitable mechanism for origin of high-$^{87}Sr/^{86}Sr$ continental basalts because of the unreasonably high geothermal gradients that would be required (HEIER *et al.*, 1965).

A rather considerable body of strontium isotopic data has been amassed on the volumetrically small, but widely distributed alkalic igneous rocks and carbonatites. Both rock types, which commonly occur together, are quite enriched in a group of trace elements including strontium.

The vast majority of the carbonatites that have been studied have $^{87}Sr/^{86}Sr$ ratios that are like those of oceanic basalts. FAURE and POWELL (1972) have recently summarized the available strontium isotopic data for carbonatite and found a mean $^{87}Sr/^{86}Sr$ ratio of 0.7034, a value identical with the mean calculated for oceanic basalts.

Strongly alkalic igneous rocks tend to have higher $^{87}Sr/^{86}Sr$ ratios than carbonatites, even when they occur together as in East Africa (BELL and POWELL, 1969; BELL and POWELL, 1970). FAURE and POWELL (1972) noted in their review of $^{87}Sr/^{86}Sr$ ratios in alkalic igneous rocks that these rocks commonly exhibit a good correlation of $^{87}Sr/^{86}Sr$ with Rb/Sr, and they suggested that this relationship was the result of the alkalic rocks being derived from an alkali-rich part of the upper mantle.

b) Andesites

Strontium isotope ratios should be particularly applicable to the testing of various theories as to the origins of andesites and related rock-types of the calc-alkaline orogenic suites, because some models invoke their genesis either directly or indirectly from the mantle, whereas other models require an input of sialic material.

The $^{87}Sr/^{86}Sr$ results that have been obtained so far on andesites and related rocks of calc-alkaline suites are quite variable (Fig. 38-B-6) and suggest that the source materials are not always the same; possibly the tectonic setting is more important in producing andesites than the actual material being fused. Strontium isotopic studies of andesites from island arcs indicate that the vast majority have $^{87}Sr/^{86}Sr$ very near 0.7035 (HEDGE, 1966; PUSHKAR, 1968; PETERMAN *et al.*, 1970; GILL, 1970; DONNELLY *et al.*, 1971; HEDGE and LEWIS, 1971). These ratios are almost perfectly coincidental with most basalts from oceanic islands and hence fit a model in which the arc volcanic rocks are derived from the mantle above the seismic zone which dips beneath the islands. However, ocean-ridge basalts tend to have somewhat lower

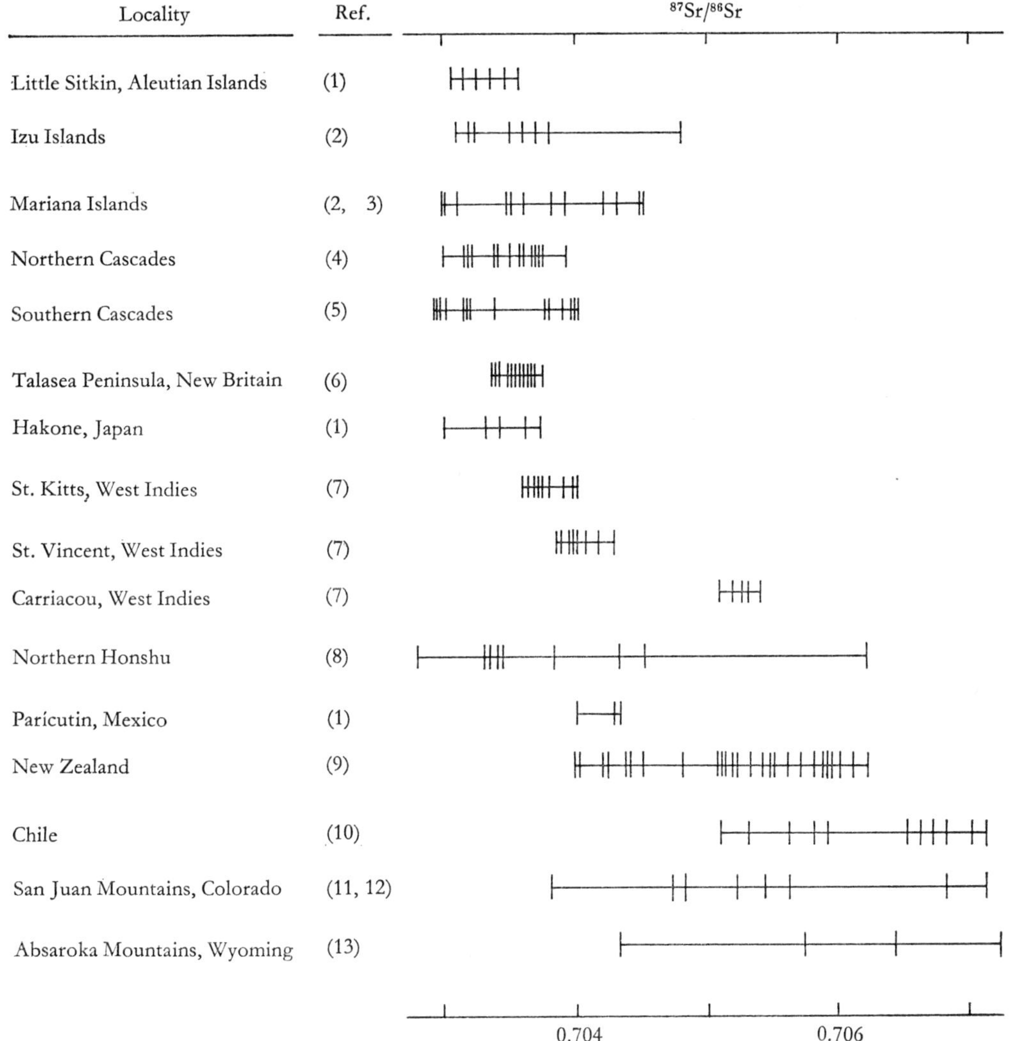

Fig. 38-B-6. $^{87}Sr/^{86}Sr$ values in andesites and associated calc-alkaline lavas. Data sources are: (1) authors, unpublished data; (2) PUSHKAR (1967); (3) HEDGE (1966); (4) PETERMAN *et al.* (1970b); (5) PETERMAN *et al.* (1970a); (6) PETERMAN *et al.* (1970); (7) HEDGE and LEWIS (1971); (8) HEDGE and KNIGHT (1969); (9) EWART and STIPP (1968); (10) PICHLER and ZEIL (1971); (11) DOE *et al* (1970); (12) DOE *et al.* (1969); (13) PETERMAN *et al.* (1970)

$^{87}Sr/^{86}Sr$ ratios ($\approx$ 0.7026), and if such rocks constitute the bulk of the oceanic crust, refusion of this crust combined with 5—10 percent of sediments would also produce andesites having the observed $^{87}Sr/^{86}Sr$ ratios (HART *et al.*, 1970; DONNELLY *et al.*, 1971).

Andesites which erupted on an old continental crust tend to have higher and more variable $^{87}Sr/^{86}Sr$ ratios than those of island arcs (EWART and STIPP, 1968; DOE *et al.*

1969; PETERMAN *et al.*, 1970), whereas those which erupted through younger continental segments have $^{87}Sr/^{86}Sr$ ratios similar to those of the island arcs (PUSHKAR, 1968; PETERMAN *et al.*, 1970b; HEDGE *et al.*, 1970). Japan shows a crude E-W progression of $^{87}Sr/^{86}Sr$ ratios in andesites — those on the east side have $^{87}Sr/^{86}Sr$ of about 0.7035 and the values generally become higher to the west (HEDGE and KNIGHT, 1969; KURASAWA, 1970). The higher $^{87}Sr/^{86}Sr$ ratios of the continental andesites seem to require the incorporation of radiogenic ^{87}Sr from the crust, but whether this incorporation comes about by the fusion of deeper crustal rocks or by assimilation of crustal Sr into mantle-derived magmas is still debatable.

c) Granites

Great many data on initial $^{87}Sr/^{86}Sr$ ratios of granitic rocks of diverse geological ages have become available as by-products of Rb-Sr dating. These data have been summarized (GAST, 1960; HURLEY *et al.*, 1962; HEDGE and WALTHALL, 1963; HURLEY, 1968) with the general result that the initial $^{87}Sr/^{86}Sr$ ratios of granites range from values like those of oceanic basalts up to ratios that are similar to estimates of the average $^{87}Sr/^{86}Sr$ ratio of the sialic crust. The distribution is strongly skewed, however, and the vast majority of granites have initial $^{87}Sr/^{86}Sr$ ratios that are like those of andesites and other volcanic rocks in a similar geological setting.

Studies of the initial $^{87}Sr/^{86}Sr$ ratios of granitic rocks of large Cenozoic batholiths (HURLEY *et al.*, 1965; FAIRBAIRN *et al.*, 1964; DOE *et al.*, 1968) give results in the range 0.705 to 0.710 and seem to indicate that these granites contain Sr from both mantle and crustal sources; resolution of the mechanisms involved, however, requires more detailed studies of strontium isotopes combined with other isotopic, geochemical, and geological parameters.

Revised manuscript received: May 197

References: Section 38-B

ALLSOPP, H. L., NICOLAYSEN, L. O., HAHN-WEINHEIMER, P.: Rb/K ratios and Sr-isotopic compositions of minerals in eclogitic and peridotitic rocks. Earth Planet. Sci. Lett. **5**, 231 (1968).

ARMSTRONG, R. L.: A model for the evolution of strontium and lead isotopes in a dynamic earth. Rev. Geophys. **6**, 175 (1968).

ARMSTRONG, R. L.: Glacial erosion and the variable isotopic composition of strontium in sea water. Nature, Phys. Sci. **230**, 132 (1971).

BELL, K., POWELL, J. L.: Strontium isotopic studies of alkalic rocks: The potassium-rich lavas of the Birunga and Toro-Ankole Regions, East and Central Equatorial Africa. J. Petrol. **10**, 536 (1969).

BELL, K., POWELL, J. L.: Strontium isotopic studies of alkalic rocks: The alkalic complexes of Eastern Uganda. Bull. Geol. Soc. Am. **81**, 3481 (1970).

BENCE, A. E., HURLEY, P. M.: Rubidium-strontium isotopic relationships in oceanic basalts (abstract). Trans. Am. Geophys. Union **48**, 251 (1967).

BERG, G. W., ALLSOPP, H. L.: Low $^{87}Sr/^{86}Sr$ ratios in fresh South African kimberlites. Earth Planet. Sci. Lett. **16**, 27 (1972).

BISCAYE, P. E., DASCH, E. J.: The rubidium, strontium, strontium isotope system in deep sea sediments: Argentine basin. J. Geophys. Res. **76**, 5087 (1971).

BONATTI, E., HONNOREZ, J., FERRARA, G.: Equatorial Mid-Atlantic Ridge: Petrologic and Sr isotopic evidence for an alpine-type rock assemblage. Earth Planet. Sci. Lett. **9**, 247 (1970).

CHAUDHURI, S., BROOKINS, D. G.: The Rb—Sr whole-rock age of the Stearns shale (Lower Permian), eastern Kansas, before and after acid-leaching experiments. Bull. Geol. Soc. Am. **80**, 2605 (1969).

CHURCH, S. E.: Strontium isotope and alkali element geochemistry of selected sediments from the northeast Pacific Ocean. Geochim. Cosmochim. Acta **35**, 1300 (1971).

COLEMAN, R. G.: The Colebrooke Schist of southwestern Oregon and its relation to the tectonic evolution of the region. U.S. Geol. Surv. Bull. **1339**, 1 (1972).

COMPSTON, W., LOVERING, J. F.: The strontium isotopic geochemistry of granulitic and eclogitic inclusions from the basic pipes at Delegate, Eastern Australia. Geochim. Cosmochim. Acta **33**, 691 (1969).

COMPSTON, W., MCDOUGALL, I., HEIER, K. S.: Geochemical comparison of the Mesozoic basaltic rocks of Antarctica, South Africa, South America and Tasmania. Geochim. Cosmochim. Acta **32**, 129 (1968).

COMPSTON, W., PIDGEON, R. T.: Rubidium-strontium dating of shales by the total-rock method. J. Geophys. Res. **69**, 3493 (1962).

DASCH, E. J.: Strontium isotope disequilibrium in a porphyritic alkali basalt and its bearing on magmatic processes. J. Geophys. Res. **74**, 560 (1969a).

DASCH, E. J.: Strontium isotopes in weathering profiles, deep-sea sediments and sedimentary rocks. Geochim. Cosmochim. Acta **33**, 1521 (1969b).

DASCH, E. J., BISCAYE, P. E.: Isotopic composition of strontium in Cretaceous-to-Recent pelagic foraminifera. Earth Planet. Sci. Lett. **11**, 201 (1971).

DASCH, E. J., DYMOND, J. R., HEATH, G. R.: Isotopic analysis of metalliferous sediment from the East Pacific Rise. Earth Planet. Sci. Lett. **13**, 175 (1971).

DASCH, E. J., GREEN, D. H.: Strontium isotope geochemistry of lherzolite inclusions and host basaltic rocks, Victoria, Australia [Abst.] Geol. Soc. America, Cordilleran Section Program 144 (1972).

DAVIES, R. D., ALLSOPP, H. L., ERLANK, A. J., MANTON, W. I.: Sr-isotopic studies on various layered intrusions in southern Africa. In: Symposium on the Bushveld Igneous Complex and Other Layered Intrusions. Geological Society of South Africa, Special Publ. **1**, 576 (1970).

DICKINSON, D. R., DODSON, M. H., GASS, I. G., REX, D. C.: Correlation of initial $^{87}Sr/^{86}Sr$ with Rb/Sr in some late Tertiary volcanic rocks of South Arabia. Earth Planet. Sci. Letters **6**, 84 (1969).

DOE, B. R.: Lead and strontium isotopic studies of Cenozoic volcanic rocks in the Rocky Mountain region — A summary. In: Cenozoic Volcanism in the Southern Rocky Mountains. Colorado School of Mines Quart. **63**, 149 (1968).

DOE, B. R., LIPMAN, P. W., HEDGE, C. E.: Radiogenic tracers and the source of continental andesites: A beginning at the San Juan volcanic field, Colorado. Bull. Oregon Dept. Geol. and Mineral Industries **65**, 143 (1969).

DOE, B. R., LIPMAN, P. W., HEDGE, C. E., KURASAWA, H.: Primitive and contaminated basalts from the southern Rocky Mountains, U.S.A. Contr. Mineral. and Petrol. **21**, 142 (1969).

DOE, B. R., TILLING, R. I., HEDGE, C. E., KLEPPER, M. R.: Lead and strontium isotope studies of the Boulder Batholith, southwestern Montana. Econ. Geol. **63**, 884 (1968).

DONNELLY, T. W., RODGERS, J. J. W., PUSHKAR, P., ARMSTRONG, R. L.: Chemical evolution of the igneous rocks of the eastern West Indies: An investigation of thorium, uranium, and potassium distributions, and lead and strontium isotopic ratios. Geol. Soc. Am. Mem. **130**, 181 (1971).

EWART, A., STIPP, J. J.: Petrogenesis of the volcanic rocks of the Central North Island, New Zealand, as indicated by a study of $^{87}Sr/^{86}Sr$ ratios, and Sr, Rb, K, U, and Th abundances. Geochim. Cosmochim. Acta **32**, 699 (1968).

FAIRBAIRN, H. W., HURLEY, P. M., PINSON, W. H.: Initial $^{87}Sr/^{86}Sr$ and possible sources of granitic rocks in southern British Columbia. J. Geophys. Res. **69**, 4889 (1964).

FAURE, G., ELLIOT, D. H.: Isotope composition of strontium in Mesozoic basalt and dolerite from Dronning Maud Land. Brit. Antarctic Survey Bull. **25**, 23 (1971).

FAURE, G., HURLEY, P. M., POWELL, J. L.: The isotopic composition of strontium in surface water from the North Atlantic Ocean. Geochim. Cosmochim. Acta **29**, 209 (1965).

FAURE, G., JONES, L. M.: Anomalous strontium in the Red Sea brines. In: DEGENS, E. T., ROSS, D. A., (eds.). Hot Brines and Recent Heavy Metal Deposits in the Red Sea. Berlin-Heidelberg-New York: Springer 1969.

FAURE, F., POWELL, J. L.: Strontium Isotope Geology. Berlin-Heidelberg-New York: Springer 1972.

GAST, P. W.: Limitations on the composition of the upper mantle. J. Geophys. Res. **65**, 1287 (1960).

GAST, P. W.: The isotopic composition of strontium and the age of stone meteorites — I. Geochim. Cosmochim. Acta **26**, 927 (1962).

GAST, P. W.: Trace element fractionation and the origin of tholeiitic and alkaline magma types. Geochim. Cosmochim. Acta **32**, 1057 (1968).

GAST, P. W., TILTON, G. R., HEDGE, C.: Isotopic composition of lead and strontium from Ascension and Gough Islands. Science **145**, 1181 (1964).

GILL, J. B.: Geochemistry of Viti Leva, Fiji, and its evolution as an island arc. Contr. Mineral. and Petrol. **27**, 179 (1970).

HAMILTON, E. I.: The isotopic composition of strontium in the Skaergaard Intrusion, East Greenland. J. Petrol. **4**, 383 (1963).

HAMILTON, E. I.: The isotopic composition of strontium in Atlantic Ocean water. Earth Planet. Sci. Lett. **1**, 435 (1966).

HART, S. R.: K, Rb, Cs, Sr, and Ba contents and Sr isotope ratios of ocean floor basalts. Phil. Trans. Roy. Soc. London, Ser. A **268**, 573 (1971).

HART, S. R., BROOKS, C., KROGH, T. E., DAVIS, G. L., NAVA, D.: Ancient and modern volcanic rocks: A trace element model. Earth Planet. Sci. Lett. **10**, 17 (1970).

HART, S. R., GLASSLEY, W. E., KARIG, D. E.: Basalts and sea floor spreading behind the Mariana Island arc. Earth Planet. Sci. Lett. **15**, 12 (1972).

HART, S. R., TILTON, G.: The isotope geochemistry of strontium and lead in Lake Superior sediments and water. In: The Earth Beneath the Continents. Geophys. Mon. Am. Geophys. Union **10**, 127 (1966).

HEDGE, C. E.: Variations in radiogenic strontium found in volcanic rocks. J. Geophys. Res. **71**, 6119 (1966).

HEDGE, C. E., HILDRETH, R. A., HENDERSON, W. T.: Strontium isotopes in some Cenozoic lavas from Oregon and Washington. Earth Planet. Sci. Lett. **8**, 434 (1970).

HEDGE, C. E., KNIGHT, R. J.: Lead and strontium isotopes in volcanic rocks from northern Honshu, Japan. Geochem. J. **3**, 15 (1969).

HEDGE, C. E., LEWIS, J. F.: Isotopic composition of strontium in three basalt-andesite centers along the Lesser Antilles Arc. Contr. Mineral. and Petrol. **32**, 39 (1971).

HEDGE, C. E., NOBLE, D. C.: Upper Cenozoic basalts with high $^{87}Sr/^{86}Sr$ and Sr/Rb ratios, southern Great Basin, Western United States. Bull. Geol. Soc. Am. **82**, 3503 (1971).

HEDGE, C. E., PETERMAN, Z. E.: The strontium isotopic composition of basalts from the Gordo and Juan de Fuca Rises, north-eastern Pacific Ocean. Contr. Mineral. and Petrol. **27**, 114 (1970).

HEDGE, C. E., PETERMAN, Z. E., DICKINSON, W. R.: Petrogenesis of lavas from Western Samoa. Bull. Geol. Soc. Am. **83**, 2709 (1972).

HEDGE, C. E., WALTHALL, F. G.: Radiogenic strontium-87 as an index of geologic processes. Science **140**, 1214 (1963).

HEIER, K. S., COMPSTON, W., MCDOUGALL, I.: Thorium and uranium concentrations, and the isotopic composition of strontium in the differentiated Tasmanian dolerites. Geochim. Cosmochim. Acta **29**, 643 (1965).

HEIER, K. S., THORESSEN, K.: Geochemistry of high grade metamorphic rocks, Lofoten-Vesterålen, North Norway. Geochim. Cosmochim. Acta **35**, 89 (1971).

HERZOG, L. F., PINSON, W. H.: Rb/Sr age, elemental and isotopic abundance studies of stony meteorites. Am. J. Sci. **254**, 555 (1956).

HILDRETH, R. A., HENDERSON, W. T.: Comparison of $^{87}Sr/^{86}Sr$ for sea-water strontium and the Eimer and Amend $SrCO_3$. Geochim. Cosmochim. Acta **35**, 235 (1971).

HOEFS, J., WEDEPOHL, K. H.: Strontium isotope studies on young volcanic rocks from Germany and Italy. Contr. Mineral. and Petrol. **19**, 328 (1968).

HUBBARD, N. J.: Some chemical features of lavas from the Manu'a Islands, Samoa. Pacific Sci. **25**, 178 (1971).

HURLEY, P. M.: Absolute abundance and distribution of Rb, K, and Sr in the earth. Geochim. Cosmochim. Acta **32**, 273 (1968).

HURLEY, P. M., BATEMAN, P. C., FAIRBAIRN, H. W., PINSON, W. H., JR.: Investigation of initial $^{87}Sr/^{86}Sr$ ratios in the Sierra Nevada plutonic province. Bull. Geol. Soc. Am. **76**, 165 (1965).

HURLEY, P. M., FAIRBAIRN, H. W., PINSON, W. H., JR.: Rb—Sr isotopic evidence in the origin of potash-rich lavas of western Italy. Earth Planet. Sci. Lett. **5**, 301 (1966).

HURLEY, P. M., HUGHES, H., FAURE, G., FAIRBAIRN, H. W., PINSON, W. H.: Radiogenic strontium-87 model of continent formation. J. Geophys. Res. **67** 5315 (1962).

HUTCHISON, R., DAWSON, J. B.: Rb, Sr and $^{87}Sr/^{86}Sr$ in ultrabasic xenoliths and host-rocks, Lashaine Volcano, Tanzania. Earth Planet Sci. Lett. **9**, 87 (1970).

KAUSHAL, S. K., WETHERILL, G. W.: ^{87}Rb—^{87}Sr age of bronzite (H group) chondrites. J. Geophys. Res. **74**, 2717 (1969).

KUDO, A. M., BROOKINS, D. G., LAUGHLIN, A. W.: Sr isotopic disequilibrium in lherzolites from the Puerco Necks, New Mexico. Earth Planet. Sci. Lett. **15**, 291 (1972).

KURASAWA, H.: Strontium and lead isotopes of volcanic rocks in Japan. In: OGATA and HAYAKAWA (eds.): Recent Developments in Mass Spectroscopy. Proc. Internat. Conf. Mass. Spectroscopy, Kyoto, Tokyo: University of Tokyo Press 1970.

LAMBERT, I. B.: The composition andevolution of the deep continental crust. In: GLOVER, E. J. (ed.): Symposium on Archaean Rocks. Geol. Soc. Australia Spec. Pub. **3**, 419 (1971).

LANPHERE, M. A.: Sr—Rb—K and Sr isotopic relationships in ultramafic rocks, south-eastern Alaska. Earth Planet. Sci. Lett. **4**, 185 (1968).

LAUGHLIN, A. W., BROOKINS, D. G., CARDEN, J. R.: Variations in the initial strontium ratios of a single basalt flow. Earth Planet. Sci. Lett. **14**, 79 (1972).
LAUGHLIN, A. W., BROOKINS, D. G., KUDO, A. M., CAUSEY, J. D.: Chemical and strontium isotopic investigations of ultramafic inclusions and basalt, Bandera Crater, New Mexico. Geochim. Cosmochim. Acta **35**, 107 (1971).
LEEMAN, W. P., MANTON, W. I.: Strontium isotopic composition of basaltic lavas from the Snake River Plain, southern Idaho. Earth Planet. Sci. Lett. **11**, 420 (1971).
LEGGO, P. J., HUTCHISON, R.: A Rb—Sr isotope study of ultrabasic xenoliths and their basaltic host rocks from the Massif Central, France. Earth Planet. Sci. Lett. **5**, 71 (1968).
MANTON, W. I., TATSUMOTO, M.: Some Pb and Sr isotopic measurements on eclogites from the Roberts Victor Mine, South Africa. Earth Planet. Sci. Lett. **10**, 217 (1971).
(M.I.T.) = Massachusetts Institute of Technology: Evidence from western Ontario of the isotopic composition of strontium in Archean Seas. M.I.T. Annual Report 145 (1965).
McDOUGALL, I., COMPSTON, W.: Strontium isotope composition and potassium-rubidium ratios in some rocks from Reunion and Rodriguez, Indian Ocean. Nature **207**, 252 (1965).
MITCHELL, R. H., CROCKET, J. H.: The isotopic composition of strontium in some South African kimberlites. Contr. Mineral. and Petrol. **30**, 277 (1971).
MOORBATH, S., BELL, J. D.: Strontium isotope abundance studies and rubidium-strontium age determinations on Tertiary igneous rocks from the Isle of Skye, North-West Scotland. J. Petrol. **6**, 37 (1965).
MUFFLER, L. J. P., DOE, B. R.: Composition and mean age of detritus of the Colorado River Delta in the Salton Trough, southeastern California. J. Sediment. Petrol. **38**, 384 (1968).
MURTHY, V. R., BEISER, E.: Strontium isotopes in ocean water and marine sediments. Geochim. Cosmochim. Acta **32**, 1121 (1968).
National Bureau of Standards: Certificate of analysis, Standard Reference Material 987 (1971).
NOBLE, D. C., HEDGE, C. E.: $^{87}Sr/^{86}Sr$ variations within individual ash-flow sheets. U.S. Geol. Surv. Profess. Paper **650**-C, 133 (1969).
O'NEIL, J. R., HEDGE, C. E., JACKSON, E. D.: Isotopic investigations of xenoliths and host basalts from the Honolulu Volcanic Series. Earth Planet. Sci. Lett. **8**, 253 (1970).
PANKHURST, R. J.: Strontium isotope studies related to petrogenesis in the Caledonian basic igneous province of Northeast Scotland. J. Petrol. **10**, 115 (1969).
PAPANASTASSIOU, D. A., WASSERBURG, G. J.: Initial strontium isotopic abundances and the resolution of small time differences in the formation of planetary objects. Earth Planet. Sci. Lett. **5**, 361 (1969).
PAUL, D. K.: Strontium isotope studies on ultramafic inclusions from Dreiser Weiher, Eifel, Germany. Contr. Mineral. and Petrol. **34**, 22 (1971).
PERRY, E. C., JR., MONSTER, J., REIMER, T.: Sulfur isotopes in Swaziland system barites and the evolution of the earth's atmosphere. Science **171**, 1015 (1971).
PETERMAN, Z. E., CARMICHAEL, I. S. E., SMITH, A. L.: Strontium isotopes in quaternary basalts of southeastern California. Earth Planet. Sci. Lett. **7**, 381 (1970a).
PETERMAN, Z. E., CARMICHAEL, I. S. E., SMITH, A. L.: $^{87}Sr/^{86}Sr$ ratios of quaternary lavas of the Cascade Range, northern California. Bull. Geol. Soc. Am. **81**, 311 (1970b).
PETERMAN, Z. E., COLEMAN, R. G., HILDRETH, R. A.: $^{87}Sr/^{86}Sr$ in mafic rocks of the Troodos Massif, Cyprus. U.S. Geol. Surv. Profess. Paper **750**-D, D157 (1971).
PETERMAN, Z. E., DOE, B. R., PROSTKA, H. J.: Lead and strontium isotopes in rocks of the Absaroka Volcanic Field, Wyoming. Contr. Mineral. and Petrol. **27**, 121 (1970).
PETERMAN, Z. E., HEDGE, C. E.: Related strontium isotopic and chemical variations in oceanic basalts. Bull. Geol. Soc. Am. **82**, 493 (1971).
PETERMAN, Z. E., HEDGE, C. E., COLEMAN, R. G., SNAVELY, P. D., JR.: $^{87}Sr/^{86}Sr$ ratios in some eugeosynclinal sedimentary rocks and their bearing on the origin of granitic magma in orogenic belts. Earth Planet. Sci. Lett. **2**, 433 (1967).
PETERMAN, Z. E., HEDGE, C. E., TOURTELOT, H. A.: Isotopic composition in sea water throughout Phanerozoic time. Geochim. Cosmochim. Acta **34**, 105 (1970).
PETERMAN, Z. E., LOWDER, G. G., CARMICHAEL, I. S. E.: $^{87}Sr/^{86}Sr$ ratios of the Talasea Series, New Britain, Territory of New Guinea. Bull. Geol. Soc. Am. **81**, 39 (1970).

PETERMAN, Z. E., WHETTEN, J. T.: $^{87}Sr/^{86}Sr$ variation in Columbia River bottom sediments as a function of provenance. Geol. Soc. Am. Mem. **135**, 29 (1972).

PICHLER, H., ZEIL, W.: The Cenozoic rhyolite-andesite association of the Chilean Andes. Bull. Volcanol. **35**, 424 (1971).

POWELL, J. L., DELONG, S. E.: Isotopic composition of strontium in volcanic rocks from Oahu. Science **153**, 1239 (1966).

POWELL, J. L., FAURE, G., HURLEY, P. M.: Strontium-87 abundance in a suite of Hawaiian volcanic rocks of varying silica content. J. Geophys. Res. **70**, 1509 (1965).

PUSHKAR, P.: Strontium isotope ratios in volcanic rocks of three island arc areas. J. Geophys. Res. **73**, 2701 (1968).

PUSHKAR, P., PETERSON, M. N. A.: Strontium isotopic analyses of three marine phillipsites from the Pacific Ocean. Earth Planet. Sci. Lett. **2**, 349 (1967).

SCHUMACHER, E.: Altersbestimmung von Steinmeteoriten mit der Rubidium-Strontium-Methode. Z. Naturforsch. **11**a, 209 (1956).

SHAW, D. M., REILLY, G. A., MUYSSON, J. R., PATTENDEN, G. E., CAMPBELL, F. E.: An estimate of the chemical composition of the Canadian Shield. Can. J. Earth. Sci. **4**, 829 (1967).

STUEBER, A. M.: Abundances of K, Rb, Sr and Sr isotopes in ultramafic rocks and minerals from western North Carolina. Geochim. Cosmochim. Acta **33**, 543 (1969).

STUEBER, A. M., MURTHY, V. R.: Strontium isotope and alkali element abundances in ultramafic rocks. Geochim. Cosmochim. Acta **30**, 1243 (1966).

SUBBARAO, K. V., HEDGE, C. E.: K, Rb, Sr, and $^{87}Sr/^{86}Sr$ in rocks from the Mid-Indian Ocean Ridge. Earth Planet. Sci. Lett. **18**, 223 (1973).

TATSUMOTO, M., HEDGE, C. E., ENGEL, A. E. J.: Potassium, rubidium, strontium, thorium, uranium, and the ratio of strontium-87 to strontium-86 in oceanic tholeiitic basalt. Science **150**, 886 (1965).

WASSERBURG, G. J., PAPANASTASSIOU, D. A.: Age of an Apollo 15 mare basalt; lunar crust and mantle evolution. Earth Planet. Sci. Lett. **13**, 97 (1971).

WASSERBURG, G. J., PAPANASTASSIOU, D. A., SANZ, H. G.: Initial strontium for a chondrite and the determination of a metamorphism or formation interval. Earth Planet. Sci. Lett. **7**, 33 (1969).

WEBSTER, R. K., MORGAN, J. W., SMALES, A. A.: Some recent Harwell analytical work on geochronology. Trans. Am. Geophys. Union **38**, 543 (1957).

WHETTEN, J. T.: Sediments from the lower Columbia River and origin of graywacke. Science **152**, 1057 (1966).

Revised manuscript received: May 1973

38-C. Abundance in Cosmos, Meteorites, Tektites, and Lunar Materials

I. Cosmos

Ross and Aller (1975) give the solar abundance of Sr as log $N_{Sr} = 2.90$, referred to log $N_H = 12.00$. Using their Si abundance of log $N_{Si} = 7.65$, the solar abundance of Sr, relative to $Si = 10^6$ atoms, is 18. This figure is consistent with the abundance recorded in chondritic meteorites (Table 38-C-1). From the data of Ross and Aller, the solar Ca/Sr atomic ratio is 2800, essentially identical with that for chondrites.

Table 38-C-1. *Strontium in chondrites* (for symbols of classes see: Chapter 4, Vol. 1 of this handbook). All data by I analysis, except Reference 2 (X)

Class (number of samples)		Range ppm Sr	Average ppm Sr	Atoms Sr/ 10^6Si	Ca/Sr atoms	Reference
Ccl	(2)	7.1–7.6	7.4	23	2,300	1
Cc2	(2)	10.8–11.7	11.3	28	2,600	1
Cc2	(3)	9–10	10	25	2,900	2
Cc3	(4)	12–17	14	29	2,600	2
CH	(12)	8–12	9.1	17	2,300	2
CH	(15)	9.3–11.1	10.0	19	2,600	3
CL	(20)	9–12	10.5	18	2,400	2
CL	(5)	10.1–11.9	11.1	19	2,500	4
CLL	(12)	10.5–11.9	11.1	19	2,400	5
Ce	(8)	6.5–8.5	7.7	14	2,700	6

References: 1. Mittlefehldt and Wetherill (1977). 2. Von Michaelis *et al.* (1969). 3. Kaushal and Wetherill (1969). 4. Gopalan and Wetherill (1968). 5. Gopalan and Wetherill (1969). 6. Gopalan and Wetherill (1970).

II. Meteorites

A large number of Sr analyses have been made in connection with Rb-Sr age determinations; these analyses have been carried out by mass-spectrometric isotope-dilution techniques (I). Other techniques used for Sr analysis include emission spectrographic (S) and X-ray fluorescence (X). Each of these techniques appear to give reliable results at the Sr concentrations found in eucrites (Table 38-C-3).

The chondritic abundances are summarized in Table 38-C-1. Strontium is an element which shows a relatively small variation in concentration between different classes of chondrites. Highest concentrations are found in Type III carbonaceous chondrites (Cc3), and lowest concentrations in enstatite chondrites (Ce). Strontium shows geochemical coherence with calcium, as indicated by the rather uniform Ca/Sr ratios in Table 38-C-1. As might be expected, most of the strontium in chondrites

Table 38-C-2. *Strontium in achondrites, excluding eucrites*

Class (name of meteorite)		Content ppm Sr	Analytical method	Reference
Ae	(Bishopville)	12.3	I	1
Ae	(Norton County)	1.4	I	2
Ab	(Johnstown)	2.1	I	3
Ao	(Chassigny)	7.2	M	4
Aop	(Haverö)	0.7	N/R	5
Aa	(Angra dos Reis)	133	I	6
Ado	(Nakhla)	58	I	7
Aor	(Frankfort)	24	X	8
Aor	(Chaves)	40	X	8
Aor	(Malvern)	59	X	8
Aor	(Binda)	33	X	9

References: 1. COMPSTON *et al.* (1965). 2. BOGARD *et al.* (1967). 3. WEBSTER *et al.* (1957). 4. MASON *et al.* (1970). 5. WÄNKE *et al.* (1972). 6. TERA *et al.* (1970). 7. GALE *et al.* (1975). 8. McCARTHY *et al.* (1972). 9. McCARTHY *et al.* (1973).

Table 38-C-3. *Strontium in eucrites (plagioclase-pigeonite achondrites), as determined by different techniques*

Meteorite	Analytical technique X	S	I
	ppm Sr	ppm Sr	ppm Sr
	Reference 1	2	3
Bereba	81	66	74.7
Cachari	84	60	
Chervony Kut		66	
Haraiya	74	80	
Jonzac		70	74.3
Juvinas	78	65	77.1
Macibini	77	66	
Moore County		74	64.1
Nuevo Laredo			80.5
Pasamonte	80	80	76.4
Serre de Mage		31	
Shergotty		35	
Sioux County	76	68	76.0
Stannern	92	97	87.7
Average	80	66	76

References: 1. McCARTHY *et al.* (1973). 2. JÉROME (1970). 3. TERA *et al.* (1970).

resides in the calcium minerals—plagioclase, diopside, and phosphates (chlorapatite and merrillite). In Modoc (L6) ALLEN and MASON (1973) recorded 11 ppm Sr in the bulk meteorite, 75 ppm in plagioclase, 87 ppm in phosphate, and 21 ppm in diopside. In Peace River (L6) GRAY *et al.* (1973) found 11.18 ppm Sr in the bulk meteorite, 92.86 ppm in plagioclase, and 75.49 ppm in phosphate. SHIMA and HONDA (1967), in selective solution experiments, showed that in Bruderheim, an L6 chondrite, practically all the Sr was contained in the HF-soluble fraction (plagioclase and pyroxene), whereas in Abee, an E4 chondrite, about half was in the sulfide fraction, which dissolved in bromine water. Thus strontium resembles calcium and magnesium in showing some chalcophile affinity in enstatite chondrites.

The relatively high concentration of strontium in the Cc3 chondrites can be ascribed, in some of them at least, to the presence of Ca-rich chondrules consisting largely of melilite and fassaite (Ca, Al-rich pyroxene). The Allende meteorite contains 14 ppm Sr, but these chondrules contain 100–250 ppm Sr (GRAY *et al.*, 1973; MARTIN and MASON, 1974). MASON and MARTIN (1974) separated melilite and pyroxene from one of these chondrules (110 ppm Sr), and found 160 ppm in the melilite and 30 ppm in the pyroxene.

Except for eucrites, the data for achondrites are much sparser than for chondrites. They are summarized in Tables 38-C-2 and 38-C-3. The calcium-rich achondrites—angrite (Aa), nakhlite (Ado), howardites (Aor), and eucrites (Ap) are notably enriched in Sr relative to the chondrites. The unique angrite Angra dos Reis has the highest Ca (17.5%) and Sr (133 ppm) concentrations of any meteorite; its Sr has the most primitive $^{87}Sr/^{86}Sr$ ratio (0.69884)[1], this primitive ratio being due to the extremely low Rb content, which has thus essentially contributed no radiogenic ^{87}Sr. The eucrites show fairly uniform Sr contents if the two with anomalously low contents—Serre de Mage and Shergotty—are excluded. Serre de Mage is unusual in being exceptionally rich in plagioclase, and Shergotty may be a unique meteorite distinct from the eucrites.

The data for the calcium-poor achondrites—aubrites (Ae), diogenites (Ab), chassignites (Ao), and ureilites (Aop)—are inadequate for drawing any wide-ranging conclusions regarding these meteorites. The great difference between the Sr contents of the two aubrites reflects the presence of plagioclase in Bishopville and its absence in Norton County. Bishopville consists almost entirely of enstatite (~85%) and plagioclase (~15%); according to COMPSTON *et al.* (1965) the plagioclase contains 77 ppm and the enstatite 1.6 ppm Sr. A plagioclase-free aubrite might be expected, therefore, to contain about 1.6 ppm Sr, and the figure for Norton County (1.4 ppm) is clearly consistent with this interpretation.

III. Tektites

The data on Sr in different groups of tektites are summarized in Table 38-C-4. The most extensive series of measurements is that of COMPSTON and CHAPMAN (1969), wherein different groups of australites and philippinites are distinguished on the basis of major-element composition. If we exclude the high Na/K australites and Darwin glass, the overall range of Sr concentration in tektites is rather limited, as fits their rather restricted bulk chemical composition. TAYLOR and SACHS (1964) noted that in the 43 australites they analyzed (none belonging to the high Na/K group), Sr did not correlate with SiO_2 (as did many major and trace elements) but did correlate positively with Ca; the Ca and Sr contents of these australites were similar to those in greywacke and distinctly different from those of granite and basalt.

[1] For comparison with initial $^{87}Sr/^{86}Sr$ ratios of chondrites a value of 0.69995 in the Guareña H6 type, as reported by WASSERBURG *et al.* (1969), can be mentioned.

Table 38-C-4. *Strontium in tektites, in order of decreasing concentration*

Class (number of samples)	Range ppm Sr	Average ppm Sr	Analytical Method	Reference
Australites, high Na/K (5)	382–420	404	I	1
Philippinites, high Ca	193–440	313	I	1
Ivory Coast (8)	255–355	303	S	2
Ivory Coast (13)	275–399	304	I	3
Australites, high Ca (10)	132–316	217	I	1
Australites, intermediate (4)	179–235	211	I	1
Australites, normal (6)	172–199	187	I	1
Martha's Vineyard (1)	—	180	S	4
Philippinites, normal (9)	155–187	168	I	1
Georgia (7)	116–186	150	S	4
Javanites (13)	117–171	147	I	1
Philippinites, low Ca/Al (6)	122–156	136	I	1
Moldavites (23)	130–156	136	X	5
Moldavites (9)	124–151	136	I	6
Indochinites (19)	99–145	121	I	1
Bediasites (10)	60–130	85	S	4
Darwin glass (8)	13–16	14	S	7

References: 1. COMPSTON and CHAPMAN (1969). 2. CUTTITTA *et al.* (1972). 3. SCHNETZLER *et al.* (1966). 4. CUTTITTA *et al.* (1967). 5. PHILPOTTS and PINSON (1966). 6. SCHNETZLER *et al.* (1969). 7. TAYLOR and SOLOMON (1964).

The high Na/K australites are very rare (only nine have been recorded), and are chemically distinct from all other australites and most other tektites in having an Na/K ratio greater than unity, and have other distinct chemical differences as well as the high Sr content. In addition, they give a fission-track age of approximately 4 million years, quite different from that of all other australites (FLEISCHER *et al.*, 1969). These authors conclude that the high Na/K australites represent a unique tektite fall, preceding that of the Australasian strewnfield, which they place at 0.7 million years ago. On the basis of this and other evidence, CHALMERS *et al.* (1976) conclude that the identification of these specimens as tektites remains in doubt.

Darwin glass, although often included with tektites, has a distinctive composition, being much higher in SiO_2 ($\sim$89%) and much lower in other major and trace elements than tektites. It is probably an impactite formed by the melting of local rocks.

The unique tektite from Martha's Vineyard (Massachusetts) has a strontium content within the range of that for the Georgia tektites. CLARKE and CARRON (1961) noted a close similarity between the Martha's Vineyard specimen and the Georgia tektites in color, density, magnetic properties, and major-element chemistry.

IV. Lunar Materials

In the extensive Rb-Sr dating of lunar materials, many specimens have been analyzed for Sr by mass-spectrometric isotope-dilution techniques. These data have been assembled and reviewed by NYQUIST (1977), and a summary is presented in Table 38-C-5.

Table 38-C-5. *Strontium in lunar materials.* (Data from review article by NYQUIST, 1977)

Mission	Material (number of samples)	Range ppm Sr	Average ppm Sr
Apollo 11	Basalts (low-K) (17)	146–218	175
	Basalts (high-K) (16)	156–184	169
	Soils (3)	163–165	164
Apollo 12	Basalts (42)	81–186	115
	Soils (14)	127–173	153
Apollo 14	Basalts (13)	143–198	172
	Soils (14)	143–187	178
Apollo 15	Basalts (29)	74–130	100
	Anorthosites (4)	172–240	191
	Soils (26)	103–142	130
Apollo 16	Anorthosites (13)	163–235	192
	Soils (15)	169–188	176
Apollo 17	Basalts (33)	121–215	169
	Soils (14)	151–208	168
Luna 16	Basalt (1)	—	437
	Soils (7)	270–306	283
Luna 20	Soils (6)	138–144	141

Relatively few Sr analyses have been made on separated phases from lunar rocks; however, the data show, as for meteorites, that Sr is concentrated in plagioclase. PHILPOTTS and SCHNETZLER (1970) analyzed individual phases from the Apollo 11 rocks, with the following results (ppm Sr): basalt 10044, bulk 167, plagioclase 541, pyroxene 62.6, opaque 29.6; basalt 10062, bulk 194, plagioclase 396, pyroxene 64.6, opaque 41.3.

The data in Table 38-C-5 show that Sr concentrations are rather uniform throughout the whole range of common lunar materials. There appears to be little difference between the highland material of Apollo 16 and the mare material of Apollo 11 and 17, in spite of the much higher plagioclase content of the Apollo 16 rocks and soils. Evidently the anorthositic plagioclase of Apollo 16 has much lower Sr concentration than the plagioclase of the mare basalt. It is interesting that the material collected by Luna 16, from a mare site, has the highest Sr content of any lunar material so far analyzed. The high Sr content of the Luna 16 basalt is not due to a high plagioclase content, but to a high Sr concentration (1,084 ppm) in the plagioclase.

A few clasts of unusual composition show very low Sr contents. The lowest Sr content in a lunar clast is 2.88 ppm in 72417, a fragment of dunite. The low Sr content is consistent with the inability of the large Sr cation to be accomodated in the structure of olivine.

Manuscript received: January 1977

38-D. Abundance in Rock-Forming Minerals; Strontium Minerals

The bulk of the strontium in the Earth's crust occurs as a trace element, dispersed in rock-forming and accessory minerals. Strontium can also be concentrated sufficiently to form its own minerals in hydrothermal deposits, sulfate, and carbonate rocks, although these are of minor importance. A summary of strontium minerals, compiled primarily from VLASOV (1964), FLEISCHER (1966), STRUNZ (1970) and ROBERTS *et al.* (1974), is given in Table 38-D-1. Recent summaries of the geochemistry of strontium have been given by VLASOV (1964), TAYLOR (1965), FAURE and POWELL (1972), and WEHMILLER (1972).

The distribution of strontium in rock-forming minerals is controlled by its diadochy with calcium and potassium. Because Sr^{2+} (1.13 Å) is intermediate in size between Ca^{2+} (0.99 Å) and K^+ (1.33 Å), the substitutional relationships are not simple. Thus Sr has a higher tendency for eight-fold or greater coordination with oxygen than Ca. Ca is able to occupy both six- and eight-fold coordinated lattice positions and K occurs in sites of up to twelve-fold coordination.

Numerous studies of *plagioclases* from mafic igneous intrusions such as the Skaergaard (WAGER and MITCHELL, 1951), the Stillwater (TUREKIAN and KULP, 1956), the layered bodies of Somalia (BUTLER and SKIBA, 1962), and the Gosse Pile intrusion, Australia (MOORE, 1971) have demonstrated trends of increasing Sr and Sr/Ca with decreasing anorthite contents. The classical explanation for this relationship is that the smaller Ca ions are preferred in the plagioclase sites and the liquid becomes progressively enriched in Sr ions. BROOKS (1968) and BERLIN and HENDERSON (1968) reinterpreted such trends in light of liquid compositions, and found that Sr is actually more readily incorporated into plagioclase than Ca. This behavior had been noted previously by WAGER and MITCHELL (1951) and by CARMICHAEL and McDONALD (1961), and was discussed by RINGWOOD (1955). BROOKS (1968) suggested that the observed trends are due to the simultaneous crystallization of clinopyroxene, which has a very strong preference for Ca over Sr; this effect predominates over that of the plagioclase, and the liquid Sr/Ca ratio increases with differentiation. BERLIN and HENDERSON (1968) felt that the effect of the clinopyroxene cannot normally be predominant, and that in most cases the residual liquid would be expected to become depleted in Sr. They felt that the observed trends must therefore be due to the Sr distribution factor increasing with decreasing anorthite content of the plagioclases. EWART and TAYLOR (1969) and PHILPOTTS and SCHNETZLER (1970) presented analytical data which showed that phenocryst/liquid distribution factors for Sr were greater than unity and generally increased with decreasing An content of the plagioclase. MOORE (1971) suggested that the controlling factor in determining the Sr/Ca ratio in plagioclase is probably the plagioclase composition, which in turn determines the lattice structure. The observed trends are then a function of the change of relative ease of Sr entry into the plagioclase lattice with change in composition.

Table 38-D-1. *Strontium minerals*

Halides	
Tikhonenkovite	$SrAlF_4(OH) \cdot H_2O$
Jarlite	$NaSr_3Al_3F_{16}$
Oxide	
Pandaite	$(Ba, Sr)_2(Nb, Ti)_2[O, OH]_7$
Carbonates	
Strontianite	$Sr[CO_3]$
Carbocernaite	$(Na, Ca, Sr, Ce)[CO_3]$
Burbankite	$(Na, Ca, Sr, Ba, Ce)_6[CO_3]_5$
Ambatoarinite	$Sr(Ce, La, Nd)O[CO_3]_3$ (?)
Stenonite	$Sr_2Al[CO_3]F_5$
Benstonite	$MgCa_6(Ba, Sr)_6[CO_3]_{13}$
Ancylite	$(Ce, La)_4(Sr, Ca)_3[CO_3]_7(OH)_4 \cdot 3H_2O$
Weloganite	$Sr_5Zr_2[CO_3]_9 \cdot 4H_2O$
Borates	
Tunellite	$Sr[B_6O_{10}] \cdot 4H_2O$
Kurgantaite	$(Sr, Ca)_2[B_4O_8] \cdot H_2O$
Veatchite	$Sr_2[B_{11}O_{16}](OH)_5 \cdot H_2O$ (monocl.: Aa)
p-Veatchite	$(Sr, Ca)_2[B_{11}O_{16}](OH)_5 \cdot H_2O$ (monocl.: $P2_1/m$)
Strontioginorite	$(Sr, Ca)_2[B_{14}O_{23}] \cdot 8H_2O$
Strontiohilgardite	$(Sr, Ca)_2[B_5O_8](OH)_2Cl$
Strontioborite	$(Sr, Ca)_4Mg_2[B_{24}O_{42}] \cdot 9H_2O$ (?)
Sulfates, phosphates, arsenates and vanadates	
Celestite	$Sr[SO_4]$
Kalistrontite	$K_2Sr[SO_4]_2$
Svanbergite	$SrAl_3[SO_4][PO_4](OH)_6$
Strontium-apatite	$(Sr, Ca)_5[PO_4]_3(OH, F)$
Böggildite	$Na_2Sr_2Al_2[PO_4]F_9$
Palermoite	$(Li, Na)_2(Sr, Ca)Al_4[PO_4]_4(OH)_4$
Goyazite	$SrAl_3[PO_4]_2(OH)_5 \cdot H_2O$
Lusungite	$(Sr, Pb)Fe_3[PO_4]_2(OH)_5 \cdot H_2O$
Belovite	$(Sr, Ce, Na, Ca)_5[PO_4]_3(OH)$
Fermorite	$(Ca, Sr)_5[(As, P)O_4]_3(F, OH)$
Delrioite	$CaSr[V_2O_6](OH)_2 \cdot 2H_2O$
Santafeite	$Na_2(Mn, Ca, Sr)_6Mn_3[(V, As)_6O_{28}] \cdot 8H_2O$
Silicates	
Lamphrophyllite	$Na_2(Sr, Ba)_2Ti_3[SiO_4]_4(OH, F)_2$
Nordite	$Na_3(Sr, Ca)Ce(Mn, Mg, Fe, Zn)_2[Si_6O_{18}]$
Brewsterite	$(Sr, Ba, Ca)[AlSi_3O_8]_2 \cdot 5H_2O$
Haradaite	$SrV[Si_2O_7]$
Yoshimuraite	$(Ba, Sr)_2TiMn_2[SiO_4]_2[PO_4, SO_4](OH, Cl)$

Trends of decreasing Sr contents with decreasing anorthite contents for plagioclases from granitic rocks have been observed by SEN *et al.* (1959) and by HALL (1967). HALL pointed out that these plagioclases contain less than about 20% An, and that the rocks contain large amounts of alkali feldspar which take up some of the Sr which would otherwise go into the plagioclase.

Table 38-D-2. *Sr in plagioclase feldspars* (for references see Table 38-D-3)

Plagioclase	Number of samples	Sr content (ppm)	
		Range	Mean
Anorthite An 90–100	15	130–1,000	289
Bytownite An 70–90	66	141–2,050	605
Labradorite An 50–70	141	180–3,000	1,013
Andesine An 30–50	111	169–3,000	906
Oligoclase An 10–30	41	105–4,250	1,002
Albite An 0–10	43	3–5,000	484

HEIER (1962) summarized the available data on Sr in plagioclases ranging in composition from albite to anorthite, and noted no direct relation between Sr and Ca. His summary table has been updated and is presented here as Table 38-D-2. EWART and TAYLOR (1969) found a strong tendency for maximum Sr concentration in the compositional range An_{40} to An_{55}, with no evidence of a simple correlation with Ca. HEIER (1962) suggested that the low Sr contents in albites might be explained by lack of availability in systems where albite forms. NEIVA (1975) found that albites from pegmatites generally have less Sr than albites from the corresponding parental granites, reflecting magmatic differentiation. HEIER (1966) reported unusually high Sr contents in albites from nepheline-albite pegmatites (Table 38-D-3).

Strontium readily substitutes for potassium in *K-feldspars*; according to HEIER (1966) the K is effectively in nine-fold coordination with oxygen. BERLIN and HENDERSON (1969) showed that Sr is enriched relative to K in crystallizing K-feldspar, consistent with the capture principle, and that the residual liquid is depleted in Sr relative to the original liquid. RHODES (1969) found that the Sr contents of K-feldspars are not influenced by ordering processes; the Sr content of orthoclase is not significantly different from that of microcline (Table 38-D-3). HEIER and TAYLOR (1959) found no apparent relationship between the Sr and Ca contents of K-feldspars from granites and pegmatites. They also observed that during differentiation K-feldspar is enriched in Sr relative to Ca (the same relationship which has already been noted for plagioclase). BERLIN and HENDERSON (1969) demonstrated this Sr/Ca enrichment through analyses of K-feldspar phenocrysts and groundmass phases.

The distribution of Sr between coexisting plagioclase and alkali feldspar has received much attention. HEIER (1962), considering data available in the literature for igneous and metamorphic rocks, found that Sr is present in the two feldspars in similar amounts. The ratio (Sr in K-feldspar/Sr in plagioclase) ranged between 0.5 and 2. SEN *et al.* (1959), HALL (1967), and NAGASAWA (1971) found that plagioclases contain more Sr than coexisting K-feldspars in granitic rocks. This is also shown by the data of NOCKOLDS and MITCHELL (1948). BERLIN and HENDERSON (1969) showed that the Sr content of plagioclases is much higher than that of coexisting alkali feldspars from porphyritic volcanic rocks regardless of crystallization history. Thus plagioclase seems to be a better solvent for Sr than alkali feldspar, at least at liquidus temperatures. Contradictory data for feldspars from metamorphic and other coarse-grained rocks (Table 38-D-4) may largely be due to subsolidus rearrangements.

The strontium concentration levels in the other common rock-forming silicate minerals are considerably lower than those found in the feldspars. This is best demonstrated by a comparison of partition coefficients which relate the concentration of Sr in a mineral phase to its concentration in the liquid in equilibrium with the mineral. WAGER and MITCHELL (1951) pointed out that such partition coefficients make allowance for varying concentrations of a trace element in different magmas, thus illustrating the relative ease of entry of a trace element into the structure of a particular mineral. Mineral/liquid Sr partition coefficients (D_{Sr}) have been measured through Sr analyses of the phenocryst and groundmass phases of various porphyritic volcanic rocks. The available data are summarized in Table 38-D-5; order of magnitude differences in the mean values are probably significant. Only the feldspars have values of D_{Sr} greater than unity.

Numerous workers (e.g., WAGER and MITCHELL, 1951) have noted the limited diadochy between Sr and Ca in the *clinopyroxenes*. This is apparently due to the fact that, although Ca is surrounded by eight oxygens, it is bonded to only six (TAYLOR, 1965). Nevertheless, clinopyroxenes from some ultramafic nodules in kimberlite pipes and basalts contain several hundred ppm Sr (Table 38-D-3). The lack of structural sites which can readily accomodate Sr in orthopyroxene and in olivine accounts for the extremely low D_{Sr} values for these two minerals. The Sr which is present (Table 38-D-3) may be in structural defects or adsorbed on grain surfaces. The relative partitioning of Sr between clinopyroxene, orthopyroxene and olivine is illustrated by the analyses of coexisting minerals from ultramafic rocks (Table 38-D-4).

The rather limited substitution of Sr for Ca in *amphiboles* has been discussed by MOXHAM (1964) and by TAUSON (1965). MOXHAM (1964) suggested that the largest ion which can occupy the Ca site should have a radius no larger than about 1.00 Å. NAGASAWA and SCHNETZLER (1971) suggested that Sr does not occupy the Ca site in hornblende but occurs in the larger, otherwise vacant site (cavity) in the structure. The Sr partition coefficient for amphibole seems to be about the same as that for clinopyroxene, although Sr analyses for the two minerals from the same rock indicate relative enrichment in the amphibole.

The low Sr concentration levels in the *micas* are probably due to the twelve-fold coordination of K in these minerals; this site is apparently too large for Sr (TAYLOR, 1965). The partition coefficient for Sr in micas (based on just three determinations) is about the same as D_{Sr} for amphiboles, although the data of DODGE and ROSS (1971) clearly show an enrichment of Sr in hornblende over coexisting biotite in granitic rocks. When biotite and muscovite coexist in metamorphic rocks there seems to be a tendency for significantly higher levels of Sr in the latter phase (Table 38-D-4).

TAYLOR (1965) stated that the entry of trace elements into crystal lattices of metamorphic minerals is controlled by the same factors which operate during crystallization of a magma (ionic size and charge, bond type, type of crystal lattice, and

Table 38-D-3. *Strontium in rock-forming minerals*

Sample, source	Number of samples	Sr content (ppm) Range	Mean (Method)	Reference
K-Feldspar				
Phenocrysts, trachytes and phonolites	9	240–2,700	1,268 (S)	Berlin and Henderson (1969)
Phenocrysts, peralkaline silicic rocks, Ethiopia	4	11.4–27.7	20.6 (I)	Dickinson and Gibson (1972)
Orthoclases, granites, Australia	26	7–663	203 (X)	Rhodes (1969)
Microclines, granites, Australia	44	9–1,430	213 (X)	Rhodes (1969)
Granites and rhyolites, Ontario and Nova Scotia, (Canada)	24	20.2–458	88.6 (I)	Fairbairn *et al.* (1960), (1961); Cormier (1972); Gibbins and McNutt (1975)
Granites, Ireland	27	120–640	404 (X)	Hall (1967); Wilson and Coats (1972)
Granites, Lone Grove pluton, Texas (U.S.A.)	10	59–225	162 (I)	Zartman (1964)
Granites and rhyolites, Australia	20	13.3–455	126 (I, X)	Pidgeon and Compston (1965); McDougall *et al.* (1966); White *et al.* (1967)
Granites, San Isabel batholith, Colorado (U.S.A.)	122	not given	561 (X)	Murray and Rogers (1973)
Cape Granite, coarsely porphyritic, S. Africa	35	105–240	166 (S)	Kolbe and Taylor (1966)
Granitic rocks, N. Portugal	29	135–2,270	799 (S)	De Albuquerque (1975)
Granites, N. Portugal	8	35–160	117 (S)	Neiva (1975)
Aplites and pegmatites, N. Portugal	16	3–150	47 (S)	Neiva (1975)
Granitoid rocks, Ural Mts. (U.S.S.R.)	17	10–300	96 (S)	Fershtater *et al.* (1969)
Granites and gneisses, Precambrian, S. Norway	39	150–1,000	471 (S)	Heier and Taylor (1959)
Granites and gneisses, SW Minnesota (U.S.A.)	16	81.1–533	368 (I)	Goldich *et al.* (1970)
Granitic gneisses, Willyama complex, Broken Hill (Australia)	14	48.01–659.6	170 (I)	Pidgeon (1967)
Pegmatites, Connecticut and Texas (U.S.A.)	8	11.5–48	27 (I)	Zartman (1964); Brookins *et al.* (1969)
Pegmatites, large, Precambrian, S. Norway	38	20–310	107 (S)	Heier and Taylor (1959)
Granodiorite and quartz monzonite, Utah (U.S.A.)	20	141–456	301 (A)	Kuryvial (1976)
Quartz diorites and granodiorites, Donegal, (Ireland)	7	450–720	619 (X)	Hall (1967)
Nepheline syenites, Stjernöy (N. Norway)	3	4,100–5,100	4,667 (S)	Heier (1966)
Metamorphic rocks, S. W. New Hampshire (U.S.A.)	20	13–1,334	286 (A)	Scotford (1973)
Metamorphic rocks, amphibolite facies, Australia and Ceylon	31	120–570	335 (X)	Virgo (1969)

Metamorphic rocks, granulite facies, Australia and Ceylon	28	146–1,537	389 (X)	Virgo (1969)
Metamorphic rocks, high-grade, Langöy, (N. Norway)	18	500–1,700	1,032 (S)	Heier (1960)
Plagioclase				
Phenocrysts, basalts and andesites	23	258–1,750	545 (I, X, S)	Ewart and Taylor (1969); Griffin and Murthy (1969); Philpotts and Schnetzler (1970); Ewart *et al.* (1973)
Phenocrysts, dacites	14	416–1,310	827 (X, S, A, I)	Dudas *et al.* (1971); Nagasawa and Schnetzleer (1971); Ewart *et al.* (1973)
Phenocrysts, rhyolites and dacite, New Zealand	12	490–880	703 (S)	Ewart and Taylor (1969)
Phenocrysts, trachytes and phonolites	5	1,600–4,000	2,380 (S)	Berlin and Henderson (1969)
Basalts and andesites, Izu-Hakone region (Japan)	38	130–640	315 (S)	Iida (1961)
Dolerite dikes, Precambrian, Canada and U.S.A.	12	163.4–606.8	316 (I)	Gates and Hurley (1973); Stueber *et al.* (1976)
Lunar basalts, Apollo 11 and Apollo 12	7	299–531	435 (I)	Compston *et al.* (1970); Schnetzler and Philpotts (1971)
Anorthosite, Nain, Labrador (Canada)	13	505–822	670 (I)	Gill and Murthy (1970)
Anorthosite massifs, Norway	7	414–1,180	915 (N/R)	Griffin *et al.* (1974)
Gabbros and pyroxenites, Gosse Pile intrusion (Central Australia)	13	403–1,337	579 (X)	Moore (1971)
Gabbros and metagabbros, two basic layered masses, Somalia	31	1,000–1,450	1,265 (X)	Skiba and Butler (1963)
Granitic rocks, Donegal (Ireland)	17	120–1,240	641 (X)	Hall (1967)
Granitic rocks, Pilbara (W. Australia)	7	0.652–439.6	192 (I)	Oversby (1976)
Granitic rocks, Isle of Skye (N. W. Scotland)	5	16.1–146	59.6 (I)	Moorbath and Bell (1965)
Albites, granites, N. Portugal	8	30–120	61 (S)	Neiva (1975)
Albites, aplites and pegmatites, N. Portugal	16	4–100	22 (S)	Neiva (1975)
Albites, pegmatites, Virginia and Connecticut (U.S.A.)	10	3.6–49.7	17.7 (I)	Brookins *et al.* (1969); Laughlin (1973)
Albites, nepheline-albite pegmatites, Stjernöy, (N. Norway)	5	1,500–5,000	3,880 (S)	Heier (1966)
Amphibolites, amphibolite facies, N. W. Adirondack Mts., New York (U.S.A.)	7	320–600	506 (S)	Engel *et al.* (1964)

Table 38-D-3 (continued)

Sample, source	Number of samples	Sr content (ppm) Range	Mean (Method)	Reference
Amphibolites, granulite facies, N. W. Adirondack Mts., New York (U.S.A.)	7	340–560	437 (S)	ENGEL *et al.* (1964)
Metamorphic rocks, amphibolite facies, Australia and Ceylon	31	38–729	286 (X)	VIRGO (1969)
Metamorphic rocks, granulite facies, Australia and Ceylon	28	33–659	207 (X)	VIRGO (1969)
Amphibolite facies rocks, world-wide distribution	34	170–1,100	420 (S)	SEN (1960)
Granulite facies rocks, world-wide distribution	26	170–1,940	1,014 (S)	SEN (1960)
Metamorphic rocks, S. W. New Hampshire (U.S.A.)	20	33–1,496	294 (A)	SCOTFORD (1973)
Amphibolite and granulite rocks, Somalia	35	250–1,100	508 (X)	SKIBA and BUTLER (1963)
Biotite				
Granite, Cape (S. Africa)	21	3–11.2	5.8 (S)	KOLBE and TAYLOR (1966)
Granites, Nova Scotia, (Canada)	20	3.3–35.4	9.83[a] (I)	FAIRBAIRN *et al.* (1960)
Granites and rhyolites, Sudbury, Ontario (Canada)	14	13.16–59.89	27.7 (I)	FAIRBAIRN*et al.* (1961) GIBBINS and MCNUTT (1975)
Granitic rocks, west coast, North America	22	3.17–45.0	17.5 (I, S)	WANLESS *et al.* (1968); MENZER (1970); DODGE and ROSS (1971)
Granitic rocks, Cape Breton Is., Nova Scotia (Canada)	25	9.90–158	67.8 (I)	CORMIER (1972)
Granitic rocks, Suriname, (South America)	17	8.2–34.0	15.1 (I)	PRIEM *et al.* (1971)
Granitic rocks, Sierra Nevada batholith, California (U.S.A.)	33	<6–28	≤ 12 (S)	DODGE *et al.* (1969)
Granodiorite and adamellite, S. E. Australia	30	4.55–22.29	10.7 (I)	WILLIAMS *et al.* (1975)
Granites, schists, and gneisses, Dalradian Series (Scotland)	31	6.7–206	32.2[a] (I)	BELL (1968)

Granites and gneisses, S. W. Minnesota (U.S.A.)	15	3.94–36.94	21.7[a] (I)	Goldich *et al.* (1970)
Alkalic intrusions, syenite to peridotite, U.S.A.	20	14.0–450	162 (I)	Zartman *et al.* (1967)
Pegmatites, North Carolina and Georgia (U.S.A.)	4	3.94–10.21	7.02 (I)	Deuser and Herzog (1962)
Mafic igneous complexes, N. E. Scotland	9	6.9–33.3	14.4 (I)	Pankhurst (1970)
Norites, Sudbury, Ontario (Canada)	6	14.3–45.1	27.8 (I)	Gibbins and McNutt (1975)
Gneisses, Manhatten and Reading Prongs, New York (U.S.A.)	12	3.5–27.5	15.0[a] (I)	Long and Kulp (1962)
Gneisses, Limpopo Belt, (S. Africa)	11	12.3–152	39.3 (I)	Van Breemen and Dodson (1972)
Gneisses and schists, epidote-amphibolite facies, Ontario (Canada)	20	19–92	46 (S)	Moxham (1964)
Paragneisses, N. W. Adirondack Mts., New York (U.S.A.)	22	15–20	17 (S)	Engel and Engel (1960)
Schists, gneisses and amphibolites, E. Alps (Austria)	13	6.0–110	32.2 (I)	Hawkesworth (1976)
Igneous and metamorphic rocks, Swiss and Italian Alps	18	0.867–10.2	3.07[a] (I)	Armstrong and Jager (1966)
Metamorphosed iron formation, Grangesberg (Sweden)	13	2–50	13 (A)	Annersten and Ekstrom (1971)
Muscovite				
Granites, Nova Scotia (Canada)	5	3.3–45.2	19.6[a] (I)	Fairbairn *et al.* (1960)
Granites and pegmatites, Dalradian Series (Scotland)	7	10.7–85.2	32.5[a] (I)	Bell (1968)
Granites, aplites and pegmatites, N. Portugal	9	3.0–10	4.3 (S)	Neiva (1975)
Granitic rocks, N. Portugal	10	5–34	14 (S)	De Albuquerque (1975)
Granodiorites, British Columbia (Canada)	8	5.8–17.0	11.1 (I)	Ryan and Blenkinsop (1971)
Pegmatites, Blue Ridge and Piedmont provinces (U.S.A.)	14	5.11–48.44	19.36 (I)	Deuser and Herzog (1962)
Pegmatites, Willyama complex, Broken Hill (Australia)	8	5.68–21.81	10.2[a] (I)	Pidgeon (1967)
Pegmatites, Wyoming, Colorado, Connecticut (U.S.A.)	9	4.91–18.9	13.3 (I)	Hills *et al.* (1968); Brookins *et al.* (1969) Vera and van Schmus (1974)

Table 38-D-3 (continued)

Sample, source	Number of samples	Sr content (ppm) Range	Mean (Method)	Reference
Pegmatites, Moines (N. Scotland)	27	4.17–126	38.8 (I)	Van Breemen *et al.* (1974)
Schists, Dalradian Series (Scotland)	2	64.4–69.8	67.1 (I)	Bell (1968)
Schists and gneiss, Hida metamorphic belt (Japan)	4	23.9–254	114 (I)	Shibata *et al.* (1970)
Schists and gneisses, E. Alps (Austria)	17	2.0–398	180 (I)	Hawkesworth (1976)
Amphibole				
Hornblende phenocrysts, rhyolites and dacites	6	21.0–36.0	28.2 (I, S)	Ewart and Taylor (1969); Nagasawa and Schnetzler (1971);
Hornblende phenocrysts, camptonites	2	481–503	492 (I)	Philpotts and Schnetzler (1970)
Kaersutites, xenoliths in dikes and breccias	10	30–1,060	697 (X)	Kesson and Price (1972)
Hornblendes, granitic rocks, California (U.S.A.)	38	3–120	38 (S)	Sen *et al.* (1959); Dodge *et al.* (1968) Dodge and Ross (1971)
Hornblendes, tonalitic rocks, N. Portugal	11	5–24	12 (S)	De Albuquerque (1974)
Hornblendes, quartz diorite and granite, Ben Nevis (Scotland)	9	5–10	7 (S)	Haslam (1968)
Hornblendes, diorite porphyry, Henry Mts., Utah (U.S.A.)	15	$<$50–300	$\leq$167 (S)	Engel (1959)
Hornblendes, ultramafic rocks, Ural Mts. (U.S.S.R.)	7	80–500	317 (S)	Borisenko (1967)
Hornblendes, amphibolites, amphibolite facies, Adirondack Mts., New York (U.S.A.)	7	26–64	46 (S)	Engel and Engel (1962)
Hornblendes, amphibolites, granulite facies, Adirondack Mts., New York (U.S.A.)	9	28–78	47 (S)	Engel and Engel (1962)
Alkali amphiboles, blueschists, California (U.S.A.)	14	$<$4–220	$\leq$37 (S)	Coleman and Papike (1968)
Hornblendes, gneisses and schists, N. W. Quebec (Canada)	20	14–77	37 (S)	Moxham (1964)
Hastingsitic amphiboles, metasomatic alkaline gneisses, Ontario (Canada)	14	33–331	149 (S)	Appleyard (1975)

Hornblendes, charnockites, Kondapalli (India)	9	10–85	36 (S)	Leelanandam (1970)
Hornblendes, skarns and hybrid rocks, Quebec and Ontario (Canada)	5	52–262	95 (S)	Moxham (1960)
Ca-amphiboles, metamorphosed iron formation, Grängesberg (Sweden)	10	11–203	37 (A)	Annersten and Ekstrom (1971)
Clinopyroxene				
Phenocrysts, andesites and dacites	17	13–35	25 (S)	Ewart and Taylor (1969); Ewart *et al.* (1973)
Phenocrysts, basalts	4	16.6–83.5	46.3 (I)	Griffin and Murthy (1969); Hart and Brooks (1974a)
Phenocrysts, volcanic rocks	5	11.6–92.8	53.1 (I)	Philpotts and Schnetzler (1970)
Augites and pigeonites, lunar basalts, Apollo 12	6	3.51–38.0	13.9 (I)	Schnetzler and Philpotts (1971)
Gabbro cumulates, Kap Edvard Holm complex (E. Greenland)	9	15–70	27 (S)	Deer and Abbott (1965)
Gabbros and pyroxenites, Gosse Pile layered intrusion (C. Australia)	11	4–44	18 (X)	Moore (1971)
Gabbros and diorites, Bushveld intrusion (S. Africa)	9	25–40	31 (S)	Atkins (1969)
Anorthosite-mangerite rocks, Quebec (Canada)	10	5–50	17 (S)	Philpotts (1966)
Teschenite sill, New South Wales (Australia)	7	125–175	139 (S)	Wilkinson (1959)
Dolerite dikes, Precambrian, North America	11	29.9–557	125 (I)	Gates and Hurley (1973); Stueber *et al.* (1976)
Hedenbergites, Ilimaussaq syenite (S. Greenland)	5	16–18	17 (S)	Larsen (1976)
Ultramafic intrusions, W. North Carolina (U.S.A.)	3	5.02–9.07	7.65 (I)	Stueber (1969)
Ultramafic flows and sills, Canadian shield	7	3.6–20.0	12.3 (I)	Hart and Brooks (1974b)
Lherzolite, Mid-Indian Ocean Ridge	1		8.9 (I)	Hart (1972)
Lherzolite and wehrlite inclusions, basalts	6	16.0–224	68.2 (I)	Stueber and Ikramuddin (1974)
Lherzolite inclusions, basalts, Victoria (Australia)	7	110–356	207 (I)	Burwell (1975)
Peridotite inclusions, basalts	15	5.79–200	73.7 (I, S, X)	Ross *et al.* (1954); Peterman *et al.* (1970) Hutchison and Dawson (1970); Paul (1971); Philpotts *et al.* (1972)
Lherzolite inclusions, basalts, Salt Lake Crater (Hawaii)	7	71.2–139.5	101 (I)	Shimizu (1974)

Table 38-D-3 (continued)

Sample, source	Number of samples	Sr content (ppm)		Reference
		Range	Mean (Method)	
Eclogite inclusions, basalts	13	52.5–123	95.2 (I)	Griffin and Murphy (1969) Shimizu (1974)
Lherzolite inclusions, kimberlite pipes, S. Africa	7	84.6–558	233 (I)	Shimizu (1974)
Peridotite and eclogite inclusions, kimberlite pipes, S. Africa	22	58.9–853	278 (I, S)	Nixon *et al.* (1963); Allsopp *et al.* (1969) Griffin and Murthy (1969); Philpotts *et al.* (1972)
Eclogite lenses, metamorphic terranes	10	30–223	84.4 (I, S)	Allsopp *et al.* (1969); Bryhni *et al.* (1969) Griffin and Murphy (1969)
Skarns, Quebec and Ontario (Canada)	23	36–123	77 (S)	Moxham (1960)
Amphibolites, N. W. Adirondack Mts. New York (U.S.A.)	7	10–16	13 (S)	Engel *et al.* (1964)
Orthopyroxene				
Hypersthene phenocrysts, andesites and dacites	7	0.80–17.5	6.36 (I, S)	Ewart and Taylor (1969); Philpotts and Schnetzler (1970); Nagasawa and Schnetzler (1971)
Norites, S. Quebec (Canada)	2	5–20	12.5 (S)	Philpotts (1966)
Bronzites, gabbros, Bushveld intrusion (S. Africa)	3	17–21	19 (S)	Atkins (1969)
Mafic pluton, Kola Peninsula (U.S.S.R.)	1		1.69 (I)	Birck and Allegre (1973)
Ultramafic intrusions, W. North Carolina (U.S.A.)	3	0.226–0.462	0.32 (I)	Stueber (1969)
Lherzolite inclusions, basalts	11	0.318–15.0	3.23 (I, X)	Peterman *et al.* (1970); Paul (1971) Philpotts *et al.* (1972); Stueber and Ikramuddin (1974)
Lherzolite inclusions, basalts, Victoria (Australia)	7	0.942–5.56	2.10 (I)	Burwell (1975)

Peridotite inclusions, kimberlite pipes, S. Africa	4	2.26–16.9	8.29 (I)	Griffin and Murthy (1969) Philpotts *et al.* (1972)
Granulite inclusion, basic pipe, Australia	1		12.1 (I)	Compston and Lovering (1969)
Amphibolites, Adirondack Mts., New York (U.S.A.)	6	<10–14	≤10 (S)	Engel *et al.* (1964)
Olivine				
Phenocrysts, basalts	5	0.459–13.9	4.05 (I)	Griffin and Murthy (1969) Philpotts and Schnetzler (1970); Hart and Brooks (1974a)
Lunar troctolitic granulite, Apollo 17	1		0.82 (I)	Bogard *et al.* (1974)
Lunar troctolitic microbreccia, Apollo 17	2	12.5–21.0	16.7 (I)	Nunes *et al.* (1974)
Fayalites, Ilimaussaq syenite (S. Greenland)	2	36–37	36.5 (S)	Larsen (1976)
Ultramafic intrusions, W. North Carolina (U.S.A.)	2	0.182–0.383	0.28 (I)	Stueber (1969)
Ultramafic intrusion, Wyoming (U.S.A.)	2	2.33–2.88	2.61 (I)	Davis (1974)
Peridotite inclusions, basalts	12	0.234–15.5	2.35 (I)	Paul (1971); Kudo *et al.* (1972); Stueber and Ikramuddin (1974); Burwell (1975)
Peridotite inclusions, kimberlite pipes, S. Africa	3	0.162–1.09	0.53 (I)	Griffin and Murthy (1969)
Peridotite lense, metamorphic terrane, Norway	1		2.16 (I)	Griffin and Murthy (1969)
Garnet				
Phenocryst, dacite, Japan	1		3.57 (I)	Philpotts and Schnetzler (1970)
Eclogite inclusions, basalts	3	1.59–1.82	1.71 (I)	Griffin and Murthy (1969)
Eclogite and peridotite inclusions, kimberlite pipes, Africa	12	0.98–33.87	10.8 (I)	Griffin and Murthy (1969); Allsopp *et al.* (1969); Philpotts *et al.* (1972) Shimizu (1974)
Eclogite inclusion, basic pipe, Australia	1		5.5 (I)	Compston and Lovering (1969)
Eclogitic rocks, gneiss massif, Bavaria (Germany)	11	2–90	31 (S)	Hahn-Weinheimer and Luecke (1963)
Eclogite lenses, metamorphic terranes	4	3.34–14.5	9.41 (I)	Griffin and Murthy (1969); Allsopp *et al.* (1969)
Gneisses, Nordfjord (Norway)	6	10–20	12 (S)	Bryhni *et al.* (1969)
Blueschists, Franciscan, California (U.S.A.)	23	<4–130	≤12 (S)	Lee *et al.* (1963)
Charnockitic granite, Pyrenees (France)	1		15.4 (I)	Vitrac-Michard and Allègre (1975)

Table 38-D-3 (continued)

Sample, source	Number of samples	Sr content (ppm) Range	Mean (Method)	Reference
Other minerals				
Glauconites, sedimentary rocks	47	3.84–163	26.8 (I)	Hurley *et al.* (1960); McDougall *et al.* (1965); Obradovich and Peterman (1968)
Phlogopites, ultramafic inclusions, kimberlite pipes, S. Africa	3	58.96–180.7	105 (I)	Allsopp *et al.* (1969)
Phlogopites, Stillwater ultramafic rocks, Montana (U.S.A.)	3	3.9–12.1	6.8 (I)	Kistler *et al.* (1969)
Phlogopite phenocrysts, phonolite and rhyodacite	2	108–121	115 (I)	Philpotts and Schnetzler (1970)
Nephelines, nepheline syenites and pegmatites, Stjernöy (Norway)	8	80–430	209 (S, I)	Heier (1966)
Leucites, leucitites and leucite basalts	5	110–1,400	482 (S, X)	Henderson (1965)
Pseudoleucites, juvites and tinguaites	7	260–2,050	783 (S, X)	Henderson (1965)
Scapolites, skarns, Grenville province (Canada)	40	120–3,700	1,870 (S)	Shaw *et al.* (1963)
Calcites, skarns, Grenville province (Canada)	19	530–6,600	2,350 (S)	Shaw *et al.* (1963)
Calcites, carbonatites, Nyanza (Kenya)	4	1,220–3,400	2,170 (S)	Barber (1974)
Melilites, Oka complex, Quebec (Canada)	5	8,400–10,900	9,200 (M)	Watkinson (1972)
Sphenes, Koksharov massif (U.S.S.R.)	12	0–6,700	2,140 (X)	Rass (1964)
Apatites, igneous environments, worldwide distribution	21	35–73,558	1,449 (S)	Cruft (1966)
Apatites, metamorphic environments, worldwide distribution	25	8–9,839	2,872 (S)	Cruft (1966)
Apatites, pegmatites, various localities	18	4.9–4,320	483 (I)	Riley (1970)
Apatites, pegmatites, E. Siberia (U.S.S.R.)	11	250–1,010	620 (X)	Shmakin and Shiryayeva (1968)
Apatites, carbonatites, Africa	4	4,540–7,680	6,351 (X)	Prins (1973)

[a] Non-radiogenic Sr only.

Table 38-D-4. *Strontium in some coexisting minerals of igneous and metamorphic rocks*

Felsic rocks	Sr content of minerals (ppm)				Analytical method	Reference
	Plagioclase	K-feldspar	Biotite	Muscovite		
Granite	53.3	120	13.2	9.01	I	PETERMAN *et al.* (1968)
Pegmatite	19.2	24.4	7.00	5.69	I	PETERMAN *et al.* (1968)
Rhyolite porphyry	65.7	76.5	4.63		I	McDOUGALL *et al.* (1966)
Granodiorite	540	216	18.7		I	NAGASAWA (1971)
Quartz monzonite	101	204	37.9		X	WHITNEY *et al.* (1976)
Gneiss	158[a]	165[a]	3.68[a]	33.2[a]	I	BAADSGAARD *et al.* (1976)
Gneiss	874	832	40.5		I	HAWKESWORTH (1976)
Schist			18.9	67.6	I	MOORBATH *et al.* (1968)

Mafic rocks	Sr content of minerals (ppm)					Analytical method	Reference
	Plagioclase	Clinopyroxene	Orthopyroxene	Olivine	Amphibole		
Gabbro, Skaergaard	1,000	20		10		S	WAGER and MITCHELL (1951)
Gabbro	280	18				X	EWART and BRYAN (1972)
Norite	495	44.3	1.69			I	BIRCK and ALLÈGRE (1973)
Dolerite	290	47.1				I	GATES and HURLEY (1973)
Basalt porphyry	1,048	43.4		0.098		I	HART and BROOKS (1974a)
Lunar basalt	396	64.6				I	PHILPOTTS and SCHNETZLER (1970)
Lunar troctolite	207			0.817		I	BOGARD *et al.* (1974)
Granulite inclusion, mafic pipe	613	80	12.1			I	COMPSTON and LOVERING (1969)
Amphibolite	420	14	12		42	S	ENGEL *et al.* (1964)
Eclogite inclusion, mafic pipe		156			482	I	COMPSTON and LOVERING (1969)

Table 38-D-4 (continued)

Ultramafic rocks	Sr content of minerals (ppm)				Analytical method	Reference
	Clinopyroxene	Orthopyroxene	Olivine	Garnet		
Lherzolite inclusion, basalt	42.1	2.48	1.13		I	DASCH and GREEN (1975)
Wehrlite inclusion, basalt	47.9		0.783		I	STUEBER and IKRAMUDDIN (1974)
Lherzolite inclusion, kimberlite pipe	139	2.26	0.162	11.8	I	GRIFFIN and MURTHY (1969)
Alpine lherzolite intrusion	5.02	0.462	0.182		I	STUEBER (1969)
Eclogite inclusion, kimberlite pipe	123			1.72	I	GRIFFIN and MURTHY (1969)
Eclogite lense, metamorphic terrane	141			10.9	I	ALLSOPP *et al.* (1969)
Peridotite lense, metamorphic terrane	95.3		2.16		I	GRIFFIN and MURTHY (1969)

[a] Non-radiogenic Sr only.

Table 38-D-5. *Phenocryst/matrix partition coefficients for Sr*

Mineral	Number of determinations	Range	Mean	Matrix (see footnote)	Reference
K-Feldspar	10	1.90 –17.3	5.92	c, d, f, g, h	1, 4, 8, 9
Plagioclase	39	1.22 –31.4	5.78	a, b, c, e, f, g, h	1, 2, 4 through 11
Clinopyroxene	21	0.0019 –0.516	0.128	a, b, d,	2, 3, 5, 6, 8, 11
Orthopyroxene	7	0.0085 –0.046	0.027	b, c	5, 8, 10
Olivine	5	0.00019–0.018	0.006	a	6, 8, 11
Amphibole	10	0.022 –0.641	0.327	b, c, e, i	5, 8, 10
Mica	3	0.081 –0.672	0.291	c, d, g	8

Matrices: a) basalt; b) andesite; c) dacite; d) rhyodacite; e) rhyolite; f) trachyte; g) phonolite; h) pitchstone; i) camptonite.

References: 1. CARMICHAEL and MCDONALD (1961); 2. BERLIN and HENDERSON (1968); 3. ONUMA *et al.* (1968); 4. BERLIN and HENDERSON (1969); 5. EWART and TAYLOR (1969); 6. GRIFFIN and MURTHY (1969); 7. HIGUCHI and NAGASAWA (1969); 8. PHILPOTTS and SCHNETZLER (1970); 9. NAGASAWA (1971); NAGASAWA and SCHNETZLER (1971); 11. HART and BROOKS (1974a).

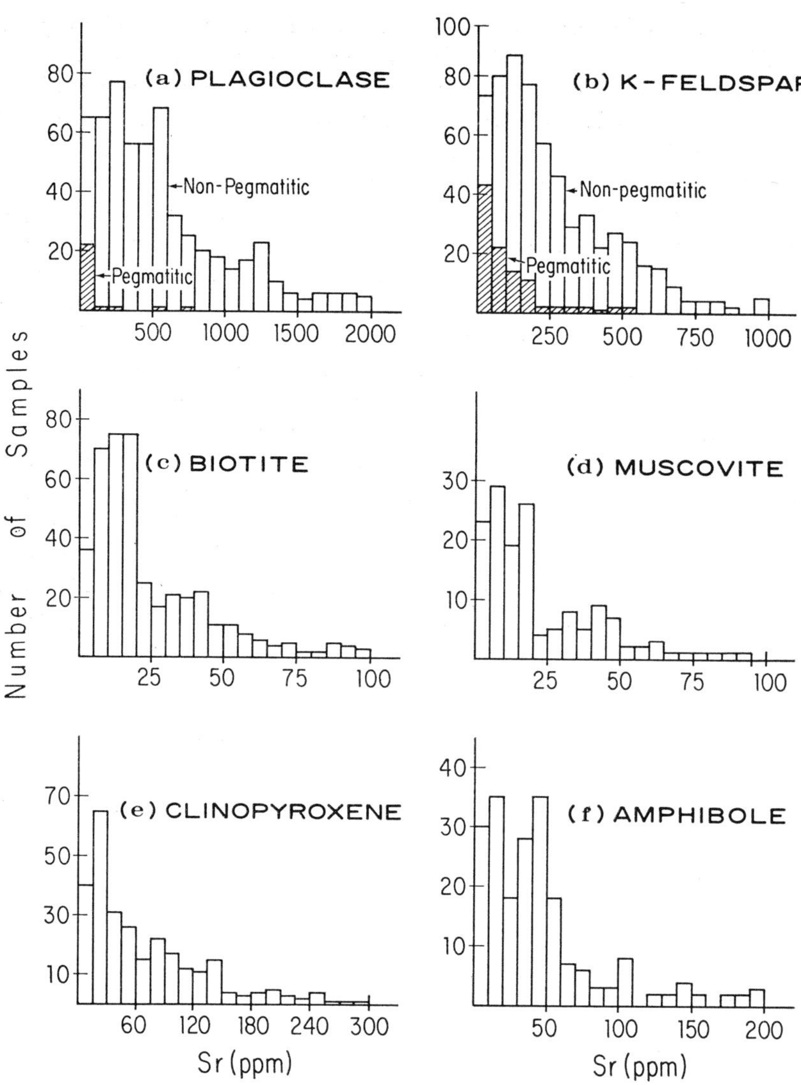

Fig. 38-D-1. (a) Distribution of Sr in 629 plagioclases (24 values above 2,000 ppm; max. 5,000 ppm). (b) Distribution of Sr in 762 K-feldspars (45 values above 1,000 ppm; max. 5,100 ppm). (c) Distribution of *total* Sr in 454 biotites (32 values above 100 ppm; max. 450 ppm). (d) Distribution of *total* Sr in 167 muscovites (19 values above 100 ppm; max. 398 ppm). (e) Distribution of Sr in 296 clinopyroxenes (14 values above 300 ppm; max. 853 ppm). (f) Distribution of Sr in 242 amphiboles (33 values above 200 ppm; max. 1,060 ppm)

trace element concentration). In general, similar Sr concentration levels are found in a given mineral occurring in igneous and in metamorphic rocks (Table 38-D-3). However, the additional influence of physical environment (temperature-pressure conditions) has been considered in studies of Sr distribution between coexisting minerals as a function of metamorphic grade.

In a study of high-grade metamorphic rocks from northern Norway, HEIER (1960) found Sr enrichment in K-feldspars over coexisting plagioclases. Systematic and significant changes in Sr partitioning between coexisting feldspars with gross changes in metamorphic grade were demonstrated by VIRGO (1968) for rocks from several areas in Australia and Ceylon. SCOTFORD (1973) showed that the partitioning of Sr between coexisting feldspars in rocks from southern New Hampshire (U.S.A.) is dependent on metamorphic grade. Both authors found that increasing metamorphic grade results in preferential entrance of Sr into the K-feldspar phase.

PEARSON and SHAW (1960) analyzed 22 samples of kyanite, sillimanite and andalusite and found that Sr was above their detection limit (10 ppm) in only two of them. HERZ and DUTRA (1964) reported less than 3 ppm Sr in ten of 18 kyanites from metamorphic rocks in Brazil; their maximum value was 19 ppm Sr.

TUREKIAN and PHINNEY (1962), with a detection limit of 4 ppm, could not find Sr in 22 biotite-garnet pairs from a metamorphic sequence in Nova Scotia, Canada. MOXHAM (1964) analyzed 20 pairs of coexisting biotite and hornblende, from gneisses and schists in the epidote-amphibolite facies of metamorphism. The distribution of Sr between the two minerals was somewhat random, although a suggestion of enrichment in biotite was indicated. A similar erratic distribution of Sr between coexisting biotite and hornblende from metamorphic rocks was reported by DEVORE (1955). ENGEL and ENGEL (1962) found no change with metamorphic grade in the Sr content of hornblendes from amphibolites.

Manuscript received: September 1976

38-E. Abundance in Common Igneous Rock Types

A very large number of plutonic and volcanic igneous rocks has been analyzed for Sr, partly because of the interest in the isotopic composition of this element (Section 38-B). The analytical techniques used include isotope dilution, X-ray fluorescence, atomic absorption spectrometry and optical spectrography. Previous summaries of the geochemistry of Sr have been given by NOLL (1934), TUREKIAN and KULP (1956), and by TUREKIAN and WEDEPOHL (1961). The Sr content of andesites was studied by TAYLOR and WHITE (1966) and that of basalts reviewed by PRINZ (1967).

The data presented here were taken from a cross-section of papers published primarily since 1965. Although the coverage is by no means complete, a sufficient number of sources are represented to give reliable estimates of the Sr contents of most major rock types. Table 38-E-1 contains data for plutonic igneous rocks arranged roughly in order of increasing silica concentration. The volcanic rocks are presented in Table 38-E-2 under a series of headings reflecting the different geologic settings in which they occur, i.e. sea floor, oceanic islands, island arcs and subduction zones, and continents. Under each heading, data are presented for specific geographic regions. Where appropriate, calc-alkaline and alkalic suites occurring together in certain areas are listed separately.

Table 38-E-3 contains average Sr concentrations for the igneous rocks included in this review. The data for plutonic rocks are arranged in the same order as in Table 38-E-1. The volcanic rocks are arranged into series in accordance with the classification of IRVINE and BARRAGAR (1971). The peralkaline rocks, comendite and pantellerite, are assigned to the potassic and sodic suites of the alkali olivine basalt series respectively. This association, although conjectural, is compatible with their Na_2O/K_2O ratios as given by NOBLE (1968).

The manner of presentation of the data for the volcanic rocks is intended to emphasize the diversity of chemical varieties that exist in a particular geologic setting. For example, even though island arcs and subduction zones are characterized by volcanics of the calc-alkaline series, alkalic suites do occur in this setting, as for example in Grenada of the lesser Antilles, Oki-Dogo Island of Japan, in the Eolian arc of the Tyrrhenian Sea, etc. These occurrences, as well as a similar dichotomy on the continents, are significant indicators of changing conditions of petrogenesis in the sub-crustal sources.

The Sr concentrations of most *ultramafic rocks* (dunite, pyroxenite, peridotite, lherzolite, harzburgite and wehrlite) are generally low and average less than 65 ppm. Only anorthosites and alkali-rich rocks (lamprophyres and kimberlites) have high Sr contents ranging up to 1,000 ppm or more, in some cases. Eclogite xenoliths in kimberlite pipes contain 116 ppm Sr on the average whereas eclogites in metamorphic rocks and xenoliths in basalt may contain several hundred parts per million.

Mafic plutonic rocks (gabbro, norite and "diabase") have average Sr contents ranging from less than 200 ppm ("diabase" and dolerite) to about 300 ppm (gabbro).

Troctolites are enriched in Sr (477 ppm) because of their high concentration of plagioclase.

The average Sr content of the *intermediate plutonic rocks* decreases from diorite (Sr = 472 ppm) to granodiorite (Sr = 457 ppm), and quartz monzonite (Sr = 271 ppm). Quartz diorites, tonalites, and trondhjemites contain about 400 ppm of Sr. Undifferentiated granitic gneisses form a very heterogeneous group of rocks whose average Sr content is about 250 ppm.

The declining trend of Sr concentrations continues with *granites* (Sr = 147 ppm) and late stage differentiates (Sr = 55 ppm). The *alkali-rich plutonic rocks* (syenites and alkali gabbros) have highly variable, but generally elevated, Sr concentrations ranging up to nearly 3,000 ppm and averaging 553 ppm for syenites and 1,367 ppm for alkali gabbros. The highest Sr concentrations occur in carbonatites with values up to nearly 4,000 ppm and an average of 2,350 ppm.

The *oceanic tholeiites* and komatiites have low Sr contents averaging 124 ppm and 108 ppm, respectively. Spilites, keratophyres, ophiolite basalts, and greenstones form a somewhat heterogeneous group of rocks related to volcanism of the sea floor whose average Sr content is 183 ppm. The Sr concentration of some of these rocks could be influenced by metamorphism. The volcanic rocks of oceanic islands are often alkali-rich and can be subdivided into three series: i) Sodic series (hawaiite, mugearite, benmorite and trachyte); ii) Potassic series (trachybasalt, trachyandesite or tristanite, and K-rich trachyte); iii) Feldspathoidal series (basanite, nepheline hawaiite, nepheline mugearite, nepheline benmorite, phonolite and other rocks rich in leucite, analcite or melilite). Many varieties of alkali-rich volcanic rocks have high Sr concentrations ranging up to several thousand parts per million. The sodic volcanic rocks of oceanic islands illustrate the pattern. The Sr content of these rocks increases from 452 ppm in ankaramites to 528 ppm in alkali olivine basalt. The Sr content continues to rise in hawaiite (749 ppm), reaches a peak in mugearite (850 ppm) and then declines through benmorite (537 ppm) to trachyte (215 ppm). This pattern is typical also of the potassic volcanics and the feldspathoidal rocks. The latter generally have the highest Sr concentrations and commonly exceed 1,000 ppm.

The *calc-alkaline series* (basalt-andesite-dacite-rhyolite) of island arcs and subduction zones also show a range of Sr concentrations from 362 ppm in basalt to 431 ppm in andesite, followed by a decline to 317 ppm in dacite, 208 ppm in rhyodacite and 115 ppm in rhyolite. Alkalic suites in this geologic setting are less common than in oceanic islands but resemble those from oceanic islands in their Sr contents where they do occur.

The calc-alkaline and alkalic series of continental areas follow the pattern outlined above. Extremely high Sr concentrations approaching 5,000 ppm have been reported from certain feldspathoidal rocks in the East African rift system, in the Italian and German alkali-rich provinces and in Wyoming and Montana. Some carbonated lavas from East Africa contain in excess of 11,000 ppm Sr.

The lowest Sr concentrations occur in highly differentiated volcanic rocks (trachyte, phonolyte, rhyolite, obsidian) although the average values do not always reflect this, due to the heterogeneity of this group of rocks. The comendites and pantellerites have uniformly low Sr concentrations ranging up to a few tens of parts per million. Ignimbrites are generally acidic in composition and, on the average, contain around 200 to 250 ppm Sr.

Table 38-E-1. *Strontium concentrations of plutonic igneous rocks*

Rock type, locality	No. of samples	Range ppm Sr	Average ppm Sr	Method	References
		I. Ultramafic Rocks			
Dunites					
Miscellaneous	13	0.6–14.7	6.0	I	1, 2, 3, 5
North Carolina	6	0.12–5.2	1.54	I	4
Pyroxenites					
Miscellaneous	10	1.63–198.9	76.3	I, X	1, 2, 4, 16, 7, 9, 37
Rhodesia and South Africa	6	0.23–60.2	42.6	I	6
Anorthosites					
U.S.A. und Canada	61	156–750	666	I, X	8, 11, 12
South Africa	5	337–1,038	619	X, I	6
Kola Pen. (U.S.S.R.)	1		257	I	10
Indian Ocean	3	220–320	260	S	17
Peridotites					
Miscellaneous	13	0.41–28.6	14.6	I, X	1, 4, 5, 7, 15
Scotland	5	7.0–50.4	36.1	X	14
North Carolina	2	1.10–2.10	1.60	I	4
South Africa	2	9.05–27.2	18.1	I	2, 13
Highwood Mtns. (Montana, U.S.A.)	1		1,784	I	1
Lherzolites					
Pindos, Greece	1		2.55	I	7
New Mexico and Montana (U.S.A.)	13	3–104	40.0	X, A	18, 19
Dreiser Weiher (Germany)	4	2.6–40.5	14.9	I	16
Victoria (Australia)	7	6.6–25.0	14.7	I	20
Harzburgites, wehrlites					
Losberg (S. Africa)	2		88	X, I	6
Dreiser Weiher (Germany)	3	60.7–93.2	75.3	I	16
Stillwater, Montana (U.S.A.)	3	14.2–38.5	14.2	I	8
Kimberlites, biotite pyroxenites					
South Africa	25	48–1,883	924	I	25
Tanzania	1		517	X	120
Bearpaw, Montana (U.S.A.)	1		845	X	24
Eclogites					
South Africa (pipe)	6	36–180	107	I	1, 13, 25, 26
Salt Lake (basalt)	2	92–566	329	I	13
Bavaria (metamorphic)	1		215	I	13
Australia (pipe)	2	117–169	143	I	27
Oregon (metamorphic)	3	58–650	329	S	28
California (metamorphic)	5	140–300	230	S	29
New Caledonia (metamorphic)	1		200	S	29

Table 38-E-1 (continued)

Rock type, locality	No. of samples	Range ppm Sr	Average ppm Sr	Method	References
Spinifex peridotites					
Yilgarn (Australia)	7	11–24	19.1	X?	30
Munro, Ontario (Canada)	3	10–38	19.7	X?	30
Barberton (S. Africa)	1		31	X?	30
II. Mafic Rocks					
Gabbros, ferrogabbros, syenogabbros, quartz gabbros					
Ophiolite, Cyprus and Greece	6	41–189	97	X, I	7, 31
Miscellaneous, Scotland	32	118–471	295	X	14
South Africa, Namibia and Sudan	25	178–860	364	X, I, S	6, 32, 36, 45
Girnar (India)	7	130–590	240	S, I	21, 22
Indian Ocean	9	99–210	140	S	17
Venezuela and Guatemala	9	132–485	280	X, I	34, 35
Kungnat (Greenland)	4	455–600	534	X	127
Miscellaneous	9	145–632	318	I, X	10, 33, 37, 38
Norites					
Scotland	21	217–235	269	X	14
Miscellaneous	9	18–328	167	I, X	2, 6 ,8, 10
Troctolites					
Scotland and Sudan	3	443–516	477	X	14, 32
"Diabases" and dolerites					
Ophiolite, Cyprus and Greece	4	98–134	113	I, X	7, 31
Precambrian of Minnesota, Montana and Wyoming (U.S.A.)	74	59–513	190	I, X	39, 40, 48
Mesozoic, eastern U.S.A.	121	78–231	140	X	43, 123
Ardnamurchin (Scotland)	97	216–268	251	X	41
Victoria Land (Antarctica)	67	75–139	116	X	46, 47
Miscellaneous	13	107–197	159	I, X	42, 44, 49
III. Diorites					
Diorites					
Yeoval (Australia)	18	316–628	485	X	37
Vorspessart and Odenwald (Germany)	29	290–870	534	X	52
Girnar (India)	9	173–740	396	S, I	21, 22
Venezuela, Guatemala, Andes	6	224–495	332	X, I	34, 35, 51
Arizona, Colorado, British Columbia	13	219–911	497	X, I	23, 50, 53
Miscellaneous	3	256–536	355	X, I	6, 33
IV. Granitic Rocks					
Quartz diorites, tonalites, trondhjemites					
Alps	48	180–1,080	370	A	59
Rhodesia and South Africa	18	222–627	366	X	44, 57
Precambrian, Minnesota (U.S.A.)	18	83–1,127	625	I	12, 54

Table 38-E-1 (continued)

Rock type, locality	No. of samples	Range ppm Sr	Average ppm Sr	Method	References
Colorado, New Mexico, California, British Columbia	18	57–601	394	X	53, 55, 58
Venezuela and Andes, South America	16	145–564	320	X	34, 51
Miscellaneous	12	134–587	368	X, I, S	31, 33, 56, 60
Granodiorites					
Alps	27	40–1,060	352	A	59
Bayrischer Wald (Germany)	12	230–760	466	X	52
Yeoval and Snowy Mtns. (Australia)	23	123–912	213	X, S	37, 64
U.S.A. and Canada	56	460–1,100	659	I, X	12, 53, 54, 55, 58, 62
Guatemala and Andes	12	128–282	245	I, X	35, 51, 61
Miscellaneous	14	357–650	473	I, X	60, 63, 67
Quartz monzonites					
Precambrian, Wisconsin and Wyoming (U.S.A.)	40	29–876	179	I, X	38, 54, 62, 68, 69
Colorado, Arizona, Nevada, California, British Columbia	26	141–392	315	X	53, 55, 65 66, 70
Granites					
South Africa, Namibia, and Sudan	53	20–490	124	X, I, S	6, 32, 45, 64, 91, 109
Australia	14	10.2–117	60	X, I, S	37, 64, 94
Guatemala, Venezuela, and Brazil	34	2.6–435	109	I, X	34, 35, 61, 93, 107, 108
Scotland	54	23–247	119	I, X	95, 100, 106
Greenland	33	10–651	252	I, X	127
Alps and Portugal	22	<10–690	204	A, S	59, 60
China, Japan, Korea	16	24–315	142	I	33, 63, 79, 96
U.S.A. and British Columbia	285	3.2–917	150	I, X, S	12, 50, 55, 62, 68, 69, 70, 75, 76, 81, 89, 90, 98, 99, 101, 102, 103, 104, 105
V. Alkalic Rocks					
Syenites					
Girnar, India	9	22–320	120	S, I	21, 22
S. Africa, Namibia, Uganda, Tanzania, Sudan	22	90–1,536	499	X, I, S	24, 32, 36, 45, 120

Table 38-E-1 (continued)

Rock type, locality	No. of Samples	Range ppm Sr	Average ppm Sr	Method	References
Montana, Colorada, Arizona, Minnesota, Wisconsin	21	50–2,845	1,422	X, I	23, 24, 54, 68, 117
Gardar, Greenland	21	5.2–770	87	I	122
Miscellaneous	11	44–794	218	X, I	63, 118, 119, 124
Alkali gabbros (shonkinite, essexite, theralite, teschenite, ijolite, jacupirangite, melteigite, jumillite, monticellite peridotite, urtite, etc.)					
Montana, Arizona	23	680–2,394	1,458	X	24, 117
Tanzania, Uganda, Spain, Australia	23	446–1,707	1,327	X	24, 120
Carbonatites					
Elk Creek, Nevada	18	300–3,500	1,800	A, S	121
Eifel, Germany	3	951–3,797	2,063	X	125
Tanzania, Uganda	7		3,889	X	24, 120
Marlsburg, Germany	99		298	X	73
Miscellaneous	2	153–288	221	X, I	51, 63

References: 1. Stueber and Murthy (1966); 2. Jahn and Shih (1974); 3. Stueber and Ikramuddin (1974); 4. Stueber (1969); 5. Flanagan (1973); 6. Davies *et al.* (1970); 7. Montigny *et al.* (1973); 8. Fenton and Faure (1969); 9. Compston and Lovering (1969); 10. Birck and Allègre (1973); 11. Papezik (1965); 12. Hart and Davis (1969); 13. Griffin and Murthy (1968); 14. Pankhurst (1969); 15. Subbarao and Hedge (1973); 16. Paul (1971); 17. Engel and Fisher (1975); 18. Laughlin *et al.* (1971); 19. Stull and McMillan (1973); 20. Burwell (1976); 21. Bose (1973); 22. Paul *et al.* (1977); 23. Jahn (1973); 24. Powell and Bell (1970); 25. Mitchell and Crocket (1971); 26. Allsopp *et al.* (1969; 27. Compston and Lovering (1969); 28. Ghent and Coleman (1973); 29. Coleman *et al.* (1965); 30. Nesbitt and Sun (1976); 31. Peterman *et al.* (1971); 32. Neary *et al.* (1976); 33. Jahn *et al.* (1976); 34. Santamaria and Schubert (1974); 35. Clemons and Long (1971); 36. Siedner (1965); 37. Gulson (1972); 38. Reed and Zartman (1973); 39. Baadsgaard and Mueller (1973); 40. Arth and Hanson (1972); 41. Holland and Brown (1972), 42. Hebeda *et al.* (1973); 43. Weigand and Ragland (1970); 44. Glikson (1976); 45. Manton (1968); 46. Compston *et al.* (1968); 47. Gunn (1966); Condie *et al.* (1969); 49. Heier *et al.* (1965), 50. Stuckless and O'Neil (1973), 51. McNutt *et al.* (1975); 52. Okrush and Richter (1969); 53. Petö (1973); 54. Arth and Hanson (1975); 55. Kistler and Peterman (1973); 56. Pidgeon and Hopgood (1975); 57. Hawkesworth *et al.* (1975); 58. Barker *et al.* (1976); 59. Gundlach *et al.* (1967); 60. de Albuquerque (1971); 61. Pushkar *et al.* (1972); 62. Condie and Lo (1971); 63. Shibata and Adachi (1974); 64. Kolbe and Taylor (1966); 65. Wasserburg and Lanphere (1965); 66. Zielinski and Lippman (1976); 67. Duchesne *et al.* (1974); 68. Van Schmus *et al.* (1975a); 69. Van Schmus *et al.* (1975b); 70. Vera and Van Schmus (1974); 71. Long (1969); 72. Clifford *et al.* (1969); 73. Hahn-Weinheimer and Ackermann (1967); 74. Hedge *et al.* (1967); 75. Hills *et al.* (1968); 76. Condie (1969); 77. Baadsgaard *et al.* (1976); 78. Jahn and Murthy (1975); 79. Fullagar and Park (1975); 80. James *et al.* (1976); 81. Johnson and Hills (1976); 82. Taylor (1975); 83. Moorbath *et al.* (1975); 84. Van Breemen *et al.* (1974); 85. Barker *et al.* (1969); 86. Fullagar and Odum (1973); 87. Goldich *et al.* (1970); 88. Spooner *et al.* (1971); 89. Riley (1970a); 90. Pearson *et al.* (1966); 91. McCarthy (1976); 92. Oversby (1976); 93. Pushkar (1968); 94. Worden and Compston (1973); 95. Pidgeon and Johnson (1974); 96. Nohda (1973); 97. Robinson *et al.* (1976); 98. Mose *et al.* (1976); 99. Whitney *et al.* (1976); 100. Pankhurst and Pidgeon (1976); 101. Mukhopadhyay *et al.* (1975); 102. Fullagar and Shiver (1973);

103. SPOONER and FAIRBAIRN (1970); 104. FULLAGAR (1971); 105. HILLS and DASCH (1972); 106. PANKHURST (1974); 107. GOMES *et al.* (1975); 108. PRIEM *et al.* (1971); 109. VAN BREEMEN and DODSON (1972); 110. MOORBATH and WELKE (1969); 111. HAMET and ALLÉGRE (1976); 112. BECKINSALE (1974); 113. FULLAGAR *et al.* (1971); 114. BOTTINO *et al.* (1970); 115. ZARTMAN and MARVIN (1971); 116. BROOKINS *et al.* (1969); 117. NASH and WILKINSON (1971); 118. VOLLMER (1976); 119. KLERKX *et al.* (1974a); 120. BELL *et al.* (1973). 121. BROOKINS *et al.* (1975); 122. VAN BREEMEN and UPTON (1972); 123. SMITH *et al.* (1975); 124. DIETRICH and HEIER (1967); 125. LLOYD and BAILEY (1969); 126. HUGHES and BROWN (1972); 127. MACDONALD *et al.* (1973).

Table 38-E-2. *Strontium concentrations of volcanic igneous rocks*

Rock type, locality	No. of samples	Range ppm Sr	Average ppm Sr	Method	References
		I. Sea Floor			
1. Oceanic tholeiite					
Atlantic Ocean	33	74–190	122	X, I, S	1, 2, 3, 4, 5
Pacific and Atlantic	24	107–166	134	X	6
Pacific Ocean	15	96–154	131	X, I	7, 8
Indian Ocean	24	69–160	143	X, S	9, 10
2. Spilite, keratophyre, ophiolite, greenstone					
Cyprus and Greece	6	70–138	114	I, X	11, 12
Lushs Bight, Rambler and Roberts Arm, Nfdl.	?	116–239	185	X	13, 121
Onverwacht, S. Africa	9	12.5–120	71	I, X	14, 19
Lesser Antilles	5	38–109	74	X	120
Europe (spilites)	39	41–627	272	X	16, 18
Norway (greenstones)	94	39–528	177	X	15
Australia, Archaean	15	81–197	130	X?	17
North America, Archaean	1,425	52–897	194	I	20, 21, 22
3. Komatiite					
Onverwacht, S. Africa	3	1.2–41	23	I	14
Rambler, Newfoundland	9		136	X	121
4. Olivine basalt, alkalic basalt, ankaramite, plagioclase basalt, trachyte					
Pacific and Indian Ocean	29	20–2,400	510	S, X	3, 9, 10, 23
Mariana basin (ol. bas.)	9	155–208	186	X	24
Skinner Cove, Nfdl.	10	122–1,151	622	X	77
		II. Oceanic Islands			
1. Atlantic Ocean (St. Helena, Cape Verde, Canary Islands, Madeira, Fernando de Noronha)					
Ankaramite	4	230–429	354	X	25, 26
Basalt	27	500–878	677	X	25, 26, 31, 46
Trachybasalt	18	540–902	682	X	25, 26
Trachyandesite	14	395–968	645	X	25, 26
Trachyte	20	9–550	321	X	25, 26, 30
Phonolitic trachyte	10	36–246	116	X	25, 26
Hawaiite	11	676–1,081	891	X	31
Mugearite	2	1,179–1,252	1,216	X	28, 31
Basanite, limburgite	194	1,000–1,450	1,073	X	27, 29
Trachybasanite	14	800–1,800	1,440	X?	29
Phonolite, phonotephrite	34	14–1,590	510	X	25, 26, 28, 29, 46
Nephelinite	3		1,335		28

Table 38-E-2 (continued)

Rock type, locality	No. of samples	Range ppm Sr	Average ppm Sr	Method	References
2. Indian Ocean (Reunion, Mauritius, Kerguelen, Crozet, Amsterdam, Saint Paul and Moheli Island)					
Tholeiite basalt	14	284–535	349	X	32, 36
Basalt	35	231–650	330	X, A	32, 34, 37
Alk. olivine basalt	58	261–930	489	X, M?	32, 33, 34, 35, 36, 39
Hawaiite	11	459–713	529	X, M?	33, 34, 36, 39
Mugearite, benmorite	3	495–738	621	X, M?	33, 34
Trachyte	2	110–280	195	M?	33
Ankaramite	6	280–645	454	X	34, 39
Basanite	26	391–1,180	915	X	35, 36. 39
Nephelinite	8	520 1,240	833	X	35, 39
3. Pacific Ocean (Hawaiian Islands, Easter Island, Samoa, Marquesas Islands, Austral Islands, Society Islands, Guadelupe, Galapagos)					
Basalt, tholeiite	81	221–751	582	X	3, 40, 41, 44, 45, 46, 116
Alk. olivine basalt	5	590–900	798	A	43
Hawaiite	11	196–1,420	818	X, A	40, 43, 116
Mugearite	7	270–1,435	816	X, A	40, 43, 116
Benmorite	3	154–250	201	X	116
Trachyte	7	12–448	90	X	40, 46, 116
Comendite	3	25–46	33	X	116
Ankaramite	2	400–506	453	X, I	40, 137
Basanite	9	460–1,275	731	X, A	40, 42
Phonolite	10	160–1,585	633	A	43
Nephelinite	5	825–1,390	1,152	A	42
III. Island Arcs and Subduction Zones					
1. Lesser Antilles, Central America and Andes, South America					
Basalt	10		465	X	55, 57, 58, 119
Andesite, basaltic andesite	143	178–856	441	X	46, 47, 48, 49, 50, 51, 52, 53, 54, 55, 56
Dacite	10	151–712	343	X	50, 51, 52, 53
Rhyodacite	8	160–364	203	X	50, 52, 56
Rhyolite, obsidian	45	80–216	115	X, I, S	52, 53, 56
Ignimbrite	7	140–307	221	X	57
Pumice	15		298	X	48
Alk. basalt	31	396–829	582	X	49, 113
Basanitoid, basanite	20	566–1,135	768	X	49, 113
Trachyandesite	2	544–633	543	X	51
2. New Zealand, Japan and Indonesia					
Basalt, tholeiite	29	121–508	288	X, A, I	46, 58, 59, 60, 61
Andesite	84	144–919	393	X, A	58, 59, 61
Dacite	3	242–315	271	X	58
Rhyolite, obsidian	31	28–147	91	X, A	58, 59, 61
Ignimbrite	5	156–210	180	X	58

Table 38-E-2 (continued)

Rock type, locality	No. of samples	Range ppm Sr	Average ppm Sr	Method	References
Trachybasalt, trachyandesite, mugearite	3		695	I	60
Trachyte	5	3.5–36	16	I	60
Olivine leucitite, tephrite, phonolite	5	764–3,600	1,434	X	61
3. Tonga, New Hebrides, Solomon Islands, Mariana Islands, Bonin Islands, Izu Island, Fiji and New Guinea					
Basalt	57	115–703	364	X	46, 50, 62, 63, 65, 66
Andesite	45	120–810	427	X, I	50, 62, 63, 64, 65, 66, 67
Dacite, rhyodacite	18	60–800	244	X, I	46, 50, 62, 64, 65, 66, 67
Rhyolite, obsidian	3	124–200	159	I	50, 66
Absarokite, shoshonite	6	711–1,331	983	X	67
High K-basalt	3		530	X	63
Trachyandesite	2	408–453	431	X	46, 63
4. Eolian Arc (Tyrrhenian Sea) and Sardinia					
Basalt	24	404–1,295	685	X, I	68, 69, 70, 71
Andesite	17	246–744	560	X	68, 69, 71
Dacite	4	499–742	596	X	69, 71
Rhyolite	8	5–430	198	X, I	68, 69, 70, 71
Pantellerite	2	13–14	13.5	I	70
Hawaiite	6	464–659	539	I	70
Mugearite	3	688–1,026	856	X, I	68, 70
Trachyte	2	183–338	261	I	70
Tephrite	3	686–1,300	1,080	X	68, 69
IV A. Continents (Calc-alkali Series)					
1. Cascade Range, Washinton and Oregon (U.S.A.)					
Basalt	25	190–880	473	X	46, 81, 82, 83
Andesite	26	265–1,490	703	X	81, 82, 83
Dacite	6	103–775	398	X	82
Rhyolite, obsidian	5	64–79	71	X	81, 83
2. Columbia River and Snake River Basalt, Washington, Oregon, Idaho (U.S.A.)					
Basalt	142	139–515	289	X, A	46, 84, 85, 86, 87, 88, 89
Latite	3	190–277	232	A	87
Obsidian, vitrophyre	5	32–90	59	A	87
3. Continental U.S.A. (California, Nevada, Arizona, New Mexico, Utah, Texas, Minnesota, Connecticut and Maine).					
Basalt	266	111–1,640	554	X, A, I, S	46, 80, 89, 92, 94, 95, 96, 97, 98, 99, 100, 102, 103, 106, 107, 108, 109, 110, 118, 137

Table 38-E-2 (continued)

Rock type, locality	No. of samples	Range ppm Sr	Average ppm Sr	Method	References
Andesite	11	93–1,371	577	X, A, I	46, 80, 100, 101, 112
Dacite, rhyodacite quartz latite	18	129–847	392	X, I	46, 80, 90, 93, 99
Rhyolite, obsidian	55	2–144	40	X, I	46, 91, 94, 117, 136
Ignimbrite	12	39–739	294	X	101
4. Transantarctic Mountains, Antarctica					
Tholeiites, basalts	46	97–378	166	I, X	73, 74
5. S. Africa, Rhodesia, Kenya and Sudan					
Basalt	31	38–983	208	I, X	76, 77, 78, 79
Andesite	31		274	X	78, 79
Dacite, rhyodacite	5	131–205	163	I	76
Rhyolite	11	23–108	90	I, X	76, 77, 79
6. Baffin Bay, Baffin Island (Greenland)					
Basalt	63	87–286	175	X	75
IV B. Continents (Alkalic Series)					
1. Wyoming and Montana (U.S.A.)					
Andesite	4	793–892	849	X	134
Shoshonite, absarokite	7	524–1,400	1,018	X	134
Rhyodacite	2	773–2,480	1,627	X	134
Phonolite	2	811–876	844	X	135
Orendite, wyomingite, madupite	17	1,716–4,855	2,625	X	135
2. Colorado, Nevada, New Mexico, Utah and Arizona (U.S.A.)					
Basalt	22	128–1,300	513	X	106, 107, 108, 109, 110
Trachybasalt	4	973–1,500	1,190	X	105, 135
Trachyandesite	7	780–1,500	1,234	X	104, 105
Rhyodacite, latite, quartz latite	10	700–1,500	899	X	104, 105
Rhyolite	3	100–365	276	X	104, 105
Basanite	18	539–1,634	1,057	A, X	99, 111, 118
3. Australia and Tasmania					
Tholeiite	3	260–392	307	X	130, 133
Alkali basalt	8	555–1,024	775	X	131, 133
Nepheline basanite	5	810–1,980	1,096	X	130, 133
Nepheline hawaiite	2	659–776	718	X	130
Nepheline mugearite	2	861–1,158	1,009	X	130
Nephelinite	4	1,018–1,583	1,330	X, I	131, 132
4. Tanzania, Uganda, Rwanda, Zaire					
Basalt	3	522–935	735	X	124
Ankaratrite, basanite	4	852–1,646	1,277	X	125, 126
Trachyandesite	2	856–1,646	1,277	X	125, 126
Latite	4		613	X	125

Table 38-E-2 (continued)

Rock type, locality	No. of samples	Range ppm Sr	Average ppm Sr	Method	References
Trachyte	8	14–1,091	235	X	124, 125, 126
Pantellerite	8	1–73	14	X	124
Absarokite, shoshonite	14		1,056	X	125
Banakite	4		1,316	X	125
Phonolite	9		1,386	X	123, 124, 126
Nephelinite	31	744–3,643	2,180	X	133, 125, 126
Leucitite	21	695–3,389	1,748	X	125
Melilitite, kalsilitite	7	2,305–3,212	2,912	X	125
Ugandite, mafurite, kivite murambite, mikenite, katungite	35	572–4,083	1,566	X	125
Carbonated lavas	5	1,605–11,149	6,113	X	125
5. South Arabia (Aden volcano), Ethiopia (Fantale), Nigeria, Namibia (Paresis Complex)					
Olivine basalt	5	405–1,020	721	X	122, 127, 128
Basanite	2	919–1,020	1,175	X	128
Hawaiite, mugearite	8	406–535	432	X	127
Trachyandesite	9	188–486	266	X	127
Trachyte	11	36–492	187	X	122, 127, 128
Phonolite	5	12–1,810	856	X	128
Rhyolite	15	6–207	84	X	122, 127, 129
Comendite	10		30	X	127, 129
6. Germany and Italy					
Tholeiite	3		593	X	119
Olivine basalt	25		1,400	X	119
Mugearite, basanite	2		2,669	X	119, 120
Tephrite	21	610–1,789	860	X	119, 120
Trachyte	12		801	X	119, 120, 121
Latite	4	348–930	633	X	120
Vesuvite	21	720–1,030	900	X	121
Leucitite	9	1,300–2,500	1,836	X	119, 120
Nephelinite	18		2,050	X	119
Ignimbrite	3	139–1,150	946	X	120

References: 1. Brooks *et al.* (1974); 2. White and Bryan (1977); 3. Engel *et al.* (1965); 4. Tatsumoto *et al.* (1965); 5. Hart and Nalwalk (1970); 6. Dasch *et al.* (1973); 7. Hart *et al.* (1974); 8. Subbarao (1972); 9. Subbarao and Hedge (1973); 10. Engel and Fisher (1975); 11. Montigny *et al.* (1971); 12. Peterman *et al.* (1971); 13. Strong (1973); 14. Jahn and Shih (1974); 15. Gale and Roberts (1974); 16. Herrmann *et al.* (1974); 17. Nesbitt and Sun (1976); 18. Herrmann and Wedepohl (1970); 19. Glikson (1976). 20. Hart *et al.* (1970); 21. Jahn and Murthy (1975); 22. Jahn *et al.* (1974); 23. Subbarao *et al.* (1973); 24. Hart *et al.* (1972); 25. Baker (1969); 26. Grant *et al.* (1976); 27. Gunn and Watkins (1976); 28. Klerkx *et al.* (1974b); 29. Ridley (1970); 30. Araña *et al.* (1973); 31. Hughes and Brown (1972); 32. Hedge *et al.* (1973); 33. Zielinski (1975); 34. Baxter (1975); 35. Baxter (1976); 36. Watkins *et al.* (1974); 37. Gunn *et al.* (1971); 38. Girod *et al.* (1976); 39. Strong (1972); 40. Lessing and Catanzaro (1964); 41. Hedge *et al.* (1972). 42. Hawkins and Natland (1975); 43. Bishop and Woolley (1973); 44. Duncan and Compston (1976); 45. Stueber and Murthy (1966); 46. Hedge (1966); 47. Hedge and Lewis (1971); 48. Gunn *et al.* (1974); 49. Arculus (1976); 50. Pushkar (1968); 51. Santamaria and Schubert (1974); 52. Pushkar *et al.* (1972); 53. McNutt *et al.* (1975); 54. Siegers and Pichler (1969); 55. Noble *et al.* (1975); 56. El-Hinnawi *et al.* (1969); 57. James *et al.* (1976). 58. Ewart and Stipp (1968);

59. SHUTO (1974); 60. NAGASAWA (1973); 61. WHITFORD (1975); 62. EWART and BRYAN (1972); 63. COLLEY and WARDEN (1974); 64. TAYLOR *et al.* (1969); 65. HEMING (1974); 66. PETERMAN *et al.* (1970a); 67. MACKENZIE and CHAPPELL (1972); 68. BARBERI *et al.* (1974); 69. KLERKX *et al.* (1974b); 70. BARBERI *et al.* (1969); 71. DUPUY *et al.* (1974); 72. STRONG (1974); 73. FAURE *et al.* (1972); 74. FAURE *et al.* (1974); 75. CLARKE (1970); 76. MANTON (1968); 77. NEARY *et al.* (1976); 78. HAWKESWORTH *et al.* (1975); 79. DAVIS and CONDIE (1977); 80. ARTH and HANSON (1975); 81. PETERMAN *et al.* (1970c); 82. CHURCH and TILTON (1973); 83. HIGGINS (1973); 84. NATHAN and FRUCHTER (1974); 85. MCDOUGALL (1976); 86. UPPULURI (1974); 87. MARK *et al.* (1975); 88. LEEMAN and MANTON (1971); 89. LEEMAN and VITALIANO (1976); 90. CONDIE and HAYSLIP (1975); 91. NOBLE *et al.* (1972); 92. STULL and MCMILLAN (1973); 93. NOBLE *et al.* (1976); 94. ROBINSON *et al.* (1976); 95. GUNN and WATKINS (1970); 96. LEEMAN (1970); 97. LOWDER (1973); 98. LIPMAN (1969); 99. STUCKLESS and O'NEIL (1973); 100. HEDGE and NOBLE (1971); 101. SCOTT *et al.* (1971); 102. CONDIE and BARSKY (1972); 103. PUSHKAR and CONDIE (1973); 104. ZIELINSKI and LIPMAN (1976); 105. ROBINSON (1972); 106. DOE *et al.* (1969); 107. LAUGHLIN *et al.* (1972); 108. BROOKINS *et al.* (1975); 109. LAUGHLIN *et al.* (1971); 110. LIPMAN and MOENCH (1972); 111. JONES *et al.* (1974); 112. LEEMAN (1974); 113. SIGURDSSON *et al.* (1973); 114. NAGLE *et al.* (1973); 115. GALE (1973); 116. BAKER *et al.* (1974); 117. BROOKINS *et al.* (1973); 118. MCKEE and NOBLE (1976); 119. HOEFS and WEDEPOHL (1968); 120. VOLLMER (1976); 121. SAVELLI (1967); 122. DICKINSON and GIBSON (1972); 123. BELL *et al.* (1973); 124. WEAVER *et al.* (1972); 125. BELL and POWELL (1969); 126. BELL and POWELL (1970); 127. CARTER and NORRY (1976); 128. GRANT *et al.* (1972); 129. SIEDNER (1965); 130. STUCKLESS and IRVING (1976); 131. KESSON (1973); 132. COMPSTON and LOVERING (1969); 133. COMPSTON *et al.* (1968); 134. PETERMAN *et al.* (1970b); 135. POWELL and BELL (1970); 136. DELONG and LONG (1976); 137. STUEBER and IKRAMUDDIN (1974).

Table 38-E-3. *Average strontium concentrations of plutonic and volcanic rocks*

Rock type	No. of Samples	Sr Content (ppm)	
		Range	Average
	I. Plutonic Rocks		
1. Ultramafic rocks			
Dunite	19	0.12–14.7	4.6
Pyroxenite	16	0.23–199	64
Anorthosite	70	156–1,038	639
Peridotite[a]	22	0.4–50.4	19
Lherzolite	25	2.6–104	27
Harzburgite, wehrlite	8	14.2–93	56
Lamprophyres	33	180–3,103	1,622
Kimberlite	27	48–1,883	906
Eclogite (pipe)	8	36–180	116
Eclogite (metamorphic)	10	58–650	255
Eclogite (basalt xenolith)	2	92–566	329
Spinifex periodtite	11	10–38	20
2. Mafic rocks			
Gabbro	101	41–860	293
Norite	30	18–328	238
Troctolite	3	463–516	477
Diabase, dolerite	376	59–513	175
3. Intermediate to granitic rocks			
Diorite	79	173–870	472
Quartz diorite, tonalite, trondhjemite	130	57–1,127	401

Table 38-E-3 (continued)

Rock type	No. of Samples	Sr Content (ppm)	
		Range	Average
Granodiorite	149	40–1,100	457
Quartz monzonite	167	29–876	271
Granitic gneiss	327	2.2–2,209	252
4. Granites etc.			
Granite	512	2.6–917	147
Late differentiate	201	2.1–778	55
5. Alkalic rocks			
Syenite	84	5.2–2,924	553
Alkali gabbro	49	446–2,195	1,367
Carbonatite	28	300–3,910	2,350
II. Volcanic Rocks			
1. Tholeiite basalt series			
Tholeiite (mid-ocean ridges)	103	69–190	124
Spilite, keratophyre, ophiolite basalt, greenstone	214	38–897	183
Komatiite	12	1.2–136	108
Tholeiite (oceanic islands)	18	221–535	329
2. Calc-alkali series			
Basalt	632	38–1,640	393
Andesite	361	93–1,490	442
Dacite	49	60–800	306
Rhyodacite, latite, quartz latite	26	160–847	330
Rhyolite	127	5–430	97
Obsidian	24	2–200	49
Ignimbrite	37	12–1,150	236
Pumice	17	161–298	283
3. Alkali olivine basalt series			
Olivine basalt	423	87–2,400	626
Hawaiite	62	196–1,420	660
Mugearite	18	40–3,337	943
Benmorite	7	154–843	537
Trachyte	87	3.5–1,750	280
Pantellerite	15	0.3–73	10
Trachybasalt	36	298–1,500	820
Trachyandesite (tristanite)	65	188–2,480	725
K-trachyte	4	3–386	150
Comendite	13	25–46	31
4. Feldspathoid basalt series			
Ankaramite	16	230–645	452
Basanite, basanitoid	220	391–1,275	717
Tephrite	24	610–1,789	887
Nepheline hawaiite	6	659–1,375	1,087
Phonolite	78	12–1,845	633
Vesuvite	21	720–1,030	900

Table 38-E-3 (continued)

Rock type	No. of Samples	Sr Content (ppm)	
		Range	Average
Nephelinite	69	520–3,643	1,830
Leucitite	35	695–3,389	1,725
Melilitite	6	2,305–3,212	2,882
Absarokite, shoshonite	21	524–1,331	1,010
Orendite, wyomingite, madupite	17	1,716–4,855	2,625
Ugandite, mafurite, kivite, murambite, mikenite, katungite	35	572–4,083	1,566
Limburgite	134	910–1,024	1,006
Carbonated lava	5	1,605–11,149	6,113

[a] One sample from the Highwood Mountains, Montana (U.S.A.) was excluded from the average.

Revised manuscript received: August 1977

38-F. Behavior in Magmatogenic Processes

The behavior of Sr in a cooling magma is determined primarily by the extent to which it is preferentially incorporated or excluded from minerals. In general, plagioclase and K-Feldspar both concentrate Sr relative to silicate liquids with which they are in equilibrium. The micas, amphiboles, pyroxenes, olivine, and garnet all discriminate against Sr. Consequently, the crystallization of olivine, pyroxene and other minerals in this group increases the Sr content of the magma, whereas formation of feldspars tends to reduce it. The evolution of the Sr concentration of cooling magmas can be predicted quantitatively by means of distribution coefficients (Section 38-D) as shown by GAST (1968) and SHAW (1970). These considerations are capable of explaining the Sr concentrations of most sequences of plutonic and volcanic igneous rocks. However, the high Sr content of alkali-rich feldspathoidal rocks may be due to the conditions of magma formation and is not entirely attributable to the effects of fractional crystallization.

Late-stage differentiates, such as pegmatites and aplites, generally are depleted in Sr compared to main-stage differentiates. Whereas granites contain 147 ppm Sr on the average, late-stage differentiates carry only about 55 ppm (Table 38-E-3). K-feldspars and mica minerals in pegmatites contain less Sr than the same minerals occurring in associated granites and related rocks (Section 38-D). Some pegmatites may contain apatite which has widely varying Sr concentrations averaging several thousand parts per million (RILEY, 1970a, b; CRUFT, 1966).

The sulfide minerals contain very little Sr, probably less than 1 ppm. The Sr of hydrothermal ore deposits resides primarily in associated barite, carbonate minerals, and fluorite. The latter may contain several 100 ppm of Sr. Barite of hydrothermal origin often exceeds 1% Sr. Isotopic studies by REESMAN (1968) indicate that the Sr of hydrothermal vein deposits was derived from the country rock and did not originate from igneous intrusives. Some hydrothermal vein deposits contain celestite and strontianite (KRIEGER, 1933).

Revised manuscript received: September 1977

38-G. Behavior during Weathering and Rock Alteration

I. Weathering Profiles

According to the information in Section 38-D, the major host minerals of the crustal strontium are feldspars. Plagioclases and K-feldspars are almost equally important. The Mg—Fe minerals usually contain less than a tenth of the strontium of coexisting feldspars in magmatic and metamorphic rocks. Feldspars (especially plagioclase low in anorthite and K-feldspar) are commonly more resistant against chemical weathering than magnesium-iron silicates. Sr is usually less mobilized during weathering than Ca. This implies that the former element is held more tightly on clay minerals; Sr is more weakly hydrated than Ca. The differential adsorption of Sr to Ca in constituents of Indian soils has been investigated by KHASAWNEH *et al.* (1968). SHORT (1961) has observed a change in the range of Sr/Ca ratios from 0.005 to 0.044 in unweathered basalt, meta-andesite and granodiorite to the range from 0.001 to 0.003 in their weathering products.

The present author has analyzed strontium in several series of weathering products of Tertiary *olivine alkali-basalts* from the Göttingen area (NW Germany) and has compared them with the Sr content of the fresh basalt. The mineral and major element composition of these samples has been described by BOLTER (1961). The original strontium content of the basalt is preserved during the early steps of decomposition (of olivine and pyroxene). It decreases with increasing plagioclase decomposition. The final weathering product of koalinite and smectite minerals only contains 20 to 30 percent of the original basaltic strontium concentration. SHORT (1961) has published detailed data on the behavior of Sr and other trace elements during weathering of basalt in Colorado (semi-arid climate). The Sr loss relative to Ti is only 25 percent and has not reached the levels reported above for German basalts. The maximum Sr is usually contained in those silt and sand fractions with the highest proportion of plagioclase. A more effective leaching of the original Sr content of gabbro from the Lizard Peninsula (U.K.) and of olivine basalt from the Nordmarka area (Norway) was observed by BUTLER (1953, 1954). The clay residue contains about 20 percent of the Sr concentration of the gabbro.

During mechanical decomposition of *granite* from the Harz Mountains (F.R. Germany), combined with slight leaching by weathering solutions, a maximum loss of about 30 percent strontium has been recorded by the present author. A higher rate of chemical decomposition can be observed in a laterite profile on granite (Eklara, India; for data on mineral and major element composition see KÖSTER, 1955). Here a constant level of 130 ppm Sr from fresh granite to kaolinite-rich material has been recorded (WEDEPOHL, unpublished; method: X). However the Sr/Ti ratio is decreasing from 0.06 to 0.014. Assuming that the less mobile element Ti has been constant throughout the profile, this decrease of the ratio indicates a loss of about 75 percent of the original Sr. There exist additional observations on the behavior of Sr during weathering of granitic rocks. SHORT (1961) has recorded a maximum loss of about

60 percent Sr relative to Ti from a granodiorite in Wyoming (USA) in its A horizon under semi-humid conditions. Slightly higher was the Sr mobilization from silt of a decomposed granodiorite gneiss in SW England (BUTLER, 1953) und from a quartz porphyry in Japan (SUWA *et al.*, 1958).

Additional Sr data on decomposition products of schists, syenite, serpentine and monzonite are reported by BUTLER (1953, 1954) and on andesite by SHORT (1961). All indicate that strontium is a rather mobile element during rock weathering, especially under decomposition of feldspar.

II. Soils

SHACKLETTE *et al.* (1971) have sampled 862 soils at a depth of about 20 cm from locations about 50 miles apart throughout the conterminous United States. Analyzed spectrographically they contain on average 240 ppm Sr, which is not much less than the 300 ppm Sr reported by VINOGRADOV (1959) from worldwide sampling. Abundant values for the United States range between 60 and 400 ppm Sr. The former authors present a map with symbols for the different Sr contents of surficial materials demonstrating that soils from the western USA contain more strontium than those from the southeastern states. The geometric mean for the western states is 210 ppm Sr and for the eastern states 51 ppm Sr.

Many soils, represented by the above listed averages, are apparently close to the average Sr concentration in the upper continental crust.

III. Bauxite

GORDON and MURATA (1952) report an average Sr concentration of 190 ppm Sr in bauxite from Arkansas (USA) which is 70 percent of the original Sr in nepheline syenite (270 ppm Sr) from which this bauxite has been formed.

IV. Rock Alteration by Sea Water

The presently largest rate of basaltic volcanism is submarine and connected with the formation of the ocean ridges. The interaction of basalt and sea water must influence the chemical composition of both. One approach to understanding the various chemical changes is to compare the more altered rim zones of basaltic pillows with presumably less altered interior portions. HART *et al.* (1974, with references to earlier investigations), in a comparison of this type in three submarine basalt pillows, observed no change of Sr in one sample but a slight and a noticeable decrease of Sr in the rim of two samples. The maximum change is about 10 percent. The sample with the slight change of the original strontium shows no apparent variation in $^{87}Sr/^{86}Sr$, but in a more altered pillow the Sr isotope ratio is higher in the margin (adsorption of Sr from sea water) than in the core. These results are in agreement with higher $^{87}Sr/^{86}Sr$ ratios in altered midocean ridge basalts reported by DASCH *et al.* (1973). Both groups of authors found the increase of the $^{87}Sr/^{86}Sr$ ratio correlated with the water content of the altered basalt. The increase of ^{87}Sr and the decrease of total Sr in pillow rims indicates a complicated exchange of Sr between sea water and rock probably at different stages of alteration.

Mottl *et al.* (1974) have reported experimental reactions between powdered fresh mid-ocean ridge basalt and sea water at 200 to 500° C (pH 3.3 to 6). Most quenched solutions from these reactions are lower in Sr than the starting sea water. They range from 0.2 to 8.5 ppm Sr (starting sea water: 7.4 ppm Sr). The three exceptions with slightly higher than sea water strontium concentrations are formed at 300° and 400° C. At 500° C the lowest Sr solution is attained.

Manuscript received: June 1977

38-H. Reactions Controlling Strontium Abundance in Natural Waters

I. Introduction

Of the total input of strontium into the ocean via the world rivers, about 80% is due to the weathering of sedimentary carbonates and sulfates (Brass, 1976, Section 38-I). The remaining 20% derives from silicates. Garrels and Mackenzie (1971) estimated the total mass of evaporites at about 20% of that of carbonates and that presently only a portion of this undergoes substantial weathering. Thus the major rock type involved in the strontium cycle is carbonate, followed by the silicates and evaporites.

In the following I will briefly discuss the evidence available on the distribution of strontium between its aqueous solutions and carbonates and sulfates. Not much attention has been given to the adsorption of strontium onto clay minerals, although this is a potentially important process. A thorough knowledge of this partitioning of strontium is necessary in order to understand the geochemical processes affecting the strontium cycle.

II. Carbonates

In this section I will discuss the distribution of strontium between aqueous solutions and the solid phases, aragonite and calcite. These two calcium carbonate phases are the most important primary carbonates formed in the oceans and lakes. Although information gained from direct precipitation measurements is of great importance, one should also consider biochemical pathways of calcium carbonate precipitation (Kinsman, 1969).

Holland *et al.* (1963, 1964a, 1964b); Kinsman and Holland (1969); Katz *et al.* (1972) have made extensive measurements of the co-precipitation of strontium with aragonite and calcite. These studies suggested that the distribution coefficient can best be described by a logarithmic law or the so-called Doerner-Hoskins (1925) relation. This implies that only surface equilibrium is maintained between precipitates and solutions. For aragonite with a distribution coefficient close to unity this causes no great problem. However, for calcite, which has a much smaller distribution coefficient (Holland *et al.*, 1964a, b), problems may arise, especially when relatively large amounts are precipitated in the experiments. The logarithmic distribution law has also been used with success in barium sulfate-strontium sulfate systems (Cohen and Gordon, 1961; Hanor, 1968). On the other hand, Usdowski (1973), in his studies of strontium co-precipitation with aragonite, calcite, and calcium sulfate, found that a linear distribution law represented his data very well. However, agreement between the data of Usdowski (1973) and those of Kinsman and Holland (1969) is generally good, so that the use of a linear or a logarithmic law yields approximately similar results.

Table 38-H-1. *Distribution coefficient k_{Sr}[a] of strontium between aqueous solutions and aragonite or calcite*[b]

Temperature (°C)	k_{Sr}	Reference
Aragonite		
15–16	1.17 ±0.04	1
24	1.14 ±0.04	2
25	1.18 ±0.10	3
30	1.15 ±0.05	1
45	1.04 ±0.03	1
60	0.97 ±0.03	1
80	0.88 ±0.03	1
96	0.85 ±0.07	1
Calcite		
25	0.14 ±0.02	2, 4
	0.15 ±0.01	3
80	0.08 ±0.008	3
96–100	0.076±0.006	4

[a] $$\frac{m^X_{Sr^{2+}}}{m^X_{Ca^{2+}}} = k^X_{Sr} \frac{m^L_{Sr^{2+}}}{m^L_{Ca^{2+}}}$$

where m = molal concentration
X = aragonite or calcite
L = liquid phase.

[b] KINSMAN and HOLLAND (1969) showed that over the temperature range 25–100° C $k^A_{Sr}/k^C_{Sr} \simeq 8$.

References: 1. KINSMAN and HOLLAND (1969); 2. JOSHI (1960); 3. USDOWSKI (1973); 4. HOLLAND *et al.* (1964b).

Data reported by KINSMAN and HOLLAND (1969) and USDOWSKI (1973) are presented in Table 38-H-1. Note that the slight difference in definition of the distribution coefficient is ignored here.

Aragonite systems at higher temperature showed some dependence on sulfate and cations other than calcium and strontium. In addition, at these higher temperatures the distribution coefficient may decrease slightly at very high Sr/Ca ratios of the solutions (KINSMAN and HOLLAND, 1969). BODINE *et al.* (1965) found that pressure effects up to 1,150 bars were negligible for the distribution coefficient with calcite. KINSMAN and HOLLAND (1969) inferred from this that no discernable pressure effect will be evident for aragonite either.

Several authors (FLÜGEL and WEDEPOHL, 1967; KINSMAN, 1969; USDOWSKI, 1973) have used the above information on the distribution coefficient of strontium in studies of carbonate diagenesis. With regard to this, KATZ *et al.* (1972) investigated the behavior of strontium during the aragonite-calcite transformation. These authors established distribution coefficients that appear much smaller than those determined from direct precipitation measurements (HOLLAND *et al.*, 1964a). This conclusion is at variance with the data of USDOWSKI (1973), who recrystallized strontium-free aragonite in strontium-containing solutions at 80° C. His value was in agreement with that of HOLLAND *et al.* (1964a). In view of the importance of these experiments with regard to conclusions about the environmental conditions of aragonite recrystallization (KINSMAN, 1969; KATZ *et al.*, 1972; USDOWSKI, 1973), it is important to carry out further experiments using a greater variety of initial conditions.

LOWENSTAM (1963a, b) has shown that skeletal material of more developed forms of organisms (especially molluscs) shows much lower strontium contents than expected. This is due to a more complex biochemical pathway of aragonite or calcite deposition. In such cases it is possible that temperature effects are more pronounced (HALLAM and PRICE, 1968; KINSMAN, 1969). It appears that even in these cases a distribution law is still closely followed (FAURE *et al.*, 1967; Section 38-I).

III. Sulfates

In the ocean, celestite ($SrSO_4$) is important because it forms the skeletal material of the planktonic species *Acantharia*. This species, although important in ocean surface waters, disappears below 400 m (see Section 38-I). WHITFIELD (1975) predicted values of the stoichiometric solubility product of strontium sulfate in seawater at 25° C and several ionic strengths. He expressed his data in terms of $pK_s = -\log K_s = -\log (m_{Sr} \cdot m_{SO_4})$, where m is the molality. The values of 4.94, 4.82, 4.70, and 4.57 were found for ionic strengths of 0.5, 0.7, 1.0, and 1.5. WHITFIELD (1975) was also able to predict the values of the solubilities of strontium sulfates in several synthetic brines which were in good agreement with those determined by DAVIS and COLLINS (1971). Little work has been done on the temperature dependence of the solubility product. However, NORTH (1974) studied the pressure depencence of the strontium sulfate solubility product at 2° C and found, for seawater of 35.2‰ salinity, the following values of K_s:

Pressure, atm.	1	250	500	750	1,000
$K_s \times 10^5$	0.95 ± 0.02	1.40 ± 0.05	2.10 ± 0.05	2.97 ± 0.06	4.35 ± 0.07

From theoretical predictions of the pressure dependence of the solubility product, NORTH (1974) concluded that possibly the phase $SrSO_4 \cdot H_2O$, and not anhydrous strontium sulfate, should be considered, when making these calculations.

In seawater, $m_{Sr^{2+}} \simeq 0.08$ mmolal and $m_{SO_4} \simeq 28.5$ mmolal, yielding a concentration product of 2.8×10^{-6}. Thus, over the entire range of temperatures and pressures common in the ocean, seawater is undersaturated with respect to celestite.

Another important aspect of the involvement of strontium in the sulfate cycle is its co-precipitation with barite in the deep ocean. HANOR (1968) considered this problem in detail. Although strontium sulfate can form a complete series of solid solutions with barium sulfate, actual analyses show only a very limited range (STARKE, 1964). The majority of barites show mole fractions of strontium less than 0.07. HANOR (1968) was able to predict the same frequency distribution as that found by STARKE (1964) on the basis of calculations made using the logarithmic distribution law.

USDOWSKI (1973) investigated the co-precipitation of strontium with gypsum at 25° C at two sodium chloride concentrations. His results are presented in Table 38-H-2. Data for anhydrite obtained by USDOWSKI (1973) are given in Table 38-H-3. WOOD and SHAW (1976) considered the geochemistry of celestites in the Yate area

Table 38-H-2. *Distribution of strontium between gypsum and solutions at 25° C* (Usdowski. 1973)

36.5 g NaCl/liter			3.6 g NaCl/liter		
Sr^L ppm	Sr^G ppm	Sr^G/Sr^L	Sr^L ppm	Sr^G ppm	Sr^G/Sr^L
7.4	440	59.5	14.0	740	52.8
14.1	690	49.0	14.1	650	46.0
14.2	600	42.2	28.2	1,300	46.0
28.2	1,370	48.5	28.2	1,300	46.0
28.0	1,490	53.2	56.6	2,500	44.2
55.2	3,600	65.3	111.0	6,200	55.9
111.0	6,700	60.4	112.0	5,900	52.7
112.0	6,100	54.5			
	Average	54.0		Average	49.0

Sr^L = equilibrium concentration in liquid phase.
Sr^G = equilibrium concentration in solid phase.

Table 38-H-3. *Distribution of strontium between anhydrite and solutions* (Usdowski, 1973)

60° C			120° C			180° C		
Sr^L ppm	Sr^A ppm	Sr^A/Sr^L	Sr^L ppm	Sr^A ppm	Sr^A/Sr^L	Sr^L ppm	Sr^A ppm	Sr^A/Sr^L
1.1	1,480	1,345	1.5	1,200	800	0.58	175	302
1.7	2,000	1,175	1.5	1,300	866	0.55	160	291
1.7	2,200	1,295	1.8	1,400	778	0.75	230	307
2.7	3,000	1,110	1.8	1,500	834	0.8	230	288
5.5	6,500	1,180	1.9	1,350	710	1.4	370	264
5.0	6,300	1,160	2.0	1,400	700	1.4	370	264
7.5	8,800	1,170	2.4	1,900	792	2.4	600	250
8.0	9,500	1,190	2.5	1,900	760	2.4	600	250
11.0	12,500	1,135	4.4	2,800	637	4.4	1,100	250
10.0	12,700	1,270	4.6	3,100	675	4.2	1,100	262
	Average	1,210	8.5	4,500	530	6.4	2,200	344
			7.5	4,800	640	5.8	2,200	380
			11.0	7,500	682	12.0	4,400	366
			11.0	8,500	773	12.0	4,300	358
			13.5	12,000	890		Average	299
			13.5	11,500	852			
				Average	744			

Run at 60° C saturated with NaCl; other runs only in $SrCl_2$ solutions.
Sr^L = strontium concentration in liquid phase.
Sr^A = strontium concentration in solid phase.

near Bristol (U.K.) and agreed with the observations of Braitsch (1971) and Usdowski (1973) that celestite forms only in the halite stage of precipitation from seawater. This is in contrast to the conclusions by Müller and Puchelt (1961), who suggested celestite precipitation as early as in the gypsum phase of precipitation.

IV. Adsorption on Clays

Relatively little work has been done on the adsorption of strontium on clay minerals. WAHLBERG *et al.* (1965a, b) studied these exchange processes in solutions which simulated conditions that may be encountered in soils. Unfortunately, no systematic studies have been carried out in solutions similar to seawater or at ionic strengths resembling that of seawater.

WAHLBERG *et al.* (1965a, b) defined the distribution coefficient of strontium as

$$K_d = \frac{(SrX_2)}{(Sr^{2+})}$$

where (SrX_2) = the amount of strontium adsorbed by the clay in m equivalents/gram; (Sr^{2+}) = normality of strontium in the equilibrium solution. These studies established that, for simple ion exchange with one competing cation, the mass action equilibrium constant

$$K_c = \left(\frac{SrX_2}{Sr^{2+}}\right)\left(\frac{Y^{2+}}{YX_2}\right)$$

where $Y = Ca^{2+}$ or Mg^{2+}, is close to unity, whereas for the exchange with Na^+ and K^+ the values are somewhat higher. Measured K_d values were generally in good agreement with calculated ones. To a good approximation, distribution coefficients calculated for more complex systems were also in agreement with observed values. Generally, distribution coefficients are constant at strontium concentrations below 10^{-3} N (0.5 mM or 45 ppm), but they decrease rapidly at higher strontium concentrations. WAHLBERG *et al.* (1965b), however, did point out that the adsorption process should not be treated as a simple ion exchange equilibrium, so that measurements in seawater-like systems are clearly necessary. McDUFF and GIESKES (1976) assumed that the adsorption density in marine sediments has become independent of the strontium concentration in interstitial waters. This assumption should be substantiated experimentally.

Revised manuscript received: December 1976

38-I. Abundance in Natural Waters

I. Introduction

As indicated in Section 38-B, the major stable isotope of strontium in nature is ^{88}Sr (82.74%), with ^{84}Sr, ^{86}Sr, and ^{87}Sr as minor components. The variation in the $^{87}Sr/^{86}Sr$ has been the subject of intensive investigation, and very high precisions in the determination of this ratio have been achived. The same is not necessarily true for the total strontium content of natural waters. This will be a brief consideration of the geochemical cycle of stable strontium and the processes that affect its distribution in the rivers, the ocean, and the interstitial waters of marine sediments.

In most studies of the geochemistry of strontium, particular attention is given to the ratio of strontium to calcium (TUREKIAN and KULP, 1956; TUREKIAN, 1964; BRASS, 1976). This is done because a large part of the strontium cycle occurs through the carbonates (75–80%, BRASS, 1976). For these reasons, we shall often refer to the Sr/Ca ratio in the aqueous phase.

II. Rivers

Studies of the strontium content of rivers are still relatively scarce, and in particular the Sr/Ca ratio in relation to the weathering terrain deserves further scrutiny. BRASS (1976) reported $^{87}Sr/^{86}Sr$ ratios distinctly lower than those of present-day sea water (0.7091) in waters draining terrains with mafic rocks (0.7036–0.7066), ratios higher than those of sea water in waters draining old granitic and comparable terrains (0.7103–0.7146), and ratios slightly higher than the sea water value in waters from young sedimentary terrains with large contributions from limestones of marine origin. STUEBER *et al.* (1973) made similar observations from underground water studies in the Scioto River basin, i.e., $^{87}Sr/^{86}Sr$ 0.708–0.709 in waters from carbonates and 0.710–0.713 in waters fom shales. The total stable strontium concentrations, however, varied widely and did not necessarily reflect the nature of the drainage basins, but rather the contact time of the source waters with the rock. ODUM (1957) gives an average Sr/Ca ratio for world rivers of 4.8×10^{-3}. TUREKIAN (1964) quotes values of the Sr/Ca weight ratio in United States rivers as ranging from 4.5×10^{-3} to 8.2×10^{-3}, which led BRASS (1976) to assume a ratio of 5.0×10^{-3} as a reasonable average of the limestone-derived ratio in rivers. The latter author then used this ratio to calculate the "average river content" of limestone-derived strontium as 55.5 μg/l, basing this on the "average river content" of limestone- (and some evaporite-) derived calcium of 11.1 mg/l. The latter is 75% of the average river value of 14.8 mg/l (GIBBS, 1972). The remainder of the calcium is derived from silicate weathering (GARRELS and PERRY, 1974). This leads to an additional strontium content of 13.0 μg/l, giving a total of 68.5 μg/l as the "average river content." With an average river flow of 3.2×10^{16} l/year, this yields a total river flux of dissolved strontium of 2.2×10^{12} grams/year.

REEDER *et al.* (1972) studied in great detail the water chemistry of the Mackenzie River drainage basin and established a Sr/Ca ratio of the Mackenzie River entering the delta of 5.8×10^{-3}, in good agreement with the above average. However, in tributary rivers flowing through limestone and dolomite terrains, these authors found Sr/Ca ratios as high as 18×10^{-3}. In addition, their strontium concentrations were generally higher than the "world average" of 68.5 μg/l, as were most of the values of BRASS (1976). For Indian rivers, SREEKUMARAN *et al.* (1970) reported strontium concentrations ranging from 80 to 460 μg/l, and SKOUGSTAD and HORR (1963) found an average Sr concentration of 110 μg/l for the major U.S. water supplies. STRADOMSKII and KONOVALOV (1968) made an intensive study of the distribution of strontium in the surface waters of rivers of the Soviet Union and calculated a total flux of about 0.5×10^{12} grams/year into the open ocean, the bulk of which reaches the Arctic Ocean (0.35×10^{12} grams/year). The above data suggest that the estimate of the river-derived strontium flux of 2.21×10^{12} grams/year is a minimum estimate, especially because it is based on a rather low "average river content" of strontium of 68.5 μg/l. Clearly it is important to obtain more data on all the major rivers of the world, especially at or near the confluence with the world ocean.

III. Lakes

Relatively few geochemical studies of the strontium cycle in lakes have been undertaken. Generally the Sr/Ca ratios reflect the rocks of the source areas of the waters. For instance, FAURE *et al.* (1967) found generally low values of dissolved strontium in Lake Superior with a rather uniform distribution (21.1–22.9 μg/l) and a low Sr/Ca ratio of $(1.53 \pm 0.04) \times 10^{-3}$. This reflects the dominance of igneous and metamorphic rocks surrounding this lake. Tributaries to this lake have more variable contents of Sr and also different Sr/Ca ratios. But, of course, the lake represents a relatively well mixed reservoir compared with the rivers that have been sampled only episodically. The data of FAURE *et al.* (1967) for Lake Huron are more variable, with higher strontium concentrations (62–113 μg/l) and higher Sr/Ca ratios (3.0–3.8×10^{-3}). Values are generally representative of the source areas of the tributaries, the higher values typically representing areas with more carbonate rocks.

The observations on the distribution coefficient of Sr/Ca (defined as $D^{\text{solid}}_{\text{liquid}} = \frac{(\text{Sr/Ca})\ \text{solid}}{(\text{Sr/Ca})\ \text{liquid}}$) in molluscs in the above lakes (FAURE *et al.*, 1967) are interesting. Shells of the pelecypod *Lampsilis* gave a distribution coefficient of 0.256 ± 0.027, a ratio much lower than the experimental value of 1.00 for aragonite determined by HOLLAND *et al.* (1964). This lower value can be understood in terms of the more complicated biochemical fractionation of strontium in these more evolved molluscs (LOWENSTAM, 1963; KINSMAN, 1969). Nevertheless, although the observed distribution coefficient is smaller than the experimental one, the Sr/Ca ratio in carbonate shells seems a primary function of the Sr/Ca ratio of the water. Similar observations were made by MÜLLER (1968, 1969) in Lake Constance.

IV. The Ocean

Only recently have more accurate determinations of the strontium concentrations of sea water been carried out (ODUM, 1951; FABRICAND *et al.*, 1966, 1967, ANDERSON

Table 38-I-1. *Strontium in ocean water* (Brass and Turekian, 1974)

Depth (m)	Sr (μg/kg)	Sr/S[a] (μg/g)
Geosecs Leg I Station 3 (51°01′N, 43°01′W)		
19	7,470	217.6
28	7,536	218.7
59	7,685	220.4
105	7,685	219.5
182	7,627	219.6
274	7,703	220.2
377	7,668	220.2
575	7,685	220.2
577	7,673	220.1
813	7,684	220.1
959	7,708	220.7
1,376	7,709	220.6
1,575	7,684	219.9
1,577	7,690	220.1
1,725	7,685	219.9
1,988	7,695	220.2
2,160	7,693	220.2
2,479	7,699	220.3
2,595	7,742	221.5
2,696	7,690	220.1
2,989	7,689	220.1
3,267	7,676	219.7
3,368	7,689	220.1
3,486	7,705	220.6
3,730	7,683	220.0
Geosecs I Pacific Test Station (28°29′N, 121°38′W)		
10	7,404	219.7
700	7,720	224.4
2,000	7,715	222.9
3,500	7,719	223.1
4,000	7,695	221.9
Scorpio Station 56 (43°16′S, 113°48′W)		
2,259	7,719	222.6
2,999	7,738	223.0
Scorpio Station 81 (35°20′S, 75°46′W)		
3,635	7,718	222.4
3,936	7,689	221.6
Geosecs Leg VI Station 68 (37°14′S, 52°00′W)		
10	7,545	220.2
189	7,539	220.7
386	7,580	221.7
632	7,682	223.9
889	7,698	223.4
1,483	7,745	223.1
1,741	7,692	221.5
2,136	7,685	220.9
2,488	7,688	221.1
2,891	7,702	221.7
3,387	7,684	221.4
3,698	7,690	221.6
4,595	7,685	221.6
4,595	7,668	221.2
4,891	7,688	221.7
5,493	7,732	223.0
5,793	7,663	221.0
Geosecs Leg VII Station 76 (57°44′S, 66°08′W)		
53	7,482	220.7
159	7,532	222.0
597	7,629	221.6
1,096	7,698	222.3
1,145	7,727	223.0
1,198	7,716	222.6
1,530	7,739	223.0
1,968	7,716	222.2
2,118	7,728	222.6
2,373	7,734	222.7
2,588	7,731	222.6
2,969	7,711	222.1
3,167	7,704	221.9
3,975	7,707	222.0
4,179	7,742	223.0
4,486	7,704	221.9

[a] S = salinity.

et al., 1970). Yet all the above authors used methods that had coefficients of variation not precise enough to detect any systematic variations in the strontium/salinity ratio of sea water. Brass and Turekian (1974) developed a mass spectrometric method with a precision of 0.2% and were the first to establish significant variations in the strontium/salinity ratio with depth. In Table 38-I-1 the data for the strontium concentrations and the strontium/salinity ratios are presented. Typically, these data show a gradual enrichment in strontium in the deep waters as they move to the North Pacific Ocean. In the deep waters of the North Pacific, an increase of about 1.5% in the strontium/salinity is observed. BRASS and TUREKIAN (1974) demonstrated clearly that this strontium increase cannot be due only to the dissolution of carbonate tests in the deep ocean because Sr/Ca ratios in carbonate tests should then be about ten times higher than those observed (TUREKIAN, 1964). The precipitation of strontium sulfate tests by acantharians in the surface waters and their almost complete dissolution below 400 meters (BISHOP *et al.*, 1977) is a much more likely explanation of the mechanism differentiating the strontium between surface and deep waters in the oceans. BERNER (1977) estimates that a substantial portion of the calcium carbonate flux to the deep ocean is supported from aragonite. This also constitutes a potential source of Sr to the deep waters, as the aragonite almost quantitatively dissolves.

The main process responsible for the removal of strontium from the ocean is the co-precipitation with carbonate tests (TUREKIAN, 1964). The residence time of strontium in the oceans, based on a river flux of 2.21×10^{12} grams/year is about 5×10^6 years, in good agreement with the estimate of GOLDBERG *et al.* (1971). A possible higher river flux, of course, may reduce this estimate. This residence time is about four times that of calcium (GARRELS and PERRY, 1974). As Sr/Ca ratios of deep-sea carbonates do not significantly change with time, at least not over the last 60×10^6 years (TUREKIAN, 1964; GIESKES, unpublished data), it appears that the strontium concentration of the ocean is in a steady state, or that it changes with a very slow time constant (TUREKIAN, 1964). This appears to be in contrast to the $^{87}Sr/^{86}Sr$ ratio that has shown substantial variation during the last 500 million years (VEIZER and COMPSTON, 1974; BRASS, 1976, Section 38-B). These variations have been explained by BRASS (1976) as being a result of variations in the relative proportion of "basaltic" and "granitic" silicate rocks due to changing patterns in global tectonics. SPOONER (1976) has invoked basalt-sea water interactions at spreading ridges as the factor explaining the discrepancy between the input $^{87}Sr/^{86}Sr$ ratio and that of sea water. In addition, changes in the isotope ratio during Phanerozoic time were explained by changes in exposed land area. Although the flux estimates of SPOONER (1976) appear to be in error, his arguments about isotope exchange with basalts obtained qualitative support from the observations of O'NIONS and PANKHURST (1976). These authors observed slight increases in the $^{87}Sr/^{86}Sr$ ratio of basalts as a result of alteration or contamination by sea water.

In adjacent seas, even at salinities half those of open ocean waters, the $^{87}Sr/^{86}Sr$ ratio, the Sr/Ca ratio, and the strontium/salinity ratio are very close to those of sea water. For instance, FAURE *et al.* (1967) found the $^{87}Sr/^{86}Sr$ ratio in the Hudson Bay to be 0.7093 ± 0.0003, both in the dissolved strontium and the strontium contained in *Mytilus edulis* shells. In addition, they established a Sr/Ca ratio of $(18.7 \pm 0.3) \times 10^{-3}$ in the water, similar to that of Atlantic Ocean surface water. BOJANOWSKI and OSTROWSKI (1968) found a strontium/salinity ratio of 212.6 μg/kg in deeper waters of the southern Baltic Sea, slightly less than that of sea water, c.f., Table 38-I-1. The surface waters, with their lower salinities, typically showed lower ratios.

V. Interstitial Waters of Marine Sediments

Studies of interstitial waters, particularly those obtained during the Deep Sea Drilling Project, have revealed distinct increases in the dissolved strontium concentrations with depth into the sediment. This is particularly true in carbonate sediments, and strontium concentrations below 200–300 m appear to be primarily a function of sedimentation rate (SAYLES and MANHEIM, 1975, GIESKES, 1975, 1976). Generally, large Sr/Ca ratios imply that much of the calcium is reprecipitated upon recrystallization of carbonate oozes, whereas the strontium is released to the interstitial waters and, in part, diffuses back into the ocean. However, recent studies in our laboratory on the Sr/Ca ratio of calcitic material in a pure carbonate ooze on the Manihiki Plateau (DSDP Leg 33, Site 317) show rather constant values of the Sr/Ca ratio with depth, i.e., 3.3×10^{-3} (or 15×10^{-4} atomic ratio). The average Sr/Ca ratio of limestones is 1.5×10^{-3} (TUREKIAN and KULP, 1956), which implies that only a small portion of these recent marine carbonates has recrystallized.

Increases in strontium concentrations are usually most pronounced in carbonate sediments (SAYLES and MANHEIM, 1975; GIESKES, 1976), but in sediments with mainly terrigenous contributions pronounced concentration gradients have also been observed (GIESKES *et al.*, 1977). On the average, strontium gradients are no more than 4–5 mg/l-100 m, resulting in a diffusive flux back into the ocean of about 0.2×10^{12} grams/year, or less than 10% of the river input. Of great interest will be studies of the strontium isotope composition of the pore fluids, especially because a more clear understanding can be reached of the origin of this strontium, i.e., whether it is derived from dissolution and/or recrystallization of carbonates or is due to submarine alteration of volcanic material in the ocean. Studies of this ratio in carbonate cores in the equatorial Pacific Ocean (LAWRENCE *et al.*, 1975) indicated a sea water origin of the dissolved strontium, but strontium ratios in drill site 323 in the Antarctic Ocean (LAWRENCE, personal communication) indicated an origin both in sea water and in volcanic materials. Clearly, studies of the strontium isotopes, in conjunction with those of the absolute strontium concentrations in interstitial waters, will be of importance in the overall geochemistry of strontium.

Strontium concentrations of larger magnitude are often found in cores above evaporites. MCDUFF *et al.* (1977) reported values as high as 1.23 g/l in one Deep Sea Drilling Project site in the Mediterranean. Generally, marine evaporites, especially the calcium sulfates, have Sr/Ca ratios similar to those of primary marine carbonates (USDOWSKI, 1973; WEDEPOHL, personal communication). However, upon diagenesis of evaporites, substantial migration of strontium and enrichment, particularly in the polyhalites, may occur (STEPNIEWSKI, 1973).

VI. Formation Waters

Studies on the distribution of strontium in ground waters are relatively rare. Among these is the already mentioned study of SKOUGSTAD and HORR (1963) on United States natural water supplies. BILLINGS *et al.* (1969) studied formation waters in the western Canada sedimentary basin and found a range of strontium values of 1–1,120 mg/l. They also found chloride values ranging from 6–190,000 mg/l. Presumably the relationship between strontium and chloride is similar to that found by CARPENTER and MILLER (1969) for formation waters in Saline County, Missouri.

Soviet workers have reported the use of dissolved strontium in ground waters as a potential indicator of oil-bearing formations. SKIBA (1967) studied in great detail the processes that lead to precipitation or dissolution of strontium salts in the supergene zone. The accumulation of strontium in the sodium chloride formation waters occurred by leaching of strontium sulfate and carbonate in the petroleum source rocks and other sediment formations. SCHEPAK and MIGOVICH (1970) found increased strontium levels in oil- and gas-bearing formations in the eastern Carpathian geosyncline. Similarly, SIVAN *et al.* (1973) noted slightly higher strontium concentrations in formation waters in the Crimea region.

Revised manuscript received: December 1976

38-K. Abundance in Common Sediments and Sedimentary Rock Types

I. Introduction

Strontium distribution in chemical and biochemical sedimentary rocks has been extensively studied because of its potential utility as a diagnostic paleoenvironmental and diagenetic tracer element. The knowledge of its distribution in terrigeneous sedimetntary rocks is to a degree the byproduct of Rb-Sr geochronological studies and, therefore, limited mostly to shales. The selective review of particular subclasses of sedimentary rocks is summarized below.

II. Terrigeneous Rocks

No analytical data are available for the matrix or clasts of the coarse (>2 mm) terrigeneous rocks.

a) Sandstones

The majority of the reported average Sr concentrations for undifferentiated *sandstones* (s. l.) are in the 40–150 ppm and for *graywackes* in the 100–400 ppm range (Table 38-K-1). The semiquantitative spectrographic data (McLaughlin, 1955; Weber and Middleton, 1961) for *arkoses* suggest a Sr content of 50–200 ppm.

From Sr isotopic studies Peterman and Whetten (1972) concluded that Sr in the Columbia River sand, a possible precursor of graywackes, was of a mixed granitic and basaltic provenance. However, neither the provenance nor the mineralogical control of Sr in older sediments is clear. Strontium substitutes diadochically for Ca^{2+} (and K^{+}) and the major controlling mineralogical phases are probably detrital and authigenic plagioclases, K-feldspars and carbonates (see Section 38-D). This may account for the higher Sr concentrations in graywackes than in quartz rich sandstones (and arkoses). However, due to this multiple mineralogical control the correlations with major elements are of low significance (e.g. Caby *et al.*, 1977). Additional work on mineral phases is needed to improve the understanding of Sr repartition in the whole rock systems.

b) Siltstones, Mudstones (Clays and Shales)

The average Sr concentrations in carbonate poor Quaternary argillaceous sediments are between 130 and 280 ppm (Table 38-K-2). Strontium appears to be derived either almost entirely from continental weathering products (Argentine Basin; Dasch, 1969b) or represents a mixed population composed of young volcanogenic and old continental components (Red Sea and Ross Sea; Boger and Faure, 1976 and Shaffer and Faure, 1976). The average Sr concentrations of 450 ppm for pelagic clays, determined spectrographically by Goldberg and Arrhenius

Table 38-K-1. *Strontium in sandstones and greywackes*

Age, location	No. of samples	Average ppm Sr	Range ppm Sr	Analytical method	Reference
Sandstones					
Eocene, Rocky Mts. region (U.S.A.)	216	138[a]		S	VINE and TOURTELOT (1973)
Jurassic-Tertiary, California (U.S.A.)	4	210	64–379	X	PETERMAN *et al.* (1967)
Permian, red bed sandstones, United Kingdom	33	47		X	COSGROVE (1973)
Paleozoic, Montagne Noire (France)	3	37	28–46	I	GEBAUER and GRÜNENFELDER (1974)
Greywackes					
Quaternary, Columbia River sands	14	361	204–512	X	PETERMAN and WHETTEN (1972)
Jurassic-Tertiary, California (U.S.A.)	41	296	33–973	X	PETERMAN *et al.* (1967)
Permian-Jurassic, New Zealand	9	376	255–471	X	EWART and STIPP (1968)
Paleozoic, New York (U.S.A.)	119	110		S	ONDRICK and GRIFFITH (1969)
Late Precambrian, Hoggar (Algeria)	35	383		A	CABY *et al.* (1977)
Early Proterozoic, Minnesota (U.S.A.)	2	171	119–222	I	PETERMAN (1966)
Archean, Wyoming (U.S.A.)	23	424	159–650	X	CONDIE (1967)
Archean, Fig Tree Group, S. Africa	22	180	41–568	X	CONDIE *et al.* (1970)
Archean, Swaziland System, S. Africa	4	45	22–98	I	ALLSOPP *et al.* (1968)

[a] Geometric mean.

Table 38-K-2. *Strontium in clays, argillaceous rocks and glauconites*

Age, location	No. of samples	Average ppm Sr	Range ppm Sr	Analytical method	Reference
Quaternary oceanic sediments					
Atlantic Ocean	51	156[a]	17–452	X	Dasch (1969b)
Argentine Basin	55	172[a]	25–367	X	Dasch (1969b)
Mid-Atlantic Ridge	24	134[a]	50–315	X	Dasch (1969b)
Red Sea, median valley	79	203[a]	112–304	X	Boger and Faure (1976)
Northern Atlantic, clays	38	141	10–460	S	Chester and Messiha-Hanna (1970)
Atlantic Ocean, clays	18	210	99–600	X	Wedepohl (1960)
Atlantic Ocean, clays (< 20% $CaCO_3$,)	17	220	93–475	S	Turekian (1964)
Pacific Ocean	6	224	76–600	X	Church (1971)
Barents Sea, silts	50	280	39–1,111	X	Wright (1974)
Barents Sea, clays	60	177	75–838	X	Wright (1974)
Ross Sea, Antarctica	40	138[a]	91–238	X	Shaffer and Faure (1976)
Tertiary, Mesozoic sedim. rocks					
Jurassic-Tertiary, California (U.S.A.)	6	359	143–723	X	Peterman *et al.* (1967)
Lias, United Kingdom	11	130	71–200	S	Catt *et al.* (1971)
Triassic, illites from sandstones, France	91	198	90–355	S	Mosser *et al.* (1972)
Paleozoic sedim. rocks					
Permian, red bed mudstones, United Kingdom	56	307		X	Cosgrove (1973)
Permian, Japan (partly metamorphosed)	56	120	23–255	X	Banno and Chappell (1969)
Permian, Kansas (U.S.A.)	13	287	84–530	I	Brookins *et al.* (1970)
Permian marls, Germany	44	195	51–680	X	Wedepohl (1964)
Carboniferous, United Kingdom	19	114	65–144	S	Curtis (1969)
Pennsylvanian, New Mexico (U.S.A.)	28	113	63–199	X, T	Mukhopadhayay and Brookins (1976)
Lower Paleozoic, Norway	95	220		X	Bjørlykke (1974)
Lower Peleozoic, Scotland (U.K.)	133	65	9–474	X	Stephens *et al.* (1975)
Silurian, Canberra (Australia)	22	24	12–43	I	Bofinger *et al.* (1970)
Silurian, Wales (U.K)	27	345	60–675	S	Spencer (1966)
Cambrian, Wales (U.K.)	25	85	60–160	S	Mohr (1959)

Proterozoic sedim. rocks					
Late Precambrian, Montana (U.S.A.)	37	52	14–121	I,X	Obradovich and Peterman (1968)
Late Proterozoic, tillites, Norway	39	97	52–227	X	Pringle (1973)
Late Precambrian, Brazil (<2 μ fraction)	21	38	18–71	I	Bonhomme (1976)
Late Precambrian, Mauritania (<2 μ fraction)	15	101	31–315	I	Clauer (1973)
Middle Precambrian, Minnesota (U.S.A.)	28	90	17–279	I	Peterman (1966)
Archean sedim. rocks					
Archean, Fig Tree Group, S. Africa	2	51	16–85	X	Condie *et al.* (1970)
Swaziland System, S. Africa	5	38	11–99	I	Allsopp *et al.* (1968)
Middle Marker, Swaziland System, S. Africa	6	64	47–75	I	Hurley *et al.* (1972)
Glauconites					
Tertiary, Belgium	8	24	11–30	I	Odin *et al.* (1974)
Cretaceous, North Carolina (U.S.A.)	4	16	13–20	I	Harris (1974)
Late Proterozoic, Northern Territory (Australia)	5	19	10–36	I	Cooper *et al.* (1971)
Middle Proterozoic, Northern Territory (Australia)	13	15	4–29	I	McDougall *et al.* (1965)

[a] Acid insoluble residue.

(1958) and EL-WAKEEL and RILEY (1961), are at variance with the results of other authors and the possibility of analytical discrepancy cannot be discounted.

The degree of retention of the inherited Sr provenance during post-depositional history of the rock is of paramount importance in the Rb-Sr dating of sedimentary rocks. CLAUER *et al.* (1975) demonstrated through isotopic studies of interstitial waters, that some isotopic exchange between clays and interstitial waters in deep sea sediments occurs within the first 1.6 m of the sedimentary column. However, this exchange, if considered for the bulk of the sediment, is probably of a small magnitude, because DASCH (1969b) was not able to detect any systematic variations either for the bulk sediment or for its size fractions. During the deep burial diagenetic stage, isotopic equilibration and elemental repartition leads to a partial homogenization of the system. PERRY and TUREKIAN (1974) observed that in the Miocene of Luisiana "strontium from interstitial waters, calcium carbonate, and adsorbed on the clay minerals, is released and then sequestered by the newly forming phases, together with strontium mobilized from the silicate minerals. The diagenetic changes involve the destruction of detrital mica and feldspar and the formation of an illite-rich clay from smectite. Diagenesis and homogenization arenot complete in the deepest part (5,323 m) of the section studied". The net increase in Sr in the shale is from 130 ppm at 1,582 to 158 ppm at 5,523 m. GEBAUER and GRÜNENFELDER (1974) concluded that complete Sr isotopic homogenization of the rock is not reached even during the anchimetamorphic (zeolite facies) stage, with detrital micas and chlorites still retaining their provenance signature, while authigenic plagioclase and illites appear to be isotopically homogenized. Whether in the process of this internal redistribution the total rock behaves as a closed or a partially open system with respect to elemental strontium cannot be ascertained on the basis of the available data.

The average Sr concentrations in the Phanerozoic argillaceous rocks are within the 20-360 ppm range (Table 38-K-2). This spread is partly due to mineralogical variations reflecting environmental and provenance factors. LEBEDEV (1967) reports an average Sr content of 73 ppm in fresh water, 247 ppm in brackish and 283 ppm in marine clays from the Mesozoic formations of the Caspian Depression and western Siberia. This may be in part due to a higher carbonate content of the latter, but the lack of mineral data precludes further generalizations.

REIMER (1972) suggested that the Sr content of shales decreases with increasing geologic age, with the bulk of the decrease observed during the Cenozoic—Late Paleozoic. This is consistent with the observation that no Precambrian assemblage in Table 38-K-2 has an average Sr concentration in excess of 100 ppm. Such an apparent secular trend may be a consequence of sampling bias, since most of the Precambrian as well as the low-Sr Phanerozoic populations were selected for dating purposes, with preferences for high Rb/Sr ratios. It is essential to discount this possibility before attempting interpretations based on the post-depositional mobility of Sr or on primary evolutionary phenomena.

The observed average Sr concentrations in a relatively homogeneous ($<2\,\mu$) *clay fraction* are within the 35–200 ppm range (Table 38-K-2) (BONHOMME, 1976; CLAUER, 1973; DASCH, 1969; MOSSER *et al.*, 1972; MUKROPADHYAY and BROOKINS, 1976) but no consistent trends have been observed for individual clay minerals. In contrast, *glauconites* have average Sr concentrations of $\sim$15–30 ppm (Tables

Table 38-K-3. *Strontium in cherts, phosphorites and evaporites*

Age, location	No. of samples	Average ppm Sr	Range ppm Sr	Analytical method	Reference
Cherts					
Quaternary, deep sea siliceous sediments, worldwide	3	230	120–390	S	El-Wakeel and Riley (1961)
Cretaceous pelagic radiolarites, Indonesia	18	265	50–900	X	Audley-Charles (1965)
Proterozoic, Ontario (Canada)	1	4		I	Faure and Kovach (1969)
Phosphorites					
Cambrian, Queensland (Australia)	54	200	46–1,000	S	Cook (1972)
Phanerozoic, worldwide	65	1,025	75–2,130	X	McArthur (1978)
Evaporites					
Gypsum:					
Quaternary, Persian Gulf, Baja California, Jarvis Island	38	1 013	210–2,820	A	Butler (1973)
Carboniferous, Spitsbergen	4	687	225–1,029	X	Holliday (1967)
Anhydrite:					
Quaternary, Persian Gulf, Baja California	41	2,125	880–5,220	A	Butler (1973)
Malm, L. Saxony (German F.R.)	41	2,460	1,400–3,400	F	Müller (1962)
Permian, Germany	34	2,040	1,300–2,900	F	Jung and Knitzschke (1960)
Permian, shaly anhydrite, Germany	20	1,558	80–3,100	X	Pundeer (1969)
Permian, German D.R.	22	1,423[a]	1,000–2,100	F	Herrmann (1961)
Permian, L. Saxony (German F.R.)	13	2,430	1,800–3,000	F	Müller (1962)
Carboniferous, Spitsbergen	3	1,421	972–1,682	X	Holliday (1967)
Halite:					
Permian, Harz Mts. (German D.R.)		100[a]		F	Herrmann (1961)
Permian, Magdeburg (German D.R.)		44[a]		F	Herrmann (1961)
Permian, Harz (Germany)	3	1,933	1,600–2,200	F	Jung and Knitzschke (1960)
Permian-Tertiary, Germany	25	3	0.7–7	F	Müller (1962)

[a] Water and alcohol insoluble residue.

38-D-3 and 38-K-2). Since the latter are again isotope dilution measurements on samples utilized for Rb-Sr dating, the quoted averages may be slightly low.

III. Chemical and Biochemical Sediments (Excluding Carbonates)

a) Ironstones

CATT *et al.* (1971) report an average Sr content of 265 ppm (range 77–1,000 ppm) from Lias ironstones of England.

b) Cherts

The wide variations in the average Sr content of cherts (4–265 ppm) (Table 38-K-3) are difficult to evaluate from the sparse data available. Since some radiolaria contain celestite in their skeleton (ODUM, 1951; Sections 38-H-III and 38-L), it is possible that radiolarites have a higher Sr content than the cherts derived from other biochemical and inorganic sources.

c) Phosphorites

Strontium substitutes for calcium in the lattice of apatite. Pure fluor-apatite should contain ~2,400 ppm of Sr (MCARTHUR, 1978). The average Sr concentration in phosphorites of various ages, but presently in submarine environments, is 1258 ppm (range 530–2,130), whereas in the onshore deposits it is 683 ppm (75–1,460) (MCARTHUR, 1978), reaching average values as low as 200 ppm in the Cambrian phosphorites of Queensland (COOK, 1972) (Table 38-K-3). Although the possibility of some primary depositional variations in Sr content cannot be discounted, the differences appear to be related mostly to the degree of post-depositional alteration (ODUM, 1957; COOK, 1972).

d) Evaporites

At near-surface temperatures the Sr partition coefficient $K = \frac{[Sr/Ca]_{gyps.}}{[Sr/Ca]_{liq.}}$ between gypsum and solution is ~0.3–0.1 and for anhydrite and solution it is ~0.5–0.2 (computed from USDOWSKI, 1973; BUTLER, 1973; ICHIKUNI and MUSHA, 1978; Section 38-H-III). Both partition coefficients decrease with increasing temperature. Consequently, marine *gypsum* should contain ~1,100 ppm of Sr at $CaCO_3$ saturation, ~2,200 at NaCl saturation and ~9,000 ppm at blödite saturation level (USDOWSKI, 1973). This is in general agreement with observations on natural gypsum (Table 38-K-3). In addition, 118 gypsum samples from the Permian of Oklahoma (U.S.A.) have an average Sr content of 785 ppm (JOHNSON and HAM, 1970 in BUTLER, 1973) and 43 samples from the Permian of the Donets Basin (U.S.S.R.) of 1,900 ppm Sr (GONCHAROV, 1967).

Anhydrite, mostly a diagenetic replacement product of gypsum, should contain from about 1,000, or more likely 2,000, to 2,400 ppm Sr (HERRMANN, 1961; BUTLER, 1973; USDOWSKI, 1973); in accord with the observed concentrations (Table 38-K-3). The general tendency towards lower than predicted Sr concentrations in gypsum and anhydrite (Table 38-K-3) may be a result of partial diagenetic equilibration

Table 38-K-4. *Strontium in carbonate rocks*

Age, location	No. of samples	Average ppm Sr	Range ppm Sr	Analytical method	Reference
LIMESTONES AND DOLOMITIC LIMESTONES					
Holocene continental deposits					
Germany, travertines (calcite)	17	1,460	90–2,200	A	Savelli and Wedepohl (1969)
Holocene shallow water sediments					
Jamaica	20	2,780	1,200–4,110	X	Land (1973a)
Red Sea	19	6,384	700–10,000	A	Milliman *et al.* (1969)
Bahamas	34	7,590	4,947–9,600	F	Till (1970)
Florida	27	9,178	6,300–11,300	F	Siegel (1961)
Florida	40	4,843	2,000–8,000	X	Stehli and Hower (1961)
British Honduras	27	6,093[a, b]	2,660–8,200	A	Billings and Ragland (1968)
Bahamas and Persian Gulf, oolites	15	9,769	9,010–10,460	A	Kinsman (1969)
Holocene deep sea sediments					
Atlantic Ocean	38	864	146–2,400	S	Chester and Messiha-Hanna (1970)
Atlantic Ocean	100	1,742[a, b]	1,019–4,020	S	Turekian (1964)
Worldwide sampling	10	1,110	200–2,400	S	El-Wakeel and Riley (1961)
Pleistocene shallow water sediments					
Florida	19	2,406	710–5,600	X	Stehli and Hower (1961)
Jamaica	32	2,823	530–7,550	X	Land (1973b)
Carribean	19	419	110–1,150	X	Land (1973b)
Tertiary shallow water sediments					
Western Pacific Ocean (Tertiary to Holocene reefs)	56	686[d]	456–1,490	A	Honjo and Tabuchi (1970)
France, non-marine (?) evaporitic sequence	53	872[a, c]	131–7,869	A	Renard (1975)
Bahamas (Cretaceous to Pleistocene)	11	287	70–402	X	Goodel and Garman (1969)
Tertiary deep sea sediments					
Pacific Ocean	9	848[d]	111–11,100	A	Honjo and Tabuchi (1970)
Northern Atlantic Ocean (Upper Cretaceous to Pleistocene)	48	564[a, c]	389–1,945	A	Renard *et al.* (1978b)
Northern Atlantic Ocean (Upper Cretaceous to Tertiary)	26	1,320[a, c]	149–1,935	A	Renard *et al.* (1978a)

Table 38-K-4 (continued)

Age, location	No. of samples	Average ppm Sr	Range ppm Sr	Analytical method	Reference
Mesozoic shallow water sediments					
Western Carpathians	612	422	15–1,969	A	Veizer and Demovič (1974)
Eastern Alps, Jurassic	137	322	55–3,200	S, A	Flügel and Wedepohl (1967)
Eastern Alps, Triassic	594	197 [c]		A	Kranz (1976)
Southern Germany	60	585 [a, c]	57–3,668	A	Veizer (1977b)
Mesozoic deep sea sediments					
Worldwide, Cretaceous to Danian, mostly chalk	213	1,229 [d]	152–8,190	A	Honjo and Tabuchi (1970)
Denmark, Maastrichtian to Danian, mostly chalk and some bryozoan limestone	73	838	454–1,069	A	Jørgensen (1975)
Northern Atlantic, Cretaceous to Lower Tertiary	76	642 [a, c]	119–1,630	A	Renard *et al.* (1978)
Israel, Cretaceous, chalk	11	562	280–942	A, S	Magaritz (1974)
Paleozoic shallow water sediments					
Kansas (U.S.A.) Late Paleozoic	261	482	14–2,000	S	Runnels and Schleicher (1956)
Australia and Western Carpathians (Paleozoic > Mesozoic)	80	408	60–1,870	X	Veizer and Compston (1974)
Illinois (U.S.A.), Pennsylvanian	91	490	240–810	S	Ostrom (1957)
United Kingdom, Carboniferous	27	326	145–484	X	Dickson and Barber (1977)
Arctic Canada	65	430 [c, e]	21–779	A	Veizer *et al.* (1978)
Worldwide sampling	337	370 [d]	31–7,120	A	Honjo and Tabuchi (1970)
United Kingdom, Phanerozoic	58	483		X	Barber (1974b)
North America, Phanerozoic calcitic oolites	52	414	230–2,250	F	Kahle (1965)
United Kingdom, Cambrian	156	618		S	Al-Hashimi (1976)
North America, mostly Paleozoic	92	315 [f]	44–1,560	S	Kulp *et al.* (1952)
Scotland, (U.K.)	183	420			Geol. Survey of Great Britain (1956) in: Graf (1960)
Proterozoic shallow water sediments					
Australia	25	410	40–1,220	X	Veizer and Compston (1976)
Archean shallow water sediments					
Ontario (Canada), stromatolitic limestones	47	295 [c]	30–1,450	A	Veizer (unpublished)

DOLOSTONES AND CALCITIC DOLOSTONES					
Holocene					
Carribean and Persian Gulf	9	814	620–970	X	Behrens and Land (1972)
Pleistocene					
Carribean	21	280	130–460	X	Land (1973a)
Tertiary					
Bahamas, Cretaceous to Pleistocene	43	115	<35–402	X	Goodel and Garman (1969)
France, non-marine (?) evaporitic sequence	18	769[a, c]	298–2,263	A	Renard (1975)
Bahamas, Neogene	31	171	65–270	A	Supko (1977)
Egypt, Eocene	19	96	63–157	X	Land *et al.* (1975)
Mesozoic					
Western Carpathians, Triassic	240	104	33–616	A	Veizer and Demovič (1974)
Eastern Alps, Triassic	194	72[c]		A	Kranz (1976)
Paleozoic					
North America, (Paleozoic>Mesozoic)	231	176	23–2,800	S	Weber (1964)
United Kingdom, Carboniferous	17	60	19–91	X	Dickson and Barber (1977)
United Kingdom, Carboniferous	19	47		X	Barber (1974)
Arctic Canada, Paleozoic	137	107[c, e]	18–637	A	Veizer *et al.* (1978)
Wisconsin (U.S.A.), Ordovician	17	37		A	Badiozamani (1973)
United Kingdom, Cambrian	132	56	60–210	S	Al-Hashimi (1976)
Proterozoic					
Australia	35	131	10–700	X	Veizer and Compston (1976)
Archaean					
Ontario (Canada)(Fe dolostones)	12	75[c]	<10–290	A	Veizer (unpublished)

[a] samples with non-carbonate fraction>50% excluded; [b] determined in acid soluble portion but concentrations given with respect to total rock; [c] determined in acid soluble portion with concentrations recalculated on 100% carbonate (insoluble residue free) basis; [d] determined in acid soluble portion; [e] samples containing celestite excluded; [f] recalculated from Sr/Ca ratios.

with meteoric waters. This is in accord with observations of BUTLER (1973) that gypsum and anhydrite at Abu Dhabi, which formed in equilibrium with continental brines, contained 200 and 600 ppm Sr respectively.

The reported Sr concentrations in *halite* cover four orders of magnitude (Table 38-K-3). It is likely that pure halite contains <100 ppm of Sr and the reported high Sr values are due to the presence of disseminated celestite and/or anhydrite. Although celestite should be directly precipitated from sea water brines only towards the end of the NaCl phase (USDOWSKI, 1973; Section 38-H-III), it is commonly present as a diagenetic mineral not only within the sulfate but also within the evaporitic carbonate facies.

Strontium concentrations in minerals of the late stage of the evaporitic sequence (*K-Mg chlorides and sulphates*) are probably in the ≤ 10 ppm range (MÜLLER, 1962). The observed or indicated presence of celestite may account for the wide spread of reported values (<1–4,500 ppm) (GONCHAROV, 1967; HERRMANN, 1961).

IV. Carbonate Rocks

a) Holocene Sediments

The experimentally determined partition coefficients $K_{Sr}^{Arag.}$ and $K_{Sr}^{Calc.}$ at 25° C are $\sim$1.1 and 0.14 respectively (KINSMAN and HOLLAND, 1969; KINSMAN, 1969; computed from USDOWSKI, 1973; Section 38-H-II). *Aragonite* and *calcite* in equilibrium with sea water should therefore contain $\sim$8,300 and $\sim$1,200 ppm Sr. The observed Sr concentrations in Recent marine carbonates cover, the whole range between these theoretical contents of the end members aragonite and calcite (Table 38-K-4). Their bulk Sr composition is determined by relative proportions of their inorganic and organic (Section 38-L) components. Carbonate sediments, with the exception of the deep sea carbonates (composed predominantly of the stable low-Mg calcitic globigerina, coccoliths and some, possibly aragonitic, pteropods) and some travertines consist of metastable aragonite and high-Mg ($>1\%$ Mg) calcite. Several European travertine deposits have been precipitated at normal surface temperatures from waters high in Sr which was derived from gypsum deposits (see Table 38-K-4).

JACOBSON and USDOWSKI (1976), on the basis of experimental results, suggested that the equilibrium concentration of Sr in marine early diagenetic *dolomite* should be about 1/2 of that for marine calcite. The penecontemporaneous Persian Gulf and Carribean dolomites (Table 38-K-4) approach this value, but the phreatic hyposaline dolomites of Jamaica (LAND, 1973b) have excessive Sr concentrations ($\sim$3,000 ppm).

b) Diagenetic Repartition of Strontium

The shallow marine metastable carbonate assemblage is, upon exposure to meteoric waters, transformed into a stable low-Mg calcitic lithified *limestone*. This transformation is a solution-precipitation process operating on a $\sim 10^5$ years time scale (GAVISH and FRIEDMAN, 1969). Because of the low $^{m}Sr/^{m}Ca$ ratio of many meteoric waters ($\sim$0.0032 vs. 0.0086 of sea water; KINSMAN, 1969) and particularly due to the $K_{Sr}^{Calcite} < 1$ the precipitated low-Mg calcite will incorporate less Sr into its lattice than was present in the dissolving metastable phases. As a consequence, pre-Quaternary limestones with stabilized mineralogy contain an order of magnitude lower Sr concentrations than their metastable precursors (~400 ppm; Table 38-K-4). VEIZER and DEMOVIČ (1974) observed that Sr distribution in stabilized limestones is of a bimodal nature and the particular Sr populations are facies related with the hypersaline micritic, dark-coloured and deep sea facies usually containing Sr values >300 ppm and the organodetrital ones <300 ppm.

In modelling the diagenetic repartition of Sr it is generally accepted that equilibrium exists between the dissolving metastable phases, the bulk aquifer water and the precipitated stable phases (MORROW and MAYERS, 1978). The process is discussed in terms of open and closed diagenetic systems by FLÜGEL and WEDEPOHL (1967) and by KINSMAN (1969). Accepting this model, the Sr content of the stabilized limestones is controlled solely by the composition of the aquifer water and the degree of openness of the diagenetic system.

In an alternative explanation it is assumed that the mineralogical transformation is achieved in a reaction zone, which is not in equilibrium with the bulk aquifer water (PINGITORE, 1976; KATZ and MATTHEWS, 1977; VEIZER, 1977a). Since the rate limiting step is the precipitation of the stable low-Mg calcite, the dissolution of the metastable phase controls the composition of the reaction zone. Consequently, the precipitated low-Mg calcite may retain some "memory" of the original chemistry. This alternative may explain the facies dependent chemical composition of limestones and their internal textural constituents. If correct it may also help to explain why partition coefficients obtained through recrystallization experiments (KATZ *et al.*, 1972; KATZ and MATTHEWS, 1977; JACOBSON and USDOWSKI, 1976) are at variance with values obtained from direct precipitation experiments or from natural observations (VEIZER, 1978; Section 38-H-II).

In contrast to shallow marine carbonates, the transformation of *deep sea low-Mg calcitic oozes* into *limestones* results in an average Sr decrease by a factor of $\lesssim 5$ (Table 38-K-4). The rate of this loss is measurable on a $\sim 10^7$ years time scale, as attested to by the relatively unaltered Sr concentrations in some Tertiary and even Mesozoic counterparts (Table 38-K-4), regardless of whether the samples were collected from deep sea cores or in outcrops on land. Despite this slow rate, the eventual decrease of Sr concentrations to >300 ppm has been observed even in deep sea cores (MATTER *et al.*, 1975), pointing out that the major diagenetic process is pressure solution followed by calcite precipitation into available pore spaces and not equilibration with meteoric waters as is the case in shallow marine carbonates.

Pre-Quaternary *dolostones* appear to contain lower Sr concentrations than their modern equivalents (Table 38-K-4), but this difference may only be a result of the paucity of data for Holocene dolomites. The penecontemporaneous and early diagenetic fine-grained dolostones, frequently of hypersaline facies, contain somewhat higher Sr concentrations (~100–800 ppm) than the late diagenetic coarse-grained dolostones (<150 ppm). VEIZER *et al.* (1978) argued that the former are replacements of originally aragonitic muds, while the latter represent replacements of already stabilized calcitic limestones in dolomitizing hyposaline solutions.

c) Secular Chemical Trends

VINOGRADOV *et al.* (1952) observed that the Sr content of undifferentiated carbonate rocks decreases with their increasing geologic age. However, no definite age trend is observed for limestones if young sequences with metastable mineralogy are excluded (KULP *et al.*, 1952; VEIZER, 1977a). In contrast, Precambrian dolostones show decreasing Sr concentrations with increasing age, which is related to progressively higher Fe^{2+} and Mn^{2+} substitution for Mg^{2+} and Ca^{2+} in the lattice positions of dolomite (VEIZER, 1977a).

Manuscript received: July 1978

38-L. Biogeochemistry

I. Calcareous Skeletons

The sparse pre-1950 analytical data on the chemical composition of biogenic systems were summarized by VINOGRADOV (1953). The pioneering work on Sr distribution in calcareous skeletons was carried out by ODUM (1951, 1957), THOMPSON and

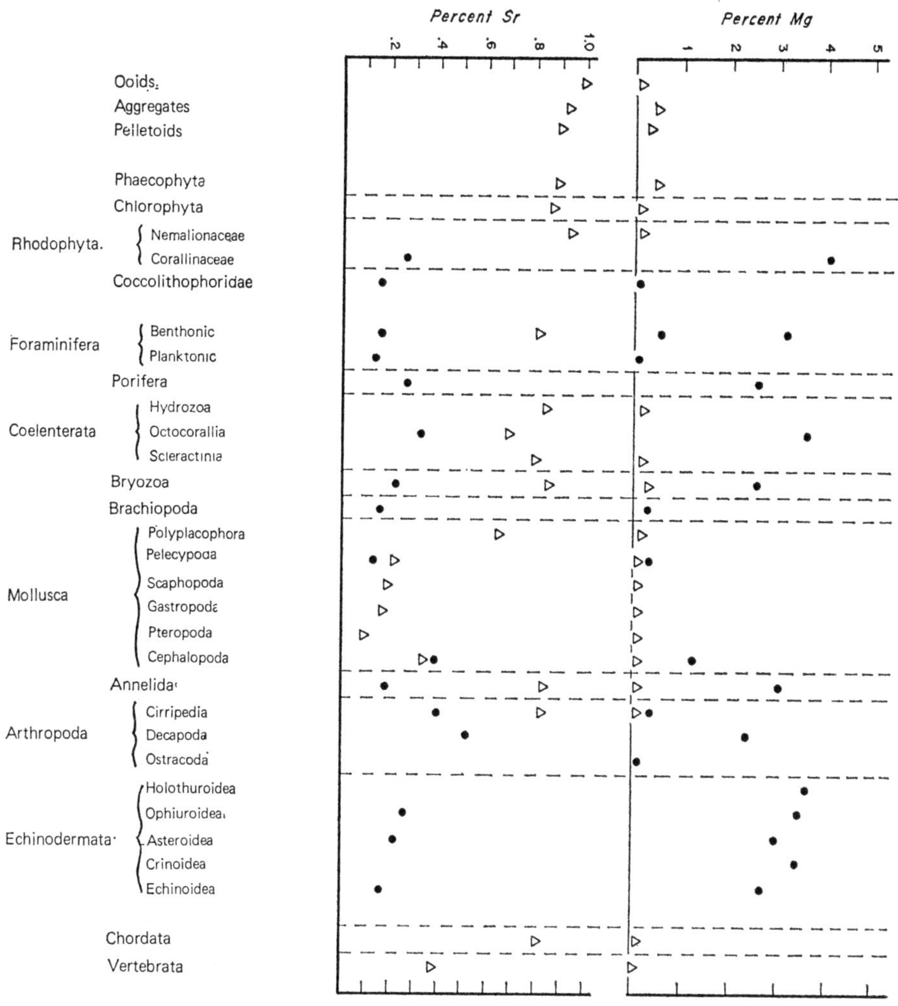

Fig. 38-L-1. Average Sr and Mg contents in the various marine carbonate components. Open triangles represent values from aragonitic carbonates; solid circles, calcite and magnesian calcite. Reproduced from MILLIMAN (1974, p. 144)

Chow (1955), Bowen (1956) and Lowenstam (1961). The work and subsequent vigorous research effort were ably summarized by Graf (1960) Lowenstam (1963b, 1964), Wolf *et al.* (1967), Dodd (1967) and particularly Milliman (1974, Chapters 4 and 5). The reader is referred to the latter for details and bibliography. Not included are the recent publications of Bender *et al.* (1975) and Lorens *et al.* (1977) on pelagic foraminifera, Goreau (1977) on corals, Buchardt (1977) on mollusks, Veizer (1974) on belemnites and Veizer and Wendt (1976) on sponges; in general confirming earlier findings.

Figure 38-L-1 from Milliman (1974) summarizes the results. The first order control of Sr distribution in modern calcareous skeletons is exercised by their mineralogical mode. Many aragonitic skeletons have Sr concentrations comparable to those predicted for inorganic aragonite (~8,000 ppm; Subsection 38-K-IV) and calcitic skeletons comparable to, or higher than, that of inorganic calcite (~1,200 ppm). The intermediate values are due in part to mixed mineralogy of the specimens, impurities, inclusions or to analytical techniques, but some are real. Thus the low Sr concentrations in aragonite of Mollusca and Vertebrata are due to biological discrimination against Sr during the transfer of fluid through cellular tissue to the site of calcification (Lowenstam, 1963a, 1963b). Buchardt and Fritz (1978) observed that the partition coefficient $K_{Sr}^{Arag.}$ for fresh water mollusks is ~0.25. Therefore, aragonite of marine mollusks should contain ~1,880 ppm Sr; in general agreement with the observed values. The second order controlling factors of Sr distribution are phylogeny, ecology, ontogeny, biochemistry and physiology of the system, but the quantitative relationships are not clear. In the present state of the art perhaps only some mollusks can be accepted as showing a Sr decrease in aragonite and increase in calcite with increasing temperature and/or decreasing salinity (Milliman, 1974). However, large differences in salinity, and correspondingly in Sr/Ca ratios of water (e.g. fresh vs. sea water), are clearly reflected in the Sr concentrations of shells. Thus the aragonitic fresh water mollusks from the Great Lakes contain only 91–441 ppm Sr (Faure *et al.*, 1967). Similarly, diagenetic alteration of shells in meteoric waters leads generally to lower Sr content.

II. Siliceous Skeletons

Odum (1951) reported that some radiolaria (Acantharia) contain *celestite*. Ash of Acantharia contained 21.8% of Sr.

III. Organic Phosphates

The *phosphatic teeth* and *bones* of Vertebrata contain 70 to 12,900 ppm of Sr (Wyckoff and Doberenz, 1968), with animal skeletons averaging 2,171 and human ones 150 ppm Sr. Flügel and Wedepohl (1967) report Sr concentrations close to 1,000 ppm in phosphates of Jurassic limestones from the Alps. Thurber *et al.* (1958) report 172 ppm of Sr (range ~40–360 ppm) in human skeletons. It is not clear if the difference in Sr content between human and animal skeletons is real or an artifact of low sampling density.

IV. Organic Matter and Coal

BOSTRÖM *et al.* (1974) report 9,300 ppm Sr in 17 samples of Recent dry organic matter from the eastern Pacific. MARTIN (1969) mentions that Sr was not concentrated by plankton collected off Panama and Colombia. BLACK and MITCHELL (1952) and YAMAMOTO (1972) report averages of 1,500 ppm Sr and 1,100 ppm Sr for dry brown algae (7 species) and dry seaweeds (59 species) respectively.

YUDOVICH *et al.* (1972) have compiled trace element data on hard and brown coal. They list a weighted mean of 240 ppm Sr in hard coal. Hard coal and brown coal ashes contain 420 and 850 ppm Sr respectively which is higher than the averages for argillaceous materials and sandstones.

Manuscript received: July 1978

38-M. Abundance in Common Metamorphic Rock Types

Table 38-M-1 presents data on the strontium distribution in a variety of metamorphic rock types. Most of the data refer to series of regional metamorphism. Some data for rocks which have undergone contact metamorphism are given at the end of this table. Most of the rock types between greenschist and granulite facies are listed approximately in the order of increasing metamorphic grade and are grouped according to the composition of their probable parental rocks. The distribution of strontium varies widely even within a single rock type, hence no meaningful Sr averages can be calculated. No systematic differences in strontium distribution are evident between ortho- and para-origin for gneisses and amphibolites, but the latter exhibit the most homogeneous distribution. For amphibolites, most of the Sr values are in the range 150–300 ppm. 1876 samples of schists, gneisses, amphibolites and granulites as listed in Table 38-M-1 contain on average 388 ppm Sr. This value deviates by only 20 percent from the overall mean abundance of 315 ppm Sr in the Canadian Shield as reported by Shaw *et al.* (1976).

The Sr content in some rocks of igneous origin appears to depend on the age of the primary material. The Sr content of Archean metavolcanics, for example, is frequently lower than that of their younger equivalents (e.g. Baragar and Goodwin, 1969). One of the main features of strontium is that its distribution in metamorphic belts varies markedly from region to region (e.g. Lambert and Heier, 1968; Shaw *et al.*, 1976). In the case of metavolcanics these regional variations are frequently interpreted (e.g. Gill and Bridgwater, 1976) as being caused by compositional heterogeneity of the source magma.

Revised manuscript received: August 1976

Table 38-M-1. *Strontium abundance in metamorphic rocks* (Number of samples in brackets)

Rock type	Origin	Range ppm Sr	Average ppm Sr	Analyt. method	References
Spilite (8)	Switzerland	41–413	247	N/R	HERRMANN *et al.* (1974)
Spilite (12)	Variscan geosyncline (W.-Germany)	41–422	266	N/R	HERRMANN *et al.* (1974)
Spilite (6)	Deccan (India)	55–600	290		VALLANCE (1974)
Meta-quartzite (6)	Pedernal Hills, New Mexico (U.S.A.)	4.1–60	22	I	MUKHOPADHYAY *et al.* (1975)
Phyllite, schist (11)	Chilean basement	19.1–178.4	63	I	MUNIZAGA *et al.* (1973)
Low grade pelite (13)	New Hampshire (U.S.A.)	110–1,700	416	S	SHAW (1954)
Schist (10)	Arahura series (New Zealand)	100–380	201	S	TAYLOR (1955)
Schist (25)	Various	24–636	232	X	TUREKIAN and KULP (1956)
Pelitic schist (16)	Connemara (Eire)	100–360	186	S	EVANS (1964)
Schist (6)	Palmer (South Australia)	582–1,138	140	X	WHITE (1966)
Al-rich schist (9)	Idaho (U.S.A.)	3–150	46	S	HIETANEN (1969)
Andesitic-dacitic schist (14)	Kongsberg (S. Norway)	10.3–376.2	148	I	O'NIONS and HEIER (1972)
Granitic schist (6)	Isua (West Greenland)	17.1–46.1	26	I	MOORBATH *et al.* (1975)
Epidiorite, schist (10)	Singhbum (India)	95–170	121	S	MUKHERJEE (1968)
Serpentine (6)	Münchberg (S. Germany)	35–75	51	S	HAHN-WEINHEIMER and ROST (1961)
Metadolerite (6)	North Carolina (U.S.A.)	211–521	338	S	WILCOX and POLDERVAART (1958)
Glaucophane metabasalt (20)	Western Crete (Greece)	120–1,400	502	X	SEIDEL (1974)
Metabasalt (83)	Archean, Norseman (W. Australia)		102	X	HALLBERG (1970)
Metabasalt (45)	Archean, Birch-Uchi (Canada)		196	X	BARAGAR and GOODWIN (1969)
Metabasalt (134)	Archean, Woods-Wabigoon (Canada)		217	X	BARAGAR and GOODWIN (1969)
Metavolcanic rock ($K_2O < 0.6\%$) (46)	Archean greenstone belts (Canada)		123	I/X	HART *et al.* (1970)

Table 38-M-1 (continued)

Rock type	Origin	Range ppm Sr	Average ppm Sr	Analyt. method	References
Metavolcanic rock (16)	Early Precambrian, Minnesota (U.S.A.)	92–1,237	455	I	HART and HANSON, (1975)
Metavolcanic rock (24)	Paleozoic, Støren (Norway)	106–528	223	X	GALE and ROBERTS, (1974)
Metabasite (10)	Lower Paleozoic, Stord (Norway)	251–351	288	X	FURNES and FAERSETH (1975)
Medium grade pelite schist (23)	New Hampshire (U.S.A.)	120–1,300	731	S	SHAW (1954)
Amphibolite facies rock (27)	Musgrave Range (Australia)		400	X	LAMBERT and HEIER (1968)
Amphibolite facies rock (15)	Cape Naturalist (Australia)		190	X	LAMBERT and HEIER (1968)
Greenschist-amphibolite (53)	Dalradian (Scotland)	83–461	235	X	VAN DE KAMP (1970)
Para-amphibolite (8)	Carolina (U.S.A.)	37–315	144	S	WILCOX and POLDERVAART (1958)
Ortho-amphibolite (8)	Carolina (U.S.A.)	397–606	478	S	WILCOX and POLDERVAART (1958)
Amphibolite (33)	Connemara (Eire)	50–880	285	S	EVANS and LEAKE (1960)
Amphibolite (32)	Western Alps (Italy)	55–840	273	S	RIVALENTI (1966)
Amphibolite (30)	British Columbia (Canada)	150–1,100	398	S	PRETO (1970)
Amphibolite (21)	Odenwald (W. Germany)	51–370	164	X	KLEMM and WEBER-DIEFENBACH (1971)
Amphibolite (32)	Inner Hebrides (Scotland)		151	X	DRURY (1973)
Ortho-amphibolite (16)	South Norway		320	X	FIELD and ELLIOT (1974)
Amphibolitic gneiss (15)	Langöy-Hinnöy (Norway)		338	X	HEIER and THORESEN (1971)
Hornblende-biotite gneiss (26)	Inner Hebrides (Scotland)		381	X	DRURY (1973)
Hornblende gneiss (17)	Inver assemblage (Scotland)		407	X	HOLLAND and LAMBERT (1975)
Hornblende-biotite gneiss (41)	East Greenland		580	X	TARNEY (1975)
Paragneiss (24)	Emeryville, N. Y. (U.S.A.)	225–490	310	S	ENGEL and ENGEL (1958)
Corundum-bearing gneiss (10)	Montana (U.S.A.)	140–500	222	S	FOSTER (1962)
Gneiss (20)	Spessart (Germany)	18–220	79	X	OKHRUSCH and RICHTER (1967)
Gneiss (22)	Willyama complex (Australia)	26–318	101	I	PIDGEON (1967)

Gneiss (254)	Lewisian (Scotland)		569	X	Sheraton (1970)
Gneiss (20)	Early Precambrian, Godthaab (Greenland)	35–485	245	X	Black *et al.* (1971)
Biotite gneiss (49)	Apsley, Ontario (Canada)		180	S	Shaw (1972)
Gneiss (254)	Drumberg area (Scotland)		569	X	Tarney *et al.* (1972)
Orthogneiss (11)	Eastern Alps (Italy)	10.6–204	69	I	Borsi *et al.* (1973)
Trondhjemitic gneiss (37)	Archean, Guyana shield (Venezuela)	90–850	485	X	Dougan (1976)
Paragneiss (18)	Gotthard Massif (Switzerland)	95–273	266	I	Nunes and Steiger (1974)
Metasedimentary gneiss (21)	Leverburgh Belt (Scotland)		405	X	Tarney (1975)
Tonalitic gneiss (13)	Archean, SW. Greenland	219–338	299	I	Pidgeon and Hopgood (1975)
Tonalitic gneiss (11)	Labrador (Canada)	200–1,113	529	X/I	Barton and Barton (1975)
Gneiss (12)	Pyrénées (France)	51–498	207	I	Vitréac-Michard and Allègre (1975)
Gneiss (10)	Uivak, Labrador (Canada)	320–860	570	X	Hurst *et al.* (1975)
Granite-gneiss (51)	Austria	40–930	132	X	Grohmann (1965)
Garnet gneiss (56)	Eastern Greenland		600	X	Tarney (1975)
High grade schist-gneiss (27)	New Hampshire (U.S.A.)	360–1,300	760	S	Shaw (1954)
Anatectic gneiss (37)	Schwarzwald (Germany)	94.2–401	203	I	Hofmann and Köhler (1973)
Migmatite, granitic gneiss (35)	Central Alps (Italy)	27–458	159	I	Hänny *et al.* (1975)
Migmatite (7)	Palmer (Australia)	127–347	240	X	White (1966)
Orthogneiss and granulite (23)	Morin, Quebec (Canada)	28–800	421	X	Philpotts (1966)
Gneiss-granulite (18)	Namaqualand (South Africa)	13–150	184	I	Clifford *et al.* (1975)
Granulite facies rocks (26)	Musgrave Range (Australia)		340	X	Lambert and Heier (1968)
Granulite facies rocks (35)	Cape Naturalist (Australia)		160	X	Lambert and Heier (1968)
Granulite (62)	Early Precambrian, Bahia (Brazil)	122–1,800	543	X	Sighinolfi (1970)
Pyroxene granulite (23)	Scourie-Kylesku (Scotland)		460	X	Bowes *et al.* (1971)
Banded granulite 20)	Lofoten-Vesterålen (Norway)		572	X	Heier and Thoresen (1971)

Table 38-M-1 (continued)

Rock type	Origin	Range ppm Sr	Average ppm Sr	Analyt. method	Reference
Stronalite (felsic granulite) (11)	Western Alps (Italy)	105–334	234		Mehnert (1975)
Granulite (8)	Vikan I Bø (Norway)	180–460	370	X	Taylor (1975)
Granulite (110)	Scourie assemblage (Scotland)		334	X	Holland and Lambert (1975)
Calcsilicate granulite (8)	Early Precambrian, Bahia (Brazil)	42–110	64	X	Sighinolfi (1974)
High-grade pyroxenite (16)	Early Precambrian, Bahia (Brazil)	22–315	104	A	Sighinolfi and Fujimori (1974)
Eclogite (17)	Münchberg Massif, (Germany)	95–360	195	S	Hahn-Weinheimer (1959)
Eclogite (12)	Kristiansund (Norway)		224	X	Griffin and Råheim (1973)
Eclogite (kyanite-bearing) (34)	Münchberg Massif (Germany)		281	X	Matthes *et al.* (1975)
Eclogite (kyanite-free) (43)	Münchberg Massif (Germany)		97	X	Matthes *et al.* (1975)
Schist, hornfels (10)	Glen Fyne complex (Scotland)	100–2,000	850	S	Nockolds and Mitchell (1948)
Pelitic hornfels (11)	Connemara (Eire)	50–340	204	S	Evans (1964)
Hornfels (7)	Red Hills (New Zealand)	22–450	135	S	Challis (1965)
Hornfels (14)	Montana (U.S.A.)	12.9–466.3	89	I	Mueller and Wooden (1976)

38-N. Behavior in Metamorphic Reactions

Only a few systematic investigations on the behaviour of strontium in series of progressive regional metamorphism have been carried out. In any event the results of these studies are beset by uncertainties of sampling and analysis. However, there are indications that strontium may be mobile in some processes of regional metamorphism. ENGEL and ENGEL (1958), for example, observed a decrease in strontium content during the granitization of paragneiss. Strontium may also be sensitive to some alkaline metasomatism, although the effects of these processes on strontium distribution may be divergent and related to the chemical nature and stage of metasomatism (e.g. YES'KOVA and YEFIMOV, 1970; BARTON and BARTON, 1975). Most of the findings, however, seem to indicate that no significant removal or introduction of strontium occurs up to amphibolite facies regional metamorphism (see e.g. SHAW, 1954; ENGEL and ENGEL, 1958; FRODESEN, 1973). Isochemical metamorphism is sometimes assumed for high-grade metamorphic conditions as well (MISRA and GRIFFIN, 1972; MATTHES *et al.*, 1975), although under these conditions the behaviour of strontium may be controlled by anatectic processes. Low Sr and Ba contents in tonalitic metamorphic suites from granite-greenstone terrains have been explained by a loss of these elements during high-grade metamorphism (HURST *et al.*, 1975). A comparison of average strontium contents in amphibolite and granulite facies terrains in Australia (LAMBERT and HEIER, 1968) indicates that regional factors play the major role. In a series of progressive metamorphism in the West Alps (Valle Strona, Italy), the author found averages of 187 ppm Sr (38 samples) and 272 ppm Sr (33 samples) in amphibolite and granulite facies rocks, respectively.

Reports of radiometric investigations of rock and mineral isochrons should be consulted for information of closed-system and open-system behaviour of Sr during regional metamorphism. It is normally assumed that strontium is more sensitive to metamorphic processes than elements like Ti and Zr (CANN, 1970; PEARCE and CANN, 1973) and REE (HERRMANN *et al.*, 1974). In a transition gabbro-amphibolite, FIELD and ELLIOT (1974) observed a much more scattered distribution of Sr in amphibolite. Evidence of a certain mobility of strontium during metamorphism is indicated by its redistribution in metamorphic minerals (GRESENS, 1967). However, metamorphic diffusion of large-radius cations like strontium is usually considered to be rather limited in distance.

Strontium may behave as a mobile element in the alteration of rocks by pneumatolytic-hydrothermal fluids. Spilitization of basaltic rocks and wall-rock metamorphism of granitoid rocks is frequently accompanied by loss and redistribution of strontium (PAMPURA, 1966; GALE and ROBERTS, 1974; HERRMANN *et al.*, 1974). Strontium, together with other trace elements (Mn, Ga, Ba), is leached out during scapolitization of gabbroic rocks (FRODESEN, 1973).

In contact metamorphism strontium may be much more mobile than in regional metamorphism. Pyrometamorphism of phyllites by a dolerite plug has been observed to cause a marked Sr depletion (SMITH, 1969). The rate of strontium exchange in contact metamorphism appears to be dependent on the chemical nature of the rocks (GHOSE, 1966) and on temperature. Progressive contact metamorphism of pelitic rocks causes a much more pronounced loss of strontium at higher temperatures (EVANS, 1964).

Revised manuscript received: August 1976

References: Sections 38-A to 38-N

ALBUQUERQUE, C. A. R. DE: Petrochemistry of a series of granitic rocks from northern Portugal. Geol. Soc. Am. Bull. **82**, 2783 (1971).

AL-HASHIMI, W. S.: Significance of strontium distribution in some carbonate rocks in the Carboniferous of Northumberland, England. J. Sediment. Petrol. **46**, 369 (1976).

ALLEN, R. O., MASON, B.: Minor and trace elements in some meteoritic minerals. Geochim. Cosmochim. Acta **37**, 1435 (1973)

ALLSOPP, H. L., NICOLAYSEN, L. O., HAHN-WEINHEIMER, P.: Rb/K ratios and Sr-isotopic compositions of minerals in eclogitic and peridotitic rocks. Earth Planet. Sci. Lett. **5**, 231 (1969).

ALLSOPP, H. L., ULRYCH, T. J., NICOLAYSEN, L. O.: Dating some significant events in the history of the Swaziland System by the Rb-Sr isochron method. Can. J. Earth Sci. **5**, 605 (1968).

ANDERSON, M. R., GASSAWAY, J. D., MALONEY, W. E.: The relationship of the strontium/chlorinity ratio to water masses in the tropical Atlantic Ocean and Caribbean Sea. Limnol. Oceanogr. **15**, 467 (1970).

ANNERSTEN, H., EKSTROM, T.: Distribution of major and minor elements in coexisting minerals from a metamorphosed iron formation. Lithos **4**, 185 (1971).

APPLEYARD, E. C.: Silica-poor hastingsitic amphiboles from metasomatic alkaline gneisses at Wolfe, eastern Ontario. Can. Mineralogist **13**, 342 (1975).

ARAÑA, V., BADIOLA, E. R., HERNAN, F. H.: Peralkaline acid tendencies in Gran Canaria (Canary Islands). Contrib. Mineral. Petrol. **40**, 53 (1973).

ARCULUS, R. J.: Geology and geochemistry of the alkali basalt-andesite association of Grenada, Lesser Antilles island arc. Geol. Soc. Am. Bull. **87**, 612 (1976).

ARMSTRONG, R. L., JAGER, E.: A comparison of K-Ar and Rb-Sr ages on alpine biotites. Earth Planet. Sci. Lett. **1**, 13 (1966).

ARMSTRONG, R. L.: A model for the evolution of strontium and lead isotopes in a dynamic earth. Rev. Geophys. **6**, 175 (1968).

ARMSTRONG, R. L.: Glacial erosion and the variable isotopic composition of strontium in sea water. Nature, Phys. Sci. **230**, 132 (1971).

ARTH, J. G., HANSON, G. N.: Quartz diorites derived by partial melting of eclogite or amphibolite at mantle depths. Contrib. Mineral. Petrol. **37**, 161 (1972).

ARTH, J. G., HANSON, G. N.: Geochemistry and origin of the early Precambrian crust of northeastern Minnesota. Geochim. Cosmochim. Acta **39**, 325 (1975).

ATKINS, F. B.: Pyroxenes of the Bushveld intrusion, South Africa. J. Petrol. **10**, 222 (1969).

AUDLEY-CHARLES, M. G.: Some aspects of the chemistry of Cretaceous siliceous sedimentary rocks from eastern Timor. Geochim. Cosmochim. Acta **29**, 1175 (1965).

BAADSGAARD, H., LAMBERT, R. ST., KRUPICKA, J.: Mineral isotopic age relationships in the polymetamorphic Amitsoq gneisses, Godthaab District, West Greenland. Geochim. Cosmochim. Acta **40**, 513 (1976).

BAADSGAARD, H., MUELLER, P. A.: K-Ar and Rb-Sr ages of intrusive Precambrian mafic rocks, southern Beartooth Mountains, Montana and Wyoming. Geol. Soc. Am. Bull. **84**, 3635 (1973).

BADIOZAMANI, K.: The Dorag dolomitization model—Application to the Middle Ordovician of Wisconsin. J. Sediment. Petrol. **43**, 965 (1973).

BÄRNINGHAUSEN, H.: Gitterkonstanten und Raumgruppe der isotypen Verbindungen $Eu(OH)_2 \cdot H_2O$, $Sr(OH)_2 \cdot H_2O$ und $Ba(OH)_2 \cdot H_2O$. Z. Anorg. Allgem. Chem. **342**, 233 (1966).

BÄRNINGHAUSEN, H., WEIDLEIN, J.: Die Kristallstruktur von Strontiumhydroxid-Monohydrat. Acta Cryst. **22**, 252 (1967).

BAKER, I.: Petrology of the volcanic rocks of Saint Helena Island, South Atlantic, Geol. Soc. Am. Bull. **80**, 1283 (1969).

BAKER, P. E., BUCKLEY, F., HOLLAND, J. G.: Petrology and geochemistry of Easter Island. Contrib. Mineral. Petrol. **44**, 85 (1974).

BANNO, S., CHAPPELL, B. W.: X-ray fluorescence analysis of Rb, Sr, Y, Pb and Th in Japanese Paleozoic slates. Geochem. J. **3**, 127 (1969).

BARAGAR, W. R. A., GOODWIN, A. M.: Andesites and Archean volcanism in the Canadian Shield. In: MCBIRNEY, A. R. (ed.), Andesite Symp., Oregon Dept. Geol. Min. Indust. **65**, 121 (1969).

BARBER, C.: The geochemistry of carbonatites and related rocks from two carbonatite complexes, South Nyanza, Kenya. Lithos **7**, 53 (1974a).

BARBER, C.: Major and trace element associations in limestones and dolomites. Chem. Geol. **14**, 273 (1974b).

BARBERI, F., BORSI, S., FERRARA, G., INNOCENTI, F.: Strontium isotopic composition of some Recent basic volcanites of the southern Tyrrhenian Sea and Sicily Channel. Contrib. Mineral. Petrol. **23**, 157 (1969).

BARBERI, F., INNOCENTI, F., FERRARA, G., KELLER, J., VILLARI, L.: Evolution of Eolian arc volcanism (southern Tyrrhenian Sea). Earth Planet Sci. Lett. **21**, 269 (1974).

BARKER, F., ARTH, J. G., PETERMAN, Z. E., FRIEDMAN, I.: The 1.7 and 1.8 b.y. old trondhjemites of southwestern Colorado and northern New Mexico: Geochemistry and depths of genesis. Geol. Soc. Am. Bull. **87**, 189 (1976).

BARKER, F., PETERMAN, Z. E., HILDRETH, R. A.: A rubidium-strontium study of the Twilight Gneiss, West Needle Mountains, Colorado. Contrib. Mineral. Petrol. **23**, 271 (1969).

BARTON, J. M., Jr., BARTON, E. S.: Age and geochemical studies of the Snyder breccia, Coastal Labrador. Can. J. Earth Sci. **12**, 361 (1975).

BAXTER, A. N.: Petrology of the older series lavas from Mauritius, Indian Ocean. Geol. Soc. Am. Bull. **86**, 1449 (1975).

BAXTER, A. N.: Geochemistry and petrogenesis of primitive alkali basalt from Mauritius, Indian Ocean. Geol. Soc. Am. Bull. **87**, 1028 (1976).

BECKINSALE, R. D.: Rb-Sr and K-Ar age determinations, and oxygen isotope data for the Glen Channel granophyre, Isle of Mull, Argyllshire, Scotland. Earth Planet. Sci. Lett. **22**, 267 (1974).

BEHRENS, E. W., LAND, L. S.: Subtidal Holocene dolomite, Baffin Bay, Texas. J. Sediment. Petrol. **42**, 155 (1972).

BELL, K.: Age relations and provenance of the Dalradian Series of Scotland. Bull. Geol. Soc. Am. **79**, 1167 (1968).

BELL, K., DAWSON, J. B., FARQUHAR, R. M.: Strontium isotope studies of alkalic rocks. The active carbonatite volcano Oldoinyo Lengai, Tanzania. Geol. Soc. Am. Bull. **84**, 1019 (1973).

BELL, K., POWELL, J. L.: Strontium isotopic studies of alkalic rocks: The potassium-rich lavas of the Birunga and Toro-Ankole regions, East and Central Equatorial Africa. J. Petrol. **10**, 536 (1969).

BELL, K., POWELL, J. L.: Strontium isotopic studies of alkalic rocks: The alkalic complexes of eastern Uganda. Geol. Soc. Am. Bull. **81**, 3481 (1970).

BENCE, A. E., HURLEY, P. M.: Rubidium-strontium isotopic relationships in oceanic basalts (abstract). Trans. Am. Geophys. Union **48**, 251 (1967).

BENDER, M. L., LORENS, R. B., WILLIAMS, D. F.: Sodium, magnesium and strontium in the tests of planktonic foraminifera. Micropaleontology **21**, 448 (1975).

BERG, G. W., ALLSOPP, H. L.: Low $^{87}Sr/^{86}Sr$ ratios in fresh South African kimberlites. Earth Planet. Sci. Lett. **16**, 27 (1972)

BERLIN, R., HENDERSON, C. M. B.: A reinterpretation of Sr and Ca fractionation trends in plagioclases from basic rocks. Earth Planet. Sci. Lett. **4**, 79 (1968).

BERLIN, R., HENDERSON, C. M. B.: The distribution of Sr and Ba between the alkali feldspar, plagioclase and groundmass phases of porphyritic trachytes and phonolites. Geochim. Cosmochim. Acta **33**, 247 (1969).

BERNER, R. A.: In: ANDERSON, N. R., MALAHOFF, A. (eds), The Fate of Fossil Fuel CO_2. New York: Plenum Press (in press).

BILLINGS, G. K., HITCHON, B., SHAW, D. R.: Geochemistry and origin of formation waters in the Western Canada sedimentary basin. 2. Alkali metals. Chem. Geol. **4**, 211 (1969).

BILLINGS, G. K., RAGLAND, P. C.: Geochemistry and mineralogy of the Recent reef and lagoonal sediments south of Belize (British Honduras). Chem. Geol. **3**, 135 (1968).

BINAS, H.: Die Struktur des Sr-Haidingerits, $SrHAsO_4 \cdot H_2O$. Z. Anorg. Allgem. Chem. **347**, 140 (1966).

BIRCK, J. L., ALLÈGRE, C. J.: ^{87}Rb-^{87}Sr systematics of Muntsche Tundra mafic pluton (Kola Peninsula, U.S.S.R.). Earth Planet. Sci. Lett. **20**, 266 (1973).

BISCAYE, P. E., DASCH, E. J.: The rubidium, strontium, strontium isotope system in deep sea sediments: Argentine basin. J. Geophys. Res. **76**, 5087 (1971).

BISHOP, A. C., WOOLLEY, A. R.: A basalt-trachyte-phonolite series from Ua Pu, Marquesas Islands, Pacific Ocean. Contrib. Mineral. Petrol. **39**, 309 (1973).

BISHOP, J. K. B., EDMOND, J. M., KETTEN, D. R., BACON, M. P., SILKER, W. B.: The chemistry, biology and vertical flux of particulate matter from the upper 400 m of the equatorial Atlantic Ocean. Deep-Sea Res. **24**, 511 (1977).

BJØRLYKKE, K.: Depositional history and geochemical composition of Lower Paleozoic epicontinental sediments from the Oslo region. Norg. Geol. Undersökelse **305**, 1 (1974).

BLACK, L. P., GALE, N. H., MOORBATH, S., PANKHURST, R. J., MCGREGOR, V. R.: Isotopic dating of very early Precambrian amphibolite facies gneisses from the Godthaab District, West Greenland. Earth Planet. Sci. Lett. **12**, 245 (1971).

BLACK, W. A. P., MITCHELL, R. L.: Trace elements in the common brown algae and in sea water. J. Marine Biol. Assoc. U.K. **30**, 575 (1952).

BODINE, M. W., HOLLAND, H. D., BORSICK, M.: Co-precipitation of manganese and strontium with calcite. Symposium: Problems of Postmagmatic Ore Deposition, Prague, **2**, 401 (1965).

BOFINGER, V. M., COMPSTON, W., GULSON, B. L.: A Rb-Sr study of the Lower Silurian State Circle Shale, Canberra, Australia. Geochim. Cosmochim. Acta **34**, 433 (1970).

BOGARD, D. D., BURNETT, D. S., EBERHARDT, P., WASSERBURG, G. J.: ^{87}Rb-^{87}Sr isochron and ^{40}K-^{40}Ar ages of the Norton County achondrite. Earth Planet. Sci. Lett. **3**, 179 (1967).

BOGARD, D. D., NYQUIST, L. E., BAUSAL, B. M., WIESMANN. H.: 76535: An old lunar rock? Abstracts, Fifth Lunar Science Conference, NASA, **70** (1974).

BOGER, P. D., FAURE, G.: Systematic variations of sialic and volcanic detritus in piston cores from the Red Sea. Geochim. Cosmochim. Acta **40**, 731 (1976).

BOJANOWSKI, R., OSTROWSKI, S.: Strontium content of the Southern Baltic waters. Acta Geophys. Polon. **16**, 351 (1968).

BOLTER, E.: Über Zersetzungsprodukte von Olivin-Feldspatbasalten. Beitr. Mineral. Petrog. **8**, 111 (1961).

BONATTI, E., HONNOREZ, J., FERRARA, G.: Equatorial Mid-Atlantic Ridge: Petrologic and Sr isotopic evidence for an alpine-type rock assemblage. Earth Planet. Sci. Lett. **9**, 247 (1970).

BONHOMME, M.: Mineralogie des fractions fines et datations rubidium-strontium dans le Groupe Bambui, MG, Brésil. Rev. Bras. Geoscienc. **6**, 211 (1976)

BORISENKO, L. F.: Trace elements in pyroxenes and amphiboles from ultramafic rocks of the Urals. Mineral. Mag. **36**, 403 (1967).

BORSI, S., DEL MORO, A., SASSI, F. P., ZIRPOLI, G.: Metamorphic evolution of the Austridic rocks to the south of the Tauern Window (Eastern Alps): radiometric and geo-petrologic data. Mem. Soc. Geol. It. **12**, 549 (1973).

BOSE, M. K.: Petrology and geochemistry of the igneous complex of Mount Girnar, Gujarat, India. Contrib. Mineral. Petrol. **39**, 247 (1973).

BOSTRÖM, K., JOENSUU, O., BROHM, I.: Plankton: its chemical composition and its significance as a source of pelagic sediments. Chem. Geol. **14**, 255 (1974).

BOTTINO, M. L., FULLAGAR, P. D., FAIRBAIRN, H. W., PINSON, W. H., JR., HURLEY, P. M.: The Blue Hills igneous complex, Massachusetts: Whole-rock Rb-Sr open system. Geol. Soc. Am. Bull. **81**, 3739 (1970).

BOWEN, H. J. M.: Strontium and barium in sea water and marine organisms. J. Marine Biol. Assoc. U.K. **35**, 451 (1956).

Bowes, D. R., Barooah, B. C.: Khoury, S. G.: Original nature of Archean rocks of North-west Scotland. Geol. Soc. Aust., Spec. Publ. **3**, 77 (1971).

Boyko, T. F.: Rare elements in granite residuum of the Azov region. Doklady Akad. Nauk SSSR **186**, 233 (1969) [Engl. Transl.]

Braitsch, O.: Salt Deposits—Their Origin and Composition. Berlin-Heidelberg-New York: Springer 1971.

Brass, G. W.: The variation of the marine $^{87}Sr/^{86}Sr$ ratio during Phanerozoic time: Interpretation using a flux model. Geochim. Cosmochim. Acta, **40**, 721 (1976).

Brass, G. W., Turekian, K. K.: Strontium distribution in Geosecs ocean profiles. Earth Planet. Sci. Lett. **23**, 141 (1974).

Brookins, D. G., Berdan, J. M., Stewart, D. B.: Isotopic and paleontologic evidence for correlating three volcanic sequences in the Maine coastal volcanic belt. Geol. Soc. Am. Bull. **84**, 1619 (1973).

Brookins, D. G., Carden, J. R., Laughlin, A. W.: Additional note on the isotopic composition of strontium in McCartys flow, Valencia County, New Mexico. Earth Planet. Sci. Lett. **25**, 327 (1975).

Brookins, D. G., Chaudhuri, S., Dulekoz, E.: Rb-Sr isotopic age of Eskridge Shale (Lower Permian), Eastern Kansas. Sediment. Geol. **4**, 103 (1970).

Brookins, D. G., Fairbairn, H. W., Hurley, P. M., Pinson, W. H.: A Rb-Sr geochronologic study of the pegmatites of the Middletown area, Connecticut. Contrib. Mineral. Petrol. **22**, 157 (1969).

Brookins, D. G., Treves, S. B., Bolivar, S. L.: Elk Creek, Nebraska, Carbonatite: Strontium geochemistry. Earth Planet. Sci. Lett. **28**, 79 (1975).

Brooks, C. K.: On the interpretation of trends in element ratios in differentiated igneous rocks, with particular reference to strontium and calcium. Chem. Geol. **3**, 15 (1968).

Brooks, K. C., Jakobsson, S. P., Campsie, J.: Dredged basaltic rocks from the seaward extensions of the Reykjanes and Snaefellsnes volcanic zones, Iceland. Earth Planet. Sci. Lett. **22**, 320 (1974).

Bryhni, I., Bollingberg, H. J., Graff, P. R.: Eclogites in quartzo-feldspathic gneisses of Nordfjord, West Norway. Norsk Geol. Tidsskr. **49**, 193 (1969).

Buchardt, B.: Oxygen isotope ratios from shell material from the Danish Middle Paleocene (Selandian) deposits and their interpretation as paleotemperature indicators. Palaeogeogr., Palaeoclimat., Palaeoecol. **22**, 209 (1977).

Buchardt, B., Fritz, P.: Strontium uptake in shell aragonite from the freshwater gastropod *Limnaea stagnalis*. Science **199**, 291 (1978).

Burwell, A. D. M.: Rb-Sr isotope geochemistry of Iherzolites and their constituent minerals from Victoria, Australia. Earth Planet. Sci. Lett. **28**, 69 (1975).

Butler, G. P.: Strontium geochemistry of modern and ancient calcium sulphate minerals. In: Purser, B. M. (ed)., The Persian Gulf. Berlin-Heidelberg-New York: Springer 1973.

Butler, J. R.: The geochemistry and mineralogy of rock weathering (1) The Lizard area, Cornwall. Geochim. Cosmochim. Acta **4**, 157 (1953).

Butler, J. R.: The geochemistry and mineralogy of rock weathering. (2) The Nordmarka area, Oslo. Geochim. Cosmochim. Acta **6**, 268 (1954).

Butler, J. R., Skiba, W.: Strontium in plagioclase feldspars from four layered basic masses in Somalia. Mineral. Mag. **33**, 213 (1962).

Caby, R., Dostal, J., Dupuy, C.: Upper Proterozoic volcanic graywackes from north-western Hoggar (Algeria)—geology and geochemistry. Precamb. Res. **5**, 283 (1977).

Cann, J. R.: Rb, Sr, Y, Zr and Nb in some ocean floor basaltic rocks. Earth Planet. Sci. Lett. **10**, 7 (1970).

Carmichael, I., McDonald, A.: The geochemistry of some natural acid glasses from the North Atlantic Tertiary volcanic province. Geochim. Cosmochim. Acta **25**, 189 (1961).

Carpenter, A. B., Miller, J. C.: Geochemistry of saline sub-surface water, Saline County (Missouri). Chem. Geol. **4**, 135 (1969).

Carter, S. R., Norry, M. J.: Genetic implications of Sr isotopic data from the Aden volcano, South Arabia. Earth Planet. Sci. Lett. **31**, 161 (1976).

CATT, J. A., GAD, M. A., LERICHE, H. H., LORD, A. R.: Geochemistry, micropaleontology and origin of the Middle Lias ironstones in northeast Yorkshire (Great Britain). Chem. Geol. **8**, 61 (1971).

CHALLIS, G. A.: High-temperature contact metamorphism at the Red Hills ultramafic intrusion — Wairau Valley — New Zealand. J. Petrol. **6**, 395 (1965).

CHALMERS, R. O., HENDERSON, E. P., MASON, B.: Occurrence, distribution, and age of Australian tektites. Smithsonian Contrib. Earth Sci. **17** (1976).

CHAUDHURI, S., BROOKINS, D. G.: The Rb-Sr whole-rock age of the Stearns shale (Lower Permian), eastern Kansas, before and after acid-leaching experiments. Bull. Geol. Soc. Am. **80**, 2605 (1969).

CHESTER, R., MESSIHA-HANNA, R. G.: Trace partitioning patterns in North Atlantic deep-sea sediments. Geochim. Cosmochim. Acta **34**, 1121 (1970).

CHURCH, S. E.: Strontium isotope and alkali element geochemistry of selected sediments from the Northeast Pacific Ocean. Geochim. Cosmochim. Acta **35**, 1300 (1971).

CHURCH, S. E., TILTON, G. R.: Lead and strontium isotopic studies in the Cascade Mountains: Bearing on andesite genesis. Geol. Soc. Am. Bull. **84**, 431 (1973).

CLARK, J. R.: The crystal structure of tunellite, $SrB_6O_9(OH)_2 \cdot 3H_2O$. Am. Mineralogist **49**, 1549 (1964).

CLARKE, D. B.: Tertiary basalts of Baffin Bay: Possible primary magma from the mantle. Contrib. Mineral. Petrol. **25**, 203 (1970).

CLARKE, R. S., CARRON, M. K.: Comparison of tektite specimens from Empire, Georgia, and Martha's Vineyard, Massachusetts. Smithsonian Misc. Collections **143**, 4 (1961).

CLAUER, N.: Utilisation de la méthode rubidium-strontium pour la datation de niveaux sédimentaires du Précambrien supérieur de l'Adrar mauritanien (Sahara occidental) et la mise en évidence de transformations précoces des minéraux argileux. Geochim. Cosmochim. Acta **37**, 2243 (1973).

CLAUER, N., HOFFERT, M., GRIMAUD, D., MILLOT, G.: Composition isotopique du strontium d'eaux interstitielles extraites de sédiments récents: un argument en faveur de l'homogénéisation isotopique des minéraux argileux. Geochim. Cosmochim. Acta **39**, 1579 (1975).

CLEMONS, R. E., LONG, L. E.: Petrologic and Rb-Sr isotopic study of the Chiquimula pluton, south-eastern Guatemala. Geol. Soc. Am. Bull. **82**, 2729 (1971).

CLIFFORD, T. N., GRONOW, J., REX, D. C., BURGER, A. J.: Geochronological and petrogenetic studies of high-grade metamorphic rocks and intrusives in Namaqualand, South Africa. J. Petrol. **16**, 154 (1975).

CLIFFORD, T. N., ROOKE, J. M., ALLSOPP, H. L.: Petrochemistry and age of the Franzfontein granitic rocks of northern South-West Africa. Geochim. Cosmochim. Acta **33**, 973 (1969).

COHEN, A. I., GORDON, L.: Co-precipitation in some binary sulfate systems. Talanta **7**, 195 (1961).

COLEMAN, R. G.: The Colebrooke Schist of southwestern Oregon and its relation to the tectonic evolution of the region. U.S. Geol. Surv. Bull. **1339**, 1 (1972).

COLEMAN, R. G., LEE, D. E., BEATTY, L. B., BRANNOCK, W. W.: Eclogite and eclogite; Their differences and similarities. Geol. Soc. Am. Bull. **76**, 483 (1965).

COLEMAN, R. G., PAPIKE, J. J.: Alkali amphiboles from the blueschists of Cazadero, California. J. Petrol. **9**, 105 (1968).

COLLEY, H., WARDEN, A. J.: Petrology of the New Hebrides. Geol. Soc. Am. Bull. **85**, 1635 (1974).

COMPSTON, W., ARRIENS, P. A., VERNON, M. J.: Rubidium-strontium chronology and chemistry of lunar material. Science **167**, 474 (1970).

COMPSTON, W., CHAPMAN, D. R.: Sr isotope patterns within the southeast Australasian strewnfield. Geochim. Cosmochim. Acta **33**, 1023 (1969).

COMPSTON, W., LOVERING, J. F.: The strontium isotopic geochemistry of granulitic and eclogitic inclusions from the basic pipes at Delegate, Eastern Australia. Geochim. Cosmochim. Acta **33**, 691 (1969).

Compston, W., Lovering, J. F., Vernon, M. J.: The rubidium-strontium age of the Bishopville aubrite and its component enstatite and feldspar. Geochim. Cosmochim. Acta **29**, 1085 (1965).

Compston, W., McDougall, I., Heier, K. S.: Geochemical comparison of the Mesozoic basaltic rocks of Antarctica, South Africa, South America and Tasmania. Geochim. Cosmochim. Acta **32**, 129 (1968).

Compston, W., Pidgeon, R. T.: Rubidium-strontium dating of shales by the total-rock method. J .Geophys. Res. **69**, 3493 (1962).

Condie, K. C.: Geochemistry of early Precambrian graywackes from Wyoming. Geochim. Cosmochim. Acta **31**, 2135 (1967).

Condie, K. C.: Petrology and geochemistry of the Laramie Batholith and related metamorphic rocks of Precambrian age, eastern Wyoming. Geol. Soc. Am. Bull. **80**, 57 (1969).

Condie, K. C., Barsky, C. K.: Origin of Quaternary basalts from the Black Rock Desert region, Utah. Geol. Soc. Am. Bull. **83**, 333 (1972).

Condie, K. C., Barsky, C. K., Mueller, P. A.: Geochemistry of Precambrian diabase dikes from Wyoming. Geochim. Cosmochim. Acta **33**, 1371 (1969).

Condie, K. C., Hayslip, D. L.: Young bimodal volcanism at Medicine Lake volcanic center, northern California. Geochim. Cosmochim. Acta **39**, 1165 (1975).

Condie, K. C., Lo, H. H.: Trace element geochemistry of the Louis Lake batholith of early Precambrian age, Wyoming. Geochim. Cosmochim. Acta **35**, 1099 (1971).

Condie, K. C., Macke, J. E., Reimer, J. O.: Petrology and geochemistry of Early Precambrian graywackes from the Fig Tree Group, South Africa. Geol. Soc. Am. Bull. **81**, 2759 (1970).

Cook, P. J.: Petrology and geochemistry of the phosphate deposits of Northwest Queensland, Australia. Econ. Geol. **67**, 1193 (1972).

Cooper, J. A., Wells, A. T., Nicholas, T.: Dating of glauconite from the Ngalia Basin, Northern Territory, Australia. J. Geol. Soc. Australia **18**, 97 (1971).

Cormier, R. F.: Radiometric ages of granitic rocks, Cape Breton Island, Nova Scotia. Can. J. Earth Sci. **9**, 1074 (1972).

Cosgrove, M. E.: The geochemistry and mineralogy of the Permian red beds of southwest England. Chem. Geol. **11**, 31 (1973).

Cruft, E. F.: Minor elements in igneous and metamorphic apatite. Geochim. Cosmochim. Acta **30**, 375 (1966).

Curtis, C. D.: Trace element distribution in some British Carboniferous sediments. Geochim. Cosmochim. Acta **33**, 519 (1969).

Cuttitta, F., Carron, M. K. Annell, C. S.: New data on selected Ivory Coast tektites. Geochim. Cosmochim. Acta **36**, 1297 (1972).

Cuttitta, F., Clarke, R. S., Carron, M. K., Annell, C. S.: Martha's Vineyard and selected Georgia tektites: new chemical data. J. Geophys. Res. **72**, 1343 (1967).

Dasch, E. J.: Strontium isotope disequilibrium in a porphyritic alkali basalt and its bearing on magmatic processes. J. Geophys. Res. **74**, 560 (1969a).

Dasch, E. J.: Strontium isotopes in weathering profiles, deep-sea sediments and sedimentary rocks. Geochim. Cosmochim. Acta **33**, 1521 (1969b).

Dasch, E. J., Biscaye, P. E.: Isotopic composition of strontium in Cretaceous-to-Recent pelagic foraminifera. Earth Planet. Sci. Lett. **11**, 201 (1971).

Dasch, E. J., Dymond, J. R., Heath, G. R.: Isotopic analysis of metalliferous sediment from the East Pacific Rise. Earth Planet. Sci. Lett. **13**, 175 (1971).

Dasch, E. J., Green, D. H.: Strontium isotope geochemistry of lherzolite inclusions and host basaltic rocks, Victoria, Australia [Abstr.] Geol. Soc. America, Cordilleran Section Program 144 (1972).

Dasch, E. J., Green, D. H.: Strontium isotope geochemistry of lherzolite inclusions and host basaltic rocks, Victoria, Australia. Am. J. Sci. **275**, 461 (1975).

Dasch, E. J., Hedge, C. E., Dymond, J.: Effect of sea water interaction on strontium isotope composition of deep-sea basalt. Earth Planet Sci. Lett. **19**, 177 (1973).

Davis, J. W., Collins, A. G.: Solubility of barium and strontium sulphates in strong electrolyte solutions. Environ. Sci. Technol. **5**, 1039 (1971).

Davis, P. A., Jr.: K, Rb, Sr and Sr isotopes in the Preacher Creek ultramafic intrusion, Wyoming. M. S.-Thesis, Miami University, Ohio (1974).

Davis, P. A., Jr., Condie, K. C.: Trace element model studies of Nyanzian greenstone belts, western Kenya. Geochim. Cosmochim. Acta **41**, 271 (1977).

Davies, R. D., Allsopp, H. L., Erlank, A. J., Manton, W. I.: Sr-isotopic studies on various layered intrusions in southern Africa. In: Symposium on the Bushveld Igneous Complex and Other Layered Intrusions. Geological Society of South Africa, Special Publ. **1**, 576 (1970).

DeAlbuquerque, C. A. R.: Geochemistry of actinolitic hornblendes from tonalitic rocks, northern Portugal. Geochim. Cosmochim. Acta **38**, 789 (1974).

DeAlbuquerque, C. A. R.: Partition of trace elements in co-existing biotite, muscovite and potassium feldspar of granitic rocks, northern Portugal. Chem. Geol. **16**, 89 (1975).

Deer, W. A., Abbott, D.: Clinopyroxenes of the gabbro cumulates of the Kap Edvard Holm complex, east Greenland. Mineral. Mag. **34**, 177 (1965).

DeLong, S. E., Long, L. E.: Petrology and Rb-Sr age of Precambrian rhyolitic dikes, Llano County, Texas. Geol. Soc. Am. Bull. **87**, 275 (1976).

Deuser, W. G., Herzog, L. F.: Rubidium-strontium age determinations of muscovites and biotites from pegmatites of the Blue Ridge and Piedmont. J. Geophys. Res. **67**, 1997 (1962).

DeVore, G. W.: The role of adsorption in the fractionation and distribution of elements. J. Geol. **63**, 159 (1955).

Dickinson, D. R., Dodson, M. H., Gass, I. G., Rex, D. C.: Correlation of initial $^{87}Sr/^{86}Sr$ with Rb/Sr in some late Tertiary volcanic rocks of South Arabia. Earth Planet. Sci. Lett. **6**, 84 (1969).

Dickinson, D. R., Gibson, I. L.: Feldspar fractionation and anomalous Sr^{87}/Sr^{86} ratios in a suite of peralkaline silicic rocks. Geol. Soc. Am. Bull. **83**, 231 (1972).

Dickson, J. A. D., Barber, C.: Chemical variation in a partially dolomitized Visean limestone bed, Isle of Man. Mineral. Mag. **41**, 145 (1977).

Dietrich, R. V., Heier, K. S.: Differentiation of quartz-bearing syenite (nordmarkite) and riebeckite-arfvedsonite granite (ekerite) of the Oslo series. Geochim. Cosmochim. Acta **31**, 275 (1967).

Dodd, J. R.: Magnesium and strontium in calcareous skeletons — a review. J. Paleontol. **41**, 1313 (1967).

Dodge, F. C. W., Papike, J. J., Mays, R. E.: Hornblendes from granitic rocks of the central Sierra Nevada Batholith, California. J. Petrol. **9**, 378 (1968).

Dodge, F. C. W., Ross, D. C.: Coexisting hornblendes and biotites from granitic rocks near the San Andreas fault, California. J. Geol. **79**, 158 (1971).

Dodge, F. C. W., Smith, V. C., Mays, R. E.: Biotites from granitic rocks of the central Sierra Nevada Batholith, California. J. Petrol. **10**, 250 (1969).

Doe, B. R.: Lead and strontium isotopic studies of Cenozoic volcanic rocks in the Rocky Mountain region — A summary. In: Cenozoic Volcanism in the Southern Rocky Mountains. Colorado School Mines Quart. **63**, 149 (1968).

Doe, B. R., Lipman, P. W., Hedge, C. E.: Radiogenetic tracers and the source of continental andesites: A beginning at the San Juan volcanic field, Colorado. Bull. Oregon Dept. Geol. and Mineral Industries **65**, 143 (1969).

Doe, B. R., Lipman, P. W., Hedge, C. E., Kurasawa, H.: Primitive and contaminated basalts from the southern Rocky Mountains, U.S.A. Contrib. Mineral. Petrol. **21**, 142 (1969).

Doe, B. R., Tilling, R. I., Hedge, C. E., Klepper, M. R.: Lead and strontium isotope studies of the Boulder Batholith, southwestern Montana. Econ. Geol. **63**, 884 (1968).

Doerner, H. A., Hoskins, W. M.: Co-precipitation of radium and barium sulfates. J. Am. Chem. Soc. **47**, 6622 (1925).

Donnay, J. D. H., Donnay, G., Cox, E. G., Kennard, O., King, M. V. (eds.): Crystal Data Determinative Tables. (2nd ed.) Am. Cryst. Assoc., Monograph No. 5 (1963).

DONNELLY, T. W., RODGERS, J. J. W., PUSHKAR, P., ARMSTRONG, R. L.: Chemical evolution of the igneous rocks of the eastern West Indies: An investigation of thorium, uranium, and potassium distributions, and lead and strontium isotopic ratios. Geol. Soc. Am. Mem. **130**, 181 (1971).

DOUGAN, T. W.: Origin of trondhjemitic biotite-quartz-oligoclase gneisses from the Venezuelan Guyana shield. Prec. Res. **3**, 317 (1976).

DRURY, S. A.: The geochemistry of Precambrian granulite facies rocks from the Lewisian complex of Tiree, Inner Hebrides, Scotland. Chem. Geol. **11**, 167 (1973).

DUCHESNE, J. C., ROELANDTS, I., DEMAIFFE, D., HERTOGEN, J., GIJBELS, R., DEWINTER, J.: Rare earth data on monzonoritic rocks related to anorthosites and their bearing on the nature of the parental magma of the anorthositic series. Earth Planet. Sci. Lett. **24**, 325 (1974).

DUDAS, M. J., SCHMITT, R. A., HARWARD, M. E.: Trace element partitioning between volcanic plagioclase and dacitic pyroclastic matrix. Earth Planet. Sci. Lett. **11**, 440 (1971).

DUNCAN, R. A., COMPSTON, W.: Sr-isotopic evidence for an old mantle source region for French Polynesian volcanism. Geol. **4**, 728 (1976).

DUNITZ, J. D., ORGEL, L. E.: Stereochemistry of ionic solids. Advan. Inorg. Chem. Radiochem. **2**, 1 (1960).

DUPUY, C., MCNUTT, R. H., COULON, C.: Determination de $^{87}Sr/^{86}Sr$ dans les andesites cénozoiques et les laves associées de Sardaigne Nord occidentale (Italie). Geochim. Cosmochim. Acta **38**, 1287 (1974).

EL-HINNAWI, E. E., PICHLER, H., ZEIL, W.: Trace element distribution in Chilean ignimbrites. Contrib. Mineral. Petrol. **24**, 50 (1969).

EL-WAKEEL, S. K., RILEY, J. P.: Chemical and mineralogical studies of deep-sea sediments. Geochim. Cosmochim. Acta **25**, 110 (1961).

ENGEL, A. E. J., ENGEL, C. G.: Progressive metamorphism and granitization of the major paragneiss, northwest Adirondack Mountains, New York. Geol. Soc. Am. Bull. **69**, 1369 (1958).

ENGEL, A. E. J., ENGEL, C. G.: Progressive metamorphism and granitization of the major paragneiss, northwest Adirondack Mountains, New York. Geol. Soc. Am. Bull. **71**, 1 (1960).

ENGEL, A. E. J., ENGEL, C. G.: Hornblendes formed during progressive metamorphism of amphibolites, northwest Adirondack Mountains, New York. Geol. Soc. Am. Bull. **73**, 1499 (1962).

ENGEL, A. E. J., ENGEL, C. G., HAVENS, R. G.: Mineralogy of amphibolite interlayers in the gneiss complex, northwest Adirondack Mountains, New York. J. Geol. **72**, 131 (1964).

ENGEL, A. E. J., ENGEL, C. G., HAVENS, R. G.: Chemical characteristics of oceanic basalts and the upper mantle. Geol. Soc. Am. Bull. **76**, 719 (1965).

ENGEL, C. G.: Igneous rocks and constituent hornblendes of the Henry Mtns., Utah. Geol. Soc. Am. Bull. **70, 951** (1959).

ENGEL, C. G., Fisher, R. L.: Granitic to ultramafic rock complexes of the Indian Ocean ridge system, western Indian Ocean. Geol. Soc. Am. Bull. **86**, 1553 (1975).

EVANS, B. W.: Fractionation of elements in the pelitic hornfelses of the Cashel-Lough Wheelaun intrusion, Connemara, Eire. Geochim. Cosmochim. Acta **28**, 127 (1964).

EVANS, B. W., LEAKE, B. E.: The composition and origin of the striped amphibolites of Connemara, Ireland. J. Petrol. **1**, 337 (1960).

EWART, A.: Geochemistry of the pantellerites of Major Islaand, New Zealand, Contrib. Mineral. Petrol. **17**, 116 (1968).

EWART, A., BRYAN, W. B.: Petrography and geochemistry of the igneous rocks from Eua, Tongan Islands. Geol. Soc. Am. Bull. **83**, 3281 (1972).

EWART, A., BRYAN, W.B., GILL, J. B.: Mineralogy and geochemistry of the younger volcanic islands of Tonga, S.W. Pacific. J. Petrol. **14**, 429 (1973).

EWART, A., STIPP, J. J.: Petrogenesis of the volcanic rocks of the central North Island, New Zealand, as indicated by a study of $^{87}Sr/^{86}Sr$ ratios, and Sr, Rb, K, U and Th abundances. Geochim. Cosmochim. Acta **32**, 699 (1968).

EWART, A., TAYLOR, S. R.: Trace element geochemistry of the rhyolitic volcanic rocks, central North Island, New Zealand. Phenocryst data. Contrib. Mineral. Petrol. **22**, 127 (1969).

FABRICAND, B. P., IMBIMBO, E. S., BREY, M. E.: Atomic absorption analyses for Ca, Li, Mg, K, Rb, and Sr at two Atlantic Ocean stations. Deep-Sea Res. **14**, 785 (1967).

FABRICAND, B. P., IMBIMBO, E. S., BREY, M. E., WESTON, J. A.: Atomic absorption analyses for Li, Mg, K, Rb, and Sr in ocean waters. J. Geophys. Res. **71**, 3917 (1966).

FAIRBAIRN, H. W., HURLEY, P. M., PINSON, W. H.: The relation of discordant Rb-Sr mineral and whole rock ages in an igneous rock to its time of crystallization and to the time of subsequent Sr^{87}/Sr^{86} metamorphism. Geochim. Cosmochim. Acta **23**, 135 (1961)

FAIRBAIRN, H. W., HURLEY, P. M., PINSON, W. H.: Initial $^{87}Sr/^{86}Sr$ and possible sources of granitic rocks in southern British Columbia. J. Geophys. Res. **69**, 4889 (1964).

FAIRBAIRN, H. W., HURLEY, P. M., PINSON, W. H., JR., CORMIER, R. F.: Age of the granitic rocks of Nova Scotia. Geol. Soc. Am. Bull. **71**, 399 (1960).

FAURE, G., BOWMAN, J. R., ELLIOT, D. H., JONES, L. M.: Strontium isotope composition and petrogenesis of the Kirkpatrick Basalt, Queen Alexandra Range, Antarctica. Contrib. Mineral. Petrol. **48**, 153 (1974).

FAURE, G., CROCKET, J. H., HURLEY, M. P.: Some aspects of the geochemistry of strontium and calcium in the Hudson Bay and the Great Lakes. Geochim. Cosmochim. Acta **31**, 451 (1967).

FAURE, G., ELLIOT, D. H.: Isotope composition of strontium in Mesozoic basalt and dolerite from Dronning Maud Land. Brit. Antarctic Survey Bull. **25**, 23 (1971).

FAURE, G., HILL, R. L., JONES, L. M., ELLIOT, D. H.: Isotope composition of strontium and silica content of Mesozoic basalt and dolerite from Antarctica. In: ADIE, R. J. (ed), Antarctic Geology and Geophysics, pp. 617–624. Oslo: Universitetsforlaget 1971.

FAURE, G., HURLEY, P. M., POWELL, J. L.: The isotopic composition of strontium in surface water from the North Atlantic Ocean. Geochim. Cosmochim. Acta **29**, 209 (1965).

FAURE, G., JONES, L. M.: Anomalous strontium in the Red Sea brines. In: DEGENS, E. T., ROSS, D. A., (eds.), Hot brines and Recent Heavy Metal Deposits in the Red Sea. Berlin-Heidelberg-New York: Springer 1969.

FAURE, G., KOVACH, J.: The age of the Gunflint Iron Formation of the Animikie Series in Ontario, Canada. Geol. Soc. Am. Bull. **80**, 1725 (1969).

FAURE, G., POWELL, J. L.: Strontium Isotope Geology. Berlin-Heidelberg-New York: Springer 1972.

FENTON, M. D., FAURE, G.: The age of the igneous rocks of the Stillwater Complex of Montanta. Geol. Soc. Am. Bull. **80**, 1599 (1969).

FERSHTATER, G. B., BORODINA, N. S., TRAYANOVA, M. V.: Lithium, rubidium, strontium and lead in the granitoids of the Urals. Geochem. Intern. **6**, 72 (1969).

FIELD, D., ELLIOT, R. B.: The chemistry of gabbro/amphibolite transitions in South Norway. Contrib. Mineral. Petrol. **47**, 63 (1974).

FLANAGAN, F. J.: 1972 values for international geochemical reference samples. Geochim. Cosmochim. Acta **37**, 1189 (1973).

FLEISCHER, M.: Index of new mineral names, discredited minerals, and changes of mineralogical nomenclature in Volumes 1–50 of "The American Mineralogist." Am. Mineralogist **51**, 1248 (1966).

FLEISCHER, R. K., PRICE, P. B., WOODS, R. T.: A second tektite fall in Australia. Earth Planet. Sci. Lett. **7**, 51 (1969).

FLÜGEL, H. W., WEDEPOHL, K. H.: Die Verteilung des Strontiums in oberjurassischen Karbonatgesteinen der Nördlichen Kalkalpen. Contrib. Mineral. Petrol. **14**, 229 (1967).

FOSTER, R. J.: Precambrian corundum-bearing rocks, Madison Range, Southwestern Montana. Geol. Soc. Am. Bull. **73**, 131 (1962).

FRODESEN, S.: Trace elements in a Precambrian gabbro intrusion, Hiåsen, Bamble area, South Norway. Norsk. Geol. Tidsskr. **53**, 1 (1973).

FULLAGAR, P. D.: Age and origin of plutonic intrusions in the Piedmont of the southeastern Appalachians. Geol. Soc. Am. Bull. **82**, 2845 (1971).

Fullagar, P. D., Lemmon, R. E., Ragland, P. C.: Petrochemical and geochronological studies of plutonic rocks in the southern Appalachians: Part 1, The Salisbury pluton. Geol. Soc. Am. Bull. **82**, 409 (1971).

Fullagar, P. D., Odom, A. L.: Geochronology of Precambrian gneisses in the Blue Ridge province of northeastern North Carolina and adjacent parts of Virginia and Tennessee. Geol. Soc. Am. Bull. **84**, 3065 (1973).

Fullagar, P. D., Park, B. K.: Rb-Sr study of granite and gneiss from Seoul, South Korea. Geol. Soc. Am. Bull. **86**, 1579 (1975).

Fullagar, P. D., Shiver, W. S.: Geochronology and petrochemistry of the Embudo Granite, New Mexico. Geol. Soc. Am. Bull. **84**, 2705 (1973).

Furnes, H., Faerseth, R. B.: Interpretation of preliminary trace element data from Lower Paleozoic greenstone sequences on Stord, West Norway. Norsk Geol. Tidsskr. **55**, 157 (1975).

Gale, G. H.: Paleozoic basaltic komatiite and ocean-floor type basalts from northeastern Newfoundland. Earth Planet. Sci. Lett. **18**, 22 (1973).

Gale, G. H., Roberts, D.: Trace element geochemistry of Norwegian Lower Paleozoic basic volcanics and its tectonic implications. Earth Planet. Sci. Lett. **22**, 380 (1974).

Gale, N. H., Arden, J., Hutchison, R.: The chronology of the Nakhla achondritic meteorite. Earth Planet. Sci. Lett. **26**, 195 (1975).

Garrels, R. M., MacKenzie, F. T.: Evolution of Sedimentary Rocks. New York: Norton and Company 1971.

Garrels, R. M., Perry, E. A.: Cycling of carbon, sulfur, and oxygen through geologic time. In: Goldberg, E. D. (ed.), The Sea, Vol. 5. New York: Wiley-Interscience 1974.

Garske, D., Peacor, D. R.: Refinement of the structure of celestite $SrSO_4$. Z. Krist. **121**, 204 (1965).

Gast, P. W.: Limitations on the composition of the upper mantle. J. Geophys. Res. **65** 1287 (1960).

Gast, P. W.: The isotopic composition of strontium and the age of stone meteorites — I. Geochim. Cosmochim. Acta **26**, 927 (1962).

Gast, P. W.: Trace element fractionation and the origin of tholeiitic and alkaline magma types. Geochim. Cosmochim. Acta **32**, 1057 (1968).

Gast, P. W., Tilton, G. R., Hedge, C.: Isotopic composition of lead and strontium from Ascension and Gough Islands. Science **145**, 1181 (1964).

Gates, T. M., Hurley, P. M.: Evaluation of Rb-Sr dating methods applied to the Matachewan, Abitibi, Mackenzie, and Sudbury dike swarms in Canada. Can. J. Earth Sci. **10**, 900 (1973).

Gavish, E., Friedman, G..M.: Progressive diagenesis in Quaternary to Late Tertiary carbonate sediments: sequence and time scale. J. Sediment. Petrol. **39**, 980 (1969).

Gebauer, D., Grünenfelder, M.: Rb-Sr whole-rock dating of late diagenetic to anchimetamorphic Paleozoic sediments in southern France (Montagne Noire). Contrib. Mineral. Petrol. **47**, 113 (1974).

Ghent, E. D., Coleman, R. G.: Eclogites from southwestern Oregon. Geol. Soc. Am. Bull. **84**, 2471 (1973).

Ghose, N. C.: Behaviour of trace elements during thermal metamorphism and or granitization of the metasediments and basic igneous rocks. Geol. Rundschau **55**, 688 (1966).

Gibbins, W. A., McNutt, R. H.: Rubidium-strontium mineral ages and polymetamorphism at Sudbury, Ontario. Can. J. Earth Sci. **12**, 1990 (1975).

Gibbs, R. J.: Water chemistry of the Amazon River. Geochim. Cosmochim. Acta **36**, 1061 (1972).

Gieskes, J. M.: Chemistry of interstitial waters of marine sediments. Annual Rev. Earth Planet. Sci. **3**, 433 (1975).

Gieskes, J. M.: Interstitial water studies. In: Schlanger *et al.* (eds.), Initial Reports of the Deep Sea Drilling Project, **33**. Washington, D. C.: U.S. Government Printing Office 1976.

Gieskes, J. M., Lawrence, J. R., Galleiski, G.: Interstitial water studies, Leg. 38. In: Talwani, M. *et al.* (eds.), Initial Reports of the Deep Sea Drilling Project, **38**. Washington, D. C.: U.S. Government Printing Office (in press).

GILL, J. B.: Geochemistry of Viti Leva, Fiji, and its evolution as an island arc. Contrib. Mineral. Petrol. **27**, 179 (1970).

GILL, J. B., MURTHY, V. R.: Distribution of K, Rb, Sr and Ba in Nain anorthosite plagioclase. Geochim. Cosmochim. Acta **34**, 401 (1970).

GILL, R. C. O., BRIDGWATER, D.: The Ameralik dykes of West Greenland, the earliest known basaltic rocks intruding stable continental crust. Earth Planet. Sci. Lett. **29**, 276 (1976).

GIROD, M., CAMUS, G., VIALETTE, Y.: Sur la présence de tholéiites à l'île de Saint-Paul (Océan Indien). Contrib. Mineral. Petrol. **33**, 108 (1971).

GLIKSON, A. Y.: Trace element geochemistry and origin of early Precambrian acid igneous series, Barberton Mountain Land, Transvaal. Geochim. Cosmochim. Acta **40**, 1261 (1976).

GOLDBERG, E. D., ARRHENIUS, G. O. S.: Chemistry of Pacific pelagic sediments. Geochim. Cosmochim. Acta **13**, 153 (1958).

GOLDBERG, E. D., BROECKER, W. S., GROSS, M. G., TUREKIAN, K. K.: Marine chemistry. In: Radioactivity in the Marine Environment. Washington, D. C.: National Acad. Sci. 1971.

GOLDICH, S. S., HEDGE, C. E., STERN, T. W.: Age of the Morton and Montevideo gneisses and related rocks, southwestern Minnesota. Geol. Soc. Am. Bull. **81**, 3671 (1970).

GOLDICH, S. S., TREVES, S. B., SUHR, N. H., STUCKLESS, J. S.: Geochemistry of the Cenozoic volcanic rocks of Ross Island and vicinity, Antarctica. J. Geol. **83**, 415 (1975).

GOMES, C. B., CORDANI, U. G., BASEI, M. A. S.: Radiometric ages from the Serra dos Carajas area, northern Brazil. Geol. Soc. Am. Bull. **86**, 939 (1975).

GONCHAROV, YU. I.: Strontium in halide strata of the Donets Permian. Lithology Mineral. Resources **1967**, 48 (1967).

GOODEL, H. G., GARMAN, R. K.: Carbonate geochemistry of Superior deep test well, Andros Island, Bahamas. Am. Ass. Petroleum Geol. Bull. **53**, 513 (1969).

GOPALAN, K., WETHERILL, G. W.: Rubidium-strontium age of hypersthene (L) chondrites. J. Geophys. Res. **73**, 7133 (1968).

GOPALAN, K., WETHERILL, G. W.: Rubidium-strontium age of amphoterite (LL) chondrites. J. Geophys. Res. **74**, 4349 (1969).

GOPALAN, K., WETHERILL, G. W.: Rubidium-strontium studies on enstatite chondrites: whole meteorite and mineral isochrons. J. Geophys. Res. **75**, 3457 (1970).

GORDON, M., MURATA, K. J.: Minor elements in Arkansas bauxite. Econ. Geol. **47**, 169 (1952).

GOREAU, T. J.: Coral skeletal chemistry; physiological and environmental regulation of stable isotopes and trace metals in *Monastrea annularis*. Proc. Roy. Soc. London, Ser. B **196**, 291 (1977).

GRAF, D. L.: Geochemistry of carbonate sediments and sedimentary carbonate rocks. Part III — Minor element distribution. Illinois State Geol. Surv. Circ. **301** (1960).

GRANT, N. K., POWELL, J. L., BURKHOLDER, F. R., WALTHER, J. V., COLEMAN, M. L.: The isotopic composition of strontium and oxygen in lavas from St. Helena, South Atlantic. Earth Planet. Sci. Lett. **31**, 209 (1976).

GRANT, N. K., REX, D. C., FREETH, S. J.: Potassium-argon ages and strontium isotope ratio measurements from volcanic rocks in northeastern Nigeria. Contrib. Mineral. Petrol. **35**, 277 (1972).

GRAY, C. M., PAPANASTASSIOU, D. A., WASSERBURG, G. J.: The identification of early condensates from the solar nebula. Icarus **20**, 213 (1973).

GRESENS, R. L.: Tectonic-hydrothermal pegmatites. Contrib. Mineral. Petrol. **16**, 1 (1967).

GRIFFIN, W. L., MURTHY V. R.: The abundance of K, Rb, Sr and Ba in some ultramafic rocks and minerals. Earth Planet. Sci. Lett. **4**, 497 (1968).

GRIFFIN, W. L., MURTHY, V. R.: Distribution of K, Rb, Sr and Ba in some minerals relevant to basalt genesis. Geochim. Cosmochim. Acta **33**, 1389 (1969).

GRIFFIN, W. L., RÅHEIM, A.: Convergent metamorphism of eclogites and dolerites, Kristiansund area, Norway. Lithos **6**, 21 (1973).

GRIFFIN, W. L., SUNDVOLL, B., KRISTMANNSDOTTIR, H.: Trace element composition of anorthosite plagioclase. Earth Planet. Sci. Lett. **24**, 213 (1974).

GROHMANN, H.: Beitrag zur Geochemie österreichischer Granitoide. Tschermaks Mineral. Petrog. Mitt. **10**, 436 (1965).

GULSON, B. L.: The high-K diorites and associated rocks of the Yeoval diorite complex, N.S.W. Contrib. Mineral. Petrol. **35**, 173 (1972).

GUNDLACH, H., KARL, F., MÜLLER, G.: Vergleichende geochemische Untersuchungen an ost- und südalpinen Graniten, Granodioriten und Tonaliten. Contrib. Mineral. Petrol. **16**, 285 (1967).

GUNN, B. M.: Modal and element variation in Antarctic tholeiites. Geochim. Cosmochim. Acta **30**, 881 (1966).

GUNN, B. M., ABRANSON, C. E., NOUGIER, J., WATKINS, N. D., HAJASH, A.: Amsterdam Island, an isolated volcano in the southern Indian Ocean. Contrib. Mineral. Petrol. **32**, 79 (1971).

GUNN, B. M., ROOBOL, M. J., SMITH, A. L.: Petrochemistry of the Pelean-type volcanoes of Martinique. Geol. Soc. Am. Bull. **85**, 1023 (1974).

GUNN, B. M., WATKINS, N. D.: Geochemistry of the Steens Mountain basalts, Oregon. Geol. Soc. Am. Bull. **81**, 1497 (1970).

GUNN, B. M., WATKINS, N. D.: Geochemistry of the Cap Verde Islands and Fernando de Noronha. Geol. Soc. Am. Bull. **87**, 1089 (1976).

HÄNNY, R., GRAUERT, B., SOPTRTJANOVA, G.: Paleozoic migmatites affected by high-grade Tertiary metamorphism in the Central Alps (Valle Bodengo, Italy). Contrib. Mineral. Petrol. **51**, 173 (1975).

HAHN-WEINHEIMER, P.: Geochemische Untersuchungen an den ultrabasischen und basischen Gesteinen der Münchberger Gneissmasse (Fichtelgebirge). Neues Jahrb. Mineral. Abhandl. **92**, 203 (1959).

HAHN-WEINHEIMER, P., ACKERMANN, H.: Geochemical investigation of differentiated granite plutons of the southern Black Forest—II. The zoning of the Marlsburg Granite pluton as indicated by the elements titanium, zirconium, phosphorus, strontium, barium, rubidium, potassium and sodium. Geochim. Cosmochim. Acta **31**, 2197 (1967).

HAHN-WEINHEIMER, P., LUECKE, W.: Garnets from the eclogites of the Muenchberger gneiss massif (N. E. Bavaria). Can. Mineralogist **7**, 764 (1963).

HAHN-WEINHEIMER, P., ROST, F.: Akzessorische Mineralien und Elemente im Serpentinit von Leupoldgrün (Münchberger Gneissmasse). Ein Beitrag zur Geochemie ultrabasischer Gesteine. Geochim. Cosmochim. Acta **21**, 165 (1961).

HALL, A.: The distribution of some major and trace elements in feldspars from the Rosses and Ardara granite complexes, Donegal, Ireland. Geochim. Cosmochim. Acta **31**, 835 (1967).

HALLAM, A., PRICE, N. B.: Environmental and biochemical control of strontium in shells of *Cardium edule*. Geochim. Cosmochim. Acta **32**, 319 (1968).

HALLBERG, J. A.: The petrology and geochemistry of metamorphosed Archean basic volcanic rocks between Coolgardie and Norseman, Western Australia. Ph. D-Thesis, Univ. W. Australia (1970).

HAMET, J., ALLÈGRE, C. J.: Rb-Sr systematics in granite from central Nepal (Manaslu): Significance of the Oligocene age and high $^{87}Sr/^{86}Sr$ ratio in Himalayan orogeny. Geol. **4**, 470 (1976).

HAMILTON, E. I.: The isotopic composition of strontium in the Skaergaard Intrusion, East Greenland. J. Petrol. **4**, 383 (1963).

HAMILTON, E. I.: The isotopic composition of strontium in Atlantic Ocean water. Earth Planet. Sci. Lett. **1**, 435 (1966).

HANOR, J. S.: Frequency distribution of compositions in the barite-celestite series. Am. Mineralogist, **53**, 1215 (1968).

HARRIS, W. B.: Rb-Sr study of Cretaceous lobate glauconite pellets, North Carolina. Geol. Soc. Am. Bull. **85**, 1475 (1974).

HART, J. G., HANSON, G. N.: Geochemistry and origin of the early Precambrian crust of northeastern Minnesota. Geochim. Cosmochim. Acta **39**, 325 (1975).

HART, S. R.: K, Rb, Cs, Sr, and Ba contents and Sr isotope ratios of ocean floor basalts. Phil. Trans. Roy. Soc. London, Ser. A **268**, 573 (1971).

HART, S. R.: The geochemistry of a lherzolite from the mid-Indian Ocean ridge. Carnegie Inst. Year Book **71**, 290 (1972).

HART, S. R., BROOKS, C.: Clinopyroxene-matrix partitioning of K, Rb, Cs, Sr and Ba. Geochim. Cosmochim. Acta **38**, 1799 (1974a).

HART, S. R., BROOKS, C.: The geochemistry and evolution of Early Precambrian mantle. Carnegie Inst. Yearbook **73**, 967 (1974b).

HART, S. R., BROOKS, C., KROGH, T. E., DAVIS, G. L., NAVA, D.: Ancient and modern volcanic rocks: A trace element model. Earth Planet. Sci. Lett. **10**, 17 (1970).

HART, S. R., DAVIS, G. L.: Zircon U-Pb and whole-rock Rb-Sr ages and early crustal development near Rainy Lake, Ontario. Geol. Soc. Am. Bull. **80**, 595 (1969).

HART, S. R., ERLANK, A. J., KABLE, E. J. D.: Sea floor basalt alteration: Some chemical and Sr isotopic effects. Contrib. Mineral. Petrol. **44**, 219 (1974).

HART, S. R., GLASSLEY, W. E., KARIG, D. E.: Basalts and sea floor spreading behind the Mariana Island arc. Earth. Planet. Sci. Lett. **15**, 12 (1972).

HART, S. R., NALWALK, A. J.: K, Rb, Cs and Sr relationships in submarine basalts from the Puerto Rico trench. Geochim. Cosmochim. Acta **34**, 145 (1970).

HART, S. R., TILTON, G.: The isotope geochemistry of strontium and lead in Lake Superior sediments and water. In: The Earth Beneath the Continents. Geophys. Mon. Am. Geophys. Union **10**, 127 (1966).

HASLAM, H. W.: The crystallization of intermediate and acid magmas at Ben Nevis, Scotland. J. Petrol. **9**, 84 (1968).

HAWKESWORTH, C. J.: Rb/Sr geochronology in the eastern Alps. Contrib. Mineral. Petrol. **54**, 225 (1976).

HAWKESWORTH, C. J., MOORBATH, S., O'NIONS, R. K.: Age relationships between greenstone belts and "granites" in the Rhodesian Archaean craton. Earth Planet. Sci. Lett. **25**, 251 (1975).

HAWKINS, J. W., JR., NATLAND, J. H.: Nephelinites and basanites of the Samoan linear volcanic chain: Their possible tectonic significance. Earth Planet. Sci. Lett. **24**, 427 (1975).

HEBEDA, E. H., BOELRIJK, N. A. I. M., PRIEM, H. N. A., VERDURMEN, E. A. TH., VERSCHURE, R. H.: Excess radiogenic argon in the Precambrian Avanavero Dolerite in western Suriname (South America). Earth Planet. Sci. Lett. **20**, 189 (1973).

HEDGE, C. E.: Variations in radiogenic strontium found in volcanic rocks. J. Geophys. Res. **71**, 6119 (1966).

HEDGE, C. E., HILDRETH, R. A., HENDERSON, W. T.: Strontium isotopes in some Cenozoic lavas from Oregon and Washington. Earth Planet. Sci. Lett. **8**, 434 (1970).

HEDGE, C. E., KNIGHT, R. J.: Lead and strontium isotopes in volcanic rocks from northern Honshu, Japan. Geochem. J. **3**, 15 (1969).

HEDGE, C. E., LEWIS, J. F.: Isotopic composition of strontium in the basalt-andesite centers along the Lesser Antilles Arc. Contrib. Mineral. Petrol. **32**, 39 (1971).

HEDGE, C. E., NOBLE, D. C.: Upper Cenozoic basalts with high $^{87}Sr/^{86}Sr$ and Sr/Rb ratios, southern Great Basin, Western United States. Bull. Geol. Soc. Am. **82**, 3503 (1971).

HEDGE, C. E., PETERMAN, Z. E.: The strontium isotopic composition of basalts from the Gordo and Juan de Fuca Rises, north-eastern Pacific Ocean. Contrib. Mineral. Petrol. **27**, 114 (1970).

HEDGE, C. E., PETERMAN, Z. E., BRADDOCK, W. A.: Age of the major Precambrian regional metamorphism in the northern Front Range, Colorado. Geol. Soc. Am. Bull. **78**, 551 (1967).

HEDGE, C. E., PETERMAN, Z. E., DICKINSON, W. R.: Petrogenesis of lavas from Western Samoa. Bull. Geol. Soc. Am. **83**, 2709 (1972).

HEDGE, C. E., WALTHALL, F. G.: Radiogenic strontium-87 as an index of geologic processes. Science **140** 1214 (1963).

HEDGE, C. E., WATKINS, N. D., HILDRETH, R. A., DOERING, W. P.: $^{87}Sr/^{86}Sr$ ratios in basalts from islands in the Indian Ocean. Earth Planet. Sci. Lett. **21**, 29 (1973).

HEIER, K. S.: Petrology and geochemistry of high-grade metamorphic and igneous rocks on Langöy, Noerthern Norway. Norg. Geol. Undersökelse **207**, 1 (1960).

HEIER, K. S.: Trace elements in feldspars — A review. Norsk Geol. Tidsskr. **42**, 415 (1962).

HEIER, K. S.: Some crystallo-chemical relations of nephelines and feldspars on Stjernöy, North Norway. J. Petrol. **7**, 95 (1966).

HEIER, K. S., COMPSTON, W., MCDOUGALL, I.: Thorium and uranium concentrations, and the isotopic composition of strontium in the differentiated Tasmanian dolerites. Geochim. Cosmochim. Acta **29**, 643 (1965).

HEIER, K. S., TAYLOR, S. R.: Distribution of Ca, Sr and Ba in southern Norwegian Precambrian alkali feldspars. Geochim. Cosmochim. Acta **17**, 286 (1959).

HEIER, K. S., THORESEN, K.: Geochemistry of high grade metamorphic rocks, Lofoten-Vesterålen, North Norway, Geochim. Cosmochim. Acta **35**, 89 (1971).

HEMING, R. F.: Geology and petrology of Rabaul Caldera, Papua, New Guinea. Geol. Soc. Am. Bull. **85**, 1253 (1974).

HENDERSON, C. M. B.: Minor element chemistry of leucite and pseudoleucite. Mineral. Mag. **35**, 596 (1965).

HERRMANN, A. G.: Zur Geochemie des Strontiums in den salinaren Zechsteinablagerungen der Stassfurt-Serie des Südharzbezirkes. Chem. Erde **21**, 138 (1961).

HERRMANN, A. G., POTTS, M. J., KNAKE, D.: Geochemistry of the rare earth elements in spilites from the oceanic and continental crust. Contrib. Mineral. Petrol. **44**, 1 (1974).

HERRMANN, A. G., WEDEPOHL, K. H.: Untersuchungen an spilitischen Gesteinen der variskischen Geosynkline in Nordwestdeutschland. Contrib. Mineral. Petrol. **29**, 255 (1970).

HERZ, N., DUTRA, C. V.: Trace elements in alkali feldspars, Quadrilatero Ferrifero, Minas Gerais, Brazil. Am. Mineralogist **51**, 1593 (1966).

HERZOG, L. F., PINSON, W. H.: Rb/Sr age, elemental and isotopic abundance studies of stony meteorites. Am. J. Sci. **254**, 555 (1956).

HIETANEN, A.: Distribution of Fe and Mg between garnet, staurolite, and biotite in aluminium rich schist in various metamorphic zones north of the Idaho batholith. Am. J. Sci. **267**, 422 (1969).

HIGGINS, M. W.: Petrology of Newberry Volcano, central Oregon. Geol. Soc. Am. Bull. **84**, 455 (1973).

HIGUCHI, H., NAGASAWA, H.: Partition of trace elements between rock-forming minerals and the host volcanic rocks. Earth Planet. Sci. Lett. **7**, 281 (1969).

HILDRETH, R. A., HENDERSON, W.T.: Comparison of $^{87}Sr/^{86}Sr$ for sea-water strontium and the Eimer and Amend $SrCO_3$. Geochim. Cosmochim. Acta **35**, 235 (1971).

HILLS, F. A., DASCH, E. J.: Rb/Sr study of the Stony Creek Granite, southern Connecticut: A case for limited remobilization. Geol. Soc. Am. Bull. **83**, 3457 (1972).

HILLS, F. A., GAST, P. W., HOUSTON, R. S., SWAINBANK, I. G.: Precambrian geochronology of the Medicine Bow Mountains, southeastern Wyoming. Geol. Soc. Am. Bull. **79**, 1757 (1968).

HOEFS, J., WEDEPOHL, K. H.: Strontium isotope studies on young volcanic rocks from Germany and Italy. Contrib. Mineral. Petrol. **19**, 328 (1968).

HOFMANN, A., KÖHLER, H.: Whole rock Rb-Sr ages of anatectic gneisses from the Schwarzwald, SW Germany. Neues Jahrb. Mineral. Abhandl. **119**, 163 (1973).

HOLLAND, H. D., BORSICK, M., MUNOZ, J., OXBURGH, U. M.: The co-precipitation of Sr^{++} with aragonite and of Ca^{++} with strontianite between 90 and 100° C. Geochim. Cosmochim. Acta **27**, 957 (1963).

HOLLAND, H. D., HOLLAND, H. J., MUNOZ, J. L.: The co-precipitation of cations with $CaCO_3$—II. The co-precipitation of Sr^{++} with calcite between 90° and 100° C. Geochim. Cosmochim. Acta **28**, 1287 (1964).

HOLLAND, H. D., KIRSIPU, T. V., HUEBNER, J. S., OXBURGH, U. M.: On some aspects of the chemical evolution of cave waters. J. Geol. **72**, 36 (1964).

HOLLAND, J. G., BROWN, G. M.: Hebridean tholeiitic magmas: A geochemical study of the Ardnamurchan cone sheets. Contrib. Mineral. Petrol. **37**, 139 (1972).

HOLLAND, J. G., LAMBERT, R. ST. J.: The chemistry and origin of the Lewisian gneisses of the Scottish mainland: The Scourie and Inver assemblages and sub-crustal accretion. Prec. Res. **2**, 161 (1975).

HOLLIDAY, D. W.: Secondary gypsum in Middle Carboniferous rocks of Spitsbergen. Geol. Mag. **104**, 171 (1967).

HONJO, S., TABUCHI, H.: Distribution of some minor elements in carbonate rocks, 1. A list of analytical values of Mg, Sr, Fe, Mn, Zn, Cu, Cr, Ni and U. Pacific Geol. **2**, 41 (1970).

HUBBARD, N. J.: Some chemical features of lavas from the Manu'a Islands, Samoa. Pacific Sci. **25**, 178 (1971).

HUGHES, D. J., BROWN, G. C.: Basalts from Madeira: A petrochemical contribution to the genesis of oceanic alkali rock series. Contrib. Mineral. Petrol. **37**, 91 (1972).

HURLEY, P. M.: Absolute abundance and distribution of Rb, K and Sr in the earth. Geochim. Cosmochim. Acta **32**, 273 (1968).

HURLEY, P. M., BATEMAN, P. C., FAIRBAIRN, H. W., PINSON, W. H., JR.: Investigation of initial $^{87}Sr/^{86}Sr$ ratios in the Sierra Nevada plutonic province. Bull. Geol. Soc. Am. **76**, 165 (1965).

HURLEY, P. M., CORMIER, R. F., HOWER, J., FAIRBAIRN, H. W., PINSON, W. H., JR.: Reliability of glauconite for age measurement by K-Ar and Rb-Sr methods. Bull. Am. Assoc. Petrol. Geologists **44**, 1793 (1960).

HURLEY, P. M., FAIRBAIRN, H. W., PINSON, W. H., JR.: Rb-Sr isotopic evidence in the origin of potash-rich lavas of western Italy. Earth Planet. Sci. Lett. **5**, 301 (1966).

HURLEY, P. M., HUGHES, H., FAURE, G., FAIRBAIRN, H. W., PINSON, W. H.: Radiogenic strontium-87 model of continent formation. J. Geophys. Res. **67**, 5315 (1962).

HURLEY, P. M., PINSON, W. H., NAGY, B., TESKA, T. M.: Ancient age of the Middle Marker horizon, Onverwacht Group, Swaziland sequence, South Africa. Earth Planet. Sci. Lett. **5**, 360 (1972).

HURST, R. W., BRIDGWATER, D., COLLERSON, K. D., WETHERILL, G. W.: 3600-m.y. Rb-Sr ages from very early Archean gneisses from Saglek Bay, Labrador. Earth Planet. Sci. Lett. **27**, 393 (1975).

HUTCHISON, R., DAWSON, J. B.: Rb, Sr and $^{87}Sr/^{86}Sr$ in ultrabasic xenoliths and host-rocks, Lashaine Volcano, Tanzania. Earth Planet. Sci. Lett. **9**, 87 (1970).

ICHIKUNI, M., MUSHA, S.: Partition of strontium between gypsum and solution. Chem. Geol. **21**, 359 (1978).

IIDA, C.: Trace elements in minerals and rocks of the Izu-Hakone region, Japan. Part II. Plagioclase. J. Earth Sci. Nagoya Univ. **9**, 14 (1961).

IRVINE, T. N., BARAGAR, W. R. A.: A guide to the chemical classification of the common volcanic rocks. Can. J. Earth Sci. **8**, 523 (1971).

JACOBSON, R. L., USDOWSKI, H. E.: Partitioning of strontium between calcite, dolomite and liquids: An experimental study under higher temperature diagenetic conditions, and a model for the prediction of mineral pairs for geothermometry. Contrib. Mineral. Petrol. **59**, 171 (1976).

JAHN, B.-M.: A petrogenetic model for the igneous complex in the Spanish Peaks region, Colorado. Contrib. Mineral. Petrol. **41**, 241 (1973).

JAHN, B.-M., CHEN, P. Y., YEN, T. P.: Rb-Sr ages of granitic rocks in southeastern China and their tectonic significance. Geol. Soc. Am. Bull. **86**, 763 (1976).

JAHN, B.-M., MURTHY, V. R.: Rb-Sr ages of the Archean rocks from the Vermilion district, northeastern Minnesota. Geochim. Cosmochim. Acta **39**, 1679 (1975).

JAHN, B.-M., SHIH, C.-Y.: On the age of the Onverwacht Group, Swaziland Sequence, South Africa. Geochim. Cosmochim. Acta **38**, 873 (1974).

JAHN, B.-M., SHIH, C.-Y., MURTHY, V. R.: Trace element geochemistry of Archean volcanic rocks. Geochim. Cosmochim. Acta **38**, 611 (1974).

JAKEŠ, P., WHITE, A. J. R.: Major and trace element abundances in volcanic rocks of orogenic areas. Geol. Soc. Am. Bull. **83**, 29 (1972).

JAMES, D. E., BROOKS, C., CUYUBAMBA, A.: Andean Cenozoic volcanism: Magma genesis in the light of strontium isotopic composition and trace-element geochemistry. Geol. Soc. Am. Bull. **87**, 592 (1976).

JÉROME, D. Y.: Composition and origin of some achondritic meteorites. Ph. D. Thesis, University of Oregon (1970).

JØRGENSEN, N. O.: Mg/Sr distribution and diagenesis of Maastrichtian white chalk and Danian bryozoan limestone from Jylland, Denmark. Bull. Geol. Soc. Denmark **24**, 299 (1975).

JOHNSON, R. C., HILLS, F. A.: Precambrian geochronology and geology of the Boxelder Canyon area, northern Laramie Range, Wyoming. Geol. Soc. Am. Bull. **87**, 809 (1976).

Jones, L. M., Walker, R. L., Stormer, J. C., Jr.: Isotope composition of strontium and origin of volcanic rocks of the Raton-Clayton district, northeastern New Mexico. Geol. Soc. Am. Bull. **85**, 33 (1974).

Joshi, M. S.: Precipitation of carbonates from sea water. Prog. Rep. U.S. AEC Contract No. At (30-1)-2266 (1960).

Jung, W., Knitzschke, G.: Kombiniert-feinstratigraphisch geochemische Untersuchungen der Anhydrite des Zechsteins 1 im SE-Harzvorland. Geologie **9**, 58 (1960).

Kahle, C. F.: Strontium in oolitic limestones. J. Sediment. Petrol. **35**, 846 (1965).

Kamp, P. C. van de: The green beds of the Scottish Dalradian series: geochemistry, origin, and metamorphism of mafic sediments. J. Geol. **78**, 281 (1970).

Katz, A., Matthews, A.: The dolomitization of $CaCO_3$: an experimental study at 252–295° C Geochim. Cosmochim. Acta **41**, 297 (1977).

Katz, A., Sass, E., Starinsky, A., Holland, H. D.: Strontium behaviour in the aragonite-calcite transformation: an experimental study at 40–98° C. Geochim. Cosmochim. Acta **36**, 481 (1972).

Kaushal, S. K., Wetherill, G. W.: ^{87}Rb-^{87}Sr age of bronzite (H group) chondrites. J. Geophys. Res. **74**, 2717 (1969).

Kaushal, S. K., Wetherill, G. W.: Rubidium 87-strontium 87 age of carbonaceous chondrites. J. Geophys. Res. **75**, 463 (1970).

Kesson, S., Price, R. C.: The major and trace element chemistry of kaersutite and its bearing on the petrogenesis of alkaline rocks. Contrib. Mineral. Petrol. **35**, 119 (1972).

Kesson, S. E.: The primary geochemistry of the Monaro Volcanics, southeastern Australia—evidence for upper mantle heterogeneity. Contrib. Mineral. Petrol. **42**, 93 (1973).

Khasawneh, F. E., Juo, A. S. R., Barber, S. A.: Soil properties influencing differential Ca to Sr adsorption. Soil Sci. Soc. Am. Proc. **32**, 209 (1968).

Kinsman, D. J. J.: Interpretation of Sr^{2+} concentrations in carbonate minerals and rocks. J. Sediment. Petrol. **39**, 486 (1969).

Kinsman, D. J. J., Holland, H. D.: The co-precipitation of cations with $CaCO_3$—IV. The co-precipitation of Sr^{2+} with aragonite between 16° and 96° C. Geochim. Cosmochim. Acta **33**, 1 (1969).

Kistler, R. W., Bateman, P. C., Brannock, W. W.: Isotopic ages of minerals from granitic rocks of the Central Sierra Nevada and Inyo Mountains, California. Geol. Soc. Am. Bull. **76**, 155 (1965).

Kistler, R. W., Obradovich, J. D., Jackson, E. D.: Isotopic ages of rocks and minerals from the Stillwater Complex, Montana. J. Geophys. Res. **74**, 3226 (1969).

Kistler, R. W., Peterman, Z. E.: Variations in Sr, Rb, K, Na and initial Sr^{87}/Sr^{86} in Mesozoic granitic rocks and intruded wall rocks in central California. Geol. Soc. Am. Bull. **84**, 3489 (1973).

Klemm, D. D., Weber-Diefenbach, K.: Geochemische Untersuchungen an Amphiboliten und Dioriten des nördlichen Odenwaldes. Neues Jahrb. Mineral. Abhandl. **116**, 80 (1971).

Klerkx, J., Deutsch, S., De Paepe, P.: Rubidium, strontium content and strontium isotopic composition of strongly alkalic basaltic rocks from the Cape Verde Islands. Contrib. Mineral. Petrol. **45**, 17 (1974).

Klerkx, J., Deutsch, S., Hertogen, J., DeWinter, J., Gijbels, R.: Comments on "Evolution of Eolian arc volcanism (southern Tyrrhenian Sea)" by F. Barberi, G. Ferrara, F. Innocenti, J. Keller and L. Villari. Earth Planet. Sci. Lett. **23**, 297 (1974).

Köster, H. M.: Beitrag zur Kenntnis indischer Laterite. Heidelberger Beitr. Mineral. Petrog. **5**, 23 (1955).

Kolbe, P., Taylor, S. R.: Major and trace element relationships in granodiorites and granites from Australia and South Africa. Contrib. Mineral. Petrol. **12**, 202 (1966).

Kranz, J. R.: Strontium—ein Fazies-Diagenese-Indikator im Oberen Wettersteinkalk (Mittel-Trias) der Ostalpen. Geol. Rundschau **65**, 593 (1976).

Kravchenko, V. B.: The crystal structure of the monoclinic modification of $SrB_2O_4 \cdot 4H_2O = Sr[B(OH)_4]_2$. Zh. Strukt. Khim. **6**, 835 (1965) (translated).

Krieger, P.: The occurence of strontianite at Sierra Mojada, Mexico. Am. Mineralogist **18**, 345 (1933).

KROGH-MOE, J.: The crystal structure of strontium diborate, $SrO \cdot 2B_2O_3$. Acta Chem. Scand. **18**, 2055 (1964).

KUDO, A. M., BROOKINS, D. G., LAUGHLIN, A. W.: Sr isotopic disequilibrium in lherzolites from the Puerco Necks, New Mexico. Earth Planet. Sci. Lett. **15**, 291 (1972).

KULP, J. L., TUREKIAN, K., BOYD, D. W.: Strontium content of limestones and fossils. Geol. Soc. Am. Bull. **63**, 701 (1952).

KURASAWA, H.: Strontium and lead isotopes of volcanic rocks in Japan. In: OGATA and HAYAKAWA (eds.): Recent Developments in Mass Spectroscopy. Proc. Internat. Conf. Mass. Spectroscopy, Kyoto, Tokyo: University of Tokyo Press 1970.

KURYVIAL, R. J.: Element partitioning in alkali feldspars from three intrusive bodies of the central Wasatch Range, Utah. Geol. Soc. Am. Bull. **87**, 657 (1976).

KUTSCHABSKY, L.: Zur Kristallstruktur des $Sr[B(OH)_4]_2$. Z. Chem. **5**, 110 (1965).

KYLE, P. R., RANKIN, P. C.: Rare earth element geochemistry of Late Cenozoic alkaline lavas of the McMurdo Volcanic Group, Antarctica. Geochim. Cosmochim. Acta **40**, 1497 (1976).

LAMBERT, I. B.: The composition and evolution of the deep continental crust. In: GLOVER, E. J. (ed.): Symposium on Archaean Rocks. Geol. Soc. Australia Spec. Pub. **3**, 419 (1971).

LAMBERT, I. B., HEIER, K. S.: Geochemical investigations of deep-seated rocks in the Australian shield. Lithos **1**, 30 (1968).

LAND, L. S.: Contemporaneous dolomitization of Middle Pleistocene reefs by meteoric water, North Jamaica. Bull. Marine Sci. **23**, 64 (1973a).

LAND, L. S.: Holocene meteoric dolomitization of Pleistocene limestones, North Jamaica. Sedimentology **20**, 411 (1973b).

LAND, L. S., SALEM, M. R. I., MORROW, D. W.: Paleohydrology of ancient dolomites: geochemical evidence. Am. Ass. Petrol. Geol. Bull. **59**, 1062 (1975).

LANPHERE, M. A.: Sr-Rb-K and Sr isotopic relationships in ultramafic rocks, south-eastern Alaska. Earth Planet. Sci. Lett. **4**, 185 (1968).

LARSEN, L. M.: Clinopyroxenes and coexisting mafic minerals from the alkaline Ilimaussaq intrusion, South Greenland. J. Petrol. **17**, 258 (1976).

LAUGHLIN, A. W.: Potassium, rubidium and strontium abundances in minerals of the Rutherford and Morefield pegmatites, Amelia, Virginia. Earth Planet. Sci. Lett. **17**, 375 (1973).

LAUGHLIN, A. W., BROOKINS, D. G., CARDEN, J. R.: Variations in the initial strontium ratios of a single basalt flow. Earth Planet. Sci. Lett. **14**, 79 (1972).

LAUGHLIN, A. W., BROOKINS, D. G., CAUSEY, J. D.: Late Cenozoic basalts from the Bandera lava field, Valencia County, New Mexico. Geol. Soc. Am. Bull. **83**, 1543 (1972).

LAUGHLIN, A. W., BROOKINS, D. G., KUDO, A. M., CAUSEY, J. D.: Chemical and strontium isotopic investigations of ultramafic inclusions and basalts, Bandera Crater, New Mexico. Geochim. Cosmochim. Acta **35**, 107 (1971).

LAWRENCE, J. R., GIESKES, J. M., BROECKER, W. S.: Oxygen isotope and cation composition of DSDP pore waters and the alteration of Layer II basalts. Earth Planet. Sci. Lett. **27**, 1 (1975).

LEBEDEV, A. B.: Trace elements in marine and fresh water clays. Geochemistry Internat. **4**, 821 (1967).

LEE, D. E., COLEMAN, R. G., ERD, R. C.: Garnet types from the Cazadero area, California. J. Petrol. **4**, 460 (1963).

LEELANANDAM, C.: Chemical mineralogy of hornblendes and biotites from the charnockitic rocks of Kondapalli, India. J. Petrol. **11**, 475 (1970).

LEEMAN, W. P.: The isotopic composition of strontium in late-Cenozoic basalts from the Basin-Range province, western United States. Geochim. Cosmochim. Acta **34**, 857 (1970).

LEEMAN, W. P.: Late Cenozoic alkali-rich basalt from the western Grand Canyon area, Utah and Arizona: Isotopic composition of strontium. Geol. Soc. Am. Bull. **85**, 1691 (1974).

LEEMAN, W. P., MANTON, W. I.: Strontium isotopic composition of basaltic lavas from the Snake River Plain, southern Idaho. Earth Planet. Sci. Lett. **11**, 420 (1971).

LEEMAN, W. P., VITALIANO, C. J.: Petrology of McKinney Basalt, Snake River Plain, Idaho. Geol. Soc. Am. Bull. **87**, 1777 (1976).

LEGGO, P. J., HUTCHISON, R.: A Rb-Sr isotope study of ultrabasic xenoliths and their basaltic host rocks from the Massif Central, France. Earth Planet. Sci. Lett. **5**, 71 (1968).

LESSING, P., CATANZARO, E. J.: Sr^{87}/Sr^{86} ratios in Hawaiian lavas. J. Geophys. Res. **69**, 9599 (1964).

LIPMAN, P. W.: Alkalic and tholeiitic basaltic volcanism related to the Rio Grande depression, southern Colorado and northern New Mexico. Geol. Soc. Am. Bull. **80**, 1343 (1969).

LIPMAN, P. W., MOENCH, R. H.: Basalts of the Mount Taylor volcanic field. New Mexico. Geol. Soc. Am. Bull. **83**, 1335 (1972).

LLOYD, F. E., BAILEY, D. K.: Carbonatite in the tuffs of the West Eifel, Germany. Contrib. Mineral. Petrol. **23**, 136 (1969).

LONG, L. E.: Whole-rock Rb-Sr age of the Yonkers Gneiss, Manhattan Prong. Geol. Soc. Am. Bull. **80**, 2087 (1969).

LONG, L. E., KULP, J. L.: Isotopic age study of the metamorphic history of the Manhattan and Reading Prongs. Geol. Soc. Am. Bull. **73**, 969 (1962).

LORENS, R. B., WILLIAMS, D. F., BENDER, M. L.: The early nonstructural chemical diagenesis of foraminiferal calcite. J. Sediment. Petrol. **47**, 1602 (1977).

LOWDER, G. G.: Late Cenozoic transitional alkali olivine-tholeiitic basalt and andesite from the margin of the Great Basin, southwest Utah. Geol. Soc. Am. Bull. **84**, 2993 (1973).

LOWENSTAM, H. A.: Mineralogy, O^{18}/O^{16} ratios, and strontium and magnesium contents of Recent and fossil brachiopods and their bearing on the history of the oceans. J. Geol. **69**, 241 (1961).

LOWENSTAM, H. A.: Sr/Ca ratio of skeletal aragonites from the recent marine biota at Palau and from fossil gastropods. In: CRAIG, H., MILLER, S. L., WASSERBURG, G. J. (eds.), Isotopic and Cosmic Chemistry. Amsterdam: North Holland 1963a.

LOWENSTAM, H. A.: Biologic problems relating to the composition and diagenesis of sediments. In: DONNELLY, T. W. (ed.), The Earth Sciences, Problems and Progress in Current Research. Rice Univ. Semi-Centennial Publ. 1963b.

LOWENSTAM, H. A.: Coexisting calcites and aragonites from skeletal carbonates of marine organisms and their strontium and magnesium contents. In: MIYAKE Y., KOYAMA, T. (eds.). Recent Researches in the Fields of Hydrosphere, Atmosphere and Nuclear Geochemistry. Tokyo: Maruzen Co. 1964.

MCARTHUR, J. M.: Systematic variations in the contents of Na, Sr, CO_3 and SO_4 in marine carbonate-fluorapatite and their relation to weathering. Chem. Geol. **21**, 89 (1978).

MCCARTHY, T. S.: Chemical interrelationships in a low-pressure granulite terrain in Namaqualand, South Africa, and their bearing on granite genesis and the composition of the lower crust. Geochim. Cosmochim. Acta **40**, 1057 (1976).

MCCARTHY, T. S., AHRENS, L. H., ERLANK, A. J.: Further evidence in support of the mixing model for howardite origin. Earth Planet. Sci. Lett. **15**, 86 (1972).

MCCARTHY, T. S., ERLANK, A. J., WILLIS, J. P.: On the origin of eucrites and diogenites. Earth Planet. Sci. Lett. **18**, 433 (1973).

MCDONALD, R., UPTON, B. G. J., THOMAS, J. E.: Potassium and fluorine-rich hydrous phase coexisting with peralkaline granite in South Greenland. Earth Planet. Sci. Lett. **18**, 217 (1973).

MCDOUGALL, I.: Geochemistry and origin of basalt of the Columbia River Group, Oregon and Washington. Geol. Soc. Am. Bull. **87**, 777 (1976).

MCDOUGALL, I., COMPSTON, W.: Strontium isotope composition and potassium-rubidium ratios in some rocks from Reunion and Rodriguez, Indian Ocean. Nature **207**, 252 (1965).

MCDOUGALL, I., COMPSTON, W., BOFINGER, V. M.: Isotopic age determinations on Upper Devonian rocks from Victoria, Australia: A revised estimate for the age of the Devonian-Carboniferous boundary. Geol. Soc. Am. Bull. **77**, 1075 (1966).

MCDOUGALL, I., DUNN, P. R., COMPSTON, W., WEBB, A. W., RICHARDS, J. R., BOFINGER, V. M.: Isotopic age determinations on Precambrian rocks of the Carpentaria region, Northern Territory, Australia. J. Geol. Soc. Australia **12**, 67 (1965).

MCDUFF, R. E., GIESKES, J. M.: Calcium and magnesium profiles in DSDP interstitial waters: Diffusion or reaction? Earth Planet. Sci. Lett. **33**, 1 (1976).

McDuff, R. E., Lawrence, J. R., Gieskes, J. M.: Interstitial water studies, Leg 42A. In: Hsü, K. *et al.* (eds.), Initial Reports of the Deep Sea Drilling Project, **42**A, Washington, D. C.: U.S. Government Printing Office (1978).

McKee, E. H., Noble, D. C.: Age of the Cardenas Lavas, Grand Canyon, Arizona. Geol. Soc. Am. Bull. **87**, 1188 (1976).

Mackenzie, D. E., Chappell, B. W.: Shoshonitic and calc-alkaline lavas from the highlands of Papua New Guinea. Contrib. Mineral. Petrol. **35**, 50 (1972).

McLaughlin, R. J. W.: Geochemical changes due to weathering under varying climatic conditions. Geochim. Cosmochim. Acta **8**, 109 (1955).

McNutt, R. H., Crocket, J. H., Clark, A. H., Caelles, J. C., Farrar, E., Haynes, S., Zentillis, M.: Initial $^{87}Sr/^{86}Sr$ ratios of plutonic and volcanic rocks of the central Andes between latitudes 26° and 29° south. Earth Planet. Sci. Lett. **27**, 305 (1975).

Magaritz, M.: Lithification of chalky limestone: a case study in Senonian rocks from Israel. J. Sediment. Petrol. **44**, 947 (1974).

Manton, W. I.: The origin of associated basic and acid rocks in the Lebombo-Nuanetsi igneous province, southern Africa, as implied by strontium isotopes. J. Petrol. **9**, 23 (1968).

Manton, W. I., Tatsumoto, M.: Some Pb and Sr isotopic measurements on eclogites from the Roberts Victor Mine, South Africa. Earth Planet. Sci. Lett. **10**, 217 (1971).

Mark, R. K., LeeHu, C., Bowman, H. R., Asaro, F., McKee, E. H., Coats, R. R.: A high $^{87}Sr/^{86}Sr$ mantle source for low alkali tholeiite, northern Great Basin. Geochim. Cosmochim. Acta **39**, 1671 (1975).

Martin, J. H.: Distribution of C, H, N, P, Fe, Mn, Zn, Ca, Sr and Sc in plankton samples collected off Panama and Colombia. Bioscience **19**, 898 (1969).

Martin, P. M., Mason, B.: Major and trace elements in the Allende meteorite. Nature **249**, 333 (1974).

Mason, B., Martin, P. M.: Minor and trace element distribution in melilite and pyroxene from the Allende meteorite. Earth Planet. Sci. Lett. **22**, 141 (1974).

Mason, B., Nelen, J. A., Muir, P., Taylor, S. R.: The composition of the Chassigny meteorite. Meteoritics **11**, 21 (1976).

(M.I.T.) = Massachusetts Institure of Technology: Evidence from western Ontario of the isotopic composition of strontium in Archean Seas. M.I.T. Annual Report 145 (1965).

Matter, A., Douglas, R. G., Perch-Nielsen, K.: Fossil preservation, geochemistry, and diagenesis of pelagic carbonates from Shatsky Rise, northwest Pacific. Joides Initial Reports of the Deep Sea Drilling Project. 891. Washington: U.S. Government Printing Office 1975.

Matthes, S., Richter, P., Schmidt, K.: Die Eklogitvorkommen des kristallinen Grundgebirges in NE-Bayern. Neues Jahrb. Mineral. Abhandl. **126**, 45 (1975).

Mehnert, K. R.: The Ivrea Zone. A model of the deep crust. Neues Jahrb. Mineral. Abhandl. **125**, 156 (1975).

Menzer, F. J., Jr.: Geochronologic study of granitic rocks from the Okanogan Range, North-Central Washington. Geol. Soc. Am. Bull. **81**, 573 (1970).

Michaelis, H. von, Ahrens, L. H., Willis, J. P.: The composition of stony meteorites II. The analytical data and an assessment of their quality. Earth Planet. Sci. Lett. **5**, 387 (1969).

Milliman, J. D.: Marine Carbonates. Berlin-Heidelberg-New York: Springer 1974.

Milliman, J. D., Ross, D. A., Ku, T.-H.: Precipitation and lithification of deep-sea carbonates in the Red Sea. J. Sediment. Petrol. **39**, 724 (1969).

Misra, S. N., Griffin, W. L.: Geochemistry and metamorphism of dolerite dikes from Austvågøy in Lofoten. Norsk Geol. Tidsskr. **52**, 409 (1972).

Mitchell, R. H., Crocket, J. H.: The isotopic composition of strontium in some South African kimberlites. Contrib. Mineral. Petrol. **30**, 277 (1971).

Mittlefehldt, D. W., Wetherill, G. W.: Rb-Sr studies of carbonaceous chondrites. (In press 1977).

Mohr, P. A.: A geochemical study of the Lower Cambrian Manganese Shale Group of the Harlech Dome, North Wales. Geochim. Cosmochim. Acta **17**, 186 (1959).

MONTIGNY, R., BOUGAULT, H., BOTTINGA, Y., ALLÈGRE, C. J.: Trace element geochemistry and genesis of the Pindos ophiolite suite. Geochim. Cosmochim. Acta **37**, 2135 (1973).

MOORBATH, S., BELL, J. D.: Strontium isotope abundance studies and rubidium-strontium age determinations on Tertiary igneous rocks from the Isle of Skye, North-West Scotland. J. Petrol. **6**, 37 (1965).

MOORBATH, S., BELL, K., LEAKE, B. E., MCKERROW, W. S.: Geochronological studies in Connemara and Murrisk, western Ireland. In: HAMILTON, E. I., FARQUHAR, R. M. (eds.), Radiometric Dating for Geologists. New York: Interscience 1968.

MOORBATH, S., O'NIONS, R. K., PANKHURST, R. J.: The evolution of early Precambrian crustal rocks at Isua, West Greenland—Geochemical and isotopic evidence. Earth Planet. Sci. Lett. **27**, 229 (1975).

MOORBATH, S., WELKE, H.: Isotopic evidence for the continental affinity of the Rockall Bank, North Atlantic. Earth Planet. Sci. Lett. **5**, 221 (1969).

MOORE, A. C.: Mineralogy of the Gosse Pile ultramafic intrusion, central Australia. I. Plagioclase. J. Geol. Soc. Australia **18**, 115 (1971).

MORROW, D. W., MAYERS, I. R.: Simulation of limestone diagenesis—a model based on strontium depletion. Can. J. Earth Sci. **15**, 376 (1978).

MOSE, D. G., RATCLIFFE, N. M., ODOM, A. L., HAYES, J.: Rb-Sr geochronology and tectonic setting of the Peekskill pluton, southeastern New York. Geol. Soc. Am. Bull. **87**, 361 (1976).

MOSSER, CH., GALL, J. C., TARDY, Y.: Geochimie des illites du grès a Voltzia (du Buntsandstein Superieur) des Vosqes du nord, France. Chem. Geol. **9**, 157 (1972).

MOTTL, M. J., CORR, R. F., HOLLAND, H. D.: Chemical exchange between sea water and mid-ocean ridge basalt during hydrothermal alteration: an experimental study. G.S.A. Abstracts with Programs **6**, 879 (1974).

MOXHAM, R. L.: Minor element distribution in some metamorphic pyroxenes. Can. Mineralogist **6**, 522 (1960).

MOXHAM, R. L.: Distribution of minor elements in coexisting hornblendes and biotites. Can. Mineralogist **8**, 204 (1964).

MÜLLER, G.: Zur Geochemie des Strontiums in ozeanen Evaporiten unter besonderer Berücksichtigung der sedimentären Coelestinlagerstätte von Hemmelte-West (Süd-Oldenburg). Geologie **11**, Beiheft 35, 1 (1962).

MÜLLER, G.: In: Recent Developments in Carbonate Sedimentology, Cent. Eur. Sem. 1967. New York: Springer 1968.

MÜLLER, G.: High strontium contents and Sr/Ca ratios in Lake Constance waters and carbonates and their sources in the drainage area of the Rhine River (Alpenrhein). Mineral. Deposita **4**, 75–84 (1969).

MÜLLER, G., PUCHELT, H.: Die Bildung von Coelestin ($SrSO_4$) aus Meerwasser. Naturwissenschaften **8**, 301 (1961).

MUELLER, P. A., WOODEN, J. L.: Rb-Sr whole-rock age of the contact aureole of the Stillwater igneous complex, Montana. Earth Planet. Sci. Lett. **29**, 384 (1976).

MUFFLER, L. J. P., DOE, B. R.: Composition and mean age of detritus of the Colorado River Delta in the Salton Trough, southeastern California. J. Sediment. Petrol. **38**, 384 (1968).

MUKHERJEE, B.: Genetic significance of trace elements in certain rocks of Singhbum, India. Mineral. Mag. **36**, 661 (1968).

MUKHOPADHYAY, B., BROOKINS, D. G.: Rb-Sr whole-rock geochronology and clay mineralogy of the Madera Formation near Albuquerque, New Mexico. J. Sediment. Petrol. **46**, 680 1976).

MUKHOPADHYAY, B., BROOKINS, D. G., BOLIVAR, S. L.: Rb-Sr whole-rock study of the Precambian rocks of the Pedernal Hills, New Mexico. Earth Planet. Sci. Lett. **27**, 283 (1975).

MUNIZAGA, F. L., AGUIRRE, F., HERVÉ, F.: Rb/Sr ages of rocks from the Chilean metamorphic basement. Earth Planet. Sci. Lett. **18**, 87 (1973).

MURRAY, M. M., ROGERS, J. J. W.: Distribution of rubidium and strontium in the potassium feldspars of two granite batholiths. Geochem. J. **6**, 117 (1973).

MURTHY, V. R., BEISER, E.: Strontium isotopes in ocean water and marine sediments. Geochim. Cosmochim. Acta **32**, 1121 (1968).

NAGASAWA, H.: Partitioning of Eu and Sr between coexisting plagioclase and K-feldspar. Earth Planet. Sci. Lett. **13**, 139 (1971).

NAGASAWA, H.: Rare-earth distribution in alkali rocks from Oki-Dogo Island, Japan. Contrib. Mineral. Petrol. **39**, 301 (1973).

NAGASAWA, H., SCHNETZLER, C. C.: Partitioning of rare earth, alkali and alkaline earth elements between phenocrysts and acidic igneous magma. Geochim. Cosmochim. Acta **35**, 953 (1971).

NAGLE, F., FINK, L. K., BOSTRÖM, K., STIPP, J. J.: Copper in pillow basalts from La Desirade, Lesser Antilles Island Arc. Earth Planet. Sci. Lett. **19**, 193 (1973).

NASH, W. P., WILKINSON, J. F. G.: Shonkin Sag laccolith, Montana. Contrib. Mineral. Petrol. **33**, 162 (1971).

NATHAN, S., FRUCHTER , . S.: Geochemical and paleomagnetic stratigraphy of the Picture Gorge and Yakima Basalts (Columbia River Group) in central Oregon. Geol. Soc. Am. Bull. **85**, 63 (1974).

National Bureau of Standards: Certificate of analysis, Standard Reference Material 987 (1971).

NEARY, C. R., GASS, I. G., CAVENAGH, B. J.: Granite association of northeastern Sudan. Geol. Soc. Am. Bull. **87**, 1501 (1976).

NEIVA, A. M. R.: Geochemistry of coexisting aplites and pegmatites and of their minerals from central northern Portugal. Chem. Geol. **16**, 153 (1975).

NESBITT, R. W., SUN, S. S.: Geochemistry of Archaean spinifex-textured peridotites and magnesian tholeiites. Earth Planet. Sci. Lett. **31**, 433 (1976).

NIXON, P. H., VON KNORRING, O., ROOKE, J. M.: Kimberlites and associated inclusions of Basutoland: A mineralogical and geochemical study. Am. Mineralogist **48**, 1090 (1963).

NOBLE, D. C.: Systematic variation of major elements in comendite and pantellerite glasses. Earth Planet. Sci. Lett. **4**, 167 (1968).

NOBLE, D. C., BOWMAN, H. R., HEBERT, A. J., SILBERMAN, M. L., HEROPOULOS, C. E., FABBI, B. P., HEDGE, C. E.: Chemical and isotopic constraints on the origin of low-silica latite and andesite from the Andes of central Peru. Geol. **3**, 501 (1975).

NOBLE, D. C., HEDGE, C. E.: $^{87}Sr/^{86}Sr$ variations within individual ash-flow sheets. U.S. Geol. Surv. Profess. Papers **650**-C, 133 (1969).

NOBLE, D. C., KORRINGA, M. K., CHURCH, S. E., BOWMAN, H. R., SILBERMAN, M. L., HEROPOULOS, C. E.: Elemental and isotopic geochemistry of nonhydrated quartz latite glasses from the Eureka Valley tuff, east-central California. Geol. Soc. Am. Bull. **87**, 754 (1976).

NOBLE, D. C., KORRINGA, M. K., HEDGE, C. E., RIDDLE, G. O.: Highly differentiated subalkaline rhyolite from Glass Mountain, Mono County, California. Geol. Soc. Am. Bull. **83**, 1179 (1972).

NOCKOLDS, S. R., MITCHELL, R. L.: The geochemistry of some Caledonian plutonic rocks: a study in the relationship between the major and trace elements of igneous rocks and their minerals. Trans. Roy. Soc. Edinburgh **61**, 533 (1948).

NOHDA, S.: Rb-Sr dating of the Yatsushiro granite and gneiss, Kyushu, Japan. Earth Planet. Sci. Lett. **20**, 140 (1973).

NOLL, W.: Geochemie des Strontiums. Chem. Erde **8**, (1933/34), 507 (1934).

NORTH, N. A.: Pressure dependence of $SrSO_4$ solubility. Geochim. Cosmochim. Acta **38**, 1075 (1974).

NUNES, P. D., STEIGER, R. H.: A U-Pb zircon, and Rb-Sr and U-Th-Pb whole-rock study of a polymetamorphic terrane in the Central Alps, Switzerland. Contrib. Mineral. Petrol. **47**, 255 (1974).

NUNES, P. D., TATSUMOTO, M., UNRUH, D. M.: U-Th-Pb and Rb-Sr systematics of Apollo 17 boulder 7 from the north massif of the Taurus-Littrow Valley. Earth Planet. Sci. Lett. **23**, 445 (1974).

NYQUIST, L. E.: Lunar Rb-Sr chronology. Phys. Chem. Earth **10**, (1977).

OBRADOVICH, J. D., PETERMAN, Z. E.: Geochronology of the Belt Series, Montana. Can. J. Earth Sci. **5**, 737 (1968).

ODIN, G. S., HUNZIKER, J. C., KEPPENS, E., LAGA, P. G., PASTEELS, P.: Analyse radiométrique de glauconies par les méthodes au strontium et à l'argon. L'Oligo-Miocène de Belgique. Bull. Soc. Belge Géol. **83**, 35 (1974).

ODUM, H. T.: Notes on the strontium content of sea water, celestite, radiolaria, and strontianite snail shells. Science **114**, 211 (1951).

ODUM, H. T.: Biogeochemical deposition of strontium. Publ. Inst. Marine Sci., Univ. Texas **4**, 39 (1957).

OKRUSCH, M., RICHTER, P.: Petrographische, geochemische und mineralogische Untersuchungen zum Problem der Granitoide im mittleren Spessartkristallin. Neues Jahrb. Mineral. Abhandl. **107**, 21 (1967).

OKRUSCH, M., RICHTER, P.: Zur Geochemie der Diorit-Gruppe. Contrib. Mineral. Petrol. **21**, 75 (1969).

ONDRICK, C. W., GRIFFITHS, J. G.: Frequency distribution of elements in Renselaer graywacke, Troy, New York. Geol. Soc. Am. Bull. **80**, 509 (1969).

O'NEIL, J. R., HEDGE, C. E., JACKSON, E. D.: Isotopic investigations of xenoliths and host basalts from the Honolulu Volcanic Series. Earth Planet. Sci. Lett. **8**, 253 (1970).

O'NIONS, R. K., HEIER, K. S.: A reconnaissance Rb-Sr geochronological study of the Kongsberg area, south Norway. Norsk Geol. Tidsskr. **52**, 143 (1972).

O'NIONS, R. K., PANKHURST, R. J.: Sr-isotope and rare earth element geochemistry of DSDP Leg 37 basalts. Earth Planet. Sci. Lett. **31**, 255 (1976).

ONUMA, N., HIGUCHI, H., WAKITA, H., NAGASAWA, H.: Trace element partition between two pyroxenes and the host lava. Earth Planet. Sci. Lett. **5**, 47 (1968).

OSTROM, M. E.: Trace elements in Illinois Pennsylvanian limestones. Illinois State Geol. Circ. **243**, 1 (1957).

OVERSBY, V. M.: Isotopic ages and geochemistry of Archaean acid igneous rocks from the Pilbara, Western Australia. Geochim. Cosmochim. Acta **40**, 817 (1976).

PAMPURA, V. D.: Behaviour of rubidium, lithium, barium and strontium in wall-rock metamorphism of granitoids (taking as example the molybdenum deposits of the eastern Transbaikal region). Geochim. Internat. **3**, 1244 (1966).

PANKHURST, R. J.: Strontium isotope studies related to petrogenesis in the Caledonian basic igneous province of Northeast Scotland. J. Petrol. **10**, 115 (1969).

PANKHURST, R. J.: The geochronology of the basic igneous complexes, northeast Scotland. Scott. J. Geol. **6**, 83 (1970).

PANKHURST, R. J.: Rb-Sr whole-rock chronology of Caledonian events in Northeast Scotland. Geol. Soc. Am. Bull. **85**, 345 (1974).

PANKHURST, R. J., MOORBATH, S., REX, D. C., TURNER, G.: Mineral age patterns in ca. 3700 m y old rocks from West Greenland. Earth Planet. Sci. Lett. **20**, 157 (1973).

PANKHURST, R. J., PIDGEON, R. T.: Inherited isotope systems and the source region prehistory of early Caledonian granites in the Dalradian series of Scotland. Earth Planet. Sci. Lett. **31**, 55 (1976).

PAPANASTASSIOU, D. A., WASSERBURG, G. J.: Initial strontium isotopic abundances and the resolution of small time differences in the formation of planetary objects. Earth Planet. Sci. Lett. **5**, 361 (1969).

PAPEZIK, V. S.: Geochemistry of some Canadian anorthosites. Geochim. Cosmochim. Acta **29**, 673 (1965).

PAUL, D. K.: Strontium isotope studies on ultramafic inclusions from Dreiser Weiher, Eifel, Germany. Contrib. Mineral. Petrol. **34**, 22 (1971).

PAUL, D. K., POTTS, P. J., REX, D. C., BECKINSALE, R. D.: Geochemical and petrogenetic study of the Girnar igneous complex, Deccan volcanic province, India. Geol. Soc. Am. Bull. **88**, 227 (1977).

PEARCE, J. A., CANN, J. R.: Tectonic setting of basic volcanic rocks determined using trace element analyses. Earth Planet. Sci. Lett. **19**, 290 (1973).

PEARSON, G. R., SHAW, D. M.: Trace elements in kyanite, sillimanite and andalusite. Am. Mineralogist **45**, 808 (1960).

PEARSON, R. C., HEDGE, C. E., THOMAS, H. H., STERN, T. W.: Geochronology of the St. Kevin Granite and neighboring Precambrian rocks, northern Sawatch Range, Colorado. Geol. Soc. Am. Bull. **77**, 1109 (1966).

PERLOFF, A., BLOCK, S.: The crystal structure of the strontium and lead tetraborates, $SrO \cdot 2B_2O_3$ and $PbO \cdot 2B_2O_3$. Acta Cryst. **20**, 274 (1966).

PERROTTA, A. J., SMITH, J. V.: The crystal structure of brewsterite. Acta Cryst. **17**, 857 (1964).

PERRY, A. E., TUREKIAN, K. K.: The effect of diagenesis on the redistribution of strontium isotopes in shales. Geochim. Cosmochim. Acta **38**, 929 (1974).

PERRY, E. A., GIESKES, J. M., LAWRENCE, J. R.: Mg, Ca, and O^{18}/O^{16} exchange in the sediment pore water system, Hole 149, DSDP. Geochim. Cosmochim. Acta **40**, 413 (1976).

PERRY, E. C., JR., MONSTER, J., REIMER, T.: Sulfur isotopes in Swaziland system barites and the evolution of the earth's atmosphere. Science **171**, 1015 (1971).

PETERMAN, Z. E.: Rb/Sr dating of middle Precambrian sedimentary rocks of Minnesota. Geol. Soc. Am. Bull. **77**, 1031 (1966).

PETERMAN, Z. E., CARMICHAEL, I. S. E., SMITH, A. L.: Strontium isotopes in quaternary basalts of southeastern California. Earth Planet. Sci. Lett. **7**, 381 (1970a).

PETERMAN, Z. E., CARMICHAEL, I. S. E., SMITH, A. L.: $^{87}Sr/^{86}Sr$ ratios of quaternary lavas of the Cascade Range, northern California. Bull. Geol. Soc. Am. **81**, 311 (1970b).

PETERMAN, Z. E., COLEMAN, R. G., HILDRETH, R. A.: $^{87}Sr/^{86}Sr$ in mafic rocks of the Troodos Massif, Cyprus. U.S. Geol. Surv. Profess. Papers **750**-D, D157 (1971).

PETERMAN, Z. E., DOE, B. R., PROSTKA, H. J.: Lead and strontium isotopes in rocks of the Absaroka Volcanic Field, Wyoming. Contrib. Mineral. Petrol. **27**, 121 (1970).

PETERMAN, Z. E., HEDGE, C. E.: Related strontium isotopic and chemical variations in oceanic basalts. Bull. Geol. Soc. Am. **82**, 493 (1971).

PETERMAN, Z. E., HEDGE, C. E., BRADDOCK, W. A.: Age of Precambrian events in the northeastern Front Range, Colorado. J. Geophys. Res. **73**, 2277 (1968).

PETERMAN, Z. E., HEDGE, C. E., COLEMAN, R. G., SNAVELY, P. D., JR.: $^{87}Sr/^{86}Sr$ ratios in some eugeosynclinal sedimentary rocks and their bearing on the origin of granitic magma in orogenic belts. Earth Planet. Sci. Lett. **2**, 433 (1967).

PETERMAN, Z. E., HEDGE, C. E., TOURTELOT, H. A.: Isotopic composition in sea water throughout Phanerozoic time. Geochim. Cosmochim. Acta **34**, 105 (1970).

PETERMAN, Z. E., LOWDER, G. G., CARMICHAEL, I. S. E.: $^{87}Sr/^{86}Sr$ ratios of the Talasea Series, New Britain, Territory of New Guinea. Bull. Geol. Soc. Am. **81**, 39 (1970).

PETERMAN, Z. E., WHETTEN, J. T.: $^{87}Sr/^{86}Sr$ variation in Columbia River bottom sediments as a function of provenance. Geol. Soc. Am. Mem. **135**, 29 (1972).

PETÖ, P.: Petrochemical study of the Similkameen Batholith, British Columbia. Geol. Soc. Am. Bull. **84**, 3977 (1973).

PHILPOTTS, A. R.: Origin of the anorthosite-mangerite rocks in southern Quebec. J. Petrol. **7**, 1 (1966).

PHILPOTTS, J. A., PINSON, W. H.: New data on the chemical composition and origin of moldavites. Geochim. Cosmochim. Acta **30**, 253 (1966).

PHILPOTTS, J. A., SCHNETZLER, C. C.: Phenocryst-matrix partition coefficients for K, Rb, Sr and Ba, with applications to anorthosite and basalt genesis. Geochim. Cosmochim. Acta **34**, 307 (1970).

PHILPOTTS, J. A., SCHNETZLER, C. C.: Potassium, rubidium, strontium, barium and rare-earth concentrations in lunar rocks and separated phases. Science **167**, 493 (1970).

PHILPOTTS, J. A., SCHNETZLER, C. C., THOMAS, H. H.: Petrogenetic implications of some new geochemical data on eclogitic and ultrabasic inclusions. Geochim. Cosmochim. Acta **36**, 1131 (1972).

PICHLER, H., ZEIL, W.: The Cenozoic rhyolite-andesite association of the Chilean Andes. Bull. Volcanol. **35**, 424 (1971).

PIDGEON, R. T.: A rubidium-strontium geochronological study of the Willyama Complex, Broken Hill, Australia. J. Petrol. **8**, 283 (1967).

PIDGEON, R. T., COMPSTON, W.: The age and origin of the Cooma granite and its associated metamorphic zones, New South Wales. J. Petrol. **6**, 193 (1965).

PIDGEON, R. T., HOPGOOD, A. M.: Geochronology of Archean gneisses and tonalites from north of the Frederikshåbs Isblink, S.W. Greenland. Geochim. Cosmochim. Acta **39**, 1333 (1975).

PIDGEON, R. T., JOHNSON, M. R. W.: A comparison of zircon U-Pb and whole-rock Rb-Sr systems in three phases of the Carn Chuinneag granite, Northern Scotland. Earth Planet. Sci. Lett. **24**, 105 (1974).

PINGITORE, N. E.: Vadose and phreatic diagenesis: processes, products and their recognition in corals. J. Sediment. Petrol. **46**, 985 (1976).

POWELL, J. L., BELL, K.: Strontium isotopic studies of alkalic rocks: Localities from Australia, Spain, and the western United States. Contrib. Mineral. Petrol. **27**, 1 (1970).

POWELL, J. L., DELONG, S. E.: Isotopic composition of strontium in volcanic rocks from Oahu. Science **153**, 1239 (1966).

POWELL, J. L., FAURE, G., HURLEY, P. M.: Strontium-87 abundance in a suite of Hawaiian volcanic rocks of varying silica content. J. Geophys. Res. **70**, 1509 (1965).

PRETO, V. A. G.: Amphibolites from the Grand Forsk quadrangle of British Columbia, Canada. Geol. Soc. Am. Bull. **81**, 763 (1970).

PRICE, R. C., COMPSTON, W.: The geochemistry of the Dunedin Volcano: Strontium isotope chemistry. Contrib. Mineral. Petrol. **42**, 55 (1973).

PRICE, R. C., TAYLOR, S. R.: The geochemistry of the Dunedin Volcano, East Otago, New Zealand. Rare Earth elements. Contrib. Mineral. Petrol. **40**, 195 (1973).

PRIEM, H. N. A., BOELRIJK, N. A. I. M., HEBEDA, E. H., VERDURMEN, E. A. TH., VERSCHURE, R. H.: Isotopic ages of the Trans-Amazonian acidic magmatism amd the Nickerie metamorphic episode in the Precambrian basement of Suriname, South America. Geol. Soc. Am. Bull. **82**, 1667 (1971).

PRIEM, H. N. A., BOELRIJK, N. A. I. M., HEBEDA, E. H., VERDURMEN, E. A. TH., VERSCHURE, R. H.: Age of the Precambrian Roraima Formation in northeastern South America: Evidence from isotopic dating of Roraima pyroclastic volcanic rocks in Suriname. Geol. Soc. Am. Bull. **84**, 1677 (1973).

PRIEM, N. H. A., BOELRIJK, N. A. I. M., HEBEDA, E. H., VERDURMEN, E. A. TH., VERSCHURE, R. H., BON, E. H.: Granitic complexes and associated tin mineralization of "Grenville" age in Rondonia, western Brazil. Geol. Soc. Am. Bull. **82**, 1095 (1971).

PRINGLE, I. R.: Rb-Sr age determinations on shales associated with Varanger ice age. Geol. Mag. **109**, 465 (1973).

PRINS, P.: Apatite from African carbonatites. Lithos **6**, 133 (1973).

PRINZ, M.: Geochemistry of basaltic rocks: Traceelements. In: HESS, H. H., POLDERVART, A. (eds.), Basalt: The Poldervaart Treatise on Rocks of Basaltic Composition, Vol. 1, p. 271. New York-London-Sydney: Interscience Publ. 1967.

PUNDEER, G. S.: Mineralogy, genesis and diagenesis of a brecciated shaly clay from the Zechstein evaporite series of Germany. Contrib. Mineral. Petrol. **23**, 65 (1969).

PUSHKAR, P.: Strontium isotope ratios in volcanic rocks of three island arc areas. J. Geophys. Res. **73**, 2701 (1968).

PUSHKAR, P., CONDIE, K. C.: Origin of the Quaternary basalts from the Black Rock desert region, Utah: Strontium isotopic evidence. Geol. Soc. Am. Bull. **84**, 1053 (1973).

PUSHKAR, P., MCBIRNEY, A. R., KUDO, A. M.: The isotopic composition of strontium in Central America ignimbrites. Bull. Volcanol. **35**, 265 (1972).

PUSHKAR, P., PETERSON, M. N. A.: Strontium isotopic analyses of three marine phillipsites from the Pacific Ocean. Earth Planet. Sci. Lett. **2**, 349 (1967).

RASS, I. T.: Some rare elements in sphene and apatite of the Koksharov ultramafic-alkalic massif (Primor'ye). Geochem. Intern. **1**, 213 (1964).

REED, J. C., JR., ZARTMAN, R. E.: Geochronoloy of Precambrian rocks of the Teton Range, Wyoming. Geol. Soc. Am. Bull. **84**, 561 (1973).

REEDER, S. W., HITCHON, B., LEVINSON, A. A.: Hydrogeochemistry of the surface waters of the Mackenzie River drainage basin, Canada. I. Factors controlling inorganic composition. Geochim. Cosmochim. Acta **36**, 825 (1972).

REESMAN, R. H.: Strontium isotopic compositions of gangue minerals from hydrothermal vein deposits. Econ. Geol. **63**, 731 (1968).

REIMER, T.: The evolution of the rubidium and strontium content of shales. Neues Jahrb. Mineral. Abhandl. **116**, 167 (1972).

RENARD, M.: Etude géochimique de la fraction carbonatée d'un faciès de bordure de dépôt gypseux (exemple du gypse ludien du Bassin de Paris). Sediment. Geol. **13**, 191 (1975).

RENARD, M., LETOLLE, R., BOURBON, M., RICHEBOIS, G.: Some trace elements in the carbonate samples recovered from holes 390, 390A, 391C, and 392A of DSDP Leg. 44. (Unpubl. manuscript) (1978)

RENARD, M., LETOLLE, R., RICHEBOIS, G.: Strontium, manganese, iron contents and oxygen isotopes in the carbonate samples recovered from site 398C of Leg 47B. (Unpubl. manuscript) (1978a).

RENARD, M., LETOLLE, R., RICHEBOIS, G.: Some trace elements, oxygen and carbon isotopes in the carbonate samples recovered from site 400A of Leg 48. (Unpubl. manuscript) (1978b).

RHODES, J. M.: On the chemistry of potassium feldspars in granitic rocks. Chem. Geol. **4**, 373 (1969).

RIDLEY, W. I.: The petrology of the Las Canadas Volcanoes, Tenerife, Canary Islands. Contrib. Mineral. Petrol. **26**, 124 (1970).

RILEY, G. H.: Isotopic discrepancies in zoned pegmatites, Black Hills, South Dakota. Geochim. Cosmochim. Acta **34**, 713 (1970).

RILEY, G. H.: Isotopic discrepancies in zoned pegmatites, Black Hills, South Dakota. Geochim. Cosmochim. Acta **34**, 713 (1970a).

RILEY, G. H.: Excess Sr^{87} in pegmatitic phosphates. Geochim. Cosmochim. Acta **34**, 727 (1970b).

RINGWOOD, A. E.: The principles governing trace element distribution during magmatic crystallization. Part I: The influence of electronegativity. Geochim. Cosmochim. Acta **7**, 189 (1955).

RIVALENTI, G.: Problema della genesi degli gneiss anfibolici della serie "dioritica-kinzigitica" delle Alpi Pennine. Periodico Mineral. (Rome) **35**, 933 (1966).

ROBERTS, W. L., RAPP, G. R., JR,. WEBER, J.: Encyclopedia of Minerals. New York: Van Nostrand Reinhold 1974.

ROBINSON, P. T.: Petrology of the potassic Silver Peak volcanic center, western Nevada. Geol. Soc. Am. Bull. **83**, 1693 (1972).

ROBINSON, P. T., ELDERS, W. A., MUFFLER, L. J. P.: Quaternary volcanism in the Salton Sea geothermal field, Imperial Valley, California. Geol. Soc. Am. Bull. **87**, 347 (1976).

ROSS, C. S., FOSTER, M. D., MYERS, A. T.: Origin of dunites and of olivine-rich inclusions in basaltic rocks. Am. Mineralogist **39**, 693 (1954).

ROSS, J. E., ALLER, L. H.: The chemical composition of the sun. Science **191**, 1223 (1976).

RUNNELS, R. T., SCHLEICHER, J. A.: Chemical composition of eastern Kansas limestones. Kansas. Geol. Surv. Bull. **119**, 81 (1956).

RYAN, B. D., BLENKINSOP, J.: Geology and geochronology of the Hellroaring Creek stock, British Columbia. Can. J. Earth. Sci. **8**, 85 (1971).

SANTAMARIA, F., SCHUBERT, C.: Geochemistry and geochronology of the southern Caribbean—northern Venezuela plate boundary. Geol. Soc. Am. Bull. **85**, 1085 (1974).

SAVELLI, C.: The problem of rock assimilation by Somma-Vesuvius magma I. Composition of Somma and Vesuvius lavas. Contrib. Mineral. Petrol. **16**, 328 (1967).

SAVELLI, C., WEDEPOHL, K. H.: Geochemische Untersuchungen an Sinterkalken. Contrib. Mineral. Petrol. **21**, 238 (1969).

SAYLES, F. L., MANHEIM, F. T.: Interstitial solutions and diagenesis in deeply buried marine sediments: Results from the Deep Sea Drilling Project. Geochim. Cosmochim. Acta **39**, 103 (1975).

SCHNETZLER, C. C., PHILPOTTS, J. A.: Alkali, alkaline earth, and rare-earth element concentrations in some Apollo 12 soils, rocks, and separated phases. Proc. Second Lunar Sci. Conf. Geochim. Cosmochim. Acta, Suppl. 2, **2**, 1101 (1971).

SCHNETZLER, C. C., PHILPOTTS, J. A., PINSON, W. H.: Rubidium-strontium correlation studies of moldavites and Ries crater material. Geochim. Cosmochim. Acta **33**, 1015 (1969).

SCHNETZLER, C. C., PINSON, W. H., HURLEY, P. M.: Rubidium-strontium age of the Bosumtwi crater area, Ghana, compared with the age of the Ivory Coast tektites. Science **151**, 817 (1966).

SCHUMACHER, E.: Altersbestimmung von Steinmeteoriten mit der Rubidium-Strontium-Methode. Z. Naturforsch. **11**a, 209 (1956).

SCOTFORD, D. M.: Strontium partitioning between coexisting K-feldspar and plagioclase as an indicator of metamorphic grade in southwestern New Hampshire. Geol. Soc. Am. Bull. **84**, 3985 (1973).

SCOTT, R. B., NESBITT, R. W., DASCH, E. J., ARMSTRONG, R. L.: A strontium isotope evolution model for Cenozoic magma genesis, eastern Great Basin, U.S.A. Bull. Volcanol. **35**, 1 (1971).

SEDLACEK, P., DORNBERGER-SCHIFF, K.: Strukturprinzip des Strontiummetavanadat, $Sr(VO_3)_2 \cdot 4H_2O$. Acta Cryst. **18**, 407 (1965).

SEIDEL, E.: Zr contents of glaucophane-bearing meta-basalts of western Crete, Greece. Contrib. Mineral. Petrol. **44**, 231 (1974).

SEN, N., NOCKOLDS, S. R., ALLEN, R.: Trace elements in minerals from rocks of the S. Californian batholith. Geochim. Cosmochim. Acta **16**, 58 (1959).

SEN, S. K.: Some aspects of the distribution of barium, strontium, iron and titanium in plagioclase feldspars. J. Geol. **68**, 638 (1960).

SHACKLETTE, H. T., HAMILTON, J. C., BOERNGEN, J. G., BOWLES, J. M.: Elemental composition of surficial materials in the conterminous United States. U.S. Geol. Surv. Profess. Papers **574**-D (1971).

SHAFFER, N. R., FAURE, G.: Regional variation of $^{87}Sr/^{86}Sr$ ratios and mineral compositions of sediments from the Ross Sea, Antarctica. Geol. Soc. Am. Bull. **87**, 1491 (1976).

SHAW, D. M.: Trace element in pelitic rocks. I, II. Geol. Soc. Am. Bull. **65**, 1151+1167 (1954).

SHAW, D. M.: Trace element fractionation during anatexis. Geochim. Cosmochim. Acta **34**, 237 (1970).

SHAW, D. M.: Development of the early continental crust. Part 1. Use of trace element distribution coefficent models for the Protoarchean crust. Can. J. Earth Sci. **9**, 1577 (1972).

SHAW, D. M., DOSTAL, J., KEAYS, R. R.: Additional estimates of continental surface Precambrian shield composition in Canada. Geochim. Cosmochim. Acta **40**, 73 (1976).

SHAW, D. M., MOXHAM, R. L., FILBY, R. H., LAPKOWSKY, W. W.: The petrology and geochemistry of some Grenville skarns. Part II: Geochemistry. Can. Mineralogist **7**, 578 (1963).

SHAW, D. M., REILLY, G. A., MUYSSON, J. R., PATTENDEN, G. E., CAMPBELL, F. E.: An estimate of the chemical composition of the Canadian Shield. Can. J. Earth. Sci. **4**, 829 (1967).

SHCHEPAK, V. M., MIGOVICH, V. I.: Strontium in formation waters of the eastern Carpathian syncline and its significance during prospecting for oil and gas. Sov. Geol. **13**, 70 (1970).

SHERATON, J. W.: The origin of the Lewisian gneisses of northwestern Scotland, with particular reference to the Drumberg area, Sutherland. Earth Planet. Sci. Lett. **8**, 301 (1970).

SHIBATA, K., ADACHI, M.: Rb-Sr whole-rock ages of Precambrian metamorphic rocks in the Kamiaso conglomerate from central Japan. Earth Planet. Sci. Lett. **21**, 277 (1974).

SHIBATA, K., NOZAWA, T., WANLESS, R. K.: Rb-Sr geochronology of the Hida metamorphic belt, Japan. Can. J. Earth Sci. **7**, 1383 (1970).

SHIMA, M., HONDA, M.: Distributions of alkali, alkaline earth and rare earth elements in component minerals of chondrites. Geochim. Cosmochim. Acta **31**, 1995 (1967).

SHIMIZU, N.: Geochemistry of ultramafic inclusions from Salt Lake Crater, Hawaii and from Southern African kimberlites. In: AHRENS, L. H., RANKANA, K., RUNCORN, S. K. (eds.), Physics and Chemistry of the Earth, **9**, 655 (1974).

SHMAKIN, B. M., SHIRYAYEVA, V. A.: Distribution of rare earth and some other elements in apatites of muscovite pegmatites, eastern Siberia. Geochem. Intern. **5**, 962 (1968).

SHORT, N. M.: Geochemical variations in four residual soils. J. Geol. **69**, 534 (1961).

SHUTO, K.: The strontium isotopic study of the Tertiary acid volcanic rocks from the southern part of northeast Japan. Sci. Rept., Tokyo Kyoiku Daigaku, Sect. C, **12**, 75 (1974).

SIEDNER, G.: Geochemical features of a strongly fractionated alkali igneous suite. Geochim. Cosmochim. Acta **29**,113 (1965).

SIEGEL, F.: Variations of Sr/Ca ratios and Mg contents in Recent carbonate sediments of the northern Florida Keys area. J. Sediment. Petrol. **31**, 336 (1961).

SIEGERS, A., PICHLER, H.: Trace element abundances in the "Andesite" Formation of northern Chile. Geochim. Cosmochim. Acta **33**, 883 (1969).

SIGHINOLFI, G. P.: Investigations into the deep levels of the continental crust: petrology and chemistry of the granulite facies terrains from Bahia (Brazil). Atti Soc. Tosc. Sci. Nat. Mem. **76 A**, 327 (1970).

SIGHINOLFI, G. P.: Geochemistry of Early Precambrian carbonate rocks from the Brazilian shield: implications for Archean carbonate sedimentation. Contrib. Mineral. Petrol. **46**, 189 (1974).

SIGHINOLFI, G. P., FUJIMORI, S.: Petrology and chemistry of diopsidic rocks in granulite terrains from the Brazilian basement. Atti Soc. Tosc. Sci. Nat. Mem. **81 A**, 103 (1974).

SIGURDSSON, H., TOMBLIN, J. F., BROWN, G. M., HOLLAND, J. G., ARCULUS, R. J.: Strongly undersaturated magmas in the Lesser Antilles Island arc. Earth Planet. Sci. Lett. **18**, 285 (1973).

SIVAN, T. P., MIGOVICH, V. I., OZERNYI, O. M., SHEREMETE'EV, S. K.: Strontium in the formation waters of the Crimea petroleum and gas bearing region. Geol. Geokhim. Goryuch. Iskop. **34**, 88 (1973).

SKIBA, N. S.: Strontium geochemistry in a supergene zone. Rasseyan. Elem. Osad. Form. Tyan-Shanya, Akad. Nauk. SSR, Inst. Geol. **1967**, 34 (1967).

SKIBA, W., BUTLER, J. R.: The use of Sr-An relationships in plagioclases to distinguish between Somalian metagabbros and countryrock amphibolites. J. Petrol. **4**, 352 (1963).

SKOUGSTAD, M. W., HORR, C. A.: Occurrence and distribution of strontium in natural water. U.S. Geol. Surv. Water Supply Papers **1496**-D, 55 (1963).

SMITH, D. G. W.: Pyrometamorphism of phyllites by a dolerite plug. J. Petrol. **10**, 20 (1969).

SMITH, R. C., II, ROSE, A. W., LANNING, R. M.: Geology and geochemistry of Triassic diabase in Pennsylvania. Geol. Soc. Am. Bull. **86**, 943 (1975).

SPENCER, D.: Factors affecting element distribution in a Silurian graptolite band. Chem. Geol. **1**, 221 (1966).

SPOONER, C. M., BERRANGÉ, J. P., FAIRBAIRN, H. W.: Rb-Sr whole-rock age of the Kanuku Complex, Guyana. Geol. Soc. Am. Bull. **82**, 207 (1971).

SPOONER, C. M., FAIRBAIRN, H. W.: Relation of radiometric age of granitic rocks near Calais, Maine, to the time of Acadian orogeny. Geol. Soc. Am. Bull. **81**, 3663 (1970).

SPOONER, E. T. C.: The strontium isotopic composition of seawater and seawater-oceanic crust interaction. Earth Planet. Sci. Lett. **31**, 167 (1976).

SREEKUMARAN, CHENGALATH, JOSEPH, K. T., PARAMESWARAN, M.: Occurrence of Na, K, Rb, Ca, and Sr in some Indian rivers. Current Sci. (India) **39**, 105 (1970).

STARKE, R.: Die Strontiumgehalte der Baryte. Freiberger Forschungsh. C**150** (1964).

STEHLI, F. G., HOWER, J.: Mineralogy and early diagenesis of carbonate sediments. J. Sediment. Petrol. **31**, 358 (1961).

STEPHENS, W. E., WATSON, S. W., PHILIP, P. R., WEIR, J. A.: Element associations and distributions through a Lower Paleozoic graptolite shale sequence in the Southern Uplands of Scotland. Chem. Geol. **16**, 269 (1975).

STEPNIEWSKI, M.: Trace elements in the Zechstein salt minerals from the Puch Bay region. Biul. Inst. Geol. Warsaw **272**, 7 (1973).

STEWART, J. W., EVERNDEN, J. F., SNELLING, N. J.: Age determinations from Andean Peru: A reconnaissance survey. Geol. Soc. Am. Bull. **85**, 1107 (1974).

STRADOMSKII, V. B., KONOVALOV, G. S.: Estimation of the distribution of stable strontium in surface waters and its loss from territory in USSR. Gidrokhim. Mater. **48**, 40 (1968).

STRONG, D. F.: Petrology of the Island of Moheli, western Indian Ocean. Geol. Soc. Am. Bull. **83**, 389 (1972).

STRONG, D. F.: Lushs Bight and Roberts Arm Groups of central Newfoundland: Possible juxtaposed oceanic and island-arc volcanic suites. Geol. Soc. Am. Bull. **84**, 3917 (1973).

STRONG, D. F.: An "off-axis" alkali volcanic suite associated with the Bay of Islands ophiolites, Newfoundland. Earth Planet. Sci. Lett. **21**, 301 (1974).

STRUNZ, H.: Mineralogische Tabellen. Leipzig: Akademische Verlagsgesellschaft 1970.

STUCKLESS, J. S., IRVING, A. J.: Strontium isotope geochemistry of megacrysts and host basalt from southeastern Australia. Geochim. Cosmochim. Acta **40**, 209 (1976).

STUCKLESS, J. S., O'NEIL, J. R.: Petrogenesis of the Superstition-Superior volcanic area as inferred from strontium-and oxygen-isotope studies. Geol. Soc. Am. Bull. **84**, 1987 (1973).

STUEBER, A. M.: Abundances of K, Rb, Sr and Sr isotopes in ultramafic rocks and minerals from western North Carolina. Geochim. Cosmochim. Acta **33**, 543 (1969).

STUEBER, A. M., BALDWIN, A. D., IKRAMUDDIN, M., PUSHKAR, P.: Sr, Ca and the isotopic composition of strontium in underground waters from the Scioto River Basin, Ohio. U.S. Nat. Tech. Inform. Serv. PB Rep. **226104/8 GA**, (1973).

STUEBER, A. M., HEIMLICH, R. A., IKRAMUDDIN, M.: Rb-Sr ages of Precambrian mafic dikes, Bighorn Mountains, Wyoming. Geol. Soc. Am. Bull. **87**, 909 (1976).

STUEBER, A. M., IKRAMUDDIN, M.: Rubidium, strontium and the isotopic composition of strontium in ultramafic nodule minerals and host basalts. Geochim. Cosmochim. Acta **38**, 207 (1974).

STUEBER, A. M., MURTHY, V. R.: Strontium isotope and alkali element abundances in ultramafic rocks. Geochim. Cosmochim. Acta **30**, 1243 (1966).

STULL, R. J., MCMILLAN, K.: Origin of lherzolite inclusions in the Malapai Hill basalt, Joshua Tree National Monument, California. Geol. Soc. Am. Bull. **84**, 2343 (1973).

SUBBARAO, K. V.: The strontium isotopic composition of basalt from the East Pacific and Chile Rises and abyssal hills in the eastern Pacific Ocean. Contrib. Mineral. Petrol. **37**, 111 (1972).

SUBBARAO, K. V., CLARK, G. S., FORBES, R. B.: Strontium isotopes in some seamounts basalts from the northeastern Pacific Ocean. Can. J. Earth Sci. **10**, 1479 (1973).

SUBBARAO, K. V., HEDGE, C. E.: K, Rb, Sr and $^{87}Sr/^{86}Sr$ in rocks from the Mid-Indian Ocean Ridge. Earth Planet. Sci. Lett. **18**, 223 (1973).

SUPKO, P. R.: Subsurface dolomites, San Salvador, Bahamas. J. Sediment. Petrol. **47**, 1063 (1977).

SUWA, K., MATSUZAWA, I., IIDA, C., YAMASAKI, K.: Mineralogical studies on the weathering of a quartz porphyry. J. Earth Sci. Nagoya Univ. **6**, 75 (1958).

TARNEY, J.: Geochemistry of Archean gneisses, with implications as to the origin and evolution of the Precambrian crust. In: The Early History of the Earth. Nato Adv. S. Inst., Leicester (Abstracts) (1975).

TARNEY, J., SKINNER, A. C., SHERATON, J. W.: A geochemical comparison of major Archean gneiss units from Northwest Scotland and East Greenland. Proc. 24th Int. Geol. Congr., Montreal **1**, 162 (1972).

TATSUMOTO, M., HEDGE, C. E., ENGEL, A. E. J.: Potassium, rubidium, strontium, thorium, uranium, and the ratio of strontium-87 to strontium-86 in oceanic tholeiitic basalt. Science **150**, 886 (1965).

TAUSON, L. V.: Distribution of trace elements during crystallization of magmas. In: AHRENS, L. H., RANKAMA, K. RUNCORN, S. K. (eds.), Physics and Chemistry of the Earth, **6**, 217 (1965).

TAYLOR, P. N.: An early Precambrian age for migmatitic gneisses from Vikan i Bo, Vesteralen, north Norway. Earth Planet. Sci. Lett. **27**, 35 (1975).

TAYLOR, S. R.: The origin of some New Zealand metamorphic rocks as shown by their major and trace element composition. Geochim. Cosmochim. Acta **8**, 182 (1955).

TAYLOR, S. R.: The application of trace element data to problems in petrology. In: AHRENS, L. H., RANKAMA, K. RUNCORN, S. K. (eds.), Physics and Chemistry of the Earth, **6**, 133 (1965).

TAYLOR, S. R., CAPP, A. C., GRAHAM, A. L., BLAKE, D. H.: Trace element abundances in andesites, II. Saipan, Bougainville and Fiji. Contrib. Mineral. Petrol. **23**, 1 (1969).

TAYLOR, S. R., SACHS, M.: Geochemical evidence for the origin of australites. Geochim. Cosmochim. Acta **28**, 235 (1964).

TAYLOR, S. R., SOLOMON, M.: The geochemistry of Darwin glass. Geochim. Cosmochim. Acta **28**, 471 (1964).

TAYLOR, S. R., WHITE, A. J. R.: Trace element abundances in andesites. Bull. Volcanol. **29**, 177 (1966).

TERA, F., EUGSTER, O., BURNETT, D. S., WASSERBURG, G. J.: Comparative study of Li, Na, K, Rb, Cs, Ca, Sr and Ba abundances in achondrites and in Apollo 11 lunar samples. Proc. Apollo 11 Lunar Sci. Confer. Geochim. Cosmochim. Acta, Suppl. **1**, **2**, 1637 (1970).

THOMPSON, T. G., CHOW, J.: The strontium-calcium ratio in carbonate-secreting marine organisms. Deep-Sea Res., Suppl. to **3**, 20 (1955).

THURBER, D. L., KULP, J. L., HODGES, E., GAST, P. W., WAMPLER, J. M.: Common strontium content of the human skeleton. Science **128**, 256 (1958).

TILL, R.: The relationship between environment and sediment composition (geochemistry and petrology) in the Bimini Lagoon, Bahamas. J. Sediment. Petrol. **40**, 367 (1970).

TUREKIAN, K. K.: The marine geochemistry of strontium. Geochim. Cosmochim. Acta **28**, 1479 (1964).

TUREKIAN, K. K., KULP, J. L.: The geochemistry of strontium. Geochim. Cosmochim. Acta **10**, 245 (1956).

TUREKIAN, K. K., PHINNEY, W. C.: The distribution of Ni, Ca, Cr, Co, Ba, and Sr between biotite-garnet pairs in a metamorphic sequence. Am. Mineralogist **47**, 1434 (1962).

TUREKIAN, K. K., WEDEPOHL, K. H.: Distribution of the elements in some major units of the Earth's crust. Geol. Soc. Am. Bull. **72**, 175 (1961).

UPPULURI, V. R.: Prineville chemical type: A new basalt type in the Columbia River group. Geol. Soc. Am. Bull. **85**, 1315 (1974).

USDOWSKI, E.: Das geochemische Verhalten des Strontiums bei der Genese und Diagenese von Ca-Karbonat- und Ca-Sulfat-Mineralen. Contrib. Mineral. Petrol. **38**, 177 (1973).

VALLANCE, T. G.: Spilitic degradation of a tholeiitic basalt. J. Petrol. **15**, 79 (1974).

VAN BREEMEN, O., AFTALION, M., ALLAART, J. H.: Isotopic and geochronologic studies on granites from the Ketilidian mobile belt of South Greenland. Geol. Soc. Am. Bull. **85**, 403 (1974).

VAN BREEMEN, O., DODSON, M. H.: Metamorphic chronology of the Limpopo Belt, southern Africa. Geol. Soc. Am. Bull. **83**, 2005 (1972).

VAN BREEMEN, O., PIDGEON, R. T., JOHNSON, M. R. W.: Precambrian and Paleozoic pegmatites in the Moines of northern Scotland. J. Geol. Soc. London **130**, 493 (1974).

VAN BREEMEN, O., UPTON, B. G. J.: Age of some Gardar Intrusive Complexes, South Greenland. Geol. Soc. Am. Bull. **83**, 3381 (1972).

VAN SCHMUS, W. R., MEDARIS, L. G., JR., BANKS, P. O.: Geology and age of the Wolf River batholith, Wisconsin. Geol. Soc. Am. Bull. **86**, 907 (1975).

VAN SCHMUS, W. R., THURMAN, E. M., PETERMAN, Z. E.: Geology and Rb-Sr chronology of Middle Precambrian rocks in eastern and central Wisconsin. Geol. Soc. Am. Bull. **86**, 1255 (1975).

VEIZER, J.: Chemical diagenesis of belemnite shells and possible consequences for paleotemperature determinations. Neues Jahrb. Geol. Palaeontol. Abhandl. **147**, 91 (1974).

VEIZER, J.: Diagenesis of pre-Quaternary carbonates as indicated by tracer studies. J. Sediment. Petrol. **47**, 565 (1977a).

VEIZER, J.: Geochemistry of lithographic limestones and dark marls from the Jurassic of southern Germany. Neues Jahrb. Geol. Palaeontol. Abhandl. **153**, 129 (1977b).

VEIZER, J.: Simulation of limestone diagenesis: a model based on strontium depletion: Discussion. Can. J. Earth Sci. **15**, 1683 (1978).

VEIZER, J., COMPSTON, W.: $^{87}Sr/^{86}Sr$ composition of seawater during the Phanerozoic. Geochim. Cosmochim. Acta **38**, 1461 (1974).

VEIZER, J., COMPSTON, W.: $^{87}Sr/^{86}Sr$ in Precambrian carbonates as an index of crustal evolution. Geochim. Cosmochim. Acta **40**, 905 (1976).

VEIZER, J., DEMOVIČ, R.: Strontium as a tool in facies analysis. J. Sediment. Petrol. **44**, 93 (1974).

VEIZER, J., LEMIEUX, J., JONES, B., GIBLING, M. R., SAVELLE, J.: Paleosalinity and dolomitization of a Lower Paleozoic carbonate sequence, Somerset and Prince of Wales Islands, Arctic Canada. Can. J. Earth Sci. **15**, 1448 (1978).

VEIZER, J., WENDT, J.: Mineralogy and chemical composition of Recent and fossil skeletons of calcareous sponges. Neues Jahrb. Geol. Palaeontol. Monatsh. **1976** (9), 558 (1976).

VERA, R. H., VAN SCHMUS, W. R.: Geochronology of some Precambrian rocks of the southern Front Range, Colorado. Geol. Soc. Am. Bull. **85**, 77 (1974).

VINE, J. D., TOURTELOT, E. B.: Geochemistry of Lower Eocene sandstones in the Rocky Mountains region. U.S. Geol. Surv. Profess. Papers **789** (1973).

VINOGRADOV, A. P.: The Elementary Chemical Composition of Marine Organisms. New Haven: Sears Found. Mar. Res. 1953. [Transl. from Russian].

VINOGRADOV, A. P.: The Geochemistry of Rare and Dispersed Chemical Elements in Soils, 2nd ed. New York: Consultants Bur. Enterpr. 1959.

VINOGRADOV, A. P., RONOV, A. B., RATYNSKII, V. I.: Izmenenije chimitscheskogo sostava karbonatnych porod russkoj platformy. Izv. Akad. Nauk. SSSR **1**, 33 (1952).

VIRGO, D.: Partition of strontium between coexisting K-feldspar and plagioclase in some metamorphic rocks. J. Geol. **76**, 331 (1968).

VIRGO, D.: Partitioning of sodium between coexisting K-feldspar and plagioclase from some metamorphic rocks. J. Geol. **77**, 173 (1969).

VITRAC-MICHARD, A., ALLÈGRE, C. J.: A study of the formation and history of a piece of continental crust by ^{87}Rb-^{87}Sr method: the case of the French Oriental Pyrénées. Contrib. Mineral. Petrol. **50**, 257 (1975).

VLASOV, K. A. (ed.): Geochemistry and Mineralogy of Rare Elements and Genetic Types of their Deposits. Vol. I. Geochemistry of Rare Elements. Translated by Z. LERMAN, Israel Program for Scientific Translations, Jerusalem, 1966 (1964).

VOLLMER, R.: Rb-Sr and U-Th-Pb systematics of alkaline rocks: the alkaline rocks from Italy. Geochim. Cosmochim. Acta **40**, 283 (1976).

WÄNKE, H., BADDENHAUSEN, H., SPETTEL, B., TESCHKE, F., QUIJANO-RICO, M., DREIBUS, G., PALME, H.: The chemistry of the Haverö ureilite. Meteoritics **7**, 579 (1972).

WAGER, L. R., MITCHELL, R. L.: The distribution of trace elements during strong fractionation of basic magma—a further study of the Skaergaard intrusion, East Greenland. Geochim. Cosmochim. Acta **1**, 129 (1951).

WAHLBERG, J. S., BAKER, J. H., VERNON, R. W., DEWAR, R. S.: Exchange adsorption of strontium on clay minerals. U.S. Geol. Surv. Bull. **1140**-C (1965a).

WAHLBERG, J. S., DEWAR, J. S., Comparison of distribution coefficients for strontium exchange from solutions containing one and two competing cations. U.S. Geol. Surv. Bull. **1140**-D (1965b).

WANLESS, R. K., LOVERIDGE, W. D.: MURSKY, G.: A geochronological study of the White Creek batholith, southeastern British Columbia. Can. J. Earth Sci. **5**, 375 (1968).

WASSERBURG, G. J., LANPHERE, M. A.: Age determinations in the Precambrian of Arizona and Nevada. Geol. Soc. Am. Bull. **76**, 735 (1965).

WASSERBURG, G. J., PAPANASTASSIOU, D. A.: Age of an Apollo 15 mare basalt; lunar crust and mantle evolution. Earth Planet. Sci. Lett. **13**, 97 (1971).

WASSERBURG, G. J., PAPANASTASSIOU, D. A., SANZ, H. G.: Initial strontium for a chondrite and the determination of a metamorphism or formation interval. Earth Planet. Sci. Lett. **7**, 33 (1969).

WATKINS, N. D., GUNN, B. M., NOUGIER, J., BAKSI, A. K.: Kerguelen: Continental fragment or oceanic island? Geol. Soc. Am. Bull. **85**, 201 (1974).

WATKINSON, D. H.: Electron microprobe analysis of melilite and garnet from the Oka complex, Quebec. Can. Mineralogist **11**, 457 (1972).

WEAVER, S. D., SCEAL, J. S. C., GIBSON, I. L.: Trace element data relevant to the origin of trachytic and pantelleritic lavas in the East African rift system. Contrib. Mineral. Petrol. **36**, 181 (1972).

WEBER, J. N.: Trace element composition of dolostones and dolomites and its bearing on the dolomite problem. Geochim. Cosmochim. Acta **28**, 1817 (1964).

WEBER, J. N., MIDDLETON, G. V.: Geochemistry of the turbidites of the Normanskill and Charny formations—1. Effect of turbidity currents on the chemical differentation of turbidites. Geochim. Cosmochim. Acta **22**, 200 (1961).

WEBSTER, R. K., MORGAN, J. W., SMALES, A. A.: Some recent Harwell analytical work on geochronology. Trans. Am. Geophys. Unions **38**, 543 (1957).

WEDEPOHL, K. H.: Spurenanalytische Untersuchungen an Tiefseetonen aus dem Atlantik. Ein Beitrag zur Deutung der geochemischen Sonderstellung von pelagischen Tonen. Geochim. Cosmochim. Acta **18**, 200 (1960).

WEDEPOHL, K. H.: Untersuchungen am Kupferschiefer in Nordwestdeutschland; ein Beitrag zur Deutung der Genese bituminöser Sedimente. Geochim. Cosmochim. Acta **28**, 305 (1964).

WEHMILLER, J.: Strontium: Element and geochemistry. In: FAIRBRIDGE, R. W. (ed.), The Encyclopedia of Geochemistry and Environmental Sciences. Vol. IVA. New York: Van Nostrand Reinhold 1972.

WEIGAND, P. W., RAGLAND, P. C.: Geochemistry of Mesozoic dolerite dikes from eastern North America. Contrib. Mineral. Petrol. **29**, 195 (1970).

WHETTEN, J. T.: Sediments from the lower Columbia River and origin of graywacke. Science **152**, 1057 (1966).

WHITE, A. J. R.: Genesis of migmatites from the Palmer region of South Australia. Chem. Geol. **1**, 165 (1966).

WHITE, A. J. R., COMPSTON, W., KLEEMAN, A. W.: The Palmer Granite—a study of a granite within a regional metamorphic environment. J. Petrol. **8**, 29 (1967).

WHITE, W. M., BRYAN, W. B.: Sr-isotope ,K, Rb, Cs, Sr, Ba, and rare-earth geochemistry of basalts from the FAMOUS area. Geol. Soc. Am. Bull. **88**, 571 (1977).

WHITFIELD, M.: The extension of chemical models for sea water to include trace components at 25 °C and 1 atm. pressure. Geochim. Cosmochim. Acta **39**, 1545 (1975).

WHITFORD, D. J.: Strontium isotopic studies of the volcanic rocks of the Sunda arc, Indonesia, and their petrogenetic implications. Geochim. Cosmochim. Acta **39**, 1287 (1975).

WHITNEY, J. A., JONES, L. M., WALKER, R. L.: Age and origin of the Stone Mountain Granite, Lithonia district, Georgia. Geol. Soc. Am. Bull. **87**, 1067 (1976).

WILCOX, R. E., POLDERVAART, A.: Metadolerite dike swarm in Bakersville-Roan Mountain area, North Carolina. Geol. Soc. Am. Bull. **69**, 1323 (1958).

WILHELMI, K.-A.: The crystal structure of strontium dichromate, $SrCr_2O_7$. Arkiv Kemi **26**, 149 (1966).

WILKINSON, J. F. G.: The geochemistry of a differentiated teschenite sill near Gunnedah, New South Wales. Geochim. Cosmochim. Acta **16**, 123 (1959).

WILLIAMS, I. S., COMPSTON, W., CHAPPELL, B. W., SHIRAHASE, T.: Rubidium-strontium age determinations on micas from a geologically controlled composite batholith. J. Geol. Soc. Australia **22**, 497 (1975).

WILSON, J. R., COATS, J. S.: Alkali feldspars from part of the Galway granite, Ireland. Mineral. Mag. **38**, 801 (1972).

WITZMAN, H., BEULICH, W.: Beitrag zur Struktur wasserfreier Strontiumborate. Naturwissenschaften **52**, 157 (1965).

WOLF, K. H., CHILINGAR, G. V., BEALES, F. W.: Elemental composition of carbonate skeletons, minerals and sediments. In: CHILINGAR, G. V., BISSEL, H. J., FAIRBRIDGE, R. W.(eds.), Carbonate Rocks B. Amsterdam: Elsevier 1967.

WOOD, M. W., SHAW, H. F.: The geochemistry of celestites from the Yate area near Bristol (U.K.). Chem. Geol. **17**, 179 (1976).

WORDEN, J. M., COMPSTON, W.: A Rb-Sr isotopic study of weathering in the Mertondale granite, Western Australia. Geochim. Cosmochim. Acta **37**, 2567 (1973).

WRIGHT, P. L.: The chemistry and mineralogy of the clay fraction of sediments from the southern Barents Sea. Chem. Geol. **13**, 197 (1974).

WYCKOFF, R. W. G., DOBERENZ, A. R.: The strontium content of fossil teeth and bones. Geochim. Cosmochim. Acta **32**, 109 (1968).

YAMAMOTO, T.: Chemical studies on the seaweeds (27). The relations between concentration factor in seaweeds and residence time of some elements in sea water. Records Oceanog. Works Japan **11**, 65 (1972).

YES'KOVA, E. M., YEFIMOV, A. F.: Distribution of rare elements in the apoeffusive alkalic metasomatites of the Urals. Geochem. Internat. **7**, 721 (1970).

YUDOVICH, YA. E., KORYCHEVA, A. A., OBRUCHNIKOV, A. S., STEPANOV, YU. V.: Mean trace-element contents in coals. Geochem. Internat. **9**, 712 (1972).

ZARTMAN, R. E.: A geochronologic study of the Lone Grove pluton from the Llano Uplift, Texas. J. Petrol. **5**, 359 (1964).

ZARTMAN, R. E., BROCK, M. R., HEYL, A. V., THOMAS, H. H.: K-Ar and Rb-Sr ages of some alkalic intrusive rocks from central and eastern United States. Am. J. Sci. **265**, 848 (1967).

ZARTMAN, R. E., MARVIN, R. F.: Radiometric age (Late Ordovician) of the Quincy, Cape Ann, and Peabody granites from eastern Massachusetts. Geol. Soc. Am. Bull. **82**, 937 (1971).

ZIELINSKI, R. A.: Trace element evaluation of a suite of rocks from Reunion Island, Indian Ocean. Geochim. Cosmochim. Acta **39**, 713 (1975).

ZIELINSKI, R. A., LIPMAN, P. W.: Trace element variations at Summer Coon volcano, San Juan Mountains, Colorado, and the origin of continental andesite. Geol. Soc. Am. Bull. **87**, 1499 (1976).

Revised manuscript of A received: July 1970

Revised manuscript of B received: May 1973

Manuscript of C received: January 1977

Manuscript of D received: September 1976

Revised manuscript of E received: August 1977

Revised manuscript of F received: September 1977

Manuscript of G received: June 1977

Revised manuscript of H, I received: December 1976

Manuscript of K, L received: July 1978

Revised manuscript of M, N received: August 1976

Yttrium 39

see Lanthanides 57–71

Zirconium 40

A	G. Bayer	(Institut für Kristallographie und Petrographie, Eidgen. Techn. Hochschule, Zürich, Switzerland)
B—O	A. J. Erlank, H. S. Smith, J. W. Marchant, M. P. Cardoso, L. H. Ahrens	(Department of Geochemistry, University of Capetown, Rondebosch, South Africa)

40-A. Crystal Chemistry

I. General

Zirconium belongs to the IVB-group transition elements of the periodic system. It has the outer electron configuration d^2s^2 and is predominantly stable in the 4-valent state as Zr^{4+}. Due to the lanthanide contraction, the atomic and the ionic radii of Zr and of Hf are very similar, much more so than for any other pair of related elements:

	Atomic radius (Å)	Ionic radius (Å) for Me^{4+} (SHANNON and PREWITT, 1969); coordination number:		
		6	7	8
Zr	1.60	0.72	0.78	0.84
Hf	1.59	0.71		0.83

Estimates of the bond character in oxides (from the differences of electronegativity) lead to 67% ionic character for the Zr—O bond and to 70% ionic character for the Hf—O bond.

The similar chemical and physical properties of Zr and Hf explain why these elements are always associated and very difficult to separate. All Zr minerals contain hafnium, usually in concentrations upto 2%. The most abundant and important Zr minerals are zircon ($ZrSiO_4$) and baddeleyite (ZrO_2).

The different absorption cross-sections of zirconium and hafnium for thermal neutrons are of importance for applications in atomic reactors. The average absorption cross-section is 105 barns for hafnium and only 0.18 barns for zirconium.

Probably all Zr-compounds have corresponding isostructural Hf-compounds, however not all of them have been synthesized so far.

II. Metallic Zirconium, Interstitial Compounds

Zirconium crystallizes in two modifications, in the hexagonal close-packed α-form (Mg-type, A 3) at room temperature, and in the body-centered cubic β-form (W-type, A 2) at high temperatures upto the melting point. The temperature for the reversible $\alpha \rightleftharpoons \beta$ transformation is 862° C and is strongly influenced by impurities and addition of other elements (GEBHARDT and SEGHEZZI, 1964).

The high solubility of zirconium in other elements is of interest because of its effect on physical properties. Both Zr and Hf form interstitial compounds with H, B, C, N, and O like the other transition elements.

Incorporation of oxygen into hexagonal zirconium results in the formation of superlattices in which the oxygens are distributed in an ordered way. The hexagonal packing of the metal atoms is maintained upto the composition Zr_2O, but ZrO has a cubic structure (STEEB and RIEKERT, 1969).

Typically, the interstitial compounds show wide composition ranges, e.g. from 35 to 50 atomic % C for the NaCl-type carbides ZrC and HfC (BENESOVSKY and RUDY, SR 1960, 89). Most of the interstitial compounds of zirconium are very hard and have high melting points, especially the carbide, boride and nitride, and also the oxide (Table 40-A-1).

Table 40-A-1. *High melting zirconium compounds*

Compound	Melting point (°C)	Structural type
ZrB_2	2,990	AlB_2-type (C 32)
ZrC	3,530	NaCl-type (B 1)
ZrN	2,980	NaCl-type (B 1)
Zr_2Si	2,110	$CuAl_2$-type (C 16)
$ZrSi_2$	1,520	$ZrSi_2$-type (C 49)
ZrO_2	2,690	fluorite-type (C 1) (high temp.)

A comprehensive compilation of data was given by KIEFFER and BENESOVSKY (1963), where also most of the structure work by NOWOTNY and co-workers can be found. Besides the many carbides, nitrides, borides and silicides, there also exists a large number of intermetallic compounds of zirconium and hafnium, e.g. LAVES-type phases like $ZrCr_2$, ZrW_2, $HfMo_2$, HfW_2 (ELLIOT, SR 1961, 161), or Al_2Cu-type compounds like Zr_2Al, Hf_2Al (NOWOTNY, SCHOB and BENESOVSKY, SR 1961, 10).

III. Oxides and Oxide Compounds

The predominant oxygen coordination of zirconium is 6-fold (octahedron) and 8-fold (cube, square antiprism, bisdisphenoid) with Me—O distances varying usually between 2.00 and 2.30 Å. There are exceptions, however, where longer interatomic distances are found or where the oxygen coordination is different (Table 40-A-2).

Table 40-A-2. *Zr-O distances in oxide compounds and silicates*

Compound	Coordination number of Zr	Distance Zr-O (Å)	Reference
K_2ZrO_3	5	1.92 (1 ×), 2.13 (4 ×)	GATEHOUSE and LLOYD (1970a)
β-$K_2Zr_2O_5$	6	2.01—2.22	GATEHOUSE and LLOYD (1970b)
$ZrTe_3O_8$	6	2.017	MEUNIER (1970)
$(K, Na)_2Zr[Si_6O_{15}]$	6	2.03—2.09	FLEET (1965)
α-$Zr[SO_4]_3$	7	2.03—2.19	BEAR and MUMME (1970)

Table 40-A-2 (continued)

Compound	Coordination number of Zr	Distance Zr-O (Å)	Reference
$Zr_3Yb_4O_{12}$	6, 7	2.03—2.59	THORNBER and BEVAN (1970)
$Na_2Zr[Si_4O_{11}]$	6	2.04—2.13	VORONKOV and PJATENKO (SR 1961, 531)
$Na_2Zr[Si_6O_{15}]$ 3 $H_2O \cdot 0.5$ NaOH	6	2.05—2.16	ILJUKHIN and BELOV (SR 1960, 498)
ZrO_2 (mcl.)	7	2.05—2.28	SMITH and NEWKIRK (1965)
ZrO_2 (tetrag.)	8	2.06 (4 ×), 2.46 (4 ×)	TEUFER (SR 1962, 480)
$KZr_2[PO_4]_3$	6	2.06	SLJUKIC *et al.* (1969)
$La_2Zr_2O_7$	6	2.09	DEISEROTH and MUELLER-BUSCHBAUM (1970)
$Zr[SO_4]_2 \cdot 7\ H_2O$	8	2.12—2.28	BEAR and MUMME (1969)
$Zr[SiO_4]$	8	2.13 (4 ×), 2.27 (4 ×)	ROBINSON *et al.* (1971)
$Zr[IO_3]_4$	8	2.197 (4 ×), 2.216 (4 ×)	LARSON and CROMER (SR 1961, 498)

Zirconium dioxide, ZrO_2, is known in three crystalline modifications under normal atmospheric pressure: in the stable monoclinic, baddeleyite structure (as mineral), and in the metastable, tetragonal and cubic fluorite-type structures at high temperatures. In addition orthorhombic, high-pressure polymorphs of ZrO_2 and of HfO_2 have been prepared at 1,600° C and 60—110 kbars (BENDELIANI *et al.*, 1967; BOCQUILLON *et al.*, 1968). The reversible transformations, especially of the monoclinic into the tetragonal form, show a strong hysteresis. The existence of pure, cubic ZrO_2 is now well established. Tetragonal and cubic ZrO_2 can be prepared also at low temperatures, e.g. by decomposition of alkoxides. They are converted however to the stable, monoclinic form on heating above 400° C.

The polymorphic forms of ZrO_2 are structurally identical to those of HfO_2, the related oxides forming solid solutions in all proportions. The stable, monoclinic modification has a peculiar 7-fold oxygen coordination which is shown in Fig. 40-A-1 (SMITH and NEWKIRK, 1965). The Zr—O distances vary between 2.051 and 2.285 Å. The coordination in the tetragonal and in the cubic fluorite-modifications is 8-fold. The coordination polyhedron of tetragonal ZrO_2 can be regarded as a combination of two distorted oxygen-tetrahedra similar to those in $Zr[SiO_4]$ (TEUFER, SR 1962, 480); this results in 4 short (2.06 Å) and 4 long (2.46 Å) Zr—O distances. A comparison of the Zr—O coordinations in the three modifications of ZrO_2, can be seen in Fig. 40-A-2.

The structural relationship between tetragonal and cubic ZrO_2 is very close. For the 7-fold oxygen coordination in monoclinic ZrO_2, some bonds have to be broken and rotated. The reconstructive monoclinic tetragonal transformation in

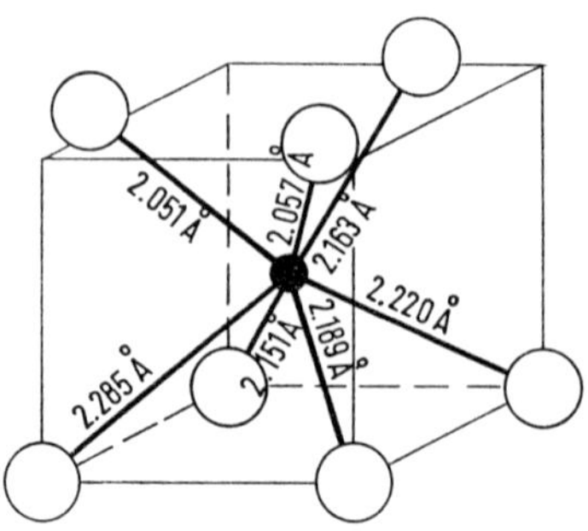

Fig. 40-A-1. ZrO_7 – polyhedron and Zr–O distances in monoclinic ZrO_2, baddeleyite (SMITH and NEWKIRK, 1965)

Table 40-A-3. *Oxide compounds of zirconium*

Compound	Coordination number of Zr	Structural type	Reference
$ZrMo_2O_8$	6, 8	—	RIMSKY *et al.* (1968)
$Zr[P_2O_7]$	6	cubic $Me^{IV}[P_2O_7]$	VOLLENKLE *et al.* (1963)
ZrO_2	7	baddeleyite	SMITH and NEWKIRK (1965)
ZrO_2	8	fluorite (tetrag. distorted)	TEUFER (SR 1962, 480)
$Zr[SiO_4]$	8	zircon	KRSTANOVIC (SR 1958, 314)
$Zr[GeO_4]$	8	scheelite	LEFEVRE and COLLONGUES (SR 1960, 457)
$ZrTiO_4$	6	α-PbO_2	NEWNHAM (1967)
$Li_2Zr[WO_4]_3$	6	wolframite	CHANG (1967)
$ZrTe_3O_8$	6	cubic $Me^{IV}Te_3O_8$	MEUNIER (1970), BAYER (1969)
$Gd_2Zr_2O_7$, $Sm_2Zr_2O_7$, $Nd_2Zr_2O_7$, $La_2Zr_2O_7$	6	pyrochlore	PEREZ Y JORBA (SR 1962, 535)
$CaZr[BO_3]_2$, $SrZr[BO_3]_2$, $BaZr[BO_3]_2$	6	dolomite	BAYER (1971)
$SrZrO_3$, $BaZrO_3$, Na_2ZrTeO_6	6	perovskite	SMITH and WELCH (SR 1960, 354), BAYER (1969)
$Ca_3Zr_2Fe_2^{[4]}Ti^{[4]}O_{12}$, $Ca_2LaZr_2Fe_3^{[4]}O_{12}$, $Ca_3Zr_2[Al_2SiO_{12}]$	6	garnet	ITO and FRONDEL (1967) MILTON *et al.* (1961)
Sr_2ZrO_4, Ba_2ZrO_4	6	K_2NiF_4	SCHOLDER *et al.* (1968)
$NaScZrO_4$	6	$CaFe_2O_4$	REID *et al.* (1968)
K_2ZrO_3	5	—	GATEHOUSE and LLOYD (1970a)

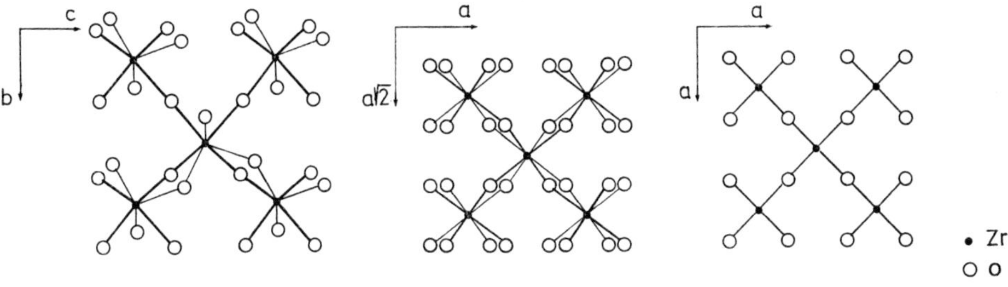

Fig. 40-A-2. Linking of Zr–O polyhedra in the three modifications of ZrO_2: a) monoclinic ZrO_2; b) tetragonal ZrO_2; c) cubic ZrO_2 (SMITH and NEWKIRK, 1965)

ZrO_2 can be inhibited by addition (10—25 mole %) of certain cubic oxides such as MgO, CaO, Sc_2O_3, Y_2O_3 and $(RE)_2O_3$, which form cubic, solid solutions with zirconia (RYSHKEWITCH, 1960). These so-called "stabilized zirconias" have a sub-stoichiometric fluorite-structure, due to the incorporation of the lower-valent cations which create random or ordered oxygen vacancies in the lattice (BARKER and WILLIAMS, 1968); typical compositions are 1 MgO:4 ZrO_2, 1 CaO:4 ZrO_2, and 1 Y_2O_3:8 ZrO_2. The stability of such non-stoichiometric, cubic zirconias depends on the kind and concentration of oxide addition and on the temperature.

There exists a large variety of oxide compounds containing Zr; a selection of compounds is presented in Table 40-A-3.

IV. Silicates

The most important and abundant mineral of zirconium is $Zr[SiO_4]$, zircon. In this orthosilicate, each Zr is surrounded by 4 oxygen atoms at 2.131 Å and by 4 additional oxygens at a distance of 2.268 Å (ROBINSON *et al.*, 1971). This 8-fold oxygen coordination can be described as a triangular dodecahedron, or alternatively as a combination of two interpenetrating ZrO_4-sphenoids of different sizes. The zircon structure is built up of chains of alternating edge-sharing SiO_4-tetrahedra and ZrO_8-dodecahedra running parallel to the z-axis (Fig. 40-A-3). These chains are joined by edge-sharing of the ZrO_8-dodecahedra. The similarity of such structural chains in zircon and in the garnet structure has been pointed out by ROBINSON *et al.* (1971). There exists also a close relationship between the zircon and the scheelite structure; the oxygen coordinations around the cations are similar, except that the scheelite structure has a denser oxygen packing which leads to a different distortion of the coordination polyhedra (BAYER, 1972). The expected high pressure transformation of $Zr[SiO_4]$ (zircon) to $Zr[SiO_4]$ (scheelite) could be experimentally verified at 900° C and 120 kbars by REID and RINGWOOD (1969); it is accompanied by an increase in density of 11%. Altered mineral varieties of zircon are quite common; partial replacement of O_4 in $[SiO_4]$ by $(OH)_4$ is found in hydrated zircons. An isotropic form of zircon with lower density, so-called "metamict zircon", is formed in the presence of radioactive atoms which decay and cause more or less complete breakdown of the zircon structure. There are indications that the relatively low stability of zircon (thermal decomposition, tendency to metamictization) is

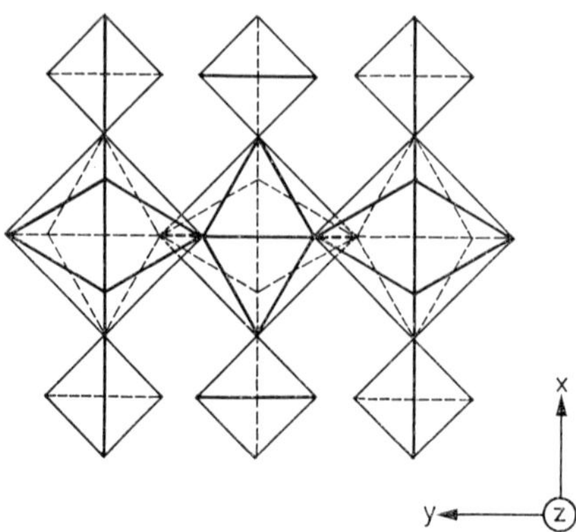

Fig. 40-A-3. Linking of ZrO_8-dodecahedra and SiO_4-tetrahedra in the zircon structure (ROBINSON *et al.* 1971)

associated with the unusually high coordination number of Zr (8) in zircon (MACHATSCHKI, 1941). For a better understanding of the zircon structure and its instability, calculations of lattice energies were carried out for a series of geometrically deformed models (SAHL and ZEMANN, 1965) and the results showed that the effective charge of silicon most probably has a value between 1^+ and 2^+.

Additional silicate minerals of zirconium are listed in Table 40-A-4, in the sequence of increasing $[SiO]_4$-tetrahedra linking. The oxygen coordination of zirconium in such compounds is usually octahedral, except for zircon, and in many silicates zirconium can be replaced partially by titanium. Various complex zirconium-rich silicates occur as primary minerals in pegmatites, the most important one of these minerals being eudialyte, $(Fe, Mn, Mg)_3Zr_3(Zr, Nb)_x(Ca, RE)_6Na_{12}[Si_9O_{27-y}(OH)_y]_2$ $[Si_3O_9]_2Cl_z$. Different formulae have been suggested for this mineral due to analytical difficulties, extensive isomorphous replacement, only partially occupied atomic positions and presence of ions and molecules in structural holes. The recently reported structure determinations (GIUSEPETTI *et al.*, 1971; GOLYSHEV *et al.*, 1971) showed that this cyclo-silicate contains 3-fold and 9-fold $[SiO_4]$-tetrahedra rings and that it has a zeolite-like structure. Eudialyte is commonly associated with its decomposition products, e.g. with the cyclosilicates catapleite and lovozerite. Such minerals are structurally related also to other zirconium-rich minerals like dalyite and wadeite. Dalyite, $K_2^{[8]}Zr^{[6]}[Si_6\,O_{15}]$, represents an interesting layer silicate in which the sheets contain 4-, 6- and 8-membered rings of $[SiO_4]$-tetrahedra; these $\overset{2}{\infty}$ $[Si_2O_5]$-sheets are joined by ZrO_6-octahedra (FLEET, 1965). Frame work silicates containing zirconium are not known so far although they are theoretically possible. A new Zr-containing silicate "tranquillityite" was recently discovered in Apollo 11 and 12 lunar samples (LOVERING *et al.*, 1971). The composition was given as $Fe_8^{2+}(Zr, Y)_2Ti_3Si_3O_{24}$, but the structure has not yet been determined.

Table 40-A-4. *Zr-containing silicate minerals*

Linking of $[SiO_4]$-tetrahedra	Composition	Coordination number of Zr	Mineral name	Reference
Separate $[SiO_4]$	$Zr[SiO_4]$	8	zircon	KRSTANOVIC (SR 1958, 314)
	$Ca_3Zr_2[Al_2SiO_{12}]$	6	kimzeyite (garnet)	MILTON *et al.* (1961)
Double $[Si_2O_7]$	$Na_2Zr[Si_2O_7]$	6	keldyshite	VORONKOV and ZAKHAROVA (1970)
	$Ca_2NaZr(F, OH, O)_2[Si_2O_7]$	6	wöhlerite	SIBARYVA and BELOV (SR 1962, 714)
Rings $[SiO_3]_n$	$BaZr[Si_3O_9]$	6	synthetic (benitoite)	BLASSE and BRIL (1970)
	$K_2Zr[Si_3O_9]$	6	wadeite	HENSHAW (SR 1955, 463)
	$Na_2Zr[Si_3O_9] \cdot 2\,H_2O$	6	catapleite	BRUNOWSKI (SR 1937, 25)
	$Na_2Zr[Si_6O_{12}(OH)_6] \cdot 0.5\,NaOH$	6	lovozerite	ILJUKHIN and BELOV (SR 1960, 498)
	$(Fe, Mn, Mg)_3Zr_3(Zr, Nb)_x (Ca, RE)_6Na_{12} [Si_9O_{27-y}(OH)_y]_2 [Si_3O_9]_2Cl_z$	6	eudialyte	GOLYSHEV *et al.* (1971) GUISEPPETTI *et al.* (1971)
Chains $[Si_2O_6]$	$Na_2(Ti, Zr)_2O_3[Si_2O_6]$	6	lorenzenite	KRAUS and MUSSGNUG (SR 1949, 283)
Double chains $[Si_6O_{15}]$	$Na_2Zr[Si_6O_{15}] \cdot 3\,H_2O$	6	elpidite	NERONOVA and BELOV (1965)
Bands $[Si_8O_{22}]$	$Na_2Zr[Si_4O_{11}]$	6	vlasovite	VORONKOV and PATJENKO (SR 1961, 531)
Sheets $[Si_2O_5]_n$	$K_2Zr[Si_6O_{15}]$	6	dalyite	FLEET (1965)

V. Halides and Sulfides

Zirconium forms a great variety of halides, oxyhalides and chalcogenides which often are structurally related to corresponding compounds of Hf and of other elements like Ge, Ti, Sn, Th, Mo, U (WELLS, 1962; KREBS, 1968). Examples are: $ZrCl_3$, $ZrBr_3$, ZrI_3 ($MoBr_3$-type); $ZrBr_4$ (SnI_4-type); K_2ZrF_6 (K_2ThF_6-type); Rb_2ZrCl_6, Cs_2ZrCl_6 (K_2PtCl_6-type); Na_3ZrF_7 (α-Na_3UF_7-type); ZrS (PbO-type); ZrS_2, $ZrSe_2$, $ZrTe_2$, (CdI_2-type). Many sulfides have composition ranges (HAHN and NESS, SR 1957, 171). Isostructural sulfide and oxide compounds exist in the case of perovskites, e.g. $BaZrS_3$ and $BaZrO_3$ (KREBS, 1968). The Zr-halides show different coordination numbers for the halogen atoms around zirconium, depending on the size of the halogen atom and on the stoichiometry. For example the coordination number is 4 in $ZrBr_4$ (BERDONOSOV *et al.*, SR 1962, 444), 6 in Rb_2ZrCl_6 and Cs_2ZrCl_6 (WELLS, 1962), 7 in Na_3ZrF_7 (HARRIS, SR 1959, 291), and 8 in ZrF_4 (ZACHARIASEN, SR 1949, 168) and in K_2ZrF_6 (BODE and TEUFER, SR 1956, 224). Recent crystal structure determinations of the complex fluorides $Na_7Zr_6F_{31}$ (BURNS, ELLISON and LEVY, 1968) and $Rb_5Zr_4F_{21}$ (BRUNTON, 1971) showed that zirconium is surrounded by eight F-ions (square antiprism) in the sodium compound, whereas the complex Rb—Zr fluoride has four crystallographically independent Zr-ions with 6-, 7-, and 8-fold fluorine coordination.

Revised manuscript received: July 1972

References: Section 40-A

Barker, W. W., Williams, L. S.: Some limitations of cubic stabilization in zirconia. J. Austral. Cer. Soc. **4**, 1 (1968).

Bayer, G.: Zur Kristallchemie des Tellurs, Telluroxide und Oxidverbindungen mit Tellur. Fortschr. Mineral. **46**, 41 (1969).

Bayer, G.: Thermal expansion anisotropy of dolomite-type borates, $Me^{2+}Me^{4+}B_2O_6$. Z. Krist. **133**, 85 (1971).

Bayer, G.: Thermal expansion of ABO_4-compounds with zircon- and scheelite-structures. J. Less-Common Metals **26**, 255 (1972).

Bear, I. J., Mumme, W. G.: The crystal chemistry of zirconium sulphates. I—III. Acta Cryst. B **25**, 1558 (1969).

Bear, I. J., Mumme, W. G.: The crystal chemistry of zirconium sulphates. VI. The structure of α-$Zr(SO_4)_2$. Acta Cryst. B **26**, 1140 (1970).

Bendeliani, N. A., Popova, S. V., Vereshchagin, L. F.: New modifications of ZrO_2 and HfO_2 produced under high pressures. Geokhimiya **6**, 677 (1967).

Blasse, G., Bril, A.: Fluorescence and structure of barium zirconium trisilicate. J. Solid State Chem. **2**, 105 (1970).

Bocquillon, G., Suse, C., Vodar, B.: Allotropy of hafnium oxide under high pressure. Rev. Inter. Hautes Temp. Refract. **5**, 247 (1968).

Brunton, G.: The crystal structure of $Rb_5Zr_4F_{21}$. Acta Cryst. B **27**, 1944 (1971).

Burns, J. H., Ellison, R. D., Levy, H. A.: The crystal structure of $Na_7Zr_6F_{31}$. Acta Cryst. B **24**, 230 (1968).

Chang, L. L. Y.: $Li_2Zr(WO_4)_3$, a wolframite-type compound. Mineral. Mag. **36**, 436 (1967).

Deiseroth, H. J., Mueller-Buschbaum, H. K.: Ein Beitrag zur Pyrochlorstruktur an $La_2Zr_2O_7$. Z. Anorg. Allgem. Chem. **375**, 152 (1970).

Fleet, S. G.: The crystal structure of dalyite. Z. Krist. **121**, 349 (1965).

Gatehouse, B. M., Lloyd, D. J.: The crystal structure of potassium metazirconate, K_2ZrO_3, and its tin analogue K_2SnO_3. J. Solid State Chem. **2**, 410 (1970a).

Gatehouse, B. M., Lloyd, D. J.: The crystal structure of beta potassium dizirconate: β-$K_2Zr_2O_5$. J. Solid State Chem. **1**, 478 (1970b).

Gebhardt, E., Seghezzi, H. D.: Reaktorwerkstoffe. Teil I. Metallische Werkstoffe. Stuttgart: B. G. Teubner 1964.

Giuseppetti, G., Mazzi, F., Tadini, C.: The crystal structure of eudialyte. Tschermaks Mineral. Petrog. Mitt. **16**, 105 (1971).

Golyshev, V. M., Simonov, V. I., Belov, N. V.: Crystal structure of eudialyte. Soviet Phys. Cryst. **16** (1), 70 (1971).

Ito, J., Frondel, C.: Synthetic zirconium and titanium garnets. Am. Mineralogist **52**, 773 (1967).

Kieffer, R., Benesovsky, F.: Hartstoffe. Wien: Springer 1963.

Krebs, H.: Grundzüge der anorganischen Kristallchemie. Stuttgart: Enke 1968.

Lovering, J. F., Wark, D. A., Reid, A. F., Warf, N. G., Keil, K., Prinz, M., Bunch, T. E.: Tranquillityite, a new silicate mineral from Apollo 11 and 12 basaltic rocks. Proc. Second Lunar Science Conference **1**, 39 (1971).

Machatschki, F.: Zur Frage der Stabilität des Zirkons. Zentr. Min. A, 38 (1941).

Meunier, G.: Les systèmes MO_2—TeO_2 ($M = Ti, Sn, Hf, Zr$). Thèse, Université de Bordeaux, 1970.

Milton, C., Ingram, B. L., Blade, L. V.: Kimzeyite, a zirconium garnet from Magnet Cove, Arkansas. Am. Mineralogist **46**, 533 (1961).

NERONOVA, N. N., BELOV, N. V.: Crystal structure of elpidite, $Na_2Zr[Si_6O_{15}] \cdot 3H_2O$. Soviet Phys.-Cryst. **9**, 700 (1965).

NEWNHAM, R. E.: Crystal structure of $ZrTiO_4$. J. Am. Ceram. Soc. **50**, 216 (1967).

REID, A. F., RINGWOOD, A. E.: Newly observed high pressure transformations in Mn_3O_4, $CaAl_2O_4$ and $ZrSiO_4$. Earth Planet. Sci. Lett. **6**, 205 (1969).

REID, A. F., WADSLEY, A. D., SIENKO, M. J.: Crystal chemistry of sodium scandium titanate, $NaScTiO_4$, and its isomorphs. Inorg. Chem. **7**, 112 (1968).

ROBINSON, K., GIBBS, G. V., RIBBE, P. H.: The structure of zircon, a comparison with garnet. Am. Mineralogist **56**, 782 (1971).

RUH, R., ROCKETT, T. J.: Proposed phase diagram for the system ZrO_2. J. Am. Cer. Soc. **53** (6), 360 (1970).

RYSHKEWITCH, E.: Oxide Ceramics. New York-London: Academic Press 1960.

SAHL, K., ZEMANN, J.: Gitterenergetische Berechnungen an Zirkon. Ein Beitrag zur Ladungsverteilung in der Silikatgruppe. Tschermaks Mineral. Petrog. Mitt. **10**, 97 (1965).

SCHOLDER, R., RAEDE, D., SCHWARZ, H.: Über Zirkonate, Hafnate und Thorate von Barium, Strontium, Lithium und Natrium. Z. Anorg. Allgem. Chem. **362**, 149 (1968).

SHANNON, R. D., PREWITT, C. T.: Effective ionic radii in oxides and fluorides. Acta Cryst. B **25**, 925 (1969).

SLJUKIC, M., MATKOVIC, B., PRODIC, B.: The crystal structure of $KZr_2(PO_4)_3$. Z. Krist. **130**, 148 (1969).

SMITH, D. K., NEWKIRK, H. W.: The crystal structure of baddeleyite (monoclinic ZrO_2) and its relation to the polymorphism of ZrO_2. Acta Cryst. **18**, 983 (1965).

STEEB, S., RIEKERT, A.: Die Struktur des Zirkoniumsuboxides Zr_2O. J. Less-Common Metals **17**, 429 (1969).

THORNBER, M. R., BEVAN, D. J. M.: Mixed oxides of the type MO_2 (fluorite)—M_2O_3. IV. Crystal structures of the high- and low-temperature forms of $Zr_3Yb_4O_{12}$. J. Solid State Chem. **1**, 536 (1970).

VOELLENKLE, H., WITTMANN, A., NOWOTNY, H.: Über Diphosphate vom Typ Me(IV)-P_2O_7. Monatsh. Chem. **94**, 956 (1963).

VORONKOV, A. A., SHUMSYATSKAYA, N. G., STRICHKOV, Y. T.: Crystal structure of a new natural modification of $Na_2ZrSi_2O_7$. Zh. Struct. Khim. **11**, 932 (1970).

WELLS, A. F.: Structural Inorganic Chemistry, 3rd ed. Oxford: Clarendon Press 1962.

Revised manuscript received: July 1972

40-B. Isotopes in Nature

The conventional abundance of the five stable isotopes of zirconium is as follows:

Isotope:	^{90}Zr	^{91}Zr	^{92}Zr	^{94}Zr	^{96}Zr
%:	51.46	11.23	17.11	17.40	2.80

^{90}Zr is especially abundant because it has a so-called "magic" number of neutrons (N=50). No natural isotopic fractionation has been reported, and none is expected at these mass numbers. MINSTER and ALLÉGRE (1976) could find no significant variation in zirconium isotopic composition which could be attributed to the decay of long-lived ^{92}Nb in samples of recent and ancient terrestrial zircons, chondrites and achondrites, and lunar soil and basalt. SHIMA and HINTENBERGER (1976) likewise found that the isotopic composition of zirconium in stony meteorites is identical to that of terrestrial reagents. The atomic weight of zirconium is 91.22.

The radioactive isotopes ^{86}Zr, ^{87}Zr, ^{88}Zr, ^{89}Zr, ^{93}Zr, ^{95}Zr, ^{97}Zr and ^{99}Zr have been prepared in the laboratory. Further details regarding these may be found in RANKAMA (1963).

Revised manuscript received: June 1977

40-C. Abundance in Cosmos, Meteorites, Tektites and Lunar Samples

I. Meteorites

For reasons which will become clear the abundance of zirconium in meteorites is discussed first, before the section dealing with cosmic abundances is presented.

The abundance of zirconium in meteorites is important for two main reasons. First, although meteorites are generally used in obtaining so-called "cosmic" abundances for non-volatile elements, the so-called "magic" ($N=50$) nuclide ^{90}Zr is particularly important for defining this portion of the abundance curve, and for testing current theories of nucleosynthesis. Secondly, uncertainty regarding the abundance of zirconium in chondrites, has led to speculation regarding fractionation between Zr and Hf in the earth and meteorites. The importance of both of these aspects has been dealt with by ERLANK and WILLIS (1964).

Zirconium is dominantly lithophile in its geochemical behaviour as shown by the analyses of SETSER and EHMANN (1964). They reported average abundances of 35, 6.3 and 1.4 ppm in chondritic meteorites, troilite from an iron meteorite, and iron meteorites respectively. Even though the chondritic analyses at least have subsequently been shown to be in error (see Table 40-C-1), the relative relationships observed are a reasonable representation of the distribution between the silicate, sulfide and metal phases of meteorites, as also shown in Table 40-C-4. The mineral zircon has been observed in mesosiderites and one chondrite (MASON and GRAHAM, 1970) and recently in the Cc1 chondrite Alais (MACDOUGALL and KERRIDGE, 1976). Note, however, the possible chalcophilic behavior of zirconium in the highly reduced enstatite chondrites, as shown by the presence of 23 ppm Zr in the troilite from the chondrite Khairpur (Table 40-C-4).

The determination of zirconium in chondrites has a history of analytical error which is summarized in Table 40-C-1. As is readily apparent the earliest determinations (>30 ppm Zr) are incorrect, and more recent determinations have steadily revised "average" abundances towards lower values. Preferred abundances for chondrites are presented in Table 40-C-2. The high value of 173 ppm Zr in the Cc1 chondrite Alais is clearly anomalous and probably related to the relative abundance of zircon in this meteorite (MACDOUGALL and KERRIDGE, 1976). But it is interesting that this meteorite is also enriched in other refractory elements such as Ta, Hf and the rare earths (GANAPATHY *et al.*, 1976). Apart from this observation, it is difficult to detect any real fractionation between the various chondrite classes as shown by analyses for any particular meteorite by different analysts. It appears that while the average abundance of zirconium in chondrites is within the range 5–10 ppm, further and more precise analyses are required as recently shown by GANAPATHY *et al.* (1976) for the important Cc1 chondrite Orgueil.

Data for achondrites (Table 40-C-3) show a distinct fractionation of zirconium between the Ca-rich and Ca-poor achondrites. Some of the former (i. e. the eucrites) have abundance levels comparable to that observed in some terrestrial tholeiitic basalts (Table 40-E-3).

The only available data for iron meteorites are given by Setser and Ehmann (1964) who reported from 1.2–1.6 ppm Zr in four samples. These analyses may be in error on the basis of the chondrite data presented by these authors (Table 40-C-2).

The distribution of zirconium within meteorites is shown in Table 40-C-4. Apart from the possible presence of zircon (Mason and Graham, 1970), it appears that much, or the bulk of, zirconium in meteorites is contained in clinopyroxenes. This

Table 40-C-1. *Comparison of zirconium averages in chondrites*

References	Method	No. of samples	ppm Zr
Hevesy and Würstlin (1934)	X	6	30–160
Pinson *et al.* (1953)	S	21	33
Merz (1962)	N/R	5	33
Setser and Ehmann (1964)	N/R	12	35
Schmitt *et al.* (1962)	N/R	4	13
Merz and Schrage (1964)	N/R	3	12
Erlank and Willis (1964)	X	8	~10
Schmitt *et al.* (1964)	S	28	9
von Michaelis *et al.* (1969)	X	45	8.4
Ehmann and Rebagay (1970)	N/R	28	6.7
Ehmann and Chyi (1974)	N/R	8	7.2
Ganapathy *et al.* (1976)	N/R	4	5.0

Table 40-C-2. *Preferred zirconium abundances in chondrites*

Name	ppm Zr	Method	Reference	Name	ppm Zr	Method	Reference
Carbonaceous chondrites Type 1 (Cc1)							
Orgueil	11	S	1	Ivuna	8.6	N/R	4
	9.4	N/R	4		8.7	Mass	8
	5.2	N/R	9	Alais	173	N/R	10
	3.1	N/R	10				
	4.0	N/R	11				
Carbonaceous chondrites Type 2 (Cc2)							
Murray	5.2	N/R	4	Nawapali	6.9	N/R	12
	8	S	1	Mighei	5.8	N/R	4
	9	X	5		10	X	5
	13	N/R	2		11	S	1
	5.9	I	13	Boriskino	6.5	N/R	4
Cold Bokkeveld	9	X	5		11	S	1
	9	S	1				
	5.3	N/R	12	Murchison	7.8	Mass	8
Renazzo	7.0	Mass	8		6.2	N/R	9
					4.6	N/R	10
Haripura	7.4	N/R	9		9.6	I	13

Table 40-C-2 (continued)

Name	ppm Zr	Method	Reference
High iron-low metal group chondrites (CHL or Cc3)			
Felix	9	S	1
	9	X	5
Mokoia	12	S	1
	10	X	5
Allende	13	N/R	4
	11.7	Mass	8
	5.9	N/R	10
	6.4	N/R	12
	7.4	I	13
	10.7	I	13
Karoonda	9.5	N/R	4
	8.0	N/R	9
Ornans	12	N/R	4
Groznaia	9	S	1
Vigarano	13	X	5
Kainsaz	6.6	X	6
Efremovka	6.0	X	6
Lance	9	X	5
Leoville	7.5	I	13
High iron group chondrites (CH)			
Allegan	8	S	1
	9	X	5
Miller	4	S	1
	14	N/R	2
Ochansk	5	S	1
	5.7	N/R	4
Pantar	8	X	5
	6.1	I	13
Pantar II	8	S	1
Pultusk	8	X	5
	10	S	1
Plainview	6	X	5
	6.8	N/R	4
	12	N/R	3
Richardton	6.1	N/R	4
	9	X	5
	7.2	I	13
Forest City	5.6	N/R	4
	7	X	5
Beardsley	7	X	5
Breit-scheid	7	S	1
	8.7	I	13
Hugoton	8	S	1
Morland	15	N/R	3
Mount Browne	7.2	N/R	4
Potter	10	S	1
Guareña	6.8	Mass	8
Coolidge	11.9	N/R	9
Canyon City	4.0	N/R	12
Ehole	7	S	1
Kesen	9	X	5
Geluks-fontein	8	X	5
Idutywa	9	X	5
Moshesh Location	9	X	5
Schaap-Kooi	8	X	5
Tulia	8	S	1
Low iron group chondrites (CL)			
Harleton	6.9	N/R	4
	10	S	1
Leedey	10	X	5
	8	S	1
Mocs	4.9	N/R	4
	12	S	1
Modoc	4	S	1
	8	X	5
	8	Mass	7
	12	N/R	2
Peace River	4.5	N/R	4
	13	S	1

Table 40-C-2 (continued)

Name	ppm Zr	Method	Reference	Name	ppm Zr	Method	Reference
Kyushu	8	X	5	Holbrook	9	X	5
	13	N/R	2	Jackals-fontein	9	X	5
Alfianello	8	X	5				
Paragould	10	S	1	Marion	11	X	5
Walters	19	S	1	Muizenberg	10	X	5
Kunashak	4.3	N/R	4	Monze	9	X	5
Chainpur	7.0	N/R	4	New Concord	9	X	5
	6.5	N/R	9	Farmington	9	X	5
Beenham, N.M.	6.8	N/R	4	Wittekrantz	10	X	5
Upper Volta	6.7	N/R	12	McKinney	11	X	5
Kiel	6.6	I	13	Bjurbole	9	X	5
Bruderheim	5.1	N/R	4	Ioka	6.3	N/R	4
	6.6	N/R	10				
	6.2	I	13	Diep River	8	X	5
Saratov	6.9	N/R	4	St.Michel	8	X	5
Colby, Wisc.	8	X	5	Barwell	7.8	N/R	4
Drake Creek	7	X	5	Wethersfield	6.8	N/R	9
Low iron-low metal group chondrites (CLL)							
Dhurmsala	8	X	5	Manbhoom	6	S	1
	5.2	N/R	9	Vavilovka	13	S	1
Jelica	6.9	N/R	4	St. Severin	6	Mass	7
Olivenza	9.2	N/R	4		10.7	I	13
Enstatite chondrites							
Ce				*Ce2*			
St. Marks	7	X	5	Atlanta	4.1	N/R	4
Cel					6	X	5
Abee	3.8	N/R	4	Daniel's Kuil	7.7	N/R	4
	7	X	5	Hvittis	8	X	5
	14	S	1				
	7.9	I	13	Pillistfer	6	X	5
Indarch	6	X	5	Blithfield	10	X	5

References: 1. Schmitt *et al.* (1964); 2. Schmitt *et al.* (1962); 3. Merz and Schrage (1964); 4. Ehmann and Rebagay (1970); 5. von Michaelis *et al.* (1969); 6. Ahrens *et al.* (1973); 7. Mason and Graham (1970); 8. Graham and Mason (1972); 9. Ehmann and Chyi (1974); 10. Ganapathy *et al.* (1976); 11. Palme (1974); 12. Ehmann *et al.* (1975); 13. Shima and Hintenberger (1976).

Table 40-C-3. *Preferred zirconium abundances in achondrites*

Name	ppm Zr	Method	Reference	Name	ppm Zr	Method	Reference
Ca-poor achondrites							
Enstatite achondrites (Ae)				*Bronzite achondrites (Ab)*			
Norton County	1.1	N/R	2	Johnstown	1.9	N/R	2
	4	X	3		4	X	3
	8	S	1		5.8	I	9
Peña Blanca Spring	6	S	1	Shalka	1.0	N/R	2
					3	X	3
Cumberland Falls	0.64	N/R	2	*Ureilite (Aop)*			
				North Haig	3.9	N/R	8
Ca-rich achondrites							
Eucrites (Ap)				Macibini	54	X	4
Binda	15	X	4	Petersburg	50	S	1
	17	N/R	2	Nuevo Laredo	67	S	1
Moore County	22	Mass	7	Stannern	87	X	4
					70	S	1
Haraiya	36	X	4		40	I	9
	39	Mass	7	*Howardites (Aor)*			
Sioux County	42	X	4	Kapoeta	59	N/R	2
	48	X	3	Malvern	36	X	5
					42	X	3
Cachari	45	X	4	Chaves	28	X	5
Juvinas	46	X	4	Frankfort	19	X	5
	53	S	1		14	X	5
Bereba	52	X	4	*Angrite (Aa)*			
Pasamonte	53	X	4	Angra dos Reis	100	Mass	6
	58	X	3				
	61	N/R	2				
	61	Mass	6				

References: 1. SCHMITT *et al.* (1964); 2. EHMANN and REBAGAY (1970); 3. VON MICHAELIS *et al.* (1969); 4. MCCARTHY *et al.* (1973); 5. MCCARTHY *et al.* (1972); 6. GRAHAM and MASON (1972); 7. ALLEN and MASON (1973); 8. EHMANN and CHYI (1974); 9. SHIMA and HINTENBERGER (1976).

Table 40-C-4. *Zirconium in separated phases and minerals from meteorites*

Phase or Mineral	Meteorite	Class	ppm Zr	Method	Reference
Total meteorite	Allende	CHL	6.4	N/R	1
White inclusions	Allende	CHL	2.4, 7.7	N/R	1
Chondrules	Allende	CHL	4.6, 3.1	N/R	1
Black matrix	Allende	CHL	6.1	N/R	1
Fusion crust	Allende	CHL	6.8	N/R	1

Table 40-C-4 (continued)

Phase or Mineral	Meteorite	Class	ppm Zr	Method	Reference
Olivine	Brenham	P	3	S	2
	Marjalahti	P	6	S	2
	Modoc	CL	<3	Mass	3
	Guareña	CH	5	Mass	3
Orthopyroxene	Winona	Chondrite[a]	6	Mass	4
	Modoc	CL	6	Mass	4
	St. Severin	CLL	22[b]	Mass	4
	Johnstown	Ab	2	Mass	4
	Guareña	CH	7	Mass	3
	Khairpur	Ce2	4	Mass	3
Clinopyroxene	Winona	Chondrite[a]	90[b]	Mass	4
	Haraiya	Ap	29	Mass	4
	Moore County	Ap	14	Mass	3
	Modoc	CL	110[b]	Mass	3
	Guareña	CH	50	Mass	3
	Allende	CHL	98	Mass	5
Plagioclase	Winona	Chondrite[a]	4	Mass	4
	Modoc	CL	7	Mass	4
	St. Severin	CLL	8	Mass	4
	Haraiya	Ap	2	Mass	4
	Guareña	CH	5	Mass	3
	Khairpur	Ce2	5	Mass	3
	Moore County	Ap	2	Mass	3
Melilite	Allende	CHL	7.8	Mass	5
Chromite	Modoc	CL	5	Mass	4
	St. Severin	CLL	2	Mass	4
	Guareña	CH	3	Mass	3
Troilite	Modoc	CL	3	Mass	4
	St. Severin	CLL	3	Mass	4
	Soroti	Iron[a]	3	Mass	4
	Guareña	CH	3	Mass	3
	Khairpur	Ce2	23	Mass	3
Phosphate	Modoc	CL	650[b]	Mass	4
	St. Severin	CLL	18	Mass	4
Metal	Modoc	CL	<3	Mass	3

[a] Precise classification unknown. [b] Possible zircon impurity.

References: 1. Ehmann *et al.* (1975); 2. Schmitt *et al.* (1964); 3. Allen and Mason (1973); 4. Mason and Graham (1970); 5. Mason and Martin (1974).

is well exemplified by the value of 98 ppm Zr in an Al-rich pyroxene (fassaite) separated from a pyroxene-melilite chondrule extracted from the Allende (CHL) chondrite. It is important when evaluating condensation models invoked for the origin of meteorites, bearing in mind the refractory nature of zirconium.

II. Cosmos

Three types of approach have been adopted to establish the "cosmic" or solar system abundance of zirconium. The first and most commonly used involves the use of chondrites. As we have seen, earlier measurements of zirconium in chondrites have been subject to analytical error. This is important since SUESS and UREY (1956) used an average of 32 ppm Zr in their classic abundance compilation (see Table 40-C-1). Furthermore, since it is generally accepted that the Cc1 chondrites may best represent primitive meteoritic material, with the specimen Orgueil being particularly important in this regard, only the most recent abundances for this meteorite have been used in presenting the atomic abundances shown in Table 40-C-5.

The second approach involves the use of modern theories of nucleosynthesis to predict cosmic abundances. Although these predictions are not entirely independent, since they initially relied to some extent on meteorite abundances, it is noteworthy that the compilation by SEEGER *et al.* (1965) is in agreement with the subsequent meteoritic estimates.

Finally, abundance measurements in the solar atmosphere have also been made for zirconium, the most recent being those incorporated in the compilation presented by ROSS and ALLER (1976). It is encouraging that the atomic abundances derived by these three types of approach, as summarized in Table 40-C-5, are in fairly close agreement.

III. Tektites

All available data for zirconium in tektites, impact glasses and possible parent materials are given in Table 40-C-6. The large range existing within certain groups is almost certainly due to analytical error. For example, since the meteorite data of PINSON *et al.* (1953) and SETSER and EHMANN (1964) are incorrect (Table 40-C-1) this probably also applies to their tektite data. The data of COHEN (1959) are consistently high. TAYLOR (1966) also obtained consistently lower values, using a different technique, for the same five samples previously analysed by TAYLOR and SACHS (1964). Nevertheless, there appear to be real differences in zirconium content between tektite types, probably indicative of differences in parent material composition.

Since the Apollo missions it is now generally accepted that tektites originate by cometary or meteoritic impact on terrestrial sedimentary rocks. Thus australites have Zr contents similar to Henbury glass produced from Henbury sub-greywacke, and the Ivory Coast tektites have Zr contents similar to glass from the Bosumtwi crater.

Although tektites are glassy bodies, the mineral baddeleyite (ZrO_2) has been observed as an inclusion in the Martha's Vineyard tektite (CLARKE and WOSINSKI, 1967).

IV. Lunar Materials

a) Lunar Rocks

A summary of what we consider to be the most reliable data for materials returned from the lunar surface is presented in Table 40-C-7. The table is arranged in order of the various Apollo missions (see TAYLOR, 1975 for a description of the landing sites) and in terms of the simple classification basalt, breccia and fines, in keeping

Table 40-C-5. *Atomic abundance of zirconium* (Si = 10^6)

Reference	Source	Atomic abundance
EHMANN and CHYI (1974)	Orgueil	15
PALME (1974)	Orgueil	12
GANAPATHY *et al.* (1976)	Orgueil	9
SEEGER *et al.* (1965)	Theoretical	15.6
ROSS and ALLER (1976)	Solar	12.6

Table 40-C-6. *Zirconium abundances in tektites, impact glasses, and possible parent materials*

Sample type	Reference	No. of samples	Method	ppm Zr	
				Range	Average
Moldavites	von HEVESY and WÜRSTLIN (1934)	3	X	110–160	140
	PREUSS (1935)		S	75–400	150
	COHEN (1959)	2	S	550, 550	550
	VOROB'EV (1960)	10	S	180–410	260
	SETSER and EHMANN (1964)	1	N/R		289
Indochinites	PINSON *et al.* (1953)	2	S	200, 200	200
	VOROB'EV (1959)	6	S	40–270	91
	SETSER and EHMANN (1964)	1	N/R		92
Australites	COHEN (1959)				550
	TAYLOR and SACHS (1964)	43	S	295–540	410
	TAYLOR (1966)	5	Mass	204–390	274
	CHAO (1963)	6	S	250–330	283
	SETSER and EHMANN (1964)	1	N/R		403
Bediasites	COHEN (1959)	2	S	550, 550	550
	CHAO (1963)	21	S	160–290	214
	SETSER and EHMANN (1964)	1	N/R		130
Other types					
Billitonite	SETSER and EHMANN (1964)	1	N/R		262
Javaite	SETSER and EHMANN (1964)	1	N/R		93
Philippinite	SETSER and EHMANN (1964)	1	N/R		201
Thailand	SETSER and EHMANN (1964)	1	N/R		381
Ivory Coast	CUTTITA *et al.* (1972)	8	S	108–120	116
Related materials					
Darwin glass	TAYLOR and SOLOMON (1962)	8	S	220–490	390
Henbury glass	TAYLOR and KOLBE (1965)	4	S	415–480	440
Henbury subgreywacke	TAYLOR and KOLBE (1965)	3	S	330–630	430
Libyan desert glass	SETSER and EHMANN (1964)	1	N/R		119
Bosumtwi crater glass	CUTTITA *et al.* (1972)	3	S	110–130	120

Table 40-C-7. *Zirconium abundances in Lunar materials* (Number of analyses in brackets)

Sample type		Mission	ppm Zr	Method	Reference
Basalts	(19)	Apollo 11	194–646	N/R, S	1, 38, 41
	(16)	Apollo 12	81–160	N/R, S, X	4, 5, 6, 38
	(4)	Apollo 14	215–250	Mass, N/R	14, 26
	(21)	Apollo 15	57–156	I, Mass, N/R, X	7, 16, 18, 22, 25, 26, 36, 40, 41
	(27)	Apollo 17	71–437	X, N/R	28, 29, 33, 35, 41, 42
Breccias	(2)	Apollo 11	328–342	X	1
	(1)	Apollo 12	175	X	6
	(18)	Apollo 14	260–1,350	N/R, X	8, 9, 11, 14
	(11)	Apollo 15	11–480	I, Mass, N/R, X	12, 20, 24, 25
	(37)	Apollo 16[a]	0.1–922	I, Mass, N/R, X	13, 15, 21, 30, 38, 42, 43
	(32)	Apollo 17	85–874	I, Mass, N/R, X	19, 29, 30, 32, 34, 35, 37, 38, 42
Fines (soils)	(2)	Apollo 11	309–318	X	1, 10
	(12)	Apollo 12	465–762	X, N/R	6, 42
	(13)	Apollo 14	790–1,117	I, Mass, N/R, S, X	2, 3, 9, 26, 42
	(31)	Apollo 15	28–1,293	I, Mass, N/R, X	25, 26, 36, 40, 42
	(27)	Apollo 16	5–228	I, Mass, N/R, X	13, 15, 17, 23, 31
	(118)	Apollo 17	158–565	I, Mass, N/R, X	19, 27, 29, 31, 32, 33, 34, 35, 37, 38, 39, 42

[a] All Apollo 16 rocks have been classed as breccias for the purposes of this table.

References: 1. Compston *et al.* [P[1]]; 2. Schnetzler *et al.* (1971); 3. Taylor *et al.* (1971); 4. Taylor *et al.* [P[2]]; 5. Willis *et al.* [P[2]]; 6. Compston *et al.* [P[2]]; 7. Schnetzler *et al.* (1972); 8. Brunfelt *et al.* [P[3]]; 9. Taylor *et al.* [P[3]]; 10. Willis *et al.* [P[3]]; 11. Philpotts *et al.* [P[3]]; 12. Willis *et al.* [A[15]]; 13. Duncan *et al.* [L[4]]; 14. Ehmann *et al.* [L[4]]; 15. Taylor *et al.* (1973); 16. Rhodes *et al.* [L[4]]; 17. Taylor *et al.* [L[4]]; 18. Chappell and Green (1973); 19. Philpotts *et al.* (1973). 20. Taylor (1973). 21. Duncan *et al.* [P[4]]; 22. Rhodes and Hubbard [P[4]]; 23. Chyi and Ehmann [P[4]]; 24. Philpotts *et al.* [P[4]]; 25. Taylor *et al.* [P[4]]; 26. Ehmann *et al.* [A[15]]; 27. Duncan *et al.* [L[5]]; 28. Duncan *et al.* [L[5]]; 29. Philpotts *et al.* [L[5]]; 30. Taylor *et al.* [L[5]]; 31. Ehmann and Chyi [P[5]]; 32. Rhodes *et al.* [P[5]]; 33. Duncan *et al.* [P[5]]; 34. Philpotts *et al.* [P[5]]; 35. Rhodes *et al.* [L[5]]; 36. Duncan *et al.* [L[6]]; 37. Ehmann *et al.* [L[6]]; 38. Ehmann *et al.* [P[6]]; 39. Rhodes *et al.* [P[6]]; 40. Duncan *et al.* [P[6]]; 41. Duncan *et al.* [P[7]]; 42. Garg and Ehmann [P[7]]; 43. Warner *et al.* [P[7]].

References designated[P[1]] to [P[7]] are from the Proceedings of the First to the Seventh Lunar Science Conferences, published annually as Supplements 1 (1970) to 7 (1976) of Geochimica et Cosmochimica Acta. References designated [L[4]] to [L[6]] comprise the abstracts presented at the Fourth (1973), Fifth (1974) and Sixth (1975) Lunar Science Conferences, and reproduced and distributed by the Lunar Science Institute, 3303 NASA Road 1, Houston, Texas 77058. References designated [A[15]] are from a special issue of abstracts devoted to the Apollo 15 samples, and also reproduced by the Lunar Science Institute (1972). Other references not designated as indicated above may be found in the bibliography at the end of this chapter.

with other chapters in this handbook. Only ranges of concentration are given, because such a classification has no real genetic significance, and because in many cases the same samples have been analysed by different investigators. For example, the Apollo 16 rocks show a range in Zr concentration from 0.1–922 ppm. This large range

Table 40-C-8. *Average zirconium abundances in Lunar rocks* (Taylor, 1975)

Maria basalts	ppm Zr	*Highland rocks*	ppm Zr
Green glass, Apollo 15	22	Anorthosite	0.5
Olivine basalt, Apollo 12	110	Anorthositic gabbro	11
Olivine basalt, Apollo 15	94	Gabbroic anorthosite	48
Quartz basalt, Apollo 15	71	Troctolite	85
Quartz basalt, Apollo 12	120	Low-K Fra Mauro basalt	480
High-K basalt, Apollo 11	560	Medium-K Fra Mauro basalt	930
Low-K basalt, Apollo 11	360		
High-Ti basalt, Apollo 17	180		
Al-rich basalt, Apollo 12	200		
Al-rich basalt, Luna 16	295		

is due to the fact that these rocks include cataclastic anorthosites (low Zr), and a variety of breccias containing several types of glass and rock, most of which are probably not locally derived. Thus Table 40-C-7 should be treated simply as a reference source in order to obtain Zr data for any particular landing site.

A more meaningful interpretation is given in Table 40-C-8, taken directly from the useful book by Taylor (1975). The maria basalts have Zr contents comparable to terrestrial basalts, with the exception of the high-K Apollo 11 basalts. Although the highland rocks are brecciated and difficult to interpret, careful study has revealed the presence of serveral different rock types. The anorthositic clan has, predictably, low Zr concentrations, but the Fra Mauro basalts (the medium high K varieties are designated by the acronym Kreep by some workers) have very high Zr contents for *basaltic rocks*, and thus have special importance.

Inter-element relationships involving volatile and/or involatile elements (e. g. K-Zr, Zr-Nb) were noticed by Erlank *et al.* (1972), highlighting the importance of the distribution of Zr in lunar materials with regard to the origin of lunar lavas, the possibility of selective volatilization of alkalies from the moon, and the question of whether the moon accreted in an inhomogeneous manner. A recent review of these problems is given by Duncan *et al.* (1976).

Fractionation between Zr and Hf in lunar materials (see Sect. 72-C) has led to suggestions (Taylor *et al.* 1972; Ehmann *et al.* 1975) that this may be due to partial reduction of Zr^{4+} to Zr^{3+}, but this interesting speculation has not yet been confirmed.

b) Lunar Minerals

Zirconium, together with other so-called "incompatible" elements, is concentrated in the residual or last crystallising fraction (mesostasis) of lunar rocks, accounting for the close inter-element relationships referred to above. A variety of Zr-rich minerals occur in this fraction, including zircon and baddeleyite, but also a selection of Fe-Ti-Zr rich minerals (some of which have high concentrations of U, REE, Nb etc). At least one does not occur terrestrially (tranquillityite). Lack of clarity and definition is due to the small sizes of these phases in the mesostasis. Peckett *et al.*

(1972) have suggested that these phases be referred to as members of the "tranquillityite group"; they and BUSCHE *et al.* (1972) have presented analyses that indicate that at least some bear resemblance to the zirkelite formula (AB_2O_5). However, WARK *et al.* (1973) have presented additional data which confirm a zirconolite structure (AB_3O_7). They have further drawn attention to the confusion with terrestrial varieties and they suggest that the name zirkelite should no longer be used.

Revised manuscript received: June 1977

40-D. Abundance in Rock-Forming Minerals, Phase Equilibria, Zirconium Minerals

I. Rock-Forming Minerals

As shown in Sect. 40-A zirconium occurs in oxides and silicates as Zr^{4+}, with the Zr-O bond being mainly ionic in character. The geochemistry of zirconium is dominated by its lithophile nature and the presence (or absence) of the commonest zirconium mineral, zircon ($ZrSiO_4$).

Before discussing the distribution of Zr in rock-forming minerals, mention should be made of the effects of zircon contamination, either due to poor technique in the separating process by which mineral fractions are obtained, or because of the presence of minute zircon inclusions (these comments obviously do not apply to microprobe analyses). The effects of zircon impurities have been demonstrated by BROOKS (1969). He points out that the presence of one grain of zircon (with 50% Zr) amongst 10^4 equal sized grains of another mineral would raise the observed concentration by 50 ppm Zr. A further example is given in Table 40-D-1 for apatite analyses presented by CRUFT (1966); 34 samples contain <2 ppm Zr, 10 contain between 2 and 312 ppm. However, CRUFT considers that the high values obtained in the latter category are due to zircon impurities. The analyses presented in Table 40-D-1 have been selected, and should be studied, bearing these comments in mind. Further analyses for zirconium in minerals, including many rare varieties not mentioned in Table 40-D-1, may be found in VLASOV (1966) and DEGENHARDT (1957).

VLASOV (1966) mentions that zirconium is able to replace several other elements, including titanium, niobium, tantalum, the rare earths, iron, calcium etc., and discusses in some detail the substitution of zirconium in minerals containing these elements as major constituents. However, some of his assertions, for example (p. 284) the replacement of calcium by zirconium in the case of pyrochlore, sphene and perovskite, are open to question and require verification. Shown below are the most recent ionic radii (for 6-fold co-ordination and appropriate oxidation state) given by WHITTAKER and MUNTUS (1970) for zirconium and the elements it most likely replaces in the commoner rock forming minerals.

Element:	Zr^{4+}	Mg^{2+}	Fe^{2+}	Y^{3+}	Ti^{4+}	Nb^{5+}	Ta^{5+}
Ionic Radius (Å):	0.80	0.80	0.86	0.98	0.69	0.72	0.72

Previous sets of ionic radii, together with data for ionization potentials and electronegativities, may be found in Sect. 40-A and in Chapt. 12 of Vol. I of this handbook.

a) Oxide Minerals

Amongst the oxides, highest concentrations of zirconium are found in those minerals containing titanium, yttrium, niobium and tantalum (Table 40-D-1). Thus ilmenite, rutile and perovskite can contain up to 0.1% Zr, presumably reflecting

Table 40-D-1. *Zirconium abundances in common rock forming minerals*

Mineral	No. of samples	Rock type and locality	ppm Zr		Method	Reference
			Range	Mean		
Oxides						
Quartz	2	Various locations	10–11	11	C	1
Spinel	2	Peridotite, Lizard area, Cornwall (England)		45	S	2
Chromite	1	Yugoslavia		42	C	1
Magnetite	3	Various locations	34–73	59	C	1
Titanomagnetite	1	Gabbro, Skaergaard intrusion (East Greenland)		9	X	3
Titanomagnetite	2	Various locations	119–418	269	C	1
Ilmenite	5	Amphibolite, Adirondack Mountains, New York (U.S.A.)	300–390	322	S	4
Ilmenite	2	Gabbro, Skaergaard intrusion (East Greenland)	319–461	390	X	3
Ilmenite	10	Megacrysts, kimberlite, Southern Africa	546–1,190	954	X	5
Ilmenite	4	Various locations	155–550	280	C	1
Ilmenite	3	Rocks and breccia, Apollo 11 samples	0.09–0.18%	0.13%	M	6
Perovskite	8	Volcanic rocks, various locations	0.07–0.30%	0.15%	M	7
Columbite	2	Lueshe carbonatite (Congo)	0.51–0.52%	0.52%	X	8
Columbite	1	Annerod (Norway)		0.30%	C	1
Pyrochlore	4	Lueshe carbonatite (Congo)	0.08–0.38%	0.22%	X	8
Pyrochlore	5	Oka, Quebec (Canada)	0.44–3.04%	1.24%	X	9
Rutile	2	Various locations	0.17–0.18%	0.18%	C	1
Xenotime	2	Alkali granites, Nigeria	2.0 –3.8%	2.9%	X	10
Silicates						
Orthopyroxene (enstatite)	3	Peridotite, Lizard area, Cornwall (England)		16	S	2
Orthopyroxene	7	Mafic granulites, Central Australia	2–4	3	X	11
Diopside	6	Kimberlites, Lesotho	8–70	42	S	12
Clinopyroxene (diopside, augite)	6	Peridotite, Lizard area, Cornwall (England)	16–45	27	S	2
Clinopyroxene	3	Gabbro, Skaergaard intrusion (East Greenland)	12–109	66	X	3
Clinopyroxene	6	Mafic granulites, Central Australia	23–58	36	X	11
Clinopyroxene	8	Lavas, Kartala (Grande Comore)	35–227	95	X	13

Table 40-D-1 (continued)

Mineral	No. of samples	Rock type and locality	ppm Zr		Method	Reference
			Range	Mean		
Clinopyroxene	7	Glaucophane schist, New Caledonia and California	40–230	101	S	14
Clinopyroxene	38	Skarn, Grenville Province (Canada)	10–68	38	S	15
Clinopyroxene	6	Charnockite, Kondapalli (India)	28–56	41	W	16
Titanaugite, aegirine-augite	3	Trachyandesite, Trachyte, Gough Island	100–1,000	650	S	17
Aegirine, hedenbergite	11	Ilimaussaq alkaline intrusion (S. Greenland)	0.16–1.35%	0.55%	M	18
Aegirine	2	Lujavrite, Lovozero Massif (U.S.S.R.)	0.11–0.27%	0.19%	X	19
Aegirine	6	Kangerdlugssuaq alkaline intrusion (East Greenland)	0.11–0.48%		X	20
Omphacite	3	Kimberlites, Lesotho	25–100	60	S	12
Pyroxene	6	Gabbro, Skaergaard intrusion (East Greenland)	<10–50		S	21
Pyroxene	14	Bushveld igneous complex (South Africa)	25–120	44	S	22
Hornblende	9	Diorite, Salem granite suite (South West Africa)	35–70	45	X	23
Hornblende	3	Monzonite, Salem granite suite (South West Africa)	33–59	48	X	23
Hornblende	2	Adamellite, Salem granite suite (South West Africa)	142–148	145	X	23
Hornblende	8	Caledonian plutonic rocks, Scotland	30–200	98	S	24
Hornblende	20	Gneiss and schist, N.W. Ontario (Canada)	85–370	159	S	25
Hornblende	24	Amphibolite, Adirondack Mountains, New York (U.S.A.)	34–150	82	S	26
Hornblende	6	Mafic granulites, Central Australia	20–52	35	X	11
Actinolite	2	Blueschist facies, California (U.S.A.)	60–80	70	S	27
Crossite-riebeckite	4	Blueschist facies, California (U.S.A.)	48–120	85	S	27
Riebeckitic-arfvedsonite	2	Ekerite, Oslo region (Norway)	0.11–0.19%	0.15%	S	28
Riebeckitic-arfvedsonite	1	Nordmarkite, Oslo region (Norway)		0.43%	S	28
Riebeckitic-arfvedsonite	14	Younger granites, Nigeria	0.10–0.64%	0.24%	C	29
Arfvedsonite	4	Kangerdlugssuaq alkaline intrusion (East Greenland)	248–2,470		X	20
Astrophyllite	1	Kangerdlugssuaq alkaline intrusion (East Greenland)		1.13%	X	20
Astrophyllite	2	Ekerite, Oslo region (Norway)	0.40–0.51%	0.46%	S	28
Amphibole	5	Skarn, Grenville Province (Canada)	43–120	66	S	15
Kaersutite	10	Basic alkaline rocks, Various locations	45–179	102	X	30
Glaucophane	9	Blueschist facies, California (U.S.A.)	60–150	87	S	27

Table 40-D-1 (continued)

Mineral	No. of samples	Rock type and locality	ppm Zr		Method	Reference
			Range	Mean		
Biotite	9	Diorite, Salem granite suite (South West Africa)	3–7	4	X	23
Biotite	4	Monzonite, Salem granite suite (South West Africa)	2–45	17	X	23
Biotite	6	Adamellite, Salem granite suite (South West Africa)	25–159	63	X	23
Biotite	16	Granite, granodiorite, tonalite, Portugal	14–125	46	S	31
Biotite	2	Kangerdlugssuaq alkaline intrusion (East Greenland)	100–115	108	X	20
Biotite	7	Meta-arkoses, French Valley Formn. (S. California)	74–210	132	S	32
Biotite	25	Paragneiss, Adirondack Mountains, New York (U.S.A.)	30–300	163	S	33
Biotite	8	Gneiss, Ontario (Canada)	75–150	103	S	34
Biotite	20	Gneiss and schist, N.W. Ontario (Canada)	125–490	218	S	25
Muscovite	2	Various locations	18–52	35	C	1
Chlorite	9	Pelitic schists, Central Vermont (U.S.A.)	23–73	43	S	35
Chloritoid	5	Pelitic schists, Central Vermont (U.S.A.)	32–150	79	S	35
Plagioclase	7	Mafic granulites, Central Australia		<5	X	11
Potassium feldspar	58	Cape granites, South Africa	<4–21		S	36
Leucite	1	Vesuvius (Italy)		11	C	1
Garnet	1	Eclogite, Robert's Victor Mine (South Africa)		64	S	37
Garnet	16	Kimberlites, Lesotho	3–130	55	S	38
Garnet	7	Glaucophane schist, New Caledonia and California	50–160	93	S	14
Garnet	8	Pelitic schists, Central Vermont (U.S.A.)	110–340	178	S	35
Garnet	10	Paragneiss, Adirondack Mountains, New York (U.S.A.)	130–250	172	S	33
Garnet	5	Gneiss, Ontario (Canada)	250–450	340	S	34
Garnet	13	Eclogite, amphibolite, Germany	41–158	75	S	37
Tourmaline (schorlite)	5	Granites, Northern Portugal	64–220	144	S	39
Tourmaline	40	Granitic rocks, South-West England	60–172	103	S	40
Tourmaline	6	Hydrothermal, South-West England	61–295	159	S	40
Kyanite	12	Various locations	tr–92		S	41
Andalusite	7	Various locations	tr–280		S	41

Table 40-D-1 (continued)

Mineral	No. of samples	Rock type and locality	ppm Zr		Method	Reference
			Range	Mean		
Sillimanite	3	Various locations	tr–180		S	41
Scapolite	40	Skarn, Grenville Province (Canada)	0–130	13	S	15
Sphene	7	Volcanic rocks, various locations	0.30–2.67%	1.26%	M	7
Sphene	7	Caledonian plutonic rocks, Scotland	100–1,500	790	S	24
Sphene	6	Granodiorite, Glen Fyne (England)	0.65–0.79%	0.74%	S	42
Sphene	6	Adamellite, Shap (England)	0.10–1.30%	0.50%	S	42
Thorite	1	Alkali Granites, Nigeria		1.1%	X	10
Apatite	10	Igneous and metamorphic rocks, various locations	2–312	74	S	43
Apatite	34	Igneous and metamorphic rocks, various locations		<2	S	43

tr = trace

References: 1. Degenhardt (1957); 2. Green (1964); 3. Brooks (1969); 4. Floyd (1972); 5. Mitchell *et al.* (1973); 6. Lovering and Ware (1970); 7. Smith (1970); 8. van Wambeke (1965); 9. Perrault (1968); 10. Heinrich (1963); 11. Woodford and Wilson (1976); 12. Carmichael and McDonald (1961); 13. Strong (1972); 14. Coleman *et al.* (1965); 15. Shaw *et al.* (1963); 16. Leelanandam (1967); 17. Le Maitre (1962); 18. Larsen (1976); 19. Gerasimovskii *et al.* (1962); 20. Brooks (1970); 21. Wager and Mitchell (1951); 22. Atkins (1969); 23. Miller (1973); 24. Nockolds and Mitchell (1948); 25. Moxham (1965); 26. Engel and Engel (1962a); 27. Coleman and Papike (1968); 28. Dietrich *et al.* (1965); 29. Borley (1963); 30. Kesson and Price (1972); 31. Albuquerque (1973); 32. Schwarcz (1966); 33. Engel and Engel (1960); 34. Wynne-Edwards and Hay (1963); 35. Albee *et al.* (1965); 36. Kolbe (1966); 37. Hahn-Weinheimer and Luecke (1963); 38. Nixon *et al.* (1963); 39. Neiva (1974); 40. Power (1968); 41. Pearson and Shaw (1960); 42. Farrand (1960); 43. Cruft (1966).

replacement of Ti^{4+} by Zr^{4+}. Concentrations of 0.5% Zr in columbite and up to 3% Zr in pyrochlore are probably related to a coupled replacement of Nb^{5+} by Zr^{4+} and other ions of lower charge. High concentrations (2–4% Zr) in xenotime are due to the fact that $ZrSiO_4$ and YPO_4 are isostructural, and reflect the coupled substitution $Y^{3+} + P^{5+} \rightleftharpoons Zr^{4+} + Si^{4+}$. However, apart from ilmenite, the other minerals mentioned above are usually present in accessory amounts in most rock types, and only assume importance in rare types such as carbonatites and in alkaline complexes.

The more common opaque minerals such as magnetite and spinels do not contain appreciable zirconium, as may be seen from Table 40-D-1.

b) Silicate Minerals

Leucocratic minerals such as the feldspar and feldspathoid groups contain, as expected, only trace amounts of zirconium and need not be further considered. The same comments apply to olivines (no olivine data are given in Table 40-D-1, but Zr data for meteoritic olivines may be found in Table 40-C-4) and orthopyroxenes.

Varying, but significant, amounts of zirconium are found in the more common varieties of clinopyroxenes, amphiboles, micas and garnets. Concentrations of around 100 ppm Zr are frequently encountered (Table 40-D-1). Calculation of averages has not been attempted because part of the variability reported may be due to zircon impurities and inclusions, especially for micas. However, the most striking feature about the distribution of zirconium in ferro-magnesian minerals concerns the enrichment of zirconium in alkali pyroxenes and amphiboles, where concentrations of 0.1 to 0.5% are commonly observed. Extreme enrichment (1.35% Zr by microprobe analysis) has been noted by LARSEN (1976) for the pyroxene aegirine and by BROOKS (1970) for the amphibole astrophyllite (1.13% Zr by X-ray fluorescence) and it is noteworthy that both minerals occur in alkaline complexes in Greenland in which zircon is rare, or a late crystallizing phase.

The degree to which zirconium substitutes in ferro-magnesian minerals will depend mainly on (i) whether or not zircon crystallizes early and (ii) the presence of suitable ions with low valency in order to achieve coupled substitution. With regard to (i) BROOKS (1969) has noted the enrichment of zirconium in early formed clinopyroxenes from the Skaergaard intrusion (i. e. before the onset of zircon crystallization); furthermore it is well known that zircon crystallization is delayed or suppressed in alkaline and agpaitic magmas. The possible presence of complexes (e. g. ZrO_4^{4-}) in magmas and their non-acceptance into silicate minerals, leading to enrichment in residual magmas (RINGWOOD, 1955) should also be considered. Further discussion of these aspects is given in Section 40-E. In the case of (ii), DEGENHARDT (1957), on the basis of increased Zr concentrations in Fe-rich pyroxenes, proposed the following type of substitution, $Zr^{4+} \rightleftharpoons Na^{+} + Fe^{3+}$.

More recently, BROOKS (1969) has suggested that the substitution is of the type $3M^{2+} \rightleftharpoons 2Na^{+} + Zr^{4+}$ (where $M = Ca^{2+}$, Fe^{2+} or Mg^{2+}). Whatever the correct situation, it seems clear that the presence of Na greatly enhances zirconium substitution in pyroxenes and amphiboles, and this is also important because the presence of this element in artificial magmas suppresses zircon crystallization (DIETRICH, 1968).

Of the other minerals listed in Table 40-D-1 and not previously discussed, the concentration of zirconium in the aluminium silicates kyanite, andalusite, sillimanite (up to 100 ppm Zr) is not understood. Mention should be made, however, of the high concentrations found in sphene, with some examples containing from 1 to 2% Zr; presumably this reflects a direct substitution of Zr^{4+} for Ti^{4+}. As mentioned earlier, apatite is considered to contain only trace amounts of zirconium (CRUFT, 1966).

II. Phase Equilibria

Since zircon is the most abundant zirconium rock-forming mineral, it is appropriate to discuss its stability, in terms of ZrO_2—SiO_2 phase relationships, in some detail. The most recent experimental data are those of BUTTERMAN and FOSTER (1967). Their conclusions, which incorporate the data of earlier investigators, are summarized in their phase diagram presented as Fig. 40-D-1. The upper limit of stability of zircon is given as 1,676 ± 7° C, which is only about ten degrees lower than the zirconia—silica eutectic which occurs at 1,687 ± 4° C (i. e. the minimum temperature of liquid formation in this system). Other features of interest are the two-liquid region (liquid immiscibility) between 41–62 wt.% SiO_2, with a lower limit of 2,250° C, and the polymorphic inversions in the ZrO_2 and SiO_2 end members, namely the well known inversions in SiO_2 at 573°, 867° and 1,470°, the inversion of monoclinic to tetragonal ZrO_2 at 1,170°, and of tetragonal to cubic ZrO_2 at 2285°. Finally, there do not appear to be any solid solution effects at the ZrO_2—end of the system.

Hydrothermal syntheses of zircon have been carried out by FRONDEL and COLLETTE (1957), MUMPTON and ROY (1961) and CARUBA *et al.* (1975). These authors show that the crystallization of zircon (both α and β phases have been obtained) requires an acidic environment, and that uranium and hafnium, but not thorium, can substitute for zirconium in zircon. The interesting work carried out by

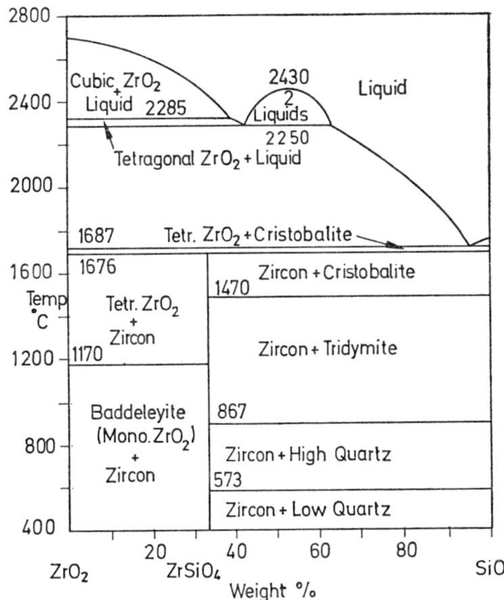

Fig. 40-D-1. Proposed phase diagram for the ZrO_2-SiO_2 system (BUTTERMAN and FOSTER, 1967)

DIETRICH (1968) on synthetic quartz-albite-orthoclase rich gels spiked with ZrO_2 shows that the addition of small amounts of $Na_2Si_2O_5$ or NaF to the mixtures suppressed, in part or completely, the crystallization of zircon in short-term runs.

Synthetic zirconium and titanium garnets have been made by ITO and FRONDEL (1967) who have demonstrated nearly complete solid solubility at 1,050° C in the system $Ca_3Fe_2Si_3O_{12}$ (andradite)—$Ca_3Zr_2Fe_2SiO_{12}$ (kimzeyite)—$Ca_3Ti_2Fe_2SiO_{12}$ (schorlomite).

Further data relating to the phase equilibria between zirconium and other elements (or oxides) may be obtained in the useful compilation provided by LEVIN *et al.* (1964); specifically Zr-O (Fig. 25), ZrO_2—Nb_2O_5 and ZrO_2—Ta_2O_5 (Figs. 373 and 374) and TiO_2—ZrO_2 (Figs. 369 and 370).

III. Zirconium Minerals

The compilation of zirconium minerals, given in Table 40-D-2, shows that most occur as oxides and silicates. Zirconium oxides are usually found in nepheline syenites and in carbonatites of alkali-rich ultramafic massifs, where they are formed by the reaction of the ultramafic rock with carbonate solutions in a silica deficient medium (VLASOV, 1966). Amongst the silicates, the commonest mineral zircon is found in nearly all rock clans. Data for 40 minor elements in zircons have been compiled by GÖRZ (1974) who notes that some of these may occur in minerals included within the zircons. With respect to Nb, Hf, Th, U, Ti, Y, REE and Al, he indicates that it is possible to recognize characteristic ranges of concentration for zircons of different origin. The other zirconium silicates, as previously discussed, occur mainly in alkali-rich and agpaitic rocks and are not normally found in rocks which contain zircon. The most famous example is perhaps the Lovozero alkali massif in the U. S. S. R. which contains a wide variety of zirconium silicates, such as eudialyte, catapleite, elpidite, zirfesite etc. (VLASOV *et al.* 1966).

From a geological viewpoint, the mineral zircon has received particular attention, especially with regard to its use for Th-U-Pb radiometric dating (see Sect. 82-B) and as indicator (utilizing crystal shape and morphology) of a magmatic or metamorphic origin for granitic rocks (POLDERVAART, 1950).

Table 40-D-2. *Zirconium minerals* (for sources see footnote)

Oxides	
Baddeleyite	ZrO_2
Belyankinite	$Ca(Ti, Zr, Nb)_6O_{13} \cdot 14H_2O$
Calzirtite	$Ca(Zr, Ca)_2Zr_4(Ti, Nb, Fe)_2O_{16}$
Oliveirite	$Zr_3Ti_2O_{10} \cdot 2H_2O$
Pseudo-armalcolite	$(Ti, Zr, Fe, Cr, Mg, Al, Ca, Si)_3O_5$
Tazheranite	$(Zr, Ca, Ti^{3+}, Ti^{4+}, Al, Fe)_4O_6$
Uhligite	$Ca_3(Ti, Al, Zr)_9O_{20}$
Zirconolite	$CaZrTi_2O_7$
Zirkelite	$(Zr, Ca, Ti, Fe, Mg, REE, U, Th)_2O_5$
Carbonates and sulfates	
Weloganite	$Sr_5Zr_2[CO_3]_9 \cdot 4H_2O$
Zircosulfate	$Zr[SO_4]_2 \cdot 4H_2O$

Table 40-D-2 (continued)

Silicates	
Armstrongite	$CaZrSi_6O_{15} \cdot (2.5H_2O)$
Calcium catapleite	$CaZrSi_3O_9 \cdot 2H_2O$
Catapleite	$Na_2ZrSi_3O_9 \cdot 2H_2O$
Dalyite	$K_2ZrSi_6O_{15}$
Elpidite	$Na_2ZrSi_6O_{15} \cdot 3H_2O$
Eudialyte	$(Na, Ca)_5(Zr, Fe, Mn)[Si_6O_{17}](O, OH, Cl)$
Giannetite	Na, Ca, Mn, Ti, Zr silicate
Hiortdahlite	$NaCa_2ZrO[Si_2O_7]F$
Keldyshite	$(Na, H)_2Zr[Si_2O_7]$
Kimzeyite	$Ca_3(Zr, Ti)_2(Al, Si)_3O_{12}$
Lavenite	$(Na, Ca, Mn)_3(Zr, Ti, Fe)O[Si_2O_7]F$
Lemoynite	$(Na, Ca)_3Zr_2Si_8O_{22}$
Lorenzenite	$Na_2(Ti, Zr)_2O_3[Si_2O_6]$
Lovozerite	$(Na, Ca)_3(Zr,Ti)Si_6(O, OH)_{18}$
Pennaite	Na, Ca, Ti, Fe, Mn, Zr silicate
Rosenbuschite	$(Na, Ca)_3(Zr, Fe, Ti)O[Si_2O_7]F$
Seidozerite	$Na_2(Zr, Ti, Mn)_2Si_2O_8F$
Sogdianovite	$(K, Na)_2Li_2(Li, Fe, Al, Ti)_{1.8}(Zr, Ti)[Si_2O_5]_6$
Tranquillityite	$Fe_8(Zr, Y)_2Ti_3Si_3O_{24}$
Vlasovite	$Na_2Zr[Si_4O_{11}]$
Wadeite	$K_2CaZrSi_4O_{12}$
Wöhlerite	$NaCa_2(Zr, Nb)O[Si_2O_7]F$
Zektzerite	$LiNa(Zr, Ti, Hf)Si_6O_{15}$
Zircon	$ZrSiO_4$
Zircophyllite	$(K, Na, Mn)_2(Mn, Fe)_7(Zr, Nb, Ti)_2(Si,Ti)_8$ $\cdot (O(OH)F)_{31}(0.9H_2O)$
Zirconium schorlomite	$Ca_3(Fe, Zr)_2(Si, Ti)_3O_{12}$
Zirfesite	Hydrous silicate of Zr and Fe^{3+}

Main references: 1. PALACHE *et al.* (1958); 2. VLASOV (1966); 3. FLEISCHER (1966); 4. AMERICAN MINERALOGIST **51**, 529 (1966); **54**, 576, 1221 (1969); **55**, 318 (1970); **57**, 1913 (1972); **58**, 140, 966, 967 (1973); **59**, 208, 633 (1974); **62**, 416 (1977).

Revised manuscript received: June 1977

40-E. Abundance in Common Igneous Rock Types; Crustal Abundance

Because zirconium is one of the more abundant trace elements in igneous rocks (DEGENHARDT, 1957, estimated the average igneous rock to contain 155 ppm Zr), a large amount of zirconium analyses for igneous rocks exist in the literature.

Some previous reviews of the geochemistry of zirconium include the following: RANKAMA and SAHAMA (1950), GOLDSCHMIDT (1954), DEGENHARDT (1957) and VLASOV (1966). A wealth of data for zirconium in igneous rocks may be found in NOCKOLDS and MITCHELL (1948), WAGER and MITCHELL (1951), NOCKOLDS and ALLEN (1953, 1954, 1956) and CHAO and FLEISCHER (1960); in addition many analyses for zirconium in mafic rocks from Czechoslovakia may be found in CAMBEL and KAMENICKÝ (1969). POLDERVAART (1956) has treated the occurence of zircon in igneous rocks, and ABSHIRE (1958) has provided a useful bibliography of zirconium references.

In general, the data contained in the references mentioned above have not been used in the tables that follow, except where no other data are available, because they have been mainly obtained by less accurate and reliable optical spectrographic and colorimetric analytical methods. Similarly many other workers have used these techniques and, where no indication of accuracy or precision have been given, have not been quoted. The most commonly and routinely reliable method currently in use is X-ray fluorescence, for which the detection limit is at the ppm level. The potentially more sensitive neutron activation, isotope dilution–mass spectrometric and spark source–mass spectrographic methods have also been applied to zirconium analysis in recent years and are useful for specific materials such as mineral separates, meteorites and lunar rocks, for which only small quantities of material may be available, because of the small amounts of material required for analysis.

An attempt has been made in this compilation to include more recent and reliable data from diverse localities and rock types. Nevertheless, it is important to note that many of the important features of the distribution of zirconium, such as the general increase in concentration in the sequence ultramafic—mafic—intermediate—acid rocks, the striking concentration of zirconium in agpaitic and peralkaline rocks, and the correlation of zirconium content with parental magma type and tectonic framework of the igneous rocks concerned, were recognized by these earlier workers. These references also include many diagrams showing the variation of zirconium with rock type, differentiation etc. and for this reason none are presented in this section. Histograms have also not been favoured; instead the raw data are presented in the tables that follow. We have also not attempted to calculate averages for the igneous rocks discussed in view of the large range in concentration encountered for specific rock types, but reference will be made to the averages estimated by DEGENHARDT (1957), VINOGRADOV (1962) and other workers.

I. Standard Reference Rocks

It has become common pratice to use standard reference rocks for calibrating or checking the accuracy of an analytical method. The first reference rocks to be used internationally were G-1, a granite, and W-1, a "diabase" or dolerite. Compilations of data for these have been issued at periodic intervals, the last complete compilation being that of FLEISCHER (1969), who also provides references for the previous compilations. All zirconium data listed in FLEISCHER (1969) and the earlier compilations are shown in Fig. 40-E-1 with the different analytical techniques indicated accordingly. The wide spread in values obtained by optical spectrographic techniques is apparent and the histogram illustrates well the comments made with regard to analytical techniques in the introduction to this section and elsewhere.

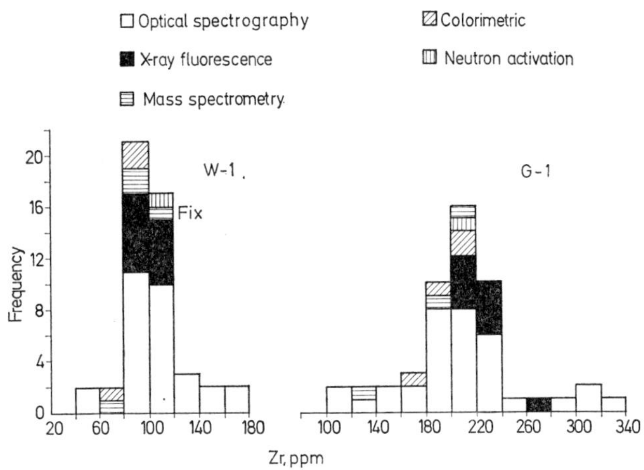

Fig. 40-E-1. Histogram of zirconium data for reference rocks W-1 and G-1

Table 40-E-1. *Zirconium values for standard reference rocks* (Data from FLANAGAN, 1973)

Recommended values		Magnitudes	
Sample	ppm Zr	Sample	ppm Zr
Basalt BR	240	Dunite DTS-1	3
Basalt BM	105	Dunite NIM-D	50
"Diabase" W-1	105	Peridotite PCC-1	7
Andesite AGV-1	225	Pyroxenite NIM-P	20
Granodiorite GSP-1	500	Basalt BCR-1	190
Granite G-1	210	Basalt JB-1	300
Granite G-2	300	Norite NIM-N	25
Granite GA	140	Granodiorite JG-1	160
Granite GR	180	Granite NIM-G	300
Granite GM	145	Syenite NIM-S	30
Granite GH	160	Syenite SY-1	3,030
Slate TB	175	Sulphide SU-1	110[a]
Limestone KH	30		

[a] Average value.

Table 40-E-1 lists the standards which are more generally used, together with recommended values, magnitudes and averages proposed by FLANAGAN (1973) for zirconium. This reference, together with that of ABBEY (1975), may be consulted for further details regarding the source and locality of these reference standards; the latter reference also contains "usable" or preferred zirconium values for some of these materials.

II. Ultramafic and Ultrabasic Rocks

Since trace element concentrations are partly dependant on mineralogy, the zirconium contents of ultramafic and ultrabasic rocks, as shown in Table 40-E-2, are expected to be low from the mineral data given in Table 40-D-1, especially for those types principally composed of olivine, orthopyroxene and plagioclase. Thus common peridotite nodules from kimberlite, composed principally of olivine and orthopyroxene (~90%), with lesser amounts of diopside and garnet, typically contain from 1–20 ppm Zr. The lower concentrations within this range are similar to those found in chondrites. Higher concentrations are encountered in pyroxenitic varieties and in pyroxenites from other localities and compare more favorably with the ultramafic rock averages of 45 ppm and 30 ppm Zr given by DEGENHARDT (1957) and VINOGRADOV (1962) respectively. Amphibole rich varieties such as those from St. Paul's rocks (MELSON *et al.* 1972) and kimberlite nodules with potassic richterite (ERLANK, 1973) contain still higher zirconium contents. Extreme examples of this are shown by metasomatised amphibole and mica rich nodules from Uganda and the Eifel region (LLOYD and BAILEY, 1975) who report from 58–738 ppm Zr in such materials.

VLASOV (1966) reports from 270–410 ppm Zr in the ultrabasic alkali rock massifs from the Kola Peninsula, with the bulk of the zirconium apparently occurring in soda-rich pyroxenes and amphiboles, but it is not clear to what extent metasomatic processes have been operative during the formation of these rocks.

The unusual Archaean ultramafic lavas (peridotitic komatiites) have been included in Table 40-E-2, even though they have been metamorphosed to greenstones, because zirconium is apparently immobile during this process (CANN, 1970; SMITH and ERLANK, 1975) and are interesting in that their zirconium contents would indicate a low zirconium content for their upper mantle source regions.

III. Mafic Rocks

A large amount of recent zirconium data occur in the literature for mafic rocks and their differentiates. Thus Table 40-E-3 is somewhat lengthy; note, however, that spilites and metabasalts (greenstones) are also included, since zirconium is essentially immobile during the processes that produced these rocks (CANN, 1970; HART *et al.*, 1974; SMITH and ERLANK, 1975).

The abundances given in Table 40-E-3 may be compared with the averages given by previous authors for basaltic rocks, viz. DEGENHARDT (1957) 110 ppm; TUREKIAN and WEDEPOHL (1961) 140 ppm; VINOGRADOV (1962) 100 ppm; PRINZ (1967) 116 ppm. CHAO and FLEISCHER (1960) noted that the large range in concentration encountered in mafic rocks was related to geographic variations and were the first to emphasize the importance of tectonic framework; their data showed that island arc basalts contain less zirconium (10–60 ppm) than other oceanic basalts (120–300 ppm), and that although zirconium increases with differentiation in all basaltic series, the rate of increase is more rapid in the alkali basalt than in the tholeiitic or calc-alkali basalt sequences. These conclusions are confirmed in this paper; the distribution of zirconium in mafic rocks is dominated by source rock (mantle) composition and the preference for zirconium to concentrate in the liquid phase during melting and crystallisation processes, explains the use of the terms "incompatible" and "residual" element by many authors.

The following *generalisations* are suggested by the data in Table 40-E-3 and the literature.

i) The dominant rock type in the oceanic crust is tholeiitic (variously referred to by such terms as oceanic tholeiite, abyssal tholeiite, MORB) with a mean concentration around 100 ppm Zr. Regional variations in zirconium concentration have not been noted (Kable, 1972), and the zirconium concentration of these rocks shows no obvious correlation with so-called depleted or undepleted mantle or mantle plumes, although inter-element ratios such as Zr/Nb do (Erlank and Kable, 1976).

ii) In oceanic basalts, zirconium increases in the sequence calc-alkali basalt, or low K-tholeiite, (10–100 ppm), oceanic tholeiite or MORB (about 100 ppm), island tholeiites (150–200 ppm), alkali basalts (200–300 ppm). Differentiates of the latter (hawaiites, trachybasalts) contain even higher concentrations. The higher concentrations in alkali basalts as compared with oceanic tholeiites is usually ascribed to the former being produced by smaller degrees of partial melting but the generally lower concentrations of zirconium in island-arc basalts versus mid-ocean ridge tholeiites has not been explained. Presumably this is related to source area compositions. These characteristics, together with variations in composition for the elements yttrium, titanium, and niobium, have been used by Pearce and Cann (1971 and 1973) for inferring the tectonic setting of volcanic rocks, as first indicated by Chao and Fleischer (1960).

iii) Nepheline basalts (usually thought to be produced by even smaller degrees of partial melting than alkali basalts) such as those from the Hawaiian and Grand Comore islands, have relatively low concentrations (100—250 ppm Zr); perhaps zirconium does not behave as an incompatible element in these circumstances, and is retained in a residual mantle phase (ilmenite?).

iv) Continental varieties of the volcanic mafic rocks discussed above do not, in general, within the limits of concentrations encountered, show any appreciable or systematic differences with their oceanic counterparts, unlike other elements such as rubidium and uranium (see appropriate chapters in this handbook). Two exceptions are (α) the high K—high Mg tholeiitic Karroo basalts from the Nuanetsi Province, Rhodesia (Cox *et al.* 1967; Vail *et al.*, 1969) which have an average of 370 ppm Zr and (β) the Archaean tholeiites from Australia (Hallberg, 1972) and South Africa (Smith and Erlank, 1975) which have mean concentrations around 60 ppm Zr. Nevertheless, there is no real evidence for a secular change in mantle (or basaltic) composition with geologic age for zirconium. Evidence for contamination by the continental crust is also equivocal (Gottfried *et al.*, 1968).

Table 40-E-2. *Zirconium abundances in ultramafic and ultrabasic rocks*

Rock type	No. of samples	Locality	ppm Zr		Method	Reference
			Range	Mean		
Peridotite mylonite	5	St. Paul's Rocks (Atlantic)	<5–34		S	1
Brown hornblende mylonite	3	St. Paul's Rocks (Atlantic)	80–90	84	S	1
Peridotite	2	Reunion	7–15	11	X	2
Peridotite	1	Garabal Hill (Scotland)		26	X	3
Garnet peridotite nodules (granular)		Kimberlite, South Africa	6–21		X	4
Garnet peridotite nodules with K-richterite	3	Kimberlite, South Africa	42–169	98	X	4
Garnet peridotite nodules (sheared)		Kimberlite, South Africa	2–10		X	4
Garnet peridotite and pyroxenite nodules	21	Matsoku kimberlite (Lesotho)	1–46	17	X	5
Peridotite nodules	15	Lashaine (Tanzania)	2.1–8.5	4.2	X	6
Pyroxenite	14	Bushveld igneous complex (South Africa)	30–55	42	S	7
Harzburgite	5	Losberg intrusion (South Africa)	21–37	28	X	8
Hortonolite dunite	5	Bushveld igneous complex (South Africa)	70–120	95	S	7
Ultramafic rocks	6	Connemara (Ireland)	18–80	40	S	9
Ultramafic rocks	4	Tiree (Scotland)		33	X	10
Pyroxenite	2	Archaean, Western Australia	15–17	16	X	11
Peridotite	7	Archaean, Western Australia	14–26	17	X	11
Peridotitic komatiite (greenstone)	32	Archaean, Barberton (South Africa)	7–38	19	X	12
Anorthosite	4	Bushveld igneous complex (South Africa)	12–25	17	S	7

References: 1. Melson *et al.* (1972); 2. Upton and Wadsworth (1972); 3. Brooks (1970); 4. Erlank (1973); 5. Gurney (1975); 6. Rhodes and Dawson (1975); 7. Liebenberg (1960); 8. Danchin and Ferguson (1970); 9. Evans (1964); 10. Drury (1973); 11. Williams and Hallberg (1973); 12. Smith and Erlank (1975).

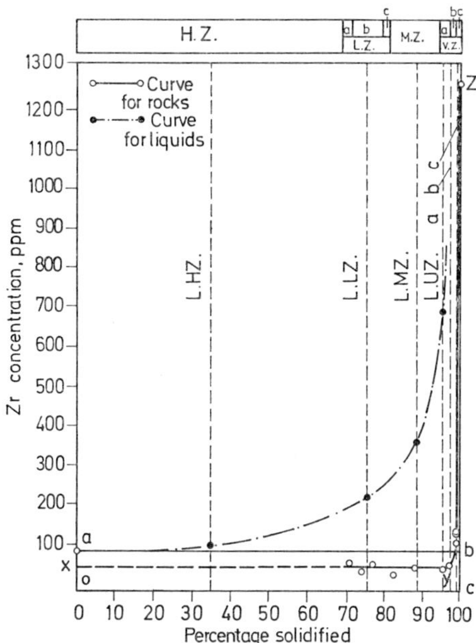

Fig. 40-E-2. Distribution of zirconium in the Skaergaard intrusion as a function of the percentage of magma solidified (lower curve shows the concentrations in the rocks; upper curve, estimated compositions of the liquids; while the horizontal line, ab, is the concentration in the chilled margin). From BROOKS (1969)

v) Plutonic rocks, as may be expected from previous discussion, have generally lower zirconium concentrations; the concentration will depend critically on the mode of origin and whether or not the rocks are cumulates and, in the latter case, as to the type of cumulate (i. e. amount of trapped or intercumulus liquid). The type example is the Skaergaard Complex, and the comments made here and previously are well illustrated in Fig. 40-E-2.

Additional reading relating to the importance of zirconium in mafic rocks may be found in SCEAL and WEAVER (1971); PEARCE and CANN (1973); BARBERI *et al.* (1975) and ERLANK and KABLE (1976).

IV. Kimberlites, Carbonatites and Alkali-Rich Lavas

The volumetrically unimportant, but geologically significant, rock types included under this heading have rather variable zirconium abundance distributions. There are genetic associations between some types (e. g. kimberlites and carbonatites). However, representatives of rift systems and island arcs indicate that the tabulation is not genetic, but based on uniqueness and generally undersaturated and highly alkaline character. Furthermore, the derivation of these rocks is not well understood, and mechanisms involving small degrees of partial melting, high-pressure eclogite fractionation, zone refining (or wall-rock reaction), magma mixing with sialic material, and liquid immiscibility, have been variously considered in the literature.

Table 40-E-3. *Zirconium abundances in mafic rocks* (including spilites and metabasalts)

Rock type	No. of samples	Locality	ppm Zr		Method	Reference
			Range	Mean		
Oceanic volcanic rocks						
Tholeiitic basalt	24	Reykjanes Ridge (Atlantic Ocean)		96	X	1
Tholeiitic basalt	18	Mid-Atlantic Ridge, 45° N	65–105	87	X	2, 3
Alkali basalt	3	Mid-Atlantic Ridge, 45° N	96–137	112	X	3
Tholeiitic basalt	30	Mid-Atlantic Ridge	62–160	119	X	3, 5, 60
Tholeiitic basalt	8	Palmer Ridge (Atlantic Ocean)	52–100	70	X	2
Tholeiitic basalt	3	Vema Fracture Zone (Mid-Atlantic Ridge)	90–160	123	S	4
Tholeiitic basalt	4	Juan de Fuca Ridge (Pacific Ocean)	121–143	133	X	3
Alkali basalt	10	East Pacific Rise	200–420	333	S	5
Tholeiitic basalt	3	East Pacific Rise	50– 82	70	X	3
Tholeiitic basalt	5	Pacific Ocean	44–150	98	S	5
Tholeiitic basalt	5	Blanco Fracture Zone (Pacific Ocean)	130–133	131	X	3
Tholeiitic basalt	3	Marianas Ocean Basin		101	X	6
Tholeiitic basalt	7	Carlsberg Ridge (Indian Ocean)	90–140	117	X	2
Spilite (Metatholeiite)	2	Carlsberg Ridge (Indian Ocean)	134–147	141	X	7
Tholeiitic basalt	4	Gulf of Aden	41–100	66	X	2
Basalts (various)	10	Iceland	102–277	180	X	3
Tholeiitic basalt	79	Hawaiian Islands		166	X, S	3, 11, 14
Alkali basalt	7	Hawaiian Islands		223	X, S	3, 14
Nepheline basalt	12	Hawaiian Islands		133	X, S	3, 14
Hawaiite	3	Hawaiian Islands	477–788	594	X	3
Alkali basalt	8	Azores Group of Islands	151–313	209	X	3
Hawaiite	4	Ascension Island	231–320	287	X	3
Trachybasalt	6	Bouvet Island	183–301	260	X	3
Picrite basalt	2	Gough Island	159–163	161	X	3
Alkali basalt	2	Gough Island	285–322	304	X	3
Trachybasalt	7	Gough Island	346–409	380	X	3

Trachybasalt	4	Tristan da Cunha	302–384	358	X	3
Basalt	7	Madeira	219–292	245	X	8
Hawaiite	11	Madeira	251–401	324	X	8
Alkali basalt	6	Fogo Island (Cape Verde)	295–403	354	X	3
Hawaiite	3	Easter Island	224–417	293	X	3
Tholeiitic basalt	3	Galapagos Islands	116–244	177	X	3
Ferrobasalt	2	Galapagos Islands	292–405	349	X	3
Alkali basalt	2	Galapagos Islands	213–228	221	X	3
Alkali basalt	5	Clarion Island	243–328	287	X	3
Alkali basalt	5	Socorro Island	248–336	290	X	3
Trachybasalt	2	Socorro Island	398–405	402	X	3
Alkali basalt	2	Pitcairn Island	294–390	342	X	3
Picrite basalt	2	Juan Fernandez Archipelago	314–393	354	X	3
Alkali basalt	4	Juan Fernandez Archipelago	238–314	266	X	3
Alkali basalt	17	Samoan Islands	232–472	318	X	3, 9
Hawaiite	7	Samoan Islands	282–510	351	X	3, 9
Picrite basalt	5	Samoan Islands	175–282	211	X	9
Alkali basalt	10	Kartala (Grande Comore)	110–240	164	X	10
Nepheline basalt	8	La Grille (Grande Comore)	145–217	181	X	10
Alkali basalt	2	Heard Island	222–261	242	X	3
Trachybasalt	2	Heard Island	333–426	380	X	3
Alkali basalt	3	Kerguelen Archipelago	178–305	253	X	3
Basalt	8	Mauritius	103–379	203	X	3
Alkali basalt	6	Prince Edward Island	204–325	278	X	3
Trachybasalt	7	Prince Edward Island	364–483	415	X	3
Alkali basalt	15	Marion Island	173–277	225	X	3
Trachybasalt	13	Marion Island	267–370	318	X	3
"Diabase", basalt	12	Troodos Massif (Cyprus)		63	X	11
Calc-alkali basalt	14	Kamchatka (Kurile Islands)		100	S	12
Calc-alkali basalt	2	Viti Levu (Fiji)	62–68	65	X	13
Calc-alkali basalt	12	Aleutian Islands		44	S	14
Calc-alkali basalt	30	Japan Arc		50	X, S	6, 11, 14
Calc-alkali basalt	3	Yalasea (New Guinea Arc)	30–40	33	X	15
Calc-alkali basalt	53	Java Arc		107	X	6

Table 40-E-3 (continued)

Rock type	No. of samples	Locality	ppm Zr		Method	Reference
			Range	Mean		
Calc-alkali basalt	7	Falcon Island (Tonga Arc)		33	X	6
Calc-alkali basalt	8	Guam (Marianas Arc)		52	X	6
Calc-alkali basalt	18	South Sandwich Islands		56	X	6, 61
Calc-alkali basalt	16	Lesser Antilles		87	X	6, 61, 62
Calc-alkali basalt	10	Erromangan (New Hebrides)		61	?	16
Calc-alkali basalt	4	Palau and Guam Islands		10	S	14
Continental volcanic rocks						
Tholeiitic basalt	9	Deccan Traps (India)		132	X	6
Olivine basalt	7	Lake Rudolf (East Africa)	100–395	175	X	17
Basalt and gabbro	49	Central Rift Valley (Kenya)	75–1,450	237	X	18
Basalt	3	Paka, Gregory Rift (Kenya)	119–150	132	X	19
Basalt	11	Afar (Ethiopia)		177	X	6
Trachybasalt	5	Roccamonfina (Italy)		148	X	20
Olivine basalt	10	Roccamonfina (Italy)		122	X	20
Basalt	14	Northern Cascades, Oregon (U.S.A.)		119	S	14
Olivine basalt	20	Snake River, Idaho (U.S.A.)		225	S	14
Olivine basalt	3	East African Rift System	169–213	192	X	21
Alkaline basalt	2	Tertiary, Aden volcanic series (S. Arabia)	110–160	135	X	22
Hawaiite	3	Tertiary, Aden volcanic series (S. Arabia)	207–275	234	X	22
Alkali basalt	5	Nandewar Mountains, N.S.W. (Australia)	127–221	193	X	23
Tholeiitic basalt	40	Tertiary, Svartenhuk (W. Greenland)	67–161	110	X	24
Tholeiitic basalt	37	Tertiary, Baffin Island	45–66	58	X	24
Tholeiitic basalt (Karroo)	37	Mesozoic, Rhodesia	161–595	370	X	25, 26
Tholeiitic basalt (Karroo)	7	Mesozoic, Swaziland	35–100	70	X	26
Tholeiitic basalt (Karroo)	21	Mesozoic (Lesotho)	72–156	110	X	27
Tholeiitic basalt	8	Mesozoic, Parana Basin (Brazil)	200–360	252	S	28
Spilite (basaltic)	14	Devonian-Carboniferous, N.W. Germany	60–210	117	X	29
Spilite (alkali basalt)	26	Devonian, S.W. England		218	X	30
Spilite (metatholeiite)	13	Devonian, S.W. England		49	X	30

Spilite (pillow lavas)	10	Precambrian to Devonian, England	116–353	202		37
Basalt	18	Ordovician, Borrowdale (N.W. England)		141	X	31
Metatholeiite (greenstone)	94	Lower Palaeozoic, Central and Western Norway	41–216	115	X	32
Basaltic komatiite (greenschist)	9	Paleozoic, Rambler area (Newfoundland)		14	X	33
Calc-alkali basalt	24	Roberts Arm group (Newfoundland)		78	X	34
Tholeiitic basalt	13	Lush's Bight group (Newfoundland)		82	X	34
Metatholeiite, andesite	46	Precambrian, Chitaldrug (India)	45–502	130	S	35
Glaucophane schist (metatholeiite)	7	Precambrian, Mona complex (Anglesey)		127	X	36
Basalt (greenstone)	155	Superior Province (Canada)		119	S	38
Metatholeiite (greenstone)	236	Archaean, Eastern Goldfields (W. Australia)		60	X	39
Metatholeiite (greenstone)	18	Barberton (South Africa)	27–131	61	X	40
Basaltic komatiite (greenstone)	42	Barberton (South Africa)	19–66	48	X	40
Basalt	15	Grossglockner (Austria)		112	X	11
Tholeiitic basalt	12	Pilanesberg complex (South Africa)	130–260	194	X	41
Hypabyssal (doleritic rocks)						
Dolerite	3	Mid-Atlantic Ridge, 22° N	82–116	96	X	3
"Diabase"	4	Dillsburg, Pennsylvania (U.S.A.)	55–70	64	S	42
Doleritic series, 48–50% SiO_2	16	Ardnamurchan, British Tertiary Province		86	X	43
Doleritic series, 50–52% SiO_2	16	Ardnamurchan, British Tertiary Province		86	X	43
Doleritic series, 52–54% SiO_2	26	Ardnamurchan, British Tertiary Province		232	X	43
Dolerite (central zone)	7	Mesozoic, Great Lake (Tasmania)	60–110	86	S	42
Dolerite (lower zone)	6	Mesozoic, Great Lake (Tasmania)	50–60	52	S	42
Dolerite	20	Mesozoic, Tasmania	110–220	157	S	44
Dolerite (quartz normative)	61	Mesozoic, Eastern North America		77	X	45
Dolerite (olivine normative)	47	Mesozoic, Eastern North America		51	X	45
Dolerite	14	Mesozoic, Karroo (South Africa)		97	S	46
Tholeiitic dolerite	54	Triassic, Piedmont, North Carolina (U.S.A.)	23–105	58	X	47
Tholeiitic dolerite	43	Mesozoic, Ferrar dolerites (Antarctica)	42–280	100	X	48
Dolerite	21	Lush's Bight group (Newfoundland)		47	X	34
Picritic dykes	6	Precambrian, Assynt district (Scotland)	34–140	82	X	49
Tholeiitic dolerite	45	Precambrian, Wyoming (U.S.A.)	57–271	128	X	50
Metadolerite, gabbro	84	Archaean, Eastern Goldfields (W. Australia)		54	X	39

Table 40-E-3 (continued)

Rock type	No. of samples	Locality	ppm Zr		Method	Reference
			Range	Mean		
Plutonic rocks						
Gabbro	19	Midatlantic ridge, 24° N	<10–140		S	51
Gabbro	5	Reunion	120–215	172	X	52
Gabbro	2	Paresis (S.W. Africa)	82–173	126	S	53
Gabbro	2	Garabal Hill (Scotland)	32–63	48	X	54
Gabbro	7	San Marcos, Southern California Batholith (U.S.A.)		40	S	14
Gabbro-dolerite	5	Alamdzhakh intrusion, Siberia (U.S.S.R.)	17–73	42	?	55
Ferrogabbro	5	Alamdzhakh intrusion, Siberia (U.S.S.R.)	55–180	98	?	55
Gabbro-diorite	2	Mount Chernaya intrusion, Siberia (U.S.S.R.)	120–230	165	?	55
Gabbro-dolerite	12	Mount Chernaya intrusion, Siberia (U.S.S.R.)	9–85	35	?	55
Gabbro, norite	13	Bushveld igneous complex (South Africa)	27–45	32	S	56
Norite, border phase	3	Bushveld igneous complex (South Africa)	40–125	88	S	56
Quartz gabbro	10	Losberg intrusion (South Africa)	52–88	68	X	57
Quartz norite	3	Losberg intrusion (South Africa)	35–51	41	X	57
Chilled margin	1	Skaergaard intrusion (East Greenland)		94	X	58
Gabbro, lower-main zone	5	Skaergaard intrusion (East Greenland)	36–68	53	X	58
Norite	8	Archaean (Western Australia)		36	X	59

References: 1. Brooks *et al.* (1974); 2. Cann (1970); 3. Kable (1972); 4. Melson and Thompson (1971); 5. Engel *et al.* (1965); 6. Pearce and Cann (1973); 7. Cann (1969); 8. Hughes and Brown (1972); 9. Hubbard (1971); 10. Strong (1972); 11. Pearce and Cann (1971); 12. Markhinin and Sapozhnikova (1962); 13. Gill (1970); 14. Chao and Fleischer (1960); 15. Lowder and Carmichael (1970); 16. Colley (1970); 17. Brown and Carmichael (1971); 18. McCall and Hornung (1972); 19. Sceal and Weaver (1971); 20. Appleton (1972); 21. Weaver *et al.* (1972); 22. Cox *et al.* (1970); 23. Abbott (1969); 24. Clarke (1970); 25. Vail *et al.* (1969); 26. Cox *et al.* (1967); 27. Cox and Hornung (1966); 28. Cordani and Vandoros (1967); 29. Herrmann and Wedepohl (1970); 30. Floyd (1972); 31. Fitton (1972); 32. Gale and Roberts (1974); 33. Gale (1973); 34. Strong (1973); 35. Naqvi and Hussain (1973); 36. Thorpe (1972); 37. Bloxam and Lewis (1972); 38. Goodwin (1972); 39. Hallberg (1972); 40. Smith and Erlank (1975); 41. Ferguson (1973); 42. Gottfried *et al.* (1968); 43. Holland and Brown (1972); 44. Tiller (1959); 45. Weigand and Ragland (1970); 46. Nockolds and Allen (1956); 47. Ragland *et al.* (1968); 48. Gunn (1966); 49. Tarney (1973); 50. Condie *et al.* (1969); 51. Thompson (1973); 52. Upton and Wadsworth (1972); 53. Siedner (1965); 54. Brooks (1970); 55. Nesterenko *et al.* (1971); 56. Liebenberg (1960); 57. Danchin and Ferguson (1970); 58. Brooks (1969); 59. Williams and Hallberg (1973); 60. Thompson *et al.* (1972); 61. Baker (1968a); 62. Baker (1968b).

The abundances presented in Table 40-E-4 should be viewed with these comments, and those for mafic rocks, in mind. The following *general observations* may be made:

i) The zirconium content of kimberlites, though variable, is not as high as might be expected, except for perovskite rich varieties (DAWSON and HAWTHORNE, 1973).

ii) Shoshonites and absarokites from island arcs have less zirconium than those from rift valleys.

iii) The undersaturated basanites, nephelinites, melilitites, leucitites etc. are generally enriched in zirconium relative to most mafic rocks, but this feature is not confined to rift valleys (RIDLEY, 1970; MOORE, 1975).

iv) Carbonatites have widely varying concentrations of zirconium; presumably this is partly dependant on mineralogy (i. e. presence of minerals such as perovskite).

Unlike mafic rocks, zircons have been reported in some of these rocks, for example large zircons have been recovered from kimberlite (AHRENS *et al.*, 1967).

V. Intermediate Rocks

Zirconium abundances in intermediate rocks are given in Table 40-E-5 and should be compared with the averages given by DEGENHARDT (1957) for andesites (120 ppm Zr) and trachytes (500 ppm Zr). CHAO and FLEISCHER's (1960) data showed that andesites (25–110 ppm) contain less zirconium than other oceanic lavas of equivalent composition such as mugearites (250–520 ppm). Data for oceanic rocks in Table 40-E-5 indicate that all intermediate rocks are enriched in zirconium as compared with possible parental mafic varieties. However, the rate of increase is such that the alkali basalt suite differentiates (trachyandesites, mugearites through to trachytes) are relatively more enriched in zirconium (commonly 400–1,000 ppm and up to 1,800 ppm in Gough Island trachytes) than tholeiitic differentiates (about 500 ppm in icelandites) and island arc andesites (30–120 ppm). The low zirconium contents of the latter are noteworthy irrespective of whether they are derived from a basaltic parent or whether they are of direct mantle derivation. The greater enrichment in the alkaline types may be related to the greater solubility of zirconium in alkali-rich magmas (see further discussion in Subsects. 40-E-VI and VIII); note that alkali pyroxenes and amphiboles in these rocks may contain substantial amounts of zirconium (e. g. LE MAITRE, 1962).

VI. Syenites

Data for syenites and related alkali-rich rocks are given in Table 40-E-6 and may be compared with the averages given by DEGENHARDT (1957) for syenites (310 ppm Zr) and nepheline syenites (680 ppm Zr). It is customary to consider syenitic rocks as being of miaskitic or agpaitic type, and discussion is given on this basis.

a) *Miaskitic rocks* are characterized by a molecular $(Na_2O + K_2O)/Al_2O_3$ ratio of unity or less than unity, with K_2O often exceeding Na_2O. Although much more abundant than the agpaitic varieties, they do not show such high zirconium concentrations probably because zircon, which is ubiquitous, crystallizes at a relatively early stage, preventing the build-up of zirconium in residual magmas. The Stjernøy nepheline syenite (HEIER, 1964) is particularly depleted, with all samples having < 80 ppm Zr.

Table 40-E-4. *Zirconium abundances in kimberlites, carbonatites and alkali-rich lavas*

Rock type	No. of samples	Locality	ppm Zr Range	ppm Zr Mean	Method	Reference
Kimberlite	10	Koffiefontein Mine (South Africa)	98–139	118	X	1
Kimberlite	10	Ebenhaezer (South Africa)	110–145	136	X	1
Kimberlite	6	Bobbejaan fissure, Bellsbank (South Africa)	223–344	278	X	1
Kimberlite	6	Main fissure, Bellsbank (South Africa)	238–367	313	X	1
Kimberlite	10	Water fissure, Bellsbank (South Africa)	102–193	140	X	1
Kimberlite	14	Lesotho	300–700	446	S	2
Kimberlite (perovskite-rich)	2	Benfontein (South Africa)	0.22–0.23%	0.23%	X	3
Kimberlite	459	Yakutiya (U.S.S.R.)		200	S	4
Kimberlite	15	Premier Mine (South Africa)	84–142	107	X	1
Lamprophyre	2	Pilanesberg complex (South Africa)	41–320	181	?	5
Shoshonite	9	Viti Levu (Fiji)	26–71	52	X	6
Shoshonite	4	Papua (New Guinea)	87–160	118	X	7
Shoshonite	5	W. Rift Valley (E. Africa)	375–460	422	X	8
Shoshonite	25	Permian volcanics, Devonshire (England)	89–876	281	X	9
Basanite	5	Las Canadas (Tenerife)	350–600	480	X, S	10
Trachybasanite	14	Las Canadas (Tenerife)	330–850	587	X, S	10
Basanite	5	Korath Range (Ethiopia)	160–250	208	X	11
Basanite	2	W. Rift Valley (E. Africa)	334–518	426	X	8
Basanite	6	Eastern Rift zone (Uganda and Ruanda)	298–462	414	X	12
Potassic lava	6	Leucite Hills, Wyoming (U.S.A.)	0.16–0.20%	0.18%	?	13
Absarokite	2	Papua (New Guinea)	160–176	168	X	7
Absarokite	9	W. Rift Valley (E. Africa)	369–503	421	X	8
Nephelinite	17	W. Rift Valley (E. Africa)	508–1,124	651	X	8
Leucitite	21	W. Rift Valley (E. Africa)	243–852	514	X	8
Leucitite, tephrite	58	Roccamonfina (Italy)		257	X	14
Olivine melilitite	28	Namaqualand (South Africa)	290–1,113	655	X	15
Melilitite	6	W. Rift Valley (E. Africa)	636–836	718	X	8
Jumillite	8	Murcia (Spain)	220–890	640	X	16
Ugandite	6	W. Rift Valley (E. Africa)	268–854	439	X	8

Mikenite	3	W. Rift Valley (E. Africa)	446–562	519	X	8
Banakite	4	W. Rift Valley (E. Africa)	418–534	479	X	8
Mafurite	6	W. Rift Valley (E. Africa)	318–816	558	X	8
Katungite	5	W. Rift Valley (E. Africa)	519–655	580	X	8
Kivite	9	W. Rift Valley (E. Africa)	280–415	373	X	8
Murambite	6	W. Rift Valley (E. Africa)	273–373	311	X	8
Carbonated lava	4	W. Rift Valley (E. Africa)	0.13–0.19%	0.15%	X	8
Carbonatite	36	Oka complex (Canada)		0.18%	?	17
Carbonatite	2	Callender Bay (Canada)	60–160	110	W	18
Carbonatite	8	Dorowa (Rhodesia)	6–80	24	S	19
Carbonatite	3	Shawa (Rhodesia)	<10–30		S	19
Carbonatite	2	Oldoinyo Dili (Kenya)	200–450	325	S	20
Carbonatite	2	Kerimasi (Kenya)	30–50	40	S	20
Carbonatite	3	West Eifel (Germany)	298–627	435	X	21
Carbonatite	25	Various locations		83	?	17
Carbonatite	2	Eastern Rift Zone (Uganda)	800–820	810	X	12

References: 1. Kable *et al.* (1975); 2. Dawson (1962); 3. Dawson and Hawthorne (1973); 4. Litinskii (1961); 5. Ferguson (1973); 6. Gill (1970); 7. Mackenzie and Chappell (1972); 8. Bell and Powell (1969); 9. Cosgrove (1972); 10. Ridley (1970); 11. Brown and Carmichael (1969); 12. Gerasimovskii *et al.* (1972); 13. Carmichael (1967); 14. Appleton (1972); 15. Moore (1975); 16. Borley (1967); 17. Gold (1963); 18. Ferguson and Currie (1972); 19. Johnson (1961); 20. Bowden (1962); 21. Lloyd and Bailey (1969).

Table 40-E-5. *Zirconium abundances in intermediate rocks*

Rock type	No. of samples	Locality	ppm Zr		Method	Reference
			Range	Mean		
Oceanic rocks						
Diorite	2	Mid-Atlantic Ridge, 45° N	200–760	480	S	1
Icelandite	3	Iceland	451–547	498	X	2
Icelandite	2	Galapagos Islands	523–540	532	X	2
Andesite	36	Aleutian Islands		78	S	3
Andesite	8	Viti Levu (Fiji)	54–115	83	X,S	4, 11
Basaltic andesite	21	Kamchatka and Kurile Islands		110	S	5
Andesite	6	North Island (New Zealand)	90–110	98	S	6
Andesite	13	Palau, Guam (Pagan Islands)		37	S	3
Andesite	10	Hakone (Japan)		64	S	3
Andesite	3	Talasea (New Guinea)	55–75	67	X	7
Andesite	7	Japan Arc	46–125	104	S	6
Andesite	12	St. Kitts (Lesser Antilles)	65–110	93	W	8
Andesite	7	Erromangan (New Hebrides)		92		9
Andesite, high-K	2	Papua (New Guinea)	124–128	126	X	10
Andesite, low-K	2	Saipan (Mariana Island)	80–95	88	S	11
Andesitic rock	10	Tonga Islands	24–40	30	X	12
Andesite, low-Si	5	Bougainville	90–105	98	S	11
Andesite, high-K	2	Bougainville		170	S	11
Andesite	2	Bougainville	110–115	113	S	11
Trachyandesite	2	Heard Island	828–857	843	X	2
Mugearite	3	Ascension Island	409–494	447	X	2
Mugearite	10	Hawaiian Islands		425	S	3
Benmoreite	2	Easter Island	577–783	680	X	2
Trachyandesite	3	Tristan da Cunha	509–534	517	X	2
Trachyandesite	3	Gough Island	510–607	549	X	2
Trachyte	4	Gough Island	0.06–0.18%	0.14%	X	2
Trachyte	5	Mauritius	0.12–0.13%	0.13%	X	2
Trachyte	2	Hawaiian Islands	0.09–0.10%	0.10%	X	2
Trachyte	2	Hawaiian Islands		850	S	3

Trachyte	2	Pitcairn Island	0.09–0.11%	0.10%	X	2
Trachyte	2	Tristan Da Cunha	838–965	876	X	2
Trachyte	5	Bouvet Island	670–690	677	X	2
Trachyte	4	Azores	369–409	386	X	2
Trachyte	2	Ascension Island	765–860	813	X	2
Trachyte	2	Kerguelen Archipelago	506–1,265	886	X	2
Continental rocks						
Trachyte	3	Paka, Gregory Rift (Kenya)	0.07–0.16%	0.11%	X	13
Trachyandesite	2	W. Rift Valley (E. Africa)	486–515	501	X	14
Trachyte	2	W. Rift Valley (E. Africa)	429–500	465	X	14
Trachyte (pantelleritic)	2	East African Rift System	0.05–0.17%	0.11%	?	15
Trachyandesite	2	Jebel Khariz (Arabia)	318–404	361	X	16
Trachyte	2	Jebel Khariz (Arabia)	323–696	510	X	16
Trachyte	4	Nandewar Mountains, N.S.W. (Australia)	345–936	668	X	17
Mugearite	2	Nandewar Mountains, N.S.W. (Australia)	327–398	363	X	17
Trachyandesite	4	Tertiary, Aden volcanic series (S. Arabia)	210–600	430	X	18
Trachyte	6	Tertiary, Aden volcanic series (S. Arabia)	700–750	717	X	18
Mugearite	2	Tertiary, Aden volcanic series (S. Arabia)		350	X	18
Doleritic series, 54–56% SiO_2	11	Ardnamurchan, British Tertiary Province		264	X	19
Doleritic series, 56–58% SiO_2	3	Ardnamurchan, British Tertiary Province		319	X	19
Doleritic series, 58–60% SiO_2	4	Ardnamurchan, British Tertiary Province		378	X	19
Doleritic series, >60% SiO_2	8	Ardnamurchan, British Tertiary Province		414	X	19
Andesite	3	Quaternary, Taupo volcanics (New Zealand)	88–110	103	S	20
Andesite	17	Northern Cascades, Oregon (U.S.A.)		136	S	3
Andesite	12	Southern Cascades, California (U.S.A.)		142	S	3
Andesite	3	Snake River, Idaho (U.S.A.)		200	S	3
Andesite, latite	24	Cenozoic, Northern Chile	170–260	205	X	21
Andesite	4	Yeoval, N.S.W. (Australia)	74–117	98	X	22
Spilite (andesitic)	5	Devonian-Carboniferous, N.W. Germany	86–250	169	X	23
Basaltic andesite	46	Ordovician, Borrowdale (N.W. England)		179	X	24
Andesite	46	Ordovician, Borrowdale (N.W. England)		227	X	24
Andesite (greenstone)	146	Superior Province (Canada)		167	S	25
Diorite	21	Precambrian, Angmagssalik area (East Greenland)		820	X	26

Table 40-E-5 (continued)

Rock type	No. of samples	Locality	ppm Zr		Method	Reference
			Range	Mean		
Diorite	7	Bushveld igneous complex (South Africa)	45–200	142	S	27
Quartz diorite, granodiorite	4	Verkhisstsk intrusion, Central Urals (U.S.S.R.)	26–140	99	X	28
Diorite, quartz diorite	7	Balsheporozhsky complex, West Zabaikalje (U.S.S.R.)		117	?	29
Diorite, quartz diorite	9	Dzhidinsky complex, West Zabaikalje (U.S.S.R.)		85	?	29
Diorite, quartz-diorite	8	Tannuolsky complex, Tuva (U.S.S.R.)		93	?	29
Shonkinite	6	Shonkin Sag Laccolith, Montana (U.S.A.)	150–190	165	X	30
Diorite	9	Salem granite suite (South West Africa)	209–382	247	X	31
Dioritic rock	65	South Germany	30–770	174	X	32
Diorite	24	Yeoval, N.S.W. (Australia)	64–136	106	X	22

References: 1. Aumento (1969); 2. Kable (1972); 3. Chao and Fleischer (1960); 4. Gill (1970); 5. Markhinin and Sapozhnikova (1962); 6. Taylor and White (1966); 7. Lowder and Carmichael (1970); 8. Baker (1968b); 9. Colley (1970); 10. Mackenzie and Chappell (1972); 11. Taylor *et al.* (1969); 12. Ewart *et al.* (1973); 13. Sceal and Weaver (1971); 14. Bell and Powell (1969); 15. Weaver *et al.* (1972); 16. Gass and Mallick (1968); 17. Abbott (1969); 18. Cox *et al.* (1970); 19. Holland and Brown (1972); 20. Ewart *et al.* (1968b); 21. Siegers *et al.* (1969); 22. Gulson (1972); 23. Herrmann and Wedepohl (1970); 24. Fitton (1972); 25. Goodwin (1972); 26. Wright *et al.* (1973); 27. Liebenberg (1960); 28. Lipova *et al.* (1957); 29. Znamenskoj *et al.* (1972); 30. Nash and Wilkinson (1971); 31. Miller (1973); 32. Okrusch and Richter (1969).

b) *Agpaitic syenites* are characterized not only by a molecular $(Na_2O + K_2O)/Al_2O_3$ ratio of >1 and $Na > K$, but also by their high contents of Zr, H_2O, F and Cl, and their exotic mineralogy (notably eudialyte). The most famous examples are the Lovozero and Ilimaussaq layered intrusions and it is readily apparent from Table 40-E-6 that some rocks from these massifs contain the highest zirconium concentrations of all rock clans. Concentrations at the 1% level are common, with the highest concentrations being recorded in the Ilimaussaq kakortokites (up to 3% Zr according to FERGUSON, 1970). These cumulate rocks, and also some of the layered rocks of the Lovozero Massif, contain most zirconium in the from of eudialyte; however other zirconium minerals (which are probably metasomatic according to VLASOV, 1966) and soda-rich amphiboles and pyroxenes may also contain substantial proportions of zirconium.

The extreme enrichment of zirconium in these rocks appears to be due to the absence of zircon (where present, it occurs as a late postmagmatic phase, VLASOV, 1966) and the alkaline and volatile-rich nature of the magmas concerned. This has led several authors to speculate that such liquids will have a low degree of polymerisation, which will encourage the formation of stable, soluble zirconium complexes. For example, RINGWOOD (1955) and GERASIMOVSKII (1966) suggest the presence of ZrO_4^{4-}, while VLASOV (1966) mentions the stability of $M_2(ZrF_6)$ and $Na_2[Zr(CO_3)_2]$ complexes in alkaline solutions. The experiments of DIETRICH (1968), as discussed in Sect. 40-D, lend some support to these suggestions, but more data are required.

Further details about these interesting agpaitic rocks may be found in GERASIMOVKII (1966) and SÖRENSON (1968).

VII. Granitic Rocks

DEGENHARDT (1957) showed that most of the zirconium in typical granitic rocks was present in zircon, and estimated the average granodiorite and granite to contain 140 ppm and 175 ppm Zr respectively. VINOGRADOV (1962) derived a somewhat higher average (210 ppm) for the average "felsic" rock (granites, granodiorites etc.). These figures may be compared with the abundances for granitic rocks given in Table 40-E-7, and it is apparent that they serve as reasonable magnitudes, even though the abundances shown exhibit a wide range of variation (CHAO and FLEISCHER, 1960, indicated a range of 50–700 ppm Zr).

In the typical and more abundant (calc-alkaline) granitic rocks the zirconium concentration does not rise to high levels. In some sequences, for example the Cape granites in South Africa (KOLBE, 1966), the zirconium concentration decreases with increasing differentiation (in the latter case from 300–30 ppm) or with age of intrusion, for example in the Verkhissetsk intrusives, U.S.S.R. (LIPOVA *et al.* 1957).

In contrast, zirconium does not show the same behavior in more alkaline granites, and increases in concentration with increasing differentiation, resulting in higher abundance levels. Thus the Younger granites from Nigeria (BOWDEN, 1966) have the following average abundances: biotite granites (235 ppm Zr), amphibole (ferrohastingsite) granites (530 ppm Zr) and riebeckite granites (1,000 ppm Zr). Zircon is relatively rare in the latter granites and zirconium is thought to be present in the alkaline amphiboles. The increase is correlated by BOWDEN (1966) with

Table 40-E-6. *Zirconium abundances in syenites and related rocks*

Rock type	No. of samples	Locality	ppm Zr		Method	Reference
			Range	Mean		
Nepheline syenite	5	Lovozero massif, first complex, Kola Peninsula (U.S.S.R.)	0.08–0.21%	0.13%	X	1
Foyaite	1	Lovozero massif, second complex, Kola Peninsula (U.S.S.R.)	0.05–1.23%	0.37%	X	1
Urtite	3	Lovozero massif, second complex, Kola Peninsula (U.S.S.R.)	0.10–0.28%	0.16%	X	1
Lujavrite	11	Lovozero massif, second complex, Kola Peninsula (U.S.S.R.)	0.10–0.44%	0.28%	X	1
Lujavrite	12	Lovozero massif, third complex, Kola Peninsula (U.S.S.R.)	0.70–1.71%	1.06%	X	1
Sodalite syenite	3	Lovozero massif, third complex, Kola Peninsula (U.S.S.R.)	0.09–0.21%	0.13%	X	1
Lujavrite	12	Ilimaussaq intrusion (South Greenland)	0.20–1.18%	0.63%	S	2
Naujite	5	Ilimaussaq intrusion (South Greenland)	0.04–0.82%	0.27%	S	2
Sodalite foyaite	5	Ilimaussaq intrusion (South Greenland)	0.08–0.69%	0.34%	S	2
Syenite	2	Ilimaussaq intrusion (South Greenland)	0.17–0.18%	0.18%	S	2
Augite syenite	2	Ilimaussaq intrusion (South Greenland)	408–655	532	S	2
Kakortokite	54	Ilimaussaq intrusion (South Greenland)	0.05–3.03%	0.92%	S	2
Ekerite	3	Oslo Province (Norway)	730–870	813	S	3
Nordmarkite	2	Kangerdlugssuag (East Greenland)	145–245	195	X	4
Nordmarkite	3	Oslo Province (Norway)	500–740	603	S	3
Umptekite	3	Spitskop (South Africa)	400–620	490	S	5
Pulaskite	1	Kangerdlugssuag (East Greenland)		167	X	4
Foyaite	1	Kangerdlugssuag (East Greenland)		355	X	4
Foyaite	2	Spitskop (South Africa)	180–260	220	S	5
Ijolite	3	Spitskop (South Africa)	120–130	125	S	5
Melteigite	1	Spitskop (South Africa)		190	S	5
Monzonite, syenite, quartz syenite	12	Dzhidinsky complex, West Zabaikalje (U.S.S.R.)		380	?	6
Nepheline syenite	12	Stjernøy (Norway)	n.d.[a]–80		S	7

Table 40-E-6 (continued)

Rock type	No. of samples	Locality	ppm Zr		Method	Reference
			Range	Mean		
Soda syenite	3	Shonkin Sag Laccolith, Montana (U.S.A.)	380–615	482	X	8
Syenite pegmatite	1	Shonkin Sag Laccolith, Montana (U.S.A.)		210	X	8
Syenite	2	Shonkin Sag Laccolith, Montana (U.S.A.)	180–195	188	X	8
Nepheline syenite	7	Callender Bay (Canada)	250–1,600	749	?	9
Nepheline syenite	53	Paleozoic, East Tuva and North Mongolia (U.S.S.R.)	260–520	410	?	10
Syenite, trachyte	72	Central Rift Valley (Kenya)	70–1,920	916	X	11
Syenite	3	Paresis (S.W. Africa)	184–340	260	S	12
Nepheline syenite	2	Pilanesberg complex (South Africa)	0.14–0.17%	0.16%	?	13
Nepheline syenite	11	Inner zone, Cortejo Filoneano ring complex, Nejoio (Angola)	0.18–0.38%	0.28%	X	14
Nepheline syenite	8	Outer zone, Cortejo Filoneano ring complex, Nejoio (Angola)	325–1,716	638	X	14
Cancranite syenite	8	Outer zone, Cortejo Filoneano ring complex, Nejoio (Angola)	373–1,344	739	X	14
Bostonite	4	Paresis (S.W. Africa)	243–640	441	S	12
Tephrite	3	Korath Range (Ethiopia)	240–320	267	X	15

[a] n.d. = not detected.

References: 1. Gerasimovskii *et al.* (1958); 2. Ferguson (1970); 3. Dietrich *et al.* (1965); 4. Brooks (1970); 5. Liebenberg (1960); 6. Znamenskoj *et al.* (1972); 7. Heier (1964); 8. Nash and Wilkinson (1971); 9. Ferguson and Currie (1972); 10. Tugarinov *et al.* (1968); 11. McCall and Hornung (1972); 12. Siedner (1965); 13. Ferguson (1973); 14. Rodrigues (1973); 15. Brown and Carmichael (1969).

Table 40-E-7. *Zirconium abundances in granitic rocks*

Rock type	No. of samples	Locality	ppm Zr Range	ppm Zr Mean	Method	Reference
Aplite	1	Mid-Atlantic Ridge, 24° N		150	S	1
Adamellite	30	Paleozoic, North Carolina (U.S.A.)	27–185	89	X	2
Granite	65	Paleozoic, East Tuva and North Mongolia (U.S.S.R.)	72–305	159	?	3
Granite	77	Bärhalde, Black Forest (Germany)	28–115	54	X	4
Granite	66	Schluchsee, Black Forest (Germany)	81–231	165	X	4
Granite	86	Malsburg, Black Forest (Germany)	128–232	194	X	4
Granodiorite	2	Garabal Hill (Scotland)	124–132	128	X	5
Pegmatite	1	Garabal Hill (Scotland)		27	X	5
Gneissic granite	4	Snowy Mountains (Australia)	200–245	221	S	6
Granodiorite	20	Snowy Mountains (Australia)	97–350	195	S	6
Leucogranite	8	Snowy Mountains (Australia)	52–138	88	S	6
Alkali granite	4	Franzfontein (South West Africa)	250–650	475	S	7
Granite (coarse to medium-grained)	26	Cape Granite (South Africa)	115–330	200	S	6
Granite (fine grained)	8	Cape Granite (South Africa)	28–235	107	S	6
Aplite, pegmatite	2	Verkhissetsk intrusion, Central Urals (U.S.S.R.)	18–49	34	X	8
Granite	7	Verkhissetsk intrusion, Central Urals (U.S.S.R.)	32–120	80	X	8
Granite	87	Precambrian, Wyoming, (U.S.A.)	16–370	142	X	9
Granite	4	Precambrian, Louis Lake Batholith, Wyoming (U.S.A.)	176–479	206	X	10
Granodiorite	17	Precambrian, Louis Lake Batholith, Wyoming (U.S.A.)	238–367	329	X	10
Quartz monzonite	4	Precambrian, Louis Lake Batholith, Wyoming (U.S.A.)	187–599	264	X	10
Granitic rock	20	Precambrian, Brazil	59–410	210	S	11
Granitic vein (concordant)	23	Lewisian, Argyllshire (Scotland)	55–587	106	X	12
Granitic vein (remobilised)	4	Lewisian, Argyllshire (Scotland)	30–264	159	X	12
Biotite pegmatite	5	Lewisian, Argyllshire (Scotland)	114–2,093	541	X	12
Granite	3	Bushveld igneous complex (South Africa)	250–500	350	S	13
Granite (Makhutso type)	20	Bushveld igneous complex (South Africa)	140–403	269	X	14

Aplite	6	Precambrian, Louis Lake Batholith, Wyoming (U.S.A.)	51–188	106	X	10
Adamellite	6	Salem granite suite (South West Africa)	256–477	380	X	15
Monzonite	4	Salem granite suite (South West Africa)	424–575	472	X	15
Amphibole granite	18	Younger series, Northern Nigeria	220–1,500	530	X	16
Biotite granite	12	Younger series, Northern Nigeria	110–420	235	X	16
Riebeckite granite	12	Younger series, Northern Nigeria	300–2,200	1,003	X	16
Granite	6	Yeoval, N.S.W. (Australia)	142–293	220	X	17
Granite	47	Kagenfels (Germany)	83–612	115	X	18
Granodiorite	3	Yeoval N.S.W. (Australia)	163–184	173	X	17
Granite, granite gneiss	55	Alps and Bohemian massif (Austria)	70–350	182	S	19
Granodiorite	25	Alps and Bohemian massif (Austria)	120–350	200	S	19
Tonalite	4	Alps and Bohemian massif (Austria)	75–300	174	S	19
Aplite	8	Alps and Bohemian massif (Austria)	10–130	63	S	19
Granodiorite	9	Southern California Batholith (U.S.A.)		140	S	20
Tonalite	8	Southern California Batholith (U.S.A.)		140	S	20
Granite	4	Southern California Batholith (U.S.A.)		120	S	20
Granite	222	Variskian, Rudny Mountains (G.D.R.)	20–210	73	?	21
Plagiogranite	25	Mainsky complex West, Sajan (U.S.S.R.)		85	?	22
Tonalite, plagiogranite	10	Tannuolsky complex, Tuva (U.S.S.R.)		68	?	22
Granodiorite, granite	10	Tannuolsky complex, Tuva (U.S.S.R.)		148	?	22
Granodiorite	15	Dzhidinsky complex, West Zabaikalje (U.S.S.R.)		250	?	22
Granite (two mica)	15	Balsheporozhsky complex, West Sajan (U.S.S.R.)		112	?	22
Tonalite, granodiorite, granite	116	Susamyrsky complex, Tjan-Shan (U.S.S.R.)		200	?	22
Porphyry granite	10	Dzhojsky complex, West Sajan (U.S.S.R.)		165	?	22
Biotite granite, alaskite granite	14	Sjutkholsky complex, Tuva (U.S.S.R.)		72	?	22
Quartz diorite, granodiorite	36	Shakhtaminsky complex, East Zabaikalje (U.S.S.R.)		276	?	22
Porphyry granite, granodiorite	93	Amudzhkono-Sretensky complex, Zabaikalje (U.S.S.R.)		300	?	22
Biotite granite	103	Kukulbejsky complex, East Zabaikalje (U.S.S.R.)		210	?	22
Leucocratic granite	26	Nerchugansky complex, East Zabaikalje (U.S.S.R.)		228	?	22

References: 1. Thompson (1973); 2. Fullagar *et al.* (1971); 3. Tugarinov *et al.* (1968); 4. Hahn-Weinheimer and Johanning (1968); 5. Brooks (1970); 6. Kolbe and Taylor (1966); 7. Clifford *et al.* (1962); 8. Lipova *et al.* (1957); 9. Condie (1969); 10. Condie and Lo (1971); 11. Herz and Dutra (1960); 12. Drury (1972); 13. Liebenberg (1960); 14. De Bruiyn and Rhodes (1975); 15. Miller (1973); 16. Bowden (1966); 17. Gulson (1972); 18. Hahn-Weinheimer *et al.* (1971); 19. Grohmann (1965); 20. Chao and Fleischer (1960); 21. Tischendorf and Lange (1972); 22. Znamenskoj *et al.* (1972).

Table 40-E-8. *Zirconium abundances in felsic volcanic rocks*

Rock type	No. of samples	Locality	ppm Zr		Method	Reference
			Range	Mean		
Oceanic volcanic rocks						
Andesitic dacite	10	Kamchatka and Kurile Islands		390	S	1
Dacite	21	Aleutian Islands		109	S	2
Latite	3	Aleutian Islands		300	S	2
Dacite	23	Hakone Lavas (Japan)		97	S	2
Dacite	13	Tonga Islands	23–51	47	X	3
Dacite	3	Talasea (New Guinea)	90–130	112	X	4
Rhyodacite	14	Aleutian Islands		152	S	2
Rhyolite	2	Talasea (New Guinea)	150–160	155	X	4
Rhyolite	2	Viti Levu (Fiji)	71–72	72	X	5
Rhyolite	3	Kamchatka and Kurile Islands		630	S	1
Soda rhyolite	3	Aleutian Islands		233	S	2
Rhyolite	3	Easter Islands	802–949	886	X	6
Dacite	2	Iceland	713–847	780	X	6
Pitchstone	13	Iceland	130–650	406	X	7
Pitchstone	7	Arran, North Atlantic	205–290	229	X	7
Comendite	2	Ascension Island	0.10–0.11%	0.11%	X	6
Comendite	2	Sardinia	0.13–0.17%	0.15%	S	8
Obsidian, comendite	3	Easter Islands	814–1,036	962	?	9
Obsidian, comendite	2	Bouvet Island	0.15–0.16%	0.15%	?	9
Pantellerite	5	Pantelleria	0.15–0.22%	0.18%	X	10
Hyalopantellerite	3	Pantelleria	0.16–0.22%	0.18%	S	8
Pantellerite	5	Mayor Island (New Zealand)	0.11–0.15%	0.13%	S,X	11
Phonolite	30	Las Canadas (Tenerife)	550–1,150	854	X,S	12
Continental volcanic rocks						
Rhyolite	7	Quaternary, North America	25–280	136	X	13
Dacite	6	Quaternary, North America	125–240	175	X	13
Rhyolitic pumice, ignimbrite	26	Quaternary, Taupo volcanics (New Zealand)	100–210	158	S	14

Pumice	4	Quaternary, North America	110–240	190	X	13
Obsidian	31	Quaternary, North America	25–355	181	X	13
Quartz latite	2	Snake River, Idaho (U.S.A.)		0.20%	S	2
Dacite	16	Northern Cascades, Oregon (U.S.A.)		142	S	2
Quartz latite	3	Northern Cascades, Oregon (U.S.A.)		258	S	2
Dacite	9	Southern Cascades, California (U.S.A.)		161	S	2
Rhyodacite	9	Northern Cascades, Oregon (U.S.A.)		189	S	2
Rhyodacite	12	Southern Cascades, California (U.S.A.)		170	S	2
Rhyolite	15	Snake River, Idaho (U.S.A.)		388	S	2
Pantellerite tuff	29	Fantale, Ethiopian rift	0.07–0.12%	0.10%	X	15
Pantellerite	6	Afar rift (Ethiopia)	0.11–0.12%	0.11%	X	16
Pantellerite	20	Fantale, Ethiopian rift	0.09–0.15%	0.11%	X	17
Obsidian	2	Lake Rudolf (E. Africa)	730–800	765	X	18
Obsidian (pantelleritic)	6	East African Rift System	0.06–0.17%	0.12%	X	19
Latite	4	W. Rift Valley, E. Africa	329–354	346	X	20
Rhyolite	4	Jebel Khariz (Arabia)	296–1,000	698	X	21
Comendite	2	Nandewar Mountains, N.S.W. (Australia)	0.08–0.14%	0.11%	X	22
Alkali rhyolite	4	Nandewar Mountains, N.S.W. (Australia)	622–903	769	X	22
Rhyolite	10	Tertiary, Aden volcanic series (S. Arabia)	550–1,450	850	X	23
Latite	11	Roccamonfina (Italy)		229	X	24
Leucite phonolite	10	Roccamonfina (Italy)		399	X	24
Rhyodacite	7	Cenozoic, Chile	120–1,200	356	S	25
Alkali rhyolite	31	Cenozoic, Chile	140–800	315	S	25
Rhyolite	9	Cenozoic, Chile	215–1,200	681	S	25
Quartz-feldspar porphyry	6	Paresis (S.W. Africa)	364–564	453	S	26
Comendite	9	Paresis (S.W. Africa)	0.07–0.13%	0.10%	S	26
Rhyolite	5	Paresis (S.W. Africa)	364–530	468	S	26
Phonolite, tinguaite	9	Nejoio (Angola)	0.11–0.54%	0.18%	X	27
Dacite	32	Ordovician, Borrowdale (N.W. England)		357	X	28

Table 40-E-8 (continued)

Rock type	No. of samples	Locality	ppm Zr		Method	Reference
			Range	Mean		
Rhyolite	7	Ordovician, Borrowdale (N.W. England)		282	X	28
Dacite	20	Roberts Arm group (Newfoundland)		201	X	29
Verite (pitchstone)	8	Almeria (Spain)	220–890	640	X	30
Dacite (greenstone)	79	Superior Province (Canada)		164	S	31
Rhyodacite (greenstone)	9	Superior Province (Canada)		177	S	31
Rhyolite (greenstone)	28	Superior Province (Canada)		169	S	31

References: 1. Markhinin and Sapozhnikova (1962); 2. Chao and Fleischer (1960); 3. Ewart *et al.* (1973); 4. Lowder and Carmichael (1970); 5. Gill (1970); 6. Kable (1972); 7. Carmichael and McDonald (1961); 8. Noble and Haffty (1969); 9. Bailey and MacDonald (1970); 10. Butler and Smith (1962); 11. Ewart *et al.* (1968 a); 12. Ridley (1970); 13. Jack and Carmichael (1968); 14. Ewart *et al.* (1968 b); 15. Gibson (1970); 16. Barberi *et al.* (1975); 17. Gibson (1972); 18. Brown and Carmichael (1971); 19. Weaver *et al.* (1972); 20. Bell and Powell (1969); 21. Gass and Mallick (1968); 22. Abbott (1969); 23. Cox *et al.* (1970); 24. Appleton (1972); 25. El-Hinnawi *et al.* (1969); 26. Siedner (1965); 27. Rodrigues (1973); 28. Fitton (1972); 29. Strong (1973); 30. Borley (1967); 31. Goodwin (1972).

alkalinity and the agpaitic coefficient and is consistent with the comments previously made in this section with regard to the greater solubility of zirconium in such magmas.

VIII. Felsic Lavas

Chao and Fleischer (1960) gave the following concentration ranges for felsic lavas: dacites (50–250 ppm Zr), latites (40–600 ppm Zr) and rhyolites (50–700 ppm Zr). Reference to Table 40-E-8 shows that much higher concentrations are reached in some felsic lavas, with concentrations of 1,000–2,000 ppm Zr being fairly common.

It will be recognized that the abundances of zirconium in felsic lavas are again related to tectonic setting and associated parental magma types. Thus, when considering the oceanic varieties, dacites and rhyolites from island arcs contain less zirconium than those with a tholeiitic parentage (e. g. Iceland) which in turn contain less zirconium than the types associated with the alkali basalt clan; the peralkaline comendites and pantellerites having the highest concentrations. Previous comments regarding the solubility of zirconium in such melts again apply and it is noteworthy that microprobe beam-scanning by Nicholls and Carmichael (1969) of pantellerites showed zirconium to be evenly distributed throughout the glassy matrix.

Similar comments may apparently be made for continental varieties, and zirconium distributions are obviously useful for interpreting evolutionary processes in Archaean rocks. Note, however, the very high concentrations reported by Rodrigues (1973) for phonolites and tinguaites (1,100–5,400 ppm Zr) from Angola.

IX. Crustal Abundance

Estimates of the abundance of elements in the earth's continental crust have traditionally involved assumptions as to the relative volumes of igneous rocks present (granitic, intermediate, mafic etc.) and the three estimates given for the average igneous rock in Table 40-E-9 have been derived using this type of approach.

Table 40-E-9. *Estimated crustal abundances for zirconium*

Source and reference	ppm Zr
Average igneous rocks	
Degenhardt (1957)	156
Vinogradov (1962)	170
Taylor (1964)	165
Average metamorphic rocks	
Lambert and Heier (1968), Australian Shield	220
Sheraton (1970), Lewisian, Scotland	202
Sighinolfi (1971), Brazilian Shield	220
Eade and Fahrig (1973), Canadian Shield	190
Shaw *et al.* (1976), Canadian Shield	240
Average sedimentary rocks	
Degenhardt (1957)	155
Vinogradov (1962)	210
Andrews-Jones (1968)	200

We have not attemped to repeat this type of calculation since we consider it more instructive to use values obtained for metamorphic rocks from the old shield areas (cratons), because these may provide a better sampling in view of the presence of granulite facies rocks commonly thought to be representative of deeper crustal levels. Our subjective assessment is that the average of 190 ppm Zr obtained by EADE and FAHRIG (1973) is especially important in view of the large number of samples and geographical coverage involved, and serves as a reasonable magnitude for the abundance of zirconium in the continental crust. Although sedimentary rocks are volumetrically unimportant, estimates for the average sedimentary rocks shown in Table 40-E-9 are in reasonable agreement with the other estimates quoted.

The oceanic crust is thought to be dominantly composed of tholeiitic basalt (abyssal tholeiite or MORB-type basalt) and, if so, then a magnitude of 100 ppm Zr for the oceanic crust may be suggested from the data given in Table 40-E-3.

Revised manuscript received: June 1977

40-F. Behavior in Magmatogenic Processes

From the discussion in Sect. 40-E, it is apparent that, for those rock suites where zirconium is incorporated in early crystallising zircon, the ability for it to become concentrated in late stage fluids is reduced. Thus, pegmatites and aplites derived from typical granitic rocks and miaskitic syenites, generally have low concentrations of zirconium (see Table 40-E-7), and extensive precipitation of zirconium-bearing minerals is confined to the pegmatite replacement stage (albitisation) where zircon or its varieties (cyrtolite, malacon) may occur (VLASOV, 1966). The high concentrations of zirconium in some biotite pegmatites from the Lewisian in Scotland (114–2,090 ppm Zr) as found by DRURY (1973) and shown in Table 40-E-7, are thought to be due to metasomatic processes.

In contrast, pegmatites associated with agpaitic nepheline syenites are, as expected, enriched in zirconium and contain a variety of zirconium minerals. VLASOV (1966) considers that most of these are formed by eudialyte (an early crystallising mineral) replacement during the final stages of pegmatite formation, and points out that they are located in the central parts of the pegmatite bodies, where they are associated with the final crystallising residue.

In hydrothermal processes, zirconium is known to concentrate in veins, generally associated with alkali rocks. VLASOV (1966) mentions the concentration of zirconium in natrolite, aegirine-apatite, fluorite, and wolframite and cassiterite veins, but does not supply quantitative information.

In summary, the behaviour of zirconium is such that it becomes concentrated in high temperature magmatic processes, and its refractory nature dictates that it is not geochemically active or prominent in gas transport or ore-deposition processes.

The behavior of zirconium in metasomatic processes is discussed in Sect. 40-N.

Revised manuscript received: June 1977

40-G. Behavior during Weathering and Alteration of Rocks

The behavior of zirconium during weathering and transportation is inadequately known, and there is disagreement about quite fundamental questions, such as the relative amounts of the metal concentrated in the resistates as opposed to the hydrolysates (RANKAMA and SAHAMA, 1950; GOLDSCHMIDT, 1954; see also Sect. 40-K). Generally speaking, zirconium tends to accumulate in weathering profiles, and soil/rock concentration ratios greater than ten have been reported (KÖSTER, 1961). Since zircon accounts for most of the zirconium present in a great many crustal rocks, and has a relatively high chemical stability (SMITHSON, 1950), it follows that this mineral has a profound influence on the geochemistry of zirconium during the weathering cycle. Many aspects of the weathering, transportation and diagenesis of zircon have been comprehensively reviewed and examined in detail by CARROLL (1953) and MARSHALL (1967). MARSHALL (1940), MARSHALL and HASEMAN (1942), HASEMAN and MARSHALL (1945) and VOGEL (1975), in making mass-balance calculations for weathering profiles, utilised the fact that zircon is practically indestructible in most soils. However, it is clear that zircon is attacked under certain conditions (BREWER, 1955; VLASOV, 1966) and CARROLL (1953) has cautioned against the indiscriminate use of zircon as a reference indicator of chemical changes in soil profiles. The consensus of opinion is that zircon is most readily attacked under alkaline conditions (CARROLL, 1953; DEGENHARDT, 1957; BLUMENTHAL, 1958; COLEMAN and ERD, 1961). DEGENHARDT (1957) specified solutions of alkali bicarbonates as effective solvents. SAXENA (1966) believed that either alkaline or acidic (e. g. carbonated, lime-rich) waters might attack zircon. This last solvent was the one favored by STROCK (1941), while BREWER (1955) envisaged the dissolution of zircon by carbonated humic acid solutions. CARROLL (1953) has shown that zircon is apparently dissolved or corroded directly, without the formation of intermediate minerals, and that the so-called "altered zircons" (as reported, for example, by GOLDICH, 1938), are probably metamict structures which have been physically damaged by radioactivity and are not chemically altered. Factors reducing the stability of zircons are metamictisation (related to radiation damage caused by unstable isotopes included in the zircon lattice), zoning, inclusions, fractures, small grain-size and, possibly, the presence of certain "corrosive" clays and organic substances (CARROLL, 1953; MARSHALL, 1967; KROGH, 1973; DAVIS *et al.*, 1976). Zirconium is undoubtedly also released during the weathering of minerals other than zircon, but the relative quantities involved are unknown. GOLDSCHMIDT (1954) mentions secondary concentrations of almost pure zirconium dioxide forming as a hardpan on certain zirconium-rich, eudialyte-bearing nepheline syenites in tropical regions. Considerable masses of zirconium may also be released during the weathering of minerals of which the metal is a minor but not insignificant component (VLASOV, 1966; HOFMEYR, 1971; see also Sect. 40-D). Generally speaking, the solubilities of naturally occurring zirconium compounds, complexes, ions etc. in typical surface waters and groundwaters are very small. It is therefore extremely unusual to find more than a minute concentration of the element in stable solution in natural waters (see Sect. 40-H), and although some ionic complexes of zirconium may survive for a time (SAXENA, 1966; DEGENHARDT, 1957), much of it is readily removed from solution by two known mechanisms: (i) under favourable circumstances zirconium is incorporated in certain clay-forming minerals in the weathering profile, and is thus accumulated. DEGENHARDT (1957) reported typical concentrations of zirconium in clay minerals (approximate values: kaolinite 80 ppm, halloysite 13 ppm, illite 100 ppm, montmorillonite 140 ppm) and noted that montmorillonite absorbed zirconium readily. However, occasional strong concentrations of zirconium occur in clay minerals such as kaolinite (over 1,000 ppm, DEGENHARDT, 1957), illite and halloysites (VLASOV, 1966). (ii) Dissolved zirconium is readly hydrolysed (KATCHENKOV, 1967; VLASOV, 1966; DEGENHARDT, 1957; GOLDSCHMIDT, 1954) to colloidal particles of oxides, hydroxides or phosphates, which may be trapped in the weathering profile or washed away. DEGENHARDT (1957) reported 50 ppm Zr in material precipitated from spring waters, and HEWETT (1963) found as much as 150 ppm Zr in manganiferous hot spring encrustations.

I. Weathering Profiles

DENNEN and ANDERSON (1962) compared the compositions of various fresh rocks with the respective weathering crusts on them. Zirconium was little changed or slightly enriched (up to twofold) in most samples and was significantly depleted in the crust of one sample only (a granite).

DEGENHARDT (1957) presented data for zirconium in two sections of weathering granite. In one the metal increased in concentration from 195 ppm in the fresh rock (5 m deep) to 306 ppm in thoroughly decomposed material at 30 cm depth. In the other profile zirconium had been leached from the bedrock (380 ppm) and redeposited in iron-rich horizons (375 and 683 ppm Zr). Zircon corrosion was strongest in the deeper parts of the profile where the bicarbonate content of percolating rain water was highest.

Zirconium is, on average, equally enriched (100 to >246 ppm) in both the silt and clay fractions of soils formed on zircon-bearing arkoses from different parts of New Zealand (MCLAUGHLIN, 1955). The depth at which the metal is most concentrated is inversely proportional to the intensity of weathering. A warmer climate did not affect the loss of zirconium, but did increase the proportion released from zircon and reprecipitated in the clay fraction.

BUTLER (1953, 1954) conducted a similar study for various Cornish and Norwegian rocks and their respective weathering products (Table 40-G-1). Zirconium commonly accumulates in the separated silt fraction of the Cornish soils and in at least one case (adamellite) the destruction of zircon has produced a relative enrichment of zirconium in the separated clay fraction. Zirconium is depleted in all the Norwegian profiles. The conflict was not reconciled.

OERTEL (1961) reported the separate mean abundances of zirconium in each of seventeen soil profiles developed on dolerite in Tasmania—these means varied from 160—450 ppm. The soil/dolerite zirconium ratio for various profiles varied from

Table 40-G-1. *Zirconium concentrations (in ppm) in various Cornish and Norwegian rocks and the silt and clay fractions of associated soils* (BUTLER, 1953, method S)

	Cornwall					Oslo Area			
	1	2	3	4	5	6	7	8	9
Fresh rock	150	—	20	300	<10	190	400	160	300
Rotted rock	—	110	—	—	—	—	—	—	—
Silt	280	250	—	330	95	150	125	70	20
Clay	130	25	30	120	15	120	55	80	<10

1. Adamellite; 2. Granodiorite gneiss; 3. Hornblende gabbro (clay from nearby profile); 4. Hornblende schist; 5. Tremolite serpentine; 6. Quartz free alkali syenite; 7. Hypersthene monzonite; 8. Altered olivine basalt; 9. Muscovite schist.

1.0—3.4, and exceeded 2.0 in seven cases and 1.25 in fourteen cases. It was concluded that pedogenic factors other than parent material have a marked effect on the zirconium soil/rock ratio and that the concentration of the element in the soil cannot be reliably predicted from an analysis of the parent rock.

II. Soils

The following estimates have been made of the abundance of zirconium in the soils of the world: MITCHELL (1944): generally 50—1,000 ppm; VINOGRADOV (1959): (presumably a worldwide estimate) 300 ppm; ANDREWS-JONES (1968): "typical" range 60 to about 2,000 ppm, "extreme" range 10—5,000 ppm, mean 300 ppm. MITCHELL (1964) reported the range of zirconium concentrations in Scottish surface soils to be from 200 to >1,000 ppm. Several authors, notably SHACKLETTE (1971) and TIDBALL (1973), have reported zirconium analyses of soils from the conterminous United States of America. Most of this data has been summarised by CONNOR and SHACKLETTE (1975). The mean zirconium content of 3,179 soil samples was 264 ppm and the observed range was from <30—2,000 ppm. The smallest and largest mean values reported by different authors (i. e., for different subsets of samples) were 120 and 460 ppm. In the subset of 862 samples described by SHACKLETTE *et al.* (1971), which were collected at depths of about 20 cm from locations about 50 miles apart throughout the forty-eight states, the mean zirconium concentration was 240 ppm and values ranged from <10 (sic) to 2,000 ppm (see their map), with 68% of all determinations falling between 125 and 400 ppm.

III. Laterites, Bauxites etc.

The mean zirconium content of 120 samples of terra rossa carbonate residuum from Missouri and Arkansas (CONNOR and SHACKLETTE, 1975) was 63 ppm, with a range of 30–150 ppm. The mean values of five subsets each of 24 samples, collected from several different geological formations, were remarkably similar—between 61 and 67 ppm.

GOLDSCHMIDT (1954) stated that zirconium concentrates strongly in laterites; FREDRICKSON (1948), CARROLL and WOOF (1951), CARROLL (1953) and DEGENHARDT

(1957) have observed corrosion of zircon in laterites. ZEISSINK (1971) presented analyses of zirconium in two lateritic profiles developed on Australian serpentinites (Table 40-G-2). Zirconium is powerfully concentrated in only one profile.

Zirconium is concentrated in bauxites and kaolinites (GOLDSCHMIDT, 1937; GORDON and MURATA, 1952). DEGENHARDT (1957) gave a mean value of 510 ppm (range 423–623) for European bauxites derived from limestone. Zirconium analyses for twenty-nine bauxite samples were given by ADAMS and RICHARDSON (1960) (Table 40-G-3). As expected, the concentration of zirconium is highest in bauxites derived from felsic or alkalic igneous rocks and lowest in those from carbonate rocks. The weathering of certain Malaysian andesite lavas (240 ppm Zr) has produced zirconium enriched bauxites (mean 417 ppm, range 350–530 ppm) (WOLFENDEN, 1965). Two bauxites from Arkansas contained 2,570 and 2,530 ppm Zr, and the parent nepheline syenite yielded values of 550, 640 and 850 ppm. Zirconium in disaggregated bauxites from Surinam is concentrated in mud particles greater than 6 μm in diameter, but in bauxites from Arkansas the metal is quite uniformly distributed in all size fractions. Zirconium in these bauxites is not related to the uranium or thorium contents, and is apparently not present as zircon. DEGENHARDT (1957) believes that zirconium in bauxites is bound to iron or aluminium compounds.

Data for zirconium in two similar kaolinite-rich drill cores from West Bengal (MUKHERJEE *et al.*, 1969) are compiled in Table 40-G-4. The authors conclude that the Adda profile is residual, while the Chaubatta profile is transported. The mean abundance of zirconium in five Malaysian kaolinites (WOLFENDEN, 1965) was 440 ppm. The range was 400–510 ppm and the concentration factor relative to the parent andesite was 1.8. DEGENHARDT (1957) lists zirconium values for two Euro-

Table 40-G-2. *Zirconium content of two Australian lateritic profiles* (ZEISSINK, 1971, method: X)

Greenvale profile		Depth	Rockhampton profile	
ppm Zr	Description		Description	ppm Zr
132		0– 5 ft		252
35		5–10		172
28	Ferruginous	10–15		177
72		15–20	Ferruginous	180
121		20–25		101
125		25–30		117
89	Montmorillonitic	30–35		122
77		35–40		107
219		40–45		86
63	Weathered serpentinite	45–50		96
68		50–55	Montmorillonitic	—
75		55–60		19
84		60–65		3
		65–70		<3
	Serpentinite	70–75	Weathered serpentinite	<3
		75–80		<3
		80–85		<3
		85–90	Serpentinite	<3

Table 40-G-3. *Zirconium analyses of bauxites derived from different rock types* (ADAMS and RICHARDSON, 1960, method W)

Source	No. of samples	ppm Zr	
		Range	Average
Carbonate rocks	11	220– 620	480
Shales	4	220–1,000	590
Mafic igneous rocks	5	200–1,320	690
Felsic igneous rocks	5	760–6,500	2,480

Table 40-G-4. *Zirconium in two kaolinite-rich boreholes from West Bengal* (MUKHER.JEE *et al* 1969, method S)

Adda borehole			Chaubatta borehole		
Depth (m)	Description	ppm Zr	Depth (m)	Description	ppm Zr
11	Clay	95	8,14	Clay	87
21	Clay	90	24	Clay	93
29	Clay	83	32	Basalt	97
31, 32[a]	Archaean gneiss	52	[b]	Fresh basalt	92

[a] Mean. [b] Obtained elsewhere.

pean kaolinites (88 and 92 ppm) and one from Arkansas (1,120 ppm). The latter sample was derived from a zircon-rich nepheline syenite, and about 78 ppm of the zirconium was readily extracted from the clay by 2N HCl.

IV. Sea Water Alteration

CANN (1970) and HART *et al.* (1974) found that the concentration of zirconium in submarine basaltic pillows was unchanged by sea water alteration.

Revised manuscript received: June 1977

40-H. Solubilities of Compounds which Control Concentrations of Zirconium in Natural Waters; Adsorption Processes; Valence States in Natural Environments

Zirconium exhibits known oxidation states of 0, II, III and IV, but only the last of these is important in nature and there is no evidence for the existence of lower oxidation states in aqueous solution (COTTON and WILKINSON, 1966). Zircon ($ZrSiO_4$), the principal zirconium mineral, is very insoluble in most natural aqueous solutions, unless artificially comminutated (BOUYOUCOS, 1921). DEGENHARDT (1957) placed 100 mg samples of powdered zircon in 250 ml aliquots of various solvents, at 20° C, and determined the concentration of zirconium in the solution after seven days (Table 40-H-1). It is probably unrealistic to extrapolate from this experiment to a natural situation. Other important solid zirconium compounds encountered in nature such as baddeleyite (ZrO_2) and hydrolysed colloidal precipitates of indefinite composition are likewise generally insoluble. The dearth of natural soluble zirconium compounds and the fact that the Zr^{4+} ion is readily hydrolysed even in fairly acidic media accounts for the very low concentrations of zirconium in most natural waters. Exceptions are probably due to concentrations of the metal in colloidal suspension, or, much less probably, in some sort of stable complex.

The forms of zirconium compounds in natural waters and associated precipitates are little known. Attempts have been made to quantify the physicochemical characteristics of zirconium in aqueous solutions (ROZZELL and ANDELMAN, 1971) but the behavior of the element in water is exceedingly complex (BILINSKI and TYREE, 1971) and apparently little progress has been made in the application of data of this sort to the natural sedimentary cycle of the element. STROCK and DREXLER (1941) proposed that zirconium in some acidic spring waters was stabilised as calcium dicarbonato zirconylate, $Ca(ZrO)_2[CO_3]_2$ or as the ion $ZrO[CO_3]_2^{2-}$. Some authors (SAXENA, 1966; RANKAMA and SAHAMA, 1950) refer loosely to "zirconates" in solution, which would be species such as ZrO_3^{2-}, $Zr_2O_5^{2-}$ or ZrO_4^{4-} but according to COTTON and WILKINSON (1966) these do not exist and neither do aqueous species such as Zr^{4+} and zirconyl ion, ZrO^{2+}. Experiments indicate that Zr^{4+} ions exist in polymers such as $[Zr_3(OH)_6Cl_3]^{3+}$ in acidic solutions rich in zirconium, but this may not hold for very dilute and nearly neutral waters.

The nature of zirconium hydrolysates is equally problematic. Geochemists have referred to these as hydrated oxides and as hydroxides (SAXENA, 1966). According to COTTON and WILKINSON (1966) the hydrolysis of tetravalent zirconium solutions yields $ZrO_2 \cdot nH_2O$, and they doubt that the hydroxide, $Zr(OH)_4$, can exist. But according to work reviewed by STUMM and BRAUNER (1975) the principle zirconium compound in sea water is the hydroxide $Zr(OH)_4^0$, or perhaps more generally, species of the form $Zr(OH)_n^{4-n}$.

Little is known about the adsorption of zirconium, although it seems very probable that such proccsses are partly responsible for the immobilisation of the metal in some weathering profiles and in certain fine-grained sediments (see Sects. 40-G and 40-K).

Radio-isotopic studies of adsorption of zirconium onto deep-sea sediments have been reported by BREWER (1975), but the results are difficult to interpret. Zirconium in some samples appears to be strongly associated with a certain size fraction, but in other samples it is not.

Table 40-H-1. *The solubility of crushed zircon* (DEGENHARDT, 1957, method C)[a]

Solvent	pH	mg/l Zr
Distilled water	7	12
1% sodium carbonate	11	32
1% sodium bicarbonate	8.7	36
1% sodium bicarbonate saturated with calcium bicarbonate at 20° C	6.7	40
0.1 N HCl	1	16
0.1 N NaOH	13	16

[a] See text for details.

Revised manuscript received: June 1977

40-I. Abundance in Natural Waters and in the Atmosphere

Literature references to zirconium in natural waters and in the atmosphere are relatively few and widely scattered. The best means of accessing this data is via Chemical Abstracts.

Zirconium analyses of *natural waters* are compiled in Table 40-I-1. These do not include unusually high values that have been recorded for *sea water*: DEGENHARDT (1957) found approximately 4 μg/l in Atlantic surface water and SASTRY *et al.* (1974) reported between 2.27 and 20.89 μg/l in tropical coastal waters off Bombay. The significance, if any, of these higher values remains uncertain. HORN and ADAMS (1966) estimated that the total masses of zirconium in the connate waters of the earth and in the oceans were, respectively, 5.86×10^8 and 56.1×10^8 tonnes. However, WEDEPOHL (pers. comm. 1977) has pointed out that a simple calculation using data from this handbook (0.03 μg Zr per litre sea water, Table 40-I-1; $1{,}349.929 \times 10^6$ km³ of sea water, Table 10-1, Volume 1, p. 298) shows that the total mass of zirconium in the oceans can apparently not be greater than 4.0×10^7 tonnes. Problems of this sort cannot be resolved at this time because there is far too little information about the distribution and abundance of zirconium in the sea (BREWER, pers. comm. 1977). The attempt by SCHUTZ and TUREKIAN (1965) to determine zirconium in sea water profiles by neutron activation analysis was frustrated by serious analytical inter-

Table 40-I-1. *Zirconium in natural waters*

Water-type	No. of samples	Origin	μg/l Zr		Method	Reference
			Range	Mean		
Sea water	—	Average ocean water	—	0.026[a]	—	TUREKIAN (1968)
Sea water	—	Average ocean water	—	0.03	—	BREWER (1975)
Sea water	—	—	0.011–0.041	0.026	L	SHIGEMATSU *et al.* (1964)
Oilfield water	823	Canada and U.S.A.	<10–20	<10	S[b]	RITTENHOUSE *et al.* (1969)
Lake water	440	Maine (U.S.A.)	0.05–22.5	2.61	S[b]	LIVINGSTONE (1963)
Spring water	1	Saratoga (U.S.A.)	—	350	—	STROCK and DREXLER (1941)
Spring water	1	Germany	—	2	C	DEGENHARDT (1957)
Tap water	1	Germany	—	3	C	DEGENHARDT (1957)
Tap water	1	New York City	—	2.5	S[b]	KLEINKOPF (1960)

[a] 35‰ salinity. [b] Concentration in water calculated from spectrographic analysis of sample of dried residue; e.g. KLEINKOPF (1960) used 50 mg of residue.

ferences. Zirconium could only occasionally be detected spectrometrically at very low levels in deep-sea interstitial brines (HENDRICKS *et al.*, 1969).

Zirconium is almost never sought in hydrogeochemical mineral prospecting surveys (MARCHANT, 1977) and no useful specific data is available from this source. The metal is apparently not regarded as a serious *water pollution* threat. MINENKO and EPIFANTSEVA (1973) stated that the concentration of zirconium reached a maximum of 1.1 mg/l in the waters of five coal mines in the USSR. BREZONIK (1976) stated that the *water supplies* of the USA generally contained only ppb amounts of zirconium, that the range of typical values was unknown and that the metal would not be regared as potentially toxic at ppm or lower concentrations.

Zirconium in the *atmosphere* ought to have very little geochemical or ecological significance. LEE *et al.* (1975) reported 0.12 $\mu g/m^3$ Zr in the air of Brisbane (Australia). SHUM and LOVELAND (1974) observed slight zirconium air pollution caused by a metallurgical processing plant in Oregon. Zirconium may be present in appreciable amounts in some volcanic gasses (WHITE and WARING, 1963).

Revised manuscript received: June 1977

40-K. Abundance in Common Sediments and Sedimentary Rocks

Although zirconium, with an estimated mean concentration of 155 ppm (DEGENHARDT, 1957) or 200 ppm (ANDREWS-JONES, 1968 and this paper), is amongst the most plentiful trace elements in sedimentary rocks, and has an abundance of the same sort of order as those of vanadium, rubidium and strontium, its distribution is, in certain respects, not well known (GRAF, 1962). Table 40-K-1 lists estimates of the mean concentration of zirconium in common sedimentary materials. The agreement between most comparable independant estimates is reasonable. Zirconium analyses for sedimentary materials from more recent references are compiled in Tables 40-K-2 and 40-K-3. There are few data, but these agree fairly well with the older estimates in Table 40-K-1.

Some details of the sedimentary cycle of zirconium remain unclear (see also Sect. 40-G). It is uncertain what proportions of eroded zirconium are transported as detrital zircon, in hydrolyzates and in solution. A few studies have been made of the relationship between zirconium in eroded source rocks and in sediments closely associated with those rocks. Thus WILLIS *et al.* (1977) found the following mean zirconium concentrations in quartz calcite feldspar sediments from the Saldanha Bay-Langebaan Lagoon complex of South Africa: Bay (deeper water), 145 ppm; intermediate depths, 36 ppm; lagoon (shallower water), 62 ppm; adjoining marsh area, 101 ppm. The granite of the area (149 ppm Zr) is clearly an important source of the zirconium in the sediments, but the relative depletion and irregular distribution of the element in the latter can apparently not be explained by any single simple mechanism.

That the mineral zircon can have an important role in the sedimentary cycle of zirconium is attested by studies of zircons in solving problems of petrogenesis, pedogenesis, provenance, correlation and radiometric dating of sedimentary formations (MARSHALL, 1967). Zircons may be of igneous, metamorphic and, to some extent, authigenic origin and are resistant enough (FREISE, 1931) to sometimes survive several cycles of erosion. Zircons can thus accumulate in high energy environments (MOORE, 1968) and may become concentrated in clastic sediments, especially sands. They are scarcer in chemical precipitates such as limestone and evaporites (VLASOV, 1966). HILL and PARKER (1970) found that most of the zirconium in samples of beach and offshore sediments from the British Isles occurred as zircon, which varied in amount with the grain size of these materials, being most abundant in well-sorted fine to very fine sands. The bulk of the zirconium in most South African shales (HOFMEYR, 1971) and in New Zealand arkoses (MCLAUGHLIN, 1955) is in detrital zircon.

However, this detrital model is often inadequate. The accompanying tabled data and other sources show that (i) zirconium is moderately concentrated in sandstones as opposed to shales. (ii) Deep-sea clays and average shales have similar amounts of zirconium, as do deep-sea and near-shore sediments, although the zirconium content of the pelagic material is more variable (CHESTER, 1965). LANGE (1970) noted that oceanic clays from different areas have similar amounts of zirconium (Mean deep-sea clays, 150; Atlantic 130; Pacific 170; Persian Gulf 188 ppm Zr). Zirconium is apparently associated with the colloidal fraction of deep-sea clays (CHESTER, 1965). (iii) Many manganese nodules are considerably enriched in zirconium. CHESTER (1965) estimated a two to five fold enrichment of zirconium in manganese nodules compared to the average pelagic sediment. GLASBY *et al.* (1971) found great variations in the abundance of zirconium in manganese nodules from the Gulf of Aden and concluded that this was not due to submarine volcanism.

These observations indicate that the weathering of zircon, or minerals in which zirconium is present in substantial amounts (VLASOV, 1966; see also Sect. 40-D), liberates more of the metal than was once thought (GOLDBERG and ARRHENIUS, 1958), although some of it is rapidly immobilised again in certain clays, either by ion-exchange absorption or by substituting for aluminium (NICHOLLS and LORING, 1962; DEGENHARDT, 1957). HOFMEYR (1971) found that clays separated from South African shales contained up to 200 ppm zirconium. Any dissolved zirconium escaping from a weathering profile would soon hydrolyse to a colloidal hydrous oxide[1], which could later be precipitated in manganese nodules or might, in time, become included in authigenic ferromanganese minerals in various sediments (CHESTER, 1965). Several authors cited by CARROLL (1953) and MARSHALL (1967), particularly SAXENA (1966), have suggested that zircon may form authigenically, for example through the reaction of silica with precipitated zirconium hydroxide. However, MARSHALL (1967) has shown that the proportion of authigenic zircon is probably quite low in most sedimentary rocks, and is largely restricted to secondary outgrowths on detrital zircons.

Factor analysis of analytical data for certain English Silurian shales (SPENCER, 1966) indicated that Zr, B and Ti have a strong coherence and were probably originally related in a finely comminutated detrital fraction. These elements were probably converted to clay-sized crystals of oxides or hydrated oxides before sedimentation.

A combination of detrital and colloidal mechanisms probably operates in many situations. IL'INA *et al.* (1970) found that Zr and Ti varied sympathetically in sandstones, siltstones and clays (Devonian and older) of the Moscow syncline, and concluded that both elements originated in terrigenous material entering the basin. Accentuated marine conditions during deposition are marked by an increase in the carbonate fraction of the sediments and a decrease in the concentrations of both Zr and Ti. Furthermore, nearshore carbonates of the basin have a larger terrigenous component and thus more zirconium than associated deep-water carbonates. The authors suggested that the zirconium in the carbonates is associated with the clay fraction. They found that the zirconium content of the sediments was dependant on facies, paleoclimate, tectonic conditions and whether the source rocks were mainly felsic or mainly mafic.

Some aspects of the distribution of zirconium remain problematic, for example the apparent concentration of this element in deep-sea carbonates compared to the

[1] See Sect. 40-H for details about the nature of precipitated colloidal zirconium compounds.

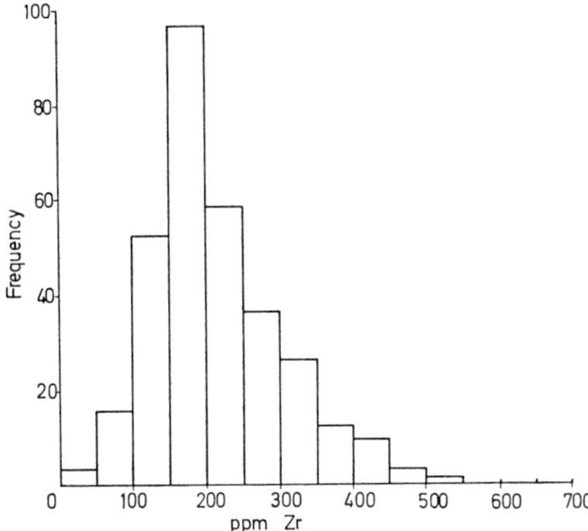

Fig. 40-K-1. Frequency distribution of zirconium concentrations in South African shales (HOFMEYR 1971)

Table 40-K-1. *Estimates of the probable mean values for the concentration of zirconium in typical sedimentary species*

Authors	Sand-stones etc.	Shales etc.	Carbon-ates	Deep sea carbon-ates	Deep sea clays	Evap-orites
RANKAMA and SAHAMA (1950)	—	120	—	—	—	—
DEGENHARDT (1957)	220	160	17–19	—	—	—
GREEN (1959)	179	137–200	20–137	—	—	—
VINOGRADOV (1962)	—	200	—	—	—	—
WEDEPOHL (1960)	—	—	—	—	150	—
HORN and ADAMS (1966)	204	142	18.1	84.4	150	0.4
This paper (Table 40-K-2)	188/387[a]	197	18[b]	—	152	—

[a] Greywackes. [b] DEGENHARDT (1957).

average carbonate rock (Table 40-K-1). WEDEPOHL (pers. comm. 1977) has suggested that the problem may be more apparent than real, in that deep-sea marls, with only 30% $CaCO_3$ are often classified as "carbonates".

LEBEDEV (1967) found that the concentration of zirconium in some western Siberian fresh-water clays was statistically higher than in marine clays of that region, but could not observe this difference in clays from the Caspian depression. DEGENHARDT (1957) gives an example of the use of zirconium analyses to distinguish stratigraphic horizons in oceanic clays. Geographical regions or sedimentary formations which are significantly enriched in zirconium relative to the mean values have been noted by ONDRICK and GRIFFITHS (1969), ANGINO and ANDREWS (1968), LANGE (1970), CONDIE *et al.* (1970) and HOFMEYR (1971). The very old Precambrian shales of South Africa (Figtree and Witwatersrand) are depleted in zirconium, perhaps because they were derived principally from ultramafic rocks (HOFMEYR, 1971;

Table 40-K-2. *Zirconium in various sedimentary materials*

Rock type	No. of samples	Locality, age	ppm Zr		Method	Reference
			Range	Mean		
Undifferentiated sedimentary materials						
Sediments	16	Persian Gulf	170–240	188	X	1
Shelf sediments	40	Washington-Oregon Coast		183	S	2
Glacial marine sediments	5	Weddell Sea	180–350	238	S	3
Arenaceous and argillaceous rocks	49	Carboniferous, County Durham (England)	60–589	264	X	4
Sands, sandstones and greywackes						
Beach and offshore sands	72	British Isles	<20–1,210		X	5
Sands	15	Gulf of Paria, West of Trinidad	195–730	418	S	6
River sediments (greywacke terrain)	4	Recent, Columbia River (U.S.A.)	160–185	169	S	7
Sandstones	412	Devonian and older, Moscow (U.S.S.R.)		160	S	8
Sandstones	262	Paleozoic to Tertiary, European U.S.S.R.		220	S	9
Sandstones	51	Carboniferous to Cretaceous, W. Germany	33–480	259	W/C	33
Siltstones	218	Paleozoic to Tertiary, European U.S.S.R.		250	S	9
Siltstones	358	Devonian and older, Moscow (U.S.S.R.)		130	S	8
Sandy seat-earth	2	Carboniferous, North Wales	205–390	298	S	10
Subgreywackes	7	Ordivician, California (U.S.A.)		155	X	11
Greywackes	24	Ordovician, New York (U.S.A.)	270–830	464	S	12
Greywackes	20	Devonian, Carboniferous, W. Germany	80–>295	266	W/C	33
Greywackes	12	Silurian, California (U.S.A.)	44–105	82	X	11
Greywackes	31	Pre-Silurian, Quebec (Canada)	280–2,000	748	S	12
Greywackes	22	Archaean, Barberton (South Africa)	103–245	150	X	13
Greywackes	17	Vosges, Western Alps	142–238	182	S	14
Greywackes	119	Troy, New York (U.S.A.)		413	S	15
Arkoses	9	New Zealand	164–270	188	?	17
Argillaceous sediments and sedimentary rocks						
Pelagic sediments	22	Pacific Ocean	96–390	180	S	18
Pelagic clays	18	Atlantic Ocean	60–270	130	X	38

Siliceous pelagic sediments	3	Atlantic, Pacific, Indian Ocean, Mediterranean Sea	140–200	167	S	19
Argillaceous pelagic sediments	12	Atlantic, Pacific, Indian Ocean, Mediterranean Sea	20–180	126	S	19
Volcanic pelagic sediments	3	Atlantic, Pacific, Indian Ocean, Mediterranean Sea	150–260	190	S	19
Volcanic muds (<5% $CaCO_3$)	3	Pacific, Deep Sea	103–172	142	S	20
Siliceous oozes (<2% $CaCO_3$)	3	Pacific, Deep Sea	156–177	168	S	20
Oozes, consolidated	7	Mid-Atlantic Ridge 43° N, 20° S	10–100	40	S	21
Red clays (<5% $CaCO_3$)	5	Pacific, Deep Sea	142–188	171	S	20
Deep sea hemipelagic sediments	12	Pacific Ocean	29–52	39	S	22
Silty clays	7	Near-shore, Louisiana (U.S.A.)	216–374	283	S	20
Clays	12	Gulf of Paria, West of Trinidad	115–350	169	S	6
Clays, globigerina oozes	9	Gulf of Mexico	19–196	116	S	20
Mudstones	6	Carboniferous, North Wales (U.K.)	160–265	223	S	10
Marlstones	6	Pierre Shale, U.S.A.		140	S	23
Marl slates, siltstones	86	Permian, S.E. Durham (England)	20–170	85	X	24
Claystones (nonmarine)	8	Late Cretaceous, U.S.A.	160–270	200	S	25
Shales, claystones	14	Pierre Shale, U.S.A.		210	S	23
Shales, claystones (marine nearshore)	32	Late Cretaceous, U.S.A.	130–200	167	S	25
Shales, claystones (marine offshore)	53	Late Cretaceous, U.S.A.	90–350	193	S	25
Shales, claystones (org.C <0.5%)	19	Pierre Shale, U.S.A.		180	S	23
Shales, claystones (org.C 0.5–1.0%)	20	Pierre Shale, U.S.A.		200	S	23
Carbonaceous shales (nonmarine)	14	Late Cretaceous, U.S.A.	150–470	243	S	25
Shales	27	Silurian, England and Wales	128–206	171	S	26
Shales	133	Ordivician to Silurian, Southern Scotland	34–466	136	X	27
Shales	16	Kimmeridge, Dorset (England)	68–167	123	X	28
Shales (composites)	50	Europe, Japan		153	X	38
Shales	17	Lower Jurassic, Yorkshire (England)	110–280	180	S	29
Clays, shales	1,179	Devonian and older, Moscow (U.S.S.R.)		170	S	8
Clays, shales	773	Paleozoic to Tertiary, European U.S.S.R.		210	S	9
Clays, shales (fresh water)	159	Jurassic, Cretaceous, U.S.S.R.		282	S	30

Table 40-K-2 (continued)

Rock type	No. of samples	Locality, age	ppm Zr Range	ppm Zr Mean	Method	Reference
Clays, shales (brackish water)	322	Jurassic, Cretaceous, U.S.S.R.		210	S	30
Clays, shales (marine)	439	Jurassic, Cretaceous, U.S.S.R.		222	S	30
Argillaceous rocks	313	Precambrian to Jurassic, S. Africa	31–888	214	X	39
Bentonite	10	Pierre Shale, U.S.A.		260	S	23
Carbonate rocks						
Calcareous pelagic sediments	10	Atlantic, Pacific, Indian Ocean, Mediterranean Sea	10–250	140	S	19
Globigerina oozes	2	Atlantic Ocean	10–30	20	W/C	33
Argillaceous carbonates	247	Paleozoic to Tertiary, European U.S.S.R.		80	S	9
Carbonate rocks	183	Scotland	<10–200		?	31
Carbonate rocks	876	Ordovician and Devonian, Moscow (U.S.S.R.)		40	S	8
Carbonate rocks	903	Paleozoic to Tertiary, European U.S.S.R.		Trace	S	9
Limestones	78	Upper Jurassic, Kimmeridge Bay (England)	6–230	115	X	32
Carbonate sediments	2	Mid-Atlantic Ridge 43° N, 20° S	60–200	130	S	21
Limestones	101	Paleozoic to Cretaceous, W. Germany	9–45	18	W/C	33
Dolostones	3	Mid-Atlantic Ridge 43° N, 20° S	10–55	29	S	21
Miscellaneous sedimentary materials						
Rock salt	2	Reyershausen (W. Germany)	0.2–0.5	0.4	C	33
Halite rocks	7	U.S.A.		<4	C	16
Polyhalite rock	1	Permian, Saldo Formation, New Mexico (U.S.A.)		22	C	16
Sylvinite	2	Stassfurth, Vienenburg (Germany)	<0.1–0.1	≤0.1	C	33
Gypsum, anhydrite	2	Permian, Lower Saxony (W. Germany)	4–26	15	C	33
Anhydrite	3	Pennsylvanian and Permian, Utah and New Mexico (U.S.A.)	13–38	29	C	16
Loess	24	Quaternary, Missouri (U.S.A.)	150–300	230	S	40
Cherts and flint	4	Various locations	12–25	19	C	33
Ironstones	5	Silurian, Clinton Formation, New York State (U.S.A.)	0–700	174	?	16
Ironstones	6	Devonian, Martin Formation, Arizona (U.S.A.)	30–150	80	?	16

Sedimentary iron ores	4	Jurassic, Lower Saxony (W. Germany)	24–335	121	C	33
Pelletal phosphorite	4	Miocene, continental shelf, S.W. Africa	17–44	29	X	34
Glauconitized phosphorite	4	Miocene, continental shelf, S.W. Africa	11–37	28	X	34
Concretionary phosphorite	3	Modern, continental shelf, S.W. Africa		<3	X	34
Glauconite	3	Continental shelf, S.W. Africa	33–55	44	X	34
Manganese nodules	7	Atlantic Ocean	211–457	385	X	35
Manganese nodules	17	Indian Ocean	59–781	301	X	35
Manganese nodules	20	Pacific Ocean	27–945	404	X	35
Manganese nodule (5% $CaCO_3$)	1	Pacific Ocean		445	S	20
Manganese nodules	4	Gulf of Aden	20–480	155	S	36
Manganese nodules	12	Loch Fyne, Argyllshire (Scotland)	40–60	53	X	37
Mean (manganese nodules)				288		

References: 1. Lange (1970); 2. White (1970); 3. Angino and Andrews (1968); 4. Hirst and Kaye (1971); 5. Hill and Parker (1970); 6. Hirst (1962); 7. Whetten (1966); 8. Il'ina *et al.* (1970); 9. Katchenkov (1967); 10. Nicholls and Loring (1962); 11. Condie and Snansieng (1971); 12. Weber and Middleton (1961); 13. Condie *et al.* (1970); 14. Rivalenti and Sighinolfi (1969); 15. Ondrick and Griffiths (1969); 16. Fleischer (1963, 1966); 17. McLaughlin (1955); 18. Goldberg and Arrhenius (1958); 19. El Wakeel and Riley (1961); 20. Young (1968); 21. Thompson (1972); 22. Bonatti *et al.* (1971); 23. Tourtelot *et al.* (1960); 24. Hirst and Dunham (1963); 25. Tourtelot (1964); 26. Spencer (1966); 27. Stephens *et al.* (1975); 28. Cosgrove (1970); 29. Gad *et al.* (1969); 30. Lebedev (1967); 31. Graf (1960); 32. Dunn (1974); 33. Degenhardt (1957); 34. Bremner (1977); 35. Willis (1970); 36. Glasby *et al.* (1971); 37. Calvert and Price (1970); 38. Wedepohl (1960); 39. Hofmeyr (1971); 40. Connor and Shacklette (1975).

Table 40-K-3. *Zirconium in South African argillaceous sedimentary rocks* (HOFMEYR, 1971, method X)

Formation	No. of samples	Range	Average
Fig Tree series, Swaziland system	23	37–130	88
Witwatersrand system	9	99–188	133
Kunjas series, Sinclair group	5	168–261	225
Malmesbury formation	9	146–271	189
Nama system	22	31–838	246
Cape system	45	126–478	237
Karroo system, carbonaceous rocks	132	102–888	255
Karroo system, non-carbonaceous rocks	61	71–307	163
Dredged Agulhas bank sediments	7	104–260	210
Totals	313	31–888	214
$<2\ \mu m$ fraction of South African argillaceous rocks	33	44–205	121

DANCHIN, 1967), although this has been disputed (CONDIE *et al.*, 1970). IL'INA *et al.* (1970) suggested that age-variations in the zirconium content of sedimentary rocks in the Moscow basin reflected the type of source rock exposed during different geological periods.

The distribution of zirconium concentrations in the Rensselaer greywacke is log-normal (ONDRICK and GRIFFITHS, 1969) and in South African shales it has a marked positive skewness, which may be a characteristic inherited from the source rocks (HOFMEYR, 1971, Fig. 40-K-1). Few strong correlations have been reported between zirconium and other elements in sediments. HOFMEYR (1971) found no correlation between zirconium and any other element in South African shales. BREMNER (1977) found a strong correlation between zirconium and yttrium and chromium in Southern African marine phosphorites. In associated glauconitic sediments zirconium correlated only with chromium.

Note added in press: There is a wealth of data for zirconium in recent oceanic sediments in "Initial Reports of the Deep Sea Drilling Project" (Nat. Sci. Foundation, USA), Vol. VI, 829 and 1131; Vol. XXIII, 923 and 975; Vol. XXXIV, 559; Vol. XXXV, 427 and 447 (1971 to 1976).

Revised manuscript received: June 1977

40-L. Biogeochemistry

A moderate number of analyses of zirconium in biological materials are scattered in many publications, which can be accessed via Chemical Abstracts. Zirconium is not known to be essential to any organism, nor are naturally-occurring compounds of the element particularly toxic (LEHNINGER, 1970). Intravenous administration of a sufficient quantity of soluble zirconium salts will kill animals (e. g. lethal dose for rabbits is 150 mg/kg body mass, FAIRHALL, 1949). Zirconium (and certain other metals) alter the taste-sensations in mice (IZUMI, 1974). In general zirconium and its compounds are not considered to constitute an industrial health hazard (INDUSTRIAL HYGIENE FOUNDATION, 1954). BARNES (1955) reported that no occupational case of systematic poisoning in man had been observed and that the oxide was practically physiologically inert. Many zirconium compounds, such as the oxide, are only slowly eliminated from lung tissue, but do no apparent serious harm there. Human dental enamel contains between 0.1 and 1.0 ppm Zr (LOSEE *et al.*, 1973).

There has been considerable interest in radioactive ^{95}Zr (bomb zirconium) in organisms. A review is given by BACHMAN and BANKS (1965).

I. Organic Matter

Most organisms contain only traces of zirconium, but the element does accumulate to a degree in certain marine species (RANKAMA and SAHAMA, 1950). Analysis of a dried Pacific *plankton* sample yielded 5.8 ppm Zr, which is about 10^5 times the zirconium content of sea water. VINOGRADOVA and KOVAL'SKIY (1962) reported from 15 to 50 ppm Zr in the ash of planktonic diatoms and from 10 to 70 ppm in the ash of zooplankton from the Black Sea. Samples of Cretaceous radiolaria from Timor contain about 50 ppm Zr, with a maximum observed value of 120 ppm (AUDLEY-CHARLES, 1965). However, if interelement concentration ratios in most simple marine organisms are considered, then zirconium is found to be neither more nor less concentrated in them than in average shale (BOSTRÖM *et al.*, 1974). GRIBOVSKAYA *et al.* (1974) found that the mass of zirconium contained in the algae *Chara intermedia* in Lake Issyk-Kul was 3.7 kg/ha.

DEGENHARDT (1957) reported less than two parts per million zirconium in a whole piglet and 60 ppm in the bones of an adult pig.

Although zirconium can apparently accumulate to a degree in some *plants* (RANKAMA and SAHAMA, 1950), the average enrichment coefficient for this metal in plants is particularly low. Because of the extreme immobility of the zirconium ion, it is unlikely that significant quantities of the element will be accumulated in plant matter (BROOKS, 1972). A survey of many species of wild Belorussian berries (SENCHUK and BORUKH, 1976) revealed only traces of zirconium. The leaves of apple trees contained 0.79 ppb Zr (SHUMILO, 1975). CONNOR and SHACKLETTE (1975) list some 400 zirconium analyses of the ashes of twenty-odd cultivated and native species of plants (vegetables and other crops, trees and bushes) from Georgia and Missouri. Reported mean values for different species varied between 2.4 and 85 ppm. For almost all species the minimum observed zirconium concentration was less than 20 ppm and only nine species produced individual analyses in excess of 100 ppm. Cabbage ash contained up to 700 ppm, beans up to 150 ppm and certain pine trees up to 150 ppm zirconium.

II. Coal

Zirconium is concentrated in certain coals. GOLDSCHMIDT (1954) reported up to 0.5% Zr in coal ash. The ranges of zirconium concentrations in samples from two Appalachian anthracites were, respectively, 202–959 and 200–300 ppm (MEDLIN *et al.*, 1975). Nevertheless, these values are apparently exceptional. YUDOVICH *et al.* (1972) reviewed trace element analyses of coals from many countries. For hard coals the values of the geometric mean zirconium concentration of coal and the enrichment factor relative to the concentration of the metal in mean continental sediments were 28 ppm Zr and 0.2 respectively. For brown coals the corresponding values were 10 ppm Zr and 0.1.

Revised manuscript received: June 1977

40-M. Abundance in Common Metamorphic Rock Types

The zirconium data for metamorphic rocks (Table 40-M-1) are grouped into four broad categories, regionally metamorphosed rocks, crustal averages, contact metamorphic rocks and metasomatic rocks. Additional data on greenstones of basaltic composition have been included in Table 40-E-3 for the reasons discussed in Sect. 40-E.

Zirconium commonly occurs as accessory zircon in metamorphic rocks. GOLDSCHMIDT (1954) has noted that relict textures of zircon may be preserved through whole cycles of metamorphism, although secondary metamorphic overgrowths have also been observed (SAXENA, 1966). MARSHALL (1967) on the other hand, considers overgrowth features to be uncommon, and has re-emphasised the use of both the shape, and U/Pb dating of zircon, for recording premetamorphic events. Many of the common metamorphic minerals however are able to contain significant amounts of the total zirconium present in the whole rock, e. g. clinopyroxene, amphibole, biotite, garnet, ilmenite and sphene (see Table 40-D-1).

DEGENHARDT (1957) noted that the zirconium content of metamorphic rocks is about equal to the amount in the presumed parent rocks, and this is confirmed in this study although in many cases data are scarce. Ortho-amphibolites and metagabbros show the same range in zirconium concentration as basalts, dolerites and gabbros from Table 40-E-3 and are indistinguishable from para-amphibolites and undifferentiated amphibolites listed in Table 40-M-1, in terms of the range and average ziconium concentration.

Meta-peridotites and ultramafic rocks have low zirconium concentrations similar to many of the values reported in Table 40-E-2, and the few data on marbles indicate low concentrations comparable to sedimentary carbonate rocks.

Average concentrations from Table 40-M-1 are: Schists (217 samples) 150 ppm Zr; gneisses (520 samples) 176 ppm Zr and granulites (166 samples) 268 ppm Zr. The data presented by LAMBERT and HEIER (1968) for granulites from the Cape Naturaliste area (Australia) have significantly higher zirconium contents than schists, gneisses and granulites from other areas.

The average zirconium in the commonest metamorphic rocks—slates, amphibolites, schists, gneisses and granulites—(1,293 samples) from Table 40-M-1 is 183 ppm Zr. This value is in good agreement with the average of EADE and FAHRIG (1973), for the Canadian Precambrian shield (190 ppm Zr) but lower than the average quoted by SHAW *et al.* (1976) of 237 ppm Zr.

The metasomatic rocks in contrast often have very high zirconium abundances (commonly 0.1–1.0% Zr), relative to both the host rock and the parent rock of the metasomatic fluids. (VLASOV, 1966; HIETANEN, 1962; YES'KOVA and YEFIMOV, 1970). Zirconium in fenites associated with carbonatites is not as high as in the fenites associated with alkali rich rocks such as the Lovozero alkali massif in the U.S.S.R. (e. g. VLASOV, 1966; CURRIE and FERGUSON, 1971).

Table 40-M-1. *Zirconium abundances in metamorphic rocks*

Rock type	No. of samples	Locality	ppm Zr		Method	Reference
			Range	Mean		
Regionally metamorphosed rocks						
Slate	5	New Hampshire (U.S.A.)	170–310	248	S	1
Slate	87	Vosges (France)	61–465	151	X	2
Meta-arkose	7	Southern California (U.S.A.)	183–607	315	S	3
Marble	6	Grenville Province, Ontario (Canada)	25–99	72	X	4
Greenschist, amphibolite	58	Dalradian Series (Scotland)	83–681	183	X	5
Greenschist	3	Pounamu Series (New Zealand)	65–160	122	S	6
Schist	8	New Hampshire (U.S.A.)	140–350	225	S	1
Schist	2	Söröy (Northern Norway)	270–310	290	X, S	7
Phyllite, mica schist	174	Precambrian, Finland	60–310	138	S	8
Sericitic phyllite	2	Precambrian, Mysore (India)	138–165	151	S	9
Micaceous schist	4	Precambrian, Mysore (India)	125–210	159	S	9
Actinolite-quartz-chlorite schist	5	Precambrian, Mysore (India)	70–125	86	S	9
Quartz-albite-biotite schist	8	Arahura series (New Zealand)	100–250	155	S	6
Schist, gneiss, quartzite	6	Clearwater County, Idaho (U.S.A.)	200–400	367	S	10
Metagabbro	16	Söröy (Northern Norway)	50–200	121	X, S	7
Ortho-amphibolite	16	Grenville Province, Ontario (Canada)	64–372	139	X	4
Ortho-amphibolite	13	Randesund gneiss (Norway)	90–280	173	X	11
Para-amphibolite	31	Grenville Province, Ontario (Canada)	52–280	148	X	4
Amphibolite	36	Waldviertel and Eastern Alps (Austria)	10–1,075	158	S	12
Amphibolite	20	Precambrian, Jaraguá (Brazil)	47–220	104	?	13
Amphibolite	15	Colton area, New York (U.S.A.)	72–330	127	S	14
Amphibolite	18	Emeryville area, New York (U.S.A.)	51–220	123	S	14
Amphibolite facies rocks	27	Musgrave range (Australia)		250	X	15
Amphibolite facies rocks	14	Cape Naturaliste (Australia)		400	X	15
Amphibolite facies rocks	20	Eyre Peninsula (Australia)		195	X	15
Amphibolite	32	Coll and Tiree (Scotland)		89	X	16
Gneiss, garnet amphibolite	6	Clearwater County, Idaho (U.S.A.)	30–200	112	S	10
Gneiss	3	New Hampshire (U.S.A.)	140–260	190	S	1
Gneiss	22	Alps and Bohemian Massif (Austria)	60–300	165	S	17

Biotite gneiss	57	Ontario (Canada)	20–500	128	S	18
Biotite-plagioclase gneiss	16	Grenville Province, Ontario (Canada)	109–960	242	X	4
Biotite-hornblende gneiss	9	Precambrian, Wyoming (U.S.A.)	13–290	156	X	19
Acid to basic gneiss	254	Drumbeg area (Scotland)	18–1,945	202	X	20
Hornblende-biotite gneiss	26	Coll and Tiree (Scotland)		169	X	16
Garnet-biotite gneiss	14	Coll and Tiree (Scotland)		125	X	16
Pyroxene gneiss	8	Tiree (Scotland)		144	X	16
Quartz-pyroxene gneiss	22	Ontario (Canada)	20–300	67	S	18
Pyroxene-scapolite gneiss	12	Grenville Province, Ontario (Canada)	52–780	164	X	4
Paragneiss (amphibolite facies)	9	Emeryville area, New York (U.S.A.)	130–200	171	S	21
Paragneiss (granulite facies)	8	Colton area, New York (U.S.A.)	100–200	176	S	21
Granitized gneiss	4	Colton area, New York (U.S.A.)	100–200	160	S	21
Granitized gneiss	23	Emeryville area, New York (U.S.A.)	150–200	161	S	21
Carbonate-silicate gneiss	21	Grenville Province, Ontario (Canada)	35–225	129	X	4
Gneiss, migmatite	24	Precambrian, Wyoming (U.S.A.)	45–284	133	X	19
Adamellite, migmatitic gneiss	5	Callander Bay (Canada)	250–880	478	W	22
Metaperidotite	4	Söröy (Northern Norway)	55–100	73	X, S	7
Ultramafic rocks (granulite facies)	4	Tiree (Sotland)		33	X	16
Pyribolite	9	Tiree (Scotland)		89	X	16
Acid granulite ($>70\%$ SiO_2)	9	Brazilian Shield		195	X	23
Sub-acid granulite (65–70% SiO_2)	11	Brazilian Shield		244	X	23
Intermediate granulite(55–65% SiO_2)	25	Brazilian Shield		296	X	23
Mafic granulite ($<55\%$ SiO_2)	17	Brazilian Shield		114	X	23
Granulite facies rocks	26	Musgrave Range (Australia)		310	X	15
Granulite facies rocks	17	Fraser Range (Australia)		115	X	15
Granulite facies rocks	30	Cape Naturaliste (Australia)		525	X	15
Granulite facies rocks	11	Eyre Peninsula (Australia)		185	X	15
Banded granulites	20	Lofoten-Vesterålen (N. Norway)		20	X	35
Eclogite	79	Eastern Alps (Austria)	22–272	98	X	34
Eclogite	6	Nordfjord (West Norway)	60–190	110	S	24
Eclogite	5	California (U.S.A.)	100–150	140	S	25
Flaser-eclogite	12	Voltri group, Pennidic Belt, Apennines (Italy)		45	X	26
Fine grained eclogite	4	Voltri group, Pennidic Belt, Apennines (Italy)		48	X	26

Table 40-M-1 (continued)

Rock type	No. of samples	Locality	ppm Zr		Method	Reference
			Range	Mean		
Crustal averages						
Average Precambrian shield	5,512	Canada	155–380[a]	240	X	27
Average Precambrian shield	5,084	Canada	<100–655[a]	190	S	28
Average Archaean rock	40	S.W. Australian Shield		220	X	15
Contact metamorphic rocks						
Hornfels	24	N.S.W. (Australia)	132–251	186	S	29
Pelitic schist	16	Connemara (Ireland)	105–380	241	S	30
Hornfels	11	Connemara (Ireland)	120–800	375	S	30
Hornfelsic xenoliths	5	Connemara (Ireland)	75–240	148	S	30
Metasomatic rocks						
Spinel skarn	4	Lespromkhoz (U.S.S.R.)	5–20	11	S	31
Monticellite skarn	22	Lespromkhoz (U.S.S.R.)	30–700	210	S	31
Metasomatic rocks including skarns	4	Clearwater County, Idaho (U.S.A.)	0.01–0.30%	0.17%	S	10
Microclinization stage metasomatite	45	Urals (U.S.S.R.)	0.03–1.48%	0.29%	?	32
Albitization stage metasomatite	47	Urals (U.S.S.R.)	0.03–1.18%	0.27%	?	32
Late hydrothermal stage metasomatite	10	Urals (U.S.S.R.)	0.13–0.39%	0.27%	?	32
Fenite	11	Callander Bay (Canada)	270–1,100	490	W	22
Fenitized monzonitic granite	9	Nejoio (Angola)	93–525	208	X	33
Fenitized granite	5	Nejoio (Angola)	273–447	350	X	33
Fenitized granodiorite	12	Nejoio (Angola)	148–314	209	X	33
Granitoid rocks	35	Ontario (Canada)	235–320	272	S	18

[a] Ranges of composite samples.

References: 1. Shaw (1956); 2. Tobschall (1975); 3. Schwarcz (1966); 4. Van de Kamp (1968); 5. Van de Kamp (1970); 6. Taylor (1955); 7. Stumpfl and Sturt (1965); 8. Lonka (1967); 9. Naqvi and Hussain (1972); 10. Hietanen (1962); 11. Ball (1966); 12. Janda *et al.* (1965); 13. De Barros Gomes *et al.* (1964); 14. Engel and Engel (1962 b); 15. Lambert and Heier (1968); 16. Drury (1973); 17. Grohmann (1965); 18. Currie (1968); 19. Condie (1969); 20. Sheraton (1970); 21. Engel and Engel (1958); 22. Currie and Ferguson (1971); 23. Sighinolfi (1971); 24. Bryhni *et al.* (1969); 25. Coleman *et al.* (1965); 26. Mottana and Bocchio (1975); 27. Shaw *et al.* (1976); 28. Eade and Fahrig (1973); 29. Joyce (1969); 30. Evans (1964); 31. Sinyakov and Nepeina (1969); 32. Yes'kova and Yefimov (1970); 33. Rodrigues (1973); 34. Richter (1973); 35. Heier and Thoresen (1970).

Revised manuscript received: June 1977

40-N. Behavior in Metamorphic Reactions

Much of our knowledge on the behavior of zirconium in metamorphic reactions has been gained from investigations of sedimentary rock units which have been subjected to progressive metamorphism, and this is conveniently discussed as follows.

I. Regional Metamorphic Processes

The effects of regional metamorphism on the geochemistry of pelitic shales that have been metamorphosed progressively from slates to schists and gneisses, has been studied by SHAW (1954). No significant change in the zirconium concentration could be detected with increasing metamorphic grade within the range of concentrations of the original pelitic sediment. ENGEL and ENGEL (1958) similarly found no detectable change in the zirconium concentration of paragneisses from the Adirondack Mountains, New York, which have been metamorphosed from amphibolite facies to granulite facies, over a 56 km zone. In contrast to the whole rocks, the zirconium content of the biotites from these samples increased systematically from 77 ppm to 190 ppm with increasing metamorphic grade, (ENGEL and ENGEL, 1960), while no systematic trend was observed for zirconium in the co-existing garnets.

Subtle changes in the zirconium concentration may not have been detected in the above studies, due to the large natural variability of zirconium in the parent materials; as has been noted by SCHWARCZ (1966) in a study of progressively metamorphosed arkosic quartzites. Reference to the literature shows that zircon is a very common accessory mineral in many metamorphic rocks. The immobile nature of zirconium may in part be due to the relative stability of zircon in the metamorphic enviroment.

The effect of low-grade metamorphism on the zirconium concentration in basaltic rocks has been investigated by a number of workers, e. g. CANN (1970); PEARCE and CANN (1971); WOOD *et al.* (1976); and SMITH and SMITH (1976). These authors have found that zirconium maintains the same degree of correlation with elements such as titanium, yttrium, niobium and phosphorus, in fresh and metamorphosed rocks. This has been interpreted to indicate that these elements have been relatively unaffected by metasomatic transport. This conclusion is particularly relevant for geochemical investigations of altered basaltic magmas, particularly in ancient greenstone belts, where zirconium can be used to characterize the primary geochemistry of the lavas with some confidence (NESBITT and SUN, 1976; HALLBERG and WILLIAMS, 1972; nnd CONDIE and HARRISON, 1976).

II. Contact Metamorphic Processes

EVANS (1964) has shown an increase of zirconium in pelitic hornfelses (average 375 ppm Zr) surrounding the Cashel Lough Wheelaun mafic and ultramafic intrusion, compared to the parent pelite (average 250 ppm Zr). In contrast, the most intensely altered xenoliths show a decrease in zirconium to below 100 ppm Zr. The relative enrichment and depletion trends for zirconium are thought to be due to anatectic melting and removal of zirconium, this process being most pronounced in the more intensely altered xenoliths.

CURRIE (1968) has also noted a similar zirconium enriched envelope of gneisses enclosing granitoid bodies from Ontario, Canada. Both the surrounding country rock gneisses and the granitoid bodies have lower zirconium concentrations than the envelope of zirconium enriched gneiss.

One of us (ERLANK, unpublished data), has studied the well-known Sea Point Contact, South Africa, where the 550 m. y. Cape granites are intrusive into the Malmesbury Formation sediments of argillaceous composition (WALKER and MATHIAS, 1946). Within the contact aureole the sediments have been converted to biotite-muscovite-cordierite hornfelses. Unlike other elements such as rubidium, cesium and fluorine which have been enriched in the hornfelses immediately adjacent to the granite, zirconium shows no detectable systematic change in concentration as the contact is approached, or in hornfels xenoliths enclosed within the granite. It should be noted that the granite has similar zirconium contents (KOLBE, 1966), and thus no concentration gradient was present at the time of intrusion. Relevant data are shown in Table 40-N-1.

Table 40-N-1. *Distribution of zirconium in rocks from the Sea Point Contact, Cape Town, South Africa.* (Granite data from KOLBE, 1966, method S; other data from ERLANK, unpublished, method X.)

Rock type	No. of samples	ppm Zr	
		Range	Average
Unmetamorphosed Malmesbury argillites	13	156–240	202
Biotite–muscovite–cordierite aureole hornfelses	24	153–238	189
Biotite–muscovite–cordierite hornfels xenoliths	8	175–204	192
Cape Granites, coarse-grained varieties	26	115–330	200

III. Fenitisation Processes

The enrichment of zirconium in fenites associated with carbonatites is not as spectacular as in the fenites associated with alkali rich rocks such as Lovozero alkali massif in the U.S.S.R. Fenites are developed from metasomatic fluids migrating through channelways in the country rock (CURRIE and FERGUSON, 1971; HIETANEN, 1962; MCKIE, 1966). The solubility of zirconium is probably enhanced by the alkali rich nature of the metasomatic fluids (DIETRICH, 1968; SINYAKOV and NEPEINA, 1969). See Subsect. 40-E-VI for further discussion on this point.

Zirconium in the fenites associated with peralkaline rocks occurs principally in independent zirconium minerals such as eudialyte, zircon and baddeleyite, and other minerals e. g. aegirine, tremolite, pyrochlore and ilmenite.

VLASOV (1966) considers that magmatic rocks release substantial amounts of dispersed zirconium during post-magmatic stages, which leads to the development of new mineral species in the metasomatic rocks. Alteration of the primary zirconium rich metasomatic minerals to biotite amphibole or albite, releases zirconium which does not migrate over large distances, before crystallizing as zircon (VLASOV, 1966). The interested reader is refered to the references already cited in this discussion of fenites for more detailed accounts on the behaviour of zirconium in metasomatic rocks.

As mentioned in Subsect. 40-E-II, high concentrations of zirconium have been reported in ultramafic nodules which contain abundant mica and Na and K-rich amphibole, and these are believed to be the result of upper mantle metasomatic processes (ERLANK, 1973 and 1976; LLOYD and BAILEY, 1975), which may be similar to the fenitisation processes discussed above.

Revised manuscript received: June 1977

40-O. Relations to Other Elements, Economic Importance

I. Relations to Other Elements

The interested reader should note that the close geochemical coherence existing between zirconium and hafnium, usually depicted as the Zr/Hf ratio, is dealt with in Chapt. 72 of this handbook.

II. Economic Importance

The main source of zirconium is from the two minerals zircon and baddeleyite. Zircon is mainly recovered from beach placer deposits, where it is associated with other heavy minerals such as ilmenite, rutile and monazite. Baddeleyite occurring with zircon is mined in the Pǫco-de-Caldos district, Brazil, from placer deposites, which originate from veins consisting essentially of baddeleyite and zircon. Huge reserves of zirconium as zircon and eudialyte occur in metasomatic rocks, nepheline and alkali syenites, and alkali granites (Vlasov, 1966).

The main producers of zircon are: Australia (New South Wales, Queensland, Western Australia), U.S.A. (Florida, South Carolina), Brazil, India, Western Africa, Sri-Lanka and Malaysia. The main source of baddeleyite is from Brazil.

Many of the industrial uses of zirconium stem from the refractory properties of zirconium oxide and zircon. Zirconia is used extensively for the manufacture of crucibles, firebricks and other materials that are subject to exceptionally high temperatures. Zirconium oxide ceramics have low coefficients of expansion, high dielectric strength and are resistant to corrosion by acids, slags and glasses.

The thermal stability, resistance to corrosion and low absorption cross section for neutrons of zirconium has led to the extensive use of this metal and its alloys in the construction of nuclear reactors. Reactor grade zirconium must be essentially free of hafnium.

Zirconium metal is used in the electronics industry, as a residual gas-getter in vacuum tubes, in rectifiers and condensers and in the manufacture of steel as a scavenger and alloying additive with iron and other metals (Lustman and Kerze, 1955). It is also utilized in explosive primers and glass to metal seals. Further applications may be found in Miller (1954).

Finally, zirconium salts have found some use as antidotes for plutonium poisoning, and are frequently used in ointments for the treatment of itching and blistering of the skin caused by the toxic chelon 3-pentadecylcatechol. This chelon occurs in the sap of *rhus toxicodendrum* (North American Poison Ivy). In practice, contact with *rhus toxicodendrum*, which is easily recognised by its three leaves growing from a woody stem, should be avoided (Blackburn, 1969).

Acknowledgment: We are grateful to R.V. Danchin, C. C. Klein and D. C. Corrigall for assisting with the compilation of data, the CSIR and the University of Cape Town for financial support, and Mrs. G. Verblun for typing the manuscript.

Revised manuscript received: June 1977

References: Sections 40-B to 40-O

ABBEY, S.: Studies in standard samples of silicate rocks and minerals. Part 4, 1974. Edition of usable values. Geol. Survey of Canada Paper **74**, 41 (1975).

ABBOTT, M. J.: Petrology of the Nandewar Volcano, N.S.W., Australia. Contr. Mineral. and Petrol. **20**, 115 (1969).

ABSHIRE, E.: Bibliography of zirconium. Supplement to Information Circular **7771**. U.S. Bur. Mines Inf. Circ. **7830** (1958).

ADAMS, J. S. A., RICHARDSON, K. A.: Thorium, uranium and zirconium concentrations in bauxite. Econ. Geol. **55**, 1653 (1960).

AHRENS, L. H., CHERRY, R. D., ERLANK, A. J.: Observations on the Th-U relationship in zircons from granitic rocks and from kimberlites. Geochim. Cosmochim. Acta **31**, 2379 (1967).

AHRENS, L. H., WILLIS, J. P., ERLANK, A. J.: The chemical composition of Kainsaz and Efremovka. Meteoritics, **8**, 133 (1973).

ALBEE, A. L., BINGHAM, E., CHODOS, A. A., MAYNES, A. D.: Phase equilibra in three assemblages of kyanite-zone pelitic schists, Lincoln Mountain Quadrangle, Central Vermont. J. Petrol. **6**, 247 (1965).

ALBUQUERQUE, C. A. R.: Geochemistry of biotites from granitic rocks, Northern Portugal. Geochim. Cosmochim. Acta **37**, 1779 (1973).

ALLEN, R. O., Jr., MASON, B.: Minor and trace elements in some meteoritic minerals. Geochim. Cosmochim. Acta **37**, 1435 (1973).

ANDREWS-JONES, D. A.: The application of geochemical techniques to mineral exploration. Mineral. Ind. Bull. **11**, 31 (1968).

ANGINO, E. E., ANDREWS, R. S.: Trace element chemistry, heavy minerals and sediment statistics of Weddel Sea sediments. J. Sedim. Petrol. **38**, 634 (1968).

APPLETON, J. D.: Petrogenesis of potassium-rich lavas from the Roccamonfina Volcano, Roman Region, Italy. J. Petrol. **13**, 425 (1972).

ATKINS, F. B.: Pyroxenes of the Bushveld intrusion, South Africa. J. Petrol. **10**, 222 (1969).

AUDLEY-CHARLES, M. G.: Some aspects of the chemistry of Cretaceous siliceous sedimentary rocks from eastern Timor. Geochim. Cosmochim. Acta **29**, 1175 (1965).

AUMENTO, F.: Diorites from the Mid-Atlantic ridge at 45° N. Science **165**, 1112 (1969).

BACHMAN, R. Z., BANKS, C. V.: Nonferrous metallurgy. II. Zirconium, hafnium, vanadium, niobium, tantalum, chromium, molybdenum, and tungsten. Anal. Chem. An. Rev. **37**, 103R (1965).

BAILEY, D. K., MACDONALD, R.: Petrochemical variations among mildly peralkaline (comendite) obsidians from the oceans and continents. Contr. Mineral. and Petrol. **28**, 340 (1970).

BAKER, P. E.: Comparative volcanology and petrology of the Atlantic island-arcs. Bull. Volcanologique **32**, 189 (1968a).

BAKER, P. E.: Petrology of Mt. Misery Volcano, St. Kitts, West Indies. Lithos **1**, 124 (1968b).

BALL, T. K.: The geochemistry of the Randesund banded gneisses. Norsk Geol. Tids. **46**, 379 (1966).

BARBERI, F., SANTACROCE, R., FERRARA, G., TREUIL, M., VARET, J.: A transitional basalt-pantellerite sequence of fractional crystallization, the Boina Centre (Afar Rift, Ethiopa). J. Petrol. **16**, 22 (1975).

BARNES, E. C.: Industrial hygiene and safety. In: LUSTMAN, B., and KERZE, F., Jr. (eds.) The Metallurgy of Zirconium. New York: McGraw-Hill Book Co. Inc. 1955.

BELL, K., POWELL, J. L.: Strontium isotopic studies of alkalic rocks: the potassium-rich lavas of the Birunga and Toro-Ankole Regions, East and Central Equatorial Africa. J. Petrol. **10**, 536 (1969).

BILINSKI, H., TYREE, S. Y. (Jr.): The rate of hydrolysis of hafnium in 1 M NaCl. In: HEM, J. D. (ed.): Nonequilibrium Systems in Natural Water Chemistry. Advances in Chemistry Series **106**, Washington: Amer. Chem. Soc. 1971.

BLACKBURN, T. R.: Equilibrium. A Chemistry of Solutions. New York: Holt, Rinehart and Winston, Inc. 1969.

BLOXAM, T. W., LEWIS, A. D.: Ti, Zr, and Cr in some British pillow lavas and their petrogenic affinities. Nature **237**, 134 (1972).

BLUMENTHAL, W. B.: The Chemical Behaviour of Zirconium. Princeton, N. J.: Van Nostrand 1958.

BONATTI, E., FISHER, D. E., JOENSUU, O., RYDELL, H. S.: Post-depositional mobility of some transition elements, phosphorus, uranium and thorium in deep sea sediments. Geochim. Cosmochim. Acta **35**, 189 (1971).

BORLEY, G. D.: Amphiboles from the younger granites of Nigeria. Part 1. Chemical classification. Mineral. Mag. **33**, 358 (1963).

BORLEY, G. D.: Potash-rich volcanic rocks from southern Spain. Mineral. Mag. **36**, 364 (1967).

BOSTRÖM, K., JOENSUU, O., BROHM, I.: Plankton: its chemical composition and its significance as a source of pelagic sediments. Chem. Geol. **14**, 255 (1974).

BOUYOUCOS, G. J.: Rate and extent of solubility of minerals and rocks under different treatments and conditions. Michigan Exper. Sta. Tech. Bull. **50** (1921).

BOWDEN, P.: Trace elements in Tanganyika carbonatites. Nature **196**, 570 (1962).

BOWDEN, P.: Zirconium in younger granites of Northern Nigeria. Geochim. Cosmochim. Acta **30**, 985 (1966).

BREMNER, J. M.: Unpublished work. University of Cape Town (1977).

BREWER, P. G.: Minor elements in sea water. In: RILEY, J. D., SKIRROW, G. (eds.): Chemical Oceanography. (2nd ed.) Vol. I. London: Academic Press 1975.

BREWER, R.: Mineralogical Examination of yellow Podzolic Soil formed on Granodiorite. Melbourne: C.S.I.R.O. 1955.

BREZONIK, P. L.: Analysis and speciation of trace metals in water supplies. In: RUBIN, A. J. (ed.), Aqueous-Environmental Chemistry of Metals. Michigan: Ann Arbor Science 1976.

BROOKS, C. K.: On the distribution of zirconium and hafnium in the Skaergaard intrusion, East Greenland. Geochim. Cosmochim. Acta **33**, 357 (1969).

BROOKS, C. K.: The concentrations of zirconium and hafnium in some igneous and metamorphic rocks and minerals. Geochim. Cosmochim. Acta **34**, 411 (1970).

BROOKS, C. K., JAKOBSSON, S. P., CAMPSIE, J.: Dredged basaltic rocks from the seaward extensions of the Reykjanes and Snaefellsnes volcanic zones, Iceland. Earth Planet. Sci. Lett. **22**, 320 (1974).

BROOKS, R. R.: Geobotany and Biogeochemistry in Mineral Exploration. New York: Harper and Row 1972.

BROWN, F. H., CARMICHAEL, I. S. E.: Quaternary volcanoes of the Lake Rudolf region: I. The basanite-tephrite series of the Korath Range. Lithos **2**, 239 (1969).

BROWN, F. H., CARMICHAEL, I. S. E.: Quaternary volcanoes of the Lake Rudolf region: II. The lavas of North Island, South Island and the Barrier. Lithos **4**, 305 (1971).

BRYHNI, I., BOLLINGBERG, H. J., GRAFF, P. R.: Eclogites in quartzo-felspathic gneisses of Nordfjord, West Norway. Norsk. Geol. Tids. **49**, 193 (1969).

BUSCHE, F. D., PRINZ, M., KEIL, K., KURAT, G.: Lunar zirkelite: a uranium-bearing phase. Earth Planet. Sci. Lett. **14**, 313 (1972).

BUTLER, J. R.: The geochemistry and mineralogy of rock weathering. (1) The Lizard area, Cornwall. Geochim. Cosmochim. Acta **4**, 157 (1953).

BUTLER, J. R.: The geochemistry and mineralogy of rock weathering. (2) The Nordmarka Area, Oslo. Geochim. Cosmochim. Acta **6**, 268 (1954).

BUTLER, J. R., SMITH, A. Z.: Zirconium, niobium and certain other trace elements in some alkali igneous rocks. Geochim. Cosmochim. Acta **26**, 945 (1962).

BUTTERMAN, W. C., FOSTER, W. R.: Zircon stability and the ZrO_2—SiO_2 phase diagram. Amer. Mineral. **52**, 880 (1967).

CALVERT, S. E., PRICE, N. B.: Composition of manganese nodules and manganese carbonates from Loch Fyne, Scotland. Contr. Mineral. and Petrol. **29**, 215 (1970).

CAMBEL, B., KAMENICKÝ, L.: Zirkonium in den basischen Gesteinen der Westkarpaten. Geol. Zborn. Slov. Akad. Vied. **20**, 213 (1969).

CANN, J. R.: Spilites from the Carlsberg Ridge, Indian Ocean. J. Petrol. **10**, 1 (1969).

CANN, J. R.: Rb, Sr, Y, Zr, and Nb in some ocean floor basaltic rocks. Earth Planet. Sci. Lett. **10**, 7 (1970).

CARMICHAEL, I. S. E.: The mineralogy and petrology of the volcanic rocks from the Leucite Hills, Wyoming. Contr. Mineral. Petrol. **15**, 24 (1967).

CARMICHAEL, I., MCDONALD, A.: The geochemistry of some natural acid glasses from the North Atlantic Tertiary volcanic province. Geochim. Cosmochim. Acta **25**, 189 (1961).

CARROLL, D.: Weatherability of zircon. Sedimentary Petrology **23**, 106 (1953).

CARROLL, D., WOOF, M.: Laterite developed on basalt at Inverell, New South Wales. Soil Sci. **62**, 87 (1951).

CARUBA, R., BAUMER, A., TURCO, G.: Nouvelles synthèses hydrothermales du zircon: substitutions isomorphiques; relation morphologie—milieu de croissance. Geochim. Cosmochim. Acta **39**, 11 (1975).

CHAO, E. C. T.: The petrographic and chemical characteristics of tektites. In: O'KEEFE, J. A. (ed.): Tektites. University of Chicago Press 1963.

CHAO, E. C. T., FLEISCHER, M.: Abundance of zirconium in igneous rocks. Geol. Congress Report of 21st Session. Norden. Part I. Geochemical Cycles. 106 (1960).

CHAPPELL, B. W., GREEN, D. H.: Chemical compositions and petrogenic relationships in Apollo 15 mare basalts. Earth Planet. Sci. Lett. **18**, 237 (1973).

CHESTER, R.: Elemental geochemistry of marine sediments. In: RILEY, J. P., SKIRROW, G. (eds.): Chemical Oceanography. Vol. 2, London and New York: Academic Press 1965.

CLARKE, D. B.: Tertiary basalts of Baffin Bay: possible primary magma from the mantle. Contr. Mineral. and Petrol. **25**, 203 (1970).

CLARKE, R. S., WOSINSKI, J. F.: Baddeleyite inclusion in the Martha's Vineyard tektite. Geochim. Cosmochim. Acta **31**, 397 (1967).

CLIFFORD, T. N., NICOLAYSEN, L. O., BURGER, A. J.: Petrology and age of the Pre-Otavi basement granite at Franzfontein, northern South-West Africa. J. Petrol. **3**, 244 (1962).

COHEN, A. J.: Moldavites and similar tektites from Georgia, U.S.A. Geochim. Cosmochim. Acta **17**, 150 (1959).

COLEMAN, R. G., ERD, R. C.: Hydrozircon from the Wind River Formation, Wyoming, U.S.A. U.S. Geol. Surv. Prof. Paper **424-C** (1961).

COLEMAN, R. G., LEE, D. E., BEATTY, L. B., BRANNOCK, W. W.: Eclogites and eclogites: their differences and similarities. Bull. Geol. Soc. Amer. **77**, 483 (1965).

COLEMAN, R. G., PAPIKE, J. J.: Alkali amphiboles from the blueschists of Cazadero, California. J. Petrol. **9**, 105 (1968).

COLLEY, H.: Andesitic volcanism in the New Hebrides. Proc. Geol. Soc. London **1662**, 46 (1970).

CONDIE, K. C.: Petrology and geochemistry of the Laramie Batholith and related metamorphic rocks of Precambrian age, Eastern Wyoming. Bull. Geol. Soc. Amer. **80**, 57 (1969).

CONDIE, K. C., BARSKY, C. K., MUELLER, P. A.: Geochemistry of Precambrian diabase dikes from Wyoming. Geochim. Cosmochim. Acta **33**, 1371 (1969).

CONDIE, K. C., HARRISON, N. M.: Geochemistry of the Archaean Bulawayan group, Midlands greenstone belt, Rhodesia. Precambrian Res. **3**, 253 (1976).

CONDIE, K. C., LO, H. H.: Trace element geochemistry of the Louis Lake batholith of early Precambrian age, Wyoming. Geochim. Cosmochim. Acta **35**, 1099 (1971).

CONDIE, K. C., MACKE, J. E., REIMER, T. O.: Petrology and geochemistry of early Precambrian graywackes from the Fig Tree Group, South Africa. Bull. Geol. Soc. Amer. **81**, 2759 (1970).

CONDIE, K. C., SNANSIENG, S.: Petrology and geochemistry of the Duzel (Ordovician) and Gazelle (Silurian) Formations, Northern California. J. Sedim. Petrol. **41**, 741 (1971).

CONNOR, J. J., SHACKLETTE, H. T.: Background geochemistry of some rocks, soils, plants, and vegetables in the Conterminous United States. U.S. Geol. Prof. Paper **574-F**, 168 (1975).

CORDANI, U. G., VANDOROS, P.: Basaltic rocks of the Parana Basin. In: BIGARELLA, J. J., BECKER, R. D., PINTO, I. D. (eds): Problems in Brazilian Gondwana Geology. Curitiba-Parana-Brazil 1967.

COSGROVE, M. E.: Iodine in the bituminous Kimmeridge shales of the Dorset coast, England. Geochim. Cosmochim. Acta **34**, 830 (1970).

COSGROVE, M. E.: The geochemistry of the potassium-rich Permian volcanic rocks of Devonshire, England. Contr. Mineral. and Petrol. **36**, 155 (1972).

COTTON, F. A., WILKINSON, G.: Advanced Inorganic Chemistry. New York: J. Wiley and Son 1966.

COX, K. G., GASS, I. G., MALLICK, D. I. J.: The peralkaline volcanic suite of Aden and Little Aden, South Arabia. J. Petrol. **11**, 433 (1970).

COX, K. G., HORNUNG, G.: The petrology of the Karroo basalts of Basutoland. Amer. Mineral. **51**, 1414 (1966).

COX, K. G., MACDONALD, R., HORNUNG, G.: Geochemical and petrographic provinces in the Karroo basalts of Southern Africa. Amer. Mineral. **52**, 1451 (1967).

CRUFT, E. F.: Minor elements in igneous and metamorphic apatite. Geochim. Cosmochim. Acta **30**, 375 (1966).

CURRIE, K. L.: Variations in the hafnium zirconium ratio of high grade metamorphic rocks from south-eastern Ontario, Canada. Earth Planet. Sci. Lett. **4**, 299 (1968).

CURRIE, K. L., FERGUSON, J.: A study of fenitization around the alkaline carbonatite complex at Callender Bay, Ontario, Canada. Can. J. Earth Sci. **8**, 498 (1971).

CUTTITTA, F., CARRON, M. K., ANNELL, C. S.: New data on selected Ivory Coast tektites. Geochim. Cosmochim. Acta **36**, 1297 (1972).

DANCHIN, R. V.: Chromium and nickel in the Fig Tree shale from South Africa. Science **158**, 261 (1967).

DANCHIN, R. V., FERGUSON, J.: The geochemistry of the Losberg intrusion, Fochville, Transvaal. Geol. Soc. S. Africa Spec. Publ. **1**, 689 (1970).

DAVIS, G. L., KROGH, T. E., ERLANK, A. J.: The ages of zircons from kimberlites from South Africa. Carnegie Inst. Wash. Year Book 1975–1976, **75**, 821 (1976).

DAWSON, J. B.: Basutoland kimberlites. Bull. Geol. Soc. Amer. **73**, 545 (1962).

DAWSON, J. B., HAWTHORNE, J. B.: Magmatic sedimentation and carbonatitic differentation in kimberlite sills at Benfontein, South Africa. J. Geol. Soc. **129**, 61 (1973).

DE BARROS GOMES, C., SANTINI, P., DUTRA, C. V.: Petrochemistry of a Precambrian amphibolite from the Jaraguá area São Paula, Brazil. J. Geol. **72**, 664 (1964).

DE BRUIYN, H., RHODES, R. C.: A new variety of Bushveld granite in the Dennilton area, Transvaal. Trans. Geol. Soc. S. Africa **78**, 89 (1975).

DEGENHARDT, H.: Untersuchungen zur geochemischen Verteilung des Zirkoniums in der Lithosphäre. Geochim. Cosmochim. Acta **11**, 279 (1957).

DENNEN, W. H., ANDERSON, P. J.: Chemical changes in incipient rock weathering. Bull. Geol. Soc. Amer. **73**, 375 (1962).

DIETRICH, R. V.: Behaviour of zirconium in certain artificial magmas under diverse P-T conditions. Lithos **1**, 20 (1968).

DIETRICH, R. V., HEIER, K. S., TAYLOR, S. R.: Studies on the igneous rock complex of the Oslo region. XX. Petrology and geochemistry of ekerite. Norske Videnskaps-Akademi i Oslo. I. Mat. Naturv. Klasse. Ny Serie **19** (1965).

DRURY, S. A.: The chemistry of some granitic veins from the Lewisian of Coll and Tiree, Argyllshire, Scotland. Chem. Geol. **9**, 175 (1972).

DRURY, S. A.: The geochemistry of Precambrian granulite facies rocks from the Lewisian complex of Tiree, Inner Hebrides, Scotland. Chem. Geol. **11**, 167 (1973).

DUNCAN, A. R., ERLANK, A. J., SHER, M. K., ABRAHAM, Y. C., WILLIS, J. P., AHRENS, L. H.: Some trace element constraints on lunar basalt genesis. Proc. Seventh Lunar Sci. Conf., Geochim. Cosmochim. Acta Suppl. **7**, 1659 (1976).

DUNN, C. E.: Identification of sedimentary cycle through Fourier analysis of geochemical data. Chem. Geol. **13**, 217 (1974).

EADE, K. E., FAHRIG, W. F.: Regional lithological and temporal variation in the abundances of some trace elements in the Canadian Shield. Canad. Geol. Surv. Paper **72–46**, 46 (1973).

EHMANN, W. D., CHYI, L. L.: Zirconium and hafnium in meteorites. Earth Planet. Sci. Lett. **21**, 230 (1974).

EHMANN, W. D., CHYI, L. L., GARG, A. N., HAWKE, B. R., MA, M.-S., MILLER, M. D., JAMES, W. D., Jr., PACER, R. A.: Chemical studies of the lunar regolith with emphasis on zirconium and hafnium. Proc. Sixth Lunar Sci. Conf., Geochim. Cosmochim. Acta Suppl. **6**, 1351 (1975).

EHMANN, W. D., REBAGAY, T. V.: Zirconium and hafnium in meteorites by activation analysis. Geochim. Cosmochim. Acta **34**, 649 (1970).

EL-HINNAWI, E. E., PICHLER, H., ZEIL, W.: Trace element distribution in Chilean ignimbrites. Contr. Mineral. and Petrol. **24**, 50 (1969).

EL WAKEEL, S. K., RILEY, J. P.: Chemical and mineralogical studies of deep-sea sediments. Geochim. Cosmochim. Acta **25**, 110 (1961).

ENGEL, A. E. J., ENGEL, C. G.: Progressive metamorphism and granitization of the major paragneiss, Northwest Adirondack Mountains, New York. Part I: Total rock. Bull. Geol. Soc. Amer. **69**, 1369 (1958).

ENGEL, A. E. J., ENGEL, C. G.: Progressive metamorphism and granitization of the major paragneiss, Northwest Adirondack Mountains, New York. Part II. Mineralogy. Bull. Geol. Soc. Amer. **71**, 1 (1960).

ENGEL, A. E. J., ENGEL, C. G.: Hornblendes formed during progressive metamorphism of amphibolites, Northwest Adirondack Mountains, New York. Bull. Geol. Soc. Amer. **73**, 1499 (1962a).

ENGEL, A. E. J., ENGEL, C. G.: Hornblendes formed during progressive metamorphism of amphibolites, Northwest Adirondack Mountains, New York. In: ENGEL, A. E.J., JAMES, H. L., LEONARD, B. F. (eds.): Petrologic Studies: a Volume to Honour A. F. BUDDINGTON, Geol. Soc. America 1962b.

ENGEL, A. E., ENGEL, C. G., HAVENS, R. G.: Chemical characteristics of oceanic basalts and the upper mantle. Bull. Geol. Soc. Amer. **76**, 719 (1965).

ERLANK, A. J.: Kimberlitic potassic richterite and the distribution of potassium in the upper mantle. In: AHRENS, L. H., DAWSON, J. B., DUNCAN, A. R., ERLANK, A. J. (eds): Extended Abstracts, International Conference on Kimberlites. University of Cape Town (1973).

ERLANK, A. J.: Upper mantle metasomatism as revealed by potassic richterite bearing peridotite xenoliths from kimberlite. EØS **57**, 597 (1976).

ERLANK, A. J., KABLE, E. J. D.: The significance of incompatible elements in Mid-Atlantic Ridge basalts from 45° N with particular reference to Zr/Nb. Contr. Mineral Petrol. **54**, 281 (1976).

ERLANK, A. J., WILLIS, J. P.: The zirconium content of chondrites and the zirconium-hafnium dilemma. Geochim. Cosmochim. Acta **28**, 1715 (1964).

ERLANK, A. J., WILLIS, J. P., AHRENS, L. H., MCCARTHY, T. S.: Inter-element relationships between the moon and stony meteorites with particular reference to some refractory elements. In: WATKINS, C. (ed.): Lunar Science III, Houston: Lunar Science Institute 1972.

EVANS, B. W.: Fractionation of elements in the pelitic hornfelses of the Cashel-Lough Wheelaun intrusion, Connemara, Eire. Geochim. Cosmochim. Acta **28**, 127 (1964).

EWART, A., BRYAN, W. B., GILL, J. B.: Mineralogy and geochemistry of the younger volcanic islands of Tonga, S. W. Pacific. J. Petrol. **14**, 429 (1973).

EWART, A., TAYLOR, S. R., CAPP, A. C.: Geochemistry of the pantellerites of Mayor Island, New Zealand. Contr. Mineral. and Petrol. **17**, 116 (1968a).

EWART, A., TAYLOR, S. R., CAPP, A. C.: Trace and minor element geochemistry of the rhyolitic volcanic rocks, Central North Island, New Zealand. Total rock and residual liquid data. Contr. Mineral. and Petrol. **18**, 76 (1968b).

FAIRHALL, L. T.: Industrial Toxicology. Baltimore: Williams and Wilkins Co. 1949.

FARRAND, M. G.: The distribution of some elements across the contacts of four xenoliths. Geol. Mag. **97**, 488 (1960).

FERGUSON, J.: The significance of the Kakortokite in the evolution of the Ilimaussaq intrusion, South Greenland, Grønlands Geol. Undersøgelse Bull. **89**, Meddelelser om Grønland **190** (1970).

FERGUSON, J.: The Pilansberg alkaline province, Southern Africa. Trans. Geol. Soc. S. Africa **76**, 249 (1973).

FERGUSON, J., CURRIE, K. L.: The geology and petrology of the alkaline carbonatite complex at Callender Bay, Ontario. Geol. Survey Canada Bull. **217** (1972).

FITTON, J. G.: The genetic significance of almandine-pyrope phenocrysts in the calc-alkaline Borrowdale volcanic group, Nothern England. Contr. Mineral. and Petrol. **36**, 231 (1972).

FLANAGAN, F. J.: 1972 values for international geochemical reference samples. Geochim. Cosmochim. Acta **37**, 1189 (1973).

FLEISCHER, M. (ed.): Data of geochemistry. U.S. Geol. Surv. Prof. Papers **440 W**, **440 Y** (1963, 1966).

FLEISCHER, M.: Index of new mineral names, discredited minerals, and changes of mineralogical nomenclature in Volumes 1–50 of the American Mineralogist, Amer. Mineral. **51**, 1247 (1966).

FLEISCHER, M.: U.S. Geological Survey standards — I. Additional data on rocks G-1 and W-1, 1965–1967. Geochim. Cosmochim. Acta **33**, 65 (1969).

FLOYD, P. A.: Geochemical characteristics of spilitic greenstones from South-West England. Nature **239**, 75 (1972).

FREDRICKSON, A. F.: Mode of occurrence of titanium and zirconium in laterites. Amer. Mineral. **33**, 374 (1948).

FREISE, F. W.: Untersuchung von Mineralen auf Abnutzbarkeit bei Verfrachtung im Wasser. Mineral. Pertol. Mitt. **41**, 1 (1931).

FRONDEL, C., COLLETTE, R. L.: Hydrothermal synthesis of zircon, thorite and huttonite. Amer. Mineral. **42**, 759 (1957).

FULLAGAR, P. D., LEMMON, R. E., RAGLAND, P. C.: Petrochemical and geochronological studies of plutonic rocks in the southern Appalachians: I. The Salisbury Pluton. Bull. Geol. Soc. Amer. **82**, 409 (1971).

GAD, M. A., CATT, J. A., LE RICHE, H. H.: Geochemistry of the Witbian (Upper Lias) sediments of the Yorkshire coast. Proc. Yorkshire Geol. Soc. **37**, 105 (1969).

GALE, G. H.: Paleozoic basaltic komatiite and ocean floor type basalts from North Eastern Newfoundland. Earth Planet. Sci. Lett. **18**, 22 (1973).

GALE, G. H., ROBERTS, D.: Trace element geochemistry of Norwegian lower palaeozoic basic volcanics and its tectonic implications. Earth Planet. Sci. Lett. **22**, 380 (1974).

GANAPATHY, R., PAPIA, G. M., GROSSMAN, L.: The abundances of zirconium and hafnium in the solar system. Earth Planet. Sci. Lett. **29**, 302 (1976).

GASS, I. G., MALLICK, D. I. J.: Jebel Khariz: an Upper Miocene strato-volcano of comenditic affinity on the South Arabian coast. Bull. Volcanologique **32**, 33 (1968).

GERASIMOVSKII, V. I.: Geochemical features of agpaitic nepheline syenites. In: VINOGRADOV, A. P. (ed.): Chemistry of the Earth's Crust, Vol. I. Jerusalem: Israel Program for Scientific Translations 1966.

GERASIMOVSKII, V. I., NESMEYANOVA, L. I., KAKHANA, M. M., KHAZIZOVA, V. D.: On regularities of zirconium and hafnium distribution in effusive rocks of East Africa rift zones [in Russian]. Geochimiya **12**, 1595 (1972).

GERASIMOVSKII, V. I., TUZOVA, A. M., SHEVALEEVSKII, I. D.: The zirconium hafnium (celtium) ratio in the Lovozero Massif rocks. Geochemistry **8**, 926 (1958).

GERASIMOVSKII, V. I., TUZOVA, A. M., SHEVALEEVSKII, I. D.: Zirconia-hafnia ratio in minerals and rocks of the Lovozero Massif. Geochemistry **6**, 585 (1962).

GIBSON, I. L.: A pantelleritic welded ash-flow tuff from the Ethiopian rift valley. Contr. Mineral. and Petrol. **28**, 89 (1970).

GIBSON, I. L.: The chemistry and petrogenesis of a suite of pantellerites from the Ethiopian rift. J. Petrol. **13**, 31 (1972).

GILL, J. B.: Geochemistry of Viti Levu, Fiji, and its evolution as an island arc. Contr. Mineral. and Petrol. **27**, 179 (1970).

GLASBY, G. P., TOOMS, J. S., CANN, J. R.: The geochemistry of manganese encrustations from the Gulf of Aden. Deep-Sea Res. **18**, 1179 (1971).

GÖRZ, H.: Microprobe studies of inclusions in zircons and compilation of minor and trace elements in zircons from the literature. Chemie der Erde **33**, 326 (1974).

GOLD, D. P.: Average chemical composition of carbonatites. Econ. Geol. **58**, 988 (1963).

GOLDBERG, E. D., ARRHENIUS, G. O. S.: Chemistry of Pacific pelagic sediments. Geochim. Cosmochim. Acta **13**, 153 (1958).

GOLDICH, S. S.: A study in rock weathering. J. Geol. **46**, 17 (1938).

GOLDSCHMIDT, V. M.: The principles of distribution of chemical elements in minerals and rocks. J. Chem. Soc. London. **1**, 655 (1937).

GOLDSCHMIDT, V.M.: Geochemistry. MUIR, A. (ed.): Oxford: Clarendon Press 1954.

GOODWIN, A. M.: Variations in tectonic styles in Canada: The Superior Province. Geol. Assoc. Canada Special Report **11** (1972).

GORDON, M., MURATA, K. J.: Minor elements in Arkansas bauxite. Econ. Geol. **47**, 169 (1952).

GOTTFRIED, D., GREENLAND, L. P., CAMPBELL, E. Y.: Variation of Nb—Ta, Zr—Hf, Th—U and K—Cs in two diabase-granophyre suites. Geochim. Cosmochim. Acta **32**, 925 (1968).

GRAF, D. L.: Geochemistry of carbonate sediments and sedimentary carbonate rocks. Part III. Minor element distribution. Illinois State Geol. Surv. Circular **301** (1960).

GRAF, D. L.: Minor element distribution in sedimentary carbonate rocks. Geochim. Cosmochim. Acta **26**, 815 (1962).

GRAHAM, A. L., MASON, B.: Niobium in meteorites. Geochim. Cosmochim. Acta. **36**, 917 (1972).

GREEN, D. H.: The petrogenesis of the high-temperature peridotite intrusion in the Lizard Area, Cornwall. J. Petrol. **5**, 134 (1964).

GREEN, J.: Geochemical table of the elements for 1959. Bull. Geol. Soc. Amer. **70**, 9 (1959).

GRIBOVSKAYA, I. F., VOROTNITSKAYA, J. E., LETUNOVA, S. V.: Migration of chemical elements in Lake Issyk-Kul. Tr. Biogeokhim. Lab., Akad. Nauk SSSR **13**, 224 (1974).

GROHMANN, H.: Beitrag zur Geochemie österreichischer Granitoide. Tschermaks Mineral. Petrogr. Mitt. **10**, 436 (1965).

GULSON, B. L.: The high-K diorites and associated rocks of the Yeoval diorite complex, N.S.W. Contr. Mineral. and Petrol. **35**, 173 (1972).

GUN, B. M.: Modal and element variation in Antarctic tholeiites. Geochim. Cosmochim. Acta **30**, 381 (1966).

GURNEY, J. J.: Unpublished data, University of Cape Town (1975).

HAHN-WEINHEIMER, P., JOHANNING, H.: Geochemical investigation on differentiated granite plutons in the Black Forest. In: AHRENS, L. H. (ed.): Origin and Distribution of the Elements. Oxford: Pergamon Press 1968.

HAHN-WEINHEIMER, P., LUECKE, W.: Garnets from the eclogites of the Muenchberger gneiss massif (N. E. Bavaria). Can. Mineral. **7**, 764 (1963).

HAHN-WEINHEIMER, P., PROPACH, G., RASCHKA, H.: Zur Genese des Kagenfels Granits. Bull. Serv. Carte Geol. Als. Lorr. **24**, 5 (1971).

HALLBERG, J. A.: Geochemistry of Archaean volcanic belts in the Eastern Goldfields region of Western Australia. J. Petrol. **13**, 45 (1972).

HALLBERG, J. A., WILLIAMS, D.: Archaean mafic and ultramafic rock associations in the Eastern Goldfields region, Western Australia. Earth Planet. Sci. Lett. **15**, 191 (1972).

HART, S. R., ERLANK, A. J., KABLE, E. J. D.: Sea floor basalt alteration: some chemical and Sr isotopic effects. Contr. Mineral. and Petrol. **44**, 219 (1974).

HASEMAN, J. F., MARSHALL, C. E.: The use of heavy minerals in studies of the origin and development of soils. Univ. Missouri Coll. Agric. Res. Bull. **387** (1945).

HEIER, K. S.: Geochemistry of the nepheline syenite on Stjernøy, North Norway. Norsk Geol. Tids. **44**, 205 (1964).

HEIER, K. S., THORESEN, K.: Geochemistry of high grade metamorphic rocks, Lofoten-Vesterålen, North Norway. Geochim. Cosmochim. Acta **35**, 89 (1971).

HEINRICH, E. W.: Xenotime and thorite from Nigeria. Amer. Mineral. **48**, 206 (1963).

HENDRICKS, R. L., REISBICK, F. B., MAHAFFEY, E. J., ROBERTS, D. B., PETERSON, N. A.: Chemical composition of sediments and interstitial brines from the Atlantic II, Discovery and Chain deeps. In: DEGENS, E. T. and ROSS, D. A. (eds.): Hot Brines and Recent Heavy Metal Deposits in the Red Sea. Berlin: Springer-Verlag 1969.

HERRMANN, A. G., WEDEPOHL, K. H.: Untersuchungen an spilitischen Gesteinen der variskischen Geosynkline in Nordwestdeutschland. Contr. Mineral. and Petrol. **29**, 255 (1970).

HERZ, N., DUTRA, C. V.: Minor element abundance in a part of the Brazilian shield. Geochim. Cosmochim. Acta **21**, 81 (1960).

HEVESY, G. V., WÜRSTLIN, K.: Die Häufigkeit des Zirkoniums. Z. Anorg. Allgem. Chem. **216**, 305 (1934).

HEWETT, D. F., FLEISCHER, M., CONKLIN, N.: Deposits of the manganese oxides. Suppl. Econ. Geol. **58**, 1 (1963).

HIETANEN, A.: Metasomatic metamorphism in western Clearwater County, Idaho, U.S. Geol. Surv. Prof. Paper **344-A** (1962).

HILL, P. A., PARKER, A.: Tin and zirconium in the sediments around the British Isles: a preliminary reconnaissance. Econ. Geol. **65**, 409 (1970).

HIRST, D. M.: The geochemistry of modern sediments from the Gulf of Paria—II. The location and distribution of trace elements. Geochim. Cosmochim. Acta **26**, 1147 (1962).

HIRST, D. M., DUNHAM, K. C.: Chemistry and petrography of the Marl Slate of S. E. Durham, England. Econ. Geol. **58**, 912 (1963).

HIRST, D. M., KAYE, M. J.: Factors controlling the mineralogy and chemistry of an Upper Visean sedimentary sequence from Rookhope, County Durham. Chem. Geol. **8**, 37 (1971).

HOFMEYR, P. K.: The abundances and distribution of some trace elements in some selected South African shales. Ph. D. Thesis, University of Cape Town (1971).

HOLLAND, J. G., BROWN, G. M.: Hebridean tholeiitic magmas: a geochemical study of the Ardnamurchan cone sheets. Contr. Mineral. and Petrol. **37**, 139 (1972).

HORN, M. K., ADAMS, J. A. S.: Computer-derived geochemical balances and element abundances. Geochim. Cosmochim. Acta **30**, 279 (1966).

HUBBARD, N. J.: Some chemical features of lavas from the Manu'a Islands, Samoa. Pacific Sci. **25**, 178 (1971).

HUGHES, D. J., BROWN, G. C.: Basalts from Madeira: a petrochemical contribution to the genesis of oceanic alkali rock series. Contr. Mineral. and Petrol. **37**, 91 (1972).

IL'INA, N. S., KATCHENKOV, S. M., FRUKHT, D. L.: Distribution of Ti and Zr in the Pre-devonian and Devonian sediments of the Moscow syncline. Geochem. Internat. **7**, 677 (1970).

INDUSTRIAL HYGIENE FOUNDATION: Review of Literature on Toxicity, Physiological Effects and Medical Uses of Zirconium and Its Compounds. Pittsburgh, Pa.: Mellon Inst. 1954.

ITO, J., FRONDEL, C.: Synthetic zirconium and titanium garnets. Amer. Mineral. **52**, 773 (1967).

IZUMI, K.: Threshold of bitter tasting and phosphorylcholinecytidyl transferase activity of tongue epithelium in mice treated with metals. Fukushima J. Med. Sci. **20** (**1–2**), 1 (1974).

JACK, R. N., CARMICHAEL, I. S. E.: The chemical fingerprinting of acid volcanic rocks. Calif. Div. of Mines and Geology, Special Report **100**, 17 (1968).

JANDA, I., SCHROLL, E., SEDLAZEK, M.: Zum Problem der geochemischen Unterscheidung von Para- und Orthoamphiboliten am Beispiel einiger Vorkommen des Waldviertels und der Ostalpen. Tschermaks Mineral. Petrog. Mitt. **10**, 552 (1965).

JOHNSON, R. L.: The geology of the Dorowa and Shawa carbonatite complexes, Southern Rhodesia. Trans. Geol. Soc. S. Africa **64**, 101 (1961).

JOYCE, A. S.: Chemical variation in a pelitic hornfels. Chem. Geol. **6**, 51 (1970).

KABLE, E. J. D.: Some aspects of the geochemistry of selected elements in basalts and associated lavas. Ph. D. Thesis, University of Cape Town 1972.

KABLE, E. J. D., FESQ, H. W., GURNEY, J. J.: The significance of the interelement relationships of some minor and trace elements in South African kimberlites. Phy. Chem. Earth **9**, 709 (1975).

KATCHENKOV, S. M.: Average contents of certain minor chemical elements in the principle types of sedimentary rocks. In: VINOGRADOV, A. P. (ed.): Chemistry of the Earth's Crust. Vol. II. Jerusalem: Israel Program for Scientific Translations 1967.

Kesson, S., Price, R. C.: The major and trace element chemistry of kaersutite and its bearing on the petrogenesis of alkaline rocks. Contr. Mineral. and Petrol. **35**, 119 (1972).

Kleinkopf, M. D.: Spectrographic determination of trace elements in lake waters of northern Maine. Bull. Geol. Soc. Amer. **71**, 1231 (1960).

Kolbe, P.: Geochemical investigation of the Cape Granite, South-Western Cape Province, South Africa. Trans. Geol. Soc. S. Africa **69**, 161 (1966).

Kolbe, P., Taylor, S. R.: Major and trace element relationships in granodiorites and granites from Australia and South Africa. Contr. Mineral. and Petrol. **12**, 202 (1966).

Köster, H. M.: Vergleich einiger Methoden zur Untersuchung von geochemischen Vorgängen bei der Verwitterung. Beitr. Mineral. Petrogr. **8**, 69 (1961).

Krogh, T. E.: A low-contamination method for hydrothermal decomposition of zircon and extraction of U and Pb for isotopic age determinations. Geochim. Cosmochim. Acta **37**, 485 (1973).

Lambert, I. B., Heier, K. S.: Geochemical investigations of deep-seated rocks in the Australian shield. Lithos **1**, 30 (1968).

Lange, J.: Geochemische Untersuchungen an Sedimenten des Persischen Golfes. Contr. Mineral. and Petrol. **28**, 288 (1970).

Larsen, L. M.: Clinopyroxenes and co-existing mafic minerals from the alkaline Ilimaussaq intrusion, South Greenland. J. Petrol. **17**, 258 (1976).

Lebedev, B. A.: Trace elements in marine and fresh water clays. Geochem. Intern. **4**, 821 (1967).

Lee, M. M., Chaudhri, M. A., Rouse, J. L., Spicer, B.: Environmental studies with the Melbourne University Cyclotron. Experientia Suppl., **24** Int. Conf. Cyclotrons, Their Appl., 1975.

Leelanandam, C.: Chemical study of pyroxenes from the charnockitic rocks of Kondapalli (Andhra Pradesh), India, with emphasis on the distribution of elements in co-existing pyroxenes. Mineral. Mag. **36**, 153 (1967).

Lehninger, A. L.: Biogeochemistry. New York: Worth 1970.

Le Maitre, R. W.: Petrology of volcanic rocks, Gough Island, South Atlantic. Bull. Geol. Soc. Amer. **73**, 1309 (1962).

Levin, E. M., Robins, C. R., McMurdie, H. F.: Phase Diagrams for Ceramists. Ohio: The Amer. Ceramic Soc. 1964.

Liebenberg, C. J.: The trace elements of the rocks of the Bushveld igneous complex. Univ. of Pretoria. Publ. **12**, 1 (1960).

Lipova, I. M., Shevaleevskii, I. D., Tuzova, A. M.: On the ratio of zirconium and hafnium in granitoids of the Verkhissetsk intrusion. Geochemistry **2**, 159 (1957).

Litinskii, V. A.: On the content of Ni, Cr, Ti, Nb and some other elements in kimberlites and the possibility of geochemical prospecting for kimberlite bodies. Geochemistry **9**, 813 (1961).

Livingstone, D. A.: Chemical composition of rivers and lakes. In: Fleischer, M. (ed.): Data of Geochem., Sixth ed. U.S. Geol. Surv. Prof. Paper **440-G** (1963).

Lloyd, F. E., Bailey, D. K.: Carbonatite in the tuffs of the West Eifel, Germany. Contr. Mineral. and Petrol. **23**, 136 (1969).

Lloyd, F. E., Bailey, D. K.: Light element metasomatism of the continental mantle, the evidence and the consequences. Phy. Chem. Earth **9**, 389 (1975).

Lonka, A.: Trace elements in the Finnish Precambrian phyllites as indicators of salinity at the time of sedimentation. Bull. Comm. Geol. Finlande **228**, 1 (1967).

Losee, F., Cutress, T. W., Brown, R.: Trace elements in human dental enamel. Trace Subst. Environ. Health **7**, 19 (1973).

Lovering, J. F., Ware, N. G.: Electron probe microanalyses of minerals and glasses in Apollo 11 lunar samples. Proc. Apollo 11 Lunar Sci. Conf. **1**, 633 (1970).

Lowder, G. G., Carmichael, I. S. E.: The volcanoes and caldera of Talasea, New Britain: Geology and Petrology. Bull. Geol. Soc. Amer. **81**, 17 (1970).

Lustman, B., Kerz, F. (eds.): The Metallurgy of Zirconium. New York-Toronto-London: McGraw Hill 1955.

MacDougall, J. D., Kerridge, J. F.: Unusual anhydrous mineral assemblage in the Alais (C 1) meteorite. Meteoritics **11**, 326 (1976).

MacKenzie, D. E., Chappell, B. W.: Shoshonitic and calc-alkaline lavas from the highlands of Papua New Guinea. Contr. Mineral. and Petrol. **35**, 50 (1972).

Marchant, J. W.: Doctoral Thesis, University of Cape Town (in prep.) (1977).

Markhinin, E. K., Sapozhnikova, A. M.: Zirconium content in volcanic rocks of Kamchatka and the Kurile Islands. Geochemistry **9**, 965 (1962).

Marshall, B.: The present status of zircon. Sedimentology **9**, 119 (1967).

Marshall, C. E.: A petrographic method for the study of soil formation process. Proc. Soil. Sci. Soc. Amer. **5**, 100 (1940).

Marshall, C. E., Hasemann, J. F.: The quantitative evaluation of soil formation and development by heavy mineral studies: a Grundy silt-loam profile. Proc. Soil. Sci. Soc. Amer. **7**, 448 (1942).

Mason, B., Graham, A. L.: Minor and trace elements in meteoritic minerals. Smithsonian Contrib. Earth Sci. **3**, 1 (1970).

Mason, B., Martin, P. M.: Minor and trace element distribution in melilite and pyroxene from the Allende meteorite. Earth Planet. Sci. Lett. **22**, 141 (1974).

McCail, G. J. H., Hornung, G.: A geochemical study of Silali volcano, Kenya, with special reference to the origin of the intermediate acid eruptives of the Central Rift Valley. Tectonopyhsics **15**, 97 (1972).

McCarthy, T. S., Ahrens, L. H., Erlank, A. J.: Further evidence in support of the mixing model for howardite orgin. Earth Planet. Sci. Lett. **15**, 86 (1972).

McCarthy, T. S., Erlank, A. J., Willis, J. P.: On the origin of eucrites and diogenites. Earth Planet. Sci. Lett. **18**, 433 (1973).

McKie, D.: Fenitization. In: Tuttle, O. F., Gittins, J. (eds.): Carbonatites. New York, London and Sydney: Interscience Publishers 1966.

McLaughlin, R. J. W.: Geochemical changes due to weathering under varying climatic conditions. Geochim. Cosmochim. Acta **8**, 109 (1955).

Medlin, J. H., Coleman, S. L., Wood, G. H., Rait, N.: Differences in minor and trace element geochemistry of anthracite in the Appalachian Basin. Geol. Soc. Amer. Abstracts with Programs, **7**, 1198 (1975).

Melson, W. G., Hart, S. R., Thompson, G.: St. Paul's rocks, Equatorial Atlantic: Petrogenesis, radiometric ages, and implications on sea-floor spreading. In: Studies in Earth and Space Series. Geol. Soc. Amer. Memoir **132**, 241 (1972).

Melson, W. G., Thompson, G.: Petrology of a transform fault zone and adjacent ridge segments. Phil. Trans. Roy. Soc. Lond. A. **268**, 423 (1971).

Merz, E.: The determination of hafnium and zirconium in meteorites by neutron activation analysis. Geochim. Cosmochim. Acta **26**, 347 (1962).

Merz, E., Schrage, E.: Activation analysis determination of the zirconium-hafnium ratios in stone meteorites and minerals. Geochim. Cosmochim. Acta **28**, 1873 (1964).

Miller, G. L.: Metallurgy of the Rarer Metals — 2. Zirconium. London: Butterworths Scientific Publications 1954.

Miller, R. McG.: The Salem granite suite, South West Africa: Genesis by partial melting of the Khomas schist. Geol. Surv. S. Africa, Memoir **64** (1973).

Minenko, O. A., Epifantseva, M. V.: Content of trace elements in mine waters. Nauchn. Tr. Permsk. Nauchno-Issled. Ugol'n. Inst. **16**, 125 (1973).

Minister, J. F., Allégre, C. J.: ^{92}Nb extinct radioactivity: a search for variations of zirconium isotopic composition. Meteoritics **11**, 336 (1976).

Mitchell, R. H., Brunfelt, A. O., Nixon, P. H.: Ilmenite association trace element studies. In: Nixon, P. H. (ed.): Lesotho Kimberlites, Lesotho National Development Corporation, Maseru 1973.

Mitchell, R. L.: The distribution of trace elements in soils and grasses. Proc. Nutrition Soc. Engl. and Scot. **1**, 183 (1944).

Mitchell, R. L.: Trace elements in soils. In: Bear, F. E. (ed.): Chemistry of the Soil. 2nd (ed.), New York: Reinhold Publishing Corp. 1964.

Moore, A. E.: Unpublished data, University of Cape Town (1975).

Moore, J. R.: Recent sedimentation in northern Cardigan Bay, Wales. Bull. Brit. Mus. (Nat. Hist.) **2**, 95 (1968).

MOTTAMA, A., BOCCHIO, R.: Superferric eclogites of the Voltri group (Pennidic Belt, Apennines). Contr. Mineral. and Petrol. **49**, 201 (1975).

MOXHAM, R. L.: Distribution of minor elements in co-existing hornblendes and biotites. Can. Mineral. **8**, 204 (1965).

MUKHERJEE, B., RAO, M. G., KARUNAKARAN, C.: Genesis of kaolin deposits of Birbhum, West Bengal, India. Clay Minerals **8**, 161 (1969).

MUMPTON, F. A., ROY, R.: Hydrothermal stability studies of the zircon—thorite group. Geochim. Cosmochim. Acta **21**, 217 (1961).

NAQVI, S. M., HUSSAIN, S. M.: Petrochemistry of Precambrian metasediments from the central part of the Chitaldrug schist belt, Mysore, India. Chem. Geol. **10**, 109 (1972).

NAQVI, S. M., HUSSAIN, S. M.: Relation between trace and major element composition of the Chitaldrug metabasalts, Mysore, India, and the Archaean mantle. Chem. Geol. **11**, 17 (1973).

NASH, W. P., WILKINSON, J. F. G.: Shonkin Sag laccolith, Montana: II. Bulk rock geochemistry. Contr. Mineral. and Petrol. **33**, 162 (1971).

NEIVA, A. M. R.: Geochemistry of tourmaline (schorlite) from granites, aplites and pegmatites from northern Portugal. Geochim. Cosmochim. Acta **38**, 1307 (1974).

NESBITT, R. W., SUN, S.: Geochemistry of Archaean spinifex-textured peridotites and magnesian and low-magnesian tholeiites. Earth Planet. Sci. Lett. **31**, 433 (1976).

NESTERENKO, G. V., ZNAMENSKIY, YE. B., AL'MUKHAMEDOV, A. I., TSYKHANSKIY, V. D.: Nb, Ta, Zr, and Hf in trap magma differentiation. Geochem. Intern. **8**, 725 (1971).

NICHOLLS, J., CARMICHAEL, I. S. E.: Peralkaline acid liquids: a petrological study. Contr. Mineral. and Petrol. **20**, 268 (1969).

NICHOLLS, G. D., LORING, D. H.: The geochemistry of some British Carboniferous sediments. Geochim. Cosmochim. Acta **26**, 181 (1962).

NIXON, P. H., VON KNORRING, O., ROOKE, J. M.: Kimberlites and associated inclusions of Basutoland: a mineralogical and geochemical study. Amer. Mineral. **48**, 1090 (1963).

NOBLE, D. C., HAFFTY, J.: Minor-element and revised major-element contents of some Mediterranean pantellerites and comendites. J. Petrol. **10**, 502 (1969).

NOCKOLDS, S. R., ALLEN, R.: The geochemistry of some igneous rock series. Geochim. Cosmochim. Acta **4**, 105 (1953).

NOCKOLDS, S. R., ALLEN, R.: The geochemistry of some igneous rock series: Part II. Geochim. Cosmochim. Acta **5**, 245 (1954).

NOCKOLDS, S. R., ALLEN, R.: The geochemistry of some igneous rock series: Part III. Geochim. Cosmochim. Acta **9**, 34 (1956).

NOCKOLDS, S. R., MITCHELL, R. L.: The geochemistry of some Caledonian plutonic rocks: a study in the relationship between the major and trace elements of igneous rocks and their minerals. Trans. Roy. Soc. Edinburgh **61**, 533 (1948).

OERTEL, A. C.: Relation between trace-element concentrations in soil and parent material. J. Soil Sci. **12**, 119 (1961).

OKRUSCH, M., RICHTER, P.: Zur Geochemie der Diorit-Gruppe. Contr. Mineral. and Petrol. **21**, 75 (1969).

ONDRICK, C. W., GRIFFITHS, J. C.: Frequency distribution of elements in Rensselaer greywacke, Troy, New York. Bull. Geol. Soc. Amer. **80**, 509 (1969).

PALACHE, C., BERMAN, H., FRONDEL, C.: The System of Mineralogy of James Dwight Dana and Edward Salisbury Dana. (7th ed.) Vol. I. and II. New York: John Wiley and Sons 1958.

PALME, H.: Zerstörungsfreie Bestimmung einiger Spurenelemente in Mond- und Meteoritenproben mit 14 MeV-Neutronen. In: KIESL, W., MALISSA, J. H. (eds.): Analyse extraterrestrischer Materiale. New York: Springer-Verlag 1974.

PEARCE, J. A., CANN, J. R.: Ophiolite origin investigated by discriminant analysis using Ti, Zr and Y. Earth Planet. Sci. Lett. **12**, 339 (1971).

PEARCE, J. A., CANN, J. R.: Tectonic setting of basic volcanic rocks determined using trace element analysis. Earth Planet. Sci. Lett. **19**, 290 (1973).

PEARSON, G. R., SHAW, D. M.: Trace elements in kyanite, sillimanite and andalusite. Amer. Mineral. **45**, 808 (1960).

PECKETT, A., PHILLIPS, R., BROWN, G. M.: New zirconium-rich minerals from Apollo **14** and 15 lunar rocks. Nature **236**, 215 (1972).

PERRAULT, G.: La composition chimique et la structure cristalline du pyrochlore d'Oka, P. Q. Can. Mineral. **9**, 383 (1968).

PHILPOTTS, J. A., SCHUMANN, S., SCHNETZLER, C. C., KOUNS, C. W., DOAN, A. S., JR., WOOD, F. M., BICKEL, A. L., LUM-STAAB, R. K. L.: Apollo 17: Geochemical aspects of some soils, basalts and breccias. EØS **54**, 603 (1973).

PINSON, W. H., AHRENS, L. H., FRANCK, M. L.: The abundances of lithium, scandium, strontium, barium and zirconium in chondrites and some ultramafic rocks. Geochim. Cosmochim. Acta **4**, 251 (1953).

POLDERVAART, A.: Statistical studies of zircon as a criterion in granitisation. Nature **165**, 574 (1950).

POLDERVAART, A.: Zircon in rocks. 2. Igneous rocks. Amer. J. Sci. **254**, 521 (1956).

POWER, G. M.: Chemical variation in tourmalines from South-West England. Mineral. Mag. **36**, 1078 (1968).

PREUSS, E.: Spektralanaytische Untersuchung der Tektite. Chemie der Erde **9**, 365 (1935).

PRINZ, M.: Geochemistry of basaltic rocks: trace elements. In: HESS, H. H., POLDERVAART, A. (eds.): Basalts. The Poldervaart Treatise on Rocks of Basaltic Composition. Vol. I. New York, London and Sydney: Interscience Publ. 1967.

RAGLAND, P. C., ROGERS, J. J. W., JUSTUS, P. S.: Origin and differentiation of Triassic dolerite magmas, North Carolina, U. S. A. Contr. Mineral. and Petrol. **20**, 57 (1968).

RANKAMA, K.: Progress in Isotope Geology. New York, London and Sydney: Interscience Publ. 1963.

RANKAMA, K., SAHAMA, T. G.: Geochemistry. Universtiy of Chicago Press 1950.

RHODES, J. M., DAWSON, J. B.: Major and trace element chemistry of peridotite inclusions from the Lashaine volcano, Tanzania. Phys. Chem. Earth **9**, 545 (1975).

RICHTER, W.: Vergleichende Untersuchungen an ostalpinen Eklogiten. Tschermaks Min. Petr. Mitt. **19**, 1 (1973).

RIDLEY, W. I.: The petrology of the Las Canadas volcanoes, Tenerife, Canary Islands. Contr. Mineral. and Petrol. **26**, 124 (1970).

RINGWOOD, A. E.: The principles governing trace-element behaviour during magmatic crystallization. Part II. The role of complex formation. Geochim. Cosmochim. Acta **7**, 242 (1955).

RITTENHOUSE, G., FULTON, R. B., GRABOWSKI, R. J., BERNARD, J. L.: Minor elements in oil-field waters. Chem. Geol. **4**, 189 (1969).

RIVALENTI, G., SIGHINOLFI, G. P.: Geochemical study of greywackes as a possible starting material of para-amphibolites. Contr. Mineral. and Petrol. **23**, 173 (1969).

RODRIGUES, B.: Processos de fenitização relacionados com a estrutura anelar do Nejoio. Ph. D. Thesis, Luanda University, (1973).

ROSS, J. E., ALLER, L. H.: The chemical composition of the sun. Science **191**, 1223 (1976).

ROZZELL, T. C., ANDELMAN, J. B.: Plutonium in the water environment. II. Sorption of aqueous plutonium on silica surfaces. In: HEM, J. D. (ed.): Nonequilibrium Systems in Natural Water Chemistry. Advances in Chemistry Series **106**. Washington: Amer. Chem. Soc. 1971.

SASTRY, V. N., KRISHNAMOORTHY, T. M., DOSHI, G. R., SARMA, T. P.: Manganese dioxide as sorber in the determination of cerium, ruthenium, zirconium, and cesium fission nuclides from water by neutron activation analysis. Indian J. Chem. **12**, 200 (1974).

SAXENA, S. K.: Evolution of zircons in sedimentary and metamorphic rocks. Sedimentology **6**, 1 (1966).

SCEAL, J. S. C., WEAVER, S. D.: Trace element data bearing on the origin of salic rocks from the Quaternary volcano Paka, Gregory Rift, Kenya. Earth Planet. Sci. Lett. **12**, 327 (1971).

SCHMITT, R. A., BINGHAM, E., CHODOS, A.: Zirconium abundances in meteorites and implications of nucleosynthesis. Geochim. Cosmochim. Acta **28**, 1961 (1964).

SCHMITT, R. A., SMITH, R. H., PERRY, K. I., OLEHY, D. A.: Abundances of the fourteen rare-earth elements plus scandium, yttrium and zirconium in meteorites. General Atomic Report GA 3687, NASA Contract NASw-579 (1962).

Schnetzler, C. C., Nava, D. F.: Chemical composition of Apollo 14 soils 14163 and 14259. Earth Planet. Sci. Lett. **11**, 345 (1971).

Schnetzler, C. C., Philpotts, J. A., Nava, D. F., Schumann, S., Thomas, H. H.: Geochemistry of Apollo 15 basalt 15555 and soil 15531. Science, **175**, 426 (1972).

Schutz, D. F., Turekian, K. K.: The investigation of the geographical and vertical distribution of several trace elements in sea water using neutron activation analysis. Geochim. Cosmochim. Acta **29**, 259 (1965).

Schwarcz, H. P.: Chemical and mineralogic variations in an arkosic quartzite during progressive regional metamorphism. Bull. Geol. Soc. Amer. **77**, 509 (1966).

Seeger, P. A., Fowler, W. A., Clayton, D. D.: Nucleosynthesis of heavy elements by neutron capture. Astrophys. J. Suppl. Series, **11**, 121 (1965).

Senchuk, G. V., Borukh, I. F.: Wild berries of Belorussia. Rastit. Resur. **12**, 113 (1976).

Setser, J. L., Ehmann, W. D.: Zirconium and hafnium abundances in meteorites, tektites, and terrestrial materials. Geochim. Cosmochim. Acta. **28**, 769 (1964).

Shacklette, H. T., Hamilton, J. C., Boerngen, J. G., Bowles, J. M.: Elemental composition of surfical materials in the conterminous United States. U. S. Geol. Surv. Prof. Paper. **574-D** (1971).

Shaw, D. M.: Trace elements in pelitic rocks. Part I: Variation during metamorphism. Part II: Geochemical relations. Bull. Geol. Soc. Amer. **65**, 1151 (1954).

Shaw, D. M.: Geochemistry of pelitic rocks. Part III: Major elements and general geochemistry. Bull. Geol. Soc. Amer. **67**, 919 (1956).

Shaw, D. M., Dostal, J., Keays, R.: Additional estimates of continental surface Precambrian shield composition in Canada. Geochim. Cosmochim. Acta **40**, 73 (1976).

Shaw, D. M., Kudo, A. M.: A test of the discriminant function in the amphibolite problem. Mineral. Mag. **34**, 423 (1965).

Shaw, D. M., Moxham, R. L., Filby, R. H., Lapkowsky, W. W.: The petrology and geochemistry of some Grenville skarns. Can. Mineral. **7**, 578 (1963).

Sheraton, J. W.: The origin of the Lewisian gneisses of Northwest Scotland, with particular reference to the Drumbeg area, Sutherland. Earth Planet. Sci. Lett. **8**, 301 (1970).

Shigematsu, T., Nishikawa, Y., Hirakai, K., Nakagawa, H.: Determination of zirconium in sea water. Nippon Kagaku Zassui **85**, 490 (1964).

Shima, M., Hintenberger, H.: Elemental and isotopic abundances of Ti, Zr and Hf in stony meteorites. Meteoritics **11**, 364 (1976).

Shum, Y. S., Loveland, W. D.: Atmospheric trace element abundances and their application in air pollution tracing. Report RLO-2227-T7-27, Oregon State Univ. Dep. Chem. (1974).

Shumilo, P. E.: Content and distribution of trace elements in apple tree organs. Tezisy Dokl. Soobshch.-Konf. Molodykh Uch. Mald. **9**, 231 (1975).

Siedner, G.: Geochemical features of a strongly fractionated alkali igneous suite. Geochim. Cosmochim. Acta **29**, 113 (1965).

Siegers, A., Pichler, H., Zeil, W.: Trace element abundances in the "Andesite" Formation of Northern Chile. Geochim. Cosmochim. Acta **33**, 882 (1969).

Sighinolfi, G. P.: Investigations into deep crustal levels: Fractionating effects and geochemical trends related to high-grade metamorphism. Geochim. Cosmochim. Acta **35**, 1005 (1971).

Sinyakov, V. I., Nepeina, L. A.: On the behaviour of trace elements in the process of formation of magnesian skarns. Geochem. Intern. **6**, 337 (1969).

Smith, A. L.: Sphene, perovskite and co-existing Fe-Ti oxide minerals. Amer. Mineral. **55**, 264 (1970).

Smith, H. S., Erlank, A. J.: Unpublished data, University of Cape Town (1975).

Smith, R. E., Smith, S. E.: Comments on the use of Ti, Zr, Y, Sr, K, P and Nb in classification of basaltic magmas. Earth Planet. Sci. Lett. **32**, 114 (1976).

Smithson, F.: The mineralogy of arenaceous deposits. Sci. Prog. **149**, 10 (1950).

Sörenson, H.: Rythmic igneous layering in peralkaline intrusions. Lithos **2**, 261 (1968).

Spencer, D.: Factors affecting element distributions in a Silurian graptolite band. Chem. Geol. **1**, 221 (1966).

STEPHENS, W. E., WATSON, S. W., PHILIP, P. R., WEIR, S. A.: Element associations and distributions through a lower Palaeozoic graptolitic shale sequence in the southern uplands of Scotland. Chem. Geol. **16**, 269 (1975).

STROCK, L. W.: Geochemical data on Saratoga mineral waters-applied in deducing a new theory of their origin Amer. J. Sci. **239**, 857 (1941).

STROCK, L. W., DREXLER, S.: Geochemical study of Saratoga mineral waters by a spectrochemical analysis of their trace elements. J. Opt. Soc. Amer. **31**, 167 (1941).

STRONG, D. F.: The petrology of the lavas of Grande Comore. J. Petrol. **13**, 181 (1972).

STRONG, D. F.: Lushs Bight and Roberts Arm Groups of Central Newfoundland: Possible juxtaposed oceanic and island-arc volcanic suites. Bull. Geol. Soc. Amer. **84**, 3917 (1973).

STUMM, W., BRAUNER, P. A.: Chemical speciation. In: RILEY, J. P., SKIRROW, G. (eds).: Chemical Oceanography. (2nd. ed.) Vol. I. London: Academic Press 1975.

STUMPFL, E. F., STURT, B. A.: A preliminary account of the geochemistry and ore mineral parageneses of some Caledonian basic igneous rocks from Sørøy, northern Norway. Norges Geol. Unders. **234**, 196 (1965).

SUESS, H. E., UREY, H. C.: Abundances of the elements. Revs. Modern Phys. **28**, 53 (1956).

TARNEY, J.: The Scourie dyke suite and the nature of the Inverian event in Assynt. In: PARK, R. G., TARNEY, J. (eds.): The Early Precambrian of Scotland and related rocks of Greenland. England: University of Keele (1973).

TAYLOR, S. R.: The origin of some New Zealand metamorphic rocks as shown by their major and trace element composition. Geochim. Cosmochim. Acta **8**, 182 (1955).

TAYLOR, S. R.: Abundance of chemical elements in the continental crust: a new table. Geochim. Cosmochim. Acta **28**, 1273 (1964).

TAYLOR, S. R.: Australites, Henbury impact glass and subgreywacke: a comparison of the abundances of 51 elements. Geochim. Cosmochim. Acta **30**, 1121 (1966).

TAYLOR, S. R.: Geochemistry of the lunar highlands. The Moon **7**, 181 (1973).

TAYLOR, S. R.: Lunar Science: A Post-Apollo View. Oxford: Pergamon Press 1975.

TAYLOR, S. R., CAPP, A. C., GRAHAM, A. L., BLAKE, D. H.: Trace element abundances in andesites. II. Saipan, Bougainville and Fiji. Contr. Mineral. and Petrol. **23**, 1 (1969).

TAYLOR, S. R., GORTON, M. P., MUIR, P., NANCE, W., RUDOWSKI, R., WARE, N. G.: Composition of Descartes region, lunar highlands. Geochim. Cosmochim. Acta **37**, 2665 (1973).

TAYLOR, S. R., KOLBE, P.: The geochemistry of Henbury impact glass. Geochim. Cosmochim. Acta **29**, 741 (1965).

TAYLOR, S. R., MUIR, P., KAYE, M.: Trace element chemistry of Apollo 14 Lunar-soil from Fra Mauro. Geochim. Cosmochim. Acta **35**, 975 (1971).

TAYLOR, S. R., KAYE, M., MUIR, P., NANCE, W., RUDOWSKI, R., WARE, N.: Composition of the lunar uplands: Chemistry of Apollo 14 samples from Fra Mauro. Proc. Third Lunar Sci. Conf., Geochim. Cosmochim. Acta Suppl. **3**, 1231 (1972).

TAYLOR, S. R., SACHS, M.: Geochemical evidence for the origin of australites. Geochim. Cosmochim. Acta **28**, 235 (1964).

TAYLOR, S. R., SOLOMON, M.: Geochemical and geological evidence for the origin of Darwin glass. Nature **196**, 124 (1962).

TAYLOR, S. R., WHITE, A. J. R.: Trace element abundances in andesites. Bull. Volcanologique **29**, 177 (1966).

THOMPSON, G.: A geochemical study of some lithified carbonate sediments from the deep-sea. Geochim. Cosmochim. Acta **36**, 1237 (1972).

THOMPSON, G.: Trace element distributions in fractionated oceanic rocks. 2. Gabbros and related rocks. Chem. Geol. **12**, 99 (1973).

THOMPSON, G., SHIDO, F., MIYASHIRO, A.: Trace element distributions in fractionated oceanic basalts. Chem. Geol. **9**, 89 (1972).

THORPE, R. S.: Ocean floor basalt affinity of Precambrian glaucophane schist from Anglesey. Nature **240**, 164 (1972).

TIDBALL, R. R.: Distribution of elements in agricultural soils of Missouri. Geol. Soc. Amer. Abstracts with Programs **5**, 358 (1973).

TILLER, K. G.: The distribution of trace elements during differentiation of the Mt. Wellington dolerite sill. Papers and Proc. Roy. Soc. Tasmania **93**, 153 (1959).

TISCHENDORF, G., LANGE, G.: Geochemistry of the Variskian granites of the western part of the Ore Mountains, Rudny, (GDR) as a function of their age. In: Granitisation, Granites and Pegmatites. First Internat. Geochem. Congress, Moscow 1971, **2**, 303 (1972) [In Russian].

TOBSCHALL, H. J.: Geochemical investigations on the composition and the depositional environment of Paleozoic marine pelites: The content of the major elements and the trace elements Ni, Cu, Zn, Rb, Sr, Y, Zr, Nb and Ba in the Steiger slates (Vosges, France). Chem. Erde Bd. **34**, 105 (1975).

TOURTELOT, H. A.: Minor element composition and organic carbon content of marine and non-marine shales of Late Cretaceous age in the western interior of the United States. Geochim. Cosmochim. Acta **28**, 1579 (1964).

TOURTELOT, H. A., SCHULTZ, L. G., GILL, J. R.: Stratigraphic variations in mineralogy and chemical composition of the Pierre shale in South Dakota and adjacent parts of North Dakota, Nebraska, Wyoming and Montana. U.S. Geol. Survey Prof. Papers **400B**, 447 (1960).

TUGARINOV, A. I., PAVLENKO, A. S., KOVALENKO, V. I.: Geochemical evidence on the origin of aplogranites. Geochem. Intern. **5**, 1156 (1968).

TUREKIAN, K. K.: Oceans. Englewood Cliffs, New Jersey: Prentice-Hall 1968.

TUREKIAN, K. K., WEDEPOHL, K. H.: Distribution of the elements in some major units of the earth's crust. Bull. Geol. Soc. Amer. **72**, 175 (1961).

UPTON, B. G. J., WADSWORTH, W. J.: Peridotitic and gabbroic rocks associated with the shield-forming lavas of Réunion. Contr. Mineral. and Petrol. **35**, 139 (1972).

VAIL, J. R., HORNUNG, G., COX, K. G.: Karroo basalts of the Tuli syncline, Rhodesia. Bull. Volcanologique **33**, 398 (1969).

VAN DE KAMP, P. C.: Geochemistry and origin of metasediments in the Haliburton-Madoc area, southeastern Ontario. Can. J. Earth Sci. **5**, 1337 (1968).

VAN DE KAMP, P. C.: The green beds of the Scottish Dalradian series: Geochemistry, origin, and metamorphism of mafic sediments. J. Geol. **78**, 281 (1970).

VAN WAMBEKE, L.: A study of some niobium-bearing minerals of the Luashe carbonatite deposit. Euratom/N.V. Hollandse Metallurgische Industria Billiton **017-61-9**ISPN (1965).

VINOGRADOV, A. P.: The Geochemistry of Rare and Dispersed Chemical Elements in Soils. (2nd ed.) New York: Consultants Bur. Enterprises 1959.

VINOGRADOV, A. P.: Average contents of chemical elements in the principal types of igneous rocks of the earth's crust. Geochemistry **7**, 641 (1962).

VINOGRADOVA, Z. A., KOVAL'SKIY, V. V.: Elemental composition of the Black Sea plankton. Dokl. Academ. Nauk SSSR **147**, 217 (1962) [English Transl.].

VLASOV, K. A.: Geochemistry and mineralogy of rare elements and genetic types of their deposits. Vol. I. Geochemistry of Rare Elements. Jerusalem: Israel Program for Scientific Translations 1966.

VLASOV, K. A., KUZ'MENKO, M. Z., ES'KOVA, E. M.: The Lovozero Alkali Massif. Edinburgh and London: Oliver and Boyd 1966.

VOGEL, D. E.: Precambrian weathering in acid metavolcanic rocks from the Superior Province, Villebon Township, South-Central Quebec. Can. J. Earth Sci. **12**, 2080 (1975).

VON MICHAELIS, H., AHRENS, L. H., WILLIS, J. P.: The composition of stony meteorites. II. Analytical data and an assessment of their quality. Earth Planet. Sci. Lett. **5**, 387 (1969).

VOROB'EV, G. G.: Composition of tektites—I. Indochinites. Meteoritika **17**, 64 (1959).

VOROB'EV, G. G.: Study of tektite content—II. Moldavites. Meteoritika **18**, 35 (1960).

WAGER, L. R., MITCHELL, R. L.: The distribution of trace elements during strong fractionation of basic magma—a further study of the Skaergaard intrusion, East Greenland Geochim. Cosmochim. Acta **1**, 129 (1951).

WALKER, F., MATHIAS, M.: The petrology of two granite-slate contacts at Cape Town, South Africa. Quart. J. Geol. Soc. Lond. **102**, 499 (1946).

WARK, D. A., REID, A. F., LOVERING, J. F., EL GORESY, A.: Zirconolite (versus zirkelite). In: CHAMBERLAIN, J. W., WATKINS, C. (eds.): Lunar Science IV. Houston: Lunar Sci. Inst. 1973.

WEAVER, S. D., SCEAL, J. S. C., GIBSON, I. L.: Trace element data relevant to the origin of trachytic and pantelleritic lavas in the East African Rift System. Contr. Mineral. and Petrol. **36**, 181 (1972).

WEBER, J. N., MIDDLETON, G. V.: Geochemistry of the turbidites of the Normanskill and Charny formations—II. Distribution of trace elements. Geochim. Cosmochim. Acta **22**, 244 (1961).

WEDEPOHL, K. H.: Spurenanalytische Untersuchungen an Tiefseetonen aus dem Atlantik. Geochim. Cosmochim. Acta **18**, 200 (1960).

WEIGAND, P. W., RAGLAND, P. C.: Geochemistry of Mesozoic dolerite dikes from eastern North America. Contr. Mineral. and Petrol. **29**, 195 (1970).

WHETTEN, J. T.: Sediments from the Lower Columbia River and origin of greywacke. Science **152**, 1057 (1966).

WHITE, D. E., WARING, G. A.: Volcanic emanations. In: FLEISCHER, M. (ed.): Data of geochemistry (6th ed.). U.S. Geol. Surv. Prof. Paper **440-K** (1963).

WHITE, S. M.: Mineralogy and geochemistry of continental shelf sediments off the Washington-Oregon coast. J. Sedim. Petrol. **40**, 38 (1970).

WHITTAKER, E. J. W., MUNTUS, R.: Ionic radii for use in geochemistry. Geochim. Cosmochim. Acta **34**, 945 (1970).

WILLIAMS, D. A. C., HALLBERG, J. A.: Archaean layered intrusions of the Eastern Goldfields Region, Western Australia. Contr. Mineral. and Petrol. **38**, 71 (1973).

WILLIS, J. P.: Investigations on the composition of manganese nodules with particular reference to certain trace elements. M.Sc. Thesis, University of Cape Town (1970).

WILLIS, J. P., FORTUIN, H. H. G., EAGLE, G. A.: A preliminary report on the geochemistry of recent sediments in Saldanha Bay and Langebaan Lagoon. Trans. Roy. Soc. S. Africa **42**, 497 (1977).

WOLFENDEN, E. B.: Geochemical behaviour of trace elements during bauxite formation in Sarawak, Malaysia. Geochim. Cosmochim. Acta **29**, 1051 (1965).

WOOD, D. A., GIBSON, I. L., THOMPSON, R. N.: Element mobility during zeolite facies metamorphism of the Tertiary basalts of eastern Iceland. Contr. Mineral. and Petrol. **55**, 241 (1976).

WOODFORD, P. J., WILSON, A. F.: Chemistry of co-existing pyroxenes, hornblendes and plagioclases in mafic granulites, Strangways Range, Central Australia. N. Jb. Mineralogie, Abhandlungen **128**, 1 (1976).

WRIGHT, A. E., TARNEY, J., PALMER, K. F., MOORLOCK, B. S. P., SKINNER, A. C.: The geology of the Angmagssalik area, East Greenland and possible relationships with the Lewisian of Scotland. In: PARK, R. S., TARNEY, J. (eds.): The Early Precambrian of Scotland and related rocks of Greenland. England: University of Keele 1973.

WYNNE-EDWARDS, H. R., HAY, P. W.: Co-existing cordierite and garnet in regionally metamorphosed rocks from the Westport Area, Ontario. Can. Mineral. **7**, 453 (1963).

YES'KOVA, E. M., YEFIMOV, A. F.: Distribution of rare elements in the apoeffusive alkalic metasomatites of the Urals. Geochem. Intern. **7**, 721 (1970).

YOUNG, E. J.: Spectrographic data on cores from the Pacific Ocean and the Gulf of Mexico. Geochim. Cosmochim. Acta **32**, 466 (1968).

YUDOVICH, Ya. E., KORYCHEVA, A. A., OBRUCHNIKOV, A. S., STEPANOV, Ym. V.: Mean trace element contents in coals. Geochem. Intern. **9**, 712 (1972).

ZEISSINK, H. E.: Trace element behaviour in two nickeliferous laterite profiles. Chem. Geol. **7**, 25 (1971).

ZNAMENSKOJ, E. B., POPOLITOV, E. I., TSYKHANSKIJ, V. D.: The behaviour of Nb, Ta, Zr and Hf in granitoids of various stages of the formation of fold belts. In: Granitisation, granites and pegmatites. First Internat. Geochem. Congress, Moscow 1971, **2**, 69 (1972) [In Russian].

Revised manuscript received: June 1977

Niobium 41

A	O. W. FLÖRKE and W. GEBERT	(Institut für Mineralogie, Ruhr-Universität Bochum, Germany)
B, C	K. H. WEDEPOHL	(Geochemisches Institut der Universität, Göttingen, Germany)
D	E. W. Heinrich	(Dept. of Geology and Mineralogy, University of Michigan, Ann Arbor, Michigan, U.S.A.)
E	K. H. WEDEPOHL	
F, G	E. W. HEINRICH	
I—M, O	K. H. WEDEPOHL	

41-A. Crystal Chemistry

I. Introduction

The electron configuration of Nb in the ground-state is Kr $4d^4$ $5s^1$ which accounts for its transition metal character. The d-level is only partly filled with four unpaired electrons; this results in a peculiar bond type and explains the high melting point (2750° K) as well as the body-centered cubic structure of the metal (PAULING, 1938 and 1949; HUME-ROTHERY *et al.*, 1951; NOWOTNY, 1959). It should be noted that even in nonmetallic compounds the influence of the d-electrons plays an important role. The five-valent state of Nb is the most stable one.

Table 41-A-1. *Valency, coordination number and ionic radii of Nb*

Valency	Coordination number	Ionic radius [Å] according to	
		WHITTAKER and MUNTUS (1970)	SHANNON and PREWITT (1969)[a]
2+	VI	0.79	0.71
3+	VI	0.78	0.70
4+	VI	0.77	0.69
5+	IV	0.40	0.32
5+	VI	0.72	0.64
5+	VII	0.74	0.66

[a] Based on $r(^{VI}O^{2-}) = 1.40$ Å.

Close relationships with Ta cause extensive mutual substitution and solid solution in their compounds (SCHRÖCKE, 1966). Nb^{5+} in oxide compounds prefers octahedral coordination, and only rarely occurs in tetrahedral coordination. 7-fold coordination in form of a pentagonal bipyramid is found in some compounds not known as minerals.

The niobium minerals are almost exclusively oxygen compounds (see Table 41-A-2).

II. Oxide Minerals with Nb in Octahedral Coordination

a) Derivatives of the α-PbO_2 Type

In ixiolite $(Nb,Ta,Mn,Fe,Sn,Ti)_2O_4$ (NICKEL *et al.*, 1963), the oxygen ions form a deformed h.c.p. The metal-ions occupy half of the octahedral holes and each oxygen ion is coordinated by three cations. The metal-oxygen octahedra mutually share two edges, thus lining up into zigzag chains along [001]. Referring to the α-PbO_2 type (ZASLAVSKI and TOLKAČEV, SR 1952, 224), the positions are statistically occupied

Table 41-A-2. *Structural properties of selected Nb minerals*

Name and composition	Structure type	Coordination number	Reference
Ixiolite $(Nb,Ta,Sn,Mn,Fe,Ti)_2O_4$	α-PbO_2	6	NICKEL *et al.* (1963)
Niobite $(Fe,Mn)(Nb,Ta)_2O_6$	α-PbO_2	6	STURDIVANT (SB 1928-32, 337)
Euxenite $(Y,Er,Ce,U,Pb,Ca)(Nb,Ta,Ti)_2(O,OH)_6$	α-PbO_2	6	ARNOTT (1950)
Polykrase $(Y, Ce,Ca,U,Th)(Ti,Nb,Ta)_2(O,OH)_6$	α-PbO_2	6	STRUNZ (1970)
Samarskite $(Y,Er,Fe,Mn,Ca,U,Zr,Th)(Nb,Ta)_2(O,OH)_6$	α-PbO_2	6	STRUNZ (1970)
Fersmite $(Ca,Ce,Na)(Nb,Ti,Fe,Al)_2(O,OH,F)_6$	α-PbO_2	6	ALEKSANDROV (SR 1960, 367)
Mossite $(Fe,Mn)(Nb,Ta)_2O_6$	Rutile	6	GOLDSCHMIDT (1926)
Pyrochlore $(Ca,Na)_2Nb_2O_6(O,OH,F)$	Pyrochlore	6	v. GAERTNER (SR 1928-32, 340)
Koppite $(Ca,Ce)_2(Nb,Fe)_2O_6(O,OH,F)$	Pyrochlore	6	BRANDENBERGER (SR 1928-32, 341)
Betafite $(Ca,U)_2(Nb,Ta,Ti)_2O_6(O,OH,F)$	Pyrochlore	6	STRUNZ (1970)
Pandaite (Ba etc.)$_2$$(Nb,Ta,Ti)_2O_6(O,OH,F)$	Fluorite	6	JÄGER *et al.* (SR 1959, 395)
Dysanalyte $(Na,Ca,Ce)(Ti,Nb,Fe)O_3$	Perowskite	6	NICKEL and MCADAM (1963)
Niobiumloparite $(Na,Ca,Ce)(Ti,Nb)O_3$	Perowskite	6	STRUNZ (1970)
Latrappite $(Ca,Na,RE)(Nb,Ti,Fe)O_3$	Perowskite	6	NICKEL (1964)
Lueshite $NaNbO_3$	Perowskite	6	SAKOWSKI-COWLEY *et al.* (1969)
Aeschynite $(RE,Ca,Fe,Mn,Th)(Nb,Ti)_2(O,OH)_6$		6	ALEKSANDROV (SR 1963, 536)
Epistolite $Na_2Ti_2Nb_2O_2(OH)_4\ [Si_2O_7]_2 \cdot 2H_2O$		6	KHALILOV *et al.* (1965)
Barsanovite $(Na,Ca,Sr,RE)_9(Fe,Mn)_2(Zr,Nb)_2\ Cl[Si_{12}O_{36}]$		6	DORFMAN *et al.* (1963)
Baotite $Ba_4(Ti,Nb)_8ClO_{16}[Si_4O_{12}]$		6	SIMONOV (SR 1961, 504)
Nenadkevichite $(Na,K)_{2-x}(Nb,Ti)_2(O,OH)_2[Si_4O_{12}] \cdot 2H_2O$		6	PERRAULT (1973)
Fergusonite $(Y,Yb)NbO_4$	Scheelite	4	KOMKOV (SR 1969, 388)

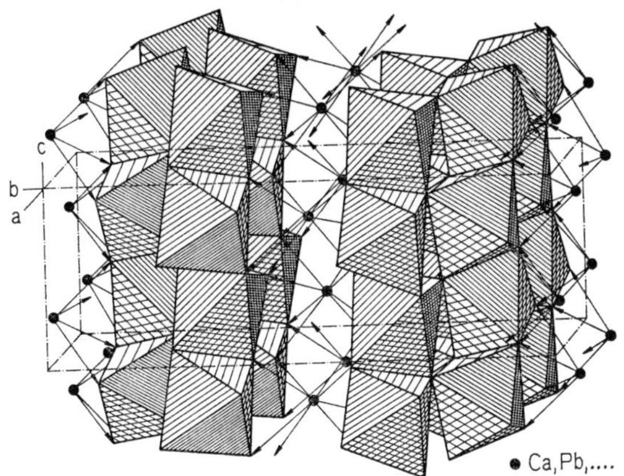

Fig. 41-A-1. Structure model of fersmite, according to ALEKSANDROV (1960)

by 5- and 3-valent ions (Nb, Ta and Fe, Mn) or by 4-valent ions (Sn, Ti). The compound $FeNbO_4$ occurs in the α-PbO_2 type at higher temperatures, and in the wolframite-type at lower temperature (LAVES *et al.*, 1963).

Niobite, (Fe,Mn) $(Nb,Ta)_2O_6$ (STURDIVANT, SB 1928-32, 337), shows continuous solid solution with tantalite thus forming the columbites. Fe and Mn can be substituted by Mg, and Nb and Ta by Ti. The ordered distribution of the cations results in a trifold a-axis as compared to ixiolite (BRANDT, SR 1942-44, 179). As in α-PbO_2, the O-O distances along common edges of the octahedra in this structure are shortened to about 2.56 Å.

The minerals of the euxenite group are related to the niobite type, e.g. euxenite $(Y,Er,Ce,U,Pb,Ca)(Nb,Ta,Ti)_2(O,OH)_6$, polykrase $(Y,Ce,Ca,U,Th)(Ti,Nb,Ta)_2$-$(O,OH)_6$, samarskite $(Y,Er,Fe,Mn,Ca,U,Th,Zr)(Nb,Ta)_2(O,OH)_6$, and fersmite $(Ca,Ce,Na)(Nb,Ti,Fe,Al)_2(O,OH,F)_6$.

The characteristic zigzag chains of octahedra still predominate, but do not build up a three-dimensional framework. They are joined by common corners to form two-dimensional "double-layers". These are interlinked with each other by the large cations (e.g. Ca,Pb) which have an 8-fold oxygen coordination. In fersmite the common edges of octahedra have O-O distances of 2.52 Å, the other O-O distances ranging between 2.71 and 3.06 Å (ALEKSANDROV, SR 1960, 367; see Fig. 41-A-1).

b) Derivatives of the Rutile Type

Mossite, $(Fe,Mn)(Nb,Ta)_2O_6$, forms a continuous solid solution with the corresponding Ta compound tapiolite; it crystallizes in the trirutile type (GOLDSCHMIDT, 1926). In contrast to Ta compounds (LAVES *et al.*, 1963), this type is not very frequently formed with Nb compounds thus reflecting a certain difference in the crystal chemistry of Nb and Ta. The tripling of the c-axis is due to an ordered distribution of the A ions and of the 5-valent B ions in the Ti positions of rutile. Compounds of the type AB_2O_6 with comparable ionic radii of A and B (between 0.60 and 0.75 Å) tend to crystallize in this structure-type (BAYER, SR 1962, 526).

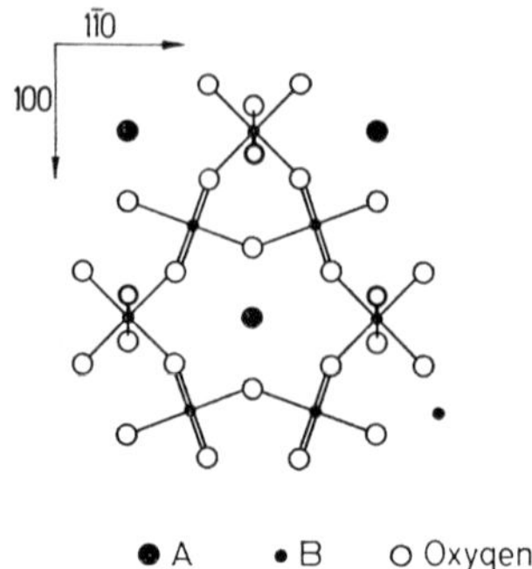

Fig. 41-A-2. The pyrochlore structure (WADSLEY, 1964)

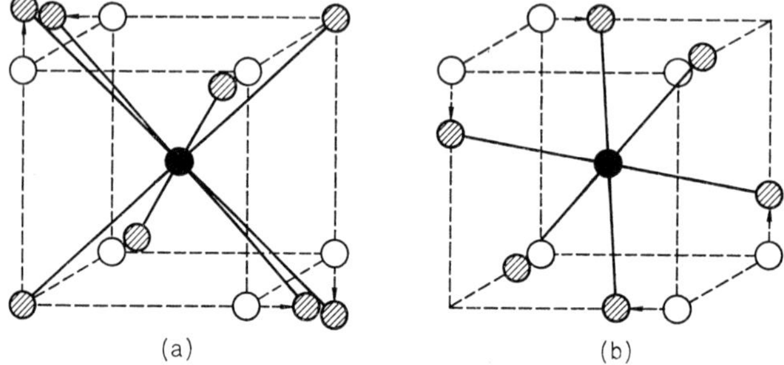

Fig. 41-A-3a and b. Derivation of BO_8 groups (a) and AO_6 groups (b) in the pyrochlore structure from ideal fluorite-type AX_8 coordination (BRANDENBERGER, 1931)

Due to an order-disorder transformation the trirutile type may change into the rutile type at high temperatures.

c) Derivatives of the Pyrochlore Type

Pyrochlore, $(Ca,Na)_2Nb_2O_6(O,OH,F)$ (v. GAERTNER, SR 1928-32, 340), forms solid solutions with the corresponding Ta compound mikrolite. The general composition of compounds crystallizing in this type is $A'A''B_2O_6X$, A′ and A″ having valencies of 1+, 2+ or 3+, and B having valencies of 5+ or 4+. As compared with the fluorite type a_0 is doubled (i.e. about 10.4 Å) (see Fig. 41-A-2).

The larger A-cations are coordinated by 8 O-ions, six of which form a distorted hexagon (2×A-O: 2.26, 6×A-O: 2.65 Å). The smaller B-cations are situated in a distorted oxygen octahedron (B-O: 1.95 Å, Fig. 41-A-3).

The BO_6-octahedra share all corners with each other, thus forming a framework with large empty holes which can be occupied by additional X anions (WADSLEY, 1964). This structure-type tolerates extensive substitution as for example in koppite $(Ca,Ce)_2(Nb,Fe)_2O_6(O,OH,F)$ (BRANDENBERGER, SR 1928-32, 341), betafite $(Ca,U)_2(Nb,Ta,Ti)_2O_6(O,OH,F)$ (STRUNZ, 1970, 191), obruchevite $(Ca,Na,Y)_2(Nb,Ta,Ti)_2O_6$

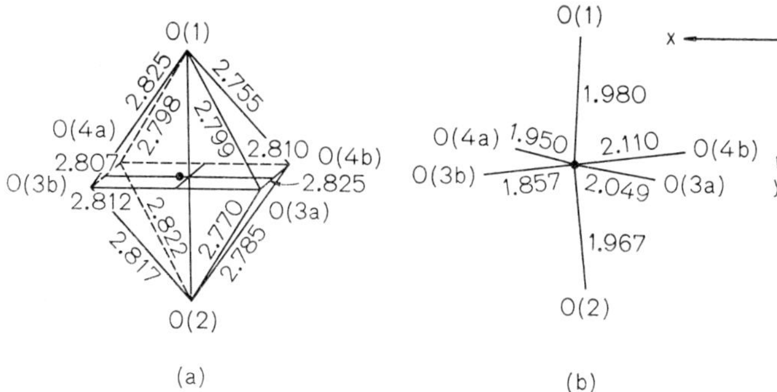

Fig. 41-A-4a and b. Projection of octahedron showing edge lengths (a) and Nb-O bond lengths (b) (SAKOWSKI-COWLEY *et al.*, 1969)

(O,OH,F), and pandaite (Ba, . . .)$_2$(Nb,Ta,Ti)$_2$O$_6$(O,OH,F) (JÄGER and NIGGLI, SR 1959, 395). Additional variability is achieved by order-disorder in A and X positions. Since many of the pyrochlore-type minerals contain radioactive elements they may be metamict (LIMA-DE-FARIA, 1971).

d) Derivatives of the Perovskite Type

Dysanalyte (Na,Ca,Ce)(Ti,Nb,Fe)O$_3$, latrappite (Ca,Na,RE)(Nb,Ti,Fe)O$_3$, niobiumloparite (Na,Ca,Ce)(Ti,Nb)O$_3$ (STRUNZ, 1970), and lueshite, NaNbO$_3$, with a common formula ABO$_3$ may be considered as close packings of A and O ions with the cubic stacking sequence, giving the A ions a 12-fold oxygen coordination. The B ions are coordinated by 6 oxygen ions. These octahedra with respect to the oxygen arrangement are remarkably regular but the central B cations are shifted from the geometrical centre by 0.13 Å towards one edge within the plane of 4 oxygens (Fig. 41-A-4). This causes two short and two long Nb-O distances in this plane (1.86 to 1.95 and 2.05 to 2.11 Å) (SAKOWSKI-COWLEY *et al.*, 1969). The NbO$_6$ octahedra form a 3-dimensional framework with all corners mutually shared. Lueshite has a b-axis twice as long as the other minerals of this group, due to slight twisting of the octahedra. The condition for substitution is that: $r_A \approx r_0$. Deformations of the ideal type depend on the size of A and B relative to O as well as on their valencies.

The sum of valencies of the A cations is counterbalanced by B in different valencies. In other related structures the balance may be obtained by coupled substitution in the A positions, e.g. (Na$_{1-x}$Cd$_{x/2}$)NbO$_3$. Nonstoichiometry in the O positions leads to other derivatives of the perovskite type: ABO$_{3-x}$. In this case the framework is changed and the B ions have coordination numbers less than 6. A non-mineral example is $Fe^{3+}Nb_{0.5}O_{2.5}$ (WADSLEY, 1967).

e) Aeschynite

Aeschynite, (RE,Ca,Fe,Mn,Th)(Nb,Ti)$_2$(O,OH)$_6$, as yet cannot be derived from a simple structure type. (Nb,Ti)O$_6$ octahedra are coupled into pairs by common edges which are shortened to 2.58 Å. By joining common corners along [001], the pairs of octahedra form kinked double chains. The interlinkage of these chains with

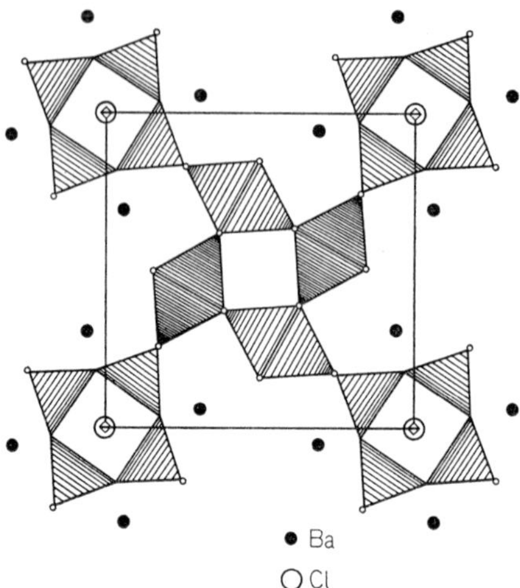

Fig. 41-A-5. Crystal structure projection of baotite (SIMONOV, 1961)

each other by common corners results in a three-dimensional framework of octahedra. The holes in this framework are occupied by the large cations in 8-fold coordination with oxygen (ALEKSANDROV, SR 1962, 536).

f) Silicate Minerals of Nb

The structure of epistolite, $Na_2Ti_2\ Nb_2O_2(OH)_4\ [Si_2O_7]_2 \cdot 2H_2O$, is built up by NbO_6 octahedra which are linked by common corners with Si_2O_7 groups and with $(Na,Ti)O_6$ octahedra as well. The structure is piled up along the c-axis by sheets of these interlinked tetrahedra and octahedra groups where the sheets are connected with each other by $(OH)^-$ and H_2O (KHALILOV *et al.*, 1965).

Barsanovite, $(Na,Ca,Sr,RE)_9(Fe,Mn)_2(Zr,Nb)_2\ Cl[Si_3O_9]_4$, has a structure similar to eudialite, but monoclinic deformed. SiO_4 tetrahedra form 3-membered rings, Zr and Nb occupy the centres of oxygen octahedra which share corners with the tetrahedra of the Si_3O_9-rings (DORFMAN *et al.*, 1963).

Baotite, $Ba_4(Ti,Nb)_8ClO_{16}[Si_4O_{12}]$, represents a ring silicate with SiO_4 tetrahedra sharing two corners with two neighboring tetrahedra thus forming a Si_4O_{12} ring with fourfold symmetry. The $(Ti,Nb)O_6$ octahedra form 4-membered rings around a 4_1-axis and are linked to one free corner of the SiO_4 tetrahedra (Fig. 41-A-5; SIMONOV, SR 1960, 504). Sharing common edges the octahedra groups form infinite columns along the c-axis.

Nenadkevichite, $(Na,K)_{2-x}(Nb,Ti)_2(O,OH)_2[Si_4O_{12}] \cdot 4H_2O$, crystallizes in Pbam with $a_0 = 7.408$, $b_0 = 14.198$, $c_0 = 7.148$ Å. The silica tetrahedra form square Si_4O_{12} rings in (100), joined by chains of NbO_6 octahedra along [100]. Large holes in this arrangement are occupied by Na and H_2O (PERRAULT *et al.*, 1973).

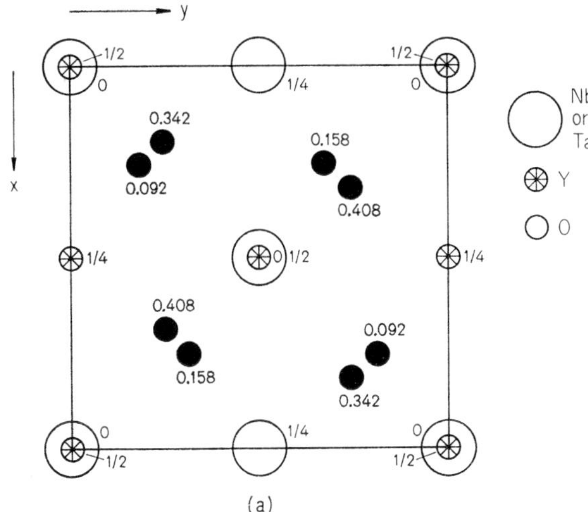

Fig. 41-A-6. Crystal structure projection of fergusonite according to WOLTEN (1967). (One half of unit cell)

III. Oxide Minerals with Nb in Tetrahedral Coordination

Fergusonite, $(Y,Yb)NbO_4$, crystallizes in the scheelite type. The NbO_4 tetrahedra (Nb-O about 1.9 Å) are somewhat flattened in the c-direction. The 3-valent cations have an 8-fold hexahedral oxygen coordination (KOMKOV, SR 1959, 388). This structure type is important for many non-mineral Nb-compounds (KELLER, SR 1962, 520) (Fig. 41-A-6).

Above 1,000° C a monoclinic polymorph is formed, which can be derived from the low-form by a shear deformation in $(3\bar{1}0)$ leaving the coordination number and the interlinkage of the NbO_4 tetrahedra unchanged but altering the Nb-O distances slightly (KOMKOV, SR 1959, 388).

IV. Oxidic Compounds with Nb in 7-fold Coordination

In non-mineral compounds Nb is reported to exist in a 7-fold coordination with oxygen. In $NaNb_6O_{15}$, three of four crystallographically different Nb ions have an octahedral 6-fold coordination with mean Nb-O distances of 1.98 Å. The fourth Nb ion has a 7-fold coordination (pentagonal bipyramid), the mean Nb-O distance being 2.03 Å (ANDERSSON, 1965).

Similar coordinations are found in $Bi_3Nb_{17}O_{47}$ with $Nb^{[6]}$-O distances of about 1.98 and $Nb^{[7]}$-O distances of about 2.05 Å (KEVE and SKAPSKI, 1971; BRAUER, SR 1941, 123; ANDERSSON, 1964, 1965 and 1967; ASTRÖM, 1966; LUNDBERG, 1965 and 1971).

Revised manuscript received: January 1974

References: Section 41-A

ANDERSSON, S.: The crystal structure of Nb_3O_7F. Acta Chem. Scand. **18**, 2339 (1964).

ANDERSSON, S.: The crystal structure of $NaNb_6O_{15}F$ and $NaNb_6O_{15}OH$. Acta Chem. Scand. **19**, 557 and 2285 (1965).

ANDERSSON, S.: Aspects on the problems of synthesis and structure of some oxide or oxide-like compounds formed by the transition elements in the groups 4, 5 and 6 of the periodic table. Arkiv Kemi **26**, 521 (1967).

ARNOTT, R. J.: X-ray diffraction on some radioactive oxide minerals. Am. Mineralogist **35**, 386 (1950).

ASTRÖM, A.: The crystal structures of $Nb_{31}O_{77}F$ and $Nb_{17}O_{42}F$. Acta Chem. Scand. **20**, 969 (1966).

DORFMAN, M. D., ILOKHIN, V. V., BUROVA, T. A.: The crystal structure of barsanovite. Dokl. Akad. Nauk SSSR **153**, 1164 (1963) [In Russian.]

GOLDSCHMIDT, V. M.: Geochemische Verteilungsgesetze der Elemente. VI. Über die Krystallstrukturen vom Rutiltypus, mit Bemerkungen zur Geochemie zweiwertiger und vierwertiger Elemente. Skrifter Norske Videnskaps-Akad. Oslo. I. Math.-Naturv. Kl. **1926**, No. 1 (1926)

HUME-ROTHERY, W., IRVING, H. M., WILLIAMS, R.: The valencies of the transition elements in the metallic state. Proc. Roy. Soc. London, Ser. A **208**, 431 (1951).

KEVE, E. T., SKAPSKI, A. C.: $Bi_3Nb_{17}O_{47}$: A potentially ferroelectric crystal structure of a tungsten bronze type. J. Chem. Soc. A **1971**, 1280 (1971).

KHALILOV, A. D., MAKAROV, E. S., MAMEDOV, K. S., PYANZINA, L. J.: Über die Kristallstruktur von Mineralien der Gruppe Murmanit — Lomonosovit. Dokl. Akad. Nauk. SSSR **162**, 179 (1965). [In Russian.]

LAVES, F., BAYER, G., PANAGOS, A., Strukturelle Beziehungen zwischen den Typen α-PbO_2 $FeWO_4$ (Wolframit) und $FeNb_2O_6$ (Columbit) und über die Polymorphie des $FeNbO_4$. Schweiz. Mineral. Petrogr. Mitt. **43**, 217 (1963).

LIMA-DE-FARIA, J.: Identification of Metamict Minerals by X-ray Powder Photographs. Lisboa (1971).

LUNDBERG, M.: The crystal structure of $LiNb_6O_{15}F$. Acta Chem. Scand. **19**, 2274 (1965)

LUNDBERG, M.: The crystal structure of $LiNb_3O_8$. Acta Chem. Scand. **25**, 3337 (1971).

NICKEL, E. H.: Latrappite — a proposed new name for the perowskite-type calcium niobate mineral from the Oka Area of Quebec. Can. Mineralogist **8**, 121 (1964).

NICKEL, E. H., MCADAM, R. C.: Niobian perovskite from Oka, Quebec; a new classification for minerals of the perovskite group. Can. Mineralogist **7**, 683 (1963).

NICKEL, E. H., ROWLAND, J. F., MCADAM, R. C.: Ixiolite — a columbite substructure. Am. Mineralogist **48**, 961 (1963).

NOWOTNY, H.: Hochschmelzende Metalle. 3. Plansee Seminar. BENESOVSKY, F. (ed). Wien. 1959.

PAULING, L.: The nature of the interatomic forces in metals. Phys. Rev. **54**, 899 (1938).

PAULING, L.: A resonating-valence-bond theory of metals and intermetallic compounds. Proc. Roy. Soc. London, Ser. A **196**, 343 (1949).

PERRAULT, G., BOUCHER, C., VICAT, J., CANILLO, E., ROSSI, G.: The crystal structure of nenadkevichite. Amer. Cryst. Ass. Winter Meeting, Progr. Abstr. **1973**, 66 (1973).

SAKOWSKI-COWLEY, A. C., LUKASZEWICZ, K., MEGAW, H. D.: The structure of sodium niobate at room temperature, and the problem of reliability in pseudosymmetric structures. Acta Cryst. B **25**, 851 (1969).

SCHRÖCKE, H.: Über Festkörpergleichgewichte innerhalb der Columbit/Tapiolit-Gruppe mit YTi (Nb,Ta) O_6, Euxenit und mit $FeNbO_4$. Neues Jahrb. Mineral. Abhandl. **106**, 1 (1966).

SHANNON, R. D., PREWITT, C. T.: Effective ionic radii in oxides and fluorides. Acta Cryst. B **25**, 925 (1969).
STRUNZ, H.: Mineralogische Tabellen, 5. Aufl. Leipzig: Akademische Verlagsges 1970.
WADSLEY, D.: In: MANDELCORN, L. (ed.) Nonstoichiometric Compounds. New York: Academic Press 1964.
WHITTACKER, E. J. W., MUNTUS, R.: Ionic radii for use in geochemistry. Geochim. Cosmochim. Acta **34**, 945 (1970).
WOLTEN, G. M.: The structure of the M-phase of $YTaO_4$, a third fergusonite polymorph. Acta Cryst. **23**, 939 (1967).

Revised manuscript received: January 1974

41-B. Isotopes in Nature

Niobium has but one natural stable isotope, ^{93}Nb. The majority of the synthetically produced isotopes, ^{90}Nb to ^{92}Nb and ^{94}Nb to ^{99}Nb, has a half-live of minutes to days. An exception is ^{94}Nb for which HOLLANDER *et al.* (1953) list a half-life of more than $5 \cdot 10^4$ years.

Revised manuscript received: March 1977

41-C. Abundance in Cosmos, Meteorites, Tektites and Lunar Materials

I. Cosmos

The cosmic abundance of niobium, based on the insufficient data on chondrites available to Suess and Urey (1956), was estimated to be 1 atom niobium per 10^6 atoms silicon. This figure was revised to 1.15 atoms Nb/10^6 atoms Si by Cameron (1968). Solar photopheric data by Müller (1968) are 5.5 times higher than the abundance reported by the latter author.

II. Meteorites

The few data on Nb concentrations in meteorites are listed in Table 41-C-1. According to this table ordinary chondrites might contain 0.4 or 0.5 ppm Nb. Wänke *et al.* (1974) report an appreciable accumulation of niobium in the refractory inclusions of the Allende Cc_3-chondrite (factor of accumulation relative to Cc_1-chondrites: 22).

A lithophilic character must be assumed for this element from the information of Table 41-C-1. Graham and Mason (1972) mention their preliminary investigations on the distribution of niobium between different minerals in two ordinary chondrites. The analysis of mineral separates showed the presence of this element in all of the concentrates except the metal phase, with no marked concentration in any of them. The only meteoritic mineral with notable concentrations of niobium as detected by

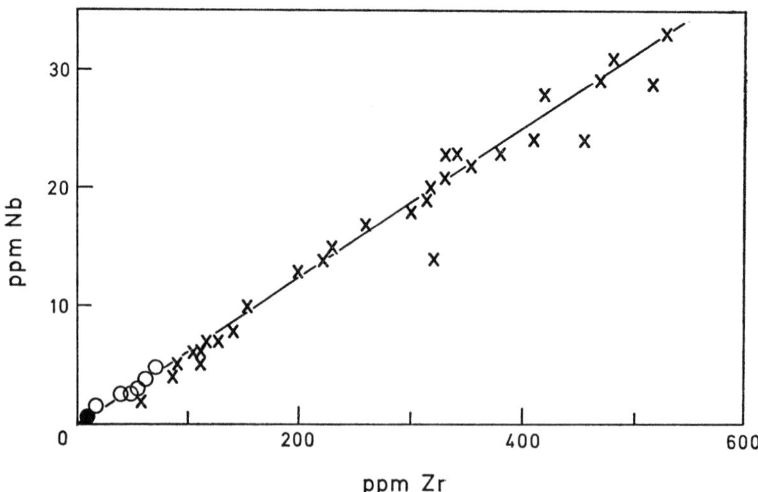

Fig. 41-C-1. Relation between Zr and Nb in extraterrestrial matter according to Graham and Mason (1972). Chondrites (filled circles); achondrites (circles); lunar materials (crosses)

Table 41-C-1. *Niobium in meteorites*

Material	Class	ppm Nb	ppm Nb	Analytical method	Reference
Chondrites					
Ivuna	Cc1	0.3		Mass	1
Murchison	Cc2	0.6		Mass	1
Renazzo	Cc2	0.5		Mass	1
Allende	Cc3	0.9		Mass	1
Modoc	CL	0.4		Mass	1
St. Michel	CL	0.3		W/S	2
Guarena	CH	0.4		Mass	1
Lake Labyrinth	CLL	2.1		W/S	2
Average chondrites		0.5		—	3
Achondrites					
Luotolax	Aor	1.5		Mass	1
Sioux Co.	Ap	2.7	2.8	Mass, X	1, 4
Stannern	Ap	4.7	6.3	Mass, X	1, 4
Bereba	Ap	2.5	4.0	Mass, X	1, 4
Pasamonte	Ap	3.6	3.5	Mass, X	1, 4
Juvinas	Ap	2.7	2.7	Mass, X	1, 4
Angra dos Reis	Aa	5		Mass	5
Stony Irons					
Admire (olivine)	P	<20		S	6
Albin (olivine)	P	<20		S	6
Brenham (olivine)	P	<20		S	6
Marjalahti (olivine)	P	<0.07		W/S	2
Springwater (olivine)	P	<20		S	6
Irons					
Casas Grandes	Om	<0.03		W/S	2
Kyancutta	Om	<0.003		W/S	2
Coolac	Og	<0.02		W/S	2
Troilite (2)	Ce, CH	1.6; 0.2		Mass	1

References: 1. Graham and Mason (1972); 2. Rankama (1948); 3. Levin *et al.* (1956); 4. Erlank *et al.* (1972); 5. Smales *et al.* (1970); 6. Lovering (1957).

El Goresy (1971) is accessory rutile from iron meteorites (1.1 to 2% Nb) and from a mesosiderite (up to 0.27% Nb).

There exists a good correlation of Nb with Ti, Zr and Ta in meteorites according to their chemical properties and their comparable refractory character. Graham and Mason (1972) report an average Ti/Nb ratio of 1,850 and a Zr/Nb ratio of 19 for meteorites. Fig. 41-C-1, according to the above mentioned authors, demonstrates the remarkably linear covariance of Nb and Zr in extra-terrestrial matter including meteorites and lunar materials.

III. Tektites

Chao (1963) reports ranges of 13 to 24 ppm Nb in 21 bediasites (average: 17 ppm Nb) and 17 to 27 ppm Nb in 6 australites (average: 22 ppm Nb; analytical method for both sets: S). These values are not characteristic for a certain unaltered terrestrial rock type.

IV. Lunar Materials

There exist more than 300 analytical data on niobium in lunar materials (Table 41-C-2). These mainly consider basalts and fines (soils). Data from X-ray and acti-

Table 41-C-2. *Niobium in lunar materials* (Number of samples in brackets; Most references are from Volume 2 of the Proceedings of 7 Lunar Science Conferences. These are marked as [P1] to [P7]. The Proceedings are Supplements 1 to 7 of Geochim. Cosmochim. Acta 1970 to 1976.)

Rock type	Range ppm Nb	Average ppm Nb	Analytical method	Reference
Apollo 11				
Basalts (5)	20–29	23	S	ANNELL and HELZ [P1]
Basalts (4)	14–25	22	X	COMPSTON *et al.* [P1]
Basalts (5)	36–47	42	Mass	MORRISON *et al.* [P1]
Basalts (6)	17.6–29	25	X	DUNCAN *et al.* [P7]
Breccias (2)	37–45	41	Mass	MORRISON *et al.* [P1]
Breccias (2)	—	19	X	COMPSTON *et al.* [P1]
Breccias (7)	14–31	23	S	ANNELL and HELZ [P1]
Fines (1)	—	18	S	ANNELL and HELZ [P1]
Fines (1)	—	18	X	COMPSTON *et al.* [P1]
Fines (1)	—	33	Mass	MORRISON *et al.* [P1]
Fines (1)	—	10	Mass	SMALES *et al.* [P1]
Apollo 12				
Basalts (3)	6.4–8.2	7.2	X	WILLIS *et al.* [P3]
Basalts (2)	9.4–11	10	Mass	MORRISON *et al.* [P2]
Basalts (8)	$<$10–16	12	S	CUTTITTA *et al.* [P2]
Basalts (4)	12–16	14	Mass	BOUCHET *et al.* [P2]
Basalts (10)	2–7	5.7	X	COMPSTON *et al.* [P2]
Basalts (3)	6–7	6.7	X	RHODES and HUBBARD [P4]
Breccia (1)	—	11	X	COMPSTON *et al.* [P2]
Fines (3)	29–44	34	X	COMPSTON *et al.* [P2]
Fines (1)	—	35	Mass	MORRISON *et al.* [P2]
Fines (2)	33–43	38	X	WILLIS *et al.* [P3]
Fines (6)	26–74	38	S	CUTTITTA *et al.* [P2]
Fines (2)	42–80	61	Mass	BOUCHET *et al.* [P2]
Fines (2)	29–30	30	S	DWORNIK *et al.* [P5]
Apollo 14				
Basalt (1)	—	64	X	HUBBARD *et al.* [P3]
Basalts (2)	29–33	31	S	ROSE *et al.* [P3]
Basalts (2)	15.7–54.5	35	X	WILLIS *et al.* [P3]
Basalt (1)	—	59	X, Mass	STRASHEIM *et al.* [P3]
Breccia (1)	—	81	X	HUBBARD *et al.* [P3]
Breccias (6)	44–64	51	S	ROSE *et al.* [P3]
Breccia (1)	—	70.7	X	WILLIS *et al.* [P3]
Fines (1)	—	65	X	HUBBARD *et al.* [P3]
Fines (6)	39–70	55	S	ROSE *et al.* [P3]
Fines (2)	61.3–63.4	62	X	WILLIS *et al.* [P3]
Fines (2)	54–62	58	X, Mass	STRASHEIM *et al.* [P3]

Table 41-C-2 (continued)

Rock type	Range ppm Nb	Average ppm Nb	Analytical method	Reference
Apollo 15				
Gabbros (4)	<10–12	≤11	S	CUTTITTA *et al.* [P4]
Basalts (5)	5.4–10	6.6	X	DUNCAN *et al.* [P6] [P7]
Basalts (11)	<10–17	≤10	S	CUTTITTA *et al.* [P4]
Basalts (2)	4.7–6	5.4	N/R	WÄNKE *et al.* [P6]
Basalts (4)	4.3–6.2	5.4	X	RHODES and HUBBARD [P4]
Basalt (1)	—	6.2	Mass	TAYLOR *et al.* [P4]
Glasses (green) (3)	1.5–2.1	1.8	Mass	TAYLOR *et al.* [P4]
Anorthositic gabbro (1)	—	4	X	DUNCAN *et al.* [P6]
Lunar highlands (average)	—	10.5/7.5	Mass	TAYLOR *et al.* [P4] [P5]
Breccias (highland) (12)	0.4–33	19	Mass	TAYLOR *et al.* [P4]
Fines (7)	12–23	16	S	CUTTITTA *et al.* [P4]
Fines (7)	9.6–33	20	Mass	TAYLOR *et al.* [P4]
Fines (2)	28–34	31	S	ROSE *et al.* [P6]
Fines (7)	14–24.8	19	X	DUNCAN *et al.* [P6]
Apollo 16				
Crystalline rocks (6)	<10–60	19	S	ROSE *et al.* [P4]
KREEP basalt (1)	—	33	Mass	MORRISON *et al.* [P4]
Gabbroic anorthosites (2)	1.3–13	7.2	N/R	WÄNKE *et al.* [P4]
Anorthositic gabbro (1)	—	2.1	Mass	MORRISON *et al.* [P4]
Breccias (4)	<10–30	≤17	S	ROSE *et al.* [P6]
Breccias (4)	<0.2–20	12	N/R	WÄNKE *et al.* [P5]
Breccias (3)	—	<10	S	ROSE *et al.* [P4]
Breccias (2)	4.1–11.8	8	N/R	WÄNKE *et al.* [P6]
Fines (10)	<10–14	≤11	S	ROSE *et al.* [P4]
Fines (2)	11–13	12	Mass	MORRISON *et al.* [P4]
Fines (6)	3.8–11.5	9.6	N/R	WÄNKE *et al.* [P4]
Fines (23)	<10–17	≤11	S	FINKELMAN *et al.* [P6]
Apollo 17				
Basalts (2)	—	21	X	RHODES *et al.* [P5]
Basalts (11)	<10–31	≤16	S	ROSE *et al.* [P5] [P6]
Basalts (5)	18.5–29.1	23.4	X	DUNCAN *et al.* [P5]
Basalt (1)	—	19	N/R	WÄNKE *et al.* [P5]
Basalts (5)	17.4–24	21	N/R	WÄNKE *et al.* [P6]
Troctolite (1)	—	1.2	X	RHODES *et al.* [P5]
Noritic breccias (4)	32–33	33	X	RHODES *et al.* [P5]
Anorthositic gabbro (1)	—	4	X	RHODES *et al.* [P5]
Fines (4)	14–16	14	N/R	WÄNKE *et al.* [P5]
Fines (10)	13.3–20.6	17.2	X	DUNCAN *et al.* [P5]
Fines (low Ti) (9)	10–13	11	S	ROSE *et al.* [P5]
Fines (high Ti) (8)	14–26	20	S	ROSE *et al.* [P5]
Fines (34)	13–21	17	X	RHODES *et al.* [P5]
Orange soil (1)	—	15	X	RHODES *et al.* [P5]

vation procedures have been given priority in averages. The niobium concentration in basalt flows is variable between the different mare areas. Nb is low in Apollo 12 and 15 basalts (mainly 5 to 7 ppm Nb), four times higher in Apollo 11 and 17 samples (mainly 19 to 25 ppm Nb) and exceeds this level in Apollo 14 basalts. Terrestrial oceanic tholeiites (5 ppm Nb) are usually as low as Apollo 12 and 15 basalts in niobium. Ca-rich achondrites have even lower Nb concentrations. Apollo 11 and 17 basalts are higher than normal tholeiites (13 ppm Nb) but not as high as terrestrial alkali olivine basalts (69 ppm Nb) (WEDEPOHL, 1975).

Except in Apollo 11, 16, and 17 fines, the niobium concentration in lunar fines exceeds that of the local magmatic rocks. Several authors assume that the so-called KREEP concentrates contain at least 85 ppm Nb (SCHONFELD and MEYER, 1972). HLAVA *et al.* (1972) found niobian rutile with 7.1% Nb_2O_5 in a lithic fragment of Apollo 14 microbreccia 14162.16. The lithic fragment has a basaltic texture and is of KREEP composition.

Anorthositic gabbro of the lunar highlands is as low as 2 to 7 ppm Nb because plagioclase usually has lower Nb concentrations than pyroxene (TAYLOR *et al.*, 1973).

A Nb-Zr correlation in meteorites comparable to that in lunar rocks has been demonstrated in Fig. 41-C-1. WÄNKE *et al.* (1975) have observed that the Zr/Nb ratio decreases from KREEP and Apollo 11, 12, 15 mare basalts (Zr/Nb: 15–19) to Apollo 17 mare basalts (Zr/Nb: 10–15).

The average total niobium concentration in the moon has been estimated as being 4.8, 3.3 and 2.2 ppm Nb respectively (WÄNKE *et al.*, 1973; GANAPATHY and ANDERS, 1974; TAYLOR and JAKES, 1974).

Revised manuscript received: March 1977

41-D. Abundance in Rock-Forming Minerals, Niobium Minerals, Phase Equilibria

I. Rock-Forming Minerals

Niobium has strong chemical relations to tantalum (Chapt. 73). In most minerals and rocks in which these elements can be analyzed they are correlated, and only exceptionally does one occur highly fractionated from the other. This association stems mainly from the identical valance states of their ions and the very close similarity of their ionic radii: $Nb^{5+} = 0.69$ Å and $Ta^{5+} = 0.68$ Å (Vol. I of this handbook, pages 390/391). During magmatic differentiation both elements accumulate; thus later differentiates are characterized by increased abundances (PARKER and FLEISCHER, 1968).

In rocks in which niobium (and tantalum) appear vicariously, the chief host minerals are pyroxene (Table 41-D-4), amphibole (Tables 41-D-1/4), biotite (Tables

Table 41-D-1. *Distribution of niobium and tantalum in granitic rocks from the Kalbinsk Massif and the eastern Sayan Mountains* (U.S.S.R.) (ZNAMENSKII *et al.*, 1957; ZNAMENSKII *et al.*, 1962) (Analytical method: W/X)

Mineral	Content of mineral in rock (wt. percent)	Content of Nb and Ta in minerals		
		Nb (ppm)	Ta (ppm)	Nb:Ta
Porphyritic biotite granodiorite, Asu-Bulak, Kalbinsk				
Biotite	13.00	96.0	7.6	12.0
Ilmenite	0.30	860.0	130.0	6.6
Zircon	0.26	70.0	110.0	—
Bulk rock		21.0	2.0	10.5
Porphyritic biotite granite, Miroliubovsk Ridge, Kalbinsk				
Biotite	8.00	130.0	18.0	7.2
Ilmenite	0.17	1,200.0	210.0	5.8
Bulk rock		15 0	2.2	6.8
Coarse-grained biotite granite, Mount Baich, Kalbinsk				
Biotite	5.6	200.0	26.0	7.7
Ilmenite	0.05	3,600.0	640.0	5.6
Bulk rock		13.0	2.0	6.5
Coarse-grained biotite granite, Chur-Chut-Su, Kalbinsk				
Biotite	8.1	80.0	9.5	8.4
Ilmenite	0.1	1,700.0	260.0	6.5
Bulk rock		10.0	1.3	7.7
Coarse-grained biotite granite, Tikhaya River, Sayan M.				
Biotite	4.5	145.0	3.9	37.2
Sphene	0.25	1,550.0	115.0	13.5
Bulk rock		11.0	<0.7	>15.7

Table 41-D-1 (continued)

Mineral	Content of mineral in rock (wt. pertent)	Content of Nb and Ta in minerals		
		Nb (ppm)	Ta (ppm)	Nb:Ua
Coarse-grained biotite granite, B. Yerma River, Sayan M.				
Biotite	8.2	37.0	<3.5	>10.6
Sphene	0.5	2,400.0	200.0	12.0
Bulk rock		14.0	1.0	14.0
Medium-grained biotite granodiorite, Zimovninskii Massif, Sayan M.				
Biotite	15.0	29	<3.3	>8.8
Sphene	0.3	1,350	155	8.7
Bulk rock		7.6	<1	>7.6
Medium-grained biotite hornblende granodiorite, Zimovninskii Massif, Sayan M.				
Biotite	8.0	12	<4	>3
Hornblende	7.4	1	n.d.	—
Sphene	0.7	705	58	12
Bulk rock		6.2	<1	>6.2

Table 41-D-2. *Tantalum and niobium in biotites of granitoids of the U.S.S.R.* (Odikadze, 1967) (Number of samples in brackets)

Rock type	Nb, ppm		Ta, ppm		Nb/Ta
	Range	Average	Range	Average	
Transcaucasian intermontane downwarp, Dzirula Massif					
Gneissoid biotite quartz diorites, Caledonian (6)	40–56	50	5–8	7	7.14
Porphyritic two-mica microcline granites, Hercynian (2)	120–150	135	8–9	8	17
Alaskites, two-mica granite dikes, Hercynian (3)	90–110	100	12–18	14	7
Unreplaced quartz-microcline pegmatites (1)	—	58	—	7	8
Albitized and greisenized pegmatites with columbite, manganotantalite and cassiterite, Hercynian (3)	25–30	27	5–8	6	4.5
Porphyroblastic microcline granites, Hercynian (7)	30–110	55	7–8	7	7.8
Average for late Hercynian granitoids (16)	—	75	—	8	9.4
Biotite anorthoclase granodiorites, late Hercynian (2)	100–120	100	4–8	6	16.6
Average for granitoids of the Dzirula massif (24)	—	78	—	7	11
Pre-Caucasus Range, Urushtenskiy, Belyagidonsk, Tyrnauz areas					
Quartz-plagioclase pegmatites, Caledonian (5)	10–40	19	2–3	2.5	7.6

Table 41-D-2 (continued)

Rock type	Nb, ppm		Ta, ppm		Nb/Ta
	Range	Average	Range	Average	
Vaza-Khokh hornblende biotite gneisses, Pre-Caledonian (3)	84–290	210	7–8	7.5	28
Biotite granites, Zdiadag area, Hercynian (9)	175–350	290	28–80	65	4.5
Biotite schlieren, Vaza-Khokh area, Hercynian (2)	3–7	5	0–6	3	1.7
Average for Belyagidonsk granitoids (14)	—	168	—	25	6.7
Porphyritic anorthoclase granites, El'dzhurta, Cimmerian (10)	60–170	138	5–21	16	8.6
Biotite-quartz-feldspar pegmatoidal injections (1)	—	72	—	6	12
Microcline granite with large porphyroblasts (1)	—	120	—	11	11
Medium-grained biotite granites, Hercynian (8)	20–36	28	3–6	5	5.6
Average Pre-Caucasus Range (38)		111		15.3	7.3
Main Range, Dar'yal, Kty-Teberda Massifs					
Gneissoid porphyroblastic biotite granites, Hercynian (2)	14–21	17	2–4	3	7
Two-mica granites, Hercynian (3)	120–170	140	10–15	13	8.6
Average for Main Range granitoids (5)	—	78	—	8	8
Granite porphyries, Neogene intrusives Caucasus (3)	15–30	21	3–12	7	3
Adzhar-Trialeti folded belt, Guriysk Syenite Massif					
Syenite pegmatites with rare earth apatite (3)	85–130	110	5–10	8	14
Average of biotites Table 41-D-2 (173)		94[a]		11.4	8.5
Average of biotites Table 41-D-1 (8)		91[a]		≤ 9.5	≥ 9.6

[a] Lovering (1972) observed a mean of 34 ppm Nb (analytical method: S) in 101 biotite samples from granites, granodiorites etc. in the Western United States. Sitnin (1966) lists the following averages for biotites: granodiorites, plagiogranites: 52 ppm Nb; unaltered granites: 80 ppm Nb; leucocratic and two-mica granites: 200 ppm Nb. The average Nb/Ta ratio in biotites of these various granitic rocks from the U.S.S.R. is 13 and in muscovites: 5.

41-D-1/2), muscovite (Table 41-D-3), sphene, ilmenite, and magnetite. Table 41-D-1 demonstrates that biotite contains 60 to 86% of the total Nb of ilmenite granites. In sphene bearing granites this Ti mineral is the main Nb host containing 35 to 86% of the total niobium.

Table 41-D-3. *Niobium and tantalum in muscovites of granitoids of the U.S.S.R.* (Odikadze, 1967) (Number of samples in brackets)

Rock type	Nb, ppm		Ta, ppm		Nb/Ta
	Range	Average	Range	Average	
Transcaucasian intermont. downwarp, Dzirula Massif					
Gneissoid biotite quartz diorites, Caledonian (4)	38–43	40	2–3	3	13
Porphyritic two-mica microcline granites, Hercynian (1)	—	85	—	7	12
Alaskites and two-mica granites, Hercynian (3)	55–70	65	5–9	8	8
Quartz-microcline pegmatites (10)	10–250	110	8–65	22	5
Albitized and greisenized quartz-microcline pegmatites with Be, Nb, Ta, Sn, etc. minerals Hercynian (17)	49–410	200	39–150	93	2.13
Average for late Hercynian granitoids (30)	—	115	—	32	3.6
Average for Dzirula Massif (34)	—	77	—	17	4.6
Pre-Caucasus Range, Urushtenskiy, Vaza-Khokh, Sadon Complexes					
Muscovite plagiogranite gneiss, Caledonian (7)	10–70	30	5–9	7	4.5
Muscovite-quartz-plagioclase greisenized pegmatites, Bol'shaya Laba R. (5)	15–30	20	3–5	4	5
Muscovite-quartz-microcline pegmatites with carburan (1)	—	5	—	8	0.62
Average of Urushtenskiy Complex (13)	—	18	—	6	3
Albitized and greisenized granites, Hercynian (3)	135–170	150	28–35	31	5
Quartz-microcline pegmatite dikes (2)	110–140	125	45–60	55	2.27
Average for Vaza-Khokh (Zdiadag) Complex (5)	—	137	—	43	3.1
Two-mica granite, Sadon, Hercynian (1)	—	60	—	2	30
Muscovite granite, Sadon, Hercynian (1)	—	60	—	6	10
Muscovite-quartz-microcline pegmatite dikes (3)	108–330	203	41–140	75	2.7
Average for the Sadon Complex (5)	—	108	—	27	4
Main Range, Kty-Teberda					
Muscovite-quartz-microcline pegmatites, Hercynian (5)	70–180	140	25–74	50	2.8
Quartz-microcline pegmatites, Hercynian, Baksan River (4)	40–160	108	38–60	49	2.2
Average for the Main Range (9)	—	124	—	50	2.5
Average for the Caucasus (66)	—	93	—	28	3.3

Table 41-D-4. *Contents of Nb and Ta in olivine, pyroxene, hornblende, feldspar, sphene, titano-magnetite, ilmenite, and perovskite from magmatic rocks in New Zealand, Germany, the NE U.S.S.R. and the Kola Peninsula (U.S.S.R.)* (Number of samples in brackets)

Mineral	Rock	Locality	ppm Nb	ppm Ta	Nb/Ta	Analytical method	Reference
Olivine (3)	Alk. basalt	Eifel (F.R.G.)	16	n.d.	n.d.	X	1
Clino-pyroxene (7)	Alk. basalt	Eifel (F.R.G.)	24	n.d.	n.d.	X	1
Pyroxene (13)	Pyroxenites	Kola (U.S.S.R.)	27	7.9	3.4	S	2
Pyroxene (7)	Pyrox. melteigites	Kola (U.S.S.R.)	70	7.4	9.4	S	2
Pyroxene (2)	Alk. pegmatites	Kola (U.S.S.R.)	93	12	7.7	S	2
Diopside (2)	Pyroxenite	NE U.S.S.R.	12	1.2	10	S	5
Aegirine (2)	Granite	NE U.S.S.R.	3.1	1.2	2.5	S	5
Diopside (3)	Diorite	NE U.S.S.R.	22	1.7	12.9	S	5
Pyroxene (3)	Pyroxenite	NE U.S.S.R.	3.3	n.d.	n.d.	S	5
Augite (1)	Volcanic breccia	Kakanui (N.Z.)	0.4	n.d.	n.d.	Mass	3
Hornblende (1)	Volcanic breccia	Kakanui (N.Z.)	12	n.d.	n.d.	Mass	3
Kaersutite (9)	Mafic alk. rocks	Australia, N. Zealand	44	n.d.	n.d.	X	4
Hornblende (2)	Pyroxenite	NE U.S.S.R.	40	5.5	7.3	S	5
Riebeckite (3)	Granite	NE U.S.S.R.	21	8.2	2.6	S	5
Hornblende (3)	Gabbro	NE U.S.S.R.	37	6.3	5.9	S	5
Hornblende (7)	Granodiorite	NE U.S.S.R.	28	2.6	10.8	S	5
Hornblende (2)	Alkali pegmatites	Kola (U.S.S.R.)	52	4.1	12.8	S	2
Hornblende (3)	Basanite etc.	Eifel (F.R.G.)	36	n.d.	n.d.	X	1
Alkali feldspar (4)	Trachyte etc.	Eifel (F.R.G.)	9	n.d.	n.d.	X	1
Plagioclase (8)	Alkali basalt	Eifel (F.R.G.)	10	n.d.	n.d.	X	1
Albite (3)	Alkali basalts etc.	Eifel (F.R.G.)	24	n.d.	n.d.	X	1
Sphene (2)	Gabbro etc.	NE U.S.S.R.	3,400	480	7.1	S	5
Sphene (2)	Granite	NE U.S.S.R.	700	640	1.1	S	5
Sphene (3)	Pyroxenite	Kola (U.S.S.R.)	1,570	110	14.1	S	2
Sphene (5)	Calc. silicate rock	Kola (U.S.S.R.)	6,070	550	11.1	S	2
Sphene (6)	Alkali pegmatite	Kola (U.S.S.R.)	4,070	190	21.3	S	2
Titano-magnetite (5)	Alkali pegmatite	Kola (U.S.S.R.)	8.4	1.6	5.3	S	2
Titano-magnetite (5)	Alkali basalt etc.	Eifel (F.R.G.)	620	n.d.	n.d.	X	1
Titano-magnetite (2)	Gabbro etc.	NE U.S.S.R.	3,480	341	10	S	5
Ilmenite (2)	Granite etc.	NE U.S.S.R.	1,250	384	3.3	S	5
Perovskite (3)	Ijolite etc.	NE U.S.S.R.	1,960	117	16.8	S	5
Perovskite (22)	Olivinite	Kola (U.S.S.R.)	3,460	407	8.5	S	2
Perovskite (14)	Pyroxenite	Kola (U.S.S.R.)	5,070	530	9.5	S	2
Perovskite (5)	Calc. silicate rock	Kola (U.S.S.R.)	8,930	660	13.6	S	2
Perovskite (18)	Alkali pegmatite	Kola (U.S.S.R.)	7,730	530	14.5	S	2

References: 1. Huckenholz (1965); 2. Kukharenko *et al.* (1965); 3. Mason and Allen (1973); 4. Kesson (1972); 5. Nekrasov (1970).

Table 41-D-5. *Niobium and tantalum contents of cassiterites*

Source of samples	No. of samples	ppm Nb	ppm Ta	Nb/Ta
1. *Siberia* (DUDYKINA, 1959)				
I. Tin-bearing pegmatite	37	6,000	5,300	1.1
II. Quartz-cassiterite				
a) Greisens and quartz-feldspar veins with topaz, tourmaline, beryl	38	5,200	1,570	3.3
b) Quartz veins	129	600	150	4
c) Transitional deposits	24	30	—	
III. Sulfide-cassiterite deposits	43	20	—	
2. *Transvaal, South Africa* (LEUBE and STUMPFL, 1963)				
a) Nieuwpoort Mine (Range of variation: 10–50 ppm Nb)	6	26		
b) Leeuwpoort Mines (Range of variation: 10–100 ppm Nb)	12	44		
3. *Tin-belt, Southeast Asia* (HOSKING, 1970)				
Tambun, Kinta		13,000	22,000	0.6
Pulai, Kinta		2,000	2,000	1
Tekka, Kinta		tr	1,000	
Kacha, Kinta		tr	1,000	
Siputeh, Kinta		tr	tr	
Sungei Way, Selangor		tr	nil	
Kota Tinggi, Johore		11,000	69,000	0.16

The close association of niobium and titanium has long been recognized (FLEISCHER *et al.*, 1952). A good correlation between Nb + Ta and Ti in rock-forming minerals has been demonstrated in diagrams by NEKRASOV (1970). Titanium minerals in which niobium proxies include titanian magnetite (Table 41-D-4), rutile, brookite, sphene and perovskite (Table 41-D-4). Niobium also substitutes in certain tin and tungsten minerals, especially cassiterite and wolframite (Tables 41-D-5, 41-D-6). In wolframites the presence of scandium has been suggested as compensating for the ionic inequality resulting from the substitution $Nb^{5+} \leftrightarrow W^{6+}$, but data by BORISENKO and LISUNOV (1958), exhibiting a large variation of Sc/Nb ratios, da not support this claim (Table 41-D-6). Maximum concentrations in zirconium minerals occur above 1% Nb (RANKAMA, 1947).

II. Niobium Minerals

Niobium (and tantalum) minerals essentially belong to only two chemical groups of minerals: oxides and hydroxides. Exceptions are the borate, behierite and several silicates (Table 41-D-7).

A large number of both rare and more common minerals contain 1–5 percent of Nb and Ta (Table 41-D-8).

The various schemes of coupled, hetero-valent isomorphism in niobium-tantalum minerals suggested by various investigators are summarized in Table 41-D-9.

Table 41-D-6. *Average content of Sc and Nb in wolframites from various rare metal deposits* (A: both Sc and Nb in detectable concentrations; B: Sc not dectected; Nb present). (According to BORISENKO and LISUNOV, 1958) (Number of samples in brackets)

Type	Deposits A	ppm Sc	ppm Nb	Deposits B	ppm Nb
Pneuma-tolytic-hydro-thermal	Zinnwald, Erzgebirge[a] (11)	1,700	3,100	Kara-Oba, Central Kazakhstan (3)	2,100
	Polyarnoe, NE Asia (22)	496	1,470	North Kounrad, Central Kazakhstan (3)	840
	Akchatau, Central Kazakhstan (54)	404	1,190	Volframovye Sopke, Central Kazakhstan (5)	1,260
	Kamenka, Altai (3)	391	559		
	Bainazar, Central Kazakhstan (10)	385	1,610		
	Etyka, E. Transbaikal (2)	326	2,100		
	Sherlova Gora, E. Transbaikal (15)	319	1,330		
	Darat, Central Kazakhstan (4)	261	1,540		
	North Kounrad, Central Kazakhstan (9)	156	1,820		
	Kara-Oba, Central Kazakhstan (35)	130	1,610		
	Volframovye Sopki, Central Kazakhstan (4)	104	1,540		
	East Kounrad, Central Kazakhstan (8)	98	1,260		
Hydro-thermal	Antonova Gora, E. Transbaikal (3)	33	840	Belukha, E. Transbaikal (2)	490
				Bukuka, E. Transbaikal (2)	1,120
				Spokoinoe, E. Transbaikal (8)	3,000
				Kalgutinskoe, Altai (4)	1,050
				Dedova Gora, E. Transbaikal (9)	98
				Antonova Gora, E. Transbaikal (20)	560

[a] According to data from LEUTWEIN (1951).

Niobium-tantalum ratios in minerals vary widely (Tables 41-D-1 to 4) even within a single species, depending largely on whether the mineral occurs in calc-alkalic or alkalic rocks and where the host rocks occur, in time, in a particular consanguineous sequence.

In granitic pegmatites, for example, early formed Nb-Ta species tend to be Nb-rich, with paragenetically later species becoming enriched in tantalum. Conversely, in differentiated alkalic igneous rock complexes, the concentration of niobium increases in the rocks of the later outer zones.

Table 41-D-7. *Minerals of niobium and tantalum* (Mainly according to PARKER and FLEISCHER, 1968, but supplemented)

Oxides, hydroxides

Ilmenorutile-strüverite series, BX_2, B_2X_4
- Ilmenorutile $(Ti,Nb,Ta,Fe)_2O_4$
- Strüverite $(Ti,Ta,Nb,Fe)_2O_4$

Tapiolite-mossite series, AB_2X_6
- Tapiolite $(Fe,Mn)(Ta,Nb)_2O_6$
- Mossite $(Fe,Mn)(Nb,Ta)_2O_6$

Perovskite series ABX_3
- Dysanalyte $(Ca,Ce,Na)(Ti,Nb,Ta)O_3$
- Latrappite $(Ca,Na)(Nb,Ti,Fe)O_3$
- Loparite $(Na,Ce,Ca)(Ti,Nb)O_3$
- Metaloparite=hydrated loparite
- Nioboloparite=niobian loparite
- Irinite=thorian loparite
- Lueshite $NaNbO_3$
- Igdloite $NaNbO_3$
- Natroniobite $NaNbO_3$ or $NaNb_2O_5OH$

Fergusonite-formanite series ABX_4
- Fergusonite $(Y,Er,Ce,Fe)(Nb,Ta,Ti)O_4$
- Risörite $(Y,Er,Ce,Fe)(Nb,Ti,Ta)O_4$
- Beta-fergusonite=polymorph of fergusonite
- Formanite $(Y,Er,U,Th,Ca)(Ta,Nb,Ti)O_4$

Stibiotantalite-stibiocolumbite series, ABX_4
- Stibiotantalite $Sb(Ta,Nb)O_4$
- Stibiobismutotantalite=bismuthian stibiotantalite
- Stibiocolumbite $Sb(Nb,Ta)O_4$
- Bismutotantalite $Bi(Ta,Nb)O_4$

Columbite-tantalite series, AB_2X_6
- Columbite $(Fe,Mn)(Nb,Ta)_2O_6$
- Ferrocolumbite $(Fe,Mn)(Nb,Ta)_2O_6$
- Manganocolumbite $(Mn,Fe)(Nb,Ta)_2O_6$
- Magnocolumbite $(Mg,Fe)(Nb,Ta)_2O_6$
- Tantalite $(Fe,Mn)(Ta,Nb)_2O_6$
- Ferrotantalite $(Fe,Mn)(Ta,Nb)_2O_6$
- Manganotantalite $(Mn,Fe)(Ta,Nb)_2O_6$
- Ixiolite $(Ta,Fe,Sn,Nb,Mn)_4O_8$
- Pseudo-ixiolite $(Ta,Nb,Mn,Fe,Sn,Ti)_2O_4$
- Wodginite $(Ta,Sn,Mn,Nb,Fe,Ti)_{16}O_{32}$
- Olovotantalite=Sn-Mn tantalate=wodginite?

Euxenite-polycrase series, AB_2X_6
- Euxenite $(Y,Ca,Ce,U,Th)(Nb,Ti,Ta)_2O_6$
- Tanteuxenite $(Y,Ca,Ce,U,Th)(Ta,Ti,Nb)_2O_6$
- Polycrase $(Y,Ca,Ce,U,Th)(Ti,Nb,Ta)_2O_6$
- Tantpolycrase $(Y,Ca,Ce,U,Th)(Ti,Ta,Nb)_2O_6$
- Kobeite $(Y,Fe,U)(Ti,Nb,Ta)_2(O,OH)_6$
- Fersmite $(Ca,Ce)(Nb,Ti)_2(O,F)_6$

Aeschynite-priorite series AB_2X_6
- Aeschynite $(Ce,Ca,Fe,Th)(Ti,Nb)_2O_6$
- Sinicite $(Ce,Nd,Th,U)(Ti,Nb)_2O_6$
- Niobo-aeschynite $(Ce,Ca,Th)(Nb,Ti)_2O_6$
- Lyndochite=niobian-thorian aeschynite(?)
- Polymignyte $(Ca,Fe,Ce)(Zr,Ti,Nb,Ta)_2O_6$
- Priorite $(Y,Er,Ca,U,Th)(Ti,Nb)_2O_6$

Table 41-D-7 (continued)

Pyrochlore-betafite-microlite series $A_{2-x}B_2X_7$
Pyrochlore $(Na,Ca,Ce)_2(Nb,Ti,Ta)_2(O,OH,F)_7$
Koppite=cerian-ferrian pyrochlore
Marignacite $(Ce,Ca,Na)_2(Nb,Ti,Ta)_2(O,OH,F)_7$
Niobozirconolite $CaZr(Ti,Nb)_2O_7$
Obruchevite $(Y,U,Ca)_2(Nb,Ti,Ta)_2(O,OH,F)_7$
Pandaite $(Ba,Sr)_2(Nb,Ti,Ta)_2(O,OH,F)_7$
Priazovite=uranian-yttrian pyrochlore
Scheteligite=$(Ca,Y,Sb,Mn)_2(Ti,Ta,Nb,W)_2O_6(O,OH)$
Yttrobetafite=intermediate between obruchevite and pyrochlore
Yttrohatchettolite $(Ca,Y,U)_2(Nb,Ta,Ti)_2(O,OH,F)_7$
Betafite $(U,Ca)_2(Nb,Ti,Ta)_2(O,OH,F)_7$
Hatchettolite $(Ca,U)_2(Nb,Ti,Ta)_2(O,OH,F)_7$
Rare-earth betafite $(RE,U,Ca)_2(Nb,Ti,Ta)_2(O,OH,F)_7$
Samiresite $(U,Pb)_2(Nb,Ti,Ta)_2(O,OH,F)_7$
Titanobetafite $(U,Ca)_2(Ti,Nb,Ta)_2(O,OH,F)_7$
Zirconium betafite $(U,Ca)_2(Ti,Nb,Zr)_2(O,OH,F)_7$
Microlite $(Ca,Na)_2(Ta,Nb,Ti)_2(O,OH,F)_7$
Bismutomicrolite=bismuthian microlite
Djalmaite=microlite
Plumbomicrolite $(Pb,Ca)(Ta,Nb)_2(O,OH,F)_7$
Rijkeboerite $(Ba,Fe,Pb,U)(Ta,Nb,Ti,Sn)_2(O,OH,H_2O)_7$
Westgrenite $(Bi,Ca)(Ta,Nb)_2O_6(OH)$
Sukulaite $Sn_2Ta_2O_7$
Samarskite series AB_2X_6
Samarskite $(Y,Fe,U)(Nb,Ti,Ta)_2(O,OH)_6$
Calciosamarskite=calcian samarskite
Vietinghofite $(Fe,U,Y)(Nb,Ti,Ta)_2O_6$
Khlopinite $(Fe,Y,U)_2(Ti,Nb,Ta)_2O_6$
Ishikawaite $(U,Fe,Y)(Nb,Ta,Ti)_2O_6$
Plumboniobite $(Ca,Pb,Fe,Y,U)Nb_2O_6$
Yttrotantalite $(Y,Fe)(Nb,Ta)O_4$ (approximately)
Hjelmite=doubtful Mn-Ca yttrotantalite
Other tantaloniobates
Thoreaulite $Sn(Ta,Nb)_2O_7$
Pisekite=complex multiple oxide
Belyankinite $Ca(Ti,Zr,Si,Nb)_6O_{13}\cdot 14H_2O$
Gerasimovskite $(Mn,Ca)_2(Nb,Ti)_5O_{12}\cdot 9H_2O$
Simpsonite $Al_4(Ta,Nb)_3(O,OH,F)_{14}$
Rankamaite $(Na,K,Pb,Li)_{5.86}$ $(Ta,Nb,Al)_{22.16}(O,OH)_{60}$

Borate

Behierite $(Ta,Nb)BO_4$

Silicates

Nenadkevichite $(Na,Ca)(Nb,Ti)(Si_2O_7)\cdot 2H_2O$
Niobolabuntsovite $(Ca,Ba)(Ti,Nb)(Si_2O_7)\cdot 2H_2O$
Shcherbakovite $(K,Na,Ba)_3(Ti,Nb)_2(Si_2O_7)_2$
Wöhlerite $NaCa_2(Zr,Nb)(Si_2O_7)(O,F)$
Karnasurtite $(Ce,La,Th)(Al,Fe)(Ti,Nb)(SiP)_2O_7(OH)_4\cdot 3H_2O$
Niocalite $Ca_4NbSi_2O_{10}(OH,F)$
Lomonosovite $Na_2(Ti,Nb)_2Si_2O_9\cdot Na_3PO_4$
Beta-lomonosovite=hydrous variety of lomonosovite
Murmanite $Na_2(Ti,Nb)_2Si_2O_9\cdot nH_2O$
Betamurmanite=weathered beta lomonosovite

Table 41-D-7 (continued)

Epistolite $(Na,Ca)_{2-x}(Nb,Ti)_2Si_2O_{9-x}(OH)_x \cdot n(H_2O,Na_3PO_4)$
Fersmanite $(Ca,Na)_2(Ti,Nb)(SiO_5)(F,OH)$
Baotite $Ba_4(Ti,Nb)_8(Si_4O_{12})ClO_{16}$
Niobophyllite $(K,Na)_3(Fe, Mn)_6(Nb,Ti)_2(Si,Al)_8(O,OH,F)_{31}$
Ilimaussite $Na_4Ba_2CeFeNb_2Si_8O_{28} \cdot 5H_2O$

Table 41-D-8. *Minerals reported to contain at certain localities 1–5 percent niobium and tantalum.* (According to PARKER and FLEISCHER, 1968)

Astrophyllite	$(K,Na)_3(Fe,Mn)_7(Ti,Zr)_2Si_8O_{24}(OH)_7$
Baddeleyite	ZrO_2
Brookite	TiO_2
Cassiterite	SnO_2
Catapleiite	$(Na_2,Ca)ZrSi_3O_9 \cdot 2H_2O$
Chevkinite	$(Ca,Ce,Th)_4(Fe,Mg)_2(Ti,Fe)_3Si_4O_{22}$
Chlorite	$(Mg,Fe)_5Al(Al,Si)_3O_{10}(OH)_8$
Elpidite	$Na_2ZrSi_6O_{12}(OH)_6$
Eudialyte	$(Ca,Na,Ce)_5(Zr,Fe)_2Si_6(O,OH,Cl)_{20}$
Freudenbergite	$Na_2(Fe,Ti)_8O_{16}$
Hematite	Fe_2O_3
Hiortdahlite	$(Ca,Na)_3ZrSi_2O_7(O,OH,F)_2$
Ilmenite	$FeTiO_3$
Keilhauite	Yttrian sphene.
Kimzeyite	$Ca_3(Zr,Ti)_2(Al,Fe,Si)_3O_{12}$
Kupletskite	$(K,Na)_2(Mn,Fe)_4TiSi_4O_{14}(OH)_2$
Labuntsovite	$(K,Na,Ba)TiSi_2(O,OH)_7 \cdot H_2O$
Låvenite	$(Na,Ca)_3ZrSi_2O_7(O,OH,F)_2$
Lorenzenite	$Na_2(Ti,Zr)_2Si_2O_9$
Manganosteenstrupine	$(Ce,La,Na,Mn)_6(Si,P)_6O_{18}(OH)$
Mosandrite	$(Na,Ca,Ce)_3TiSi_2O_8(O,F)$
Perovskite	$CaTiO_3$
Rutile	TiO_2
Sphene	$CaTiSiO_5$
Thortveitite	$(Sc,Y)_2Si_2O_7$
Titanolåvenite	$(Na,Ca)_3(Ti,Zr)Si_2O_7(O,OH,F)_2$
Triplite	$(Mn,Fe)_2(PO_4)F$
Tritomite	$(Ce,La,Y,Th)_5(Si,B)_3(O,OH,F)_{13}(?)$
Tundrite	$Na_3(Ce,La)_4(Ti,Nb)_2[SiO_4]_2[CO_3]_3O_4(OH) \cdot 2H_2O$
Vernadite	$MnO_2 \cdot nH_2O(?)$
Vinogradovite	$Na_5Ti_4AlSi_6O_{24} \cdot 3H_2O$
Wolframite	$(Mn,Fe)WO_4$
Yttrotitanite	Yttrian sphene
Zircon	$ZrSiO_4$
Zirkelite (zirconolite)	$(Ca,Th,Ce)Zr(Ti,Nb)_2O_7$

According to the data reported by KUKHARENKO *et al.* (1965), the Nb/Ta ratio in pyrochlore from different carbonatites of the Kola Peninsula varies between 21.8, 20.5, and 7.0 to 6.0. Baddeleyite from the same deposits has a much smaller range of Nb/Ta: 2.6 to 3.9.

Table 41-D-9. *Schemes of isomorphous substitution in niobium-tantalum-bearing minerals.* (According to PARKER and FLEISCHER, 1968)

Scheme	Example
Ti	
$(Nb, Ta)^{5+}+Na^{1+} \leftrightarrow Ti^{4+}+Ca^{2+}$	Loparite, perovskite
$(Nb, Ta)^{5+}+2Na^{1+}+RE^{3+} \leftrightarrow Ti^{4+}+3Ca^{2+}$ [a]	Sphene, irinite
$(Nb, Ta)^{5+}+(Fe, Al)^{3+} \leftrightarrow 2Ti^{4+}$	Biotite
$2(Nb, Ta)^{5+}+Fe^{2+} \leftrightarrow 3Ti^{4+}$	Rutile, tapiolite
Zr	
$(Nb, Ta)^{5+}+Na^{1+} \leftrightarrow Zr^{4+}+Ca^{2+}$	Wöhlerite, zircon
$2(Nb, Ta)^{5+}+Fe^{2+} \leftrightarrow 3Zr^{4+}$	Polymignyte
$2(Nb, Ta)^{5+}+Ca^{2+} \leftrightarrow 2(Ti, Zr)^{4+}+(Th, U)^{4+}$	Zirconium betafite
$(Nb, Ta)^{5+}+RE^{3+} \leftrightarrow 2Zr^{4+}$ [a]	Zircon
$(Nb, Ta)^{5+}+Fe^{3+} \leftrightarrow 2Zr^{4+}$	Zircon, baddeleyite
W	
$(Nb, Ta)^{5+}+Sc^{3+} \leftrightarrow W^{6+}+Fe^{2+}$	Wolframite
Sn	
$2(Nb, Ta)^{5+}+Fe^{2+} \leftrightarrow 3Sn^{4+}$	Cassiterite
Al, Si, Fe	
$(Nb, Ta)^{5+}+2(Mn, Mg)^{2+} \leftrightarrow 3Fe^{3+}$	Iron oxides
$(Nb, Ta)^{5+}+2Mg^{2+} \leftrightarrow Al^{3+}+2Fe^{3+}$	Pyroxenes, amphiboles
$(Nb, Ta)^{5+}+Na^{1+}+Al^{3+} \leftrightarrow Ca^{2+}+Fe^{3+}+Si^{4+}$	Aegirine
$(Nb, Ta)^{5+}+Be^{2+} \leftrightarrow Al^{3+}+Si^{4+}$	Muscovite
$(Nb, Ta)^{5+}+Li^{1+} \leftrightarrow 2Al^{3+}$	Muscovite
Others	
$(Nb, Ta)^{5+}+Li^{1+} \leftrightarrow Mg^{2+}+Ti^{4+}$	Biotite, lepidolite
$(Nb, Ta)^{5+}+(Ce, La)^{3+} \leftrightarrow 2Th^{4+}$	Thorite
$(Nb, Ta)^{5+}+Si^{4+} \leftrightarrow P^{5+}+Ce^{4+}$	Monazite(?)
$(Nb, Ta)^{5+}+2Si^{4+} \leftrightarrow 2P^{5+}+RE^{3+}$ [a]	Erikite

[a] RE = rare-earth elements.

III. Phase Equilibria

The compounds TiO_2 and NbO_2 form a continuous series of mixed crystals with a rutile structure up to 75% NbO_2. The comparable rutile structure of $(Ti_{1-x}\ Ta_x)O_2$ extends up to $x=0.5$. At $x=0.5$ the compound is $Ti^{3+}Ta^{5+}O_4$ (RÜDORF and LUGINSLAND, 1964). Non-stochiometric phases with an α-PbO_2 structure form between 48 and 60 mol % Nb_2O_5 in the system Fe_2O_3—Nb_2O_5 (WEITZEL and SCHRÖCKE, 1973). At higher than 60 mol % Nb_2O_3 a rutile structure is stable.

Phase equilibria within the columbite tapiolite series and of this group with euxenite $Y(Ti, Nb, Ta)_2O_6$ have been studied by SCHRÖCKE (1966). The same author gives details about the heterotypic mixed crystals formed between the compounds columbite, tapiolite, ferberite, and hübnerite (SCHRÖCKE, 1961). Beside large sections governed by continuous series of mixed crystals there exist ranges of unmixing in the systems $FeWO_4$—$FeNb_2O_6$ and $FeNbO_4$—$MnWO_4$.

Revised manuscript received: March 1977

41-E. Abundance in Common Igneous Rock Types; Crustal Abundance

According to the information from Sect. 41-D, biotite, amphibole and Ti minerals are the major host phases of niobium in magmatic rocks. The existence of good correlations between Nb+Ta and Ti in rock-forming minerals has been proved by NEKRASOV (1970; Figs. 4 and 6) and others.

I. Standard Reference Rocks

Numerous niobium determinations in rocks and minerals are results of optical spectrography, spectrophotometry and X-ray spectrometry. Less abundant are data from spark source mass spectrometry. The limit of detection for optical and X-ray spectrometry is close to 10 ppm. Some authors use chemical preconcentration in combination with these procedures. Mass spectrometry and neutron activation can approach levels of Nb concentrations lower than 10 ppm. FLANAGAN (1973) has compiled the recommended averages for magmatic reference rocks which are listed in Table 41-E-1.

Table 41-E-1. *Recommended values and magnitudes for selected international reference rocks* (FLANAGAN, 1973)

Symbol	Rock type	ppm Nb[a]	Symbol	Rock type	ppm Nb[b]
G-1	Granite (USGS)	23.5	GA	Granite (Nancy)	*13*
G-2	Granite (USGS)	13.5	GH	Granite (Nancy)	*85*
GSP-1	Granodiorite (USGS)	29.0	NIM-G	Granite (S. Africa)	*50*
AGV-1	Andesite (USGS)	15.0	GM	Granite (ZGI)	17[a]
BCR-1	Basalt (USGS)	13.5	NIM-S	Syenite (S. Africa)	*3*
W-1	"Diabase" (USGS)	9.5	SY-1	Syenite (Canada SSC)	150[a]
			BM	Basalt (spilite) (ZGI)	*10*
			NIM-N	Norite (S. Africa)	*2*
			DTS-1	Dunite (USGS)	<3[a]
			NIM-D	Dunite (S. Africa)	*3*
			PCC-1	Peridotite (USGS)	<2[a]
			NIM-P1	Pyroxenite (S. Africa)	*3*

[a] Recommended value; [b] Magnitude (italics), except values marked with a.

II. Granitic Rocks and Their Effusive Equivalents

Data on niobium and tantalum in calc alkalic granites, granodiorites, quartz diorites and comparable rocks are listed in Tables 41-E-2 and 41-E-3. The range of

Table 41-E-2. *Niobium and tantalum contents of selected granites (excluding alkalic granites), quartz monzonites, granodiorites, and volcanic rocks of similar chemical composition* (Analytical methods are listed with the references in the footnote.)

Rock type, location	Number of analyses	ppm Nb	ppm Ta	Nb/Ta	Reference
Plutonic rocks					
Granite, granodiorite, Kalba (U.S.S.R.)	3	n.d.	2.0[a]	n.d.	1
Kulinda, Baikal region (U.S.S.R.)	1	n.d.	16.[a]	n.d.	1
Central Kazakhstan (U.S.S.R.)	1	n.d.	8.2[a]	n.d.	1
Susamyr batholith (U.S.S.R.)	55	27.0	2.5	11.0	2
Alaskite granite, Slyudyanka (U.S.S.R.)	9	2.4	<0.8[b]	>3	3
Granite, Mamsk-Chuesk, Bodaibensk (U.S.S.R.)	1	0.9	<0.4	>2.3	3
Granite, Mamsk-Oronskii complex (U.S.S.R.)	1	1.2	<0.5	>2.4	3
Granite, Konkudero-Mamakanskii (U.S.S.R.)	11	1.0	<0.5	>2.	3
Granite, Kalbin, E. Kazakhstan (U.S.S.R.)	14	17.	2.1	8.	3
Granite, Korosten, Ukraine (U.S.S.R.)	7	10.	<1.1	>9.	3
Granite, Urik-Oka, E. Sayan Mtns. (U.S.S.R.)	4	12.	<0.8	>15.	3
Granite, Kaibsk, Kazakhstan (U.S.S.R.)	20	16.	1.1	15.	3
Quartz monzonite, Rubidoux Mtn., Calif. (U.S.A.)	2	5.4	0.42	13	4
Granite, G-1 standard, Rhode Island (U.S.A.)	1	23.5	1.5	15.7	5
Granite, Fennoscandia	12	7.	5.8[c]	1.2	6
Granite (rapakivi), Finland	9	39.	1.7[c]	23.	6
Granodiorite, granite, etc. Kokchetav uplift (U.S.S.R.)	99	25.4	n.d.	n.d.	7
Granodiorite, granite, etc. Kokchetav uplift (U.S.S.R.)	31	20.	n.d.	n.d.	7
Alaskite, leucogranite, etc., Kokchetav uplift (U.S.S.R.)	68	37.	n.d.	n.d.	7
Granite and microgranite, S. Rhodesia	6	96[a]	n.d.	n.d.	8
Granite, Paraguay	1	10	n.d.	n.d.	9
Aplite, Paraguay	2	20	n.d.	n.d.	10
Granite aplite, Montana (U.S.A.)	4	13	n.d.	n.d.	10
Quartz monzonite, Montana (U.S.A.)	4	7	n.d.	n.d.	10
Quartz porphyry, Idaho (U.S.A.)	1	28	n.d.	n.d.	10
Granite porphyry, Idaho (U.S.A.)	1	28	n.d.	n.d.	10
Alaskite, Idaho (U.S.A.)	1	15	n.d.	n.d.	10
Granite, Idaho (U.S.A.)	1	20	n.d.	n.d.	10
Granodiorite, California (U.S.A.)	2	20	n.d.	n.d.	10
Quartz monzonite, California (U.S.A.)	5	20	n.d.	n.d.	10
Granite, California (U.S.A.)	2	15	n.d.	n.d.	10
Granodiorite, Johnston County, Okla (U.S.A.)	5	20	n.d.	n.d.	10
Average	373	23.5	(2.6 extrapolated)		
Average, analyses with both Nb, Ta reported	146	18.1	2.0	9.1	

Table 41-E-2 (continued)

Rock type, location	Number of analyses	ppm Nb	ppm Ta	Nb/Ta	Reference
Volcanic rocks					
Rhyolite, S. Rhodesia	3	110[a]	n.d.	n.d.	8
Pitchstone, Iceland	31	23	n.d.	n.d.	11
Rhyolite porphyry, Paraguay	1	10	n.d.	n.d.	9
Rhyolite and quartz latite, Oregon (U.S.A.)	3	30	n.d.	n.d.	10
Rhyolite, eastern Idaho (U.S.A.)	7	56	n.d.	n.d.	12
Rhyolite, northwestern Wyoming (U.S.A.)	8	30	n.d.	n.d.	13
Rhyolite, New Mexico (U.S.A.)	12	33	n.d.	n.d.	10
Rhyodacite-dacite, quartz latite, New Mexico (U.S.A.)	15	29	n.d.	n.d.	10
Rhyolite, Nevada (U.S.A.)	15	26	n.d.	n.d.	14
Quartz latite, Craters of the Moon, Idaho (U.S.A.)	2	300[a]	n.d.	n.d.	10
Quartz latite porphyry, Nevada (U.S.A.)	9	18	n.d.	n.d.	10
Average		28			

[a] Not included in averages; [b] Figures shown as < are averaged at half value; [c] Some figures given by RANKAMA as less than a certain value were averaged at half that value.

References: 1. KUZ'MENKO (1966); 2. GERASIMOVSKIY and KARPUSHINA (1965) (W/C); 3. ZNAMENSKII (1964); 4. GOTTFRIED (1965 written communication) (W/C); 5. FLANAGAN (1973); 6. RANKAMA (1944, 1948) (W/S); 7. PODOL'SKII and SERYKH (1964) (W/C); 8. COX *et al.* (1965); 9. ECKEL (1959); 10. Unpublished results U.S. Geol. Survey (W/C); 11. CARMICHAEL and MACDONALD (1961) (S); 12. HAMILTON (1966) (S); 13. HAMILTON (1963) (S); 14. CORNWALL (1962).

variation is large. Some proportion of this scattering and the great variations in the Nb/Ta ratio are probably due to insufficient accurate measurements as GERASIMOVSKIY and KARPUSHINA (1965) demonstrated by a check of earlier results on the Susamyr batholith. PARKER and FLEISCHER (1968) list several weaknesses and limitations with respect to the presently available Nb and Ta abundance data.

Alkalic granites often contain appreciably higher concentrations of Nb and Ta than the common calc-alkalic granites. Therefore they are listed in a separate compilation (Table 41-E-4).

The alkalic granites are often products of metasomatic albitization or greisen formation. BUGAYETS and NARSEYEV (1964) have proved that in plutons of Kazakhstan metasomatic alteration has released niobium from the rock-forming minerals which has been redeposited as columbite and samarskite. Comparable statements have been published by KALININ and GOLDIN (1967) on metasomatically altered granites and granodiorites from the Pechora region, Urals, mentioning that only the most altered "apogranites" contain independent Nb-Ta minerals (fergusonite, samarskite). GERASIMOVSKIY and KARPUSHINA (1967) have observed abnormally high Nb and Ta concentrations (up to 1,750 ppm Nb and 150 ppm Ta) in some younger intrusive granites in Nigeria. Their rocks enriched in niobium and tantalum have usually low

Table 41-E-3. *Niobium and tantalum contents of granitic rocks (excluding alkali granites) and volcanic rocks of similar chemical composition not listed in the compilation by* Parker and Fleischer (1968) (Analytical methods are listed with references in footnote)

Rock type	Number of analyses	ppm Nb	ppm Ta	Nb/Ta	Reference
Plutonic rocks					
Granodiorite, quartz diorite, granite, Gissar Pluton, Central Tadzhikistan (U.S.S.R.)	92	11.8	1.8	6.6	1
Granite, Devonian-Triassic, NE U.S.S.R.	7	20	2.4	8.3	2
Granitic rocks, Jurassic-Cretaceous, NE U.S.S.R.	54	8.5	1.4[a]	6.1	2
Granitic rocks, SW and S Africa	4	13	n.d.	n.d.	3
Granitic rocks, Franzfontein (SW Africa)	11	≤20	n.d.	n.d.	5
Granitic rocks, Minas Gerais (Brazil)	39	25	n.d.	n.d.	4
Average granitic rocks, this table	207	14.2	1.7[b]	8.4	
Average granitic rocks, Table 41-E-2	373	23.5	2.6[c]	9.1	
Average quartzofeldspathic rocks[d], Canadian Shield	292	26.4	n.d.	n.d.	6
Volcanic rocks					
Obsidians (comendites), Easter Island	3	121	n.d.	n.d.	7
Obsidians, pantellerites, East African Rift	6	225	n.d.	n.d.	8

[a] 21 samples; [b] 153 samples; [c] Extrapolated; [d] Granite, granitic gneiss, pegmatite, rhyolite, arkose, sandstone.

References: 1. Mogarovskiy and Mel'nichenko (1968) (W/C); 2. Nekrasov (1970) (S); 3. Edge and Ahrens (1963) (W/S); 4. Herz and Dutra (1960) (S); 5. Clifford *et al.* (1969) (S); 6. Shaw *et al.* (1976) (X); 7. Baker *et al.* (1974); 8. Weaver *et al.* (1972) (X).

Table 41-E-4. *Niobium and tantalum contents of some sodic and alkalic granites and rhyolites* (Mainly compiled by Parker and Fleischer, 1968. Analytical methods are listed with references in footnote.) (Number of samples in brackets)

Rock type, location	ppm Nb	ppm Ta	Nb/Ta	Reference
Plutonic rocks				
Alkalic granite, Kola Peninsula (U.S.S.R.)	14	1.3	10.7	1
Alkalic granite, Kola Peninsula (U.S.S.R.)	30	2.3	13	1
Alkalic granite, Kiev, Kola Peninsula (U.S.S.R.)	26	2.5	10.4	1

Table 41-E-4 (continued)

Rock type, location	ppm Nb	pp Ta	Nb/Ta	Reference
Alkalic granite, Kiev, Kola Peninsula (U.S.S.R.)	35	3.5	10	1
Soda granite, Lunyo (Uganda) (1)	350	n.d.	n.d.	2
Alkalic granite, Liruei (Nigeria)	80	6.5	12	3
Alkalic granite, Amo (Nigeria)	70	5.1	14	3
Alkalic granite, Mount Tremont, N. Hampshire (U.S.A.)	150	n.d.	n.d.	3
Hastingsite granite, Jackson Falls, N. Hampshire (U.S.A.)	50	8.1	6.2	3
Alkalic granite, Kungnät (Greenland)	275	n.d.	n.d.	3
Alkalic granite, Rockall, N. Atlantic	150	4.3	35	3
Biotite granite (Conway), Redstone, N. Hampshire (U.S.A.)	50	9.2	5.4	3
Alaskite, Ognitskii complex (U.S.S.R.) (3)	17.3	1.6	10.8	4
Riebeckite granite, Ognitskii complex (U.S.S.R.) (3)	13.7	1.0	13.7	4
Riebeckite granite, Ognitskii complex (U.S.S.R.) (4)	22.7	1.0	22.7	4
Albite granite, Attu Islands, Aleutians (1)	30	n.d.	n.d.	5
Riebeckite (?) granite, Wichita Mountains, Okla. (U.S.A.) (20)	71	n.d.	n.d.	5
Riebeckite granite, Brewster County, Tex. (U.S.A.) (1)	250	n.d.	n.d.	5
Riebeckite granite, Mount Rosa, Colo. (U.S.A.) (3)	267	n.d.	n.d.	5
Riebeckite granite, Conway, N. Hampshire (U.S.A.) (2)	225	n.d.	n.d.	5
Albite granite (metasomatized), Dogdan, Kazakhstan (U.S.S.R.) (100)	140	n.d.	n.d.	6
Albite granite (metasomatized), Man'Khambro, Urals (U.S.S.R.) (39)	70	24	2.9	7
Average (186)	114			
Average, analyses with both Nb and Ta reported (58)	59	3.9[a]	15[a]	
Volcanic rocks				
Alkali rhyolite, soda rhyolite, New Mexico (U.S.A.) (20)	61	n.d.	n.d.	5
Alkali rhyolite, Oregon (U.S.A.)	30	n.d.	n.d.	5
Riebeckite rhyolite, Brewster County, Tex. (U.S.A.) (7)	260	n.d.	n.d.	5
Riebeckite rhyolite, Jeff Davis County, Tex. (U.S.A.) (2)	35	n.d.	n.d.	5
Average	97			

[a] Excluding the data from Man'Khambro, Urals (U.S.S.R.).

References: 1. Gerasimovskiy and Karpushina (1965) (W/C); 2. Von Knorring (1960) (S); 3. Butler and Smith (1962) (Nb: X; Ta: N/R). 4. Kovalenko *et al.* (1964) (W/S); 5. Unpubl. data U.S. Geol. Surv. (W/C); 6. Bugayets and Narseyev (1964) (S); 7. Kalinin and Goldin (1967) (W, S).

Ti/Nb + Ta and K/Na ratios and high concentrations of lithium and fluorine. The average Nb and Ta concentrations in alkalic granites listed in Table 41-E-4 are 4 to 5 times higher than in common calc-alkali granites of Tables 41-E-2 and 3. The niobium and tantalum accumulation in alkalic rocks is treated in Subsect. 41-E-III and some speculations about the transport of these elements are included in Sect. 41-F.

Some averages for Nb and Ta abundances and for Nb/Ta ratios in common calc-alkali granitic rocks and comparable species are listed in Table 41-E-3. From them an overall average of 22 ppm Nb and 2.1 ppm Ta (Nb/Ta: 10.5) can be computed. This niobium figure is close to those published by RANKAMA (1948), TUREKIAN and WEDEPOHL (1961) and VINOGRADOV (1962). But the new Ta average is lower than those reported by the above mentioned authors (4.2; 3.9; 3.5 ppm Ta respectively).

MOGAROVSKIY and MEL'NICHENKO (1968) observed that in genetically related series of granitic rocks in Central Tadzhikistan, Tien Shan (U.S.S.R.), the Nb and Ta content increases from early to later intrusive phases.

Rhyolites, quartz latites, rhyodacites and obsidians are often higher in Nb than granitic intrusives.

III. Alkalic Rocks

Niobium and tantalum concentrations increase on average in the following sequence: calc-alkalic granitic rocks < alkalic granitic rocks < syenites, trachytes (Table 41-E-8) < nepheline syenites, phonolites (Tables 41-E-5 and 6). The niobium increase in this sequence is: 22 < 59–114 < 100 < 208 ppm Nb. Alkalic rocks are highly variable in these elements from species to species (Tables 41-E-5 and 7) and between localities (Table 41-E-6). The nepheline syenites with the highest niobium and tantalum accumulations are those from the Lovozero complex, Kola Peninsula (U.S.S.R.).

In syenites and nepheline syenites the Nb/Ta ratios range between 11 and 13. These values are slightly higher than the average ratio of calc-alkalic granitic rocks (9.1). Even between the different rock types of the Lovozero complex the Nb/Ta ratio does not scatter more than from 10.3 to 15.3 (GERASIMOVSKIY *et al.*, 1966).

The average niobium and tantalum content is higher in agpaitic than in miaskitic nepheline syenites. Nb and Ta minerals are not typical of agpaitic rock varieties

Table 41-E-5. *Niobium and tantalum contents of some nepheline syenites, phonolites, and additional nepheline-bearing rocks* (Compiled by PARKER and FLEISCHER, 1968; supplemented by the present author. Number of analyses in brackets) (Analytical methods are listed with the references in footnote.)

Rock type, location	ppm Nb	ppm Ta	Nb/Ta	Reference
Miaskite, Vishnevye Mts. (U.S.S.R.)	250	21	12	1
Nepheline syenite, Borsuksai (U.S.S.R.)	410	20	20	2
Nepheline syenite, Il'men Mts. (U.S.S.R.)	110	8	13	2
Nepheline syenite, hydrosodalite syenite, Lovozero (U.S.S.R.) (18)	320	29	11	3
Foyaite, Lovozero (U.S.S.R.) (20)	702	49.5	14.4	3

Table 41-E-5 (continued)

Rock type, location	ppm Nb	ppm Ta	Nb/Ta	Reference
Lujavrite, Lovozero (U.S.S.R.) (21)	607	59	10.3	3
Amphibole lujavrite, Lovozero (U.S.S.R.) (7)	760	52	14.6	3
Eudialyte lujavrite, Lovozero (U.S.S.R.) (16)	1,053	94	11.2	3
Porphyritic lujavrite, Lovozero (U.S.S.R.) (5)	440	33	13.3	3
Porphyritic lujavrite, Lovozero (U.S.S.R.)	980	85	11.5	3
Poikilitic sodalite syenite, Lovozero (U.S.S.R.) (6)	490	32	15.3	3
Phonolite, Lovozero (U.S.S.R.)	399	37.5	10.6	3
Pseudoleucite phonolite, Lovozero (U.S.S.R.)	530	50	10.6	3
Nepheline syenite, Karasyrsk (U.S.S.R.)	160	<10	n.d.	4
Nepheline syenite, Tatarsk (U.S.S.R.)	140	n.d.	n.d.	4
Nepheline syenite, Karsk (U.S.S.R.)	100	n.d.	n.d.	4
Foyaite, Khibina (U.S.S.R.)	140	n.d.	n.d.	5
Khibinite, Khibina (U.S.S.R.)	56	n.d.	n.d.	5
Rischorrite, Khibina (U.S.S.R.)	56	n.d.	n.d.	5
Nepheline syenite, Kishingar (India)	44	4	11	6
Nepheline syenite, Shonkin Sag, Mont. (U.S.A.)	25	n.d.	n.d.	7
Aegirine syenite, Shonkin Sag, Mont. (U.S.A.)	42	n.d.	n.d.	7
Aplitic syenite, Shonkin Sag, Mont. (U.S.A.)	43	n.d.	n.d.	7
Transition rocks, Shonkin Sag, Mont. (U.S.A.)	20	n.d.	n.d.	7
Mafic phonolite, Shonkin Sag, Mont. (U.S.A.)	22	n.d.	n.d.	7
Nepheline syenite, White Mountain, plutonic-volcanic series, N. H. (U.S.A.)	179	n.d.	n.d.	7
Sericitized syenite, Bearpaw Mts., Mont. (U.S.A.)	400	n.d.	n.d.	8
Sericitized syenite dike, Bearpaw Mts., Mont. (U.S.A.)	60	n.d.	n.d.	8
Nepheline syenite, Iron Hill, Color. (U.S.A.)	150	n.d.	n.d.	9
Nepheline syenite, Arkansas (U.S.A.)	200	n.d.	n.d.	10
Nepheline syenite, Arkansas (U.S.A.)	100	n.d.	n.d.	10
Pulaskite, Arkansas (U.S.A.)	90	n.d.	n.d.	10
Feldspathic trachyte, Magnet Cove, Ark. (U.S.A.)	200	n.d.	n.d.	11
Pseudoleucite syenite, Magnet Cove, Ark. (U.S.A.)	74	n.d.	n.d.	11
Nepheline syenite, Magnet Cove, Ark. (U.S.A.)	99	n.d.	n.d.	11
Pseudoleucite syenite, Magnet Cove, Ark. (U.S.A.)	96	n.d.	n.d.	11
Nepheline syenite, Magnet Cove, Ark. (U.S.A.)	87	n.d.	n.d.	11
Tinguaite, Magnet Cove, Ark. (U.S.A.)	220	n.d.	n.d.	11
Sodalite trachyte, Magnet Cove, Ark. (U.S.A.)	200	n.d.	n.d.	11
Nepheline syenite, Monchique (Portugal)	420	0.8	525	12
Phonolite, Unterwiesenthal (Germany)	200	0.8	250	12
Phonolite, Kaiserstuhl (Germany) (5)	400	n.d.	n.d.	13

Table 41-E-5 (continued)

Rock ype, location	ppm Nb	ppm Ta	Nb/Ta	Reference
Tinguaite, Kaiserstuhl (Germany) (10)	285	n.d.	n.d.	13
Nepheline syenite, Greenland (6)	73	n.d.	n.d.	14
Nepheline syenite, Wheatland County, Mont. (U.S.A.) (3)	44	n.d.	n.d.	15
Nepheline syenite, Marathon County, Wis. (U.S.A.) (3)	175	n.d.	n.d.	15
Nepheline syenite, Fremont County, Colo. (U.S.A.) (3)	107	n.d.	n.d.	15
Phonolite, Chouteau County, Mont. (U.S.A.) (4)	20	n.d.	n.d.	15
Phonolite, Meagher County, Mont. (U.S.A.) (2)	75	n.d.	n.d.	15
Leucite phonolite, Sweetwater County, Wyo. (U.S.A.) (17)	58	n.d.	n.d.	15
Phonolite, Tenerife (Canary Islands) (29)	357	n.d.	n.d.	16
Essexite, Ilimaussaq (Greenland)	440	n.d.	n.d.	6
Jacupirangite, Magnet Cove, Ark. (U.S.A.)	20	n.d.	n.d.	11
Ijolite, Magnet Cove, Ark. (U.S.A.)	50	n.d.	n.d.	11
Ijolite, Magnet Cove, Ark. (U.S.A.)	150	n.d.	n.d.	11
Ijolite, Magnet Cove, Ark. (U.S.A.)	20	n.d.	n.d.	11
Melteigite, Magnet Cove, Ark (U.S.A.)	140	n.d.	n.d.	11
Ijolite-melteigite, Lovozero massif (U.S.S.R.) (2)	1,016	90	11.3	3
Ijolite-urtite, Lovozero massif (U.S.S.R.) (11)	572	54	10.6	3
Ijolite, Khibina (U.S.S.R.)	210	n.d.	n.d.	17
Ijolite, Iron Hill, Color. (U.S.A.)	250	n.d.	n.d.	9
Essexite, Kaiserstuhl (Germany) (3)	128	n.d.	n.d.	13
Tephrite, Kaiserstuhl (Germany) (5)	115	n.d.	n.d.	13
Essexite porphyry, monchiquite, Kaiserstuhl (Germany) (5)	105	n.d.	n.d.	13
Bergalite, Kaiserstuhl (Germany) (2)	430	n.d.	n.d.	13
Mondhaldeite, gauteite, shonkinite, Kaiserstuhl (Germany) (5)	140	n.d.	n.d.	13
Olivine nephelinite, Kaiserstuhl (Germany) (5)	75	n.d.	n.d.	13
Shonkinite, Shonkin Sag, Mont. (U.S.A.) (2)	15	n.d.	n.d.	13
Shonkinite, Shonkin Sag, Mont. (U.S.A.)	13	n.d.	n.d.	13
Shonkinite, Paraguay (2)	65	n.d.	n.d.	18
Shonkinite, Chouteau County, Mont. (U.S.A.) (2)	20	n.d.	n.d.	15
Nephelinite, Monaro (SE Australia) (3)	124	n.d.	n.d.	19
Nephelinite, Maymecha-Kotuy (U.S.S.R.) (6)	197	11	17.9	21
Shoshonite, Devonshire (UK) (25)	17	n.d.	n.d.	20

References: 1. Es'kova (1959) (W/C); 2. Kuz'menko (1966); 3. Gerasimovskiy *et al.* (1966); 4. Es'kova (1960) (W/C); 5. Borodin (1955); 6. Gerasimovskiy and Karpushina (1965) (W/C); 7. Gottfried *et al.* (1961) (W/C); 8. Pecora (1962); 9. Temple and Grogan (1965); 10. Gordon *et al.* (1958) (S); 11. Erickson and Blade (1963); 12. Rankama (1944, 1948) (W/S); 13. Van Wambeke *et al.* (1964); 14. Upton (1964); 15. Unpubl. data U.S. Geol. Surv. (W/C); 16. Ridley (1970) (S, X); 17. Borodin (1955); 18. Eckel (1959); 19. Kesson (1973) (X); 20. Cosgrove (1972) (X); 21. Gladkikh and Victorova (1967) (W/C).

Table 41-E-6. *Niobium and tantalum contents of nepheline syenites and phonolites from different massifs.* (Data summarized from Table 41-E-5; Number of analyses averaged in brackets)

Massifs represented	ppm Nb	ppm Ta	Nb/Ta
Lovozero (U.S.S.R.) (10)	628	52	12
Kishingar (India) (1)	44	4	11
Vishnevye Mts. (U.S.S.R.) (1)	250	21	12
Borsuksai (U.S.S.R.) (2)	410	20	21
Il'men Mts. (U.S.S.R.) (1)	110	8	13
Shonkin Sag, Mont. (U.S.A.) (11)	25	n.d.	n.d.
White Mountain plutonic-volcanic series, N. H. (U.S.A.) (1)	179	n.d.	n.d.
Bearpaw Mts., Mont. (U.S.A.) (2)	230	n.d.	n.d.
Iron Hill, Color. (U.S.A.) (1)	150	n.d.	n.d.
Arkansas bauxite region (U.S.A.) (3)	130	n.d.	n.d.
Magnet Cove, Ark. (U.S.A.) (7)	140	n.d.	n.d.
Karasyrsk (U.S.S.R.) (1)	160	<10	n.d.
Tatarsk (U.S.S.R.) (1)	140	n.d.	n.d.
Koisk (U.S.S.R.) (1)	100	n.d.	n.d.
Khibina (U.S.S.R.) (3)	85	n.d.	n.d.
Kaiserstuhl (Germany) (10)	285	n.d.	n.d.
Tugtutôq (Greenland) (6)	73	n.d.	n.d.
Wheatland County, Mont. (U.S.A.) (3)	44	n.d.	n.d.
Marathon County, Wis. (U.S.A.) (3)	175	n.d.	n.d.
Fremont County, Color. (U.S.A.) (3)	107	n.d.	n.d.
Average	208		
Average excluding Lovozero massif	140		

Table 41-E-7. *Niobium and tantalum contents in rocks from Ilimaussaq (SW Greenland)* (According to HAMILTON, 1964; HANSEN, 1968)

	ppm Nb	ppm Ta	Nb/Ta
Lava (basaltic)	17	<8	
Augite syenite	4–13		
Augite syenite	210	5	43
Foyaite	560	25	23
Foyaite	750		
Sodalite foyaite	780	n.d.	
Naujaite (poikilitic sodalite syenite)	615		
Naujaite	700	49	14
Lujavrite	380–700		
Lujavrite	700	41	17
Medium- to coarse-grained lujavrite in Kvanefjeld area	1,050	16	64
Medium- to coarse-grained lujavrite in Kvanefjeld area	630	12	51
Albititic vein with epistolite	1,750	n.d.	
Porphyritic dyke unaffected	<35	<8	
Porphyritic dyke sheared	1,120	66	17
Porphyritic dyke sheared	1,600	16	98
Porphyritic dyke sheared	1,800	25	74
Porphyritic dyke sheared	3,900	57	68
Porphyritic dyke sheared	4,200	8	512

Table 41-E-8. *Niobium and tantalum contents of some syenites and trachytes* (Compiled by PARKER and FLEISCHER, 1968, supplemented by the present author. Analytical methods are listed with the references in the footnote; Number of analyses in brackets)

Rock type, location	ppm Nb	ppm Ta	Nb/Ta	Reference
Biotite syenite, Vishnevye Mts. (U.S.S.R.)	170	16	11	1
Aegirine syenite, Vishnevye Mts. (U.S.S.R.)	130	12	11	1
Nordmarkite, Greenland	105	8.7	12	2
Syenite, Kola Peninsula (U.S.S.R.)	210	18	12	2
Trachyte, Guadalupe Island	120	n.d.	n.d.	3
Trachyte, Guadalupe Island	140	n.d.	n.d.	3
Trachyte, Easter Island (1)	150	n.d.	n.d.	4
Trachyte, Hocheifel (West Germany) (5)	95	n.d.	n.d.	5
Trachyte, Magnet Cove, Ark. (U.S.A.)	100	n.d.	n.d.	6
Trachyte, Magnet Cove, Ark. (U.S.A.)	20	n.d.	n.d.	6
Biotite syenite, Il'men Mts. (U.S.S.R.)	77	n.d.	n.d.	7
Syenite, Chouteau County, Mont. (U.S.A.) (4)	20	n.d.	n.d.	8
Syenite, Brewster County, Tex. (U.S.A.)	88	n.d.	n.d.	8
Syenite, Fremont County, Color. (U.S.A.) (4)	25	n.d.	n.d.	8
Trachyte, Meagher County, Mont. (U.S.A.) (2)	20	n.d.	n.d.	8
Trachyte, Brewster County, Tex. (U.S.A.) (6)	100	n.d.	n.d.	8
Potash trachyte, Uganda	400[a]	n.d.	n.d.	9
Trachyte, East African Rift (5)	222	n.d.	n.d.	10
Trachyte, Kusnetskiy, Maymecha-Kotny (U.S.S.R.) (7)	75	$\leq$12	$\geq$6	11
Average	96			
Average with both Nb and Ta reported	104	13	(11)	

[a] Not included in average.

References: 1. ES'KOVA (1959) (W/C); 2. GERASIMOVSKIY and KARPUSHINA (1965) (W/C); 3. ENGEL *et al.* (1965) (S); 4. Baker and BUCKLEY (1974); 5. HUCKENHOLZ (1965) (X); 6. ERICKSON and BLADE (1963); 7. ES'KOVA (1966); 8. Unpubl. data U.S. Geol. Surv. (W/C); 9. SUTHERLAND (1965); 10. WEAVER *et al.* (1972) (X); 10. GLADKIKH and VIKTOROVA (1967).

where these elements are correlated with Ti and Zr. Pyrochlore commonly occurs in miaskitic nepheline syenites (GERASIMOVSKIY and KARPUSHINA, 1966).

An unusual niobium accumulation and Nb to Ta fractionation occurs in late hydrothermal dykes in agpaitic roof rocks of the Ilimaussaq complex in SW Greenland (Table 41-E-7). Here the most important minerals are pyrochlore and certain species of the epistolite-murmanite group.

Meaningful averages beside those listed are difficult to obtain because of the large spread of values.

IV. Gabbroic, Basaltic and Andesitic Rocks

Niobium and tantalum concentrations in *gabbros* are listed in Table 41-E-9.

There is a reasonably large number of reliable niobium data on *basalts* from different geotectonic origin available (Table 41-E-10). Information on the tantalum distribution in these rocks is rare.

Table 41-E-9. *Niobium and tantalum contents of some gabbroic rocks* (Number of samples in brackets)

Location	ppm Nb	ppm Ta	Nb/Ta	Analytical method	Reference
Skaergaard (Greenland) (9)	n.d.	0.92	n.d.	N/R	1
San Marcos, S. Calif. (U.S.A.) (6)	4.7	0.19	25	W/C, N/R	2
Chernogorsk, Siberia (U.S.S.R.)	9.8	1	10	W/C	3
Kokchetav uplift (U.S.S.R.)	7	n.d.	n.d.	W/C	4
Selennyakh, Yakutiya (U.S.S.R.) (14)	6.5	1.3	5	S (W/C)	5
Victoria Land (Antarctica) (6)	<60	n.d.	n.d.	S	6
Magnet Cove[a], Ark. (U.S.A.)	20	n.d.	n.d.		7
Mean gabbro (23)	6.8	0.95	7.2		

[a] Metagabbro.

References: 1. Atkins and Smales (1960); 2. Gottfried *et al.* (1961); Gottfried and Dinnin (1965); 3. Gerasimovskiy and Karpushina (1965); 4. Podol'skii and Serykh (1964); 5. Nekrasov (1970); 6. Hamilton (1965); 7. Erickson and Blade (1963).

In mafic effusive rocks a regular increase of Nb from oceanic (abyssal) tholeiitic basalts to tholeiitic basalts to alkali olivine basalts is observable. The means for these three groups, calculated from data of Table 41-E-10, are 5, 10.7 and 76 ppm Nb respectively. They are comparable to averages reported by Wedepohl (1975) on smaller sets of samples. Low-K tholeiites from volcanic arcs are even lower in Nb than oceanic tholeiites. The Y/Nb ratio has been used by Pearce and Cann (1973) to successfully determine the petrologic character of basaltic types. This ratio ranges from 0.1 to 1 for alkalic basalt, from 1 to 2.5 for transitional species and is higher than 2.5, or can be even higher than 10, for tholeiitic basalts. Because of low mobilization of Nb, Y, Zr and P during metamorphism these elements can be used to analyze the petrographic character of metabasalts, spilites etc. (Pearce and Cann, 1973; Floyd and Winchester, 1975; Smith and Smith, 1976). Erlank and Kable (1976) have demonstrated that the abundances of Zr and Nb are unaffected by seawater alteration of basalt and thus the Zr/Nb ratio is useful for petrologic considerations about the source and formation of oceanic basaltic magmas. These authors observed different Zr/Nb ratios ranging from 6.4 to 110 in different latitudes of sampling the Mid-Atlantic Ridge. The different ratios probably reflect heterogeneity in the source areas of the melts. The average niobium concentration in *gabbros* (6.8 ppm Nb) is between that of oceanic and normal tholeiites (Table 41-E-9). Shaw *et al.* (1976) list an average abundance of ≤ 12.5 ppm Nb in basalts, gabbros and greenstones of the Canadian Shield.

The Nb/Ta ratio of gabbros is 9.7, which is very close to a ratio of 9.1 in calc alkaline granites (Table 41-E-2). There are too few data to calculate meaningful Nb/Ta ratios for the different basaltic types.

The increase of Nb concentration with alkalinity in basaltic rocks is probably due to a decrease of the partial melt fraction formed in the mantle source of the basaltic magmas. Nb (and Ta) must be highly fractionated into the magma during partial melting because mantle rocks are low in these elements (Subsect. 41-E-V).

Table 41-E-10. *Niobium and tantalum contents of some basaltic rocks* (Number of samples in brackets)

Rock type	Location	ppm Nb	ppm Ta	Nb/Ta	Analytical method	Reference
Oceanic tholeiite						
Oceanic tholeiite (3)	Eastern Iceland	13.7	2	6.9	X, N/R	1
Oceanic tholeiite (> 2)	Reykjanes etc. (Iceland)	10.5	n.d.	n.d.	X	1
Oceanic tholeiite (4)	Mid-Atlantic Ridge 45° N	9.2	n.d.	n.d.	X	2
Oceanic tholeiite	Typical MORB	3.3	n.d.	n.d.	X	2
Oceanic tholeiite (7)	Ridges etc., Indian Ocean, Pacific Ocean	≤15.8	n.d.	n.d.	W/C	3
Oceanic tholeiite (32)	Atlantic Ridge, Carlsberg Ridge	6.9	n.d.	n.d.	X	4
Oceanic tholeiite (10)	Atlantic, Pacific etc. floor	<30	n.d.	n.d.	S	5
Mean oceanic tholeiite (72)	3 oceans	5.0	n.d.	n.d.	X	6
Tholeiitic basalt						
Tholeiite basalt (3)	Easter Island	33	n.d.	n.d.		7
Tholeiitic basalt (23)	Blosseville (E Greenland)	10.4	n.d.	n.d.	X	8
Tholeiitic basalt (20)	Karroo, Rhodesia, Swaziland	38	n.d.	n.d.	X, S	9
"Diabase" (3)	Dillsburg, Pa (U.S.A.)	13.9	0.92	15.1	W/C, N/R	10
"Dolerite" (15)	Great Lake, Tasmania (Australia)	6.3	0.48	13.1	W/C, N/R	10
Metabasalt (55)	Cliefden, NSW (Australia)	≤4	n.d.	n.d.	X	11
"Trapp" (9)	Deccan (India)	10	n.d.	n.d.	X	6
"Trapp" (20)	Hawaii	14	n.d.	n.d.	X	6
"Trapp" etc. (28)	Siberian Platform	2.6	0.48	5.4		12
Mean tholeiite (176)		10.7				This paper
Calc alkali basalt, mean (60)	Volcanic arcs	2.5	n.d.	n.d.	X	6
Low-K tholeiite, mean (46)	Volcanic arcs	1.5	n.d.	n.d.	X	6

Table 41-E-10 (continued)

Rock type	Location	ppm Nb	ppm Ta	Nb/Ta	Analytical method	Reference
Alkali olivine basalt						
Alkali olivine basalt (3)	S. E. Australia	67	n.d.	n.d.	X	13
Alkali olivine basalt (3)	East African Rift	63	n.d.	n.d.	X	14
Alkali olivine basalt (4)	Dunedin (New Zealand)	63	n.d.	n.d.	X	15
Alkali olivine basalt (7)	Eifel (F. R. Germany)	65	n.d.	n.d.	X	16
Basanite, tephrite, alk. basalt (14)	Kuznetskiy, Maymechakotuy (U.S.S.R.)	60	8	7.5	W/C	17
Alkali olivine basalt (4)	Guadalupe, East Pacific	75	n.d.	n.d.	S	5
Alkali olivine basalt (10)	East Pacific Rise	72	n.d.	n.d.	S	5
Alkali olivine basalt (12)	Azores, Madeira	84	n.d.	n.d.	X	6
Alkali olivine basalt (14)	Tristan da Cunha	113	n.d.	n.d.		18
Basanite (3)	S. W. Australia	81	n.d.	n.d.	X	13
Hawaiite (4)	Eifel (F. R. Germany)	69	n.d.	n.d.	X	16
Mugearite (3)	Eifel (F. R. Germany)	84	n.d.	n.d.	X	16
Hawaiite, mugearite (6)	Dunedin (New Zealand)	83	n.d.	n.d.	X	15
Hawaiite, mugearite (6)	Easter Island	38	n.d.	n.d.		7
Mean alkali basalt (93)		76				This paper

References: 1. Wood *et al.* (1976); 2. Erlank and Kable (1976); 3. Gladkikh and Chernycheva (1966); 4. Cann (1970); 5. Engel *et al.* (1965); 6. Pearce and Cann (1973); 7. Baker *et al.* (1974); 8. Brooks *et al.* (1976); 9. Cox *et al.* (1967); 10. Gottfried *et al.* (1968); 11. Smith and Smith (1976); 12. Znamenskii *et al.* (1965); 13. Kesson (1973); 14. Weaver *et al.* (1972); 15. Price and Chappell (1975); 16. Huckenholz (1965); 17. Gladkikh and Viktorova (1967); 18. Baker *et al.* (1964)

The number of data on niobium concentrations in *andesites* is very small. TAYLOR (1969) in his compilation on the trace element chemistry reports an average of 4.3 ppm Nb for andesites. TAYLOR *et al.* (1969) have observed such concentrations in normal andesites from Saipan, Bougainville and Fiji, but they got only 0.3 ppm Nb in low-K andesites and 11 ppm Nb (method: Mass) in a high-K andesite. GLADKIKH and VICTOROVA (1967) report appreciably higher concentrations (49 ppm Nb) from some Devonian andesites in the U.S.S.R.

V. Ultramafic Rocks

The number of reliable Nb and Ta data on common mantle rocks is very small.

The 2 dunites, 1 peridotite and 1 pyroxenite in the sets of internationally exchanged standard reference rocks (Table 41-E-1) contain from <2 to 3 ppm Nb. These figures have not yet received the status of recommended values by FLANAGAN (1973) because of limited analytical data. For 1 dunite and 1 peridotite only maximum values of 0.1 ppm Nb are reported.

GERASIMOVSKIY and KARPUSHINA (1965; analytical method: W/C) obtained 5.5 ppm for a peridotite, 7 ppm Nb for a dunite and 5.4 ppm Nb for a pyroxenite from the Kola Peninsula (U.S.S.R.). These rocks are reported to contain 0.49, 0.55 and 0.62 ppm Ta and an average Nb/Ta ratio of 10.8. NEKRASOV (1970) has analyzed 4 pyroxenites from Yakutiya (U.S.S.R.) with a chemically monitored spectrographic method. His averages are 4 ppm Nb, 0.8 ppm Ta and a Nb/Ta ratio as low as 5.

NESBITT and SUN (1976) report corrected X-ray fluorescence values on 7 Archaean spinifex-textured peridotites from the Yilgarn Block (SW Australia), on 3 comparable rocks from Munro Township (Canada) and one sample from the Barberton Mountains (S. Africa). The average niobium concentrations for these 3 groups are: 0.7 ppm, 1.2 ppm and 1 ppm Nb respectively. The Zr-Nb correlation in these ultramafic rocks and related high magnesian and low-magnesian tholeiitic basalts is close to the line for meteorites and lunar rocks of Fig. 41-C-1 (with slightly lower Nb per Zr concentrations).

RHODES and DAWSON (1975) have analyzed 7 garnet peridotites and 8 garnet-free peridotites from the Lashaine ancaramitic volcano in northern Tanzania. These xenoliths contain on average: 1.4 and 1.2 ppm Nb respectively (analytical method: X).

It might tentatively be assumed that the most abundant *ultramafic rocks of the upper mantle contain 1.5 ppm Nb.*

Data on alkalic peridotites including kimberlites are not as scarce as those on the above listed rocks. KUDRYAVTSEV (1964) and BURKOV and PODPORINA (1965) report averages of 94 and 160 ppm Nb for 349 and 57 analyses respectively on kimberlites from Yakutia (U.S.S.R.). DAWSON (1962) lists a higher figure of 240 ppm Nb (analytical method: S) for 14 kimberlites from Basutoland.

KABLE *et al.* (1975) give a lower average of 81.3 ppm Nb for 60 kimberlites (ranging from 32 to 277 ppm Nb) from the Premier mine and the fissure and pipe materials from the Kimberley area (S. Africa). Considering additional data from Lesotho and Russia, for which KABLE *et al.* (1975) give the references, an overall mean of about 100 ppm Nb in kimberlites is to be expected.

VI. Crustal Abundance

The average niobium concentration in magmatic rocks of the upper continental crust has been computed in Table 41-E-11 using the figures on relative rock abundances of Table 7–8 in Vol. I of this handbook.

Table 41-E-11. *Niobium concentration in magmatic rocks of the upper continental crust*

Magmatic upper crustal rocks	Rock abundance in vol. %	ppm Nb
Granites, quartz monzonites	44	22 (Granites, quartz monzonites; Granodiorites; Quartz diorites)
Granodiorites	34	
Quartz diorites	8	
Diorites	1	
Gabbros, basalts	13	10
Syenites, alkalic rocks	<0.5	100
Peridotites	<0.5	1.5
Average	~ 100	20

We get an average abundance in these rocks of 20 ppm Nb. The same figure has already been published by GOLDSCHMIDT (1937), VINOGRADOV (1962) and TAYLOR (1964) on the basis of much less analytical information.

SHAW *et al.* (1976) got an overall mean abundance of 26 ppm Nb in the Precambrian shield of Canada. In this investigation a large proportion of metamorphic rocks has been considered.

As will be mentioned in Sect. 41-M granulite facies rocks as typical representatives of the lower continental crust seem to be depleted in Nb relative to normal portions of the upper continental crust.

If it is tentatively assumed that a large proportion of the oceanic crust is represented by oceanic basalts, it might contain about 5 ppm Nb or slightly more.

In the total earth's crust a mean between 10 and 15 ppm Nb is to be expected.

Revised manuscript received: March 1977

41-F. Behavior in Magmatogenic Processes

Although various discrete niobium and/or tantalum minerals may occur as accessory species in a wide variety of ordinary igneous rocks, including those of both alkalic and calc-alkalic series, relatively high abundances of niobium-tantalum minerals are found primarily in:

1. Alkalic granites (see Subsect. 41-E-II).
2. Metasomatically altered granites (albitized or greisenized) (see Subsect. 41-E-II).
3. Granite pegmatites.
4. Carbonatites and late hydrothermal veins of alkalic complexes.

In both calc-alkalic and alkalic consanguineous series or multistage massifs, total Nb and Ta contents generally increase from early to later stages of the host rock. In early series Nb and Ta occur chiefly camouflaged in rock-forming biotite, amphibole and Ti minerals, rarely as Nb-mineral accessories. In later stage units Nb and Ta occur both vicariously and as discrete accessories; in the latest stage rocks camouflaged Nb-Ta may be accompanied by abundant Nb-Ta minerals. Some authors assume that in calc-alkalic sequences the Nb : Ta ratio decreases with increasing magmatic stage and in alkalic sequences it increases. But the general validity of such a rule has still to be confirmed by additional and more reliable data.

The solubility of Nb_2O_5 in steam at 500 °C and 1,050 atm is 28 ppm (Morey, 1957).

I. Pegmatites

All three major structural categories of pegmatites can contain niobium-tantalum species: homogeneous (unzoned); zoned; and complex (zones plus replacement units). In unzoned pegmatites, columbite of relatively uniform composition is the characteristic mineral. In zoned pegmatites several generations of columbite-tantalite may occur, successive crops of which are progressively richer in tantalite. This general sequence is assumed by Goldschmidt (1954), but Fersman (1931/1952) states that the trend is opposite. In complex pegmatites the several columbite-tantalite generations may be followed by such tantalum-rich species as wodginite, stibiotantalite, bismutotantalite, simpsonite and microlite.

Fersman (1931) has estimated the average niobium concentration in granitic pegmatites as 10 ppm Nb. Beus (1966) in his general survey of pegmatitic muscovites states that Ta concentrations $\gg 20$ ppm are abundant in muscovites from rare-metal pegmatites which contain minerals of the tantalite-columbite group. High Nb concentrations in muscovite ($\gg 200$ ppm Nb) are diagnostic of columbite-bearing pegmatites of the U.S.A. and the U.S.S.R. A correlation between Nb and Be in pegmatitic muscovite and the occurrence of columbite and beryl has already been observed by Heinrich (1962).

II. Carbonatites

Niobium bearing carbonatites are widespread in eastern Canada (Ontario), Brazil[1] (HEINRICH, 1961), in Uganda, Zaire, Congo and the U.S.S.R. (DE KUN, 1961; PECORA, 1956). They can contain 0.X% Nb[2] and reserves in the range X00,000 t Nb. Several details on Nb minerals in these deposits etc. are compiled by PECORA (1956). In carbonatites the initial species is usually pyrochlore, which may display considerable variation in composition and physical properties. This can be followed by the replacement of its Ca-Na by Ba or Sr, followed by its replacement by fersmite and eventually by columbite (HEINRICH, 1966).

Results of experiments by ALEKSANDROV (1967) on reactions of $Nb_2O_5 \cdot nH_2O$ (n = 2–5) with solutions of $KHCO_3$ and K_2CO_3 at temperatures from 50 to 450 °C and under pressures of 500 to 1,000 atm indicate that niobium forms carbonate complexes in carbonate solutions at elevated temperatures. The experimental data demonstrate that this element could have been transported in nature in the form of polynuclear complexes in concentrations up to about 6 gNb/liter.

Niobium forms stable fluoride complexes; hence high concentration of fluorine in solution inhibits precipitation of niobium. Generally high fluorine concentrations (as fluoride) in carbonatites are antipathetic loci for accumulation of niobium minerals.

[1] The partly decomposed carbonatite of Araxà, Minas Gerais (Brazil) contains about 2% Nb_2O_5 with reserves of about $3 \cdot 10^8$ t. The 1969 production of 3,570 t Nb at this mine has contributed 60 percent to the world production (GIES and KASPER, 1972).

[2] GOLD (1966) assumes that "typical" carbonatites contain 560 ppm Nb+Ta.

Revised manuscript received: March 1977

41-G. Behavior during Weathering and Alteration of Rocks

Niobium and tantalum are either released during the weathering cycle to solutions through the destruction of host species as biotite, amphibole (and Ti minerals) of magmatic or metamorphic rocks or remain behind within resistate minerals such as ilmenite, sphene, zircon etc. Residual minerals will be accumulated in placers conglomerates and sands (see Sect. 41-K). The main placer mineral is columbite often being associated with cassiterite in Malaysia, Indonesia, Nigeria, Congo, Tanganyika. In 1968 24% of the world production of Nb and Ta was contributed from placer deposits (GIES and KASPER, 1972). Both the dissolved elements and the residual species tend to be concentrated in clays, laterites and bauxites. However, significant amounts of niobium reach the oceans to be deposited with pelagic clays (Sect. 41-K).

PACHADZHANOV (1963), who compared the behavior of niobium under humid environments with that under arid environments before being deposited on the Russian platform, found higher average concentrations in clays formed under humid conditions (20.1 ppm Nb) than in clays from arid environments (13.3 ppm Nb, analytical method: S). He ascribes the difference to higher residual concentration owing to greater leaching, under humid conditions, of such ions as sodium, potassium, calcium, magnesium and silicon. The Nb : Ta ratio is higher in the arid environment clays (14.8) owing to the greater mobility of the niobium and its easier removal in the humid environment (Nb/Ta: 8.4).

Niobium is chemically unique inasmuch as most of its compounds are almost insoluble in acid and alkaline media. Hence, in the absence of complexing agents (see 41-F), the soluble niobium compounds require high concentrations of acid or alkali for their formation and stabilization (GRIMALDI and BERGER, 1961). Organic complexes of Nb seem to be more stable than those of Ta.

Analyses of twenty lateritic soils from West Africa show an average Nb content of 24 ppm; four from within a few miles of a niobium deposit contain 79 to 87 ppm (GRIMALDI and BERGER, 1961). As the aluminum content of lateritic soils increases, the following depletion sequence applies: Si > Nb > Al = Fe. Thus, in general, high enrichments in niobium are not to be expected in lateritic soils, and unusually high niobium contents of laterites and bauxites stem from provenance enrichment (derivation from nepheline syenites, nearby niobian carbonatites etc.).

The Russian bauxites, investigated by PACHADZANOV (1964), are representatives of a variety of source rocks. All 35 varieties contain slightly more than average crustal abundances of Nb and Ta. The average Nb in these bauxites is 28 ppm and the average Ta: 4.5 ppm (Nb/Ta: 6.2). The Nb content varies from 4.5 to 91.8 ppm Nb. Four red bauxites from Hungary in the same report contain 35.6 ppm Nb and 5.1 ppm Ta and 2 bauxites from France: 61 ppm Nb. Nb + Ta is correlated with the Ti concentration in all these bauxites. The three elements occur mainly together in gibbsite, boehmite, diaspor and clay minerals and to a lesser degree in Ti phases.

Arkansas bauxite, derived from nepheline syenite with 130 ppm Nb, contains as much as 500 ppm Nb (range: 200 to 1,000 ppm Nb; analytical method: S) (GORDON and MURATA, 1952). In another report on Arkansas bauxite FLEISCHER *et al.* (1952) mention an average of 450 ppm Nb (46 samples). They found that in these deposits Nb mainly occurrs in ilmenite. The figures given by PEARSON (1955) on bauxite from Nyasaland (500 ppm Nb) derived from alkalic rocks are comparable with the data on Arkansas bauxite.

Investigations of niobium contents of soils and stream sediments near four pyrochlore carbonatites of the Feira district in Zambia revealed that little or no niobium enters the drainage system in solution. The pattern of geochemical anomalies derives from mechanical dispersion of residual grains of pyrochlore and niobian magnetite (WATTS *et al.*, 1963). The residual accumulation of Nb (up to 14% Nb_2O_5) during weathering of the Araxà carbonatite (Brazil) has been mentioned in Sect. 41-F.

The weathering alteration of the complex oxide minerals of Ti, Nb and Ta ($A_xB_yX_z$) has been studied by VAN WAMBEKE (1970). Such species (e.g. euxenite, samarskite), which are invariably metamict, alter chiefly by selective leaching of A-site atoms (Na, RE, Ca) and the partial replacement of O or F by (OH) or H_2O. These minerals because of both physical and chemical instabilities, normally do not accumulate in placer deposits. Exceptions are metastable accumulations, derived from nearby bedrock sources under cold-climate physical weathering environments (Pleistocene) and concentrated in stream sediments behind special physiographic or structural barriers (Bear Valley, Idaho) (HEINRICH, 1958).

803 samples of soils or other regoliths taken from locations about 50 miles apart throughout the conterminous United States contain a geometric mean of 12 ppm Nb (analytical method: S) (SHACKLETTE *et al.*, 1971). A map with the distribution of values and a frequency curve is given in this report.

Revised manuscript received: March 1977

41-I. Abundance in Natural Waters

CARLISLE and HUMMERSTONE (1958) have determined the abundance of niobium in unfiltered surface sea water from 40 km SE Plymouth to be 0.015 ppb Nb. In filtered surface water from Plymouth Sound niobium was analyzed at a concentration of 0.005 ppb while at this station unfiltered water reached a level of 0.05 to 0.1 ppb. The method was radiometrically checked chromatography. The oceanic residence time for niobium is calculated to be 300 years (GOLDBERG, 1965). This value is comparable to the residence time of Th, Cr and of several RE elements.

Revised manuscript received: March 1977

41-K. Abundance in Common Sediments and Sedimentary Rock Types

Data on the abundances of niobium and tantalum in various types of sedimentary rocks are listed in Table 41-K-1.

Values from X-ray analyses are more reliable than the older spectrographic determinations. There is enough information on niobium in shales and clays as the most abundant sedimentary rocks. Two sets of data from South Africa (see Fig. 41-K-1) and the U.S.S.R. (670 samples) are listed in Table 41-K-1. A one to one average of these sets might be representative for argillaceous rocks. It contains 17 ppm Nb. The Ta figure must be taken from the Russian material exclusively (2 ppm Ta). The Nb and Ta concentrations in the most abundant detrital rocks are comparable to the average composition of continental upper crustal rocks (Table 41-E-11 and a Nb/Ta ratio of ~10). The figure of 20 ppm Nb listed by

Table 41-K-1. *Niobium and tantalum in some sedimentary rocks and manganese nodules*

Material	Number of samples	ppm Nb	ppm Ta	Nb/Ta	Reference
Argillaceous rocks, South Africa (Swaziland to Karroo age)	316	15.7			1
Clay and sand sediments		20			2
Clays and shales	10	10	<0.9		3
Deep sea clays and oozes	5	16	<4		3
Shales and clays from arid environment, Russian Platform	199	13.3	0.9	14.8	4
Shales and clays from humid environment, Russian Platform	155	20.1	2.4	8.4	4
Shales and clays, average for Russian Platform	354	18.3	2	9.1	4
Siltites, Montana, Idaho (U.S.A.)	31	5.5			5
Pelitic rocks, Montana, Idaho (U.S.A.)	79	5.8			5
Manganese nodule, outer shell, Challenger	1	35			3
Manganese nodule, Albatross	1	147			3
Manganese nodule, core, Indian Ocean	1	7.6	2.3	3.3	6
Manganese nodule, outer shell, Indian Ocean	1	29	0.17	170	6
Manganese nodules		20			7
Manganese nodules, Atlantic-, Indian- and Pacific Ocean	44	36			8

References: 1. Hofmeyr (1971) (X); 2. Goldschmidt (1937) (S); 3. Rankama (1944, 1948) (W/S); 4. Pachadzhanov (1963) (S); 5. Harrison and Grimes (1970) (S, semi-quantitative method with many samples below sensitivity); 6. Pachadzhanov *et al.* (1963) (S); 7. Korkisch in: Manheim (1965); 8. Willis (1970) (X).

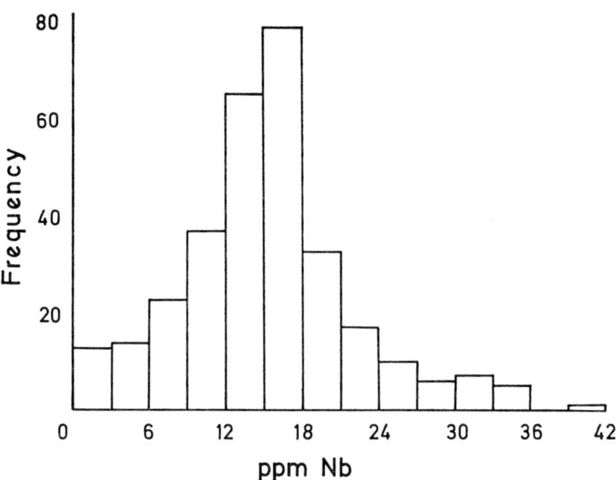

Fig. 41-K-1. Frequency distribution of niobium concentrations in 316 argillaceous rocks from South Africa from early Precambrian to Triassic age according to HOFMEYR (1971)

GOLDSCHMIDT (1937) on clay and sand sediments is close. PACHADZHANOV (1963) and HOFMEYR (1971) mention that there exist good correlations between Nb (Ta) and Ti and between Nb (Ta) and Zr in argillaceous rocks from Russia and South Africa. According to this author Ta is slightly enriched in the silt-sand fraction and Nb in the clay fraction.

Numerous and reliable data on Nb in manganese nodules from the ocean bottom have been contributed by WILLIS (1970). His averages for the Atlantic, Indian and Pacific Ocean are 41, 32 and 38 ppm Nb respectively which do not indicate systematic differences between the oceans. In pelagic clays manganese is diluted by an average factor of about 40 relative to manganese nodules. With this factor we might estimate that the manganese oxide fraction in pelagic clays cannot contribute more than 1 ppm Nb to the bulk niobium content. Therefore deep sea and nearshore deposited clays should contain comparable niobium concentrations.

The most highly enriched sedimentary accumulations occur as:

Eluvial placers: (i) Of pyrochlore over deeply weathered carbonatites: Mrima Hill, Kenya, has 100 mt averaging 0.7% Nb_2O_5 (HEINRICH, 1966); (ii) Of columbite-tantalite over deeply weathered granitic pegmatites: Sao Joao del Rei district, Minas Gerais, Brazil (HEINRICH, 1964).

Alluvial (stream) placers: (i) Of columbite-tantalite: Petaca district, New Mexico (HEINRICH, 1961); (ii) Of euxenite and other radioactive multiple oxide Nb-Ta species: Bear Valley, Idaho (HEINRICH, 1958) (Sect. 41-G).

One of the most unusual occurrences of a niobium-bearing mineral is that of labuntsovite, (K,Ba,Na,Ca,Mn) (Ti,Nb) $(Si,Al)_2(O,OH)_7 \cdot H_2O$, in the Green River continental evaporite formation of southwestern Wyoming (MILTON and EUGSTER, 1959).

Revised manuscript received: March 1977

41-L. Biogeochemistry

Land plants *(Sphagnum, Salix, Picea obovata L., Carex vesicaria, Ribes nigrum)* as analyzed by TYUTINA *et al.* (1959) (analytical method: W/C; S) contain a low niobium concentration of less than 0.4 ppm in dry tissues. Several plants *(Rubus arcticus L., Chamaenerium angustofolium L., Vaccinium myrtillus L., Rubus chamaemorus L.)* sampled by these authors near and over a niobium deposit in the Srednii Timan region, Komi A.S.S.R. (U.S.S.R.), have a great capacity for extracting Nb from the soil (Table 41-L-1). TYUTINA *et al.* (1959) concluded that an increase from the above mentioned normal level to a range from 7 to 10 ppm Nb serves as an indicator anomaly for a nearby occurring Nb deposit. They regularly observed a higher niobium concentration in leaves than in branches and twigs. The authors checked the niobium content in the roots of *Chamaenerium angustofolium* and found it to be about 2 or 3 times as high as in the leaves.

Table 41-L-1. *Niobium content of selected plants in a Nb mineralized area of the Komi ASSR (U.S.S.R.)* (According to TYUTINA *et al.*, 1959)

Plant	Percentage of plants with a certain range of Nb contents in dry leaves						Total number of samples
	≤0.4 ppm	0.5 to 0.7 ppm	0.8 to 1ppm	1.1 to 2.4 ppm	1.1 to 3.5 ppm	1.1 to 8.4 ppm	
Betula pubescens Ehr. and *B. verrucosa* Ehr.	56	33	1.5	9.5	—	—	40
Sphagnum Sp.	70	30	—	—	—	—	20
Vaccinium myrtillus L.	70	13	—	17	—	—	15
Chamaenerium angustifolium L.	57	7	8	—	28	—	20
Picea obovata L.	79	21	—	—	—	—	15
Rubus arcticus L.	21	15	14	—	—	50	20
Carex vesicaria L.	85	15	—	—	—	—	15
Ribes nigrum L.	85	15	—	—	—	—	10
Salix Sp.	63	22	15	—	—	—	20

According to RATYNSKIY and GLUSHNEV (1967) Nb has a smaller affinity for the organic portion of brown and bituminous coal than Ge, W, Ga and Be but a higher affinity than Mo, Sc, Y, La, Zn and Pb.

A few species of marine animals have been analyzed for niobium. *Mytilus edulis* for instance contains less than 1 ppb Nb in dry weight (CARLISLE, 1958). But some ascidians *(Molgula manhattensis, Styela plicata)* actively incorporate niobium (CARLISLE, 1958) and tantalum (KOKUBU and HIDAKA, 1965). In dry matter of pooled material from more than hundred individuals the first author found 25 to 75 ppm Nb (method: chromatography radiometrically checked). KOKUBU and HIDAKA (1965) for their dry material report averages of 250 ppm Nb and 230 ppm Ta (method: W/C). It is interesting to note that niobium (plus tantalum) and vanadium exclude each other and are separately specific for certain species of these animals (CARLISLE, 1958). The niobium (and tantalum) appears to be organically bound.

Carbonaceous shales of early Precambrian to Triassic age from South Africa contain slightly more Nb than non-carbonaceous pelitic rocks (HOFMEYR, 1971). The average concentrations for 132 and 184 samples are 19.9 and 12.6 ppm Nb respectively.

The presence of citric, tartaric and oxalic acids is known to produce a marked increase in the water solubility of Nb and Ta compounds because of chelation effects.

Revised manuscript received: March 1977

41-M. Abundance in Common Metamorphic Rock Types

Data on the niobium concentration in metamorphic rock series are scarce. In the Canadian Precambrian shield mica rich gneisses, garnetiferous gneisses, schists and argillites from Northern Quebec (16 samples) and Baffin Island (28 samples) contain average concentrations of 10 and 28 ppm Nb respectively (SHAW *et al.*, 1976; analytical method: X). The total mean for this group of rocks is 21 ppm Nb. Quartzofeldspathic rocks including granitic gneiss and granite from all sampling areas of the Canadian shield analyzed by the same authors have the following regional averages: 9 (86), 45 (101), 4.1 (47) and 35 (64) ppm Nb (number of samples in brackets). Considering the number of samples per set an average of 26 ppm Nb can be computed for this rock class. The authors mention that several samples were below or very close to the detection limit of their X-ray method (~2 ppm Nb) and therefore the means are not yet well-established. The partial data show regional variation. SW Quebec which contains anomalously high Ti, Y and Zr also has the highest content of Nb. The group averages of this investigation are comparable to Nb in granitic rocks (Tables 41-E-2 and 3) but slightly higher than the mean Nb in argillaceous sediments (Table 41-K-1).

Granulite facies rocks from the Lewisian complex of the Inner Hebrides (Scotland) are depleted in Nb relative to medium and low grade gneisses and schists (DRURY, 1973; analytical method: X). This author reports means of 4, 6 and 7 ppm Nb for 8 pyroxene gneisses, 26 migmatitic hornblende gneisses and 14 garnet biotite gneisses respectively from Tiree and Coll.

Metabasaltic tuffs from the Scottish Dalradian series are even lower in niobium than the average of tholeiitic basalts. These greenschists and amphibolites contain from 2 to 17 ppm Nb with a mean of 6 ppm Nb (VAN DE KAMP, 1970; analytical method: X).

Revised manuscript received: March 1977

41-O. Economic Importance

The melting point of metallic Nb is high (2,360° C). Its resistance against chemical attack is large. About 3% of the U.S. production has been used as metallic Nb and 18% for magnetic alloys. In the same year 79% of this metal went into the manufacture of stainless steel. 0.5 to 1% Nb addition to a steel improves its resistance against corrosion remarkably. Small concentrations (1 to 2%) of very pure Nb_2O_5 improve the quality of alnico magnets through the control of crystal formation. In nuclear reactors pure Nb metal is installed as fuel-element cladding. Nb-W-Hf high temperature alloys are used for rocket and aircraft engines. In addition production of niobium carbides for abrasives and of pure metallic niobium for vacuum getters has to be mentioned.

The U.S. production of niobium has been used for industrial equipment and pipelines for the manufacture of engines, cars and drilling apparatus.

The mining of niobium minerals has shifted from placer deposits to pyrochlore accumulations in alkalic rocks and carbonatites within recent years. The latter are mainly mined at Araxa (Brazil), Oka (Canada) and Lueshe (Zaire). In 1972 Araxa has produced 65% of the western world production. Additional large producers are Canada (22%), Nigeria (13%), Zaire, Malaysia, Australia, and the U.S.S.R.

A classification scheme of Nb and Ta deposits has been published by KUZMENKO (1961) and reproduced by PARKER and FLEISCHER (1968).

There is some fluctuation in the rate of world production—1962: 1,770 t Nb; 1970: 9,400 t Nb; 1972: 5,700 t Nb (excluding eastern countries).

The largest reserves occur in Brazil, Canada, Nigeria and Kenya. At present the minimum grade to be open-pit-mined is about 0.2% Nb_2O_5 (GOCHT, 1974).

Revised manuscript received: March 1977

References to Sections 41-B to 41-G, 41-I to 41-M, 41-O

ALEKSANDROV, I. V.: Niobium in the carbonate solutions and some considerations on the migration of rare elements under hydrothermal conditions. Geochem. Intern. **4**, 558 (1967).

ATKINS, D. H. F., SMALES, A. A.: The determination of tantalum and tungsten in rocks and meteorites by neutron activation analysis. Anal. Chim. Acta **22**, 462 (1960).

BAKER, P. E., BUCKLEY, F., HOLLAND, J. G.: Petrology and geochemistry of Easter Island. Contrib. Mineral. Petrol. **44**, 85 (1974).

BAKER, P. E., GASS, I. G., HARRIS, P. G., LEMAITRE, R. W.: The volcanological report of the Royal Society Expedition to Tristan da Cunha, 1962. Phil. Trans. Roy. Soc. London, Ser. A **256**, 439 (1964).

BEUS, A.A.: Distribution of tantalum and niobium in muscovites of granitic pegmatites. Geochem. Intern. **3**, 974 (1966).

BORISENKO, L. F., LISUNOV, N. V.: The distribution of scandium and niobium in wolframites. Geochem, (6), 735 (1958).

BORODIN, L. S.: Features of the concentration of niobium in nepheline syenites. Dokl. Akad. Nauk SSSR **103**, 1061 (1955).

BROOKS, C. K., NIELSEN, T. F. D., PETERSEN, T. S.: The Blosseville coast basalts of East Greenland—their occurrence, composition and temporal variations. Contrib. Mineral. Petrol. **58**, 279 (1976).

BUGAYETS, A. N., NARSEYEV, V. A.: On the distribution of niobium in some granitic plutons of Kazakhstan. Geochemistry **1**, 27 (1964).

BURKOV, V. V., PODPORINA, E. K.: Rare elements in kimberlitic rocks. Dokl. Akad. Nauk SSSR **163**, 197 (1965).

BUTLER, J. R., SMITH, A. Z.: Zirconium, niobium and certain other trace elements in some alkali igneous rocks. Geochim. Cosmochim. Acta **26**, 945 (1962).

CAMERON, A. G. W.: A new table of abundances of the elements in the solar system. In: AHRENS, L. H. (ed.), Origin and Distribution of the Elements. Oxford: Pergamon Press 1968.

CANN, J. R.: Rb, Sr, Y, Zr and Nb in some ocean floor basaltic rocks. Earth Planet. Sci. L. **10**, 7 (1970).

CARLISLE, D. B.: Niobium in ascidians. Nature **4613**, 933 (1958).

CARLISLE, D. B., HUMMERSTONE, L. G.: Niobium in sea-water. Nature **181**, 1002 (1958).

CARMICHAEL, I., MACDONALD, A.: The geochemistry of some natural acid glasses from the North Atlantic Tertiary volcanic province. Geochim. Cosmochim. Acta **25**, 189 (1961).

CHAO, E. C. T.: Petrographic and chemical characteristics of tektites. In: O'KEEFE, J. A. (ed.), Tektites. Chicago: Univ. Chicago Press 1963.

CLIFFORD, T. N., ROOKE, J. M., ALLSOPP, H. L.: Petrochemistry and age of the Franzfontein granitic rocks of northern South-West Africa. Geochim. Cosmochim Acta **33**, 973 (1969).

CORNWALL, H. R.: Calderas and associated volcanic rocks near Beatty, Nye County, Nevada. In: Petrologic Studies (Buddington Vol.), New York, Geol. Soc. Am. 357 (1962).

COSGROVE, M. E.: The geochemistry of the potassium-rich Permian volcanic rocks of Devonshire, England. Contrib. Mineral. Petrol. **36**, 155 (1972).

COX, K. G., JOHNSON, R. L., MONKMAN, L. J., STILLMAN, C. J., VAIL, J. R., WOOD, D. N.: The geology of the Nuanetsi igneous province. Phil. Trans. Roy. Soc. London, Ser. A **257**, 71 (1965).

COX, K. G., MACDONALD, R., HORNUNG, G.: Geochemical and petrographic provinces in the Karroo basalts of southern Africa. Am. Mineralogist **52**, 1451 (1967).

DAWSON, J. B.: Basutoland kimberlites. Bull. Geol. Soc. Am. **73**, 545 (1962).

DE KUN, N.: Die Niobkarbonatite von Afrika. Neues Jahrb. Mineral. Monatsh. 124 (1961).

DRURY, S. A.: The geochemistry of Precambrian granulite facies rocks from the Lewisian complex of Tiree, Inner Hebrides, Scotland. Chem. Geol. **11**, 167 (1973).

DUDYKINA, A. S.: Paragenetic associations of element-admixtures in cassiterites of different genetic types of tin-ore deposits. Akad. Nauk SSR, Inst. Geol. Rudnykh Mestorozhd. Petrog., Mineral., i Geokhim. **28**, 111 (1959).

ECKEL, E. B.: Geology and mineral resources of Paraguay—a reconnaissance. U.S. Geol. Surv. Profess. Papers **327** (1959).

EDGE, R. A., AHRENS, L. H.: The niobium contents of some South African granitic and alkali rocks. Trans Geol. Soc. S. Africa **66**, 109 (1963).

EL GORESY, A.: Meteoritic rutile, a niobium-bearing mineral. Earth Planet. Sci L. **11**, 359 (1971).

ENGEL, A. E. J., ENGEL, C. G., HAVENS, R. G.: Chemical characteristics of oceanic basalts and the upper mantle. Bull. Geol. Soc. Am. **76**, 719 (1965).

ERICKSON, R. L., BLADE, L. V.: Geochemistry and petrology of the alkalic igneous complex at Magnet Cove, Arkansas. U.S. Geol. Surv. Profess. Papers **245** (1963).

ERLANK, A. J., KABLE, E. J. D.: The significance of incompatible elements in Mid-Atlantic ridge basalts from 45° N with particular reference to Zr/Nb. Contrib. Mineral. Petrol. **54**, 281 (1976).

ERLANK, A. J., WILLIS, J. P., AHRENS, L. H., GURNEY, J. J., MCCARTHY, T. S.: Interelement relationships between the moon and stony meteorites with particular reference to some refractory elements. In: WATKINS, C. (ed.), Lunar Science III. Revised Abstracts of Papers Third Lunar Sci. Conf. 238 (1972).

ES'KOVA, E. M.: The geochemistey of niobium and tantalum in nepheline syenite massifs in the Vishnevye Mountains. Geochemistry **2**, 158 (1959).

ES'KOVA, E. M.: The geochemistry of niobium and tantalum in nepheline syenites of the USSR. 21. Internat. Geol. Congr. (Copenhagen) Papers of Soviet Geologists **1**, 101 (1960).

ES'KOVA, E. M.: Niobium. In: Geochemistry of Rare Elements. Vol. 1 (Moscow, Izdatel' stvo Nauka) Engl. Transl. New York: Davey & Co. 1966.

FERSMAN, A. E.: Les Pegmatites I. Les Pegmatites Granitiques. Louvain et Bruxelles 1952 (Transl. of the 1931 original Russian text).

FLANAGAN, F. J.: 1972 values for international geochemical reference samples. Geochim. Cosmochim. Acta **37**, 1189 (1973).

FLEISCHER, M., MURATA, K. J., FLETSCHER, J. D., NARTEN, P. F.: Geochemical association of niobium (columbium) and titanium and its geological and economic significance. U.S. Geol. Surv. Circ. 225 (1952).

FLOYD, P. A., WINCHESTER, J. A.: Magma type and tectonic setting discrimination using immobile elements. Earth Planet. Sci. L. **27**, 211 (1975).

GANAPATHY, R., ANDERS, E.: Bulk compositions of the moon and the earth, estimated from meteorites. Proc. Fifth Lunar Sci. Conf. Geochim. Cosmochim. Acta, Suppl. 5, **2**, 118 (1974).

GERASIMOVSKIY, V. I., KARPUSHINA, V. A.: On the relationship of Nb and Ta in igneous rocks. Geochem. Intern. **2**, 575 (1965).

GERASIMOVSKIY, V. I., KARPUSHINA, V. A.: Geochemistry of niobium and tantalum in nepheline syenites. Geochem. Intern. **7**, 609 (1966).

GERASIMOVSKIY, V. I., KARPUSHINA, V. A.: Niobium and tantalum content in Nigerian granites. Geochem. Intern. **4**, 599 (1967).

GERASIMOVSKIY V. I., VOLKOV, V. P., KOGARKO, L. N., POLYAKOV, A. I., SAPRYIKINA, T. V., BALASHOV, Y. A.: Geochemistry of the Lovozero Massif. Akad. Nauk SSSR, Moscow 1966.

GIES, H., KASPER, R.: Zur Geochemie und Lagerstättenkunde von Niob und Tantal II. Erzmetall **25**, 632 (1972).

GLADKIKH, V. S., CHERNYSHEVA, V. I.: Rare elements in suboceanic mafic extrusives. Geochem. Intern. **3**, 786 (1966).

GLADKIKH, V. S., VICTOROVA, M. Y.: Distribution of niobium and tantalum in the vol. canic rocks of the Kuznetskiy Alatan and Maymecha-Kotuy province. Geochem. Intern. **4**, 321 (1967).

GOCHT, W.: Handbuch der Metallmärkte. Berlin-Heidelberg-New York: Springer 1974.

GOLD, D. P.: The average and typical composition of carbonatites. Int. Min. Ass. 4th New Delhi Pap. Proc. (Mineral. Soc. India) p. 88 (1966).

GOLDBERG, E. D.: In: RILEY, V. P., SKIRROW, G. (eds.), Chemical Oceanography. Vol. 1. London: Acad. Press: 1965.

GOLDSCHMIDT, V. M.: Geochemische Verteilungsgesetze der Elemente. 9. Die Mengenverhältnisse der Elemente und der Atomarten. Norske Vidensk.-Akad. Oslo, Math.-Narurw. Kl. **4**, 1 (1937).

GOLDSCHMIDT, V. M.: Geochemistry. Oxford: Oxford University Press 1954.

GORDON, M., MURATA, K. J.: Minor elements in Arkansas bauxite. Econ. Geol. **47**, 169 (1952).

GORDON, M., TRACEY, J. I., ELLIS, M. W.: Geology of the Arkansas bauxite region. U.S. Geol. Surv. Profess. Papers **299** (1958).

GOTTFRIED, D., DINNIN, J. I.: Distribution of tantalum in some igneous rocks and coexisting minerals of the southern California batholith. U.S. Geol. Surv. Profess. Papers **525-B**, B96 (1965).

GOTTFRIED, D., GREENLAND, L. P., CAMPBELL, E. Y.: Variation of Nb-Ta, Zr-Hf, Th-U and K-Cs in two diabase-granophyre suites. Geochim. Cosmochim. Acta **32**, 925 (1968).

GOTTFRIED, D., JENKINS, L., GRIMALDI, F. S.: Distribution of niobium in three contrasting comagmatic series of igneous rocks. U.S. Geol. Surv. Profess. Papers **424-B**, B-256 (1961).

GRAHAM, A. L., MASON, B.: Niobium in meteorites. Geochim. Cosmochim. Acta **36**, 917 (1972).

GRIMALDI, F. S., BERGER, I. A.: Niobium contents of soils from West Africa. Geochim. Cosmochim. Acta **25**, 71 (1961).

HAMILTON, E.: The geochemistry of the northern part of the Ilimaussaq intrusion, S. W. Greenland. Meddr. Grønland **162**, 1 (1964).

HAMILTON, W.: Petrology of rhyolite and basalt, north-western Yellowstone Plateau. U.S. Geol. Surv. Profess. Papers **475C**, 78 (1963).

HAMILTON, W.: Diabase sheets of the Taylor Glacier region, Victoria Land, Antarctica. U.S. Geol. Surv. Profess. Papers **456B**, 31 (1965).

HAMILTON, W.: Geology and petrogenesis of the Island Park Caldera of rhyolite and basalt, eastern Idaho. U.S. Geol. Surv. Profess. Papers **504C** (1966).

HANSEN, J.: Niobium mineralization in the Ilimaussaq alkaline complex, South-west Greenland. XXIII. Intern. Geol. Congr. **7**, 263 (1968).

HARRISON, J. E., GRIMES, D. J.: Mineralogy and geochemistry of some belt rocks, Montana and Idaho. U.S. Geol. Surv. Bull. **1312-O** (1970).

HEINRICH, E. W.: Mineralogy and Geology of Radioactive Raw Materials. New York: McGraw-Hill 1958.

HEINRICH, E. W.: Types of Nb-Ta deposits in the western hemisphere (abs.). Fortschr. Mineral. **39**, 134 (1961).

HEINRICH, E. W.: Geochemical prospecting for beryl and columbite. Econ. Geol. **57**, 616 (1962).

HEINRICH, E. W.: Tin-tantalum-lithium pegmatites of the São João del Rei district, Minas Gerais, Brazil. Econ. Geol. **59**, 982 (1964).

HEINRICH, E. W.: The Geology of Carbonatites. Chicago: Rand McNally & Co. 1966.

HERZ, N., DUTRA, C. V.: Minor element abundance in a part of the Brazilian shield. Geochim. Cosmochim. Acta **21**, 81 (1960).

HLAVA, P. F., PRINZ, M., KEIL, K.: Niobian rutile in an Apollo 14 KREEP fragment. Meteoritics **7**, 479 (1972).

HOFMEYR, P.K.: The abundances and distribution of some trace elements in some selected South African shales. Ph. D.-Thesis, Capetown (1971).

HOLLANDER, J. M., PERLMAN, I., SEABORG, G. T.: Table of isotopes. Rev. Mod. Phys. **25**, 469 (1953).

HOSKING, K. F. G.: The primary tin deposits of south-east Asia. Minerals, Sci. and Eng. **2** (4), 24 (1970).

HUCKENHOLZ, H. G.: Die Verteilung des Niobs in den Gesteinen und Mineralen der Alkalibasalt-Assoziation der Hocheifel. Geochim. Cosmochim. Acta **29**, 807 (1965).

KABLE, E. J. D., FESQ, H. W., GURNEY, J. J.: The significance of the inter-element relationships of some minor and trace elements in South African kimberlites. Phys. Chem. Earth **9**, 709 (1975).

KALININ, Y. P., GOLDIN, B. A.: Distribution of tantalum and niobium in apogranites of the Man'khambo granite-granodiorite massif (southern Pechora region, Urals). Geochem. Intern. **4**, 270 (1967).

KESSON, S.: The major and trace element chemistry of kaersutite and its bearing on the petrogenesis of alkaline rocks. Contrib. Mineral. Petrol. **35**, 119 (1972).

KESSON, S. E.: The primary geochemistry of the Monaro alkaline volcanics, southeastern Australia—evidence for upper mantle heterogeneity. Contrib. Mineral. Petrol. **42**, 93 (1973).

KNORRING, O. VON: Some geochemical aspects of a columbite-bearing soda granite from South-East Uganda. Nature **188**, 204 (1960).

KOKUBU, N., HIDAKA, T.: Tantalum and niobium in ascidians. Nature **205**, 1028 (1965).

KOVALENKO, V. I., KRINBERG, I. A., MIRONOV, V. P., SELIVANOVA, G. I.: Behavior of U, Th, Nb and Ta during albitization of granitoids of the Ognitskii Complex (eastern Sayan Mtns.). Geochem. Intern. **5**, 868 (1964).

KUDRYAVTSEV, V. A.: Distribution of niobium in Yakutian kimberlites. Akad. Nauk SSSR Sibirskoye Otdeleniye, Inst. Geokhim. Trudy p. 142 (1964) [in Russian].

KUKHARENKO, A. A., ORLOVA, M. P., BULAKH, A. G., BAGDASAROV, E. A., RIMSKAYA-KORSAKOVA, O. M., NEVEDOV, E. I., IL'INSKII, G. A., SERGEEV, A. S., ABAKUMOVA, N. B.: Caledonian complex of ultrabasic, alkalic rocks and carbonatites of the Kola Peninsula and Northern Karelia. Moscow: Izdatel'stvo "Nedra" 1965.

KUZ'MENKO, M. V.: The geochemistry of tantalum and niobium. Intern Geol. Rev. **3**, 9 (1961).

KUZ'MENKO, M. V.: Tantalum. In: Geochimistry of Rare Elements. Vol. 1 (Russian Edition Moscow, Izdatel'stvo Nauka) Engl. Transl. D. Davey: New York 1966.

LEUBE, A., STUMPFL, E. F.: The Rooiberg and Leeuwpoort tin mines, Transvaal, South Africa. Econ. Geol. **58**, 527 (1963).

LEUTWEIN, F.: Die Wolframit-Gruppe. Minerallagerstättenkundliche Untersuchungen. Freiberger Forschungsh. **8** (1951).

LEVIN, B. Y., KOZLOVSKAIA, S. V., STARKOVA, A. G.: The average chemical composition of meteorites. Meteoritika **14**, 38 (1956).

LOVERING, J. F.: Differentiation in the iron-nickel core of a parent meteorite body. Geochim. Cosmochim. Acta **12**, 238 (1957).

LOVERING, T. G.: Distribution of minor elements in biotite samples from felsic intrusive rocks as a tool for correlation. U.S. Geol. Surv. Bull. **1314-D** (1972).

MANHEIM, F. T.: Manganese-iron accumulations in the shallow marine environment. Narragansett Marine Laboratory, Occ. Publ. **3**, 217 (1965).

MASON, B., ALLEN, R. O.: Minor and trace elements in augite, hornblende and pyrope megacrysts from Kakanui, New Zealand. New Zealand J. Geol. Geophys. **16**, 935 (1973).

MILTON, C., EUGSTER, H. P.: Mineral assemblages of the Green River Formation. In: Researches in Geochemistry. New York: John Wiley 1959.

MOGAROVSKIY, V. V., MEL'NICHENKO, A.K.: Distribution of niobium and tantalum in the granitoids of the Gissar Pluton, Central Tadzhikistan. Geochem. Intern. **5**, 893 (1968).

MOREY, G. W.: The solubility of solids in gases. Econ. Geol. **52**, 225 (1957).

MÜLLER, E. A.: The solar abundances. In: AHRENS, L. H. (ed.), Origin and Distribution of the elements. Oxford: Pergamon Press 1968.

NEKRASOV, I. Y.: Distribution of tantalum and niobium in magmatic and postmagmatic rocks and minerals of the northeastern USSR. Geochem. Intern. **7**, 378 (1970).

NESBITT, R. W., SUN, S.-S.: Geochemistry of Archaean spinifex-textured peridotites and magnesian and low-magnesian tholeiites. Earth Planet. Sci. L. **31**, 433 (1976).

ODIKADZE, G. L.: Distribution of tantalum, niobium, tin, and fluorine in micas from granitoids of the Greater Caucasus and the Dzirula crystalline massif. Geochem. Intern. **4**, 754 (1967).

PACHADZHANOV, D. N.: Geochemistry of niobium and tantalum in clays. Geochemistry **10**, 963 (1963).

PACHADZHANOV, D. N.: Geochemical bond between tantalum, niobium, and titanium in bauxites. Geochem. Intern. **1**, 889 (1964).

PACHADZHANOV, D. N., BANDURHIN, G. A., MIGDISOV, A. A., GIRIN, YU. P.: Data on the geochemistry of manganese nodules from the Indian Ocean. Geochemistry (5), 520 (1963).

PARKER, R. L., FLEISCHER, M.: Geochemistry of niobium and tantalum. U.S. Geol. Surv. Profess. Papers **612** (1968).

PEARCE, J. A., CANN, J. R.: Tectonic setting of basic volcanic rocks determined using trace element analyses. Earth Planet. Sci. L. **19**, 290 (1973).

PEARSON, T. G.: The chemical background of the aluminum industry. London, Royal Inst. Chem., Lectures, Mon. and Repts. **3**, 1 (1955).

PECORA, W. T.: Carbonatites: a review. Bull. Geol. Soc. Am. **67**, 1537 (1956).

PECORA, W. T.: Carbonatite problem in the Bearpaw Mountains, Montana. In: Petrologic Studies (Buddington Vol.) New York: Geol. Soc. Am., 1962.

PODOL'SKII, A. M., SERYKH, V. I.: Niobium in the intrusive rocks of the Kokchetav uplift (Kazakhstan). Geochem. Intern. **5**, 983 (1964).

PRICE, R. C., CHAPPELL, B. W.: Fractional crystallisation and the petrology of Dunedin Volcano. Contrib. Mineral. Petrol. **53**, 157 (1975).

RANKAMA, K.: On the geochemistry of tantalum. Finlande Comm. Géol. Bull. **133**, 1 (1944).

RANKAMA, K.: On the geochemistry of columbium. Science **106**, 13 (1947).

RANKAMA, K.: On the geochemistry of niobium. Ann. Acad. Sci. Fennicae, Ser. A III **13**, 57 pp. (1948).

RATYNSKIY, V. M., GLUSHNEV, S. V.: Relationships in the distribution of metals in coals. Dokl. Akad. Nauk SSSR **177**, 236 (1967).

RHODES, J. M., DAWSON, J. B.: Major and trace element chemistry of peridotite inclusions from the Lashaine volcano, Tanzania. Phys. Chem. Earth **9**, 545 (1975).

RIDLEY, W. I.: The petrology of the Las Canadas volcanoes, Tenerife, Canary Islands. Contrib. Mineral. Petrol. **26**, 124 (1970).

RÜDORF, W., LUGINSLAND, H.-H.: Untersuchungen an ternären Oxiden der Übergangsmetalle. II. Die Systeme TiO_2—NbO_2 und TiO_2—$TiTaO_4$. Z. Anorg. Allgem. Chem. **334**, 125 (1964).

SCHONFELD, E., MEYER, C.: The abundances of components of the lunar soils by a least-squares mixing model and the formation age of KREEP. Cosmochim. Acta, Suppl. 3, **2**, 1397 (1972).

SCHRÖCKE, H.: Heterotype Mischbarkeit zwischen Wolframit- und Columbitgruppe. Beitr. Mineral. Petrog. **8**, 92 (1961).

SCHRÖCKE, H.: Über Festkörpergleichgewichte innerhalb der Columbit-Tapiolitgruppe sowie der Columbit-Tapiolitgruppe mit YTi(Nb, Ta) O_6, Euxenit und mit $FeNbO_4$. Neues Jahrb. Mineral. Abhandl. **106**, 1 (1966).

SHACKLETTE, H. T., HAMILTON, J. C., BOERNGEN, J. G., BOWLES, J. M.: Elemental composition of surficial materials in the conterminous United States. U.S. Geol. Surv. Profess. Papers **574-D** (1971).

SHAW, D. M., DOSTAL, J., KEAYS, R. R.: Additional estimates of continental surface Precambrian Shield composition. Geochim. Cosmochim. Acta **40**, 73 (1976).

SITNIN, A. A.: Tantalum and niobium concentration in granitoid micas of the USSR. Geochem. Intern. **3**, 843 (1966).

SMALES, A. A., MAPPER, D., WEBB, M. S. W., WEBSTER, R. K., WILSON, J. D.: Elemental composition of lunar surface material. Proc. First Lunar Sci. Conf. Geochim. Cosmochim. Acta, Suppl. 1, **2**, 1575 (1970).

SMITH, R. E., SMITH, S. E.: Comments on the use of Ti, Zr, Y, Sr, K, P and Nb in classification of basaltic magmas. Earth Planet. Sci. L. **32**, 114 (1976).

SUESS, H. E., UREY, H. C.: Abundances of the elements. Rev. Mod. Phys. **28**, 53 (1956).

SUTHERLAND, D. S.: Potash-trachytes and ultra-potassic rocks associated with the carbonatite complex of the Toror Hills, Uganda. Mineral. Mag. **35**, 363 (1965).

TAYLOR, S. R.: Abundance of chemical elements in the continental crust—A new table. Geochim. Cosmochim. Acta **28**, 1273 (1964).

TAYLOR, S. R.: Trace element chemistry of andesites and associated calc-alkaline rocks. In: MCBIRNEY, A. R. (ed.), Proceedings of the Andesite Conference. Oregon Dept. Geol. Mineral. Ind. Bull. **65**, 43 (1969).

TAYLOR, S. R., CAPP, A. C., GRAHAM, A. L., BLAKE, D. H.: Trace element abundances in andesites II. Saipan, Bougainville and Fiji. Contrib. Mineral. Pertrol. **23**, 1 (1969).

TAYLOR, S. R., GORTON, M. P., MUIR, P., NANCE, W., RUDOWSKI, R., WARE, N.: Lunar highlands composition: Apennine Front. Proc. Fourth Lunar Sci. Conf. Geochim. Cosmochim. Acta, Suppl. 4, **2**, 1445 (1973).

TAYLOR, S. R., JAKEŠ, P.: The geochemical evolution of the moon. Proc. Fifth Lunar Sci. Conf. Geochim. Gosmochim. Acta, Suppl. 5, **2**, 1287 (1974).

TEMPLE, A. K., GROGAN, R. M.: Carbonatite and related alkalic rocks at Powderhorn, Colorado. Econ. Geol. **60**, 672 (1965).

TUREKIAN, K.K., WEDEPOHL, K. H.: Distribution of the elements in some major units of the earth's crust. Geol. Soc. Am. Bull. **72**, 175 (1961).

TYUTINA, N. A., ALESKOVSKII, V. B., VASIL'EV, P. I.: An experiment in biogeochemical sampling and the method of determination of niobium in plants. Geochemistry **6**, 668 (1959),

UPTON, B. G. J.: The geology of Tugtutôq and neighboring islands, South Greenland. Parts III, IV Medd. om Grönland **169** (1964).

VAN DE KAMP, P. C.: The green beds of the Scottish Dalradian series: geochemistry, origin, and metamorphism of mafic sediments. J. Geol. **78**, 281 (1970).

VINOGRADOV, A. P.: Average content of chemical elements in the principal types of igneous rocks of the earth's crust. Geochemistry **7**, 641 (1962).

WÄNKE, H., BADDENHAUSEN, W., DREIBUS, G., JAGOUTZ, E., KRUSE, H., PALME, H., SPETTEL, B., TESCHKE, F.: Multielement analyses of Apollo 15, 16 and 17 samples and the bulk composition of the moon. Proc. Fourth Lunar Sci. Conf. Geochim. Cosmochim. Acta, Suppl. 4, **2**, 1461 (1973).

WÄNKE, H., BADDENHAUSEN, H., PALME, H., SPETTEL, B.: On the chemistry of the Allende inclusions and their origin as high-temperature condensates. Earth Planet. Sci. L. **23**, 1 (1974).

WÄNKE, H., PALME, H., BADDENHAUSEN, H., DREIBUS, G., JAGOUTZ, E., KRUSE, H., PALME, C., SPETTEL, B., TESCHKE, F., THACKER, R.: New data on the chemistry of lunar samples: Primary matter in the lunar highlands and the bulk composition of the moon. Proc. Sixth Lunar Sci. Conf. Geochim. Cosmochim. Acta, Suppl. 6, **2**, 1313 (1975).

WAMBEKE, L. VAN: The alteration processes of the complex titano-niobo-tantalates and their consequences. Neues Jahrb. Mineral. Abhandl. **112**, 117 (1970).

WAMBEKE, L. VAN, BRINCK, J. W., DEUTZMANN, W., GONFIANTINI, R., HUBAUX, A., METAIS, D., OMENETTO, P., TONGIORGI, E., VERFAILLIE, G., WEBER, K., WIMMENAUER, W.: Les roches alcalines et les carbonatites du Kaiserstuhl. European Atomic Energy Comm. EUR. **1827 d-e** (1964).

WATTS, J. T., TOOMS, J. S., WEBB, J. S.: Geochemical dispersion of niobium from pyrochlore-bearing carbonatites in Northern Rhodesia. Inst. Min. Met. Bull. **681** (Trans. 72), 729 (1963).

WEAVER, S. D., SCEAL, J. S. C., GIBSON, I. L.: Trace element data relevant to the origin of trachytic and pantelleritic lavas in the East African Rift system. Contrib. Mineral. Petrol. **36**, 181 (1972).

WEDEPOHL, K. H.: The contribution of chemical data to assumptions about the origin of magmas from the mantle. Fortschr. Mineral. **52**, 141 (1975).

WEITZEL, H., SCHRÖCKE, H.: Nichtstöchiometrische Mischphasen mit α-PbO_2-Struktur und mit verwandten Strukturen. I. $FeNbO_4$ im System Fe_2O_3-Nb_2O_5. Neues Jahrb. Mineral. Abhandl. **119**, 3, 285 (1973).

WILLIS, J. P.: Investigations on the composition of manganese nodules with particular reference to certain trace elements. Ph.D-Thesis, University of Capetown (1970).

WOOD, D. A., GIBSON, I. L., THOMPSON, R. N.: Elemental mobility during zeolite facies metamorphism of the Tertiary basalts of eastern Iceland. Contrib. Mineral. Petrol. **55,** 241 (1976) .

ZNAMENSKII, E. B.: The average contents of niobium and tantalum in granitic rocks. Akad. Nauk SSSR Geokhim. Konf. Khimiya Zemnoi Kory. Trudy **2,** 301 (1964).

ZNAMENSKII, E. B., KONYSOVA, V. V., KRINBERG, I. A., POPOLITOV, E. I., FLEROVA, K. V., TSYKHANSKII, V. D.: Distribution of titanium, niobium, and tantalum in granitic rocks containing sphene. Geochemistry **1962,** 919 (1962).

ZNAMENSKII, E. B., NESTERENKO, G. V., TSYKHANSKII, V. D., KONYSOVA, V. V.: Abundance of niobium and tantalum in traps. In: Problemij Geokhimii (Vinogradov Vol.) Akad. Nauk, p. 410, 1965.

ZNAMENSKII, E. B., RODIONOVA, L. M., KAKHANA, M. M.: Distribution of niobium and tantalum in granites. Geochemistry (3), 267 (1957).

Revised manuscript received: March 1977

Molybdenum 42

A	H. T. Evans, Jr.	(U.S. Geological Survey, Reston, Virginia, U.S.A.)
B—O	S. Landergren and	(Källängsvägen 8, Lidingö, Sweden)
	F. T. Manheim	(Dept. of Marine Sciences, University of South Florida, St. Petersburg, Florida, U.S.A.)

42-A. Crystal Chemistry

Molybdenum and tungsten are transition metal elements in the fifth and sixth rows, respectively, of group VIB of the periodic table. Thus, their chemistry is complex and encompasses a wide range of oxidation states. In nature, however, they are almost exclusively hexavalent in their oxygen compounds, and are otherwise found only in the two sulfide species molybdenite and tungstenite (except for the rare species hemusite, Cu_6SnMoS_8). The two elements are separated by the lanthanum series of elements, the result being that through the lanthanide contraction phenomenon they have nearly identical atomic size. With respect to oxygen, the radii of Mo^{6+} and W^{6+} are 0.42 and 0.41 Å, respectively, for tetrahedral coordination (SHANNON and PREWITT, 1969). Consequently, the chemical behavior of the two elements is very similar, and isomorphism between corresponding compounds is common, sometimes leading to solid solution systems. The most characteristic crystal chemical properties of molybdenum are treated here, but reference is also made to the chapter on tungsten (Section 74-A) for further information.

I. Molybdenum Sulfides

The common ore mineral of molybdenum is molybdenite, $Mo^{[6]}S_2$. The crystal structure is hexagonal and contains infinite double sulfide layers stacked on one another with interlayer van der Waals contacts between sulfur atoms. The crystals are consequently very soft and foliar, very similar in mechanical properties to graphite (both are excellent lubricants), but with a metallic luster that has a distinctly bluish color. The layer is an unusual one in which the two triangular sulfur sheets are exactly superposed, providing interstices occupied by Mo with trigonal-prism coordination geometry (see Fig. 42-A-1). The S—S distance parallel to c is 2.99 Å (parallel to a, 3.16 Å), and there is no tendency toward the formation of S_2 groups. Similar layers and coordination are found in sulfide compounds of Nb, Ta and W, but only the last occurs in nature as the rare mineral tungstenite. A theoretical treatment by HUISMAN *et al.* (1971) gives the bonding conditions under which these elements are more stable in trigonal-prism than octahedral coordination.

The molybdenite-type structure (Fig. 42-A-1) was first established by DICKINSON and PAULING (SB 1913—28, 214), who found the Mo—S bond length to be 2.23 Å; no further refinement has appeared since. The layered nature of the structure suggests the possibility of stacking sequences other than the common two-unit hexagonal one (commonly denoted 2H). BELL and HERFERT (SR 1957, 155) synthesized a rhombohedral three-layer form (denoted 3R) and this was shown by SEMILETOV (SR 1961, 199) to be composed of the same type of layers as the hexagonal form. The 3R type was first demonstrated to exist in natural material by TRAILL (1963). TAKEUCHI and NOWACKI (SR 1964, 68) refined the crystal structure of a 3R crystal by Fourier methods and found a Mo—S bond of length 2.41 ± 0.2 Å. The possible stacking polytypes of molybdenite have been exhaustively investigated to as many as seven repeat layers

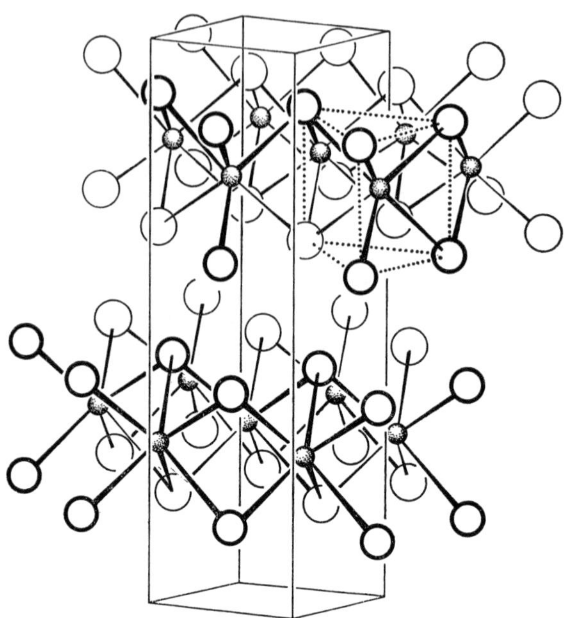

Fig. 42-A-1. The crystal structure of molybdenite, $Mo^{[6]}S_2$ (2H). The trigonal-prismatic coordination of S atoms around a molybdenum atom is shown by dotted lines, the hexagonal unit cell by full lines

(TAKEUCHI and NOWACKI, SR 1964, 68; WICKMAN and SMITH, 1970), but a broad search for such polytypes from more than 80 localities by FRONDEL and WICKMAN (1970) showed the presence only of the types 2H (about 80 percent), 3R, and mixtures of the two. The stability conditions of the two forms in terms of temperature and composition have been studied by CLARK (1970).

II. Oxomolybdenum Species

Under strongly oxidizing and weathering conditions, various oxomolybdate species are found, the principal one being wulfenite, $Pb^{[8]}[Mo^{[4]}O_4]$. This compound belongs to the scheelite group, whose structure is described in Section 74-A. Wulfenite contains nearly regular tetrahedral $[MoO_4]^{2-}$ groups in which the Mo—O bond has been found to be 1.772 ± 0.006 Å in length (LECIEJEWICZ, 1965). As discussed by HURLBUT (1955), crystals of wulfenite have long been considered to belong to the polar tetrahedral crystal class C_4 because of its morphological, etching, and piezoelectric properties, although no crystal structure evidence has yet been found for any lower symmetry than C_{4h} (space group $I4_1/a$).

Tetrahedral MoO_4 groups are present in the crystal structure of the minerals powellite, $Ca^{[8]}[Mo^{[4]}O_4]$ (scheelite group structure, approximately refined by ALEKSANDROV *et al.*, 1968), lindgrenite, $Cu_3^{[4+2]}(OH)_2[Mo^{[4]}O_4]_2$ (structure by CALVERT and BARNES, SR 1957, 249), and possibly in weathering products such as ferrimolybdite, $Fe_2[MoO_4]_3 \cdot 7$—$8\ H_2O$, and umohoite, $UO_2[MoO_4] \cdot 4\ H_2O$.

Among minerals, octahedral oxygen coordination for Mo has been found in the rare species koechlinite, $Bi_2^{[3+4]}Mo^{[6]}O_6$ (ZEMANN, SR 1956, 449; VAN DEN ELZEN and RIECK, 1973), in which layers of corner-linked MoO_6 octahedra alternate with pseudotetragonal $(BiO)_n$ layers similar to those in BiOCl. In its solution chemistry, molybdenum(6+) forms tetrahedral orthomolybdate $[MoO_4]^{2-}$ ions in alkaline solutions but tends strongly toward higher coordination and more condensed species in more acid ranges, ending finally in the precipitation of hydrated oxide compounds. Representatives of the latter types are almost unknown in nature, but the mineral ilsemannite may be related to such systems. This earthy material has an unknown structural basis, but it is essentially a hydrous oxide of acid origin, with some of the Mo reduced to 4+ or 5+, producing the characteristic blue color found in the synthetic "molybdenum blue" complexes obtained by partial reduction of polymolybdate and molybdenum oxide compounds.

Manuscript received: November 1973

References: Section 42-A

ALEKSANDROV, V. B., GORBATYI, L. V., ILYUKHIN, V. V.: Crystal structure of powellite $CaMoO_4$. Kristallografiya **13**, 512 (1968); Soviet Phys.-Cryst. **13**, 414 (1969).

BELL, R. E., HEFERT, R. E.: Preparation and characterization of a new crystalline form of molybdenum disulphide. J. Am. Chem. Soc. **79**, 3351 (1957).

CLARK, A. H.: Compositional differences between hexagonal and rhombohedral molybdenite. Neues Jahrb. Mineral., Monatsh. **1970**, 33.

FRONDEL, J. W., WICKMAN, F. E.: Molybdenite polytypes in theory and occurrence. II. Some naturally occurring polytypes of molybdenite. Am. Mineralogist **55**, 1857 (1970).

HUISMAN, R., DE JONGE, R., HAAS, C., JELLINEK, F.: Trigonal-prismatic coordination in solid compounds of transition metals. J. Solid State Chem. **3**, 56 (1971).

HURLBUT, C. S.: Wulfenite symmetry as shown on crystals from Yugoslavia. Am. Mineralogist **40**, 857 (1955).

LECIEJEWICZ, J.: A neutron crystallographic investigation of lead molybdenum oxide, $PbMoO_4$. Z. Krist. **121**, 158 (1965).

SHANNON, R. D., PREWITT, C. T.: Effective ionic radii in oxides and fluorides. Acta Cryst. **B 25**, 925 (1969).

TRAILL, R. J.: A rhombohedral polytype of molybdenite. Can. Mineralogist **7**, 524 (1963).

VAN DEN ELZEN, A. F., RIECK, G. D.: Redetermination of the structure of Bi_2MoO_6 koechlinite. Acta Cryst. **B 29**, 2436 (1973).

WICKMAN, R. E., SMITH, D. K.: Molybdenite polytypes in theory and occurrence. I. Theoretical considerations of polytypism in molybdenite. Am. Mineralogist **55**, 1843 (1970).

Manuscript received: November 1973

42-C. Abundance in Cosmos, Meteorites, Lunar Materials and Tektites

I. Cosmos

Molybdenum is a minor constituent in the cosmos. Some abundance estimates are reviewed in Table 42-C-1. GOLES (1969) has mentioned the problems of speculations on cosmic abundances. Our knowledge is limited to our solar system.

Table 42-C-1. *Cosmic abundance of molybdenum*
(Cosmic abundance units: C.A.U. = atoms per 10^6 atoms Si)

Material	Abundance (C.A.U.)	References
Cosmos	19	BROWN (1949)
Cosmos	2.42	SUESS and UREY (1956)
Cosmos	2.52	CAMERON (1959)
Solar atmosphere	10	GOLES (1969)
Meteorites (chondrites)	3.70	CASE *et al.* (1973)

II. Meteorites

Several modern analyses of the concentration in meteorites are available at present. However, estimates of cosmic abundance based on meteorite analyses are reliable only if the proportion of metal-sulfide-silicate phases is considered in the calculations. This very important feature in the geochemistry of molybdenum was first recognized by KURODA and SANDELL (1954) in their pioneer work. For older references the reader is referred to the latter publication. These authors analyzed the three phases separately in two composite meteorite samples. The differences in the concentrations reflect the different siderophile, chalcophile and lithophile tendencies in the distribution of molybdenum, as may be seen from Table 42-C-2.

Table 42-C-2. *Distribution of molybdenum in metal-, sulfide-, and silicate phases in ordinary chondrites* (KURODA and SANDELL, 1954; analytical method: W/C); (MASON and GRAHAM, 1970; analytical method: Mass)

	Metal	Silicate	Sulfide
Composite Sample I Arriba, Beenham, Cope, Cotesfield, Morland, Plainview, Roy (4 CL; 3 CH)	7.8	0.70	5.9
Composite Sample II Alamogordo, Estacado, Gladstone, Gruver, Kingfisher, Ladder Creek, Melrose (4 CH; 3 CL)	8.2	0.50	5.9
CL-, CLL-chondrites (Modoc, St. Severin)	8.5	n.d.	3.0

Table 42-C-3. *Molybdenum contents in iron meteorites.* (Number of samples in brackets.)

Name and class	Analytical method	Range, ppm Mo	Average, ppm Mo	References
Toluca, Om	I		6.4	1
Canyon Diablo, Om	I		6.3	1
Grant, Of	I		7.4	1
Sikhote Alin, Ogg	I		6.9	1
Santa Luzia, Ogg	I		6.2	1
Charkas, Om	I		7.2	1
Yardymly, Og	I		7.1	1
H (11)	N/R	2.7 - 7.8	6.5	2
Ogg (4)	N/R	5.9 - 8.1	7.3	2
Og (7)	N/R	4.9 - 8.1	6.3	2
Om (18)	N/R	4.1 - 8.5	7.5	2
Of (12)	N/R	3.2 -10.7	6.6	2
Off (4)	N/R	4.8 - 8.5	6.9	2
D (6)	N/R	2.2 -24.5	10.9	2
O (Breccia) (2)	N/R	5.9 - 6.6	6.3	2
Granular metabolite (2)	N/R	6.4 - 8.1	7.3	2
Odessa	N/R	6.2 - 6.5 (metal) 2.25- 2.27 (troilite)	—	3
Mean of siderites			7.07	

References: 1. Murthy (1963); 2. Smales *et al.* (1967); 3. Imamura and Honda (1976).

Three collections of analyses of iron meteorites by Murthy (1963), Smales *et al.* (1967) and Imamura and Honda (1976) are assembled in Table 42-C-3. The concentrations in the two sets of analyses are fairly similar and within the same range, as are the analyses of Kuroda and Sandell (1954). A somewhat higher average content and greater concentration range should be noted among the nickel-rich ataxites.

Table 42-C-4 presents analyses of carbonaceous enstatite and ordinary chondrites. It is interesting to note the good agreement between these analyses and those performed by Kuroda and Sandell (1954). Imamura and Honda (1976) have used the partitioning of Mo between meteoritic minerals to calculate effective temperatures

Table 42-C-4. *Molybdenum concentrations in chondrites.* (Analytical method: N/R)

Name and class		Metal (ppm Mo)	Silicate (ppm Mo)	Troilite (ppm Mo)	Total (ppm Mo)	References
Carbonaceous chondrites						
Ivuna	Cc1				1.2, 1.5	Case *et al.* (1973)
Orgueil	Cc1				1.5, 1.3	Case *et al.* (1973)
Cold Bokkeveld	Cc1				1.5	Case *et al.* (1973)
Mighei	Cc2				1.8	Case *et al.* (1973)
Murray	Cc2				1.6, 1.6	Case *et al.* (1973)
Grosnaja	Cc2				1.8	Case *et al.* (1973)
Vigarano	Cc3				2.4	Case *et al.* (1973)

Table 42-C-4 (continued)

Name and class		Metal (ppm Mo)	Silicate (ppm Mo)	Troilite (ppm Mo)	Total (ppm Mo)	References
Felix	Cc3				2.1, 2.3	CASE *et al.* (1973)
Lancé	Cc3				1.7	CASE *et al.* (1973)
Ornans	Cc3				1.9	CASE *et al.* (1973)
Warrenton	Cc3				1.7	CASE *et al.* (1973)
Allende	Cc3	4.0	2.2	—	2.2	IMAMURA and HONDA (1976)
Mean	Cc				1.8	
Ordinary chondrites						
Sharps	CL				1.8	CASE *et al.* (1973)
Bremerförde	CL				1.3	CASE *et al.* (1973)
Prairie Dog Creek	CL				1.7	CASE *et al.* (1973)
Clovis Nr. 1	CL				1.5	CASE *et al.* (1973)
Mezö-Madaras	CL				(3.7)	CASE *et al.* (1973)
Khokar	CL				1.3	CASE *et al.* (1973)
Barratta	CL				1.6	CASE *et al.* (1973)
Holbrook	CL				4.6±2.2	CASE *et al.* (1973)
	CL	0.15	0.76	9.5	2.6	IMAMURA and HONDA (1976)
Bruderheim	CL	6.38	0.31	3.5	1.1	CASE *et al.* (1973)
	CL	6.44	0.37		1.26	IMAMURA and HONDA (1976)
Parnalee	CL				0.82	CASE *et al.* (1973)
Hamlet	CL				1.0	CASE *et al.* (1973)
Fayetteville:						
dark	CH				1.6	CASE *et al.* (1973)
light	CH				2.0	CASE *et al.* (1973)
Chainpur	CHL				1.4	CASE *et al.* (1973)
Ngawi	CHL				1.4	CASE *et al.* (1973)
Plainview	CH				2±0.6	CASE *et al.* (1973)
Krymka	CH				0.95	CASE *et al.* (1973)
Kesen	CH	4.3	0.91	2.7	1.7	IMAMURA and HONDA (1976)
Yonozu	CH	4.1	0.23	2.6	1.1	IMAMURA and HONDA (1976)
Beardsley	CH	4.5	0.36 0.40	—	1.2	IMAMURA and HONDA (1976)
Richardton	CH	4.04 4.06	0.42 0.36	3.6	1.2	IMAMURA and HONDA (1976)
Peace River	CL	5.2 6.6	0.71 0.69	5.1	1.31	IMAMURA and HONDA (1976)
Potter	CL	5.1	0.54	—	1.2	IMAMURA and HONDA (1976)
St. Séverin	CLL	3.8 3.8	0.63 0.66	3.5	0.99	IMAMURA and HONDA (1976)
Mean CL, CH, CHL, CLL					1.41	
Enstatite chondrites						
Abee	Ce	4.37 4.25	(0.63) (0.66)	2.7	—	IMAMURA and HONDA (1976)
Indarch	Ce	2.9 3.3	1.2	9.2	—	IMAMURA and HONDA (1976)

of equilibration, arriving at values ranging from greater than 720° C for the Kesen chondrite to 920° C for the Holbrook and Bruderheim stones.

According to LIPSCHUTZ (1971), no reliable analyses of achondrites existed at his writing.

III. Tektites

One analysis of a tektite from Muldoon, Texas reported by KURODA and SANDELL (1954) gave 0.5 ppm Mo and four determinations by TAYLOR (1968) of australites yielded 0.4, 0.4, 0.6, and 1.0 ppm Mo.

IV. Lunar Materials

There appear to be very few analyses for molybdenum in lunar materials. Although much care has been taken to avoid contamination, the reliability of the molybdenum concentrations are nevertheless open to question (Table 42-C-5).

Table 42-C-5. *Molybdenum concentrations in materials from the Lunar expeditions—Apollo 11, Luna 16 and Luna 20*

Type of sample	ppm Mo	Analytical method	References
Apollo 11:			
Basalt (A)	0.4, 0.4, 0.4	N/R, I	1
Basalt (B)	0.4, 0.7	N/R, I	1
Breccia (C)	0.4, 0.7	N/R, I	1
Soil (D)	0.7	N/R, I	1
Basalt (A)	0.03	N/R	2
	0.055	N/R	2
	0.10	N/R	2
Basalt (B)	0.24	N/R	2
	0.16	N/R	2
	0.03	N/R	2
Breccia (C)	0.20	N/R	2
	0.03	N/R	2
Soil (D)	0.35	N/R	2
LUNA 16:			
Soil and rock	7 ± 1	I/S	3
LUNA 20:			
Soil	26 ± 2	I/S	3
Anorthosite	2.5	I/S	4
Soil	32 (?)	I/S	4
Soil	27 (?)	I/S	4
Soil	5.3	I/S	4
Basalt	0.24	I/S	4

References: 1. MORRISON *et al.* (1970); 2. TUREKIAN and KHARKAR (1970); 3. JEROME and PHILIPPOT (1973); 4. VINOGRADOV (1973).

The analyses by MORRISON *et al.* (1970) seem to be rather uniform with an average content of about 0.5 ppm Mo. Some differences between the analyses by TUREKIAN and KHARKAR (1970), JEROME and PHILIPPOT (1973) and VINOGRADOV (1973) cannot be explained yet. It should be noted that VINOGRADOV (1973) himself questioned two of his soil analyses. In fact, the "soil" contains fine lunar rock material and a minor fraction of meteoritic matter.

Judging from the lunar rocks analyzed so far, the molybdenum concentrations seem to be of the same magnitude as corresponding terrestrial rocks—or somewhat lower.

Revised manuscript received: February 1978

42-D. Abundance in Rock-Forming Minerals, Molybdenum Minerals

I. Valence State

Molybdenum occurs naturally in oxidation states ranging from Mo^{3+} to Mo^{6+}, with the latter the most stable under oxidizing conditions, and neutral-mildly acid conditions at high temperatures. Molybdenum compounds possess coordination numbers 4 and 6, and disproportionate to yield complex mixtures in which the element may exhibit several valences. At high concentrations, molybdate ions aggregate in solution to form polymolybdates and heteropolyanions. These forms are controlled by concentration, temperature, pH and Eh, and are considered in Sects. 42-F, 42-H and 42-L.

Under reducing conditions and in acid medium, Mo^{4+} is the main stable form, often as the ubiquitous sulfide, MoS_2. Mo^{5+} has been indicated as occurring in ilsemannite ($Mo_3O_8 \cdot nH_2O$?). At low temperatures, Mo^{3+} possesses a small field of stability in aqueous medium (Figs. 42-H-1, 42-H-2), but this field disappears at higher temperatures. Valencies lower than Mo^{4+} have not been recorded in solid naturally occurring species, including meteorites, and no oxide below MoO_2 was obtained in laboratory experiments (Brewer, 1953; Samsonov, 1973). However, laboratory experiments have confirmed Mo_2S_3 rather than MoS_2 as the lowest sulfide which is in equilibrium with metallic Mo in meteoritic associations excluding water (McCabe, 1955; see also Imamura and Honda, 1976, and Pouillard and Perrot, 1975).

II. Rock-Forming Minerals

Molybdenum concentrations in the major rock-forming silicates and quartz are close to the concentrations in the crust as a whole (1 ppm). Mo is most enriched in tungsten minerals. It is preferentially accumulated in minerals containing Ti^{4+} and Fe^{3+} such as sphene, ilmenite, titanomagnetite and biotite. This isomorphism can be accounted for in part on the basis of similarity in ionic radii (for Mo^{6+} (0.69, 0.62 Å, Goldschmidt, Pauling see Table 12-8 in Vol. I of this handbook) and Mo^{4+} (0.68, 0.66 Å) in comparison with Fe^{3+}(0.67 Å) and Ti^{4+}(0.64, 0.68 Å). Kuroda and Sandell (1954) suggested diadochic replacement of Mo for Fe^{2+} and Mg^{2+}, especially in ferriferous olivine, and Wager and Mitchell (1948) proposed limited Si—Mo substitution. Goldschmidt (1954) emphasized the close crystallochemical relationship between Mo and W in terms of atomic and ionic radii and coordination system, but pointed out important differences in thermodynamic behavior, especially in the presence of sulfide. Mo can be coprecipitated with fine-grained sulfides such as FeS, ZnS, PbS, and CuS, but tends to be expelled from crystal lattices as a discrete MoS_2 phase during growth of crystals. In contrast, WS_2 is almost unknown in rock associations, being much more soluble.

Table 42-D-1. *Molybdenum in separated mineral phases of granitic rocks from eastern Sibiria* (Transbaikal, U.S.S.R.). (References in footnote)

	Mineral	% in rock	Mo in mineral (ppm)	Proportion of Mo in rock (%)	Total Mo (ppm)
Leucocratic granite, (mean 3 samples) Gudzhir complex Reference 1	Feldspars	65.5	2.4	62.7	
	Quartz	32.83	1.8	15.43	
	Biotite	0.83	1.9	0.67	
	Titanomagnetite	0.07	36.0	1.1	
	Sphene	0.13	48.0	2.2	
	Σ	99.36		82.17	2.56
Porphyritic granodiorite, Amanan complex, near Lakom'ya intrusion Reference 1	Feldspars	66.3	2.1	63.8	
	Quartz	23.4	1.6	17.0	
	Hornblende	7.4	4.0	13.2	
	Biotite	2.8	1.0	1.3	
	Sphene	0.15	10.0	0.6	
	Σ	100.05		95.80	2.20
Granodiorite, Amudzhikan complex ,Karenga intrusion Reference 1	Plagioclase	60.7	1.0	32.1	
	Potash feldspar	4.8	0.6	1.5	
	Quartz	22.4	1.0	11.6	
	Biotite	7.7	9.3	38.0	
	Hornblende	3.4	4.8	5.8	
	Sphene	1.0	20.0	10.5	
	Σ	100.0		99.5	1.90
Hornblende-biotite granodiorite, Casimur River Reference 2 (Analytical method: S)	Plagioclase	55.3	0.95	43.0	
	K-feldspar	9.3	0.60	5.0	
	Quartz	18.7	0.30	5.0	
	Biotite	13.0	3.0	32.5	
	Amphibole	2.77	4.5	10.0	
	Sphene	0.19	43.0	7.0	
	Σ	99.3		97.5	1.20
Porphyritic hornblende-biotite granodiorite, Unda River Reference 2 (Analytical method: S)	Plagioclase	56.69	0.80	32.0	
	K-feldspar	11.47	0.71	6.0	
	Quartz	20.0	0.30	—	
	Biotite	8.4	7.3	44.0	
	Amphibole	1.88	5.0	6.0	
	Sphene	0.35	58.0	14.0	
	Σ	98.64		102.0	1.40
Porphyritic biotite granodiorite, Unda River, Kavykuchi-Undinskiye village Reference 2 (Analytical method: S)	Plagioclase	53.3	0.79	25.0	
	K-feldspar	15.4	1.21	11.0	
	Quartz	21.2	0.30	—	
	Biotite	8.06	7.6	36.0	
	Amphibole	0.76	5.6	2.0	
	Sphene	0.43	83.0	21.0	
	Σ	99.15		95.0	1.70

Table 42-D-1 (continued)

	Mineral	% in rock	Mo in mineral (ppm)	Proportion of Mo in rock (%)	Total Mo (ppm)
Porphyritic biotite granite,	Plagioclase	42.6	0.81	34.0	
Unda River	K-feldspar	22.0	0.94	21.0	
Reference 2	Quartz	26.8	0.3	—	
(Analytical method: S)	Biotite	6.48	3.1	20.0	
	Amphibole	0.78	5.0	4.0	
	Sphene	0.40	39.0	16.0	
	Σ	98.84		95.0	1.0
Leucocratic biotite granite,	Plagioclase	27.4	0.80	27.5	
Kudikan	K-feldspar	36.5	0.80	37.5	
Reference 2	Quartz	31.3	0.3	—	
(Analytical method: S)	Biotite	3.97	63.0	31.0	
	Σ	99.21		96.0	0.8
Leucocratic biotite granite,	Plagioclase	37.3	0.62	25.5	
Alenguy River	K-feldspar	27.3	0.81	24.5	
Reference 2	Quartz	30.0	0.30	—	
(Analytical Method: S)	Biotite	4.17	4.2	20.0	
	Sphene	0.09	33.0	3.0	
	Σ	98.94		73.0	0.9
Leucocratic granite,	Plagioclase	29.4	1.15	48.0	
Unda River near Alenguy	K-feldspar	35.8	0.97	50.0	
Reference 2	Quartz	32.0	0.30	—	
(Analytical method: S)	Biotite	1.44	9.5	20.0	
	Σ	98.69		118.0	0.7
Biotite granite, medium	Plagioclase	38.8	0.79	42.5	
grained, Gazimur River	K-feldspar	17.0	0.75	18.5	
Reference 2	Quartz	37.4	0.30	—	
(Analytical method: S)	Biotite	6.68	3.8	36.0	
	Σ	99.89		97.0	0.7

References: 1. Tauson *et al.* (1970). 2. Kozlov and Roshchupkina (1965).

Studies of separated rock-forming minerals have shown that in melanocratic types of granitoids, such as granodiorite, sphene, amphibole and biotite carry most of the Mo, with biotite a principle carrier (Kuroda and Sandell, 1954; Tauson, 1961). In granites in which total dark minerals are 6% or less, most of the Mo is contained in feldspars (Kozlov and Roshchupkina, 1965; Tauson *et al.*, 1970), Table 42-D-1. The former authors found that pure quartz samples from Transbaikal granites contained Mo less than 0.3 ppm (their detection limit) and that granites with 5 ppm Mo or more always contained large flakes of molybdenite upon prolonged grinding and centrifugation in separation liquids. Ranges of Mo in sphene from sialic rocks (Tables 42-D-1 and 2) are confirmed by Sen *et al.* (1950), who found 20–40 ppm Mo in sphene from Caledonian granite, granodiorite and tonalite.

Tabele 42-D-2. *Molybdenum, tin and tungsten concentrations in biotite granites of eastern Siberia* (Transbaikal, U.S.S.R.) IVANOVA and BUTUZOVA, 1968) (Analytical method: S)[a]

Mineral	Rock	Sn (ppm)	W (ppm)	Mo (ppm)
Plagioclase	Alenuyev granite,	0.6	0.9–1.1	0.68–0.70
K-feldspar	Konduy massif	0.6	1.0–1.2	0.58–0.64
Quartz		2.0	1.4	0.2–0.2
Biotite		10.0–18.0	5.0	1.0
Sphene		450	15.0–16.0	52.0–106

[a] Utilizing halogen volatilizers.

Table 42-D-3. *Molybdenum concentrations in coexisting minerals of mafic rocks* (Analytical method: S)

Mineral	Rock	ppm Mo	References
Hypersthene Magnetite	Olivine norite, S. California batholith (U.S.A.)	2 20	SEN (1959)
Hypersthene Magnetite	Quartz-biotite norite, S. California batholith (U.S.A.)	3 30	SEN (1959)
Hornblende Magnetite	Granite, coarse grained, S. California batholith (U.S.A.)	3 100	SEN (1959)
Hornblende Biote	Epidote-amphibolite schists (20)	25.0–29.4	MOXHAM (1965)

From the data in Tables 42-D-3 to 42-D-5, iron and titaniferous minerals are the dominant carriers of Mo in mafic rocks. Considerations of ionic size (Mg^{2+}0.78 Å; Fe^{3+}0.67 Å; Mo^{4+}0.68 Å) suggest that molybdenum should be preferentially enriched in fayalite rather than Mg-rich (forsterite) olivine. However, information on partitioning of Mo among mafic rocks in a manner similar to that performed by Soviet authors for granitic rocks is very limited (LYAKHOVICH, 1972). That the major ore minerals are normally grossly undersaturated in Mo may be seen from the large concentrations (100 ppm Mo) found in magnetites from granitic rocks (Table 42-D-5).

The first major study on the distribution of Mo in sulfide and many rare minerals was conducted by I. and W. NODDACK (1931) as a part of the search for their newly discovered element rhenium. The pioneering geochemists conducted an investigation of trace metals including Mo on some 1,600 mineral specimens from around the world. The highest molybdenum concentrations occurred in minerals of 4–6 valent metals such as W, U, V, Ge, Sn, and As. They report 800 and 900 ppm Mo for a scheelite and a huebnerite and 200 and 300 ppm Mo for an enargite and proustite respectively.

A broad review of common sulfides (FLEISCHER, 1955) indicated that chalcopyrite, followed by pyrite contained the largest concentrations of Mo. This has been confirmed by studies of cogenetic minerals shown in Table 42-D-6. However, studies of several hundred samples of pyrite, chalcopyrite and sphalerite from the Western U.S. (ROSE, 1967) also demonstrate the great variability characteristic of Mo distribution, owing to its complex linkage with pH, Eh, temperature, and coexisting mineral species. ROSE was unable to link trace metals to age or genetic parameters.

Table 42-D-4. *Molybdenum concentrations in mafic rock-forming silicates.* (Analytical method: S except where noted, Numbers of samples in brackets)

Mineral	Rock type, locality	Range, ppm Mo	References
Olivine (5)	Gabbros, Skaergaard intrusion (Greenland)	3–30	WAGER and MITCHELL (1951)
Hypersthene, amphibole	Mafic and intermediate rocks (gabbro, basalt, norite etc.)	0.2–3.0	WAGER and MITCHELL (1951); NOCKOLDS and MITCHELL (1948); SEN *et al.* (1959); KURODA and SANDELL (1954) C; ANNERSTEN and EKSTRÖM (1971)
Pyroxene	Mafic rocks (gabbro, basalt etc.)	0.2–4.5	KURODA and SANDELL (1954) C; HOWIE (1955); GERASIMOVSKIY *et al.* (1965), REKHARSKII (1970)
Hornblende, amphibole	Granitic rocks	1–5.6	RABINOVICH *et al.* (1958) C; KOZLOV and ROSHCHUPKINA (1965) C; REKHARSKII (1970); TAUSON *et al.* (1970) C

Table 42-D-5. *Molybdenum concentrations in rock-forming oxides.* (Analytical method: S and C, number of samples in brackests)

Mineral	Rock, locality	Range, ppm Mo	References
Magnetite	Gabbros, Skaergaard (Grennland) (5)	3–5	WAGER and MITCHELL (1951)
	Charnockite series, intermediate-mafic rocks, India	5–150 (mean 20)	HOWIE (1955)
	Mafic rocks (5) S. California batholith	20	SEN *et al.* (1959)
	Granitic rocks, S. California b.	20–100	SEN *et al.* (1959)
	Exhalations, Mt. Katmai (Alaska)	200	ZIES (1924, 1929, cited by WEDEPOHL, 1960)
	Iron ores, chiefly Sweden (average)	7	FRIETSCH (1970)
	All sources; bimodal distribution (210)	5, 20	UZKUT (1974)
Ilmenite	Gabbros	3	WAGER and MITCHELL (1951); LYAKHOVICH and BALANOVA (1969) C; IVANOVA (1963)
	Granites	1.8–328 (mean 28)	RABINOVICH *et al.* (1958); LYAKHOVICH and BALANOVA (1969) C; REKHARSKII (1970); NOCKOLDS and MITCHELL (1948); GERASIMOVKIY *et al.* (1965) C; TAUSON *et al.* (1970); KOSALS and MAZUROV (1970)

On the basis of a very large statistical sample of pyrites (550) from Czechoslovakia, CAMBEL and JARKOVSKY (1967) showed clearly that sedimentary minerals have the largest concentrations of Mo (mean 26 ppm), whereas hydrothermal and low temperature metamorphic pyrites had a mean of 14 ppm. Among sedimentary sulfides, the primary enrichment of Mo in host rock appears to be a governing factor. Thus Mo-rich upper Cambrian Swedish alum shales show a mean of 170 ppm Mo in pyrites, whereas the relatively Mo-poor Middle Cambrian alum shales yield only a total of 7 ppm (ARMANDS, 1973). Likewise, Mo-rich Pleistocene shales from the Red Sea yield pyrites averaging 170 ppm Mo (MANHEIM, 1974), whereas KEITH and DEGENS (1959) detected <2 ppm Mo in pyrites and marcasites from marine and freshwater carboniferous shales from Pennsylvania (U.S.A.).

Sphalerites and galena generally show low Mo concentrations, but, again, the metallogenetic province and specific associations are important. For example, in the Red Sea hot brine deposits, BERTINE and TUREKIAN (1973) and WOO (1976) show significant Mo in marmatite (Zn, Fe)S, under conditions where sulfide associated with Cu + Zn approaches total sulfide concentrations. Here, sulfidic iron is present largely in zinc and copper minerals (e.g. sphalerite and chalcopyrite) and the excess Fe forms low-Mo oxide phases such as hematite or ferrous montmorillonite (BISCHOFF, 1969).

For Mo in iron and manganese oxides, see Sect. 42-K.

III. Molybdenum Minerals, Phase Equilibria

Few minerals having molybdenum as a major constituent have been reported, and of these only molybdenite, ferrimolybdite, molybdeniferous scheelite (powellite) and wulfenite are found in significant concentrations. Molybdenite dominates Mo mineral phases overwhelmingly, and is found in many pegmatite and pneumatolytic deposits in small quantities or disseminations. Ferrimolybdite occurs as a yellow oxidation product in the Climax molybdenum deposit, Colorado, whereas ilsemannite is most noticeable as the molybdenum blue forming in acid rhyolite exhalative volcanics in the Valley of the Ten Thousand Smokes, Alaska. See Table 42-D-7 for a list of minerals.

Phase equilibria in the ternary system Mo—Fe—S in relation to natural minerals have been studied by GROVER *et al.* (1975). These authors present diagrams for the occurrence of MoS_2, $Mo_{2.06}S_3$ and a phase $\sim FeMo_4S_6$ at temperatures between 500 and 750° C.

Table 42-D-6. *Molybdenum concentration in sulfide minerals.* (References and analytical methods in footnote; value in parentheses is number of samples)

Mineral	Source	Range, ppm Mo	Average, ppm Mo	Reference
Sphalerite	Central Mining District, Utah (12)	3–70		1
Sphalerite	Bingham District, Utah (58)	<2–7	<2	1
Sphalerite	Plio-Pleistocene shales, Red Sea		<10	2
Sphalerite	N.W. Illinois Pb—Zn district (23)		<30	3

Table 42-D-6 (continued)

Mineral	Source	Range, ppm Mo	Average, ppm Mo	Reference
Sphalerite (dominant)	Black Kuroko ore, Japan (7)	tr–900	200	4
Sphalerite-maimatite	Red Sea hot brine area (Atlantis II Deep)		528	5
Chalcopyrite	Bingham District, New Mexico (porphyry copper) (56)	2–1,500	200 [a]	1
Chalcopyrite	Singhbhum migmatites, Bihar (India) (23)		44	6
Pyrite (dominant)	Yellow Kuroko ore, Japan (7)	500–2,300	1,200	4
Marcasite, pyrite	Co-occurring in N.W. Illinois Pb-Zn district (23)	–30 –30		3
Pyrite and marcasite	Paleozoic shales, Pennsylvania (22)		$\leqq$3	8
Pyrite	Alpine lead-zinc deposits (3) (6)		10	9
Pyrite	Alum shale, Upper Cambrian, S. Sweden (39)		170	10
Pyrite	Pliocene-Pleistocene Mo-rich shales, Red Sea (7)	30–500	178	2
Pyrite	17 Central Asian ore deposits (U.S.S.R.)	5.1–68	30	14
Pyrite	Graphitic slates, quartz sulfide ores, West Carpathian Mts. (Czechoslovakia) (30)	<20–90	51	11
Pyrite	Czechoslovakian ore deposits (550)	<3–2,500	28.6	11
Pyrite	Norway	5–20	12	7
Pyrite	Poland (77)	1–1,000	41	12
Pyrite, pyrrhotite	Coexisting minerals, W. Carpathian mountains, Czechoslovakia (17)		37 63	11
Pyrite Pyrrhotite	Coexisting minerals, Hel'pa, central Czechoslovakia (75)		20 20	11
Pyrrhotite	Norway (3)	20–40	30	7
Pyrrhotite	Poland (4)	1–15	6.5	13
Galena	U.S.A. (128) (Mo detected in 53 of 128 samples)		23	12
Galena	N.W. Illinois, Pb-Zn deposits (with fluorspar)		<10	3

[a] Three >1,000 values counted as 1,000.

References: 1. Rose (1967) (S); 2. Manheim and Siems (1974) (S); 3. Bradbury *et al.* (1970) (S); 4. Lambert and Sato (1974) (S); 5. Bertine (1970) (N/R); Woo (1976) (M); 6. Talapatra (1968) (S); 7. Noddack and Noddack (1931) (C/X); 8. Keith and Degens (1959) (S); 9. Hegemann (1949) (S); 10. Armands (1973) (S); 11. Cambel and Jarkovsky (1965) (S); 12. Mosier *et al.* (1976) (S); 13. Michalek (1958) (S); 14. Badalov *et al.* (1966).

Table 42-D-7. *Molybdenum minerals*

Oxides	
Ilsemannite	$Mo_3O_8 \cdot nH_2O$(?)
Molybdite[a]	MoO_3
Sulfides, selenides	
Molybdenite	MoS_2
Jordisite	MoS_2 (amorph.)
Castaingite	Mo_2S_5
Hemusite	Cu_6SnMoS_8
Femolite	Mo_5FeS_{11}
Drysdallite	$MoSe_2$
Molybdates etc.	
Powellite	$Ca[MoO_4]$[b]
Wulfenite	$Pb[MoO_4]$[b]
Chillagite	$Pb[(Mo,W)O_4]$
Lindgrenite	$Cu_3[(OH)MoO_4]_2$
Ferrimolybdite	$Fe_2^3\ [MoO_4]_3 \cdot 7H_2O$
(molybdenum ocher)	$Fe_2O_3 . 3MoO_3 \cdot 7.5H_2O$)
Koechlinite	Bi_2MoO_6
Betpakdalite	$CaFe_2H_8[AsO_4]_2[MoO_4]_5 . 10H_2O$
Sedovite	$U[MoO_4]_2$
Moluranite	$H_4U(UO_2)_3[MoO_4]_7 \cdot 18H_2O$
Iriginite	$(UO_2)Mo_2O_7 \cdot 3H_2O$
Umohoite	$UO_2[MoO_4] \cdot 4H_2O$
Mourite	$UMo_5O_{12}(OH)_{10}$
Calcurmolite	$Ca(UO_2)_3[MoO_4]_3(OH)_2 \cdot 11H_2O$
Cousinite	$MgU_2Mo_2O_{13} \cdot 6H_2O$(?)
Betpakdalite	$CaFe_2H_8[AsO_4]_2[MoO_4]_5 \cdot 10H_2O$

[a] SCHALLER (1907) showed that the yellow mineral earlier known as molybdite was a hydrated ferric molybdate and not molybdic oxide. PALACHE *et al.* (1951) relegate the name to a synonym for ferrimolybdite, but CECH and POVONDRA (1963), cited by FLEISCHER (1964) resurrected the original designation upon finding naturally occurring MoO_3 in Krupka, NW Bohemia.

[b] Complete isomorphous series between Mo^{+6} and W^{+6} are known.

Revised manuscript received: February 1978

42-E. Abundance in Common Igneous Rocks and in the Earth's Crust

I. Standard Rocks and Analytical Techniques

The low concentrations of Mo present in most igneous rocks require special analytical techniques to permit reliable quantitative evaluation. One of the most successful and widely used methods has been the spectrophotometric determination of the brown thiocyanate complex after preliminary chemical precipitation or extraction of Mo (SANDELL and GOLDICH, 1943; KURODA and SANDELL, 1954; CHAN and RILEY, 1966). The thiocyanate complex had earlier been used successfully for determining low concentrations of Mo in natural waters (KURODA, 1939). Other

Table 42-E-1. *Recommended values for standard reference rocks.* Data are from FLANAGAN (1973) unless otherwise noted. Parentheses indicate approximate value

Rock Type	Symbol[a]		Mo (ppm)	Rock Type	Symbol[a]		Mo (ppm)
Granite	USGS	G-1	6.5[b]	"Diabase"	USGS	W-1	0.57[b]
			5.3[c]		SS	4984	0.8[c]
Granite	SS	4979	4.4[c]	Lujavrite	NIM	1	(2)
	SS	4980	5.5[c]	Norite	NIM	N	(2)
	SS	4983	1.2[c]				
Granite	USGS	G-2	0.36	Peridotite	USGS	PCC-1	(0.21)
Granite	CRPG	GA	(1)	Pyroxenite	NIM	P	(1)
	CRPG	GH	(4)	Dunite	SS	4975	0.07[c]
	CRPG	GR	(18)				
	NIM	G	(2)	Dunite (Chrysotile)	NIM	D	(2)
	ZGI	GM	1.1				
Granodiorite	USGS	GSV-1	0.90	Dunite	USGS	DTS-1	0.2
Granodiorite	GSJ	J-G-1	(2)				
Gabbro-diorite	SS	4982	1.6[c]	Basalt	USGS	BCR-1	1.1
					ZGI	BM	0.6
Syenite	SSC	SY-1	(4)		SS	4978	1.4[c]
	NIM-	S	0.3		CRPG	Br	3
Andesite	USGS	AGV-1	2.3	Slate	ZGI	TB	(1)
				Sulfide Ore	SSC	SV-1	(9)

[a] Including an abbreviation of the organization which has preparated the sample:

Abbreviations: SSC Canadian Assoc. Applied Spectroscopy; GSJ Geological Survey of Japan; NIM National Inst. Metall. Research, Joahnnesburg, S. Africa; USGS U.S. Geological Survey; ZGI Zentral. Geol. Institut, DDR (German Dem. Republic); CRPG Centre de Recherche Petrogr. et Geochimiques, France; SS U.S. National Bureau of Standards "Radium Series".

[b] FLEISCHER (1965, 1969) gives detailed lists of data on USGS G-1 and W-1.

[c] BERTINE (1970).

organic complexes such as dithiol have also been employed, preceded by precipitation of Mo with $Fe(OH)_3$ (ISHIBASHI, 1953).

The normal limit for determination of Mo by spectrographic techniques is 2–10 ppm, but special variants permit lower limits. NODDACK and NODDACK (1939) employed a chemical pre-enrichment system with a spectrographic end determination. IVANOVA and BUTUZOVA (1968 and references cited) took advantage of the volatilizing effect of halogens in the form of AgCl, $CuCl_2$ and teflon at high temperatures to lower spectrographic detection limits to 0.2 ppm Mo.

More recently, low concentrations of Mo in meteorites, natural waters and other phases have been determined by neutron activation (KHARKAR *et al.*, 1968; BERTINE and TUREKIAN, 1973; IMAMURA and HONDA, 1976). The daughter of ^{99}Mo, ^{99m}Tc is measured with suitable procedures for separating interfering radioisotopes and monitoring extraction yields. CALVERT and PRICE (1977) determined Mo in sediments and manganese concretions by X-ray fluorescence; however, the method is not as sensitive as C, N/R and W/S at low concentrations.

II. Magmatic Rocks

Table 42-E-2 reviews the frequency distribution of molybdenum in 255 analyses of magmatic rocks by several authors (BERTINE, 1970; PATTERSON, 1952; WAGER and MITCHELL, 1953; NOCKOLDS and ALLEN, 1954; MATHIAS, 1957; HERZ and DUTRA, 1960; CARMICHAEL and MCDONALD, 1961; and KURODA and SANDELL, 1954). As may be seen, 73% fall below 2 ppm. Some analyses are semi-quantitative or approximate numbers. This is especially true of spectrographic determinations on a low ppm level. Aside from the pioneering work of KURODA and SANDELL (1954), the great bulk of modern analyses at low Mo levels have been performed by Soviet scientists, who have analyzed thousands of rocks, as well as their separated mineral phases (Section 42-D), and associated elements such as W and Sn. The first extensive studies on Mo in rocks were performed by VON HEVESY and HOBBIE (1933), but their data appear to be too high and now have only historical interest.

Table 42-E-2. *Frequency distribution of molybdenum in magmatic rocks* (see text)

Conc. class, ppm Mo	Number of analyses	Frequency distribution in per cent
0– 1	132	51.8
1– 2	54	21.2
2– 3	18	7.1
3– 4	19	7.4
4– 5	7	2.7
5– 6	10	3.9
6– 7	3	1.2
7– 8	3	1.2
8– 9	1	0.4
9–10	0	0
>10	8	3.1
	255	100.0

Tabel 42-E-3. *Mo in molybdeniferous granitoid rocks chiefly of eastern Transbaikal (T) and N. Caucasus (C) (U.S.S.R.)*

Rock type, locality	Range, ppm Mo	Mean, ppm Mo	Method	Reference
Leucocratic granites, Gudzhir		2.4	N/R	1
Intrusive granites, Gudzhir		3.3	N/R	1
Granodiorites, Amanan complex		2.0	N/R	1
Porphyritic granites, Amudzhikan-Bretensk		1.6	N/R	1
Granosyenites, Khulugli		2.1	N/R	1
Gneissoid biotite granites, Talochi-Zimovya		1.7	N/R	1
Migmatites, Lakom'ya-Khulugli	0.02–	2.8	N/R	1
Paleozoic granites, Verkhneundinsk massif	0.3–2.2	0.7		2
U. Jurassic granites, Tsaganoluyevsk massif	0.7–3.4	1.4		2
U. Jurassic granites, Kukul'bey	0.6–3.0	1.2		2
Granites, Durulguyev, Taptenai, Sakhanai and Zun-Undur massifs		1.08		3
Granitoids, E. Transbaikal		*2.4*		11
Granodiorites of all ages and areas, E. Transbaikal		*2.0*		4
Granite, Shakhtaminsk		1.8		7
Granites of all ages and areas, E. Transbaikal		*1.6*		4
Granitoids, Upper Paleozoic Undino-Gozimur region		1.96	N/R	1
Granites, El'dzhurta		1.02		12
Granites and granodiorites, chief mountain range of Caucasus		1.0		5
Quartz diorite intrusives (small)		0.41		6
Granodiorite intrusives (small)		2.09		6
Granitoids, Kazakhstan		1.7		8
Granitoids, Central Tien Shan		0.6		9
Granitoids, N. Tien Shan		1.2		10

Referenzes: 1. Kozlov and Roshchupkina (1965); 2. Ivanova and Butuzova (1968); 3. Ivanova (1963); 4. Kanishchev and Menaker (1974); 5. Studenikova and Glinkina (1967); 6. Studenikova and Glinkina (1964); 7. Tauson *et al.* (1968); 8. Vinogradov *et al.* (1958); 9. Tauson (1961); 10. Rekharskii and Krutetskaya (1960); 11. Rabinovich *et al.* (1958); 12. Lyakhovich (1972).

Extraordinarily detailed studies of granitoid rocks from the Transbaikal, Northern Caucasus, Kirghiz, and other regions have been reported by Soviet workers (Vinogradov *et al.*, 1958; Tauson *et al.*, 1968, 1970; Ivanova and Butuzova, 1968; Studenikova and Glinkina, 1967, and references cited).

A number of authors (Herz and Dutra, 1960; Studenikova and Glinka, 1967; Tauson *et al.*, 1968; Kanishchev and Menaker, 1974) have shown that the con-

Table 42-E-4. *Mo in selected granitoid rocks and gross averages.* (Number of samples in brackets)

Source	ppm Mo	References
Granites, U.S.A., different regions (11)	2.2	SANDELL and GOLDICH (1943)
Granites, U.S.A., different regions (36)	1.3	KURODA and SANDELL (1954)
Granites, South Africa (5)	0.9	KURODA and SANDELL (1954)
Granites, Japan (19)	1.8	ISHIMORI (1951)
Granites, Precambrian, Minas Gerais (Brazil) (4)	2.7	HERZ and DUTRA (1960)
Estimated mean of rhyolitic volcanics Taupo, New Zealand	2.6	EWART *et al.* (1968, in: MACDONALD and BAILEY, 1973)
Estimated world average, granitic rocks	1.0	KURODA and SANDELL (1954)
Estimated world average, high-Ca granitic rocks	1.0	TUREKIAN and WEDEPOHL (1961)
Estimated world average, low-Ca granitic rocks	1.3	TUREKIAN and WEDEPOHL (1961)
Estimated world average, granitic rocks	1.0	VINOGRADOV (1962)
Estimated world average, granitic rocks	2.0, 2.1	TAYLOR (1964); UZKUT (1974)
Estimated unmineralized granitic rocks	1.0	This Report
Estimated total granitic rocks	2.0	This Report

Table 42-E-5. *Molybdenum in rocks of the northern Caucasus* (STUDENIKOVA and GLINKINA, 1967). (Method: C)

	Precambrian	M. Paleozoic	U. Paleozoic	Jurassic	Quarternary
Granite (alaskite to granodiorite)	0.8–1.6	1.1–1.2	0.5–2.9 1.2–4.4	1.4 1.0	
Tonalite		1.5	1.0		
Gabbrodiorite, quartz diorite		<0.2–3.1	0.2–0.63		
"Diabase"		2.4–3.0		0.8	
Andesite			0.8–1.2		1.9
Porphyry, basalt		0.9–2.7		1.2	<0.2–3.1

Overall mean: Intrusives: 1.2 ppm Mo; Mafic effusives: 2.4–2.9 ppm Mo.

centration of Mo in *granites* is rather uniform and lacks any significant relationship with age (see Table 42-E-5). Even in highly mineralized areas, Mo is rarely enriched in granitic country rock, but rather is concentrated near fissures, fractures, and other zones of weakness (see 42-F).

Overall statistics show that, whereas a great majority of granites have values comparable to those of mafic rocks; granitoids are nearly exclusive hosts for Mo enrichments and ores and, therefore, their overall averages should approach their

Table 42-E-6. *Mo abundance in intermediate, mafic and other igneous rocks*

Rock type, locality	No. of samples	Range, ppm Mo	Average, ppm Mo	References (Anal. method)
Trachytes, Hawaii, Marquesas, Pitcairn Island	7	2–8	4.6	NOCKOLDS and ALLEN (1954) (S)
Liparites, N. Caucasus (U.S.S.R.)	15		0.70	LYAKHOVICH (1972)
Vitroandesites, N. Caucasus (U.S.S.R.)	5		0.7	LYAKHOVICH (1972)
Basalts (submarine), Hawaiian shelf	10		1.4	LISITSINA *et al.* (1975)
Basalts, N. Ireland	4	0.3–1	0.83	PATTERSON (1952) (S)
Basalts, Hawaii	4	<1–2		WAGER and MITCHELL (1953) (S)
Tinguaite, olivine tephrite, basalt and foyaite hybride, Messum Complex (S. Africa)	6	3–10		MATHIAS (1957)
"Diabase"-pegmatites, Siberia	15		1.0	LYAKHOVICH (1972)
Basalts, U.S.S.R.	3	0.9–4.5	2.8	VINOGRADOV *et al.* (1958) (W/S)
Diorites, U.S.S.R.	3	0.9–1.3	1.05	VINOGRADOV *et al.* (1958) (W/S)
"Diabases", U.S.S.R.	4	0.94–4.2	2.1	VINOGRADOV *et al.* (1958) W/S)
Gabbro-diabases	3	0.8–3.5	1.9	VINOGRADOV *et al.* (1958) (W/S)
Dunite, U.S.S.R.	1		0.23	VINOGRADOV *et al.* (1958) (W/S)
Andesites and andesitic basalts, Uksichan volcano, Kamchatka (U.S.S.R.)	39		≦10	OGORODOV *et al.* (1967)
Pitchstones, Eigg and Arran (U.K.), Iceland	36	1.1–1.9	1.8	CARMICHAEL and McDONALD (1961) (S)
Glass in vitroandesite	2		1.8	LYAKHOVICH (1972)
Comendites and pantellerites (alkali-oversaturated obsidians)	15		13.0	Data from various authors, cited by MACDONALD and BAILEY (1973)
Aegerine-nepheline phonolite			15.0	CARMICHAEL and McDONALD (1961)
Carbonatites		4–100	50	HIGAZY (1954), GOLD (1963), DEANS (1966), UZKUT (1974) etc.

parent sediments, or about 1.8 ppm. Large enrichments of Mo in peralkaline obsidian glasses along with high concentrations of Cl and F have been reported (Table 42-E-6).

In *mafic rocks*, molybdenite segregations and other large concentrations of molybdenum are very rare. LYAKHOVICH (1972) has shown that Mo is enriched in the glassy portion of vitroandesitic porphyries rather than in phenocrysts. Very large enrichments (13 ppm) in peralkaline obsidian glasses have been reported (MAC-

Table 42-E-7. *Mo averages in intermediate, mafic and other igneous rocks*

Rock type	Range, ppm Mo	Average, ppm Mo	Reference (Analyt. method)
Ultramafic rocks	0.1–1.4	0.4	KURODA and SANDELL (1954) (C)
Ultramafic rocks		0.3	TUREKIAN and WEDEPOHL (1961)
Ultramafic rocks		0.2	VINOGRADOV (1962)
Basalts	0.2–7.3	1.1	KURODA and SANDELL (1954) (C)
Basalts		1.5	TUREKIAN and WEDEPOHL (1961)
Basalts		1.4	VINOGRADOV (1962)
Basalts		1.0	TAYLOR (1964)
Basalts, Transbaikal (2,225 samples)		1.1	KANISHCHEV and MENAKER (1974)
Intermediate rocks	0.1–1.5	0.7	KURODA and SANDELL (1954) (C)
Gabbros and related rocks	0.1–1.5	0.6	KURODA and SANDELL (1954) (C)
Intermediate rocks		1.0	VINOGRADOV (1962)
Diorite-metamorphites, Transbaikal (2,860 samples)		2.1	KANISHCHEV and MENAKER (1974)

DONALD and BAILEY, 1973), along with high concentrations of Cl and F, and resistate elements such as W and Zr.

Mo abundances and averages of intermediate, mafic and ultramafic rocks are listed in Tables 42-E-6 and 7.

III. Earth's Crust

Considerations in developing averages for the crust are discussed in VINOGRADOV (1962), TAYLOR (1964), and RONOV and YAROSHEVSKII (1972) and references cited. Averages for abundant magmatic rock types, listed in Table 42-E-8, are those of RONOV and YAROSHEVSKII (1972). It should be noted that misleading figures will be obtained if detailed rock proportions and Mo concentrations are not utilized in favor of approximations useful for other purposes. One must also make allowance

Table 42-E-8. *Averages for abundant magmatic rocks* according to RONOV and YAROSHEVSKII (1972)

Rock type	ppm Mo
Granite (unmineralized)	1.1
Granitoids (total)	1.8
Intermediate rocks	0.8
Basalts	1.2
Ultrabasic (ultramafic) rocks	0.3
Peralkaline glasses	3–15 (?)
Continental crust	*1.2*

Table 42-E-9. *Mo abundance in the earth's crust*

	ppm Mo	Reference
Continental crust	1.5	TAYLOR (1964)
Continental crust, total crust	1.2	RONOV and YAROSHEVSKII (1972)
Earth's crust	1.5	GOLDSCHMIDT (1937)
Earth's crust	1.0	MASON (1958); this paper
Earth's crust	1.1	VINOGRADOV (1962)

for the fact that some of the best documented areas have been intensively and precisely studied because they are regions of widespread Mo mineralization (e.g. E. Transbaikal, U.S.S.R.). This will affect total granite averages from the area to a significant extent. The greatest uncertainties in crustal estimates probably lie in metamorphic rocks and sediments, and no overall study of Mo distribution incorporating an adequate data base has yet been performed.

Crustal values of different authors are compiled in Table 42-E-9.

Revised manuscript received: February 1978

42-F. Processes Connected with Magmatism

The behavior of molybdenum under conditions of higher temperature and pressure in the earth's crusts is less well known than for several other anion-forming metallic elements. Extensive studies include those of KORENBAUM (1970), whose monograph depicts the thermodynamic stability fields for aqueous molybdates, molybdenum halogenides, and reduced Mo species at temperatures from 25° to 400° C, as well as characterizing the occurrences of economic Mo deposits; GROVER *et al.* (1975), who provide experimental investigations of the Fe—Mo—S system at temperatures from 500° to 750° C; MOH *et al.* (1974), who report on the Mo—S system at elevated temperatures. UZKUT (1974) offers an extensive review. Unfortunately, the detailed work of KORENBAUM (1970) chiefly utilizes older and apparently incorrect free energy data for key Mo phases MoO_4^{2-} and MoS_2.

I. Magmatic Gases

KRAUSKOPF (1964), ARUTYUNYAN (1969), and UZKUT (1974) discuss gaseous transport of molybdenum. Several volatile species are listed in Table 42-F-1. A number of species have low sublimation or boiling points. Because of hydrolysis or decom-

Table 42-F-1. *Melting* (M.P.) *and boiling points* (B.P.) *of undissociated molybdenum species*

Component	MP °C	BP °C	Remarks	Reference
MoO_2		1,000[b]	Decomposes to MoO_2^+ in air. At 1,000° C partly sublimes and disproportionates to MoO_3 and Mo	SAMSONOV (1973)
MoO_3	795	650[b] 1,257		SAMSONOV (1973)
Mo_2S_3	1,740		Decomposes in air	MOH *et al.* (1974)
MoS_2	1,900		Decomposes in air	MOH *et al.* (1974)
MoS_3		695[b]		MOSER and BEER in: DUVAL (1963)
		780[b]		HERSTEIN in: DUVAL (1963)
MoF_6	17.5	35	Decomposes in water to HF and molybdic acid	UZKUT (1974)
$MoCl_4$	110[a]		Stable to 1,500° C in gas form; hydrolyzes to oxychloride and trioxide (the former is stable in 4.7–7N HCl only)	UZKUT (1974)
MoO_2Cl_2		[b]		WEAST (1976)

[a] decomposes. [b] sublimes.

position in water some could occur above the critical temperature of water. Their presence in most hydrothermal conditions is excluded. Further, volatilization of the halogen species requires high halogen content and acidity. Suitable conditions occur mainly in the granitic-pegmatitic-pneumatolytic stage of magmatism, where most silicates have been crystallized, leaving chiefly volatiles such as HF, HCl, H_3BO_3, H_2O and associated phases.

Lowering of pressure and/or temperature with consequent loss of halogens and hydrolysis to aqueous molybdenum-halogen complexes, and loss of halogens through precipitation of halogen-containing phases such as mica, topaz and fluorite can trigger molybdenum deposition.

II. Magmatism

Experimental petrologic studies on molybdenum phases are relatively few. GROVER *et al.* (1975), and MOH *et al.* (1974) studied Fe--Mo—S and Mo—S systems, respectively. GROVER *et al.* (1975) found only narrow ranges of miscibility and solid solution for sulfides. Even at 950° C, solid solution among phases such as MoS_2, FeS, Fe_2Mo_2—$FeMo_4S_6$ are generally less than 1%.

Granitic melts are able to sustain only very limited amounts of Mo in solid solution, excesses over about 3 ppm being attributable to discrete MoS_2 flakes (KOZLOV and ROSHCHUPKINA, 1965). RAMDOHR (1940) calls MoS_2 a rare mineral in granites. Several authors report the occurrence of molybdenite in granitic rocks as mentioned in Section 42-E.

III. Metasomatism and Hydrothermal Transport

The systems relevant to metasomatism involve H_2S and halogens. Figure 42-F-1, modified from KORENBAUM (1970), illustrates qualitatively the pertinent relationships. Updating of data using newer free energy values is required to interpret the information quantitatively, as indicated in 42-H.

At relatively low (hydrothermal) temperatures and neutral pH values, molybdenum remains soluble in the form of thiomolybdates even at relatively high ΣS. With higher temperatures, the thiomolybdate field retreats toward basic pH values (Fig. 42-F-1). Molybdenite precipitation thus takes place under more acid conditions whereas molybdates, and polyanions predominate under neutral regimes.

In the presence of halogens, there is a complex interplay as mentioned earlier. Fluorides can occur above the critical temperature of water (373° C). No treatment appears to be available for the important chloride system, although experiments have demonstrated increasing volatilities with temperature (KHITAROV *et al.*, 1967). The specific influence of fluoride on the sulfide system is relatively small, except in extending the MoS_2 field toward acid values at the expense of MoO_2^+, especially at low temperatures.

IV. Deposition of Ores

Most major molybdenum mineralizations are associated with granitic-pegmatitic intrusions followed by pneumatolytic and hydrothermal activity in the region 200–300° C and lower. High temperature, tin-tungsten deposits in greisens such as

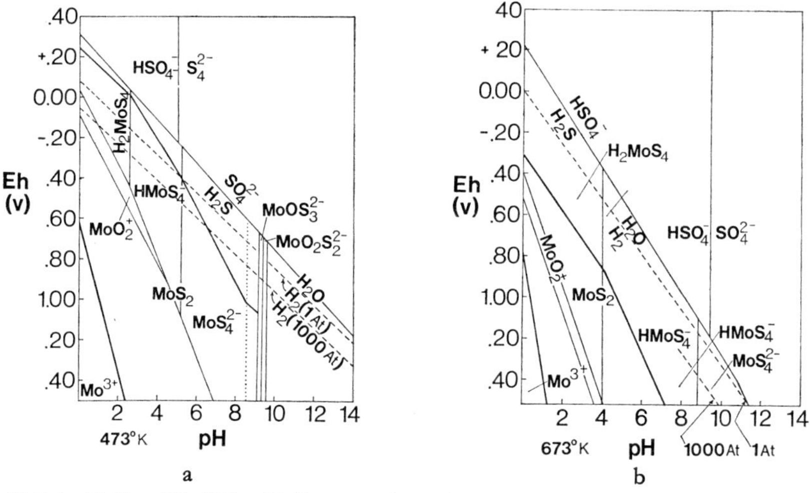

Fig. 42-F-1. (a) Simplified Eh-pH diagram of molybdenum and sulfur species at $\Sigma Mo = 10^{-6}$ and $\Sigma S = 1$ molar at 200° C. Modified from Korenbaum (1970). (b) Same for 400° C

the tin deposits of Bolivia, Indonesia, Cornwall, and the Pyrenees, are often very low in molybdenum, whereas scheelites from contact metamorphic zones may have significant Mo. Frequently a major deposit may show a variety of temperature regimes and parageneses. In the Vitim River Mesozoic molybdenite deposits (Yablonovyi mountain range, Transbaikal, U.S.S.R.), early high temperature-pneumatolytic mineralization includes alkali feldspars, wolframite, molybdenite, pyrite, ilmenorutile, quartz, muscovite, and fluorite. The zone of pneumatolytic influence extends about 200 m (Pokalov, 1964).

Molybdenite and copper-molybdenum parageneses are stable under acid conditions in which lead and especially zinc sulfides remain soluble. As pH values increase toward neutral regimes, molybdenum becomes soluble as thiomolybdates and molybdates and may be remobilized. In mildly acid conditions (pH 3–5), ferrimolybdite forms a common oxidation product of sulfide ores; whereas at neutral pH values lead molybdate, wulfenite, may accompany lead-zinc deposits associated with significant molybdenum concentrations.

V. Geochemical Prospecting

Mo is a particularly valuable pathfinder element for Mo ores as well as copper ores containing significant Mo because of its low and relatively constant background concentrations in most rocks, mobility under earth surface conditions, and role as a plant nutrient (Vinogradov, 1955; Vinogradov and Malyuga, 1957, Malyuga, 1964; Hawkes and Webb, 1962; Levinson, 1971; Brooks, 1972). Mo is enriched in the A horizon of soil profiles (Fig. 42-F-2); this relationship has been used to locate orebodies. Whereas Bright (1974) found a Mo dispersion halo at only 300 m from the giant Henderson molybdenum deposit in Colorado for a 20 ppm Mo threshhold, Hansuld (1967) and Tooms *et al.* (1965) found dispersion halos at 300–400 m around ores for 5 ppm anomaly threshholds, and Tauson *et al.* (1970)

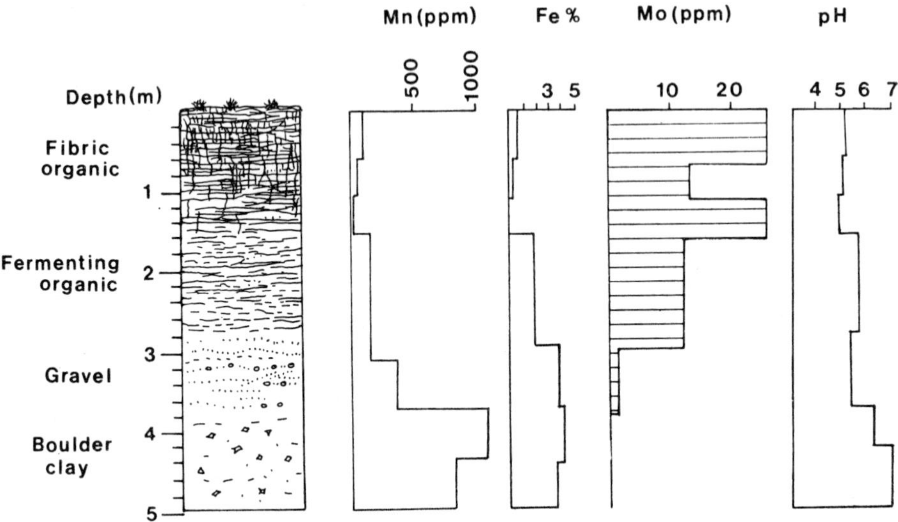

Fig. 42-F-2. Distribution of Mo, and other parameters with depth in soil profile in mineralized area, Ontario, Canada. (Modified from HORSNAIL and ELLIOTT, 1971)

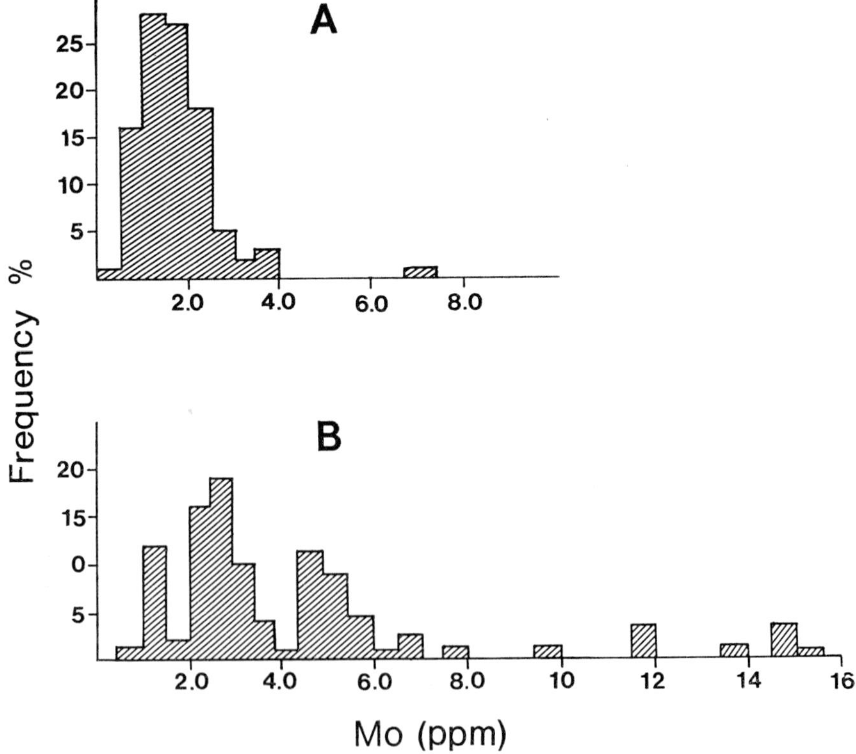

Fig. 42-F-3. Distribution of Mo in rock samples from (A) unmineralized control area and (B) Shakhtama ore field (Eastern Transbaikal), Siberia, from TAUSON *et al.* (1968)

discerned influence of ore deposits over a 17 km² area through use of several hundred precision determinations of Mo to the sub-ppm level and statistical analysis (Fig. 42-F-3). Prospecting by plant ash analysis and analysis of stream and ground waters and stream sediments has detected ore deposits at up to 10 km or even, more (CHAFFEE and TESSIN, 1970; CHAFFEE, 1975; ERICKSON and MARRANZINO, 1961; HUFF and MARRANZINO, 1961; WEBB, 1965; HUFF, 1970; NOWLAN, 1976; BELL, 1976).

Revised manuscript received: February 1978

42-G. Behavior during Weathering and Rock Alteration

I. Weathering of Magmatic Rocks

Much of the Mo incorporated in igneous rock-forming minerals is in reduced or partially reduced valence forms; i.e. Mo^{5+} or Mo^{4+}. In granitic rocks, concentrations higher than about 2 ppm may be largely attributable to MoS_2, whereas in mafic rocks excess Mo is chiefly associated with iron-containing minerals (see Sections 42-D, 42-E). Consequently, under oxidizing surficial conditions, a slow oxidation to more soluble Mo^{6+} or molybdates will take place, and net leaching is to be expected. However, differential accumulation of Mo may be anticipated by uptake on iron, manganese, or aluminium oxides. The net results may be poorly predictable, owing to the complexity and sensitivity of Mo reactions to pH and Eh conditions, as well as the heterogeneous distribution of Mo-enriched minerals, which may yield enriched weathered zones. Some typical sequences of weathering suites in Table 42-G-1 indicates the difficulty in interpreting isolated analyses.

The problem of scatter in isolated samples has been partly overcome by the availability of information from geochemical prospecting studies, where large statistical collections of data reveal regional trends not only around ore bodies, but data on analytical background over given rock sequences.

Weathering-alteration profiles for Mo-enriched hydrothermal ore deposits may be classified as in Table 42-G-2, modified from Mikhailov (1964).

Table 42-G-1. *Weathering sequences of igneous rocks.*
(From Bertine and Turekian (1973), and Bertine (1970) (Method: N/R)

Source	Description	ppm Mo
Sleeping Giant Basalt, S.G. State Park, Connecticut (U.S.A.)	Fresh gray doleritic basalt	0.64
	Weathered chloritic basalt	0.36
Hong Kong Granite	Fresh, pink granite	6.1
	Weathered, iron stained granite	4.0
	Highly weathered, kaolinitic granite	6.1
Littleton Formation, New Hampshire (U.S.A.)	Fresh medium to coarse sillimanite gneiss	2.0
	Weathered, illitic-vermiculitic gneiss	0.46
Kinsman Formation, New Hampshire (U.S.A.)	Fresh quartz monzonite	4.8
	Weathered illitic-vermiculitic quartz monzonite	3.7
Elberton, Georgia (U.S.A.)	Fresh biotite granodiorite	1.9
	Slightly weathered granodiorite	1.6
	Highly weathered granodiorite (2)	1.3, 0.8
	Red saprolite, much iron oxide	2.3

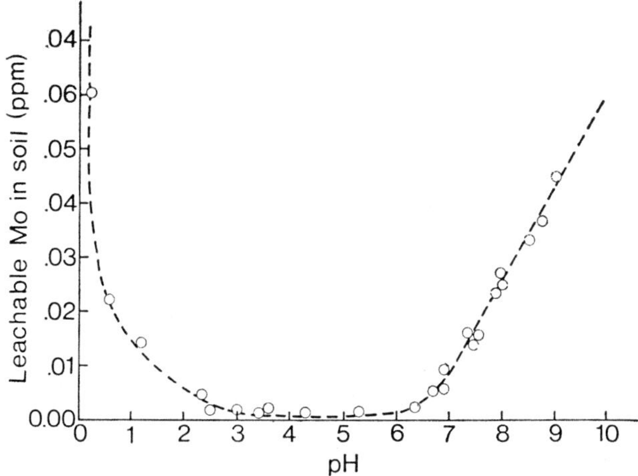

Fig. 42-G-1. Leachable Mo in Lanna soil from Sweden (redrawn from KARLSSON, 1961). Total Mo is 1.6 ppm. Leaching solutions are mixtures of HCl and KOH in the presence of 0.001 M $CaCl_2$

Table 42-G-2. *Alteration characteristics of hydrothermal deposits of molybdenum ore* (modified from MIKHAILOV, 1964)

Zones	1	2	3	4
Parent minerals	MoS_2, minor FeS_2, $CuFeS_2$	FeS_2, $CuFeS_2$, MoS_2	FeS_2, $CuFeS_2$, PbS, ZnS, MoS_2	FeS_2, UO_2, MoS_2, $CuFeS_2$, PbS, ZnS
pH range	6–8	4–7.5	2.5–6	5–8
Secondary minerals	Powellite	Molybdeniferous limonite, ferri-molybdite, ilse-mannite, Cu-Mn oxides, eosite	Molybdeniferous limonite, jarosite, wulfenite, ilse-mannite, Fe-Mn oxides, ferrimolyb-dite	Urano-Mo minerals, powel-lite, wulfenite, ferrimolybdite
Mobility of Mo	High	Precipitation and enrichment of molybdates	Enrichment of Mo in oxides	Variable

II. Soils

The molybdenum content of soils frequently resembles that of their parent rocks, and averages between 1 and 3 ppm (Table 42-G-3). The world mean of 2.0 ppm (VINOGRADOV, 1959) looks to be reasonable for total Mo concentration in soils. Soils containing large concentrations of organic matter as humus may take up appreciable amounts of Mo from ground water, especially where pH within the soils is low (below 5). Mo tends to be concentrated in humus-rich layers in the lower "A" or upper "B" horizon where pH also reaches a minimum (TIMOFEEVA, 1965). Mo has a unique position among the nutrient microelements in that it is least soluble at low pH, and molybdenum fertilization may not be successful without liming to raise pH

Table 42-G-3. *Molybdenum in soils.* (Total Mo in dry matter unless otherwise indicated. Number of samples in brackets)

Area	Range, ppm Mo	Average, ppm Mo	Anal. method	Reference
Conterminous U.S.A. (844)	<3–7	<3	S	SHACKLETTE *et al.* (1971)
Florida (citrus soils)	0.2–0.4	0.3	C	STEWART and LEONARD (1952)
U.S.A. (400)	0.6–3.5	2.0	C	ROBINSON *et al.* (1951)
Hawaii		27	C	FUJIMOTO and SHERMAN (1951)
Hawaii	trace–13		C	SHERMAN and KAPTEYN (1966)
New Zealand (8)	0.41–1.28	0.7	C	PERRIN (1946) in: KARLSSON (1961)
New Zealand (17)	0.2–9.7		C	KIDSON (1954) in: KARLSSON (1961)
Australia (500)		2.0	C	ROBINSON and ALEXANDER (1953)
Argentina (199 surface samples)	0.2–24	2.0	C	TRELLES and AMATO (1950)
Argentina (38 subsurface samples)	–8	1.5	C	TREELES and AMATO (1950)
Somerset (U.K.) (72 Mo-excess soils)	2–224	39	C	LEWIS (1943)
Thuringia (G.D.R.) (3 developed on Mo-rich alum shale)	10–30	20	S	LEUTWEIN (1951)
Aberdeenshire, Scotland (161)	1–20	2	C/S	SWAINE (1951)
Norway, Sweden	0.32–46		C	KARLSSON (1961)
Eastern Elbrus area (U.S.S.R.) (57 including soil forming rocks)		2.2	?	RZHAKSINSKAYA (1965) in: REISENAUER (1968)
Russian plain (52)	1–12	2.6	C	VINOGRADOV and VINOGRADOVA (1948) in: VINOGRADOV (1959)
Sierra Leone (lateritic latosol and red loam on amphibolite)	1–5	2.6	C/S	TOOMS *et al.* (1965)
Average from many countries (1,000)		2.0		VINOGRADOV (1959)
Average for Argentina, Russia, Scotland and U.K.	0.2–5	2.0		SWAINE (1955)

(KARLSSON, 1961). The range of solubility of Mo at varying pH in clayey soils from Sweden is illustrated by Fig. 42-G-1. High molybdenum concentrations may also be found in some arid or desert soils where much of the anomalous Mo is in a form easily extractable by plants.

Larger and world-wide compilations of trace metal information including total Mo in soils exist in the earlier literature (Table 42-G-3). In more recent time, soils are normally analyzed for "available" Mo, or that portion of the total which may be extracted by plant roots via pore solutions. Thus, Mo in plants generally follows extractable, rather than total Mo. See Table 42-G-1 and also extensive references in SWAINE (1955), VINOGRADOV (1959), KARLSSON (1961)[1], REISENAUER (undated—1968?), and references cited. See also WILLIAMS (1966), RISH (1965), SAUCHELLI (1969) and typical recent references such as BISHOP (1974) and MACLEAN and LANGILLE (1973) in the specialized agronomy literature.

[1] KARLSSON (1961) reports 0.38 ppm Mo in 0.3 M oxalate extract at pH 3.3 from 5 Swedish agricultural soils, containing 1.9 ppm Mo. A 0.05 M Ca lactate extract was as low as 0.04 ppm Mo on average.

Revised manuscript received: February 1978

42-H. Solubility and Stability of Naturally Occurring Molybdenum Species at Low Temperature and Pressure

In natural oxygenated waters at the earth's surface, the chief stable molybdenum species are those of Mo^{6+} H_2MoO_4, $HMoO_4^-$, and MoO_4^{2-}. The former, as well as amphoteric species such as $Mo(OH)_6$, $MoO(OH)_5^-$, and MoO_2^{2-} are significant only under conditions of higher acidity, such as volcanic springs or pools formed from oxidative leaching of sulfide rocks. In neutral to moderately alkaline pH ranges in sea or fresh water, the dominant phase is molybdate (MoO_4^{2-}). Higher oxypolymers begin to form significantly only at Mo concentrations of 10^{-4} molar and higher, and are theoretically distributed according to pH as follows (Pourbaix *et al.*, 1963, cited by Uzkut, 1974. The data are subject to ΔG_f corrections cited below.

	pH Range
MoO_4^{2-}	6
$Mo_3O_{11}^{4-}$	4.5–6.0
$Mo_6O_{21}^{6-}$	1.5–4.5
$Mo_{12}O_{41}^{10-}$	0.9–1.5
$Mo_{24}O_{48}^{12-}$	–0.9

Under reducing conditions, Mo^{4+} dominates, but Mo^{6+} persists at higher pH values. See Sect. 42-L for biologically significant species.

I. Thermodynamic Properties and Solubilities of Mo Species

Thermodynamic properties of selected molybdates are shown in Table 42-H-1. Many of the values on thermodynamic properties of Mo species have been compiled or derived in a monograph by Korenbaum (1970). It is important to note that many of the author's measurements are based on assumptions or fragmentary data, and that Mo data in general show unusually poor agreement. In particular, recent ΔG_f values for the critical MoO_4^{2-} ion converge on $\sim$ —200 kcal/mole, rather than data cited by Garrels and Christ (1965) and Latimer (1952). Figure 42-H-1 shows some of the basic Eh-pH relationships for molybdenum species at earth-surface temperatures.

II. Adsorption and Precipitation under Oxidizing Conditions

Adsorption, or coprecipitation, of Mo has been documented on a variety of phases: iron and manganese oxides, aluminum oxides, UO_2—MoO_4 oxide complexes, as well as with the molybdates of certain cations: Pb (wulfenite), Cu, Bi, and Ca (powellite). (See Section 42-D.) Uptake conditions are strongly influenced by pH, concentration ranges, and associated ionic components. Some apparently contradictory reports (e.g. Krauskopf, 1956; Kharkar *et al.*, 1968) may be ascribed to variable experimental conditions.

Table 42-H-1. *Solubility and thermodynamic properties of selected molybdates* (chiefly from URUSOV *et al.*, 1967). Free energy (ΔG^0), enthalpy (ΔH_f^0), and entropy (S^0), are given for 25° C. K_s is the solubility (activity) product derived from ΔG_f^0 using ion free energies from GARRELS and CHRIST (1965) and ΔG_f^0 $(MoO_4)^-$ = 199.9 kcal/m from WAGMAN *et al.* (1969). Solubilities are from ANONYMOUS (1973)

Species	ΔG_f^0 (kcal/m)	ΔH_f^0 (kcal/m)	S^0 (cal/deg)	Log K_s	Solubility (g/100 ml)
Ag_2MoO_4	–197.7 [a]	–206.6 [a]	—	–25.4	0.00386
$CaMoO_4$	–344.0	–369.5	29.3	– 8.7	0.0050 [c] to 0.013
$SrMoO_4$	–345.1	–369.8	34.7	– 8.8	0.003
$BaMoO_4$	–347.6	–372.4	36.3	–10.0	0.0055 to 0.02
$MnMoO_4$	–260.6	–284.7	31.5	– 4.6	—
	–267.5 [a]	–290.7	—	– 9.7	—
$FeMoO_4$	–231.4 [e]	–255.3	30.9	– 8.2	0.0076
	–233.0			– 9.4	—
	–234.8 [a]	—	33.4	–10.7	—
$CoMoO_4$	–224.0 [a]	–247	32.8	– 8.3	—
$NiMoO_4$	–226.5 [a]	–251.3	29.5	–11.1	—
$CuMoO_4$	–195.5	–210.8	31.5	– 8.1	0.038
$PbMoO_4$	–225.9	–249.9	39.7	–14.8	1.2×10^{-5} to $4\cdot10^{-5}$
$(Fe)_2(MoO_4)_3 \cdot 7H_2O$	–675 [b]	—	—	–27.5 –52 [d]	

[a] From references cited in KARAPETYANTS and KARAPETYANTS (1970).
[b] TITLEY (1963). A K_s closer to that of ([d]) is obtained by ignoring combined H_2O.
[c] Minimum solubility for $CaMo_4$ of 3 mg/l was found at pH 3.2 (VINOGRADOV, 1957). Widely circulated free energy values for molybdate (–218.8 and –205.4 cal/mole) do not agree with more recent data and yield $CaMO_4$ solubilities that are too large. Likewise, the important data for MoS_2 are uncertain.
[d] FOLLETT and BARBER (1967).
[e] WAGMAN *et al.* (1969).

A number of authors have shown that coprecipitation of Mo on ferric hydroxide achieves a maximum effectiveness at pH 3–4, and decreases with either higher or lower pH (JONES, 1974; ISHIBASHI *et al.*, 1962; REISENAUER *et al.*, 1962; KARLSSON, LE GENDRE and RUNNELS, 1975; KIM and ZEITLIN, 1968). The accumulation 1961; of negatively charged molybdate ions by positively charged $Fe(OH)_3$ colloids offers a ready means of precipitation of Mo.

ISHIBASHI *et al.* (1962) attributed the markedly lower concentrations of W and V than Mo in sea water (0.1, 2 and 11 μg/l respectively) to the much greater co-precipitation of V and W by iron oxides at pH ranges typical for sea water (7.8–8.3). However, as noted below and in Sect. 42-K, it appears that manganese oxides are more effective than iron in removing Mo from sea water.

Experiments by TANAKA (1958) and CHAN and RILEY (1966), showed that, whereas precipitation of Mo by manganese oxides reached maximum at pH levels

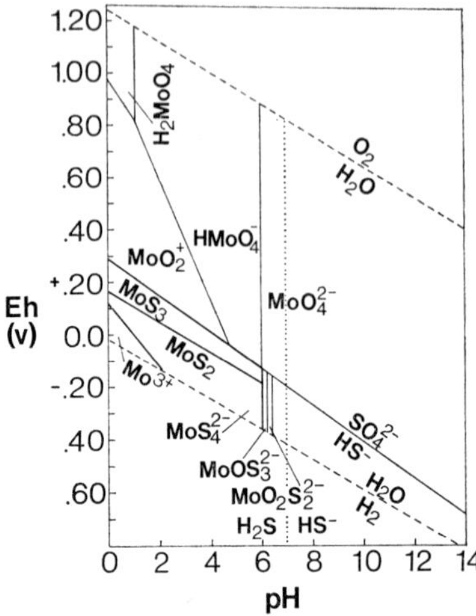

Fig. 42-H-1. Simplified diagram of stability of Mo species in the presence of sulfide. Modified from KORENBAUM (1970). Standard conditions; $\Sigma Mo = 10^{-6}$; $\Sigma S = 10^{-4}$ M. Fields for sulfomolybdic acid and its dissociation at the left of the diagram are omitted. Heavy lines designate field of stability of solid species. Note that correction for ΔG_{MoO_4} will shift fields to left

similar to those for Fe oxides (Fig. 42-H-2), its effectiveness extends to much higher pH levels. Moreover, studies of ferromanganese oxides, oxidized soils (BOYLE *et al.*, 1966), and sediments (42-K) show that Mo nearly always correlates with Mn, rather than Fe. A few notable exceptions (TROBISCH and SCHILLING, 1963) deserve special study.

The manner of uptake of Mo on manganese oxides is problematical since, at neutral pH values, the surface charge of Mn oxide colloids ought to be negative. CALVERT and PRICE (1977) have discussed this issue and cite literature to conclude that uptake of other ions may reverse the charge. In our view, substitution of Mo^{6+} for Mn^{4+} probably occurs in the MnO_2 moiety of ferromanganese oxide minerals such as birnessite, psilomelane, or todorokite.

The solubility of calcium molybdate and other cation molybdates is too great to remove significant Mo from the ocean or most surface waters; powellite ($CaMoO_4$) and molybdiferous scheelite occurrences being typical of high-temperature deposits. However, in the oxidized zone of molybdeniferous sulfide ore deposits, species such as ferrimolybdite ($Fe_2(MoO_4)_3 \cdot 7H_2O$) and wulfenite ($PbMoO_4$) are well known. According to TITLEY (1963) (see Fig. 42-H-1), ferrimolybdite is stable only at pH values less than 5.

At pH values of 4 to 5, aluminum hydroxide could serve as a collector for Mo, and at pH values greater than 9.5, precipitation of calcium and magnesium hydroxides and carbonates from sea water has been shown to remove significant Mo (KIM and ZEITLIN, 1968). However, there is no evidence that the Mo concentrations are retained through aging processes or recrystallization. Bauxites (Sect. 42-K) and brucite, as in algal brucite ($Mg(OH)_2$), (MILLIMAN, 1974) do not show significant concentrations of Mo.

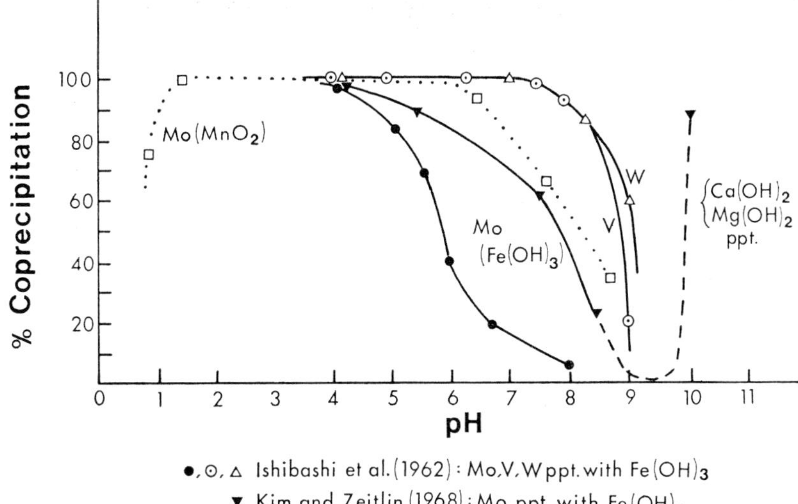

Fig. 42-H-2. Coprecipitation of Mo by manganese and iron oxides under varying pH conditions

III. Adsorption and Precipitation under Reducing Conditions

In the presence of H_2S at neutral or basic pH values, Mo forms soluble thiomolybdates in the series $MoO_2S_2^{2-}$, $MoOS_3^{2-}$, MoS_4^{2-}, by gradual replacement of oxygen in MoO_4^{2-} (Fig. 42-H-2). At pH 6 in simulated natural solutions, some yellow-amber MoS_3 is formed, but remains in colloidal solution for more than a month (Korolev, 1958). In the presence of iron (FeS), however, experiments by Korolev (1958) and Bertine (1972) showed that Mo is brought down as the trisulfide with yields up to 98%. Precipitation efficiency increases with increasing iron content and decreasing pH and *decreasing* Mo in the precipitate. Free energy calculations indicate that MoS_2 and FeS_2 should be the stable phases.

In summary, acidification of thiomolybdate or Mo—H_2S systems results in precipitation of MoS_3. The reaction can proceed at neutral solutions only in the presence of iron sulfide as a collector (see also Arutyunian, 1966). Figure 42-H-1 and equations of Table 42-H-2 illustrate the Mo—S relationships.

It is important to note that Figure 42-H-1, modified from Korenbaum (1970), represents relative positions of stability fields only; a ΔG value for MoO_4^{2-} intermediate between that utilized by Korenbaum (1970) and the value of Wagman *et al.* (1969) used in the equation of Table 42-H-2 was employed to calculate thiomolybdate boundaries (see the technical report 1960, cited in Garrels and Christ, 1965).

Unlike vanadium and nickel, no molybdenum porphyrin compounds have been identified in natural environments, nor have other specific Mo-organic complexes yet been described, with exception of the heteropolyanions referred to in Sect. 42-L. *Sorption of Mo on peat* and other organic matter has been ascribed to reduction of

Table 42-H-2. *Reactions of potential importance in low-temperature natural aquatic systems* including data on G_r^0 energies and reaction constants)

Reaction						Data
$HMoO_4^-$		$\rightleftarrows H^+$	$+ MoO_4^{2-}$			$\Delta G_r^0 = 5.1$ $K = 10^{-3,7}$
$MoO_2S_2^{2-}$	$+ 2H_2O$	$\rightleftarrows MoO_4^{2-}$	$+ 2HS^-$	$+ 2H^+$		$\Delta G_r^0 = 6.31$ $K = 10^{-4,6}$
$MoOS_3^{2-}$	$+ H_2O$	$\rightleftarrows MoO_2S_2^{2-}$	$+ HS^-$	$+ H^+$		$\Delta G_r^0 = 4.51$ $K = 10^{-3,3}$
MoS_4^{2-}	$+ H_2O$	$\rightleftarrows MoOS_3^{2-}$	$+ HS^-$	$+ H^+$		$\Delta G_r^0 = 7.41$ $K = 10^{-5,4}$
MoS_3	$+ HS^-$	$\rightleftarrows MoS_4^{2-}$	$+ H^+$			$\Delta G_r^0 = 74.8$ $K = 10^{-59,8}$
MoS_3	$+ 4H_2O$	$\rightleftarrows MoO_4^{2-}$	$+ 3HS^-$	$+ 5H^+$		$\Delta G_r^0 = 96.0$ $K = 10^{-70,4}$
MoS_3	$+ 4H_2O$	$\rightleftarrows HMoO_4^-$	$+ 3H_2S$	$+ H^+$		$\Delta G_r^0 = 68.6$ $K = 10^{-50,4}$
MoS_2	$+ HS^-$	$\rightleftarrows MoS_3$	$+ H^+$	$+ 2e^-$		$\Delta G_r^0 = 0.59$

$$Eh = 0.0128 + \frac{0.059}{2} \log \frac{(H^+)}{(HS^-)}$$

MoO_4^{2-} to a cationic form (MoO_2^+); whereas only 6% uptake of molybdate in distilled water was reported by KHARKAR *et al.* (1968), 50% adsorption occurred in one hour at pH 4, and 90% adsorption was found at pH 1 (SZILAGYI, 1967; BERTINE, 1972). This pH-sensitive behavior has implications for occasional high concentrations of Mo in peat, lignite and coal, resins and amber.

Revised manuscript received: February 1978

42-I. Natural Waters and Atmospheric Precipitation

Major contributions toward clarifying the distribution of Mo in natural waters have been made by Japanese workers studying Mo in hot springs (KURODA, 1939, 1940a; KIMURA *et al.*, 1955); ocean waters (ISHIBASHI *et al.*, 1955; SUGAWARA and OKABE, 1960; and lake and stream water and atmospheric precipitation (SUGAWARA *et al.*, 1961; SUGAWARA, 1967; and publications cited in these works). Mean Mo values are: rivers 0.5 μg/l, ocean 11.0 μg/l, brines 1.7 μg/l, and thermal waters 20μg/l.

I. Fresh Surface Waters and Atmospheric Precipitates

In Japan, the Mo concentration of *atmospheric precipitates* ranges from 0.02 to 0.11 μg/l with a mean of 0.06 μg/l (300 samples), according to SUGAWARA (1967). Snow samples analyzed by SUGAWARA *et al.* (1961 Method: C), are somewhat higher in concentration (0.12–0.30 and 0.20 μg/l; range and mean for 4 samples). The Mo in Japanese atmospheric precipitation was largely from windblown dust, and incorporates minimal industrial pollutant input. A partial estimate of the solid phases, after removing estimated sea salt contributions, yields the following components: SiO_2: 42%, Al_2O_3: 5.0%, Fe_2O_3: 7.8%, MgO: 9.8%, CaO: 23.2% and Mo: 14 ppm. Low-boiling Mo-halogen compounds of volcanic origin are virtually excluded owing to the instability and hydrolysis of these species in the presence of water (see data of NEMOTO *et al.*, 1957).

The concentration of Mo in *streams* from regions relatively uninfluenced by industrialization ranges from 0.1 to 1 μg/l, with an estimated mean of 0.5 μg/l (Table 42-I-1). Highest concentrations (10 μg/l) were obtained in rivers draining arid regions, such as the Amu Darya and Syr Darya (Aral Sea basin), and Kura and Terek (Caspian Sea basin) (KONOVALOV *et al.*, 1968). However, many of these rivers have relatively low runoff and discharge into interior basins. Much more significant contributions appear to arise from industrialization-urbanization, including liming and molybdenum fertilization of soils, monoculture of leguminous molybdenum-rich plants such as alfalfa and soy beans, chlorination of municipal waters, and unknown sources. Thus, total dissolved runoff of Mo to the oceans has nearly doubled in recent time (Table 42-I-1).

Contribution of Mo to the oceans by particulate detritus in rivers and streams is less certain than the dissolved fraction, due to the lack of data on the composition of particulates in major rivers such as the Hwang Ho and only a few isolated data on others such as the Mississippi, and also due to lack of knowledge of anthropogenic and industrial influences (MEADE, 1969; TUREKIAN, 1971). It is clear that the mean of 7.4 ppm in U.S. river particulates (BERTINE, 1970; BERTINE and TUREKIAN, 1973) is affected by human influences. Very high estimates for Mo in river particulates (ASTON and CHESTER, 1977) do not agree with data reported here. Based on the 1.2 ppm Mo in suspended matter of the Black Sea drainage basin (VOLKOV, 1975), total sediment transport of 18.5×10^9 tons and world surface runoff of 35.5×10^{15} tons (LISITSIN,

Table 42-I-1. *Mo in rivers, streams, lakes, and miscellaneous waters.* (Data in parentheses refer to total dissolved solids and total particulate matter in respective columns, in ppm)

Source	Dissolved Mo (μg/l)		Suspended Mo (μg/l)	Analyt. method	Reference
	Range	Average			
Rivers					
Amazon (Brazil)	0.41–0.44	0.42 (34)	4[a]; 0.7	N/R S	BERTINE (1970) MARTIN *et al.* (1973) MARTIN and MAYBECK (1976)
Congo (Africa)	0.36–0.52	0.44 (242)	4[a]	S	MARTIN *et al.* (1973) MARTIN and MAYBECK (1976)
Rhone, Pont Avignon (France)		5.03 (94)	1.8, 6.1[a] (296)	N/R	BERTINE (1970)
Rhine			9[a]	S	MARTIN *et al.* (1973)
Po, N.W. Ferrara (Italy)	1.3 –1.10	1.20 (210)	—	N/R	BERTINE (1970)
Mekong			2	S	MARTIN *et al.* (1973) MARTIN and MAYBECK (1976)
French rivers (5)			4	S	MARTIN *et al.* (1973)
Mississippi, Baton Rouge, Louisiana (U.S.A.)		<1.0 (221)		W/S	DURUM and HAFTY (1963)
Mississippi, Arkansas (U.S.A.)		2.44	0.94; 5.1[a] (185)	N/R	BERTINE (1970)
Susquehanna Marietta, Pa. (U.S.A.)		0.71 (187)	0.84; 16[a] (54)	N/R	BERTINE (1970)
Brazos, Highway 54, Texas (U.S.A.)		0.68 (189)	2.6; 2.7 (954)	N/R	BERTINE (1970)
Colorado, near Fruita (Grand Junction), Colorado (U.S.A.)		26 (∼1,500)			VOEGELI and KING (1969)
Colorado, Yuma, Arizona (U.S.A.)		6.9 (853)		W/S	DURUM and HAFTY (1963)
Columbia River below Dulles Dam, Washington (U.S.A.)		2.1 (191)		W/S	DURUM and HAFTY (1963)
Japanese rivers (10)	0.2 –0.5	0.32		C	SUGAWARA *et al.* (1961)
Japanese rivers (43)		0.60 (34)		C	SUGAWARA (1967)
Rivers of Black Sea drainage (5)	0.1–2.5	0.7	1.2	C	VOLKOV (1975)
Principal rivers, U.S.S.R.[c] (55)	0.2–10.6	1.1	0.1	C	KONOVALOV *et al.* (1968)

Table 42-I-1 (continued)

Source	Dissolved Mo (µg/l)		Suspended Mo (µg/l)	Analyt. method	Reference
	Range	Average			
North American rivers		0.35		W/S	DURUM and HAFFTY (1963)
U.S. rivers (10)	0.26–2.44	1.0 (223)		N/R	BERTINE (1970)
U.S. rivers (20)			0.95, 7.4 [a] (129)	N/R	BERTINE (1970) BERTINE and TUREKIAN (1973)
Patuxent River basin, Maryland (32 samples)	0.2–1.0	0.6 [d] (78)		S	HEIDEL and FRENIER (1965)
U.S. rivers (est.) [b]		0.68			HEM (1970)
Estimated world rivers (prior to industrial inputs) [c]		0.5	0.62		This report
Estimated world rivers (current), adjusted in volume		0.87	≥1.0		This report
Lakes etc.					
Castle Lake, California		<0.1		C	BACHMANN and GOLDMAN (1964)
Lake Superior		1.2 (94)	—	N/R	BERTINE (1970)
Japan (10)	0.05-1.2	0.49	—	C	SUGAWARA *et al.* (1961)
Maine (U.S.) (459)	<0.05–30	0.22 [f]	—	S	KLEINKOPF (1960)
Public water supply of the 100 largest U.S. cities	< 0.1–68	3.1 [g] (215)	—	S	DURFOR and BECKER (1965)

[a] In ppm of solid matter.
[b] Sum of inorganic constituents reported (excluding bicarbonate).
[c] Detectable soluble Mo (>0.2 µg/l) was found in 13 of 55 Soviet rivers, and particulate Mo (>0.1 µg/l) was found in 5 of 55 samples. No adjustment made for volume.
[d] Higher values, probably attributable to sea water admixture, occur where salinity increases significantly.
[e] The estimates of average Mo were made assuming undetectable =0. Takes into account the fact that the highest Mo values are obtained from rivers having relatively low discharge.
[f] Calculated assuming that <1–0.5.
[g] Mean of 3.1 µg/l Mo for 200 analyses. Water treatment (chlorination) and filtration systems may add more to Mo content than natural sources.

1974), we estimate a total suspended contribution of 0.62 µg/l Mo, roughly equivalent to the dissolved contribution.

Lakes show molybdenum values roughly comparable to those of streams, and are influenced by the Mo concentrations of surrounding soils. Thus, among the lakes

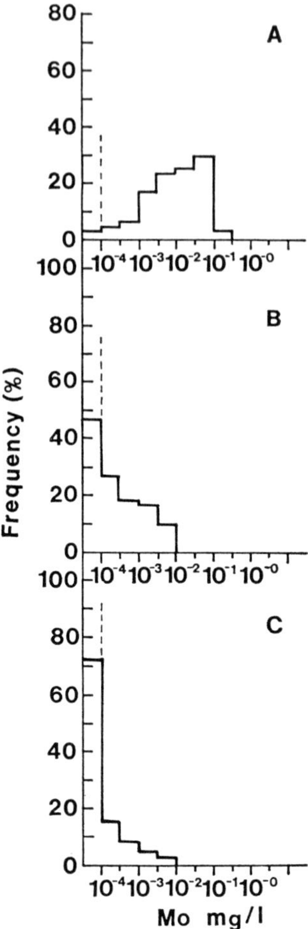

Fig. 42-I-1 A–C. Distribution of Mo in natural waters from Bulgaria (PENTCHEVA, 1967). Several hundred analyses are included in (A) Nitrogen-rich thermal waters; (B) Saline waters with concentrations up to 265 g/l total dissolved solids; and (C) Fresh surface and ground waters. Method: W/S

studied by SUGAWARA *et al.*, 1961 (Table 42-I-1), Lake Kizaki-ko with 0.4–0.6 μg Mo/l drains soils with 0.5 to 1.8 ppm Mo, whereas Lake Aoki-ko with 1.2 μg/l Mo, drains soils having 3.5 to 9.5 ppm Mo.

In meromictic (stratified) lakes, Mo may increase in stagnant deeper layers. This is explained by SUGAWARA *et al.* (1961) as being due to redissolution of Fe—Mn—Al hydroxide floccules at depth. However, where stagnation is accompanied by formation of H_2S and ferrous sulfide, Mo is sharply decreased in lower layers by capture in bottom sediment.

The mobility index (VISTELIUS and SARMANOV, 1947)[1] of Mo in surface waters of Bulgaria has been calculated for surface waters of Bulgaria by PENTCHEVA (1967).

[1] Mobility index is given as $\frac{A}{B} \cdot 10^{-4}$ where A is concentration in water, and B is ppm Mo in the earth's crust.

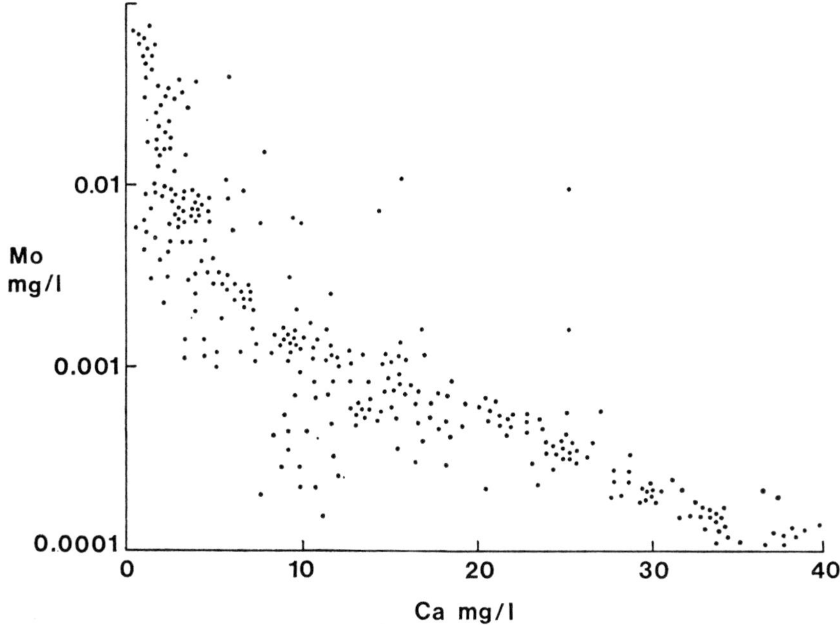

Fig. 42-I-2. Distribution of Ca and Mo in nitrogen-rich thermal ground waters from Bulgaria (PENTCHEVA, 1967)

She found a value of 0.55, higher than all other metals except (some) alkali and alkali earths (See also Fig. 42-I-1).

Thus, with the possible exception of Re, whose solubility as the rhenate, ReO_4^{2-} exceeds that of MoO_4^{2-} (LEVINSON, 1971), Mo is probably the most mobile of all metallic elements aside from alkalis and alkaline earths; this behavior helps account for its preferential accumulation in ocean water (see below).

Where waters are associated with molybdenum ores or molybdenum-enriched deposits, concentrations may be much higher than in normal surface waters. Thus, waters in the vicinity of the Climax molybdenum deposit and other Colorado-Utah-Rocky Mountain deposits reach 3,800 µg/l (VOEGELI and KING, 1969), and waters draining beneficiating plants may be even higher in Mo (RUNNELS *et al.*, 1975). See also 42-L for discussion of environment and toxicity factors.

II. Ground Waters

Normal ground waters having low saline content, resemble surface waters in molybdenum content, except in mineralized areas. This conclusion is based chiefly on studies of well and other waters during geochemical prospecting by hydrochemical analysis. Note also Fig. 42-I-1.

Under *thermal conditions*, Mo increasingly tends to resemble tungsten, uranium, germanium, and vanadium in its behavior; all of these elements are characterized by anionic mobility as molybdates and polyions (but not thiomolybdates and halogen

Table 42-I-2. *Mo in selected hot springs and other thermal waters.* Temperatures at discharge point, unless otherwise indicated

Site and remarks	T° C	Total diss. solids (g/l)	Range, µg Mo/l	Avg., µg Mo/l	Analyt. method	Reference
Haukadalur, Iceland	100 (>180 ?)[a]	1.0	60	3	S	White (1965)
Hveravellir, Iceland	90.5	1.1		30	S	White *et al.* (1963)
Amedee Hot Spring, California (U.S.A.)	93	0.8		60	S	White (1965)
40 Geysers, Yellowstone Park, Wyoming (U.S.A.)	90	0.50–2.5	<1.7–70	30	S	Rowe *et al.* (1973)
Old Faithful, Yellowstone Park, Wyoming (U.S.A.)	90	1.37		40	S	Rowe *et al.* (1973)
Arima, Japan	94	0.072		60	C	Ikeda (1955)
Geothermal brine well, Imperial Co., Calif. (U.S.A.)	~200 (>300)[a]	0.33		<50		White (1965)
300 Bulgarian N-rich thermal waters (see Fig. 42-I-1 and 2)	55 (mean)	1.0	0.05–250	35	CS	Pentcheva (1967)
Geothermal waters, North-Central Nevada (U.S.A.)					NA	Bowman *et al.* (1974)
Leach	"cold"			0.2		
Beowawe	78		0.3			
Beowawe	200–250[a]		10–20	15		
Brady	"hot"		0.10–14			
Seasonal variations at Yunohanazawa thermal spring; pH 2.6–2.9	35–51.6	0.8		7.0	C	Kuroda (1940)[a]
		0.11		8.0		
		0.18		14.0		
		0.7		10.0		
		0.13		16.0		
		0.7		10.0		
Mean		0.44		10.8		
32 Japanese warm springs in volcanic rocks	11.5–31.7			25.0	C	Oana and Ikeda (1955)

[a] Original temperature (calculated).
[b] Semiquantitative.
[c] Waters are alkaline with pH to 10.5, of sodium sulfate-bicarbonate type with total dissolved solids to 1 g/l; variable nitrogen, methane and occasionally H_2S; high silica and fluorine concentrations (from Pentcheva, 1967). See also text.

complexes) at higher temperatures. Hence, Mo is enriched in thermal fluids (Table 42-I-2). Note also inverse relationship between Ca and Mo (Fig. 42-I-1; PENTCHEVA, 1967).

Only a few data are available on Mo in saline ground waters and oil field brines, and we may accept PENTCHEVA's (1967) mean of 1.7 μg/l Mo for Bulgarian brines, ranging to 265 g/l in total dissolved salt, as a magnitude. Interstitial waters from the metalliferous sediments of the Red Sea hot brine area showed 30–300 μgMo/l (HENDRICKS *et al.*, 1969), but these may possibly have been affected by leaching of Mo from very labile and Mo-rich host sediments during sample extraction. Interstitial waters analyzed by PILIPCHUK and VOLKOV (1974 and references cited) are subject to similar considerations.

III. Ocean and Sea Water

Data from authors whose methods are tested and/or show reproducibility and internal consistency vary between 9 and 13 μg/l Mo for open ocean waters (Table 42-I-3). We have chosen 11 μg/l as an average. Mo behaves like a conservative constituent with respect to local evaporative increases in salinity (e.g. SUGAWARA *et. al.*, 1961; Pacific Ocean) and dilution in estuarine waters (e.g. United Kingdom, HEAD and BURTON, 1970; British Columbia, BERRANG and GRILL, 1974; Japanese bays, SUGAWARA *et al.*, 1961). However, JONES (1974) reported relative increases in Mo/S ratios that he interpreted as due to desorption of Mo from river-borne particulate matter near the coast of Wales. Similar relationships appear in data of SUGAWARA and OKABE (1960) from Japanese bays.

MANKOVSKAYA and SOVGA (1973) found a systematic decrease in Mo in the lower parts of Caribbean Sea profiles. The Black Sea as a whole is even more strongly

Table 42-I-3. *Mo concentrations in oceanic and other marine waters* (selected data). Numbers of samples are indicated by parentheses

Site	Range, μg Mo/l	Average, μg Mo/l	Analyt. method	Reference
Pacific Ocean				
Off Japan (surf)	10–11	10.6	C	ISHIBASHI *et al.* (1955)
N. Pacific (21)	8.9–11.6	9.9[a]	C	SUGAWARA and OKABE (1960)
Central Pacific (30)	9.8–11.6	10.3[a]	C	SUGAWARA and OKABE (1960)
Off San Francisco		12.2	C	WEISS and LAI (1963)
Off Hawaii		12.0	C	KIM and ZEITLIN (1968)
Off New Zealand	10.8–11.0	11.0	C/A	KIM and ALEXANDER (1976)
Indian Ocean (5)	11.4–12.2	11.5[a]	C	SUGAWARA and OKABE (1960)
Antarctic				
(Southern Ocean) (5)	9.8–10.6	10.2[a]	C	SUGAWARA and OKABE (1960)
Marginal Seas				
Irish Sea		10.3	C	CHAN and RILEY (1966)
Irish Sea		11[a]	C	JONES (1974)
Caribbean Sea (60)	12.5–13.5	13.2	C	MANKOVSKAYA and SOVGA (1973)

[a] Adjusted for salinity of 35‰.

depleted in Mo, both in upper oxygenated waters $\frac{Mo(\mu g/l)}{Sal(o/oo)} = 0.14\text{–}0.26$ and in the lower anoxic (H_2S) layers (Mo and Mo/Sal = 2–1 μg/l and 0.11–0.05 respectively (PILIPCHUK and VOLKOV, 1974) contrasted with 11 μg/l and 0.32 for the open ocean. The Mo depletion is caused by coprecipitation of Mo with FeS and incorporation in bottom sediments of the Black Sea (PILIPCHUK and VOLKOV, 1974). Similar depletion is expected in other stagnant water bodies, including coastal fjords and certain other areas (BERRANG and GRILL, 1974).

Additional data on Mo in marine and brackish waters are reported in ERNST and HÖRMANN (1936), BARDET *et al.* (1938), BLACK and MITCHELL (1952), ZHAVORONKINA (1960), YOUNG *et al.* (1959), BROOKS (1965), BACHMAN and GOLDMAN (1966), BELYAEV (1966), KOGAN (1967), PILIPCHUK and VOLKOV (1967), PILIPCHUK (1972), OVSYANII (1972), SKOPINTSEV (1975), and MAY *et al.* (1975).

IV. Ocean Budget

Residence times for Mo in the ocean have been calculated by a number of authors, and range from 2×10^5 years (TUREKIAN, 1971) to 7×10^5 (GOLDBERG *et al.*, 1971). See also SUGAWARA (1968), WEDEPOHL (1968), BERTINE and TUREKIAN (1973). Our revised estimates are 7.8×10^5 for pre-industrial ocean, and 13.5×10^5 years at present, including industrial input to rivers. The data are based on ocean Mo = 11 μg/l, stream Mo (pre-industrial) = 0.5 μg/l, and volume of ocean and stream discharge, respectively 1.37×10^{24} g and 38.8×10^{18} g/year, including subsurface discharge to the oceans (L'VOVICH, 1973, cited in LISITSIN, 1974).

Revised manuscript received: February 1978

42-K. Abundance in Sediments and Sedimentary Rocks

More geochemical information exists on the concentration of Mo in sediments and sedimentary rocks than in any other group of rocks. This is so because the metal is strongly enriched in two disparate regimes. These are oxidizing conditions characterized by the precipitation of manganese oxides, and highly reducing conditions accompanied by accumulation of organic matter and precipitation of iron sulfide. Moreover, though generally below a concentration of about 3 ppm, Mo in soils has received much attention owing to its dual role as an essential plant nutrient and potential toxic substance.

Table 42-K-1. *Mean concentration of Mo in sedimentary materials*

	ppm Mo
Soils	2.0
Continental and shallow marine sediments and sedimentary rocks (non-anoxic environments):	
Recent sediments	1.5
Shales and sandstones	
Shales	0.7–2.0
Sandstones	0.3
Carbonates	0.4
Deep sea pelagic sediments:	
Altantic	5
Pacific and Indian Oceans	4–18
Ferromanganese concretions and ores:	
Freswater	26
Brackish and shallow marine	114
Deep sea	380
Iron ores	35
Manganese oxide ores	200
Anoxic sediments and black shales:	
Recent deposits	26
Black (bituminous) shales and other sediments	70
Phosphorites	11

I. Continental and Shallow Marine Sediments and Sedimentary Rocks

a) Recent Sediments

Freshwater sediments of lakes and streams strongly reflect the composition of their surrounding soils and rock particles eroded from local terrains. They consequently approach earth crustal values (1 ppm) in the absence of significant manganese oxide enrichment, organic accumulations, or influence of local ore deposits. Sediments transported to the sea by streams have been suggested to lose, rather than gain, molybdenum (JONES, 1974). See also Section 42-I for the composition of particulate matter in streams and Section 42-H for discussion of Mo uptake. Data on recent non-anoxic sediments (Fig. 42-K-1) include the following sources: SUGAWARA *et al.* (1961), LANDERGREN and MANHEIM (1962), NOWLAN (1976), CRECELIUS (1969), MANHEIM (1961), EMELYANOV (1976), EMELYANOV and POSTEL'NIKOV (1976), VOLKOV and SEVASTYANOV (1968), ISAEVA (1960), STRAKHOV and NESTEROVA (1968 and LISITSINA *et al.* (1975).

b) Sedimentary Rocks

In contrast to the profusion of data on bituminous sediments, few special investigations report accurate and useful data on "normal" terrigenous clays,sands, or chemical sediments such as carbonates or evaporites. In compiling averages, we have tended to avoid isolated high Mo values unless the data are particularly well documented, in view of the known problems of dealing with data close to the detection limit of analytical techniques. HIRST (1974) showed that parallel methods (X-ray fluorescence and optical spectrography) may give values differing by a factor of 3 at low Mo levels. Another problem is how much importance to attach to analyses of composites that may represent a very large number of samples.

Extraordinary samples of this kind are the two composites: 7,614 shales with 0.74 ppm Mo, and 6,107 sandstones with 0.4 ppm Mo from the Russian Platform (VINOGRADOV *et al.*, 1958). These data were cited as a footnote, with no further explanation in a classical paper devoted mainly to Mo and W in igneous rocks. We know the authors as very competent in geochemical judgment and their detection limit for Mo was one of the lowest yet obtained (<0.1 ppm). Yet these values are significantly lower than data from dozens of other studies including composites from common sediments.

A similar example is provided by the authors of another outstanding body of information, KURODA and SANDELL (1954). Their composite of Japanese Paleozoic shales at 0.5 ppm does not agree with the European composite of 5 ppm; but the latter contains several bituminous shales. Here, we are faced with a possible problem of representativeness, for a single shale with 100 ppm Mo could raise the average of 36 samples otherwise containing 1 ppm Mo to nearly 4 ppm. Questions of this kind can only be resolved with new data by sensitive and accurate methods, and with appropriate strategies to solve the problems of representativeness in space and time. In the interim, no better than a range of values can be offered (0.7–2 ppm).

We choose 0.3 and 0.7 ppm respectively for sandstone and graywacke averages. Shales are a much more difficult question. For many elements, shale concentrations balance those of average igneous rocks (WEDEPOHL, 1968). Weathering and leaching of Mo-sulfide or organic-associated Mo-enriched phases ought to leave largely detrital residues in normal shales, and hence Mo of about 1 ppm. Yet many higher values are present in the literature. Some are probably due to analytical over-statement; others due to sampling or inclusion of organic-rich specimens in averages. Also, thin Mo-enriched shelf sediments (VINE, 1966) are over-represented in the literature with respect to thick Mo-poor geosynclinal sequences.

Other uncertainties remain in the proportion of black (bituminous) shales, with an estimated average Mo concentration of 70 ppm, to total shales. The proportion of pelagic sediments to total world sediments, and the suggestion that significant portions of the oceans may be floored by metal-rich sediments representing stagnation regimes in the Atlantic and elsewhere (e.g. LANCELOT *et al.*, 1972; EMERY and UCHUPI, 1974; TUCHOLKE and MOUNTAIN, 1977) is another problem. In an illustrative calculation, if the proportion of carbonates, sandstones plus graywackes and arkoses, and shales are 8%, 15%, and 77%, respectively (WEDEPOHL, 1968), and Mo for the three phases is given at 0.4, 0.46, and 1.0 ppm, respectively, then total sedimentary rocks (less deep ocean and bituminous rocks) would be 0.89 ppm Mo. At an average of 70 ppm, the percentage of black shales as a portion of total shales, needed to bring sedimentary rocks to 1.0 ppm Mo, would be 0.2%; for 1.2 ppm, it would be 0.6%.

Carbonates tend to converge on 0.4 ppm Mo, with a significant part of this value attributable to clays, organic-related phases, and other extraneous matter.

II. Pelagic Sediments

Deep sea sediments are distinct from "normal" terrigenous and chemical sediments in that they: (i) tend to be enriched in a characteristic suite of metals: Mn(Fe), Ni, Cu, Co, Mo, Pb, Zn and others: (ii) are generally slowly deposited under oxidizing conditions; (iii) have low organic carbon content; and (iv) are rarely represented in fossil sediments and sedimentary basins. Virtually all the *excess* Mo in terrigenous sediments is associated with ferromanganese oxides, and with Mn rather than with Fe (CRONAN and TOOMS, 1969; CRONAN, 1969a; BEZRUKOV and ANDRSHCHENKO, 1974; GLASBY *et al.*, 1974; CALVERT and PRICE, 1977; and references cited in these publications). The ferromanganese oxides in sediments may be in the form of micronodules (EL WAKEEL and RILEY, 1961), grain coatings and replacements, and finely disseminated phases.

CRONAN (1969a) indicated that the relationship between Mo and Mn is stronger in ferromanganese concretions than in sediments, but that the covariance for Mo/Mn is significant, and greater than for all other elemental combinations except Ni/Mn. CALVERT and PRICE (1977) reported relationships for Mo/Mn in Pacific pelagic sediments resembling those for "δMnO_2" in nodules. Their regression equation for sediments is Mo (ppm) $= 20.7 \times$ Mn(%).

Data for Atlantic pelagic clays tend to converge on 5 ppm Mo (Fig. 42-K-1). The 1.6 ppm Mo for 80–90% pure carbonate ooze (BERTINE and TUREKIAN, 1973) appears to represent a good value for Atlantic biogenic deposits.

Data on molybdenum concentrations in Pacific pelagic sediments (Fig. 42-K-1) are more divergent, with ranges from 3 to 45 ppm. Part of the differences can be resolved by taking into account manganese and micronodule concentrations, but more work is clearly needed to investigate whether the differences represent real phase and regional variations or analytical differences. Based on a regression line on ferromanganese concretions of Mo ppm $= 20 \times$ Mn(%), pelagic clay with an Mn concentration of 0.7% would have a Mo concentration of 14 ppm. In contrast, a number of Soviet investigations (SKORNYAKOVA, 1976; GLAGOLEVA *et al.*, 1975; LISITSINA *et al.*, 1975; VOLKOV *et al.*, 1976) and some Western workers have obtained values of approximately half this range. Curiously, the analytical data agree much

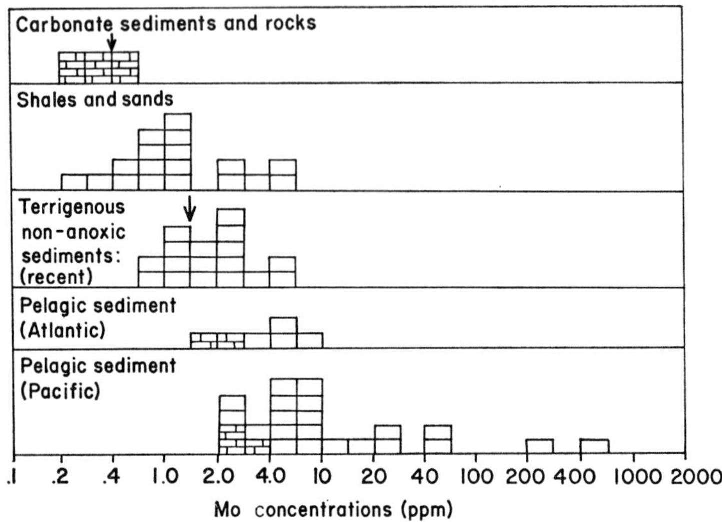

Fig. 42-K-1. Distribution of Mo in sediments and sedimentary rocks from oxidizing or presumed oxidizing environments. In general each box represents an average of values from a given area. Midpoints between values on scale are, respectively, 1.5, 3, 15, 30, etc. Values falling exactly on boundary line are assigned alternately to either higher or lower side. Arrows refer to arithmetic means of data

References to recent non-anoxic sediments and pelagic sediments are given in the text. References to shales and sandstones: Rader and Grimaldi (1961), Vinogradov *et al.* (1958), Kuroda and Sandell (1954), Degens *et al.* (1957), Zav'yalov (1965), Strakhov *et al.* (1968), Rekharskii and Krutetskaya (1960), Studenikova and Glinkina 1964, 1967); references to carbonates: Chichilo and Whittaker (1958 cited in Graf, 1960), Blumer and Erlenmeyer (1950).

better for ferromanganese concretions than for pelagic sediments. The data shown in Fig. 42-K-1 for pelagic sediments include: (Atlantic) Landergren and Manheim (1962); Wedepohl (1960); Bertine (1970; Bertine and Turekian (1973) and Lange (1974). (Pacific)—Kuroda (1940b); Kuroda (1943); Skornyakova (1976); Goldberg and Arrhenius (1958); Wedepohl (1960); Landergren and Manheim (1962); Landergren (1964); Lange (1974); Calvert and Price (1977); Lisitsina *et al.* (1975); Glagoleva *et al.* (1975); Cronan (1969a, b); El Wakeel and Riley (1961); Swanson *et al.* (1967); Boström and Peterson (1969); Volkov *et al.* (1976) and Strakhov (1976). (Indian Ocean and miscellaneous sources not included in Fig. 42-K-1) — Pushkina (1971); Strakhov *et al.* (1968); Baturin *et al.* (1967); and Pilipchuk (1972).

III. Ferromanganese Concretions and Ores

The larger relationships of Mo to *iron oxides* are clear from studies cited in 42-H, and empirical data (Fig. 42-K-3). $Fe(OH)_3$ is a powerful *adsorbent* and *precipitant* of Mo at pH values between 2.5 and 4 (Fig. 42-H-1), but Mo is lost at higher pH and on

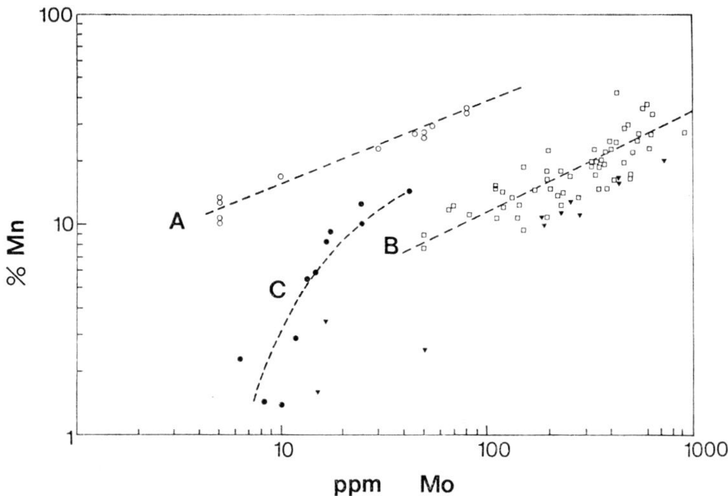

Fig. 42-K-2. Relationship of molybdenum and manganese in ferromanganese concretions from various environments. (A) Estuarine nodules, Loch Fyne, Scotland (CALVERT and PRICE, 1970); (B) Open ocean nodules from Marsden Squares 530 to approximately 400 (southern Pacific, Atlantic, and Indian Ocean, S. of 25° S. latitude), compiled in MONGET *et al.* (1976). Samples marked "x" refer to phosphatic nodules, chiefly around Agulhas Bank, southern Africa; (C) Black Sea nodules (SEVASTYANOV and VOLKOV, 1967)

aging of oxide phases. Thus, sedimentary oxides such as goethite and hematite are relatively low in Mo, as has been shown by LANDERGREN (1948), JAMES (1966), STRAKHOV *et al.* (1968) and FRIETSCH (1970). Magnetite is generally a metastable phase in sediments, and occasional higher concentrations of Mo, along with other group IV–VI elements such as Sn and W, usually reflect a Mo- and halogen-enriched province in which magnetites crystallized under higher temperature and pressure regimes. The consistent absence of, or negative correlation of Mo with Fe in sediments and oxides suggests that iron phases are not the primary limitors of Mo concentration in the ocean as suggested by ISHIBASHI *et al.* (1962) and HEM (1977). Occasional high concentrations in iron-rich materials require further explanation.

A much different relationship is found for molybdenum and *manganese oxides*. Although maximum precipitation of Mo with MnO_2 is obtained at pH 3.5 (Fig. 42-H-1), Mn oxides retain effectiveness for Mo uptake at higher pH levels than do Fe oxides. Unlike iron, Mn and Mo consistently show significant degrees of correlation in sediments and oxide phases as noted by the correlation coefficients in Table 42-K-2. These correlations apply also to microlayers.

Manganese oxides of carbonate origin, such as the massive South Ukrainian deposits of Nikopol' and Chiatura tend to reflect the low Mo concentrations in carbonates including rhodochrosite and siderite, plus minor uptake from ambient surface and ground waters.

In contrast, oxides precipitated from solutions in primary form may show values up to 1,500 ppm Mo. Owing to the enrichment of Mo in fluids associated with rocks at higher temperature (42-I; Table 42-I-2), we would normally anticipate higher Mo

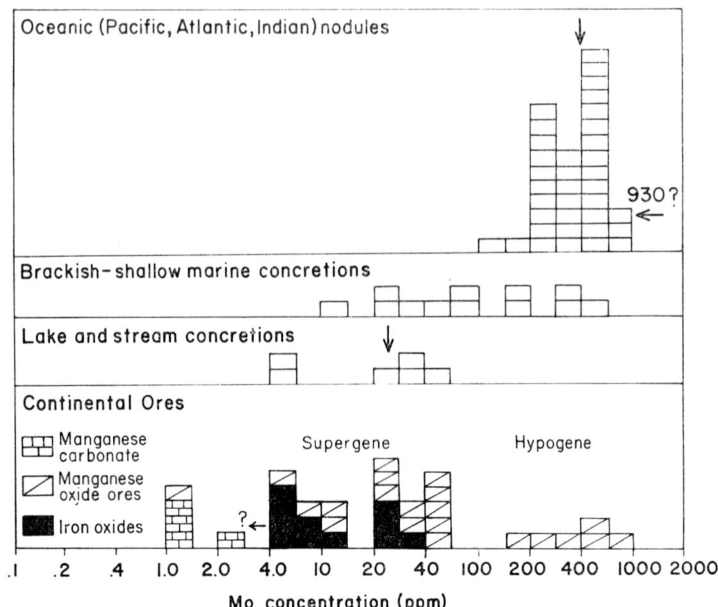

Fig. 42-K-3. Distribution of Mo in ferromanganese concretions and ores. Notes on plotting as in Fig. 42-K-1

References include: fresh water bodies—ENGEL and SHARP (1958), LAKIN *et al.* (1963), GORHAM and SWAINE (1965), MANHEIM (1965), ROSSMAN (1973, cited in CALVERT and PRICE 1977), DEAN (1970), NOWLAN (1976); brackish and shallow marine—SEVASTYANOV and VOLKOV (1967), CALVERT and PRICE (1977), STRAKHOV *et al.* (1968), MANHEIM (1965), WINTERHALTER (1966), GRILL *et al.* (1968), CALVERT and PRICE (1970); Pacific, Atlantic and Indian ocean nodules—RILEY and SINHASENI (1958), WILLIS and AHRENS (1962), MERO (1962; 1965), VOLKOV *et al.* (1976), SKORNYAKOVA (1976), GLASBY (1973), RAAB (1972), CRONAN (1972); CRONAN and TOOMS (1969), WILLIS and AHRENS (1962), MANHEIM (1972), SUMMERHAYES and WILLIS (1967), GLASBY *et al.* (1974), BEZRUKOV and ANDRUSHCHENKO (1974); iron ores—LANDERGREN (1948), JAMES (1966), STRAKHOV *et al.* (1968), KISVARSANYI and PROCTOR (1967), FRIETSCH (1970); manganese ores—HEWETT and FLEISCHER (1960,) STRAKHOV *et al.* (1968), HEWETT (1968), FRIETSCH (1970).

Table 42-K-2. *Correlation coefficients* of *elements with Mn in ferromanganese concretions and ores*

	Mo	Ni	Ba	Cu	Co	Pb	Zn	Fe
Pacific Ocean nodules (CRONAN, 1969 a, b)	+0.513	+0.446	+0.425	+0.262	+0.251	+0.242	—	−0.276
Pacific Ocean nodules (STRAKHOV *et al.*, 1968)	+0.506	+0.366	—	+0.344	0.0	−0.261	+0.190	−0.517
Indian Ocean nodules (GLASBY *et al.*, 1974)	+0.39	+0.59	—	+0.34	−0.09	+0.02	+0.49	+0.11
Pelagic sediments (CRONAN, 1969 a, b)	+0.490	+0.631	—	+0.277	+0.174	+0.101	—	+0.155
S. Ukrainian manganese ores (STRAKHOV *et al.*, 1968)	+0.318	+0.598	—	+0.233	+0.046	+0.061	+0.163	−0.487

in oxide deposits created by precipitation of ascending solutions (hypogene) than the reverse (supergene). However, enrichments may also occur in supergene deposits (HEWETT and FLEISCHER, 1960, HEWETT *et al.*, 1963).

Although the actual mode of uptake of Mo in manganese oxides is not known, pH, mineral species, and accompanying ions are very important after the Mo concentration in precipitating fluids. These problems are elucidated by ferromanganese concretions and crusts developed in surface waters of the earth's crust. The influence of Mo concentrations of ambient waters can be seen in Fig. 42-K-2, showing the distribution of Mo and Mn in concretions from various aqueous regimes. It may be noted that for Mn normalized at 20%, interpolated values of Mo for the three regimes—South Pacific (deep ocean), Black Sea (reduced salinity, Mo-depleted), and Loch Fyne (freshwater influence?)—are approximately 300, 90, and 20 ppm Mo, respectively. The latter is not far from the general mean of Mo in freshwater concretions. Assigning appropriate Mo values in water of 11, 3.1, and 1 μg/l to the three environments (ocean, Black Sea and lake), we obtain the partition coefficient ($Mo_{concr.}/Mo_{water}$)=K, listed in Table 42-K-3.

Table 42-K-3. *Partition of Mo between ferromanganese concretions and waters*

	Concretion ppm Mo	Water (μg/l)	Partition coefficient (K)
Deep Sea	380	11	35
Black Sea (surf)	90	3.1	29
Loch Fyne [a]	25	1.0	25

[a] Including interstitial water supply.

The relative constancy of the K values is quite surprising, because the supply of soluble Mn in the environments undoubtedly varies, especially in interstitial waters, and would be expected to affect the proportion of Mo to Mn.

Other influences on Mo uptake in ferromanganese oxides include mineralogy. CALVERT and PRICE (1977) recently suggested that Mo is preferentially taken up per unit Mn concentration in todorokites (Al, K, Ca, Mo, Mn^{2+}) ($Mn_3O_7 \cdot nH_2O$), though not necessarily in shallow marine or continental forms, with respect to birnessite (Na, Ca)$Mn_7O_{14} \cdot 2H_2O$. δ-MnO_2, a weakly crystallized oxide that is associated with increased Fe, Co, and Ti (CRONAN and TOOMS, 1969; CRONAN, 1976; BEZRUKOV and ANDRUSHCHENKO, 1974; BURNS and BURNS, 1977) is depleted in Mo.

Eh-pH relationships with Mo uptake have not been defined on the basis of observational data (as opposed to theoretical data), owing in part to the difficulties in assessing conditions in microenvironments. Given the fact that at least in ocean waters dissolved Mo exceeds Mn by a factor of 10 or more (BREWER, 1976), ferromanganese oxides at neutral or higher pH values are clearly inefficient, if consistent absorbers of Mo (GLASBY, 1973).

Greatest molybdenum concentrations are found in the borderland off Baja California (720 ppm) according to CRONAN (1972) and CRONAN and TOOMS (1969). The authors also show least Mo of 8 geographic provinces in the Pacific Ocean in the

North Pacific (180 ppm). With the exception of the latter, however, Mo/Mn variations show closer agreement, averaging 22 (ppm Mo/% Mn), as do nodules from the Atlantic and Indian Oceans.

In the nodules themselves, RAAB (1972) demonstrated higher Mo concentrations in bottoms of nodules, presumably influenced by interstitial waters of underlying sediments, than in tops, precipitated from seawater. However, no significant differences between nodule cores and rinds, or within discrete annular layers could be found.

More rapidly formed shallow marine and freshwater nodules have lower Mo concentrations, presumably because their interstitial water metal sources have higher Mn concentrations (ELDERFIELD, 1976, CALVERT and PRICE, 1977, and references cited), and therefore Mo/Mn ratios.

In summary, the controls on Mo uptake ander earth surface pH conditions are dominated by the concentration of Mo in precipitating solutions. Subsidiary influences are exerted by accessory element associations and oxide mineralogy. Influence of pH and Eh are as yet poorly defined by field or experimental data.

IV. Anoxic Sediments and Organic Residue-rich Rocks (Black Shales)

Molybdenum is best known for its enrichment in anoxic or organic rich sediments. The general physicochemical conditions applicable are discussed in 42-H; it will suffice here to state that uptake proceeds principally through two major mechanisms. The first is coprecipitation of molybdenum sulfides with FeS or other metal sulfides. The occurrence of molybdenum sulfide in sediments was early established by observation of molybdenite-jordisite (MoS_2) and later castaingite (Mo_2S_5) in the Permian Kupferschiefer of Germany (GOLDSCHMIDT, 1954; RAMDOHR, 1969, and literature cited). The second is fixation of Mo by adsorption, reduction and other mechanisms in organic phases (see 42-H; 42-L), especially organic-rich (black) shales. The form of Mo-organic association is not fully explained, for although Mo-adsorption and reduction to Mo^{5+} is well documented at low pH, specific Mo-organic complexes other than enzymes are not known by us to be documented in naturally occurring sediments.

a) Recent Deposits

Molybdenum enrichment in sediments is a sensitive indicator of anoxic environmental conditions. Early distinctions were made between enrichments of Mo in sedimentary rocks interpreted to have been deposited in fully anoxic (including the sediment-water interface) conditions: Vollfaulschlamm, sapropel; and Mo-poor, organic-enriched sediments laid down beneath an oxic sediment-water interface: Halbfaulschlamm, gyttja (SCHNEIDERHÖHN *et al.*, 1949; LEUTWEIN, 1951).

Studies of Mo in recent anoxic sediments have shown that, since the bulk of the Mo is extracted from sea water, the sediment-water interface must be permeated with sulfur in order to permit maximum uptake of Mo from bottom waters. Within a gives anoxic basin, such as the Baltic Sea, Mo tends to be enriched in the

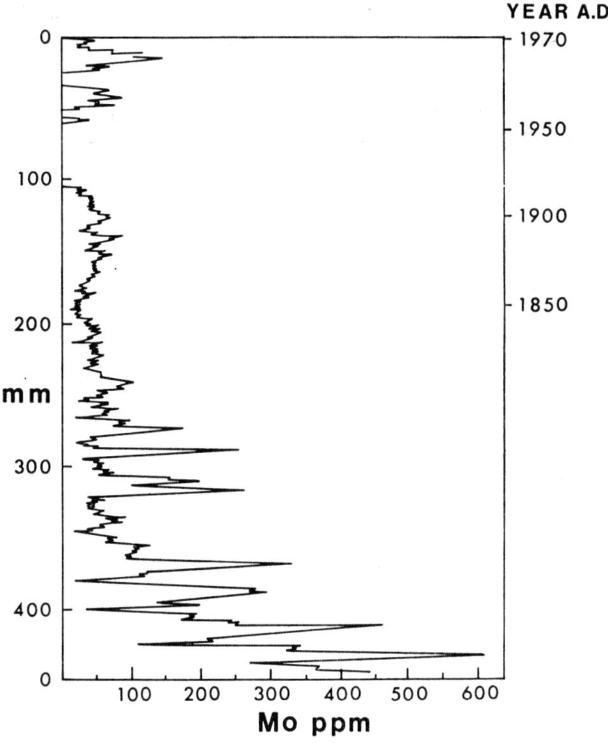

Fig. 42-K-4. Distribution of molybdenum in micro-sampled layers of sediment from the Gotland Deep, Central Baltic Sea (HALLBERG, 1974). Dates are by ^{210}Pb analysis. The apparent "O" data above 100 mm are reflected in the original figure redrawn here, and are not clarified in the author's report. The data show the markedly greater interpreted intensity of stagnation in the recent past, than in the post-1850 period. A cyclicity of 10–20 years is also reflected in the molybdenum variations

centers or most H_2S-rich portions of the area (MANHEIM, 1961; GROSS, 1967). Micro-sampling of sediment layers in the Central Baltic (Gotland Deep) has shown that Mo, or the (Mo+Cu/Zn ratio faithfully records the intensity and duration of reducing bottom conditions, insofar as intermittent oxidizing conditions are not sufficiently strong to re-oxidize and leach out metalliferous bands (HALLBERG, 1974) Fig. 42-K-4. The mechanics of such Mo recycling has been studied in Japanese lakes by SUGAWARA *et al.*, (1961).

Recent Mo-accumulating anoxic environments are mainly marine. Not separated in the histogram in Fig. 42-K-5. they can be divided geographically into 3 categories:

(i) Continental margin upwelling zones such as off Southwest Africa (CALVERT, 1976), Peru-Chile, and other coastal bays or offshore basins off the west coast of Central and South America (BERTINE and TUREKIAN, 1973).

(ii) Semi-enclosed bays, fjords and estuaries that are well supplied with nutrients and develop stratified water conditions preventing overturn and aeration of bot-

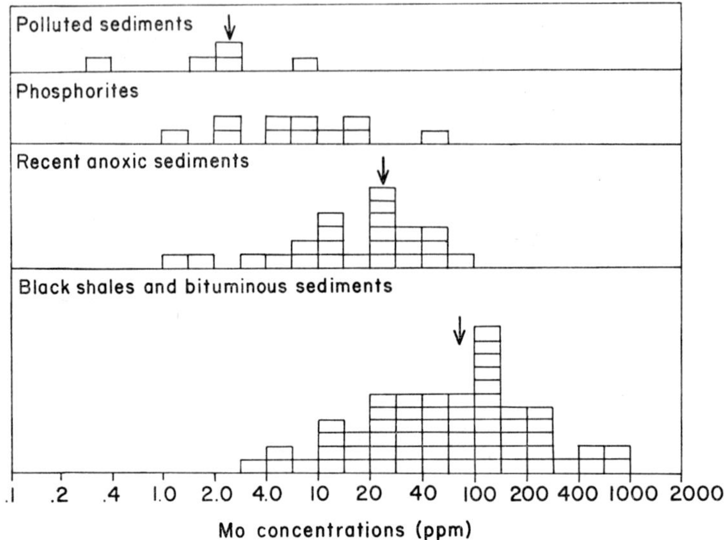

Fig. 42-K-5. Distribution of Mo in anoxic, organic-rich and related sediments. Notes on plotting are similar to Fig. 42-K-1. References for polluted sediments are CLARK and HILL (1958), BERROW and WEBBER (1972), MACKAY *et al.* (1972), PERKINS and GILCHRIST (1973), and MANHEIM (1973). Other references are given in the text

toms. Examples include Saanich Inlet, British Columbia (GROSS, 1967), Oslo- and Bunnefjord (DOFF, 1970), Mljet, Adriatic Sea (SEIBOLD, 1958), Kyrkfjarden in the Stockholm Fjord (LANDERGREN and MANHEIM, 1962); and Lake Nitinat, a fjord in Vancouver Island, British Columbia (CRECELIUS, 1969).

(iii) Landlocked seas of wide geographic extent and variable depth. Such areas are exemplified by the Black and Baltic Seas. (MANHEIM, 1961; LANDERGREN and MANHEIM, 1962; VARENTSOV, 1975; EMELYANOV, 1976; PILIPCHUK and VOLKOV, 1967; PILIPCHUK, 1972; HIRST, 1974; CALVERT, 1976; PILIPCHUK and VOLKOV, 1974.)

A highly unusual occurrence of apparent reducing sediments in a ponded sediment basin in the Mid-Atlantic Rift zone was reported by TUREKIAN and BERTINE (1971). Mo and U were found enriched in reducing sediments beneath an oxidized zone characterized by high manganese content (Fig. 42-K-5). It is possible that the Mo and U uptake were caused by hydrothermal solutions containing H_2S; such phenomena have been observed very recently in sea floor studies of the Galapagos Rift zone in the Central East Pacific Ocean (BALLARD and VON HERZEN, 1977, personal communication).

Most recent environments do not yield Mo concentrations exceeding 50 ppm. An exception is the Central Baltic Sea whose sediments were shown by HALLBERG (1974, Fig. 42-K-4) to contain zones with Mo excess of 400 ppm, rivalling some of the classical Mo and V-enriched black shales. However, in general, the recent occurrences do not fully illuminate the origin of hundreds of organic-enriched "black shales" found among fossil sediments both on the continent, and in deeper layers beneath the sea floor (Mediterranean: BATURIN *et al.*, 1967; PILIPCHUK, 1972; Red

Sea: MANHEIM, 1974; Cretaceous black shales beneath the western N. Atlantic: LANCELOT *et al.*, 1969; EMERY and UCHUPI, 1974; TUCHOLKE and MOUNTAIN, 1977).

Mo has been only infrequently analyzed in sediments affected by domestic sewage or industrial pollutants. To date, only minor enrichments have been reported, even in sediments containing industrial pollutants at levels approximating 1,500 ppm each of V, Zn, and Pb, and 1,000 ppm Cr (MANHEIM, 1973; Fig. 42-K-5). Hence, unless H_2S permeates the water column as well as underlying sediments, or special Mo-rich contaminants are involved, Mo concentrations in organic-rich sediments of pollutant origin tend to follow the ranges typical of "Halbfaulschlamm and gyttja" rather than sapropels.

b) Black (Bituminous) Shales and Other Sedimentary Rocks

The amount of analytical data on molybdenum-enriched bituminous rocks is large. A histogram of averages is provided in Fig. 42-K-5, drawn from the following publications: VINE and TOURTELOT (1969); RADER and GRIMALDI (1961); TOURTELOT (1962); DAVIDSON and LAKIN (1961); FISCHER (1961); LANDIS (1962): BREGER and SCHOPF (1955); HYDEN and DANILCHIK (1962); SUNDIUS (1941); BAIN (1960); LANDERGREN and MANHEIM (1962); ANKINOVICH and ANKINOVICH (1968); REKHARSKII and KRUTETSKAYA (1960); LANCELOT *et al.* (1972); LE RICHE (1959); GAMALEYEV and KHAMRABAYEV (1958); COUDERC (1951); BLUMER and ERLENMEYER (1950); BRANDENSTEIN *et al.* (1960); BITTERLI (1960); BORCHERT and KREJCI-GRAF (1959); TITZE in: BORCHERT and KREJCI-GRAF (1959); ADYSHEV *et al.* (1963); STRAHL (1959); TOURTELOT (1970); TIKHOMIROVA (1960); SCHNEIDERHÖHN *et al.* (1949); SCHROEDER in: BROCKKAMP (1944); FESSER (1958); SCHMITZ (1963); KREJCI-GRAF and WICKMAN (1960); MARMO (1960); PELTOLA (1960); WEDEPOHL (1964); KNITZSCHKE (1961); EISENHUTH and KAUTZSCH (1954); MESSER (1955); KAUTZSCH (1942); DEANS (1950); DUNHAM (1961); TOURTELOT *et al.* (1960); VINE (1966); KURODA and SANDELL (1954); HARANCZYK (1964); PRYAKINA (1958); MANHEIM (1974); and LEUTWEIN (1951).

Literature reviews and original information of wide scope for sediments are found in GOLDSCHMIDT *et al.* (1948), SCHNEIDERHÖHN *et al.* (1949), LEUTWEIN (1951), BORCHERT and KREJCI-GRAF (1959), KRAUSKOPF (1955), KHRUSHCHOV and SHCHERBINA (1965), WEDEPOHL (1965), VINE and TOURTELOT (1969), FISCHER and OHL (1970) and UZKUT (1974).

Some facts and generalizations on Mo accumulations in fossil bituminous sediments include the following: threshhold concentrations of organic carbon for major increase in trace concentrations of Mo typically occur between 2 and 6%. Beyond this level, correlations between Mo and C_{org}, and also sulfide sulfur are positive, but subject to a high degree of scatter. Mo:V ratios increase from 1/20 or less to 1/3 with increasing Mo concentration. Some shale bands are thin and discontinuous (e.g. Red Sea Pleistocene, Permian Kupferschiefer), whereas others, such as the Devonian Chattanooga Shale, the Scandinavian Cambro-Ordovician Alum Shales, and the Pierre Shale of Late Cretaceous age, extend up to 1000 m.

One of the most famous of the "black shales" is the (Mansfelder) Kupferschiefer (WEDEPOHL, 1965, 1968, 1971, and references cited). WEDEPOHL (1965) indicates maximum Mo values on the order of 1,500 ppm. This author has emphasized the relationship betwen V and Mo-enriched black shales and redbed country rocks (by inference, arid, oxidizing terrains). Under highly oxidizing conditions, the solubility of molybdates and vanadates enhances transport of these metals from the continents to marine basins having limited detrital sedimentation. Evaporitic conditions resulting in temporary salinity stratification and stagnation may then provide environments suitable for precipitating the Mo, V, and other metals. Such redox-driven processes have been demonstrated to occur diagenetically on a microscale. Bleached, organic-influenced reduction zones in redbed suites such as the Permian Rotliegendes sandstones have been shown by HARTMANN (1963) and other authors cited in WEDEPOHL (1971) to provide sites for heavy metal enrichment. The mechanism is also supported by observed increases in Mo content of river water in arid regions (Sect. 42-I).

Studies of a complete stratigraphic sampling (37 samples) of black coaly shales of Pennsylvanian (Carboniferous) age in Indiana gave the following correlation coefficients of Mo with other constituents: Ni 0.78, Co and Y 0.56, C_{org} 0.53, Cu 0,49, V 0.44, and Cr 0.31. Correlations with major and other minor constituents were negative (VINE, 1966). Arithmetic means were C_{org} 34.8%, V 1,200 ppm, Ni 320 ppm, Mo 400 ppm, Mn 84 ppm. The author points out that the coaly shales contain much larger concentrations of molybdenum than purer coals themselves, even on an ash basis and explains this by the "differing availability of molybdenum in a fresh-water coal swamp and in brackish or marine waters adjacent to the strand line".

Mo and Mn are closely correlated in deep sea pelagic and other oxidizing sediments, but are generally negatively correlated in reducing sediments. In fact, some of the well-known, Mo-enriched black shales have Mn concentrations as little as 100 ppm. Notable exceptions to this generality include the Baltic Sea and Norwegian fjords, where Mn is accumulated as calcic rhodochrosite, and in the Kupferschiefer among fossil Mo-rich rocks, where Mn is found in calcite and dolomite, and extrapolates to 6,000 ppm for pure authigenic carbonate (WEDEPOHL, 1964).

The form of Mo in black shales remains incompletely understood. WEDEPOHL (1964) and VOLKOV and FOMINA (1974) showed that living organic matter could contribute less than 0.2 ppm Mo or less than 1% of Mo in the Kupferschiefer and Black Sea deep sediments, respectively. Only traces of Mo in organic-solvent soluble bitumens have been obtained by CRECELIUS (1969) and HYDEN (1961), excluding its presence in porphyrins in the rocks studied. ARMANDS (1973) showed that only a small proportion of Mo in Upper Cambrian Swedish alum shales resided in pyrites and shales containing comparable concentrations of Mo. ARMANDS (1973) also found complex factor analysis associations of Mo with various phases. LE RICHE (1959) and RAZUMNAYA (1957) reported association of Mo with organic fractions. VOLKOV and FOMINA (1974) obtained one set of data with 5.6% of Mo in deepwater Black Sea sediments in the "humic acid" (NaOH soluble, H_2SO_4 insoluble) fraction, and 82% in the "fulvic acid" fraction, for a total of nearly 90% in the combined "humates". However, simultaneous studies showed nearly all of the Mo in the sediments in the pyrite fraction. The sulfides contained Mo in a form soluble in 0.1 N NaOH-H_2SO_4, and therefore apparently associated with the fulvic acid fraction. Similar studies need to be repeated in fossil sediments. Conclusions from all data are that a considerable portion of Mo may reside in finely disseminated MoS_2 within iron sulfide and other metal sulfides, or as minute coatings and impregnations in various phases, organic and inorganic.

A major gap in knowledge is the total role that bituminous sedimentary rocks play in the molybdenum budget of sediments as a whole, and in the earth's crust. VINE (1966) and WEDEPOHL (1964) have pointed out that the literature may give a misleading picture of overall metal concentrations because of the natural focus of attention and selection on "interesting" samples. To our knowledge, there has been no attempt to estimate the proportion of the geologic column comprised by bituminous sediments and their component metals.

V. Phosphorites and Miscellaneous Sediments

Phosphorites are known to be enriched in uranium and molybdenum (KRAUSKOPF, 1955, and references cited) and other elements in a manner rather similar to black shales. The mode of a specific phase involved is disputed (CALVERT, 1976), but fine coatings and impregnations of MoS_2 in sulfides, organic matter, and phosphate itself may be involved. Phosphorites richest in organic carbon have highest Mo; thus the Western U.S. phosphates have greater concentrations than the Florida deposits or many Russian phosphorites.

References for phosphorites in Fig. 42-K-5 include CLARK and HILL (1958), LANDIS (1962), BLOKH and KOCHENOV (1964) and CALVERT (1976).

VI. Rates of Mo Accumulation in Marine Sediments

Several authors have calculated the rates of accumulation of molybdenum per unit area of sea floor. These data are listed in Table 42-K-4. Rather consistent values between 1 and 4 μg Mo/cm²/1,000 years with a mean of about 2.3 μg Mo/cm²/1,000 years for pelagic sediments are obtained from data of BERTINE (1970), BERTINE and TUREKIAN (1973) and LANGE (1974).

MERO (1965) estimated a total tonnage of 17×10^{11} tons of nodules for the Pacific Ocean; ZENKEVICH and SKORNYAKOVA (1961) estimated about 20-fold less. The former author also estimated an annual accumulation of 3×10^3 tons of Mo (method not stated). New data in SKORNYAKOVA and ZENKEVICH (1976) permit calculation of a mean nodule concentration of about 1 g/cm² based on 110 grab and bottom photo stations in Pacific basins and 0.6 g/cm² for the entire Pacific. Data from other oceans are sketchy but GLASBY (1973) estimated 0.25 g/cm² for the Indian Ocean.

For a molybdenum concentration of 380 ppm in nodule material, accumulation will be 0.14 μg Mo/cm²/1,000 years. Considering crusts in areas of exposed rocks, we conclude that nodule accumulation contributes to loss of Mo from the ocean in a propostion of 1/10 or less than that withdrawn by sediments.

Anoxic sediment facies play a significant role in the total Mo budget, as pointed out by BERTINE and TUREKIAN (1973) and CALVERT (1976). Accumulation rates vary extremely widely (Table 42-K-4), but an estimate of the total Mo accumulation in the past 1,000 years shows that, although the three anoxic areas make up only a fraction of 1% of total ocean sediment (or 0.36% of deep pelagic sediment area), they remove 30% of the Mo taken up by pelagic sediments and nearly 10 times as much as Pacific nodules. This implies that, when substantially larger anoxic areas existed in the past, Mo concentrations and residence times in the oceans should have been decreased.

Table 41-K-4. *Rates of accumulation of Mo in marine sediments*

Region	Area (cm²) × 10^{15}	Rate of accumul. μg Mo/cm²/1,000 yr	Mo in 10^6 t/1,000 yr
Black Sea [a]	3.1	514	1.6
Baltic Sea [b]	2.4	30	0.07
SW Africa	0.14	1,783	0.25
Subtotal	5.64		1.92
Deep sea sediment >3,000 m depth in Pacific, Atlantic, and Indian Oceans	2,770	2.3	6.4
Pacific nodules >3,000 m depth	1,478	0.14	0.21

[a] Depth greater than 100 m.
[b] Central Baltic less Gulfs of Finland and Bothnia.

Revised manuscript received: February 1978

42-L. Biogeochemistry

Molybdenum was discovered by SCHEELE in the 18th Century (SCHEELE, 1778), but it was not until DEMARCAY (1900) that its presence was noted in plants. The biological significance of Mo was first detailed by BORTELS (1930), who concluded that the element was an essential biocatalyst for nitrogen fixation in plants through the agency of bacteria such as *Azotobacter* and *Clostridium*. A similar function occurred for certain algae (BORTELS, 1940). Results of many workers confirmed BORTELS' concepts and showed that higher plants also require Mo (SCHARRER and SCHROPP, 1934; ARNON and STOUT (1939); SCHARRER, 1955; KARLSSON, 1961 and references therein; REISENAUER, undated, probably 1968).

Molybdenum in organisms is largely present in the form of heteropolymolybdates such as phosphomolybdates. These occur in flavoprotein enzymes and nitrate reductases of plants and organisms, as well as the N-fixing enzyme, nitrogenase. The latter incorporates a molybdenum-iron protein containing 1–2 atoms of Mo in a unit having a molecular weight of 200,000–270,000 (EVANS and BARBER, 1977). The role of Mo as a catalyst in biochemical reactions is due to its tendency to form labile O- and S- bridged polynuclear complexes and its ready ability to change oxidation states between Mo^{5+} and Mo^{6+}. However, this behavior also makes clear-cut definition of chemical species difficult.

Extensive data on Mo concentrations in a variety of biological materials were given by TER MEULEN (1932), TER MEULEN and RAVENSWAAY (1935). BERTRAND (1940, 1943) first reported Mo in animal species, especially marine forms. Data on marine species have been compiled by VINOGRADOV (1953). Data on Mo concentrations in land plants have been compiled by VINOGRADOV (1954). Data on Mo in biological species are given in Tables 42-L-1 to 42-L-3.

I. Plants

The requirement for soluble Mo in soils varies with the plants, physico-chemical conditions, and the form and concentration of nitrogen, in view of the aforementioned function of Mo in N-fixation. According to KARLSSON (1961), the highest Mo requirement for leguminous plants occurs when N_2 forms the main N supply. Lesser concentrations are required for nitrate N and least is needed where ammonia-N is present. In otherwise well-balanced nutrient solutions, 0.001 ppm Mo in the root-milieu is adequate for higher plants, but 0.0001 ppm Mo is always deficient. 0.1 ppm Mo in dry matter signifies the lower limit for healthy herbs and vegetables. pH values below 5 may give rise to Mo deficiency regardless of Mo concentrations in the soil, owing to the insolubility of Mo in acid media (KARLSSON, 1961). Contrarily, the highest concentrations of Mo in hay are obtained after lime treatment of soil (SKULMOWSKY and WIERCINSKI, 1966; KARLSSON, 1961). Highest Mo concentrations are found in plants having enhanced nitrogen-fixing ability such as legumes and alders

Table 42-L-1. *Molybdenum in land plants* (dry matter). (Asterisk indicates analysis on ashed basis. Number of samples is in parentheses)

Material, source	Range, ppm Mo	Mean, ppm Mo	Anal. method	Reference
Grains: wheat, buckwheat corn, rye, oats, barley	0.2–1.3	0.77		VINOGRADOV (1954); SHIMP *et al.* (1957)
Grasses (except legumes)				
Timothy	0.6–1.1	0.8		KARLSSON (1961)
May (153), Poland	0.21–1.36			SKULMOWSKI and WIERCINSKI (1966)
Hay, Sweden				
from clay-rich soil (176)	0.1– 7.8	0.91		KARLSSON (1961)
from humus-rich soil (73)	0.1–17.7	2.7		
all samples (1,263)	0.07–17.7	1.5		
Grasses above ground (32)		34*		CANNON (1960)
Legumes (soy beans, beans, lentils, peas, vetch, clover, alfalfa, lupine, wisteria, and mimosa)	0.8–10	4.0		VINOGRADOV (1954)
Peanut meal		5.5		KARLSSON (1961)
Legumes		100*		CANNON (1960)
Vegetables (non-legume)				
Kale, carrots, sugar beets, tomatoes, potatoes	0.09–0.67	0.4		VINOGRADOV (1954) KARLSSON (1961)
Trees and shrubs				
Red spruce, balsam fir, hemlock, white pine, white birch, red maple, aspen			N/R	YOUNG and GUINN (1966)
Needles/leaves		4.1		
Twigs		3.6		
Bark		7.2		
Wood (roots, branches and trunk)		0.49		
Deciduous trees (leaves) (118)		7*	S	CANNON (1960)
Deciduous trees (twigs)	3–30*		S	WARREN (1962)
Conifer needles		5*	S	CANNON (1960)
Alders	15–100*		S	WARREN (1962)
Mesquite, catclaw, ironwood	9–15*	11*	C	CHAFFEE (1976)
Other categories				
Herbs		19*	C	CANNON (1960)
Lichens	10–50*		S	LEROY and KOSKOY (1962)
Mushrooms (2)	1.0–1.5	1.2		BERTRAND (1940)
Average plants		0.5	C	VINOGRADOV (1954)
Average plant ash		10	C	VINOGRADOV cited in MALYUGA (1964)
Average vegetation ash		20		TKALICH (1970)
Average vegetation ash (unmineralized soil)		13		CANNON (1960)

(*alnus*), as noted in Table 42-L-1. According to YOUNG and GUINN (1966), Mo values in wood, whether root, trunk, or branch, average only 1/10 those of leaves and bark of trees. This would imply correlation of Mo with ash content (leaves 12% compared with wood at 1%: POLIKARPOCHKIN and POLIKARPOCHKINA, 1964). However, the former authors' absolute values are significantly higher than would correspond with those of CANNON (1960) (Table 42-L-1 for plants from unmineralized terrains).

Nearly all land plants accumulate Mo beyond their natural physiological requirement in proportion to increases in soil solutions. As has been noted in Sect. 42-F, this behavior lends itself to use of plants in prospecting for metallic ores.

The role of molybdenum in marine plants is still poorly understood. Although the molybdenum content of sea water is more than 10 times that of mean fresh waters, the data in Table 42-L-2 shows no increase in Mo content as a function of

Table 42-L-2. *Molybdenum in lake and marine plants.* (Number of samples in brackets)

Organisms	Dry matter		Ash	Reference
	Range, ppm Mo	Mean, ppm Mo	Mean, ppm Mo	
Phytoplankton (diatoms)[a]	0.20–4.0	0.38	2.6	SUGAWARA *et al.* (1961)
Brown algae[b]	0.14–1.16	0.38	6.0	BERTRAND (1940); BLACK and MITCHELL (1952); YAMAMOTO *et al.* (1968); FUGE and JAMES (1973)
Sargassum weed[c]	0.07–5.30	0.34	2.0	YAMAMOTO *et al.* (1968)
Green algae[d]	0.08–0.96	0.31	1.2	YAMAMOTO *et al.* (1968); BERTRAND (1940); BLACK and MITCHELL (1952)
Red algae[e]		0.29	2.0	YAMAMOTO *et al.* (1968)
Fresh water plants[f]	0.18–1.13	0.55	2.4	TER MEULEN (1932); YAMAMOTO *et al.* (1968)

[a] *Microcystis* sp., *Melosira* sp., unidentified sp. Ariake Bay and Lake Suwa-ko, Japan; does not include high value of 50.6 ppm in *Rhizosolania* sp.

[b] *Pelvetia canaliculata*, *Laminaria digitata*, *Laminaria cloustonii*, *Eisenea bicyclis*, *Padine arborescens*, *Neurocarpus prolifera*, *Ishige foliaceae*, *Ishige okamurai*, *Fucus vesiculosus* (14), *Fucus serratus* (4).

[c] *Hizikia fusiforme*, *Cygtophyllum sisymbroides*, *Sargassum finggoldianum*, S. *serratifolium*, *S. patens*, *S. hemiphyllum*, *S. tortile*, *S. thunbergii* (12).

[d] *Ulva sublitoralis*, *Letterstedtia japonica*, *Enteromorpha compressa*, *Ulva fasciata*, *Ulva latissima*, *Monostroma nitidum*, *Chaetomorpha spiralis*, *Chaetomorpha crassa*, *Codium fragile*, *Fucus serratus*, *Fucus vesiculosus*, *Ascophyllum nodosum*.

[e] *Galaxaura fastigiata*, *Gelidium amensii*, *Acanthopeltis japonica*, *Grateloupia turuturu*, *Grateloupia filicini*, *Hypnea charoides*, *Gymnogongrus flabelliformis*, *Polysiphonia urceaolata*, *Ceratodictyon spongiosum*, *Chondrus ocellatus*, *Gloiopeltis tenas*.

[f] *Pleurotaenium nodulosum*, *Valliseneria spiralis*, *Var. asiatica*, *Hydrilla verticillata*, *Myriophyllum spicatum*, *Azolla sp.*

dry matter of marine plants relative to land plants. However, blue-green algae, which might be expected to yield enhanced Mo concentrations because of their special nitrogen-fixing ability, are poorly represented in existing data, as is phytoplankton in general. Until confirming information is available, isolated high values for Mo such as in *Rhizosolenia* (Table 42-L-2) must be regarded with caution. In summary, there is no evidence to data that organic matter as a separate entity contributes significantly to the molybdenum in Mo-enriched marine sediments.

II. Animals and Toxicity Relationships

Animals have significantly greater Mo concentrations than do non-leguminous plants (Table 42-L-3), which might be related to their greater proportion of protein.

Table 42-L-3. *Molybdenum in dry matter of animals and animal tissues with special emphasis on marine species.* (Content in ash marked by asterisk. Shell excluded from larger forms.)

Organism and part	Range, ppm Mo	Mean, ppm Mo	Reference
Sponges (*Halichondria* sp., *Dysidea etheria*, *Ficulum ficus*)	0.2–1.3	0.73	NODDACK and NODDACK (1939); BOWEN and SUTTON (1951); BERTRAND (1943)
Coelenterates (*Cyanea capillata*, *Metridium dianthus*, *Anemonia sulcata*	0.7–18	10.6	NICHOLLS *et al.* (1959); NODDACK and NODDACK (1939); BERTRAND (1943)
Echinoderms [a]	0.02–4.7	2.0	BERTRAND (1943); NODDACK and NODDACK (1939)
Ctenophores (*Beroë cucumis*)		2.1	NICHOLLS *et al.* (1959)
Molluscs [b] (cephalopods, oyster, mussel, gastropod, pteropod)	0.3–13.4	3.3	NICHOLLS *et al.* (1959); BERTRAND (1943)
Arthropods [c]	0.6–3.2	2.0, 13*	NICHOLLS *et al.* (1959); BERTRAND (1943)
Bryozoans (*Plumatella fungosa*)		136 (?)	BERTRAND (1943)
Tunicates (*Ciona intestinalis*, *Ascidia mentula*, *Salpa fusiformis*)	0.77–5	2.2	NODDACK and NODDACK (1939); NICHOLLS *et al.* (1959)
Worms [d]	0.65–3.5	1.9	BERTRAND (1943)
Fishes [e]	0.12–8.5	2.2	NODDACK and NODDACK (1939); BERTRAND (1943); KARLSSON (1961)
Birds (pelican, cormorant)	0.5–4.1	1.9	BERTRAND (1943)
Mammals (dog and parts; sheep wool)	0.06–1.4	0.8	BERTRAND (1943); TAUCINS and SVILANE (1965)

[a] *Asterias marthasterias*, sp., *rubens; Echinides paracentrotus*, *Brissopsis lvrifera*, *Stichopus tremulus*, *Cucumaria lefevre*.
[b] *Ommastrephes illicebrosa*, *Loligo*, *Mytilus edulis*, *Crassostrea* sp., *Patella vulgata*, *Helix* sp., *Limacina retroversa*, *Clione limacina*.
[c] *Centropages typicus*, *Calanus finmarchicus*, *Euphausia krohnii Carcinus*.
[d] *Sagitta elegans*, *Berebeis cultrifera*, *Arenicola marina*.
[e] *Ctenolabrus supestris*, *Squalus acanthius*, *Scyllum caniculata*, *Scomber Triagla*,, *Galus merlangus*, carp ovaries,, cod liver, herring meal.

However, according to BRAITHWAITE (1973), the role of Mo in animal nutrition has not been well defined.

Whereas land plants can accept relatively large quantities of excess molybdate, pasture containing 20 ppm Mo in dry material was discovered by FERGUSON *et al.* (1943 and ref. cited) to be the cause of "teartness" or molybdenosis, an illness of grazing animals. 10 ppm Mo in dry forage, or less, is the normal toxic threshold for ruminants (KARLSSON, 1961; KUBOTA *et al.* 1961, 1963). Mo excess has the biochemical effect of causing copper deficiency and hence is counteracted by copper. Cases of high Mo intake without apparent harm are known (KARLSSON, 1961).

Toxicity and function of molybdenum in human physiology is poorly documented. Mo has been indicated to have low toxicity, on the order of that of potassium. The U.S. Environmental Protection Agency's current quality criteria for water (BECK, 1976) does not mention molybdenum. On the other hand, earlier reports (ANONYMOUS, 1972; FWPCA, 1968) suggest a limit of 5 μg/l Mo for irrigation water, 10 μg/l for water supplying pastureland, and 50 μg/l for intermittent use. A recommended maximum permissible concentration in drinking waters for the USSR in 1965 was 0.5 ppm (ASMANGULYAN, 1965). One of the most extensive efforts to document the environmental significance of molybdenum is through the "Molybdenum Project" sponsored by the U.S. National Science Foundation-RANN program for the watershed region of the giant Climax molybdenum deposit and surrounding mineralized areas (RUNNELS *et al.*, 1975).

III. Fuels

a) Petroleum

Mo is found in the ash of crude petroleum over a wide range, from 1 to 7,500 ppm as may be seen in Table 42-L-4 and Fig. 42-L-1[1]. Highest concentrations of Mo are found in asphaltic or pyrobitumen residues such as elaterite, liverite, gilsonite, and wurtzilite (ERICKSON *et al.*, 1954) and the Athabasca tar sands which contain 0.2% Mo in oil ash (SCOTT *et al.*, 1954).

Statistical correlation of rank coefficient on 102 analyses of trace metals in American petroleums (HYDEN, 1961) showed that molybdenum correlated most closely with chromium in terms of basic crude oil characteristics. SOUTHWICK (cited in HYDEN, 1961) separated crude oil samples chromatographically and by solvent extraction and concluded that chromium was not combined organically. Though no information for Mo was reported, the above correlation, as well as other studies from organic-rich sediments (see Sect. 42-K) suggest lack of solvent extractability or other indices of natural metal-organic associations for Mo in non-living natural materials. This observation is contradicted by the ability of Mo to form organometallic complexes (LARSON, 1965), and by the report of GLEIM *et al.* (1975), indicating that 70% of the Mo is in the organic soluble fraction of the Athabasca tar sand. Further studies are necessary.

b) Coals and Miscellaneous Organic Phases

The distribution of Mo in a compilation of German coals by LEUTWEIN and RÖSLER (1956) is shown in Fig. 42-L-2. As is noted, the concentration of Mo in

[1] For additional information see: GOTT and ERICKSON (1952); KRAYUSHKIN *et al.* (1964, and references cited).

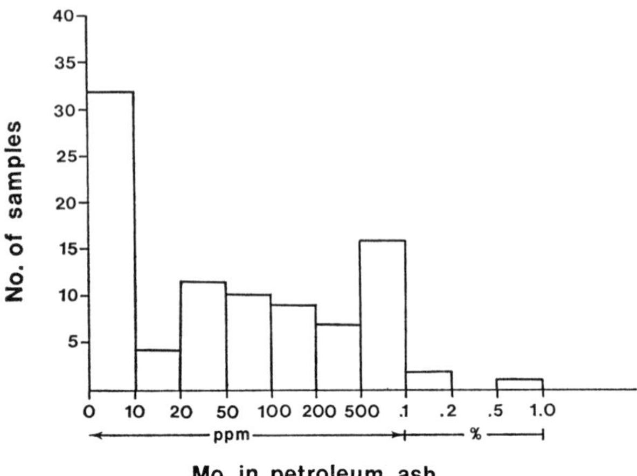

Fig. 42-L-1. Mo in petroleum ash (HYDEN, 1961), based on semiquantitative spectrographic analyses

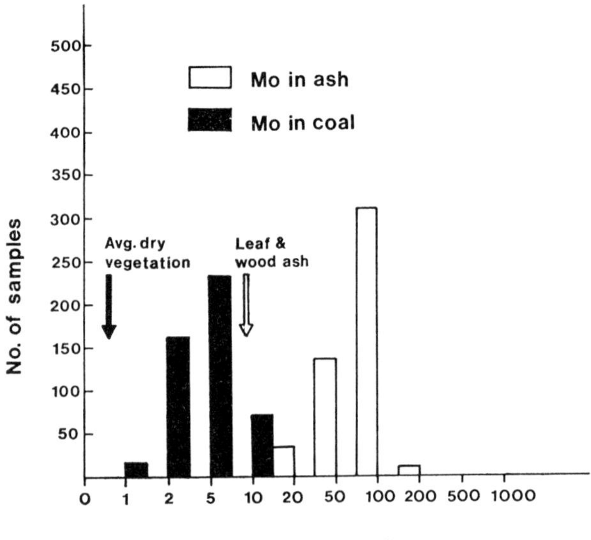

Fig. 42-L-2. Mo in coal and coal ash from Germany (LEUTWEIN and RÖSLER, 1956). Data for living vegetation summarized from Table 42-L-1

wood and average vegetation (the starting point for coal formation) falls at the lowest range for the fossil fuels. Given some reduction in initial volume of coal, the relationships suggest the role of post-depositional uptake of metal from ground waters or other ambient fluids.

Data from a large collection of American coals (SWANSON *et al.*, 1976) showed average values of Mo somewhat less than the above, ranging between 1.5 and 3 ppm in whole coal with no significant correlation with coal rank.

Table 42-L-4. *Mo in fossil fuels.* (*N* and *A* indicate natural or ash state respectively. Numbers in parentheses signify total sample)

Source	State	Range, ppm Mo	Mean, ppm Mo	Anal. meth.	Reference
A. Petroleum					
Italian light oils (4)	*N*	0.008–0.053	0.027	N/R	COLOMBO *et al.* (1964)
Italian asphalts (4)	*N*	2.8–10.1	4.5	N/R	COLOMBO *et al.* (1964)
Italian light oils (38)	*A*	<0.1–171	19	S	COLOMBO and SIRONI (1961)
U.S. crude oils (102)	*A*	<10–7,500	305	S	HYDEN (1961)
U.S. crude oils	*A*		150	S	ERICKSON *et al.* (1954)
Venezuelan Boscan	*A*		4.6	A	GLEIM *et al.* (1975)
Misc. world crude oils	*A*	3.3–7.3	5.1		GLEIM *et al.* (1975)
Russian oils	*A*		<10	S	KATCHENKOV (1948)
Oil ash, McMurray tar sands, Alberta	*A*		2,000	S	SCOTT *et al.* (1954)
B. Coal					
E. Interior basin (U.S.A.)	*N*				ZUBOVIC *et al.* (1961)
American coals, all ages and locations	*N*	0.2–30	3	S	SWANSON *et al.* (1976)
British coals	*N*	1–2		C	HOSTON and AUBREY (1950)
Swedish Kolm (woody-coaly segregation in black shale)	*A*		130	C	WELLS and STEVENS (1931)
Handlova-Novaky basin (22), Czechoslovakia	*A*	0–65	27	S	MECHAČEK (1972)
Paleozoic and Mesozoic coals, Central and E. Germany (514)	*A*	10–150	35	S	LEUTWEIN and RÖSLER (1956)
U-rich lignite, E. Montana and S. Dakota (U.S.A.)	*A*		3,000	S	DENSON and GILL (1956)
Worldwide sampling					
high rank (24) [a]	*A*		24	S	YUDOVICH *et al.* (1972)
brown coal (12)	*A*		15		

[a] Number of sets.

MECHAČEK (1972), studying coals from the Handlova-Novaky coal basin of Czechoslovakia, concluded that there was no consistent relationship between Mo and ash content, petrographic coal type (vitrain, durain, fusain), or position in vertical profile. On the other hand, a number of elements, including Mo, are influenced by the presence of hydrothermal solutions, and by the geologic provenance of the coal region (DENSON and GILL, 1956; BREGER and SCHOPF, 1955); GOLDSCHMIDT and PETERS, 1933).

In a broad survey of trace metal enrichment in sediments, KRAUSKOPF (1955) concluded that Mo was the second or third most enriched element with respect to crustal abundance (after Ag and As) in asphalt and petroleum ash, coal, black shale and phosphorite. There appears to be no current reason to alter KRAUSKOPF's conclusion that enrichments in phosphorite are associated with organic-rich phases, rather than a phosphate affinity for Mo.

ZULFUGARLY (1964) suggested that Mo as well as Cd may be volatilized during the ashing of coal. No specific data were cited, but MoO_3 with a boiling point of 1155° C or Mo chlorides could presumably be mobile under ashing conditions or in the presence of molten halogen salts.

Revised manuscript received: February 1978

42-M/N
Abundance in Common Metamorphic Rock Types; Behavior in Metamorphic Processes

An important, but only partial influence on the distribution of molybdenum in metamorphic rocks is the composition of parent rocks, whether igneous or sedimentary. Regional metamorphism causes a short-range increase in irregularity of Mo distribution due to new mineral phases and segregations. No consistent depletion or enrichment is found in typical metamorphic sequences such as those listed in Table 42-M/N-1.

Table 42-M/N-1. *Molybdenum in rocks of the regionally metamorphic Thompson Formation, E. Minnesota (U.S.A.)* according to KURODA and SANDELL (1954) Analytical method: C

Rock type	ppm Mo
Slate	0.9
Slate-graywacke	0.2
Phyllite	1.2
Phyllite	0.2
Schist	0.3
Staurolite chist	0.2
Granite-gneiss	0.5

On a larger scale, regional metamorphism tends to smooth Mo concentrations across variable rock types over distances of less than a kilometer to dozens of kilometers. This tendency is a reflection of the increased mobility of Mo at higher temperatures and dispersion of Mo in response to concentration gradients. It can only be documented by large statistical distributions of data points, as have been accumulated by Soviet authors in such areas as E. Transbaikalia and N. Kazakhstan (KANISHEV and MENAKER, 1974; STUDENIKOVA and GLINKINA, 1967; TAUSON *et al.*, 1968). Thus, averages of several thousand samples of diorites, gabbros to granodiorites in E. Transbaikalia ranged in concentration from 1.1 to 2 ppm Mo. Somewhat higher concentrations (2.1 ppm Mo) in diorite-metamorphites representing several km^3 of rock were indicated to be influenced by mobilization of Mo from sediments and other phases.

Changes are most pronounced in metasediments that contain more interstitial fluids and interconnected pore spaces than mafic igneous rocks. According to GOLOVNYA (1970), cited in UZKUT (1974), progressive metamorphism corresponding to loss of water and volatile constituents is associated with decreasing Mo concentrations.

CAMBEL and JARKOVSKY (1965) report 17 ppm Mo from highly metamorphosed sulfide ore deposits in Male Karpaty (ČSSR).

Revised manuscript received: February 1978

42-O. Economic Importance

The bulk of the world's reserves of molybdenum occur in the Western Cordillera of North and South America in the form of clustered porphyry deposits. Historically, (since 1931) the great deposit at Climax, Colorado, has dominated world production. As of 1972 more than 150 million tons of ore containing 0.05 to 0.5 percent Mo had been mined, but Climax Molybdenum Co.'s annual report indicated further reserves of about 500 million tons (King *et al.*, 1973).

The Climax deposit is connected with Tertiary (Miocene) volcanism and granitic and monzonitic porphyric intrusions into Precambrian granites, gneisses, and crystalline schists, as well as Paleozoic quartzites, conglomerates and sandstones and limestones. Molybdenite is associated with quartz, pyrite, fluorite, topaz and lesser tungsten, titanium and tin minerals. These occur in fracture fillings, veinlets and minute flakes throughout the intensely fractured stockworks of a large dome-like structure. In addition to the Climax deposit, other major porphyry deposits in hydrothermally altered zones are found at Urad-Henderson, Colorado, the Interior Belt of British Columbia; Questa, New Mexico; in Sonora, Mexico, Chile (Chuquicamata); and Peru (Toquepala). In the latter two deposits the molybdenum occurs as byproducts in copper ore. Elsewhere, porphyry molybdenum occurs significantly in intrusive deposits in the Alpine belt, and Ural Altai Mountains of the USSR.

Statistical study of 27 economic and subeconomic molybdeniferous prophyry deposits in the western North American Cordillera found them to consist of quartz diorite-quartz monozonite intrusives into Late Cretaceous sediments and meta sediments (Lowell and Guilbert, 1970). Stocks average $\geq$1.0 km in diameter and are oval to pipelike with a depth of about 1.7 km up to 3 km. Their locations and dimensions are chiefly controlled by regional faulting. Typically, 70% of the ore body is located in the intrusive stock, and 30% is in pre-ore metasedimentary country rocks with 45% of the ore hypogene and 35% supergene, depending on depth below surface and ground water patterns. Average assay in ore is 0.015% Mo.

Prior to the discovery of the large porphyry deposits, molybdenum was found in many smaller quartz vein deposits and in skarn and contact metamorphic deposits. Here other molybdenum minerals such as wulfenite, powellite and Mo-bearing scheelite may occur. Intensive prospecting in the USSR has revealed major deposits of this type in the Caucasus, central and eastern Kazakhstan and E. Transbaikal (Siberia) (Gorokhov, 1963).

Sedimentary formations, especially marine manganese concretions, contain vast quantities of this metal having molybdenum concentrations of about 0.04% or more, as noted in Sect. 42-K. However, whereas such concentrations would be within minable range as a byproduct from sulfide ores, in sediments the Mo is finely dispersed and to date has not lent itself to benefication and economic recovery. The abundant supply of Mo also has provided little incentive for such recovery.

Table 42-O-1. *Production* (J. T. KUMMER, U.S. Bur. Mines, Wash., D.C.) *and projected reserves* (KING *et al.*, 1973) *of undiscoverd deposits in known districts only*

	1976 production $\times 10^3$ tons	Identified resources $\times 10^6$ tons	Hypothetical resources $\times 10^6$ tons
United States	50.8	15.85	500
Canada	13.4	4.53	50
Chile	9.8	2.50	
U.S.S.R.	8.8	2.29	
China	1.80	0.45	
Peru	0.75	0.59	
Bulgaria	0.12	0.04	
Japan	0.11	0.04	
Total	85.40		

Leading producers of molybdenum and world resources are listed in Table 42-O-1. Ultimate (hypothetical) resources are estimated at more than 10 times identified resources (KING *et al.*, 1973).

The leading use of molybdenum is as an alloy in steelmaking to confer hardness and corrosion resistance, especially from acid attack; other applications include fertilizers, lubricants, catalysts, pigments, and chemical and laboratory uses. Price as of April, 1976 is \$ 2.92/pound of molybdenum concentrate.

The authors acknowledge the aid and criticism of V. E. SWANSON, R. U. KING, C. E. HOLMES, J. HEM, M. BOTHNER, E. JENNY and M. FLEISCHER of the U. S. Geological Survey, K. K. TUREKIAN, Yale University, F. D. SAYLES, Woods Hole Oceanographic Institution, K. BERTINE, Scripps Institute of Oceanography, and K. H. WEDEPOHL, Editor. Publication authorized by Director, U. S. Geological Survey.

Revised manuscript received: February 1978

References: Sections 42-B to 42-O

ADYSHEV, M. M., SHABALIN, V. V., KALMURZAYEV, K. E.: Trace elements in the Cambrian rocks of the Dzhetym-Too Range (Central Tien-Shan). Dokl. Akad. Nauk SSSR **151**, 170 (1963).

ANDERS, E. (ALPEROVITCH): Contribution to the problems of naturally occurring technetium. AEC Contract NYO 6139. Univ. Chicago (1954).

ANKINOVICH, S. G., ANKINOVICH, E. A.: Conditions of formation of metalliferous shales of the lower Paleozoic in southern Kazakhstan [Russ.]. In: STRAKHOV, N. M. (ed.), Geokhimiya Osadochnykh Porod i Rud. Moscow: Izdat. Nauka Moscow 1968.

ANNERSTEN, H., EKSTRÖM, T.: Distribution of major and minor elements in coexisting minerals from a metamorphosed iron formation. Lithos **4**, 185 (1971).

Anonymous: Permissible limits of Mo in natural waters and other materials. U.S. Environmental Protection Agency, Division of Water Supply and Pollution Control, Narragansett, Rhode Island (1972).

Anonymous: Aqueous solutions of molybdenum compounds for catalytic applications. Climax Molybdenum Co., Bulletin Cdb-16 (1973).

ARMANDS, G.: Geochemical studies of uranium, molybdenum and vanadium in Swedish alum shales. Contributions in Geology, Acta Univ. Stockholm., Stockholm **27**, 1 (1973).

ARNON, D. I., STOUT, P. R.: Molybdenum as an essential element for higher plants. Plant Physiol. **14**, 599 (1939).

ARUTYUNYAN, L. A.: Stability of water-soluble forms of molybdenum in sulfur-bearing solutions at high temperatures [Russ.]. Geokhimiya **4**, 479 (1966).

ARUTYUNYAN, L. A.: Possibilities of migration of molybdenum in the form of haloid compounds. Intern. Geol. Rev. **2**, 1200 (1969).

ASMANGULYAN, T. A.: Maximum permissible concentration of molybdenum in surface waters [Russ.]. Gigiena i Sanit. **30**, 6 (1965).

ASTON, S. R., CHESTER, R.: Estuarine sedimentary processes. In: BURTON, J. D., LISS, P. S. (eds.), Estuarine Chemistry. London-New York: Academic Press 1977.

BACHMANN, R. W., GOLDMAN, C. P.: The determination of microgram quantities of molybdenum in natural waters. Limnol. Oceanog. **2**, 143 (1964).

BADALOV, S. T., BESUTOVA, S. M., GODUNOVA, L. J., SHODIEV, F. S.: On the geochemistry of rhenium and molybdenum in endogene sulfide deposits of Central Asia [Russ.]. Geokhimiya **3**, 64 (1966).

BAIN, G. W.: Patterns to ores in layered rocks. Econ. Geol. **55**, 695 (1960).

BAINBRIDGE, K. T., NIER, A. O.: Relative isotopic abundances of the elements. Preliminary Rept. Nuclear Sci. Ser., Natl. Res. Council, Washington, D.C. **9** (1950).

BARDET, J., TCHAKIRIAN, A., LAGRANGE, R.: Recherche spectrographique des elements existant a l'etat traces dans l'eau de mer. Compt. Rend. **206**, 450 (1938).

BATURIN, G. N., KOCHENOV, A. U., SHIMKUS, K. M.: Uranium and rare metals in the sediments of the Black and Mediterranean Seas [Russ.]. Geokhimiya **1**, 41 (1967).

BECK, E. (ed.): Proposed Quality Criterial for Water. U.S. Environmental Protection Agency, Washington, D.C. 1976.

BELL, H. III: Geochemical reconnaisance using heavy minerals from small streams in Central South Carolina. U.S. Geol. Surv. Bull. **1401**, 23 (1976).

BELYAEV, L. I.: On the distribution and composition of heavy microelements in Black Sea waters [Russ.]. Tr. Morsk. Gidrofiz. Inst. Akad. Nauk USSR **37**, 199 (1966).

BERRANG, P. G., GRILL, E. V.: The effect of manganese oxide scavenging on molybdenum in Saanich Inlet, British Columbia. Mar. Chem. **2**, 125 (1974).

BERROW, M. L., WEBBER, J.: Trace elements in sewage sludges. Sci. Food Agric. J. **23**, 93 (1972).

BERTINE, K. K.: The marine geochemical cycle of chromium and molybdenum. Ph.D. Dissertation, Yale University, New Haven, Conn. (1970).

BERTINE, K. K.: The deposition of molybdenum in anoxic waters. Mar. Chem. **1**, 43 (1972).

BERTINE, K. K., TUREKIAN, K. K.: Molybdenum in marine deposits. Geochim. Cosmochim. Acta **37**, 1415 (1973).

BERTRAND, D.: Contribution a l'étude de la diffusion du molybdène chez les animaux. Bull. Soc. Chim. Biol. Paris **22**, 60 (1940).

BERTRAND, D.: Le molybdène et le cuivre dans la série animale. Bull. Soc. Chim. Biol. Paris **25**, 197 (1943).

BEZRUKOV, P. L., ANDRUSHCHENKO, P. F.: Iron-manganese nodules from the Indian Ocean. Intern. Geol. Rev. **15**, 342 (1974).

BISCHOFF, J. L.: Red Sea geothermal brine deposits: Their mineralogy, chemistry, and genesis. In: DEGENS, E. T., ROSS, D. A. (eds.), Hot Brines and Recent Heavy Metal Deposits in the Red Sea. Berlin-Heidelberg-New York: Springer 1969.

BISHOP, H. G.: Pasture investigations in the sandy forest country of Northwest Queensland. Part I. Nutritional Requirements. Queensland J. Agric. Anim. Sci. **31** (4), 329 (1974).

BITTERLI, P.: Bituminous Posidonienschiefer (Lias Epsilon) of Mont Terri, Jura Mountains. Bull. Ver. Schweiz. Petrol. Geol. Ing. **26**, 41 (1960).

BLACK, W. A. P., MITCHELL, R. L.: Trace elements in the common brown algae and in sea water. J. Marine Biol. Assoc. U.K. **30**, 575 (1952).

BLOKH, A. M., KOCHENOV, A. V.: Elemental composition of bone phosphate of fossil fish deposits [Russ.]. Geol. Mestorozhd. Redkikh Elementov, Nedra, Moscow **24** (1964).

BLUMER, M., ERLENMEYER, H.: Analysen schweizerischer Sedimentgesteine. Helv. Chim. Acta **33**, 45 (1950).

BORCHERT, H., KREJCI-GRAF, K.: Spurenmetalle in Sedimenten und ihren Derivaten. Bergbauwissenschaften **6**, 205 (1959).

BORTELS, H.: Molybdän als Katalysator bei der biologischen Stickstoffbindung. Arch. Mikrobiol. **1**, 333 (1930).

BOSTRÖM, K., PETERSON, M. N. A.: The origin of aluminium-poor ferromanganoan sediments in areas of high heat flow on the East Pacific rise. Marine Geol. **7**, 427 (1969).

BOWEN, V. T., SUTTON, D.: Comparative studies of mineral constituents of marine sponges. I. Genera *Dysidea*, *Chondrill*, *Terpios*. Sears Foundation J. Marine Res. **10**, 153 (1951).

BOWMAN, H. R., HEBERT, A. J., WOLLENBERG, H. A., ASARO, F.: Trace, minor, and major elements in geothermal waters and associated rock formations (North-Central Nevada). 2nd U.N. Symposium for the Development and Use of Geothermal Resources **1**, 699 (1975).

BOYLE, R. W., TUCKER, W. M., LYNCH, J., FRIEDRICH, G., ZIADDIN, M., SHAFITULLAH, M., CARTER, M., BYGRAVE, K.: Geochemistry of Pb, Zn, Cu, As, Sb, Mo, Sn, W, Ag, Ni, Co, Cr, Ba, and Mn in the waters and stream sediments of the Bathurst-Jacques River District, New Brunswick. Can. Dept. Mines Tech. Surv. Geol. Surv., Can. Paper **65-42**, (1966).

BRADBURY, J. C., HESTER, N. C., RUCH, R. R.: Analyses of some Illinois rocks for gold. U.S. Geol. Surv. Indust. Minerals Note **44** (1970).

BRAITHWAITE, E.: Molybdenum, the pervasive element. New Scientist **59**, 678 (1973).

BRANDENSTEIN, M., JANDA, I., SCHROLL, E.: Seltene Elemente in österreichischen Kohlen und Bitumengesteinen. Tschermaks Mineral. Petrog. Mitt. **7**, 260 (1960).

BREGER, I. A., SCHOPF, J. M.: Germanium and uranium in coalified wood from Upper Devonian black shale. Geochim. Cosmochim. Acta **7**, 287 (1955).

BREWER, L.: The thermodynamic properties of the oxides and their vaporization processes. Chem. Rev. **52**, 1 (1953).

BREWER, P. G.: Minor elements in sea water. In: RILEY, J. P. and SKIRROW, G. (eds.), Chemical Oceanography (2nd ed.) London, New York: Academic Press 1975.

BRIGHT, M. J.: Primary geochemical dispersion associated with the Henderson molybdenum deposit, Colorado. Econ. Geol. **69**, 1177 (1974).

BROCKAMP, B.: Zur Paläogeographie und Bitumenführung des Posidonienschiefers im deutschen Lias. Arch. Lagerstättenforsch. **77**, 7 (1944).

BROOKS, R. R.: Trace elements in New Zealand coastal waters. Geochim. Cosmochim. Acta **29**, 1369 (1965).

BROOKS, R. R.: Geobotany and Biogeochemistry in Mineral Exploration. New York: Harper and Row 1972.

BROWN, H. S.: The composition of meteoritic matter and the origin of meteorites. Science **109**, 251 (1949).

BURNS, R. G., BURNS, V. M.: Mineralogy. In: GLASBY, G. P. (ed.), Marine Manganese Deposits. Amsterdam: Elsevier 1977.

CALVERT, S. E.: The mineralogy and geochemistry of near-shore sediments. In: RILEY, J. P., CHESTER, R. (eds.), Chemical Oceanography (2nd ed.) London, New York: Academic Press **6**, 187 (1976).

CALVERT, S. E., PRICE, N. B.: Composition of manganese nodules and manganese carbonates from Loch Fyne, Scotland. Contrib. Mineral. Petrol. **29**, 215 (1970).

CALVERT, S. E., PRICE, N. B.: Chemical variation in ferromanganese nodules and associated sediments from the Pacific Ocean. Marine Chemistry **5**, 43 (1977).

CAMBEL, B., JARKOVSKY, J.: Rare elements in pyrite deposits of the western Carpathians in connection with genetic problems of mineralizations. In: KHITAROV, N. I. (ed.), Problems of Geochemistry. Moscow: Izdat. Akad. Nauk 1965.

CAMBEL, B., JARKOVSKY, J.: Geochemie der Pyrite einiger Lagerstätten der Tschechoslowakei. Vydavatel'stvo Slovenskej Akadémie Vied Bratislava 1967.

CAMERON, E. N.: A revised table of abundance of the elements. Astrophys. J. **129**, 676 (1959).

CANNON, H. L.: Botanical prospecting for ore deposits. Science **132**, 591 (1960).

CARMICHAEL, I., MCDONALD, A.: The geochemistry of some natural acid glasses from the North Atlantic volcanic province. Geochim. Cosmochim. Acta **25**, 189 (1961).

CASE, D. R., LAUL, J. L., PELLY, I. Z., WECHTER, M. A., SCHMIDT-BLEEK, F., LIPSCHUTZ, M. F.: Abundance of 13 trace elements in primitive carbonaceous and unequilibrated ordinary chondrites. Geochim. Cosmochim. Acta **37**, 19 (1973).

CECH, F., POVONDRA, P.: Natural occurrences of molybdenum trioxide, MoO_3 in Krupka [Czech] Acta Univ. Carolinae, Geologica **1**, 1 (1963).

CHAFFEE, M. A.: Geochemical exploration techniques applicable in the search for copper deposits. U.S. Geol. Surv. Profess. Papers **907-B** (1975).

CHAFFEE, M. A.: Geochemical exploration technique based on distribution of selected elements in rocks, soils, and plants, Mineral Butte Copper Deposit, Pima County, Ariz. U.S. Geol. Surv. Bull. **1278-D** (1976).

CHAFFEE, M. A., TESSIN, T. D.: An evaluation of geochemical sampling in the search for concealed "porphry" copper-molybdenum deposits on sediments in southern Arizona. 3rd Geochemical Exploration Symposium, Toronto, Canadian Inst. Mining and Metallurgy Special Vol. **11**, 401 (1971).

CHAN, K. M., RILEY, J. P.: The determination of molybdenum in natural waters, silicates, and biological materials. Anal. Chim. Acta **36**, 220 (1966).

CHICHILO, P., WHITTAKER, C. W.: Trace elements in agricultural limestones of the Atlantic Coast regions. Agron. J. **50**, 131 (1968).

CLARK, L. J., HILL, W. L.: Occurrence of manganese, copper, zinc, molybdenum, and cobalt in phosphate fertilizers and sewage sludge. J. Assoc. Offic. Agr. Chemists **41**, 631 (1958).

COLOMBO, U., SIRONI, G.: Geochemical analysis of Italian oils and asphalts. Geochim. Cosmochim. Acta **25**, 24 (1961).

COLOMBO, U., SIRONI, G., FASOLO, G. B., MALVANO, R.: Systematic neutron activation technique for the determination of trace metals in petroleum. Anal. Chem. **36**, 802 (1964).

COUDERC, J. M.: Les schistes bitumineux du Lias en Wurttemberg du Sud. Inst. Petroleum, Oil Shale and Channel Coal. **2**. 38 (1951).

CRECELIUS, E. A.: Mo enrichment in the sediments of an anoxic fjord. Trans. Am. Geophy. Union **50**, 708 (1969).

CRONAN, D. S.: Average abundances of Mn, Fe, Ni, Co, Cu, Pb, Mn, Mo, V, Cr, Ti and P in Pacific pelagic clays. Geochim. Cosmochim. Acta **2**, 33 (1969a).

CRONAN, D. S.: Interelement associations in pelagic deposits. Chem. Geol. **5**, 99 (1969b).

CRONAN, D. S.: Regional geochemistry of ferromanganese nodules in the world ocean. In: HORN, D. R. (ed.), Ferromanganese Deposits on the Ocean Floor. Washington, D.C.: Office of IDOE, National Science Foundation 1972.

CRONAN, D. S.: Manganese nodules and other ferromanganese oxide deposits. In: RILEY, J. P., CHESTER, R. (eds.), Chemical Oceanography (2nd ed.), Vol. **5**, p. 137. London-New York: Academic Press 1976.

CRONAN, D. S., TOOMS, J. S.: The geochemistry of manganese nodules and associated pelagic deposits from the Pacific and Indian Oceans. Deep-Sea Res. **66**, 335 (1969).

DAVIDSON, D. F., LAKIN, H. W.: Metal content of some black shales of the western United States. U.S. Geol. Surv. Profess. Papers **424-C**, 329 (1961).

DEAN, W. E.: Iron-manganese oxidate crusts in Oneida Lake, New York. Proc. Conf. Great Lakes Res. **13**, 217 (1970).

DEANS, T.: The Kupferschiefer and associated lead-zinc mineralization in the Permian of Silesia, Germany and England. 18th Intern. Geol. Congress **7**, 340 (1950).

DEANS, T.: Economic mineralogy of African carbonatites. In: TUTTLE, S. D. and GITTINS, J. (eds.), Carbonatites. New York: Wiley 1966.

DEGENS, E. T., WILLIAMS, E. G., KEITH, M. L.: Environmental studies of carboniferous sediments. Part I: Geochemical criteria for differentiating marine and freshwater shales. Bull. Am. Assoc. Petrol. Geologists **41**, 2427 (1957).

DEMARCAY, E.: Sur la presence dans les vegetaux du vanadium, du molybdene, et du chrome. Compt. Rend. Acad. Sci. Paris **130**, 91 (1900).

DENSON, N. M., GILL, J.: Uranium-bearing lignite and its relation to volcanic tuffs in eastern Montana and North and South Dakota. U.S. Geol. Surv. Profess. Papers **300**, 413 (1956).

DOFF, D. H.: The geochemistry of recent oxic and anoxic sediments of Oslo Fjord, Norway. Thesis, Edinburg Univ. (1970).

DUNHAM, K.: Black shale, oil, and sulfide ore. Advan. Sci. **18**, 284 (1961).

DURFOR, C. N., BECKER, E.: Public water supplies of the 100 largest cities in the United States. U.S. Geol. Surv. Water Supply Papers **1812** (1965).

DURUM, W. H., HAFTY, J.: Implications of the minor element content of some major streams of the world. Geochim. Cosmochim. Acta **27**, 1 (1963).

DUVAL, C.: Inorganic Thermogravimetric Analysis. Amsterdam, New York: Elsevier 1963.

EISENHUTH, K. H., KAUTZSCH, E.: Handbuch für den Kupferschieferbau. Leipzig: Fachbuchverlag 1954.

ELDERFIELD, H.: Hydrogenous material in marine sediments excluding manganese nodules. In: RILEY, J. P., CHESTER, R. (eds.), Chemical Oceanography (2nd ed.), Vol. 2. London-New York: Academic Press 1976.

EL WAKEEL, S. K., RILEY, J. P.: Chemical and mineralogical studies of deep-sea sediments. Geochim. Cosmochim. Acta **25**, 110 (1961).

EMELYANOV, E. M.: Trace elements in the sediments [Russ.]. In: GUDELIS, V. K. and EMELYANOV, E. M. (eds.), Geologiya Baltiiskogo Morya., Vilnius, Izdat. Mokslis. **12**, 288 (1976).

EMELYANOV, E. M., PUSTEL'NIKOV, O. S.: Suspended matter, its composition and material balance of sediments in waters of the Baltic Sea [Russ.]. In: GUDELIS, V. K. and EMELYANOV, E. M. (eds.), Geologiya Baltiiskogo Morya., Vilnius, Izdat. Mokslis. (1976).

EMERY, K. O., UCHUPI, E.: Western North Atlantic Ocean. Am. Assoc. Petrol. Geol. Mem. **17** (1972).

ENGEL, C.G., SHARP, R.P.: Chemical data on desert varnish. Bull. Geol. Soc. Am. **69**, 487 (1958).

ERICKSON, R. L., MARRANZINO, A. P.: Hydrochemical anomalies in 4-Mile Canyon, Eureka County, Nevada. U.S. Geol. Surv. Profess. Papers **424-B**, 291 (1961).

ERICKSON, R. L., MYERS, A. T., HORN, C. A.: Association of uranium and other metals with crude oil, asphalt and petroliferous rock. Am. Assoc. Petrol. Geol. Bull. **38**, 2200 (1954).

ERNST, T., HÖRMANN, H.: Bestimmung von Vanadium, Nickel und Molybdän in Meerwasser. Nachr. Ges. Wiss. Göttingen, Math.-Phys. Kl. **1**, 205 (1935/1936).

EVANS, H. J., BARBER, L. E.: Biological nitrogen fixation for food and fiber production. Science **196**, 332, (1977).

EWART, A., TAYLOR, S. R., CAPP, A. C.: Geochemistry of the pantellerites of Mayor Island, New Zealand. Contrib. Mineral. Petrol. **17**, 116 (1968).

FESSER, H.: Untersuchungen am nordwestdeutschen Posidonienschiefer und seiner organischen Substanz. Beih. Geol. Jahrb. **58**, 221 (1958).

FISCHER, R. P.: Geochemistry and geology. In: BUSCH, P. M. (ed.), Vanadium, a Materials Survey. U.S. Bur. Mines Inform. Circ. **8060**, 26 (1961).

FISCHER, R. P., OHL, J. P.: Bibliography on the geology and resources of vanadium. U.S. Geol. Surv. Bull. **1316** (1970).

FLANAGAN, F. J.: 1972 values for international geochemical reference samples. Geochim. Cosmochim. Acta **37**, 1189 (1973).

FLEISCHER, M.: Minor elements in some sulfide minerals. Econ. Geol. 50th Anniv. Vol., Pt. II, 970 (1955).

FLEISCHER, M.: New mineral names. Am. Mineral. **49**, 1497 (1964).

FLEISCHER, M.: Summary of new data on G-1 and W-1 1962—1965. Geochim. Cosmochim. Acta **29**, 1263 (1965).

FLEISCHER, M.: Additional data on G-1 and W-1 1965—1967. Geochim. Cosmochim. Acta **33**, 65 (1969).

FOLLETT, R. F., BARBER, S. A.: Molybdate phase equilibria in soils. Soil Sci. Soc. Am. Proc. **31**, 26 (1967).

FRIETSCH, R.: Trace elements in magnetite and hematite mainly from northern Sweden. Sveriges Geol. Undersök. Ser. C **646**, 3 (1970).

FUGE, R., JAMES, K. H.: Trace metal concentrations in brown seaweeds, Cardigan Bay, Wales. Marine Chemistry **1**, 281 (1973).

FUJIMOTO, G., SHERMAN, G. D.: Molybdenum content of typical soils and plants of the Hawaiian Islands. Agron. J. **43**, 424 (1951).

FWPCA: Report of the Committee on Water Quality Criteria. Fed. Water Pollut. Control./ Admin., U.S. Dept. Interior (1968).

GAMALEYEV, I. E., KHAMRABAYEV, I.: Vanadium and molybdenum in Silurian deposits of the Mal'guzar, Naratau and Tandy Mountains [Russ.]. Akad. Nauk Uzbek SSR. Uzbek Geol. Zhur **2**, 47 (1958).

GARRELS, R. M., CHRIST, C. L.: Solutions, Minerals, and Equilibria. New York: Harper & Row 1965.

GERASIMOVSKIY, V. I., PAVLENKO, L. I., NESMEYANOVA, L. I.: Geochemistry of molybdenum in nepheline syenites. Geochemistry **2**, 2 (1965).

GLAGOLEVA, M. A., VOLKOV, I. I., SOKOLOV, V. S., YAGODINSKAYA, T. A.: Chemical elements in Pacific Ocean sediments in a section from the Hawaiian Islands to the coast of Mexico. Litologiya i Poleznyi Iskopaemye **10**, 547 (1975).

GLASBY, C. P.: Mechanism of enrichment of the rarer elements in marine manganese nodules. Mar. Chem. **1**, 105 (1973).

GLASBY, G. P., TOOMS, J. S., HOWARTH, R. J.: Geochemistry of manganese concretions from the Northwest Indian Ocean. New Zealand J. Sci. **17**, 387 (1974).

GLEIM, W. K. T., GATSIS, J. G., PERRY, C. J.: The occurrence of molybdenum. In: YEN, T. F. (ed.), The Role of Trace Metals in Petroleum. Ann. Arbor Sci. Publ. **161**, 1975.

GOLD, D. P.: Average chemical composition of carbonatites. Econ. Geol. **58**, 988 (1963).

GOLDBERG, E. D., ARRHENIUS, G. O. S.: Chemistry of Pacific pelagic sediments. Geochim. Cosmochim. Acta **13**, 153 (1958).

GOLDBERG, E. D., BROECKER, W. S., GROSS, M. G., TUREKIAN, K. K.: Marine Chemistry. In: Radioactivity in the Marine Environment. Nat. Acad. Sci. **5**, 137 (1971).

GOLDSCHMIDT, V. M.: Geochemistry. London: Clarendron Press 1954.

GOLDSCHMIDT, V. M., KREJCI-GRAF, K., WITTE, H.: Spurenmetalle in Sedimenten. Nachr. Akad. Wiss. Göttingen **2**, 35 (1948).

GOLDSCHMIDT, V. M., PETERS, C.: Über die Anreicherung von seltenen Elementen in Steinkohlen. Nachr. Ges. Wiss. Göttingen **4**, 371 (1933).

GOLES, C. G.: Cosmic abundances. In: WEDEPOHL, K. H. (ed.), Handbook of Geochemistry I, **5**, 116 (1969).

GOLOVNYA, S. V.: Behavior of molybdenum during progressive metamorphosis of arenaceous-argillaceous rocks [Russ.]. In: FEODOTIEV, K. M. (ed.), Ocherki Geokhim. Rtuti Molibdena i Sery v Gidrotermicheskie Protessy. Moscow: Izdat. Nauka 1970.

GORHAM, E., SWAINE, D. J.: The influence of oxidizing and reducing conditions upon the distribution of some elements in lake sediments. Limnol. Oceanogr. **10**, 268 (1965).

GOROKHOVA, V. N.: The distribution of rhenium in the molybdenites of the Kadzharan and Aigedzor copper-molybdenum deposits. In: Renii. Moscow: Izdat. Nauka (Cited in Uzkut, 26, 1961).
GRAF, D. L.: Geochemistry of carbonate sediments and sedimentary rocks. Part III. Illin. Geol. Survey Circ. **301** (1960).
GRILL, E. V., MURRAY, J. W., MACDONALD, R. D.: Todorokite in manganese nodules from a British Columbia fjord. Nature **219**, 358 (1968).
GROSS, M. G.: Concentrations of minor elements in diatomaceous sediments of a stagnant fjord. In: LAUFF, G. H. (ed.), Estuaries. Am. Assoc. Adv. Sci., Washington, D.C. 1967.
GROVER, B., KULLERUD, G., MOH, G. H.: Phasengleichgewichtsbeziehungen im ternären System Fe-Mo-S in Relation zu natürlichen Mineralien und Erzlagerstätten. Neues Jahrb. Mineral. Abhandl. **124**, 246 (1975).
HALLBERG, R.: Paloredox conditions in the eastern Gotland basin during the recent centuries. Merentutkimuslait. Julk. **238**, 3 (1974).
HANSULD, J. A.: Eh and pH in geochemical prospection. Geol. Surv. of Canada Paper **66-54**, 172 (1967).
HARANCZYK, C.: Imvestigations of copper-bearing Zechstein shales from the Wroclaw monocline (Lower Silesia). Acad. Polon. Sci. Bull. Ser. Sci. Geol. Geogr. Warsaw **12**, 13 (1964).
HARTMANN, M.: Einige geochemische Untersuchungen an Sandsteinen aus Perm und Trias. Geochim. Cosmochim. Acta **27**, 459 (1963).
HAWKES, H. E., WEBB, J. S.: Geochemistry in Mineral Exploration. New York: Harper and Row 1962.
HEAD, P. C., BURTON, J. D.: Molybdenum in some ocean and estuarine waters. J. Marine Biol. Assoc. U.K. **50**, 439 (1970).
HEGEMANN, F.: Die Herkunft des Mo, V, As und Co im Wulfenit der alpinen Pb-Zn-Lagerstätten. Heidelberger Beitr. Mineral. Petrog. **1**, 690 (1949).
HEIDEL, S. G., FRENIER, W.: Chemical quality of water and trace elements in the Patuxent River Basin. Maryland Geol. Survey Rept. of Invest **1**, (1965).
HEM, J.: Reactions of metal ions of surfaces to hydrous iron oxide. Geochim. Cosmochim. Acta **41**, 527 (1977).
HENDRICKS, R. L., REISBICK, F. B., MAHAFFEY, F. J., ROBERTS, D. B., PETERSON, M. N. A.: Chemical composition of sediments and interstitial brines from the Atlantis II, Discovery, and Chain Deeps. In: Degens, E. T., Ross, D. A. (eds.), Hot Brines and Recent Heavy Metal Deposits in the Red Sea. Berlin-Heidelberg-New York: Springer 1969.
HERSTEIN, B.: U.S. Dept. Agr. Bull. **150** (1913) (cited in DUVAL, 1963).
HERZ, N., DUTRA, C. W.: Minor element abundance in a part of the Brazilian Shield. Geochim. Cosmochim. Acta **21**, 81 (1960).
HEVESY, G. VON, HOBBIE, R.: Die Ermittlung des Molybdän- und Wolframgehaltes von Gesteinen. Z. Anorg. Allgem. Chem. **212**, 134 (1933).
HEWETT, D. F., FLEISCHER, M.: Deposits of the manganese oxides. Econ. Geol. **55**, 1 (1960).
HEWETT, D. F., FLEISCHER, M., CONKLIN, N.: Deposits of the manganese oxides: Supplement. Econ. Geol. **58**, 1 (1963).
HIGAZY, R. A.: The Tundulu carbonatite ring-complex in southern Nyasaland. Geol. Surv. Nyasaland Mem. **2** (1954).
HIRST, D. M.: Geochemistry of sediments from eleven Black Sea cores. In: DEGENS, E. T., ROSS, D. A. (eds.), The Black Sea. Am. Assoc. Petrol. Geol. Mem. **20**, 430 (1974).
HORSNAIL, R. F., ELLIOTT, I. L.: Some environmental influences on the secondary dispersion of molybdenum and copper in western Canada. Geochem. Exploration, Canadian Inst. Mining and Metallurgy Spec. Vol. **11**, 161 (1971).
HOSTEN, L., AUBREY, K. V.: The distribution of minor elements in vitrain. J. Soc. Chem. Ind., London **69**, 41 (1950).
HOWIE, R. A.: The geochemistry of the charnockite series of Madras, India. Trans. Royal Soc. Edinburg **62**, 725 (1955).
HUFF, L. C.: A geochemical study of alluvium-covered copper deposits in Pima County, Ariz. U.S. Geol. Surv. Bull. **1312-C** (1970).

HUFF, L. C., MARRANZINO, A. P.: Geochemical prospecting for copper deposits hidden beneath alluvium in the Pima district, Arizona. U.S. Geol. Surv. Profess. Papers **424-B**, 308 (1961).

HYDEN, H. J.: Distribution of uranium and other metals. U.S. Geol. Surv. Bull. **1100-B**, 17 (1961).

HYDEN, H. H., DANILCHIK, W.: Uranium in some rocks of Pennsylvanian age in Oklahoma, Kansas, and Missouri. U.S. Geol. Surv. Bull. **1147-B** (1962).

IKEDA, N.: Chemical studies on the hot springs of Arima. VII: Investigations on the Tenmangu-no-yu Spring, Arima area [Japanese]. J. Chem. Soc. Japan **76**, 1079 (1955).

IMAMURA, K., HONDA, M.: Distribution of tungsten and molybdenum between metal, silicate, and sulphide phases of meteorites. Geochim. Cosmochim. Acta **40**, 1073 (1976).

ISAEVA, A. B.: Molybdenum in sediments of the Okhatsk Sea [Russ.]. Dokl. Akad. Nauk SSSR **131**, 3 (1960).

ISHIBASHI, M.: Studies on minute elements in sea water. Records Oceanog. Japan N.S. **1**, 88 (1953).

ISHIBASHI, M., FUJINAGA, T., KAWAMOTO, T.: Fundamental investigation on the dissolution and deposition of molybdenum, tungsten and vanadium in the sea. Records Oceanog. Works Japan, Spec. No. **6**, 215 (1962).

ISHIBASHI, M., SHIGEMATSU, T., NAKAGAWA, Y.: Quantitative determination of tungsten and molybdenum in sea water. Mem. Chem. Res. Lab. Kyoto Univ. **32**, 199 (1955).

ISHIMORI, T.: Molybdenum content in Japanese volcanic rocks. (Japanese). Bull. Chem. Soc. Japan **24**, 251 (1951).

IVANOVA, G. F.: Content of tin, tungsten and molybdenum in granites enclosing tin-tungsten deposits. [Russ.] Geokhimiya **5**, 492 (1963).

IVANOVA, G. F., BUTUZOVA, YE. G.: Distribution of tungsten, tin and molybdenum in granites of eastern Transbaikalia. Geochemistry **6**, 572 (1968).

JAMES, H. L.: Chemistry of the iron-rich sedimentary rocks. U.S. Geol. Surv. Profess. Papers **440-W** (1966).

JÉROME, D. Y., PHILIPPOT, J. C.: Chemical composition of Luna 20 soil and rock fragments. Geochim. Cosmochim. Acta **37**, 909 (1973).

JONES, P. G. W.: Molybdenum in a near-shore and estuarine environment. North Wales Est. Coastal Mar. Sci. **2**, 185 (1974).

KANISHCHEV, A. D., MENAKER, G. I.: Mean content of 15 ore-forming elements in the earth's crust in Transbaikalia. Geochem. Intern. **11**, 137 (1974).

KARAPETYANTS, M. KH., KARAPETYANTS, M. L.: Thermodynamic Constants of Inorganic and Organic Compounds. Ann Arbor: Humphrey Science Publishers 1970.

KARLSSON, N.: Om molybden i svensk vegetation och mark samt några darmed sammanhangande frågor. Statens Lantbrukskem. Kontrollanstalt Medd. **23**, (1961).

KATCHENKOV, S. M.: The elementary composition of petroleum ash [Russ.]. Dokl. Akad. Nauk SSSR **62** (3), 361 (1948).

KAUTZSCH, E.: Untersuchungsergebnisse über die Metallverteilung im Kupferschiefer. Arch. Lagerstättenforsch. **74** (1942).

KEITH, M. L., DEGENS, E. T.: Geochemical indicators of marine and fresh water sediments. In: ABELSON, P. H. (ed.), Researches in Geochemistry. New York: Wiley 1959.

KHARKAR, D. P., TUREKIAN, K. K., BERTINE, K. K.: Stream supply of dissolved Ag, Mo, Sb, Se, Cr, Co, Rb and Cs to the oceans. Geochim. Cosmochim. Acta **32**, 285 (1968).

KHITAROV, N. I., ARUTYUNYAN, L. A., LEBEDEV, E. B.: Experimental study of separation of molybdenum from granitic melts at water vapor pressures up to 3000 atmospheres. Geochem. Intern. **8**, 730 (1967).

KHITAROV, N. I., ARUTYUNYAN, L. A., MALININ, S. D.: On the possibility of migration of molybdenum in the vapor phase of molybdate salt-ions at elevated temperatures. Geochem. Intern. **2**, 98 (1967).

KHRUSHCHOV, N. A., SHCHERBINA, V. V.: Molybdenum [Russ.]. In: Metally v Osadochnykh Tolshchakh. Moscow: Nauka 1965.

KIDSON, E. B.: Molybdenum content of Nelson soils. New Zealand J. Sci. Technol. **36** A, 38 (1954).

KIM, C. H., ALEXANDER, P. W.: Use of long chain alkydamines for preconcentrated determination of Mo, W, and Re by AA spectroscopy. II: Mo in soil, sediments, and natural water. Zelanda **23**, 229 (1976).

KIM, Y. A., ZEITLIN, H.: The determination of molybdenum in seawater. Limnol. Oceanog. **13**, 534 (1968).

KIMURA, K., YOKOYAMA, Y., IKEDA, N.: Geochemical studies on the minor constituents in mineral springs of Japan. Assoc. Inst. Hydrol. Sci. Assemblie Gen. Rome, Pub. **37**, 200 (1955).

KING, R. U., SHAWE, D. R., MACKEVETT, E. M. Jr.: Molybdenum. U.S. Geol. Surv. Profess. Papers **820**, 425 (1973).

KISVARSANYL, G., PROCTOR, P. D.: Trace element content of magnetites and hematites, SE-Missouri, iron metallogenic province, USA. Econ. Geol. **62**, 449 (1967).

KLEINKOPF, M. D.: Spectrographic determination of trace elements in lake waters of northern Maine. Geol. Soc. Am. Bull. **71**, 1231 (1960).

KNITZSCHKE, G.: Vererzung der Hauptmetalle und Spurenelemente des Kupferschiefer in der Sangerhauser und Mansfelder Mulde. Z. Angew. Geol. **7**, 349 (1961).

KOGA, A.: Chemical studies on Beppu spa. IX Trace elements, Pt. 6. Determination and distribution of molybdenum [Japanese]. J. Chem. Soc. Japan **79**, 461 (1958).

KOHMAN, P.: Search for new natural radioactivity. Conf. Nuclear Processes in Geologic Settings **1**, 10. Washington, Nat. Sci. Found.-Nat. Res. Council (1953).

KONOVALOV, G. S., IVANOVA, A. A., KOLESNIKOV, T. K.: Trace elements, dissolved and particulate, in the chief rivers of the USSR [Russ.]. In: STRAKHOV, N. M. (ed.), Geokhimiya Osadochnykh Porod i Rud. Moscow: Izdat. Akad. Nauk 1968.

KORENBAUM, S. A.: Fiziko-khimicheskie usloviya kristallizatsii mineralov volframa i molybdena v gidrotermalnykh sredakh. Moscow: Izdat. Nauka 1970.

KOROLEV, D. F.: The role of iron sulfides in the accumulation of Mo in sedimentary rocks of the reduced zone. Geochemistry **4**, 452 (1958).

KOSALS, Y. A., MAZUROV, M. P.: Behavior of Mo, W, Sn, Nb, and Ta during formation of the Bitu-Dzhida granitic massif (SW part of Cisbaikalia). Geochem. Intern. **7**, 506 (1970).

KOZLOV, V. D., ROSHCHUPKINA, O. S.: The distribution of Mo in the Paleozoic granitoids of the Undino-Gazimur region (eastern Transbaikalia). Geochemistry **2**, 1066 (1965).

KRAUSKOPF, K. B.: Sedimentary deposits of rare metals. Econ. Geology 50th Anniv. Volume, Part I, 411 (1955).

KRAUSKOPF, K. B.: Factors controlling the concentrations of thirteen rare metals in sea water. Geochim. Cosmochim. Acta **9**, 1 (1956).

KRAUSKOPF, K. B.: Possible role of volatile metal compounds in ore genesis. Econ. Geol. **59**, 22 (1964).

KRAYUSHKIN, V. A., KAZAKOV, S. B., DRABINA, Y. M.: V, Ni and other trase elements in the petroleums of the Bitkovsk field. Geochemistry **4**, 822 (1964).

KREJCI-GRAF, K., WICKMAN, F. E.: Ein geochemisches Profil durch den Lias-alpha. Geochim. Cosmochim. Acta **18**, 259 (1960).

KUBOTA, J., LAZAR, V. A., LANGON, L. W., BEEMS, K. C.: The relationship of soils to molybdenum toxicity in cattle in Nevada. Soil Sci. Soc. Am. Proc. **25**, 227 (1961).

KUBOTA, J., LEMON, E. R., ALLAWAY, W. H.: The effect of soil moisture content upon the uptake of molybdenum, copper, and cobalt by alsike clover. Soil Sci. Soc. Am. Proc. **27**, 679 (1963).

KURODA, K.: Vanadium, chromium, and molybdenum contents of the hot springs of Japan. [Jap.] Bull. Chem. Soc. Japan **14**, 307 (1939).

KURODA, K.: Radium, vanadium, chromium and molybdenum contents of the hot springs of Yunohanazawa, and their seasonal variations. [Japanese] Bull. Chem. Soc. Japan **15**, 65 (1940a).

KURODA, K.: Vanadium, chromium, and Molybdenum contents of deep sea deposits. [Japan.] J. Chem. Soc. Japan **61**, 1060 (1940b).

KURODA, K.: Chemical investigations of deep-sea deposits. IX: Vanadium, chromium, and molybdenum content of deep sea deposits. [Japan.] J. Chem. Soc. Japan **63**, 496 (1943).

KURODA, P. K., SANDELL, E. B.: Geochemistry of molybdenum. Geochim. Cosmochim. Acta **6**, 35 (1954).

LAKIN, H. W., HUNT, C. B., DAVIDSON, D. F., ODA, U.: Variation in minor element content of desert varnish. U.S. Geol. Surv. Profess. Papers **475-B**, 28 (1963).

LAMBERT, I. B., SATO, T.: The Kuroko and associated ore deposits of Japan. Econ. Geol. **69**, 1215 (1974).

LANCELOT, Y., HATHAWAY, J. C., HOLLISTER, C. D.: Lithology of sediments from the western North Atlantic. Leg 11, Deep Sea Drilling Project. In: HOLLISTER, C. D., EWING, J. I., Initial Reports of the Deep Sea Drilling Project **11**, 901 (1972).

LANDERGREN, S.: On the geochemistry of Swedish iron ores and associated rocks. A study on iron-ore formation. Sveriges Geol. Undersökn., Ser. C **496** (1948).

LANDERGREN, S.: On the geochemistry of deep-sea sediments. Rept. Swedish Deep-Sea Exped. **10** (5) (1964).

LANDERGREN, S., MANHEIM, F. T.: Über die Abhängigkeit der Verteilung von Schwermetallen von der Fazies. Fortschr. Geol. Rheinland Westfalen **10**, 173 (1962).

LANDIS, E. R.: Uranium and other trace metals in Devonian and Mississippian black shales in the Central Midcontinent Area. U.S. Geol. Surv. Bull. **1107-E** (1962).

LANGE, J.: Geochemische Untersuchungen an pelagischen Sedimenten des Atlantischen und Pazifischen Ozean (DSDP, Leg I–VII). Doctoral Diss., Univ. Göttingen 1974.

LARSON, M. L.: Organic complexes of molybdenum. Climax Molybdenum Co. Bull. Cdb–**9**, Suppl. 2 (1965).

LE GENDRE, G. R., RUNNELS, D. D.: Removal of dissolved molybdenum from waste water by precipitation of ferric iron. Environ. Sci. Tech. **9**, 744 (1975).

LERICHE, H. H.: The distribution of trace elements in the Lower Lias of southern England. Geochim. Cosmochim. Acta **16**, 101 (1959).

LEROY, L. W., KOSKOY, M.: The lichen: a possible plant medium for mineral exploration. Econ. Geol. **57**, 107 (1962).

LEUTWEIN, F.: Geochemische Untersuchungen an den Alaun- und Kieselschiefern Thüringens. Arch. Lagerstättenforsch. **82**, 1 (1951).

LEUTWEIN, F., RÖSLER, H. J.: Geochemische Untersuchungen an paläozoischen und mesozoischen Kohlen Mittel- und Ostdeutschlands. Freiberger Forschungsh. C **19** (1956).

LEVINSON, A. A.: Introduction to Exploration Geochemistry. Calgary: Applied Publ. Ltd. 1971.

LEWIS, A. H.: The teart pastures of Somerset. II. Relation between the soil and teartness. J. Agr. Sci. **33**, 52 (1943).

LIPSCHUTZ, M. E.: Molybdenum. In: MASON, B. A. (ed.), Handbook of Elemental Abundances in Meteorites. New York: Gordon and Breach 1971.

LISITSIN, A. P.: Osadkoobrazovanie v Okeanakh. Moscow Nauka 1974.

LISITSINA, N. A., BUTUZOVA, G. YU., VOLKOV, I. I., GLAGOLEVA, M. A., SOKOLOV, V. S.: Influence of the Hawaiian vulcanism on sediment accumulations [Russ.]. In: PEIVE, A. V. (ed.), Problemy Litologii i Geokhimii Osadochnykh Porod i Rud. Moscow: Izdat. Nauka 1975.

LOWELL, J. D., GUILBERT, J. M.: Lateral and vertical alteration — mineralization and zoning in porphyry ore deposits. Econ. Geol. **65**, 373 (1970).

L'VOVICH, M. I.: How much fresh water is there on earth? [Russ]. Priroda **5** (1973).

LYAKHOVICH, V. V.: Partition of rarer elements between porphyritic phenocrysts and vitreous groundmass in a vitroandesite. Geochem. Intern. **10**, 793 (1972).

LYAKHOVICH, V. V., BALANOVA, T. T.: On the average content of W, Mo, Sn, Ta, Nb, and Zn in sphene and ilmenite from granitic rocks. Geochemistry **7**, 281 (1969).

MACDONALD, R., BAILEY, D. K.: The chemistry of the peralkaline oversaturated obsidians. U.S. Geol. Surv. Profess. Papers **440-N** (1973).

MACKAY, D. W., HALCROW, W., THORNTON, T.: Sludge dumping in the Firth of Clyde. Marine Pollution Bull. **3**, 7 (1972).

MACLEAN, K. S., LANGILLE, W. M.: Heavy metal studies of crops and soils in Nova Scotia. Commun. Soil Sci. Plant Analy. **4**, 495 (1973).

MALYUGA, D. P.: Biogeochemical Methods of Prospecting. Consultants Bur. New York 1964.

MANHEIM, F. T.: A geochemical profile in the Baltic Sea. Geochim. Cosmochim. Acta **25**, 52 (1961).

MANHEIM, F. T.: Manganese-iron accumulations in the shallow marine environment. In: SCHINK, D. R., CORLESS, J. T. (eds.) Marine Geochemistry. Narragansett Marine Lab. Occ. Publ. **3**, 217 (1965).

MANHEIM, F. T.: Composition and origin of manganese nodules and pavements on the Blake Plateau. In: HORN, D. R. (ed.), Ferromanganese Deposits on the Ocean Floor. U.S. Nat. Sci. Found. (IDOE) **106** (1972).

MANHEIM, F. T.: Analyses of polluted sediments from dump sites in the vicinity of Hudson Canyon. Unpublished data (1973).

MANHEIM, F. T.: Red sea geochemistry. In: WHITMARSH, R. B., WESER, O. E., ROSS, D. A.: Initial Reports of the Deep Sea Drilling Project, Washington, D. C. **23**, 975 (1974).

MANHEIM, F. T., SIEMS, D.: Chemical analyses of Red Sea sediments. In: WHITMARSH, R. B., WESER, O. E., ROSS, D. A.: Initial Reports of the Deep Sea Drilling Project, Washington, **23** B, 023 (1974).

MAN'KOVSKAYA, L. I., SOVGA, E. E.: On the composition and distribution of copper, zinc, and molybdenum in the Caribbean Sea [Russ.]. Morskie Gidrofizicheskie Issledovaniya 200 (1973).

MARMO, V.: On the sulfide and sulfide-graphite schists of Finland. Bull. Comm. Géol. Finlande **190** (1960).

MARTIN, J. M., KULBICKI, G., DE GROOT, A. J.: Terrigenous supply of radioactive and stable elements to the ocean. In: INGERSON, E. (ed.), Proc. Symposium on Hydrogeochemistry and Biogeochemistry, Tokyo, Sept. 7–9, 1970. Washington, D. C.: Clarke Co. 1973.

MASON, B., GRAHAM, A. C.: Minor and trace elements in meteoritic minerals. Smithsonian Contr. in Earth Sci. **3**, 1 (1970).

MATHIAS, M.: The geochemistry of the Messum Igneous Complex, South-West Africa. Geochim. Cosmochim. Acta **12**, 12 (1957).

MAY, S., PICCOT, D., PINTA, G.: Analyse systématique des éléments traces dans l'eau de mer par activation neutronique. J. Radioanal. Chem. **27**, 353 (1975).

MCCABE, C. L.: Sulphur pressure measurements of molybdenum sesquisulfide in equilibrium with molybdenum. J. Metals **7**, 61 (1955).

MEADE, R. H.: Landward transport of bottom sediments in estuaries of the Atlantic coastal plain. J. Sediment. Petrol. **39**, 222 (1969).

MECHAČEK, E.: Mikroelemente in Flözen des Kohlenbeckens von Handlova-Novaky. Geol. Sbornik Geol. Carpathica **23**, 311 (1972).

MERO, J.: Ocean-floor manganese nodules. Econ. Geol. **57**, 747 (1962).

MERO, J.: The Mineral Resources of the Sea. Amsterdam-London-New York: Elsevier Publ. 1965.

MESSER, E.: Kupferschiefer, Sanderz und Kobaltrücken im Richelsdorfer Gebirge. Dissert. Clausthal (1955).

MICHALEK, Z.: Molybdenum in iron sulphide minerals. Bull. Acad. Polon. Sci. Ser. Sci. Chim. Géol. Géograph **6**, 777 (1958).

MICHAILOV, A. S.: Geochemistry of molybdenum in the oxidizing zone [Russ.]. Geokhimiya **11**, 1171 (1964).

MILLIMAN, J. D.: Marine Carbonates. Berlin-Heidelberg-New York: Springer 1974.

MOH, G. H., UDUBASA, G., HÜLLER, R.: Hochtemperatur-Phasengleichgewichte im System Molybdän-Schwefel. Metall. **28**, 804 (1974).

MONGET, J. M., MURRAY, J. W., MASCLE, J.: A world-wide complication of published, multi-component analyses of ferromanganese concretions. Manganese Nodule Project Tech. Report **12**, Washington, D.C.: National Science Foundation (IDOE) 1976.

MORRISON, G. H., GERARD, J. T., KASHUBA, A. T., GANGADHARAM, E. V., ROTHENBERG, A. M., POTTER, N. M., MILLER, G. B.: Multielement analysis of lunar soil and rocks. Science **167**, 505 (1970).

MOSIER, E. L., ANTWEILER, J. C., NISHI, J. M.: Spectrochemical determination of trace elements in galena. U.S. Geol. Surv. J. Res. **3**, 625 (1976).

MOXHAM, R. L.: Distribution of minor elements in coexisting hornblendes and biotites. Can. Mineralogist **8**, 204 (1965).

MURTHY, V. R.: Isotopic anomalies of molybdenum in some iron meteorites. J. Geophys. Res. **67**, 905 (1962).

MURTHY, V. R.: Elemental and isotopic abundance of molybdenum in some meteorites. Geochim. Cosmochim. Acta **27**, 1171 (1963).

NEMOTO, T., HAYAKAWA, M., TAKAHASHI, K., OANA, S.: Report on the geological, geophysical, and geochemical studies of Showa-Shinzan, Usu Volcano [Jap.] Geol. Surv. Japan Rept. **170** (1957).

NICHOLLS, G. D., CURL, H. C., BOWEN, V. T.: Mo in plankton. Limnol. Oceanogr. **4**, 472 (1959).

NOCKOLDS, S. R., ALLEN, R.: The geochemistry of some igneous rock series: Part II. Geochim. Cosmochim. Acta **5**, 245 (1954).

NOCKOLDS, S. R., MITCHELL, R. L.: The geochemistry of some Caledonian plutonic rocks: a study in the relationship between major and trace elements of igneous rocks and their minerals. Trans. Roy. Soc. Edinburgh **61**, 533 (1948).

NODDACK, I., NODDACK, W.: Die Geochemie des Rheniums. Z. Physik. Chemie **154 A**, 207 (1931).

NODDACK, I., NODDACK, W.: Die Häufigkeit der Schwermetalle in Meerestieren. Ark. Zool. **32 A**, 1 (1939).

NOWLAN, G. A.: Concretionary manganese-iron oxides in streams and their usefulness as a sample medium for geochemical prospecting. J. Geochem. Exploration **6**, 193 (1956).

OGORODOV, N. V., KOZHEMYAKA, N. N., VAZHEERSKAYA, A. A., OGORODOVA, A. A.: Volcano Uksichan in the central ridge of Kamchatka [Russ.]. In: NABOKO, S. I. (ed.): Vulkanzm i Geokhemiya Ego Produktov. Moscow: Izdat. Nauka 1967.

OVSYANII, E. I.: Vanadium and molybdenum in Black Sea water. In: Voprosy Fiziki Morya. Sevastopol, Izdat., MGI 1972.

PALACHE, C., BERMAN, H., FRONDEL, C.: The System of Mineralogy of James Dwight and Edward Salisbury Dana (7th ed.), Vol. I and II. New York: John Wiley and Sons 1951.

PATTERSON, E. M.: A petrochemical study of the Tertiary lavas of north-east Ireland. Geochim. Cosmochim. Acta **2**, 283 (1952).

PELTOLA, E.: On the black schists in the Outokumpu region in Eastern Finland. Bull. Comm. Géol. Finlande **192**, 107 p. (1960).

PENTCHEVA, E. N.: Sur les particularités hydrochimiques de certains oligoéléments des eaux naturelles. Annali di Idrologia **5**, 90 (1967).

PERKINS, E. J., GILCHRIST, J. R., ABBOTT, O. J., HALCROW, W.: Trace metals in the Solway Firth sediments. Mar. Pollut. Bull. **4**, 59 (1973).

PERRIN, D. D.: The determination of molybdenum in soils. New Zealand J. Sci. Technol. **27** A, 396 (1946).

PILIPCHUK, M. F.: Some questions of the geochemistry of molybdenum in the Mediterranean Sea [Russ.]. Litologiya i Poleznye Iskopaemye **2**, 25 (1972).

PILIPCHUK, M. F., VOLKOV, I. I.: Molibden v vode Chernogo i Azovskogo morei. [Russ.]. Geokhimiya **8**, 977 (1967).

PILIPCHUK, M. F., VOLKOV, I. I.: Behavior of molybdenum in processes of sediment formation and diagenesis in Black Sea. In: DEGENS, F. T. and ROSS, D. A. (eds.), The Black Sea: Geology, Chemistry, and Biology. Am. Assoc. Petroleum Geologists, Memoir **20**, 542 (1947).

POKALOV, V. T.: Conditions of Formation of Endogene Molybdenum Deposits in the USSR [Russ.]. Moscow: NEDRA 1964.

POLIKARPOCHKIN, V. V., POLIKARPOCHKINA, R. T.: Biogeokhimicheskie Poiski Poleznykh Isokopaemykh. Moscow: Nauka 1964.

POUILLARD, G., PERROT, P.: Free enthalpy of formation and region of stability of molybdenum III and molybdenum IV sulfide [French]. Compt. Rend., Ser. C **281**, 143 (1975).

PRYAKHINA, Yu. A.: Carbonate concretions in the Maikop deposits of the central Fore-Caucasus [Russ.]. Izv. Akad. Nauk, Ser. Geol. **18** (1959).

PUSHKINA, Z. V.: Distribution of trace elements in sediments of the Indian Ocean south of Ceylon. Litologiya i Poleznye Iskopaemye **4**, 34 (1971).

RAAB, W.: Physical and chemical features of Pacific deep sea manganese nodules and their implications to the genesis of nodules. In: HORN, D. R. (ed.), Ferromanganese Deposits on the Ocean Floor. Washington, D.C.: National Science Foundation (IDOE) 1972.

RABINOVICH, A. V., MURAVEVA, A. N., ZHDANOVA, N. V.: Mo content of certain rocks and minerals in the intrusives of eastern Transbaikal. Geochemistry 2, 155 (1958).

RADER, L. F., GRIMALDI, F. S.: Chemical analyses for selected minor elements in Pierre Shale. U.S. Geol. Surv. Profess. Papers **391-A** (1961).

RAMDOHR, P.: Die Erzmineralien in gewöhnlichen magmatischen Gesteinen. Abhandl. Preuß. Akad. Wiss. 2, 1 (1940).

RAMDOHR, P.: The Ore Minerals and their Intergrowths. Oxford: Pergamon Press 1969.

RAZUMNAYA, F. G.: Study of the distribution of molybdenum and vanadium in carbonaceous siliceous shales by centrifuge methods [Russ.]. Sovrem. Metody Mineral. Issled. Gornykh Porod. Rud. i Mineralov 1957.

REISENAUER, H. M.: The role of molybdenum in plants and soils: a supplemental bibliography with abstracts. Climax Molybdenum Co. (undated, probably 1968).

REISENAUER, H. M., TABIKH, A. A., STOUT, P. R.: Molybdenum reactions with soils, and the hydrous oxides of iron, aluminum and titanium. Soil Sci. Soc. Am. Proc. **26**, 23 (1962).

REKHARSKII, V. I.: Molybdenum in intrusive rock minerals [Russ.]. In: FEODOTIEV, K. M. (ed.), Ocherki Geokhimii, Rtuti, Molibdena Sery Gidroterm. Protsesse, Moscow: Nauka 1970.

REKHARSKII, V. I., KRUTETSKAYA, O. V.: Molybdenum in rocks of the southwestern ridges of the N. Tyan Shan [Russ.]. Tr. IGEM **46** (1960).

RILEY, J. P., SINHASENI, P.: Chemical composition of three manganese nodules from the Pacific Ocean. J. Marine Res. **17**, 466 (1958).

RISH, M. A.: Trace elements in desert soils and serozems of Uzbekistan [Russ.]. Geogr. Klassifik Pochv Azii 152 (1965).

ROBINSON, W. O., ALEXANDER, L. T.: Molybdenum content of soils. Soil Sci. Soc. Am. Proc. **75**, 287 (1953).

ROBINSON, W. O., EDGINGTON, G., ARMIGER, W. H., BREEN, A. V.: Availability of molybdenum as influenced by liming. Soil Sci. Soc. Am. Proc. **72**, 267 (1951).

RONOV, A. B., YAROSHEVSKII, A. A.: Earth's crust geochemistry. In: FAIRBRIDGE, R. W. (ed.), Encyclopedia of Geochemistry and Environmental Sciences **4** A. New York-London: Van Nostrand-Reinhold 1972.

ROSE, A. W.: Trace elements in sulfide minerals from the Central District, New Mexico, and the Bingham District, Utah. Geochim. Cosmochim. Acta **31**, 547 (1967).

ROSSMAN, R.: Lake Michigan manganese nodules. Thesis, Univ. Mich. Ann Arbor (1973).

ROWE, J. J., FOURNIER, R. O., MOREY, G. W.: Chemical analysis of thermal water in Yellowstone National Park, Wyoming 1960—1965. U.S. Geol. Surv. Bull. **1303** (1973).

RUNNELS, D. D., CHAPPELL, W. R., MEGLEN, R. R.: The molybdenum project: geochemical aspects. In: FREEDMAN, J., GSA Spec. Paper **155**, 61 (1975).

RZHAKSINSKAYA, M. V.: Trace elements in the soils and plants of the eastern Elbrus area [Russ.]. Nauchn. Dokl. Vysshei Shkoly, Biol. Nauki: Moscow 1965.

SAMSONOV, G. V.: The Oxide Handbook. IFI/Plenum Press 1973.

SANDELL, E. B., GOLDICH, S. S.: The rarer metallic constituents of some American igneous rocks. J. Geol. **51**, 99 and 167 (1943).

SAUCHELLI, V.: Trace Elements in Agriculture. New York: Van Nostrand-Reinhold 1969.

SCHALLER, W. T.: The chemical composition of molybdic ocher. Am. J. Sci. **23**, 297 (1907).

SCHARRER, K.: Biochemie der Spurenelemente (3rd rev. ed.). Berlin and Hamburg: Paul Parey 1955.

SCHARRER, K., SCHROPP, W.: Wasser- und Sandkulturversuche über die Wirkung des Molybdat- und Wolframat-Ions. Z. Pflanzenernähr. Düng. Bodenk. **34** A, 312 (1934).

SCHEELE, C. W.: Forsök med Blyerts, Molybdaena. Kungliga Svenska Akad. Handl. **39**, 247 (1778).

SCHMITZ, H. H.: Untersuchungen am nordwestdeutschen Posidonienschiefer und seiner organischen Substanz. Beihefte Tb. **58**, 1 (1963).

SCHNEIDERHÖHN, H., CLAUS, E., LEUTWEIN, F., PRELL, G., SCHINZINGER, A., SPITZ, W.: Das Vorkommen von Spurenmetallen in deutschen Sedimentgesteinen. Neues Jahrb. Mineral. Monatsh. **A**, 50 (1949).

SCHROEDER, A. in: BROCKKAMP, B.: Zur Paläogeographie und Bitumenführung des Posidonienschiefers im deutschen Lias. Arch. Lagerstättenforsch. **77**, 33 (1944).

SCOTT, J., COLLINS, G. A., HODGSON, G. W.: Trace metals in the McMurray oil sands and other Cretaceous reservoirs of Alberta. Can. Mining Met. Bull. **57**, 34 (1954).

SEIBOLD, E., MÜLLER, G., FESSER, H.: Chemische Untersuchungen eines Sapropels aus der mittleren Adria. Erdöl Kohle **11**, 296 (1958).

SELIG, H.: Search for long-lived natural beta radioactivities. U.S. Atomic Energy Agency Rept. NYO **6626** (1954).

SEN, S., NOCKOLDS, S. R., ALLEN, R.: Trace elements in minerals from rocks of the S. Californian batholith. Geochim. Cosmochim. Acta **16**, 58 (1959).

SEVASTYANOV, V. G., VOLKOV, I. I.: Redistribution of chemical elements during the oxidation process proceeding in the bottom deposits of the oxygen zone of the Black Sea. Tr. Inst. Okeanol. Akad. Nauk, SSSR **83**, 115 (1967).

SHACKLETTE, H. T., HAMILTON, J. C., BOERNGEN, J. G., BOWLES, J. M.: Elemental composition of surficial materials in the conterminous United States. U.S. Geol. Surv. Profess. Papers. **574-D** (1971)

SHERMAN, G. D., KAPTEYN, R. J.: Reexamination of the molybdenum content of Hawaiian soils. Agron. J. **58**, 358 (1966).

SHIMP, N. F., CONNOR, J., PRICE, A. L., BEAR, F. E.: Spectrochemical analysis of soils and biological materials. Soil Sci. Soc. Am. **83**, 51 (1957).

SKOPINTSEV, B. A.: Formirovanie Sovremennogo Khimicheskogo Sostava vod Chernogo Morya. Leningrad, Gidrometeoizdat 1975.

SKORNYAKOVA, N. S.: Chemical composition of ferromanganese nodules of the Pacific [Russ.]. In: BEZRUKOV, P. L. (ed.), Zhelezo-Margantsevyie Konkretsii Tikhogo Okeana, 7, 190. Moscow: Izdat. Nauka 1976.

SKORNYAKOVA, N. S., ZENKEVICH, N. S.: Regularities of the space distribution of ferromanganese nodules [Russ.]. In: BEZRUKOV, P. L. (ed.), Zhelezo-Margantsevyie Konkretsii Tikhogo Okeana, **11**, 37. Moscow: Izdat. Nauka 1976.

SKULMOWSKI, J., WIERCINSKI, J.: Determination of the trace element content in feeds by a polarographic method. III Ann. Univ. Mariae Curie-Sklodowska Lublin, Poland, Sect. DD, Med. Vet. **20**, 179 (1966).

SMALES, A. A., MAPPER, D., FOUCHÉ, K. F.: The distribution of some trace elements in iron meteorites as determined by neutron activation. Geochim. Cosmochim. Acta **31**, 673 (1967).

STEWART, I., LEONARD, C. D.: Molybdenum deficiency in Florida citrus. Nature **170**, 714 (1952).

STRAHL, E. O.: An investigation of the relationships between selected minerals, trace elements, and organic constituents of several black shales. Penn. State Univ. Tech. Rept. NYO-7908 (1959).

STRAKHOV, N. M.: Problemy Geokhimii Sorremennogo Okeanskogo Litogeneza. Moscow: Izdat. Nauka 1976.

STRAKHOV, N. M., NESTEROVA, I. L.: On the influence of vulcanism on the geochemistry of marine deposits, with special reference to the Okhotsk Sea. In: STRAKHOV, N. M. (ed.), Geokhimiya Osadochnykh Porod i Rud. Moscow: Izdat. Nauka 1968.

STRAKHOV, N. M., SHTERENBERG, L. E., KALINENKO, V. V., TIKHOMIROVA, E. S.: Geokhimiya Osadochnogo Margantsovorudnogo Protsessa. Moscow: Izdat. Nauka 1968.

STUDENIKOVA, Z. V., GLINKINA, M. I.: Evolution of the molybdenum and tungsten contents in rocks of the geosynclinal zone of the northern Caucasus and their relationship to the formation of ore deposits. In: VINOGRADOV, A. P. (ed.), Chemistry of the Earth's Crust, Vol. 2. Jerusalem: Israel Program for Scientific Transl. 1967.

SUESS, H. E., UREY, H. C.: Abundances of the elements. Rev. Mod. Phys. **28**, 53 (1956).

SUGAWARA, K.: Migration of elements through phases of the hydrosphere and atmosphere. In: VINOGRADOV, A. P. (ed.), Chemistry of the Earth's Crust, Vol. 2. Jerusalem: Israel Program for Scientific Transl. 1967.

SUGAWARA, K.: Secular change of the chemical composition of sea water. In: AHRENS, L. H. (ed.), Origin and Distribution of the Elements. Oxford: Pergamon Press 1968.

SUGAWARA, K., OKABE, S.: Geochemistry of molybdenum in natural waters (I). J. Earth Sci., Nagoya Univ. **8**, 93 (1960).

SUGAWARA, K., OKABE, S., TANAKA, M.: Geochemistry of molybdenum in natural waters (II). J. Earth Sci., Nagoya Univ. **9**, 114 (1961).

SUMMERHAYES, C. P., WILLIS, J. P.: Geochemistry of manganese deposits in relation to environment on the sea floor around southern Africa. Marine Geol. **18**, 159 (1975).

SUNDIUS, N.: Oil shale and shale oil industry. Sveriges Geol. Undersökn., Ser. C **41**, 42 (1941).

SWAINE, D. J.: The distribution of trace elements in soils. Thesis, Univ. Aberdeen (1951).

SWAINE, D. J.: The Trace-Element Content of Soils. Commonwealth Bureau of Soil Sci., Tech. Comm. 48, Farnham Royal, Bucks, UK, 1955.

SWANSON, V. F., MEDLEN, J. H., HATCH, J. R., COLEMAN, S. L., WOOD, G. H. JR., WOODRUFF, S. D., HILDEBRAND, R. T.: Collection, chemical analysis, and evaluation of coal samples in 1975. U.S. Geol. Surv. Open-File Rept. **76–468** (1976).

SWANSON, V. F., PALACAS, J. G., COVE, A. H.: Geochemistry of deep sea sediment along the 160° Meridian of the N. Pacific Ocean. U.S. Geol. Surv. Profess. Papers **575-B**, 137 (1967).

SZILAGYI, M.: Sorption of molybdenum by humus preparations. Geochemistry **12**, 1489 (1967).

TALAPATRA, A. K.: Sulfide mineralization association with migmatization in the southeastern part of the Singhbum shear zone, Bihar, India. Econ. Geol. **63**, 156 (1968).

TANAKA, M.: Méthode de précipitation à pH constant des oxydes hydrates. Mikrochim. Acta **1988**, 209 (1958).

TAUCINS, E., SVILANE, A.: Content of major and trace elements in sheep wool as a function of their levels in fodder [Russ.]. Tr. Lab. Biokhim. Fiziol. Zhivotn. Inst. Biol. Akad. Nauk Latv. SSR **4**, 247 (1965).

TAUSON, L. V.: Geokhimiya Redkikh Elementov v Granitoidakh. Moscow: Izdat. Akad. Nauk, SSSR 1961.

TAUSON, L. V., PETROVSKAYA, S. G., SANIN, B. P.: Endogenic aureoles of dispersed molybdenum in the Shakhtaminsk mining area. [Russ.] Dokl. Akad. Nauk, SSSR **182** (4) 990 (1968).

TAUSON, L. V., SHEREMET, E. M., ANTIPIN, V. S.: Trends in the distribution of molybdenum in the granitoids of northeastern Transbaikalia. Geochem. Intern. **7**, 637 (1970).

TAYLOR, S. R.: Abundance of chemical elements in the continental crust: a new table. Geochim. Cosmochim. Acta **28**, 1273 (1964).

TAYLOR, S. R.: Geochemistry of Australian impact glasses and tektites (australites). In: AHRENS, L. H. (ed.), Origin and Distribution of the Elements, p. 533. Oxford-London: Pergamon Press 1968.

TER MEULEN, H.: Distribution of molybdenum. Nature **130**, 966 (1932).

TER MEULEN, H., RAVENSWAY, H. J.: The molybdenum content in leaves. Proc. Neth. Acad. Sci., Sect. Sci. **38**, 7 (1935).

TIKHOMIROVA, E. S.: The problem of the geochemical mobility of elements during the formation of sulfide concretions in shale-bearing deposits of the Volga and Baltic basins. Dokl. Akad. Nauk, Earth Sci. Sect. **135** (6) 1098 (1960).

TITLEY, S. R.: Some behavioral aspects of molybdenum in the supergene environment. Trans. AIME **226**, 199 (1963).

TKALICH, S. M.: Phytogeochemical Method of Prospecting for Economic Deposits [Russ.]. Leningrad: Nedra 1970.

TOOMS, J. S., ELLIOTT, I., MATHER, A. L.: Secondary dispersion of molybdenum from mineralization, Sierra Leone. Econ. Geol. **60**, 1478 (1965).

TOURTELOT, E. B.: Selected annotated bibliography of minor-element content of marine black shales and related sedimentary rocks, 1930—1965. U.S. Geol. Surv. Bull. **1292** (1970).

TOURTELOT, H. A.: Preliminary investigation of the geological setting and chemical composition of the Pierre Shale, Great Plains region. U.S. Geol. Surv. Profess. Papers **390** (1962).

TOURTELOT, H. A., SCHULZ, L. G., GILL, J. R.: Stratigraphic variation in mineralogy and chemical composition of the Pierre Shale in South Dakota and adjacent parts of North Dakota, Nebraska, Wyoming, and Montana. U.S. Geol. Surv. Profess. Papers **400-B** (1960).

TRELLES, R. A., AMATO, F. D.: As, V and Mo in the soils and certain strata of the Argentine Republic. Anales. Soc. Cent. Argentina **149**, 93 (1950).

TROBISCH, S., SCHILLING, G.: Über die Bindung des Molybdäns in Thüringer Ackerböden. Chem. Erde **23**, 91 (1963).

TUCHOLKE, B. E., MOUNTAIN, G. S.: Depositional patterns and depositional environment of Cretaceous black clays in the western North Atlantic. GSA Abstr. with Programs **9**, 1207 (1977).

TUREKIAN, K. K.: Rivers, tributaries, and estuaries. In: HOOD, D. W. (ed.), Impingement of Man on the Oceans. New York: Wiley Publ. 1971.

TUREKIAN, K. K., BERTINE, K. K.: Deposition of molybdenum and uranium along the major ocean ridge systems. Nature **229**, 250 (1971).

TUREKIAN, K. K., KHARKAR, D. P.: Major and trace elements and cosmic ray-produced radioisotope in lunar samples. Science **167**, 523 (1970).

TUREKIAN, K. K., WEDEPOHL, K. H.: Distribution of the elements in some major units of the earth's crust. Bull. GSA. **72**, 75 (1961).

URUSOV, V. S., IVANOVA, G. F., KHODAKOVSKII, I. L.: Energetic and thermodynamic properties of molybdates and tungstates and some features of the geochemistry. Geochemistry **10**, 950 (1967).

UZKUT, I.: Zur Geochemie des Molybdäns. Clausthaler Hefte zur Lagerstättenkunde u. Geochemie der Mineralischen Rohstoffe (Heft 12). Berlin: Borntraeger, 1974.

VARENTSOV, I. M., BLASHCHISHIN, A. I.: Iron and manganese concretions [Russ.]. In: GUDELIS, V. K. and EMELYANOV, E. M. (eds.), Geologiya Baltiiskogo Morya. Vilnius Izdat. Mokslis 1976.

VINE, J. D.: Element distribution in some shelf and eugeosynclinal black shales. U.S. Geol. Surv. Bull. **1214-E** (1966).

VINE, J. D., TOURTELOT, E. B.: Element distribution in some Paleozoic black shales and associated rocks. U.S. Geol. Surv. Bull. **1214**-G (1969).

VINOGRADOV, A. P.: The Elemental Composition of Marine Organisms. New Haven: Sears Foundation for Marine Research 1953.

VINOGRADOV, A. P.: Prospecting for ores using plants and soils [Russ.]. Tr. Biogeokhim. Lab. **10**, 3 (1954, 1955).

VINOGRADOV, A. P.: The Geochemistry of Rare and Dispersed Chemical Elements in Soils (2nd ed.). New York: Consultants Bureau 1959.

VINOGRADOV, A. P.: Average contents of chemical elements in some major units of the earth's crust. Geochemistry **7**, 641 (1962).

VINOGRADOV, A. P.: Preliminary data on lunar soil collected in the "Luna 20" automatic station. Geochim. Cosmochim. Acta **37**, 721 (1973).

VINOGRADOV, A. P., MALYUGA, D. P.: Geokhimicheskie Poiski Rudnykh Mestorozhdenie SSSR. Moscow: Gosgeoltekhizdat 1957.

VINOGRADOV, A. P., VINOGRADOVA, K. G.: Molybdenum in the soils of the USSR. [Russ.]. Dokl. Akad. Nauk SSSR **62**, 657 (1948).

VINOGRADOV, A. P., VAINSHTEIN, E. E., PAVLENKO, L. I.: Tungsten and molybdenum in igneous rocks. Geochemistry **5**, 399 (1958).

VINOGRADOV, V. I.: Solubility of the secondary molybdenum minerals in the weak solution of H_2SO_4 and Mo_2CO_3. Geochemistry **3**, 279 (1957).

VISTELIUS, A. B., SARMANOV, O. V.: Dokl. Akad. Nauk SSSR **47**, 4 1947 (cited in PENTCHEVA, 1967).

VOEGELI, P. T., KING, R. U.: Occurrence and distribution of molybdenum in the surface water of Colorado. U.S. Geol. Surv. Water Supply Papers **1935 N** (1969).

VOLKOV, I. I.: Chemical elements in river runoff and forms in which they are supplied to the sea. In: PEIVE, A. V. (ed.), Problemy Litologii i Geokhimii Osadochnykh Porod i Rud (Strakhov 75th Anniv. ed.) Moscow, Izdat. Nauka 1975.

VOLKOV, I. I., FOMINA, L. S.: Influence of organic material and processes of sulfide formation on distribution of some trace elements in deep water sediments of Black Sea. In: DEGENS, E. T. and ROSS, D. A. (eds.), The Black Sea: Geology, Chemistry, and Biology. Memoir A.A.P.G. **20**, 456 (1974).

VOLKOV, I. I., FOMINA, L. A., YAGODINSKAYA, T. A.: Chemical composition of ferromanganese concretions of the Pacific Ocean from Wake Island to the Mexican coast [Russ.]. In: VOLKOV, I. I. (ed.), Biogeokhimiya Diageneza Osadkov Okeana. Moscow, Nauka 1976.

VOLKOV, I. I., SEVAST'YANOV, V. F.: Redistribution of chemical elements during diagenesis [Russ.]. In: STRAKHOV, N. M. (ed.), Geokhimiya Osadochnykh Porod i Rud. Moscow: Izdat. Akad. Nauka 1968.

WAGER, L. R., MITCHELL, R. L.: The distribution of Cr, V, Ni, Co and Cu during the fractional crystallization of a basic magma. Internat. Geol. Congr. Report of the 18th Session, Great Britain, Part II (1948).

WAGER, L. R., MITCHELL, R. L.: The distribution of trace elements during strong fractionation of basic magma—a further study of the Skaergaard intrusion, East Greenland. Geochim. Cosmochim. Acta **1**, 129 (1951).

WAGER, L. R., MITCHELL, R. L.: Trace elements in a suite of Hawaiian lavas. Geochim. Cosmochim. Acta **3**, 217 (1953).

WAGMAN, D. D., EVANS, W. H., PARKER, V. B., HALOW, I., BAILEY, S. M., SCHUMM, R. H.: Selected values of chemical thermodynamic properties. Nat. Bur. Stand. Tech. Note **270-4** (1969).

WARREN, H. V.: Background data for biogeochemical prospecting in British Columbia. Trans. Roy Soc. Can. Sect. III **56**, 21 (1962).

WEAST, R. C. (ed.): Handbook of Chemistry and Physics. Cleveland: CRC Press 1976.

WEBB, J. S.: Geochemistry applied to society [Russ.]. In: VINOGRADOV, A. P. (ed.), Problemy Geokhimii. Moscow: Izdat. Nauka 1965.

WEDEPOHL, K. H.: Spurenanalytische Untersuchungen an Tiefseetonen aus dem Atlantik. Ein Beitrag zur Deutung der geochemischen Sonderstellung von pelagischen Tonen. Geochim. Cosmochim. Acta **18**, 200 (1960).

WEDEPOHL, K. H.: Untersuchungen am Kupferschiefer in Nordwestdeutschland: Ein Beitrag zur Deutung der Genese bituminöser Sedimente. Geochim. Cosmochim. Acta **28**, 305 (1964).

WEDEPOHL, K. H.: Untersuchungen an Proben von Kupferschiefer aus Nordwestdeutschland und Diskussion seiner Bildungsbedingungen. Freiberger Forschungsh. **C 193**, 107 (1965).

WEDEPOHL, K. H.: Chemical fractionation in the sedimentary environment. In: AHRENS, L. H. (ed.), Origin and Distribution of the Elements. Oxford-New York: Pergamon Press 1968.

WEDEPOHL, K. H.: "Kupferschiefer" as a prototype of syngenetic sedimentary ore deposits. Soc. Mining Geol. Japan, Spec. Issue **3**, 268 (1971).

WEISS, H. V., LAI, M. G.: The co-crystallization of ultra-micro quantities of molybdenum with α-benzoinoxine. Talanta **8**, 72 (1963).

WELLS, R. C., STEVENS, R. F.: Further studies of the Kolm. Nat. Acad. Sci. J. **21**, 409 (1931).

WETHERILL, G. W.: Isotopic composition and concentration of molybdenum in iron meteorites. J. Geophys. Res. **69**, 4403 (1964).

WHITE, D. E.: Metal contents of some geothermal brines. Symposium on Problems of Postmagmatic Ore Deposition, Vol. 2, Prague, Czechoslovakia 1965.

WHITE, D. E., HEM, J. D., WARING, G. A.: Chemical composition of subsurface waters. In: FLEISCHER, M. (ed.), Data of geochemistry (6th ed.). U.S. Geol. Surv. Profess. Papers **440-F** (1963).

WILLIAMS, J. H.: Molybdenum deficiency. Gr. Brit. Min. Agric. Fisheries and Food Tech. Bull. **21**, 119 (1966).

WILLIS, J. P., AHRENS, L. H.: Some investigations on the composition of manganese nodules, with particular reference to certain trace elements. Geochim. Cosmochim. Acta **26**, 751 (1962).

WINTER, R. G.: Search for double beta decay in cadmium and molybdenum. Physics Rev. **99**, 88 (1955).

WINTERHALTER, B.: Iron-manganese concretions from the Gulf of Bothnia and the Gulf of Finland [Fin.]. Geoteknillisia Julkaisuja **69**, 5 (1966).

Woo, C. C.: U.S. Geol. Survey, Woods Hole, Mass.: personal communication (1976).

Yamamoto, T., Fujita, T., Shigematsu, T., Ishibashi, M.: Chemical studies on the seaweeds (23). Molybdenum content in seaweeds. Rec. Oceanogr. Works, Japan **9**, 209 (1968).

Young, F. G., Smith, D. G., Langille, W. M.: The chemical composition of sea water in the vicinity of the Atlantic provinces of Canada. J. Fisheries Res. Board, Can. **16**, 7 (1959).

Young, H. E., Guinn, V. P.: Chemical elements in completely mature trees of 7 species in Maine. Tappi **49**, 190 (1966).

Yudovich, Ya. E., Korycheva, A. A., Obruchnikov, A. S., Stepanov, Yu., V.: Mean trace-element contents in coals. Geochem. Intern. **9**, 712 (1972).

Zav'yalov, V. A.: Molybdenum in middle Franskikh rocks of the Pritiman [Russ.]. In: Gulvaeva, L. A. (ed.), Mikroelementary v Kaustobiolitakh i Osadochnykh Porodakh. Moscow: Izdat. Nauka 1965.

Zenkevich, N., Skornyakova, N. S.: Iron and manganese on the sea floor [Russ.]. Priroda **3**, 47 (1961).

Zhavoronkina, T. K.: Spectrographic method for determining a series of microelements in seawater [Russ.]. Tr. Morsk. Gidrofiz. Inst. **19**, 38 (1960).

Zies, E. G.: The fumarolic incrustations in the Valley of the Ten Thousand Smokes. Nat. Geogr. Soc., Wash., D.C. Contr. Tech. Papers Katmai Ser. **1**, 157 (1924).

Zubovic, P., Stadnichenko, T., Sheffey, N. B.: Geochemistry of minor elements in coals of the Northern Great Plains coal province. U.S. Geol. Surv. Bull. **1117-A** (1961).

Zul'fugarly, D. I.: Verbreitung der Spurenelemente in Kaustobiolithen, Organismen, Sedimentgesteinen und Schichtwässern. Leipzig: VEB Deutscher Verlag für Grundstoffindustrie 1964.

Revised manuscript received: February 1978

Ruthenium 44

and the other Platinum Group Elements 45, 46, 76, 77

see Platinum 78

Rhodium 45

and the other Platinum Group Elements 44, 46, 76, 77

see Platinum 78

Palladium 46

and the other Platinum Group Elements 44, 45, 76, 77

see Platinum 78

Silver 47

A A. J. Frueh (Department of Geological Sciences, McGill University, Montreal 2, Canada)

B-M, O E. A. Vincent (Department of Geology and Mineralogy, Parks Road, Oxford, England)

47-A. Crystal Chemistry

Silver is most commonly found in nature as a sulphide or sulphosalt, or as a minor element in a sulphide. It is also to be found (especially in arid regions) combined with a halogen but this is usually a product of the alteration of the sulphide.

Ag usually forms two collinear *sp* bonds or 4 tetrahedral sp^3 bonds and only rarely is found in octahedral sites, except in ionic crystals. However, in the sulphides and sulphosalts Ag is seldom co-ordinated in regular polyhedra but rather in a quite distorted array. When the polyhedra are regular, the Ag is usually located close to two anions rather than in an equidistant position. The physical properties of these compounds (sectility, conductivity etc.) seem to indicate that an appreciable metallic nature must be present in the bonding. Many of these compounds have high temperature phases where the Ag atoms show great mobility, as evidenced by positional disorder and high self-diffusion rates.

Silver sulphide has three known stable polymorphs, and the changes between the phases are rapid, making it impossible to quench in the higher temperature phase. Ag_2S-III, the phase stable from room temperature up to 176.3° C (Kracek, 1946), is monoclinic $P2_1/n$ (9), but can be described as consisting of a very slightly distorted body-centered cubic array of S, with one $Ag_{(I)}$ displaced slightly from the center of one of the cube faces and the other $Ag_{(II)}$ just off the cube edge (Frueh, 1958; Sadanaga and Sueno, 1967). This results in the following interatomic distances and angles.

	Distance	S—Ag—S angle
$Ag_{(II)}$—S	2.54	137.7°
	2.57	106.9°
	2.69	113.8°
	2.99	
$Ag_{(I)}$—S	2.48	
	2.59	157.2°

Ag_2S-II, stable from 176.3° C to 586° C, has a structure based on a body-centered cubic lattice, with the S atoms on the nodes of the lattice and the 4 Ag atoms distributed statistically in the interstitial sites as follows (Rahlfs, *S. B.*, 1936, 110):

$$(000;\ \tfrac{1}{2}\,\tfrac{1}{2}\,\tfrac{1}{2}+)$$

1.5 Ag atoms in $\tfrac{1}{2}\,0\,\tfrac{1}{2},\ \tfrac{1}{4}\,\tfrac{1}{2}\,0,\ 0\,\tfrac{1}{4}\,\tfrac{1}{2},\ \tfrac{1}{2}\,0\,\tfrac{3}{4},\ \tfrac{3}{4}\,\tfrac{1}{2}\,0,\ 0\,\tfrac{3}{4}\,\tfrac{1}{2}$

2.5 Ag atoms in $\tfrac{1}{2}\,0\,0,\ 0\,\tfrac{1}{2}\,0,\ 0\,0\,\tfrac{1}{2}$

$u\ u\ 0,\ u\ u\ 0,\ u\ u\ 0,\ u\ u\ 0$

$u\ 0\ u,\ u\ 0\ u,\ u\ 0\ u,\ u\ 0\ u$ where $u=\tfrac{5}{8}$

$0\ u\ u,\ 0\ u\ u,\ 0\ u\ u,\ 0\ u\ u$

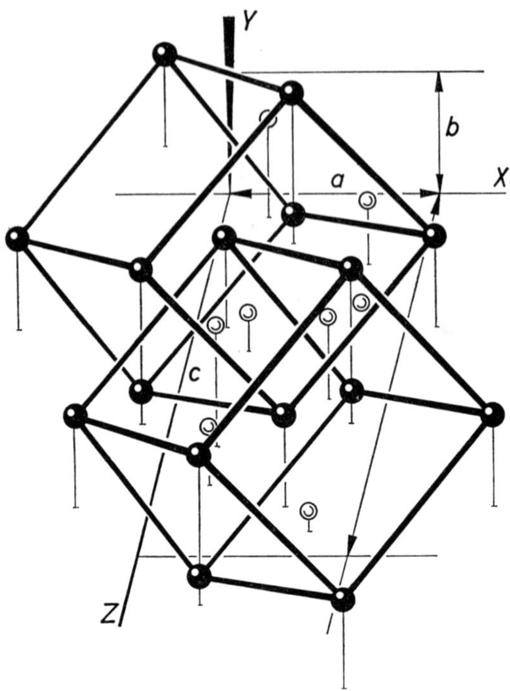

Fig. 47-A-1. The structure of acanthite, Ag_2S-III, dark circles represent sulfur atoms, light circles silver atoms. Projection inclined 20° to *b* axis

Ag_2S-I, stable from 586° C to the melting point 838° C, is probably based upon a face-centered lattice of S atoms with dynamically disordered Ag atoms in the interstices (FRUEH, 1961).

Other sulphides and sulphosalts have the following inter-atomic distances and angles —

	Composition	Ag—S distance	S—Ag—S
Stromeyerite (FRUEH, *S. R.*, 1955, 412)	CuAgS	2.40 Å	180°
Smithite (HELLNER and BURZLAFF, 1964)	$AgAsS_2$		
	Ag_I	2.51 Å, 2.56, 2.68, 2.84	
	Ag_{II}	2.24, 2.28, 2.31	
	Ag_{III} (distorted octahedra)	2.67 (2), 2.82 (2), 3.03 (2)	
	Ag_{IV} (distorted octahedra)	2.57 (2), 2.69 (2), 3.13 (2)	
Miargyrite (KNOWLES, 1964)	$AgSbS_2$	2.47 Å, 2.36 Å	175°
Proustite (ENGEL and NOWACKI, 1966)	Ag_3AsS_3	2.45 Å	162°39′
Pyrargyrite (ENGEL and NOWACKI, 1966)	Ag_3SbS_3	2.45 Å	161°09′
Marrite (WUENSCH, 1967)	PbAgAsS	2.47 Å, 2.52 Å	153°, 89°
		2.68 Å	116°

Ag_2Se also exhibits polymorphism; above 133°, the structure has a body-centered array of Se atoms with the same statistical distribution of Ag atoms as described above for Ag_2S-II (RAHLFS, *S. B.*, 1936, 114). The room-temperature structure has not yet been determined. In eucarite (CuAgSe) the Ag atoms lie in an almost planar-centered array with each Ag having four other Ag atoms at 2.96 Å. Above and below this plane it has one close Se at 2.67 Å and 4 at 3.59 Å. Thus the cohesive forces holding this structure together must in part depend upon Ag—Ag metallic bonds (FRUEH, *S. R.*, 1957, 114).

Ag_2Te also occurs as three polymorphs. From room temperature to 145° C Ag_2Te-III forms a monoclinic structure, where the Ag atoms have two structurally independent positions both of which are tetrahedrally co-ordinated to 4 Te atoms at the following distances and angles (FRUEH, *S. R.*, 1959, 240):

Ag_I—Te	2.91 Å	95°	104°
	3.01 Å	97°	
	2.96 Å	119°	
	2.86 Å		
Ag_{II}—Te	2.91 Å	96°	100°
	2.84 Å	93°	
	3.01 Å	95°	
	3.04 Å		

Ag_2Te-II, stable between 145° C and 802° C, is a face-centered cubic srtucture, with the Te atoms at the nodes of the lattice and the Ag atoms distributed amongst the octahedral holes, the tetrahedral holes and the 16-fold positions at $(\frac{1}{3}\ \frac{1}{3}\ \frac{1}{3})$ (RAHLFS, *S. B.*, 1936, 115). The structure of Ag_2Te-I, stable between 802° C and the melting temperature (~950°), has not been carefully determined, but it is suggested that it consists of a body-centered array of Te atoms with randomly distributed Ag atoms (FRUEH, 1961).

Other tellurides containing Ag have the following interatomic distances.

Mineral	Composition	Ag—Te
Sylvanite (TUNNELL, *S. R.*, 1952, 100)	$AgAuTe_4$	2.69 (2), 2.96 (2), 3.20 (2)
Petzite (FRUEH, *S. R.*, 1959, 154)	Ag_3AuTe	2.90 (2), 2.95 (2)

The halides of silver, cerargyrite $Ag^{[6]}Cl^{[6]}$ and bromyrite, $Ag^{[6]}B^{[6]}$, are characteristic ionic compounds forming a NaCl (B 1) type structure (BARTH, *S. B.*, 1925, 111). These compounds are photosensitive and their solid state chemistry is discussed in almost every text on the photographic process. $Ag^{[4]}I^{[4]}$ exists in three different modifications: miersite a cubic sphalerite-like modification (B 3-type) stable below 137° C (BARTH and LUNDE, *S. B.*, 1925, 113); iodyrite a hexagonal wurtzite-like modification (B 4-type) stable up to 146° C (DERSELB, *S. B.*, 1923, 112); and a high-temperature cubic α form (B 23-type) above 146° C (STROCK, *S. B.*, 1934).

Revised manuscript received: December 1967

References: Section 47-A

ENGEL, P., NOWACKI, W.: Verfeinerung der Kristallstruktur von Proustit und Pyrargyrit. Neues Jahrb. Mineral., Monatsh. 181 (1966).

FRUEH, A. J.: The crystallography of silver sulphide. Z. Krist. **110**, 136 (1958).

FRUEH, A. J.: Use of zone theory in problems of sulphide mineralogy Part III. Polymorphism of Ag_2Te and Ag_2S. Am. Mineralogist **46**, 654 (1961).

HELLNER, E., BURZLAFF, H.: Die Struktur des Smithits $AgAsS_2$. Naturwissenschaften **51**, 35 (1964).

KNOWLES, C. R.: A redetermination of the structure of miargyrite $AgSbS_2$. Acta Cryst. **17**, 847 (1964).

KRACEK, F. C.: Phase relations in the system Sulfur-Silver and the transitions in silver sulfide. Trans. Am. Geophys. Union **27**, 364 (1946).

SADANAGA, R., SUENO, S.: X-ray study on the α—β transition of Ag_2S. J. Mineral. Soc. Japan **5**, 124 (1967).

WUENSCH, B. S., NOWACKI, W.: The crystal structure of marrite, $PbAgAsS_3$. Z. Krist. **125**, 459 (1967).

Revised manuscript received: December 1967

47-B. Isotopes in Nature

There are two stable isotopes of silver, their abundances (as given by STROMINGER *et al.*, 1958, and quoted by other authors) being

^{107}Ag	51.4%
^{109}Ag	48.6%

Variations in the ^{107}Ag/^{109}Ag ratio reported by various researchers from meteorites and terrestrial materials are discussed by BOYLE (1968). The more recent measurements suggest that there is probably little or no significant variation in the ratio, either among terrestrial materials or between terrestrial materials and meteorites. SHIELDS *et al.* (1962) give the figure 1.07597 for the ^{107}Ag/^{109}Ag ratio in terrestrial materials. CROUCH and TURNBULL (1962) give the absolute isotopic abundance ratio as 1.0733 ± 0.0043 and the atomic weight of natural silver (^{12}C = 12) as 107.8694 ± 0.0026.

While both isotopes may be largely primordial in origin, they are also both stable fission products of ^{235}U, ^{107}Ag being derived from ^{107}Pd (half-life 7.5×10^6 years) and ^{109}Ag from ^{109}Pd (half -life 13.6 hours). BOYLE (1968) points out that it is unlikely, bearing in mind the fission yields and the observed isotopic ratio, that the silver of ore deposits has been formed directly by the natural fission of uranium; the two stable isotopes could possibly be the products of ^{235}U fission deep within the Earth over long periods of geological time, having become well-mixed and homogenised in the materials of the upper crust.

Revised manuscript received: May 1974

47-C. Abundance in Cosmos, Meteorites, Lunar Materials and Tektites

I. Cosmos

Green (1959), following Suess and Urey (1956), gives the cosmic abundance of silver as 0.00052 ppm (0.52 ppb) or 0.26 atoms Ag per 10^6 atoms Si ("cosmic abundance units").

More recently Cameron (1968), bearing in mind developments in the theories of nucleosynthesis and leaning heavily upon element abundances in Type I carbonaceous chondrites, has given the appreciably higher estimate of 0.5 atoms Ag per 10^6 atoms Si.

II. Meteorites

Most of the older analyses in the literature, and abundance estimates based upon them, are to be viewed with caution; many reported values for silver obtained by older analytical methods appear high in comparison with recent determinations made by the neutron activation analysis and mass spectrometric isotope dilution methods.

The most complete and self-consistent body of data for the silver content of iron meteorites is that provided by Smales *et al.* (1967); their values are reproduced in Table 47-C-1 a, and lead to an estimate of about 20 ppb as the average Ag content of iron meteorites, the coarse and medium octahedrite groups on the whole containing the most. Chakraburtty *et al.* (1964) have also determined silver in a number of iron meteorites, by neutron activation and by isotope dilution; their values are given in Table 47-C-1b. At the levels of Ag concerned, agreement between the results of the two methods is reasonably satisfactory; Chakraburtty *et al.* cannot explain the wide discrepancies in the cases of the Bristol and Sandia Mountains samples. In the case of the Canyon Diablo iron, the results of Chakraburtty *et al.* are appreciably higher than those of Smales *et al.* (due perhaps to differences in the individual samples analysed); agreement in the case of the Toluca meteorite is good. Dews and Newbury (1966), using isotope dilution analysis, report values between 41 and 93 ppb for the Canyon Diablo iron, their average being 58 ppb.

In the case of the chondritic meteorites, following a few determinations made by Schindewolf and Wahlgreen (1960), Greenland (1967) published results for silver and other trace elements in a series of samples representing the various main types of chondrite, and his data are further supplemented by those of Keays *et al.* (1971) who analysed a number of L (low iron)-type hypersthene-bronzites, and of Laul *et al.* (1973) who analysed 10 enstatite chondrites, 11 low iron-low metal (LL) and 22 high-iron chondrites. Laul and his co-workers (1972) have also provided data for achondrites. All these workers used neutron activation analysis for silver, and their results are assembled in Table 47-C-2.

Table 47-C-1a. *Silver content of iron meteorites* (Smales *et al.*, 1967) (Method: N)

Meteorite	Type	Ag ppb	Meteorite	Type	Ag ppb
La Primitiva	Da	30	Glorieta Mountain	Om	20
Nedagolla	Da	<10	Gun Creek	Om	30
Otumpa	Da	20	Henbury	Om	10
Tombigbee River	Da	10	Kingston	Om	20
Coahuila	H	<10	La Caille	Om	<10
San Martin	H	<10	Mount Edith	Om	<10
Tocopilla	H	<10	Puquios	Om	20
Uwet	H	<10	Rhine Villa	Om	<10
Walker County	H	40	Sams Valley	Om	20
Arispe	Ogg	50	Toluca	Om	10
Gladstone	Ogg	40	Boogaldi	Of	<10
São Julião de Moreira	Ogg	10	Carlton	Of	10
Seeläsgen	Ogg	30	Gibeon	Of	<10
Bendego	Og	30	Grand Rapids	Of	<10
Bennett County	Og	<10	Huizopa	Of	<10
Billings	Og	10	Lockport	Of	10
Bischtübe	Og	30	Moonbi	Of	50
Bohumilitz	Og	30	Obernkirchen	Of	<10
Canyon Diablo	Og	40	Otchinjau	Of	10
Cranbourne	Og	30	Bacubirito	Off	<10
Pan de Azucar	Og	70	Ballinoo	Off	40
Wichita County	Og	50	Mount Magnet	Off	10
Youndegin	Og	50	Salt River	Off	<10
Breece	Om	50	Kopjes Vlei	Gr. Met.	50
Campbellsville	Om	20	Murnpeowe	Gr. Met.	20
Canton	Om	<10	Barranca Blanca	O. Brecc.	40
Cape York	Om	<10	Santa Rosa	O. Brecc.	10
Caperr	Om	<10	Weekeroo Station	O. Brecc.	<10
Chinautla	Om	<10	Babb's Mill	Dr	<10
Clark County	Om	<10	Hoba	Dr	<10
Colfax	Om	70	Illinois Gulch	Dr	<10
Cumpas	Om	30	San Cristobal	Dr	110
Descubridora	Om	<10	Santa Catharina	Dr	30
			Smithland	Dr	<10

Table 47-C-1b. *Silver content of iron meteorites* (Chakraburtty *et al.*, 1964)

Meteorite	Type	Ag, ppb (Method: N/R)	Ag, ppb (Method: I)
Pinon	D	12	18
Sandia Mountains	H	33	79
Canyon Diablo	Og	73	85
Odessa	Og	146	98
Toluca	Om	131	83
Grant	Of	48	82
Bristol	Off	28	181

Table 47-C-2. *Silver content of stone meteorites* (Method: N/R)

Meteorite	Type	Ag ppb	Reference
a) Chondrites:			
Ivuna	Cc 1	360	GREENLAND (1967)
Orgueil	Cc 1	420	GREENLAND (1967)
Mighei	Cc 2	150	GREENLAND (1967)
Murray	Cc 2	190	GREENLAND (1967)
Chainpur	Cc 3 (CHL)	120	GREENLAND (1967)
		57	LAUL *et al.* (1973)
Felix	Cc 3 (CHL)	280	GREENLAND (1967)
Lancé	Cc 3 (CHL)	570	GREENLAND (1967)
Mokoia	Cc 3 (CHL)	120	GREENLAND (1967)
Vigarano	Cc 3 (CHL)	120	GREENLAND (1967)
Warrenton	Cc 3 (CHL)	160	GREENLAND (1967)
Abee	Ce 1	500	GREENLAND (1967)
		316	LAUL *et al.* (1973)
Indarch	Ce 1	410	GREENLAND (1967)
		377	LAUL *et al.* (1973)
Adhi-Kot	Ce 1	220	LAUL *et al.* (1973)
St. Sauveur	Ce (Intermediate)	301	LAUL *et al.* (1973)
St. Mark's	Ce (Intermediate)	189	LAUL *et al.* (1973)
Atlanta	Ce 2	14.4	LAUL *et al.* (1973)
Blithfield	Ce 2	58	LAUL *et al.* (1973)
Hvittis	Ce 2	74	GREENLAND (1967)
		167	LAUL *et al.* (1973)
Khairpur	Ce 2	230	GREENLAND (1967)
		81	LAUL *et al.* (1973)
Pillistfer	Ce 2	52	GREENLAND (1967)
		133	LAUL *et al.* (1973)
Chainpur	CHL	57	LAUL *et al.* (1973)
Parnallee	CL	356	LAUL *et al.* (1973)
Hamlet	CLL	98	LAUL *et al.* (1973)
Kelly	CLL	29	LAUL *et al.* (1973)
Soko-Banja	CLL	40	LAUL *et al.*, (1973)
Olivenza	CLL	57	LAUL *et al.* (1973)
Umbala	CLL	78	LAUL *et al.* (1973)
Jelica	CLL	56	LAUL *et al.* (1973)
Manbhoom	CLL	85	LAUL *et al.*, (1973)
Ottawa	CLL	99	LAUL *et al.* (1973)
St. Mesmin	CLL	554	LAUL *et al.* (1973)
Barratta	CL	104[a]	KEAYS *et al.* (1971)
Beardsley	CL	120	SCHINDEWOLF and WAHLGREN (1960)
Bruderheim	CL	58	GREENLAND (1967)
		62	KEAYS *et al.* (1971)
Farmington	CL	68[a]	KEAYS *et al.* (1971)
Forest City	CL	130	SCHINDEWOLF and WAHLGREN (1960)
Fukutomi	CL	37[a]	KEAYS *et al.* (1971)
Hamlet	CL	177; 78	KEAYS *et al.* (1971)
Holbrook	CL	40	SCHINDEWOLF and WAHLGREN (1960)
Hessle	CL	60	SCHINDEWOLF and WAHLGREN (1960)
Homestead	CL	155[a]	KEAYS *et al.* (1971)

Table 47-C-2 (continued)

Meteorite	Type	Ag ppb	Reference
Khohar	CL	94; 199	KEAYS *et al.* (1971)
Krymka	CL	115[b]	KEAYS *et al.* (1971)
Mezö-Madaras	CL	1,350; 861	KEAYS *et al.* (1971)
Mocs	CL	170	GREENLAND (1967)
Modoc	CL	120	SCHINDEWOLF and WAHLGREN (1960)
		107[b]	KEAYS *et al.* (1971)
Peace River	CL	84	GREENLAND (1967)
Tennasilm	CL	54[a]	KEAYS *et al.* (1971)
Walters	CL	89	GREENLAND (1967)
Allegan	CH	42	GREENLAND (1967)
Allegan (> 200 mesh)	CH	22	LAUL *et al.* (1973)
Pantar	CH	33	GREENLAND (1967)
Pantar (Light)		33	LAUL *et al.* (1967)
Bremervörde	CH	47	LAUL *et al.* (1967)
Sharps	CH	100	LAUL *et al.* (1967)
Tieschitz (> 100 mesh)	CH	98	LAUL *et al.* (1973)
Bath	CH	33	LAUL *et al.* (1973)
Bielokrynitschie	CH	98	LAUL *et al.* (1973)
Ochansk	CH	93	LAUL *et al.* (1973)
Tysnes Island (Light)	CH	170	LAUL *et al.* (1973)
Cangas de Onis	CH	71	LAUL *et al.* (1973)
Pultusk	CH	56	LAUL *et al.* (1973)
Richardton	CH	51	LAUL *et al.* (1973)
Cape Girardeau	CH	240	LAUL *et al.* (1973)
Charsonville A	CH	166	LAUL *et al.* (1973)
Charsonville B	CH	49	LAUL *et al.* (1973)
Doroninsk	CH	38	LAUL *et al.* (1973)
Kernouvé	CH	67	LAUL *et al.* (1973)
Lancon	CH	50	LAUL *et al.* (1973)
Nanjemoy	CH	194	LAUL *et al.* (1973)
Queen's Mercy	CH	32	LAUL *et al.* (1973)
Salles (Light)	CH	53	LAUL *et al.* (1973)
Supuhee (Bulk)	CH	151	LAUL *et al.* (1973)
Supuhee (Light)	CH	76	LAUL *et al.* (1973)
Supuhee (Intermediate)	CH	101	LAUL *et al.* (1973)
Supuhee (Dark)	CH	81	LAUL *et al.* (1973)
Trenzano	CH	198	LAUL *et al.* (1973)
b) Achondrites			
Béréba	Ap	98	LAUL *et al.* (1972)
Constantinople	Ap	30	LAUL *et al.* (1972)
Jonzac	Ap	37	LAUL *et al.* (1972)
Juvinas	Ap	21	LAUL *et al.* (1972)
Sioux County	Ap	5.5	LAUL *et al.* (1972)
Stannern I	Ap	2,100	LAUL *et al.* (1972)
Stannern II	Ap	40	LAUL *et al.* (1972)
Serra de Mage	Ap	3.3	LAUL *et al.* (1972)
Bialystok	Aor	19	LAUL *et al.* (1972)
Frankfort	Aor	17	LAUL *et al.* (1972)
Kapoeta (Light)	Aor	3.1	LAUL *et al.* (1972)
Kapoeta (Dark)	Aor	7.3	LAUL *et al.* (1972)

Table 47-C-2 (continued)

Meteorite	Type	Ag ppb.	Reference
Molteno (Light)	Aor	214	LAUL *et al.* (1972)
Washougal	Aor	3.8	LAUL *et al.* (1972)
Nakhla	Ado	40	LAUL *et al.* (1972)
Lafayette	Ado	58	LAUL *et al.* (1972)
Zagami	Ap ("shergottite")	37	LAUL *et al.* (1972)
Shergotty	Ap ("shergottite")	263	LAUL *et al.* (1972)
Angra dos Reis	Aa	18	LAUL *et al.* (1972)
Pesyanoe (Light)	Ae	13	LAUL *et al.* (1972)
Pesyanoe (Dark)	Ae	74	LAUL *et al.* (1972)

[a] Mean of 2 analyses.
[b] Mean of several analyses.

There appear to be no reliable figures for the concentration of silver in individual phases of meteorites. GOLDSCHMIDT (1954) gives estimates, based upon analyses by NODDACK and NODDACK, and by GOLDSCHMIDT and PETERS, of concentrations of 2—5 ppm Ag in the iron phase, and 18 ppm in troilite. These estimates appear high in the light of the more recent analyses of meteorites quoted above, and modern work on the individual mineral phases of the meteorites is clearly called for.

III. Lunar Materials

Analyses for silver in rocks and soil collected on the Apollo 11 mission were reported in the Proceedings of the Apollo 11 Lunar Science Conference (1970) by four groups of investigators, three using neutron activation analysis and the fourth spark-source mass spectrometry. The reported results are summarized in Table 47-C-3.

Despite careful precautions taken in the laboratories concerned, the rather wide and random discrepancies noted in the levels of Ag found by different groups of analysts suggest that at any rate some of the samples may have become contaminated

Table 47-C-3. *Silver in Apollo 11 Lunar surface samples.* (Values in ppb)

Material type				Method	Reference
A (Crystalline rock, grain size $<{}^{1}/_{2}$ mm)	B (Crystalline rock, grain size >1 mm)	C (Breccia)	D (Soil)		
0.69; 2.27	1.42	23.6	35.4; 53.3	N/R	GANAPATHY *et al.* (1970)
2.3 ± 0.5	4.9 ± 1.8	16.0 ± 5.6	27.1 ± 6.2	N/R	HASKIN *et al.* (1970)
100; 400; 600	70; 20	200; 10	—	Mass	MORRISON *et al.* (1970)
200; 200	650; 80; 80	650;	120	N/R	TUREKIAN and KHARKAR (1970)
64; 52	320; 62; 130	360; 360	120		

Table 47-C-4. *Silver in Apollo 12 Lunar surface samples* (Method: N/R)

Sample	Type	Ag, ppb	Reference
12002	Porphyritic basalt	0.81	ANDERS *et al.* (1971)
12008	Porphyritic vitrophyre	1.17	ANDERS *et al.* (1971)
12017	Porphyritic basalt	1.45	ANDERS *et al.* (1971)
12020	Porphyritic basalt	0.98	ANDERS *et al.* (1971)
12040	Ophitic basalt	0.41	ANDERS *et al.* (1971)
12051	Ophitic basalt	0.81	ANDERS *et al.* (1971)
12065	Porphyritic basalt	1.37	ANDERS *et al.* (1971)
Average		1.00	ANDERS *et al.* (1971)
12022	Porphyritic basalt	11	SMALES *et al.* (1971)
12002	Porphyritic basalt	30	BRUNFELT *et al.* (1971)
12009	Porphyritic vitrophyre	<1	KHARKAR and TUREKIAN (1971)
12020	Porphyritic basalt	<1	KHARKAR and TUREKIAN (1971)
12021	Porphyritic basalt (crushed and homogenised)	274	KHARKAR and TUREKIAN (1971)
12052	Porphyritic basalt	<1	KHARKAR And TUREKIAN (1971)
12063	Porphyritic basalt	<1	KHARKAR and TUREKIAN (1971)
12017	Glassy coating on porphyritic basalt	2.81	ANDERS *et al.* (1971)
12010	Breccia	3.7	LAUL *et al.* (1971)
12013	Breccia	1.7	LAUL *et al.* (1971)
12073	Breccia	2.7	LAUL *et al.* (1971)
12001	Soil	57	KHARKAR and TUREKIAN (1971)
12044	Soil	79	KHARKAR and TUREKIAN (1971)
12070	Soil	79	KHARKAR and TUREKIAN (1971)
12070	Soil	26	SMALES *et al.* (1971)
12070	Soil	140—290	BRUNFELT *et al.* (1971)
12025	Soil, core 1.7 cm	28	LAUL *et al.* (1971)
12028	Soil, core 13.2 cm	301	LAUL *et al.* (1971)
12028	Soil, core 18.9 cm	(140)[a]	LAUL *et al.* (1971)
12028	Soil, core 21.2 cm	3.6	LAUL *et al.* (1971)
12028	Soil, core 37.2 cm	7.2	LAUL *et al.* (1971)
12032	Soil	3.2	LAUL *et al.* (1971)
12033	Soil	12	LAUL *et al.* (1971)
12037	Soil	5.1	LAUL *et al.* (1971)
12057	Soil and chips	(43)[a]	LAUL *et al.* (1971)
12070	Soil	(46)[a]	LAUL *et al.* (1971)

[a] Values in parentheses considered doubtful by the authors concerned, because of possibility of sample contamination etc.

by silver before analysis, possibly from gaskets used in the pressure seals of the boxes in which the samples were brought back from the moon. This possibility would tend to place more reliance on the lower levels of Ag reported by GANAPATHY *et al.* and HASKIN *et al.*, rather than those of MORRISON *et al.* (by spark-source mass spectrometry) and TUREKIAN and KHARKAR, which are very much higher. In any event, the silver contents of the Apollo 11 rocks are not accurately known.

Similar inter-laboratory discrepancy appears in the case of the silver content of the sample returned by the Russian unmanned probe Luna 16. VINOGRADOV (1971) reports 50 and 70 ppb Ag in samples from two different levels in the core, while LAUL *et al.* (1972) found 1,340 and 540 ppb respectively in samples from similar levels; the latter authors again ascribe their high figures to sample contamination at some stage.

Silver contents of Apollo 12 samples were similarly reported in the Proceedings of the Second Lunar Science Conference (1971). Again, there are discrepancies between the results obtained by different research groups, and some contamination of the samples likewise cannot be ruled out. The figures are given in Table 47-C-4; all analyses were made by radiochemical neutron activation methods. In general, it appears that the lunar regolith is preferentially enriched in silver compared with the rocks themselves.

IV. Tektites

There appear to be extremely few published analyses for silver in tektites. COHEN (1959), using emission spectrography, reported <1 ppm Ag in five samples. TAYLOR and SACHS (1960), also using emission spectrography, failed to detect silver in fourteen australites, their detection limit being given as 0.5 ppm.

Until an adequate number of analyses by more sensitive methods becomes available, no meaningful estimate can be made of the silver content of tektites.

Revised manuscript received: May 1974

47-D. Abundance in Rock-Forming and Other Minerals; Silver Minerals

I. Valence State, Behavior

Silver, together with copper and gold, is a member of Group IB of the Periodic Table, with the outer electron configuration $4d^{10}\,5s^1$. Loss of the $5s$ electron leads to the oxidation state $+1$, the commonest valence of silver in its compounds and its most stable ionised condition. Compounds in which silver has valence $+2$ or even $+3$ are known, probably by the loss of electrons to the s and p orbitals of the fifth quantum level to form bonds. Such silver compounds are complexes and only stable under strongly oxidizing conditions, and in nature, aside from the occurrence of the native metal and alloys, only the Ag^{+1} state is likely to be encountered.

In compounds of univalent silver, with the exception of sulfides etc., the element forms bonds of moderately ionic character, but covalent bonding characterises compounds in which the element occurs in higher valence states. Ionic radii, therefore, cannot be defined satisfactorily in general terms but will vary effectively depending upon the anionic environment of the site occupied by silver in any particular crystal structure. While the radius of the Ag atom in the metal is well-defined at 1.44 Å (Pauling, 1960; Zemann, 1966), estimates of the radius of the Ag^+ ion range from 0.97 Å (Wyckoff, 1948) through 1.13 Å (Goldschmidt, 1934; Zemann, 1966) to 1.26 Å (Pauling, 1927; Ahrens, 1952). Ahrens (1952) gives the radius of the Ag^{2+} ion as 0.89 Å. Whittaker and Muntus (1970) in their critical reassessment of ionic radii, list the following values for various coordination patterns with O^{2-}:

Ion	Coordination pattern	Radius, Å
Ag^+	II	0.75
	IV	1.10
	V	1.20
	VI	1.23
	VII	1.32
	VIII	1.38
Ag^{3+}	IV square	0.73

II. Rock-Forming Minerals

Silver is not an essential or major constituent of any rock-forming mineral in the usual sense, and the available data on its distribution at trace levels in such materials leave much to be desired, both in quality and quantity. Boyle (1968) has assembled analyses from the literature and has supplemented these by careful optical

Table 47-D-1. *Silver contents of rock-forming minerals* (Method: S)

Mineral (and number of analyses)	Ag content (or range) ppb	Average Ag content, ppb	Reference
Quartz		<50	BOYLE (1968)
K-feldspar (3)		<50	BOYLE (1968)
Albite (9)	70—1,000	290	BOYLE (1968)
Labradorite (1)	50		BOYLE (1968)
Bytownite (1)	<50		BOYLE (1968)
Muscovite (3)	<50—50		BOYLE (1968)
Biotite (3)	<50—<1,000		BOYLE (1968)
Biotite (19)	100—5,000	660	DE VORE (1955)
Chlorite (8)	100—6,000	980	DE VORE (1955)
Hornblende (4)	50—80	63	BOYLE (1968)
Hornblende (23)	200—3,000	850	DE VORE (1955)
Augite (1)	<1,000		BOYLE (1968)
Aegirine (acmite) (3)	<50—<1,000		BOYLE (1968)
Olivine (3)		<1,000	BOYLE (1968)
Garnet (3)		<1,000	BOYLE (1968)
Garnet (6)	200—15,000	4,950	DE VORE (1955)
Nepheline (2)	<50—50		BOYLE (1968)
Sodalite (2)	70—100	85	BOYLE (1968)
Zircon (3)	80—<1,000		BOYLE (1968)
Tourmaline		<1,000	BOYLE (1968)

spectrographic determinations carried out in the laboratories of the Geological Survey of Canada. Table 47-D-1 summarises these data; the reader is referred to BOYLE's monograph for details.

BOYLE finds silver to be present in nearly all silicates, but seldom exceeding 0.5 ppm (500 ppb); there are no obvious relationships between Ag and any other elements in the common silicates. Apart from the determinations made by the chemists of the Geological Survey of Canada and reported by BOYLE, DE VORE (1955) reports a number of spectrographic determinations of silver in amphiboles, biotites, garnets and chlorites from metamorphic rocks. As will be seen from Table 47-D-1, these minerals have very variable silver contents, but the distribution and manner of occurrence of the element in his materials are not specifically discussed by DE VORE. BOYLE suggests that there might be limited replacement of Na by Ag in some soda-rich silicates such as albite and sodalite, as well as in zeolites; he finds no evidence to support TAYLOR's (1965) suggestion that Ag^{2+} replaces Fe^{2+}, Ca^{2+} and Na^{+} in silicates.

III. Other Minerals

BOYLE (1968) has tabulated in summary form most of the available data, many of those of better quality again emanating from the laboratories of the Geological Survey of Canada. The following information is mainly drawn from BOYLE's work.

Table 47-D-2. *Silver content of common tellurides and selenides* (Boyle, 1968; Sources: Hintze, 1889—1939; Doelter *et al.*, 1911—1931; Palache *et al.*, 1944; Chukhrov *et al.*, 1960—1965; Laboratories, Geological Survey of Canada)

Mineral	Range recorded in % Ag
Tellurides	
Altaite, PbTe	0.43—1.28
Wehrlite, $Bi_{2+x}Te_{3-x}$ (?)	0.48—4.37
Melonite, $NiTe_2$	0.08—0.86
Calaverite, $AuTe_2$	0.40—3.23
Krennerite, $AuTe_2$	0.46—5.87
Montbrayite, Au_2Te_3	0.22—0.55
Nagyagite, $Pb_5Au(Te, Sb)_4S_{5-8}$	up to 1.12
Selenides	
Berzelianite, Cu_2Se	3.51—8.50
Umangite, Cu_3Se_2	0.45—0.55
Penroseite, $(Ni, Cu, Pb)Se_2$	0.59—7.78
Clausthalite, PbSe	traces to small amounts
Klockmannite, CuSe	up to 0.73
Wittite, $Bi_6Pb_5(Se, S)_{14}$	up to 0.19
Selenokobellite, $Pb_2(Bi, Sb)_2(S, Se)_5$	0.36—0.44 (Ödman, 1941)

a) Native Elements

As would be expected, silver reaches high concentrations in native copper (0.1 to 4.0%) and gold (0.1 to 20.0%). Recorded silver contents of native tellurium range from 0.2 to 1.7%, of native antimony 0.01 to 1.0%, and of native iron from 0.0001 to 0.001%. Other native metals contain but traces or minor amounts of silver.

b) Tellurides and Selenides

Silver is also frequently concentrated in minerals of these groups: see Table 47-D-2.

c) Sulfides, Arsenides, etc.

Very variable and often high concentrations of silver have been recorded in minerals of these groups, the case of argentiferous galena being perhaps the best-known. Apart from the inherent difficulties of making really accurate routine analyses for silver, the variability of the figures reported undoubtedly reflects the widespread presence in many sulfide host minerals of admixed intergrowths and exsolution bodies of silver mineral phases which may go undetected unless careful ore-microscopic or X-ray examinations are made. On the other hand, some argentiferous sulfides are optically and structurally homogeneous and the silver may be assumed to substitute for elements like copper and lead in the crystal structure. Boyle's summary is given in Table 47-D-3 and he states (Boyle, 1968, p. 32) that it includes material from the literature to the end of 1965 as well as many analyses done in the laboratories of the Geological Survey of Canada. This table certainly gives the best possible estimates

Table 47-D-3. *Silver content of some common sulfides, antimonides, bismuthides, arsenides, and sulfosalts.* (Boyle, 1968; Sources: Hintze, 1889—1939; Doelter *et al.*, 1911—1931; Palache *et al.*, 1944; Fleischer, 1955; Chukhrov *et al.*, 1960—1965; Laboratories, Geological Survey of Canada)

Mineral	Range in ppm Ag (additional references)
Arsenides and antimonides	
Niccolite, NiAs	10—100 (higher values recorded probably due to impurities)
Safflorite, (Co, Fe)As_2	0.5—100 (higher values recorded probably due to impurities)
Skutterudite, (Co, Ni)As_3	Up to 4% Ag but probably due to impurities
Smaltite-chloanthite, (Co, Ni)As_{3-x}	1—1,000 (values up to 4% recorded but probably due to impurities)
Rammelsbergite, $NiAs_2$	1—10
Aurostibite, $AuSb_2$	Up to 1% Ag or greater
Sulfides and sulfide-arsenides	
Digenite, $Cu_{2-x}S$	10—> 1,000
Chalcocite, Cu_2S	<1—> 1,000 (contents up to 1% Ag recorded but purity uncertain)
Bornite, Cu_5FeS_4	<1—1,000
Galena, PbS	<1—8,000 (values up to 6.4% Ag recorded but probably due to impurities)
Sphalerite, ZnS	3—3,500 (values up to 1.5% Ag recorded but probably due to impurities)
Hawleyite, βCdS	~ 50
Chalcopyrite, $CuFeS_2$	5—3,300
Stannite, Cu_2FeSnS_4	Isolated values up to 1% Ag recorded. High values probably due to impurities
Pyrrhotite, $Fe_{1-x}S$	1—210 (values up to 5,500 ppm recorded but probably due to impurities)
Pentlandite, (Fe, Ni)$_9S_8$	1—50 (average for Sudbury ores—3.4 ppm) (Hawley, 1962)
Covellite, CuS	Traces to 500 ppm
Cinnabar, HgS	<1—10
Stibnite, Sb_2S_3	Traces to 500 ppm
Molybdenite, MoS_2	<1—> 100
Bismuthinite, Bi_2S_3	Traces to 3,000 ppm. Pure material from Quebec and Bolivia show only <5—70 ppm Ag
Pyrite, FeS_2	<1—500 (values up to 1,500 ppm have been recorded on isolated samples)
Cobaltite, (Co, Fe)AsS	1—100
Gersdorffite, NiAsS	1—100
Ullmannite, NiSbS	10—1,000

Table 47-D-3 (continued)

Mineral	Range in ppm Ag (additional references)
Arsenopyrite, $FeAsS$	<1—400
Molybdenite, MoS_2	0.1—100 (GOLDSCHMIDT and PETERS, 1932)
Sulfosalts	
Wittichenite, Cu_3BiS_3	100—1,500
Germanite, (Cu, Ge) (S, As)	50—100
Famatinite, $Cu_3(Sb, As)S_4$	Up to 1,900
Geocronite, $Pb_5(Sb, As)_2S_8$	0.5 to 1.5% (KUZNETSOV, 1957)
Beegerite, $Pb_6Bi_2S_9$	Up to 15.40%, but purity uncertain
Bournonite, $PbCuSbS_3$	Up to 1.69%
Meneghinite, $CuPb_{13}Sb_7S_{24}$	Up to 1,100 ppm
Boulangerite, $Pb_{2-5}Sb_{2-4}S_{5-11}$	Traces to 600; 4,000 to 5,000 (KUZNETSOV, 1959)
Chalcostibite, $CuSbS_2$	Traces to 1,000
Cosalite, $Pb_2Bi_2S_5$	3,200—7,500 (contents up to 1.67% reported)
Selenocosalite, $Pb_2Bi_2(S, Se)_5$	1.02 to 1.20% (ÖDMAN, 1941) (values up to 3.74% reported)
Kobellite, $Pb_2(Bi, Sb)_2S_5$	Up to 3.58%
Selenokobellite, $Pb_2(Bi, Sb)_2(S, Se)_5$	3,600—4,400 (ÖDMAN, 1941)
Franckeite, $Pb_5Sn_3Sb_2S_{14}$	Up to 1%
Cylindrite, $Pb_3Sn_4Sb_2S_{14}$	3,900—6,200
Jamesonite, $Pb_4FeSb_6S_{14}$	40—13,400
Semseyite, $Pb_9Sb_8S_{21}$	1,300—5,600
Cuprobismutite, $CuBiS_2$(?)	Up to 1%
Zinkenite, $Pb_6Sb_{14}S_{27}$	2,300—5,200
Plagionite, $Pb_5Sb_8S_{17}$	Traces to 1,800
Berthierite, $FeSb_2S_4$	Taces to 100

of the ranges of silver contents of the minerals concerned, and a search by the present author of the more recent literature suggests that few, if any, of the figures need modification at present.

Among the commonest sulfide ore minerals other than galena, pyrite seems almost invariably to contain traces of silver, and also, to an even greater degree, does sphalerite. Pyrrhotites generally show much lower contents of the element. In chalcopyrite, there may be appreciable substitution of Ag for Cu in the structure. BADALOV and BADALOVA (1967) report the following average silver contents for associated minerals in the Karamazar and Western Uzbekistan ore deposits, U.S.S.R. (Table 47-D-4). The ranges reported by these authors vary widely, and the minerals

Table 47-D-4. *Silver contents of ore minerals from hypogene deposits of Uzbekistan* (BADALOV and BADALOVA, 1967)

Mineral	Karamazar ores		Western Uzbekistan ores	
	(No. of samples)	Average Ag content, ppm	(No. of samples)	Average Ag content, ppm
Arsenopyrite	(12)	318	(61)	140
Pyrite	(32)	141	(54)	178
Chalcopyrite	(110)	927	(6)	192
Sphalerite	(124)	364	(3)	75
Galena	(311)	1,126	(23)	386
Tetrahedrite	(21)	15,097	—	—
Stibnite	—	—	(4)	27

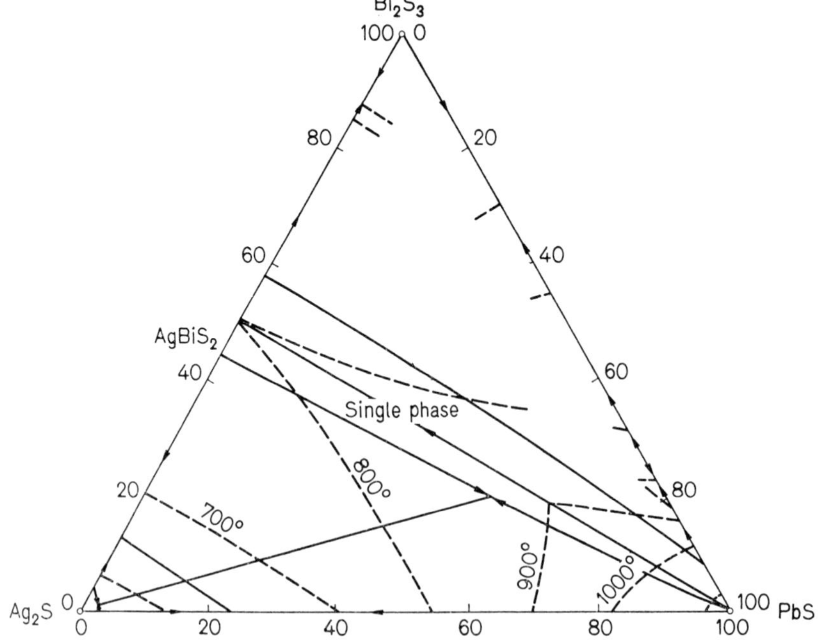

Fig. 47-D-1. Phase relations in the system $Ag_2S—Bi_2S_3—PbS$ (VAN HOOK, 1960)

come from a variety of detailed geological settings: nevertheless these data indicate the kind of distribution pattern that may be expected in other common sulfide mineral associations.

Sulfides such as pyrite exhibit bonding which, although dominantly covalent, takes on a considerable metallic character, so that minor amounts of other metals, including silver, may be expected to substitute for the iron. At the same time, as a chalcophile element silver will form covalent bonds with sulfur, so that it may occur

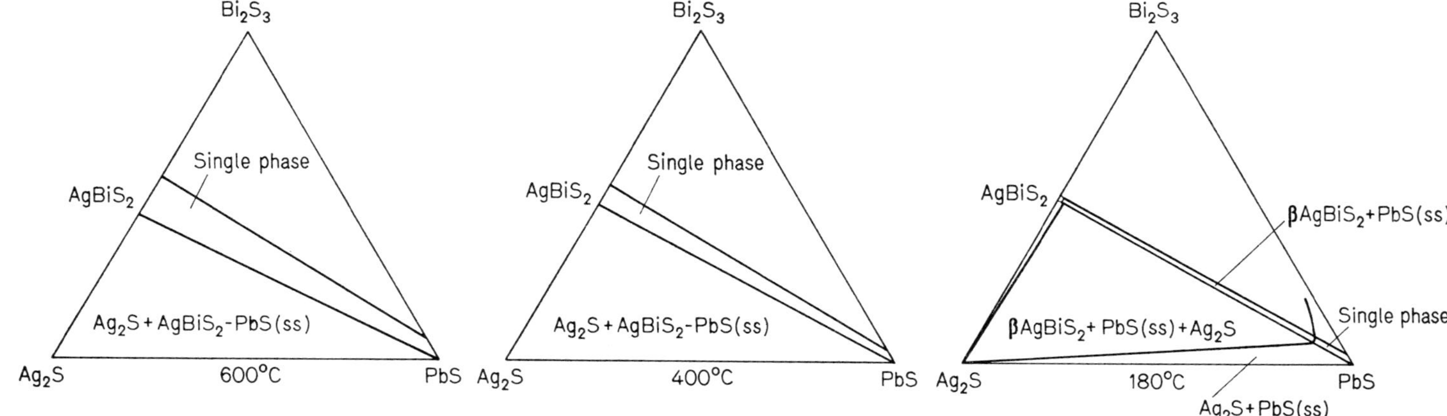

Fig. 47-D-2. Subsolidus equilibria at 600°, 400° and 180° C in the system Ag_2S—Bi_2S_3—PbS (Van Hook, 1960)

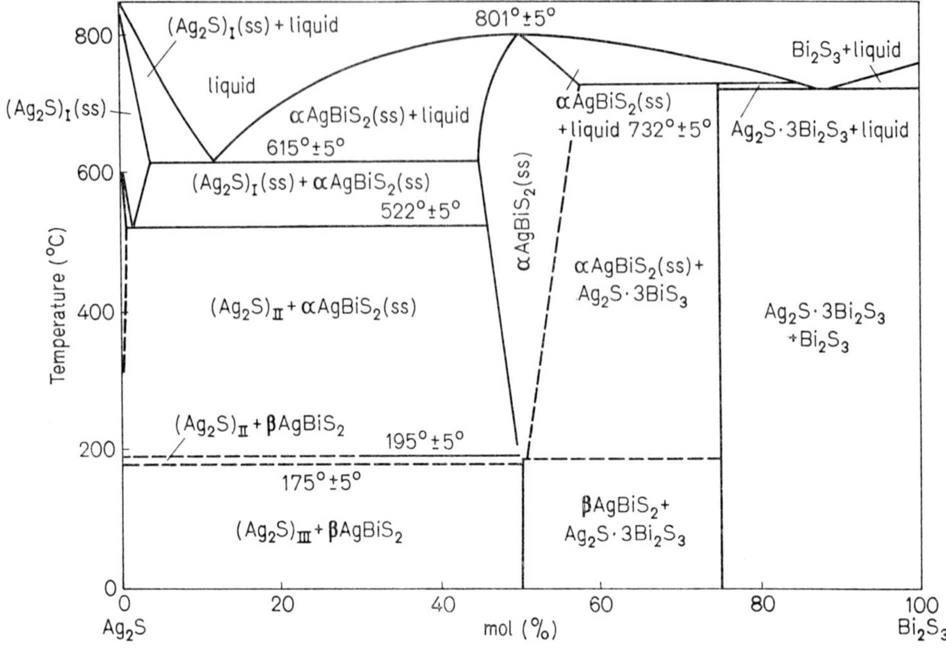

Fig. 47-D-3. The system Ag_2S—Bi_2S_3 (VAN HOOK, 1960)

in pyrite and arsenopyrite both as a substituent for iron in the structure and, to a limited extent, as an interstitial substituent as well, despite rather large differences in the atomic radii of the elements.

In optically and structurally homogeneous zinc sulfide minerals, silver may apparently substitute for some of the zinc. GAUDIN *et al.* (1951) suggest the rapid ion exchange of Ag for Zn, followed by a slow solid-state diffusion reaction, the Ag finally replacing Zn in structural sites in the rather open sphalerite lattice, or perhaps residing in interspaces.

Modern experimental work shows that Ag_2S has a maximum solubility in PbS of around 0.4 mol-%. A great many argentiferous galenas carry microscopic intergrowths or exsolution bodies of silver-rich phases like acanthite, matildite or miargyrite, and laboratory studies in the Ag_2S—Bi_2S_3—PbS plane in the system Ag—Bi—Pb—S by VAN HOOK (1960) indicate extensive ternary solid solution at higher temperatures between matildite ($AgBiS_2$) and galena with Bi_2S_3 on the one side and Ag_2S on the other (Fig. 47-D-1). The degree of solid solution either one side of the $AgBiS_2$—PbS join (especially on the Ag_2S side) diminishes with falling temperature (Fig. 47-D-2). VAN HOOK's diagram for the $AgBiS_2$—PbS join is shown in Fig. 47-D-3; cubic solid solutions persist over most of the compositional range down to temperatures of about 200° C, much $AgBiS_2$ unmixing at 195° C, which is the α—β inversion temperature of matildite. Even so, it is seen that as much as 10% of the $AgBiS_2$ component remains in solid solution in galena at about 170° C, and still several percent at room temperature. From these results it appears that the presence

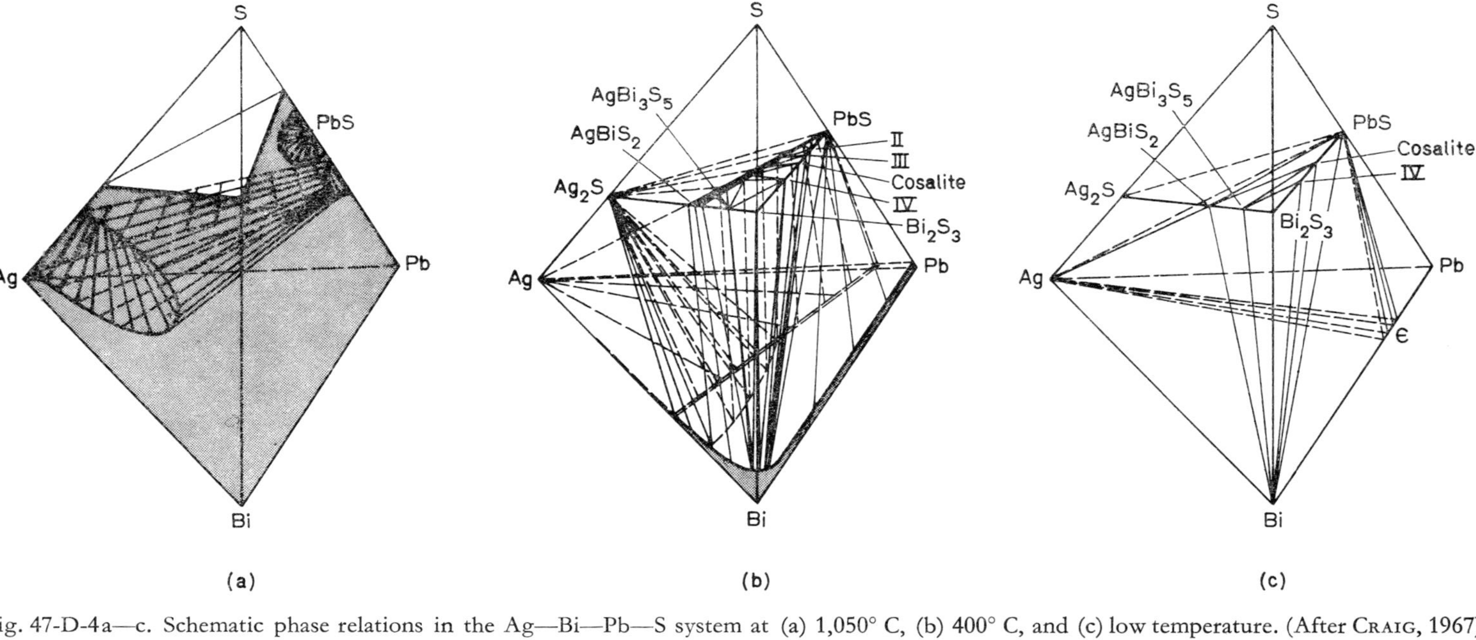

Fig. 47-D-4a—c. Schematic phase relations in the Ag—Bi—Pb—S system at (a) 1,050° C, (b) 400° C, and (c) low temperature. (After CRAIG, 1967)

of bismuth greatly increases the solubility of Ag in PbS, even at low temperatures, while VAN HOOK further suggests that the presence of antimony leads to similar relationships involving the phase miargyrite ($AgSbS_2$), possibly explaining the common occurrence of Ag, Bi and Sb together in natural galenas; GODOVIKOV (1967) also confirms that $AgSbS_2$ forms solid solutions in galena along with $AgBiS_2$ and considers that the presence of significant amounts of antimony in galena is a useful criterion for discovering galena with high silver content.

A very detailed investigation of the phase relationships and mineral assemblages in the whole quaternary Ag—Bi—Pb—S system has been reported by CRAIG (1967), whose results for the Ag_2S—Bi_2S_3—PbS plane are in broad agreement with those of VAN HOOK (1960). CRAIG draws attention to the occurrence in the system of the rare sulfosalt pavonite, $AgBi_3S_5$, the composition of which is indicated in VAN HOOK's diagram for the Ag_2S—Bi_2S_3 join (Fig. 47-D-3), but not in his other diagram.

Fig. 47-D-4 shows CRAIG's schematic phase relations in the quaternary system at various temperatures. At 1,050° C, a large two-liquid field (unshaded in the diagram) extends some way towards the Ag—Bi—Pb base of the diagram; for clarity, tie lines are omitted in this field. A large field of metal-rich homogeneous liquids (shaded) extends upwards from the Ag—Bi—Pb base of the diagram to meet the two-liquid sulfur-rich field, and enclosed within this is a further two-liquid volume extending across from the Ag—Bi—S face to span the Ag—Pb—S face; tie-lines are indicated within this two-liquid volume. This large field of homogeneous liquid extends through the central portion of the system separating the two regions of liquid immiscibility. There is also a smaller field of PbS and liquid.

At 400° C, reduction and disappearance of the two-liquid fields, and the appearance of the phases indicated, lead to the diagram shown in Fig. 47-D-4(b), in which the Ag_2S—Bi_2S_3—PbS plane may be compared with VAN HOOK's diagram in Fig. 47-D-2, and in which the tie-lines between ternary phases and liquid sulfur are omitted for clarity. At low temperature the diagram has the form shown in Fig. 47-D-4(c); the reader is referred to CRAIG's paper for details.

The crystal chemistry of argentiferous galena is lucidly reviewed by BOYLE (1968, pp. 37—39) and the reader is referred to this source for further details.

d) Halides, Carbonates, Sulfates, Oxides, etc.

There are few reliable analyses for the silver contents of minerals of these groups, and certainly insufficient for meaningful general averages or even ranges to be quoted. Again, most of the available data are summarised in BOYLE's monograph, and the examples given in Table 47-D-5 are taken from that source. Except where otherwise stated in the table, the analyses were made in the laboratories of the Geological Survey of Canada by emission spectrographic methods.

High silver contents are encountered in certain copper and lead minerals of these groups (e.g. marshite, percylite, phosgenite, malachite, anglesite, pyromorphite, bindheimite) where Ag probably substitutes fairly readily for Cu and Pb in the crystal structure; the reasons for high silver contents sometimes recorded in hydrous oxides

Table 47-D-5. *Silver content of some halides, carbonates, sulfates, phosphates, oxides etc.* (BOYLE, 1968); (Method: S)

Mineral	Location	Ag ppm (range)
Halides		
Halite, NaCl	Various	<0.5
Sylvite, KCl, and carnallite, $KMgCl_3 \cdot 6H_2O$	Südharz district (Germany) (HERRMANN, 1958)	0—0.02
Fluorite, CaF_2	Various	1.0
Carbonates		
Calcite, $CaCO_3$	Various, Canada	<1.0
Siderite, $FeCO_3$	Various, Canada	<1.0—50
Smithsonite, $ZnCO_3$	Keno Hill, Y.T. (Canada)	1—50
Cerussite, $PbCO_3$	Keno Hill, Y.T. (Canada)	1—50
Malachite, $Cu_2(CO_3)(OH)_2$	Various, Canada	50—1,000
Sulfates		
Barite, $BaSO_4$	(vein) Keno Hill, Y.T. (Canada), Rudnany (Czech.) (BERNARD, 1961)	<1—5 10—50
Anhydrite, $CaSO_4$	Walton, N.S. (Canada)	<0.05—0.4
Gypsum, $CaSO_4 \cdot 2H_2O$	Walton, N.S. (Canada)	<0.05—0.4
Phosphates		
Apatite, $Ca_5(PO_4)_3(OH, F, Cl)$	Various	<0.1
Pyromorphite, $Pb_5(PO_4)_3Cl$	Keno Hill, Y.T. (Canada)	100—1,000
Oxides		
Cuprite, Cu_2O	Chile, Arizona, Mexico	<1—300
Haematite, Fe_2O_3	Walton, N.S. (Canada) Rudnany (Czech.) (BERNARD, 1961)	<5 up to 100
Limonite, $Fe_2O_3 \cdot nH_2O$	Keno Hill, Y.T. (Canada) (in oxidized zones)	100—1,000
	Keno Hill, precipitated from springs	<0.5
	Walton, N.S. (Canada) (in oxidized zones)	0—22
Pyrolusite, MnO_2	Thüringen (Germany) (FISCHER, 1958—59)	3
Wad, and various manganese oxides	U.S.A. (HEWETT *et al.*, 1963)	30—300
Magnetite, Fe_3O_4	Various	1—5
Chromite, $FeCr_2O_4$	Various	<1—5

such as limonite and wad are not so clear, but in the case of wad, they may stem from the adsorption of positive silver ions on the surface of negatively-charged hydrous manganese oxide sols (see BOYLE, 1968, pp. 40—46).

IV. Silver Minerals

Silver minerals are listed in Table 47-D-6, among the commonest being native silver, acanthite, pyrargyrite, proustite, tetrahedrite-tennantite, chlorargyrite and argentojarosite.

Table 47-D-6. *Silver Minerals*

Native element and alloys	
Argentian gold (electrum)	(Au, Ag)
Silver	Ag
Moschellandsbergite	Ag_2Hg_3
Silver amalgam	Ag, Hg
Sulfides, antimonides, arsenides, tellurides, etc.	
Dyscrasite	Ag_3Sb
Novakite	$(Cu, Ag)_4As_3$
Argentite	αAg_2S
Acanthite	βAg_2S
Aguilarite	Ag_4SeS
Naumannite	Ag_2Se
Crookesite	$(Cu, Tl, Ag)_2Se$
Eucairite	CuAgSe
Hessite	Ag_2Te
Petzite	Ag_3AuTe_2
Stromeyerite	$Ag_{1-x}CuS$
McKinstryite	$Cu_{0.8}Ag_{1.2}S$
Sternbergite	$AgFe_2S_3$
Argentopyrite	$AgFe_2S_3$
Empressite	AgTe
Stuetzite	$Ag_{5-x}Te_3$
Muthmannite	(Ag, Au)Te
Sylvanite	$(Au, Ag)Te_4$
Sulfosalts	
Polybasite	$(Ag, Cu)_{16}Sb_2S_{11}$
Pearceite	$(Ag, Cu)_{16}As_2S_{11}$
Polyargyrite (?)	$Ag_{24}Sb_2S_{15}$
Argyrodite	Ag_8GeS_6
Canfieldite	Ag_8SnS_6
Stephanite	Ag_5SbS_4
Pyrargyrite	Ag_3SbS_3
Proustite	Ag_3AsS_3
Pyrostilpnite	Ag_3SbS_3
Stylotypite	$(Ag, Cu, Fe)_3SbS_3$
Xanthoconite	Ag_3AsS_3
Tetrahedrite-tennantite	$(Cu, Fe, Ag)_{12}(Sb, As)_4S_{13}$
Samsonite	$Ag_4MnSb_2S_6$
Lengenbachite	$Pb_6(Ag, Cu)_2As_4S_{13}$
Diaphorite	$Pb_2Ag_3Sb_3S_8$
Freieslebenite	$Pb_3Ag_5Sb_5S_{12}$
Owyheeite	$Pb_5Ag_2Sb_6S_{15}$
Schirmerite	$PbAg_4Bi_4S_9$
Miargyrite	$AgSbS_2$
Aramayoite	$Ag(Bi, Sb)S_2$
Matildite	$AgBiS_2$
Pavonite	$AgBi_3S_5$

Table 47-D-6 (continued)

	Smithite	$AgAsS_2$
	Trechmannite	$AgAsS_2$(?)
	Benjaminite	$Pb(Cu, Ag)Bi_2S_4$
	Fizelyite	$Pb_5Ag_2Sb_8S_{18}$
	Ramdohrite	$Pb_3Ag_2Sb_6S_{13}$
	Andorite	$PbAgSb_3S_6$
	Hutchinsonite	$(Pb, Tl)_2(Cu, Ag)As_5S_{10}$
	Betechtinite	$(Cu, Fe)_{11}(Pb, Ag)S_7$(?)
	Marrite	$AgPbAsS_3$
Halides		
	Chlorargyrite	AgCl
	Huantajayite	(Na, Ag)Cl
	Ostwaldite	Colloidal-like AgCl
	Bromargyrite	AgBr
	Miersite	(Ag,Cu)I
	Iodargyrite	AgI
	Boleite	$Pb(Cu, Ag)Cl_2(OH)_2 \cdot H_2O$
Basic sulfates		
	Argentojarosite	$AgFe_3(SO_4)_2(OH)_6$
	Argentian plumbojarosite	$(Pb, Ag)Fe_{3-6}(SO_4)_{2-4}(OH)_{6-12}$

Native silver is seldom found pure, but usually contains appreciable Au, Hg or other elements including As, Sb, Bi, Te, Cu, Fe, Zn, Pb, Co, Ni, Pt, and Ir. Silver and gold form a continuous series of alloys, from silver through aurian silver to argentian gold ("electrum" with $>20\%$ Ag) and gold. Silver amalgam ("arguerite"; Ag, Hg) containing up to 15—20% Hg, is known from various Canadian ore deposits (Boyle, 1968, p. 19), and from other localities throughout the world (Palache *et al.*, 1944).

Acanthite, βAg_2S, has monoclinic structure and is the stable form of silver sulfide, inverting from the high-temperature body-centred cubic form, αAg_2S (argentite) at 173° C (Frueh, 1958); natural silver sulfide is thus always acanthite. The mineral, most commonly of hypogene origin, generally contains some copper.

Proustite and pyrargyrite ("Ruby Silver"), Ag_3AsS_3 and Ag_3SbS_3, form a complete solid solution series at least down to 300° C. Boyle (1968) remarks that substitution of other elements seldom amounts to more than about 500 ppm.

Tetrahedrite and tennantite, $(Cu, Fe, Zn, Ag)_{12}Sb_4S_{13}$—$(Cu, Fe, Zn, Ag)_{12}As_4S_{13}$, are usually hypogene minerals exhibiting wide ranges of substitution of various metallic elements for Cu, As and Bi for Sb, and Se and Te for S. Palache *et al.* (1944) give the probable maximum silver content of minerals of this group as 14—18%, but remark that the estimate is based partly on older analyses of dubious accuracy and which may not always relate to pure materials.

Argentojarosite, $AgFe_3(SO_4)_2(OH)_6$, is formed in supergene conditions in the weathering of silver deposits containing pyrite, and is usually associated with limonite etc. Although first described from, and apparently common at the Tintic Standard mine, Utah, Boyle (1968) remarks that argentojarosite may well prove to be more common in some deposits than hitherto suspected.

Revised manuscript received: May 1974

47-E. Abundance in Common Igneous Rock Types

VINOGRADOV (1962), in his critical consideration of the average contents of elements in principal rock types, placed special stress upon accuracy of analytical methods used by various investigators, as well as upon numbers of samples analysed, and gives the following estimates for the silver contents of major groups of igneous rocks:

Ultramafic rocks (dunites, etc.)	50 ppb Ag
Mafic rocks (basalts, gabbros etc.)	100 ppb Ag
Intermediate rocks (diorites, andesites, etc.)	70 ppb Ag
Felsic rocks (granites, granodiorites, etc.)	50 ppb Ag

TUREKIAN and WEDEPOHL (1961) give similar average estimates. Like VINOGRADOV, these authors base their estimates largely on the comprehensive research into the silver contents of Japanese rocks by HAMAGUCHI and KURODA (1959), whose averages are reproduced in Table 47-E-1. BOYLE (1968, p. 49) also reports the silver contents of a less comprehensive series of Canadian igneous rocks, most of his figures ranging between about 0.1 and 2 ppm (100—2,000 ppb), appreciably above the levels in most of the Japanese rocks.

Silver has been determined by various investigators in the reference silicate rock samples distributed by the United States Geological Survey, and the published results, mainly as summarised by FLANAGAN (1969), together with figures for the Canadian "C.A.A.S. Syenite—1" sample, are given in Table 47-E-2.

As far as the limited data on the silver contents of igneous rocks at present allow, it must be concluded that there is probably no strong correlation between the silver content and overall chemical composition of the rock, except that silver, as a chalcophile element, may tend to be more concentrated in rocks carrying appreciable traces of sulfide minerals — in general, the more mafic igneous rocks. This is to some extent borne out by the behavior of silver in a suite of samples from the differentiated

Table 47-E-1. *Average silver contents of Japanese igneous rocks* (HAMAGUCHI and KURODA, 1959; analytical method: W/S)

Rock type	(No. of samples averaged)	Range ppb Ag	Average ppb Ag
Granites	(7)	30— 42	37
Granodiorites	(27)	20— 90	51
Liparites	(4)	25— 90	49
Diorites	(8)	20— 85	52
Andesites etc.	(24)	30—190	80
Basalts	(8)	46—190	100
"Diabases"	(5)	80—140	120
Gabbros	(5)	68—140	110
Ultramafic rocks	(5)	30— 80	60

Table 47-E-2. *Silver contents of some "standard" reference igneous rock samples*

Rock type and designation	(Number of analyses averaged)	Ag content ppb	Method	Reference
Granite, G-1,	(9)	42	N/R	MORRIS and KILLICK (1960)
U.S.G.S.	(3)	46	N/R	VINCENT and ADAMS (unpubl.)
	(3)	52	N/R	GREENLAND and FONES (1971)
		40	Mass	TAYLOR (1965)
Granite, G-2,	(2)	40	N/R	BRUNFELT and STEINNES (1968)
U.S.G.S.	(3)	49	N/R	GREENLAND and FONES (1971)
	(2)	40	S	LE RICHE (1967)
	(6)	53	S	CHAMP (1968)[a]
Granodiorite,	(2)	84	N/R	BRUNFELT and STEINNES (1968)
GSP-1, U.S.G.S.	(3)	100	N/R	GREENLAND and FONES (1971)
	(2)	20	S	LE RICHE (1967)
	(6)	82	S	CHAMP (1968)[a]
Andesite, AGV-1,	(2)	94	N/R	BRUNFELT and STEINNES (1968)
U.S.G.S.	(3)	110	N/R	GREENLAND and FONES (1971)
	(2)	60	S	LE RICHE (1967)
	(6)	110	S	CHAMP (1968)[a]
Basalt ("diabase"),	(6)	57	N/R	MORRIS and KILLICK (1960)
W-1, U.S.G.S.	(3)	70	N/R	VINCENT and ADAMS (unpubl.)
	(3)	81	N/R	GREENLAND and FONES (1971)
		30	Mass	TAYLOR (1965)
Basalt, BCR-1,	(2)	36	N/R	BRUNFELT and STEINNES (1968)
U.S.G.S.	(3)	36	N/R	GREENLAND and FONES (1971)
	(2)	20	S	LE RICHE (1967)
	(6)	54	S	CHAMP (1968)[a]
Peridotite, PCC-1,	(2)	9.1	N/R	BRUNFELT and STEINNES (1968)
U.S.G.S.	(2)	5	N/R	GREENLAND and FONES (1971)
	(2)	10	S	LE RICHE (1967)
	(6)	50	S	CHAMP (1968)[a]
Dunite, DTS-1,	(2)	10.4	N/R	BRUNFELT and STEINNES (1968)
U.S.G.S.	(15)	8	N/R	GREENLAND and FONES (1971)
	(2)	20	S	LE RICHE (1967)
	(6)	53	S	CHAMP (1968)[a]
Syenite rock-1, C.A.A.S.		Range 300—5,700 Median value 800	S	WEBBER (1965); SINE *et al.* (1969)

[a] Analyses made in laboratories of Geological Survey of Canada, Ottawa, and reported by FLANAGAN (1969).

Skaergaard basic layered intrusion of east Greenland (VINCENT and ADAMS, unpublished work). The individual analyses (made by a neutron activation method) are set out in Table 47-E-3. So far as the limited number of analyses allows of interpretation, there seems little or no systematic variation in the silver content of average rocks of the intrusion with progressive fractionation. It seems likely that the element will exist largely or mostly in the form of uncharged atoms and not in the form of an

Table 47-E-3. *Silver contents of rocks from the differentiated Skaergaard Intrusion, East Greenland.* (Analyses: VINCENT and ADAMS, unpublished results) (Analytical method: N/R)

Rock No.	Description[a]	Average ppb Ag
Late differentiates		
E.G. 5259	Transgressive salic granophyre (T. Gr)	220
4489	Transgressive hedenbergite granophyre (UBGγ)	56
Layered Series Rocks		
4332	Melanogranophyre (UBGγ)	71
4328	Fayalite ferrodiorite (UZc) (carries ~ 2% wt. iron sulfides)	35
5196	Melanocratic hortonolite ferrodiorite (UZb) (carries ~ 0.4% wt. copper sulfides)	700
5181	Average hortonolite ferrodiorite (UZa) (carries ~ 0.03% wt. copper sulfides)	460
5051	Melanocratic "olivine-free" gabbro (MZ)	120
5052	Average "olivine-free" gabbro (MZ)	70
5086	Average hypersthene-olivine-gabbro (LZb)	160
5087	Average hypersthene-olivine-gabbro (LZb)	79
Marginal Rocks		
4443	Coarse olivine gabbro (WBG)	39
4526	Gabbro-picrite (NBG)	65
4507	Chilled marginal gabbro (SBG)	110

[a] See WAGER and BROWN (1968). UBG = Upper Border Group; UZ = Upper Zone; MZ = Middle Zone; LZ = Lower Zone; WBG = Western Border Group; NBG = Northern Border Group; SBG = Southern Border Group.

ion which will at all readily substitute for any major ion in the common rock-forming silicates. From Table 47-E-3, however, it will be seen that the two Upper Zone rocks, 5181 and 5196, carrying microscopic traces of copper and copper-iron sulfide minerals, contain appreciably more silver than the remainder of the rocks analysed, and it seems reasonable to suppose that the element is strongly concentrated in the accessory copper sulfide phases. It is striking, and may be significant, that the fayalite ferrodiorite 4328, which carries about 2% of accessory iron sulfide minerals (mainly marcasite replacing pyrrhotite: WAGER *et al.*, 1957), has the lowest silver content of any Skaergaard rock analysed. Apparently, in this case, the chalcophile character of silver does not extend to its preferential concentration in iron sulfide, and its close coherence is with the other group I B metal, copper. As remarked earlier, pyrite often contains appreciable traces of silver and, for example, BOYLE (1968, p. 49) remarks that the relatively gold- and silver-enriched quartz-feldspar porphyries in the Yellowknife Greenstone Belt in Canada carry appreciable (probably) syngenetic pyrite in which nearly all the gold and silver is concentrated.

GREENLAND and FONES (1971) investigated the distribution of silver at intervals along a drill-core (no. 5123) of the Great Lake dolerite ("diabase") sheet in Tasmania. The average silver content of the sheet is given as 86 ppb; the individual analyses are

Table 47-E-4. *Silver in drill core no. 5123 of the Great Lake dolerite sheet, Tasmania* (Greenland and Fones, 1971) (Analytical method: N/R and fire-assay separation)

Depth in drill core (feet)	Rocks	Ag content, ppb	Average ppb Ag
50	Granophyre	100	110
100		120	
200	Central zone (silicic)	100	92
300	dolerites	100	
400		96	
500		110	
600		84	
700		62	
800	Lower zone (mafic)	110	78
900	dolerites	70	
1,000		86	
1,100		120	
1,200		55	
1,300		32	
1,400		120	
1,500		62	
1,600		77	
1,735		57	
Lower chilled marginal contact (drill core 5084)		140	

quoted in Table 47-E-4. The authors point to a steady increase in the concentration of silver in the rocks with increasing differentiation (though such an increase is, judging from their figures, on a modest scale). The higher silver content of the chilled marginal rock (core no. 5084; Table 47-E-4) is possibly to be attributed to deuteric alteration; Bramall and Dowie (1936) also found enhanced silver contents in some pneumatolytically reddened facies of the granitic rocks of the Malvern Hills, England.

The behavior of silver in the Great Lake dolerite contrasts with that in some of the Siberian traps investigated by Nesterenko *et al.* (1969), where the element was strongly concentrated in the most mafic rocks and its abundance closely correlated with the presence of sulfides, as in the copper sulfide-bearing horizons of the Skaergaard intrusion. In the Great Lake sheet, Cu, Au and Ir, but not Ag, become concentrated with sulfide minerals which appear only very late in the crystallisation sequence. Greenland and Fones suggest that the silicate crystal/liquid distribution coefficient for Ag is much greater than those of Cu and Au in the absence of sulfide crystallisation, so that in the residual silicate liquids, Cu and Au become enriched relative to Ag and thus later sulfides are relatively depleted in silver. The observed distribution of silver in the Great Lake intrusion could be accounted for by the overall low abundance of sulfides together with a high silicate crystal/liquid distribution coefficient relative to gold and copper. Greenland and Fones (1971) also suggest the possibility of a diadochic replacement of Ag for K in some lattice sites in the silicate minerals, but offer no concrete data in support.

Revised manuscript received: May 1974

47-F. Behavior in Magmatogenic Processes and Ore Deposition

I. Magmatogenic Processes

The association of small amounts of silver with subordinate copper and iron sulfide phases in igneous rocks, presumably representing the preferential entry of the element into an immiscible sulfide liquid or matte, has been mentioned above in Section 47-E. The behavior of the element during magmatic differentiation has been little studied, and then only in mafic rock sequences, where there is a marked if modest tendency for silver to become enriched in the later magmatic fractions.

Krauskopf (1957) calculates that at 600° C, silver, in common with other heavy metals, is more volatile as chloride or fluoride than as native metal; he gives the vapor pressure of the chloride (alone) as $10^{-6.0}$ atmospheres at this temperature and a model gas composition, and of the metal (alone) as $10^{-10.6}$ atmospheres. Both silver and gold and their relevant compounds, have among the lowest volatilities of the common heavier metals of ore deposits, yet, remarks Krauskopf, they are common in low-temperature deposits and their observed behavior bears no relation at all to their volatilities.

From Krauskopf's work, it seems unlikely then that gas transport is an important process in the formation of magmatic concentrations of silver or of silver ore deposits. So far as information exists, pegmatites, while they may show some enrichment in the element compared with the enclosing rocks, are also on the whole poorer in silver than veins and other types of deposits.

II. Ore Deposits

An excellent review and classification of silver deposits, with a wealth of examples and discussions on their genesis and associations, is given by Boyle (1968) and the following summary leans heavily upon that source, to which the reader is referred for details.

The chief sources of silver are in hypogene veins, stockworks, etc. The element may be locally enriched in some cupriferous shales and sandstones, in which silver is generally at low levels of concentration and is won mainly from the copper and lead concentrates. Although silver enters accessory sulfide mineral phases in igneous rocks, it is very seldom enriched to an economically worth-while degree, and then only as a by-product of the extraction of copper or lead.

As far as generalisation is possible, silver appears to be most frequently associated in its deposits with lead and gold, but may also be associated with a wide range of other, mainly chalcophile, elements often including bismuth. Boyle states that on a statistical basis silver tends to occur in veins with carbonates or barite as the gangue mineral, while gold is more generally associated with a quartz gangue. Appreciable secondary enrichment of silver can occur in near-surface parts of veins, due to

Table 47-F-1. *Types of silver-bearing ore deposits and examples*

1. *Shale deposits, Kupferschiefer type*
   Kupferschiefer (Germany, Poland)
   White Pine, Michigan (U.S.A.)
   Zambian (N. Rhodesian) Copperbelt
   El Boléa (Mexico)

---

2. "*Red Bed*" *sandstone deposits*
   Uranium-vanadium deposits of south-western U.S.A. (Permo-Carbinoferous to Tertiary in age e.g. Silver Reef, Utah)
   Copper deposits in Carboniferous sandstones of Nova Scotia and New Brunswick, (Canada)
   Cambrian sandstone of Laisvall area, (Sweden) (lead-bearing)
   Copper-bearing Tertiary sandstones of Corocoro, (Bolivia)
   Copper-bearing Permo-Carboniferous sediments of Dzhezkazgan, Kazakhstan (U.S.S.R.)

---

3. *Silver-bearing skarn deposits*
   Åmmeberg, (Sweden)
   Western U.S.A. (e.g. Bingham district, Utah)
   Chihuahua and Zacatecas districts, (Mexico)
   Tetreault, Quebec, and New Calumet, Ottawa River, deposits (Canada)

---

4. *Lodes, veins, stockwerks etc., essentially of epigenetic deposition in sedimentary rocks*
   Kongsberg, (Norway)
   Harz Mountains and Freiberg, Saxony, (Germany)
   Jáchymov, (Czechoslovakia)
   Laurion, (Greece)
   Broken Hill and Mount Isa, (Australia)
   Santa Eulalia, Chihuahua, (Mexico)
   Coeur d'Alene, Idaho (U.S.A.)
   In Canada: Thunder Bay, Ontario, Ainsworth, Bluebell and Sullivan mines, British Columbia, Keno Hill-Galena Hill area, Yukon

---

5. *Silver-gold veins and lodes in or associated with volcanic flow rocks*
   Yellowknife, Red Lake, Porcupine, Kirkland Lake (Canada)
   Black Mountains, Arizona, De Lamar, Idaho, Rawhide, Nevada (U.S.A.)
   Kalgoorlie (Australia)
   Transylvania (Roumania)
   Hawaki (New Zealand)
   El Oro (Mexico)
   Comstock Lode, Nevada (U.S.A.)
   Guanajuato (Mexico)
   Cripple Creek, Colorado (U.S.A.)

---

6. *Silver deposits in complex geological environments*
   Cobalt, Ontario (Canada)
   Great Bear Lake, N.W.T. (Canada)
   Potosí (Bolivia)

---

7. *Miscellaneous sources of silver*
   a) Nickel-copper ores of Sudbury type (Canada)
   b) Massive sulfide deposits of iron, copper, lead and zinc, e.g. Rammelsberg (Germany); Cerro de Pasco (Peru)
   c) Lead-zinc deposits in carbonate rocks, e.g. Pine Point — Mississippi Valley type (U.S.A.)

Table 47-F-1 (continued)

d) Native copper deposits (e.g. Michigan, U.S.A.)
e) Porphyry copper deposits (in the broad sense) (e.g. Butte, Montana; Bingham, Utah; Ely, Nevada; Ajo, San Manuel and Ray, Arizona, U.S.A.)
f) Gold deposits and placers. All contain some silver
g) Eluvial and alluvial deposits formed in appropriately mineralised areas.

supergene oxidation processes. In shale and sandstone deposits, silver is most frequently associated with the highest concentrations of copper and, occasionally, with uranium.

A classification of silver-bearing ore deposits, with examples, is given in Table 47-F-1.

As regards the origin of these deposits it seems improbable that most can be accounted for by any reasonable "magmatic" hydrothermal process, and most syngenetic and epigenetic deposits are regarded by BOYLE (1968) as owing their origin to precipitation from surface waters, or from ground, connate or metamorphic waters, and to concentration by diffusion. This author quotes one of his earlier papers (BOYLE, 1963b):

"The writer holds the opinion that the elements in veins and similar epigenetic deposits came from their enclosing country rocks and that they are concentrated mainly by diffusion processes either in groundwater systems or as a result of the interaction of metamorphic reactions and the dilatant effects imposed by structures". And, later: "The key, therefore, to the formation of epigenetic mineral deposits in the deep zones of the earth's crust is to be found in the proper timing and intensity of two geological events — mobilisation of the elements in the country rocks (a metamorphic event), and the dilatancy of structures (a tectonic event)".

Concentration by diffusion of ions or complexes through a stationary aqueous phase filling pore spaces and fractures in rocks appears a feasible process provided there is a sufficient source of the element and the appropriate concentration gradient can be maintained. BOYLE (1968) calculates that during the 200 million years or so since the copper-silver bearing sandstones of Carboniferous and Triassic age were formed, a silver ion or "molecule" could have travelled by diffusion a distance of around 4 kilometres at 25° C, so that the silver in red-bed deposits could have had its ultimate source at that kind of distance. In the analogous case where gaseous water or carbon dioxide fills the rocks at higher temperatures and pressures, a similar diffusion mechanism could obtain, and may be important in certain metamorphic and skarn-type deposits. Under these conditions, BOYLE estimates that a silver ion or "molecule" of a silver compound might diffuse over a distance around 1.75 kilometres per million years at 300—500° C. A sizeable silver deposit could form by such a mechanism in about a million years, drawing silver from rocks containing only a few tenths of a part per million of the element over this kind of distance.

Dry gaseous diffusion along grain boundaries, through pore spaces, etc., is also considered briefly by BOYLE; both this and mechanisms involving diffusion of ions through crystal lattices seem in general less likely to be important except, in the latter case, where the silver may migrate through silicate and sulfide mineral lattices to the grain boundaries, along which extensive migration does then take place.

Revised manuscript received: May 1974

47-G. Behavior during Weathering and Alteration of Rocks

I. Soils

Apart from the work of BOYLE and his co-workers at the Geological Survey of Canada, and summarised by BOYLE (1968), there appears to be a paucity of information on the silver content of soils. SWAINE (1955) reported a range of < 0.01—5 ppm (10—5,000 ppb) Ag for normal soils, and LOUNAMAA (1956) reports average silver contents from 0.3—0.5 ppm (300—500 ppb) for Finnish soils developed on a variety of different rock types.

BOYLE (1968) reports the results of several investigations of the silver distribution in Canadian soil profiles in regions where some of the soils are developed on or near metalliferous mineral deposits. As an example, figures for silver in the soils of the Bathurst-Newcastle area, New Brunswick (PRESANT, 1971) are reproduced in Table 47-G-1. Higher silver contents in bedrock or underlying mineral deposits are reflected in higher silver contents in the overlying soils, and the element tends to concentrate in the surface A_0 horizon, where it is possibly bound with organic matter.

Based chiefly upon his own work on Canadian cases, BOYLE (1968, p. 70—73) discusses the general factors governing the mobility and fixation of silver in soils. The oxidation potential, Eh, affects the behavior of silver only indirectly — but extremely profoundly — in such matters as the hydrolysis and precipitation of iron and manganese (manganese sols in particular displaying strong adsorptive capacity for silver), production of the S^{2-} ion, and the oxidation of organic matter. Since the Ag^+ ion represents the only normally stable oxidation state, the range of Eh values found in soils will have a minimal direct effect upon the mobility and concentration of silver. Soil pH exerts many effects, both directly and indirectly. Silver salts are generally more soluble and more mobile in an acid environment; in alkaline media the oxide or hydrated oxide may be precipitated directly, but under these conditions the concomitant behavior of iron, manganese, aluminium and organic matter is likely to be far more important in controlling what happens to the silver. For these reasons, BOYLE concludes that silver is rather immobile in soils where the pH is greater than > 4. Humic substances adsorb silver strongly, perhaps forming chelated complexes, leading to enrichment of the element in the A_0 soil horizon, as well as in peats and organic-rich materials of related kinds. Some complexing anions such as SO_4^{2-}, NO_3^- and HCO_3^-, and some organic acids, render silver soluble and increase its mobility in soils, while others like PO_4^{3-}, Cl^-, Br^-, I^-, CrO_4^{2-} and AsO_4^{3-} cause the precipitation of insoluble silver compounds and inhibit mobility. The presence of H_2S or S^{2-} will precipitate silver sulfide. The thiosulfate ion, $(S_2O_3)^-$, on the other hand leads to the soluble complex $Ag(S_2O_3)_2^{3-}$, while excess of Cl^- in solution may similarly lead to the formation of the soluble and mobile complexes $AgCl_2^-$ or $AgCl_3^{2-}$.

Table 47-G-1. *Silver contents of soils from the Bathurst-Newcastle mineralized area, New Brunswick, Canada.* (Data from PRESANT, 1971)

Horizon (Podzols developed mainly on glacial till)	Range ppb Ag	Mean of all horizons (26 profiles) ppb Ag	Mean of normal horizons (14 profiles) ppb Ag	Mean of horizons over Brunswick No. 6 deposit (2 profiles) ppb Ag	Mean of horizons over Nigadoo deposit (5 profiles) ppb Ag	Mean of horizons over Middle River deposit (2 profiles) ppb Ag
A_0	200—90,000	5,700	1,600	45,600	5,200	3,400
A_2	100—36,000	2,400	400	18,600	400	1,800
B_1	200— 6,700	1,800	300	3,900	4,600	4,100
B_2	100—24,000	3,500	300	19,500	4,100	6,100
C	100—16,000	2,700	300	5,300	9,000	Not detected

Notes: Brunswick No. 6 deposit is a massive sulfide body containing essentially Pb, Zn, Cu, Fe, Ag, S. Nigadoo is a sulfide vein deposit containing essentially Pb, Zn, Cu, Fe, As, Ag, S. Middle River deposit is a massive sulfide body containing essentially Pb, Zn, Cu, Fe, Ag, S.

One of the most important reactions governing the geochemical behavior of silver under low P—T conditions concerns the ratio of Fe^{2+} to Fe^{3+} in solution:

$$Fe^{2+} + Ag^{+} \rightleftharpoons Ag + Fe^{3+}.$$

In high Fe^{2+} concentrations, silver will be precipitated, while in the presence of much Fe^{3+} in solution, the element remains ionized and mobile. Soluble manganous salts probably also play a similar role:

$$2\,Ag^{+} + Mn^{2+} + 4\,OH^{-} \rightarrow 2\,Ag + MnO_2 + 2\,H_2O.$$

II. Weathering Products Other Than Soil

Silver tends to become concentrated with gold in the eluvial lateritic "banket" deposits representing the detritus of weathered gold and silver deposits, known from Brazil and other parts of Central and South America and elsewhere. While a high proportion of the native element particles in such deposits is certainly residual in origin, some may be formed by later chemical action involving percolating solutions.

Solution and re-precipitation processes may also play a significant part in the further concentration of silver and gold in placer deposits of various kinds.

Revised manuscript received: May 1974

47-H. Compounds Controlling Silver Concentration in Natural Waters

In natural waters, silver may occur in a number of more or less soluble forms; a comprehensive listing is given by BOYLE (1968, pp. 80—81). Of these, the most important are: the various chloride complexes with Na and K (e.g., $Na[AgCl_2]$); complex sulfide and polysulfide ions and complex hydrosulfides (e.g., AgS^-, $[Ag(S_4)_2]^{3-}$, $Ag_2S \cdot nH_2S$, etc.); and soluble organic compounds. The soluble thiosulfate complex ion $[Ag(S_2O_3)_2]^{3-}$ may be locally important but decomposes readily in acid media; silver could also possibly migrate as thionate or polythionate.

In addition, silver may possibly occur in colloidal forms of such compounds as the halide, sulfide, telluride, etc. Similarly, silver may be strongly associated with humic complexes in natural waters as in soils, the metal either being strongly adsorbed

Table 47-H-1. *Experimental adsorption and desorption of silver in aqueous suspensions* (KHARKAR *et al.*, 1968)

Material and concentration (mg/l)	% ^{110}Ag adsorbed from distilled water	% adsorbed ^{110}Ag desorbed in sea water
Montmorillonite		
1,000	33 ± 3.2[a]	26.5 ± 10.2
10,000	60 ± 3.7	30.7 ± 3.0
	59.5 ± 7.1	35.0 ± 5.0
Illite		
1,000	13.9 ± 3.0	
	28.0 ± 4.3	
10,000	74.6 ± 4.4	25.9 ± 10.6
	83.0 ± 8.8	
Kaolinite		
1,000	13.0 ± 1.4	
10,000	13.0 ± 0.5	
Ferric oxide (Baker)		
1,000	5.0 ± 3.6	
Manganese dioxide (Baker)		
1,000	81.4 ± 12.0	96.1 ± 3.6
	89.8 ± 25.0	
Ferric hydroxide (freshly precipitated at pH 7)		
1,000	59.0 ± 10.2	17.0 ± 28.6

[a] The errors quoted in this table are counting errors only.

or even forming chelated compounds of some kind. KRAUSKOPF (1956) demonstrated experimentally that silver is strongly adsorbed on to plankton, and this source may at times account for an appreciable proportion of the silver found in natural waters.

As regards the association of silver with suspended inorganic particulate matter in natural waters, the experimental results of KHARKAR *et al.* (1968) are important. These authors studied the adsorption and desorption of various trace elements on and off the clay minerals montmorillonite, illite, and kaolinite, reagent ferric oxide and manganese dioxide, and freshly-precipitated ferric hydroxide at pH ~7. Their results for silver (obtained using the radioisotope ^{110}Ag) are given in Table 47-H-1, which also shows the proportions of silver subsequently desorbed on treatment of the materials with sea water. While natural conditions will obviously vary, the suspended load of most rivers will generally be within the ranges covered by the experiments, which indicate that under conditions resembling those in most streams (say ~1,000 mg/liter suspended matter), montmorillonite and illite will probably adsorb about 20—30% of the silver present in solution, and kaolinite much less. Ferric hydroxide adsorbs about 60% of the silver (while ferric oxide shows negligible adsorption), and manganese dioxide about 80—90%.

On contact with sea water, a proportion of the silver adsorbed on to these materials will be released into solution once more (desorbed) because of the displacement of silver by sodium and magnesium ions present at high concentration in sea water.

Revised manuscript received: May 1974

47-I. Abundance in Natural Waters

Boyle (1968) has tabulated the data available up to 1965 and concludes that the average silver content of all fresh waters (springs, streams, rivers and lakes) is about 0.2 ppb Ag, although considerable variations either side of this average may be encountered.

I. Spring Waters

Spring waters may be slightly higher in silver content, in general, than stream, river or lake waters, while acid waters (other than hot springs) tend to carry more silver than neutral or alkaline waters. Chloride, sulfate and bicarbonate waters, as well as waters carrying much dissolved iron and manganese, tend to be relatively enriched in silver. Waters leaching metalliferous mineral deposits may of course carry abnormally high amounts of silver, up to 10 or 100 times the usual values.

II. Streams

Kharkar *et al.* (1968) have made a study of the stream supply of silver and other trace elements to the oceans. Silver shows a relatively narrow range of concentrations in the rivers these authors studied (Table 47-I-1), and in common with other trace elements the concentrations of dissolved silver in the stream water appear to be independent of its concentration on particulate matter, and also of the suspended load of the stream. Based upon adsorption — desorption experiments using ^{110}Ag as tracer, and estimated ratios of various clay mineral species in the transported sediment load of the rivers, Kharkar *et al.* conclude that the sea water-desorbable silver carried in this way amounts to only an additional 5—10% above the figures given in Table 47-I-1, so that the greatest proportion of the silver is carried in solution.

III. Sea Water

Using their data for rivers, and previously available figures for the silver contents of sea waters, Kharkar *et al.* estimate the mean oceanic residence time of the soluble silver supply at 40,000 years.

Boyle (1968) also tabulates various estimates of the silver content of sea water samples, concluding that the normal range is about 0.15 to 2.9 ppb. The most reliable values are probably those of Schutz and Turekian (1965), determined by neutron activation analysis. The average of a large number of analyses (from over 100 sample locations covering the Atlantic, Indian, Antarctic and Pacific Oceans, the Gulf of Mexico and the Sea of Labrador) is 0.29 μg Ag/liter (0.29 ppb or 0.00029 ppm). Variations in regional average silver concentrations recorded by these authors are reproduced in Table 47-I-2; wide variations are observed within each oceanic body, together with a tendency for silver content to increase with depth in regions of high phosphate and high organic productivity.

Table 47-I-1. *Silver in stream waters* (KHARKAR *et al.*, 1968)(Method: N/R)

River	Dissolved solids (mg/liter)	µg Ag/liter (~ ppb Ag) (average)
Mississippi, Minneapolis, Minn. (U.S.A.) (July 1965)	247	0.24
Susquehanna, Marietta, Pa. (U.S.A.) (June 1966)	186	0.37
Mad, Blue Lake, Calif. (U.S.A.) (November 1966)	317	0.26
Klamath, Klamath Glenn, Calif. (U.S.A.) (November 1966)	273	0.55
Russian, Calif. Highway 116, Calif. (U.S.A.) (November 1966)	298	0.10
Eel, U.S. Highway 101, Calif. (U.S.A.) (November 1966)	286	0.17
Brazos, U.S. Highway 59, Texas (U.S.A.) (June 1966)	189	0.38
Housatonic, Conn. (U.S.A.) (July 1965)		0.39
Connecticut, Conn. (U.S.A.) (July 1965)		0.17
Neuse, N. Carolina (U.S.A.) (November 1965)		0.37
Rhone, Avignon (France) (June 1966)	94	0.38
Amazon, Santarem (Brazil) (January 1965)	34	0.23
Average		0.30

Table 47-I-2. *Average regional silver contents of sea waters* (SCHUTZ and TUREKIAN, 1965) (Method: N/R)

Area	Average Ag content, µg/liter (~ ppb)
Caribbean	0.25
Gulf of Mexico	0.16
Sea of Labrador	0.13
N.W. Atlantic	0.19
N.E. Atlantic	0.25
S.W. Atlantic	0.18
S.E. Atlantic	0.64
Indian Ocean	0.69
Central Pacific	0.34
East Pacific	0.23
Antarctic	0.42

IV. Brines

Abnormally high concentrations of silver have been recorded in the remarkable metalliferous brines of the Red Sea deeps (DEGENS and ROSS, 1969) and the wells of the Salton Sea geothermal area, California (SKINNER *et al.*, 1967), as well as in the associated sediments. Presumably the element forms abundant complexes with alkali chlorides in these waters.

HENDRICKS *et al.* (in DEGENS and ROSS, 1969) report many analyses of the Red Sea brines; the silver contents (determined both by optical emission spectrography and atomic absorption spectrophotometry) mostly range from 0.001 to 0.5 mg Ag/liter of brine, with an average around 0.06 mg /liter (~60 ppb).

SKINNER *et al.* (1967) report 0.8 ppm and 2 ppm Ag (800 and 2,000 ppb) in two reservoir brines from the Salton Sea geothermal field, and from 1 to 7 *weight per cent* Ag in the sulfide-rich scale deposited from pipes discharging the hot brines.

Revised manuscript received: May 1974

47-K. Silver in Sediments and Sedimentary Rocks

Silver is present in variable and very low concentrations in ordinary shales, sandstones, limestones and evaporites; the highest amounts in sedimentary rocks are found in black shales, sulfidic "schists", alum shales, and phosphorites, and concentrations up to 10 ppm or more have been reported. BOYLE (1968) and other investigators have found the silver in these latter types of rock to be mainly associated with carbonaceous matter and very fine-grained sulfides, particularly pyrite. In sedimentary rocks carrying both pyrite and arsenopyrite, BOYLE suggests the interesting generalisation that silver is preferentially concentrated in the former mineral and gold in the latter. Much of the silver in sedimentary rocks may also be associated with potassium-rich clay minerals and hydrated iron and manganese oxides.

Data on silver contents of sedimentary rocks are rather sparse and of variable quality; they are summarised by BOYLE (1968, Table 15). Of more value is this author's listing of the silver contents of Canadian sedimentary rocks, reproduced in Table 47-K-1, since this spans a wide range of rock types and it may be assumed that the spectrographic analyses performed in the laboratories of the Geological Survey of Canada are of uniform quality.

Table 47-K-1. *Silver content of some Canadian sedimentary rocks* (BOYLE, 1961, 1963, 1965, 1968 and unpublished)(Method: S)

Rock type	Locality	Range ppm Ag	Average ppb Ag
Arenaceous rocks			
Sandstone	Walton area, N.S. (Cheverie Formation)	0.03—0.90	250
Sandstone and conglomerate	Walton area, N.S. (Triassic)	0.10—0.18	50
Quartzite and sandstone	Walton area, N.S. (Horton Bluff Formation)	<0.05—1.4	220
Thick- and thin-bedded grey quartzites	Keno Hill, Y.T.	0.13—1.30	310
Siliceous white quartzites	Keno Hill, Y.T.	0.17—0.37	250
Calcareous quartzites	Keno Hill, Y.T.	0.24—0.54	360
Argillaceous rocks			
Grey and green shale	Walton area, N.S. (Cheverie Formation)	0.05—0.90	190

Table 47-K-1 (continued)

Rock type	Locality	Range ppm Ag	Average ppb Ag
Red shale and argillite	Walton area, N.S. (Cheverie Formation)	0.05—0.29	110
Red and buff shale	Walton area, N.S. (Windsor Group, Tennycape Formation)	0.15—0.23	190
Black shale and argillite	Walton area, N.S. (Horton Bluff Formation)	<0.03—4.0	320
Carbonaceous and phosphatic shale	Walton area, N.S. (Horton Bluff Formation)	0.19—1.2	430
Albert bituminous shale	Albert mines, New Brunswick		70
Pyritiferous and arseno-pyritiferous slate	Ovens area, Nova Scotia		2,200
Pyritiferous shale	Kettle Point Formation, Kettle Point and Florence, Ontario		120
Graphitic tuff, slate, and schist	Yellowknife, N.W.T.	0.33—3.4	1,140
Graphitic argillites, schists and phyllites	Keno Hill, Y.T.	0.17—1.60	500
Quartz-sericite schist	Keno Hill, Y.T.	0.12—1.1	440
Calcareous rocks and evaporites			
Limestones	Keno Hill, Y.T.	0.10—0.27	150
Limestone	Walton area, N.S. (Macumber Formation)	<0.1—0.3	100
Limestone conglomerate	Walton area, N.S. (Macumber Formation)	<0.1—0.3	100
Calcareous concretions	Walton area, N.S. (Horton Bluff Formation)		<50
Dolomite, arenaceous and argillaceous dolomite	Walton area, N.S. (Horton Bluff Formation)	0.04—0.5	170
Calcareous anhydrite	Walton area, N.S. (Windsor evaporites)	0.05	<50
Anhydrite	Walton area, N.S. (Windsor evaporites)	0.05—0.4	<100
Gypsum	Walton area, N.S. (Windsor evaporites)	0.05—0.4	<100

Carbonate sediments appear generally to carry very little silver; LURYE (1957) found silver concentrated in the same horizons within a limestone-dolomite succession in the U.S.S.R. as lead, zinc, barium, manganese and iron.

In oxide-hydroxide sediments, silver may occur adsorbed on to colloidal iron and manganese compounds, and traces of the element are reported in many such rocks. MERO (1965) reports an average of 3 ppm Ag (Method: X) in 5 samples of manganese nodules from the floor of the Pacific Ocean, the highest value being 6 ppm.

TUREKIAN (1968) reports 22—97 ppb silver (Method: N/R) in different levels of the Lamont core V16—36 from the southern Mid-Atlantic Ridge, where the rate of accumulation of abyssal clay is estimated at about 0.1 $g/cm^2/1,000$ years. This author states that the rate of silver supplied in solution by streams, per unit area of ocean bottom, is about 50 times greater than the rate of removal of the element by the sediment at the site of the core. Silver is clearly not removed homogeneously in the oceans, and most of the element is removed in near-shore marine muds by biological agencies and trapping as silver sulfide during the production of H_2S in the sediment by the bacterial reduction of sulfate ions in the interstitial water.

A few sedimentary rocks carry abnormally high concentrations of silver, occasionally (as in the case of the famous Permian Kupferschiefer beds of Germany) attaining the status of low-grade silver ores; examples are listed by BOYLE (1968, Table 18). The possible syngenetic origin of such silver concentrations is indicated by the paleogeographical control of its abundance. The element seems frequently to occur along with copper in these materials, perhaps generally in sulfide minerals of various kinds, occasionally as native metal.

Revised manuscript received: May 1974

47-L. Biogeochemistry

I. Plants and Animals

Silver has been detected in a very wide range of plant and animal materials; some of the values recorded are given in Tables 47-L-1 and 47-L-2, after BOYLE (1968).

Most plants appear to concentrate silver well above the levels recorded in the substrates upon which they grow. SHACKLETTE (1965), for example, found the silver content of bryophytes to be fifteen times higher than that of the corresponding soil

Table 47-L-1. *Silver contents of plant materials*

Material	Ag content ppm	Reference
Algae		
Seaweeds	0.7—1.5	HABER (1927)
Seaweeds (oven-dried)	up to 0.7	BLACK and MITCHELL (1952)
Fungi		
Dried mushrooms	up to 100	RAMAGE (1930)
Dried mushrooms (gills)	up to 500	RAMAGE (1930)
Bryophyta		
Mosses (growing on silicic and ultramafic rocks)	1—6, in plant ash	LOUNAMAA (1956)
Mosses (growing on serpentinites)	up to 10	LOUNAMAA (1956)
Mosses and liverworts	Average 9, in plant ash	SHACKLETTE (1965)
Gymnosperma and Angiosperma		
Twigs and leaves (oven-dried) of trees in Nigerian Pb-Zn belt	up to 2.7	WEBB and MILLMAN (1950—51)
Twigs, conifer needles etc. Watson Bar Creek, British Columbia (Canada)	0.1—1.4 (up to 30 ppm in ash)	WARREN and DELAVAULT (1950)
Needles and twigs of conifers	up to 10	LOUNAMAA (1956)
Leaves and twigs of deciduous trees	up to 10	LOUNAMAA (1956)
Stems and leaves of dwarf shrubs	up to 30	LOUNAMAA (1956)
Herbs	up to 100	LOUNAMAA (1956)
Grasses	up to 30	LOUNAMAA (1956)

Table 47-L-2. *Silver in some marine and fresh-water animals* (dry matter) (Boyle, 1968)

Material		Ag ppm; range
Porifera	Marine sponges	Up to 1.0
Coelenterata	Jellyfishes, anemones, corals, etc.	2.0—6.0
Platyhelminthes-Annelida	Worms	Present in some species
Echinodermata	Starfishes, sea-urchins, sea lilies, etc.	1.0—4.0
Mollusca	Oysters, clams, snails, mussels, etc. (soft parts) (Concentrations often in liver, kidney and blood)	
Arthropoda	Crustacea: lobsters, crabs, etc. (Ag found both in tissues and blood) (Gills of river crustaceans take up Ag)	Up to 2.0
Chordata	Vertebrata: class Pisces (fishes) (Ag found in flesh, blood, entrails, etc.)	Up to 12

but that the degree of enrichment was less in the case of vascular plants. The limited data available also indicate very considerable enrichment of the element by higher fungi, while marine algae show an enrichment factor over sea water of as much as 2,500. In the higher plants, too, silver appears frequently to become preferentially concentrated in seeds, nuts and fruits.

Similarly, silver becomes preferentially concentrated in a wide range of marine, freshwater and terrestrial animals, and this may in turn explain in part the unusual concentrations of the element often encountered in black (organic-rich) shales and other sediments. Table 47-L-2 lists the ranges of silver contents of various marine and freshwater animals given by Boyle (1968). Noddack and Noddack (1939) estimated the enrichment factor for silver in marine animals compared with sea water to be 22,000.

As already mentioned in Sections 47-G and 47-H, silver is often very enriched in humic materials, a little by adsorption, but most of it probably bonded in silver humates, silver-organic acid complexes, organic sulfides, etc.

II. Coal

Peats and coals frequently contain detectable amounts of silver, particularly where the vegetable matter has grown upon mineralised country rocks. Most coal ashes contain less than 10 ppm Ag, but higher concentrations are found locally, especially sometimes in pyritic coals. Although the distribution of silver in coals is erratic, Boyle (1968) remarks that high silver contents are commonly recorded in low-ash coals, in sulfide(pyrite)-enriched parts of seams and adjacent rocks, and near the top and bottom of individual seams. Some of the silver is probably bound as metal-organic complexes; much of it is associated with pyrite and other sulfides, and a little may be adsorbed on to the coaly substances.

III. Oil

There are few significant data on the abundance and mode of occurrence of silver in oil shales and natural hydrocarbons generally. Up to 100 ppm Ag have been recorded in ash from Russian and other oils by KATCHENKOV (1952) and from oils from the western U.S.A. by HYDEN (1961). GOLDSCHMIDT (1954) reported 5—10 ppm Ag in bituminous marls, but BOYLE (1968) reports less than 0.1 ppm in most of the petroliferous Albert Shales of New Brunswick.

Some of the silver in these natural oils presumably has its origin in the plants and animals giving rise to the petroleum; further contributions could come from absorption and adsorption processes as the petroleum migrates through sedimentary rocks to its final reservoir, and from contact with connate and other waters containing silver.

Revised manuscript received: May 1974

47-M. Abundance in Metamorphic Rocks

Few determinations of silver in metamorphic rocks have been published; again BOYLE (1968) has summarised the available data to 1965, which are reproduced in Table 47-M-1. While it is dangerous to draw generalisations from very limited data, again silver appears in metamorphic rocks to be concentrated in any copper or iron sulfides present, and in ferromagnesian minerals. BOYLE (1968, p. 59) states that during granitization silver does not become concentrated during pegmatite formation, but is strongly enriched in late-stage veins and other deposits characteristic of the latest stages of the metamorphic and granitization process. As a chalcophile element, silver probably follows sulfur, arsenic and antimony during regional and contact metamorphism, along with gold, lead, zinc, cadmium, bismuth and copper, whose rates of diffusion appear to be about the same as that of silver. Apparently mainly on the basis of ionic radius and charge, HIGAZY (1952) suggests that Ag^{+} may substitute to some extent for K^{+} (and Ca^{2+}) in skarn rocks from the Dalradian of County Donegal, Ireland; BOYLE remarks that an association of silver with potassium is also indicated by the frequent enrichment of potassium in the wall-rock alteration zones of gold-silver veins and lead-zinc-silver deposits. Contact skarn bodies in the Whitehorse copper belt (Yukon, Canada) on the other hand show no correlation between silver and potassium or sodium; the skarns (200—1,500 ppb) are somewhat en-

Table 47-M-1. *Silver content of metamorphic rocks*

Rock type	Locality	Range ppm Ag	Average ppb Ag	Reference
Serpentinite	Norway	0.3—0.9		LUNDE and JOHNSON (1928)
Serpentinite	Leka (Norway)		30	GOLDSCHMIDT (1954)
Biotite-epidorite	Co. Donegal (Ireland)		930	HIGAZY (1952)
Biotite-skarn	Co. Donegal (Ireland)		3,700	HIGAZY (1952)
Quartzite	Co. Donegal (Ireland)		6,500	HIGAZY (1952)
Mica schist	Co. Donegal (Ireland)		1,860	HIGAZY (1952)
Eclogite	Norway	0.0—0.3		LUNDE and JOHNSON (1928)
Various crystalline schists	Malvern Hills (England)		600	BRAMMALL and DOWIE (1936)
Banded schist and gneiss (reddened)	Malvern Hills (England)	0.6—33.0	7,300	BRAMMALL and DOWIE (1936)
Sandstone in contact with red granite	Malvern Hills (England)	3—16	10,000	BRAMMALL and DOWIE (1936)

Table 47-M-1 (continued)

Rock Type	Locality	Range ppm Ag	Average ppb Ag	Reference
Amphibolites and hornblende schists (some reddened)	Malvern Hills (England)	<1.0—7.0	2,750	BRAMMALL and DOWIE (1936)
Pyroxenite-hornblendite	Malvern Hills (England)	<1.0—2.6	1,350	BRAMMALL and DOWIE (1936)
Granodiorite derived from above rock	Malvern Hills (England)		20,100	BRAMMALL and DOWIE (1936)
Clay schists and quartzite	Schwarzburger saddle, Thüringen (Germany)	0.01—0.3	110	FISCHER (1958—59)
Marble	Carrara (Italy)		200	WAGONER (1901)
Canadian metamorphic rocks				
Amphibolite (amphibolite facies)	Yellowknife greenstone belt (N.W.T.)	0.6—1.4	~1,000	BOYLE (1961)
Amphibolite (epidote amphibolite facies)	Yellowknife greenstone belt (N.W.T.)	0.3—1.2	~700	BOYLE (1961)
Chlorite schist (greenschist facies)	Yellowknife greenstone belt (N.W.T.)		~300	BOYLE (1961)
Keewatin greenstones	Cobalt district	0.19—0.30	250	
Argillites and slates	Sedimentary area, Yellowknife	0.3—2.0	~1,000	BOYLE (1961)
Meta-greywacke	Sedimentary area, Yellowknife	0.3—1.2	~750	BOYLE (1961)
Quartz-mica schists	Sedimentary area, Yellowknife	0.9—1.2	~1,000	BOYLE (1961)
Meta-conglomerate and -greywacke	Cobalt district		<50	
Skarn	Keno Hill area (Y.T.)	0.1—0.15	100	BOYLE (1965)
Skarn	Whitehorse area (Y.T.)	0.10—0.45	200	BOYLE (1968)
Crystalline limestone	Whitehorse area (Y.T.)		~50	BOYLE (1968)

riched in silver by comparison with the crystalline limestone from which they were derived (~50 ppb), but the chief concentration of the element (up to 60 ppm — 60,000 ppb) is found in the magnetite-bornite ores.

In the gold-silver deposits of the Yellowknife greenstone belt of Canada, the associated chlorite schist zones appear to be depleted in silver compared with the metavolcanic rocks in which they occur; BOYLE (1968, p. 61) suggests that the silver lost during the alteration of the volcanics to chlorite schists probably migrated into dilatant zones where it became concentrated in the gold-quartz veins and lenses.

Revised manuscript received: May 1974

47-O. Relation to Other Elements; Economic Importance

I. Relation to Other Elements

In nature, silver is most commonly associated with the other sub-group IB elements, copper and gold. Like gold, it does not easily form simple ions and is a rare, widely-dispersed element exploitable commercially in a wide range of types of ore deposit; it may be also associated with many other heavier metals and chalcophile

Table 47-O-1. *Ag/Au ratios of a variety of mineral deposits* (BOYLE, 1968)

Type of deposit	Ag/Au Ratio (range)	Remarks
Shale deposits (Kupferschiefer type)	40—150	
Disseminations, veins, etc. in sandstones (red-bed type)	100—300	Variable
Conglomerates (Rand type)	0.05—0.20	Witwatersrand Bullion ranges between 0.064—0.173. On channel samples of the Ventersdorp Contact Reef and Main Reef, the Ag/Au ratios range from 0.073—1.0
Disseminated lead-zinc deposits in carbonate rocks (Mississippi Valley-Pine Point type)	10—1,000	Few data
Skarn-type deposits	0.1—200	Variable
Massive sulfides (Ni—Cu Sudbury type)	15—31	SHCHERBINA (1956) gives 15 for Norilsk, and HAWLEY (1962) gives 31 for Sudbury ores
Porphyry copper type	10—800	Few data
Polymetallic massive sulfides (Flin Flon-Noranda-Bathurst type)	1—170	Average about 40
Polymetallic veins, mantos, etc. (Keno Hill-Sullivan-Coeur d'Alene type)	50—> 10,000	
Gold-quartz veins, lodes etc. (Precambrian, Palaeozoic, and Mesozoic age)	0.08—0.73	Average about 0.28
Gold-quartz veins, lodes, etc. (Tertiary age)	3—200	
Native silver-Co—Ni—As—Bi veins	100—> 10,000	
Hot spring siliceous sinters	> 1	
Gold placers (all types and ages)	< 1	

elements, with all of which silver may under certain conditions exhibit some degree of geochemical coherence. Because of its atomic configuration and properties, silver is more closely associated with gold than with any other element, and the Ag:Au ratio in ore deposits has attracted attention from various investigators. As might be expected, the data on this topic are conflicting and few, if any, valid generalisations can be drawn from them.

It has been suggested at various times that the Ag:Au ratio might provide a useful geothermometer in ore deposits. Certainly, there is evidence that deeper-seated deposits tend to have lower Ag:Au ratios than those formed nearer to the surface, but there are exceptions and inconsistencies even within a single deposit. Boyle (1968) also remarks that Ag:Au ratios seem to depend in a crude way on regional metallogenic peculiarities in certain types of deposit, but if all types of deposit are considered, no particular relationship is apparent.

Table 47-0-1 lists the ranges of Ag:Au ratios reported in various types of mineral deposits, as compiled by Boyle (1968).

II. Economic Importance

Only about 10 per cent of all silver mined is won from deposits primarily exploited for the metal; 90 per cent or more represents a by-product of copper, lead-zinc and gold mining. The average annual production of the metal (1960—64) according to Boyle (1968) totalled 245×10^6 ounces (troy), or 7.62×10^6 kg, distributed as in Table 47-O-2.

Table 47-O-2. *Average annual production of silver*, 1960—64

Mexico	1.31×10^6 kg
U.S.A.	1.09×10^6 kg
Peru	1.09×10^6 kg
Canada	0.96×10^6 kg
U.S.S.R.	0.81×10^6 kg
Australia	0.50×10^6 kg
Europe (Sweden, Yugoslavia, Germany)	0.47×10^6 kg
Asia	0.43×10^6 kg
Bolivia and other S. and Central American countries	0.37×10^6 kg
Africa (Congo and S. Africa)	0.31×10^6 kg
Others	0.28×10^6 kg

Revised manuscript received: May 1974

References: Sections 47-B to 47-M, 47-O

ANDERS, E., GANAPATHY, R., KEAYS, R. R., LAUL, J. C., MORGAN, J. W.: Volatile and siderophile elements in lunar rocks: Comparison with terrestrial and meteoritic basalts. Geochim. Cosmochim. Acta, Suppl. **2**, 1021 (1971).

BADALOV, S. T., BADALOVA, R. P.: Some regularities of distribution of gold and silver in the principal ore minerals of hypogene deposits of Karamazar and western Kazakhstan. Geochem. Intern. **4**, 660 (1967).

BERNARD, J. H.: Mineralogie und Geochemie der Siderit-Schwerspat-Gänge mit Sulfiden im Gebiet von Rudnany. Geologicke Prace, **58**, 1 (1961).

BLACK, W. A. P., MITCHELL, R. L.: Trace elements in the common brown algae and in sea water. J. Marine Biol. Assoc. U.K., **30**, 575 (1952).

BOYLE, R. W.: The geology, geochemistry and origin of the gold deposits of the Yellowknife district. Geol. Surv. Canad. Mem. **310** (1961).

BOYLE, R. W.: Geology of the barite, gypsum, manganese, and lead-zinc-copper-silver deposits of the Walton-Cheverie area, Nova Scotia. Geol. Surv. Can. Paper **62—25** (1963a).

BOYLE, R. W.: Diffusion in vein genesis; Symposium — Problems of Postmagmatic Ore Deposition (Prague), **1**, 377 (1963b).

BOYLE, R. W.: Geology, geochemistry and origin of the lead-zinc-silver deposits of the Keno Hill-Galena Hill area, Yukon Territory. Geol. Surv. Can. Bull. **111** (1965).

BOYLE, R. W.: The geochemistry of silver and its deposits. Geol. Surv. Can. Bull. **160** (1968).

BRAMMALL, A., DOWIE, D. L.: The distribution of gold and silver in the crystalline rocks of the Malvern Hills. Mining Mag. (London) **24**, 260 (1936).

BRUNFELT, A. O., HEIER, K. S., STEINNES E.: Determination of 40 elements in Apollo 12 materials by neutron activation analysis. Geochim. Cosmochim. Acta Suppl. **2**, 1281 (1971).

BRUNFELT A. O., STEINNES E.: (Silver contents of U.S.G.S. standard rocks). Personal communication reported in FLANAGAN, F. J. Geochim. Cosmochim. Acta **33**, 81 (1969).

CAMERON, A. G. W.: A new table of abundances of the elements in the solar system. In: AHRENS, L. H. (Ed.): Origin and Distribution of the Elements. Oxford: Pergamon Press 1968.

CHAKRABURTTY, A. K., STEVENS, C. M., RUSHING, H. C., ANDERS, E.: Isotopic composition of silver in iron meteorites. J. Geophys. Res. **69**, 505 (1964).

CHUKROV, F. V., *et al.*: Mineral Reference Book. Vol. 1: Elements, Sulfides, etc.; Vol. 2: Simple Oxides. Acad. Sci. U.S.S.R., "IGEM", Moscow (1960—65).

CLEVENGER, G. H., CARON, M. H.: The treatment of manganese-silver ores. U.S. Bur. Mines, Bull. **226** (1925).

COHEN, A. J. : Moldavites and similar tektites from Georgia, U.S.A. Geochim. Cosmochim. Acta **17**, 150 (1959).

CRAIG, J. R.: Phase relations and mineral assemblages in the Ag—Bi—Pb—S system. Miner. Deposita **1**, 278 (1967).

CROUCH, E. A. C., TURNBULL, A. H.: The absolute isotopic abundance ratio and the atomic weight of natural silver. J. Chem. Soc. Pt. **1**, 161 (1962).

DEGENS, E. T., ROSS, D. A. (Eds.): Hot Brines and Recent Heavy Metal Deposits in the Red Sea. Berlin-Heidelberg-New York: Springer-Verlag 1969.

DE VORE, G. W.: The role of adsorption in the fractionation and distribution of elements. J. Geol. **63**, 159 (1955).

DEWS, J. R., NEWBURY, R. S.: The isotopic composition of silver in the Canyon Diablo meteorite. J. Geophys. Res. **71**, 3069 (1966).

DOELTER, C., *et al.*: Handbuch der Mineralchemie. 4 volumes. Dresden and Leipzig 1911 to 1931.
FISCHER, K. W.: Zur Geochemie der Edelmetalle. Wiss. Z. Hochsch. Arch. Bauwesen Weimar, **2**, 85 (1958—59).
FLANAGAN, F. J.: U.S. Geological Survey standards — II — First compilation of data for the new U.S.G.S. rocks. Geochim. Cosmochim. Acta **33**, 81 (1969).
FLEISCHER, M.: Minor elements in some sulphide minerals. Econ. Geol. 50th Anniv. vol. 970 (1955).
FLEISCHER, M., RICHMOND, W. E., EVANS, H. T.: Studies on the manganese oxides. V, Ramsdellite, MnO_2, an orthorhombic dimorph of pyrolusite. Am. Mineralogist **47**, 47 (1962).
FOLEY, L. L.: Gold and silver in manganese ore, Polk county, Arkansas. Econ. Geol. **55**, 1757 (1960).
GANAPATHY, R., KEAYS, R. R., LAUL, J. C., ANDERS, E.: Trace elements in Apollo 11 lunar rocks: Implications for meteorite influx and origin of moon. Geochim. Cosmochim. Acta, Suppl. **1**, 1117 (1970).
GAUDIN, A. M., SPEDDEN, H. R., CORRIVEAU, M. P.: Adsorption of silver ion by sphalerite. Mining Eng. **3**, 780 (1951).
GODOVIKOV, A. A.: Über die Silber-, Wismut- und Antimonbeimengungen im Bleiglanz. Z. Angew. Geol. **13**, 125 (1967).
GOLDSCHMIDT, V. M.: MUIR, A. (Ed.): Geochemistry. Oxford: Clarendon Press 1954.
GOLDSCHMIDT, V. M., PETERS, CL.: Zur Geochemie der Edelmetalle. Nachr. Akad. Wiss. Goettingen, Math.-Physik. Kl. IIa **4**, 377 (1932).
GREEN, J.: Geochemical table of the elements for 1959. Bull. Geol. Soc. Am. **70**, 1127 (1959).
GREENLAND, L.: The abundance of Se, Te, Ag, Pd, Cd and Zn in chondritic meteorites. Geochim. Cosmochim. Acta **31**, 849 (1967).
GREENLAND, L. P., FONES, R.: Geochemical behavior of silver in a differentiated tholeiitic dolerite sheet. Neues Jahrb. Mineral. Monatsh. **9**, 393 (1971).
HABER, F.: Das Gold im Meerwasser. Z. Angew. Chem. **40**, 303 (1927).
HAMAGUCHI, H., KURODA, R.: Silver content of igneous rocks. Geochim. Cosmochim. Acta **17**, 44 (1959).
HARTLEY, W. N., RAMAGE, H.: The wide dissemination of some of the rarer elements, and the mode of their association in common ores and minerals. J. Chem. Soc. **71**, 533 (1897).
HASKIN, L. A., ALLEN, R. O., HELMKE, P. A., PASTER, T. P., ANDERSON, M. R., KOROTEV, R. L., ZWEIFEL, K. A.: Rare earths and other trace elements in Apollo 11 lunar samples. Geochim. Cosmochim. Acta., Suppl. **1**, 1213 (1970).
HAWLEY, J. E.: The Sudbury ores: Their mineralogy and origin. Can. Mineralogist **7**, 1 (1962).
HERRMANN, A. G.: Geochemische Untersuchungen an Kalisalzlagerstätten im Südharz. Freiberger Forschungsh. C **43** (1958).
HEWETT, D. F., FLEISCHER, M., CONKLIN, N.: Deposits of the manganese oxides. Econ. Geol. (Suppl.) **58**, 1 (1963).
HIGAZY, R. A.: Behavior of the trace elements in a front of metasomatic-metamorphism in the Dalradian of Co. Donegal. Geochim. Cosmochim. Acta **2**, 170 (1952).
HINTZE, C.: Handbuch der Mineralogie. 6 vols. Berlin and Leipzig 1889—1939.
HYDEN, H. J.: Distribution of uranium and other metals in crude oils. U.S. Geol. Surv. Bull. **1100-B** (1961).
KATCHENKOV, S. M.: On some general regularities of the accumulation of mineral elements in petroleum and hard coals. Dokl. Akad. Sci. USSR. **86**, 805 (1952).
KEAYS, R. R., GANAPATHY, R., ANDERS, E.: Chemical fractionations in meteorites, IV. Abundances of 14 trace elements in L-chondrites: implications for cosmothermometry. Geochim. Cosmochim. Acta **35**, 337 (1971).
KHARKAR, D. P., TUREKIAN, K. K.: Analyses of Apollo 11 and Apollo 12 rocks and soils by neutron activation. Geochim. Cosmochim. Acta, Suppl. **2**, 1301 (1971).
KHARKAR, D. P., TUREKIAN, K. K., BERTINE, K. K.: Stream supply of dissolved Ag, Mo, Sb, Se, Cr, Co, Rb and Cs to the oceans. Geochim. Cosmochim. Acta **32**, 285 (1968).

KRAUSKOPF, K. B.: Factors controlling the concentrations of thirteen rare metals in sea-water. Geochim. Cosmochim. Acta **9**, 1 (1956).

KRAUSKOPF, K. B.: The heavy metal content of magmatic vapour at 600° C. Econ. Geol. **52**, 786 (1957).

KUZNETSOV, K. F.: Geocronite in ores of the Ekaterino-Glagodatsk deposits. Dokl. Akad. Nauk SSSR. **114**, 880 (1957).

KUZNETSOV, K. F.: Rare and trace elements in ores of some polymetallic deposits of the Nerchinsk-Zavodskaya Group. Inst. Min. Geochem. and Cryst. Chem., Rare Elements, Trans. **2**, 49 (1959).

LAUL, J. C., GANAPATHY, R., ANDERS, E., MORGAN, J. W.: Chemical fractionations in meteorites — IV. Accretion temperatures of H-, LL-. and E-chondrites, from abundance of volatile trace elements. Geochim. Cosmochim. Acta **37**, 329 (1973).

LAUL, J. C., GANAPATHY, R., MORGAN, J. W., ANDERS, E.: Meteoritic and non-meteoritic trace elements in Luna 16 samples. Earth Planet. Sci. Letters **13**, 450 (1972).

LAUL, J. C., KEAYS, R. R., GANAPATHY, R., ANDERS, E., MORGAN, J. W.: Chemical fractionations in meteorites — V. Volatile and siderophile elements in achondrites and ocean ridge basalts. Geochim. Cosmochim. Acta **36**, 329 (1972).

LAUL, J. C., MORGAN, J. W., GANAPATHY, R., ANDERS, E.: Meteoritic material in lunar samples: characterization from trace elements. Geochim. Cosmochim. Acta, Suppl. **2**, 1139 (1971).

LE RICHE, H. H.: (Silver contents of U.S.G.S. standard rocks). Personal communication reported in FLANAGAN, F. J.: Geochim. Cosmochim. Acta **33**, 81 (1969).

LOUNAMAA, J.: Trace elements in plants growing wild on different rocks in Finland. Ann. Botan. Soc. Zoo. Botan. Fennic. "Vanamo", **29**, 1 (1956).

LUNDE, G.: Über das Vorkommen des Platins in norwegischen Gesteinen und Mineralen. Z. Anorg. Allgem. Chem. **161**, 1 (1927).

LUNDE, C., JOHNSON, M.: Vorkommen und Nachweis der Platinmetalle in norwegischen Gesteinen, II. Z. Anorg. Allgem. Chem. **172**, 167 (1928).

LURYE, A. M.: Certain regularities in the distribution of elements in sedimentary rocks of the Northern Bayaldyr district in Central Karatau. Geochemistry (English translation) **5**, 470 (1957).

MERO, J. L.: The Mineral Resources of the Sea. Amsterdam: Elsevier 1965.

MORRIS, D. F. C., KILLICK, R. A.: Silver and thallium contents of rocks. Geochim. Cosmochim. Acta **19**, 139 (1960).

MORRISON, G. H., GERARD, J. T., KASHUBA, A. T., GANGADHARAM, E. V., ROTHENBERG, A. M., POTTER, N. M., MILLER, G. B.: Elemental abundances of lunar soil and rocks. Geochim. Cosmochim. Acta, Suppl. **1**, 1383 (1970).

NESTERENKO, G. V., BELYAYEV, YU. I., PHI, P. H.: Silver in the evolution of mafic rocks. Geochem. Intern. **6**, 119 (1969).

NODDACK, I., NODDACK, W.: Die Häufigkeiten der Schwermetalle in Meerestieren. Arkiv. Zool. **32 A**, 1 (1939).

ÖDMAN, O. H.: Geology and ores of the Boliden deposit, Sweden. Sveriges Geol. Undersok. Arsbok Ser. C. no. 438, **35**, 1 (1941).

PALACHE, C., BERMAN, H., FRONDEL, C.: Dana's System of Mineralogy, Vol. I: Elements, Sulfides, Sulfosalts, Oxides. New York: John Wiley 1944.

PALACHE, C., BERMAN, H., FRONDEL, C.: Dana's System of Mineralogy, Vol. II: Halides, Nitrates, etc. New York: John Wiley 1951.

PRESANT, E. W.: Geochemistry of iron, manganese, lead, copper, zinc, arsenic, antimony, silver, tin and cadmium in the soils of the Bathurst area, New Brunswick. Geol. Surv. Can. Bull. **174** (1971).

RAMAGE, H.: Mushrooms — mineral content. Nature **126**, 279 (1930).

SCHINDEWOLF, U., WAHLGREN, M.: The rhodium, silver and indium content of some chondritic meteorites. Geochim. Cosmochim. Acta **18**, 36 (1960).

SCHUTZ, D. F., TUREKIAN, K. K.: The investigation of the geographical and vertical distribution of several trace elements in sea water using neutron activation analysis. Geochim. Cosmochim. Acta **29**, 259 (1965).

SHACKLETTE, H. T.: Element content of bryophytes. U.S. Geol. Surv. Bull. **1198-D** (1965).

SHCHERBINA, V. V.: Complex ions and the transfer of elements in the supergene zone. Geochemistry (English translation), **5**, 486 (1956).

SHIELDS, W. R., GARNER, E. L., DIBELER, V. H.: Absolute isotopic abundance of terrestrial silver. J. Res., Nat. Bur. Std. A, Physics and Chemistry **66 A**, 1 (1962).

SINE, N. M., TAYLOR, W. O., WEBBER, G. R., LEWIS, C. L.: Third report of analytical data for CAAS sulphide ore and syenite rock standards. Geochim. Cosmochim. Acta **33**, 121 (1969).

SKINNER, B. J., WHITE, D. E., ROSE, H. J., MAYS, R. E.: Sulfides associated with the Salton Sea geothermal brines. Econ. Geol. **62**, 316 (1967).

SMALES, A. A., MAPPER, D., FOUCHE, K. F.: The distribution of some trace elements in iron meteorites, as determined by neutron activation. Geochim. Cosmochim. Acta **31**, 673 (1967).

SMALES, A. A., MAPPER, D., WEBB, M. S. W., WEBSTER, R. K., WILSON, J. D., HISLOP, J. S.: Elemental composition of lunar surface material (Part 2). Geochim. Cosmochim. Acta, Suppl. **2**, 1253 (1971).

STROMINGER, D., HOLLANDER, J. M., SEABORG, G. T.: Table of isotopes. Rev. Mod. Phys. **30**, 585 (1958).

SUESS, H. E., UREY, H. C.: Abundances of the elements. Rev. Mod. Phys. **28**, 53 (1956).

SWAINE, D. J.: The trace-element content of soils. Commonwealth Bur. Soil Sci. Tech. Commun. **48** (1955).

TAYLOR, S. R.: Geochemical application of spark source mass spectrography. Nature **205**, 34 (1965).

TAYLOR, S. R.: The application of trace element data to problems in petrology. Physics and Chemistry of the Earth, **6**, 133 .Oxford: Pergamon Press 1965.

TAYLOR, S. R., SACHS, M.: Trace elements in australites. Nature **188**, 387 (1960).

TUREKIAN, K. K.: Deep-sea deposition of barium, cobalt and silver. Geochim. Cosmochim. Acta **32**, 603 (1968).

TUREKIAN, K. K., KHARKAR, D. P.: Neutron activation analysis of milligram quantities of Apollo 11 lunar rocks and soil. Geochim. Cosmochim. Acta, Suppl. **1**, 1659 (1970).

TUREKIAN, K. K., WEDEPOHL, K. H.: Distribution of the elements in some major units of the earth's crust. Bull. Geol. Soc. Am. **72**, 175 (1961).

VAN HOOK, H. J.: The ternary system Ag_2S—Bi_2S_3—Pb S. Econ. Geol. **55**, 759 (1960).

VINCENT, E. A., ADAMS, D. B.: Silver in standard rock samples and in rocks of Skaergaard intrusion. Unpublished work, 1961.

VINOGRADOV, A. P.: Average contents of chemical elements in the principal types of igneous rocks of the earth's crust. Geochemistry (English translation) **7**, 641 (1962).

VINOGRADOV, A. P.: Preliminary data on lunar ground brought to Earth by automatic probe "Luna-16". Geochim. Cosmochim. Acta, Suppl. **2**, 1 (1971).

WAGER, L. R., BROWN, G. M.: Layered Igneous Rocks. Edinburgh and London: Oliver and Boyd 1968.

WAGER, L. R., VINCENT, E. A., SMALES, A. A.: Sulphides in the Skaergaard intrusion, East Greenland. Econ. Geol. **52**, 855 (1957).

WAGONER, L : The detection and estimation of small quantitites of gold and silver. Trans. Inst. Mining. Eng. (Am.) **31**, 798 (1901).

WARREN, H. V., DELAVAULT, R. E.: Gold and silver content of some trees and horsetails in British Columbia. Bull. Geol. Soc. Am. **61**, 123 (1950).

WEBB, J. S., MILLMAN, A. P.: Heavy metals in vegetation as a guide to ore. Trans. Inst. Min. Met. **60**, 473 (1950—51).

WEBBER, G. R.: Second report of analytical data for CAAS syenite and sulphide standards. Geochim. Cosmochim. Acta **29**, 929 (1965).

WHITTAKER, E. J. W., MUNTUS, R.: Ionic radii for use in geochemistry. Geochim. Cosmochim. Acta **34**, 945 (1970).

ZEMANN, J.: Kristallchemie, Sammlung Göschen Bd. 1220, 1220a. Berlin: de Gruyter 1966.

Revised manuscript received: May 1974

Cadmium 48

A	B. Brehler	(Mineralogisch-Kristallographisches Institut, Technische Universität, Clausthal-Zellerfeld 1, Germany)
B—M	H. Wakita and	(Department of Chemistry, Oregon State University, Corvallis, Oregon, U.S.A.)
O	R. A. Schmitt	(Department of Chemistry, Oregon State University, Corvallis, Oregon, U.S.A.)

48-A. Crystal Chemistry

In nature cadmium is exclusively known as occurring in compounds. In some cases it replaces other elements in their minerals, especially zinc. The Cd-minerals (greenockite, β-CdS; monteponite, CdO; otavite, $CdCO_3$; hawleyite, α-CdS) originate mainly from weathering of Cd-bearing zinc minerals.

In all known compounds the valency of Cd is two. The ionic radius of Cd^{2+} has been found to be 1.03 Å (GOLDSCHMIDT, 1926; see Chapter 12 of Volume I of this handbook) and 0.97 Å (PAULING, 1927; AHRENS, 1952; see Chapter 12 of Volume I of this handbook); PAULING and HUGGINS (1934) derived the tetrahedral covalent radius as 1.48 Å.

The observed coordination numbers of Cd in its compounds are usually four and six, in a few cases also five, seven, eight, nine and twelve. Cd-compounds are often isotypic with the corresponding compounds of Zn^{2+}, Mg^{2+}, Fe^{2+}, Co^{2+}, Ni^{2+} and in some cases of Ca^{2+}. Some cadmium chalcogen compounds and silicates have applications in electronics.

Elementary cadmium crystallizes in a very distorted hexagonal close-packing; the axial ratio c/a is 1.885 (STENZEL and WEERTS, SB 1928—32, 164), instead of 1.633 as in the ideal hexagonal close-packings. The distance from each atom to its six neighbors in (00.1) is 2.96 Å, to the other six neighbors 3.28 Å (HULL and DAVEY, SB 1913—28, 42).

In the following lists of known crystal-structures of cadmium compounds, the following are omitted: all organic compounds, the large number of alloys, and all compounds whose number of cadmium atoms is smaller than a quarter of the total number of cations.

In Table 48-A-1 the halogenides and hydroxides of cadmium are listed.

In CdO the cadmium atoms are surrounded by six oxygens. The sulfide, selenide and telluride have fourfold coordinations; they can form both zincblende and wurtzite structures.

From the lattice energies calculated by DÄWERITZ (1971) from the Born-Landé equation, it can be concluded that the stability of the zincblende structure increases in the order CdS-CdSe-CdTe. In Table 48-A-2 the cadmium chalcogenides are listed.

In all structurally known double halogenides, Cd is octahedrally surrounded. $KCdF_3$ (BRISI, SR 1952, 166), $RbCdF_3$ (COUSSEINS and PINA PEREZ, 1968), $CsCdCl_3$ and $CsCdBr_3$ (NÁRAY-SZABÓ, SR 1947—48, 454) crystallize in the perovskite-type. $KCdCl_3$ (BRANDENBERGER, SR 1947—48, 438) and $RbCdCl_3$ (MAC GILLAVRY, NIJEVELD, DIERDORP and KARSTEN, SB 1939, 115) are isotypic to $(NH_4)CdCl_3$ (BRASSEUR and PAULING, SB 1938, 79). Rb_2CdCl_4 (SEIFERT and KOKNAT, 1968) and some other compounds are isotypic to K_2NiF_4.

Table 48-A-1. *Cadmium halogenides and hydroxides*

Compound	Structure-type	Cd-coordination	Reference
$Cd^{[8]}F_2^{[4]}$	fluorite	8 F Cd—F: 2.33 Å	KOLDERUP (SB 1913—28, 188); HAENDLER and BERNARD (SR 1951, 144)
$^2_\infty Cd^{[6]}Cl_2^{[3\,py]}$	$CdCl_2$	6 Cl Cd—Cl: 2.74 Å	PAULING (SB 1913—28, 774)
$^2_\infty Cd^{[6]}Br_2^{[3\,py]}$	$CdCl_2$	6 Br	FERRARI and GIORGIO (SB 1928—32, 246)
$^2_\infty Cd^{[6]}Br_2^{[3\,py]}$	"Wechsel-struktur"[a, b]	6 Br	BIJVOET and NIEUWENKAMP (SB 1933—35, 280); HÄGG, KIESSLING and LINDÉN (SR 1947—48, 490)
$^2_\infty Cd^{[6]}I_2^{[3\,py]}$[c]	CdI_2 CdI_2 (2. Modif.)	6 I Cd—I: 2.99 Å	BOZORTH (SB 1913—28, 189); HASSEL (SB 1933—35, 282)
$^2_\infty Cd^{[6]}(OH)_2$	CdI_2[d]	6 OH	NATTA (SB 1913—28, 195 and 775)
γ-$Cd^{[6]}(OH)_2$[e]		6 OH Cd—(OH): 2.16—2.39 Å	FEITKNECHT (SR 1957, 239); GLEMSER, HAUSCHILD and RICHERT (SR 1957, 239); DE WOLFF (1966, 1967)
$^2_\infty Cd^{[6]}OHCl$[f]	CdOHCl	3 Cl + 3 OH Cd—Cl: 2.69 Å (3×) Cd—OH: 2.34 Å (3×)	HOARD and GRENKO (SB 1933—35, 373)

[a] This structure is a random layer lattice of C6 and C19 type layers. With heating, it transforms rapidly into the C19 type (HÄGG, KIESSLING, LINDÉN, SR 1947—48, 490).

[b] A C6 type in some samples with enclosed 6—8 layers of the C27 type was found by PINSKER and LAPIDUS (SR 1947—48, 490) and PINSKER (SR 1947—48, 491). An X-ray study of the polytypes was given by MITCHELL (1962). Dislocations in $CdBr_2$ are discuseds by AGRAWAL and TRIGUNAYAT (1970).

[c] More than 100 polytypes of CdI_2 are known (e.g., MITCHELL, SR 1956, 252; JAIN, CHADHA and TRIGUNAYAT, 1970; PRASAD and SRIVASTAVA, 1970). In the system $CdBr_2$—CdI_2, one intermediate phase with the composition CdBrI appears. It was stated by HÄGG and LINDÉN (SR 1947—48, 490) that CdBrI posses a "Wechselstruktur"; it cannot be decided whether the I and Br ions form separate layers, or are divided statistically.

[d] During the precipitation of $Cd(OH)_2$, the "α-Form" will grow first, which is a random layer structure (FEITKNECHT, SB 1938, 89). MITCHELL (1966) studied 50 crystals of $Cd(OH)_2$ by the Weissenberg method; he observed no evidence for structural polytypism. All crystals were shown to be type 2H, $[(A\,\gamma\,B)]_n$.

[e] In this structure, the $Cd(OH)_6$-octahedra occur in pairs, sharing a face parallel to (010). All octahedra share edges with neighbors in the c-direction. Finally, the infinite double strings thus formed share corners with four neighboring strings.

[f] There are several hydroxyhalogenides known i.e., $CdCl_{0.67}(OH)_{1.33}$; $CdBr_{0.6}(OH)_{1.4}$; $CdI_{0.5}(OH)_{1.5}$ (FEITKNECHT *et al.*, SB 1937, 49; SR 1942—44, 217; SR 1945—46, 171). They have layer structures with replacement of (OH) and halogen atoms.

Table 48-A-2. *The chalcogenides of cadmium*

Compound	Structure-type	Cd-coordination	Reference
$Cd^{[6]}O^{[6]}$ [a] monteponite	NaCl	6 O Cd—O: 2.35 Å	DAVEY and HOFFMANN (SB 1913—28, 120)
α-$Cd^{[4]}S^{[4]}$ [b] hawleyite	sphalerite	4 S Cd—S: 2.52 Å	ULRICH and ZACHARIASEN (SB 1913—28, 129); TRAILL and BOYLE (1955) [c]
β-$Cd^{[4]}S^{[4]}$ [b, d] greenockite	wurtzite	4 S	BRAGG (SB 1913—28, 129); SMITH (SR 1955, 403) (lattice dimensions)
α-$Cd^{[4]}Se^{[4]}$ [e]	sphalerite	4 Se Cd—Se: 2.6 Å	GOLDSCHMIDT (SB 1913—28, 136)
β-$Cd^{[4]}Se^{[4]}$ cadmoselite	wurtzite	4 Se	GOLDSCHMIDT (SB 1913—28, 136)
$Cd^{[4]}Te^{[4]}$(I) [f]	sphalerite	4 Te Cd—Te: 2.80 Å	ZACHARIASEN (SB 1913—28, 136)
$Cd^{[4]}Te^{[4]}$(II)	wurtzite	4 Te	SEMILETOV (SR 1955, 81)
High pressure forms:			
$Cd^{[6]}S^{[6]}$ (β-CdS, 18 kb)	NaCl	Cd—S: 2.78 Å	KABALKINA and TROITSKAYA (1963)
$Cd^{[6]}Se^{[6]}$ (> 32 kb)	NaCl		MARIANO and WAREKOIS (1963); ONODERA (1969)
$Cd^{[6]}Te^{[6]}$ (high pressure)	NaCl	Cd—Te: 2.98 Å	SMITH and MARTIN (1963); ONODERA (1969)
$Cd^{[4+2]}Te^{[4+2]}$ (> 90 kb)	analog to β-Sn		SMITH and MARTIN (1963)

[a] CdO forms solid solutions with CaO (NATTA and PASSERINI, SB 1928—32, 227).

[b] Both α-CdS and β-CdS form solid solutions with HgS (RITTNER and SCHULMAN, SR 1942—44, 282).

[c] Isometric CdS has been found as a mineral. CdS was precipitated from salt solutions under several different conditions by HARTMANN (1966).

[d] β-CdS forms a continuous series of solid solutions with ZnS (BALLENTYNE and ROY, SR 1961, 251). There are also some other mixed sulfides (Cd, Zn, Fe, Mn)S (SKINNER and BETHKE, SR 1961, 251; SKINNER, SR 1961, 252). β-CdS also forms solid solutions with CdSe and CdTe (VITRIKHOVSKIJ and MIZECKAJA, SR 1959, 76).

[e] CdSe forms solid solutions with ZnSe, InAs and In_2Se_3 (GORJUNOVA, KOMOVIĆ and FRANK-KAMENECKIJ, SR 1955, 80).

[f] Cubic CdTe forms solid solutions with ZnTe (KOLOMIEC and MALKOVA, SR 1958, 71) and HgTe (GAVRIŠČAK, SR 1960, 83; WOLLEY and RAY, SR 1960, 234).

In oxyspinels Cd^{2+} has the coordination numbers four (e.g., $CdFe_2O_4$, GORTER, SR 1950, 242) and six (e.g., $CdIn_2O_4$, SKRIBLJAK, DASGUPTA and BISWAS, SR 1959, 377).

Table 48-A-3 shows the cadmium silicates of known structure.

Table 48-A-3. *Cadmium silicates*

Compound	Remarks	Reference
$Cd_3^{[6]}O[SiO_4]$	similar to $Ca_3O[SiO_4]$	DENT GLASSER and GLASSER (1964); DENT GLASSER (1965)
$Cd_3^{[8]}Al_2^{[6]}[SiO_4]_3$	garnet structure	GENTILE and ROY (1960)
$Cd_3^{[8]}V_2^{[6]}[SiO_4]_3$	garnet structure	MILL (1964)
$Cd_2^{[6]}[SiO_4]$	related to Na_2SO_4(V) Cd—O: 2.15—2.35 Å (4×) 2.61 Å (2×)	DENT GLASSER and GLASSER (1964)
$Li_2^{[4]}Cd^{[4]}[SiO_4]$ [a]	related to Li_3PO_4	TARTE et CAHAY (1970)
$Cd^{[6]}[SiO_3]$	related to β-$CaSiO_3$	DENT GLASSER and GLASSER (1964)
$Na_4^{[4]}Cd_2^{[5]}[Si_3O_{10}]$	Cd—O: 2.13—2.27 Å	SIMONOV, EGOROV-TISMENKOV and BELOV (1968)

[a] The compounds of Zn, Mg, Mn, Co and Fe are isostructural.

Table 48-A-4. *Complex cadmium oxygen compounds*

Compound	Remarks	Reference
	Coordination number 4	
α-$Cd^{[4]}SO_4$	isotypic with $HgSO_4$ Cd—O: 2.11 Å, 2.18 Å (2×) 2.21 Å	KOKKOROS and RENTZEPERIS (1963)
$Cd^{[4]}[B^{[4]}B^{[3]}O_7]$	Cd—O: 2.18—2.22 Å	IHARA and KROGH-MOE (1966)
	Coordination number 5	
$Ba^{[7]}Cd^{[5]}O_2$	(trig. bipyramidal) Cd—O: 2.23 Å (2×), 2.29 Å (2×), 2.62 Å	SCHNERING (1962)
	Coordination number 6	
$Cd^{[6]}[CO_3]$ [a] otavite	calcite type	ZACHARIASEN (SB 1913—28, 338); RAMDOHR and STRUNZ (1941)
$Cd^{[6]}Ti^{[6]}O_3$	ilmenite type	BARTH and POSNJAK (SB 1933—35, 379)
$Cd_2^{[6]}Mn_3^{[6]}O_8$	isotypic with Mn_5O_8 Cd—O: 2.16—2.52 Å av. 2.35 Å	OSWALD, WAMPETICH (1970)
β-$Cd^{[6]}[SO_4]$ [b]	isotypic with $CrVO_4$ Cd—O: 2.23 Å	COING-BOYAT (SR 1961, 441)

[a] $CdCO_3$ forms a continuous series of solid solutions with $MnCO_3$, and shows limited solubility in series with $CaCO_3$ (OFTEDAHL and VEGARD, SR 1947—48, 494). Phase relations in the system $CaCO_3$—$CdCO_3$ were investigated by CHANG and BRICE (1971).

[b] Isotypic with the Mg, Ni, Co(II), Mn(II) sulfates.

Table 48-A-4 (Continued)

Compound	Remarks	Reference
α-$Cd^{[6]}[CrO_4]$[c] (low temperature)	isotypic with $CrVO_4$	BRANDT (SR 1942—44, 181)
β-$Cd^{[6]}[CrO_4]$ (high temperature)	isotypic with α-$MnMoO_4$	MÜLLER, WHITE and ROY (1969)
β-$Cd^{[6]}[U^{[6]}O_4]$		KOVBA, POLUNINA, IPPOLITOVA, SIMANOV and SPICYN (SR 1961, 403)
$Cd^{[6]}[Sb_2^{[6]}O_6]$ $Cd^{[6]}[As_2^{[6]}O_6]$	isotypic with $PbSb_2O_6$	MAGNÉLI (SR 1940—41, 156)
$Cd^{[6]}Ta_2^{[6]}O_6$	columbite type	ISMAILZADE (SR 1958, 316)
$A_2^ICd_2^{[6]}[SO_4]_3$ $A^I = (NH_4)$, K, Rb, Tl	langbeinite type	GATTOW and ZEMANN (1958)
$Cd_3^{[6]}[Cl \| B_7O_{13}]$[d] cadmium borate	Cd: 4O + 2Cl	JONA (SR 1959, 502)
$Cd^{[6]}Sn^{[6]}(OH)_6$	6 OH isotypic with stottite	GIGLIO and NOVALES (1964)
	Coordination number 7	
$Cd_2^{[7]}SnO_4$	isotypic with Sr_2PbO_4	TRÖMEL (1967, 1969)
	Coordination number 8	
$Cd^{[8]}[MoO_4]$	scheelite type	BROCH (SB 1928—32, 450) SWANSON, GILFRICH and COOK (SR 1956, 451)
$(CdU_2)^{[8]}O_6$	fluorite type[e]	RÜDORFF, KEMMLER and LEUTNER (1962)
$Cd_2^{[8]}[Nb_2^{[6]}O_7]$[f]	pyrochlore type	BYSTRÖM (SR 1945—46, 117)
$Cd_2^{[8]}[Ta_2^{[6]}O_7]$	pyrochlore type Cd—O: 2.50 Å (6×) 2.25 Å (2×)	COOK and JAFFE (SR 1952, 254); ALESHIN and ROY (1962)
$Cd_2^{[8]}[Re_2^{[6]}O_7]$	pyrochlore type Cd—O: 2.66 Å, 2.21 Å	DONOHUE, LONGO, ROSENSTEIN and KATZ (1965)
$Cd_3^{[8]}Fe_2^{[6]}[SnO_4]_3$	garnet type	GELLER, BOZORTH, GILLEO and MILLER (SR 1960, 350)
$Cd_6^{[8]}X^{[6]}[GeO_4]_3$ X = Fe, Al, Cr, Se, Ga	garnet type	STRUNZ, FREIGANG and CONTAG (SR 1960, 456); TAUBER, WHINFREY and BANKS (SR 1961, 415)
$Cd_2^{[6]}Bi_2O_4Br_2$(I, II)	isotypic Cd—O: 2.28 Å (4×) Cd—Br: 3.44 Å (4×)	SILLÉN (SR 1939, 159)
$Cd_2^{[6]}Bi_2O_4I_2$	isotypic Cd—O: 2.32 Å Cd—I: 3.52 Å	SILLÉN (SR 1947—48, 307)

[c] Oxygen in distorted cubic close packing. The metal distribution is closely related to that of the spinels. Isotypic with the Ni, Co, Cu, Zn compounds.

[d] See also FOUASSIER, LEVASSEUR, JOUBERT, MULLER, HAGENMULLER (1970).

[e] The cations are statistically distributed.

[f] $Cd_2Nb_2O_7$ forms solid solutions with $Sr_2Nb_2O_7$ and $NaBiNb_2O_7$ (ISMAILZADE, SR 1958, 371) and with $NaNbO_3$ (LEWIS and WHITE, SR 1956, 298). $Cd_2MM'O_5F_2$ (with M = Ta(V) or Nb(V) and M' = Fe(III), Cr(III), Al, Ga, Se, In) crystallizes also in the pyrochlore type (PANNETIER and LUCAS, 1970).

Table 48-A-4 (Continued)

Compound	Remarks	Reference
	Coordination number 12	
$Cd^{[12]}Ti^{[6]}O_3$	perovskite type (monoclinic distorted)[g]	Barth and Posnjak (SB 1933—35, 379); Lebedev, Venevstsev and Shdanov (1970)
	Coordination numbers 4 and 6	
$X^{[6]}X^{[4]}[Mo_3^{[6]}O_8]$ X = Cd, Mg, Mn, Co, Ni, Zn		Mac Carroll, Katz and Ward (SR 1957, 333)
	Coordination numbers 5 and 6	
$Cd^{[6]}Cd_2^{[5]}[AsO_4]$	$Cd^{[6]}$—O: 2.19—2.50 Å $Cd^{[5]}$—O: 2.11—2.39 Å	Engel and Klee (1970)
	Coordination numbers 6 and 8	
$Cd_2^{[8]}Cd^{[6]}Ge^{[6]}[GeO_4]_3$ (high pressure phase)	garnet like structures $Cd^{[8]}$(I)—O: 2.45 Å (av.) $Cd^{[8]}$(II)—O: 2.41 Å (av.) $Cd^{[6]}$—O: 2.24 Å (av.)	Prewitt and Sleight (1969)
$Cd^{[8]}Cd^{[6]}[Sb_2^{[6]}O_7]$	isotypic with weberite	Byström (SR 1945—46, 117)
	Coordination numbers 7 and 9	
$Cd_3^{[6+3]}Cd_3^{[7]}(OH)[PO_4]_3$ $Cd_2^{[6+3]}Cd_3^{[7]}(Y)[XO_4]_3$ X = P, As, V Y = F, Cl, Br	apatite type Cd (I): 9 O, Cd (II): 6 O + 1 (OH)	Klement and Zureda (SR 1940—41, 206); Klement and Haselbeck (1965); Engel (1968 and 1970)

[g] Some other cadmium compounds (e.g., $Cd^{[12]}Sn^{[6]}O_3$) crystallize also in the perovskite structure (Náray-Szabó, SR 1947—48, 454).

In its complex oxygen compounds, in most cases cadmium has coordination numbers 6, 8 and 12, and in a few cases also 4, 5, 7 and 9. In Table 48-A-4 some complex oxygen compounds are listed.

Table 48-A-5 gives some information about the structurally known hydrated complex compounds of cadmium.

There are some binary cadmium chalcogene compounds. Some are isotypic with the normal spinel type e.g., $Cd^{[4]}In_2^{[6]}S_4$. A large group has the thiogallate structure (Hahn, Frank, Klingler, Störger and Störger, SR 1955, 414) e.g., $Cd^{[4]}Al^{[4]}S_4$ and $Cd^{[4]}$ $Ga^{[4]}S_4$.

Some high pressure modifications have been found by Range, Becker and Weiss (1968a, **b** and 1969). $Cd^{[4]}Al_2^{[6]}S_4$ crystallizes at 400° C and at a pressure above 48 kb in the normal spinel type. At different temperatures and pressures four modifications of $CdIn_2S_4$ are known. One of these crystallizes in a NaCl defect structure, the other one with a fourfold coordination of cadmium.

Table 48-A-5. *Hydrated complex compounds of cadmium*

Compound	Cd-coordination	Reference
$Cd^{[6]}(H_2O)[SO_4]$	$2\,OH_2+4\,O$ Cd—O: 2.21—2.36 Å	BRÉGEAULT and HERPIN (1970)
$Cd^{[6]}(H_2O)_2[SeO_4]$	$2\,OH_2+4\,O$	KOKKOROS (SR 1940—41, 196); HIRIYANA and SAKURI (SR 1949, 251)
$Cd^{[6]}(H_2O)_2[SO_4]\cdot{}^2/_3\,H_2O$	$2\,OH_2+4\,O$	LIPSON (SB 1936, 182)
$Cd^{[4+4]}(H_2O)_3[H_3IO_6]$	$2\,OH_2+2\,O$ $(+2\,OH_2+2\,O)$ Cd—O: 2.23 Å (av.) 2× —OH_2: 2.27 Å; 2.41 Å Cd—OH_2: 2.50 Å (av.) 2× —O: 2.64 Å; 2.79 Å	BRAIBANTI, TIRIPICCHIO, BIGOLI and PELLING-HELLI (1970)
$Cd^{[8]}(H_2O)_4[NO_3]_2$	$4\,OH_2+4\,O$ Cd—OH_2: 2.26 Å (2×) 2.33 Å (2×) Cd—O: 2.44 Å (2×) 2.59 Å (4×)	MATKOVIĆ and RIBAR (1963); MATKOVIĆ, RIBAR, ZELENKO and PETERSON (1966)
$(NH_4)_2Cd^{[6]}(H_2O)_6[SO_4]_2$	$6\,OH_2$ Cd—OH_2: 2.30 Å (4×) 2.24 Å (2×)	HOFMANN (SB 1928—32, 436); MONTGOMERY and LINGAFELTER (1966)
$Cd^{[6]}(H_2O)_6[ClO_4]_2$	$6\,OH_2$	WEST (SB 1933—35, 452);
$Cd^{[6]}(H_2O)_6[BF_4]_2$	$6\,OH_2$	MOSS, RUSSEL and SHARP (SR 1961, 304)

Some arsenide and phosphide compounds of cadmium and the elements of the fourth group of the periodic system crystallize in the chalcopyrite type, e.g., $Cd^{[4]}Ge^{[4]}As$ (GOODMAN, SR 1957, 38), and $Cd^{[4]}Sn^{[4]}P_2$ (BUEHLER and WERNICK, 1971).

Revised manuscript received: February 1972

References: Section 48-A

AGRAWAL, V. K., TRIGUNAYAT, G. C.: Arcing phenomenon in single crystals of cadmium bromide. Acta Cryst. **A 26**, 426 (1970).

ALESHIN, E., ROY, R.: Crystal chemistry of pyrochlore. J. Am. Ceram. Soc. **45**, 18 (1962).

BRAIBANTI, A., TIRIPICCHIO, A., BIGOLI, F., PELLINGHELLI, M. A.: Crystal and molecular structure of cadmium trihydrogenhexaoxoiodate. (VII) trihydrate. Acta Cryst. **B 26**, 1069 (1970).

BRÉGEAULT, J.-M., HERPIN, P.: Étude structurale du sulfate de cadmium monohydraté $CdSO_4$. H_2O. Bull. Soc. Franç. Minéral. Cristall. **93**, 37 (1970).

BUEHLER, E., WERNICK, J. H.: Concerning growth of single crystals of the II-IV-V diamond-like compounds $ZnSiP_2$, $CdSiP_2$, $ZnGeP_2$, and $CdZnP_2$ and standard enthalpies of formation for $ZnSiP_2$ and $CdSiP_2$. J. Crystal Growth **8**, 324 (1971).

CHANG, L. L. Y., BRICE, W. R.: Subsolidus phase relations in the system calcium carbonate-cadmium carbonate. Am. Mineralogist **56**, 338 (1971).

COUSSEINS, J.-C., PINA PEREZ, C.: Fluorures doubles de cadmium et d'éléments metalliques monovalents. Rev. Chim. Minerale **5**, 147 (1968).

DÄWERITZ, L.: Relative stability of zincblende and wurtzite structure in $A^{II}B^{IV}$ compounds. Kristall u. Technik **6**, 101 (1971).

DENT GLASSER, L. S.: Silicates M_3SiO_5. II. Relationships between Sr_3SiO_5, Cd_3SiO_5 and Ca_3SiO_5. Acta Cryst. **18**, 455 (1965).

— GLASSER, F. P.: The preparation and crystal data of the cadmium silicates $CdSiO_3$, Cd_2SiO_4, and Cd_3SiO_5. Inorg. Chem. **3**, 1228 (1964).

DONOHUE, P. C., LONGO, J. M., ROSENSTEIN, R. D., KATZ, L.: The preparation and structure of cadmium rhenium oxide, $Cd_2Re_2O_7$. Inorg. Chem. **4**, 1152 (1965).

ENGEL, G.: Einige Apatite des Cadmiums. Z. Anorg. Allgem. Chem. **362**, 273 (1968).

— Cadmiumapatite sowie die Verbindungen Cd_2XO_4F mit $X = P$, As und V. Z. Anorg. Allgem. Chem. **378**, 49 (1970).

— KLEE, W. E.: Die Kristallstruktur des Cadmiumorthoarsenates. Z. Krist. **132**, 332 (1970).

FOUASSIER, C., LEVASSEUR, A., JOUBERT, J. C., MULLER, J., HAGENMULLER, P.: Les systèmes B_2O_3—MO—MS. Boracites M—S (M = Mg, Mn, Fe, Cd) et sodalites M—S (M = Co, Zn). Z. Anorg. Allgem. Chem. **375**, 202 (1970).

GATTOW, G., ZEMANN, J.: Über Doppelsulfate vom Langbeinit-Typ, $A_2^+B_2^{2+}(SO_4)_3$. Z. Anorg. Allgem. Chem. **293**, 233 (1958).

GENTILE, A. L., ROY, R.: Isomorphism and crystalline solubility in the garnet family. Am. Mineralogist **45**, 701 (1960).

GIGLIO, M., NOVALES, H.: $ZnSn(OH)_6$ und $CdSn(OH)_6$. Naturwissenschaften **51**, 56 (1964).

HARTMANN, H.: Über die Polymorphie unterschiedlich präparierter Cadmiumsulfide. Kristall u. Technik **1**, 267 (1966).

IHARA, M., KROGH-MOE, J.: The crystal structure of cadmium diborate, $CdO \cdot 2B_2O_3$. Acta Cryst. **20**, 132 (1966).

JAIN, R. K., CHADHA, G. K., TRIGUNAYAT, G. C.: Crystal structures of four new polytypes of cadmium iodide. Acta Cryst. **B 26**, 1785 (1970).

KABALKINA, S. S., TROITSKAYA, Z. V.: Cadmium sulfide structures at high pressures up to 90 kilobars. Dokl. Akad. Nauk SSSR **151**, 1068 (1963).

KLEMENT, R., HASELBECK, H.: Apatite und Wagnerite zweiwertiger Metalle. Z. Anorg. Allgem. Chem. **336**, 113 (1965).

KOKKOROS, P. A., RENTZEPERIS, P. J.: The crystal structure of the anhydrous cadmium and mercuric sulfates. Z. Krist. **119**, 234 (1963).

LEBEDEV, V. M., VENEVTSEV, YU. N., SHDANOV, G. S.: High-temperature x-ray investigation of the perovskite modification of $CdTiO_3$. Soviet Phys.-Cryst. **15**, 318 (1970).

MARIANO, A. N., WAREKOIS, E. P.: High-pressure phases of some compounds of group II—VI. Science **142**, 672 (1963).

MATKOVIĆ, B., RIBAR, B.: The crystal structure of cadmium nitrate tetrahydrate. Croat. Chem. Acta **35**, 147 (1963).

— — ZELENKO, B., PETERSON, S. W.: Refinement of the structure of $Cd(NO_3)_2 \cdot 4H_2O$. Acta Cryst. **21**, 719 (1966).

MILL, B. V.: Hydrothermal synthesis of garnets containing V^{3+}, In^{3+} and Sc^{3+}. Dokl. Akad. Nauk SSSR **156**, 914 (1964).

MITCHELL, R. S.: Single crystal x-ray study of structural polytypism in cadmium bromide. Z. Krist. **117**, 309 (1962).

— Comments on an x-ray study of cadmium hydroxide single crystals. Z. Krist. **123**, 459 (1966).

MONTGOMERY, H., LINGAFELTER, E. C.: The crystal structure of tutton's salts. IV. Cadmium ammonium sulfate hexahydrate. Acta Cryst. **20**, 728 (1966).

MÜLLER, O., WHITE, W. B., ROY, R.: X-ray diffraction study of the chromates of nickel, magnesium and cadmium. Z. Krist. **130**, 112 (1969).

ONODERA, A.: High pressure transition in cadmium selenide. Rev. Phys. Chem. Japan **39**, 65 (1969).

OSWALD, H. R., WAMPETICH, M. J.: Die Kristallstrukturen von Mn_5O_8 und $Cd_2Mn_3O_8$. Helv. Chim. Acta **50**, 2023 (1970).

PANNETIER, J., LUCAS, J.: Niobates et tantalates de cadmium oxyfluores contenant de cations trivalents. Compt. Rend., Ser. C **270**, 1013 (1970).

PAULING, L., HUGGINS, M. L.: Covalent radii of atoms and interatomic distances in crystals containing electron-pair-bonds. Z. Krist. (A) **87**, 205 (1934).

PRASAD, R., SRIVASTAVA, O. N.: A new hexagonal polytype 32H of cadmium iodide, its structure and growth. Z. Krist. **131**, 376 (1970).

PREWITT, C. T., SLEIGHT, A. W.: Garnet-like structures of high-pressure cadmium germanate and calcium germanate. Science **163**, 386 (1969).

RAMDOHR, P., STRUNZ, H.: Isomorphie von Otavit mit Kalkspat. Zentr. Mineralogie A, 97/98 (1941).

RANGE, K.-J., BECKER, W., WEISS, A.: Über Hochdruckphasen des $CdAl_2S_4$, $HgAl_2S_4$, $ZnAl_2Se_4$, $CdAl_2Se_4$ und $HgAl_2Se_4$ mit Spinellstruktur. Z. Naturforsch. **23**b, 1009 (1968a).

— — — Über Hochdruckphasen des $CdIn_2Te_4$ und $HgIn_2Te_4$ mit NaCl-Struktur. Z. Naturforsch. **23**b, 1261 (1968b).

— — — Das Verhalten von $CdIn_2Se_4$ bei hohen Drucken. Z. Naturforsch. **24**b, 1654 (1969).

RÜDORFF, W., KEMMLER, S., LEUTNER, H.: Ternäre Uran (V)-oxyde [1]. Angew. Chem. **74**, 429 (1962).

SCHNERING, H. G. VON: Die Kristallstruktur des $BaCdO_2$. Z. Anorg. Allgem. Chem. **314**, 144 (1962).

SEIFERT, H. J., KOKNAT, F. W.: Neue Alkalichlorocadmate (II) in den Systemen CsCl/$CdCl_2$ und RbCl/$CdCl_2$. Z. Anorg. Allgem. Chem. **357**, 314 (1968).

SIMONOV, M. A., EGOROV-TISMENKO, YU. K., BELOV, N. V.: New trisilicate radical $[Si_3O_{10}]$ in the structure of $Na_4Cd_2[Si_3O_{10}]$. Dokl. Acad. Nauk SSSR **179**, 1329 (1968).

SMITH, P. L., MARTIN, J. E.: High pressure structures of cadmium sulfide and cadmium telluride. Phys. Letters **6**, 42 (1963).

TARTE, P., CAHAY, R.: Synthesis and structure of a new family of $Li_2X(II)SiO_4$ and $Li_2X(II)GeO_4$ compounds structurally related to lithium phosphate, [X(II) = magnesium, zinc, manganese, cobalt, iron and cadmium]. Compt. Rend. Ser. C **271**, 777 (1970).

TRAILL, R. J., BOYLE, R. W.: Hawleyite, isometric cadmium sulphide, a new mineral. Am. Mineralogist **40**, 555 (1955).

TRÖMEL, M.: Kristallstrukturdaten für Ca_2SnO_4 und Cd_2SnO_4. Naturwissenschaften **54**, 17 (1967).

— Die Kristallstruktur der Verbindungen vom Sr_2PbO_4-Typ. Z. Anorgan. Allgem. Chem. **371**, 237 (1969).

WOLFF, P. M. DE: The crystal structure of γ-$Cd(OH)_2$. Acta Cryst. **21**, 432 (1966).

— The crystal structure of γ-$Cd(OH)_2$. Corrigenda. Acta Cryst. **22**, 441 (1967).

Revised manuscript received: February 1972

48-B. Isotopes in Nature

Cadmium has the atomic number of 48, with eight stable isotopes between mass number 106 and 116. The relative abundances of the isotopes are listed in Table 48-B-1 given by LELAND and NIER (1958).

Table 48-B-1. *Stable isotopes of cadmium*

Mass number	Relative abundance (%)	Mass number	Relative abundance (%)
106	1.22	112	24.07
108	0.88	113	12.26
110	12.39	114	28.86
111	12.75	116	7.58

Revised manuscript received: September 1969

48-C. Abundance in Cosmos, Meteorites, Tektites and Lunar Materials

I. Cosmic Abundance

The cosmic abundance data of cadmium are summarized in Table 48-C-1.

Table 48-C-1. *The cosmic abundances of cadmium*

Cadmium abundance		Author
$N_{Si}=10^6$	$\log N_H=12.00$	
0.89		SUESS and UREY (1956)
0.11		VINOGRADOV (1956)
0.89		CAMERON (1959)
	1.5	GOLDBERG *et al.* (1960)
0.14		AHRENS and TAYLOR (1961)
0.81		CLAYTON *et al.* (1961)
	1.5	ALLER (1961)
	1.5	ALLER (1962)
	1.7	ALLER (1965)
	1.9	ALLER (1968)
	2.6	GREVESSE *et al.* (1968)
	1.6	MÜLLER (1968)
1.2		GOLES (1968)

II. Meteorites and Tektites

The cadmium contents of iron and stony-iron meteorites are shown in Table 48-C-2. Cadmium in seven iron meteorites ranges from ≤ 0.04 to 0.056 ppm and the average given by SCHMITT *et al.* (1963) is 0.027 ± 0.012 ppm.

Table 48-C-2. *Cadmium in iron and stony-iron meteorites*

Material	Cd (ppm)	Method	Author
a) Irons:			
Arispe, Ogg	0.022	(N/R)	SCHMITT *et al.* (1963)
Bristol, Of	0.056	(N/R)	SCHMITT *et al.* (1963)
Canyon Diablo, Om	0.020	(N/R)	SCHMITT *et al.* (1963)
Deep Springs, D	0.018	(N/R)	SCHMITT *et al.* (1963)
Odessa, Og	≤ 0.04	(N/R)	SCHMITT *et al.* (1963)
Sandia Mountains, H	0.0085	(N/R)	SCHMITT *et al.* (1963)
Sikhote-Alin, H or Ogg	0.022	(N/R)	SCHMITT *et al.* (1963)
b) Stony-Irons:			
Brenham, P (olivine phase)	0.13	(N/R)	SCHMITT *et al.* (1963)
Estherville, M (nonmetal phase)	<0.004	(N/R)	SCHMITT *et al.* (1963)
Thiel Mountains, P	0.33	(N/R)	SCHMITT *et al.* (1963)
Veramin, M	0.056	(N/R)	SCHMITT *et al.* (1963)

The cadmium contents of chondrites, achondrites and tektites are shown in Table 48-C-3. Since the agreement of duplicate determinations was unsatisfactory for some chondrites, GREENLAND (1967) suggested the possibility of heterogeneous distribution of Cd in meteoritic samples.

Table 48-C-3. *Cadmium in chondrites, achondrites and tektites*

Material	Cd (ppm)	Method	Author
a) *Chondrites:*			
Abee, Ce_1	3.3	(N/R)	SCHMITT *et al.* (1963)
Allegan, CH	0.015	(N/R)	SCHMITT *et al.* (1963)
Allegan, CH	0.037	(N/R)	GREENLAND (1967)
Chainpur, CHL	0.039	(N/R)	GREENLAND (1967)
Ehole, CH	0.015	(N/R)	GREENLAND (1967)
Felix, CHL	0.34	(N/R)	GREENLAND (1967)
Holbrook, CL	0.047	(N/R)	SCHMITT *et al.* (1963)
Hvittis, Ce_2	0.11	(N/R)	GREENLAND (1967)
Indarch, Ce_1	1.7	(N/R)	SCHMITT *et al.* (1963)
Indarch, Ce_1	<0.01	(N/R)	GREENLAND (1967)
Ivuna, Cc_1	0.91	(N/R)	SCHMITT *et al.* (1963)
Ivuna, Cc_1	1.7	(N/R)	GREENLAND (1967)
Khairpur, Ce_2	0.12	(N/R)	GREENLAND (1967)
Kyushu, CL	0.087	(N/R)	SCHMITT *et al.* (1963)
Lancé, CHL	0.17	(N/R)	GREENLAND (1967)
Mandhoom, CL	0.053	(N/R)	SCHMITT *et al.* (1963)
Mighei, Cc_2	1.2	(N/R)	SCHMITT *et al.* (1963)
Mighei, Cc_2	0.011	(N/R)	GREENLAND (1967)
Miller, CH	0.12	(N/R)	SCHMITT *et al.* (1963)
Mokoia, CHL	0.52	(N/R)	SCHMITT *et al.* (1963)
Mokoia, CHL	0.012	(N/R)	GREENLAND (1967)
Murray, Cc_2	0.59	(N/R)	SCHMITT *et al.* (1963)
Murray, Cc_2	0.33	(N/R)	GREENLAND (1967)
Orgueil, Cc_1	1.1	(N/R)	SCHMITT *et al.* (1963)
Orgueil, Cc_1	0.40	(N/R)	GREENLAND (1967)
Pantar, CH	0.023	(N/R)	GREENLAND (1967)
Peace River, CL	0.073	(N/R)	GREENLAND (1967)
Pillistfer, Ce_2	0.082	(N/R)	GREENLAND (1967)
Richardton, CH	0.020	(N/R)	SCHMITT *et al.* (1963)
St. Mark's Ce	0.042	(N/R)	SCHMITT *et al.* (1963)
Vigarano, CHL	0.039	(N/R)	GREENLAND (1967)
Walters, CL	0.080	(N/R)	GREENLAND (1967)
Warrenton, CHL	0.052	(N/R)	GREENLAND (1967)
b) *Achondrites:*			
Johnstown, Ab	0.33	(N/R)	SCHMITT *et al.* (1963)
Juvinas, Ap	0.065	(N/R)	SCHMITT *et al.* (1963)
La Fayette, Ado	0.18	(N/R)	SCHMITT *et al.* (1963)
Nakhla, Ado	1.8	(N/R)	SCHMITT *et al.* (1963)
Norton County, Ac	≤ 0.010	(N/R)	SCHMITT *et al.* (1963)
Stannern, Ap	0.022	(N/R)	SCHMITT *et al.* (1963)
c) *Tektites:*			
Indochinite, Dalat, Province de Lang Bian, Sud Annam	0.14, 0.13	(N/R)	TANNER (1968)
Indochinite, Nord du Cambodge	0.07, 0.17	(N/R)	TANNER (1968)

Revised manuscript received: September 1969

III. Lunar Materials

The cadmium contents of lunar materials from lunar maria, lunar uplands and lunar highlands are shown in Tables 48-C-4 to 48-C-6. Most references are taken from the Proceedings of the Lunar Science Conferences published in Geochim. Cosmochim. Acta, Supplements.

Table 48-C-4. *Cadmium in lunar mare materials* (Analytical method N/R)

Materials		No. of rocks	Cd (ppb) Range	Geometric mean[cc]	Reference
Apollo 11					
Mare basalts		5	2.6–10	4.5	1, 2
High K basalt 10017		1	44, 68	56	2, 3
Breccias		2	78, 106	92	1
Anorthosites[a]		3	9, 17, 83	36	4, 5
Soil		1	40–50	45	1, 3
Apollo 12					
Mare basalts		9	1.1–3.3	1.4	2, 6
Breccias		2	19, 34	25	5
Breccia 12013		1	24–300	71[b]	7
Soils		5	10–50	27	5, 6, 8
Core 1.7 cm		1		70	5
Soils 13.2 cm		1		22,000	5
18.9 cm		1		53	5
21.2 cm		1		48	5
37.2 cm		1		49	5
Apollo 15					
Mare basalts		7	0.7–2.1	1.5	9, 10, 11
Mare gabbro		1		1.5	11
Highland basalt 15256[c] (impact melted)		1		104	10
Valley and rille breccias		2	16, 36	24	9, 11
Highland breccias[c]		7	0.6–180	8	9, 10, 11
Anorthosite 15415[c]		1		0.6	10
Valley and rille soils		12	20–50	34	10, 11, 12
Highlands soils[c]		5	50–90	62	9, 11
Glass sphere		1		4	11
Apollo 17					
Mare basalts		7	1.1–≤7	2.1	13, 18, 19
Valley breccia		3	1.0–1.9	1.3	20
Highland-massif breccias[aa]		41	0.6–27	5	13, 20
Sculptured Hills breccia[bb]		1		60	13
Dunite		2	0.4–0.9	0.6	20
Valley soils		5	25–210	50	13
Highland-massif soils		9	35–320	55	13, 14, 15, 16
Sculptured Hills[bb] trench soils	0–1 cm			39	14
	ca. 5 cm			40	14
	ca. 10 cm			38	14
	15–25 cm			46	14
Glass, orange		1		320	18

Table 48-C-4 (continued)

Materials	No. of Rocks	Cd (ppb)		Reference
		Range	Geometric mean[cc]	
Luna 16				
Soil	1	200, 10,900		17

[a] >90% plagioclase. [b] mass-weighted average. [c] at stations 6 and 7. [aa] at stations 2, 3, 6 and 7. [bb] at station 8. [cc] For *replicate analysis* of one rock or soil, arithmetic means are given.

References: 1. Ganapathy, R., Keays, R. R., Laul, J. C., Anders, E.: Suppl. 1, **2**, 1117 (1970); 2. Anders, E., Ganapathy, R., Keays, R. R., Laul, J. C., Morgan, J. W.: Suppl. 2, **2**, 1021 (1971); 3. Wakita, H., Schmitt, R. A., Rey, P.: Suppl. 1, **2**, 1685 (1970). 4. Wakita, H., Schmitt, R. A.: Lunar anorthosites: Rare-earth and other elemental abundances. Science **170**, 969 (1970); 5. Laul, J. C., Morgan, J. W., Ganapathy, R., Anders, E.: Suppl. 2, **2**, 1139 (1971); 6. Baedecker, P. A., Schaudy, R., Elzie, J. L., Kimberlin, J., Wasson, J. T.: Suppl. 2, **2**, 1037 (1971); 7. Laul, J. C., Keays, R. R., Ganapathy, R., Anders, E.: Abundance of 14 trace elements in lunar rock 12013,10. Earth Planet. Sci. Lett. **9**, 211 (1970); 8. Wakita, H., Rey, P., Schmitt, R. A.: Suppl. 2, **2**, 1319 (1971); 9. Baedecker, P. A., Chou, C. L., Grudewicz, E. B., Wasson, J. T.: Suppl. 4, **2**, 1177 (1973); 10. Morgan, J. W., Krähenbühl, U., Ganapathy, R., Anders, E.: Suppl. 3, **2**, 1361 (1972); 11. Ganapathy, R., Morgan, J. W., Krähenbühl, U., Anders, E.: Suppl. 4, **2**, 1239 (1973); 12. Baedecker, P. A., Chou, C. L., Wasson, J. T.: Suppl. 2, **2**, 1343 (1972); 13. Morgan, J. W., Ganapathy, R., Higuchi, H., Krähenbühl, U., Anders, E.: Suppl. 5, **2**, 1703 (1974); 14. Baedecker, P. A., Chou, C L., Sundberg, L. L., Wasson, J. T.: Suppl. 5, **2**, 1625 (1974); 15. Laul, J. C., Schmitt, R. A.: Chemical composition of boulder-2 rocks and soils, Apollo 17, Station 2. Earth Planet. Sci. Lett., **23**, 206 (1974); 16. Laul, J. C., Hill, D. W., Schmitt, R. A.: Suppl. 5, **2**, 1047 (1974); 17. Laul, J. C., Ganapathy, R., Morgan, J. W., Anders, E.: Meteoritic and non-meteoritic trace elements in Luna 16 samples. Earth Platent. Sci., Lett. **13**, 450 (1972). 18. Chou, C.-L., Boynton, W. V., Sundberg, L. L., Wasson, J. T.: Suppl. 6, **2**, 1701 (1975); 19. Boynton, W. V., Baedecker, P. A., Chou, C.-L., Robinson, K. L., Wasson, J. T.: Suppl. 6, **2**, 2241 (1975); 20. Higuchi, H., Morgan, J. W.: Suppl. 6, **2**, 1625 (1975).

Table 48-C-5. *Cadmium in lunar uplands materials*[b] (Analytical method: N/R)

Materials	No. of rocks	Cd (ppb)		Reference
		Range	Geometric mean[a]	
Apollo 14				
Basalt	1		4.5	1
Pseudo basalt 14310	1	2.6, 8.4	5.5	1, 2
Breccias	6	7–280	49	1, 2, 3
Soils	7	80–460	130	1, 2
Yellow glass	1		16	1
Green glass	1		16	1

[a] See Table 48-C-1. [b] See also Apollo 15 and 17 for some highland materials.

References: 1. Morgan, J. W., Krähenbühl, U., Ganapathy, R., Anders, E.: Suppl. 3 **2**, 1361 (1972). 2. Baedecker, P. A., Chou, C. L., Wasson, J. T.: Suppl. 2, **2**, 1343 (1972) 3. Boynton, W. V., Baedecker, P. A., Chou, C.-L., Robinson, K. L., Wasson, J. T.: Suppl. 6, **2**, 2241 (1975).

Table 48-C-6. *Cadmium in lunar highlands material*[a] (Analytical method: N/R)

Materials	No. of rocks	Cd (ppb)		References
		Range	Geometric mean[b]	
Apollo 16				
Basalts	6	4–16	7	1, 2, 3
Breccias	9	2.7–290	27	1, 2, 3
Anorthosites	3	37–330	100	1, 3
Anorthosite 61016 light	1		180	3
dark	1		8	3
Anorthositic gabbros	2	≤0.4, 0.9		1, 3
Gabbroic anorthosite	1		61	3
Soils	17	16–140	66	1, 3
Glasses	3	1.5–5.2	1.7	2
Luna 20				
Breccia	1		48	4
Soil	1	1,700[a], 19,400[a]		4, 5

[a] High abundances may be due to the contamination from the auto-probe.
[b] See Table 48-C-1

References: 1. Krähenbühl, R., Ganapathy, R., Morgan, J. W., Anders, E.: Suppl. 4, **2**, 1325 (1973); 2. Ganapathy, R., Morgan, J. W., Higuchi, H., Anders, E.: Suppl. 5, **2**, 1659 (1974); 3. Baedecker, P. A., Chou, C.-L., Sundberg, L. L., Wasson, J. T.: Suppl. 5, **2**, 1625 (1974); 4. Morgan, J. W., Laul, J. C., Krähenbühl, U., Ganapathy, R., Anders, E.: Luna 20 soil: Abundance of 17 trace elements. Geochim. Cosmochim. Acta **37**, 953 (1973); 5. Laul, J. C., Schmitt, R. A.: Chemical composition of Luna 20 rocks and soil and Apollo 16 soils. Geochim. Cosmochim. Acta **37**, 927 (1973).

Manuscript received: September 1974

48-D. Abundance in Rock-Forming Minerals, Cadmium Minerals

I. Rock-Forming Minerals

The cadmium contents of rock-forming minerals and of ore-forming minerals are shown in Tables 48-D-1 and 48-D-2, respectively. The distribution of cadmium among co-existing minerals from the same rock was studied by VINCENT and BILEFIELD (1960), BILEFIELD and VINCENT (1961) and BUTLER and THOMPSON (1967).

Table 48-D-1. *Cadmium in rock-forming minerals. (Method in brackets)*

Material	Cd (ppm)	Author
Plagioclase, An_{40}, Skaergaard (Greenland)	0.14, 0.15 (N/R)	VINCENT and BILEFIELD (1960)
Albite, NBS-99, standard	0.016 (P) (2 detns.)	WAHLER (1968)
Biotite from biotite-granite, N. Nigeria (5 samples)	0.16–4.8 (P) (av. 1.51)	BUTLER and THOMPSON (1967)
Riebeckite from granite, N. Nigeria (6 samples)	0.88–5.8 (P) (av. 2.17)	BUTLER and THOMPSON (1967)
Arfvedsonite from arfvedsonite-granite, N. Nigeria	2.2 (P)	BUTLER and THOMPSON (1967)
Pyroxene, $Ca_{34}Mg_{35}Fe_{31}$, Skaergaard (Greenland)	0.42, 0.44 (N/R)	VINCENT and BILEFIELD (1960)
Olivine, Fa_{60}, Skaergaard (Greenland)	0.37 (N/R)	VINCENT and BILEFIELD (1960)
Aegirine from riebeckite granite, N. Nigeria (2 samples)	1.2, 1.5 (P)	BUTLER and THOMPSON (1967)
Glauconite, Glen Allen Mine, Prieska (S. Africa)	0.03 (S)	BROOKS and AHRENS (1961)
Titaniferous magnetite, Skaergaard (Greenland) (2 samples)	0.28, 0.31 (N/R)	VINCENT and BILEFIELD (1960)
Ilmenite, Skaergaard (Greenland) (3 samples)	0.12, 0.12, 0.14 (N/R)	VINCENT and BILEFIELD (1960)
Apatite, Skaergaard (Greenland) (2 samples)	0.14, 0.15 (N/R)	VINCENT and BILEFIELD (1960)

FLEISCHER (1955) and IVANOV (1964 summarized the distribution of cadmium in ore deposits. FLEISCHER (1955) mentioned that the cadmium content of sphalerite seemed to be independent of the condition of formation, while IVANOV (1964) concluded that high-temperature skarn-type lead zinc deposits and low-temperature

Table 48-D-2. *Cadmium in ore-forming minerals. (Method in brackets)*

Material	Cd (ppm)	Author
Hemimorphite, Eastern Transbaikaliya, U.S.S.R. (8 samples)	up to 65 (P)	Kulikova (1962), Kulikova (1963)
Sulfide ore-1, CAAS standard, Falconbridge, Ontario, Canada	10 (S)	Moxham (1967)
Sulfide ore-1, CAAS, Falconbridge, Ontario, Canada	0.340 (P) (2 detns.)	Wahler (1968)
Sulfide ore-1, CAAS, Falconbridge, Ontario, Canada	<2 (S)	Champ (1968b)
Zinc ore, NBS-113 standard	0.41 (C)	Huffman (1962)
Zinc ore, NBS-113	4750 (A)	Nakagawa and Harms (1968)
Mimetite, Eastern Transbaikaliya, U.S.S.R. (5 samples)	80—100 (P)	Kulikova (1962)
Bindheimite, Eastern Transbaikaliya, U.S.S.R. (3 samples)	10 (P)	Kulikova (1963)
Chalcopyrite, Yakutiya, U.S.S.R. (2 samples)	30, 60 (P)	Ivanov (1961)
Sphalerite, Kristineberg, Sweden (6 samples)	1,700—2,400 (P) (av. 2,000)	Hybbinette (1945)
Sphalerite, Karamazar, U.S.S.R. (5 samples)	2,000—2,600 (P, W)	Badalov and Enikeev (1959)
Sphalerite, Leninogrsk, U.S.S.R.	1,400—4,800	Litvinovich and Loginova (1960)
Sphalerite, Yakutiya, U.S.S.R. (9 samples)	70—2,300 (P) (av. 1,350)	Ivanov (1961)
Sphalerite, Central City, Colorado, U.S.A.	2,000—6,400 (S)	Sims and Barton (1961)
Sphalerite, Central Tadzhikstan, U.S.S.R. (43 samples)	5,900—18,500 (W) (av. 1,370)	Mogarovskii and Rosseiken (1961)
Sphalerite, (iron-rich) Eastern Transbaikaliya, U.S.S.R. (59 samples)	1,000—4,500 (P)	Meituv (1962)
Sphalerite (iron-poor), Eastern Transbaikaliya, U.S.S.R. (7 samples)	2,800—7,900 (P)	Meituv (1962)
Sphalerite, Mo-Ba deposits, Vietnam	5,100	Zdenck (1962)
Sphalerite, Eastern Transbaikaliya, U.S.S.R.	500—3,500 (P)	Kulikova (1963)
Sphalerite, Pb-Zn deposits, U.S.S.R. (150 samples)	1,700—8,000 (S) (av. 4,300)	Balakishieva (1964)
Sphalerite, Northern Alpine, 23 deposits (200 samples)	450—5,200 (S)	Fruth (1966)
Marmatite, Karamazar, U.S.S.R. (3 samples)	800—1,900 (P, W) (av. 1,400)	Badalov and Erikeev (1959)
Cleiophane, Karamazar, U.S.S.R. (7 samples)	1,200—2,700 (P, W)	Badalov and Erikeev (1959)

Table 48-D-2 (Continued)

Material	Cd (ppm)	Author
Boulangerite, Yakutiya, U.S.S.R. (4 samples)	500—3,000 (S)	IVANOV (1961)
Boulangerite, Eastern Transbaikaliya, U.S.S.R. (3 samples)	100—500 (P)	KULIKOVA (1962)
Galena, Yakutiya, U.S.S.R. (7 samples)	0—180 (P)	IVANOV (1961)
Galena, Eastern Transbaikaliya, U.S.S.R. (more than 11 samples)	10—500 (P)	KULIKOVA (1962), KULIKOVA (1963)
Bournonite, Eastern Transbaikaliya, U.S.S.R.	50—100 (P)	KULIKOVA (1962)
Tetrahedrite, Yakutiya, U.S.S.R. (5 samples)	500—1,000 (S)	IVANOV (1961)
Stannite, Yakutiya, U.S.S.R. (4 samples)	100—200 (P)	IVANOV (1961)
Cerussite (black), Eastern Transbaikaliya, U.S.S.R. (more than 11 samples)	0—100 (P)	KULIKOVA (1962), KULIKOVA (1963)
Cerussite (colorless), Eastern Transbaikaliya, U.S.S.R. (more than 10 samples)	up to 50 (P)	KULIKOVA (1962), KULIKOVA (1963)
Smithonite, Eastern Transbaikaliya, U.S.S.R. (more than 13 samples)	80—2,100 (P)	KULIKOVA (1962), KULIKOVA (1963)
Smithonite, Mo-Ba deposits, Vietnam	3,710	ZDECK (1962)
Franckeite, Yakutiya, U.S.S.R.	23 (S)	IVANOV (1961)

silver deposits had the most cadmium-rich sphalerites; next were low-temperature lead-zinc deposits; intermediate-temperature sulfide deposits had the least cadmium.

II. Cadmium Minerals

The common cadmium minerals are greenockite-CdS, octavite-$CdCO_3$ and cadmium oxide-CdO. Significant amounts of naturally occurring cadmium are found only in association with zinc ores, in which the amount of cadmium varies considerably. Usually, $0.1 \sim 0.5\%$ cadmium is present in zinc blende (sphalerite) and calamine (zinc spar, smithsonite).

Greenockite (CdS) occurs usually as an earthy coating on zinc minerals especially sphalerite. Rarely crystals are found in amygdaloidal cavities in basic igneous rocks.

Cadmium oxide (CdO) at Genarutta, near Iglesias, Sardinia, occurs as a coating over calamine (PALACHE *et al.*, 1944, 1951).

Cadmoselite (CdSe) occurs as xenomorphic grains and occasionally as very small crystals (0.1 mm) of a hexagonal pyramid and pedion (BURÝANOVA, 1960).

Revised manuscript received: September 1969

48-E. Abundance in Common Igneous Rock Types

The cadmium contents of granitic, intermediate, basic, and ultrabasic rocks are presented in Tables 48-E-1 to 48-E-4, respectively.

Table 48-E-1. *Cadmium in granitic rocks. (Method in brackets)*

Material	Cd (ppm)	Author
Granite, Germany (composite of 14 samples)	0.2 (S)	PREUSS (1940)
Granite, Minnesota, U.S.A. (composite of 5 samples)	0.13 (C)	SANDELL and GOLDICH (1943)
Granite, central Texas, U.S.A. (composite of 5 samples)	0.1 (C)	SANDELL and GOLDICH (1943)
Granite, southern Aitkin County, Minnesota, U.S.A.	0.15 (C)	SANDELL and GOLDICH (1943)
Granite, Granite Mountain, Texas, U.S.A.	0.2 (C)	SANDELL and GOLDICH (1943)
Granite, Cape peninsula area, South Africa (2 samples)	0.10, 0.12 (S)	BROOKS *et al.* (1960)
Granite, White Mts., New Hampshire, U.S.A.	<0.1 (P)	BUTLER and THOMPSON (1967)
Granite, G-1 standard, Rhode Island, U.S.A.	5 (P)	SMYTHE and GATEHOUSE (1955)[a]
Granite, G-1, Rhode Island, U.S.A.	≈0.05 (S)	BROOKS *et al.* (1960)
Granite, G-1, Rhode Island, U.S.A.	0.06 (N/R) (av. of 3 detns.)	VINCENT and BILEFIELD (1960), BILEFIELD and VINCENT (1961)
Granite, G-1, Rhode Island, U.S.A.	0.08 (P) (av. of 4 detns.)	CARMICHAEL and MCDONALD (1961 a)
Granite, G-1, Rhode Island, U.S.A.	<0.3 (C)	HUFFMAN (1962)
Granite, G-1, Rhode Island, U.S.A.	0.08 (P) (av. of 4 detns.)	STANTON *et al.* (1962)
Granite, G-1, Rhode Island, U.S.A.	<0.2 (I)	BROWN and WOLSTENHOLME (1964)
Granite, G-1, Rhode Island, U.S.A.	0.6 (A)	NAKAGAWA and HARMS (1968)
Granite, G-1, Rhode Island, U.S.A.	0.003 (P) (2 detns.)	WAHLER (1968)
Granite, G-2 standard, Rhode Island, U.S.A.	0.010 (P) (2 detns.)	WAHLER (1968)

[a] This value is abnormally high compared with the other data on the same rock.

Table 48-E-1 (Continued)

Material	Cd (ppm)	Author
Granite, G-2, Rhode Island, U.S.A.	<2 (C)	Champ (1968a)
Granite, GR standard, France	0.001 (P)	Wahler (1968)
Granite, NBS-4983 standard	0.013 (P) (3 detns.)	Wahler (1968)
Granodiorite, unknown	0.1±0.05 (C)	Sandell (1939)
Granodiorite, St. Cloud, Minnesota, U.S.A.	0.1 (C)	Sandell and Goldich (1943)
Granodiorite, GSP-1 standard, Silver Plume, Colorado, U.S.A.	0.016 (P) (3 detns.)	Wahler (1968)
Granodiorite, GSP-1. Silver Plume, Colorado, U.S.A.	2 (C)	Champ (1968a)
Acid granophyre, Skaergaard, Greenland (5 samples)	0.44—0.59 (N/R) (av. 0.51)	Vincent and Bilefield (1960), Bilefield and Vincent (1961)
Hornblende-biotite-granite, Amo N. Nigeria	0.31 (P)	Butler and Thompson (1967)
Albite-riebeckite-granite, R. Kaffo, N. Nigeria (2 samples)	0.25, 0.22 (P)	Butler and Thompson (1967)
Riebeckite-biotite-granite, with astrophyllite, Amo, N. Nigeria	0.17 (P)	Butler and Thompson (1967)
Riebeckite-granite, D. Jirana, Gamawa, N. Nigeria	0.51 (P)	Butler and Thompson (1967)
Riebeckite-granite, Afu Hill, N. Nigeria	1.56 (P)	Butler and Thompson (1967)
Riebeckite-astrophyllite-granite, Mt. Tremont, White Mts., New Hampshire, U.S.A.	0.28 (P)	Butler and Thompson (1967)
Biotite-granite, Afu Hill, N. Nigeria (7 samples)	<0.05—0.50 (P)	Butler and Thompson (1967)
Biotite-granite, White Mts., New Hampshire, U.S.A. (2 samples)	<0.1 (P)	Butler and Thompson (1967)
Biotite-granite, Bujang Melaka, Perak, Malaya (2 samples)	<0.1—0.19 (P)	Butler and Thompson (1967)
Lepidolite-biotite-granite, Afu Hill, N. Nigeria (2 samples)	0.08—1.13 (P)	Butler and Thompson (1967)
Riebeckite-aegirine-granite, R. Kwoya, N. Nigeria	0.41 (P)	Butler and Thompson (1967)
Arfvedsonite-fayalite-granite-porphyry, N. Nigeria (3 samples)	0.30, 0.39, 0.70 (P)	Butler and Thompson (1967)
Quartz monzonite, unknown	1.4—1.8 (A)	Nakagawa and Harms (1968)
Pitchstone, E. Iceland (14 samples)	0.11—0.29 (P) (av. 0.20)	Carmichael and McDonald (1961b)

Table 48-E-1 (Continued)

Material	Cd (ppm)	Author
Pitchstone, Arran (7 samples)	0.05—0.14 (P) (av. 0.08)	CARMICHAEL and McDONALD (1961 b)
Pitchstone, Eigg (3 samples)	0.15—0.34 (P) (av. 0.24)	CARMICHAEL and McDONALD (1961 b)
Rhyolite, E. Paresis, S.W.Africa (2 samples)	0.05, 0.05 (S)	BROOKS and AHRENS (1961)
Na-rhyolite, Northern Nigeria (6 samples)	0.21—0.48 (P)	BUTLER and THOMPSON (1967)
Pantellerite, Pantelleria	0.48 (P)	BUTLER and THOMPSON (1967)
Obsidian, Ascension Is.	0.29 (P)	BUTLER and THOMPSON (1967)
Obsidian, Caucasus, U.S.S.R.	0.22 (P)	BUTLER and THOMPSON (1967)
Glass from pitchstone, E. Iceland (5 samples)	0.10—0.25 (P) (av. 0.21)	CARMICHAEL and McDONALD (1961 b)
Glass from pitchstone, Arran (3 samples)	0.05—0.14 (P) (av. 0.11)	CARMICHAEL and McDONALD (1961 b)
Glass from pitchstone, Eigg	0.14 (P)	CARMICHAEL and McDONALD (1961 b)

An average for abundant granitic rock types is expected between 0.1 and 0.2 ppm Cd.

Table 48-E-2. *Cadmium in intermediate rocks. (Method in brackets)*

Material	Cd (ppm)	Author
Andesite, AGV-1 standard, Lake County, Oregon, U.S.A.	0.017 (P) (2 detns.)	WAHLER (1968)
Andesite, AGV-1, Lake County, Oregon, U.S.A.	<2 (C)	CHAMP (1968a)
Basic hedenbergite granophyre, Skaergaard, Greenland	0.27, 0.28 (N/R)	VINCENT and BILEFIELD (1960), BILEFIELD and VINCENT (1961)
Syenite, Otjiwarongo, S.W. Africa (2 samples)	0.04, 0.04 (S)	BROOKS and AHRENS (1961)
Syenite, S-1, CAAS standard, Bancroft Area, Ontario, Canada	8 (S)	MOXHAM (1967)
Syenite, S-1, CAAS, Bancroft Area, Ontario, Canada	0.093 (P) (2 detns.)	WAHLER (1968)
Syenite, S-1, CAAS, Bancroft Area, Ontario, Canada	n.d. —<2 (S)	CHAMP (1968b)
Syenite, Gamawa, N. Nigeria (3 samples)	0.32, 0.23 0.24 (P)	BUTLER and THOMPSON (1967)
Foyaite, Paresis, S.W. Africa (2 samples)	0.04, 0.04 (S)	BROOKS and AHRENS (1961)
Gabbro-diorite, NBS-4982 standard	0.012 (P)	WAHLER (1968)
Msusule-tonalite, T-1 standard	0.079 (P) (2 detns.)	WAHLER (1968)

Table 48-E-3. *Cadmium in mafic rocks. (Method in brackets)*

Material	Cd (ppm)	Author
Mafic rock, Minnesota, U.S.A. (composite of 5 samples)	0.2 (C)	SANDELL and GOLDICH (1943)
Basalt, Mt. Wellington, Tasmania	7 (P)	SMYTHE and GATEHOUSE (1955)[a]
Basalt, Paresis, S.W. Africa	0.03 (P)	BROOKS and AHRENS (1961)
Basalt, Columbia Plateau, U.S.A.	0.33 (N/R)	SCHMITT *et al.* (1963)
Basalt, NBS-4978 standard, Columbia River, U.S.A.	0.019 (P) (2 detns.)	WAHLER (1968)
Basalt, NBS-4985 standard, Deccan trap, India	0.006 (P) (2 detns.)	WAHLER (1968)
Basalt, Kilauea Iki-22, Hawaii, U.S.A.	0.5 (N/R)	SCHMITT *et al.* (1963)
Basalt, BCR-1 standard, Columbia River, Oregon, U.S.A.	0.067 (P)	WAHLER (1968)
Basalt, BCR-1, Columbia River, Oregon, U.S.A.	<2 (C)	CHAMP (1968a)
Basalt, BCR-1, Columbia River, Oregon, U.S.A.	0.081, 0.20, 0.20 (N/R)	REY *et al.* (1970)
Basalt (Dolerite), Mt. Wellington, Tasmania (2 samples)	6, 1.5 (P)	SMYTHE and GATEHOUSE (1955)[a]
Basalt (Dolerite), Cape Town, S. Africa	0.30, 0.30 (S)	BROOKS *et al.* (1960)
Basalt (Dolerite), O.F.S., S. Africa (14 samples)	0.01—0.25 (S) (av. 0.10)	BROOKS and AHRENS (1961)
Basalt, Cook County, Minnesota, U.S.A.	0.25 (C)	SANDELL and GOLDICH (1943)
Basalt ("Diabase"), W-1 standard, Centerville, Va., U.S.A.	4 (P)	SMYTH and GATEHOUSE (1955)[a]
Basalt ("Diabase"), W-1, Centerville, Va., U.S.A.	0.33 (N/R) (av. of 4 detns.)	VINCENT and BILEFIELD (1960, 1961)
Basalt ("Diabase"), W-1, Centerville, Va., U.S.A.	0.07, 0.09 (S)	BROOKS *et al.* (1960)
Basalt ("Diabase"), W-1, Centerville, Va., U.S.A.	0.2, 0.4, 0.4 (C), 0.3 (P) (av. of 4 detns.)	CARMICHAEL and MCDONALD (1961a)
Basalt ("Diabase"), W-1, Centerville, Va., U.S.A.	<0.3 (C)	HUFFMAN (1962)
Basalt ("Diabase"), W-1, Centerville, Va., U.S.A.	0.28 (P) (av. of 8 detns.)	STANTON *et al.* (1962)
Basalt ("Diabase"), W-1, Centerville, Va., U.S.A.	0.22 (P) 0.27 (N/R)	BUTLER and THOMPSON (1967)
Basalt ("Diabase"), W-1, Centerville, Va., U.S.A.	<0.2 (I)	BROWN and WOLSTENHOLME (1964)
Basalt ("Diabase"), W-1, Centerville, Va., U.S.A.	0.6 (A)	NAKAGAWA and HARMS (1968)

[a] See footnote Table 48-E-1.

Table 48-E-3 (Continued)

Material	Cd (ppm)	Author
Basalt ("Diabase"), W-1, Centerville, Va., U.S.A.	0.016 (P) (2 detns.)	WAHLER (1968)
Olivine basalt, Devils Gap, Akaroa, New Zealand	0.04, 0.05 (S)	BROOKS *et al.* (1960)
Fayalite ferrogabbro, Skaergaard, Greenland	0.28, 0.27 (N/R)	VINCENT and BILEFIELD (1960), BILEFIELD and VINCENT (1961)
Hortonolite ferrogabbro, Skaergaard, Greenland (3 samples)	0.11—0.34 (N/R)	VINCENT and BILEFIELD (1960) BILEFIELD and VINCENT (1961)
Gabbro, Germany (composite of 11 samples)	<0.03 (S)	PREUSS (1940)
Gabbro (olivine-free), Skaergaard, Greenland	0.08, 0.09 (N/R)	VINCENT and BILEFIELD (1960) BILEFIELD and VINCENT (1961)
Gabbro, Duluth, Minnesota, U.S.A.	0.1 (C)	SANDELL and GOLDICH (1943)
Olivine gabbro, Skaergaard Greenland (2 samples)	0.10—0.20 (N/R)	VINCENT and BILEFIELD (1960) BILEFIELD and VINCENT (1961)
Gabbro picrite, Skaergaard, Greenland (2 samples)	0.08, 0.09 (N/R)	VINCENT and BILEFIELD (1960) BILEFIELD and VINCENT (1961)
Chilled marginal gabbro, Skaergaard, Greenland (4 samples)	0.13 (N/R) (av. of 4 detns.)	VINCENT and BILEFIELD (1960) BILEFIELD and VINCENT (1961)
Norite, Merensky Reef, Lydenburg Tvl., S. Africa	~0.05 (S)	BROOKS *et al.* (1960)

There is very little available data on cadmium in ultrabasic rocks. SMYTHE and GATEHOUSE (1955) reported no detectable amounts of cadmium in serpentine and pyroxenite. VINOGRADOV (1962) gives Cd values in dunites by radioactivation analysis as 0.03—0.05 ppm.

Table 48-E-4. *Cadmium in ultrabasic rocks. (Method in brackets)*

Material	Cd (ppm)	Author
Peridotite, PCC-1 standard, Sonoma County, Calif., U.S.A.	<0.001 (P)	WAHLER (1968)
Peridotite, PCC-1, Sonoma County, Calif., U.S.A.	<2 (C)	CHAMP (1968a)
Dunite, DTS-1 standard, Twin Sisters, Washington, U.S.A.	0.005 (P)	WAHLER (1968)
Dunite, DTS-1, Twin Sisters, Washington, U.S.A.	<2 (C)	CHAMP (1968a)

Revised manuscript received: September 1969

48-F. Behavior in Magmatogenic Processes

Cadmium has a strong affinity for sulfur and is a typical chalcophile element (GOLDSCHMIDT, 1958). This explains its preferred accumulation in magmatic sulfides (IVANOV, 1961).

Owing to its weaker oxyphile character, cadmium becomes nearly quantitatively enriched in hydrothermal rocks and minerals (RANKAMA and SAMAHA, 1961). The highest concentration of cadmium is found in sphalerite. Cadmium is also present in galena, boulangerite and bournonite. In oxidized zones cadmium is found in smithsonite, sauconite, manganese oxides and limonite. Nine samples of sauconite from Eastern Transbaikaliya (U.S.S.R.) contain 10—40 ppm Cd (KULIKOVA, 1962).

The observations on the Skaergaard intrusion show a four-fold increase in cadmium in the granophyres derived largely from residual magmatic fractions compared to the original parent mafic magma (VINCENT and BILEFIELD, 1960).

Revised manuscript received: September 1969

48-G. Behavior during Weathering and Alteration of Rocks

During weathering, cadmium goes into solution containing sulfate and chloride.

Under conditions of strong oxidation cadmium forms oxidized minerals, such as CdO and $CdCO_3$ (GOLDSCHMIDT, 1958).

Data on cadmium in soils may be found in the following reports: MALYUGA (1941), VINOGRADOV (1945, 1959), SWAINE (1955), GREEN (1959) and BOWEN (1966).

The average content of cadmium in soils probably lies between 0.1 and 0.5 ppm. The character of the parent rock is the main factor determining cadmium content in soils (VINOGRADOV, 1959).

Revised manuscript received: September 1969

48-H. Solubilities of Cadmium Compounds in Natural Waters, and Valence State in Natural Environments

In aqueous solution cadmium ions form a large number of complex ions with ammonium, cyanide and halide ions. All cadmium compounds are soluble in excess aqueous halide solution. Table 48-H-1 lists solubilities of cadmium compounds of some interest in natural environments.

Table 48-H-1. *Solubilities of cadmium compounds*

Cadmium compound	Solubility
$Cd(OH)_2$	in cold water 0.26 mg per 100 cc
CdS	in cold water 0.13 mg per 100 cc
	solubility product: 3.6×10^{-29}

The only important valence state of cadmium compounds in the natural environment is +2. The most important factor which controls the cadmium ion concentration in natural waters is the pH and oxidation potential. The standard electrode potentials (E^0) of the following reactions are

$$Cd^{2-} + 2e^- = Cd \qquad E^0 = -0.4020 \text{ V (after Latimer, 1952)}$$
$$Cd^{2-} + S + 2e^- = CdS \qquad E^0 = 0.31 \text{ V.}$$

Revised manuscript received: September 1969

48-I. Abundance in Natural Waters

MULLIN and RILEY (1954, 1956) reported that the average cadmium concentration in 37 surface sea water samples of the Irish Sea and the English channel was 0.11 ppb, with values ranging from 0.024 to 0.25 ppb, which depended upon the sample source and season. Three samples from Japan contained 0.08—0.17 ppb cadmium (ISHIBASHI *et al.*, 1962). BROOKS (1960) obtained a value of ≥ 0.02 ppb.

The cadmium content of spring water from 72 springs in California was spectrographically determined by SILVEY (1967). The greatest concentration was 20 ppb and the average was 8 ppb. There are very few data on cadmium in river or lake water. MALYUGA (1941) reported that cadmium was found only in one sample of water from the Urov river.

SILVEY (1967) reported that cadmium was not detected in any of the samples from 65 streams in California, and in five from 82 well water and oil-field brine samples that contained 5 to 70 ppb. UDODOV and PARILOV (1961) reported the average cadmium content of 3,490 samples of natural waters of Siberia was 3 ppb with values ranging from 0.1 to 207 ppb.

LOGINOVA (1959) obtained 12 ppm to 1,000 ppm cadmium in the dry residues obtained from natural waters with a low mineral content.

Revised manuscript received: September 1969

48-K. Abundance in Sedimentary Rocks

The cadmium contents of sedimentary rocks are shown in Table 48-K-1.

Table 48-K-1. *Cadmium in sedimentary rocks. (Method in brackets)*

Material	Cd (ppm)	Author
Bituminous Shale, (Kupferschiefer) Mansfeld, Germany	500	CISSARZ (1930)
Shale, Paleozoic, Germany (composite of 36 samples)	0.3 (S)	PREUSS (1940)
Bentonite, mostly Montana and S. Dakota, U.S.A. (10 samples)	$<$0.3—11.0 (C) (av. 1.4)	TOURTELOT *et al.* (1964)
Marlstone, South Dakota, U.S.A. (8 samples)	0.4—10.0 (C) (av. 2.6)	TOURTELOT *et al.* (1964)
Shale and Claystone, mostly Montana and S. Dakota, U.S.A. (66 samples)	$<$0.3—8.4 (C) (av. 1.0)	TOURTELOT *et al.* (1964)
Plastic clay, NBS-99 standard	0.016 (A) (2 detns.)	WAHLER (1968)
Glass sand, NBS-81 standard	0.014 (A) (2 detns.)	WAHLER (1968)

The cadmium content of 84 samples from the bituminous Pierre Shale averages 1.4 ppm and ranges from <0.3 to 11 ppm. The median value of 0.8 ppm probably represents the cadmium content of the average shale (TOURTELOT *et al.*, 1964).

TUREKIAN and WEDEPOHL (1961) suggested that the cadmium content of limestone is 0.035 ppm.

The cadmium contents of Recent marine sediments are shown in Table 48-K-2. A high content of cadmium is observed in manganese nodules.

Table 48-K-2. *Cadmium in marine sediments. (Method in brackets)*

Material	Cd (ppm)	Author
Diatomaceous ooze, S. Atlantic Ocean (5 samples)	0.17—0.86 (C)	MULLIN and RILEY (1956)
Globigerina ooze, Atlantic Ocean (3 samples)	0.28—0.52 (C)	MULLIN and RILEY (1956)
Calcareous ooze, N. Atlantic Ocean	0.57 (C)	MULLIN and RILEY (1956)
Radiolarian ooze, Pacific and Atlantic Oceans (4 samples)	0.13—0.98 (C)	MULLIN and RILEY (1956)
Pelagic clay, Pacific Ocean	0.56 (C)	MULLIN and RILEY (1956)
Green mud, N. Atlantic Ocean	0.27 (C)	MULLIN and RILEY (1956)
Manganese nodule, Pacific Ocean (2 samples)	5.1—8.4 (C)	MULLIN and RILEY (1956)
Organic mud, Irish Sea (2 samples)	0.36—0.42 (C)	MULLIN and RILEY (1956)
Mud, Lutzholm Bay, Antarctica	0.12 (S)	BROOKS *et al.* (1960)

Revised manuscript received: September 1969

48-L. Biogeochemistry

The biological role of cadmium is not clear. The cadmium content of many biological materials is summarized by BOWEN (1966). Cadmium in coal ashes was reported by GOLDSCHMIDT (1937) and SLUTZER (1940).

The cadmium contents of some plants are shown in Table 48-L-1.

Table 48-L-1. *Cadmium in plants. (Method in brackets)*

Material	Cd (ppm)	Author
Spinach (dried) (3 samples)	1.0—1.2 (C)	SHIRLEY *et al.* (1949)
Lettuce (dried) (3 samples)	0.4—0.5 (C)	SHIRLEY *et al.* (1949)
Alfalfa leaf meal (2 samples)	0.1—0.2 (C)	SHIRLEY *et al.* (1949)
Peat	3—30	SALMI (1950)
Peat (ash) (50 samples)	3—30	GORDON (1952)
Corn leaves	1.4 (S)	SHIMP *et al.* (1957)
Kale (dried) (4 samples)	0.38 (N/R)	BOWEN (1967)

The cadmium contents of marine animals have been reported by NODDACK and NODDACK (1940), MALYUGA (1941), MULLIN and RILEY (1956), ISHIBASHI *et al.* (1962) and BOWEN (1966).

Some of their data are listed in Table 48-L-2.

Table 48-L-2. *Cadmium in marine animals. (Method in brackets)*

Material	Cd (ppm)	Author
Marine algae (6 samples)	0.1—1.0 (P)	MALYUGA (1941)
Marine algae (12 samples)	0.13—2.1 (C)	MULLIN and RILEY (1956)
Sea flowering plants	0.23 (P)	MALYUGA (1941)
Oyster (soft parts)	2.9—6.8	BOYLE and LYNCH (1968)
Clam	<0.5—1.3	BOYLE and LYNCH (1968)
Mollusca (shell only)	0.003—0.08 (C),	MULLIN and RILEY (1956)
(soft parts)	0.83—38 (C)	
Echinodermata (12 samples)	0.24—15 (C)	MULLIN and RILEY (1956)
Crustacea (5 samples)	0.15—1.3 (C)	MULLIN and RILEY (1956)
Protozoa (2 samples)	2.2 (C)	MULLIN and RILEY (1956)
Porifera	1.9 (C)	MULLIN and RILEY (1956)
Coelenterata	1.2 (C)	MULLIN and RILEY (1956)

The distribution of cadmium in marine animal organisms shows extremely low concentrations in calcareous shell, about 1.5 ppm in muscle and high concentrations, up to 500 ppm, in the digestive glands and renal organs of molluscs (MULLIN and RILEY, 1956).

Data on cadmium in human tissue can be found in reports by SHIMP *et al.* (1957), STICH (1957), WESTERMARK and SJÖSTRAND (1960), BOWEN (1963, 1966) and WESTER (1965). Some of these values are given in Table 48-L-3.

Table 48-L-3. *Cadmium in human tissues. (Method in brackets)*

Material	Cd (ppm)	Author
Kidney	1,470 (S)	SHIMP *et al.* (1957)
Kidney	1,000 (S)	STITCH (1957)
Kidney	21—446 (N/R)	WESTERMARK and SJÖSTRAND (1960)
Liver	50 (S)	SHIMP *et al.* (1957)
Heart	0.01—0.03 (N/R)	WESTER (1965)
Bone	4 (S)	STITCH (1957)

It is noted that the highest concentrations of cadmium are found in the kidney. PULIDO *et al.* (1960) found from 3 to 12 ppb Cd (A) in urine. Cadmium in blood, plasma, red cells and serum is summarized by BOWEN (1963, 1966).

Revised manuscript received: September 1969

48-M. Abundance in Common Metamorphic Rocks

The cadmium contents of metamorphic rocks are shown in Table 48-M-1.

Table 48-M-1. *Cadmium in metamorphic rocks. (Method in brackets)*

Material	Cd (ppm)	Author
Eclogite, Kimberley, S. Africa (2 samples)	0.04, 0.04 (S)	Brooks and Ahrens (1961)
Eclogite, Roberts Victor mine, S. Africa (2 samples)	0.02, 0.04 (S)	Brooks and Ahrens (1961)
Eclogite, Roberts Victor mine, S. Africa	0.026 (N/R)	Schmitt *et al.* (1963)
Eclogite, Delegate, N.S.W., Australia	0.24 (N/R)	Schmitt *et al.* (1963)
Garnet schist, W. Coast, Tasmania	1 (P)	Smythe and Gatehouse (1955)[a]
Grey gneiss, W. Ongul I., Lutzow-Holm Bay, Antarctica (2 samples)	0.12, 0.16 (S)	Brooks *et al.* (1960)

[a] See footnote Table 48-E-1.

The behavior of cadmium in metamorphic reactions is not well-known.

Manuscript received: September 1969

48-O. Relations between Cadmium and Other Elements, and Cadmium's Economic Importance

Crystallochemical relations between cadmium and other elements are governed by the size relationships of the radii (GOLDSCHMIDT, 1958). The similarities in charge and ionic radius between Cd^{2+} (0.97 Å) and Ca^{2+} (0.99 Å) are so close, that both elements might form the octahedral structure in silicate with possible Cd^{2+}—Ca^{2+} substitution. SANDELL and GOLDICH (1943) suggested that the higher cadmium concentration in mafic compared to granitic rocks is presumably due to replacement of Ca^{2+} by Cd^{2+}. Similarly, GOLDSCHMIDT (1958) has mentioned the detection of cadmium in Norwegian bytownite rocks as a component of calcic feldspar and in calcic augites from basalts, as due to substitution.

On the other hand, VINCENT and BILEFIELD (1960) have demonstrated in their work on the Skaergaard intrusion that cadmium cannot enter into the plagioclase and apatite structures. BROOKS and AHRENS (1961) suggested that a high degree of covalency is part of the Cd—O bond and is a probable reason why the Cd—Ca association is virtually nonexistent.

Cadmium is recovered almost entirely as a by-product in the treatment of zinc-bearing ores. Twenty countries produced cadmium in 1963 with a total output of about 26.3 million pounds. The United States, U.S.S.R., Canada, Japan and Belgium, in ranked order, accounted for 81 percent of the total. Based upon calculated zinc reserves and assigned cadmium contents, the world reserve of cadmium in ore is estimated at 900 million pounds.

The principal uses of cadmium are for electroplating, various kinds of alloys (that is, low-melting-point alloys), nuclear reactor control rods and containers for uranium alloys and storage batteries. Another small but important application is the use of cadmium sulfide crystals in photovoltaic cells, radiation detection devices and photo-sensitive elements (SCHROEDER, 1965). Cadmium fumes or dusts and solutions of the compounds are quite toxic.

Revised manuscript received: September 1969

References: Sections 48-B to 48-M, 48-O

AHRENS, L. H., and M. FLEISCHER: Report on trace constituents in granite G-1 and diabase W-1. U.S. Geol. Surv. Bull. **1113**, 83 (1960).

—, and S. R. TAYLOR: Spectrochemical analysis. Reading, Mass.: Addison-Wesley 1961.

ALLER, L. H.: The abundance of the elements: New York: Interscience Publ., Inc. 1961.

— Abundances of elements in stars and nebulas (the abundance of certain elements in solar atmosphere). AFSOR 3024, Technical Note No. 2 (1962).

— The abundance of elements in the solar atmosphere. Advances in Astronomy and Astrophysics, vol. 3. New York: Academic Press 1965.

— The chemical composition of the sun and the solar system. Proc. Anstno. Soc. Australia **1**, 133 (1968).

BADALOV, S. T., and M. P. ENIKEEV: Geochemistry of cadmium in the Almalyk and Altyn-Topkan mineralized areas of the Karamazar region. Geochemistry (English translation) No. 4, 406 (1959).

BALAKISHIEVA, B. A.: Geochemistry of cadmium, manganese, and iron in sphalerites. Izv. Akad. Nauk. Azerb. SSR, Ser. Geol.-Geogr. Nauk **1964** (6), 27; Chem. Abstr. **62**, 15907h (1965).

BILEFIELD, L. I., and E. A. VINCENT: Determination of cadmium in rocks by neutron-activation analysis. Analyst **86**, 386 (1961).

BOWEN, H. J. M.: U. K. Atomic Energy Authority Report, AERE-R 4196 (1963).

— Trace elements in biochemistry. New York: Academic Press 1966.

— The determination of antimony, cadmium, cesium, iridium and silver in biological material by radioactivation. Analyst **92**, 118 (1967).

BOYLE, R. W., and J. J. LYNCH: Speculation of zinc, cadmium, lead, copper and sulfur in Mississippi Valley and similar types of lead-zinc deposits. Econ. Geol. **63**, 421 (1968).

BROOKS, R. R.: The use of ion-exchange enrichment in the determination of trace elements in sea water. Analyst **85**, 745 (1960).

—, L. H. AHRENS, and S. R. TAYLOR: The determination of trace elements in silicate rocks by a combined spectrochemical-anion exchange technique. Geochim. Cosmochim. Acta **18**, 162 (1960).

— — Some observations on the distribution of thallium, cadmium and bismuth in silicate rocks and the significance of covalency on their degree of association with other elements. Geochim. Cosmochim. Acta **23**, 100 (1961).

BROWN, R., and W. A. WOLSTENHOLME: Analysis of geological samples by spark source mass spectrometry. Nature **201**, 598 (1964).

BUR'YANOVA, E. Z.: Mineralogy and geochemistry of cadmium in the sedimentary rocks of Tuva. Geochemistry (English translation) No. 2, 209 (1960).

BUTLER, J. R., and A. J. THOMPSON: Cadmium and zinc in some alkali acidic rocks. Geochim. Cosmochim. Acta **31**, 97 (1967).

CAMERON, A. G. W.: A revised table of abundances of the elements. Astrophys. J. **129**, 676 (1959).

CARMICHAEL, I., and A. MCDONALD: The colorimetric and polarographic determination of some trace elements in the standard rocks G-1 and W-1. Geochim. Cosmochim. Acta **22**, 87 (1961a).

— — The geochemistry of some natural acid glasses from the North Atlantic Tertiary volcanic province. Geochim. Cosmochim. Acta **25**, 189 (1961b).

CHAMP, W. H. (1968a): Quoted in FLANAGAN, F. J.: U. S. Geological Survey standards II. First compilation of data for the new U.S.G.S. rocks. Geochim. Cosmochim. Acta **33**, 81 (1969).

— (1968b): Quoted in SINE, N. M., W. O. TAYLOR, G. R. WEBBER and C. L. LEWIS: Third report of analytical data for CAAS sulfide ore and syenite rock standards. Geochim. Cosmochim. Acta **33**, 121 (1969).

CISSARZ, A.: Average composition of Mansfeld copper shale. Metall u. Erz **27**, 316 (1930); Chem. Abstr. **24**, 4245 (1930).

CLAYTON, D. D., W. A. FOWLER, T. E. HULL, and B. A. ZIMMERMAN: Neutron capture chains in heavy element synthesis. Ann. Phys. **12**, 331 (1961).

FLEISCHER, M.: Minor elements in some sulfide minerals. Econ. Geol. 50th Anniv., 970 (1955).

FRUTH, I.: Spurengehalte der Zinkblenden verschiedener Pb—Zn-Vorkommen in den nördlichen Kalkalpen. Chem. Erde **25**, (2) 105 (1966).

GOLDBERG, L., E. A. MÜLLER, and L. H. ALLER: The abundances of the elements in the solar atmosphere. Astrophys. J., Suppl. Ser. **5**, 1 (1960).

GOLDSCHMIDT, V. M.: The principles of distribution of chemical elements in minerals and rocks. J. Chem. Soc. **1937**, 655.

— Geochemistry. Oxford: University Press 1958.

GOLES, G. G.: Cosmic abundances, nucleosynthesis and cosmic chronology. Handbook of Geochemistry. Part I. Berlin-Heidelberg-New York: Springer 1968.

GORDON, M.: Trace elements in peat. Torfnachrichten **3**, 12 (1952); Chem. Abstr. **47**, 4533i (1953).

GREEN, J.: Geochemical table of the elements for 1959. Bull. Geol. Soc. Am. **70**, 1127—1184 (1959).

GREENLAND, L.: The abundances of selenium, tellurium, silver, palladium, cadmium and zinc in chondritic meteorites. Geochim. Cosmochim. Acta **31**, 849 (1967).

GREVESS, N., G. BLANQUET, and A. BOURY: Proc. I.A.G.C. Symposium on the origin and distribution of the elements, Paris (AHRENS, L. H., ed.). Oxford: Pergamon Press 1968.

HERTEL, L.: X-ray and microchemical control studies in geochemical analysis of single minerals. Geol. Rundschau **55**, (2) 355 (1966); Chem. Abstr. **65**, 16685f (1966).

HUFFMAN, C., JR.: Ion-exchange separation and spectrophotometric determination of cadmium. U. S. Geol. Surv. Profess. Papers **450**-E, 126 (1962).

HYBBINETTE, A.: Microchemical analysis of sphalerite from Kristineberg, Sweden. Ind. Eng. Chem. Anal. Ed. **17**, 654 (1945).

ISHIBASHI, M., T. SHIGEMATSU, M. TABUSE, Y. NISHIKAWA, and S. AIDA: Determination of Cd in sea water. Nippon Kagaku Zasshi **83**, 295 (1962).

IVANOV, V. V.: Geochemistry of cadmium in the deposits of the Deputatskoye group. Geochemistry (English translation) No. 2, 168 (1961).

— Distribution of cadmium in ore deposits. Geochemistry (English translation) No. 8, 757 (1964).

KULIKOVA, M. F.: Trace elements in the oxidized zone of polymetallic deposits of Eastern Transbaikaliya. Geochemistry (English translation) No. 2, 176 (1962).

— Cadmium in the oxidized zones of the Nerchinskii Zavod and Klichkin lead-zinc deposits (Eastern Transbaikaliya). Geochemistry (English translation) No. 9, 897 (1963).

LATIMER, W. M.: The oxidation states of the elements and their potentials in aqueous solutions, 2nd ed. Englewood Cliffs, N.J.: Prentice-Hall, Inc. 1952.

LELAND, W. T., and A. O. NIER: The relative abundances of the zinc and cadmium isotopes. Phys. Rev. **73**, 1206 (1948).

LITVINOVICH, A. N., and M. D. LOGINOVA: Elements in small amounts in ores of the Leninogorsk mining field. Akad. Nauk Kazakh. SSR **9**, 92 (1960); Chem. Abstr. **55**, 21994h (1961).

LOGINOVA, L. G.: Spectrographic determination of tin, zinc, cadmium, antimony, lead and arsenic in natural waters. Zh. Analit. Khim. **14**, 217 (1959).

MALYUGA, D. P.: On cadmium in organisms. Compt. rend. (Doklady) Akad. Sci. U.S.S.R. (English translation) **31**, 145 (1941).

MEITUV, G. M.: Geochemistry of rare elements in the lead-zinc deposit of the Klichkinskii region (Eastern Transbaikaliya). Geochemistry (English translation) No. 7, 694 (1962).

MOGAROVSKII, V. V., and L. V. ROSSEIKIN: Geochemistry of rare and dispersed elements at the Maikhura tin-tungsten deposit (Central Tadzhikstan). Geochemistry (English translation) No. 6, 561 (1961).

MOXHAM, R. L.: Quoted in SINE, N. M., W. O. TAYLOR, G. R. WEBBER and C. L. LEWIS, Third report of analytical data for CAAS sulfide ore and syenite rock standards. Geochim. Cosmochim. Acta **33**, 121 (1969).

MÜLLER, E. A.: Proc. I.A.G.C. Symposium on the origin and distribution of the elements, Paris (L. H. AHRENS, ed.) Oxford: Pergamon Press 1968.

MULLIN, J. B., and J. P. RILEY: Cadmium in sea-water. Nature **174**, 42 (1954).

— — The occurrence of cadmium in seawater and in marine organisms and sediments. J. Marine Res. **15**, 103 (1956).

NAKAGAWA, H. M., and T. F. HARMS: Atomic absorption determination of cadmium in geologic materials. U. S. Geol. Surv. Profess. Papers **600-D** (1968).

NODDACK, I., u. W. NODDACK: Die Häufigkeiten der Schwermetalle in Meerestieren. Arkiv Zool. **32** A, 1 (1940).

PALACHE, C., H. BERMAN, and C. FRONDEL: The system of mineralogy. 7th ed., vol. I. New York: John Wiley & Sons, 1944.

— — — The system of mineralogy. 7th ed., vol. II. New York: John Wiley & Sons, 1951.

PREUSS, E.: Beiträge zur spektralanalytischen Methodik. II. Bestimmung von Zn, Cd, Hg, In, Tl, Ge, Sn, Pb, Sb und Bi durch fraktionierte Destillation. Z. Angew. Mineral. **3**, 8 (1940).

PULIDO, P., K. FUWA, and B. L. VALLEE: Determination of cadmium in biological materials by atomic absorption spectrophotometry. Anal. Biochem. **14**, 393 (1966).

RANKAMA, K., and T. G. SAHAMA: Geochemistry. Chicago: Chicago University Press 1950.

REY, P., H. WAKITA, and R. A. SCHMITT: Radiochemical neutron activation analysis of In, Cd and the 14 Rare Earth Elements plus Y in rocks. Submitted for publication (1970).

SALMI, M.: The trace element in peat. Geol. tutkimuslaitos. Geotek. julkaisuja **51**, 20 (1950); Chem. Abstr. **45**, 6977a (1951).

SANDELL, E. B.: Determination of cadmium in silicate rocks. Ind. Eng. Chem. Anal. Ed. **11**, 364 (1939).

—, and S. S. GOLDICH: The rarer metallic constituents of some American igneous rocks II. J. Geol. **51**, 167 (1943).

SCHMITT, R. A., R. H. SMITH, and D. A. OLEHY: Cadmium abundances in meteoritic and terrestrial matter. Geochim. Cosmochim. Acta **27**, 1077 (1963).

SCHROEDER, H. J.: Cadmium, mineral facts and problems. U. S. Bur. Mines Bull. **630**, 165 (1965).

SHIMP, N. F., J. CONNOR, A. L. PRINCE, and F. E. BEAR: Spectrochemical analysis of soils and biological materials. Soil Sci. **83**, 51 (1957).

SHIRLEY, R. L., E. J. BENNE, and E. J. MILLER: Cadmium in biological materials and foods. Anal. Chem. **21**, 300 (1949).

SILVEY, W. D.: Occurrence of selected minor elements in the waters of California. U. S. Geol. Surv. Water Supply Papers **1535-L** (1967).

SIMS, P. K., and P. B. BARTON, JR.: Some aspects of the geochemistry of sphalerite, Central City district, Colorado. Econ. Geol. **56**, 1211 (1961).

SLUTZER, O.: Geology of coal. Chicago: Chicago University Press 1940.

SMYTHE, L. E., and B. M. GATEHOUSE: Polarographic determination of traces of copper, nickel, cobalt, zinc and cadmium in rocks, using rubeanic acid and 1-nitroso-2-naphthol. Anal. Chem. **27**, 901 (1955).

STANTON, R. E., A. J. MCDONALD, and I. CARMICHAEL: The determination of some trace elements in silicate rocks. Analyst **87**, 134 (1962).

STITCH, S. R.: Trace elements in human tissue. I. A semi-quantitative spectrographic survey. Biochem. J. **67**, 97 (1957).

SUESS, H. E., and H. C. UREY: Abundances of the elements. Rev. Mod. Phys. **28**, 53 (1956).

SWAINE, D. J.: The trace-element content of soils. Commonwealth Bur. Soil Sci. Tech. Commun. No. 48 (1955).

TANNER, J. T.: Cadmium abundances in tektites by neutron activation. U.S.A.E.C. Report No. NYO-844-75, 7 (1968).

TOURTELOT, H. A., C. HUFFMAN, JR., and L. F. RADER: Cadmium in samples of the Pierre shale and some equivalent stratigraphic units, Great Plains region. U. S. Geol. Surv. Profess. Papers **475-D**, 73 (1964).

TUREKIAN, K. K., and K. H. WEDEPOHL: Distribution of the elements in some major units of the earth's crust. Bull. Geol. Soc. Am. **72**, 175 (1961).

UDODOV, P. A., and YU. S. PARILOV: Certain regularities of migration of metals in natural waters. Geochemistry (English translation) No. 8, 763 (1961).
VINCENT, E. A., and L. I. BILEFIELD: Cadmium in rocks and minerals from the Skaergaard intrusion, East Greenland. Geochim. Cosmochim. Acta **19**, 63 (1960).
VINOGRADOV, A. P.: A chemical study of the biosphere. Pedology (USSR) **1945**, 348; Chem. Abstr. **40**, 1893 (1946).
— Regularity of distribution of elements in the earth's crust. Geokhimiya No. 1, 6 1956).
— The geochemistry of rare and dispersed chemical elements in soils, 2nd ed. New York: Consultants Bureau, 1959.
— Average contents of chemical elements in the principal types of igneous rocks of the earth's crust. Geochemistry (English translation) No. 7, 641 (1962).
WAHLER, W.: Pulse-polarographische Bestimmung der Spurenelemente Zn, Cd, In, Tl, Pb und Bi in 37 geochemischen Referenzproben nach Voranreicherung durch selektive Verdampfung. Neues Jahrb. Mineral. Abhandl. **108**, 36 (1968).
WESTER, P. O.: Concentration of 17 elements in subcellular fractions of beef heart tissue determined by neutron activation analysis. Biochim. Biophys. Acta **109**, 268 (1965).
WESTERMARK, T., and B. SJÖSTRAND: Activation analysis of cadmium in small biopsy samples. Intern. J. Appl. Radiation Isotopes **9**, 78 (1960).
WILSKA, S.: Quantitative spectral analysis of trace elements in water. Acta Chem. Scand. **5**, 1368 (1951).
ZDENCK, J.: Mineralogy of cadmium in the Mo-Ba deposits in Vietnam. Casopsis Mineral. Geol. **7**, 132 (1962); Chem. Abstr. **63**, 2763d (1965).

Revised manuscript received: September 1969

Indium 49

A	A. WITTMANN	(Institut für Mineralogie, Kristallographie und Strukturchemie der Technischen Hochschule, Wien, Austria)
B-C-II	T. A. LINN, JR. and R. A. SCHMITT	(Kennecott Research Center, Salt Lake City, Utah 84111, U.S.A.) (Department of Chemistry, Oregon State University, Corvallis, Oregon 97331, U.S.A.)
C-III	R. A. SCHMITT and M. WAKITA	(Department of Chemistry Faculty of Science, University of Tokyo, Japan)
D-M, O	T. A. LINN, JR. and R. A. SCHMITT	

49-A. Crystal Chemistry

I. Introduction

The element indium exhibits a distinctly chalcophilic behavior in the minerals of the earth's crust (GOLDSCHMIDT, 1954). It occurs in solid solutions in sulfide minerals with tetrahedral metal-sulfur bonds, such as stannite (Cu_2FeSnS_4), chalcopyrite ($CuFeS_2$), and sphalerite (ZnS). The first indium mineral described in the literature is roquésite, $CuInS_2$ (PICOT and PIERROT, 1963); it has the chalcopyrite structure and has already been synthesized by HAHN *et al.* (SR 1953, 19). Indite (GENKIN and MURAVEVA, 1963) corresponds to the spinel-type phase $FeIn_2S_4$ (HAHN and KLINGLER, SR 1950, 246), whereas dzhalindite, $In(OH)_3$ (GENKIN and MURAVEVA, 1963), being isostructural with $Sc(OH)_3$ (SCHUBERT and SEITZ, SR 1947—1948, 278), represents a supergene alteration product of indite. Sakuraiite, (Cu, Zn, Fe, $Ag)_3InS_4$ (KATO, 1965), has the stannite structure and is the indium analogue of kesterite (Cu_2ZnSnS_4) and briartite (Cu_2(Fe, Zn)GeS_4). Finally, it is noteworthy that native indium could be found in granites from East Transbaikalia being closely associated with lead (IVANOV, 1964).

Indium has the electronic structure [Kr] $4d^{10}\,5s^2\,5p^1$; in compounds the atoms are present commonly in the trivalent state. The existence of a few compounds containing formally monovalent atoms can be regarded as evidence for the inertness of the paired 5s-electrons. There are known InCl, InBr and InI, the bromide (STEPHENSON and MELLOR, SR 1950, 185) and iodide (JONES and TEMPLETON, SR 1955, 320) having the Tl I-structure, though they are much less stable than the Tl^+-compounds. In_2Cl_3 probably consists of equal numbers of In^+ and In^{3+} atoms, according to In_3^+-$[In^{3+}Cl_6]$ (CLARK *et al.*, 1958). The crystal structures of In_2O and In_2S have not been investigated yet.

II. Elementary Indium and Pnictides

The crystal structure of elementary In can be considered as a tetragonally distorted cubic close-packed arrangement of indium atoms with $c/a = 1.076$, the interatomic distances being 3.24 (4×) and 3.37 (8×) Å, respectively (HULL and DAVEY, SB 1913—1928, 44; DWYER and MELLOR, SB 1928—1932, 173; ZINTL and NEUMAYR, SB 1933—1935, 202; cf. PEARSON, 1958 and 1967).

The structure type of the semiconducting III-V compounds and their energy gap between valence band and conduction band have been described by GORJUNOVA (1971). The high-pressure polymorphs of InP and InAs, obtained at 130—135 and 100—103 kbar respectively (MINOMURA and DRICKAMER, 1962), have rock-salt structure (JAMIESON, 1963).

Table 49-A-1. *Structural data of synthetic indium oxygen compounds*

Compounds	Structure type	In-O distances in Å (CN = 6, unless stated)	Reference
a) Oxides and hydroxides			
In_2O_3	bixbyite; (RE_2O_3), C-type		GOLDSCHMIDT *et al.* (SB 1913—1928, 261); ZACHARIASEN (SB 1913—1928, 263); SCHUSTERIUS and PADUROW (SR 1955, 375); ROTH and SCHNEIDER (SR 1960, 331);
		2.10—2.23	BETZL *et al.* (1963); MAREZIO (1966); HASE *et al.* (1967)
	corundum	2.07, 2.27	CHRISTENSEN *et al.* (1967)
InOOH	InOOH	2.15, 2.20	CHRISTENSEN *et al.* (1964); SCHWARZMANN *et al.* (1965)
$In(OH)_3$	$Sc(OH)_3$	2.17	SCHUBERT and SEITZ (SR 1947—1948, 278); MOELLER *et al.* (SR 1947—1948, 280); CHRISTENSEN *et al.* (1967)
b) Complex oxides			
$LiInO_2$	α-$LiFeO_2$	2.15, 2.33	HOPPE and SCHEPERS (SR 1958, 336); HOPPE and RÖHRBORN (1961)
$MeInO_2$ (Me = Na, K)	α-$NaFeO_2$		HOPPE and SCHEPERS (SR 1958, 336); HOPPE (SR 1959, 347)
$NiFe_{2-x}In_xO_4$ (x = 1.5—1.875)	spinel		MAXWELL and PICKART (SR 1954, 457)
$In^{[4]}(MeNi)^{[6]}O_4$ (Me = Al, Ga, Cr)	spinel		SCHMITZ-DUMONT and KASPER (1965)
$In^{[4]}(InCd)^{[6]}O_4$	spinel (inverse)	(4) 2.13 (6) 2.19 (In, Cd)—O	SKRIBLJAK *et al.* (SR 1959, 377)
$CaIn_2O_4$	$CaFe_2O_4$	2.06—2.22	REID (1967)
$Cu_2In_2O_5$	$Cu_2In_2O_5$	2.04—2.30	BERGERHOFF and KASPER (1968)
$MeInO_3$ (Me = La, Nd, Sm)	perovskite, orthorhombic		PADUROW and SCHUSTERIUS (SR 1955, 385); GELLER (SR 1957, 315); ROTH (SR 1957, 314)
Ba_2InNbO_6	perovskite, cubic		GALASSO and DARBY (SR 1962, 509)

$(InSb)O_4$	rutile		Bayer (1963 and 1969)
$InMeO_4$ (Me = Nb, Ta)	wolframite		Keller (SR 1962, 522); Liebertz (1972)
Me_2InSbO_6(Me = Mg, Mn, Cd)	Mg_3TeO_6		Bayer (1968 and 1969)
c) Borates			
$InBO_3$	calcite		Goldschmidt and Hauptmann (SB 1928—1932, 403); Levin *et al.* (SR 1961, 491)
d) Silicates and germanates			
$Ca_3In_2Si_3O_{12}$	garnet		Mill (1964); Ito and Frondel (1968); Ito (1968)
$Sr_3In_2(OH)_{12}$	hydrogarnet		Ito (1968)
$In_2Si_2O_7$	thortveitite		Ito (1968)
$Pb_2In_2Si_2O_9$	melanotekite	(6) and (4)	Ito (1968)
$Be_3In_2Si_6O_{18}$	beryl (bazzite)		Ito (1968)
$NaInSi_2O_6$	aegirine	2.06—2.21	Christensen and Hazel (1967); Ito (1968)
$Me_3In_2Ge_3O_{12}$ (Me = Ca, Cd)	garnet		Mill (1964)
e) Phosphates			
$InPO_4$	$CrVO_4$; $NiSO_4$	1.98, 2.19	Mooney *et al.* (SR 1954, 503); Mooney (SR 1956, 307)
$InPO_4 \cdot 2\,H_2O$	strengite	2.08—2.28	Mooney-Slater (SR 1961, 464)
f) Sulfates and selenates			
$In(OH)SO_4$	closely related to $Fe(OH)SO_4$		Johansson (SR 1962, 623); Johansson (1963)
$In(OH)SO_4 \cdot 2\,H_2O$	$In(OH)SO_4 \cdot 2\,H_2O$	2.14—2.19	Johansson (SR 1961, 454)
$MeIn(SO_4)_2 \cdot 12\,H_2O$ (Me = NH_4, Rb, Cs, Tl)	alum		Haussühl (SR 1961, 446)
$CsIn(SeO_4)_2 \cdot 12\,H_2O$	alum		Haussühl (SR 1961, 446)

III. Oxygen Compounds

Except for the extremely rare mineral dzhalindite ($In(OH)_3$), all known oxygen compounds of indium have been prepared synthetically (Table 49-A-1). A close relationship of the crystal chemistry of In to that of scandium can be observed (FRONDEL, 1968) which is often displayed in complete solid solutions of related compounds such as In_2O_3—Sc_2O_3 (SCHNEIDER *et al.*, 1961; PETRŮ *et al.*, 1966) and several silicates (ITO and FRONDEL, 1968; ITO, 1968; SHANNON and PREWITT, 1969 and 1970). In an octahedral oxygen coordination, with the assumption that $r[O^{2-}] = 1.40$ Å, the effective ionic radius for In^{3+} (0.790 Å) is not considerably larger than that of Sc^{3+} (0.730 Å).

The crystal structure determination of cubic In_2O_3, which crystallizes in the C-type of rare earth sesquioxides (or bixbyite type), has been refined repeatedly (BETZL *et al.*, 1963; MAREZIO, 1966; HASE *et al.*, 1967). A remarkable agreement between experimental and theoretical atomic parameters could be attained, the latter from calculations of the lattice energy (GASHUROV and SOVERS, 1970; v. MERTENS and ZEMANN, 1966). A new rhombohedral form of In_2O_3 was prepared by thermal synthesis at about 370° C; this probably metastable form changes to the cubic modification at 1205° C (CHRISTENSEN *et al.*, 1967). Fig. 49-A-1 shows the crystal structure of $In(OH)_3$ (SCHUBERT and SEITZ, 1948; CHRISTENSEN *et al.*, 1967); the distorted octahedral $[In(OH_6]$-groups are connected at the corners to form a structure which resembles the ReO_3-type. $Cu_2In_2O_5$ represents a proper structure type (Fig. 49-A-2) and consists of octahedral $[In(OH)_6]$-groups sharing two opposite edges; neighboring chains lie parallel to [001] and are linked at corners by bridging oxygen atoms, thus forming a 3-dimensional network (BERGERHOFF and KASPER, 1968).

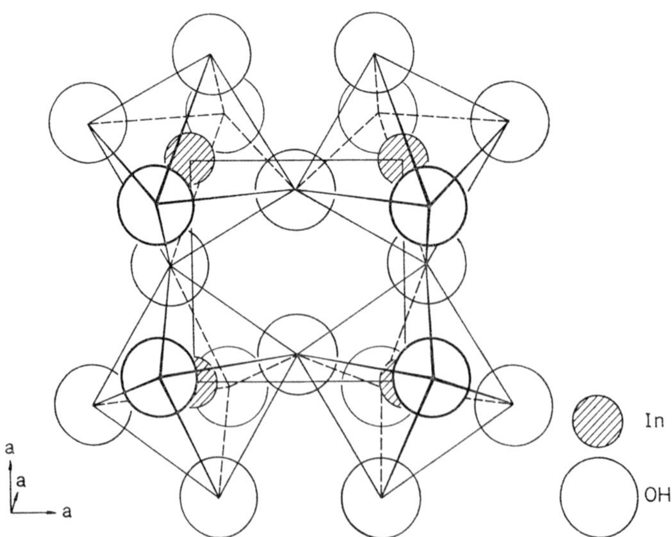

Fig. 49-A-1. Crystal structure of $In(OH)_3$; arrangement of distorted $[In(OH)_6]$-octahedra parallel to (100)

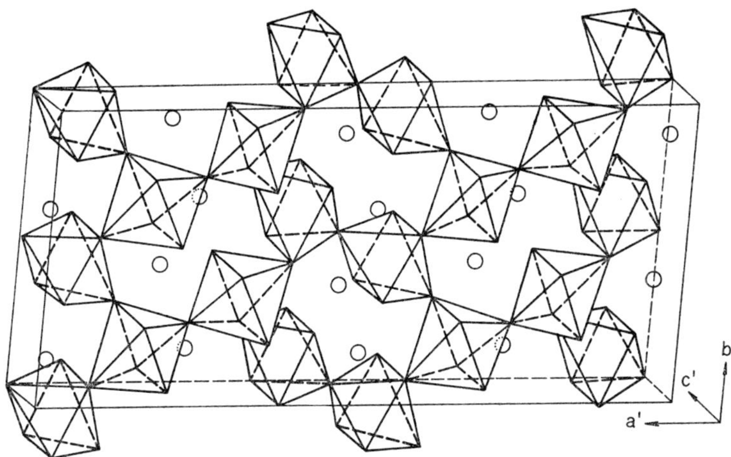

Fig. 49-A-2. Pseudo-orthorhombic unit cell of $Cu_2In_2O_5$ (o Cu atoms)

In contrast to the numerous oxygen compounds with octahedral oxygen coordination, only a few spinel-like phases are reported wherein the indium atoms are surrounded tetrahedrally. As can be summarized from Table 49-A-1, the interatomic In—O distances for the predominant distorted octahedral oxygen coordination range from 2.04 to 2.33 Å (average value from 12 reliable determinations: 2.172 Å), whereas the smaller tetrahedral In—O distance in $In^{[4]}(InCd)^{[6]}O_4$ is 2.13 Å (Skribljak *et al.*, SR 1959, 377). Assuming an oxygen radius of 1.40 Å in octahedral coordination, the radius for In^{3+} should decrease slightly from 0.79 Å to 0.77 Å.

In the high-pressure pyrochlore phase $In_2Ge_2O_7$, the indium atoms have an 8-fold oxygen coordination (Shannon and Sleight, 1968).

IV. Sulfides, Selenides and Tellurides

In general, crystal structures of indium sulfides consist of distorted close-packings of sulfur atoms, containing the metal atoms in tetrahedral and/or octahedral sites respectively (Table 49-A-2). Detailed studies on the very complex ZnS—In_2S_3 system revealed the existence of at least 11 hexagonal phases $Zn_mIn_{2n}S_{m+3n}$; their c-axes are integer multiples of 3.1 Å, thus corresponding approximately with an In—S slab (Gnehm *et al.*, 1969; Boorman and Sutherland, 1969).

Semiletov (1960 and 1961) determined the two hexagonal α- and β-In_2Se_3 polymorphs which are obviously not identical with the rhombohedral α and β phases (for avoidance of confusion written as α' and β') studied by Osamura *et al.* (1966) (Fig. 49-A-3). The β'-form contains the indium atoms in octahedral holes only, whereas in the α'-form indium alternately occupies tetrahedral and octahedral sites between neighboring close-packed Se-layers. In_6S_7 (Hogg and Duffin, 1967) and In_6Se_7 (Hogg, 1971) are structurally related, the latter being of lower symmetry. Additional structural data of selenides and tellurides are listed in Table 49-A-2. It should be mentioned that InTe (Schubert *et al.*, 1953 and 1955), crystallizing in a TlSe structure, contains

Table 49-A-2. *Structural data of synthetic indium sulfides, selenides and tellurides*

Compound	Structure type	In-X distances in Å (CN in brackets)	Reference
a) Sulfides			
InS	SnS, herzenbergite	(6) 2.56—3.84	Schubert *et al.* (SR 1954, 176 and 382); Duffin and Hogg (1966)
α-In_2S_3	defect zinc blende type	(6), (4)	Hahn and Klingler (SR 1949, 178)
β-In_2S_3	stable >300° C, defect spinel type, tetragonal	(6) 2.54—2.68 (4) 2.44, 2.48	Hahn and Klingler (SR 1949, 178); Rooymans (SR 1959, 163); King (SR 1962, 245); Steigmann *et al.* (1965)
In_6S_7	In_6S_7	(6) average 2.89	Schubert *et al.* (SR 1954, 176 and 382); Duffin and Hogg (1966); Hogg and Duffin (1967)
b) Complex sulfides			
$NaInS_2$	α-$NaFeO_2$	(6) 2.63	Hoppe (SR 1959, 401); Hoppe *et al.* (SR 1961, 172)
$MeInS_2$ (Me = Cu, Ag)	chalcopyrite	(4)	Hahn *et al.* (SR 1952, 274 and 1953, 19)
$Me^{[4]}In_2^{[6]}S_3$ (Me = Ca, Cd, Hg)	spinel	(6)	Hahn and Klingler (SR 1950, 246)
$In^{[4]}(InMe)^{[6]}S_3$ (Me = Mg, Fe, Co, Ni)	spinel (inverse)	(6), (4)	Hahn and Klingler (SR 1950, 246)
$Zn^{[6]}In^{[6]}In^{[4]}S_3$	$ZnIn_2S_3$	(6) 2.67, (4) 2.40	Lappe *et al.* (SR 1962, 246); Donika *et al.* (1967)
$Zn_2^{[4]}(ZnIn)^{[4]}In^{[6]}S_6$	$Zn_3In_2S_6$	(6) 2.62 (4) 2.41 (In, Zn)-S	Zhitar *et al.* (1966); Donika *et al.* (1967)
$Bi_2In_3^{[6]}In^{[4]}S_9$	$Bi_2In_4S_9$	(6) 2.63 (4) 2.48	Chapuis *et al.* (1972)
c) Selenides			
In_2Se	In_2Se	(2) 2.72 In-Se-chains	Schubert *et al.* (SR 1954, 176 and 389); Man and Semiletov (1965)
InSe	GaS	(4) 2.51 In-Se (3 ×) 3.16 In-In (1 ×)	Damon and Redington (SR 1954, 176); Schubert *et al.* (SR 1954, 389); Semiletov (SR 1958, 142)

In_6Se_7	In_6Se_7	(6) average 2.94	HOGG (1971)
α-In_2Se_3	SeInSeInSe-layer structure, hexagonal	(6) 2.95, (4) 2.51	SEMILETOV (SR 1961, 170)
β-In_2Se_3	stable >200° C, superstructure of wurtzite type	(4) 2.51	HAHN (SR 1953, 167); HAHN and FRANK (SR 1957, 135); MIYAZAWA and SUGAIKE (SR 1957, 135); SEMILETOV (SR 1960, 161 and 1961, 170)
α′-In_2Se_3 (l) β′-In_2Se_3 (h)	15-layer structures, rhombohedral	(6) 2.87, (4) 2.69 (6) 2.87	OSAMURA *et al.* (1966)
d) Complex selenides			
$NaInSe_2$	α-$NaFeO_2$	(6) 2.76	HOPPE *et al.* (SR 1961, 172)
$MeInSe_2$ (Me = Cu, Ag)	chalcopyrite	(4)	HAHN *et al.* (SR 1952, 274 and 1953, 19)
$MeIn_2Se_4$ (Me = Zn, Hg)	Al_2CdS_4	(4)	HAHN *et al.* (SR 1955, 414)
e) Tellurides			
InTe	TlSe	(8) 3.57, (4) 2.78	SCHUBERT *et al.* (SR 1953, 191 and 1955, 198); SUGAIKE (1957)
InTe(II)	NaCl; high-pressure phase >28 kbar, >150° C	(6) 3.09	BANUS *et al.* (1963); SCLAR *et al.* (1964); DENEKE and RABENAU (1965); GELLER *et al.* (1965)
In_3Te_4	anti-As_3Sn_4; high-pressure phase, ~ 35 kbar, 550° C	(6) 3.03	GELLER *et al.* (1965)
α-In_2Te_3	9-layer-packing of Te atoms, cubic	(4)	INUZUKA and SUGAIKE (SR 1954, 177); SUGAIKE (1957); ZASLAVSKIJ and SERGEEVA (1960); ŽUZE *et al.* (SR 1961, 173)
β-In_2Te_3	stable >550° C, defect sphalerite type	(4)	ZASLAVSKIJ and SERGEEVA (1960); ŽUZE *et al.* (SR 1961, 173); THOMASSEN *et al.* (1963)
f) Complex tellurides			
$MeInTe_2$ (Me = Cu, Ag, Au)	chalcopyrite	(4)	HAHN *et al.* (SR 1952, 274 and 1953, 19); ZALAR and CADOFF (SR 1962, 181)
$MeIn_2Te_4$(Me = Cd, Zn, Hg)	Al_2CdS_4	(4)	HAHN *et al.* (SR 1955, 414); THOMASSEN *et al.* (1963)

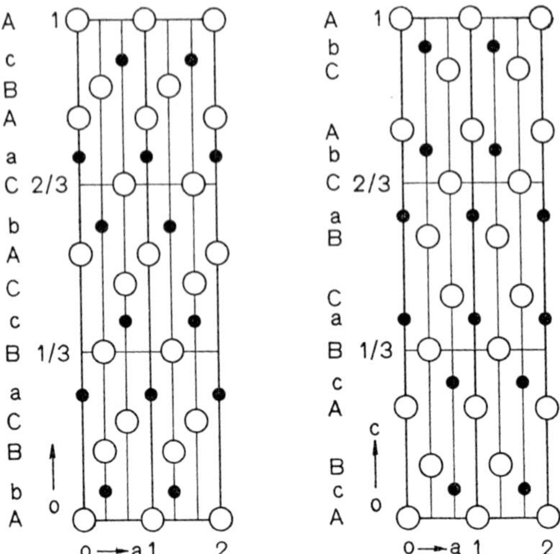

Fig. 49-A-3. The crystal structures of α'-In_2Se_3 and β'-In_2Se_3 projected down [110] (OSAMURA *et al.*, 1966)

2 types of indium atoms and may be written correctly as $In^{[8]}(In^{[4]}Te_2)$; one set of In atoms occupies the tetrahedral sites in the chain-like links of tellurium tetrahedra sharing opposite edges, the other set occurs between the infinite chains in positions of 8-coordination.

Average values of interatomic distances in tetrahedral coordination are: In-S = 2.42 Å; In—Se = 2.55 Å; In—Te = 2.78 Å; and in octahedral coordination (of high pressure phases): In—S = 2.68 Å; In—Se = 2.88 Å; and In—Te = 3.06 Å. The information on In—Te distances is as yet insufficient.

Revised manuscript received: January 1973

References: Section 49-A

BANUS, M. D., HANNEMAN, R. E., STRONGIN, M., GOOEN, K.: High-pressure transitions in $A^{III}B^{VI}$ compounds: indium telluride. Science **142**, 662 (1963).

BAYER, G.: $InSbO_4$ and $ScSbO_4$, new compounds with the rutile structure. Z. Krist. **118**, 158 (1963).

BAYER, G.: Isostructural oxide compounds of tellurium and antimony of the type Me_3XO_6. Naturwissenschaften **55**, 33 (1968).

BAYER, G.: Zur Kristallchemie des Tellurs. Fortschr. Mineral. **46**, 41 (1969).

BERGERHOFF, G., KASPER, H.: Die Kristallstruktur des Kupfer-Indium-Oxids, $Cu_2In_2O_5$. Acta Cryst. B **24**, 388 (1968).

BETZL, M., HASE, W., KLEINSTÜCK, K., TOBISCH, J.: Messung der kohärenten Streuamplitude von In für thermische Neutronen und Bestimmung der Strukturparameter von In_2O_3. Z. Krist. **118**, 473 (1963).

BOORMAN, R. S., SUTHERLAND, J. K.: Subsolidus phase relations in the $ZnS—In_2S_3$ system: 600 to 1080° C. J. Materials Sci. **4**, 658 (1969).

CHAPUIS, G., GNEHM, CH., KRÄMER, V.: Crystal structure and crystal chemistry of bismuth indium sulphide, $Bi_2In_4S_9$. Acta Cryst. B **28**, 3128 (1972).

CHRISTENSEN, A. N., BROCH, N. C., VON HEIDERSTAM, O., NILSSON, Å.: Hydrothermal investigation of the system $In_2O_3—H_2O—Na_2O$ and $In_2O_3—D_2O—Na_2O$. The crystal structure of rhombohedral In_2O_3 and of $In(OH)_3$.. Acta Chem. Scand. **21**, 1046 (1967).

CHRISTENSEN, A. N., GRØNBAEK, R., RASMUSSEN, S. E.: The crystal structure of InOOH. Acta Chem. Scand. **18**, 1261 (1964).

CHRISTENSEN, A. N., HAZEL, R. C.: The crystal structure of $NaIn(SiO_3)_2$. Acta Chem. Scand. **21**, 1425 (1967).

CLARK, R. J., GRISWOLD, E., KLEINBERG, J.: Some observations on lower halides of indium. J. Am. Chem. Soc. **80**, 4764 (1958).

DENEKE, K., RABENAU, A.: Über die Natur der Phase In_3SbTe_2. Z. Anorg. Allgem. Chem. **333**, 201 (1964).

DONIKA, F. G., KIOSSE, G. A., RADAUTSAN, S. I., SEMILETOV, S. A., ZHITAR, V. F.: (The crystal structure of $Zn_3In_2S_6$). Kristallografiya **12**, 854 (1967).

DUFFIN, W. J., HOGG, J. H. C.: Crystalline phases in the system $In—In_2S_3$. Acta Cryst. **20**, 566 (1966).

FRONDEL, C.: Crystal chemistry of scandium as a trace element in minerals. Z. Krist. **127**, 121 (1968).

GASHUROV, G., SOVERS, O. J.: Theoretical calculation of structural parameters of C-type sesquioxides. Acta Cryst. B **26**, 938 (1970).

GELLER, S., JAYARAMAN, A., HULL, G. W., JR.: Crystal chemistry and superconductivity of pressure-induced phases in the In—Te system. J. Phys. Chem. Solids **26**, 353 (1965).

GENKIN, A. D., MURAVEVA, I. V.: (Indite and dzhalindite, new indium minerals). Zap. Vses. Mineralog. Obshchestva **92**, 445 (1963); cf. Am. Mineralogist **49**, 493 (1964).

GNEHM, C., NITSCHE, R., WILD, P.: New phases in the system Zn—In—S. Naturwissenschaften **56**, 86 (1969).

GOLDSCHMIDT, V. M.: In: Geochemistry. MUIR, A. (ed.). Oxford: Clarendon Press 1954.

GORJUNOVA, N. A.: Halbleiter mit diamantähnlicher Struktur. Leipzig: B. G. Teubner 1971 (Transl. from Russian).

HASE, W., KLEINSTÜCK, K., SCHULZE, G. E. R.: Kristallstrukturuntersuchungen an Sesquioxiden mit α-Mn_2O_3-Struktur. Z. Krist. **124**, 428 (1967).

HOGG, J. H. C.: The crystal structure of In_6Se_7. Acta Cryst. B **27**, 1630 (1971).

HOGG, J. H. C., DUFFIN, W. J.: The crystal structure of In_6S_7. Acta Cryst. **23**, 111 (1967).

HOPPE, R., RÖHRBORN, H.-J.: Zur Kristallstruktur von $LiInO_2$. Naturwissenschaften **48**, 452 (1961).

ITO, J.: Synthetic indium silicate and indium hydrogarnet. Am. Mineralogist **53**, 1663 (1968).

ITO, J., FRONDEL, C.: Syntheses of the scandium analogues of aegirine, spodumene, andradite and melanotekite. Am. Mineralogist **53**, 1276 (1968).

IVANOV, V. V.: Native indium. In: VLASOV, K. A. (ed.): Geochemistry, Mineralogy and Genetic Types of Deposits of Rare Elements, **2**, 568 Moscow, Nauka Publ. 1964.

JAMIESON, J. C.: Crystal structures at high pressures of metallic modifications of compounds of indium, gallium and aluminum. Science **139**, 845 (1963).

JOHANNSON, G.: On the structures of some hydroxo salts of Al^{3+}, Ga^{3+}, In^{3+} and Tl^{3+}. Svensk Kem. Tidskr. **75**, 41 (1963).

KATO, A.: (Sakuraiite, a new mineral.) Chigaku Kenkyu, Sakurai Vol. 1—5 (1965); cf. Am. Mineralogist **53**, 1421 (1968).

LIEBERTZ, J.: Gitterkonstanten von $InNbO_4$ und $InTaO_4$. Acta Cryst. B **28**, 3100 (1972).

MAN, L. I., SEMILETOV, S. A.: (Preliminary electron-diffraction study of the structure of the semiconducting compound In_2Se). Kristallografiya **10**, 407 (1965).

MAREZIO, M.: Refinement of the crystal structure of In_2O_3 at two wavelengths. Acta Cryst. **20**, 723 (1966).

MERTENS, H.-E. v., ZEMANN, J.: Elektrostatische Gitterenergien für den Bixbyit-Typ. Acta Cryst. **21**, 467 (1966).

MILL, B. W.: (Hydrothermal synthesis of V^{3+}-, In^{3+}- and Sc^{3+}-containing garnets). Dokl. Akad. Nauk SSSR **156**, 814 (1964).

MINOMURA, S., DRICKAMER, H. G.: Pressure induced phase transitions in silicon, germanium and some III-V compounds. J. Phys. Chem. Solids **23**, 451 (1962).

OSAMURA, K., MURAKAMI, Y., TOMIIE, Y.: Crystal structure of α- and β-indium selenide, In_2Se_3. J. Phys. Soc. Japan **21**, 1848 (1966).

PETRŮ, F., HÁJEK, B., KÁLALOVÁ, E., DOLEŽALOVÁ, J.: Über das Mischkristallsystem Scandiumoxid-Indiumoxid. Z. Chem. **6**, 190 (1966).

PICOT, P., PIERROT, R.: La roquésite, premier minéral d'indium: $CuInS_2$. Bull. Soc. Franç. Minéral. Crist. **86**, 7 (1963).

REID, A. F.: Calcium indate, an isotype of calcium ferrite and sodium scandium titanate. Inorg. Chem. **6**, 631 (1967).

SCLAR, C. B., CARRISON, L. C., SCHWARTZ, C. M.: High-pressure metallic indium telluride: preparation and crystal structure. Science **143**, 352 (1964).

SCHMITZ-DUMONT, O., KASPER, H.: Die Lichtabsorption des 2-wertigen Kupfers, Nickels und Kobalts sowie des 3-wertigen Chroms. Z. Anorg. Allgem. Chem. **341**, 252 (1965).

SCHNEIDER, S. J., ROTH, R. S., WARING, J. L.: Solid state reactions involving oxides of trivalent cations. J. Res. Nat. Bur. Std. A **65**, 345 (1961).

SHANNON, R. D., PREWITT, C. T.: Effective ionic radii in oxides and fluorides. Acta Cryst. B **25**, 925 (1969).

SHANNON, R. D., PREWITT, C. T.: Revised values of effective ionic radii. Acta Cryst. B **26**, 1046 (1970).

SHANNON, R. D., SLEIGHT, A. W.: Synthesis of new high-pressure pyrochlore phases. Inorg. Chem. **7**, 1649 (1968).

STEIGMANN, G. A., SUTHERLAND, H. H., GOODYEAR, J.: The crystal structure of β-In_2S_3. Acta Cryst. **19**, 967 (1965).

SUGAIKE, S.: Synthesis, crystal lattices and some electrical properties of indium tellurides and selenides. Mineral. J. (Japan) **2**, 63 (1957).

SCHWARZMANN, E., GLEMSER, O., MARSMANN, H.: IR- und ^1H-Kernresonanzmessungen an InO(OH). Naturwissenschaften **52**, 344 (1965).

THOMASSEN, L., MASON, D. R., ROSE, G. D., SARACE, J. C., SCHMITT, G. A.: The phase diagram for the pseudo-binary system CdTe—In_2Te_3. J. Electrochem. Soc. **110**, 1127 (1963).

ZASLAVSKIJ, A. I., SERGEEVA, V. M.: (On the polymorphism of In_2Te_3.) Fiz. Tverd. Tela **2**, 2872 (1960).

ZHITAR, V., OKSMAN, YA., RADAUTSAN, S., SMIRNOV, V.: Some photo-dielectric and luminescent properties of new semi-conducting single crystals of $Zn_3In_2S_6$ phase. Physica Status Solidi **15**, K105 (1966).

Revised manuscript received: January 1973

49-B. Isotopes in Nature

Indium (atomic number 49) consists of two isotopes of mass numbers 113 and 115. The natural isotopic abundances of ^{113}In and ^{115}In are 4.33 and 95.67 atomic percent respectively (SUNDERMAN and TOWNLEY, 1960; LEDERER *et al.*, 1967). Indium-115 is naturally radioactive, decaying with a 5×10^{14}-year half-life to stable ^{15}Sn.

Revised manuscript received: October 1970

49-C. Abundance in Cosmos, Meteorites, Tektites and Lunar Materials

I. Cosmic Abundance

The cosmic abundance of indium has been based on solar abundance and the available data are summarized in Tables 49-C-1 and 49-C-2.

Table 49-C-1. *Indium cosmic abundances relative to hydrogen*

Indium abundance[a]	Reference
0.75	ALLER (1961)
1.16	GOLDBERG *et al.* (1960)
1.28	ALLER (1965)
1.45	ALLER (1965)
1.51	ALLER (1968)
1.68	GREVESSE *et al.* (1968)

Atomic abundances based on silicon ($Si = 10^6$) are given in Table 49-C-2.

Table 49-C-2. *Indium cosmic abundances relative to silicon*

Indium abundance[b]	Reference
0.11	SUESS and UREY (1956)
0.084	VINOGRADOV (1956)
0.098	AHRENS and TAYLOR (1961)
0.22	CAMERON (1968)
0.80	UREY (1967)

The indium content of sunspot spectra was reported as 0.75 ± 0.2 ($\log N_{In}$ based on $\log N_H = 12.0$) by DUBOV and KHROMOVA (1966). A cosmic mass abundance of 0.24 ppb indium was given by GREEN (1959).

II. Meteorites

The indium content of iron meteorites are shown in Table 49-C-3, and the abundances of indium in the various classes of stony meteorites are given in Table 49-C-4.

[a] $\log N_{In}$ based on $\log N_H = 12.0$.
[b] $\log N_{In}$ based on $\log N_{Si} = 6.0$.

Table 49-C-3. *Indium in iron meteorites*

Material	No. of samples	In (ppb) range	In (ppb) mean	Method	Reference
Hexahedrite (H)	5	0.5–10	3	(N/R)	SMALES *et al.* (1967)
Hexahedrite (H)	2	0.36–0.39	0.38	(N/R)	TANDON and WASSON (1967a)
Ataxite, Ni-poor (Da)	4	4–10	10	(N/R)	SMALES *et al.* (1967)
Octahedrite, coarsest (Ogg)	4	4–11	9	(N/R)	SMALES *et al.* (1976)
Octahedrite, coarsest (Ogg)	1	—	10.6	(N/R)	TANDON and WASSON (1967a)
Octahedrite, coarse (Og)	10	1–14	11	(N/R)	SMALES *et al.* (1967)
Octahedrite, medium (Om)	20	0.5–30	12	(N/R)	SMALES *et al.* (1967)
Octahedrite, medium (Om)	2	14.8–16.1	15.5	(N/R)	TANDON and WASSON (1967a)
Octahedrite, fine (Of)	10	0.3–<10	—		SMALES *et al.* (1967)
Octahedrite, finest (Off)	4	1.1–<10	—		SMALES *et al.* (1967)
Ataxite, Ni-rich (Dr)	6	0.4–41	14	(N/R)	SMALES *et al.* (1967)

Table 49-C-4. *Indium in achondrites and chondrites*

Meteorite class	No. determined	In (ppb) range	In (ppb) mean	Method	Reference
Achondrites:					
Ae	2	0.34–0.40	0.37 ± 0.04	(N/R)	SCHMITT and SMITH (1968)
Ap	2	0.24–1.0	0.5 ± 0.1	(N/R)	SCHMITT and SMITH (1968)
Chondrites:					
Ce	4	4–76	35	(N/R)	FOUCHÉ and SMALES (1967)
Ce_1	1	135	135	(N/R)	AKAIWA (1966)
Ce_1	2	90–93	92	(N/R)	SCHMITT and SMITH (1968)
Ce_2	3	0.24–2.9	1.1	(N/R)	SCHMITT and SMITH (1968)
CH	9	0.5–8.2	2.3	(N/R)	FOUCHÉ and SMALES (1967)
CH	3	0.1–2.1	0.8	(N/R)	SCHMITT and SMITH (1968)
CH	1	0.5	0.5	(N/R)	TANDON and WASSON (1967b)
CL	9	0.5–1.3	0.7	(N/R)	FOUCHÉ and SMALES (1967)
CL	26	0.05–55	6	(N/R)	TANDON and WASSON (1968)
CL	5	0.3–14	5	(N/R)	SCHMITT and SMITH (1968)
CL	10	0.07–26	7	(N/R)	KEAYS *et al.* (1971)
CLL	1	74	74	(N/R)	SCHMITT and SMITH (1968)
CHL	3	23–32	27	(N/R)	SCHMITT and SMITH (1968)
CHL	1	31	31	(N/R)	AKAIWA (1966)
CHL	1	27	27	(N/R)	WAKITA and SCHMITT (1970)
Cc_1	2	64–80	75	(N/R)	SCHMITT and SMITH (1968)
Cc_1	2	82–88	85	(N/R)	FOUCHÉ and SMALES (1967)
Cc_1	1	115	115	(N/R)	AKAIWA (1966)
Cc_2	2	47–64	55	(N/R)	SCHMITT and SMITH (1968)
Cc_2	2	46–49	48	(N/R)	FOUCHÉ and SMALES (1967)
Cc_2	2	23–50	36	(N/R)	AKAIWA (1966)

LARIMER and ANDERS (1967) reported the averages of indium content in various types of chondritic meteorites. Their data are summarized in Table 49-C-5.

Table 49-C-5. *Indium in chondritic meteorites*

Chondrite type	In (ppb)
Carbonaceous, I	220
Carbonaceous, II	100
Carbonaceous, III	43
Ordinary	0.4
Enstatite, I	140
Enstatite, II	3

TAYLOR and SACHS (1960) attempted to determine indium in tektites, but found less than their detection limit (1 ppm In) in 14 australites. No other data are available concerning indium in tektites.

Revised manuscript received: October 1970

III. Lunar Materials

The indium contents of lunar materials from lunar maria, lunar uplands and lunar highlands are shown in Tables 49-C-6 to 49-C-8. Most references are taken from the Proceeding of the Lunar Science Conferences published in Geochim. Cosmochim. Acta Supplements.

Table 49-C-6. *Indium in lunar mare materials* (*Analytical method*: *N*/*R*)

Material	No. of rocks	In (ppb)		Reference
		Range	Geometric mean[f]	
Apollo 11				
Mare basalts	5	1.0–4.4	2.6	1, 2, 3, 4, 5
Breccias	3	4–6	5	3, 4, 6
Anorthosites[a]	3	2.7, 2.7, 6.6	3.6	7, 8
Soil	indium values show contamination from returned lunar rock box			
Apollo 12				
Mare basalts	12	0.4–2.0	1.4	2, 9, 10
Breccias	2	6.5–7.0	7	8
Breccia 12013	1	2.8–6.6	3.7[b]	11
Soils	3	2.0–10	4	8, 10
Core samples 19 cm depth	1		9	8
37 cm depth	1		9	8

Table 49-C-6 (continued)

Apollo 15				
Mare basalts	8	0.3–0.6	0.5	12, 13, 14
Mare gabbro	1		0.5	14
Highland basalt 15256[c] (impact melted)	1		7	14
Valley and rille breccia	1		2.6	13
Highland breccias[c]	6	0.3–9	1.1	13, 14
Anorthosite 15415[c]	1		0.2	12
Valley and rille soils	6	3.2–11	6	15
Highland soils[c]	2	4.4, 5.1	5	13
Glass sphere	1		2	14
Apollo 17				
Highland-massif breccias[d]	6	0.2–1	0.3	16
Valley soils	2	2.1, 4.6	3.1	19
Highland-massif soils[d]	5	1.4–4.0	2.7	17, 18, 19
Sculptured hills[e] 0–1 cm	1		4.5	19
trench soils ca. 5 cm	1		2.8	19
ca. 10 cm	1		2.6	19
15–25 cm	1		2.9	19
Luna 16				
Mare basalt	1		12	20
Soil	1	3.0, 1.9, 7	4	20, 21

[a] >90% plagioclase. [b] Mass weighted average. [c] At stations 6 and 7. [d] At station 2, 3, 6 and 7. [e] At station 8. [f] For one rock or soil, arithmetic means are given.

References: 1. Ganapathy, R., Keays, R. R., Laul, J. C., Anders, E.: Suppl. 1, **2**, 1117 (1970). 2. Anders, E., Ganapathy, R., Keays, R. R., Laul, J. C., Morgan, J. W.: Suppl. 2, **2**, 1021 (1971). 3. Wakita, H., Schmitt, R. A., Rey, P.: Suppl. 1, **2**, 1685 (1970). 4. Wasson, J. T., Baedecker, P. A.: Suppl. 1, **2**, 1741 (1970). 5. Wänke, H., Rieder, R., Baddenhausen, H., Spettel, B., Teschke, F., Quijano-Rico, M., Balacescu, A.: Suppl. 1, **2**, 1719 (1970). 6. Smales, A. A., Mapper, D., Webb, M. S. W., Webster, R. K., Wilson, J. D.: Suppl. 1, **2**, 1575 (1970). 7. Wakita, H., Schmitt, R. A.: Lunar anorthosites. Rare-earth and other elemental abundances. Science **170**, 969 (1970). 8. Laul, J. C., Morgan, J. W., Ganapathy, R., Anders, E.: Suppl. 2, **2**, 1139 (1971). 9. Baedecker, P. A., Schaudy, R., Elzie, J. L., Kimberlin, J., Wasson, J. T.: Suppl. 2, **2**, 1037 (1971). 10. Wakita, H., Rey, P., Schmitt, R. A.: Suppl. 2, **2**, 1319 (1971). 11. Laul, J. C., Keays, R. R. Ganapathy, R., Anders, E.: Abundance of 14 trace elements in lunar rock 12013,10. Earth Planet. Sci. Letters **9**, 211 (1970). 12. Morgan, J. W., Krähenbühl, U., Ganapathy, R., Anders, E.: Suppl. 3, **2**, 1361 (1972). 13. Baedecker, R. A., Chou, C.-L., Grudewicz, E. B., Wasson, J. T.: Suppl. 4, **2**, 1177 (1973). 14. Ganapathy, R., Morgan, J. W., Krähenbühl, U., Anders, E.: Suppl. 4, **2**, 1239 (1973). 15. Baedecker, P. A. Chou, C.-L., Wasson, J. T.: Suppl. 3, **2**, 1343 (1972). 16. Laul, J. C., Schmitt, R. A.: Chemical composition of boulder-2 rocks and soils, Apollo 17, Station 2. Earth Planet. Sci. Letters **23**, 206 (1974). 17. Laul, J. C., Hill, D. W., Schmitt, R. A.: Suppl. 5, **2**, 1047 (1974). 18. Morgan, J. W., Ganapathy, R., Higuchi, H., Krähenbühl, U., Anders, E.: Suppl. 5, **2**, 1703 (1974). 19. Baedecker, P. A., Chou, C.-L., Sundberg, L. L., Wasson, J. T.: Suppl. 5, **2**, 1625 (1974). 20. Vinogradov, A. P.: Preliminary data on lunar soil collected by the Luna 20 unmanned spacecraft. Geochim. Cosmochim. Acta **37**, 721 (1973). 21. Laul, J. C., Ganapathy, R., Morgan, J. W., Anders, E.: Meteoritic and non-meteoritic trace elements in Luna 16 samples. Earth Planet. Sci. Letters **13**, 450 (1972).

All Supplements listed are those of Geochim. Cosmochim. Acta.

Table 49-C-7. *Indium in lunar upland materials* (*Analytical method*: *N/R*)

Material	No. of rocks	In (ppb) Range	Geometric mean	Reference
Apollo 14				
Pseudo basalt 14310	1	20, 130		1, 2
Breccias	2	2.7, 3.4	3.0	1
Soils	3	39, 41, 120	57	2
Yellow glass	1		3.6	1
Green glass	1		8	1

References: 1. Morgan, J. W., Laul, J. C., Krähenbühl, U., Ganapathy, R., Anders, E.: Geochim. Cosmochim Acta Suppl. 3, **2**, 1377 (1972). 2. Baedecker, P. A., Chou, C.-L., Wasson, J. T.: Geochim. Cosmochim. Acta Suppl. 2, **2**, 1343 (1972).

Table 49-C-8. *Indium in lunar highlands materials*[a] (*Analytical method*: *N/R*)

Material	No. of rocks	In (ppb) Range	Geometric mean[b]	Reference
Apollo 16				
Basalt	1		0.16	1
Anorthosite 61016 light	1		250	1
Anorthosite 61016 dark	1		4	1
Anorthositic gabbro	1		7	1
Gabbroic anorthosite	1		15	1
Breccia	1		6	1
Soils	8	1.5–18	10	1
Luna 20				
Breecias	2	<30, <400		2
Soil	1	4, 10	7	3, 4

[a] Also see Apollo 15 and 17 for the highland materials.
[b] See Table 49-C-6.

References: 1. Baedecker, P. A., Chou, C.-L., Sundberg, L. L., Wasson, J. T.: Geochim. Cosmochim. Acta Suppl. 5, **2**, 1625 (1974). 2. Vinogradov, A. P.: Preliminary data on lunar soil collected by the Luna 20 unmanned spacecraft. Geochim. Cosmochim. Acta **37**, 721 (1973). 3. Morgan, J. W., Laul, J. C., Krähenbühl, U., Ganapathy, R., Anders, E.: Luna 20 soil: Abundance of 17 trace elements. Geochim. Cosmochim. Acta **37**, 953 (1973). 4. Laul, J. C., Schmitt, R. A.: Chemical composition of Luna 20 rocks and soil and Apollo 16 soils. Geochim. Cosmochim. Acta **37**, 927 (1973).

Manuscript received: September 1974

49-D. Abundance in Rock-forming Minerals and Ore Minerals; Indium Minerals

I. Rock-forming and Ore Minerals

Indium is rare in the common minerals, being either an inclusion in the principal mineral or occupying a lattice site in the mineral crystal. Tables 49-D-1 and 49-D-2 list abundances of indium in the principal rock-forming minerals and the principal ore-forming minerals.

Table 49-D-1. *Indium in rock-forming minerals*

Material	No. of samples	In (ppb) range	In (ppb) mean	Method	Reference
Quartz from alaskite, Yakutiya (U.S.S.R.)	—	—	10	(C)	1
Plagioclases and K-feldspars	9	—	< 20	(S)	2
Feldspars from granite, Yakutiya (U.S.S.R.)	—	—	15	(C)	1
Feldspars (plagioclases) from gabbro, Skaergaard intrusion (East Greenland)	2	3.0—3.4	3.2	(N/R)	3
Feldspars (plagioclases) from gabbro, Skaergaard intrusion (East Greenland)	2	—	4 ± 1	(N/R)	4
Albite from alaskite, Yakutiya (U.S.S.R.)	—	—	20	(C)	1
Tourmalines, Deputaskoe (U.S.S.R.)	3	3,000—13,000	7,000	(C)	5
Chlorite, Deputaskoe (U.S.S.R.)	1	—	8,000	(C)	5
Muscovites (Finland and U.S.A.)	2	< 20—4,500	3,000	(S)	6
Muscovites from granite, Yakutiya (U.S.S.R.)	2	50—80	65	(C)	1
Biotites from granite, Saxony (Eastern Germany)	2	250—320	290	(S)	7
Biotites	2	—	< 20		6
Biotites from granites, Yakutiya (U.S.S.R.)	2	490—1,800	1,100	(C)	1
Augite (diopside)	—	—	100 ± 4	(N/R)	8
Pyroxenes from gabbros, Skaergaard intrusion (East Greenland)	9	170—1,080	700	(N/R)	3
Pyroxenes from gabbros, lamprophyres (Greenland, U.S.A., Finland, and Canada)	4	< 20—310	91	(S)	6

Table 49-D-1 (continued)

Material	No. of samples	In (ppb) range	mean	Method	Reference
Olivines, Skaergaard intrusion (East Greenland)	2	55—57	56	(N/R)	3
Serpentinites	4	—	15	(S)	7
Amphiboles (U.S.A.)	2	< 20—5,800	3,000	(S)	6
Garnets from gneisses (Greenland and U.S.A.)	3	26—180	100	(S)	6
Epidote	1	—	590	(S)	6
Cassiterites (U.S.S.R.)	9	16—30 ppm	—	(C)	9
Cassiterites (U.S.S.R.)	186	10—100 ppm	—		9
Cassiterite	1	< 1 ppm	—		10
Cassiterites	72	10—1,000 ppm	—		11
Cassiterites	—	X—100 ppm	—	(S)	12
Magnetites from gabbro, Skaergaard intrusion (East Greenland)	2	150—160	160	(N/R)	3
Magnetites from lamprophyre	2	71—120	95	(S)	6
Magnetites from gabbro, Skargaard intrusion (East Greenland)	2	—	90 ± 5	(N/R)	13
Magnetite-ilmenites from gabbro, Skaergaard intrusion (East Greenland)	2	—	150 ± 4	(N/R)	13
Ilmenites from gabbro, Skaaergaard intrusion (East Greenland)	2	280—290	290	(N/R)	3
Ilmenites	2	—	194 ± 4	(N/R)	13
Pyrrhotites, Deputaskoe (U.S.S.R.)	6	1—50 ppm	28 ppm	(C)	5
Pyrite, Deputaskoe (U.S.S.R.)	1	—	2.5 ppm	(C)	5
Siderites, Deputaskoe (U.S.S.R.)	2	—	4 ppm	(C)	5
Mangano-siderites, Deputaskoe (U.S.S.R.)	2	2—6 ppm	4 ppm	(C)	5
Calcites, North Kantau (U.S.S.R.)	3	320—1,800	—	(P)	14
Aragonites, Karkhona (U.S.S.R.)	6	220—1,600	—	(P)	14

X: less than limit of detection.

References: 1. Ivanov (1963). 2. Shaw (1957). 3. Wager *et al.* (1958). 4. Smales *et al.* (1957). 5. Ivanov and Rozbianskaya (1961). 6. Shaw (1952). 7. Voland (1969). 8. Onuma *et al.* (1968). 9. Ivanov and Lizunov (1960). 10. Ahrens (1948). 11. Itzikson and Rusanov (1946). 12. Brewer and Baker (1936). 13. Smales *et al.* (1957). 14. Kulikova (1966).

Indium concentrates in sulfide minerals, especially those having tetrahedral coordination about the principal metal ion. Note the especially high concentrations in cassiterites, sphalerites, stannites, and chalcopyrites.

Table 49-D-2. *Indium in ore-forming minerals*

Material	No. of samples	In (ppm or ppb) range	mean	Method	Reference
Wolframites, Deputaskoe (U.S.S.R.)	2	3—50 ppm	26 ppm	(C)	1
Wolframites, Northern Yakutiya (U.S.S.R.)	17	10—100 ppm	—		1
Sphalerites, (U.S.S.R.)	7	200 ppb to 100 ppm	—	(P)	2
Sphalerites, Eastern Transbaikaliya (U.S.S.R.)	—	10—5,000 ppm	—		2
Sphalerites, Eastern Transbaikaliya (U.S.S.R.)	48	30—3,300 ppm	—	(P)	3
Sphalerites, Eastern Transbaikaliya (U.S.S.R.)	9	—	20 ppm	(P)	3
Sphalerites, Karamazar and Chatkal (U.S.S.R.)	51	70—330 ppm	—		4
Sphalerites, Deputaskoe (U.S.S.R.)	12	20—3,100 ppm	800 ppm	(C)	1
Sphalerites, Northern Yakutiya (U.S.S.R.)	85	10—4,700 ppm	—		1
Sphalerites	24	X—800 ppm	—		5
Sphalerites	—	70—120 ppm	—		6
Sphalerites	75	10—700 ppm	—		7
Chalcopyrites, Eastern Transbaikaliya (U.S.S.R.)	11	5—100 ppm	—	(P)	3
Chalcopyrites, Karamazar and Chatkal (U.S.S.R.)	10	30—315 ppm	—		4
Chalcopyrites, Deputaskoe (U.S.S.R.)	5	570—1,500 ppm	—	(C)	1
Chalcopyrites, Northern Yakutiya (U.S.S.R.)	58	50—750 ppm	—		8
Chalcoyprite	1	—	4.7 ppm	(S)	9
Stannites, Deputaskoe (U.S.S.R.)	2	800—900 ppm	850 ppm	(C)	1
Stannites, Northern Yakutiya (U.S.S.R.)	27	50—1,500 ppm	—		1
Stannites	2	—	$>$ 100 ppm		5
Galenas, Kansay (U.S.S.R.)	10	0.3—70 ppm	—	(P)	2
Galenas, Eastern Transbaikaliya (U.S.S.R.)	—	10—50 ppm	—		10
Galenas, Eastern Transbaikaliya (U.S.S.R.)	11	1—34 ppm	—	(P)	3
Galenas, Karamazar and Chatkal (U.S.S.R.)	25	—	5 ppm		4
Galenas, Deputaskoe (U.S.S.R.)	6	5—10 ppm	5.3 ppm	(C)	1
Galena	1	—	910 ppb	(S)	9

X = less than limit of detection.

References: 1. Ivanov and Rozbianskaya (1961). 2. Kulikova (1966). 3. Meituv (1962). 4. Badalov (1961). 5. Anderson (1953). 6. Cambi and Maletesta (1936). 7. Oftedal (1940). 8. Ivanov and Lizunov (1959). 9. Shaw (1952). 10. Kulikova (1962).

II. Indium Minerals

Although rare in occurrence, three indium minerals have been reported (Fleischer, 1966) and are listed in Table 49-D-3.

Table 49-D-3. *Indium minerals*

Mineral	Formula	Reference
Dzhalindite (or Jalindite)	$In(OH)_3$	Genkin and Murav'eva (1963)
Indite	$FeIn_2S_4$	Genkin and Murav'eva (1963)
Roquesite	$CuInS_2$	Picot and Pierrot (1963)

Genkin and Murav'eva (1963) described the more recently discovered indium minerals, which were found in cassiterite from tin ore deposits in the far eastern USSR. Roquesite was reported by Picot and Pierrot (1963), who determined the mineral composition from electron probe measurements.

Revised manuscript received: October 1970

49-E. Abundance in Common Igneous Rocks

The indium contents of the principal igneous rock types are given in Tables 49-E-1 through 49-E-4.

Table 49-E-1. *Indium in ultramafic rocks*

Material	No. of samples	In (ppb)		Meth-od	Re-fer-ence
		range	mean		
Dunites, average	—	—	13	(—)	1
Dunites, Urals, Tagil'sk massif (U.S.S.R.)	1	—	20	(C)	2
Dunites, Canwell glacier, Alaska (U.S.A.)	1	—	6	(N/R)	3
Peridotites (Alaska; Venezuela; Puerto Rico)	3	5—33	15	(N/R)	3
Peridotites, Miyamori, Iwate, and Toba, Mie (Japan)	2	23—26	25	(N/R)	4
Peridotites, average	—	—	13	(—)	1
Peridotites, Kola Peninsula, Monchegora (U.S.S.R.)	(composite)	—	40	(C)	2
Pyroxenites, average	—	—	13	(—)	1
Pyroxenites, Urals, Tagil'sk massif (U.S.S.R.)	1	—	50	(C)	2
Pyroxenites, Kola Peninsula, Monchegora (U.S.S.R.)	(composite)	—	60	(C)	2
Garnet-pyroxenites, Kakanui (New Zealand)	1	—	14 ± 2	(N/R)	3

References: 1. Vinogradov (1962). 2. Ivanov (1963). 3. Schmitt and Smith (1968). 4. Hamaguchi *et al.* (1967).

Table 49-E-2. *Indium in mafic rocks*

Material	No. of samples	In (ppb) range	In (ppb) mean	Method	Reference
Basalt, Rhön (Germany)	1	—	180	(S)	1
Basalts, Thüringer Wald (East Germany)	6	40—210	93	(S)	1
Basalts, Aleutian Islands and Greenland	3	130—320	—	(S)	2
Basalts (Japan)	(composite of 14)	—	83 ± 4	(N/R)	3
Basalts	(composite of > 200)	—	67 ± 6	(N/R)	4
Basalts (East Pacific Rise; Guadalupe Island)	4	65—94	76	(N/R)	4
Basalts	—	—	60 ± 3	(N/R)	5
Spilites, Elbe Valley (East Germany)	3	15—50	27	(S)	1
Quartz-spilites, Elbe Valley (East Germany)	6	100—170	127	(S)	1
Olivine basalt (1921), Kilauea (Hawaii)	1	—	84 ± 4	(N/R)	3
Olivine-basalts (dolerites), Thüringer Wald (East Germany)	3	35—56	47	(S)	1
Olivine-basalt, Rhön (Germany)	1	—	80	(S)	1
Nepheline-basalts, Lausitz and Erzgebirge (East Germany)	4	25—40	34	(S)	1
Leucite-basalts, Lausitz and Erzgebirge (East Germany)	2	40—60	50	(S)	1
Gabbros (Finland, Canada, and U.S.A.)	6	—	15	(S)	2
Basalts and gabbros, average	—	—	220	(—)	6
Gabbros, Urals (U.S.S.R.)	(composite of 3)	40—90	60	(C)	7
Gabbros, Skaergaard intrusion (East Greenland)	18	50—62	—	(N/R)	8
Lamprophyres (East Germany)	4	40—80	59	(S)	1
Lamprophyres (Canada and Finland)	2	17—71	44	(S)	2
Eclogites	6	—	51	(S)	1

References: 1. Voland (1969). 2. Shaw (1952). 3. Hamaguchi *et al.* (1967). 4. Schmitt and Smith (1968). 5. Onuma *et al.* (1968). 6. Vinogradov (1962). 7. Ivanov (1963). 8. Wager *et al.* (1958).

Table 49-E-3. *Indium in intermediate rocks*

Material	No. of samples	In (ppb) range	In (ppb) mean	Method	Reference
Diorites, Yakutiya (U.S.S.R.)	—	—	50	(C)	Ivanov (1963)
Quartz-diorites, Yakutiya (U.S.S.R.)	1	—	<50	(C)	Ivanov (1963)
Hornblende-diorites, Yakutiya (U.S.S.R.)	(composite)	—	30	(C)	Ivanov (1963)
Hornblendite, Kaalamo (Finland)	—	—	53	(S)	Shaw (1952)
Andesites (Japan)	(composite)	—	50 ± 4	(N/R)	Hamaguchi *et al.* (1967)
Diorites, Yakutiya (U.S.S.R.)	—	—	96	(C)	Ivanov (1963)
Monzonites, Yakutiya (U.S.S.R.)	1	—	130	(C)	Ivanov (1963)
Syenites, Yakutiya (U.S.S.R.)	4	50—380	139	(C)	Ivanov (1963)
Nepheline-syenites (U.S.S.R.)	3	30—36	33	(S)	Voland (1969)
Phonolites, Lausitz, Erzgebirge (East Germany)	2	40—45	43	(S)	Voland (1969)

Table 49-E-4. *Indium in granitic rocks*

Material	No. of samples	In (ppb) range	In (ppb) mean	Method	Reference
Granodiorites, Mongolia	4	45—80	60	(S)	1
Granodiorites, Lausitz (East Germany)	3	35—85	53	(S)	1
Granodiorites, Urals, Kasakhstan, Tuva and Yakutiya (U.S.S.R.)	8	<20—140	94	(C)	2
Granodiorites, Kitashidara, Aichi (Japan)	1	—	27 ± 1	(N/R)	3
"Granites" ($<60\%$ SiO_2)	(composite of 85)	—	83 ± 8	(N/R)	4
Granites (60—70% SiO_2)	(composite of 191)	—	67 ± 6	(N/R)	4
Granites ($>70\%$ SiO_2)	(composite of 213)	—	57 ± 6	(N/R)	4
Granites (Japan)	(composite of 20)	—	42 ± 3	(N/R)	3
Granites, Shimoina, Nagana (Japan)	—	—	47 ± 3	(N/R)	3
Granites, Yakutiya (U.S.S.R.)	4	20—87	47	(C)	2

Table 49-E-4 (continued)

Material	No. of samples	In (ppb) range	In (ppb) mean	Meth-od	Re-fer-ence
Granites (East Germany)	11	13—22	18	(S)	1
Granites, Far East (U.S.S.R.)	(composite of 18)	—	290	(C)	2
Granites, Eastern Transbaikaliya (U.S.S.R.)	(composite of 8)	—	140	(C)	2
Granites Eastern Kasakhstan (U.S.S.R.)	(composite of 9)	—	530	(C)	2
Granites, Ukraine (U.S.S.R.)	(composite of 8)	—	200	(C)	2
Granites, Gornyi Altai (U.S.S.R.)	(composite of 18)	—	160	(C)	2
Granites, Western Tuva (U.S.S.R.)	(composite of 71)	—	20	(C)	2
Granites, Northern Caucasus	(composite of 18)	—	130	(C)	2
Granites, Urals	(composite of 96)	—	20	(C)	2
Granites, Lausitz (East Germany)	3	18—45	28	(S)	1
Granites, Thuringia (East Germany)	8	5—50	23	(S)	1
Granites, Niederbobritz (East Germany)	4	11—60	30	(S)	1
Granites, Saxony (East Germany)	3	17—60	34	(S)	1
Granites, Western Erzgebirge (East Germany)	20	14—250	42	(S)	1
Granites	10	< 20—2,000	260	(S)	5
"Tin-Granites", Erzgebirge (East Germany)	19	13—340	170	(S)	1
Granite-gneiss, Yakutiya (U.S.S.R.)	1	—	210	(C)	2
Alaskites, Yakutiya (U.S.S.R.)	3	24—50	36	(C)	2
Aplites (granitic), Yakutiya (U.S.S.R.)	3	X—33	—	(C)	2
Aplitic granites	27	—	18	(S)	1
Rhyolites (quartz-porphyries), Erzgebirge (East Germany)	35	11—85	43	(S)	1
Rhyolites (quartz-porphyries), Northwest Saxony (East Germany)	2	58—70	64	(S)	1
Rhyolites (pyroxene-poor quartz porphyries), Northwest Saxony (East Germany)	3	95—230	230	(S)	1
Pyroxene-rich quartz porphyry, Northwest Saxony (East Germany)	1	—	640	(S)	1

X = less than limit of detection.

References: 1. VOLAND (1969). 2. IVANOV (1963). 3. HAMAGUCHI *et al.* (1967). 4. SCHMITT and SMITH (1968). 5. SHAW (1952).

Ivanov (1963) and Voland (1969) noted that the indium content in granites decreased with decrease in total iron content (Figs. 49-E-1 and 49-E-2), and thus the Fe/In ratio remained relatively constant in spite of considerable varaition in indium content.

Some additional data on indium concentrations in igneous rocks of special interest are given in Tables 49-E-5 and 49-E-6.

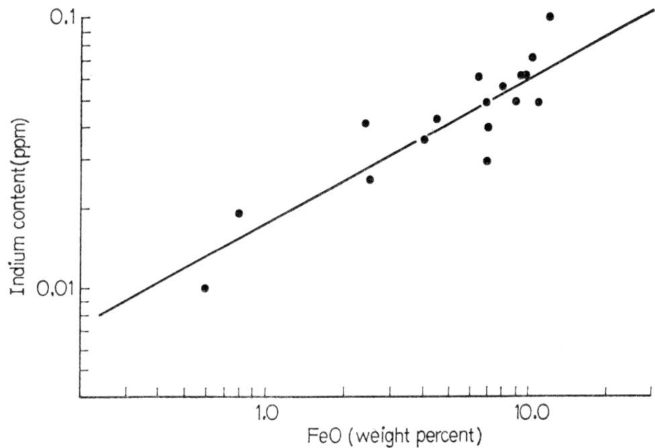

Fig. 49-E-1. Correlation between indium and FeO content of ultramafic and mafic rocks (Voland, 1969)

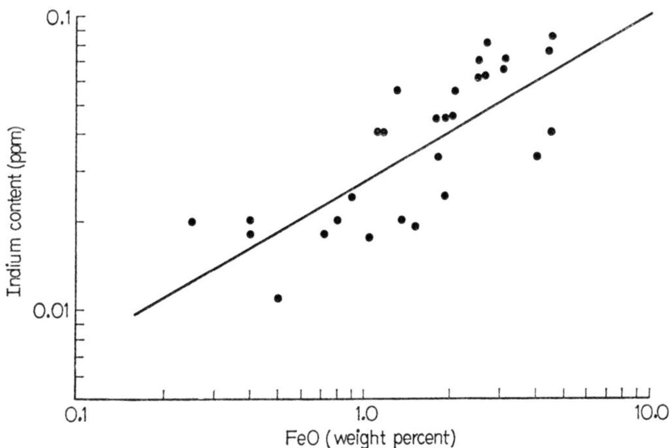

Fig. 49-E-2. Correlation between indium and FeO content of granitic rocks of norma evolution (Voland, 1969)

Table 49-E-5. *Indium in rocks of the Skaergaard intrusion, East Greenland*

Material	Collection No.	No. of samples	In (ppb) range	In (ppb) mean	Method	Reference
Chilled marginal gabbro, east margin	EG-1825	7	59—62	61 ± 2	(N/R)	1
Chilled marginal gabbro	EG-1825	9	52—62	61	(N/R)	2
Chilled marginal gabbro, south margin	EG-4507	11	50—58	54 ± 3	(N/R)	1
Chilled marginal gabbro	EG-4507	8	50—58	54	(N/R)	2
Fayalite ferrogabbro	EG-4327	7	150—174	160 ± 10	(N/R)	1
Fayalite ferrogabbro	EG-4327	7	151—170	170	(N/R)	2
Fayalite ferrogabbro	EG-4328	8	164—194	180 ± 20	(N/R)	1
Fayalite ferrogabbro	EG-4328	12	170—194	180	(N/R)	2
Acid granophyre	EG-3058	2	89—93	91 ± 2	(N/R)	1
Acid granophyre	EG-3058	2	89—93	91	(N/R)	2
Hypersthene olivine gabbro	EG-5086	2	59—62	60 ± 2	(N/R)	1
Hypersthene olivine gabbro	EG-5086	2	59—62	60	(N/R)	2
Hortonolite ferrogabbro	EG-5181	3	77—83	79	(N/R)	2

References: 1. Smales *et al.* (1957). 2. Wager *et al.* (1958).

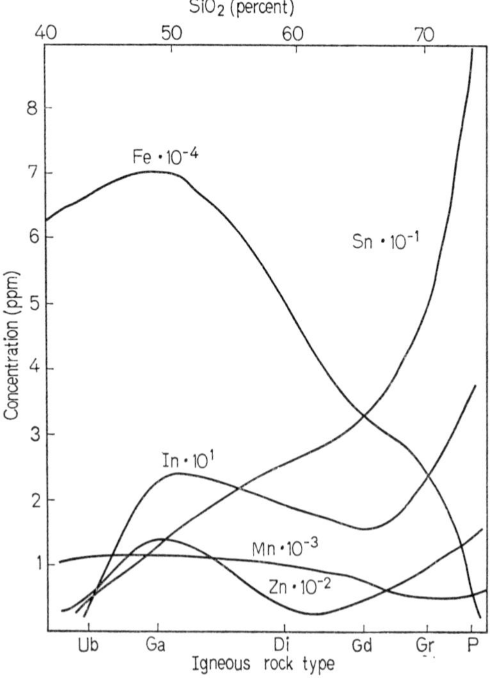

Fig. 49-E-3. Variation diagram of In, Fe, Sn, Mn and Zn (Smith, 1963). *Ub* Ultramafic; *Ga* Gabbro; *Di* Diorite; *Gd* Granodiorite; *Gr* Granite; *P* Pegmatite

Table 49-E-6. *Indium in U.S.G.S. silicate rock standards*

Material	No. of samples	In (ppb) range	mean	Method	Reference
DTS-1 dunite	2	2.6—2.8	2.7	(N/R)	1
PCC-1 peridotite	2	4.1—4.5	4.3	(N/R)	1
BCR-1 basalt	2	107—110	109	(N/R)	1
W-1 basalt ("diabase")	2	67.4—67.5	67	(N/R)	1
W-1 basalt ("diabase")	—	55—59	57	(N/R)	2
W-1 basalt ("diabase")	—	—	42	(S)	3
W-1 basalt ("diabase")	7	61—70	64 ± 4	(N/R)	4
W-1 basalt ("diabase")	—	—	50	(S)	5
W-1 basalt ("diabase")	—	—	55	(N/R)	6
W-1 basalt ("diabase")	—	—	<200	(mass. spec.)	7
W-1 basalt ("diabase")	—	—	25	(C)	8
AGV-1 andesite	2	42.6—44.5	44	(N/R)	1
STM-1 nepheline-syenite	2	85.2—87.9	87	(N/R)	1
GSP-1 granodiorite	2	49.7—57.1	53	(N/R)	1
G-1 granite	2	23.5—24.5	24	(N/R)	1
G-1 granite	—	25—27	26	(N/R)	2
G-1 granite	—	—	42	(S)	3
G-1 granite	8	24—29	25 ± 2	(N/R)	4
G-1 granite	—	—	25	(N/R)	6
G-1 granite	—	—	<200	(mass spec.)	7
G-1 granite	—	—	20	(C)	8
G-1 granite	—	—	22	(S)	5
G-2 granite	2	27.0—29.1	28	(N/R)	1

References: 1. Johansen and Steinnes (1966). 2. Hamaguchi *et al.* (1967). 3. Brooks and Ahrens (1961). 4. Smales *et al.* (1957). 5. Voland (1969). 6. Pierce and Peck (1961). 7. Brown and Wolstenholme (1964). 8. Ivanov and Cholodov (1966).

Indium remains in magma until the latest stages of crystallization. Some concentration of indium at earlier stages (in basalts and gabbros) has been observed, and the trend in the abundance data is shown by the variation diagram for indium (Fig. 49-E-3). Distribution of indium abundances in common magmatic rocks is presented in frequency diagrams (Fig. 49-E-4 and Fig. 49-E-5).

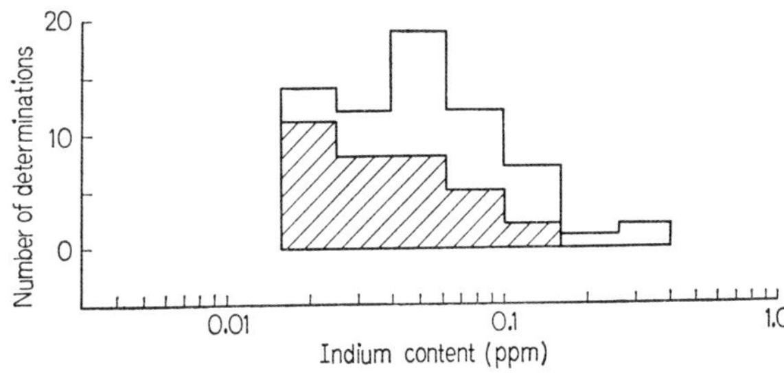

Fig. 49-E-4. Frequencies of distribution of indium in mafic and granitic volcanic rocks; 62 values (Ivanov, 1963). ▭ Mafic; ▨ Granitic

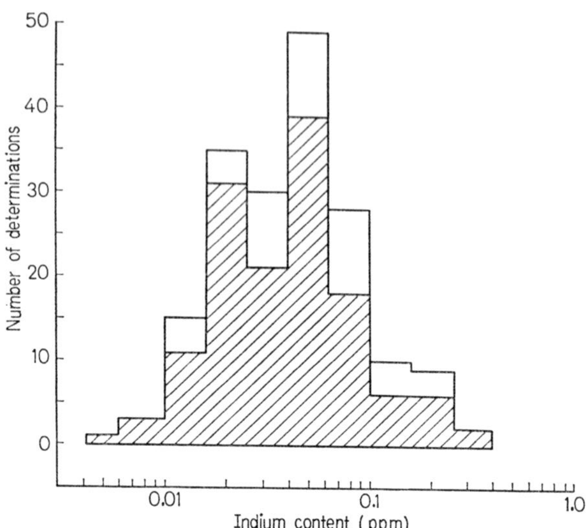

Fig. 49-E-5. Frequencies of distribution of indium in mafic and granitic volcanic rocks; 179 values (VOLAND, 1969). ▭ Mafic; ▨ Granitic

Revised manuscript received: October 1970

49-F. Behavior in Magmatogenic Processes

Among the principal ore-forming elements in the early stages of ore deposition, indium closely resembles iron, manganese and tin (IVANOV and LIZUNOV, 1959). Indium is concentrated in silicates with high iron content, and in later stages of sulfide formation, indium concentrates in the sulfides of copper and zinc and in iron-containing sulfides and carbonates (IVANOV and ROZBIANSKAYA, 1961).

Indium has strongly chalcophile properties and is concentrated by sulfide minerals, especially in products of hydrothermal and pneumatolytic processes (RANKAMA and SAHAMA, 1950[1]). Indium is also enriched in pegmatites containing tin, especially cassiterite (SHAW, 1952). Considerable amounts of indium are found in sphalerites which form at intermediate temperatures and, in general, sphalerites are the most important ore-minerals of indium (OFTEDAHL, 1940).

The variation diagram for indium in the Skaergaard intrusion follows that of manganese (WAGER *et al.*, 1957). Variation diagrams for Fe, Mn, Sn, and Zn are compared with that for In in Fig. 49-E-3 (after SMITH, 1963).

[1] Editor's note: The accumulation of In by pneumatolytic processes is confirmed by a concentration of 1450 ppb In in 7 greisen samples from the Erzgebirge area (Saxony, East Germany) (VOLAND, 1969). Indium is generally high in rocks of tin provinces (VOLAND, 1969).

Revised manuscript received: October 1970

49-G. Behavior during Weathering and Alteration of Rocks

During ordinary weathering, indium is oxidized to In^{3+} (SHAW, 1952) which disperses following the Fe^{3+} and Mn^{4+} ions, and usually precipitates under conditions which form the hydrous oxides (KULIKOVA, 1962). Relatively high indium contents were observed in limonite, and electrodialysis experiments showed that indium was present both as free hydroxide and as hydroxide bonded in the structure of hydrous iron oxide (KULIKOVA, 1964).

On oxidation, the primary sulfide ores are depleted of indium by dissolution in slightly acidic or alkaline environments. Such depletion is accelerated where complex ions can form as in the presence of sulfate, chloride or organic acid anions, e.g., humates, carbonates, oxalates, etc. (KULIKOVA, 1966).

Deposition of indium usually occurs in the oxidates following iron and manganese, in some hydrolysate sediments (SHAW, 1957), and in bauxites with aluminium (GOLDSCHMIDT, 1958). Indium found in some marine sediments may have accumulated due to the insolubility of borates of trivalent indium (KULIKOVA, 1962).

Revised manuscript received: October 1970

49-H. Solubilities of Compounds which Control Indium Concentrations in Natural Waters; Adsorption Processes; Valence States in Natural Environments

In neutral aqueous solutions indium forms a number of insoluble compounds. Nearly all are soluble in acidic solutions or in concentrated solutions of the associated anion. Table 49-H-1 lists four compounds of some interest in natural environments.

Table 49-H-1. *Insoluble compounds of indium*

Compound	Solubility product (K_{sp})
$In(OH)_3$	1.3×10^{-34}
In_2S_3	(10^{-40})[a]
$In_2(CO_3)_3$	(10^{-15})[a]
$InPO_4$	(10^{-25})[a]

[a] Estimated values.

Indium forms a number of complex ions in aqueous solutions, especially with halide ions. Low concentrations of hydroxide ion permit indium complexation due to incomplete hydrolysis. Several organic anions (e.g., oxalate ion) are capable of complexing indium (SUNDERMAN and TOWNLEY, 1960).

The electrochemical half-cell (reduction) potentials of indium species in aqueous solution are shown in Table 49-H-2 (after LATIMER, 1952).

Table 49-H-2. *Standard potentials for indium*

Half-cell reactions	ε° (vs. N.H.E.)
$In^{3+} + 3e^- = In$	-0.342
$In(OH)_3(s) + 3e^- = 3OH^- + In^{3+}$	-1.0

The compounds of trivalent indium are the most stable, and only the trivalent ion is stable in aqueous solutions. Compounds of lower valent indium undergo oxidation-reduction in the presence of water to form In^{3+} and elemental indium (BUSEV, 1962).

Revised manuscript received: October 1970

49-I. Abundance in Natural Waters

There are very few data on indium concentrations in natural waters. GOLDBERG (1961) assumed that sea water contains less than 20 ppb indium. Earlier information indicated that no indium was detectable in sea water (GOLDSCHMIDT, 1958), and SHAW (1952) reported no detectable amounts of indium in ocean salts.

CHOW and SNYDER (1969) recently reported measurements of the indium content of Pacific Ocean water at various depths below the surface, and at depths in core material from Mohole experiments. The indium profile to a depth of 3571 meters indicated a uniform indium concentration of 0.004 ppb (micrograms per liter of sea water) by isotope dilution mass spectrometry. MATTHEWS and RILEY (1970) report a range of about 0.0003 to 0.0001 ppb In of decreasing indium concentration with depth from an Atlantic (sea water) sampling station (neutron activation analysis). Core samples at depths from 33.3 to 149.2 meters below the sea floor contained from 40 to 20 ppb indium. The concentration of indium in sea water appears to be controlled by pH through the solubility product of indium hydroxide and the hydroxo-indium complex $[In(OH)_4^-]$.

Revised manuscript received: October 1970

49-K. Abundance in Common Sediments and Sedimentary Rocks

Table 49-K-1. *Indium in sediments and sedimentary rocks*

Material	No. of samples	In (ppb) range	In (ppb) mean	Method	Reference
Sediments					
Pelagic clays	6	—	73	(S)	SHAW (1957)
Pelagic clays Atlantic (4) Pacific (2)	6	< 20—280	74	(S)	SHAW (1952)
Pelagic clay, Atlantic	1	—	74	(N)	MATTHEWS and RILEY (1970)
Kaolines	12	—	69	(S)	VOLAND (1969)
Kaolines	2	< 20—120	70	(S)	SHAW (1952)
Siliceous ooze, Pacific	1	—	< 20	(S)	SHAW (1952)
Diatom ooze, Atlantic	1	—	120	(N)	MATTHEWS and RILEY (1970)
Globigerina ooze, Pacific	1	—	< 20	(S)	SHAW (1952)
Globigerina ooze, Atlantic	1	—	23	(N)	MATTHEWS and RILEY (1970)
Sedimentary rocks					
Shales, argillites, etc. (U.S.A., Finland, U.S.S.R.)	7	< 20—180	57	(S)	SHAW (1952)
Shales	16	X—230	54	(S)	SHAW (1957)
Shales, Paleozoic (Europe)	(composite of 36)	—	48 ± 5	(N/R)	SCHMITT and SMITH (1968)
Shales (North America)	(composite of 40)	—	70 ± 7	(N/R)	SCHMITT and SMITH (1968)
Shale, Thuringia (East Germany)	1	—	70	(S)	VOLAND (1969)
Slates (U.S.A.)	2	—	37	(S)	SHAW (1952)
Shales (East Germany)	16	—	62	(S)	VOLAND (1969)
Cherts (East Germany)	3	—	10	(S)	VOLAND (1969)
Greywackes (U.S.A. and Canada)	3	33—230	104	(S)	SHAW (1952)
Carbonaceous sedimentary rocks	3	—	72	(S)	SHAW (1957)
Limestone, Harz (Germany)	1	—	9	(S)	VOLAND (1969)
Bauxite	—	—	170	(S)	SHAW (1952)

X = less than limit of detection.

VINOGRADOV (1962) reported an average of 50 ppb indium in sedimentary rocks (clays and shales) which is comparable with the data listed in Table 49-K-1 for sediments.

Revised manuscript received: October 1970

49-L. Biogeochemistry

The data on indium in biological materials and organisms are extremely sparse in the present literature. Certain coal ashes were reported to contain upto 2 ppm indium (GOLDSCHMIDT, 1937). KOCH and ROESMER (1962) reported 16 ppb indium (N/R) in dried mammalian muscle. Indium in mammalian blood was studied by WOLSTENHOLME (1964) who found less than 70 ppb indium (spark source mass spectroscopy) in dried blood plasma.

Revised manuscript received: October 1970

49-M. Abundance in Common Metamorphic Rocks

The literature is lacking much data on indium in metamorphic rocks. Those that are available are summarized in Table 49-M-1.

Table 49-M-1. *Indium in metamorphic rocks*

Material	No. of samples	In (ppb) range	In (ppb) mean	Method	Reference
Phyllites and schists, New York State (U.S.A.)	11	$<$ 20—460	110	(S)	Shaw (1952)
Gneisses, Wyoming (U.S.A.), Greenland	11	$<$ 20—1,900	495	(S)	Shaw (1952)
Ortho-gneisses, Erzgebirge (East Germany)	9	—	56	(S)	Voland (1969)
Para-gneisses, Erzgebirge (East Germany)	22	—	60	(S)	Voland (1969)
Gneisses, Erzgebirge (East Germany)	6	—	110	(S)	Voland (1969)
Mica-schists, etc., Erzgebirge (East Germany)	29	—	104	(S)	Voland (1969)
Amphibolites (East Germany)	14	—	110	(S)	Voland (1969)
Skarns	8	—	3,800	(S)	Voland (1969)
Granulites, Saxony (East Germany)	3	—	23	(S)	Voland (1969)
Eclogites (East Germany)	6	—	51	(S)	Voland (1969)

Because several sets of data in Table 49-M-1 are on gneisses and mica-schists from a tin province (Erzgebirge), values higher than 100 ppb In are probably abnormal.

Revised manuscript received: October 1970

49-O. Concluding Notes

Table 49-O-1 summarizes the ranges and average abundances of indium in different geochemical materials. Since many of the data in this table are only approximate, these values should be treated with caution in correlative studies.

Table 49-O-1. *Abundances of indium in various cosmic and terrestrial matter*

Type of Matter	In (ppb)	
	range	average
Iron meteorites	0.3—41	10
Carbonaceous chondrites	46—112	70
Enstatite chondrites	0.2—150	70
Ordinary chondrites	0.05—55	0.5
Igneous rocks		
Ultramafic rocks	5—60	20
Basalts and gabbros	15—320	70
Granites	10—2,000	50
Shales	30—230	60
Pelagic clays	$<$ 20—280	70

SHAW (1957) has reviewed the geochemistry of Ga, In and Tl, all members of Group III of the periodic table of the chemical elements. All three elements are geochemically dispersed in that they form separate minerals only in very rare cases and usually occur in combination with minerals of other elements. Their distributions in nature are obviously determined by their geochemical affinities, which are dependent on some chemical properties such as ionic radii, electronegativity, electronic orbitals for complexing, etc. Elements, with similar behavior and affinity, which govern the uptake of In are principally Fe, Zn, Sn, Pb and Cu. The chalcophilic property of In greatly exceeds the siderophilic and lithophilic tendencies.

Although In has been observed in iron meteorites by SMALES *et al.* (1967) using the sensitivite N/R technique (see Table 49-C-3), the possibility of In incorporation in small sulfide inclusions cannot be ruled out. Since siderophilic properties have been observed for both Ga and Ge in iron meteorites (WASSON, 1967) and also for Sn in iron meteorites (GOLDSCHMIDT, 1958), it may be reasonably postulated, from the general periodic behavior of chemical elements, that In also has siderophilic properties. Indium appears to behave similarly to As, Sb and Ag in iron meteorites (SMALES *et al.*, 1967).

A close similarity for the ionic radii of In^{3+} (0.81 Å) and Fe^{2+} (0.74 Å) may be partially responsible for In^{3+}—Fe^{2+} diadochism in igneous rocks and minerals

(SHAW, 1952, 1957). BREWER and BAKER (1936) suggested that In tends to associate with Ag, Cd, Sn and Sb; the first two elements, Ag and Cd, immediately precede In in the Periodic Table and they terminate the second transition series in sequence and progressively add successive electrons to the $5s\,5p$ subshell; thus some similar chemical properties might be expected for these elements.

ANDERSON (1953) has discussed the occurrence of In in sulfides and sulphosalts. From a study of sulfide structural characteristics he concluded that: a) sulfides, the best collectors of In, show a tetrahedral coordination; b) the In—S tetrahedral bond length is greater than the bond lengths of Zn—S, Cu—S, and Fe—S; moreover, the In—S bond length approaches most closely the Sn—S length in stannite, a very efficient collector of In; and c) indium rejection occurs largely in minerals with small metal-sulfur bond lengths.

Indium achieved economic importance in 1924, some 60 years after its discovery. The chief commercial source of indium is the processing of the zinc mineral sphalerite and other zinc-lead ores. Producers are found in the United States, Canada, Japan, and several European countries. The United States, the chief consumer, uses about 20 tons per year and has estimated reserves of 320 tons. The principal uses of indium are in the manufacture of electronic semiconductor devices, sleeve bearings, special alloys for dental casts, fusible safety plugs and links, and solders (SCHROEDER, 1965). Recent research has been directed toward specialized uses of indium in the field of electronics.

Revised manuscript received: October 1970

References: 49-B to 49-M, 49-O

AHRENS, L. H.: Evidence of geological age against decay of Sn-115 to In-115 by electron capture. Nature **162**, 413 (1948).

— TAYLOR, S. R.: Spectrochemical Analysis, 2nd ed. Reading, Massachusetts: Addison-Wesley Publishing Co., Inc. 1961.

AKAIWA, H.: Abundances of selenium, tellurium, and indium in meteorites. J. Geophys. Res. **71**, 1919 (1966).

ALLER, L. H.: The Abundance of the Elements. New York: Interscience Publishers, Inc. 1961.

— The abundance of elements in the solar atmosphere. Advances in Astronomy and Astrophysics. New York: Academic Press 1965.

— The chemical composition of the sun and the solar system. Proc. Australian Astron. Soc. **1**, 133 (1968).

ANDERSON, J. S.: Observations on the geochemistry of indium. Geochim. Cosmochim. Acta **4**, 225 (1953).

BADALOV, S. T.: The geochemical relationship between indium and silver in zinc-silver-lead deposits. Geochemistry (English Transl.). **10**, 1005 (1961).

BREWER, F. M., BAKER, E.: Arc spectrographic determination of indium in minerals and the association of indium with tin and silver. J. Chem. Soc. 1286 (1936).

BROOKS, R. R., AHRENS, L. H.: The determination of indium and thallium in G-1, W-1 and other silicate rocks by a new technique. Geochim. Cosmochim. Acta **23**, 145 (1961).

BROWN, R., WOLSTENHOLME. W. A.: Analysis of geological samples by spark source mass spectrometry. Nature **201**, 598 (1964).

BUSEV, A. I.: The Analytical Chemistry of Indium. Oxford: Pergamon Press 1962.

CAMBI, L., MALESTESTA, L.: Germanium, gallium, and indium in Sardinian sphalerite. Rend. Inst. Lombardo Sci. **69**, 369 (1936).

CAMERON, A. G. W.: A new table of abundances of the elements in the solar system. Proc. I.A.G.C. Symposium on the Origin and Distribution of the Elements, Paris (L. H. AHREN., ed.) Oxford: Pergamon Press 1968.

CHOW, T. J., SNYDER, C. B.: Indium content of sea water. Earth and Planet. Sci. Lett. **7**, 221 (1970).

DUBOV, E. E., KHROMOVA, T. P.: Determination of the content of certain elements in the sun from sunspot spectra. III. The indium abundance. Izv. Krym. Astrofiz. Observ. **35**, 186 (1966).

FLEISCHER, M.: Index of new mineral names, discredited minerals, and changes of mineralogical nomenclature in volumes 1—50 of the American Mineralogist. Am. Mineralogist **51**, 1248 (1966).

FOUCHÉ, K. F., SMALES, A. A.: The distribution of trace elements in chondritic meteorites. I. Gallium, germanium and indium. Chem. Geol. **2**, 5 (1967).

GENKIN, A. D., MURAV'EVA, I. V.: Indite and jalindite, new indium minerals. Zap. Vses. Mineralog. Obshchestva **92**, 445 (1963).

GOLDBERG, E. D.: Marine geochemistry. Ann. Rev. Phys. Chem. **12**, 24 (1961).

GOLDBERG, L., MÜLLER, E. A., ALLER, L. H.: Abundances of the elements in the solar atmosphere. Astrophys. J., Suppl. Ser. **5**, 1 (1960).

GOLDSCHMIDT, V. M.: The principles of distribution of chemical elements in minerals and rocks. J. Chem. Soc. 655 (1937).

— Geochemistry. Oxford: Oxford University Press 1958.

GREVESSE, N., BLANQUET, G., BOURY, A.: Proc. I.A.G.C. Symposium on the Origin and Distribution of the Elements, Paris (L. H. AHRENS, ed.). Oxford: Pergamon Press 1968.

GREEN, J.: Geochemical table of the elements for 1959. Bull. Geol. Soc. Am. **70**, 1127 (1959).
HAMAGUCHI, G., TOMURA, K., ONUMA, N., HIGUCHI, H., SUDA, K.: Determination of indium in rocks by neutron activation analysis. Bunseki Kagaku **16**, 1233 (1967).
IRVING, H., VAN R. SMIT, J., SALMON, L.: Determination of indium in cylindrite by neutron activation analysis and other methods. Analyst **82**, 549 (1957).
ITZIKSON, M. I., RUSANOV, A. K.: Indium in the tin-ore deposits of the Far East. Dokl. Akad. Nauk SSSR **53**, 631 (1946).
IVANOV, V. V.: Indium in some igneous rocks of the USSR. Geochemistry (English Transl.) **12**, 115 (1963).
— CHOLODOV, V. N.: Indij. Sammelwerk: Metally v. osadočnych t olščach Nauka, Moskva 187 (1966).
— LIZUNOV, N. V.: Indium in some tin deposits of Yakutia. Geochemistry (English Transl.) **4**, 416 (1959).
— — On certain pecularities of the distribution of indium in endogenetic ore deposits. Geochemistry (English Transl.) **1**, 53 (1960).
— ROZBIANSKAYA, A. A.: Geochemistry of indium in cassiterite-silicate-sulfide ores. Geochemistry (English Transl.) **1**, 71 (1961).
JOHANSEN, O., STEINNES, E.: Determination of indium in standard rocks by neutron activation analysis. Talanta **13**, 1177 (1966).
KEAYS, R. R., GANAPATHY, R., ANDERS, E.: Chemical fractionations in meteorites. IV. Abundances of fourteen trace elements in L-chondrites; implications for cosmothermometry. Geochim. Cosmochim. Acta. **35**, 337 (1971).
KOCH, R. C., ROESMER, J.: Application of activation analysis to the determination of trace-element concentrations in meat. J. Food Sci. **27**, 309 (1962).
KULIKOVA, M. F.: Behavior of indium in the oxidized zones of some polymetallic deposits of Eastern Transbaikaliya. Geochemistry (English Transl.) **7**, 707 (1962).
— Distribution of trace elements in the oxidized zone of some lead-zinc deposits of Middle Asia. Geochemistry (English Transl.) **5**, 981 (1964).
— Geochemistry of gallium and indium in the oxidized zone of lead-zinc deposits in Soviet Central Asia. Geochemistry (English Transl.) **5**, 982 (1966).
LARIMER, J. W., ANDERS, E.: Chemical fractionations in meteorites. II. Abundance patterns and their interpretation. Geochim. Cosmochim. Acta **31**, 1239 (1967).
LATIMER, W. M.: Oxidation States of the Elements and their Potentials in Aqueous Solutions, 2nd ed. New Jersey: Prentice-Hall, Inc. 1952.
LEDERER, C. M., HOLLANDER, J. M., PERLMAN, I.: Table of Isotopes, 6th ed. New York: John Wiley & Sons, Inc. 1967.
MATTHEWS, A. D., RILEY, J. P.: Occurence of indium in seawater and some marine sediments. Nature **225**, 1242 (1970).
MEITUV, G. M.: Geochemistry of rare elements in the lead-zinc deposit of the Klichkinskii region (Eastern Transbaikaliya). Geochemistry (English Transl.) **7**, 694 (1962).
OFTEDAL, I.: Untersuchungen über die Nebenbestandteile von Erzmineralen norwegischer zinkblendeführender Vorkommen. Skr. Norske Vid.-Akad. Oslo 1. Mat.-Naturv. Kl. No. 8 (1940).
ONUMA, N., HIGUCHI, H., WAKITA, H., NAGASAWA, H.: Trace element partition between two pyroxenes and the host lava. Earth Planet. Sci. Lett. **5**, 47 (1968).
PICOT, P., PIERROT, R.: Roquesite. Bull. Soc. Franc. Mineral. Crist. **86**, 7 (1963).
PIERCE, T. B., PECK, P. F.: A rapid method for determining indium by neutron activation. Analyst **86**, 580 (1961).
PREUSS, E.: Beiträge zur spektranalytischen Methodik. II. Bestimmung von Zn, Cd, Hg, In, Tl, Ge, Sn, Sb, Pb und Bi durch fraktionierte Destillation. Z. Angew. Min. **3**, 8 (1940).
RANKAMA, K., SAHAMA, T. G.: Geochemistry. Chicago: Chicago University Press 1950.
SCHINDEWOLF, U., WAHLGREN, M.: The rhodium, silver and indium content of some chondritic meteorites. Geochim. Cosmochim. Acta **18**, 36 (1960).
SCHMITT, R. A., SMITH, R. H.: Indium abundances in chondritic and achondritic meteorites, and in terrestrial rocks. Proc. I.A.G.C. Symposium on the Origin and Distribution of the Elements, Paris (L. H. AHRENS, ed.). Oxford: Pergamon Press 1968.
SCHROEDER, H. J.: Indium, mineral facts and problems. U.S. Bur. Mines Bull. **630** (1965).

SHAW, D. M.: The geochemistry of indium. Geochim. Cosmochim. Acta **2**, 185 (1952).
— The geochemistry of gallium, indium, thallium — A review. Phys. Chem. Earth **2**, 164 (1957).
SMALES, A. A., MAPPER, D., FOUCHÉ, K. F.: The distribution of some trace elements in iron meteorites, as determined by neutron activation. Geochim. Cosmochim. Acta **31**, 673 (1967).
— VAN R. SMIT, J., IRVING, H. M.: Determination of indium in rocks and minerals by radioactivation. Analyst **82**, 539 (1957).
SMITH, F. G.: Physical Geochemistry. Reading, Massachusetts: Addison-Wesley Publishing Co., Inc. 1963.
SUESS, H., UREY, H. C.: Abundances of the elements. Rev. Mod. Phys. **28**, 53 (1956).
SUNDERMAN, D. N., TOWNLEY, C. W.: The radiochemistry of indium. NAS-NS-3014, U.S. At. En. Comm., 1960.
TANDON, S. N., WASSON, J. T.: Neutron activation determination of indium in meteorites. Radiochim. Acta **8**, 184 (1967a).
— — Indium variations in a petrologic suite of L-group chondrites. Science **158**, 259 (1967b).
— — Gallium, germanium, indium and iridium variations in a suite of L-group chondrites. Geochim. Cosmochim. Acta **32**, 1087 (1968).
TAYLOR, S. R.: Abundance of chemical elements in the continental crust: A new table. Geochim. Cosmochim. Acta **28**, 1273 (1964).
— SACHS, M.: Trace elements in australites. Nature **188**, 387 (1960).
UREY, H. C.: The abundance of the elements with special reference to the problem of the iron abundance. Quart. J. Roy. Astron. Soc. **8**, 23 (1967).
VINOGRADOV, A. P.: Regularity of distribution of elements in the earth's crust. Geokhimiya **1**, 6 (1956).
— Average contents of chemical elements in the principal types of igneous rocks of the earth's crust. Geochemistry (English Transl.). **7**, 641 (1962).
VOLAND, B.: Die Verteilung des Indiums in Eruptivgesteinen. Ein Beitrag zur Geochemie des Indiums. Freiberger Forschungsh. C **246**, 67 (1969).
WAGER, L. R. VAN R. SMIT, J., IRVING, H.: Indium content of rocks and minerals from the Skaergaard intrusion, East Greenland. Geochim. Cosmochim. Acta **13**, 81 (1958).
WAKITA, H., SCHMITT, R. A.: Rare earth and other elemental abundances in the Allende meteorite. Nature **227**, 478 (1970).
WASSON, J. T.: The chemical classification of iron meteorites. I. A study of iron meteorites with low concentrations of gallium and germanium. Geochim. Cosmochim. Acta **31**, 161 (1967).
WOLSTENHOLME, W. A.: Analysis of dried blood plasma by spark source mass spectrometry. Nature **203**, 1284 (1964).

Revised manuscript received: October 1970

Tin 50

A	G. Bergerhoff	(Anorganisch-Chemisches Institut, Universität Bonn, Germany)
B—M	H. Hamaguchi	(Department of Chemistry, University of Tokyo, Hongo, Tokyo, Japan)
	and R. Kuroda	(Faculty of Engineering, University of Chiba, Chiba City, Japan)
O	K. H. Wedepohl	(Geochemisches Institut der Universität, Göttingen, Germany)

50-A. Crystal Chemistry

The small number of minerals containing more than sparse amounts of tin can be divided into three groups according to the crystal chemistry of tin: I. the elementary state, II. compounds with the oxidation number two of tin, III. compounds with the oxidation number four of tin (Table 50-A-1).

I. Tin crystallizes in two modifications: α-tin (grey tin) and β-tin (white tin). Though the transformation point is 18°C, only white tin is known as a mineral due to the low transformation rate. α-Tin has the ideal diamond structure (BIJL and KOLKMEIJER, SB 1913—1928, 54; Sn—Sn: 4×2.80 Å), but in β-tin the coordination tetrahedron is much flattened and each atom receives two further neighbors at a greater distance, forming altogether a strongly deformed octahedron (Fig. 50-A-1).

II. In the bivalent state tin tends to form pyramidal coordinations due to the electron configuration $(5s)^2(5p)^2$ of the tin(II)-ion (DONALDSON, 1967). This is evident in the crystal structure of herzenbergite, SnS, the rock-salt structure of which is strongly deformed (Fig. 50-A-2). Teallite, $PbSnS_2$, forms an isotypic structure. Lead and tin are probably randomly distributed. In ottemannite, Sn_2S_3, one tin atom in the bivalent state is also pyramidally coordinated, the other tetravalent atoms forming double chains of octahedra as in NH_4CdCl_3.

In so far as one can speak of a distinct "valence state" in the sulfo-salts, in colusite, $Cu_{12}(As,Sn,V)_4S_{13}$, bivalent tin will substitute the pyramidally coordinated arsenic and tetravalent tin the tetrahedrally coordinated copper in the isotypic structure of tennantite (= binnite), $Cu_{12}As_4S_{13}$.

III. In the tetravalent state, too, tin forms only a few independent minerals: berndtite, SnS_2, cassiterite, SnO_2, nordenskiöldite, $CaSn[BO_3]_2$, pabstite, $BaSn[Si_3O_9]$, stokesite, $CaSn[Si_3O_9]\cdot 2H_2O$. All of them have well-known structures as specified in Table 50-A-1. Moreover there is an isomorphous series between SnO_2 and TiO_2 at temperatures above 1350°C (PADUROW, 1956). As in many synthetic compounds, such as $Zn_2^{[4]}Sn^{[6]}O_4$ (POIX, 1965), $Fe[Sn^{[6]}(OH)_6]$ (STRUNZ and CONTAG, SR 1960, 341), tin forms an octahedral coordination polyhedron, sometimes slightly deformed, as established for SnO_2 by BAUR (SR 1956, 263) (Fig. 50-A-3). The ionic radius for Sn^{4+} is given by GOLDSCHMIDT (1923): r = 0.74 Å. This being the case, one can understand the incorporation of small amounts of tin into structures such as vonsenite (or ludwigite) forming hulsite $(Fe, Mg, Sn)_2Fe(O)_2[BO_3]$, into spinel forming limaite $(Al, Zn, Sn)_3O_4$, into corundum forming nigerite $(Al, Fe)_{12}(Sn, Fe, Zn, Mg)_3O_{22}(OH)_2$, into α-$PbO_2$ forming ixiolite (orthorhombic) or wodginite (monoclinic), $(Ta, Nb, Sn, Mn, Fe)_2O_4$, and into PbS forming franckeite, $Pb_5Sn_3Sb_2S_{14}$.

There are also some examples where tin goes into tetrahedral holes. Stannite, Cu_2FeSnS_4, and kusterite, $Cu_2(Zn_{0.77}Fe_{0.23})SnS_4$ belong to the ZnS family forming a superstructure with ordered distribution of the metals. Other variations described by RAMDOHR (1944) and CLARINGBULL and HEY (1955) have been thoroughly characterized by MOH and OTTEMANN (1962).

Among structures of oxides a synthetic garnet, $Ca_3^{(8)}Sn_2^{(6)}(Ga_2Sn)^{(4)}O_{12}$ is known. Here there are only two octahedral holes per twelve oxygen atoms. Therefore every third tin atom should be in tetrahedral holes (GELLER, 1967; REINEN, 1969).

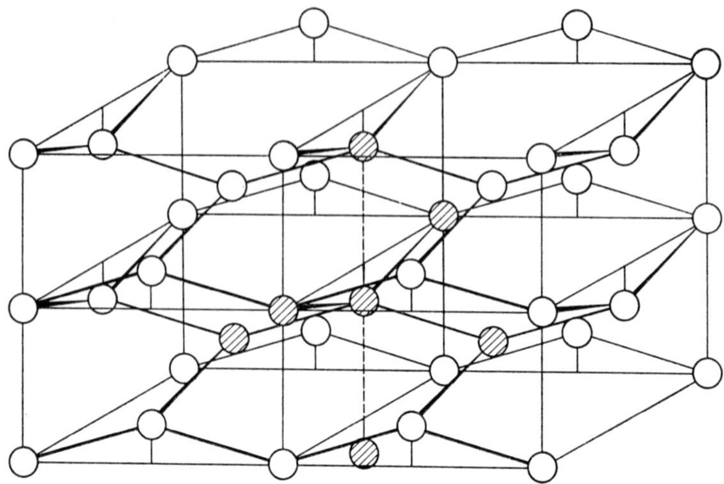

Fig. 50-A-1. Crystal structure of β-tin (white tin). One tin atom and its six neighbours are hatched

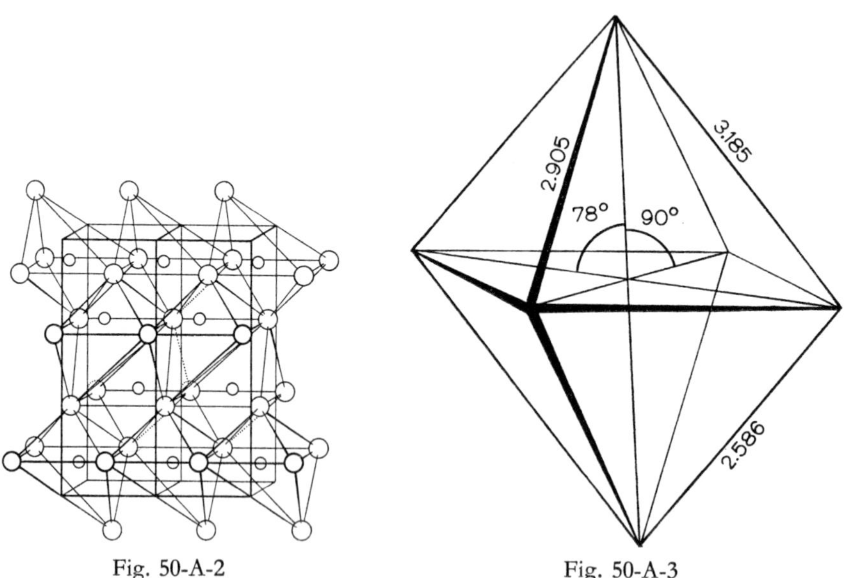

Fig. 50-A-2 Fig. 50-A-3

Fig. 50-A-2. Crystal structure of herzenbergite, SnS (small circles: *Sn*; large circles: *S*)

Fig. 50-A-3. Oxygen coordination polyhedron around Sn in cassiterite, SnO_2 (distances between O-atoms in Å)

Table 50-A-1. *Tin-containing minerals and their structures*

	Name	Composition	Distances in Å:		Structure type	Literature (related references are given in brackets)
I.	White tin	β-Sn	Sn-Sn:	4×3.016 2×3.175	pressed diamond structure	MARK and POLYANI (SB 1913—1928, 21, 54)
II.	Herzenbergite	SnS	Sn-S:	2.62; 2×2.68; 2×3.27; 3.39	deformed rocksalt (Herzenbergite and Teallite)	HOFMANN (SB 1933—1935, 14, 253)
	Teallite	$PbSnS_2$				
	Ottemannite	$SnSnS_3$	Sn^{2+}-S: Sn^{4+}-S:	2×2.644; 2.741 2.497; 2.552; 2×2.543; 2×2.611	related to NH_4CdCl_3	MOOTZ and PUHL (1967)
	Colusite	$Cu_{12}(As,Sn,V)_4S_{13}$	As-S: Cu-S:	2.246 2.337	tennantite (=binnite)	BERMAN and GONYER (1939); MURDOCH (SR 1953, 451); (WUENSCH, TAKÉUCHI and NOWACKI, 1966)
III.	Franckeite	$Pb_5Sn_3Sb_2S_{14}$	details unknown		related to lengenbachite	NUFFIELD (SR 1947—1948, 345) (NOWACKI, 1967)
	Berndtite	SnS_2	Sn-S:	6×2.55	cadmium-iodide	OFTEDAL (SB 1913—1928, 214)
	Cassiterite	SnO_2	Sn-O:	6×2.054	rutile	BAUR (SR 1956, 263)
	Nordenskiöldite	$CaSn[BO_3]_2$	details unknown		dolomite	EHRENBERG and RAMDOHR (SB 1933—1935, 406) (STEINFINK and SANS, SR 1959, 419)

Table 50-A-1 (Continued)

Name	Composition	Distances in Å:	Structure type	Literature (related references are given in brackets)
Pabstite	$BaSn[Si_3O_9]_2$	details unknown	benitoite	Gross, Wainright, and Evans (1965); (Zachariasen, SB 1928—1932, 128, 526)
Stokesite	$CaSn[Si_3O_9]\cdot 2H_2O$	Sn-O: 2×2.013 2×2.033 2×2.055	chain-silicate	Vorma (1963)
Hulsite	$(Fe, Mg, Sn)^{[6]}_2 Fe^{[6]}(O)_2[BO_3]$	Fe-O: 1.96—2.31	vonsenite	Schaller (1954, 1961); (Takéuchi, SR 1956, 374)
Limaite	$(Al, Zn, Sn)_3O_4$	details unknown	spinel	Cotelo Neiva, Rimsky, and Sandréa (1955)
Nigerite	$(Al, Fe)_{12}(Sn, Fe, Zn, Mg)_3O_{22}(OH)_2$	details unknown	related to corundum	Bannister, Hey, and Stadler (SR 1947—1948, 351)
Ixiolite	$(Ta, Nb, Sn, Mn, Fe)_2O_4$	orthorhombic; details unknown	α-PbO_2	Nickel, Rowland, and McAdam (1963a); (Laves, Bayer, and Panagos, 1963; Zaslavsky and Tolkacev, SR 1952, 224)
Wodginite	$(Ta, Sn, Mn, Fe, Nb)_2O_4$	monoclinic; details unknown	α-PbO_2	Nickel, Rowland, and McAdam (1963b)
Stannite	Cu_2FeSnS_4	Sn-S: 4×2.42	separate type; related to zinc-blende	Brockway (SB 1933—1935, 96, 440)
Kusterite	$Cu_2(Zn_{0.77}Fe_{0.23})SnS_4$	details unknown	separate type; related to zinc-blende	Ivanov and Pyatenko (1959)

Revised manuscript received: June 1969

50-B. Isotopes in Nature

Tin has the atomic number of fifty, with its ten stable isotopes between mass numbers 112 and 124. This is the largest number of stable isotopes for any element. The relative abundances of the isotopes are listed in Table 50-B-1. For information on the isotopic composition of tin see De Laeter and Jeffery (1967).

Table 50-B-1. *Tin isotopes*

	Mass number	Relative abundance (%)	Mass number	Relative abundance (%)
Sn (Z = 50)	112	0.95	118	24.01
	114	0.65	119	8.58
	115	0.34	120	32.97
	116	14.24	122	4.71
	117	7.57	124	5.98

Tin isotopes (Z = 50) are among those called proton "magic numbers". Correspondingly, there is a sudden increase in the value of the binding energy per nucleon at ^{120}Sn (Z = 50, N = 70) in the curve of average binding energy per nucleon versus the mass number. The stability of nuclei with proton or neutron magic numbers is also related to high natural abundances.

The observed thermal neutron cross sections are unusually small for nuclei with magic neutron numbers, and nuclei with 50 protons (tin) and 82 protons (lead) have small capture cross sections as compared with their neighbors (Flowers, 1952).

Revised manuscript received: September 1967

50-C. Abundance in Meteorites, Tektites, Cosmos and Lunar Materials

I. Meteorites, Tektites and Cosmos

Tin contents of chondrites and iron meteorites are listed in Tables 50-C-1 and 50-C-2, respectively. The mean value given by ONISHI and SANDELL (1957) is

Table 50-C-1. *Tin in chondrites and tektites*

Material		Sn (ppm)	Author
a) Chondrites:			
Composite (Arriba, Beenham, Cope, Cotesfield, Morland, Plainview, Roy)[a]		1 (C)	ONISHI and SANDELL (1957)
Composite (Alamogordo, Estacado, Gladstone, Gruver, Kingfisher, Ladder Creek, Melrose)[b]		1 (C)	ONISHI and SANDELL (1957)
Abee, Ce_1		0.80, 0.96 (av. 0.88) (C)	SHIMA (1964)
		1.9, 2.4 (av. 2.14) (N/R)	HAMAGUCHI *et al.* (1968)
		0.75 (av. of 2)	DE LAETER and JEFFERY (1967)
Bluff, CL		0.3 (C)	SHIMA (1964)
Boriskino, Cc_2		0.79, 0.97 (av. 0.88) (N/R)	HAMAGICHU *et al.* (1968)
Bruderheim, CL		0.85 (C)	SHIMA (1964)
Chainpur, CHL		0.33 (N/R)	HAMAGUCHI *et al.* (1968)
Cocklebiddy, CH		1.25 (I) (av. of 2)	DE LAETER and JEFFERY (1967)
Cold Bokkeveld, Cc_2		0.82 (N/R)	HAMAGUCHI *et al.* (1968)
Ehole, CH	1.00	0.86 (C)	SHIMA (1964)
Ehole, CH[c]	0.72		
Estacado, CH[d]		1 (C)	ONISHI and SANDELL (1957)
Estacado, CH		1.09 (C)	SHIMA (1964)
Forest City, CH		0.08 (C)	SHIMA (1964)
		0.24 (N/R)	HAMAGUCHI *et al.* (1968)
Hat Creek		<0.1 (C)	SHIMA (1964)
Holbrook, CL		<0.07	SHIMA (1964)
Homestead, CH	0.20	0.5 (C)	SHIMA (1964)
Homestead, CH	0.30		
Homestead (coarse parts)	0.10		
Homestead (fine parts)	0.79		
Indarch, Ce_1		1.5 (N/R)	HAMAGUCHI *et al.* (1968)
Ladder Creek, CL		1.01 (C)	SHIMA (1964)
Lake Brown, CL		1.9	DE LAETER and JEFFERY (1967)
Lancé, CHL		0.88 (N/R)	HAMAGUCHI *et al.* (1968)

Table 50-C-1 (continued)

Material	Sn (ppm)		Author
Leedey, CL	0.74	0.78 (C)	SHIMA (1964)
Leedey, CL	0.82		
Mighei, Cc_2		0.84, 0.50 (av. 0.67) (N/R)	HAMAGUCHI *et al.* (1968)
Modoc, CL		0.1 (C)	SHIMA (1964)
		0.28 (N/R)	HAMAGUCHI *et al.* (1968)
Murray, Cc_2		0.84, 1.22, 1.09 (av. 1.05) (C)	SHIMA (1964)
		0.86, 1.1 (av. 0.97) (N/R)	HAMAGUCHI *et al.* (1968)
Ness County, CL		0.08 (C)	SHIMA (1964)
Orgueil, Cc_1		1.5, 1.9, 1.5 (av. 1.6) (N/R)	HAMAGUCHI *et al.* (1968)
Potter, CL		0.5 (C)	ONISHI and SANDELL (1957)
Richardton, CH	0.54	0.46 (C)	SHIMA (1964)
Richardton, CH	0.38		
Rochester, CH		0.07 (C)	SHIMA (1964)
Tieshitz, CHL		0.39, 1.1 (av. 0.75) (N/R)	HAMAGUCHI *et al.* (1968)
Tulia, CH		0.07 (C)	SHIMA (1964)
Woolgorong, CL		<0.1 (I)	DE LAETER and JEFFERY (1967)
b) Tektites			
Australite		0.95	DE LAETER and JEFFERY (1967)
Philippinite		0.77 (N/R)	SCHMIDT (1970)[1]
Moldavite		0.88 (N/R)	SCHMIDT (1970)[1]

Samples for separating two phases:

Sample	Sn	Author
[a] Metal phase (11%)	5 ppm (C)	ONISHI and SANDELL (1957)
[b] Metal phase (13%)	5 ppm (C)	ONISHI and SANDELL (1957)
Nonmagnetic phase of composites a and b (5.7% FeS)	0.4 ± 0.2 ppm (C)	ONISHI and SANDELL (1957)
[c] Magnetic phase (29.8%)	2.1 ppm (C)	SHIMA (1964)
Nonmagnetic phase (70.2%)	0.14 ppm (C)	
[d] Metal phase of Estacado, Haven and Plainview	6 ppm (C)	ONISHI and SANDELL (1957)

1 ± 0.5 ppm and that by SHIMA (1964) 0.43 ± 0.10 ppm for chondrites. However, tin concentration varies widely, so that the values quoted may be taken as providing an approximate figure. All the recent data are much lower than the average value of 40 ppm given by the NODDACKS (1934) and also lower than the few values reported by GOLDSCHMIDT and PETERS (1933). It is quite certain either that these early workers were not able to eliminate all contamination from laboratory materials, or the spectrographic methods employed were seriously affected by a matrix effect which can cause an analytical error of a factor of ten.

[1] Annual Report: Max-Planck-Instit. f. Kernphysik, Heidelberg, Fed. Rep. Germany (Reference added by the Exec. Editor).

Table 50-C-2. *Tin in iron meteorites*

Material	Sn (ppm)	Author
Ainsworth, Ogg	1 (C)	ONISHI and SANDELL (1957)
Altonah, Of	1 (C)	ONISHI and SANDELL (1957)
Arispe, Ogg	3 (C)	ONISHI and SANDELL (1957)
Aroos, Og	5.23 (av. of 3 detns.) (C)	SHIMA (1964)
Boguslavka, H	0.2 (I)	DE LAETER and JEFFERY (1967)
Canyon Diablo, Og[a]	5.17 (av. of 6 detns.) (C)	SHIMA (1964)
Canyon Diablo, Og	7; 10 (C)	ONISHI and SANDELL (1957)
Canyon Diablo, Og	6.3 (C)	WINCHESTER and ATEN JR. (1957)
Canyon Diablo, Og	4.6 (I)	DE LAETER and JEFFERY (1967)
Chinge, D	0.1 (I)	DE LAETER and JEFFERY (1967)
Clark Country, Om	<0.2 (C)	SHIMA (1964)
Coahuila, H	0; 0 (C)	ONISHI and SANDELL (1957)
Costilla Peak, Om	0 (C)	ONISHI and SANDELL (1957)
Coya Norte, H	0 (C)	ONISHI and SANDELL (1957)
Deep Springs, D	0.55 (C)	SHIMA (1964)
Deport, Og	6 (C)	ONISHI and SANDELL (1957)
Duchesne, Of	3 (C)	ONISHI and SANDELL (1957)
Edmonton, Off-Of	5 (C)	ONISHI and SANDELL (1957)
Goose Lake, Om-Og	7 (C)	ONISHI and SANDELL (1957)
Grant, Of	<0.1 (C)	SHIMA (1964)
Haig, Og-Om	0.1 (I)	DE LAETER and JEFFERY (1967)
Henbury, Om	5; 4 (C)	ONISHI and SANDELL (1957)
Henbury, Om	<1.0 (C)	WINCHESTER and ATEN JR. (1957)
Henbury, Om	0.15 (I) (av. of 2 detns.)	DE LAETER and JEFFERY (1967)
Kumerina, Of	0.7	DE LAETER and JEFFERY (1967)
Locust grove, H (Ni-poor)	8 (C)	WINCHESTER and ATEN JR. (1957)
Mayodan, H	0.2 (C)	SHIMA (1964)
Milly Milly, Om	0.15 (I) (av. of 2 detns.)	DE LAETER and JEFFERY (1967)
Mount Joy, Ogg	1 (C)	ONISHI and SANDELL (1957)
Mt. Dooling, Om	0.5 (I)	DE LAETER and JEFFERY (1967)
Mt. Magnet, D	1.2 (I)	DE LAETER and JEFFERY (1967)
Mt. Stirling, Og	5.3 (I)	DE LAETER and JEFFERY (1967)
Muonionalusta, Of	10.7 (C)	WINCHESTER and ATEN JR. (1957)
Negrillos, H	<0.3 (C)	SHIMA (1964)
Odessa, Og	5.74 (av. of 2 detns.) (C)	SHIMA (1964)
Perryville, Off (?)	0 (C)	ONISHI and SANDELL (1957)
Sacramento Mountains, Om	0 (C)	ONISHI and SANDELL (1957)
Sandia Mountains, H	0 (C)	ONISHI and SANDELL (1957)
Sao Juliao de Moreira, Ogg (H)[b]	5 (C)	WINCHESTER and ATEN JR. (1957)
Sardis Georgia, Og	4.13 (av. of 2 detns.) (C)	SHIMA (1964)
Sierra Gorda, H	<0.1 (C)	SHIMA (1964)
Sikhote-Alin, H or Ogg	0.303 (C)	SHIMA (1964)

Table 50-C-2 (continued)

Mineral		Sn (ppm)	Author
Sikhote-Alin, H or Ogg		0.3 (I)	De Laeter and Jeffery (1967)
Spearman, Om		1 (C)	Onishi and Sandell (1957)
Tamarugal (El Inca), Om		<0.8 (C)	Winchester and Aten Jr. (1957)
Tocopilla (Cerros de Buey Muerto), H		20.2 (C)	Winchester and Aten Jr. (1957)
Toluca, Om		5.3 (I) (av. of 2 detns.)	De Laeter and Jeffery (1967)
Toluca, Om[c]		7.6 (C)	Winchester and Aten Jr. (1957)
Toluca, Om		7.72 (C)	Shima (1964)
Treysa, Om		<0.5 (C)	Shima (1964)
Wallapi, Of		1.5 (C)	Onishi and Sandell (1957)
Warburton Ranges, D		0.2 (I)	De Laeter and Jeffery (1967)
Williamstown, Om		<0.2 (C)	Shima (1964)
Willow Creek, Om		1 (C)	Onishi and Sandell (1957)
Yorndegin III, Og		7.6 (I)	De Laeter and Jeffery (1967)
Troilites:			
Odessa		<1 (C)	Shima (1964)
Stony iron meteorites:			
Bencubbin I	Metal	0.2 (I)	De Laeter and Jeffery (1967)
	Stone	0.1 (I)	
Brenham	Metal 1	n.d.	Shima (1964)
	Metal 2	0.8 (C)	Shima (1964)
	Stone 1	0.53 (C)	Shima (1964)
	Stone 2	0.6 (C)	Shima (1964)
	Troilite	<0.1 (C)	Shima (1964)
Admire	Metal 1	n.d.	Shima (1964)
	Metal 2	2.35 (C)	Shima (1964)
	Stone 1	0.77 (C)	Shima (1964)
	Stone 2	0.3 (C)	Shima (1964)
	Troilite	<0.1 (C)	Shima (1964)

[a] Noddacks (1931) found 50 ppm.
[b] Classified by Krantz as hexahedrite breccia.
[c] Noddacks (1931) found 60 ppm.

Winchester and Aten Jr. (1975) estimated a weighted average of iron meteorites to be 6.7 ppm Sn considering the abundances of different classes. The simple arithmetical average of all iron meteorites (7.2 ppm) does not differ markedly from their weighted mean. Onishi and Sandell (1957) gave 2 ppm Sn as the average. This difference in the average may reflect the fact that the tin concentration varies considerably in iron meteorites.

On the basis of a comparison of the data for chondrites and iron meteorites, and a thermodynamic calculation, Shima (1964) mentioned that tin can be expected to be siderophile at high temperatures and lithophile at low temperatures.

Previous calculations for the abundance of tin in primitive solar materials have been done by Suess and Urey (1956), Cameron (1959), Aller (1961, 1965).

SUESS and UREY determined the atomic abundances by semiempirical methods, based on earlier chemical analysis data of chondritic meteorites. CAMERON relied upon nucleosynthesis arguments for determining cosmic abundances and filled tin into the abundance curve by interpolation. CAMERON (1967) gave a revised value. ALLER's data were derived from astrophysical data of solar atmospheric elemental abundances.

Atomic abundance data of tin in various types of the chondrites, which have been used as reference models for cosmic abundance determinations were experimentally obtained by HAMAGUCHI *et al.* (1968) and are listed in Table 50-C-3 together with the calculated primordial solar abundances of tin.

Table 50-C-3. *Cosmic and atomic abundance data for tin relative to* $Si = 10^6$

Source	Sn
SUESS and UREY (1956)	1.33
CAMERON (1967)	4.22
ALLER (1961)	1.1
ALLER (1965)	{3.6 5.6
HAMAGUCHI *et al.* (1968)	
Chondrites (CH)	0.34
Chondrites (CL)	0.32
Carbonaceous chondrites (Cc_1)	3.5
Carbonaceous chondrites (Cc_2)	1.4
Enstatite chondrites (Ce_1)	2.6

Revised manuscript received: September 1967

II. Lunar Materials

The tin content of lunar materials is shown in Table 50-C-4. Most references are taken from the Proceedings of the Lunar Science Conferences in Geochimica et Cosmochimica Acta Supplements 1 to 4. Lunar basalts seem to contain less tin than terrestrial ones.

Table 50-C-4. *Tin in lunar materials (Analytical method: Mass)*

Materials	Sn (ppm)	Reference
Apollo 11		
Type A rock (fine-grained vesicular basalt)		
10057,32	0.6[a]	1
10072,24	0.4[a]	1
Type B rock (coarse-grained basalt)		
10058,26	1.2[a]	1
Type C rock (breccia)		
10056,20	0.3[a]	1
Soil (fine fraction of type D)		
10084,55	0.7[a]	1

Table 50-C-4 (continued)

Materials	Sn (ppm)	Reference
Apollo 12		
Porphyritic basalt		
12002,139	0.23	2
12022,46	0.39	2
12052,47	0.16	2
12063,66	0.12	2
Ophitic basalt		
12038,27	0.10	2
Fines		
12070,93	0.30	2
Apollo 14 (Fra Mauro)		
Breccia		
14047,32	0.4	3
Magmatic rock		
14072	0.3	3
Basalt		
14321	0.2	3
Apollo 15		
Basalt (highly vesicular)		
15016,42		
Total rock	0.19	4
Pyroxene separate	0.22	4
Plagioclase separate	0.4	4
Green glass		
15401,27	0.11	4
15421,24	0.16	4
15426,38	0.12	4
Soils		
15471,35	0.24	4
15261,69	0.6	4
15271,86	0.23	4
15401,27	0.28	4
15301,89	0.39	4
Highland breccias	0.12–0.29	4

[a] Approximate value.

References: 1. Morrison, G. H., Gerard, J. T., Kashuba, A. T., Gangadharam, E. V., Rothenberg, A. M., Potter, N. M., Miller, G. B.: Proc. Apollo 11 Lunar Sci. Conf., **2**, 1383 (1970); 2. Taylor, S. R., Rudowski, R., Muir, P., Graham, A., Kaye, M.: Proc. 2nd Lunar Sci. Conf., **2**, 1083 (1971); 3. Taylor, S. R., Kaye, M., Muir, P., Nance, W., Rudowski, R., Ware, N.: Proc. 3rd Lunar Sci. Conf., **2**, 1231 (1972); 4. Taylor, S. R., Gorton, M. P., Muir, P., Nance, W., Rudowski, R., Ware, N.: Proc. 4th Lunar Sci. Conf., **2**, 1445 (1973)

Manuscript received: September 1975

50-D. Abundance in Rock-forming Minerals (I), Tin Minerals (II), Phase Equilibria (III)

I.

Tin is rare in the early magmatic sulfides, being enriched particularly in pneumatolytic and hydrothermal formations. It has been assumed that at low temperatures the affinity of tin for sulfur increases whereas at intermediate to high temperatures it occurs almost exclusively in oxygen-bearing compounds, often cassiterite. In Table 50-D-1 are listed the tin contents of various rock-forming minerals from different localities.

Table 50-D-1. *Tin in rock-forming minerals*

Mineral	Sn (ppm)	Author
4 Magnetites, Kiruna	10 (av.) (S)	Noddacks (1931)
2 Magnetites, Altai-sayan Mt.	16, 39	Vakhrushev (1962)
2 Magnetites, Dzhida, W. Transbaykalia	56.0, 30.0 (S)	Petrova and Legeydo (1965)
9 Olivines, Fichtelgebirge (Germany)	5 (av.) (S)	Noddacks (1931)
4 Olivines	4 (av.) (2.4—8.1) (S)	Hellwege (1956)
2 Augites, Harzburg, Radautal Harz, (Germany)	< 0.01, 0.01 (S)	Ottemann (1941)
3 Orthopyroxenes (in metamorphic rocks)	< 0.8, 0.8, 2 (S)	De Vore (1955b)
7 Pyroxenes	18 (av.) (S)	Hellwege (1956)
4 Hornblendes, Sulitjelma	2 or less (S)	Noddacks (1931)
7 Hornblendes (epidote amphibolite facies), Laramie Range, Wyoming	2.4 (av.) (S)	De Vore (1955a)
10 Hornblendes (granulite facies), Laramie Range, Wyoming	6 (av.) (S)	De Vore (1955a)
14 Hornblendes (amphibolite facies), Laramie Range, Wyoming	0.8 (av.) (S)	De Vore (1955a)
10 Hornblendes (in metamorphic rocks)	6 (av.) (0.8—24) (S)	De Vore (1955b)
Hornblendes (contact metamorphic origin)	11 to 110 (S)	Hellwege (1956)
Plagioclase, Harzburg (Germany)	2.1 (S)	Ottemann (1941)
2 Anorthites	0.1, 0.4 (S)	Hellwege (1956)
Oligoclase	5.2 (S)	Hellwege (1956)
2 Albites, Wurmberg, Sommerklippen Harz (Germany)	45, 50 (S)	Ottemann (1941)
5 Albites	15.0 (av.) (3.1—44.0) (S)	Hellwege (1956)
2 Orthoclases, Schwarzwald, Wurmberg Harz (Germany)	2.0, 0.3 (S)	Ottemann (1941)
Feldspar, Brocken (Germany)	10 (S)	Ottemann (1941)
6 Potash feldspars	0.9 (av.) (0.3—2.0)	Hellwege (1956)
70 Potash feldspars, S.W. Cape, S. Africa	1.5—11 (S)	Kolbe (1965)
35 Potash feldspars (in coarsely porphyr. granites), Cape, S. Africa	2.7 (av.) (1.5—11) (S)	Kolbe and Taylor (1966)

Table 50-D-1 (Continued)

Material	Sn (ppm)	Author
9 Potash feldspars (in medium-grained granites), Cape, S. Africa	3.1 (av.) (1.8—4.0) (S)	KOLBE and TAYLOR (1966)
9 Potash feldspars (in fine-grained granites), Cape, S. Africa	3.8 (av.) (2.3—10) (S)	KOLBE and TAYLOR (1966)
Biotite, East-Pool-Mine, Cornwall	39 (S)	BROWN (1934)
2 Biotites, Wurmberg, Brocken	300, 230 (S)	OTTEMANN (1941)
4 Biotites (mostly pegmatitic), various localities	32, 49, 65, 102 (S)	AHRENS and LIEBENBERG (1950)
6 Biotites (epidote-amphibolite-facies), Laramie Range, Wyoming	1.3 (av.) (S)	DE VORE (1955a)
4 Biotites (amphibolite-facies), Laramie Range, Wyoming	7 (av.) (S)	DE VORE (1955a)
12 Biotites (granulite-facies), Laramie Range, Wyoming	4 (av.) (S)	DE VORE (1955a)
16 Biotites (in metamorphic rocks)	12 (av.) (< 0.8—30) (S)	DE VORE (1955b)
3 Biotites (in basic rocks)	7 (av.) (2.0—12.0) (S)	HELLWEGE (1956)
4 Biotites (in granites)	230 (av.) (150—300) (S)	HELLWEGE (1956)
Biotite, Karaoba, Kazakhstan		
in granite	250	GANEEV *et al.* (1961)
in pegmatitic schlieren	1,300	GANEEV *et al.* (1961)
in pegmatitic vein	360	GANEEV *et al.* (1961)
Biotites (in granitic rocks), Lesser Caucasus	30—40	MUSTAFAEV (1964)
22 Biotites	3.2—6.5 (S)	Kolbe (1965)
21 Biotites (in Cape granites), S. Africa	4.4 (av.) (3.2—6.5) (S)	KOLBE and TAYLOR (1966)
37 Muscovites (mostly from pegmatites), various localities	12—3500 (S)	AHRENS and LIEBENBERG (1950)
4 Muscovites (in metamorphic rocks)	12 (av.) (2.6—30) (S)	DE VORE (1955b)
Coarse-flake muscovite (in greisen)	580	GANEEV *et al.* (1961)
Fine-flake muscovite (in greisen)	1,240	GANEEV *et al.* (1961)
4 Muscovites	4.8—7.0 (S)	KOLBE (1965)
4 Muscovites (in Cape granites), S. Africa	6.0 (av.) (4.8—7.0) (S)	KOLBE and TAYLOR (1966)
13 Lepidolites, Africa and U.S.A.	4—700 (S)	AHRENS and LIEBENBERG (1950)
3 Phlogopites, various localities	16, 4, 4 (S)	AHRENS and LIEBENBERG (1950)
Zinnwaldite, Zinnwald	3,500 (S)	AHRENS and LIEBENBERG (1950)
3 Zinnwaldites, W. Bohemia	300, 400, 400 (P)	HOFFMAN and TRDLICKA (1963)
Sphene (in leucocratic granite), Dzhida, W. Transbaykalia	1,000 (S)	PETROVA and LEGEYDO (1965)
2 Sphenes (in porphyritic granites), Dzhida, W. Transbaykalia	322, 440 (S)	PETROVA and LEGEYDO (1965)
Ilmenite, Dzhida, W. Transbaykalia		
in leucocratic granite	200 (S)	PETROVA and LEGEYDO (1965)
in porphyritic granite	46.5 (S)	
Allanite (in leucocratic granite), Dzhida, W. Transbaykalia	150 (S)	PETROVA and LEGEYDO (1965)

It is of importance to note how the tin is distributed in monomineralic fractions separated from magmatic rocks. In Table 50-D-2 the distribution of tin in coexisting minerals of a suite of granitic rocks from the Dzhida area, U.S.S.R., is given. As can be seen, most of the tin, 60 to 70% and more, is bound in the dark minerals and sphene. In the hornblende-biotite-sphene assemblages, sphene contains on average 40%, and hornblende and biotite together 25 to 30% of the total tin present in the rock. As the content of dark minerals in the rocks decreases, the content of tin in the sphene increases. Also, tin enters preferentially into the amphiboles rather than into the associated biotites.

As can be seen in Table 50-D-2, the principal tin-bearing mineral in these granitic rocks is sphene, the others being ilmenite, magnetite and allanite. The preferential entry of tin into biotite, hornblende, and sphene may be explained by the crystal-chemical affinity of tin to replace titanium and trivalent iron as seen in Table 50-D-3.

Table 50-D-2. *Tin content of minerals of the Dzhida granitic rocks* (PETROVA and LEGEYDO, 1965)

Mineral	Mineral fraction (wt.-%)	Sn in mineral ppm	Amnt. of Sn in mineral recalc. to gram of rock (μg)	Percentage of total Sn in mineral	Concentration coefficient C_{Sn} (mineral)/ C_{Sn} (rock)
Quartz-diorite (Sn content 5.4 ppm)[a]:					
K-feldspar	21.3	1.0	0.21	3.9	0.18
Plagioclase	49.5	2.0	0.99	18.3	0.37
Quartz	1.4	0.5	0.07	1.3	0.09
Biotite	7.6	6.0	0.46	8.5	1.1
Hornblende	13.7	7.0	0.96	17.8	1.3
Sphene	1.9	120	2.28	42.3	22.2
Other accessory minerals	4.6	—	—	—	—
Total	100.0		4.97	92.1	
Granodiorite (Sn content 5.4 ppm)[a]:					
K-feldspar	16.1	1.4	0.22	4.08	0.26
Plagioclase	43.8	2.3	1.00	18.50	0.42
Quartz	17.1	1.0	0.01	0.20	0.18
Biotite	9.2	5.5	0.50	9.25	1.0
Hornblende	13.1	10.0	1.31	24.30	1.8
Sphene	0.5	400	2.00	37.00	74.0
Other accessory minerals	0.2	—	—	—	—
Total	100.0		5.04	93.33	
Quartz-syenite (Sn content 4.0 ppm)[a]:					
K-feldspar	29.3	1.0	0.29	7.25	0.25
Plagioclase	48.2	1.7	0.82	20.50	0.42
Quartz	13.4	0.5	0.07	1.75	0.12
Biotite	2.4	4.0	0.09	2.25	1.0
Hornblende	4.2	6.8	0.29	7.25	1.7
Sphene	0.7	310	2.17	54.30	77.5
Other accessory minerals	1.8	—	—	—	—
Total	100.0		3.73	93.30	

Table 50-D-2 (Continued)

Mineral	Mineral fraction (wt.-%)	Sn in mineral ppm	Amnt. of Sn in mineral recalc. to gram of rock (μg)	Percentage of total Sn in mineral	Concentration coefficient C_{Sn}(mineral)/ C_{Sn}(rock)
Porphyritic granite (Sn content 3.8 ppm)[a]:					
K-feldspar	32.0	1.3	0.42	11.06	0.3
Plagioclase	35.7	1.9	0.68	17.9	0.5
Quartz	20.6	0.5	0.10	2.63	0.13
Biotite	6.7	5.0	0.34	8.95	1.3
Hornblende	3.2	10.3	0.33	8.70	2.7
Sphene	0.4	440	1.76	46.31	116
Other accessory minerals	1.4	—	—	—	—
Total	100.0		3.63	95.55	
Leucocratic granite (Sn content 3.8 ppm)[a]:					
K-feldspar	34.3	1.1	0.38	9.70	0.3
Plagioclase	29.6	1.5	0.44	11.30	0.4
Quartz	32.7	0.7	0.23	6.00	0.18
Biotite	1.7	10.0	0.17	4.30	2.5
Sphene	0.5	475	2.38	61.00	122
Other accessory minerals	1.2				
Total	100.0		3.60	92.30	

[a] All values obtained spectrochemically.

Table 50-D-3. *Properties of tin, titanium and iron important for their crystal-chemical behavior*

	Sn (IV)	Ti (IV)	Fe (III)	Author
Ionic radius (Å)	0.71	0.68	0.64	AHRENS (1952)
	0.67	0.64	0.67	BELOV and BOKIY (1959)
Coordination number	6	6	6	
Ionization potential, eV	40.57	44.66	—	GREEN (1959)
			55.42	KAY and LEBBY (1949)
Electronegativity, Kcal/g atom	235	260	245	POVARENNYKH (1955)

[a] The values were taken from K. A. VLASOV (chief editor) "Geochemistry and mineralogy of rare elements and genetic types of their deposits" vol. I, Izd. "Nauka", 1964. (Translated by Israel Program for Scientific Translations. Jerusalem 1966).

II. Tin Minerals

Tin minerals are comparatively few.

Cassiterite (SnO_2) is the most important ore of tin and the most common of all tin minerals. The cassiterite occurs in pneumatolytic and high-temperature hydrothermal veins or metasomatic deposits that are genetically closely associated with highly siliceous igneous rocks, usually granite or rhyolithe. Cassiterite is a typical resistate, representing the chemically unchanged residue from weathering. Therefore, it becomes concentrated in bands and layers of varying thickness, forming economically valuable deposits such as those found in Malaya, Indonesia, Bolivia, U.S.S.R., Congo etc. Iron is usually present in substituting for tin, up to about Fe:Sn=1:6. Titanium, tantalum and niobium are often present in small amounts, and substitute for Sn up to (Ta, Nb):Sn=1:30. Small amounts of zinc, lead (probably due to admixture of other phases), tungsten and manganese also have been known. For detailed minor element investigations of large numbers of cassiterites see SCHRÖCKE (1955), FESSER (1968) etc.

Tin occurs in many sulfide ores in the form of relatively pure tin sulfides or incorporated with a number of other metals in more complex sulfides. The economic importance of tin from such sources has not been adequately explored, although in the U.S.S.R. some deposits of this type are actually being mined.

Stannite (Cu_2FeSnS_4) occurs in tin-bearing veins, associated principally with chalcopyrite, sphalerite, tetrahedrite, pyrite, cassiterite, wolframite and quartz. Oriented inclusions of chalcopyrite and sphalerite have been noted, and an oriented deposition of stannite twinned crystals on tetrahedrite has been reported. Stannite shows little variation in composition. Deviations from the formula may well be due to inclusions of chalcopyrite and sphalerite.

Canfieldite (Ag_8SnS_6) is an end-member of the argyrodite series. Argyrodite is a silver germanium sulfide, Ag_8GeS_6. Tin replaces germanium, and a series exists to the sulfostannate, canfieldite Ag_8SnS_6. In Bolivia, argyrodite and canfieldite occur at Aullagas, accompanying pyrargyrite, stephanite, and sphalerite.

Cylindrite is a lead, tin, antimony sulfide, possibly $Pb_3Sn_4Sb_2S_{14}$. It occurs in Bolivia in the Trinacria and other mines at Poopó, with franckeite, pyrite and sphalerite, and at the Porvenir and Francisca mines, Huanuni, with franckeite.

Teallite, a lead tin sulfide $PbSnS_2$, is sometimes found in large amounts in the silver-tin veins of Bolivia. Wurtzite, cassiterite, sphalerite, pyrite and sometimes stannite and franckeite are associated with the teallite.

Franckeite is a common mineral in the silver-tin veins of Bolivia, sometimes found in large amounts. Wurtzite, cassiterite, cylindrite, galena, pyrite, and feather-ore accompany franckeite at the Trinacria and other mines, Poopó. It also occurs abundantly at Llallagua along with pyrite, pyrrhotite and marcasite. Franckeite is a lead and tin antimony sulfide, possibly possessing the composition $Pb_5Sn_3Sb_2S_{14}$. The iron and zinc reported are apparently due to impurities.

Herzenbergite is an incompletely described species, presumably the same as artificial SnS. It occurs intimately associated with cassiterite, pyrite and quartz.

Two new natural tin sulfides Sn_2S_3 and SnS_2 have also been found (MOH and BERNDT, 1964).

The phase relations in the binary system tin-sulfur were studied by ALBERS and co-workers (ALBERS, HAAS, VINK and WASSCHER, 1961; ALBERS and SCHOL, 1961) and MOH (1962/1963). Their results are illustrated in Fig. 50-D-1.

The phase relations in the iron-tin-sulfur and copper-tin-sulfur systems were also given by MOH (1962/1963). Many investigations on the phase relations in the binary systems containing tin have been reported in the literature. The phase relations in the binary iron-tin system (Fig. 50-D-2) are reported in: "Metals Handbook" (The Japan Institute of Metals, Maruzen Co. Ltd. 1965, Tokyo).

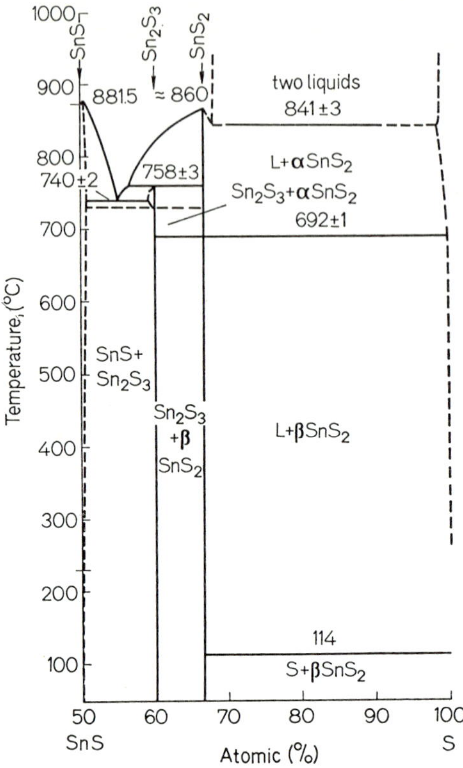

Fig. 50-D-1. Phase relations in the SnS—S portion of the system Sn—S. (Taken from MOH (1962/1963). The compound Sn_3S_4 earlier reported to occur in this system did not appear)

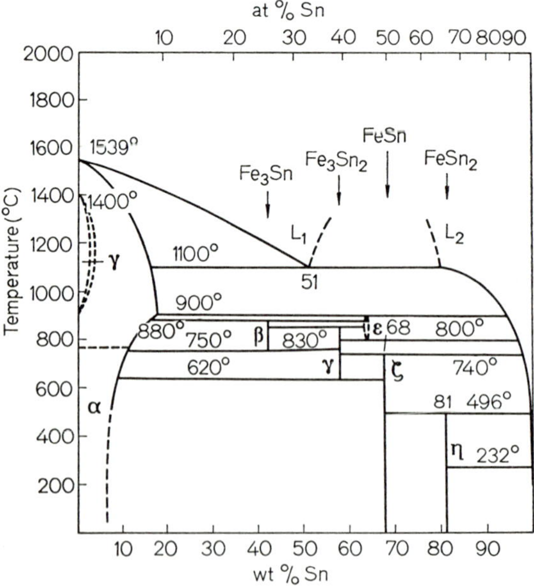

Fig. 50-D-2. Phase relations in the system Fe—Sn. Traces of carbon alter the phase relations significantly

Tin minerals are compiled in Table 50-D-4.

Table 50-D-4. *Tin minerals* (mainly according to FLEISCHER, M.: 1975 Glossary of Mineral Species. Bowie, Maryland: Mineralogical Record Inc. 1975)

Stannides	
Niggliite	PtSn
Stannopalladinite	$(Pd, Cu)_3Sn_2$ (?)
Sulfides	
Stannite	Cu_2FeSnS_4
Hexastannite (stannoidite)	$Cu_5(Fe, Zn)_2SnS_8$
Mawsonite	$Cu_7Fe_2SnS_{10}$
Kösterite	$Cu_2(Zn, Fe)SnS_4$
Colusite	$Cu_3(As, Sn, V, Fe)S_4$
Cernyte	Cu_2CdSnS_4
Herzenbergite	SnS
Ottemannite	Sn_2S_3
Berndtite	SnS_2
Teallite	$PbSnS_2$
Montesite	$PbSn_4S_5$ (?)
Franckeite	$Pb_5Sn_3Sb_2S_{14}$
Oxides	
Cassiterite	SnO_2
Varlamoffite (hydrocassiterite?)	$(Sn, Fe)(O, OH)_2$
Nigerite	$(Zn, Mg, Fe)(Sn, Zn)_2(Al, Fe)_{12}O_{22}(OH)_2$
Ixiolite	$(Ta, Nb, Sn, Mn, Fe)_2O_4$ (orthorhomb.)
Wodginite	$(Ta, Nb, Sn, Mn, Fe)_{16}O_{32}$ (mon.)
Yttrotantalite	$(Y, Fe, U, Ce, Zr)(Ta, Nb, Ti, Sn)O_4$
Rijkeboerite	$Ba(Ta, Nb)_2(O, OH)_7$
Thoreaulite	$Sn(Ta, Nb)_2O_6$
Borate, silicates	
Nordenskiöldine	$CaSn[BO_3]_2$
Malayaite	$CaSn[O\|SiO_4]$
Pabstite	$Ba(Sn, Ti)[Si_3O_9]$

III. Phase Equilibria

The phase relations in the binary systems tin-sulfur and tin-iron are illustrated in Fig. 50-D-1 and 50-D-2 respectively. Phase relations in the systems Sn—Fe—S—O and Cu—Fe—Zn—Sn—S are published by MOH, G. H. (Chem. d. Erde **33**, 244, 1974 and **34**, 1, 1975).

Revised manuscript received: September 1967 (Subsect. 50-D-I and II);
Manuscript received: November 1977 (Subsect. 50-D-III and Table 50-D-4, provided by the Exec. Editor)

50-E. Abundance in Common Igneous Rocks

The average content of tin in the upper lithosphere was estimated by CLARKE and WASHINGTON (1924) to be 6 ppm Sn. The NODDACKS (1930) confirmed this value. In 1957 ONISHI and SANDELL reported 2 ± 1 ppm Sn as an average for the crustal abundance of tin. GOLDSCHMIDT (1954) assumed a value of 40 ppm for the Paleozoic European shales as the average tin content of the rocks of the upper lithosphere. HAMAGUCHI and his collaborators (1964) have established a neutron activation method for tin and reported an activation value of 3.1 ppm Sn as the average tin content of the Earth's crust. TAYLOR (1964)compiled a table of element abundances in the continental crust, calculated on the basis of a 1:1 mixture of granite and basalt

Table 50-E-1. *Average tin content of different types of igneous rocks (ppm)*

Rock type	ONISHI and SANDELL (1957)	HAMAGUCHI and collaborators (1964)
Silicic rocks	3.5	3.6
Intermediate rocks	1.3	1.5
Mafic rocks	1.2	0.9
Ultramafic rocks	0.5	0.35

abundances and gave 2 ppm Sn as the crustal average for tin. HORN and ADAMS (1966) have currently made a new approach to geochemical balancing, utilizing a high-speed digital computer. The computed abundance for tin in igneous rocks was 2.5 ppm Sn. Based on these consideration we may conclude that the values given by CLARKE and WASHINGTON (1924) and the NODDACKS are somewhat high and that GOLDSCHMIDT's average is certainly too high. At the present time we may take 2 to 3 ppm Sn as a reasonably certain average of tin for igneous rocks.

Tin values currently reported in main types of igneous rocks agree reasonably well, although older values are sometimes markedly high. It is certain that the distribution of tin in rocks is subject to more regional variations due to tin mineralization than is the case with nearly all other metals. Judging from the abundance pattern of tin reported recently one can think of tin as being distributed as a "trace element" rather evenly throughout the crustal rocks.

Averages of tin for different igneous rock types have been obtained by ONISHI and SANDELL (1957) and HAMAGUCHI and collaborators (1964), respectively. Their results are summarized in Table 50-E-1. Agreement between the experimentally obtained averages given by ONISHI and SANDELL and HAMAGUCHI *et al.* is surprisingly good in spite of the fact that sources of rocks are widely different between the two groups of workers. As to the average figure for tin in granitic rocks, KOLBE (1965) reported 2.4–4.4 ppm Sn for 42 granites of South Africa, 2.0–4.1 ppm for 35 granites of Australia; PETROVA and LEGEYDO (1965) gave 2.5 ppm Sn for 8 granites and

Table 50-E-2. *Tin in granitic rocks*

Rock Locality	Sn (ppm)	Author
15 Granites, South Norway	1 (S/X)	NODDACKS (1931)
8 Granites (composite), U.S.A. and Canada	2.5 (C)	ONISHI and SANDELL (1957)
18 Granites (composite), various countries	4 (C)	ONISHI and SANDELL (1957)
Granites etc. (composite of 9), California	2 (C)	ONISHI and SANDELL (1957)
11 Granitic rocks (composite), Minnesota	3, 2.5 (C)	ONISHI and SANDELL (1957)
6 Granites, Liano region, Texas	4 (C)	ONISHI and SANDELL (1957)
12 Granitic rocks (composite), Japan and China	3.5 (C)	ONISHI and SANDELL (1957)
14 Granite-gneisses (composite), Canada, U.S.A., and elsewhere	2.5 (C)	ONISHI and SANDELL (1957)
Quartz monzonite, Warman, Mich.	1, 1.5 (C)	ONISHI and SANDELL (1957)
Granite, Miyako, Japan	2.0 (N/R)	HAMAGUCHI *et al.* (1964)
Yakushima, Japan	6.1 (N/R)	HAMAGUCHI *et al.* (1964)
Takakuma, Japan	7.3 (N/R)	HAMAGUCHI *et al.* (1964)
Matsuyama, Japan	7.6 (N/R)	HAMAGUCHI *et al.* (1964)
Aikawa, Japan	2.6 (N/R)	HAMAGUCHI *et al.* (1964)
Denshoji, Japan	4.3 (N/R)	HAMAGUCHI *et al.* (1964)
20 Granites (composite), Japan	2.7 (N/R)	HAMAGUCHI *et al.* (1964)
Fine migmatite, Hidaka, Japan	1.5 (N/R)	HAMAGUCHI *et al.* (1964)
Granodiorite, Dando, Japan	1.7 (N/R)	HAMAGUCHI *et al.* (1964)
Takato, Japan	1.7 (N/R)	HAMAGUCHI *et al.* (1964)
Matsuyama, Japan	2.4 (N/R)	HAMAGUCHI *et al.* (1964)
5 Granodiorites (composite), Japan	1.9 (N/R)	HAMAGUCHI *et al.* (1964)
Granitic rocks, Mekhmana, Lesser Caucasus	<1	MUSTAFAEV (1964)
Granitic rocks, Daliday, Lesser Caucasus	15 (av.)	MUSTAFAEV (1964)
8 Leucocratic granites, Dzhida, W. Transbaykalia	2.5 (av.) (3.0-1.1) (S)	PETROVA and LEGEYDO (1965)
27 Leucocratic granites, Dzhida, W. Transbaykalia	2.4 (av.) (3.3-1.6) (S)	PETROVA and LEGEYDO (1965)
Microgranosyenite, Burgultag River	3.8 (S)	PETROVA and LEGEYDO (1965)
42 Granites, S. W. Cape, S. Africa	2.4—4.4 (S)	KOLBE (1965)
35 Granites, Snowy Mountains, Australia	2.0—4.1 (S)	KOLBE (1965)
12 Porphyritic granites, Dzhida, W. Transbaykalia	2.7 (av.) (4—1.4) (S)	PETROVA and LEGEYDO (1965)
10 Granodiorites, Dzhida W. Transbaykalia	6.0 (av.) (8.2—3.9) (S)	PETROVA and LEGEYDO (1965)
4 Gneissic granites, Snowy Mts., Australia	3.7 (av.) (3.5—3.8) (S)	KOLBE and TAYLOR (1966)
20 Granodiorites, Snowy Mts., Australia	3.0 (av.) (2.0—4.1) (S)	KOLBE and TAYLOR (1966)
8 Leucogranites, Snowy Mts., Australia	3.5 (av.) (2.9—4.1) (S)	KOLBE and TAYLOR (1966)

Table 50-E-2 (Continued)

Rock Locality	Sn (ppm)	Author
17 Coarsely porphyritic granites, Cape, S. Africa	3.0 (av.) (2.4—3.9) (S)	KOLBE and TAYLOR (1966)
9 Medium grained granites, Cape, S. Africa	3.2 (av.) (2.8—4.2) (S)	KOLBE and TAYLOR (1966)
8 Fine-grained granites, Cape, S. Africa	3.4 (av.) (3.0—4.2) (S)	KOLBE and TAYLOR (1966)
Granite, G-1, Westerley, U.S.A.	<2 (S)	TUREKIAN (1957)
	5 (S)	CHODOS (1957)
	2.3 (av.) (S)	ONISHI and SANDELL (1957)
	5.1 (av.) (S)	HERZ and DUTRA (1960)
	8.8 (av.) (S)	BROOKS, AHRENS and TAYLOR (1960)
	<5 (S)	HÜGI (1961)
	<20 (S)	CLARKE and SWAINE (1962)
	3.3 (av.) (C)	CARMICHAEL and MCDONALD (1961a)
	3.4 (N/R)	HAMAGUCHI *et al.* (1962)
	5.0 (S)	INGAMELLS and SUHR (1963)
	3.4 (av.)[a] (N/R)	HAMAGUCHI *et al.* (1964)
	3.7 (av.)[b] (N/R)	HAMAGUCHI *et al.* (1964)
	5 (I)	BROWN and WOLSTENHOLME (1964)
	4.5 ± 0.4 (S)	SCHROLL and WENINGER (1965)
	2.3 (I)	TAYLOR (1965)
(recommended value)	4	FLEISCHER (1965)
Aplite, Churulgum River	4.5 (S)	PETROVA and LEGEYDO (1965)
Aplite, Utukhtuy River	1.6 (S)	PETROVA and LEGEYDO (1965)
24 Greisens	7800—780 (av.) (S)	GOLDSCHMIDT and PETERS (1933)
Greisen, Cinovec, W. Bohemia	1300 (P)	HOFFMAN and TRDLICKA (1963)
7 Pitchstones, Arran	1.2 (av.) (0.2—1.8) (C)	CARMICHAEL and MCDONALD (1961b)
12 Pitchstones, Iceland	2.6 (av.) (0.8—4.5) (C)	CARMICHAEL and MCDONALD (1961b)
3 Pitchstones, Eigg	1.1, 1.6, 2.5 (C)	CARMICHAEL and MCDONALD (1961b)

[a] Opening-up based on H_2SO_4-HF digestion.
[b] Opening-up based on Na_2O_2 fusion.

2.4 ppm Sn for 27 granites, all from the Dzhida region, U.S.S.R., as the averages. All these recent figures provide further justification for the average concentration of tin in granitic rocks reported by ONISHI and SANDELL and HAMAGUCHI *et al.* Higher tin averages reported by GOLDSCHMIDT and PETERS (1933) for granitic rocks are probably too high.

Table 50-E-3. *Tin in intermediate rocks*

Rock locality	Sn (ppm)	Author
Diorite, Harz	0.1 (S)	OTTEMANN (1941)
Diorite, Harz	0.01 (S)	OTTEMANN (1941)
6 Andesites, composite, Japan	1 (C)	ONISHI and SANDELL (1957)
Andesite, Uchimura, Japan	1.5 (N/R)	HAMAGUCHI *et al.* (1964)
Nijosan, Japan	1.8 (N/R)	HAMAGUCHI *et al.* (1964)
Yatsugatake, Japan	1.7 (N/R)	HAMAGUCHI *et al.* (1964)
15 Andesites, composite, Japan	2.0 (N/R)	HAMAGUCHI *et al.* (1964)
Diorite, Takanuki, Japan	0.7 (N/R)	HAMAGUCHI *et al.* (1964)
4 Gabbro-diorites, Modonkul' River, W. Transbaykalia	7.5 (av.) (10.0—3.8) (S)	PETROVA and LEGEYDO (1965)
10 Quartz-bearing and quartz diorites, Modonkul' and Khuldat Rivers, W. Transbaykalia	5.5 (av.) (8.7—3.2) (S)	PETROVA and LEGEYDO (1965)
24 Nepheline syenites	8—39 (av.) (S)	GOLDSCHMIDT and PETERS (1933)
4 Monzonites, U.S.A.	10 (av.) (S)	BILLINGS and RABBITT (1947)
25 Monzonites and quartz monzonites, Shabartayev, W. Transbaykalia	3.5 (av.) (6.3—2.4) (S)	PETROVA and LEGEYDO (1965)
32 Quartz syenites and granosyenites, Modonkul', Ulekchin, and Ulyatuy Rivers, W. Transbaykalia	2.9 (av.) (5.0—1.2) (S)	PETROVA and LEGEYDO (1965)

Table 50-E-4. *Tin in mafic rocks*

Rock Locality	Sn (ppm)	Author
Basaltic and gabbroic rocks	6 (av.)	VINOGRADOV (1956)
Basalts, composite of 11, mostly North America	2 (C)	ONISHI and SANDELL (1957)
Basalts, composite of 6, mostly U.S.A.	1.5 (C)	ONISHI and SANDELL (1957)
Basalts, weighted composite of Greenstone flows, Mich.	1 (C)	ONISHI and SANDELL (1957)
Basalts, weighted composite of Kearsarge flow, Mich.	1 (C)	ONISHI and SANDELL (1957)
8 Basaltic rocks, composite, Keweenawan, Mich.	1.5 (C)	ONISHI and SANDELL (1957)
7 Olivine basalts, composite, Oregon	0, 0.5 (C)	ONISHI and SANDELL (1957)
Gabbro, Duluth, Minnesota	2 (C)	ONISHI and SANDELL (1957)
Basaltic rocks	1.5 (av.)	TUREKIAN and WEDEPOHL (1961)
Basalt, Matsue, Japan	1.2 (N/R)	HAMAGUCHI *et al.* (1964)
Basalt, Ajiro, Japan	1.1 (N/R)	HAMAGUCHI *et al.* (1964)
Diabase, Tanzawa, Japan	0.5 (N/R)	HAMAGUCHI *et al.* (1964)
Tholeiitic basalt W-1, ("diabase") Centerville, U.S.A.	<2 (S)	TUREKIAN (1957)
	2.8 (av.) (C)	ONISHI and SANDELL (1957)
	8.7 (av.) (S)	BROOKS, AHRENS and TAYLOR (1960)

Table 50-E-4 (Continued)

Rock Locality	Sn (ppm)	Author
	< 5 (S)	HÜGI (1961)
	< 20 (S)	CLARKE and SWAINE (1962)
	2.5 (av.) (C)	CARMICHAEL and McDONALD (1961a)
	3.4 (av.) (N/R)	HAMAGUCHI *et al.* (1962)
	2.5 (S)	INGAMELLS and SUHR (1963)
	2.3 ± 0.2 (S)	SCHROLL and WENINGER (1965)
	2 (I)	BROWN and WOLSTENHOLME (1964)
	1.0 (I)	TAYLOR (1965)
(recommended value)	3	FLEISCHER (1965)
Norite, Lydenbury, Tvl., S. Africa	1.3 (S)	BROOKS, AHRENS and TAYLOR (1960)

Table 50-E-5. *Tin in ultramafic rocks*

Rock Locality	Sn (ppm)	Author
Ultramafic rocks, composite of 13, U.S.A. and Canada	0.5 (C)	ONISHI and SANDELL (1957)
Peridotite, Tysfjord, Norway	0.9 (S)	BROOKS, AHRENS and TAYLOR (1960)
Lherzolite, Pyrenees, France	0.3 (S)	BROOKS, AHRENS and TAYLOR (1960)
Hortonolite dunite, Lydenburg, Tvl. S. Africa	1.3 (S)	BROOKS, AHRENS and TAYLOR (1960)
Ultramafic rocks	0.5 (av.)	TUREKIAN and WEDEPOHL (1961)
Serpentinite, Tosayama, Japan	0.2 (N/R)	HAMAGUCHI *et al.* (1964)
Peridotite, Hidaka, Japan	0.1 (N/R)	HAMAGUCHI *et al.* (1964)
Peridotite, unknown locality, Japan	0.6 (N/R)	HAMAGUCHI *et al.* (1964)
Cortlandite, Abukuma, Japan	0.5 (N/R)	HAMAGUCHI *et al.* (1964)

The abundance values of tin in intermediate and mafic rocks are not established as well as those for granitic rocks, because of the scarcity of reliable data. However, the averages given by ONISHI and SANDELL and HAMAGUCHI and coworkers may be taken as probable for these rock types.

Rather consistent values were reported by several authors for the content of tin in ultramafic rocks, although the data are too few to draw a general conclusion. Approximately 0.8 ppm as the average value of tin can be obtained from three determinations reported by BROOKS, AHRENS and TAYLOR (1960). This is reasonably consistent with those given by ONISHI and SANDELL and HAMAGUCHI and coworkers. Individual analyses for tin in the main types of igneous rocks are tabulated in Tables 50-E-2 to 50-E-5.

Revised manuscript received: September 1967

50-F. Behavior during Processes Connected with Magmatism

Tin becomes enriched in the later products of magmatic differentiation. Tin (IV) behaves chiefly as a network former in silicate melts so that tin-oxygen complexes do not enter appreciably into the major rock-forming minerals, but accumulate in residual magmas until a concentration is reached which is sufficient to precipitate cassiterite as a separate mineral. Some common accessory minerals, such as apatite, sphene, ilmenite, monazite, rutile, magnetite, allanite and xenotime, also carry much of the tin present in igneous rocks. GANEEV and coworkers (1961) investigated the distribution of tin along with other elements in the granite massifs of central Kazakhstan. Two stages of mineralization, i.e. a pegmatite stage and a pneumatolytic hydrothermal phase, are apparent in these massifs. During the pneumatolytic hydrothermal stage various greisen facies and veins with rare metal mineralization are formed and marked by the abundance of micas. The tin content in the biotite from the granite is 250 ppm. Early-formed coarse-flake muscovite in the greisen contains 580 ppm Sn, while later-formed fine-flake muscovite is as high as 1,240 ppm Sn. Much lower values of tin were found on the most recent micas, gilbertite and sericite, down to 62 and 25 ppm, respectively. It is evident that the conversion of biotite to muscovite, which is very characteristic of granites containing rare metal mineralization, raises tin to a great extent. A relation between high tin content in the biotites of granites and pegmatites and tin mineralization is also indicated.

AHRENS and LIEBENBERG (1950) also showed that the tin content of pegmatite micas ranges from <4 ppm to 4,000 ppm Sn and is of value as a possible aid for prospecting rapidly for tin.

IVANOVA (1963) revealed that in granites, containing tin mineralization, the tin content ranges from 16 to 32 ppm, i.e. 5 to 10 times higher than the average tin content in common granitic rocks (3 ppm Sn). A distinct enrichment in tin and tungsten was also observed in greisenized granites, the contents of both metals increasing with increasing degree of alteration.

Revised manuscript received: September 1967

50-G. Behavior during Weathering and Rock Alteration

There is little information about the geochemical behavior of tin during weathering processes. The most important tin ore, cassiterite, is strongly resistant to weathering. However, some tin is dissolved in natural waters and its fate is determined by several factors, including oxidation-reduction conditions. HIRST (1962) thinks that the contribution of non-detrital tin removed from the solution by adsorption, co-precipitation or direct precipitation, may represent a high proportion of the total tin content of 15 sediments of the Gulf of Paria (see Table 50-K-1). Tin in solution behaves in different ways according to its oxidation state. According to GOLDSCHMIDT (1934) divalent tin has the ionic potential of 2.16 so that it belongs to a group of elements, which go into solution as cations when subjected to weathering and subsequent transportation. On the other hand, quadrivalent tin has the ionic potential of 5.14 and, therefore, is believed to precipitate promptly into hydrolysates. Divalent tin ions are not thought of as being important in weathering processes, because being a strong reducing agent, it can only be present in reducing environments like sapropel facies.

$$Sn^{4+} + 2e = Sn^{2+} \qquad \varepsilon^\circ = 0.15 \text{ V}.$$

However, an acidic environment must prevail for the divalent tin to survive in the form of stable ions. But weakly basic conditions usually dominate throughout the sapropel facies. Therefore, stannic tin may be expected to be the only stable ionic species in the weathering cycle.

GOLDSCHMIDT and PETERS (1933) found that tin becomes enriched in aluminum-rich resistates giving as much as 40 to 80 ppm Sn for bauxites. Once weathered, tin does not become concentrated in sedimentary iron ores, because tin goes into solution together with iron to a very limited extent, so that almost all the tin tends to remain in the residues as a hydrolysis product along with aluminum. GOLDSCHMIDT and PETERS (1933) found 8 ppm Sn in Bohnerz, Württemberg, and 8 ppm Sn in Salzgitter iron ore (goethite).

Revised manuscript received: September 1967

50-H. Solubilities of Tin Compounds Which Control Concentrations in Natural Waters

Solubility product computations of several tin compounds in water are given in Tables 50-H-1.

Table 50-H-1. *Solubility products of tin compounds*

Compound	°C	Solubility product[a]	Reference
$Sn(OH)_2$	25	$5 \cdot 10^{-26}$	PRYTZ, M.: Z. anorg. allgem. Chem. **174**, 371 (1928)
SnS	25	$8 \cdot 10^{-29}$	BARBER, H. H.: Semimicro quantitative analysis, 2nd. ed. 1953
$Sn(OH)_4$	25	10^{-56}	LATIMER, W. M.: Oxidation potentials. 1950

[a] Ion concentration expressed in gram · ion/litre.

Revised manuscript received: September 1967

50-I. Abundance in Natural Waters and in the Atmosphere

A high content of tin and several other elements is usually observed in waters of mineralized zones and in streams fed from these areas. UDODOV and PARILOV (1961) analyzed about 4,500 water samples spectrographically for 22 elements including tin in many regions of Siberia, including a variety of ore deposits. Tin was found in 340 of 4,278 samples analyzed, averaging 0.09 ppb. Taking the ratio of average tin content in water to its CLARKE in rocks, these authors concluded that the migration capacity of tin is very slight, about the same as that of Ba, Ti and Zr.

$$\frac{\text{Average tin in natural waters (in wt. \%)}}{\text{Average tin in rocks (Chapter 50-E) (0.0003 wt. \%)}} = 3 \cdot 10^{-5}.$$

Table 50-I-1. *Tin in terrestrial waters*

Sample	Sn (ppb)	Author
Formation water from salt deposits, Langensalza	120 (S)	HERRMANN (1961a)
Formation water from salt deposits, Kirchheilingen II	180 (S)	HERRMANN (1961a)
Formation waters from salt deposits, Southern Harz (8 samples)	150 (av.) (S)	HERRMANN (1961b)
19 Bulgarian saline underground waters (80—2,610) m deep	0.0×10^3 to 0.67×10^3 (S)	PENTCHEVA (1965)
Municipal waters from 42 U.S.A. cities	1.1 and 2.2 for 2 samples	DURUM and HAFFTY (1961)
175 finished municipal waters	0.8—30 (in 32 samples)	TAYLOR (1964)

KLEINKOPF (1960) carried out spectrographic analyses for trace elements in 439 lake and stream waters from Maine in the U.S. Of the samples analysed 336 had less than 0.01 ppb Sn with an estimated average value of 0.03 ppb Sn for the Maine lakes.

The abundance of tin in hot springs and sea waters is listed in Tables 50-I-2 and 50-I-3, respectively. The data relating to hot springs are exclusively from Japan. Tin in hot springs ranges mostly from 0.1 to 0.5 ppb and does not show a marked dependence on the temperature and pH of the hot springs. IKEDA (1955a) worked out the extensive spectrographic determination of 40 elements in a hot springs of Moto-yu, Nasu, Japan. He found 0.1 ppm Sn in the residue recovered from the evaporation of the spring water and 5 ppm Sn in the sintered deposit of the same spring. It is evident that the tin in the deposit becomes enriched by a factor of 50 relative to the water.

There is little reliable information about the concentration of tin in sea water. Recent neutron activation determination of tin in the Pacific Ocean ranges from 0.30 to 1.22 ppb, thus involving variation by a factor of 4 (HAMAGUCHI *et al.* 1964). A geometric mean may provide a tentative average of tin in sea water, coming out 0.72 ppb.

Table 50-I-2. *Tin in hot springs* (IKEDA, 1955b)

Locality	Temp. (°C)	pH	Sn (ppb)
Nasu, 6 samples	34—92	1.5—6.4	0.1∼0.5 (S)
Nikko, 6 samples	55—69	6.0—6.6	∼0.5∼1 (S)
Hakone, 7 samples	33—77	2.0—8.0	<0.1∼0.5 (S)
Kusatu, 7 samples	57—65	1.5—1.6	∼0.5 (S)
Various, 7 samples	16—91	1.2—8.2	∼0.1∼0.5 (S)

Table 50-I-3. *Tin in sea water*

Locality	Sn (ppb)	Author
Pacific Ocean:		
30° 00' N 147° 30' E, 500 m deep	0.85 (N/R)	HAMAGUCHI *et al.* (1964)
30° 00' N 144° 30' E, 500 m deep	1.22 (N/R)	HAMAGUCHI *et al.* (1964)
30° 00' N 142° 30' E, 500 m deep	0.87 (N/R)	HAMAGUCHI *et al.* (1964)
30° 00' N 139° 30' E, 500 m deep	0.30 (N/R)	HAMAGUCHI *et al.* (1964)

Sizeable amounts of tin occur in the air of American cities, especially where many industrial factories are located. Amounts from 0.003 to 0.3 $\mu g/m^3$ were found in 60.6 per cent of 754 samples taken from 22 cities (TABOR and WARREN, 1958).

Of 312 samples from 6 major industrial centers 17 percent had 0.03—0.3 $\mu g/m^3$ whereas suburban areas were considerably less. No information is available about the tin content of air in the absence of contamination.

Revised manuscript received: September 1967

50-K. Abundance in Common Sedimentary Deposits

Goldschmidt and Peters (1933) found 40 ppm Sn in the composite of thirty-six European Paleozoic shales. It should be mentioned that the value obtained by Onishi and Sandell (1957) using a colorimetric method on the same sample is smaller by

Table 50-K-1. *Tin in common sedimentary rock types and marine sediments*

Material	Sn (ppm)	Author
7 Shales, composite, North America	11 (C)	Onishi and Sandell (1957)
24 Shales (Paleozoic and Mesozoic), composite, Japan	2.5 (C)	Onishi and Sandell (1957)
36 European shales (Paleozoic), composite	5 (C)	Onishi and Sandell (1957)
Carboniferous shales	3.2 (S)	Degens, Williams and Keith (1957)
Shale, Lower shale bed, Table Mt. Series, Cape Town	19 (S)	Brooks, Ahrens and Taylor (1960)
Shales	5 (S)	Wedepohl (1961) quoted in: Turekian and Wedepohl (1961)
8 Pelagic clays, composite, Pacific Ocean	1.5 (C)	Onishi and Sandell (1957)
Ocean mud, Lutzholm Bay, 500 m deep	3.8 (S)	Brooks, Ahrens and Taylor (1960)
Sediments from Atlantic, Pacific, and Indian Oceans, and Mediterranean Sea:		
7 Pelagic calcareous sediments	18 (av.) (S)	El Wakeel and Riley (1961)
9 Pelagic argillaceous sediments	20 (av.) (S)	El Wakeel and Riley (1961)
2 Pelagic siliceous sediments	12 (av.) (S)	El Wakeel and Riley (1961)
2 Volcanic muds	5, 25 (S)	El Wakeel and Riley (1961)
2 Near shore muds (less than 2 μ fraction)	20, 23 (S)	El Wakeel and Riley (1961)
Sediment, 58 m deep, Baltic Sea	<4 (S)	Manheim (1961)
Sediment, Baltic Sea	6 (S)	Manheim (1961)
Non-pelagic manganese-iron nodules, Baltic Sea	1∼3 (av.) (S)	Manheim (1961)
15 sediments (non-detrital fraction only), Gulf of Paria	0.5∼2 (S)	Hirst (1962)
10 Pelagic clays, Pacific Ocean (range 2.9—7.5 ppm)	4.9 (av.) (N/R)	Hamaguchi *et al.* (1964)
3 Pelagic clays, Japan Sea (range 3.9—4.3 ppm)	4.1 (av.) (N/R)	Hamaguchi *et al.* (1964)
4 Volcanic muds, Pacific Ocean (range 0.8—1.8 ppm)	1.4 (av.) (N/R)	Hamaguchi *et al.* (1964)
Bottom sediment, Atlantis II Deep, Red Sea	20 (S)	Miller *et al.* (1966)
3 globigerina oozes, Pacific Ocean (range 1.4—1.6 ppm)	1.5 (av.) (N/R)	Hamaguchi *et al.* (1964)

a factor of eight than that of GOLDSCHMIDT and PETERS. The weighted average of thirty-six European shales, twenty-four Japanese Paleozoic and Mesozoic shales is 4 ppm (ONISHI and SANDELL, 1957). TUREKIAN and WEDEPOHL (1961) report a weighted average of 6 ppm Sn in shales. There is no representative and reliable information regarding the tin abundance in sandstones and carbonate rocks. HAMAGUCHI *et al.* (1964) determined tin by neutron activation in different types of oceanic sediments. They obtained values ranging between 0.8 to 7.5 ppm Sn for all sedimentary types (as listed in Table 50-K-1). This is considerably lower than the values reported by EL WAKEEL and RILEY (1961). The lower values appear to be less likely influenced by contamination. No significant difference is found in tin contents of pelagic clays of the Pacific Ocean and the Japan Sea. Contrary to GOLDSCHMIDT's statement, that it is possible for tin to precipitate from sea water along with carbonate sediment, globigerina ooze does not show any marked tendency for tin to become enriched. As a whole, pelagic sediments do not show any enrichment of tin compared with igneous rocks.

Revised manuscript received: September 1967

50-L. Biogeochemistry

There is no definite information about the biological role of tin.

The concentration of tin in coals and coal ashes, has been determined by GOLDSCHMIDT and PETERS (1933); HORTON and AUBREY (1953); LEUTWEIN and RÖSLER (1956); and ZUBOIC *et al.* (1960, 1964, 1966). There is no clear information available about the form in which tin exists in coals. According to LEUTWEIN and RÖSLER (1956), tin occurs in coals bound to organic residues. This assumption may be further supported by the fact reported by TEREBENINA and ANGELOVA (1962) that tin is mostly enriched in the benzene and ethanol extract of coals. The tin content of coals generally tends to increase with increasing concentration of ash in coals.

Data on tin in soils can be found in reports; by ALLISON and GADDUM (1940), MITCHELL (1948), VINOGRADOV (1950), LOUNAMAA (1956), FONTANA and ALFIERI (1961), PINTA and OLLAT (1961), SAROSIEK and KLYS (1962), CHAO-LUEN FANG *et al.* (1963), and SCHROEDER *et al.* (1964). The variation in the concentration of tin in soils is large, in part possibly because the abundance of tin is related to the concentration in the country rock.

In plants, tin is not always detected. GLAZOVSKAYA (1964) reports that tin does not show a tendency to become enriched in the ashes of some common plants relative to soil whereas SAROSIEK and KLYS (1962) found that several plant species, Calluna vulgaris, Gnaphalium sylvaticum, Sempervivum soboliferum, Silene inflata, Tanacetum vulgare, and Quercus sessilis accumulated tin, the average tin content in ashes of these plants being as high as 46 ppm.

Other information has been reported on tin in plants (GORDON, 1952; LOUNAMAA, 1956; HANNA and GRANT, 1962; SCHROEDER *et al.*, 1964); in human tissues (KEHOLE, CHOLAK and STORY, 1940; GOFMAN *et al.*, 1961; STOROZHEVA, 1963; SCHROEDER *et al.*, 1964); in fresh and processed foods (SCHROEDER *et al.*, 1964).

Tin contents of marine plankton (NICHOLLS, 1959) are highly variable, ranging from <1 ppm to 90 ppm in the ash (Table 50-L-1). An average for tin in terrestrial and marine plants is given by VINOGRADOV (1954) as 0.5 ppm of fresh tissues. Additional data are presented in Table 50-L-1.

Tin contents of land fishes are listed in Table 50-L-2.

Table 50-L-1. *Tin in marine organismus*

Material	Tin content	Author
a) Marine plankton (ppm in ash):		
Coelenterata		
Cyanea capillata	4 (av.) (S)	NICHOLLS (1959)
Ctenophora		
Beröe cucumis	7 (S)	NICHOLLS (1959)

Table 50-L-1 (Continued)

Material	Tin content	Author
Mollusca, Pteropoda		
Limacina retroversa	< 1 (av.) (S)	NICHOLLS (1959)
Clione limacina	20 (S)	NICHOLLS (1959
Mollusca, Cephalopoda		
Ommastrephes illicebrosa	3 (av.) (S)	NICHOLLS (1959)
Arthropoda, Copepoda		
Centropages typius and C. hamatus	50 (av.) (S)	NICHOLLS (1959)
Calanus finmarchicus	< 1 (av.) (S)	NICHOLLS (1959)
Calanus finmarchicus with ~ 50% neritic phytoplankton	90 (av.) (S)	NICHOLLS (1959)
Arthropoda Euphausiacea		
Euphausia Krohnii	< 1 (av.) (S)	NICHOLLS (1959)
Chaetognatha		
Sagitta elegans	20 (av.) (S)	NICHOLLS (1959)
Tunicata		
Salpa fusiformis	8 (av.) (S)	NICHOLLS (1959)
Sea salt	0.09 (S)	NODDACKS (1939)
Chaetoceras curvisetus (diatomaceous plankton), Black Sea	100	VINOGRADOV and KOVALJSKII (1962)
b) Marine animals (in ppm of dry matter, unless otherwise mentioned):		
Halichondria	1.7 (S)	NODDACKS (1939)
Coelenterata		
Cyanea capillata	32 (S)	NODDACKS (1939)
Metridium dianthus	15 (S)	NODDACKS (1939)
Vermes		
Lineus longissimus (Nemertea)	0.02[a] (S)	WEBB (1937)
Nephthys (Polychaeta)	— (S)	WEBB (1937)
Echinodermata		
Stichopus tremulus	6.2 (S)	NODDACKS (1939)
Brissopsis lyrifera (shell)	1.6 (S)	NODDACKS (1939)
Asterias rubens	7.2 (S)	NODDACKS (1939)
Asterias rubens	800[a] (S)	WEBB (1937)
Paracentrotus lividus (Ovary)	800[a] (S)	WEBB (1937)
Mollusca		
Helix pomatia	250[a] (S)	WEBB (1937)
Aeolidia papillosa	150[a] (S)	WEBB (1937)
Tunicata		
Ciona intestinalis	3.5 (S)	NODDACKS (1939)
Pisces		
Squalus acanthias (eviscerated)	2 (S)	NODDACKS (1939)
Ctenolabrus rupestris (whole)	4.7 (S)	NODDACKS (1939)
Arca, shell, Gaybu	0.8 (S)	WEDEPOHL (1955)
Pernambuco beech		
Sea urchin (shell), Bahia beech	2.6 (S)	WEDEPOHL (1955)
Algae (calcareous), Lithotamnium	0.7 (S)	WEDEPOHL (1955)
San Carlos-Fernando Poo		
Metapenaeus joyneri, Tokyo Bay	2.9 (N/R)	HAMAGUCHI *et al.* (1964)
Watasenia scintillans, Tokyo Bay	1.6 (N/R)	HAMAGUCHI *et al.* (1964)
Meretrix meretrix cusoria, Tokyo Bay	0.7 (N/R)	HAMAGUCHI *et al.* (1964)

[a] Expressed as % of weight of ash cation.

Table 50-L-2. *Tin in fresh-water fishes*

Material	Sn content	Author
Ca 250 Plecoglossus altivelis (ayu), Japan	1—10 ppm (in ash)[a] (S)	MORITSUGU and KOBAYASHI (1962)
Hypomesus olidus, Lake Kasumigaura, Japan	0.3 (dry basis) (N/R)	HAMAGUCHI et al. (1964)

[a] A few samples had values of 100 ppm or more.

Revised manuscript received: September 1967

50-M. Abundance in Common Metamorphic Rock Types

Little information is available about the abundance and distribution of tin in metamorphic rocks. Systematic investigation of the behavior of tin during metamorphic reaction is also lacking.

The few available data are quoted in Table 50-M-1 and 50-M-2. Detailed information on tin in metamorphic minerals is furnished by DE VORE (1955a, b).

Table 50-M-1. *Tin in metamorphic rocks*

Rocks	Sn (ppm)	Author
Grey gneiss, W. Ongul I., Lützow-Holm Bay, Antarctica	21.5 (S)	BROOKS, AHRENS and TAYLOR (1960)
Limonite-quartz-sericite schist, Quadrilátero, Ferrîferro	7.1, 5.6 (S)	HERZ and DUTRA (1960)
Hornfels, Hidaka, Japan	3.9 (N/R)	HAMAGUCHI *et al.* (1964)
Slate, Okaya, Nagano, Japan	2.6 (N/R)	HAMAGUCHI *et al.* (1964)
Eclogite, Almeklovdalen, Norway	0.7 (S)	BROOKS, AHRENS and TAYLOR (1960)

Appreciable concentrations of tin can be accumulated in minerals such as axinite, epidote, vesuvianite, garnet and datolite:

Table 50-M-2. *Tin in metamorphic minerals*

5 Epidotes	40 (av.) (trace to 170) (S)	HELLWEGE (1956)
7 Axinites	200 (av.) (4.0—600) (S)	HELLWEGE (1956)
4 Vesuvianites	300 (av.) (0.0—130) (S)	HELLWEGE (1956)
5 Garnets	20 (av.) (1.0—81) (S)	HELLWEGE (1956)
5 Datolites	10 (av.) (3.7—27) (S)	HELLWEGE (1956)

Revised manuscript received: September 1967

References: Sections 50-B to 50-M

AHRENS, L. H., and W. R. LIEBENBERG: Tin and indium in mica as determined spectroscopically. Am. Mineralogist **35**, 571 (1950).

ALBERS, W., C. HAAS, H. J. VINK, and J. D. WASSCHER: Investigation on SnS. J. Appl. Phys. **32**, 2220 (1961).

—, and K. SCHOL: The P—T—X phase diagram of the system Sn-S. Philips Res. Repts. **16**, 329 (1961).

ALLER, L. H.: The abundances of the elements. New York: Interscience Publ. Inc. 1961.

— Advances in astronomy and astrophysics, ed. by Z. KOPAL, vol. 3, p. 1—25, 1965.

ALLISON, R. V., and L. W. GADDUM: The trace element content of some important soils. A comparison. Proc. Soil Sci. Soc. Florida **2**, 68 (1940).

BILLINGS, M. P., and J. C. RABBITT: Chemical analyses and calculated modes of the Oliverian magma series, Mt. Washington Quadrangle, New Hampshire. Bull. Geol. Soc. Am. **58**, 573 (1947).

BORCHERT, H., u. J. DYBEK: Zur Geochemie des Zinns. Chem. Erde **20**, 137 (1960).

BROOKS, R. R., L. H. AHRENS, and S. R. TAYLOR: The determination of trace elements in silicate rocks by a combined spectrochemical-anion exchange technique. Geochim. Cosmochim. Acta **18**, 162 (1960).

BROWN, J. C.: Lagerstätten- und erzmikroskopische Untersuchung der Zinnerzgänge der East-Pool-Mine bei Redruth in Cornwall. Neues Jahrb. Mineral., Abt. A, Bl.-Bd. **68**, 331 (1934).

BROWN, R., and W. A. WOLSTENHOLME: Analysis of geological samples by spark source mass spectrometry. Nature **201**, 598 (1964).

CAMERON, A. G. W.: A revised table of abundances of the elements. Astrophys. J. **129**, 676 (1959).

— A new table of abundances of the elements in the solar system. Proc. Paris symposium on the origin of the elements. 1967 (in press).

CARMICHAEL, I., and A. McDONALD: The colorimetric and polarographic determination of some trace elements in the standard rocks G-1 and W-1. Geochim. Cosmochim. Acta **22**, 87 (1961a).

— — The geochemistry of some natural acid glasses from the North Atlantic Tertiary Volcanic Province. Geochim. Cosmochim. Acta **25**, 189 (1961b).

CHAO-LUEN FANG, TA-CHUAN SUNG, and BING YEH: Trace elements in the soils of North-Eastern China and Eastern Inner Mongolia. T'u Jang Hsueh Pao **11**, 130 (1963); Chem. Abstr. **60**, 9049a (1964).

CHODOS, A. A.: Quoted in AHRENS and FLEISCHER 1960.

CLARKE, C., and D. J. SWAINE: The content of several trace elements in the standard rocks of G-1 and W-1. Geochim. Cosmochim. Acta **26**, 511 (1962).

CLARKE, F. W., and H. S. WASHINGTON: The composition of the earth's crust. U.S. Geol. Surv., Profess. Papers **127** (1924).

DEGENS, E. T., E. G. WILLIAMS, and M. L. KEITH: Environmental studies of carboniferous sediments, part I; Geochemical criteria for differentiating marine and fresh water shales. Bull. Am. Assoc. Petrol. Geologists **41**, 2427 (1957).

DE LAETER, J. R., and P. M. JEFFÉRY: Tin: its isotopic and elemental abundance. Geochim. Cosmochim. Acta **31**, 969 (1967).

DE VORE, G. W.: The role of adsorption in the fractionation and distribution of elements. J. Geol. **63**, 159 (1955a).

— Crystal growth and the distribution of elements. J. Geol. **63**, 471 (1955b).

DURUM, W. H., and J. HAFFTY: Occurrence of minor elements in water. Circ. U.S. Geol. Surv. No. 445 (1961).

EL WAKEEL, S. K., and J. P. RILEY: Chemical and mineralogical studies of deep-sea sediments. Geochim. Cosmochim. Acta **25**, 110 (1961).

FESSER, H.: Spurenelemente in bolivianischen Zinnsteinen. Geol. Jahrb. **85**, 605 (1968).

FLEISCHER, M.: Summary of new data on rock samples G-1 and W-1, 1962—1965. Geochim. Cosmochim. Acta **29**, 1263 (1965).

FLOWERS, B. H.: The nuclear shell model. In: Progress in nuclear physics, ed. by O. R. FRISCH, vol. 2, p. 241. New York: Academic Press 1952.

FONTANA, P., and L. ALFIERI: Spectrographic determination of trace elements in soils of the Province of Piacenza. Ann. Fac. Agrar., Univ. Cattolica Sacro Cuore **1**, 53 (1961).

GANEEV, I. G., D. N. PACHADZHANOV, and L. A. BORISENOK: Geochemistry of gallium, tin and some other elements in the process of greisenization. Geochemistry **9**, 830 (1961).

GLAZOVSKAYA, M. A.: Biological cycle of elements in various landscape zones of the Urals, Fiz., Khim., Biol. i. Mineralog. Pochv. SSSR, Akad. Nauk SSSR, Dokl. k VIII-mu [Vośmomu] Mezhdunar. Kongr. Pochvovedov, Bucharest 1964, p. 148—157; Chem. Abstr. **62**, 4562 h (1965).

GOFMAN. J. W., O. DELALLA, G. JOHNSON, E. L. KOVICH, O. LOWE, W. MARTIN, D. L. PILUSO, R. K. TANDY, F. UPHAM, R. WEITZEL, and D. WILBUR: Chemical elements of the blood of man, UCRL-9897, Lawrence Radiation Laboratory Report, University of California, Berkeley, Fall, 1961.

GOLDSCHMIDT, V. M.: Drei Vorträge über Geochemie. Geol. Fören. Stockholm Förh. **56**, 385 (1934).

— Geochemistry, ed. by A. MUIR. Oxford: Clarendon Press 1954.

—, u. C. PETERS: Zur Kenntnis der Troilitknollen der Meteoriten, ein Beitrag zur Geochemie von Chrom, Nickel und Zinn. Nachr. Ges. Wiss. Göttingen, Jahresber. Geschäftsjahr III, **36** — IV, **37**, 278 (1933).

GORDON, M.: Trace elements in peats. Torfnachrichten **3**, 12 (1952).

HAMAGUCHI, H., R. KURODA, N. ONUMA, K. KAWABUCHI, T. MITSUBAYASHI, and K. HOSOHARA: The geochemistry of tin. Geochim. Cosmochim. Acta **28**, 1039 (1964)

— — T. SHIMIZU, I. TSUKAHARA, and R. YAMAMOTO: Values for trace elements in G-1 and W-1 with neutron activation analysis-II. Mo, Sn, Ta, W. Geochim. Cosmochim. Acta **26**, 503 (1962).

— N. ONUMA, Y. HIRAO, H. YOKOYAMA, S. BANDO, and M. FURUKAWA: The abundances of arsenic, tin, and antimony in chondritic meteorites. Geochim. Cosmochim. Acta (in press).

HANNA, W. J., and C. L. GRANT: Spectrochemical analysis of the foliage of certain trees and ornamentals for 23 elements. Bull. Torrey Botan. Club **89**, 293 (1962).

HEIDE, F.: Rare elements in tektites. Forsch. Fortschr. **12**, 232 (1936); Chem. Abstr. **31**, 6145[1] (1937).

HELLWEGE, H.: Zum Vorkommen des Zinns als Spurenelement in Mineralien. Hamburger Beitr. angew. Mineral. u. Kristallphysik **1**, 73 (1956).

HERRMANN, A. G.: Über das Vorkommen einiger Spurenelemente in Salzlösungen aus dem deutschen Zechstein. Kali Steinsalz **3**, 209 (1961a).

— The effect of Cu-, Sn-, Pb- and Mn-containing oil field waters on the Stassfurt Series of the Southern Harz Region. Neues Jahrb. Mineral., Monatsh. 52 (1961); Chem. Abstr. **55**, 12183d (1961b).

HERZ, N., and C. V. DUTRA: Minor element abundance in a part of the Brazilian Shield. Geochim. Cosmochim. Acta **21**, 81 (1960).

HIRST, D. M.: Geochemistry of modern sediments from the Gulf of Paria II. The location and distribution of trace elements. Geochim. Cosmochim. Acta **26**, 1147 (1962).

HOFFMAN, V., and Z. TRDLICKA: Sn isomorphism in zinnwaldite from the deposit of Cinovec, Western Bohemia. Casopis Mineral. Geol. **8**, 244 (1963).

HORN, M. K., and J. A. S. ADAMS: Computer-derived geochemical balances and element abundances. Geochim. Cosmochim. Acta **30**, 279 (1966).

HORTON, L., u. K. AUBREY: Gehalte der Aschen der Vitrainkohlen aus dem Flöz Barnsley (Yorkshire). Zentr. Mineral., Tl. II, 31 (1953).

HÜGI, TH.: Quoted in FLEISCHER and STEVENS 1962 as written communication.

IKEDA, N.: On the sinter deposit of the Moto-yu Spring, Yumoto, Nasu. Nippon Kagaku Zassi **76**, 1195 (1955a).

— Determination of minute quantities of tin in hot spring waters. Nippon Kagaku Zassi **76**, 1011 (1955b).

INGAMELLS, C. O., and N. H. SUHR: Chemical and spectrochemical analysis of standard silicate samples. Geochim. Cosmochim. Acta **27**, 897 (1963).

IVANOVA, G. F.: Content of tin, tungsten, and molybdenum in granites and their relation to the Sn-W deposits present in granites. Geokhimiya **470**, No. 5 (1963); Chem. Abstr. **59**, 3672d (1963)

KEHOE, R. A., J. CHOLAK, and R. V. STORY: A spectrochemical study of the normal ranges of concentration of certain trace metals in biological materials. J. Nutr. **19**, 579 (1940).

KLEINKOPF, M. D.: Spectrographic determination of trace elements in Lake Waters of Northern Maine. Bull. Geol. Soc. Am. **71**, 1231 (1960).

KOLBE, P.: The use of an oxygen jet in the spectrochemical determination of trace amounts of Pb, Tl, Ga, Cu and Sn in some silicate rocks and minerals. Geochim. Cosmochim. Acta **29**, 153 (1965).

—, and S. R. TAYLOR: Major and trace element relationships in granodiorites and granites from Australia and South Africa. Contrib. Mineral. and Petrol. **12**, 202—222 (1966).

LEUTWEIN, F., u. H. J. RÖSLER: Geochemische Untersuchungen an paläozoischen und mesozoischen Kohlen Mittel- und Ostdeutschlands. Freiberger Forschungsh. C **19**, 1 (1956); Chem. Abstr. **50**, 17377a (1956).

LOUNAMAA, J.: Trace elements in plants growing wild on different rocks in Finland. A semiquantitative spectrographic survey. Ann. Botan. Soc. Zool. Botan. Fennicae Vanamo **29**, 4 (1956).

MANHEIM, F. T.: A geochemical profile in the Baltic Sea. Geochim. Cosmochim. Acta **25**, 52 (1961).

MILLER, A. R., C. D. DENSMORE, E. T. DEGENS, J. C. HATHAWAY, F. T. MANHEIM, P. F. MCFARLIN, R. POCKLINGTON, and A. JOKELA: Hot brines and recent iron deposits in deeps of the Red Sea. Geochim. Cosmochim. Acta **30**, 341 (1966).

MITCHELL, R. L.: The spectrographic analysis of soils, plants and related material. Techn. Comm. Bur. Soil Sci. No. 44, 1948.

MOH, G.: Sulfide systems containing Sn. Carnegie Inst. Wash. Year Book **62**, 197 (1962/1963).

—, and F. BERNDT: Two new natural tin sulfides Sn_2S_3 and SnS_2. Neues Jahrb. Mineral. Monatsh. **3**, 94 (1964).

MORITSUGU, M., and J. KOBAYASHI: Trace metals in bio-materials. I Geographical difference of metals in ayu. Biol. J. Okayama Univ. **11**, 393 (1962).

MUSTAFAEV, G. V.: Distribution of tin in granitic rocks and minerals of the Dalidag and Mekhmana Massifs (Lesser Caucasus). Dokl. Akad. Nauk Azerb. SSR **20**, 57 (1964); Chem. Abstr. **62**, 12933g (1965).

NICHOLS, C. D.: Spectrographic analyses of marine plankton. Limnol. Oceanog. **4**, 472 (1959).

NODDACK, I., u. W. NODDACK: Die Häufigkeit der chemischen Elemente. Naturwissenschaften **18**, 757 (1930).

— — Die Geochemie des Rheniums. Z. Physik. Chem. (Leipzig) A **154**, 207 (1931).

— — Die geochemischen Verteilungskoeffizienten der Elemente. Svensk Kem. Tidskr. **46**, 173 (1934).

— — Die Häufigkeiten der Schwermetalle in Meerestieren. Arkiv Zool. **32** A, 1 (1939).

ONISHI, H., and E. B. SANDELL: Meteoritic and terrestrial abundance of tin. Geochim. Cosmochim. Acta **12**, 262 (1957).

OTTEMANN, J.: Untersuchung zur Verteilung von Spurenelementen, insbesondere Zinn, in Tiefengesteinen und einigen gesteinsbildenden Mineralien des Harzes. Z. Angew. Mineral. **3**, 142 (1941).

PALACHE, C., H. BERMAN, and C. FRONDEL: The system of mineralogy, 7th ed., vol. I. New York: John Wiley & Sons, Inc. 1944.

PENTCHEVA, E.: The distribution of rare and dispersed elements in Bulgarian saline underground waters. Compt. Rend. Acad. Bulgare Sci. **18**, 149 (1965); Chem. Abstr. **62**, 15899b (1965).

PETROVA, Z. I., and V. A. LEGEYDO: Geochemistry of tin in the magmatic process. Geokhimiya **4**, 482 (1965).

PINTA, M., et C. OLLAT: Recherchés physico-chimiques des èléments-traces dans les sols tropicaux-I. 1. Etude de quelques sols du Dahomey. Geochim. Cosmochim. Acta **25**, 14 (1961).

SAROSIEK, J., and B. KLYS: The tin content in plants and soil in Sudety. Acta Soc. Botan. Polan. **31**, 737 (1962); Chem. Abstr. **60**, 16451h (1964).

SCHRÖCKE, H.: Zur Geochemie erzgebirgischer Zinnerzlagerstätten. Neues Jahrb. Mineral., Abhandl. **87**, 416 (1955).

SCHROEDER, H. A., J. J. BALASSA, and I. H..TIPTON: Abnormal trace metals in man: Tin. J. Chronic Diseases **17**, 483 (1964).

SCHROLL, E., u. M. WENINGER: Eine empfindliche spektrochemische Analysenmethode zur Bestimmung von Germanium und Zinn unter Verwendung sulfidierender thermochemischer Reagenzien. Mikrochim. Acta (2) 378 (1965).

SHIMA, M.: The distribution of germanium and tin in meteorites. Geochim. Cosmochim. Acta **28**, 517 (1964).

STOROZHEVA, N. N.: Content of Pb and Sn in healthy and carious teeth. Stomatologiya **42**, 44 (1963); Chem. Abstr. **59**, 988g (1963).

SUESS, H. E., and H. C. UREY: Abundances of the elements. Rev. Mod. Phys. **28**, 53 (1956).

TABOR, E. C., and W. V. WARREN: Distribution of certain metals in the atmosphere of some American cities. A.M.A. Arch. Ind. Health **17**, 145 (1958).

TAYLOR, S. R.: Abundance of chemical elements in the continental crust: A new table. Geochim. Cosmochim. Acta **28**, 1273 (1964)

— Geochemical application of spark source mass spectrometry. Nature **205**, 34 (1965).

TEREBENINA, A., and G. ANGELOVA: Composition of the ashes of alcohol-benzene extracts of some Bulgarian coals. Compt. Rend. Acad. Bulgare Sci. **15**, 495 (1962); Chem. Abstr. **58**, 10005b (1963).

TUREKIAN, K. K.: Additional trace-element analyses of standard granite G-1 and standard diabase W-1. Science **126**, No. 3277, 745 (1957).

—, and K. H. WEDEPOHL: Distribution of the elements in some major units of the earth's crust. Geol. Soc. Am. Bull. **72**, 175 (1961).

UDODOV, P. A., and YU. S. PARILOV: Certain regularities of migration of metals in natural waters. Geochemistry No. 8, 763 (1961).

VAKHRUSHEV, V. A.: Trace elements in accessory magnetite as an index for a genetic separation of the Altai-Sayan granites. Dokl. Akad. Nauk SSSR **147**, 707 (1962); Chem. Abstr. **58**, 8778h (1963).

VINOGRADOV, A. P.: Geochemistry of rare and dispersed chemical elements in soil. Moscow-Leningrad: Acad. Sci. U. S. S. R. 1950 [in Russian].

— The elementary chemical composition of marine organisms. Memoir. Sears Foundation for Marine Research, No. II. Sears Foundation for Marine Research, Yale University, New Haven, 1953.

— Report of biogeochemical laboratory, vol. 3 and Selected papers of VERNADSKY, ed. by VINOGRADOW, vol. 1, p. 364, 1954.

VINOGRADOV, V. A.: The regularity of distribution of chemical elements in the earth's crust. Geokhimiya **1**, 1 (1956).

VINOGRADOV, Z. A., and V. V. KOVALJSKII: Elemental composition of Black Sea plankton. Dokl. Akad. Nauk SSSR **147**, 1458 (1962); Chem. Abstr. **58**, 14441a (1963).

WEBB, D. A.: Studies on the ultimate composition of biological material. II. Spectrographic analysis of marine invertebrates with special reference to the chemical composition of their environment. Sci. Proc. Roy. Dublin Soc. **21**, 505 (1937).

WEDEPOHL, K. H.: Schwermetallgehalte der Kalkgerüste einiger mariner Organismen. Nachr. Akad. Wiss. Goettingen, Math.-Physik. Kl. IIa Math.-Physik. Chem. Abt. **5**, 79 (1955).

WINCHESTER, J. W., and A. H. W. ATEN JR.: The content of tin in iron meteorites. Geochim. Cosmochim. Acta **12**, 57 (1957).

ZUBOVIC, P., T. STADNICHENKO, and N. B. SHEFFEY: Comparative abundance of the minor elements in coals from different parts of the United States, in Short Papers in the Geological Sciences. U.S. Geol. Surv., Profess. Papers **400-B**, B 87—B 88 (1960).

— — — Distribution of minor elements in coal beds of the eastern interior region. U.S. Geol. Surv., Bull. **1117-B**, (1964).

— — — Distribution of minor elements in coals of the appalachian region. U.S. Geol. Surv., Bull. **1117-C**, (1966).

Revised manuscript received: September 1967

50-O. Economic Importance

A copper alloy of tin (bronze) has been used by man for more than 5,000 years.

At present metallic tin is mainly consumed for the coating of cans and the production of alloys. In 1971 the U.S.A., Japan and Great Britain used 45% of their tin import or production in the can industry. An electrolytical process introduced in the late thirties, which provides a thin coating in the one micron range, has caused a decrease of tin consumption. The most common tin alloys are those of lead and copper. Soldering metals usually contain tin and lead. The application of tin-antimony and tin-aluminum alloys is increasing. Most present day bronce contains about 6 percent tin. Some steels have about 0.1% Sn. The use of tin containers is decreasing.

Tin oxide is a common white pigment in enamels and glass. The chemical industries consume about 5 percent of the tin production (catalysts, pesticides etc.).

Cassiterite is the common tin ore mineral whose deposits are mainly restricted to certain stanniferous crustal belts. In 1972 71% of the world tin production came from placer deposits originating primarily from tin-bearing granites. The lower limit of economic tin concentrations in placers ranges between 0.01 and 0.04% Sn. Vein and greisen deposits must contain between 0.2 and 1% Sn to be workable.

The potential world reserves of this metal are estimated to be about $3 \cdot 10^7$ t Sn and the proven reserves $7.9 \cdot 10^6$ t Sn (Gocht, W.: Handbuch der Metallmärkte. Berlin-Heidelberg-New York: Springer 1974). In 1972 the world production was $2.3 \cdot 10^5$ t. About $5.5 \cdot 10^4$ t Sn per year are recycled. The consumption increases at a rate of about one percent per year. Therefore the view on the future availability of this metal cannot be very optimistic.

In 1972 the main producers were Malaysia ($7.7 \cdot 10^4$ t), Bolivia ($3.2 \cdot 10^4$ t), China ($2.3 \cdot 10^4$ t), Thailand and Indonesia ($2.2 \cdot 10^4$ t each), Australia and the U.S.S.R. ($1.2 \cdot 10^4$ t each). Great Britain, which has contributed much tin early in history, only produces $3.3 \cdot 10^3$ t. The major reserves occur in SE Asia. Details on the world tin resources are compiled by Gocht (1974 l.c.) and Sainsbury, C. L. (U.S. Geol. Surv. Bull. **1301**, 1969).

Manuscript received: October 1977

Antimony 51

A	V. Kupčík	(Mineralogisch-Kristallographisches Institut der Universität Göttingen, Germany)
B-M, O	H. Onishi	(Japan Atomic Energy Research Institute Tokai-mura, Ibaraki-ken, Japan)

51-A. Crystal Chemistry

From a crystallochemical point of view, antimony is closely related to arsenic and bismuth. The electronegativity of antimony differs only slightly from that of bismuth but distinctly from that of arsenic. Antimony usually appears with a formal valence of 3+ and occasionally 5+. It shows amphoteric behavior .With metal it forms alloys and intermetallic ordered compounds, which may be called antimonides. In minerals antimony occurs with sulfur in complex polyanions; here its valence is 3+.

The 5+ valence state in sulfur compounds can be accepted for stibioluzonite, Cu_3SbS_4, and isochemical stibioenargite (Gaines, SR 1957, 349). The lattice constants and space group indicate the isotypy to chalkopyrite and that means a tetrahedral coordination for Sb. This observation is coherent with the formal valence rule. Synthetic $Na_3SbS_4 \cdot 9H_2O$ also contains $[Sb^{[4]}S_4]^{-3}$ tetrahedral groups. It is possible to synthetize the compound Sb_2S_5. In nature these cases are of no major importance (Gmelin, 1949/50).

In complex anions with oxygen, antimony is either in the 3+ or 5+ state. The abundant minerals occur in polymetallic hydrothermal vein deposits and their oxidation zones.

The electronic structure is $5s^2p^3$. The ionic radius of Sb^{3+} is 0.90 Å according to Goldschmidt and 0.76 Å according to Ahrens (see Table 12-8 in Volume I of this handbook). Pauling (1960) reports an ionic radius for Sb^{+5} of 0.62 Å; the covalent radius for octahedrally-coordinated antimony is 1.41 Å and the covalent radius for tetrahedrally-coordinated antimony is 1.36 Å. The numerous antimony minerals may be grouped as follows:

1. metallic antimony, alloys and antimonides;
2. sulfides and sulfosalts;
3. sulfo-oxides and sulfo-halogenides;
4. oxygen-containing compounds.

I. Antimony Alloys (i.e. Antimonides)

Metallic antimony is isostructural with arsenic and bismuth ($Sb^{[3+3]}$) having interatomic distances of 3×2.91 Å and 3×3.36 Å. Its structural geometry is that of a trigonal antiprism (Barrett *et al.*, 1963). Arsenic and antimony are completely miscible. Crystalline AsSb is called stibarsenic (Strunz, 1970).

In limited concentrations antimony can form alloys with silver. Up to 11% Sb causes no structural change, while between 10 and 16% Sb the hexagonal γ-phase, allargentum, is formed (Ramdohr, in Strunz, 1970). Stoichiochemically-ordered phases may be named antimonides. At a concentration ratio of 1:2 they have pyrite structures, such as aurostibite, $Au^{[6]}Sb_2^{[3Au+1Sb]}$, or geversite, $PtSb_2$; at a ratio of 1:1 the compounds have the structure of niccolite, such as breithauptite, $Ni^{[6]}Sb^{[6]}$, and arite, $Ni^{[6]}(As, Sb)^{[6]}$.

Additional detailed information on comparable structures is not yet available.

Table 51-A-1. *Structural data of chain-like sulfosalts*

Name	Chemical formula	Number of independent Sb-atoms in the asymmetric unit and interatomic distances					Type of Sb-S chains (see Fig. 51-A-2)	Reference
Stibnite	$[Sb^{[3+3]}Sb^{[3+2]}S_3]^1_\infty$	2	2.52 2×2.54 2×3.17 3.11	2.46 2×2.68 2×2.85			C	BAYLISS and NOWACKI (1972)
Berthierite	$Fe^{[6]}[Sb_2^{[3+2]}S_4]^1_\infty$	2	2.48 2×2.58 2×2.93	2.43 2×2.48 2×3.24			C	BUERGER and HAHN (SR 1955, 418)
Bournonite	$Cu^{[4]}[Pb^{[6]}Sb^{[3+1(2)]}S_3]^1_\infty$	2	2.49 2.49 2.52	2.39 2.39 2.45			D	EDENTHALTER and NOWACKI (1970)
Freieslebenite	$Pb^{[6]}Ag^{[3+1]}Sb^{[3+2]}S_3$	1	2.46 2.54 2.72 2.94 2.94				D	HELLNER (SR 1957, 32)
Jamesonite	$Fe^{[6]}Pb_4^{[7]}[Sb_4^{[2+3]}Sb_2^{[3+2]}S_{14}]^1_\infty$	3	2.41 2.52 2.85 2.90 2.97	2.43 2.59 2.59 3.04 3.07	2.44 2.57 2.81 2.94 3.18		D	NIIZEKI and BUERGER (SR 1957, 346)
Kobellite	$(Cu, Fe)^{[3+1]}Pb_6^{[8]}(Bi, Sb)_7^{[5-6]}S_{17,5}$	7 with variable Bi/Sb ratio		2.47 2×2.71 2×2.96		for Sb:Bi 3:1 polyhedron	—	MIEHE (1971)
Livingstonite	$Hg^{[2+4]}[Sb_4^{[3+2]}S_8]^1_\infty$	4	2.49 2.54	2.44 2.47	2.47 2.55	2.54 2.59	C	NIIZEKI and BUERGER

			2.70 2.94 2.95	2.52 3.11 3.15	2.62 2.95 2.96	2.66 2.88 2.98		(SR 1957, 347)
Miargyrite	$Ag^{[3+1]}Ag^{[3+1+1]}[Sb_2^{[3+3]}S_4]^1_\infty$	2	2.53 2.58 2.58 3.14 3.25 3.28	2.48 2.49 2.54 3.26 3.29 3.34			C	KNOWLES (SR 1964, 21)
Plagionite	$Pb_5^{[6-8]}Sb_8^{[4-5]}S_{15}$	4	2.47 2.52 2.68 2.93 2.97	2.44 2.55 2.58 3.03 3.12	2.45 2.53 2.54 3.14 3.16	2.46 2.46 2.49 3.27 3.46	galena type slabs, see text	WUENSCH (1973)
Semseyite	$Pb_9^{[6-8]}Sb_8^{[5-6]}S_{21}$	4	2.44 2.53 2.57 3.08 3.10 3.41	2.40 2.55 2.56 3.05 3.10 3.60	2.38 2.47 2.55 3.12 3.13 3.62	2.40 2.49 2.51 3.16 3.45	galena type slabs, see text	KOHATSU (1973)
Vrbaite	$Tl_4^{[2+7]}Hg_2^{[2]}[Hg^{[4]}As_8^{[3p]}Sb_2^{[3+2]}S_{20}]^2_\infty$	1	2.48 2.52 2.55 3.19 3.29	double chain with change of the sequence As—Sb				OHMASA and NOWACKI (1971)
Wolfsbergite	$Cu_2^{[4]}[Sb_2^{[3+2]}S_4]^1_\infty$	1	2.44 2×2.57 2×3.11				C	HOFMANN (SB 1933—35, 75)
Zinckenite	$Pb Sb_2 S_4$	2+1	Pb—Sb 2.40 2×2.85 2×2.81	 2.48 2×2.76 2×2.90				TAKEDA and HORIUCHI (1971)

II. Sulfides and Sulfosalts

Stibnite, Sb_2S_3 (see Table 51-A-1), has a typical sulfosalt-structure ("spießglanz"). A wide range of miscibility of Sb_2S_3 with Bi_2S_3 is possible. $(Sb, Bi)_2S_3$ crystals with a Sb to Bi ratio from 11:9 to 11:13 are called horobetsuite (HAYASE, 1959). The diadochy with As is limited. Getchellite, $AsSbS_3$ (GUILLERMO and WUENSCH, 1974), is structurally unrelated to stibnite. The relation to orpiment, As_2S_3, is based only on the similarity of lattice constants. Atoms occupy four statistically non-equivalent positions with a ratio from 0.25:0.75 to 0.65:0.35. This compound has a complicated layer structure with 8-membered rings which are normal to the plane of the layers and stacked above one another.

Many attempts have been made to group the large number of minerals called sulfosalts (118 according to NOWACKI, 1969, 1970). The first attempts were made by HOFMANN (1933—1935). The ratio $\phi = \Sigma\, S / \Sigma\, (As, Sb, Bi)$ has been used for a classification which completely lacks structural considerations. HELLNER (1957, 1958a and 1958b) suggested a structurally-oriented system using the galena structure as a reference. Another classification was recently published by NOWACKI (1969), who distinguishes 6 groups according to different values of ϕ. These groups are subdivided according to the linkage of the trigonal $(As, Sb, Bi)S_3$ pyramids. This system is appropriate to classify the arsenic compounds in which the dominance of the three shortest bonds is unquestioned. It is useful (with restrictions) for antimony, where additional larger bonds must be considered in some cases, but because of the frequent occurrence of a nearly octahedrally coordination it is often inapplicable to Bi sulfosalts.

Additional classifications of sulfosalts are given by TAKEUCHI and SADANAGA (1968) and POVARENNICH (1963); the latter classifies this group from a chemical point of view.

It seems useful to divide the sulfosalts into two groups according to the ratio $\alpha = \Sigma\, Me / \Sigma\, Pb + (As, Sb, Bi)$ where Me represents the metals and Pb + (As, Sb, Bi) the "chain-forming" elements.

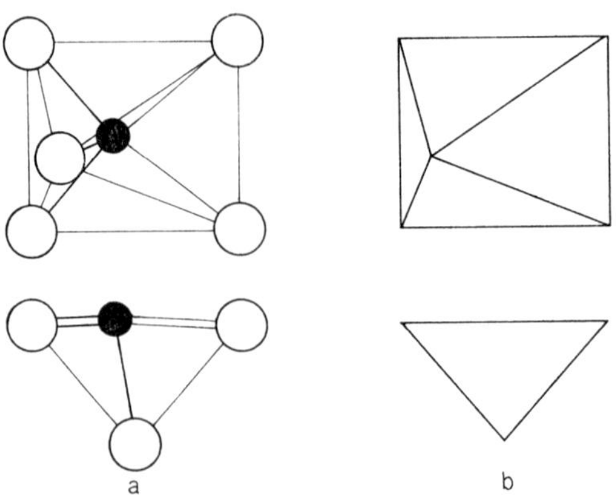

Fig. 51-A-1. a Typical $[SbS_5]$ pyramid. b Schematic presentation

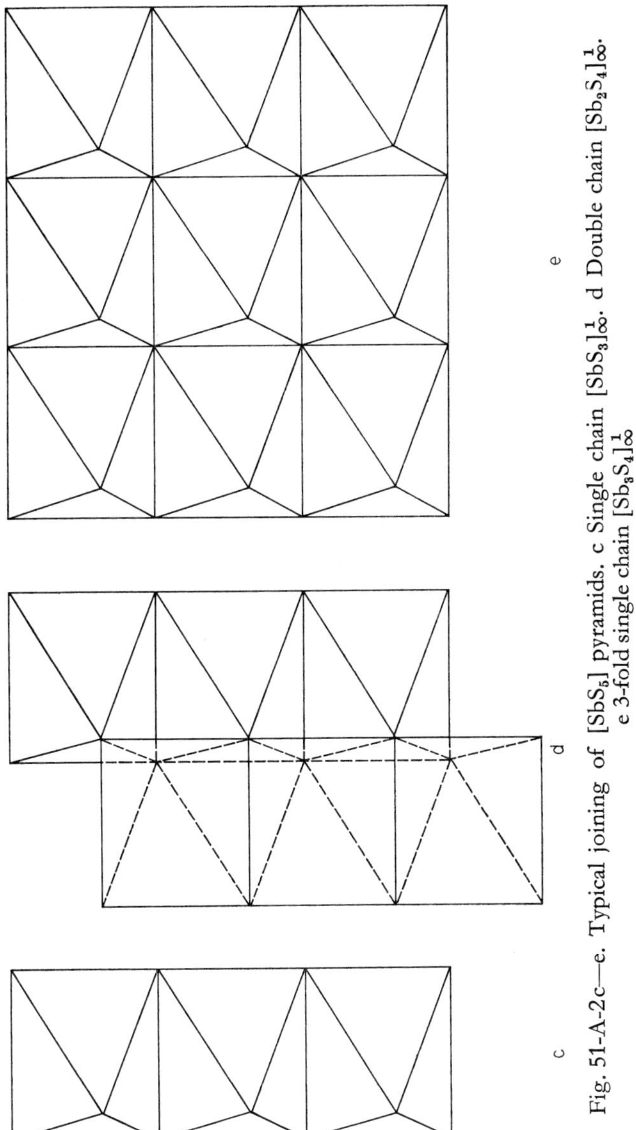

Fig. 51-A-2c—e. Typical joining of $[SbS_5]$ pyramids. c Single chain $[SbS_3]^1_\infty$. d Double chain $[Sb_2S_4]^1_\infty$. e 3-fold single chain $[Sb_3S_4]^1_\infty$

With a ratio $\alpha \leqq 1$, typical chain structures are formed which are called "spießglanzes". Stibnite, Sb_2S_3, belongs to this group. If this ratio is >1, no chains occur. These structures are of greater variety and the majority is controlled by the other cations, for example Cu, Ag, Tl, etc.

The coordination polyhedron of antimony in the "spießglanz" type can be described as a trigonal pyramid with an antimony atom at the top. By additional bonds, this arrangement is often modified to a distorted tetragonal pyramid. Chemically there are 3p bonds at right angles to each other (average angle 93.3° with a

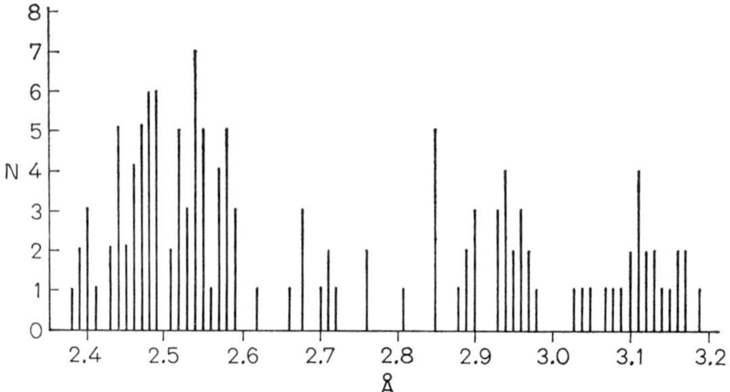

Fig. 51-A-3. Frequency of interatomic distances in the chain-type sulfosalts

mean bond length of 2.56 Å). The larger bonds have d_γ character and a bond length between 2,87 and 3.20 Å (see Fig. 51-A-1). A prominent characteristic of the sulfosalts is the occurrence of one lattice parameter being n×4 Å (along the needle axis). Most common are:

a) infinite $^1_\infty SbS_3$ chains in which basal corners of tetragonal pyramids are linked,

b) joints of infinite twin chains which are displaced at a distance of 2 Å across the 2_1 axis. These and other types of chains are shown in Fig. 51-A-2.

MeS_x polyhedra can be linked at the edges of the pyramids occurring in the above-mentioned chains. The edges in question are those which run parallel to the 4 Å lattice parameter. The compatibility of an edge of the PbS_x polyhedron with the 4 Å edge is limited. Considering the important role of lead in "spießglanzes", this limitation is the origin of an often-observed doubling or multiplication of the lattice parameters in the chain direction, and therefore of occasionally distinct structural differences between related Sb- and Bi-sulfosalts. In the case of Bi this compatibility is very good. Isomorphism between Sb and Bi is possible but is not very common (see kobellite; Miehe, 1972).

Of interest is the plagionite-group with the general formula $Pb_{3+2n}Sb_8S_{15+2n}$ and the minerals (fulöpite ($Pb_3Sb_8S_{15}$), plagionite ($Pb_5Sb_8S_{17}$), heteromorphite ($Pb_7Sb_8S_{19}$), and semseyite ($Pb_9Sb_8S_2$). The structures are composed of galena-type slabs which run parallel to 112 and $\bar{1}1\bar{2}$, or alternatively parallel to c and which are extended indefinitely to $[\bar{1}\bar{1}2]$ and [110] respectively.

For a compilation of bond lengths see Table 51-A-1 and Fig. 51-A-3.

Finite chains, $Sb_{12}S_{24}$, and isolated groups, Sb_2S_4, occur in andorite, Pb_4Ag_4-$Sb_{12}S_{24}$, as the result of an ordered sequence of Ag and Sb atoms in a 6-fold superperiod (Tokonami, 1973).

In the remaining group of sulfosalts the usual coordination around antimony can be observed, but the formation of chains is rare. More common are isolated SbS_3 groups. The failure of the valence rule in tetrahedrite, $Cu_{12}Sb_4S_{13}$, may be due to the behavior of copper atoms. No analogous isomorphic bismuth compounds are known in this group, but in some cases there is an unlimited miscibility with corresponding

Table 51-A-2. *Structural data of sulfosalts containing isolated $(SbS_3)^{3-}$ groups*

Name	Chemical formula	Interatomic distances Sb—S (Å)	References
Pyrargyrite	$Ag_3^{[2]}[Sb^{[3]}S_3^{[2Ag+1Sb]}]$	3 × 2.46	ENGEL and NOWACKI (1966)
Pyrostilprite	$Ag_3[Sb^{[3]}S_3]$	2.40; 2.48; 2.49	KUTOGLU (1968)
Tetrahedrite	$Cu_6^{[4]}Cu_6^{[3]}S[SbS_3]_4$	3 × 2.446	WUENSCH (1963)
Skinnerite	$Cu_3[Sb^{[3]}S_3]$	2.47; 2 × 2.44 (at 170° C)	MAKOVICKÝ and SKINNER (1972)
Stephanite	$Ag_5S[Sb^{[3]}S_3]$	2.47; 2.47; 2.48	RIBAR and NOWACKI (1970)

arsenic sulfosalts ($Cu_{12}As_4S_{13}$ and $Cu_{12}Sb_4S_{13}$). Only limited Bi diadochy is observed. The new mineral skinnerite, Cu_3SbS_3, is an exception (MAKOVICKY and SKINNER, 1972) and it is nearly isostructural with wittichenite, Cu_3BiS_3 (MATZAT, 1972; KOCMAN and NUFFIELD, 1973). The distribution and the coordination of the Sb atoms is the same, but the Cu atoms are distributed more or less statistically. Data are listed in Table 51-A-2.

The minerals gudmindite, $Fe^{[3Sb+3S]}Sb^{[S+3Fe]}S^{[3Fe+Sb]}$, and the isostructural ullmanite, NiSbS, crystallize in the arsenopyrite type (EO_7) with discrete Sb-S groups.

III. Oxysulfides and Sulfohalogenides

The only known natural oxysulfide is kermesite, Sb_2S_2O. There exists no analogous arsenic or bismuth mineral. Although formed of infinite sheets, the structure has great similarity to sulfosalts, since these sheets consist of two types of chains, $^1_\infty\,[Sb_2^{[3+2]}S_3]$ and $^1_\infty\,[Sb_2^{[3O+1S]}O_2S]$. The sheets are held together by weak bonding (smallest interatomic distance $>$ 3.4 Å) (see Fig. 51-A-4). The following interatomic distances have been observed:

Sb_I —S: 2.40 Å, Sb_I—O: 1.99, 2.07, 2.34 Å
Sb_{II} —S: 2.39 Å, Sb_{II}—O: 1.99, 2.05, 2.26 Å
Sb_{III} —S: 2.37, 2.61, 2.64, 3.07, 3.09 Å
Sb_{IV} —S: 2.35, 2.71, 2.75, 2.91, 3.02 Å.

Within $^1_\infty\,[Sb_2O_2S]$ chains, very short O—O_1 distances of 2.26 Å are noteworthy (KUPČÍK, 1967).

SbSI (not found in nature) has a structure formed by $^1_\infty\,[Sb_2^{[3S+2I]}S_2Cl_2]$ chains. The coordination of the antimony atom is similar to that in sulfosalts, the only difference being the two longer bonds in the coordination pyramid formed by iodine atoms. Interatomic distances are Sb—S 2.48 Å, 2.67 Å (2×); Sb—Cl 3.11 Å (2×). The substance is centrosymmetrical only above 22° C. Below this temperature it is piezoelectric and ferroelectric (OKA *et al.*, 1965).

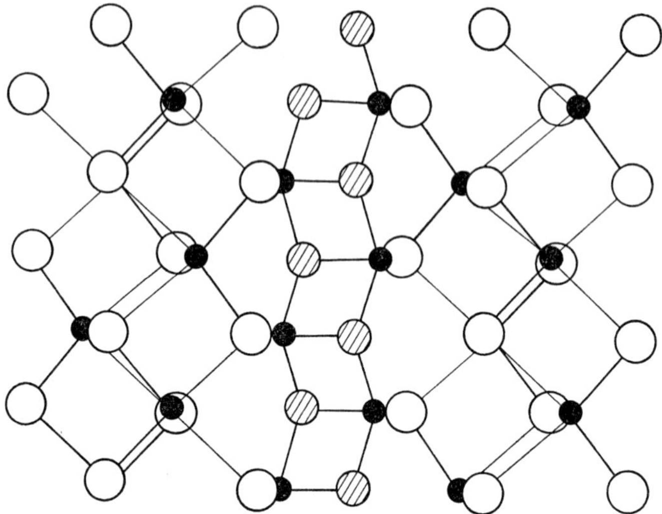

Fig. 51-A-4. Kermesite, Sb_2S_2O. (Only one layer shown). Small black circles: Sb Medium hatched circles: O. Large white circles: S

IV. Oxygen-Containing Compounds

Compared with sulfides, much less reliable data are known about antimony oxides. One reason for this lack of information is that most of the oxide minerals are poorly crystallized antimony ockers, which in addition are intermixed and intergrown. Some structural relations have been studied from powder data, but positional atomic parameters and, therefore, interatomic distances are lacking. Furthermore, in those minerals which supposedly have a pyrochlore structure, some atomic positions must be assumed to be statistically occupied. Only on the following structures has reliable information been recorded: velantinite (BUERGER, SR 1940—50, 34; MIEHE, 1973; SVENSSON, 1974), senarmontite (BOZORTH, SB 1913—1926, 245), schafarzikite (ZEMANN, SR 1951, 287), melanostibianite Mn $(Sb^{+5}_{0.5}Fe^{+3}_{0.5})O_3$ (MOORE, 1967), and manganostibite Mn_5SbAsO_{12} (MOORE, 1970). The structures of sendborgite $NaSbBe_4O_7$ (AMINOFF, SB 1933—35, 381; PAULING, 1960, p. 382), stibiotantalite Sb(Ta, Nb)O_4 (DIHLSTRÖM, SB 1938, 113), and Sb_2O_4 (NATTA and BACCAREDA, SB 1933—35, 359) are principally known, but the information is less reliable.

GRÜNDER *et al.* (1962) have mentioned the gaps in our present knowledge about the crystal structures of Sb^{+3} and Sb^{+5} and their hydrated oxides. The information about the probable structural relations within this group of antimony minerals (taken from the above paper, but changed and completed) is presented in Table 51-A-3. Subdivision of this group into double oxides and oxidic salts has been omitted.

Valentinite, Sb_2O_3, consists of infinite chains $^1_\infty(Sb^{[3+2]}O^{[2+1]}_{\frac{3}{2}})$ (see Fig. 51-A-5). According to MIEHE (1973) in his refinement of BUERGER (SR 1940—50, 34), valentinite is triclinic pseudorhombic, but this fact has no further crystallochemical consequences. The coordination about Sb may be described as a distorted square-

Table 51-A-3. *Crystal-chemical relationships in the group of antimony oxide minerals*

Name	Formula	Structure type	Reference
Sb^{+3}-compounds			
Senarmontite	Sb_2O_3	Arsenolite	BOZORTH (SR 1913—26)
Valentinite	Sb_2O_3		BUERGER (SR 1940—1950, 34), MIEHE (1973), SVENSSON (1974)
Schafarzikite	$FeSb_2O_4$	Trippkeite	ZEMANN (SR 1951, 271)
Synthetic	$NiSb_2O_4$	Trippkeite	STAHL (SR 1942—44, 174)
	$CoSb_2O_4$	Trippkeite	STAHL (SR 1942—44, 174)
Haematostibite	$8(Mn, Fe)O \cdot Sb_2O_3$	Unknown	
Sb^{+5}-compounds			
Byströmite	$MgSb_2O_6$	Trirutile	MANSON and VITALIANO (SR 1952, 257)
Tripuhyite	$FeSb_2O_6$	Trirutile	MANSON and VITALIANO (SR 1952, 257)
Ordonezite	$ZnSb_2O_6$	Trirutile	MANSON and VITALIANO (SR 1952, 257)
Swedenborgite	$NaSbBe_4O_7$		AMINOFF (SB 1933—35, 381), PAULING, ibid. 382)
Derbylite	$6FeO \cdot 5TiO_2 \cdot Sb_2O_5$	Unknown	
Katoptrite	$Mn_{14}Sb_2(Al, Fe)_4O_{21}[SiO_4]_2$	Unknown	
Yeatmanite	$(Mn, Zn)_{16}Sb_2O_{13}[SiO_4]_4$	Unknown	
Chapmanite	$SbFe_2\ OH\ (SiO_4)_2$	Unknown	
Manganostibite	$Mn_7Sb^{+5}As\ O_{12}$		MOORE (1970)
Melanostibite	$MnSb^{+5}_{0.5}Fe^{+3}_{0.5}O_3$	Ilmenite	MOORE (1967)
Sb^{+3} and Sb^{+5}-compounds			
Bindheimite	$Pb_{1-2}Sb_{2-1}(O, OH, H_2O)_7$	Pyrochlore	GÄRTNER (SB 1928—32, 340), MACHATSCHKI, ibid. 340)
Partzite	$Cu_{1-2}Sb_{2-1}(O, OH, F)_7$	Pyrochlore	
Stetefeldite	$Ag_{1-2}Sb_{2-1}(O, OH, H_2O)_7$	Pyrochlore	
Stibiconite	Sb_3O_6OH	Pyrochlore	
Anhydro-stibiconite	Sb_6O_{13}	Pyrochlore	
Romeite	$(Ca, Na, H)Sb_2O_6(O, OH, F)$	Pyrochlore	
Stibiotantalite	$Sb(Ta, Nb)O_4$	Stibio-tantalite	DIHLSTRÖM (SB 1938, 113)
Cervantite	Sb_2O_4	Stibio-tantalite	
Stibioniobite	$Sb(Nb, Ta)O_4$	Stibio-tantalite	
Synthetic	$SbBiO_4$	Stibio-tantalite	
Synthetic	Sb_2O_4	Pyrochlore	NATTA and BACCAREDA (SB 1933—35, 359)

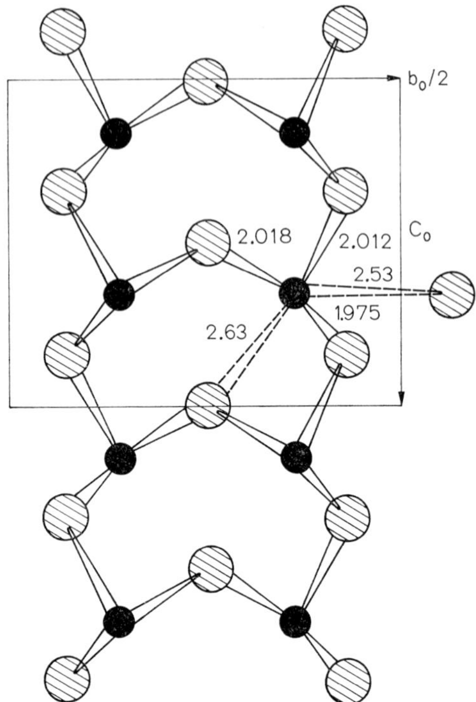

Fig. 51-A-5. Crystal structure of valentinite, Sb_2O_3. Small black circles: Sb. Large white circles: O

pyramid. The Sb—O distances are 1.975, 2.012, 2.018, 2.630 and 2.53 Å. The angles of the three shortest bonds with one another are 91.92°, 80.14°, and 89.24°. An independent refinement of the valentinite structure carried out by SVENSSON (1974) did not reveal triclinic symmetry.

The structure of senarmontite, $(Sb_2^{[3+3]}O_3^{[3+2]})_{2k}$, is commonly assumed to consist of separate Sb_4O_6 "molecules" (see Fig. 51-A-6). However, this view is contradicted by the difference in interatomic distances within one "molecule" ($3 \times 2.2_2$ Å) and those between different "molecules" ($3 \times 2.3_7$ Å). The coordination polyhedron is a trigonal antiprism forming a framework with rather close complexes of Sb_4O_6 (BOZORTH, SB 1913—26, 264).

Sb_2O_4 (NATTA, BACCAREDA, SB 1933—35, 359) probably has the pyrochlore structure, $Sb^{+3[6]}Sb^{+5[6+2]}O_4$, and is not identical with cervantite, Sb_2O_4, which is isostructural with stibiotantalite, $Sb^{[6]}(Ta, Nb)^{[6]}O_4$. All interatomic distances are not very precisely known.

Schafarzikite, $FeSb_2O_4$ (ZEMANN, SR 1951, 287), is isotopic with $NiSb_2O_4$, $CoSb_2O_4$ and some other synthetic compounds of this type. It has the same structure as trippkeite. The coordination about Sb is trigonal pyramidal with one bond distance being significantly shorter (1.87 Å and 2×2.20 Å). Polyhedra are probably connected in chains $^1_\infty(Fe^{[6]}Sb_2^{[3]}O_4)$. In oxide compounds isomorphism between antimony

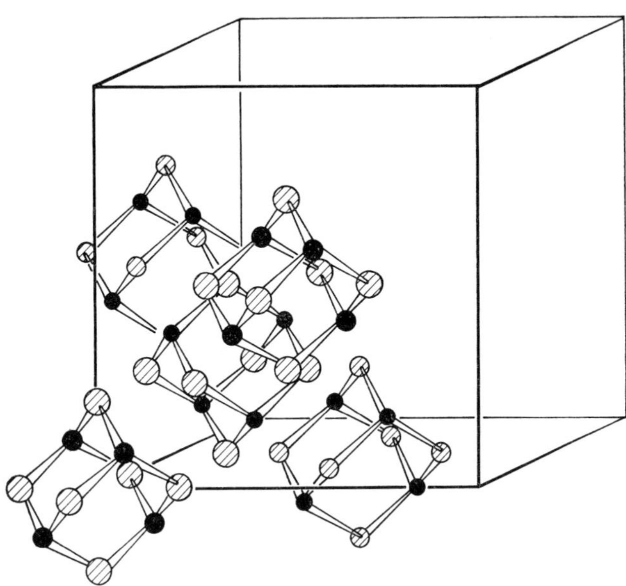

Fig. 51-A-6. Crystal structure of senarmontite, Sb_4O_6. Small black circles: Sb. Large white circles: O

and arsenic is more likely than between antimony and bismuth although the latter possibility cannot be completely excluded. An isomorphous replacement of pentavalent antimony by iron and manganese is possible because of the common octahedral coordination. The distribution of Sb and Mn or Fe is completely random in melanostibite, $Mn(Sb^{5+}_{0.5}Fe^{3+}_{0.5})O_3$ (MOORE, 1967), and manganostibite, Mn_5SbAsO_{12} (MOORE, 1970).

Revised manuscript received: March 1974

References: Section 51-A

BARRETT, C. S., CUCKA, P., HAEFNER, K.: The crystal structure of antimony at 4.2, 78 and 298° K. Acta Cryst. **16**, 451 (1963).

BAYLISS, P., NOWACKI, W.: Refinement of the crystal structure of stibnite, Sb_2S_3. Z. Krist. **135**, 308 (1972).

EDENTHARTER, A., NOWACKI, W., TAKEUCHI, Y.: Verfeinerung der Kristallstruktur von Bournonit $(SbS_3)_2Cu_2^{IV}Pb^{VII}Pb^{VIII}$ und von Seligmannit $(AsS_3)_2Cu_2^{IV}Pb^{VII}Pb^{VIII}$. Z. Krist. **131**, 397 (1970).

ENGEL, P., NOWACKI, W.: Die Verfeinerung der Kristallstruktur von Proustit, Ag_3AsS_3, und Pyrargyrit, Ag_3SbS_3. Neues Jahrb. Mineral. Monatsh. **1966**, 181 (1966).

GIUSEPETTI, G., TADINI, C.: Riesame della struttura della nadorite, $PbSbO_2Cl$. Periodico di Mineralogia **42**, 335 (1973).

GMELIN: Handbuch der Anorganischen Chemie, Bd. Antimon (4. ed.). Weinheim: Verlag Chemie 1949/50.

GRÜNDER, W., PÄTZOLD, H., STRUNZ, H.: Sb_2O_4 als Mineral (Cervantit). Neues Jahrb. Mineral. Monatsh. **1962**, 93 (1962).

GUILLERMO, T. R., WUENSCH, B. J.: The crystal structure of getchelite $AsSbS_3$. Acta Cryst. **B 29**, 2536 (1973).

HAYASE, K.: Minerals of bismuthinite-stibnite series with special reference to horobetsuite from the Horobetsu mine, Hokkaido, Japan. Mineral. J. Japan **1**, 189 (1958).

HELLNER, E.: A structural scheme for sulfide minerals. J. Geol. **66**, 503 (1957).

HELLNER, E.: Über ein strukturelles Einteilungsprinzip für sulfidische Erze. Naturwissenschaften **45**, 38 (1958a).

HELLNER, E.: Über komplex zusammengesetzte Spießglanze. Z. Krist. **110**, 169 (1958b).

HOFMANN, W.: Strukturelle und morphologische Zusammenhänge bei Erzen vom Formeltyp ABC_2. Fortschr. Mineral. **17**, 422 (1933).

HOFMANN, W.: Ergebnisse der Strukturbestimmung komplexer Sulfide. Z. Krist. **92**, 161 (1935).

KOCMAN, V., NUFFIELD, E. W.: The crystal structure of wittichenite Cu_3BiS_3. Acta Cryst. **B 29**, 2528 (1973).

KOHATSU, I. I.: The crystal chemistry of the homologous series $Pb_{3+2n}Sb_8S_{15+2n}$ (the plagionite group). (Ph-D Thesis, M.I.T., 1973).

KUPČÍK, V.: Die Kristallstruktur des Kermesits, Sb_2S_2O. Naturwissenschaften **54**, 114 (1967).

KUTOGLU, A.: Die Struktur des Pyrostilpnits (Feuerblende) Ag_3SbS_3. Neues Jahrb. Mineral. Monatsh. **1968**, 145 (1968).

MAKOVICKÝ, E., SKINNER, B. J.: The crystal structure of high-temperature modification of skinnerite. Proceedings of the winter meeting of the american crystallographic association, Albuquerque (New Mexico), April 1972.

MATZAT, E.: Die Kristallstruktur des Wittichenits, Cu_3BiS_3. Tschermaks Mineral. Petrog. Mitt. **18**, 312 (1972).

MIEHE, G.: Crystal structure of Kobellite. Nature Phys. Sci. **231**, 133 (1971).

MIEHE, G.: Private communication (1973).

MOORE, P. B.: Contributions to Swedish mineralogy; II. Melanostibite and manganostibite, two unusual antimony minerals. The identity of ferrostibian with långbanite. Arkiv Mineral. Geol. **4**, 449 (1967).

MOORE, P. B.: Manganostibite: A novel cubic close-packed structural type. Am. Mineralogist **55**, 1489 (1970).

NOWACKI, W.: Zur Klassifikation und Kristallchemie der Sulfosalze. Schweiz. Mineral. Petrog. Mitt. **49**, 109 (1969).

NOWACKI, W.: Zur Klassifikation der Sulfosalze. Acta Cryst. **B 26**, 286 (1970).

OHMASA, M., NOWACKI, W.: The crystal structure of vrbaite $Hg_3Tl_4As_8Sb_2S_{20}$. Z. Krist. **134**, 360 (1971).

OKA, A., KIKUCHI, A., MORI, T., SAWAGUCHI, E.: Atomic parameters in ferroelectric SbSI. J. Phys. Soc. Japan **21**, 405 (1965).

PAULING, L.: The Nature of the Chemical Bond and the Structure of Molecules and Crystals (3. ed.). New York: Cornell University Press 1960.

POVARENNYCH, A. S.: Grundsätze einer kristallchemischen Klassifikation der Sulfide. Geologie **12**, 377 (1963).

RAMDOHR, P.: in STRUNZ, H.: Mineralogische Tabellen (5. ed.). Leipzig: Akad. Verlagsgesellschaft 1970.

RIBAR, B., NOWACKI, W.: Die Kristallstruktur von Stephanit, $SbS_3SAg_5^{III}$. Acta Cryst. **B 26**, 201 (1970).

STRUNZ, H.: Mineralogische Tabellen (5. ed.). Leipzig: Akad. Verlagsgesellschaft 1970.

SVENSSON, GH.: The crystal structure of orthorhombic antimony trioxide, Sb_2O_3. Acta Cryst. **B 30**, 458 (1974).

TAKEDA, H., HORIUCHI, H.: Symbolic addition procedure applied to zinkenite. Structure determination. J. Mineral. Soc. Japan **10**, 283 (1971).

TAKEUCHI, Y., SADANAGA, R.: Structural principles and classification of sulfosalts. Z. Krist. **130**, 346 (1969).

TOKONAMI, M.: Private communication (1973).

WUENSCH, B. J.: The crystal structure of tetrahedrite, $Cu_{12}Sb_4S_{13}$. Z. Krist. **119**, 437 (1964).

WUENSCH, B. J.: The crystal structure of plagionite, $Pb_5Sb_8S_{15}$ (to be published, Acta Cryst. 1974).

Revised manuscript received: March 1974

51-B. Isotopes in Nature

Antimony (atomic number 51) consists of two isotopes of mass numbers 121 and 123. The natural isotopic abundances of ^{121}Sb and ^{123}Sb are 57.25 and 42.75 atomic %, respectively (WHITE and CAMERON, 1948; LEIPZIGER, 1963).

Radioactive ^{125}Sb (half-life 2.7 years) was detected in fallout, and its concentration in air and soil was determined (KASAI and YANASE, 1965).

Revised manuscript received: February 1967

51-C. Abundances in Meteorites

The antimony content of iron and stony-iron meteorites is shown in Table 51-C-1.

Table 51-C-1. *Antimony in iron and stony-iron meteorites*

Material	Sb (ppm)	Author
Irons:		
Altonah, Of	0.03 (N/R)	SMALES *et al.* (1958)
Bethany, Of	0.01	SMALES *et al.* (1958)
Bristol, Of	0.46	SMALES *et al.* (1958)
Canyon Diablo, Og-Ogg	0.8 (C)	ONISHI and SANDELL (1955)
Canyon Diablo, Og-Ogg	0.31 (N/R)	SMALES *et al.* (1958)
Canyon Diablo, Og-Ogg	0.35 (N/R)	HAMAGUCHI *et al.* (1961b)
Carbo, Om	0.19 (N/R)	SMALES *et al.* (1958)
Casas Grandes, Om	0.07	SMALES *et al.* (1958)
Duchesne, Of	0.04	SMALES *et al.* (1958)
Goose Lake, Om-Og	0.5	SMALES *et al.* (1958)
Henbury, Om	0.5 (C)	ONISHI and SANDELL (1955)
Henbury, Om	0.43 (N/R)	SMALES *et al.* (1958)
Henbury, Om	0.45 (N/R)	HAMAGUCHI *et al.* (1961b)
Spearman, Om	0.13 (N/R)	SMALES *et al.* (1958)
Tazewell, Off	0.78	SMALES *et al.* (1958)
Coahuila, H	0.58	SMALES *et al.* (1958)
San Martin, H	0.09	SMALES *et al.* (1958)
Sandia Mountains, H	0.09	SMALES *et al.* (1958)
Deep Springs, D	0.15	SMALES *et al.* (1958)
Stony-irons:		
Admire, P; silicate phase	0.02 (N/R)	HAMAGUCHI *et al.* (1961b)
Admire, P; metal phase	0.26	HAMAGUCHI *et al.* (1961b)
Bolivia, P	0.22 (N/R)	SMALES *et al.* (1958)
Imilac, P	0.23	SMALES *et al.* (1958)

Antimony in 11 octahedrites ranges from 0.01 to 0.8 ppm Sb; the average is 0.29 ppm. Three hexahedrites contain an average of 0.25 ppm. The average for 14 iron meteorites (octahedrites and hexahedrites) is 0.28 ppm Sb.

The inner and outer parts of an inclusion, presumably troilite, from the Canyon Diablo iron contained 0.64 and 0.11 ppm Sb, respectively (N/R; SMALES, MAPPER, MORGAN, WEBSTER, and WOOD, 1958).

The antimony content of chondrites and achondrite is shown in Table 51-C-2.

The average for ordinary chondrites may be taken as 0.1 ppm Sb. The average for 3 carbonaceous chondrites is 0.12 ppm. The atomic ratio of Sb/Si in ordinary chondrites is 0.12 ppm. The atomic ratio of Sb/Si in ordinary chondrites is 1.3×10^{-7}, using Si = 18%. The average weight ratio of Sb/As in ordinary chondrites is 0.1/2 = 0.05.

Table 51-C-2. *Antimony in chondrites and one achondrite*

Material	Sb (ppm)	Author
Composite I of 7 chondrites (CH and CL)	0.1 (C)	ONISHI and SANDELL (1955)
Metal phase of composite I	0.4	ONISHI and SANDELL (1955)
Composite II of 7 chondrites (CH and CL)	0.1	ONISHI and SANDELL (1955)
Metal phase of composite II	0.6	ONISHI and SANDELL (1955)
Forest City, CH	0.12 (N/R)	HAMAGUCHI *et al.* (1961b)
Modoc, CL	0.12	HAMAGUCHI *et al.* (1961b)
Modoc, CL	0.12 (N/R)	HAMAGUCHI *et al.* (1965)
Bjurböle, CL	0.064 (N/R)	ESSON *et al.* (1965)
Château Renard, CL	0.075	ESSON *et al.* (1965)
Ochansk, CH	0.14	ESSON *et al.* (1965)
Chandrakapur, C	0.073	ESSON *et al.* (1965)
Boriskino, Cc	0.13 (N/R)	HAMAGUCHI *et al.* (1965)
Mighei II, Cc	0.16	HAMAGUCHI *et al.* (1965)
Murray, Cc	0.06	HAMAGUCHI *et al.* (1965)
Nuevo Laredo, A	0.01 (N/R)	HAMAGUCHI *et al.* (1961b)

Revised manuscript received: February 1967

51-C-II. Lunar Materials

Data on the antimony content of lunar materials are compiled in Table 51-C-3. As shown, different research groups have analyzed soil (fines) of the same sample number for antimony. Only two or three values are available for each sample. The Luna 20 soil was probably contaminated with antimony prior to the analyses. The USGS basalt BCR-1 analyzed at the same time gave antimony values of 560 and 720 ppb Sb, which were in reasonable agreement. The antimony values for soil sample 12,070 do not agree but those for soil samples 10,084 and 14,163 differ by a factor of 2 to 3 only.

KREEP is a component rich in K, rare earth elements (REE) and P. The meteoritic (C1 chondrites, Cc_1) components of Apollo 11, 12 and 15 soils are estimated as being approximately 1.8, 1.7, and 1.7% respectively (MORGAN *et al.*, 1972; see footnote of Table 51-C-3); C1 chondrites contain an average of 0.14 ppm Sb (Table 51-C-2).

According to MORGAN *et al.* (1973; see footnote of Table 51-C-3), the average Apollo 15 mare basalt contains 0.098 ppb Sb and Apollo 15 pure KREEP basalt 0.34 ppb Sb.

Eighteen samples of separates 1—2 mm from Apollo 16 soils have been analyzed for antimony by neutron activation (KRÄHENBÜHL *et al.*, 1973; see footnote of Table 51-C-3). The highest antimony content is found in magnetic fractions (2 samples: 25 and 31 ppb Sb) and contrasts with magnetic fractions of Apollo 14 soil (Table 51-C-3).

The Apollo 16 rock that shows the highest antimony content (6.9 ppb) is also highest in Ge, Zn, Cd, Tl, and Se. Goethite (FeOOH) stains were observed on this rock.

Table 51-C-3. *Antimony in lunar materials* (for references see footnote)

Material (no. of samples)	Sb (ppb)	Analytical method	Reference
Apollo 11 (Mare Tranquillitatis)			
A, basalt (1)	6	N/R	Haskin *et al.* [P1]
A, basalt (3)	5—10 (av. 8)	N/R	Morrison *et al.* [P1] [S]
B, basalt (1)	7	N/R	Haskin *et al.* [P1]
B, basalt (2)	5, 10 (av. 8)	N/R	Morrison *et al.* [P1] [S]
C, breccia (1)	9	N/R	Haskin *et al.* [P1]
C, breccia (2)	5, 5 (av. 5)	N/R	Morrison *et al.* [P1] [S]
D, soil < 1 mm (3), 10084	5, 5, 2.2	N/R	Haskin *et al.* [P1]; Morrison *et al.* [P1] [S]; Morgan *et al.* [P3]
D, soil (1), 10084	< 60	N/R	Travesi *et al.* [P2]
Apollo 12 (Oceanus Procellarum)			
A, basalt (2)	8, 10 (av. 9)	N/R	Morrison *et al.* [P2]
AB, basalt (1)	4	N/R	Morrison *et al.* [P2]
AB, basalt (1)	9	N/R	Smales *et al.* [P2]
B, basalt (1)	32	N/R	Morrison *et al.* [P2]
Basalt (4)	15—130 (av. 60)	N/R	Brunfelt *et al.* [P2]
C, breccia (1)	3	N/R	Morrison *et al.* [P2]
Soil (2)	35, 130	N/R	Morrison *et al.* [P2]
Soil (3), 12070	9, < 50, 40	N/R	Morrison *et al.* [P2]; Travesi *et al.* [P2]; Brunfelt *et al.* [P2]
Soil < 1 mm (1)	0.77	N/R	Morgan *et al.* [P3]
KREEP concentrates from soil (4)	0.43—3.1 (av. 1.5)	N/R	Morgan *et al.* [P3] [G1]
Apollo 14 (Fra Mauro)			
Crystalline rock (2), 14310	4, 4.5	N/R	Brunfelt *et al.* [P3]; Morgan *et al.* [P3]
Plagioclase from above rock	< 10	N/R	Brunfelt *et al.* [P3]
Pyroxene from above rock	90	N/R	Brunfelt *et al.* [P3]
Basalt (1)	0.64	N/R	Morgan *et al.* [P3]
Breccia (2)	< 30, 24	N/R	Brunfelt *et al.* [P3]
Breccia (11),	0.46—15 (9 samples: 0.78—2.9)	N/R	Morgan *et al.* [P3]
Breccia (2), 14306	1.4, 12	N/R	Ganapathy *et al.* [P4]
Soil < 1 mm (1), 14163	3, 5.7	N/R	Brunfelt *et al.* [P3] Morgan *et al.* [P3]
Fine fraction of 14163	10	N/R	Brunfelt *et al.* [P3]

Table 51-C-3 (continued)

Material (no. of samples)	Sb (ppb)	Analytical method	Reference
Plagioclase from 14163	1300	N/R	BRUNFELT *et al.* [P3]
Glass from 14163 (2)	10, 110	N/R	BRUNFELT *et al.* [P3]
Soil < 1 mm (5)	2.4—3.6 (av. 2.9)	N/R	MORGAN *et al.* [P3]
Magnetic fractions 1—2 mm of soils (4)	1.8—2.4 (av. 2.1)	N/R	MORGAN *et al.* [P3]
Norite fractions 1—2 mm of soils (5)	1.0—2.9 (av. 1.8)	N/R	MORGAN *et al.* [P3]
Glass 1—2 mm from soils (3)	0.53—0.87 (av. 0.69)	N/R	MORGAN *et al.* [P3]
Apollo 15 (Hadley-Apennine)			
Basalt (5)	0.04, 0.07, 0.13, 0.43, 1.5 ?	N/R	MORGAN *et al.* [P3]; GANAPATHY *et al.* [P4]
Anorthosite (1)	0.07	N/R	MORGAN *et al.* [P3]; GANAPATHY *et al.* [P4]
Gabbro (2)	0.02, 0.24	N/R	GANAPATHY *et al.* [P4]
Glass (1)	4.7	N/R	GANAPATHY *et al.* [P4]
Breccia (16)	0.04—2.2	N/R	GANAPATHY *et al.* [P4]
Soil < 1 mm (8)	0.85—4.8 (av. 2.1)	N/R	LAUL and SCHMITT [G2]; MORGAN *et al.* [P3]; GANAPATHY *et al.* [P4]
Soil < 1 mm (1)	< 30	N/R	BRUNFELT *et al.* [P3]
Apollo 16 (Descartes-Cayley)			
Crystalline rock, igneous (5)	0.15, 0.21, 0.53, 1.7, 3.1	N/R	KRÄHENBÜHL *et al.* [P4]
Crystalline rock, metaclastic (3)	0.88, 4.3, 6.9	N/R	KRÄHENBÜHL *et al.* [P4]
Breccia (7)	0.04, 0.33, 0.37, 0.41, 1.9, 3.7, 3.9	N/R	KRÄHENBÜHL *et al.* [P4]
Soil < 1 mm, (9)	0.73—4.2 (av. 1.9)	N/R	KRÄHENBÜHL *et al.* [P4]
Soil < 1 mm (1)	4.5		LAUL and SCHMITT [G2]
Luna 16 (Mare Fecunditatis)			
Basalt	av. 350	Mass (?)	VINOGRADOV [G3]
Soil	av. 850	Mass (?)	VINOGRADOV [G3]
Soil < 0.125 mm (2)	3.3, 4.2	N/R	MORGAN *et al.* [P3]
Luna 20 (between Mare Fecunditatis and Mare Crisium)			
Anorthosite (1)	0.68	N/R	GANAPATHY *et al.* [P4]
Breccia (1)	1.5	N/R	MORGAN *et al.* [G4]

Table 51-C-3 (continued)

Material (no. of samples)	Sb (ppb)	Analytical method	Reference
Soil (2)	$<9{,}000$, <200	Mass (?)	VINOGRADOV [G3]
Soil (1), 22001, 9	760	N/R	LAUL and SCHMITT [G2]
Soil (1), 22001, 10	9.8	N/R	MORGAN *et al.* [G4]

References: [P1] Proc. Apollo 11 Lunar Sci. Conf., Geochim. Cosmochim. Acta, Supplement 1, 2 (1970). [S] Science **167**, No. 3918 (1970). [P2] Proc. Second Lunar Sci. Conf., Geochim. Cosmochim. Acta, Supplement 2, **2** (1971). [P3] Proc. Third Lunar Sci. Conf., Geochim. Cosmochim. Acta, Supplement 3, **2** (1972). [P4] Proc. Fourth Lunar Sci. Conf., Geochim. Cosmochim. Acta, Supplement 4, **2** (1973) [G1] MORGAN, J. W., GANAPATHY, R., LAUL, J. C., ANDERS, E.: Geochim. Cosmochim. Acta **37**, 141 (1973). [G2] LAUL, J. C., and SCHMITT, R. A.: Geochim. Cosmochim. Acta **37**, 927 (1973). [G3] VINOGRADOV, A. P.: Geochim. Cosmochim. Acta **37**, 721, (1973). [G4] MORGAN, J. W., KRÄHENBÜHL, U., GANAPATHY, R., ANDERS, E.: Geochim. Cosmochim. Acta **37**, 953 (1973).

Very approximate average antimony contents of lunar rocks and soils are as follows:

Apollo 11. Basalts (type A): 8 ppb Sb; basalts (type B): 7 ppb Sb; breccias (type C): 6 ppb Sb.

Apollo 12. Basalts: 8 ppb Sb.

Apollo 14. Breccias, variable; soils: 3 ppb Sb.

Apollo 15. Basalts: 0.1 ppb Sb; breccias: variable; soils: 2 ppb Sb.

Apollo 16. Crystalline rocks: 1 ppb Sb; breccias: variable; soils: 2 ppb Sb.

Luna 16. Soils: 4 ppb Sb.

Thus the antimony content of lunar rocks and soils is much lower than that of terrestrial basalts (see Section 51-E) and of C1 chondrites.

Manuscript received: February 1974

51-D. Abundances in Rock-Forming Minerals (I), Antimony Minerals (II)

I. Rock-Forming Minerals

The distribution of antimony among co-existing minerals of gabbros from the Skaergaard intrusion were studied by ESSON *et al.* (1965): Table 51-D-1.

Table 51-D-1. *Antimony in rock-forming minerals*

Material	Sb (ppm)	Reference
Plagioclase (3 samples) from gabbro, Skaergaard, Greenland	0.24; 0.032; 0.036 (N/R)	ESSON *et al.* (1965)
Pyroxene (3 samples) from gabbro, Skaergaard, Greenland	0.11; 0.026; 0.032	ESSON *et al.* (1965)
Olivine (3 samples) from gabbro, Skaergaard, Greenland	0.21; 1.17; 1.38	ESSON *et al.* (1965)
Titaniferous magnetite (3 samples) from gabbro, Skaergaard, Greenland	0.083; 0.026; 0.11	ESSON *et al.* (1965)
Ilmenite (2 samples) from gabbro, Skaergaard, Greenland	0.27; 0.15	ESSON *et al.* (1965)

There is apparently a marked preferential entry of antimony into early magnesian olivines. Ilmenite appears to be the next most favorable host for antimony, presumably due to the substitution of antimony (III) for iron (II).

In rock-forming minerals antimony (ionic radii, Sb^{3+} 0.76 and Sb^{5+} 0.62 Å) can probably substitute for iron (Fe^{2+} 0.74 and Fe^{3+} 0.64 Å) (ESSON *et al.* 1965).

II. Antimony Minerals

Minerals of which antimony is a major constituent include the element, antimonides, sulfides, sulfosalts (complex sulfides), oxides, and a few antimonates and antimonites. A list of selected antimony minerals is given in Table 51-D-2. The most

Table 51-D-2. *Selected antimony minerals (Mostly from* PALACHE *et al. 1944, 1951)*

Name	Composition	Manner of occurrence
Antimony	Sb	Occurs in veins, with Ag, Sb, and As ores
Allemontite	AsSb	Mostly in veins
Dyscrasite	Ag_3Sb	Found as a vein mineral in Ag deposits

Table 51-D-2 (Continued)

Name	Composition	Manner of occurrence
Breithauptite	NiSb	Occurs in calcite veins
Stibnite	Sb_2S_3	Formed at low temperatures in hydrothermal vein or replacement deposits and in hot springs
Ullmannite	NiSbS	Occurs in veins
Gudmundite	FeSbS	A hydrothermal mineral formed at a relatively late period in sulfide deposits
Polybasite	$(Ag, Cu)_{16}Sb_2S_{11}$	Occurs in Ag veins formed at low to moderate temperatures
Stephanite	Ag_5SbS_4	Found in many Ag deposits, usually in small amounts, as one of the last vein minerals to form
Pyrargyrite	Ag_3SbS_3	Formed at low temperatures as one of the last Ag minerals to crystallize in the sequence of primary deposition
Pyrostilpnite	Ag_3SbS_3	
Tetrahedrite	$(Cu, Fe)_{12}Sb_4S_{13}$ or Cu_3SbS_3	Occurs in hydrothermal veins of Cu, Pb, Zn, and Ag minerals
Samsonite	$Ag_4MnSb_2S_6$	
Geocronite	$Pb_5(Sb, As)_2S_8$	
Meneghinite	$Pb_{13}Sb_7S_{23}$	
Bournonite	$PbCuSbS_3$	Occurs in hydrothermal veins formed at moderate temperatures
Diaphorite	$Pb_2Ag_3Sb_3S_8$	
Freieslebenite	$Ag_5Pb_3Sb_5S_{12}$	
Boulangerite	$Pb_5Sb_4S_{11}$	Occurs in hydrothermal vein deposits formed at low or at moderate temperatures
Miargyrite	$AgSbS_2$	Found in hydrothermal vein deposits formed at low temperatures
Aramayoite	$Ag (Sb, Bi)S_2$	
Chalcostibite	$CuSbS_2$	
Franckeite	$Pb_5Sn_3Sb_2S_{14}$	
Jamesonite	$Pb_4FeSb_6S_{14}$	Found in druses and hydrothermal veins formed at low to moderate temperatures
Andorite	$PbAgSb_3S_6$	
Plagionite	$Pb_5Sb_8S_{17}$	
Semseyite	$Pb_9Sb_8S_{21}$	
Zinkenite	$Pb_6Sb_{14}S_{27}$	Found in vein deposits formed at low to moderate temperatures
Berthierite	$FeSb_2S_4$	
Vrbaite	$TlAs_2SbS_5$	
Livingstonite	$HgSb_4S_7$	
Senarmontite	Sb_2O_3	A secondary mineral formed by the oxidation of stibnite, native Sb, and other Sb minerals
Valentinite	Sb_2O_3	A secondary mineral formed by the oxidation of stibnite, native Sb, and other Sb minerals
Stibiotantalite	$SbTaO_4$	Occurs in pegmatites
Bindheimite	$Pb_2Sb_2O_6 (O, OH)$	
Romeite	$(Ca, Fe, Mn, Na)_2 (Sb, Ti)_2O_6(O, OH, F)$	
Swedenborgite	$Na_4Be_4SbO_7$	
Nadorite	$PbSbO_2Cl$	

important ore mineral of antimony is stibnite. Other important minerals are complex sulfides such as stephanite, pyrargirite, and tetrahedrite. Chemical processes in the formation of stibnite were discussed by TUNELL (1964). For additional information see chapter 80.

Phase diagrams are given for the binary systems of Sb—As, Sb—Se, Sb—Si, Sb—Sn, Sb—Te, Sb—Ti, Sb—Tl, Sb—Zn, and Sb—Zr (HANSEN and ANDERKO, 1958). The melting curve of antimony at pressures up to 70 kbars was determined (KLEMENT, JAYARAMAN, and KENNEDY, 1963). The 1-atm. isobaric phase relations were determined for the system As—Sb (SKINNER, 1965).

Revised manuscript received: February 1967

51-E. Abundance in Common Igneous Rocks

In the work of ONISHI and SANDELL (1955), the level of antimony in igneous rocks was found to be only slightly higher than the limit of the analytical method (C). Consequently, composite samples were analyzed to obtain a rough average.

Table 51-E-1. *Antimony in ultrabasic rocks*

Material	Sb (ppm)	Author
Composite of 13 samples, U.S.A. and Canada; peridotite (4), dunite (3), bronzitite (1), harzburgite (2), pyroxenite (1), unspecifid ultrabasic (2)	0.1 (C)	ONISHI and SANDELL (1955)
Composite of 4 samples, U.S.A.; serpentinites (2), dunite (1), lherzolite (1)	0.1	ONISHI and SANDELL (1955)

Table 51-E-2. *Antimony in basic rocks*

Material	Sb (ppm)	Author
Composite of balsalts, mostly North America (11 samples)	0.3 (C)	ONISHI and SANDELL (1955)
Composite of basalts, mostly U.S.A. (6 samples)	0.9	ONISHI and SANDELL (1955)
Composite of basalts, Minnesota and Wisconsin, U.S.A. (5 samples)	0.1	ONISHI and SANDELL (1955)
Composite of basalts, Michigan, U.S.A. (20 samples)	0.1	ONISHI and SANDELL (1955)
Composite of olivine basalts, Oregon, U.S.A. (7 samples)	0.1	ONISHI and SANDELL (1955)
Composite of basalts, Japan (5 samples)	1.4	ONISHI and SANDELL (1955)
Composite of gabbros, U.S.A. and Canada (4 samples)	0.2	ONISHI and SANDELL (1955)
Composite of gabbros, Minnesota, U.S.A. (5 samples)	0.1	ONISHI and SANDELL (1955)
Composite of basaltic rocks, Japan (7 samples)	0.038 (N/R)	HAMAGUCHI *et al.* (1961a)
Rocks of the layered series (mainly gabbros), Skaergaard intrusion, Greenland	0.025—0.19 (av. 0.11) (N/R)	ESSON *et al.* (1965)
Chilled olivine gabbro, Skaergaard intrusion[a]	0.16	ESSON *et al.* (1965)

[a] This rock is taken to represent perhaps the best estimate of the composition of the original magma (ESSON *et al.*, 1965).

The antimony contents of ultrabasic rocks and basic rocks are compiled in Tables 51-E-1 and 51-E-2, respectively.

In Table 51-E-2, two of the basalt composites contain what appear to be abnormally high amounts (0.9 and 1.4 ppm) of antimony. Their omission in the computation of a basaltic average is probably justified. A tentative average for basic rocks (basalt, and gabbro) is 0.1—0.2 ppm Sb.

The antimony contents of intermediate rocks and granitic rocks are shown in Tables 51-E-3 and 51-E-4, respectively.

Table 51-E-3. *Antimony in intermediate roocks*

Material	Sb (ppm)	Author
Composite of diorites and quartz diorites, western Canada (10 samples)	0.2 (C)	ONISHI and SANDELL (1955)
Composite of andesites, Japan (6 samples)	0.2	ONISHI and SANDELL (1955)
Granophyres, Skaergaard intrusion, Greenland (2 samples)	0.12; 0.21 (N/R)	ESSON *et al.* (1965)
Andesite	0.18 (spark source mass spectrography)	TAYLOR (1965)

Table 51-E-4. *Antimony in granitic rocks*

Material	Sb (ppm)	Author
Composite of granites, Germany (14 samples)	0.3 (S)	PREUSS (1940)
Composite of granites, U.S.A. and Canada (37 samples)	0.3 (C)	ONISHI and SANDELL (1955)
Composite of granites, U.S.A. and elsewhere (18 samples)	0.2	ONISHI and SANDELL (1955)
Composite of granitic rocks, Minnesota, U.S.A. (11 samples)	0.1	ONISHI and SANDELL (1955)
Composite of granites, Llano region, Texas, U.S.A. (5 samples)	0.1	ONISHI and SANDELL (1955)
Composite of granitic rocks, mostly Japan (12 samples)	0.2	ONISHI and SANDELL (1955)
Composite of granite-gneisses, Canada, U.S.A., and elsewhere (14 samples)	0.3	ONISHI and SANDELL (1955)
Composite of silicic rocks, U.S.A. (7 samples)	0.6	ONISHI and SANDELL (1955)
Composite of granitic rocks, Japan (20 samples)	0.24 (N/R)	HAMAGUCHI *et al.* (1961a)
Acid granophyres, Skaergaard intrusion, Greenland (4 samples)	0.09—0.44 (av. 0.20) (N/R)	ESSON *et al.* (1965)
Granodiorites, Mayo Lake, Keno Hill-Galena Hill area, Yukon, Canada (4 composite samples)	<1.0—1.0 (av. <1.0)	BOYLE (1965)

From Tables 51-E-3 and 51-E-4 a value of 0.2 ppm Sb is given as the average for intermediate rocks and granitic rocks. It seems that granitic rocks contain somewhat more antimony than do basaltic and gabbroic rocks, but the accuracy of the data does not warrant a positive conclusion.

The average content of antimony in igneous rocks may be taken as 0.2 ppm Sb, based on the averages for granitic rocks, basalts, and gabbros. The average weight ratio of Sb/As in igneous rocks is 0.2/1.5 = approximately 0.1. GOLDSCHMIDT's (1937) estimate of 1 ppm Sb for the crustal rocks is certainly too high.

In order to judge reliability of the analysis of silicate rocks for antimony, the results of antimony determinations in granite G-1 and diabase W-1 are listed in Table 51-E-5.

Table 51-E-5. *Summary of antimony determinations in G-1 and W-1*

Rock	Sb (ppm)	Method	Author
G-1	0.6	C	WARD and LAKIN (1954)
G-1	0.28	N/R	HAMAGUCHI *et al.* (1961 a)
G-1	0.1	Spark source mass spectrography	BROWN and WOLSTENHOLME (1964)
G-1	0.30	N/R	ESSON *et al.* (1965)
G-1	0.35	Spark source mass spectrography	TAYLOR (1965)
W-1	1.2	C	WARD and LAKIN (1954)
W-1	0.95	N/R	HAMAGUCHI *et al.* (1961 a)
W-1	0.3	Spark source mass spectrography	BROWN and WOLSTENHOLME (1964)
W-1	1.03	N/R	ESSON *et al.* (1965)
W-1	0.89	Spark source mass spectrography	TAYLOR (1965)

With the exception of the values obtained by BROWN and WOLSTENHOLME, the figures are in good agreement; the ranges are 0.28—0.6 and 0.89—1.2 ppm Sb for G-1 and W-1, respectively. The averages (= arithmetic means) are 0.4 and 1.0 ppm Sb for G-1 and W-1, respectively. These averages are recommended when G-1 and W-1 are used as standard samples.

The present state of the analytical methods for the determination of traces of antimony in silicate rocks will be briefly discussed below. Unfortunately, the sensitivities of the available colorimetric methods are slightly higher than the average antimony content of igneous rocks. It is desirable to have a colorimetric method that enables a larger size of sample, e.g. 5 g, to be handled. The sensitivity of the ordinary emission spectrographic analysis is too low to permit the determination of antimony in silicate rocks. However, PREUSS (1940) developed a method using an auxiliary furnace technique and increased the spectrographic sensitivity down to ~0.3 ppm Sb. Neutron activation analysis is probably the most useful method. Spark sources mass spectrography is a very promising method.

We do not have a method for the separate determination of antimony (III) and antimony (V) in silicate rocks. The development of such a method is very significant, because it will give a clue to the elucidation of the manner of occurrence of antimony in silicate rocks.

Revised manuscript received: February 1967

51-F. Behavior in Volcanic Processes (Abundance in Sulfur and Sulfide Minerals)

GEILMANN and BILTZ (1931) found 9 ppm and 3 ppm Sb in 2 samples of volcanic sulfur from Papandajan, Java. Of 47 volcanic sulfur samples from Japan, only one sample contained spectrographically detectable amounts of antimony (IWASAKI and NASHIZAWA, 1937; the lower limit of detection was not mentioned).

FLEISCHER (1955) summarized data on the minor element content of sulfide minerals. Table 51-F-1 gives data on antimony in galena, sphalerite, chalcopyrite, pyrite, and arsenopyrite. A paper by TIMASHEVA (1963) was not seen except in abstract. Galena usually shows high contents of antimony.

The relation between the sequence of hypogene mineralization and distribution of antimony in the Keno Hill-Galena Hill area, Yukon, Canada was studied by BOYLE (1965). Most of the antimony was bound in various sulfosalts or complex sulfides [freibergite (= tetrahedrite containing much silver), jamesonite, boulangerite, etc.] of the second and third stages of mineralization. Small amounts of antimony were also present in arsenopyrite, pyrite, galena, sphalerite, and chalcopyrite of these stages.

Table 51-F-1. *Antimony in sulfide minerals*

Material	Sb (ppm)	Author
Galenas	up to 10,000	FLEISCHER (1955)
Galenas, northeast Japan	< 25—5,000 (S)	TAKAHASHI (1963)
Galenas, Keno Hill-Galena Hill area, Yukon, Canada (15 samples)	2,400—6,600 (C)	BOYLE (1965)
Galenas, U.S.S.R., Romania, and Bulgaria (5 samples)	500—5,000 (av. 3,000) (S)	KRAVCHENKO *et al.* (1966)
Sphalerites	up to 5,000	FLEISCHER (1955)
Sphalerites, northeast Japan	< 25—1,000 (S)	TAKAHASHI (1963)
Sphalerites, Keno Hill-Galena Hill area, Yukon, Canada (11 samples)	< 5—3,800 (C)	BOYLE (1965)
Chalcopyrites, northeast Japan	all < 25 (S)	TAKAHASHI (1963)
Chalcopyrite, Galena Hill, Yukon, Canada	25 (C)	BOYLE (1965)
Pyrites	up to 1,000	FLEISCHER (1955)
Pyrites, northeast Japan	mostly < 25; max. 2,500 (S)	TAKAHASHI (1963)
Pyrites from deposits, Keno Hill-Galena Hill area, Yukon, Canada (24 samples)	< 10—400 (C)	BOYLE (1965)
Pyrites from metamorphic rocks, Keno Hill-Galena Hill area, Yukon, Canada (9 samples)	5—45; 135	BOYLE (1965)
Arsenopyrites, Keno Hill-Galena Hill area, Yukon, Canada (6 samples)	170—1,160	BOYLE (1965

Revised manuscript received: February 1967

51-G. Abundance in Soils

The behavior of antimony during weathering of rocks remains to be explored. Therefore, this section summarizes data on antimony in soils.

Ward and Lakin (1954) reported 2.3—9.5 ppm Sb (C) in soils from Idaho and North Carolina, U.S.A. However, it is not clear if these soils came from non-mineralized areas. In the Nyeba lead-zinc district of Nigeria, soils which were 30—120 m distant from the Ameri lode contained 1—5 ppm Sb, and soils which were 30—90 m distant from the Palm Wine lode contained 1—2 ppm Sb (C) (Hawkes, 1954).

Residual soils of the Keno Hill-Galena Hill area, Yukon, Canada were studied by Boyle (1965). The soils which probably had not been influenced by mineralization contained 1 to 3 ppm Sb; a nonweighted average was $1._9$ ppm. The average arsenic content of these soils was as high as 16 ppm As, and the weight ratio of Sb/As was 0.1_2.

If the above results are representative of soils, antimony is concentrated in soils compared to igneous rocks.

Revised manuscript received: February 1967

51-H. Solubilities of Compounds which Control Concentrations in Natural Waters; Adsorption Processes, and Valence States in Natural Environments

The solubilities of Sb_2O_3 and Sb_2O_5 in water are 1.31 mg and 8.77 mg per 100 ml of water at 35° C, respectively. The solubility of Sb_2S_3 is 0.175 mg per 100 ml of water at 18° C, and that of Sb_2S_5 is 0.7 mg per 100 ml of water at 26° C.

Antimony sulfides, Sb_2S_3 and Sb_2S_5, are soluble in sodium sulfide solution. Stibnite occurs in the siliceous muds and sinters of Steamboat Springs, Nevada, U.S.A. (Brannock, Fix, Gianella, and White, 1948; White, Thompson, and Sandberg, 1964). The equilibrium relations in the system Sb_2S_3—Na_2S—H_2O at 100 and 200° C and 100 bars are shown in a paper by Tunell (1964).

According to Goldschmidt (1937 and 1954), sedimentary iron oxide ores contain antimony in amounts roughly one-fifth by weight of arsenic. Because Goldschmidt and Peters (1934) found 70—700 ppm As (S) in these ores, the enrichment of antimony in them was indicated. This enrichment was probably caused by the adsorption of antimony on ferric hydroxide.

Standard electrode potentials (ε°) of some geochemically important reactions at 25° C are as follows. All the data are from Latimer (1952).

	ε°
$Sb_2O_3 + 6H^+ + 6e^- = 2Sb + 3H_2O$	0.152 v
$SbO^+ + 2H^+ + 3e^- = Sb + H_2O$	0.212 v
$Sb_2O_5 + 6H^+ + 4e^- = 2SbO^+ + 3H_2O$	0.581 v

Krauskopf (1955) showed fields of stability of some antimony compounds as functions of pH and oxidation-reduction potential.

Antimonic acid is known only in solution, but it gives antimonates of the type $K[Sb(OH)_6]$.

It seems that the ratio of Sb^{5+}/Sb^{3+} is fairly small in magmas (Esson, Stevens, and Vincent, 1965).

Revised manuscript received: February 1967

51-I. Abundance in Natural Waters

The antimony content of river and ground waters is shown in Table 51-I-1.

Table 51-I-1. *Antimony in river and ground waters*

Material	Sb (ppb)	Author
River waters:		
Skellefte River, Sweden	0.05 (N/R)	LANDSTRÖM and WENNER (1965)
Ume River, Sweden	0.07	LANDSTRÖM and WENNER (1965)
Ångerman River, Sweden	0.08	LANDSTRÖM and WENNER (1965)
Ground waters:		
Near Skelleftå, Sweden (2 locations)	0.01; 0.02 (N/R)	LANDSTRÖM and WENNER (1965)
Brån, Sweden (2 locations)	0.03; 0.2	LANDSTRÖM and WENNER (1965)
Sollefteå, Sweden (3 locations)	0.06; 0.04; 0.5	LANDSTRÖM and WENNER (1965)

Table 51-I-2. *Antimony in hot spring waters*

Locality	Sb (ppm)	Author
Umnak Island, Alaska, U.S.A. (9 springs)	< 0.07—0.93	BYERS and BRANNOCK (1949)
Ouray Springs, Colorado, U.S.A.	0.5	WHITE, HEM, and WARING (1963)
Boiling Springs, Idaho, U.S.A.	0.1	WHITE, HEM, and WARING (1963)
Steamboat Springs, Nevada, U.S.A.	0.07—0.4	BRANNOCK *et al.* (1948), WHITE and BRANNOCK (1950), WHITE, HEM, and WARING (1963)
Yellowstone Park, Wyoming, U.S.A.	0.1	WHITE, HEM, and WARING (1963)
Kamchatka, U.S.S.R. (2 springs)	0.1 (S); 0.6	WHITE, HEM, and WARING (1963)
Hot springs of Hokkaido, Japan	max. about 0.8 (S)	NISHIMURA (1958)
Nasu, Tochigi Prefecture, Japan	0.0001 (S)	IKEDA (1955a)
Arima, Hyogo Prefecture, Japan	about 0.0005 (S)	IKEDA (1955b)
Tokaanu, New Zealand (2 springs)	0.34; 0.68 (C)	RITCHIE (1961)
Wairakei, New Zealand (8 springs)	0.085—0.12	RITCHIE (1961)
Geyser Valley, New Zealand (2 springs)	0.12; 0.14	RITCHIE (1961)
Ohaki, New Zealand	0.90	RITCHIE (1961)
Orakei-Korako, New Zealand (4 springs)	0.017—0.048	RITCHIE (1961)
Waiotapu, New Zealand (3 springs)	0.034—0.22	RITCHIE (1961)
Waimangu, New Zealand	0.038	RITCHIE (1961)
Whakarewarewa, New Zealand	0.036	RITCHIE (1961)
Rotorua, New Zealand	0.008	RITCHIE (1961)

Table 51-I-3. *Antimony in sea water*

Area	Depth (m)	Sb (ppb)	Author
Caribbean	3—3,000	0.20—0.31 (av. 0.26) (N/R)	SCHUTZ and TUREKIAN (1965)
Gulf of Mexico,			
25° 03′ N, 87° 05′ W	10	0.58 (av. 0.46)	SCHUTZ and TUREKIAN (1965)
25° 03 N, 87° 05′ W	3,000	0.34	
N.W. Atlantic	5—4,500	0.12—0.43 (av. 0.24)	SCHUTZ and TUREKIAN (1965)
N.E. Atlantic	8—800	0.14—0.30 (av. 0.21)	SCHUTZ and TUREKIAN (1965)
S.W. Atlantic,			
19° 57′ S, 34° 03′ W	4,000	1.1 (av. of 3 = 0.53,	SCHUTZ and TUREKIAN (1965)
21° 49′ S, 35° 43′ W	10	0.30 av. of 2 = 0.25)	
21° 49′ S, 35° 43′ W	100	0.19	
Irish Sea	surface	0.13—0.40 (av. 0.24)(C)	PORTMANN and RILEY (1966)
Indian Ocean,			
24° 57′ S, 79° 33′ E	surface to 1,200	0.31 (av. 0.37) (N/R)	SCHUTZ and TUREKIAN (1965)
25° 41′ S, 101° 56′ E		0.42	
Central Pacific	surface to 4,350	0.18—1.2 (av. 0.51)	SCHUTZ and TUREKIAN (1965)
East Pacific	14—1,300	0.16—0.38; 1.75 (av. of all = 0.51; 0.26[a])	SCHUTZ and TUREKIAN (1965)
Antarctic	14—1,300	0.18—0.26 (av. 0.24)	SCHUTZ and TUREKIAN (1965)

[a] The high value of 1.75 ppb was excluded.

A sample of Searles Lake brine, California, U.S.A. contained 5 ppm Sb (WHITE, HEM, and WARING, 1963).

Data on the antimony content of hot spring waters (including some mineral spring waters) are listed in Table 51-I-2.

The highest concentration of antimony in hot spring waters is close to 1 ppm Sb.

The distribution of antimony and other trace elements in sea water was investigated by SCHUTZ and TUREKIAN (1965) using neutron activation analysis. Their data are summarized in Table 51-I-3. It is noted that the early estimate by NODDACK and NODDACK (1939), i.e., <0.5 ppb Sb, is concordant with the recent data.

The average for sea water may be taken as 0.3 ppb Sb (μg per l). If we take the average chlorine content as 19,000 ppm (mg per l), we obtain the atomic ratio of $Sb/Cl = 5 \times 10^{-9}$. The weight ratio of Sb/As in sea water is $0.3/2 = 0.15$.

Revised manuscript received: February 1967

51-K. Abundance in Common Sedimentary Rocks

The antimony content of Mansfeld copper shale (Kupferschiefer), Germany is 0.0012% and 0.03% Sb according to NODDACK and NODDACK (1931) and CISSARZ and MORITZ (1933), respectively. A new analysis is desirable.

Table 51-K-1. *Antimony in shales and slates*

Material	Sb (ppm)	Author
Composite, Europe (36 samples)	3 (S)	PREUSS (1940)
Nyeba district, Nigeria (3 samples)	(all) 4 (C)[a]	HAWKES (1954)
Composite, Wisconsin, U.S.A. (3 samples)	0.1 (C)	ONISHI and SANDELL (1955)
Composite, Minnesota, U.S.A. (4 samples)	0.2	ONISHI and SANDELL (1955)
New York and Pennsylvania, U.S.A. (5 samples)	0.3—2.4 (av. 0.8)	ONISHI and SANDELL (1955)
Composite, North America (7 samples)	2.6	ONISHI and SANDELL (1955)
Japan (5 samples)	0.2—0.9 (av. 0.4)	ONISHI and SANDELL (1955)
Non-weighted composite of Japanese (24 samples) and European (36) shales	2.3	ONISHI and SANDELL (1955)
Oil shale (tasmanite), Tasmania	0.2	ONISHI and SANDELL (1955)
Slate, Granville, New York, U.S.A.	1.7 (C)	ONISHI and SANDELL (1955)
Slate, Powlet, Vermont, U.S.A.	0.5	ONISHI and SANDELL (1955)
Slate, Cuyuna Range, Minnesota, U.S.A.	0.4	ONISHI and SANDELL (1955)

[a] These samples are high in lead (67—114 ppm).

Shales are strongly variable in antimony. Some contain no more than igneous rocks, whereas others contain as much as a few parts per million. At present the average for shales may be taken as 1—2 ppm Sb. The antimony content of shales as a whole may be as much as ten times greater than that of igneous rocks. Although antimony and arsenic do not vary together in shales, the average weight ratio of Sb/As in shales is approximately 0.1.

The antimony content of various sedimentary rocks and sediments other than shales is given in Table 51-K-2.

Because little antimony may be expected to enter quartz, calcite, or dolomite, sandstones and limestones should be low in antimony, and what is found should be

Table 51-K-2. *Antimony in various sedimentary deposits other than shales*

Material	Sb (ppm)	Author
Composite of limestones, mostly U.S.A. (9 samples)	0.3 (C)	ONISHI and SANDELL (1955)
Composite of sandstones, Germany (23 samples)	1 (S)	PREUSS (1940)
Sandstone, Minnesota, U.S.A., washed with HCl	0.0 (C)	ONISHI and SANDELL (1955)
Bauxites, British Guiana and Haiti (2 samples)	1.9; 2.3	ONISHI and SANDELL (1955)
Pebble phosphate, Florida, U.S.A.	1.3	ONISHI and SANDELL (1955)
Phosphate rock, Tennessee Brown, Natl. Bur. Standards Sample 56a	0.2	ONISHI and SANDELL (1955)
Composite of pelagic clays, Pacific Ocean, depth of sea 4096—5794 m (4 samples)	1.4	ONISHI and SANDELL (1955)
Composite of pelagic clays, Pacific Ocean, depth of sea 4293—5844 m (4 samples)	0.7	ONISHI and SANDELL (1955)
Composite of globigerina ooze, Pacific Ocean, depth of sea 2140—4540 m (6 samples)	0.2	ONISHI and SANDELL (1955)

present largely in the shaly or an accessory component. At least some bauxites are relatively high in antimony. Phosphate rock is evidently variable in antimony.

Pelagic clays are clearly enriched in antimony compared to igneous rocks. An average of 1 ppm Sb is indicated, roughly one-tenth of the arsenic content. Globigerina ooze is very low in antimony.

Revised manuscript received: February 1967

51-L. Biogeochemistry

Antimony in marine animals was determined spectroscopically by NODDACK and NODDACK (1939). The samples were taken from Gullmarfjord, Sweden. Their results are shown in Table 51-L-1. Although the data obtained by the NODDACKS need confirmation, they should serve as good indications of magnitude.

Table 51-L-1. *Antimony in marine animals*

Material (dried samples)	Sb (ppm)
Tunicata	
Ciona intestinalis (whole)	0.1
Porifera	
Halichondria (whole)	0.08
Coelenterata	
Cyanea capillata (whole)	0.16
Metridium dianthus (whole)	0.23
Echinodermata	
Stichopus tremulus (eviscerated)	0.24
Brissopsis lyrifera (test)	0.18
Asterias rubens (eviscerated)	0.1
Pisces	
Ctenolabrus rupestris (whole)	0.2
Squalus acanthius (eviscerated)	0.2

Data on antimony in other biological materials are few. Using neutron activation analysis, WESTER (1965) reported 0.0007—0.01 ppm Sb in fresh (wet) samples of conductive tissue of 4 hearts from healthy cattle.

Due to the insufficient sensitivity of the analytical method, the content of antimony in coal is not fully known (Table 51-L-2).

Table 51-L-2. *Antimony in coals*

Material	Sb (ppm)	Author
Ash of coal, Virginia, U.S.A.	av. $<$ 40 (S)	HEADLEE and HUNTER (1953)
Ash of coal, Germany	max. ca. 3,000 (S)	OTTE (1953)
Bituminous coals, New South Wales, Australia	av. $<$ 30 in coal or $<$ 200 in ash	CLARK and SWAINE (1962), SWAINE (1962)

Revised manuscript received: February 1967

51-M. Abundance in Common Metamorphic Rocks

The literature on this subject is scarce. Table 51-M-1 gives the antimony content of metamorphic rocks. BOYLE (1965) considers that, in the rocks he studied, most of the antimony is probably bound in sulfide minerals.

Table 51-M-1. *Antimony in metamorphic rocks*

Material	Sb (ppm)	Author
Rocks of Keno Hill-Galena Hill area, Yukon, Canada		
Graphitic argillites, schists, and phyllites (29 composite samples)	1.0—7.0 (av. 3.1)	BOYLE (1965)
Quartz-sericite schists (8 composite samples)	1—1.5 (av. 1)	BOYLE (1965)
Thick- and thin-bedded gray quartzites (27 composite samples)	1.0—6.0 (av. 1.5)	BOYLE (1965)
Siliceous white quartzites (9 composite samples)	<1—1 (av. <1)	BOYLE (1965)
Calcareous quartzites (3 composite samples)	<1—1 (av. <1)	BOYLE (1965)
Skarn (6 composite samples)	1 (av. 1)	BOYLE (1965)

Revised manuscript received: February 1967

51-O. Concluding Notes

Table 51-O-1 summarizes the average abundances of antimony in different cosmic and crustal materials. Many of the figures in this table are approximate, and it is hoped that they will be improved or confirmed by future work.

Table 51-O-1. *Average abundances of Sb in different materials from cosmos and earth's crust*

Material	Sb (ppm)
Iron meteorites	0.3
Ordinary chondrites	0.1
Igneous rocks	0.2
Basalts and gabbros	0.1—0.2
Granitic rocks	0.2
Sea water	0.0003 (=0.3 ppb)
Shales	1—2
Pelagic clays	1

It may be of interest of refer here to the geochemical balance of antimony. Because sedimentary material as a whole is richer in antimony than igneous rocks, significant amounts of antimony must have escaped from the interior of the earth in the water of hot springs and as volcanic exhalations. As shown in Table 51-I-92, many hot springs contain appreciable amounts of antimony. It appears that the geochemical balance of antimony is similar to that of arsenic.

Revised manuscript received: February 1969

References: Sections 51-B to 51-O

BOYLE, R. W.: Geology, geochemistry, and origin of the lead-zinc-silver deposits of the Keno Hill-Galena Hill area, Yukon territory. Geol. Surv. Canada Bull. 111 (1965).

BRANNOCK, W. W., P. F. FIX, V. P. GIANELLA, and D. E. WHITE: Preliminary geochemical results at Steamboat Springs, Nevada. Trans. Am. Geophys. Union **29**, 211 (1948).

BROWN, R., and W. A. WOLSTENHOLME: Analysis of geological samples by spark source mass spectrometry. Nature **201**, 598 (1964).

BYERS JR., F. M., and W. W BRANNOCK: Volcanic activity on Umnak and Great Sitkin Islands, 1946—1948. Trans. Am. Geophys. Union **30**, 719 (1949).

CISSARZ, A., u. H. MORITZ: Untersuchungen über die Metallverteilung in Mansfelder Hochofenprodukten und ihre geochemische Bedeutung. Metallwirtschaft **12**, 131 (1933).

CLARK, M. C., and D. J. SWAINE: Trace elements in coal. I. New South Wales coals. C.S.I.R.O. Div. Coal Res. Tech. Commun. **45** (1962).

ESSON, J., R. H. STEVENS, and E. A. VINCENT: Aspects of the geochemistry of arsenic and antimony, exemplified by the Skaergaard intrusion. Mineral. Mag. **35**, 88 (1965).

FLEISCHER, M.: Minor elements in some sulfide minerals. Econ. Geol. **50**, 970 (1955).

GEILMANN, W., u. W. BLITZ: Über die Zusammensetzung vulkanischen Schwefels vom Papandajan (West-Java). Z. Anorg. Allgem. Chem. **197**, 422 (1931).

GOLDSCHMIDT, V. M.: Geochemische Verteilungsgesetze der Elemente. IX. Die Mengenverhältnisse der Elemente und der Atom-Arten. Skrifter Norske Videnskaps-Akad. Oslo: I, Mat.-Naturv. Kl. No. 4 (1937; published 1938).

— Geochemistry. Oxford: University Press 1954.

—, u. C. PETERS: Zur Geochemie des Arsens. Nachr. Ges. Wiss. Göttingen, Jahresber. Geschäftsjahr, Math.-physik. Kl. IV, N.F. **1**, 11 (1934).

HAMAGUCHI, H., R. KURODA, K. TOMURA, M. OSAWA, K. WATANABE, N. ONUMA, T. YASUNAGA, K. HOSOHARA, and T. ENDO: Values for trace elements in G-1 and W-1 with neutron activation analysis. Geochim. Cosmochim. Acta **23**, 297 (1961 a).

—, T. NAKAI, and T. ENDO: Determination of arsenic and antimony in meteorites by neutron activation. Nippon Kagaku Zasshi **82**, 1485 (1961 b).

—, N. ONUMA, M. FURUKAWA, Y. HIRAO, and S. BANDO: Tin, antimony, and arsenic in carbonaceous chondrites. Discussion Meeting on Geochemistry, Tokyo, October, 1965.

HANSEN, M., and K. ANDERKO: Constitution of binary alloys. New York: McGraw-Hill Book Co. 1958.

HAWKES, H. E.: Geochemical prospecting investigations in the Nyeba lead-zinc district, Nigeria. U.S. Geol. Surv., Bull. **1000-B**, 51 (1954).

HEADLEE, A. J. W., and R. G. HUNTER: Elements in coal ash and their industrial significance. Ind. Eng. Chem. **45**, 548 (1953).

IKEDA, N.: Chemical studies on the hot springs of Nasu. VIII. Nippon Kagaku Zasshi **76**, 713 (1955 a).

— Chemical studies on the hot springs of Arima. VI. Nippon Kagaku Zasshi **76**, 842 (1955 b).

IWASAKI, I., and N. NASHIZAWA: Geochemical investigations of volcanoes in Japan. VI. Spectrographic detection of elements in volcanic sulfur in Japan. J. Chem. Soc. Japan **58**, 173 (1937).

KASAI, A., and Y. YANASE: ^{125}Sb contents in airborne dust and soil. Radioisotopes (Tokyo) **14**, 518 (1965).

KLEMENT JR., W., A. JAYARAMAN, and G. C. KENNEDY: Phase diagrams of arsenic, antimony, and bismuth at pressures up to 70 kbars. Phys. Rev. **151**, 632 (1963).

KRAUSKOPF, K. B.: Sedimentary deposits of rare metals. Econ. Geol. **50**, 411 (1955).

KRAVCHENKO, A. F., A. KH. TIMCHENKO, and A. A. GODOVIKOV: (Electrophysical properties of galena from different deposits.) Dokl. Akad. Nauk SSSR **167**, 172 (1966).

LANDSTRÖM, O., and C. G. WENNER: Neutron-activation analysis of natural water applied to hydrogeology. Aktiebolaget Atomenergi AE-**204** (1965).

LATIMER, W. M.: The oxidation states of the elements and their potentials in aqueous solutions, 2nd ed. New Jersey: Prentice-Hall 1952.

LEIPZIGER, F. D.: Some new upper limits of isotopic abundance by mass spectrometry. Appl. Spectry. **17**, 158 (1963).

NISHIMURA, M.: Chemical investigations of hot springs in Japan. XXVII and XXVIII. Spectrographic studies on minor metallic constituents in hot spring waters of Hokkaido. 1. and 2. Nippon Kagaku Zasshi **79**, 172 and 183 (1958).

NODDACK, I., u. W. NODDACK: Die Geochemie des Rheniums. Z. Physik. Chem. (Leipzig) A **154**, 207 (1931).

— — Die Häufigkeit der Schwermetalle in Meerestieren. Arkiv Zool. A **32**, No. 4 (1939).

ONISHI, H., and E. B. SANDELL: Notes on the geochemistry of antimony. Geochim. Cosmochim. Acta **8**, 213 (1955).

OTTE, M.-U.: Spurenelemente in einigen deutschen Steinkohlen. Chem. Erde **16**, 237 (1953).

PALACHE, C., H. BERMAN, and C. FRONDEL: The System of Mineralogy, 7th ed., vol. I. New York: John Wiley & Sons 1944.

— — — The System of Mineralogy, 7th ed., vol. II. New York: John Wiley & Sons 1951.

PORTMANN, J. E., and J. P. RILEY: The determination of antimony in natural waters with particular reference to sea water. Anal. Chim. Acta **35**, 35 (1966).

PREUSS, E.: Beiträge zur spektralanalytischen Methodik. II. Bestimmung von Zn, Cd, Hg, In, Tl, Ge, Sn, Pb, Sb und Bi durch fraktionierte Destillation. Z. Angew. Mineral. **3**, 8 (1940).

RITCHIE, J. A.: Arsenic and antimony in some New Zealand thermal waters. New Zealand J. Sci. **4**, 218 (1961).

SCHUTZ, D. F., and K. K. TUREKIAN: The investigation of the geographical and vertical distribution of several trace elements in sea water using neutron activation analysis. Geochim. Cosmochim. Acta **29**, 259 (1965).

SKINNER, B. J.: The system arsenic-antimony. Econ. Geol. **60**, 228 (1965).

SMALES, A. A., D. MAPPER, J. W. MORGAN, R. K. WEBSTER, and A. J. WOOD: Some geochemical determinations using radioactive and stable isotopes. Proc. U.N. Intern. Conf. Peaceful Uses At. Energy 2nd, Geneva, 1958, **2**, 242 (1958).

SWAINE, D. J.: Trace elements in coal. II. Origin, mode of occurrence, and economic importance. C.S.I.R.O. Div. Coal Res. Tech. Commun. **45** (1962).

TAKAHASHI, K.: Geochemical study on minor elements in sulfide minerals. Sulfide minerals from inner northeast Japan province. Chishitsu Chosasho Hokoku (Rept. Geol. Surv. Japan) No. 199 (1963).

TAYLOR, S. R.: Geochemical analysis by spark source mass spectrography. Geochim. Cosmochim. Acta **29**, 1243 (1965).

TIMASHEVA, E. E.: (Rare and trace elements in sulfide minerals of the Transcarpathian deposits.) Tr. Ukr. Nauchn.-Issled. Geol. Razved. Inst. **1963**, 386; Chem. Abstr. **61**, 11769 (1964).

TUNELL, G.: Chemical processes in the formation of mercury ores and ores of mercury and antimony. Geochim. Cosmochim. Acta **28**, 1019 (1964).

WARD, F. N., and H. W. LAKIN: Determination of traces of antimony in soils and rocks. Anal. Chem. **26**, 1168 (1954).

WESTER, P. O.: Trace elements in the conductive tissue of beef heart determined by neutron activation analysis. Aktiebolaget Atomenergi AE-**191** (1965).

WHITE, D. E., and W. W. BRANNOCK: The sources of heat and water supply of thermal springs, with particular reference to Steamboat Springs, Nevade. Trans. Am. Geophys. Union **31**, 566 (1950).

—, J. D. HEM, and G. A. WARING: Data of geochemistry, 6th ed. Chemical composition of subsurface waters. U.S. Geol. Surv., Profess. Papers **440-F** (1963).

WHITE, D. E., G. A. THOMPSON, and C. H. SANDBERG: Rocks, structure, and geologic history of Streamboat Springs thermal area, Washoe County, Nevada. U.S. Geol. Surv., Profess. Papers **458-B** (1964).

WHITE, J. R., and A. E. CAMERON: The natural abundance of isotopes of stable elements. Phys. Rev. **74,** 991 (1948).

Revised manuscript received: February 1969

51-D. Abundances in Rock-Forming Minerals (I) Antimony Minerals (II)

I. Rock-Forming Minerals

The distribution of antimony among co-existing minerals of gabbros from the Skaergaard intrusion were studied by ESSON *et al.* (1965): Table 51-D-1.

Table 51-D-1. *Antimony in rockforming minerals*

Material	Sb (ppm)	Reference
Plagioclase (3 samples) from gabbro, Skaergaard (Greenland)	0.24; 0.032; 0.036 (N/R)	ESSON *et al.* (1965)
Pyroxene (3 samples) from gabbro, Skaergaard (Greenland)	0.11; 0.026; 0.032	ESSON *et al.* (1965)
Olivine (3 samples) from gabbro, Skaergaard (Greenland)	0.21; 1.17; 1.38	ESSON *et al.* (1965)
Titaniferous magnetite (3 samples) from gabbro, Skaergaard (Greenland)	0.083; 0.026; 0.11	ESSON *et al.* (1965)
Ilmenite (2 samples) from gabbro, Skaergaard (Greenland)	0.27; 0.15	ESSON *et al.* (1965)

There is apparently a marked preferential entry of antimony into early magnesian olivines. Ilmenite appears to be the next most favorable host for antimony, presumably due to the substitution of antimony (III) for iron (II).

In rock-forming minerals antimony (ionic radii, Sb^{3+} 0.76 and Sb^{5+} 0.62 Å) can probably substitute for iron (Fe^{2+} 0.74 and Fe^{3+} 0.64 Å) (ESSON *et al.* 1965).

II. Antimony Minerals

Minerals of which antimony is a major constituent include the element, antimonides, sulfides, sulfosalts (compley sulfides), oxides, and a few antimonates and antimonites. A list of antimony minerals is given in Table 51-D-2. The most important ore mineral of antimony is stibnite. Other important minerals are complex sulfides such as stephanite, pyrargyrite, and tetrahedrite. Chemical processes in the formation of stibnite were discussed by TUNELL (1964). For additional information see Chapter 80.

Phase diagrams are given for the binary systems of Sb—As, Sb—Se, Sb—Si, Sb—Sn, Sb—Te, Sb—Ti, Sb—Tl, Sb—Zn, and Sb—Zr by HANSEN and ANDERKO (1958). The melting curve of antimony at pressures up to 70 kbars was determined by KLEMENT *et al.* (1963). The 1-atm. isobaric phase relations were determined for the system As—Sb (SKINNER, 1965).

Table 51-D-2. *Antimony minerals* (mainly according to STRUNZ, H.: Mineralogische Tabellen (5. ed.) Leipzig: Akademische Verlagsgesellschaft, 1970)

Name	Formula
Native metal, alloys, antimonides	
Antimony	Sb
Dyscrasite	Ag_3Sb
Stibarsen (allemontite)	AsSb
Aurostibite	$AuSb_2$
Horsfordite	Cu_5Sb
Arite	Ni(As, Sb)
Breithauptite	NiSb
Stibiopalladinite	Pd_5Sb_2
Geversite	$PtSb_2$
Sulfides, sulfosalts	
Stibnite (antimonite)	Sb_2S_3
Billingsleyite	$Ag_7(As, Sb)S_6$
Antimonpearceite	$(Ag, Cu)_{16}(Sb, As)_2S_{11}$
Polybasite	$(Ag, Cu)_{16}Sb_2S_{11}$
Samsonite	$Ag_4MnSb_2S_6$
Stephanite	Ag_5SbS_4
Aramayoite	$Ag(Sb, Bi)S_2$
Miargyrite	$AgSbS_2$
Pyrargyrite	Ag_3SbS_3 (trig.)
Pyrostilpnite	Ag_3SbS_3 (mon.)
Getchellite	$AsSbS_3$
Chalcostibite	$CuSbS_2$
Tetrahedrite	$Cu_{12}Sb_4S_{13}$
Stibioluzonite	Cu_3SbS_4 (tetrag.)
Stibioenargite	Cu_3SbS_4 (orthorhomb.)
Berthierite	$FeSb_2S_4$
Gudmundite	FeSbS
Livingstonite	$HgSb_4S_8$
Gerstleyite	$(Na, Li)_4As_2Sb_8S_{17} \cdot 6H_2O$
Hauchecornite	$(Ni, Co)_9(Bi, Sb)_2S_8$
Ullmannite	NiSbS
Nakaseite	$Pb_4Ag_3CuSb_{12}S_{24}$
Freieslebenite	$PbAgSbS_3$
Andorite	$Pb_4Ag_4Sb_{12}S_{24}$
Ramdohrite	$Pb_6Ag_4Sb_{10}S_{23}$
Diaphorite	$Pb_2Ag_3Sb_3S_8$
Owyheeite	$Ag_2Pb_5Sb_6S_{15}$
Teremkovite	$Ag_2Pb_7Sb_8S_{20}$
Fizelyite	$Pb_5Ag_2Sb_8S_{18}$
Geocronite	Pb_5SbAsS_8
Kobellite	$Pb_5Bi_8S_{17}$
Bournonite	$PbCuSbS_3$
Jamesonite	$Pb_4FeSb_6S_{14}$
Twinnite	$Pb(Sb, As)_2S_4$
Veenite	$Pb_2(Sb, As)_2S_5$
Guettardite	$Pb_9(Sb, As)_{16}S_{33}$
Sterryite	$Pb_{12}(Sb, As)_{10}S_{27}$
Sorbyite	$Pb_{17}(Sb, As)_{22}S_{50}$
Tintinaite	$Pb_5(Sb, Bi)_8S_{17}$
Fueloeppite	$Pb_3Sb_8 S_{15}$

Table 51-D-2 (continued)

Name	Formula
Meneghinite	$Pb_{13}CuSb_7S_{24}$
Boulangerite	$Pb_5Sb_4S_{11}$
Plagionite	$Pb_5Sb_8S_{17}$
Zinckenite	$Pb_6Sb_{14}S_{27}$
Robinsonite	$Pb_7Sb_{12}S_{25}$
Semseyite	$Pb_9Sb_8S_{21}$
Heteromorphite	$Pb_7Sb_8S_{19}$
Dadsonite	$Pb_{11}Sb_{12}S_{29}$
Playfairite	$Pb_{16}Sb_{18}S_{43}$
Madocite	$Pb_{17}(Sb, As)_{16}S_{41}$
Launayite	$Pb_{22}Sb_{26}S_{61}$
Franckeite	$Pb_5Sn_3Sb_2S_{14}$
Cylindrite	$Pb_3FeSn_4Sb_2S_{14}$
Vrbaite	$Tl_4Hg_3Sb_2As_8S_{20}$
Oxysulfide	
Kermesite	Sb_2S_2O
Oxygen-containing compounds	
Senarmontite	Sb_2O_3 (cub.)
Valentinite	Sb_2O_3 (orthorhomb.)
Cervantite	Sb_2O_4
Onoratoite	$Sb_8O_{11}Cl_2$
Stibioniobite (stibiocolumbite)	$Sb(Nb, Ta)O_4$
Stibioconite	$Sb_3O_6(OH)$
Stibiotantalite	$Sb(Ta, Nb)O_4$
Stetefeldtite	$Ag_{1-2}Sb_{2-1}(O, OH)_7$
Romeite	$(Ca, Fe, Na, H)_2Sb_2O_6(O, OH, F)$
Partzite	$Cu_{1-2}Sb_{2-1}(O, OH, F)_7$
Schafarzikite	$FeSb_2O_4$
Tripuhyite	$FeSb_2O_6$
Derbylite	$Fe_3Ti_3SbO_{11}OH$
Bystroemite	$MgSb_2O_6$
Haematostibite	$(Mn, Fe)_8Sb_2O_{13}$ (?)
Manganostibite	$(Mn, Fe)_7SbAsO_{12}$
Melanostibite	$Mn(Sb, Fe)O_3$
Swedenborgite	$NaSbBe_4O_7$
Nadorite	$PbSbO_2Cl$
Bindheimite	$Pb_{1-2}Sb_{2-1}(O, OH, H_2O)_{6-7}$
Ordonezite	$ZnSb_2O_6$
Silicates	
Chapmanite	$SbFe_2(OH)[SiO_4]_2$
Katoptrite	$(Mn, Mg, Fe^{2+})_{14}Sb_2(Al, Fe)_4[Si_2O_{29}]$
Parwelite	$(Mn, Mg)_5Sb[(Si, As)_2O_{10-11}]$
Yeatmanite	$(Mn, Zn)_{16}Sb_2[Si_4O_{29}]$

Revised manuscript received: February 1967 and October 1977 (Table 51-D-2)

51-O. Average Abundance, Economic Importance

I. Average Abundance

Table 51-O-1 summarizes the average abundances of antimony in different cosmic and crustal materials. Many of the figures in this table are approximate, and it is hoped that they will be improved or confirmed by future work.

Table 51-O-1. *Average abundances of Sb in different materials from cosmos and earth's crust*

Material	Sb (ppm)
Iron meteorites	0.3
Ordinary condrites	0.1
Igneous rocks	0.2
Basalts and gabbros	0.1—0.2
Granitic rocks	0.2
Sea water	0.0003 (= 0.3 ppb)
Shales	1—2
Pelagic clays	1

It may be of interest to refer here to the geochemical balance of antimony. Because sedimentary material as a whole is richer in antimony than igneous rocks, significant amounts of antimony must have escaped from the interior of the earth in the water of hot springs and as volcanic exhalations. As shown in Table 51-I-2, many hot springs contain appreciable amounts of antimony. It appears that the geochemical balance of antimony is similar to that of arsenic.

II. Economic Importance

A lead-antimony alloy, antimonial lead, is used for storage batteries. Lead-antimony-tin alloys are used as bearing metal, type metal, etc. Antimony(III) oxide is being increasingly used in flame retardants.

In 1974 world mine production of antimony was about 78,000 short tons (Minerals Yearbook, 1974). The chief producing countries are: Republic of South Africa, Bolivia, People's Republic of China and USSR.

The chief source of antimony is the ore mineral stibnite.

Revised manuscript received: February 1969 (I), September 1977 (II)

Tellurium 52

A	J. ZEMANN	(Mineralogisches Institut der Universität Wien, Austria)
B—K, O	F. LEUTWEIN	(Centre de Recherches Pétrographiques et Géochimiques, 54-Vandoeuvre-Nancy, France)

52-A. Crystal Chemistry

Tellurium in the ground state has the electron configuration $[Kr](4d)^{10}(5s)^2(5p)^4$; the electronegativity is about 2.1. The generally accepted effective ionic radii for the 2−, 4+ and 6+ oxidation states are: r(Te^{2-}) = 2.21 Å, r(Te^{4+}) = 0.70 Å, r(Te^{6+}) = 0.56 Å (for compilations of ionic radii, see this Handbook, Vol. I, p. 391, and SHANNON and PREWITT, 1969). With the exception of the tellurides of the metals with lowest electronegativities, Te compounds with predominantly ionic character seem to be rare, however; mostly the bonds seem to be complex with large covalent and metallic contributions.

I. Native Tellurium and Tellurides

Native tellurium, a semiconductor, contains screw chains $^{1}_{\infty}Te^{[2Te]}$ with Te—Te = 2.84 Å within the chains and closest Te—Te contacts of 3.50 Å between different chains (CHERIN and UNGER, 1967).

The *tellurides* of the alkali metals and of the alkali earth metals are largely ionic; from them an effective radius for Te^{2-} can be derived. The Te^{2-} ion can act as acceptor of hydrogen bridges ($Na_2Te \cdot 5H_2O$; BEDLIVY and PREISINGER, 1965). The telluride minerals, however, are either semiconductors or conductors. Altaite, PbTe (PbS-type), and coloradoite, HgTe (metacinnabarite-type), are examples with simple chemical formulae and simple structures. In compounds with a tellurium-to-metal ratio larger than 1, there is a tendency for strong Te—Te interactions, and also for the formation of polytellurium groups; a simple example is frohbergite, $Fe^{[6Te]}[Te_2]$, which crystallizes in the marcasite-type (TENGNÈR, SB 1938, 166; the Te—Te distance in the Te_2-group measures 2.92 Å which compares well with the Te—Te bond length in native tellurium).

The most important telluride minerals are the gold, silver, and gold-silver tellurides. For the phase diagram for Au—Ag—Te, see CABRI (1965a), and for the Ag—Te system, KRACEK *et al.* (1966). The atomic arrangements in calaverite, $AuTe_2$, krennerite, $(\underline{Au}, Ag)Te_2$, and sylvanite, $(Au_{1+x}Ag_{1-x})Te_4$, bear resemblances insofar as in all of them the Au, Ag atoms are surrounded by six Te in the form of a distorted octahedron (with two trans Au, Ag—Te bonds shorter than the rest). These (Au, Ag)Te_6 "octahedra" are joined via edges to form $^{2}_{\infty}(Au, Ag)^{[6]}Te_2$ sheets. In calaverite and sylvanite the result is a structure which is somewhat similar to the CdI_2-type (flat sheets of octahedra); in krennerite, the sheets are corrugated. In all three cases, however, there are strong Te—Te bonds between different "sheets" so that the similarity to layer structures is only formal. Table 52-A-1 lists the Au, Ag tellurides. The chemical bond in calaverite, krennerite and sylvanite has been discussed by TUNELL and PAULING (1952), and BARANOVA *et al.* (1971) have discussed the structures of the Au, Ag tellurides from the view point of arrangements of "molecules" built of atoms with shortest distances.

Tetradymite, Bi_2Te_2S, has a sheet structure $^{2}_{\infty}Bi_2^{[3S,\,3Te]}Te_2^{[3Bi]}S^{[6Bi]}$ with the sequence Te—Bi—S—Bi—Te within one sheet (HARKER, SB 1933—1935, 287; cf. ibid. p. 28). The Te—Te distances between different $^{2}_{\infty}Bi_2Te_2S$-sheets compare with the Te—Te distances between different chains in native tellurium. Bi_2Te_3, tellurobismutite, crystallizes in the same structure type (FRONDEL, 1940).

Table 52-A-1. *Crystal chemical data of gold-, silver- and gold-silver -tellurides*

Chemical formula Structural formula Mineral name	Remarks	Reference
$\sim Au_2Te_3$ (with minor amounts of Bi, Sb, and Pb) — Montbrayite	not synthesized in the system Au-Te, but with minor additions of Sb, or Pb+Bi +Sb; structure only very approximately known, partly unusual interatomic distances	BACHECHI (1971, 1972)
$AuTe_2$ (with up to 3 weight-% Ag) $Au^{[2+4Te]}Te_2$ Calaverite	natural and synthesized; indication of $^{1}_{\infty}Te^{[2\,Te]}$-chains with Te-Te = 3.19 Å (next Te-Te ≧3.47 Å), $Au\text{-}Te_{min}$ = 2.68 Å (2 ×) (next ≧2.97 Å)	TUNELL and PAULING (1952)
(Au, Ag) Te_2 (with 3.5 to 6.0% Ag) $(\underline{Au}, Ag)_3^{[2+4Te]}[Te_3][Te_2]Te$ Krennerite	natural and synthesized; Te-Te = 3.02—3.08 Å in [Te_3] and [Te_2]-groups, next ≧3.28 Å; Au-Te = 2.63—2.67 Å in $AuTe_2$-groups, next ≧2.87 Å	TUNELL and MURATA (SR 1950, 114)
$Au_{1+x}Ag_{1-x}Te_4$ (x = 0.0—∼0.5) $(Au, Ag)^{[2+4Te]}[Te_2]$ Sylvanite	natural and synthesized; Te-Te = 2.87 Å in Te_2-groups, next ≧3.55 Å; (Au, Ag)-Te = 2.67—2.69 Å in (Au, Ag) Te_2-groups, next ≧2.75 Å	TUNELL (SR 1940—1941, 72)
$AuAg_3Te_2$ $Au^{[2Te,6Ag]}\,Ag_3^{[4Te,2Au,2Ag]}$ $Te_2^{[1Au,6Ag]}$ Petzite[a]	natural and synthesized; Au-Te = 2.54 Å (2 ×), Ag-Te = 2.90—2.95 Å (4 ×); $Au\text{-}Ag_{min}$ = 3.09 Å, $Ag\text{-}Ag_{min}$ = 3.10 Å (next Ag-Ag ≧3.27 Å)	FRUEH (SR 1959, 154), MESSIEN and BAIWIR (1966)
Ag_2Te (low temp. modification)[b] $Ag^{[4Te,6Ag]}Ag^{[4Te,4Ag]}Te^{[8Ag]}$[c] Hessite	natural and synthesized, Ag-Ag = 2.83—3.13 Å (next≧3.73 Å); Ag-Te = 2.85—3.04 Å (next ≧3.95 Å)	FRUEH (SR 1959, 240)
$Ag_{5-x}Te_3$ (x = 0.06—0.12)[d] — Stuetzite[e]	natural and synthesized; structure not known with certainty[f]	IMAMOV and PINSKER (1966)
Ag_9Te_5, "γ-phase" — —	synthetic only; structure unknown	
AgTe — Empressite	not synthesized; structure unknown	

For footnotes see page 52-A-3.

Footnotes to Table 52-A-1:

[a] There exists a high temperature modification of $AuAg_3Te_2$ from the melting point (735 °C) down to 319°C with Te in b.c.c. arrangement, Au,Ag disordered, and a intermediate modification from 319 °C to 210°C with an unindexed powder pattern (cf. CABRI, 1965; FRUEH, 1959a; LLABRES and MESSIEN, 1968).

[b] There exists a high temperature modification from the melting point (960° C) down to ~690° C with Te in b.c.c. arrangement and Ag disordered, and an intermediate modification between ~ 690° C and 145° C with Te in f.c.c. arrangement and Ag disordered (cf. CABRI, 1965a; KRACEK *et al.*, 1966; FRUEH, 1961; LLABRES and MESSIEN, 1968).

[c] In the original paper of FRUEH (1959b; SR 1959, 240), it is erroneously stated that half of the Ag atoms are surrounded by 4 Te atoms and half by 5 Te atoms; the correct coordination of Te around Ag is already given by FRUEH in his article 47-A in this handbook.

[d] According to STUMPFL and RUCKLIDGE (1968), the homogenity range of natural stuetzite extends down to $Ag_{4.7}Ag_3$. According to KRACEK *et al.* (1966), $Ag_{5-x}Te_3$ shows a high temperature modification between ~ 280 °C and 420° C.

[e] For literature on stuetzite-empressite terminological problems, see THOMPSON *et al.* (1951), DONNAY *et al.* (1956), HONEA (1964), CABRI (1965b), HONEA (1965), STUMPFL and RUCKLIDGE (1968).

[f] IMAVOV and PINSKER (1966) have published a structure for a hexagonal silver telluride which, according to lattice constants and extinctions, seems to be identical with stuetzite. However, they have derived a formula Ag_7Te_4 (= $Ag_{5.25}Te_3$) from their electron diffractograms, and the interatomic distances and coordinations given are partly unusual. Therefore, a redetermination seems to be desirable.

Tellurides are sometimes isotypic with the corresponding sulfides and selenides (e.g. PbS—PbSe—PbTe; Sb_2S_3—Sb_2Se_3—Sb_2Te_3), and sometimes heterotypic (e.g. low Ag_2S — low Ag_2Se — low Ag_2Te; CuS—CuSe—CuTe). Limited substitution of Te for S in sulfides seems to be common. In sulfosalts tellurium can, however, play a role similar to antimony. A clear example is $Ba[TeS_3]\cdot 2H_2O$ which contains pyramidal $[TeS_3]$-groups (GERL and SCHÄFER, 1972). Microprobe analyses of Te-bearing tetrahedrites suggest that Te replaces Sb in this mineral (SPRINGER, 1969; LEVY, 1967) and that, therefore, such tetrahedrites contain TeS_3 pyramids. The structural role of Te in the complex Pb—Au—Sb tellurium sulfide nagyagite is unknown at present; the chemical composition of this mineral is given by STUMPFL (1970).

II. Tellurium Oxides, Tellurites and Tellurates

At ordinary pressure, TeO_2 exists in two modifications: tellurite, $^2_\infty\,Te^{[4O]}O_2^{[2Te]}$, and paratellurite, $^3_\infty\,Te^{[4O]}O_2^{[2Te]}$. In both crystal species, the oxygen coordination around Te(IV) is similar to a trigonal bipyramid with one equatorial corner unoccupied (Fig. 52-A-1a). In tellurite, these TeO_4-groups partly have O—O edges in common; in paratellurite, they are joined exclusively via oxygen "corners" to build a 3-dimensional framework. The higher oxides Te_2O_5 and TeO_3, as well as the modifications of $Te(OH)_3$, are known artificially only.

In the list of Te(IV) minerals (Table 52-A-2), the 4-coordination of oxygen around tellurium changes continuously via a (3+1)-coordination (Fig. 52-A-1b) into a 3-coordination (Fig. 52-A-1c); for a detailed discussion of this problem, see ZEMANN (1971) and PERTLIK and ZEMANN (1971). The single tellurite groups can polymerize, for example, to $[Te_2O_5]$ groups, but also to more complicated polyanions. A rationali-

Table 52-A-2. *Bond lengths and bond angles in TeO_2 minerals and tellurite minerals*

No.	Chemical formula (Mineral name)	Te-O bond lengths (Å)				O-Te-O bond angles (°)						Reference
		$Te—O_c$	$Te—O_a$	$Te—O_b$	$Te—O_d$	$O_a—Te—O_c$	$O_b—Te—O_c$	$O_a—Te—O_b$	$O_a—Te—O_d$	$O_b—Te—O_d$	$O_c—Te—O_d$	
1	tetr. TeO_2 (paratellurite)	2.08	1.90	1.90	2.08	85	88	102	88	85	169	Leciejewicz (SR 1961, 362), Lindqvist (1968)
2	$Zn_2[Te_3O_8]$[a] [Te(1)]	2.10	1.83	1.83	2.10	84	93	106	93	84	175	Hanke (1966)
3	$UO_2[Te_3O_7]$[b]	2.15	1.78	2.02	2.16	89	68	96	83	91	157	Galy and Meunier (1971)
4	orthorhomb. TeO_2 (tellurite)	2.07	1.88	1.93	2.20	90	90	101	78	90	168	Beyer (1967)
5	(Mn, Ca, Zn) $[Te_2O_5]$ (denningite)	2.04	1.84	1.87	2.36	90	85	98	88	76	160	Walitzi (1965)
6	$Fe(OH)[Te_2O_5]$ (mackayite)	1.98	1.90	1.90	2.37	92	92	96	71	87	162	Pertlik (1969)
7	$Fe_2[TeO_3]_2SO_4 \cdot 3\,H_2O$ (poughite) [Te(2)]	1.89	1.87	1.91	2.38	105	104	88	82	68	170	Pertlik (1971)
8	$Zn_2[Te_3O_8]$[a] [Te(2)]	1.98	1.88	1.93	2.41	94	80	98	83	77	156	Hanke (1966)
9	$\{Fe_2[TeO_3]_3 \cdot H_2O\} \cdot x\,H_2O$[c] (emmonsite) [Te(1)]	1.81	1.79	1.86	2.50	94	90	98	81	71	160	Pertlik (1972)
10	$\{Fe_2[TeO_3]_3 \cdot H_2O\} \cdot x\,H_2O$[c] (emmonsite) [Te(2)]	1.87	1.84	1.87	2.52	92	91	96	84	74	164	Pertlik (1972)
11	$H_3Fe_2[TeO_3]_4Cl$ (rodalquilarite) [Te(2)]	1.87	1.88	1.91	2.54	88	99	97	76	89	162	Dusausoy and Protas (1969)
12	$Fe(OH)[TeO_3] \cdot H_2O$ (sonoraite) [Te(2)]	1.89	1.86	1.91	2.57	94	92	96	81	70	161	Donnay *et al.* (1970)
13	$Fe_2[TeO_3]_2SO_4 \cdot 3\,H_2O$ (poughite) [Te(1)]	1.99	1.85	1.91	2.70	91	88	97	89	72	160	Pertlik (1971)
14	$Fe(OH)[TeO_3] \cdot H_2O$ (sonoraite) [Te(1)]	1.89	1.89	1.89	2.74	96	100	94	96	87	166	Donnay *et al.* (1970)
15	$\{Fe_2[TeO_3]_3 \cdot H_2O\} \cdot x\,H_2O$[c] (emmonsite) [Te(3)]	1.85	1.81	1.97	2.75	96	93	100	71	69	154	Pertlik (1972)
16	$\{(Zn, Fe)_2[TeO_3]_3\}\,Na_xH_{2-x} \cdot y\,H_2O$ (zemannite)	1.90	1.90	1.92	2.91	94	97	97	73	69	160	Matzat (1967)
17	$H_3Fe_2[TeO_3]_4Cl$ (rodalquilarite) [Te(1)]	1.90	1.86	1.91	2.93	97	90	102	68	65	146	Dusausoy and Protas (1969)
18	$Cu[TeO_3] \cdot 2\,H_2O$ (teineite)	1.88	1.82	1.88	2.98	98	96	102	59	106	150	Zemann and Zemann (SR 1961, 635)

[a] The atomic arrangement in $Zn_2[Te_3O_8]$ is very similar to that in spiroffite, $(Mn, Zn)_2[Te_3O_8]$.
[b] Identical with, or very similar to, cliffordite, UTe_3O_8 (?).
[c] $x = 0$—1.

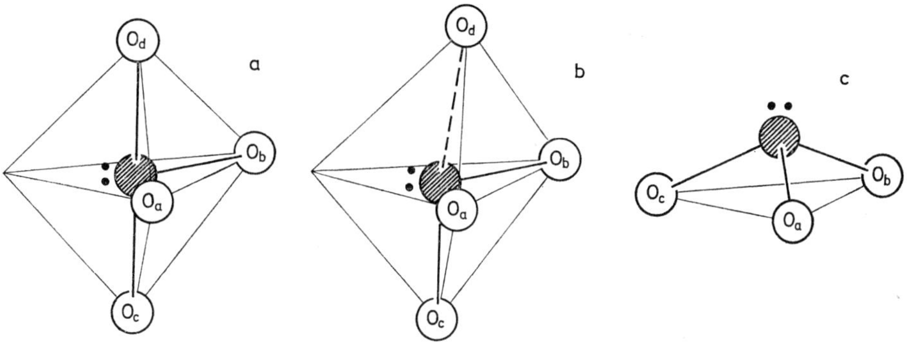

Fig. 52-A-1. Oxygen coordinations around Te(IV). The lettering of the oxygens corresponds to that in Table 52-A-2. Schematic positions of lone electron pairs are indicated by two dots

zation of the stereochemistry of Te(IV) towards oxygen can be obtained by placing a lone electron pair with a large space-requirement at the "free" corner of the coordinations (for a survey article of this class of minerals, see ZEMANN, 1968).

Some few rare tellurium minerals have been described to be tellurates (magnolite, Hg_2TeO_4; montanite, $Bi_2TeO_6 \cdot 2H_2O$), but recent confirmations of the oxidation state of tellurium are lacking. Artificial tellurates usually show an octahedral oxygen coordination (e.g. $Mg_3[TeO_6]$: SCHULZ and BAYER, 1971; $K[TeO(OH)_5] \cdot H_2O$: RAMAN, 1964; mon. $Te(OH)_6$: LINDQVIST, 1970; cub. $Te(OH)_6$: COHEN-ADDAD, 1971); the Te—O distances measure $\sim 1.90 \pm 0.05$ Å. Polymerization of two tellurate octahedra at one O—O edge leads to binuclear complexes, as in Na_2K_4-$[Te_2^{[6]}O_8(OH)_2] \cdot 14H_2O$ (LINDQVIST, 1969), and further polymerization at O—O edges leads to infinite chains, as in $^1_\infty K[Te^{[6]}O_3(OH)]$ (LAMMERS, 1964). LAMMERS and ZEMANN (1965) have reported the polymerization of tellurate octahedra via corners to infinite chains in $^1_\infty K[Te^{[6]}O_2(OH)_3]$. Only occasionally do $[TeO_4]^{2-}$ tetrahedra seem to occur, as in $NaDy[TeO_4]_2$, which crystallizes in the scheelite type (SCHIEBER, 1965), and, according to PATRY (SB 1936, 173) and TARTE and LEYDER (1971), in $KTeO_4$.

For a compilation of information on this class of compounds see also BAYER (1969).

III. Halogenides

Tellurium halogenides do not occur naturally. TeF_4 has a chain structure, $^1_\infty [Te^{[5]}F_4]$, the coordination around Te(IV) being approximately an octahedron with one corner unoccupied (EDWARDS and HEWAIDY, 1968). $K[TeF_5]$ contains TeF_5^{1-} groups of the same shape (EDWARDS and MOUTY, 1969), and $TeCl_4$ contains $[(TeCl_3)_4Cl_4]$ units in which each tellurium has three close Te neighbors (Te—Cl $\sim$ 2.31 Å) and three at a considerably greater distance (Te—Cl $\sim$ 2.93 Å) (BUSS and KREBS, 1970). In the halogenides with higher atomic number, there exist $[Te^{[6]}Hal_6]^{2-}$ complexes, as in $K_2[Te^{[6]}Br_6]$ (BROWN, 1964; cf. for spectra also GREENWOOD and STRAUGHAN, 1966; ADAMS and MORRIS, 1967).

Manuscript received: September 1972

References: Section 52-A

ADAMS, D. M., MORRIS, D. M.: Vibrational spectra of halides and complex halides. Part III. Hexahalogenotellurates. J. Chem. Soc. (A) **1967**, 2067 (1967).

BACHECHI, F.: Crystal structure of montbrayite. Nature Phys. Sci. **231**, 67 (1971).

BACHECHI, F.: Synthesis and stability of montbrayite, Au_2Te_3. Am. Mineralogist **57**, 146 (1972).

BARANOVA, R. V., IMAMOV, R. M., PINSKER, Z. G.: Crystal chemistry of A^{I}-B^{VI} compounds. Certain structural characteristics and geometric analysis of ordering mechanism for silver and gold tellurides. Soviet Phys.-Cryst. (English Transl.) **16**, 99 (1971).

BAYER, G.: Zur Kristallchemie des Tellurs. Telluroxide und Oxidverbindungen mit Tellur. Fortschr. Mineral. **46**, 42 (1969).

BEDLIVY, D., PREISINGER, A.: Die Kristallstruktur von $Na_2S \cdot 5H_2O$, $Na_2Se \cdot 5H_2O$ und $Na_2Te \cdot 5H_2O$. Z. Krist. **121**, 131 (1965).

BEYER, H.: Verfeinerung der Kristallstruktur von Tellurit, dem rhombischen TeO_2. Z. Krist. **124**, 228 (1967).

BROWN, I. D.: The crystal structure of K_2TeBr_6. Can. J. Chem. **42**, 2758 (1964).

BUSS, B., KREBS, B.: Kristallstruktur von $TeCl_4$: Tetramere in festen Chalkogen(IV)-halogeniden. Angew. Chem. **82**, 446 (1970).

CABRI, L. J.: Phase relations in the Au—Ag—Te system and their mineralogical significance. Econ. Geol. **60**, 1569 (1965a).

CABRI, L. J. P.: Discussion of "Empressite and stuetzite redefined" by R. M. HONEA. Am. Mineralogist **50**, 795 (1965b).

CHERIN, P., UNGER, PH.: Two-dimensional refinement of the crystal structure of tellurium. Acta Cryst. **23**, 670 (1967).

COHEN-ADDAD, C.: Étude du composé $Te(OH)_6$ par diffraction des neutrons. Bull. Soc. Franç. Mineral. Crist. **94**, 172 (1971).

DONNAY, G., KRACEK, F. C., ROWLAND, W. R., JR.: The chemical formula of empressite. Am. Mineralogist **41**, 722 (1956).

DONNAY, G., STEWART, J. M., PRESTON, H.: The crystal structure of sonoraite, $Fe^{3+}Te^{4+}$-$O_3(OH) \cdot H_2O$. Tschermaks Mineral. Petrog. Mitt. [3] **14**, 27 (1970).

DUSAUSOY, Y., PROTAS, J.: Determination et etude de la structure cristalline de la rodalquilarite, chlorotellurite acide de fer. Acta Cryst. B **25**, 1551 (1969).

EDWARDS, A. J., HEWAIDY, F. I.: Fluoride crystal structures. Part IV. Tellurium tetrafluoride. J. Chem. Soc. (A) **1968**, 2977 (1968).

EDWARDS, A. J., MOUTY, M. A.: Fluoride crystal structures. Part V. Potassium pentafluorotellurate (IV). J. Chem. Soc. (A) **1969**, 703 (1969).

FRONDEL, C.: Redefinition of tellurobismuthite and vandiestite. Am. J. Sci. **238**, 880 (1940).

FRUEH, A. J., JR.: The crystallography of petzite, Ag_3AuTe_2. Am. Mineralogist **44**, 693 (1959a).

FRUEH, A. J. ,JR.: The structure of hessite, Ag_2Te—III. Z. Krist. **112**, 44 (1959b).

FRUEH, A. J., JR.: Use of zone theory in problems of sulfide mineralogy. Part III. Polymorphism of Ag_2Te and Ag_2S. Am. Mineralogist **46**, 654 (1961).

GALY, J., MEUNIER, G.: A propos de la cliffordite UTe_3O_8. Le système UO_3—TeO_2 à 700° C. Structure cristalline de UTe_3O_9. Acta Cryst. B **27**, 608 (1971).

GERL, H., SCHÄFER, H.: Die Kristallstruktur von $BaTeS_3 \cdot 2H_2O$. Z. Naturforsch. **27** b, 1421 (1972).

GREENWOOD, N. N., STRAUGHAN, B. P.: Metal-halogen vibrations of simple octahedral ions containing tin, selenium, and tellurium. J. Chem. Soc. (A) **1966**, 962 (1966).

HANKE, K.: Die Kristallstruktur von $Zn_2Te_3O_8$. Naturwissenschaften **53**, 273 (1966).

HONEA, R. M.: Empressite and stuetzite redefined. Am. Mineralogist **49**, 325 (1964).

HONEA, R. M.: Reply to discussion of "Empressite and stuetzite redefined" by R. M. HONEA. Am. Mineralogist **50**, 802 (1965).

IMAMOV, R. M., PINSKER, Z. G.: Determination of the crystal structure of the hexagonal phase in the silver-tellurium system. Soviet Phys.-Cryst. (English Transl.) **11**, 182 (1966).

KRACEK, F. C., KSANDA, C. J., CABRI, L. J.: Phase relations in silver-tellurium system. Am. Mineralogist **51**, 14 (1966).

LAMMERS, P.: Darstellung und Kristallstruktur von $KTeO_3(OH)$. Naturwissenschaften **23**, 552 (1954).

LAMMERS, P., ZEMANN, J.: Über ein neues Kaliumtellurat und seinen Strukturtyp. Z. Anorg. Allgem. Chem. **334**, 225 (1965).

LÉVY, C.: Contribution à la minéralogie des sulfures de cuivre du type Cu_3XS_4. Mém. Bur. Rech. Géol. Minières, **54**, 178 pp. (1967).

LINDQVIST, O.: Refinement of the structure of α-TeO_2. Acta Chem. Scand. **22**, 977 (1968).

LINDQVIST, O.: The crystal structure of the tellurate $Na_2K_4[Te_2O_8(OH)_2]\ (H_2O)_{14}$. Acta Chem. Scand. **23**, 3062 (1969).

LINDQVIST, O.: The crystal structure of telluric acid, $Te(OH)_6$(mon). Acta Chem. Scand. **24**, 3178 (1970).

LLABRES, G., MESSIEN, P.: Sur les structures cristallines des variétés α des sulfures d'argent et des sulfurides mixtes d'argent et d'or. Bull. Soc. Roy. Sci. Liège **37**, 329 (1968).

MATZAT, E.: Die Kristallstruktur eines unbenannten zeolithartigen Telluritminerals, $\{(Zn, Fe)_2[TeO_3]_3\}Na_xH_{2-x} \cdot y\, H_2O$. Tschermaks Mineral. Petrog. Mitt. [3] **12**, 108 (1967).

MESSIEN, P., BAIWIR, M.: Structuie cristalline des tellurure et séléniure mixtes d'argent et d'or β-Ag_3AuTe_2—β-Ag_3AuSe_2. Bull. Soc. Roy. Sci. Liège **35**, 234 (1966).

PERTLIK, F.: Hydrothermalsynthese, Formel und Struktur von Mackayit, $Fe(OH)[Te_2O_5]$. Tschermaks Mineral. Petrog. Mitt. [3] **13**, 219 (1969).

PERTLIK, F.: Die Kristallstruktur von Poughit, $Fe_2[TeO_3]_2[SO_4] \cdot 3H_2O$. Tschermaks Mineral. Petrog. Mitt. [3] **15**, 279 (1971).

PERTLIK, F.: Der Strukturtyp von Emmonsit, $\{Fe_2[TeO_3]_3 \cdot H_2O\} \cdot xH_2O$ (x = 0—1). Tschermaks Mineral. Petrog. Mitt. [3] **18**, 157 (1972).

PERTLIK, F., ZEMANN, J.: Übergang zwischen den Koordinationszahlen 3 und 4 von Sauerstoff um 4-wertiges Tellur. Österr. Akad. Wiss., Math.-Naturw. Kl., Anz. **1971**, 175 (1971).

RAMAN, S.: Crystal structure of $KTeO(OH)_5 \cdot H_2O$. Inorg. Chem. **3**, 634 (1964).

SCHIEBER, M. M.: Growth of rare earth scheelites by the flux method. Inorg. Chem. **4**, 762 (1965).

SCHULZ, H., BAYER, G.: Structure determination of Mg_3TeO_6. Acta Cryst. B **27**, 815 (1971).

SHANNON, R. D., PREWITT, C. T.: Effective ionic radii in oxides and fluorides. Acta Cryst. B **25**, 925 (1969).

SPRINGER, G.: Electronprobe analyses of tetrahedrite. Neues Jahrb. Mineral. Monatsh. **1969**, 24 (1969).

STUMPFL, E. F.: New electron probe and optical data on gold tellurides. Am. Mineralogist **55**, 808 (1970).

STUMPFL, E. F., RUCKLIDGE, J.: New data on natural phases in the system Ag—Te. Am. Mineralogist **53**, 1513 (1968).

TARTE, P., LEYDER, F.: Sur la coordinence tétraédrique de tellure dans les metatellurates alcalins K_2TeO_4, Rb_2TeO_4 et Cs_2TeO_4. Compt. Rend., Série C, **273**, 852 (1971).

THOMPSON, R. M., PEACOCK, M. A., ROWLAND, J. F., BERRY, L. G.: Empressite and "stuetzite". Am. Mineralogist **36**, 458 (1951).

TUNELL, G., PAULING, L.: The atomic arrangement and bonds of the gold-silver ditellurides. Acta Cryst. **5**, 375 (1952).

WALITZI, E. M.: Die Kristallstruktur von Denningit, $(Mn, Ca, Zn)Te_2O_5$. Ein Beispiel für die Koordination um vierwertiges Tellur. Tschermaks Mineral. Petrog. Mitt. [3] **10**, 241 (1965).

ZEMANN, J.: The crystal chemistry of the tellurium oxide and tellurium oxosalt minerals. Z. Krist. **127**, 319 (1968).

ZEMANN, J.: Zur Stereochemie des Te(IV) gegenüber Sauerstoff. Monatsh. Chem. **102**, 1209 (1971).

Manuscript received: September 1972

52-B. Isotopes in Nature

Tellurium (atomic number 52) has 8 stable isotopes whose relative abundances (BAINBRIDGE and NIER, 1950) are given in the following table:

Table 52-B-1. *Isotopes of Te and their relative abundances*

Te isotope	%
120	0.089
122	2.46
123	0.87
124	4.61
125	6.99
126	18.71
128	31.79
130	34.49

If the antimony isotope ^{123}Sb (relative abundance: 42.75%) were radioactive, forming ^{123}Te under β-decay, the half-life would be extremely long, more than 10^{12} to 10^{14} years (RANKAMA, 1954). INGHRAM and REYNOLDS (1950) found that ^{130}Te is unstable, forming ^{130}Xe at the end of two β-decay processes — but the half-life-time is about 10^{21} a.

Revised manuscript received: January 1971

52-C. Abundance in Cosmos, Meteorites and Lunar Samples

I. Cosmic Abundance

Direct spectral observations of tellurium in solar or stellar atmospheres are lacking. CAMERON (1967) calculated the cosmic abundance of this element by using carbonaceous chondrites of type 1 (for comparison, the values for S and Se are also given). Relative to $Si = 10^6$, he found:

$$S = 5.06 \cdot 10^5$$
$$Se = 70.1$$
$$Te = 6.76$$

In these meteorites, tellurium is about ten times rarer than selenium. According to SUESS and UREY (1956), the ratio of Te/S in the cosmos is 13×10^6 that of Se/Te 14.

II. Abundance in Meteorites

In the older investigations on tellurium in meteorites by NODDACK and NODDACK (1933), a value (17 ppm Te) has only been reported for troilite. GOLDSCHMIDT has calculated the average concentration of tellurium in meteorites as 0.73 ppm. Recent investigations on the abundance of tellurium in meteorites with reliable and sensitive methods have been published by DU FRESNE (1960), SCHINDEWOLF (1960), GOLES and ANDERS (1960, 1962), REED (1963) and GREENLAND (1967).

The average tellurium concentration in 20 ordinary and enstatite chondrites has been reported by DU FRESNE (1960) (using a spectrophotometric method) to be 1.34 ppm Te (7.9 ppm Se). SCHINDEWOLF's mean value for 4 ordinary chondrites (using neutron activation) is lower, 0.61 ppm Te; REED (1963) obtained a value of 0.64 ppm for 11 chondrites and KIESL *et al.* (1970) a value of 0.81 ppm Te for 16 chondrites by neutron activation.

In 1962, GOLES and ANDERS (1962) published 23 analyses of meteorites by neutron activation. The range for Te in 7 bronzite and hypersthene chondrites is upto 0.51 ppm; in three enstatite chondrites, they found 2.1 ppm and in two carbonaceous chondrites 1.9 ppm. For iron meteorites, the values cover a wide range of 17—90 ppm Te, whereas in five troilites 1.2—5.0 ppm Te were found.

In 1967, GREENLAND published a very complete study on several trace elements in chondritic meteorites and these results are given in the following two tables; the analyses were made by neutron activation.

To obtain a better comparison with stellar spectrographic data, Table 52-C-2 gives the results calculated in atom-percentage for $Si = 10^6$.

Table 52-C-1. *Analytical results (data in ppm) on some chondritic meteorites by N/R.* (After GREENLAND, 1967)

Meteorite	Se	Te
Carbonaceous type I		
Ivuna	26	1.0
Orgueil	27	0.94
Carbonaceous type II		
Mighei	14	1.2
Murray	10	3.3
Carbonaceous type III		
Chainpur	9.0	1.4
Felix	11	0.44
Lancé	9.3	0.15
Mokoia	9.6	0.70
Vigarano	6.5	0.82
Warrenton	11	0.90
Enstatite type A		
Abee	11	3.9
Indarch	34	2.6
Enstatite type B		
Daniei's Kuil	19	0.21
Hvittis	10	0.79
Khairpur	24	0.74
Pillistfer	13	0.18
Hypersthene-bronzite		
Allegan (H)	6.1	—
Bruderheim (L)	6.4	0.76
Ehole (H)	6.4	0.93
Mocs (L)	7.3	0.89
Pantar (H)	4.8	0.61
Peace River (L)	6.4	0.82
Walters (L)	8.2	0.60

(H) indicates a high-iron olivine bronzite chondrite and (L) indicates a low-iron olivine hypersthene chondrite.

Table 52-C-2. *Atomic abundance relative to* $Si = 10^6$ (GREENLAND, 1967)

Material	Se	Te
Carbonaceous type I chondrites	89	2.0
Carbonaceous type II chondrites	32	3.8
Carbonaceous type III chondrites	21	1.1
Enstatite type A chondrites	48	4.2
Enstatite type B chondrites	32	0.56
Ordinary chondrites	13	0.96
Cosmic (SUESS and UREY, 1956)	68	4.7

III. Lunar Samples

Only a limited amount of data has been published on tellurium in lunar rocks. Ganapathy *et al.* (1970) report 8 to 13 ppm Te in four Apollo 11 samples, and 72 and 73 ppm Te in two lunar breccias from the same sampling locality (by neutron activation analysis). Lunar fines and breccias are mainly enriched in relatively volatile elements. Vinogradov (1971) has published three values of 150 to 200 ppm Te on lunar regolith from the Russian automatic station Luna 16 (mass spectrometric analysis).

Revised manuscript received: January 1971

52-D. Abundance in Rock-Forming Minerals; Tellurium Minerals

In natural environments, tellurium has few lithophile tendencies; it is mainly restricted to sulfides and to low-temperature supergene minerals. In rock-forming silicates it has not yet been detected probably because of the limit of sensitivity of the methods applied.

The ionic radii of tellurium in its various ionization states are very different from those of sulfur and even of selenium. It forms anions in minerals (tellurides, tellurites and tellurates), and its redox potentials are widely different from those of the analogous sulfur components. Therefore, it is not surprising that there exists a rather large number (about 40) of independent minerals of this element. The tendency for isomorphic replacement of the far more abundant sulfur in its compounds is smaller for Te than for Se.

Some tellurium minerals, e.g. sylvanite, were exploited a long time before the detection of tellurium (1798) in important gold ores. Tellurium minerals are rare and are often only of microscopic size. Because of the grain size and the presence of intergrowths, analyses made without a microprobe are often of minor value as a basis for mineral formulae. STUMPFL (1970) found by microprobe, that sylvanite has often more gold than its formula indicates. He also states, that a correct distinction of krennerite and calaverite is only possible by microprobe.

The supergene tellurium minerals tellurites and tellurates have a high proportion of ionic bonding. In the group of hypogene Te minerals, metallic luster, proportions of metallic bonds (often accompanied by infra-red transparency and photoelectric effects) are frequent. Some minerals are typical intermetallic compounds.

In the following table tellurium minerals are listed mainly after SINDEEVA (1964).

The ionic radius of tellurium and the physico-chemical properties of H_2Te, H_2TeO_3, H_2TeO_4 are so different from those of the far more abundant sulfur compounds that there is only a very limited camouflage or diadochy to be expected. There are far more tellurium than selenium minerals. There is, surprising, a great number of gold-tellurium minerals (8), followed by silver, bismuth, copper and mercury ores. Only one lead telluride is known, but no zinc, cadmium, manganese or tin-tellurium compounds are known as natural minerals at this time.

Phase Equilibria

Selenium and tellurium form homogeneous melts in all proportions and below the solidus, form homogeneous mixed crystals in the hexagonal rhombohedral form of selenium. Te and S melts are completely miscible; in the solid state miscibility is rather limited — in β-sulfur maximum contents of 2% Te and in α-sulfur only 0.5% Te have been reported.

Table 52-D-1. *Tellurium minerals*

1. Native tellurium

Native tellurium. Often contains traces of iron and gold, and no traces of selenium.

Seleniferous tellurium. Under laboratory conditions, Se and Te are able to form homogeneous mixed crystals, though the atomic radii are rather different. The natural mineral usually has a higher Te content ($^2/_3$ Te, $^1/_3$ Se).

2. Tellurides

Hessite	Ag_2Te
Empressite	$AgTe_3$
Petzite	Ag_3AuTe_2
Stützite	$Ag_{(5-\alpha)}Te_3$
Muthmannite	$(Au, Ag)Te$
Calaverite	$AuTe_2$
Krennerite	$(Au, Ag)Te_2$
Sylvanite	$(Au, Ag)Te_4$
Montbrayite	Au_2Te_3
Altaite	$PbTe$
Coloradoite	$HgTe$
Frohbergite	$FeTe_2$
Melonite	$NiTe_2$
Tellurobismuthite	Bi_2Te_3
Wehrlite	$Bi_{(2+\alpha)}Te_{(3-\alpha)}$
Hedleyite	Bi_7Te_3
Tetradymite	Bi_2Te_2S
Csiklovaite	Bi_2TeS_2
Joseïte A	$Bi_{(4+\alpha)}Te_{(1-\alpha)}S_2$
Joseïte B	$Bi_{(4+\alpha)}Te_{(2-\alpha)}S$
Gruenlingite	Bi_4TeS_3
Oruetite	Bi_8TeS_4 perhaps
Rickardite	$Cu_{(4-\alpha)}Te_2$
Weissite	Cu_2Te
Nagyagite	$Pb_5Au_{0.7}Sb_{1.1}(Te_{2.1}, S_{5.4})_{7.5}$

3. Sulfosalts

Goldfieldite	$Cu_6Sb_2(S, Te)_3$
Arsenotellurite	$Te_2As_2S_7$

The existence of neither minerals is yet confirmed on the base of very detailed studies.

4. Oxides

Tellurite	TeO_2 (orthorhomb.)
Paratellurite	TeO_2 (tetrag.)

5. Tellurites and tellurates

Montanite	$Bi_2TeO_4(OH)_4$	(?)	
Teineite	$CuTeO_3 \cdot 2H_2O$		
Emmonsite	$Fe_2(TeO_3)_3 \cdot 2\,H_2O$		
Mackayite	$Fe(OH)(Te_2O_5)$		
Blakeite	$Fe_2(TeO_3)_3$	(?)	
Schmitterite	$UO_2\ TeO_3$		
Magnolite	Hg_2TeO_4	(?)	chemical and crystallographic data lacking
Ferrotellurite	$FeTeO_4$	(?)	
Lead tellurate	$PbTeO_4$	(?)	(DOMEYKO, 1875), possibly identical with a mineral later described as dunhamite (FAIRBANKS, 1946)
Denningite	$(Mn, Ca, Zn)Te_2O_5$		
Spiroffite	$(Mn, Zn, Ca)_2Te_3O_8$		
Moctezumite	$PbUTe_2O_8$		
Poughite	$Fe_2[TeO_3]_2[SO_4] \cdot 3H_2O$		

Revised manuscript received: January 1971

52-E. Abundance in Common Igneous Rock Types

The average content of tellurium in the earth's crust — the clarke — is mainly based on extrapolations or on analyses of composite samples of "igneous rocks"; rather contraversial results have been obtained. NODDACK and NODDACK (1930) report 0.8 ppm Se and 0.01 ppm Te as average values. More recently, VINOGRADOV (1954) made the following estimate of the order of magnitude of the crustal abundance: 0.01 ppm selenium and 0.001 ppm tellurium. SINDEEVA (1964) analyzed a large number of different rock-types for Se and Te — but the Te-contents were always under the limit of detection (0.01 ppm)[1]. The tellurium content of the whole continental earth's crust (~16 km) was estimated by SCHTSCHERBINA (1937) to be 2×10^{11} t.

TUREKIAN and WEDEPOHL (1961) give 0.002 ppm as average content of all magmatic rocks of the upper continental crust.

[1] GANAPATHY *et al.* (1970) analyzed a pyroxene gabbro from the Adirondacks and obtained a value of 10 ppbTe(N/R).

Revised manuscript received: January 1971

52-F. Behavior in Magmatogenic Processes

The presence of traces of tellurium and of gaseous H_2Te in volcanic exhalations is often mentioned; in addition, Te might be transported as gaseous $TeO(OH)_2$ (GLEMSER *et al.* 1964). Volcanic sulfur as well as exhalation-crusts from the Liparian islands can contain traces of this element. In general, traces of 8—15 ppm, and rarely even 200—250 ppm in seleniferous volcanic sulfur are mentioned.

Tellurium has not been detected in pegmatites, but in hydrothermal sulfides, even of higher origin, and in skarn deposits, tellurium minerals are well known (SINDEEVA, 1964).

In sulfides from high temperature quartz, wolframite cassiterite veins, contents of selenium and traces of tellurium have been reported.

Tellurium has not been detected in tin and molybdenum greisens, but hydrothermal copper-molybdenum deposits (characterized by several ore-forming phases) often contain tellurium, especially in the younger molybdenite, either in isomorphous form or as tellurium minerals. The average content of 107 molybdenites from the Kadzharan deposit, USSR, is reported to be 33 ppm Te (KARAMYAN, 1962).

The deposits of magmatic pyrrhotite-pentlandite with chalcopyrite may have considerable amounts of Se and Te — though the concentrations are usually low. The recovery of these elements from this type of ore is relatively easy, because Se and Te accumulate in dusts or anodic slugs of the copper electrolysis plants.

Another important group of telluriferous deposits are the sedimentary pyrite masses of high temperature sources. Deposits are known where the amount of tellurium equals that of selenium or exceeds it. In general, pyrite, pyrrhotine and chalcopyrite are the richest, and sphalerite the poorest hosts of Te and Se. Selenium minerals are in general absent. Tellurium mostly appears in the form of fine-grained separate minerals — tetradymite, altaite, hessite and others. As a general tendency, younger deposits, of Mesozoic or Tertiary age, are richer in these elements than older ones (SINDEEVA, 1964).

From a mineralogical point of view, the most interesting tellurium deposits are the sulfide-bearing gold veins. (From the economical point of view, the two deposits cited before are by far the more important). They may have very different ages and they are not always of subvolcanic origin. Deposits of this type are not very abundant, but those from Roumania (here subvolcanic), Calgoorlie (Australia), Cripple Creek (Nevada), Beresovskoye in the Central Ural, Dzhelambet (Kazakstan), Kirovskoye in the Amur-region (for detailed description, see SINDEEVA, 1964) and from the early Precambrian gold belt of Ontario-Quebec (Canada) are known.

Kirovskoye is a vein deposit with a complex mineral composition. The gangue is quartz with rare carbonates. The common sulfides are pyrite, arseno-pyrite, bismuthinite, chalcopyrite, sphalerite, galena, tetraedrite, pyrrotite and gold. Rarer minerals are boulangerite, jamesonite, scheelite and molybdenite and tellurides of gold, silver and bismuth. This evidently represents mineralization in several phases of catathermal to mesothermal character. The host rocks are granitic gneisses and sediments of Jurassic age, which are cut by a post-middle Jurassic granodiorite and its differentiates. A table of the Se and Te contents is given by SINDEEVA (1964).

Table 52-F-1. *Contents of selenium and tellurium (by W, C) in minerals of the Kirovskoye deposit.* (After SINDEEVA, 1964)

Minerals	Percent Te	Percent Se
Bismuthinite	0.3275—0.2600	0.3275—0.2800
Arsenopyrite	0.0240—0.0001	0.019—0.0006
Chalcopyrite	0.0016—0.0008	0.003—0.0008
Pyrite	upto 0.0066	upto 0.0030
Jamesonite	upto 0.0018	upto 0.0005

Table 52-F-2. *Contents of tellurium (ppm) in sulfide-minerals of some pyrite deposits of the Ural.* After VLASSOV (1969). (Analytical method not stated)

Ore deposit	Pyrite	Chalcopyrite	Sphalerite	Tetraedrite
III — Internationala	6—220	20—387	<1—9	n.d.
Krasnogwardeskoë	6— 39	nil—236	>1	20—860
Lewicha	tr —50	9—6,600	nil—1	n.d.
Sibaï	1— 32	88	nil—7	n.d.
Karabasch	2—1,350	400	n.d.	1—900

Table 52-F-3. *Content of tellurium (ppm) in sulfide minerals of some copper-molybdenum deposits of the USSR.* After VLASSOV (1964), (W, C)

Ore deposit	Pyrite	Chalcopyrite	Molybdenite	Sphalerite	Galena
Almalyk	5—510	5—20	10—12	5—30	50—520
Kadscharan	5—12	22	30	—	—
Dasstakert	22	9—31	n.d.	n.d.	40

It is interesting to note that the relative enrichment of Te widely exceeds that of selenium; no selenium-minerals have been found in this deposit.

The very important group of hydrothermal lead-zinc deposits (of vein type) is of no significance as regards the Se and Te contents. Traces of both elements are often found here — more often Te than Se — but the absolute quantities are low. At Freiberg/Saxony, the author found traces of 2 to 20 ppm of Te especially in tetraedrite and the bismuthiferous galena; selenium was present in even lower concentrations (unpublished spectrographic data).

Tellurium minerals are formed from high to low thermal conditions. Tetradymite is mostly a high thermal mineral, appearing even in skarn deposits. Sylvanite and nagyagite are known only from mesothermal veins.

Schtscherbina (1937) inferred from microscopic observations as well as from thermodynamic calculations that there exists a series of different thermodynamic affinities of certain metals for tellurium in preference to sulfur:

Cu—Pb—Ni—Bi—Hg—Ag—Au

(sequence of increasing affinity). Gold tellurides may exist in the presence of sulfides of all other metals of this sequence. Melonite (NiTe) may only co-exist with sulfides of Cu, Pb and all other tellurides, and not with sulfides of the metals to the right of nickel. Copper-tellurides may co-exist with all tellurides of this sequence, but not in presence of the sulfides of these metals. (It must be noted that this rule holds under favorable thermodynamic conditions and only for minerals of the same paragenesis *sensu stricto.*) From Schtscherbina's conclusions, it follows that $Ag_2Te + PbS$ (hessite + galena) is a stable paragenesis, but not $PbTe + Ag_2S$ (altaite + argentite), which is in good agreement with microscopic observations.

Revised manuscript received: January 1971

52-G. Behavior during Weathering and Alteration of Rocks

In the oxidation zone of ore deposits, tellurium shows little tendency for migration. The tellurides are — in contrast to sulfides and selenides — stable even at rather high oxidation potentials. The oxidation potential for hexavalent sulfur (SO_4^{2-}) is under natural conditions very easily attained. No Te^{6+}-minerals are well known; they are to be expected only under exceptional climatic conditions. The stable oxidized compounds are mostly tellurites. Because of their stability tellurides, even under oxidising conditions, accumulate in placerdeposits together with gold and other heavy minerals. Examples of such deposits are reported by SINDEEVA (1964). In the Angora region in Siberia, several important gold-tellurium placers are known but are not at present exploited. Here the heavy minerals are: gold, scheelite, cassiterite, bismuthinite, native silver, pyrite, chalcopyrite. Tellurium minerals are hessite and joseite. This list shows, that only the special climatic conditions of this tundra region may allow the existence of readily oxidizable sulfides. It is interesting that the primary sources of these placers are not yet known.

The natural migration of Te may only occur in zones of a high oxidation potential and of leaching above the so-called "iron cap", especially in arid regions. Tellurium is fixed under such conditions either, as native tellurium or seleno-tellurium or, as in deposits of Colorado and Montana (USA), it is oxidized to TeO_2 which has a strong amphoteric character and may migrate under acid and alkaline conditions. So tetradymite gives the bismuth tellurite montanite ($Bi_2TeO_4 \cdot 2H_2O$) and coloradoite gives magnolite (Hg_2TeO_4). Finally sulfuric acid replaces tellurium in the metallic compounds and we find such iron tellurites as emmonsite and mackayite.

Revised manuscript received: January 1971

52-H. Solubilities of Compounds which Control Concentration of Tellurium in Waters

Under earth's surface conditions, tellurium may exist as native tellurium and its field of existence under various pH-Eh conditions is more extensive than that of native sulfur. Tellurates (TeO_4^{2-}) are only to be expected in strong alkaline solutions under extremely oxidizing conditions. The most abundant form of Te will be the tetravalent state (TeO_3^{2-}). The behavior of tellurium-ions plotted in a pH-Eh diagram (Fig. 52-H-1) has recently been investigated by DYACHKOVA and KHODAKOVSKIY (1968) and the following solubility products were found:

	25° C	150° C	300° C
PbTe	38.60	31.1	29.2
Ag_2Te	64.00	47.2	39.2
ZnTe	25.87	22.9	23.4
CdTe	34.40	28.9	28.1

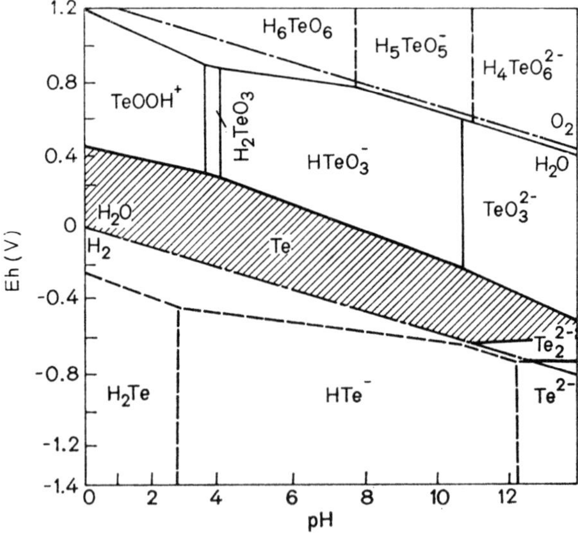

Fig. 52-H-1 a

Fig. 52-H-1 a—c. Stability fields of various compounds of tellurium in relation to Eh and pH at Σ Te $= 10^{-7}$ at: a = 25° C, 1 atm.; b = 150° C, 5 atm.; c = 300° C, 85 atm. After DYACHKOVA and KHODAKORSKIY (1968)

Revised manuscript received: January 1971

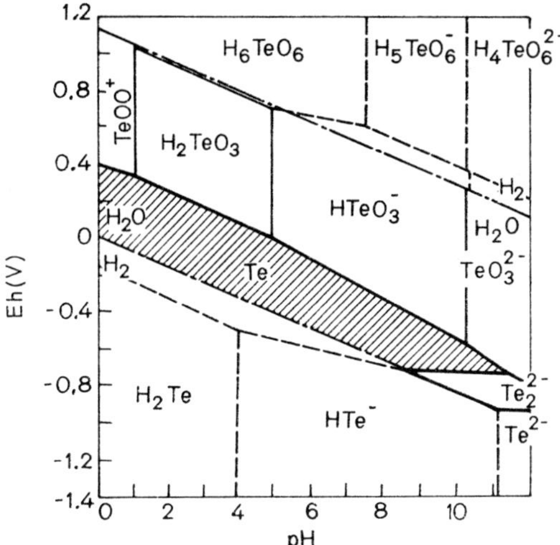

Fig. 52-H-1b

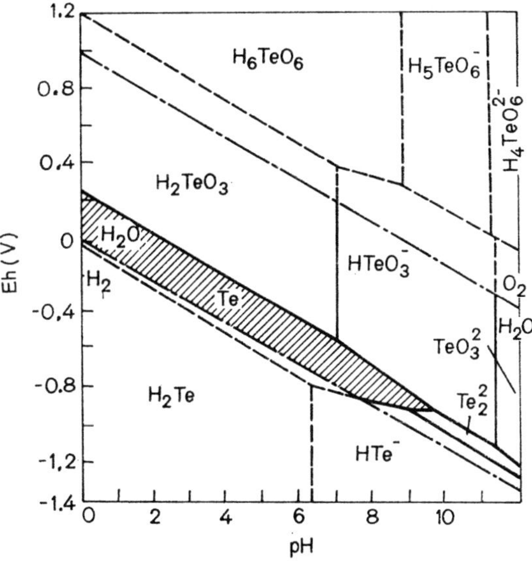

Fig. 52-H-1c

52-I. Abundance in Natural Waters

The tellurium content of ocean and fresh waters is not yet known. SINDEEVA (1964) mentions that in the waters of an abandoned mine in Siberia, 0.4 ppm Te was found (W, C). Unpublished data of the author (1966) on 6 samples of Elbe river waters upstream from Hamburg and seawater from near Helgoland showed only in the waters of the outflow of the Hamburg harbour 0.05 ppm Te (but 1 ppm Se); in all other samples the Te content was less than 0.01 ppm Te (W, C).

Revised manuscript received: January 1971

52-K. Abundance in Common Sediments, Sedimentary Rock Types and Sedimentary Ore Deposits

Adequate data on the tellurium content of rocks of the exogenic cycle are virtually lacking. In graywackes and shales its content is less than 0.1—1 ppm Te. LAKIN *et al.* (1963) report a range of 5—125 ppm Te (average: 48 ppm Te) in 12 manganese nodules from the Pacific and Indian Oceans. In samples of several sedimentary pyrite-nodules, SINDEEVA (1964) reported 2 ppm Te; in sedimentary iron, manganese and phosphorite-ores, SINDEEVA detected no tellurium. However, in the sedimentary sulfur deposits of Vetroyamovskoye, there exists 30 ppm Se and 6 ppm Te. In a systematic analysis of salt minerals (sylvinite, carnallite from Solikamsk, szaybelyite, ulexite, colemanite and hydroborazite from Inder), from potassium deposits of Ukrainia, no Te could be detected (limit of detection: 0.5 ppm).

In the Kupferschiefer marls from an euxinic environment (Eisleben) NODDACK and NODDACK (1936) found 0.02 ppm Te by X-ray analysis.

In contrast to selenium, tellurium has not yet been found in soils, nor plants, coal or lignite-ashes. [Unpublished data of LEUTWEIN (1948) from coals and soils of Eastern Germany — limit of detection 0.1—1 ppm (S-W method).]

Revised manuscript received: January 1971

52-O. Relations to Other Elements; Economic Importance, etc.

Many additional analytical investigations involving Te are required as the geochemical cycle of this element is only little known. However, it does appear that the cosmic abundance is higher than the abundance in the earth's crust. The chemical behavior and ionic radii of S, Se and Te are so different, that a camouflage of Te in the far more abundant sulfur minerals can not at present be detected. Though very rare (crustal abundance about 0.01 ppm), Te forms a rather large number of separate minerals but the crystallochemical relationships between Te and Se are not close and the ratio of Se to Te in different rocks is not well known. In pyrite, pyrrhotite, chalcopyrite and pentlandite ore deposits the content of these elements is very low, but nevertheless such deposits provide an estimated 80% of the world's production. Here the ratio of Se to Te is about 10:1.

Tellurium is an important element for the electrotechnical and electronics industries but technical applications in metallurgy and chemical industries appear to be less important. The famous telluride deposits of gold, silver, and other metals are of only minor economic importance. Both metals are accumulated during the metallurgical processes in the copper and pyrite smelters as by-products. For tellurium the main sources are flue-dust and anode-slugs of electrolysis plants but complete information on the world annual production is not available. The main producers are the United States and Canada and for 1958, SINDEEVA (1964) gives the respective annual amounts produced as 77 tons and 20 tons respectively.

Revised manuscript received: January 1971

References: Sections 52-B to 52-K, 52-O

BAINBRIDGE, K. T., NIER, A. O.: Relative isotopic abundances of the elements. Preliminary Report No. 9, Nucl. Sci. Series, Nat. Research Council U.S., Washington, D. C. (1950).

BOULADON, J., PICOT, P.: Sur les minéralisations en cuivre des ophiolites de Corse, des Alpes françaises et de Ligurie. Bull. du B.R.G.M., p. 23 (1968).

CHUKROV, F. V.: Discovery of a bismuth telluride in Manka Deposit (S. Altai). Notes of the All-Russian Min. Soc., No. 2, **76** (1947).

DOMEYKO, I.: Über die Entdeckung von Tellur-Mineralien in Chile. Compt. Rend. **81** (1875).

DU FRESNE, A.: Selenium and tellurium in meteorites. Geochim. Cosmochim. Acta **20**, 141 (1960).

DYACKKOVA, I. B., KHODAKOVSKIY, J. L.: Thermodynamic equilibria in the systems S—H_2O, Se—H_2O, Te-H_2O in the 25—300° temperature range and their geochemical interpretation (Geokhimija, 1968). Cited after Geochemistry International **5**, 1108 (1968).

FAIRBANKS, E. E.: The punched card identification of ore minerals. Econ. Geol. **41**, 761 (1946).

FERSMAN, A. E.: Geochemie. Vol. 4. Leningrad: Staatl. Wiss. Tech. Verlag chemischer Literatur 1939 (Russ.)

GANAPATHY, R., KEAYS, R. R., LAUL, J. C., ANDERS, E.: Trace elements in Apollo 11 lunar rocks: Implication for meteorite influx and origin of moon. Geochim. Cosmochim. Acta, Proceedings of the Apollo 11 Lunar Science Conference, Part 2, Suppl. 1, Vol. 34 (1970).

GARRELS, R. M.: Mineral Equilibria. New York: Harpers Geoscience Series 1960.

GLEMSER, v. O., HAESELER, R. v., MÜLLER, A.: Über gasförmiges $TeO(OH)_2$. Z. Anorg. Allgem. Chem. **329**, 53 (1964).

GMELIN's Handbuch der anorganischen Chemie, VIII. ed., Syst. No. 11, Tellurium. Weinheim: Verlag Chemie 1940.

GOLDSCHMIDT, V. M.: Geochemische Verteilungsgesetze und kosmische Häufigkeit der Elemente. Naturwissenschaften **18**, 999 (1930).

— BARTH, T., HOLMSEN, D., LUNDE, G., THOMASSEN, L., ULRICH, F., ZACHARIASSEN, W.: Geochemische Verteilungsgesetze der Elemente I—IX. Skrifter Norske Videnskaps-Akad. Oslo, I. Math. Naturw. Kl. 1923—1937.

GOLES, G. G., ANDERS, F.: I content of meteorites and their ^{129}I—^{129}X ages. J. Geophys. Res. **65**, 4181 (1960).

— — Abundance of iodine, tellurium and selenium in meteorites. Geochim. Cosmochim. Acta **26**, 723 (1962).

GREENLAND, L.: The abundances of Se, Te, Ag, Pd, Ce and Zn in chondritic meteorites. Geochim. Cosmochim. Acta **31**, 849 (1967).

INGHRAM, M. G., REYNOLDS, I. H.: Double β-decay of ^{130}Te. Phys. Rev. **78**, 822 (1950).

KARAMYAN, K. A.: Correlation between rhenium, selenium and tellurium in the molybdenites of the Kadzkaran copper-molybdenum deposit. Geochemistry **2**, 194 (1962).

KIESL, W., GRASS, F., BÖCKL, R., PONTA, U.: Cosmochemical abundances of trace elements in meteorites I. J. Radioanal. Chem. **6**, 447 (1970).

LAKIN, H. W., THOMPSON, C. E., DAVIDSON, D. F.: Tellurium content of marine manganese oxides and other manganese oxides. Science **142**, 1568 (1963).

NODDACK, I., NODDACK, W.: Die Häufigkeit der chemischen Elemente. Naturwissenschaften **18**, 757 (1930).

— — Die geochemischen Verteilungskoeffizienten der Elemente. Svensk. Kem. Tidskr. **46**, 173 (1934).

RAMDOHR, P.: Die Erzmineralien und ihre Verwachsungen, 2 ed. Berlin: Akademie-Verlag 1955.

RANKAMA, K.: Isotope Geology. London: Pergamon Press 1954.

— SAHAMA, T.: Geochemistry. Chicago: The University of Chicago Press 1950.

REED, G. W.: Heavy elements in the Pantar meteorite. J. Geophys. Res. **68**, 3531 (1963).

SCHINDEWOLF, U.: Selenium and tellurium content of stony meteorites by neutron activation. Geochim. Cosmochim. Acta **19**, 134 (1960).

SCHTSCHERBINA, W. W.: Principal features of the geochemistry of tellurium [Russ.]. Bull. Acad. USSR, Ser. Geol. 980 (1937).

SINDEEVA, N. D.: Mineralogy and Types of Deposits of Selenium and Tellurium. Engl. transl. (ed. E. Ingerson.) New York: Interscience Publishers 1964.

STUMPFL, E. F.: New electronprobe and optical data on gold-tellurides. Am. Mineralogist **55**,808 (1970).

SUESS, H. E., UREY, H. C.: Abundances of the elements. Rev. Mod. Phys. **28**, 53 (1956).

TUREKIAN, K. K., WEDEPOHL, K. H.: Distribution of the elements in some major units of the earth's crust. Geol. Soc. Am. Bull. **72**, 175 (1961).

VINOGRADOV, A. P.: Geochemie seltener und nur in Spuren vorhandener chemischer Elemente im Boden (1954). German translation (Acad. Verlag, Berlin) of the Russian original, 1950.

— Trace-elements and research problems: microelements in plants of various systematic position. Agrochimija, No. 8, 2—30 (1965) [Russ.]. (Mikroelementy i zadatchi na'uky.)

— Introduction to the Geochemistry of the Oceans. Ed. Nauka (Sciences) Moscow 1967 [Russ.].

— Preliminary data on lunar ground supplied by the automatic station "Luna 16". Geochimija **3**, 261 (1971).

VLASSOV, V. A.: Geochemistry and Mineralogy of the Genetic Types of Deposits of Rare Elements, vol. 1, chapt. Tellurium, p. 586. Moskow 1964 [Russ.]

Revised manuscript received: January 1971

Iodine 53

A	B. Brehler	(Mineralogisch-Kristallographisches Institut, Technische Universität, Clausthal-Zellerfeld 1, Germany)
B—M, O	R. Fuge	(Geology Department, University College of Wales, Aberystwyth, Wales, U.K.)

53-A. Crystal Chemistry

Iodine occurs as a minor constituent of various minerals but only rarely forms separate minerals such as marshite (CuI), miersite (α-AgI), jodargyrite (β-AgI), lautarite ($Ca[IO_3]_2$), salesite ($Cu(OH)[IO_3]$), and bellingerite ($Cu_3[IO_3]_6 \cdot 2\,H_2O$).

Iodine is the halogen with the lowest electronegativity (2.5; PAULING, 1960, see Chapter 12 of Volume I of this handbook).

The outer electron shell has the structure $5s^2\,5p^5$. Iodine is known in the oxidation states -1, $+1$, $+3$, $+5$ and $+7$. The ionic radius of I^- has been found to be 2.20 Å (GOLDSCHMIDT, 1926, see Chapter 12 of Volume I of this handbook), and PAULING and HUGGINS (1934) report a covalent tetrahedral radius of 1.28 Å.

I. Elementary Iodine

Iodine forms crystals with I_2 molecules (HARRIS, MACK and BLAKE, SB 1913—28, 760); at 110° K, the I—I-distance in the molecule is 2.715 Å. This molecule forms a two-dimensional network with I ... I (intermolecular): 3.50 Å. The bonding angles of approximately 90° can be explained in a first approximation as being formed by orthogonal iodine 5 p orbitals (VAN BOLHUIS *et al.*, 1967).

II. Interhalogen Compounds

The crystal structures of several compounds of the form XX'_n (where X and X' are different halogen atoms and $n = 1, 3, 5$ or 7) are known.

$I^{[7]}F_7$ forms a molecular structure with I—F: 1.81—1.97 Å (BURBANK, SR 1962, 446).

α-ICl is built up of two sets of non-equivalent ICl molecules with bond lengths of 2.37 Å and 2.44 Å. The short distance between the iodines (3.08 Å) indicates strong interactions between the molecules (BOSWIJK, VAN DER HEIDE, VOS and WIEBENGA, SR 1956, 238).

In I_2Cl_6, planar molecules $I_2^{[4Cl,\,pl]}Cl_6$ are packed in layers, the shortest I—Cl distance being 2.38 Å (BOSWIJK and WIEBENGA, SR 1954, 390).

IBr is the only known bromine-iodine interhalogen compound. The structure is similar to that of I_2, I—Br: 2.52 Å. The shortest intermolecular distances I—Br are 3.18 and 3.78 Å (SWINK and CARPENTER, 1968a and 1968b).

III. Binary Iodides

Most of the elements form crystalline iodides with a wide range of bonding character; several examples are listed in Table 53-A-1. BaHI, SrHI and CaHI crystallize in the PbFCl-type (EHRLICH and KULKE, SR 1956, 213).

Table 53-A-1. *Structures of binary iodides*

Coordination number of iodine	Structure-type	Compounds, interatomic distances	Reference
8	$Cs^{[8]}Cl^{[8]}$	CsI	CLARKE and DUANE, (SB 1913—28, 107)
		TlI Tl—I: 3.66 Å	BARTH and LUNDE (SB 1913—28, 113)
6	$Na^{[6]}Cl^{[6]}$	LiI	WYCKOFF (SB 1913—28, 101)
		KI	DUANE and CLARKE (SB 1913—28, 106)
		NH_4I[a]	VEGARD (SB 1913—28, 109)
		RbI[b] Rb—I: 3.62 Å	DAVEY (SB 1913—28, 107)
4	β-$Zn^{[4]}S^{[4]}$	AgI[c], iodyrite	WILSEY (SB 1913—28, 112)
	α-$Zn^{[4]}S^{[4]}$	CuI, marshite	WYCKOFF and POSNJAK (SB 1913—28, 110)
		AgI, miersite[aa]	WILSEY (SB 1913—28, 112)
3	$^{2}_{\infty}Cd^{[6]}I_2^{[3py]}$	PbI_2 Pb—I: 3.12 Å	VAN ARKEL (SB 1913—28, 191)
		CdI_2[bb] Cd—I: 2.99 Å	BOZORTH (SB 1913—28, 189)
		TiI_2, VI_2	KLEMM and GRIMM (SR 1947—48, 259)
		MgI_2, CaI_2	BLUM (SB 1933—35, 281)
2	$^{2}_{\infty}Bi^{[6]}I_3^{[2]}$	BiI_3 Bi—I: 3.1 Å	BRAEKKEN (SB 1928—32, 294)
		SbI_3 Sb—I: 2.87 Å (3 ×) 3.32 Å (3 ×)	TROTTER and ZOBEL (1966)
	$^{2}_{\infty}Hg^{[4]}I_2^{[2]}$	HgI_2[bb] (red form)	BIJVOET, CLAASSEN and KARSSEN (SB 1913—28, 189)
		AuI Au—I: 2.62 Å (2 ×)	JAGODZINSKI (SR 1959, 318)
1 (molecules)	planar molecules BI_3	BI_3 B—I: 2.10 Å	RING, DONNAY and KOSKI (SR 1962, 438) LAURITA and KOSKI (1959)
		AsI_3 As—I: 2.56 Å I—As—I: 102.0°	TROTTER (1965)

Table 53-A-1 (continued)

2/1	$^1_\infty Nb^{[6]}I_2{}^{[2]}I_2{}^{[1]}$	α-NbI_4 Nb—I[2]: 2.76 Å (av.) (2 ×) 2.90 Å (av.) (2 ×) Nb—I[1]: 2.69 Å (av.) (2 ×)	Dahl and Wampler (SR 1962, 443)
3/4	$Sr^{[7]}I^{[3]}I^{[4]}$	SrI_2 Sr—I (I): 3.31 Å (av.) (3 ×) Sr—I (II): 3.85 Å (av.) (4 ×)	Rietschel and Bärnighausen (1969)
5/4	$Pb^{[7+2]}Cl^{[5]}Cl^{[4]}$	BaI_2	Döll and Klemm (SB 1939, 80)

[a] Below −17.6° C, NH_4I crystallizes in the CsCl-type.

[b] Under high pressures RbI crystallizes in the CsCl-type (Verescagin and Kabalkina, SR 1957, 215).

[c] AgI crystallizes above 146° C in the α-AgI-type. In this structure iodine forms a cubic body-centred lattice and the Ag atoms are statistically distributed over all the holes in this lattice (Strock, SB 1933—35, 232; Hoshino, SR 1957, 216).

[aa] In miersite, 20% of the Ag atoms are replaced by Cu (Aminoff, SR 1913—28, 189).

[bb] There also exists a metastable modification; its structure unit is a Hg_4I_{10} group consisting of four corner linked HgI_4 tetrahedra (Schwarzenbach, 1969).

IV. Complex Iodides

Several complex iodides exist and examples are given in Table 53-A-2.

Table 53-A-2. *Complex iodides*

Compound	Interatomic distances, structure type	Reference
Cs $Pb^{[6]}I_3$	Pb—I: 3.01—3.42 Å	Møller (SR 1959, 321)
β-$Cu^{[4]}Hg^{[4]}I_4$		Hahn, Frank and Klingler (SR 1955, 337)
Zn_3PI_3	(disordered sphalerite type)	Suchow and Witzen (SR 1962, 467)
$Sr(OH_2)_6[Pb^{[6]}I_6]\cdot H_2O$	Pb—I: 3.14—3.20 Å	Ferrari, Braibanti and Lanfredi (SR 1961, 336)

V. Oxyiodides and Related Compounds

Oxychlorides, oxybromides and oxyiodides are closely related.

BiOI (Sillén, SR 1947—48, 312) and SnOI (Kruse, Asprey and Morosin, SR 1961, 338) are isotypical to BiOCl (see Section 17-A). In iodolaurionite, Pb(OH)I (Malćić and Zivadinović, SR 1960, 298), the iodine is coordinated by 4 I, 5 Pb and 5 OH; the interatomic distances are I—I: 4.16—4.35 Å; Pb—I: 3.42—3.68 Å.

VI. Polyhalides

Iodine forms a series of compounds with univalent polyhalide anions containing one, two or three different halogen species. Examples are:

$NH_4[I_3]$ (MOONEY, SB 1933—35, 321; CHEESMAN and FINNEY, 1970). The iodine forms a nearly linear group with I—I: 2.79 Å and 3.11 Å. The short I—N distance of 3.62 Å may be evidence for hydrogen bonding between NH_4^+ and I_3^-.

$K[ICl_4] \cdot H_2O$ (MOONEY, SB 1938, 95; ELEMA *et al.*, 1963) has a slightly pyramidal ICl_4^- anion with I—Cl: 2.42—2.60 Å.

$Cs[BrI_2]$ (CARPENTER, 1966); in the nearly linear triatomic anion, bond lengths are Br—I: 2.91 Å, I—I: 2.78 Å.

$NH_4[ClIBr]$ (MOONEY, SB 1937, 76; MIGCHELSON and VOS, 1967) consists of linear halogen groups with the iodine in the centre.

VII. Iodates and Related Acids

The IO_3 group forms a trigonal pyramid; in some compounds the iodine atoms are surrounded by three oxygen atoms of other IO_3 groups, building a very distorted octahedron. Table 53-A-3 gives structural information on iodates and related acids.

Table 53-A-3. *Structures of iodates and related acids*

Compound	Interatomic distances	Reference
$Cu(OH)[IO_3]$	I—O: 1.78 Å (1 ×); 1.82 Å (2 ×); 2.50 Å (1 ×); 2.69 Å (2 ×)	GHOSE (SR 1962, 644)
α-HIO_3	I—O (intraionic): 1.80; 1.81; 1.89 Å I—O (interionic): 2.45; 2.70, 2.90 Å	ROGERS and HELMHOLTZ (SR 1949, 231); GARRETT (SR 1954, 393)
HI_3O_8 [a]	HIO_3: I—O: 1.80 Å (av.) I—OH: 1.90 Å I_2O_5: I—O: 1.79 Å (av.) I—OI: 1.96 Å (av.)	FEIKEMA and VOS (1966)
$LiIO_3$	I—O: 1.81 Å (3 ×) 2.89 Å (3 ×)	ROSENZWEIG and MOROSIN (1966); DE BOER *et al.* (1966)
$NaIO_3$	I—O: 1.80—1.83 Å (3 ×) 2.84—3.30 Å (3 ×)	MAC GILLAVRY, PANTHALEON, VAN ECK (SR 1947—48, 363)
α-$RbIO_3$	I—O: 1.81 Å (3 ×) 2.75 Å (3 ×)	ALCOCK (1972)
$K_2[IO_3][IO_2(OH)]Cl$ [b]	I—O: 1.77—1.85 Å I—OH: 1.93 Å I—Cl: 3.08 Å 3.05 Å	MANOTTI LANFREDI *et al.* (1972)
$Sr(IO_3)_2 \cdot H_2O$ [c]	I—O: 1.79—1.83 Å (3 ×) 2.85—3.22 Å (4 ×)	MANOTTI LANFREDI *et al.* (1972)

[a] HI_3O_8 consists of $IO_2(OH)$ and I_2O_5 units which have strong intermolecular relations. Many intermolecular distances I ... O are as short as 2.6 Å. I_2O_5 consists of two IO_3 pyramids which have one oxygen atom in common.

[b] $K_2H(IO_3)_2Cl$ consists of anions IO_3^-, molecules $IO_2(OH)$ and of Cl^- and K^+.

[c] Each iodine atom together with four oxygen atoms forms a distorted pentagonal bipyramid.

VIII. Periodates

Only a few crystal structures of periodates are known (Table 53-A-4). A review on iodine(VII) oxy-acids and the periodates has been given by SIEBERT (1967).

In the periodates, iodine has the coordination number 4 (distorted tetrahedron) and 6 (distorted octahedron).

Tetrahedral $[IO_4]$-groups have been found in the compounds $MeIO_4$. Compounds with a six-fold coordination of iodine have large differences in composition and structure. In the $[IO_6]$-octahedron a variable number of oxygen atoms may be replaced by OH, and moreover two octahedra are able to share edges or faces; sharing corners is not known.

Table 53-A-4. *Structures of periodates*

Compound	Structure type, interatomic distances	Reference
a) Coordination number 4		
$Na[IO_4]$	$CaWO_4$ I—O: 1.775 Å	KIRKPATRICK and DICKINSON (SB 1913—26, 372) KÁLMÁN and CRUICKSHANK (1970)
$K[IO_4]$	$CaWO_4$	HYLLERAS (SB 1913—26, 373)
b) Coordination number 6		
	Isolated octahedra	
$I(OH)_5O$	I—O: 1.78 Å I—OH: 1.89 Å	FEIKEMA (SR 1961, 499) FEIKEMA (1966)
$Cd(OH_2)_3[I(OH)_3O_3]$	I—O: 1.86 Å (av.) I—OH: 1.95 Å (av.)	BRAIBANTI *et al.* (1970)
	Edge-shared double octahedra	
$K_4[I_2(OH)_2O_8]\cdot 8\,H_2O$	I—O: 1.81 Å (av.) I—OH: 1.98 Å I—OI: 2.00 Å (av.)	FERRARI *et al.* (1965); SIEBERT and WEDEMEYER (1965)
	Face-shared double octahedra	
$K_4[I_2O_9]$	I—O: 1.77 Å I—OI: 2.01 Å	BREHLER *et al.* (1968)

In all known iodate(VII) structures the observed I—OH-distances are longer than the I—O-distances. Also the distances of the shared oxygen to iodine are longer than those of the unshared oxygen. In the I—OH bonds the double bond character decreases compared with I—O. The difference between bonds with shared and with unshared atoms is normal.

For additional information about the crystal chemistry of iodine see WELLS (1962).

Revised manuscript received: June 1973

General Introduction

A study of the geochemistry of iodine was included by CORRENS (1956) in his general review of the geochemistry of the halogens. Most of the data which he utilised was from the work of VON FELLENBERG and LUNDE (1926, 1927); the results quoted by these authors, who used classical methods for the analyses, have been criticised by recent workers.

GOLES and ANDERS (1962) using a neutron activation method analysed several samples of meteorites which had previously been analysed by VON FELLENBERG (1927). Comparison of the results show generally that the earlier values are five to thirty-five times higher than those of the later workers. The range of values quoted by GOLES and ANDERS (1962) are in general agreement with the values for meteorites quoted by several other workers (REED and ALLEN, 1966; GOLES *et al.*, 1967).

In view of their results for meteorites, GOLES and ANDERS have questioned the validity of the results quoted by VON FELLENBERG and LUNDE for terrestrial materials. This view appears to have been partly confirmed by the results obtained by YOSHIDA *et al.* (1971), based mainly on Japanese volcanic rocks. However, values quoted by KURODA and CROUCH (1962) do not differ markedly from those of similar rock types given by VON FELLENBERG and LUNDE (1926, 1927). This is also true of some values quoted in other works of recent origin (KOGARKO and GULYAYEVA, 1965; BECKER and MANUEL, 1972).

While stressing the need for caution in use of the results of these early workers, the author has chosen to include much of their data for terrestrial materials.

Revised manuscript received: March 1973

53-B. Isotopes in Nature

There are twenty-four known isotopes of iodine, ranging from ^{117}I to ^{130m}I and ^{130}I to ^{139}I; eighteen of these isotopes have half-lives of less than one day. The only stable isotope is ^{127}I and the atomic weight of iodine is 126.9044 (HEATH, 1971).

Iodine-129, which has a half-life of 1.7×10^7 y (HEATH, 1971), was formed in the primordial solar system. This isotope decays to the radiogenic ^{129}Xe (REYNOLDS, 1960) and the occurrence of excess ^{129}Xe in meteorites has been used by several workers as a dating method to estimate formation times and cooling rates etc. (JEFFERY and REYNOLDS, 1961; HOHENBERG *et al.*, 1967; ALEXANDER and MANUEL, 1968). In addition, excess ^{129}Xe has been found in terrestrial materials (BOULOS and MANUEL, 1972).

The supply of primordial ^{129}I has long been exhausted; however, this isotope is continually produced by spontaneous fission of uranium and also by spallation reactions of xenon in the upper atmosphere (EDWARDS, 1962). EDWARDS has estimated that these natural processes contribute a steady-state concentration of $\geqq 10^{-14}$ g of ^{129}I per gram ^{127}I. In addition, comparatively large quantities of ^{129}I are produced during nuclear fission fallout from nuclear weapons and reactors (EDWARDS, 1962).

EDWARDS and REY (1969) analysed iodine from some brines and iodine derived from other sources; most of these samples were collected prior to the advent of the nuclear age. Their results (Table 53-B-1) indicate a ratio of $^{129}I/^{127}I$ of about 5×10^{-12}, which agrees fairly well with their estimated value of 2.2×10^{-12}. As expected, the lowest values were obtained for the geologically oldest sample, the Michigan brine.

SRINIVASAN *et al.* (1971) determined the quantity of ^{129}Xe in a natural iodyrite sample from Broken Hill, Australia. On the basis of their determination, they have estimated that the pre-nuclear age equilibrium ratio of $^{129}I/^{127}I$ was 2.2—3.3×10^{-15}.

Table 53-B-1. ^{129}I *in iodine derived from various sources* (from EDWARDS and REY, 1969)

Sample Source	Age of deposit	Year produced or sampled	$^{129}I/^{127}I$
Brine, Long Beach, California, U.S.A.	Pliocene	1930	1×10^{-11} 1×10^{-11} 5×10^{-12}
Chile nitrate	—	1939	5×10^{-12} $4{\cdot}5 \times 10^{-12}$
Merck I_2, source not known	—	1920	5×10^{-12}
Brine, Michigan	Silurian	1966	2×10^{-12} 3×10^{-12}

^{131}I, which has a half-life or 8.07 days, has been detected in vegetation by PENDLETON and LLOYD (1970) and in the atmosphere, along with ^{132}I (half-life 2.3 hours) and ^{133}I (half-life 20.9 hours), by HATTORI and NAGAHARA (1967); these isotopes were derived from nuclear fallout.

Revised manuscript received: March 1973

53-C. Abundance in Meteorites, Tektites and Lunar Samples

I. Meteorites

a) Stones

1. Chondrites

The iodine content of chondritic meteorites ranges from less than 10 ppb in some ordinary chondrites to almost one ppm in some Cc_1 chondrites (Table 53-C-1). Iodine is classified by LARIMER and ANDERS (1967) as a strongly depleted element in ordinary and Ce_2 chondrites (Table 53-C-2). In common with other strongly depleted elements, iodine is depleted relative to Cc_1 chondrites in all chondrite classes.

Table 53-C-1. *Iodine in chondrites* (All methods: N/R)

Class	Meteorite	Number of samples	I (ppb)	Reference
Cc_1	Orgueil	1	230	REED and ALLEN (1966)
		1	400	GOLES *et al.* (1967)
	Ivuna[a]	3	920	REED and ALLEN (1966)
Cc_2	Mighei	2	310	GOLES and ANDERS (1962)
		1	550	REED and ALLEN (1966)
		1	480	GOLES *et al.* (1967)
	Murray	3	230	GOLES and ANDERS (1962)
		1	300	GOLES *et al.* (1967)
Ce_1	Abee	2	145	GOLES and ANDERS (1962)
		1	300	MERRIHUE (1965)
		1	<180	REED and ALLEN (1966)
	Indarch	3	270	GOLES and ANDERS (1962)
		1	470	REED and ALLEN (1966)
		1	310	GOLES *et al.* (1967)
Ce_i	St. Marks	2	82	GOLES and ANDERS (1962)
Ce_2	Hvittis	1	74	REED and ALLEN (1966)
		2	53	GOLES *et al.* (1967)
CHL	Lancé	1	110	REED and ALLEN (1966)
		1	170	GOLES *et al.* (1967)
	Felix	1	260	GOLES *et al.* (1967)
	Karoonda	1	220	REED and ALLEN (1966)
CLL	Chainpur	1	200	GOLES *et al.* (1967)
CL	Bruderheim	1	16	GOLES and ANDERS (1962)
		1	7.8	MERRIHUE (1965)
		2	{ 450 { <74	REED and ALLEN (1966)

Table 53-C-1 (continued)

Class	Meteorite	Number of samples	I (ppb)	Reference
		2	6.5	GOLES *et al.* (1967)
		2	22.5	CLARK *et al.* (1967)
	Mocs	1	50	GOLES and ANDERS (1962)
		2	14	GOLES *et al.* (1967)
	Peace River	1	13	GOLES *et al.* (1967)
		2	17.5	CLARK *et al.* (1967)
	Bjurböle	1	16	TURNER (1965)
	Ergeo	1	90	GOLES and ANDERS (1962)
	Harleton	2	65	REED and ALLEN (1966)
	Holbrook	1	30	REED and ALLEN (1966)
	Walters	1	14	GOLES *et al.* (1967)
CH	Richardton	1	76	REYNOLDS (1960)
		3	28	GOLES and ANDERS (1962)
		1	23	MERRIHUE (1965)
	Allegan	2	{120 69	REED and ALLEN (1966)
	Beardsley	2	63	GOLES and ANDERS (1962)
	Ehole	1	9	GOLES *et al.* (1967)
	Miller	1	67	REED and ALLEN (1966)
	Pantar (l)[b]	1	4.4	TURNER (1965)
	(l)[b]	1	<200	REED and ALLEN (1966)
	(l)[b]	1	19	GOLES *et al.* (1967)
	Pantar (d)[b]	1	60	TURNER (1965)
	(d)[b]	1	<550	REED and ALLEN (1966)
	Plainview[c]	2	49	GOLES and ANDERS (1962)

[a] GOLES *et al.* (1967) quote a value of 11.2 ppm for Ivuna and suggest that this high value is due to contamination of the sample. However, REED (1971) has suggested this value may have been due to an unusual iodine concentration in the aliquot.
[b] l and d — light and dark regions.
[c] Plainview is a "find".

Table 53-C-2. *The relative abundance of iodine in the various classes of chondrites* (from LARIMER and ANDERS, 1967)

Class	Absolute concentration of iodine (ppb)[a]	Atoms/10^6 Si atoms	Depletion factors relative to the Cc_1 meteorites
Cc_1	745 (4)	1.4	
Cc_2	363 (3)	0.63	0.45
CHL	165 (2)	0.25	0.18
Ordinary chondrites	41 (13)	0.058	0.041
Ce_1	295 (3)	0.39	0.28
Ce_2	78 (2)	0.10	0.072

The values for ordinary chondrites are geometric means, all others are arithmetic means.
[a] Values derived from essentially the same data as LARIMER and ANDERS (1967), i.e. from GOLES and ANDERS (1962) and REED and ALLEN (1966). Figures in parentheses—number of samples.

Table 53-C-3. *Leachable iodine in chondrites* (All methods: N/R)

Meteorite	Class	Iodine (ppb)	Leachable iodine as % of total[a]	Reference
Orgueil	Cc_1	230	65	Reed and Allen (1966)
Ivuna	Cc_1	500	64	Reed and Allen (1966)
		1,040	77	Reed and Allen (1966)
		1,210	79	Reed and Allen (1966)
Mighei	Cc_2	550	71	Reed and Allen (1966)
Murray	Cc_2	300	35	Goles and Anders (1962)
		230	39	Goles and Anders (1962)
Indarch	Ce_1	470	49	Reed and Allen (1966)
Hvittis	Ce_2	74	45	Reed and Allen (1966)
Lancé	CHL	110	35	Reed and Allen (1966)
Karoonda	CHL	220	21	Reed and Allen (1966)
Bruderheim	CL	16	60	Goles and Anders (1962)
Harleton	CL	76	24	Reed and Allen (1966)
		54	40	Reed and Allen (1966)
Holbrook	CL	30	40	Reed and Allen (1966)
Richardton	CH	33	54	Goles and Anders (1962)
Allegan	CH	69	62	Reed and Allen (1966)
Miller	CH	67	24	Reed and Allen (1966)

[a] Reed and Allen (1966) leached the samples in hot water, whereas Goles and Anders (1962) used a hot, slightly alkaline solution of CH_3COONH_4 and NH_4OH.

From the work of Goles and Anders (1962) and Reed and Allen (1966), it appears that appreciable quantities of iodine occur in a water-soluble phase (Table 53-C-3). Goles and Anders (1962) have observed the chalcophilic character of iodine in iron meteorites and this led them to suggest tentatively that some of the iodine in chondrites could be present in water-soluble sulfides such as oldhamite (CaS).

Multiple analyses of individual chondrites for iodine reveal less scatter than do other halogens and this, according to Dodd (1969), suggests the occurrence of iodine in a major and/or a homogeneously distributed minor phase.

Hohenberg and Reynolds (1969) have found evidence of the concentration of iodine at grain boundaries and Reed (1971) suggests that iodine may be present in a water-soluble phase occurring on grain surfaces.

Goles and Anders (1962) observed that iodine in the CL meteorite Bruderheim occurred in the finest fractions following crushing and sieving.

Most of the ordinary chondrites analysed for iodine are of the higher petrological types; however, Dodd (1969) has pointed out that the CLL-3 chondrite Chainpur has several times as much iodine as the CL-6 type chondrites. He suggests from this evidence that iodine, like bromine, shows fractionation from the lower to higher (equilibrated) petrologic types of ordinary chondrites.

2. Achondrites

Clark *et al.* (1967) determined the iodine content of achondrites and their values, recorded in Table 53-C-4, show a range similar to that found for chondrites.

Table 53-C-4. *Iodine in achondrites* (from CLARK *et al.*, 1967) (Method: N/R)

Meteorite	Class	Number of determinations	I (ppb)
Cumberland Falls	Ae	1	460
Norton County	Ae	2	100
Pena Blanca Spring	Ae	2	22
Shallowater	Ae	2	180
Johnstown	Ab	1	25
Shalka	Ab	2	120
Lafayette	Ado	2	100
Nakhla	Ado	1	180
Kapoeta (d)[a]	Aor	2	34
Kapoeta (l)[a]	Aor	2	40
Juvinas	Ap	2	39
Moore County	Ap	2	140
Pasamonte	Ap	4	110
Petersburg	Ap	2	50
Sioux County	Ap	2	14
Stannern	Ap	2	880

[a] l and d—light and dark regions.

Table 53-C-5. *Iodine in iron meteorites and in separated phases of iron meteorites* (All methods: N/R)

Sample	I (ppb)	Reference
Og, Odessa	130	GOLES *et al.* (1967)
Og, Sardis	99	GOLES *et al.* (1967)
Troilite, Sardis	3,590	GOLES and ANDERS (1962)
Metal, Toluca	295	GOLES and ANDERS (1962)
Troilite, Toluca	1,030	GOLES and ANDERS (1962)
Metal, Canyon Diablo	28	GOLES and ANDERS (1962)
Troilite, Canyon Diablo	62	GOLES and ANDERS (1962)
Metal, Grant	11	GOLES and ANDERS (1962)
Troilite, Grant	24	GOLES and ANDERS (1962)
Troilite, Soroti	50	GOLES and ANDERS (1962)
Apatite, Mt. Stirling	1,700	REED and ALLEN (1966)

b) Irons

Data obtained for iodine in iron meteorites is summarised in Table 53-C-5.

GOLES *et al.* (1967) quote two analyses of octahedrites and GOLES and ANDERS (1962) analysed separated troilite and metal phases of several iron meteorites. From these analyses, they concluded that iodine is weakly chalcophilic in iron meteorites, the ratio of I (troilite)/I (metal) for a given meteorite ranging from 2.2 to 6.1.

Table 53-C-6. *Iodine in tektites and impact glasses* (from BECKER and MANUEL, 1972) (Method: N/R)

Sample	I (ppm)
Tektites	
Lee County, Texas	0.17
Australite	0.22
Moldavite	0.19
Philipinite	0.26
Thailand	0.56
Impact Glasses	
Aouelloul	1.4
Meteor Crater, Arizona	2.9
Monturaqui	73.0
Wabar	1.1

Table 53-C-7. *Iodine in lunar samples* (from REED and JOVANOVIC, 1971)(Method: N/R)

Sample		I (ppb)
Apollo 11		
Soil	10084,1	74
Rock	10017,22	4.7
Apollo 12		
Soil	12070,61	35
Breccia	12034,23	10
Rock	12052,18	8.7
Rock	12052,49	56
Rock	12022,109	(508)[a]
Rock	12022,69	14
Rock	12021,84	19

[a] this value may be unreliable.

REED and ALLEN (1966) found comparatively large amounts of iodine in an apatite from the Mt. Stirling meteorite. REED (1971) suggests that this occurrence of high iodine in a phosphate inclusion reflects the fact that the metal is a less favourable host.

II. Tektites

The only data available for iodine in tektites is that quoted for five samples from various localities by BECKER and MANUEL (1972); the average value for these is 280 ppb. These values are given in Table 53-C-6 along with values quoted by the same authors for various impact glasses.

BECKER and MANUEL (1972) comment that their values for tektites are lower than those they obtained for impact glasses and obsidians (see Table 53-E-2).

III. Lunar Samples

The iodine contents of lunar samples have been reported by REED and JOVANOVIC (1970a, 1971); however, these authors expressed doubt as to the validity of the results in their earlier work (REED and JOVANOVIC, 1970b) and for this reason only those values quoted in the 1971 paper are recorded here (Table 53-C-7). REED and JOVANOVIC (1971) have observed that their values for iodine in lunar materials are within the same range as values quoted for ordinary chondrites and they also found that almost all of the iodine is soluble in hot water.

Revised manuscript received: March 1973

53-D. Abundance in Rock-Forming Minerals; Iodine Minerals

Most of the available data for iodine in minerals is that quoted by von Fellenberg and Lunde (1926, 1927). (for comments on the reliability of this data, see Introduction to this Chapter); these values are given, along with some of more recent origin, in Table 53-D-1.

Table 53-D-1. *Iodine in rock-forming minerals*

Mineral	Source	Iodine (ppb)	Method	Reference
Biotite	Kragerö (Norway)	500	W	von Fellenberg and Lunde (1926, 1927)
Muscovite	Pegmatite, Halvorsröd, Raade, Östfold (Norway)	690	W	von Fellenberg and Lunde (1926, 1927)
Phlogopite	Bamle, Ödegaarden (Norway)	630	W	von Fellenberg and Lunde (1926, 1927)
Hornblende	Kragerö (Norway)	160	W	von Fellenberg and Lunde (1926, 1927)
Hypersthene	Soggendal, Ekersund (Norway)	940	W	von Fellenberg and Lunde (1926, 1927)
Microcline perthite	Halvorsröd, Raade, Östfold (Norway)	370	W	von Fellenberg and Lunde (1926, 1927)
Albite	Halvorsröd, Raade, Östfold (Norway)	150	W	von Fellenberg and Lunde (1926, 1927)
Labradorite	Soggendal, Ekersund (Norway)	440	W	von Fellenberg and Lunde (1926, 1927)
Smoky quartz	Halvorsröd, Raade, Östfold (Norway)	330	W	von Fellenberg and Lunde (1926, 1927)
Olivine	Jackson County, New Jersey (U.S.A.)	70	C	Kuroda and Crouch (1962)
Sodalite	Brevik (Norway)	900	W	von Fellenberg and Lunde (1926, 1927)
Sodalite	Pegmatite, Lovozero alkalic massif (U.S.S.R.)	900	W	Kogarko and Gulyayeva (1965)
Eudialyte	Pegmatite, Lovozero alkalic massif (U.S.S.R.)	1,200	W	Kogarko and Gulyayeva (1965)
Eudialyte	Phase III rocks, Lovozero alkalic massif (U.S.S.R.)	800	W	Kogarko and Gulyayeva (1965)
Scapolite	Risör (Norway)	230	W	von Fellenberg and Lunde (1926, 1927)

Table 53-D-1 (continued)

Mineral	Source	Iodine (ppb)	Method	Reference
Ilmenite	Blaafjell, Ekersund (Norway)	770	W	VON FELLENBERG and LUNDE (1926, 1927)
Fluorspar	Halvorsröd, Raade, Ösfold (Norway)	550	W	VON FELLENBERG and LUNDE (1926, 1927)
Apatite (yellow)	Bamle, Ödegaarden (Norway)	180	W	VON FELLENBERG and LUNDE (1926, 1927)
Apatite	Mt. Stirling, iron meteorite	1,700	N/R	REED and ALLEN (1966)

It appears likely that many of the mineral samples analysed by VON FELLENBERG and LUNDE are from pegmatites and may, therefore, be relatively enriched in iodine due to enrichment of that element in the residual liquids of magmas.

The relatively large ionic radius of iodine, 2.20 Å (WEAST, 1971), would appear to make it unlikely that there is any marked replacement of the OH^- or F^- groups in minerals. CORRENS (1956) comments on the fact that the iodine contents of all minerals are very similar, there being no enrichment in any group. He has suggested that iodine in rock-forming minerals is most likely to be present as fluid inclusions.

It appears that there is somewhat of an enrichment of iodine in the chlorine-containing minerals eudialyte and sodalite from the Lovozero alkali complex (KOGARKO and GULYAYEVA, 1965). These authors are of the opinion that at least some of the iodine in these minerals is due to substitution of the iodide ion for the chloride ion.

A summary of iodide and iodate minerals is given in Table 53-D-2.

Table 53-D-2. *Iodine minerals* (mainly from HEY, 1950 and FLEISCHER, 1966)

Mineral	Formula
Iodides	
Marshite	Cu I
Miersite (var. cuproiodargyrite, Ag<Cu)	(Ag, Cu) I; (Ag>Cu)
Iodargyrite	Ag I
Iodembolite	Ag (Cl, Br, I)
Coccinite	$Hg\ I_2$
Hopingite	$Hg\ I_2$ cubic[a]
Iodates	
Bellingerite	$Cu\ (IO_3)_2$
Salesite	$Cu\ IO_3\ (OH)$
Lautarite	$Ca\ (IO_3)_2$
Dietzeite	$Ca_2\ (IO_3)_2CrO_4$
Schwartzembergite	$Pb_5IO_3Cl_3O_3$

[a] SPENCER (1952).

Revised manuscript received: March 1973

53-E. Abundance in Common Igneous Rocks

The distribution of iodine in igneous rocks is likely to be controlled by several factors, including its large ionic radius (CORRENS, 1956), its possible occurrence in melts in non-ionic form (KOGARKO and GULYAYEVA, 1965) and its possible chalcophilic character (RANKAMA and SAHAMA 1950).

The figures for iodine content in intrusive rocks are recorded in Table 53-E-1 but from these few data very little information can be derived. It is apparent that there is little variation in iodine content with differing silica content, the values for ultramafic rocks being very similar to those for the granites. One interesting fact, which has emerged from the work of BECKER *et al.* (1968), is the comparative enrichment of iodine in carbonatites.

The data for iodine in extrusive rocks, mainly derived from the work of YOSHIDA *et al.* (1971), are recorded in Table 53-E-2; there appears to be little variation in the iodine content of the various rock types. The values for obsidians quoted by BECKER

Table 53-E-1. *Iodine in intrusive rocks (including carbonatites)*

Sample	Source	I (ppb)	Method	Reference
Granite	Llano County, Texas (U.S.A.)	140	C	KURODA and CROUCH (1962)
Granite (G-1)	Westerley, Rhode Island (U.S.A.)	170	C	KURODA and CROUCH (1962)
Granite (N.B.S. 789)	Chelmsford (U.S.A.)	150	C	KURODA and CROUCH (1962)
Granite	—	360	W	SCHNEIDER and MILLER (1965)
Granite	Fredrikshald (Norway)	200	W	VON FELLENBERG and LUNDE (1926, 1927)
Quartz syenite	Vermontville, New York (U.S.A.)	250	C	KURODA and CROUCH (1962)
Nepheline syenite	Bancroft, Ontario (Canada)	330	C	KURODA and CROUCH (1962)
Larvikite	Larvik (Norway)	300	W	VON FELLENBERG and LUNDE (1926, 1927)
Porphyrite (2)	—	200—620	W	SCHNEIDER and MILLER (1965)
Hydrothermally-altered porphyrite (2)	—	320—2,720	W	SCHNEIDER and MILLER (1965)
Basalt ("diabase")	Comerset County, New Jersey (U.S.A.)	160	C	KURODA and CROUCH (1962)

Table 53-E-1 (continued)

Sample	Source	I (ppb)	Method	Reference
Olivine gabbro	Cripple Creek, Colorado (U.S.A.)	140	C	KURODA and CROUCH (1962)
Labradorite rock	Ekersund (Norway)	230	W	VON FELLENBERG and LUNDE (1926, 1927)
Eclogite	California (U.S.A.)	60	N/R	BECKER *et al.* (1968)
Eclogite	Germany	180	N/R	BECKER *et al.* (1968)
Peridotite (PCC-1)	Cazaddero, California (U.S.A.)	110	N/R	BECKER *et al.* (1968)
Peridotite	New Guinea	70	C	YOSHIDA *et al.* (1971)
Peridotite, interior	Nodule, Hualalai (Hawaii)	8.7	N/R	BECKER *et al.* (1968)
Peridotite, lava crust	Nodule, Hualalai (Hawaii)	300	N/R	BECKER *et al.* (1968)
Mica peridotite	Vermont (U.S.A.)	660	N/R	BECKER *et al.* (1968)
Peridotite with serpentine	Vermont (U.S.A.)	300	N/R	BECKER *et al.* (1968)
Harzburgite	Red-Hill (New Zealand)	78	C	YOSHIDA *et al.* (1971)
Dunite	Dun Mountain (New Zealand)	130	C	YOSHIDA *et al.* (1971)
Dunite	North Carolina (U.S.A.)	150	N/R	BECKER *et al.* (1968)
Dunite (DTS-1)	Twin Sisters, Washington (U.S.A.)	110	N/R	BECKER *et al.* (1968)
Dunite (NBS-4975)		80	N/R	BECKER *et al.* (1968)
Carbonatite	Tanganyika	770	N/R	BECKER *et al.* (1968)
Carbonatite	Arkansas (U.S.A.)	<2,000	N/R	BECKER *et al.* (1968)
Carbonatite with koppite	Norway	450	N/R	BECKER *et al.* (1968)
Carbonatite with perovskite	Arkansas (U.S.A.)	1,000	N/R	BECKER *et al.* (1968)

and MANUEL (1972), however, show much higher values than do other groups of rocks. These values for obsidians from the American continent are much higher than those of similar rock types from Japan recorded by YOSHIDA *et al.*

The values for Japanese volcanic rocks from YOSHIDA *et al.* (1971) range from 11 to 320 ppb with a mean value of 88 ppb; the distribution of iodine in these samples (Fig. 53-E-1) is a lognormal one.

YOSHIDA *et al.* found that there was no correlation between the distribution of iodine and any of the other halogens. However, these workers did find evidence of a geographical variation of I/Br ratios, the samples from the Circum-Japan Sea province having higher values for this ratio than other provinces.

Table 53-E-2. *Iodine in volcanic rocks (number of samples in brackets)*

Rock type	Locality	Range (ppb I)	I (ppb)	Method	Reference
Basalt	Daltenberg, Linz, Siebengebirge (Germany)		310	W	von Fellenberg and Lunde (1926, 1927)
Basalts (4)	Tokyo Metr. (Japan)	29—130	82	C	Yoshida *et al.* (1971)
Basalt	Sakasagawa, Itô, Shizuoka Pref. (Japan)		84	C	Yoshida *et al.* (1971)
Basalt	Fuji-san, Yamanashi Pref. (Japan)		32	C	Yoshida *et al.* (1971)
Basalt	Tokachi-dake, Hokkaidô (Japan)		72	C	Yoshida *et al.* (1971)
Basalts (3)	Aichi Pref. (Japan)	95—140	112	C	Yoshida *et al.* (1971)
Basalts (2)	Karatsu, Saga Pref. (Japan)	29; 46	38	C	Yoshida *et al.* (1971)
Basalts (2)	Shimane Pref. (Japan)	18; 42	30	C	Yoshida *et al.* (1971)
Basalt	Satsuma-iwô-jima, Kagoshima Pref. (Japan)		120	C	Yoshida *et al.* (1971)
Basalt (BCR-1)	Columbia River basalt, Bridal Veil, Washington (U.S.A.)		160	N/R	Becker and Manuel (1972)
Trachyandesitic basalt	Shidara, Aichi Pref. (Japan)		70	C	Yoshida *et al.* (1971)
Nepheline basalt	Hamada, Shimane Pref. (Japan)		45	C	Yoshida *et al.* (1971)
Basalt (dolerite)	Shidara, Aichi Pref. (Japan)		96	C	Yoshida *et al.* (1971)
Andesites (3)	Kogoshima Pref. (Japan)	42—82	59	C	Yoshida *et al.* (1971)
Andesite	Bonin Islands, Tokyo Metr. (Japan)		230	C	Yoshida *et al.* (1971)
Andesite	Hakone, Kanagawa Pref. (Japan)		91	C	
Andesites (2)	Gumma Pref. (Japan)	250; 260	255	C	Yoshida *et al.* (1971)
Andesite bomb	Tokachi-dake, Hokkaidô (Japan)		150	C	Yoshida *et al.* (1971)
Andesite	Yôtei-zan, Hokkaidô (Japan)		22	C	Yoshida *et al.* (1971)
Andesite	Nishino-shima, Oki, Shimane Pref. (Japan)		64	C	Yoshida *et al.* (1971)
Andesite (AGV-1)	Guano Valley, Oregon (U.S.A.)		270	N/R	Becker and Manuel (1972)
Glassy andesite	Satsuma iwô-jima, Kagoshima Pref. (Japan)		130	C	Yoshida *et al.* (1971)
Dacitic andesite	Sukomogawa, Hakone, Kanagawa Pref. (Japan)		70	C	Yoshida *et al.* (1971)
Dacitic andesite	Sambe-yama, Shimane Pref. (Japan)		32	C	Yoshida *et al.* (1971)
Trachyandesite	Sulphur Islands, Tokyo Metr. (Japan)		260	C	Yoshida *et al.* (1971)

Trachytes (2)	Shimane Pref. (Japan)	36; 80	58	C	YOSHIDA *et al.* (1971)
Dacite	Usu, Hokkaidô (Japan)	34; 51	43	C	YOSHIDA *et al.* (1971)
Dacite	Yake-yama, Akita Pref. (Japan)		220	C	YOSHIDA *et al.* (1971)
Dacite	Akagi-yama, Gumma Pref. (Japan)		66	C	YOSHIDA *et al.* (1971)
Liparites (3)	Tokyo Metr. (Japan)	11—73	49	C	YOSHIDA *et al.* (1971)
Liparite	Shimoda, Shizuoka Pref. (Japan)		140	C	YOSHIDA *et al.* (1971)
Rhyolite	Oki-Dôgo, Shimane Pref. (Japan)		26	C	YOSHIDA *et al.* (1971)
Alkali rhyolite	Madara-jima, Saga Pref. (Japan)		320	C	YOSHIDA *et al.* (1971)
Obsidian	Iceland		320	W	VON FELLENBERG and LUNDE (1926, 1927)
Obsidian	Hokkaidô (Japan)		20	C	YOSHIDA *et al.* (1971)
Obsidian	Imari, Saga Pref. (Japan)		57	C	YOSHIDA *et al.* (1971)
Obsidian	Hime-Shima, Ôita Pref. (Japan)		65	C	YOSHIDA *et al.* (1971)
Obsidian	Oregon (U.S.A.)		730	N/R	BECKER and MANUEL (1972)
Obsidian	Arizona (U.S.A.)		730	N/R	BECKER and MANUEL (1972)
Obsidian	Utah (U.S.A.)		540	N/R	BECKER and MANUEL (1972)
Obsidian	Mexico		1,900	N/R	BECKER and MANUEL (1972)
Glassy rock	Odi-Dôgo, Shimane Pref. (Japan)		110	C	YOSHIDA *et al.* (1971).

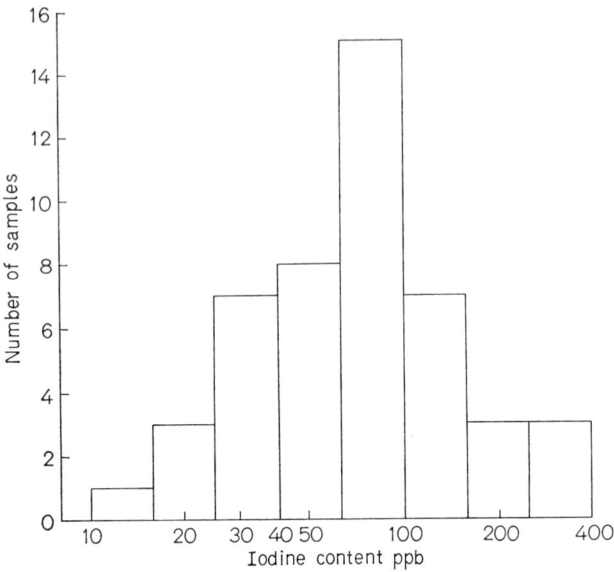

Fig. 53-E-1. Distribution of iodine in 47 Japanese volcanic rocks (from YOSHIDA *et al.*, 1971)

YOSHIDA *et al.* also carried out some work on the leachable iodine content of volcanic rocks; they found that in general the quantity of leachable iodine was small, being normally less than 20% of the total.

KOGARKO and GULYAYEVA (1965) determined iodine in several rocks from three intrusive phases of the Lovozero alkali massif and found the distribution of iodine to be fairly uniform, ranging from 300 to 1,300 ppb, the weighted average for the massif being 500 ppb. This uniformity of the distribution of iodine is in marked contrast to the other halogens (KOGARKO and GULYAYEVA, 1965) and they concluded that the distribution of iodine in these rocks was not markedly influenced by chemical composition. However, they do note the higher concentration of iodine in the silica-deficient, alkali, chlorine- and bromine-rich tawites and poikilitic sodalite syenites of Phase III (1,000 and 1,200 ppb I).

In view of the values for iodine recorded in the Tables 53-E-1 and 53-E-2, the average values for the various rock types would appear to be lower than the value of 500 ppb quoted by TUREKIAN and WEDEPOHL (1961). It is probable that the average value for most igneous rock groups is within the range 80—150 ppb, while the average value for alkalic rocks is possibly somewhat higher.

Revised manuscript received: March 1973

53-F. Behavior in Magmatogenic Processes (Volcanic Gases, Sublimates, and Hot Spring Waters)

In residual magmatic liquids iodine appears to be relatively enriched (see pegmatite minerals of Table 53-D-1).

HONDA *et al.* (1966) studied the iodine and chlorine abundance of gases from eight active volcanoes in Japan and found the iodine content varied appreciably both geographically and with temperature (Fig. 53-F-1). The increased iodine content with temperature conflicts with the evidence of BASHARINA (1965) who found in a study of gases from Kamchatka volcanoes that iodine decreased markedly with increasing temperature, being undetectable in some high temperature gases.

The I/Cl ratio of volcanic gases was observed by HONDA *et al.* (1966) to vary within the range 10^{-3} to 10^{-5} (Table 53-F-1), this ratio is different for different volcanoes, but is characteristic of each volcano. HONDA *et al.* (1966) found that the I/Cl ratio for the fumaroles at Showashinzan remained constant over several hours despite marked fluctuations in the iodine content.

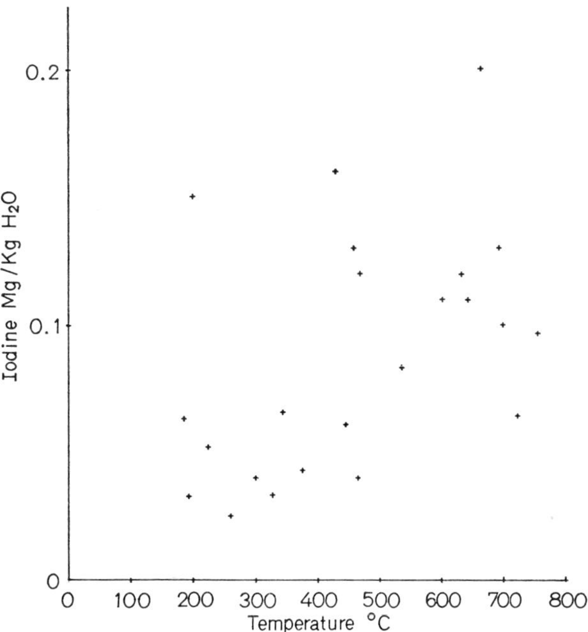

Fig. 53-F-1. A plot of iodine content *versus* temperature for volcanic gases of Showashinzan (from HONDA *et al.*, 1966)

Table 53-F-1. *Iodine in volcanic gases, sublimates and hot spring waters associated with volcanic areas*

Sample	Temperature (°C)	ppm I (in water)	I/Cl × 10^4	Reference
Gases:				
Fumerole, Showashinzan (Japan)	759—187	0.025—0.20 (C) Mean (23) 0.088	0.66—2.9 1.25	HONDA *et al.* (1966)
Fumerole, Iwatesan (Japan)	206	0.053—0.19 (C) Mean (3) 0.13	0.17—0.59 0.44	HONDA *et al.* (1966)
Fumerole, Issaikyoyama (Japan)	213—141	0.030, 0.087 (C) Mean (2) 0.059	2.9, 4.1 3.5	HONDA *et al.* (1966)
Fumerole, Nasudake (Japan)	489—97	0.16—3.9 (C) Mean (17) 0.88	0.77—13.0 8.5	HONDA *et al.* (1966)
Fumerole, Yakedake (Japan)	168; 163	0.58, 0.93 (C) Mean (2) 0.76	3.4—7.4 5.4	HONDA *et al.* (1966)
Fumerole, Kuji-Ioyama (Japan)	508—96	0.13—1.30 (C) Mean (10) 0.87	1.2—9.7 2.6	HONDA *et al.* (1966)
Fumerole, Kirishima Ioyama (Japan)	220—121	0.036—0.84 (C) Mean (8) 0.45	0.35—3.9 2.5	HONDA *et al.* (1966)
Fumerole, Miharayama (Japan)	392—388	0.045—1.5 (C) Mean (4) 1.19	2.2—7.8 6.0	HONDA *et al.* (1966)
Sheveluch, Kamchatka (U.S.S.R.)	240 260 360	1.2 2.1 0.2	2.9 4.2 0.5	BASHARINA (1965)
Eliuchevskoi volcano, Kamchatka (U.S.S.R.)	220 360	12.8 0.2	19.0 0.48	BASHARINA (1965)
Sublimates:				
Alkali chloride, Miyake-zima volcano, Tokyo Metr. (Japan)		—	0.01	YOSHIDA *et al.* (1965)
Alkali chloride, Iwō-zima volcano, Kagoshima Pref. (Japan)		—	0.01; 0.02	YOSHIDA *et al.* (1965)
Ammonium chloride, Syōwa-sinzan, Hokkaido (Japan)		—	0.05; 0.06	YOSHIDA *et al.* (1965)
Ammonium chloride, Mt. Iwate, Iwate Pref. (Japan)		—	0.01	YOSHIDA *et al.* (1965)
Ammonium chloride, Miyake-zima volcano Tokyo Metr. (Japan)		—	0.00—0.56	YOSHIDA *et al.* (1965)
Ammonium chloride, Tokati-dake. Hokkaido (Japan)		—	0.044	YOSHIDA *et al.* (1965)

Table 53-F-1 (continued)

Sample	Temperature	ppm I (in water)	I/Cl × 10^4	Reference
Hot spring waters:				
Nasudake area (Japan)		0.0035—0.095 Mean (10) 0.045	0.34—41 6.37	HONDA *et al.* (1966)
Kuju-Ioyama area (Japan)		0.0026—0.23 Mean (9) 0.047	0.51—5.6 1.71	HONDA *et al.* (1966)
Kirishima-Ioyama area (Japan)		0.000—0.069 Mean (5) 0.019	0.66—2.9 1.59	HONDA *et al.* (1966)

In a study of hot spring waters associated with the volcanic areas studied, HONDA *et al.* (1966) found the I/Cl ratio to be very similar to that of the gases to which they are genetically related (Table 53-F-1).

Alkali and ammonium chlorides occurring as volcanic sublimates were found by YOSHIDA *et al.* (1965) (Table 53-F-1) to be low in iodine and this is in accord with their observations of sublimates produced by heating volcanic rocks.

Revised manuscript received: March 1973

53-G. Behavior during Rock-Weathering; Soils

There is little information on the behavior of iodine during weathering of rocks. According to GOLDSCHMIDT (1954) and RANKAMA and SAHAMA (1950), weathering of rocks results in the release of much of their iodine which forms soluble compounds.

KONOVALOV (1959) found that rivers draining Tertiary marine sediments have higher iodine contents than rivers draining other areas and this was considered to be due to iodine being easily leached from the marine sediments.

There is a very marked increase in the iodine content of soils as compared to the rocks from which they derive (GOLDSCHMIDT, 1954; VINOGRADOV, 1959) (Tables 53-G-1 and 53-G-2). Many authors have suggested that much of the iodine in soils is derived from atmospheric sources (GOLDSCHMIDT, 1954; RANKAMA and SAHAMA, 1950; VINOGRADOV, 1959), while another major source of soil iodine is that supplied by plant remains (SHACKLETTE and CUTHBERT, 1967).

A fairly large volume of data has been recorded on the iodine content of soils, quite a large percentage of this data deriving from the U.S.S.R. Many of the workers have noted the strong correlation of iodine content with that of humus (SINITSKAYA

Table 53-G-1. *Iodine content of soils*

Sample locality	Soil type	ppm I (Method)	Reference
Russian Plain	Tundra	12.0	VINOGRADOV (1959)
	Podzolic	2.5	
	Grey forest	2.6	
	Chernozem, Steppe and chesnut	5.3	
	Serozems	2.5	
	Subtropic red	10.0	
Moldavia	Eluvial sands and suspensions	0.85	IRINEVICH *et al.* (1970)
	Clays	3.0	
	Chernozems	3.7	
	Loess-like clay	1.5—3.2	
	Silty loams	2.0	
Sverdlovsk	Dark grey and podzol chernozems	2.6	PROSKURYAKOVA *et al.* (1969b)
	Meadow chernozems	2.8	
	Peat humus	6.7	
Armenia	Chernozems	3.8	SAFRAZBEKYAN (1970)
	Chesnut	2.1	
Norway	Coastal area	18.0 (N/R)	LÅG and STEINNES (1971)
	Inland area	7.3 (N/R)	
U.S.A.	From downwind side of busy road	3.0 (C)	VOUGHT *et al.* (1970)
	From upwind side of busy road	4.5 (C)	

Table 53-G-2. *Distribution of iodine in soil profiles* (from VINOGRADOV, 1959)

Soil type	Horizon	ppm I (dry soil)	C (%)
Mountain tundra	A_0	12.7	17.8
	B	1.8	5.48
	C	3.0	2.50
Loamy podzolic	A_1	4.7	—
	B	1.8	—
	C	0.56	—
		2.2	—
Grey forest	A	6.7	3.73
	A_1	4.2	1.0
	B	2.0	0.39
	C	1.2	0.03
		0.67	0.0
Chernozem	A_1	6.3	6.3
	A_2	6.8	4.15
	B	2.2	1.83
	C	5.5	3.69
Light chesnut	A	6.2	1.81
	B	9.8	0.0
	C	3.5	0.0
		1.8	0.0
Red soil	A	11.8	10.8
	B	7.2	0.0

1969; PROSKURYAKOVA *et al.*, 1969b; IRINEVICH *et al.*, 1970). Low humus soils such as grey forest or podzolic soils are generally iodine-poor, while high humus soils such as the chernozems and peaty soils are generally iodine-rich.

Silty and clay soils appear to be enriched in iodine (PROSKURYAKOVA *et al.*, 1969a) and DE *et al.* (1971) have found that the clay fractions of soil fix iodide, a feature which is most marked for illite. VINOGRADOV (1959) suggests that the iodine content of soils depends primarily on their content of fine fractions and only secondly on their organic content.

Several workers have commented on the low percentage of water-soluble iodine in soils (MAGOMEDOVA, 1970; POKATILOV, 1968); however, VINOGRADOV (1959) claims that the greater part of soil iodine is water-soluble.

An acid medium in soils favours leaching of iodine (SAZONOV, 1969), whereas carbonate horizons in soils tend to act as natural barriers to iodine migration (ZYRIN and IMADI, 1967).

LÅG and STEINNES (1971) found in a study of Norwegian soils that the iodine content decreases with increasing distance from the sea. This was also found by IRINEVICH *et al.* (1970) to occur in Moldavian soils and by ZYRIN and IMADI (1967) in soils of the Russian plain and the Crimea.

A marked increase of the iodine content of soils along the upwind side of a main road was observed by VOUGHT *et al.* (1970) and was considered to be due to contamination from automobile exhaust gases.

In the vertical soil profile, the upper layers of the A horizon are richer in iodine than lower layers (VINOGRADOV, 1959) and this feature is again closely connected with the organic content of the soils. In water-rich environments, however, the iodine content is frequently greater in the lower soil horizons, a feature often encountered in podzolic soils (VINOGRADOV, 1959).

Revised manuscript received: March 1973

53-H/I. Chemistry and Abundance in Atmosphere (I) and Natural Waters (II)

I. Atmosphere

Several workers have commented on the enrichment of iodine in the atmosphere compared to other halogens (GOLDSCHMIDT, 1954; RANKAMA and SAHAMA, 1950). The chemical form of atmospheric iodine has been the subject of much discussion; RANKAMA and SAHAMA (1950) suggested that it is present in a free state, whereas GOLDSCHMIDT (1954) thought that an equilibrium existed between free iodine vapor and iodine adsorbed on dust particles, and CORRENS (1956) concluded that some gaseous iodine exists in the atmosphere.

It has been generally accepted that the oceans are a major source of atmospheric iodine; other sources are volcanic gases (MIYAKE and TSUNOGAI, 1963) and rotting bio-materials (CHAMBERLAIN and CHADWICK, 1966). It has also been observed (VOUGHT *et al.*, 1970) that some iodine in urban atmosphere may be derived from combustion of fossil fuels and this has been corroborated by SCHROLL and KRACHSBERGER (1970) who found higher iodine contents in the atmosphere of Vienna than in that of the neighboring countryside, especially in winter.

The high I/Cl ratio of the atmosphere has been explained by MIYAKE and TSUNOGAI (1963) as resulting from the action of ultraviolet light on iodide ions occurring in the surface layers of the ocean; this action causes the conversion of the iodide to elemental iodine which is then lost to the atmosphere.

Another possible mechanism for iodine injection into the atmosphere is that organically-bound iodine occurring in the surface film of the ocean is incorporated into spray droplets which are carried into the atmosphere (DUCE, 1967).

In a series of papers (DUCE *et al.*, 1963, 1965, 1966; WINCHESTER and DUCE, 1966, 1967; LININGER *et al.*, 1966), the iodine content of aerosols and precipitation in Hawaii, Alaska (during winter) and Massachusetts were reported (Table 53-H/I-1). These workers concluded that iodine was enriched, relative to chlorine, in all samples of aerosols and that the I/Cl ratio increases with increasing altitude and distance from the sea. They comment also that there is a great similarity in the iodine content of aerosols in the three areas studied.

From a series of studies conducted over Hawaii, DUCE *et al.* (1963, 1965) concluded that over half of the atmospheric iodine was in gaseous form.

PASLAWSKA and OSTROWSKI (1968) found a large decrease in the iodine content of air with increasing distance from the Baltic Sea, 65—80% of the iodine being lost at 3.5 km from the sea. They also found that the iodine content of air was affected by wind direction, and was lower during and immediately after precipitation.

Table 53-H/I-1. *Iodine in the atmosphere and some natural waters*

Sample	µg/l I (except aerosols)	Method	Reference
Aerosols			
Hawaii	≈2 ng/m³	(N/R)	Lininger *et al.* (1966)
Barrow, Alaska	0.3—9.4 ng/m³	(N/R)	Duce *et al.* (1966)
Cambridge, Massachusetts (U.S.A.)	2—10 ng/m³	(N/R)	Lininger *et al.* (1966)
Rain water			
Hawaii	1.01—15.3 Mean (85) 4.94	(N/R)	Duce *et al.* (1966)
Alaska	0.60—6.2 Mean (31) 1.77	(N/R)	Duce *et al.* (1966)
Japan	Mean (300) 1.8	(W)	Sugawara (1967)
River water			
Japan	Mean (43) 2.2	(W)	Sugawara (1967)
U.S.S.R.	3.3—42.4 Mean (25) 11.0	(?)	Konovalov (1959)
Moscow region, U.S.S.R.	4—18	(?)	Obukhova and Golobov (1969)
Opole region, Poland	1.23—9.79	(?)	Ewy *et al.* (1968)
Saale River, Germany	4.54	(?)	Correns (1956)
Potomac River, Washington (U.S.A.)	2.2	(C)	Vought *et al.* (1970)
Lake water			
Freshwater, Central Kazakhstan	Mean (26) 160	(C)	Mun and Basilevich (1963)
Saline water, Central Kazakhstan	Mean (38) 680	(C)	Mun and Basilevich (1963)
Thermal and mineral springs			
Japan (average composition)	730	(W)	Sugawara (1967)
Saratoga Springs, New York (U.S.A.)	310—3,760 Mean 1,270	(?)	Correns (1956)
Beppu spa, Kyushu (Japan)	30—290 Mean (89) 96	(?)	Kikkawa and Shiga (1966)
Shirahama, Wakayama Pref. (Japan)	123—158 Mean (7) 140	(?)	Kikkawa and Shiga (1966)
Obama, Nagasaki Pref. (Japan)	210—329 Mean (17) 277	(?)	Kikkawa and Shiga (1966)
Subsurface brines and formation water			
Oilfield, Andarko basin, Oklahoma (U.S.A.)	23—1,400 Mean: 370	W	Collins (1969)

Table 53-H/I-1 (continued)

Sample	µg/l I	Method	Reference
Oilfield, Mississippi and Alabama (U.S.A.)	2—65	W	Collins *et al.* (1967)
Oil and gas field, California (U.S.A.)	21—23 Mean: 22.5	?	White *et al.* (1963)
Oilfield, Poland	Mean: 199	?	White *et al.* (1963)
Gasfield, Japan	Mean: 132	?	White *et al.* (1963)
Devonian limestone, Michigan (U.S.A.)	Mean: 8	?	White *et al.* (1963)
Lower Devonian sandstone, Michigan (U.S.A.)	Mean: 40	?	White *et al.* (1963)
Salt deposit, Searles Lake California (U.S.A.)	25, 29 Mean: 27	?	White *et al.* (1963)
Sedimentary basin, W. Canada	<1—39 Mean: 15	W	Billings *et al.* (1969)

II. Natural Waters

a) Rain Water (for data see Table 53-H/I-1)

From a study of the halogen content of rainwater over Hawaii, Duce *et al.* (1963, 1965) found that the iodine content, in common with the other halogens, decreases with increasing altitude. However, the iodine content decreases to a lesser extent than chlorine and thus the I/Cl ratio increased with altitude. The I/Cl ratio of the Hawaiian samples is 500—1,000 times greater than that of sea water. Sugawara (1967) quotes an "enrichment coefficient" of iodine against chlorine of 2,400.

Dean (1963) found that about 40% of the iodine in New Zealand rain-water was organically-bound.

b) Surface Drainage and Ground Water (for data see Table 53-H/I-1)

From the recorded analyses of rain and river waters, it is apparent that there is a similarity in their iodine contents. Correns (1956) considered that an appreciable quantity of the halogens in river waters resulted from atmospheric precipitation.

Sugawara (1967) has calculated the iodine input of the Japanese land-surface *via* precipitation, dry fallout and thermal spring waters. A comparison with the iodine content of Japanese rivers shows that over 50% of the iodine input is not washed out *via* the surface drainage. It would appear that much of this "excess" iodine is absorbed by plants (Shacklette and Cuthbert, 1967) and by soil organic matter (Vinogradov, 1959).

In his study of the micronutrients of Russian rivers, Konovalov (1959) found that waters draining areas of Tertiary marine deposits had comparatively high iodine contents and noted high iodine contents in rivers draining iodine-rich chernozem soils.

The iodine content of ground water in the Russian plain and Crimea areas has been shown by ZYRIN and IMADI (1967) to be highest in the coastal areas and lowest in areas of black or grey forest soils.

RYKHILOV (1970) found no regular seasonal variation in the distribution of iodine in two Russian rivers, but KAMENEV (1968) found much higher iodine contents in Russian rivers in summer than in spring or winter.

It has been shown by VOUGHT *et al.* (1970) that sewage outfall into rivers results in an increase in their iodine content and similarly they suggest that waters containing plant debris may be expected to have higher iodine contents.

From a study of fresh water and salt lakes in Central Kazakhstan, MUN and BAZILEVICH (1963) found that saline waters contained 3—5 times as much iodine as fresh waters. In addition, these workers observed that soda-rich bicarbonate waters contained more iodine than the chloride-sulfate waters.

SUGAWARA and TERADA (1957) found that over 90% of the total iodine in river waters was in the form of the iodide ion.

c) Thermal and Mineral Spring Waters (for data see Table 53-H/I-1)

The iodine content of thermal and mineral waters varies over a wide range and the chemistry of hot spring waters can be greatly modified by secondary processes such as mixing with ground or sea water. However, KIKKAWA and SHIGA (1966) have pointed out that there is a basic similarity in iodine content of hot spring waters from three regions of Japan.

d) Sea Water

1. Abundance and Distribution

Iodine occurs in sea water essentially in the ionic states of iodide and iodate (SUGAWARA and TERADA, 1957, 1958; JOHANNESSON, 1958), but smaller amounts may exist in organo-iodine compounds (SUGAWARA and TERADA, 1957) especially in the surface film (DUCE, 1967).

The distribution of iodine in sea water has been fairly extensively studied by several workers. One of the first of these studies was that of SUGAWARA and TERADA (1957) who reported on the iodine distribution of the Western Pacific Ocean. They obtained an average value of 39 μg/l, and also observed that the percentage of iodide to total iodine varied from 0 to 100%.

BARKLEY and THOMPSON (1960) obtained values for Pacific and Arctic Ocean waters which were higher than those of SUGAWARA and TERADA; their average value for surface waters was 60 μg/l and for deeper waters 63 μg/l. MIYAKE and TSUNOGAI (1966) also obtained values which were higher than those of SUGAWARA and TERADA, 50 μg/l for the surface waters and 60 μg/l for deeper waters.

The ratio iodate/total iodine was found by BARKLEY and THOMPSON to be fairly constant for Pacific Ocean waters, being generally in the range $\frac{1}{3}$ to $\frac{2}{3}$. The Arctic Ocean samples, however, showed a marked variation, with all of the iodine of the surface layers being in the form of iodide, while in waters below 500 m iodate was the only iodine species. MIYAKE and TSUNOGAI (1966) observed that the iodide content of surface waters was approximately 10 μg/l and decreased with depth.

Table 53-H/I-2. *Iodine distribution in Pacific Ocean water* (from TSUNOGAI, 1971)(Method: C)

Depth (m)	Number of samples	Iodide (μg/l)	Iodate (μg/l)	Total Iodine (μg/l)
(i) Data for samples obtained from stations where surface temperature is greater than 15 °C				
0—100	32	9.4	33.5	42.9
100—200	19	6.0	39.0	44.9
200—500	30	3.2	45.9	49.1
500—1,000	27	3.2	47.3	50.5
1,000—2,000	24	2.8	51.5	54.3
>2,000[a]	43	1.8	52.4	54.2
(ii) Data for samples obtained from stations where surface water is colder than 15 °C				
0—100	15	4.3	41.0	45.3
100—200	6	3.7	44.4	48.1
200—500	12	2.9	45.9	48.9
500—1,000	12	2.4	52.0	54.4
1,000—2,000	11	1.9	51.9	53.8
>2,000[a]	14	2.0	52.0	54.1

[a] Excluding bottom water.

In their recent study of the iodine content of surface Pacific Ocean waters (Table 53-H/I-2), TSUNOGAI and HENMI (1971) found an average value of 47 μg/l, while for waters in excess of 1,000 m an average value of 52 μg/l was obtained. The iodide content of sea water was generally greatest in surface waters, with a mean value of 9.3 μg/l. Higher iodide contents were found in warm waters than in cold, the highest values being found in equatorial waters which averaged 16.5 μg/l, with the highest value 26.6 μg/l. However, TSUNOGAI and HENMI found that in surface Antarctic sea water, the total iodine content is present as the iodide ion.

From work on iodine in deep ocean waters, TSUNOGAI (1971) has found that the bottom-most layers of these deep waters are much enriched in iodide, the bottom 30 m containing 5.3 μg/l, whereas waters in regions above this contained only 1.6 μg/l.

2. The Chemistry of Iodine in Seawater

The iodate ion is the thermodynamically stable state of iodine in normal sea water (SILLÉN, 1961; TSUNOGAI and SASE, 1969) and several works on iodine in sea water have confirmed that the iodate species is the most abundant in this medium. However, the iodide ion is almost always present in significant amounts, particularly in surface layers (TSUNOGAI and HENMI, 1971).

The oxidation of iodide to iodate is accelerated in the surface layers of the sea by the action of sunlight (TSUNOGAI and SASE, 1969). Also, due to solar radiation, iodide is possibly selectively lost to the atmosphere as iodine vapor (MIYAKE and TSUNOGAI, 1963).

The iodide ion is the form in which iodine is utilised by plankton (SUGAWARA and TERADA, 1961) and algae (SHAW, 1962). TSUNOGAI and SASE (1969) have found that iodate can be reduced to iodide by organisms utilising the enzyme nitrate reductase and they have suggested that it is the action of such organisms which results in the constant supply of the iodide ion in surface waters.

e) Subsurface Brines

Many subsurface brines are enriched in iodine, particularly oilfield brines, which can be used as commercial sources (COLLINS, 1966). The enrichment of iodine in oilfield brines is thought to be linked with the high iodine content of marine organisms and organic matter (COLLINS, 1969; BENTOR, 1969). It is possible also that desorption from clays is important (KOZIN, 1960; HITCHON *et al.*, 1971). WHITE *et al.* (1963) have found that oilfield brines rich in sodium and chloride are characterised by a high I/Cl ratio, whereas those with high calcium contents have a lower I/Cl ratio.

Revised manuscript received: March 1973

53-K. Abundance in Sediments and Sedimentary Rocks

I. Sedimentary Rocks

a) Clastic Rocks (for data see Table 53-K-1)

BECKER *et al.* (1972) suggest that previous estimates for iodine abundance are too high for sandstones and too low for shales and limestones. They found sandstones to have very low iodine contents, apart from one containing argillaceous material. This high iodine content of an argillaceous limestone may, they suggest, reflect adsorption of iodine by clays.

The high iodine content of Kimmeridge shale samples from England was shown by COSGROVE (1970) to correlate strongly with the content of organic material. Several of the samples analysed were found to contain remains of marine plants and COSGROVE (1970) suggested that the high iodine content may be derived from this source.

A similar source has been suggested for the high iodine content of shales and limestones from the Andarko basin, Oklahoma (COLLINS *et al.*, 1971). These samples, some of which have been found to contain algal remains, are associated with oil field brines which contain up to 1,400 ppm iodine (COLLINS, 1969).

A correlation of iodine with carbonaceous material in shales and clays has been found by ITKINA and LYGALOVA (1964), but these workers comment that this correlation is not as marked as that of bromine with organic carbon.

b) Evaporites and Saltpetre (Niter)

There is little information on the iodine content of evaporite deposits, most of that which is available was summarized by CORRENS (1956); from this data it is apparent that the iodine concentration is very small.

The Chile saltpetre deposits have long been known to contain appreciable quantities of iodine, much of which is present as iodate (GOLDSCHMIDT, 1954; CORRENS, 1956), and the origin of these deposits has been the subject of much discussion (GOLDSCHMIDT, 1954). More recently, JOHANNESSON and GIBSON (1962) and CLARIDGE and CAMBELL (1968) have found iodate-rich nitrate deposits in Antarctica. The latter workers have suggested that these deposits are genetically related to those of the Atacama desert of Northern Chile. CORRENS (1956) and CLARIDGE and CAMBELL (1968) have suggested an atmospheric origin for the nitrate deposits. Much of the iodine could have been derived from the oceans and been oxidized in the upper atmosphere (CLARIDGE and CAMBELL, 1968) while volcanic exhalation may also have contributed some iodine.

The virtually unique nitrate deposits (smaller ones exist in California) are thought by MUELLER (1968) to be due to the unique "climatic asymmetry" of the Atacama region. According to this worker the nitrates originated as brine solutions in the

Table 53-K-1. *Iodine in sedimentary rocks*

Sample (number of samples)	ppm I	Method	Reference
Sandstones			
Sandstone, Keuper	0.8	(W)	von Fellenberg (1924)
Sandstone, Keuper	0.02		Wilke-Dörfurt (1927)
Sandstones(2)(Switzerland)	0.51	(W)	von Fellenberg (1924)
Quartzite, Taunus (Germany)	0.5—0.8		Cauer (1930)
Sandstones(3), Bavly formation, Bashkiria (U.S.S.R.)	0.5	(W)	Itkina and Lygalova (1964)
Sandstones(8), Carboniferous, Bashkiria (U.S.S.R.)	1.5	(W)	Itkina and Lygalova (1964)
Sandstone (white), Klondyke, Mo. (U.S.A.)	0.068	(N/R)	Becker *et al.* (1972)
Sandstone (red), Potsdam, New York (U.S.A.)	0.14	(N/R)	Becker *et al.* (1972)
Greywacke, Eifel Mts. (Germany)	0.084	(N/R)	Becker *et al.* (1972)
Argillaceous sandstone, Portageville, New York (U.S.A.)	37.6	(N/R)	Becker *et al.* (1972)
Shales and Clays			
Shale, Hunsrück (Germany)	0.2—2.2		Cauer (1930)
Shale, Taunus (Germany)	0.3—0.8		Cauer (1930)
Flysch shale	0.2	(W)	von Fellenberg (1924)
Shale, Neocomian	0.78	(W)	von Fellenberg (1924)
Clays and argillites(23), Bavly formation, Bashkiria (U.S.S.R.)	0.6	(W)	Itkina and Lygalova (1964)
Clays and argillites(33), Carboniferous (U.S.S.R.)	1.7	(W)	Itkina and Lygalova (1964)
Clays and argillites(7), Domanik formation (U.S.S.R.)	1.7	(W)	Itkina and Lygalova (1964)
Shales(7), carbonaceous, Carboniferous (U.S.S.R.)	6.8	(W)	Itkina and Lygalova (1964)
Marls(15), Domanik formation, Bashkiria (U.S.S.R.)	1.2	(W)	Itkina and Lygalova (1964)
Shales(16), Kimmeridge, Dorset (England)	17.0	(X)	Cosgrove (1970)
Shales(4), Paleozoic, Kingfisher Co., Oklahoma (U.S.A.)	7.08	(N/R)	Collins *et al.* (1971)
Argillaceous shale, Rochester, New York (U.S.A.)	13.0	(N/R)	Becker *et al.* (1972)
Calcareous shale, Lima, New York (U.S.A.)	38.0	(N/R)	Becker *et al.* (1972)
Pyritic argillite, Transvaal (South Africa)	7.0	(N/R)	Becker *et al.* (1972)
Argillites, from freshwater origin	0.9		Gulyayeva and Itkina (1962a)
Carbonate rocks			
Limestones (Switzerland)	0.25—0.75	(W)	von Fellenberg (1924)
Dolomite, Triassic	1.0	(W)	von Fellenberg (1924)

Limestone, Upper Jurassic, Württemberg (Germany)	6.6		Wilke-Dörfurt (1927)
Limestones(6), Domanik formation (U.S.S.R.)	1.3	(W)	Itkina and Lygalova (1964)
Argillaceous limestones(15), Domanik formation (U.S.S.R.)	2.2	(W)	Itkina and Lygalova (1964)
Limestones(6), Paleozoic, Kingfisher Co., Oklahoma (U.S.A.)	4.2	(N/R)	Collins *et al.* (1971)
Limestone (gray), Buffalo, New York (U.S.A.)	3.0	(N/R)	Becker *et al.* (1972)
Argillaceous limestone, Trenton falls, New York (U.S.A.)	8.0	(N/R)	Becker *et al.* (1972)
Cherty limestone, Le Roy, New York (U.S.A.)	23.0	(N/R)	Becker *et al.* (1972)
Dolomitic limestone, Rochester, New York (U.S.A.)	5.6	(N/R)	Becker *et al.* (1972)
Chalk, Dover (England)	29.0	(N/R)	Becker *et al.* (1972)
Limestone	2.67	(C)	Grimaldi and Schnepfe (1971)
Phosphates			
Podolian phosphate	0.8		Wilke-Dörfurt (1927)
Phosphate, Tunis	2.88		Wilke-Dörfurt (1927)
Phosphate, Algeria	2.87		Wilke-Dörfurt (1927)
Phosphate, Tennessee (U.S.A.)	2.6		Wilke-Dörfurt (1927)
Phosphates(3), Florida, (U.S.A.)	28.7		Wilke-Dörfurt (1927)
Phosphate, Staffel, Lahn (Germany)	93.5		Wilke-Dörfurt (1927)
Phosphorite, Limburg (Germany)	280.0		Wilke-Dörfurt (1927)
Evaporites			
Anhydrites	0.0155—0.1365		Roeber (1938)
Rock salt	0—0.051		Roeber (1938)
Halites(4), differing geological ages	<2		Moore (1971)
Sylvine	upto 0.036		Roeber (1938)
Carnallite	upto 0.029		Roeber (1938)
Chile nitrate deposit	≈ 400		Tower and Brewer (1964)

semi-arid Andes. These brines accumulated in desert basins where more normal evaporite deposits occur and nitrate deposits were concentrated on more arid higher ground due to capillary migration.

II. Recent Sediments

a) Marine Sediments

Marine and oceanic sediments are very much richer in iodine than sedimentary rocks (SHISHKINA and PAVLOVA, 1965) (Table 53-K-2). The degree of enrichment of iodine in these sediments increases with increasing organic content and the degree of fineness of the sediments (SHISHKINA and PAVLOVA, 1965). BOJANOWSKI and PASLAWSKA (1970) suggests that the biological accumulation of iodine plays an important role in the accumulation of iodine in the sediments of the Baltic Sea.

Table 53-K-2. *Iodine in Recent sediments* (number of samples in brackets)

Sample	ppm I	Method	Reference
Marine sediments[a]			
Calcareous ooze (7), Indian Ocean	39.8	(W)	SHISHKINA and PAVLOVA (1965)
Calcareous ooze (4), Atlantic Ocean	37.5	(W)	SHISHKINA and PAVLOVA (1965)
Calcareous ooze (8), Mediterranean Sea	16.3	(W)	SHISHKINA and PAVLOVA (1965)
Red clays (4), Pacific Ocean	28.8	(W)	SHISHKINA and PAVLOVA (1965)
Gray clays (4), Pacific Ocean	247.2	(W)	SHISHKINA and PAVLOVA (1965)
Gray clays (2), Indian Ocean	112.8	(W)	SHISHKINA and PAVLOVA (1965)
Gray clays (4), Red Sea	35.5	(W)	SHISHKINA and PAVLOVA (1965)
Gray clays (2), Baltic Sea	101.1	(W)	SHISHKINA and PAVLOVA (1965)
Deep Sea sediments (4), Atlantic Ocean	36.7	(N/R)	BENNETT and MANUEL (1968)
Muds (5), Southern Baltic Sea	200.0	(C?)	BOJANOWSKI and PASLAWSKA (1970)
Muddy sands (3), Southern Baltic Sea	33.0	(C?)	BOJANOWSKI and PASLAWSKA (1970)
Sands (2), Southern Baltic Sea	2.9	(C?)	BOJANOWKI and PASLAWSKA (1970)
Sediments (8), South-western Barents Sea	405.0	(X)	PRICE *et al.* (1970)
Lake sediments[a]			
Muds, freshwater (12), Central Kazakhstan (U.S.S.R.)	6.1	(W)	MUN and BASILEVICH (1964)
Muds, saline lakes (19), Central Kazakhstan (U.S.S.R.)	5.3	(W)	MUN and BASILEVICH (1964)

[a] All superficial sediments.

From studies of superficial sediments in the Barents Sea, PRICE *et al.* (1970) found that the iodine content of the sediments depended greatly on the organic carbon content (Fig. 53-K-1), but very little on the degree of fineness of the sediment.

Organic-rich gray clays contain the highest iodine contents of the oceanic sediments (SHISHKINA and PAVLOVA, 1965), whereas the coarse-grained sands are lowest in iodine content (BOJANOWSKI and PASLAWSKA, 1970).

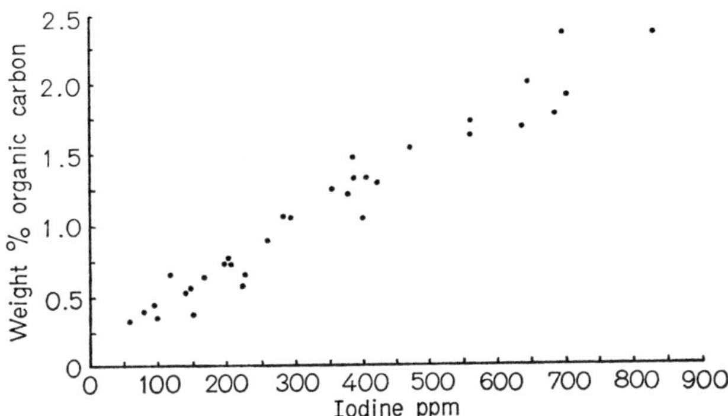

Fig. 53-K-1. A plot of iodine content *versus* organic carbon content for superficial sediments of the Southwestern Barents Sea (from PRICE *et al.*, 1970)

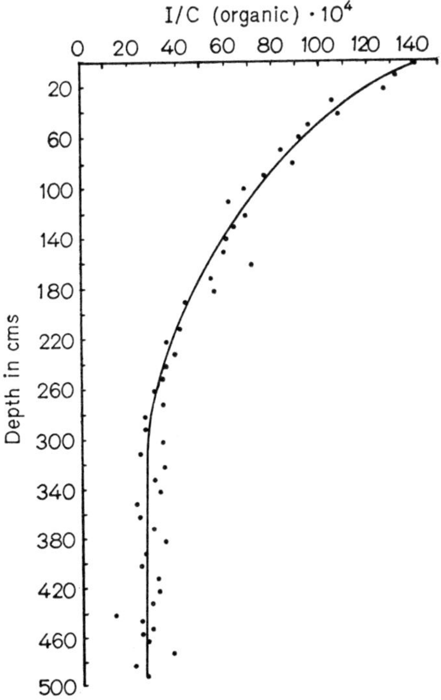

Fig. 53-K-2. The variation in I/C ratio with depth in sediments of a core sample from the Gulf of California (from PRICE *et al.*, 1970)

Several studies on the iodine content of marine sediments have described the distribution of the element in core samples; these investigations reveal that the iodine content descreases with depth from the surface. Shishkina and Pavlova (1965) found that at a depth of about 7 m the iodine content was halved. Bojanowski and Paslawska (1970) found that the decrease of iodine with depth parallelled that of organic carbon. However, Price *et al.* (1970) found that the ratio of I/C shows a marked decrease, there being a loss of iodine during burial of the sediments (Fig. 53-K-2).

The pore fluids of the sediments are enriched in iodine relative to normal sea water and can contain up to 38% of the iodine of the sediments (Shishkina and Pavlova, 1965); the iodine content of the pore liquids increases with depth of burial (Shishkina and Pavlova, 1965).

The degree of binding of iodine in dried marine sediments has been investigated by Walters and Winchester (1971) by studies of nuclear recoil reactions following neutron capture. They found that most of the iodine was surface-bound, being adsorbed on the surface or held in covalent bonds in association with carbon; of this surface-held fraction, 23% was extracted with organic solvents. These workers also found that some iodine was bound within the sediment grains, i.e. in lattice sites or inclusions.

b) Lake Sediments

In their study of lacustrine muds and pore solutions, Mun and Bazilevich (1964) found that iodine accumulates in the muds of both fresh and saline water lakes (Table 53-K-2). They found that as in marine sediments, there is a direct correlation of iodine content with organic carbon, in addition lacustrine muds show a decrease of iodine content with depth of burial, possibly due to release of iodine from decaying organic matter. The iodine contents of pore solutions show an increase with depth.

A correlation of iodine content of lake sediments with organic carbon content was found in several lakes of Northern England and Scotland by Pennington and Lishman (1971).

Sugawara *et al.* (1958) found that in some Japanese lake waters the iodide ion is captured by suspended floccula of hydroxides. These suspended particles tend to sink and accumulate in bottom sediments where a reducing environment can release the iodine into the lower layers of the lake water, or into the sediment pore liquids.

Revised manuscript received: March 1973

53-L. Biogeochemistry

Iodine has been described by GOLDSCHMIDT (1954) as a biophile element. Much of its geochemistry is connected with its involvement in biological processes (GOLDSCHMIDT, 1954). Its enrichment in organic-rich sediments and soils has been dealt with in Sections 53-G and 53-K. That it accumulates in peat and coal can be seen from data in Table 53-L-1.

Table 53-L-1. *Iodine content of plant groups and coals* (data on plants from SHACKLETTE and CUTHBERT, 1967)(Method: C)

Plant	Number of samples	ppm I (dry matter)	
		Range	Mean
Marine brown algae	4	55.0—8,800.0	2,488.7
Marine red algae	4	200.0—565.0	382.5
Marine flowering plants	2	50.0—55.0	52.5
Garden vegetables (dicots)	12	2.8—10.0	6.9
Fleshy fungi ("mushrooms")	8	5.3—9.5	6.7
Garden vegetables (monocots)	8	3.4—10.4	6.3
Lycopodium (club-moss)	3	4.8—7.5	6.0
Grasses	6	4.3—7.1	5.5
Ferns	7	3.8—7.1	5.4
Epiphytes (Spanish-moss)	5	4.0—7.0	5.0
Forbs, high saline soil	4	4.0—6.0	5.0
Green algae	3	3.0—6.2	4.5
Coniferous trees	11	2.9—6.9	3.9
Mosses and liverworts	5	3.1—4.0	3.3
Deciduous trees	22	1.1—6.2	2.7
Equisetum (horsetail)	3	2.0—3.5	2.7
Lichen	1	—	1.8

Coals	ppm I	Method	Reference
Coal, delta, lake, swamp facies (U.S.S.R.)	13.7 (160 in ash)	(W)	GULYAYEVA and ITKINA (1962b)
Coal, near-shore marine facies (U.S.S.R.)	9.5 (142 in ash)	(W)	GULYAYEVA and ITKINA (1962b)
"Coal", Kimmeridge, Dorset (England)	72.0	(X)	COSGROVE (1970)
Coals, Ruhr and Saxony (Germany, D.G.R.)	0.85—11.17		WILKE-DÖRFURT and ROMERSPERGER (1930)
Bitumen, marine (U.S.S.R.)	6.7 (587 in ash)	(W)	GULYAYEVA and ITKINA (1962b)

Crude petroleum is sometimes found to be enriched in iodine; KREJCI-GRAFF *et al.* (1968) found 0.1—0.8 ppm in samples from the Vienna basin. However, COLLINS and EGLESON (1967) found less than 0.05 ppm iodine in a petroleum sample from Oklahoma which was associated with brines containing upto 1,400 µg/l iodine. MAKSIMOVA (1962) found iodo-hydrocarbons in crude oils and solid bitumens.

Iodine is an essential element for mammals (BOWEN, 1966); it is concentrated mainly in the thyroid gland and to a lesser extent in bile, hair, and pituitary, salivary and mammary glands, and ovaries (UNDERWOOD, 1971; VOUGHT, 1970). Iodine is present in the thyroid hormones of man and its deficiency causes malfunction of the thyroid gland. Such deficiency can be associated with problems of lack of growth and reproduction, feeble-mindedness, deaf-mutism and with the development of goitre due to excessive enlargement of the thyroid gland (UNDERWOOD, 1971). As a precaution against the occurrence of thyroid deficiences in humans, iodine is added to table-salt in many countries. In addition, it is added to animal feeds and mineral licks (CHILEAN IODINE EDUCATIONAL BUREAU, 1956). An account of the role of iodine in human and animal metabolism, and the distribution and extent of goitrous areas of the world is given by UNDERWOOD (1971).

Its distribution in, and uptake by plants and animals has been fairly well studied in connection with its primary association with endemic goitre in man and animals (UNDERWOOD, 1971).

Many marine algae concentrate iodine greatly (VINOGRADOV, 1953; SHAW, 1962), values recorded for the brown algae (VINOGRADOV, 1953; SHACKLETTE and CUTHBERT, 1967) being among the highest for plant samples (Table 53-L-1). However, as pointed out by SHAW (1962) there is a very marked species variation, some of which is attributed by VINOGRADOV (1953) to the variation in depth of habit. SHAW (1962) states that only the iodide ion is taken up by marine algae and this is also said to occur with planktonic material (SUGAWARA and TERADA, 1961).

In the marine environment, iodine is also accumulated by diatoms, sponges, coelenterates and annelids (VINOGRADOV, 1953).

In terrestrial plants there are marked variations in iodine content. According to SHACKLETTE and CUTHBERT (1967) this variation is species-characteristic and appears to be largely unaffected by the variation in soil type. These authors have suggested that atmospheric iodine absorbed by leaves may be an important source of the element for terrestrial plants. Analyses of Spanish moss, a plant which has no roots and must rely on direct absorption from the atmosphere and rain water for its iodine intake, showed it to have similar iodine contents to that of other terrestrial plants (Table 53-L-1). In addition, CHAMBERLAIN and CHADWICK (1953) showed that plant leaves can take up radioactive iodine.

VOUGHT *et al.* (1970) have also suggested that much of the iodine taken in by terrestrial plants is derived from direct absorption from the atmosphere and precipitation, and they suggest further that atmospheric iodine is an important source of iodine for humans. The iodine content of edible matter of foodstuffs ranges from 0.3 to 1.3 ppm I (dry matter) (SENGUPTA and PAL, 1971).

It has been shown by SELEZNEV and TYURYUKANOV (1971), that pea and wheat plants more readily assimilate inorganic iodine than that which is bound in peat or humus-rich soil. However, these authors found that soil bacteria readily convert these iodo-organic compounds to lower molecular weight forms which can more easily be taken up by plants.

PORTYANKO *et al.* (1970) found that sunflower plants grown on high iodine-containing soils were richer in iodine than those grown on lower iodine-containing soils. PORTYANKO and KOSTINA (1968) showed that various parts of plants were very different in their iodine contents. Among the vegetative organs, leaves are highest, while of the generative organs, stamens and pistils are highest. ZYRIN and BYKOVA (1960) found that the roots of terrestrial plants contain 5 to 20 times as much iodine as the parts of the plant above the ground.

The iodine content of dried rice plants grown on volcanic ash soils in Japan was found by TENSHO and YEN (1970) to be between 1.3 and 341 ppm; the iodine content of the soil varied between 14 and 23 ppm iodine. The very high iodine and bromine contents are thought to be a cause of the "Reclamation-Akagare" disease in the plants.

Information on iodine in dry animal tissue has been compiled in Table 53-L-2.

Table 53-L-2. *Iodine content of dry animal tissue* (from compilation of BOWEN, 1966, except where indicated)

Group	ppm I	Group	ppm I
Annelida	160	Pisces	1
Mollusca	4	Mammalia	0.43
Echinodermata	6	Coelenterata	15
Crustacea	1	Corals[a] (Mean of 3)	37,900
Insecta	0.9		

[a] From COLLINS (1969)

Revised manuscript received: March 1973

53-M. Abundance in Metamorphic Rocks

Very few data are available for the iodine content of metamorphic rocks. KURODA and CROUCH (1962) record one value of 40 ppb, using colorimetry, for a biotite gneiss from De Grasse, New York. In Table 53-E-1 two eclogites have been quoted as containing 60 and 180 ppb I (N/R).

In the older literature, the author has been able to find only two analyses from the work of VON FELLENBERG and LUNDE (1927). Both values, 70 ppb and 550 ppb, were obtained by classical methods and are for samples of marble.

Revised manuscript received: March 1973

53-O. Economic Importance; Relationship to Other Elements

I. Economic Importance

The major commercial source of iodine is the Chile nitrate deposit (CHILEAN IODINE EDUCATIONAL BUREAU, 1956). Some iodine-rich brines are also commercial sources (HEILBRON, 1946; SHCHEPAK and GAVRILENKO, 1968), particularly oilfield brines (COLLINS, 1966). KHOLODOV (1966) has suggested that some phosphorite deposits can also be rich sources of iodine.

Some workers have suggested that iodine can be used as an indicator element for hydrothermal deposits (SCHNEIDER and MILLER, 1965), and the iodine and bromine contents of oilfield brines have been used by BOJARSKI (1966) as an indicator of hydrocarbon deposits.

CHITAEVA *et al.* (1971) found iodine to be enriched in the supergene zone of a chalcopyrite deposit. BARRINGER (1965) has suggested the use of remote sensing of iodine vapor from airborne and space surveys to detect mineral deposits.

The major uses of commercial iodine are found in medicine, where the element is put to both curative and preventative uses. It is used externally as an antiseptic and internally to treat several ailments mainly concerned with thyroid deficiencies. It is added to table-salt in many countries as a precaution against goitre (CHILEAN IODINE EDUCATIONAL BUREAU, 1956) and is added to animal feeds and mineral licks for prevention of goitre and for ensuring high milk yields in cows, etc.

One rather specialised use of iodine in medicine and the biological sciences is as radio-iodine for treating the thyroid gland and as a tracer for identifying the metabolic pathways and functions of iodine in the body.

Among its many other uses can be numbered its importance as a catalyst in organic chemistry, its use in photography as KI, in the dyeing industry, and its use in photo-electric cells (HEILBRON, 1946).

II. Relationship to Other Elements (see Table 53-O-1)

Iodine, being a severely depleted element in chondritic meteorites, tends to correlate with other severely depleted elements such as tellurium and bromine (REED and ALLEN, 1966).

In general, in igneous rocks iodine shows little variation with silica content and little correlation with any other halogens (YOSHIDA *et al.*, 1971); however, in common with chlorine and bromine, iodine appears to be enriched in undersaturated rocks (KOGARKO and GULYAYEVA, 1965).

HONDA (1966) has shown that the I/Cl ratio of volcanic gases varies for different volcanoes but is characteristic for each volcano. In additon, this worker has shown that the I/Cl ratios of hot spring waters associated with volcanic areas are very

Table 53-O-1. *Tentative estimates of average Br/I and Cl/I ratios*

Source	Br/I	Cl/I
Carbonaceous and enstatite chondrites	10—30	1,500—8,000
Ordinary chondrites	7	700
Igneous Rocks	2—8	1,000—2,000
Volcanic gases	2—10	100—2,000
Soils	0.2—2	10—25
Sedimentary rocks (and sediments)	0.2—10	10—1,000
Rain water	3	600
Sea water	1,300	3.9×10^5

similar to the gases with which they are genetically related. The I/Cl ratio of some oilfield brines is particularly high (WHITE *et al.*, 1963), while the I/Cl ratio of the atmosphere is several hundred times that of sea water.

Iodine, a biophile element (GOLDSCHMIDT, 1954), shows a strong correlation with the carbonaceous content of soils, sediments and sedimentary rocks.

Revised manuscript received: March 1973

References: Sections 53-A to 53-M, 53-O

Alcock, N. W.: The crystal structure of α-rubidium iodate. Acta Cryst. **B 28**, 2783 (1972).

Alexander, E. C., Manuel, O. K.: Xenon in the inclusions of Canyon Diablo and Toluca iron meteorites. Earth Planet. Sci. Lett. **4**, 113 (1968).

Barkley, R. A., Thompson, T. G.: The total iodine and iodate-iodine content of sea water. Deep Sea Res. Oceanog. Abstr. **7**, 24 (1960).

Barringer, A. R.: Developments towards the remote sensing of vapors as an airborne and space exploration tool. U.S. Clearinghouse Fed. Sci. Tech. Inform. A.D. 614032, 279 (1965).

Basharina, L. A.: Gases of Kamchatka volcanoes. Bull. Volcanol. **28**, 95 (1965).

Becker, V. J., Bennett, J. H., Manuel, O. K.: Iodine and uranium in ultrabasic rocks and carbonatites. Earth Planet. Sci. Lett. **4**, 357 (1968).

Becker, V. J., Bennett, J. H., Manuel, O. K.: Iodine and uranium in sedimentary rocks. Chem. Geol. **9**, 133 (1972).

Becker, V. J., Manuel, O. K.: Chlorine, bromine, iodine and uranium in tektites, obsidians and impact glasses. J. Geophys. Res. **77**, 6353 (1972).

Bennett, J. H., Manuel, O. K.: On iodine abundances in deep-sea sediments. J. Geophys. Res. **73**, 2302 (1968).

Bentor, Y. K.: On the evolution of subsurface brines in Israel. Chem. Geol. **4**, 83 (1969).

Billings, G. K., Hitchon, B., Shaw, D. R.: Geochemistry and origin of formation waters in the western Canada sedimentary basin, 2. Alkali metals. Chem. Geol. **4**, 211 (1969).

Bojanowski, R., Paslawska, S.: On the occurrence of iodine in bottom sediments and interstitial waters of the southern Baltic Sea. Acta Geophys. Polon. **18**, 277 (1970).

Bojarski, L.: (I and Br as hydrochemical indicators of hydrocarbons in the Mesozoic and Paleozoic deposits of northern Poland.) Kwart. Geol. **10**, 177 (1966).

Boulos, M. S., Manuel, O. K.: Extinct radioactive nuclides and production of xenon isotopes in natural gas. Nature Phys. Sci. **235**, 110 (1972).

Bowen, H. J. M.: Trace Elements in Biochemistry. London-New York: Academic Press 1966.

Braibanti, A., Tiripicchio, A., Bigoli, F., Pellinghelli, M. A.: Crystal and molecular structure of cadmium trihydrogenhexaoxoiodate (VII) trihydrate. Acta Cryst. **B 26**, 1069 (1970).

Brehler, B., Jacobi, H., Siebert, H.: Kristallstruktur und Schwingungsspektrum von $K_4J_2O_9$. Z. Anorg. Allgem. Chem. **362**, 301 (1968).

Carpenter, G. B.: The crystal structure of CsI_2Br. Acta Cryst. **20**, 330 (1966).

Cauer, H.: Über das Vorkommen von Jod in Gesteinen, Erden und Wässern und seine Beziehung zum Kropf. J. Landwirtschaft **77**, 251 (1930).

Chamberlain, A. C., Chadwick, R. C.: Deposition of airborne radio-iodine vapour. Nucleonics **11**, 22 (1953).

Chamberlain, A. C., Chadwick, R. C.: Transport of iodine from atmosphere to ground. Tellus **18**, 226 (1966).

Cheesman, G. H., Finney, A. J. T.: Refinement of the structure of ammonium triiodide, NH_4I_3. Acta Cryst. **B 26**, 904 (1970).

Chilean Iodine Educational Bureau: Geochemistry of Iodine. London: Shenval Press 1956.

Chitaeva, N. A., Miller, A. D., Grosse, Y. I.: (Iodine distribution in the supergene zone of the Gaisk chalcopyrite deposit.) Geokhimiya 696 (1971).

Claridge, G. G., Cambell, I. B.: Origin of nitrate deposits. Nature **217**, 428 (1968).

Clark, R. S., Rowe, M. W., Ganapathy, R., Kuroda, P. K.: Iodine, uranium and tellurium contents in meteorites. Geochim. Cosmochim. Acta **31**, 1605 (1967).

Collins, A. G.: Here's how producers can turn brine disposal into profit. Oil Gas J. **64**, 112 (1966).

Collins, A. G.: Chemistry of some Andarko Basin brines containing high concentrations of iodide. Chem. Geol. **4**, 169 (1969).

Collins, A. G., Bennett, J. H., Manuel, O. K.: Iodine and algae in sedimentary rocks associated with iodine-rich brines. Bull. Geol. Soc. Am. **82**, 2607 (1971).

Collins, A. G., Egleson, G. C.: Iodide abundance in oilfield brines in Oklahoma. Science **156**, 934 (1967).

Collins, A. G., Zelinski, W. F., Pearson, C. A.: Bromide and iodide in oilfield brines in some Tertiary and Cretaceous formations in Mississippi and Alabama. Amer. Chem. Soc., Div. Water Waste Chem., Preprints **7**, 166 (1967).

Correns, C. W.: The geochemistry of the halogens. In: Ahrens, L. H., Rankama, K., Runcorn, S. K. (Eds.): Physics and Chemistry of the Earth, vol. 1. London: Pergamon Press 1956.

Cosgrove, M. E.: Iodine in the bituminous Kimmeridge shales of the Dorset Coast, England. Geochim. Cosmochim. Acta **34**, 830 (1970).

Dean, G. A.: The iodine content of some New Zealand drinking waters with a note on the contribution from sea spray to the iodine in rain. New Zealand J. Sci. **6**, 208 (1963).

De Boer, J. L., van Bolhuis, F., Olthof-Hazekamp, R., Vos, A.: Reinvestigation of the crystal structure of lithium iodate. Acta Cryst. **21**, 841 (1966).

De Rao, S. K., N. S. S., Tripathi, C. M., Rai, C.: Retention of iodide by soil clays. Indian J. Agr. Chem. **4**, 43 (1971).

Dodd, R. T.: Metamorphism of the ordinary chondrites. Geochim. Cosmochim. Acta **33**, 161 (1969).

Duce, R. A.: Iodine in the atmosphere. Amer. Chem. Soc., Div. Water and Waste Chem., Preprints **7**, 79 (1967).

Duce, R. A., Wasson, J. T., Winchester, J. W., Burns, F.: Atmospheric iodine, bromine and chlorine. J. Geophys. Res. **68**, 3943 (1963).

Duce, R. A., Winchester, J. W., Nahl, T. W. van: Iodine, bromine and chlorine in the Hawaiian marine atmosphere. J. Geophys. Res. **70**, 1775 (1965).

Duce, R. A., Winchester, J. W., Nahl, T. W. van: Iodine, bromine and chlorine in winter aerosols and snow from Barrow, Alaska. Tellus **18**, 238 (1966).

Edwards, R. R.: Iodine-129: Its occurrence in nature and its utility as a tracer. Science **137**, 851 (1962).

Edwards, R. R., Rey, P.: Terrestrial occurrence and distribution of iodine-129. U.S. At. Energy Comm., NYO-3624-3 (1967).

Elema, R. J., de Boer, J. L., Vos, A.: The refinement of the crystal structure of $KICl_4 \cdot H_2O$. Acta Cryst. **16**, 243 (1963).

Ewy, Z., Pyska, H., Styczynski, H.: Iodine content in milk and water in the Opole [Poland] area.) Roczniki Nauk Rolniczychy, Ser. B **90**, 399 (1968); Chem. Abstr. 70-105255 k (1970).

Feikema, Y. D.: The crystal structures of two oxy-acids of iodine. I. A study of orthoperiodic acid, H_5IO_6, by neutron diffraction. Acta Cryst. **20**, 765 (1966).

Feikema, Y. D., Vos, A.: The crystal structures of two oxy acids of iodine. II. An x-ray diffraction study of anhydro-iodic acid, HI_3O_8. Acta Cryst. **20**, 769 (1966).

Fellenberg, T. von: Untersuchungen über das Vorkommen von Jod in der Natur. VII. Biochem. Z. **152**, 153 (1924).

Fellenberg, T. von: Untersuchungen über das Vorkommen von Jod in der Natur. XI. Zur Geochemie des Jods, II. Biochem. Z. **187**, 1 (1927).

Fellenberg, T. von, Lunde, G.: Untersuchungen über das Vorkommen von Jod in der Natur. X. Beitrag zur Geochemie des Jods. Biochem. Z. **175**, 162 (1926).

Fellenberg, T. von, Lunde, G.: Contribution à la geochimie de l'iode. Norsk. Geol. Tidsskr. **9**, 48 (1927).

Ferrari, A., Braibanti, A., Tiripicchio, A.: The crystal structure of tetrapotassium dihydrogendecaoxodiiodate (VII) octahydrate. Acta Cryst. **19**, 629 (1965).

Fleischer, M.: Index of new mineral names, discredited minerals and changes of mineralogical nomenclature in vols. 1—50 of the American Mineralogist. Am. Mineralogist **51**, 1248 (1966).

Goldschmidt, V. M.: Geochemistry. Oxford: University Press 1954.

Goles, G. G., Anders, E.: Abundances of iodine, tellurium and uranium in meteorites. Geochim. Cosmochim. Acta **26**, 723 (1962).

Goles, G. G., Greenland, L. P., Jérome, D. Y.: Abundances of chlorine, bromine and iodine in meteorites. Geochim. Cosmochim. Acta **31**, 1771 (1967).

Grimaldi, F. S., Schnepfe, M. M.: Determination of iodine in the p.p.m. range in rocks. Anal. Chim. Acta **53**, 181 (1971).

Gulyayeva, L. A., Itkina, E. S.: Halogens in marine and freshwater sediments. Geochemistry 610 (1962a).

Gulyayeva, L. A., Itkina, E. S.: The halogens and vanadium, nickel and copper in coals. Geochemistry 398 (1962b).

Hattori, M., Nagahara, T.: γ-Ray spectra of the radioactive fallout with a Ge (Li) detector. J. Radiat. Res. **8**, 125 (1967).

Heath, R. L.: Table of isotopes. In: Weast, R. C. (Ed.): Handbook of Chemistry and Physics, 52nd ed. Cleveland, Ohio: The Chemical Rubber Co. Ltd. 1971.

Heilbron, I. M. (Chief ed.): Thorpe's Dictionary of Applied Chemistry, Vol. 7, 4th ed. London: Longmans 1946.

Hey, M. H.: An Index of Mineral Species and Varieties Arranged Chemically. London: British Museum (Natural History) 1950.

Hitchon, B., Billings, G. K., Klovan, J. E.: Geochemistry and origin of formation waters in the western Canada sedimentary basin — III. Factors controlling chemical composition. Geochim. Cosmochim. Acta **35**, 567 (1971).

Hohenberg, C. M., Podosek, F. A., Reynolds, J. H.: Xenon — iodine dating: Sharp isochronism in chondrites. Science **156**, 233 (1967).

Hohenberg, C. M., Reynolds, J. H.: The preservation of the iodine-xenon record in meteorites. J. Geophys. Res. **74**, 6679 (1969).

Honda, F., Mitzutani, Y., Sugiura, T., Oana, S.: A geochemical study of iodine in volcanic gases. Bull. Chem. Soc. Japan **39**, 2690 (1966).

Irinevich, A. D., Rabinovich, I. Z., Fil'kov, V. A.: (Iodine in Moldavian soils.) Pochvovedenie 58 (1970).

Itkina, E. S., Lygalova, V. W.: Geochemistry of iodine and bromine in the Carboniferous horizon and in the Domanik and Bavly formations of some petroliferous areas of Bashkiria. In: Gulyayeva, L. A. (Ed.): The Geochemistry of Oil and Oil Deposits. Acad. Sci. U.S.S.R. 1962; Israel Program for Scientific Translations, Jerusalem, 1964.

Jeffery, P. M., Reynolds, J. H.: Origin of excess ^{129}Xe in stone meteorites. J. Geophys. Res. **66**, 3582 (1961).

Johannesson, J. K.: Oxidised iodine in sea water. Nature **182**, 251 (1958).

Johannesson, J. K., Gibson, G. W.: Nitrate and iodate in Antarctic salt deposits. Nature **194**, 567 (1962).

Kàlmàn, A., Cruickshank, D. W. J.: Refinement of the structure of $NaIO_4$. Acta Cryst. **B 26**, 1782 (1970).

Kamenev, V. F.: The content of iodine and other halogens in the soil, water, and plants within the Ulan-Ude area. Vop. Pitaniya **27**, 77 (1968); Chem. Abstr. **69**-89613x (1968).

Kholodov, V. N.: (Rare elements in phosphorite deposits.) Geokhim. Mineral. Genet. Tipy. Mestorozhd. Redk. Elem., Akad. Nauk S.S.S.R., Gos. Geol. Kom. S.S.S.R., Inst. Mineral. Geokhim. Kristallo-khim. Redk. Elem. **3**, 685 (1966); Chem. Abstr. **68**-23194j (1968).

Kikkawa, K., Shiga, S.: Relations between halogen contents of hot spring water (I). Spec. Contrib. Geophys. Inst., Kyoto Univ. 173 (1966).

Kogarko, L. N., Gulyayeva, L. A.: Geochemistry of the halogens in the alkalic rocks of the Lovozero massif (Kola Peninsula). Geochemistry Internat. 729 (1965).

Konovalov, G. S.: Removal of microelements by the main rivers of the U.S.S.R. Dokl. Acad. Sci. S.S.S.R. **129**, 912 (1959).

KOZIN, A. N.: Geochemistry of bromine and iodine of formation waters of the Kuybyshev area of the Volga. Petrol. Geol. **4**, 110 (1960).

KREJCI-GRAF, K., ERNST, W., HUBER, W., KRAUS, F., STRADLER, G., WERNER, H.: Geochemistry of the Vienna Basin, II. Boron and iodine. Chem. Erde **27**, 143 (1968).

KURODA, P. K., CROUCH, W. H.: On the chronology of the formation of the solar system, 2. Iodine in terrestrial rocks and the xenon-129/136 formation interval of the Earth. J. Geophys. Res. **67**, 4863 (1962).

LÅG, J., STEINNES, E.: Mercury and iodine distribution in Norwegian forest soils by neutron activation analysis. Nucl. Tech. Environ. Pollut., Proc. Symp. 1970, 429 (1971).

LARIMER, J. W., ANDERS, E.: Chemical fractionations in meteorites, II. Abundance patterns and their interpretation. Geochim. Cosmochim. Acta **31**, 1239 (1967).

LAURITA, W. G., KOSKI, W. S.: Iodine nuclear quadrupole resonance spectrum of boron triiodide. J. Am. Chem. Soc. **81**, 3179 (1959).

LININGER, R. L., DUCE, R. A., WINCHESTER, J. W., MATSON, W. R.: Chlorine, bromine, iodine and lead in aerosols from Cambridge, Massachusetts. J. Geophys. Res. **71**, 2457 (1966).

MAGOMEDOVA, L. A., ZYRIN, N. G., SALMANOV, A. B.: (Iodine in the soils and rocks of mountainous Dagestan.) Agrokhimiya 117 (1970).

MAKSIMOVA, S. N.: Organic halogen and sulphur compounds in petroleum and in solid bituminous formations. In: GULYAYEVA, L. A. (Ed.): The Geochemistry of Oil and Oil Deposits. Acad. Sci. U.S.S.R. 1962; Israel Program for Scientific Translations, Jerusalem, 1964.

MANOTTI LANFREDI, A. M., PELLINGHELLI, M. A., TIRIPICCHIO, A., TIRIPICCHIO CAMELLINI, M.: The crystal structure of strontium di-iodate(V)monohydrate. Acta Cryst. **B 28**, 679 (1972).

MANOTTI LANFREDI, A. M., PELLINGHELLI, M. A., TIRIPICCHIO, A.: The crystal structure of dipotassium hydrogen diiodate(V)chloride. Acta Cryst. **B 28**, 1822 (1972).

MERRIHUE, C.: Xenon and krypton in the Bruderheim meteorite. J. Geophys. Res. **71**, 263 (1966).

MIGCHELSEN, T., VOS, A.: Reinvestigation of the crystal structure of ammonium bromochloroiodate(I), NH_4BrICl, at 140° K. Acta Cryst. **22**, 812 (1967).

MITCHELL, K. A. R.: The use of outer d orbitals in bonding. Chem. Rev. **69**, 157 (1969).

MIYAKE, Y., TSUNOGAI, S.: Evaporation of iodine from the ocean. J. Geophys. Res. **68**, 3989 (1963).

MIYAKE, Y., TSUNOGAI, S.: Problème de l'iode dans les océans. La Mer. Bull. Soc. franco-japonaise d'océanogr. **4**, 65 (1966).

MOORE, G. W.: Geologic significance of the minor-element composition of marine salt deposits. Econ. Geol. **66**, 187 (1971).

MUELLER, G.: Genetic histories of nitrate deposits from Antarctica and Chile. Nature **219**, 1131 (1968).

MUN, A. I., BAZILEVICH, Z. A.: Iodine in surface brines and waters of Central Kazakhstan. Geochemistry 528 (1963).

MUN, A. I., BAZILEVICH, Z. A.: Some characteristics of the distribution of iodine in lake muds. Geochemistry 451 (1964).

OBUKHOVA, V. A., GOLOBOV, A. D.: Iodine content of soils, natural waters, and forages of the Moscow-Oka river flatland. Agrokhimiya 106 (1969).

PASLAWSKA, S., OSTROWSKI, S.: The influence of the Baltic Sea on the iodine content of air in the inshore zone. Acta Geophys. Polon. **16**, 329 (1968).

PAULING, L., HUGGINS, M. L.: Covalent radii of atoms and interatomic distances in crystals containing electron-pair bonds. Z. Krist. **87**, 205 (1934).

PENDLETON, C. R., LLOYD, R. D.: Environmental levels of radio-activity in Utah following Operation Pinstripe. Health Data Rep. **11**, 65 (1970).

PENNINGTON, W., LISHMAN, J. P.: Iodine in lake sediments in Northern England and Scotland. Biol. Rev. Cambridge Phil. Soc. **46**, 279 (1971).

POKATILOV, Y. G.: Content of various forms of iodine in soils of geochemical areas of the Barguzin intermountain basin of the Buryat ASSR. Mikroelem. Sib. 18 (1968).

PORTYANKO, V. F., KOSTINA, A. E.: (Iodine content in the vegetative and reproductive organs of plants.) Fiziol Rast. **15**, 135 (1968).

PORTYANKO, V. F., KOSTINA, A. E., DULOVA, M. K., PORTYANKO, V. V.: Distribution of chlorine and iodine in plants. Fiziol Rast. **17**, 169 (1970).

PRICE, N. B., CALVERT, S. E., JONES, P. G. W.: The distribution of iodine and bromine in the sediments of the southwestern Barents Sea. J. Marine Res. **28**, 22 (1970).

PROSKURYAKOVA, G. F., IVANOV, N. A., RESHETNIKOV, A.: Iodine level in the soils of the 1st section of the scientific research station of the Sverdlovsk Agricultural Institute. Tr. Sverdlovsk Sel'Skokhoz. Inst. **15**, 333 (1969b).

PROSKURYAKOVA, G. F., IVANOV, N. A., SALANGINA, N. D.: (Iodine level in soils of the Sverdlovsk district.) Tr. Sverdlovsk. Sel'Skokhoz. Inst. **15**, 328 (1969a).

RANKAMA, K., SAHAMA, T. G.: Geochemistry. Chicago: Chicago University Press 1950.

REED, G. W.: In: MASON, B. (Ed.): Handbook of Elemental Abundances in Meteorites. New York-Paris-London: Gordon & Breach Science Publishers 1971.

REED, G. W., ALLEN, R. O.: Halogens in chondrites. Geochim. Cosmochim. Acta **30**, 779 (1966).

REED, G. W., JOVANOVIC, S.: Halogens, mercury, lithium and osmium in Apollo 11 samples. Proc. Apollo 11 Lunar Science Conf. **2**. Suppl. I. Geochim. Cosmochim. Acta 1487 (1970a).

REED, G. W., JOVANOVIC, S.: Halogens, mercury, lithium and osmium in Apollo 11 samples. Geochim. Cosmochim. Acta **34**, 1370 (1970b).

REED, G. W., JOVANOVIC, S.: The halogens and other trace elements in Apollo 12 samples and the implications of halides, platinum metals, and mercury on surfaces. Proc. Second Lunar Science Conf. **2**, 1261 (1971).

REYNOLDS, J. H.: Determination of the age of the elements. Phys. Rev. Letters **4**, 8 (1960).

RIETSCHEL, E. TH., BÄRNIGHAUSEN, H.: Die Kristallstruktur von Strontiumjodid SrJ_2. Z. Anorg. Allg. Chem. **368**, 62 (1969).

ROEBER, J.: Der Jodgehalt der deutschen Salzlagerstätten. Jahrbuch des Halleschen Verbandes f. d. Erforsch. d. mitteldeutschen Bodenschätze, N.F. **16**, 129 (1938).

ROSENZWEIG, A., MOROSIN, B.: A reinvestigation of the crystal structure of $LiIO_3$. Acta Cryst. **20**, 758 (1966).

RYKHILOV, G. P.: Level of fluorine, bromine and iodine in the water of the Irtysch and Om rivers. Mikroelem. Sib. 22 (1970).

SAFRAZBEKYAN, E. A.: (Iodine level in chesnut soils and chernozems in Armenia.) Biol. Zh. Arm. **23**, 89 (1970).

SAZONOV, N. N.: Contents and migration of iodine in central Yakutia soils. Pochvy Merzlotn. Obl. 124 (1969); Chem. Abstr. **72**-135213d (1970).

SCHNEIDER, L. A., MILLER, A. D.: Determination of microamounts of iodine in silicate rocks by a kinetic method. Zh. Analit. Khim. **20**, 92 (1965).

SCHROLL, E., KRACHSBERGER, H.: Geochemistry of impurities in atmospheric precipitation in the Vienna metropolitan area. Radex Rundsch. 334 (1970).

SCHWARZENBACH, D.: The crystal structure and one-dimensional disorder of the orange modification of HgI_2. Z. Krist. **128**, 97 (1969).

SELEZNEV, Y. M., TYURYUKANOV, A. N.: Factors of changing iodine-compound forms in soils. Biol. Nauki **14**, 128 (1971); Chem. Abstr. **76**-87514r (1972).

SENGUPTA, S. R., PAL, B.: Iodine and fluorine contents of foodstuffs. Indian J. Nutr. Diet. **8**, 66 (1971).

SIEBERT, H.: Chemie der Überjodsäuren und der Perjodate. Fortschr. Chem. Forsch. **8**, 470 (1967).

SIEBERT, H., WEDEMEYER, H.: Die Kristallstruktur des Dikaliumperjodat-hydrates. Angew. Chem. **77**, 507 (1965).

SHACKLETTE, H. T., CUTHBERT, M. E.: Iodine content of plant groups as influenced by variation in rock and soil type. In: CANNON, H. E., DAVIDSON, D. F. (Eds.): Relation of Geology and Trace Elements to Nutrition. U.S. Geol. Soc. Spec. Paper **90**, 31 (1967).

SHAW, T. I.: Halogens in algae. In: LEWIN, R. A. (ed.): Physiology and Biochemistry of Algae. New York: Academic Press 1962.

Shchepak, V. M., Gavrilenko, K. S.: (Formation waters as an industrial raw material.) Zakhidnikh Obl. Ukr 263 (1968); Chem. Abstr. **72**-81689t (1970).

Shishkina, O. V., Pavlova, G. A.: Iodine distribution in marine and oceanic bottom muds and in their pore fluids. Geochemistry Intern. 559 (1965).

Sillén, L. G.: Physical chemistry of sea water. In: Sears, M. (Ed.): Oceanography. Am. Assoc. Advan. Sci., Publ. **67**, 549 (1961).

Sinitskaya, G. I.: (Iodine content of the Zeya-Bureya Plain soils). Uch. Zap. Dal'nevost. Gos. Univ. **27**, 72 (1969); Chem. Abstr. **76**-79190s (1972).

Spencer, L. J.: 19th list of new mineral names. Mineral. Mag. **29**, 974 (1952).

Srinivasan, B., Alexander, E. C., Manuel, O. K.: Iodine-129 in terrestrial ores. Science **173**, 327 (1971).

Sugawara, K.: Migration of the elements through phases of the hydrosphere and atmosphere. In: Vinogradov, A. P. (Ed.): Chemistry of the Earth's Crust, 2. Acad. Sci. U.S.S.R.; Israel Program for Scientific Translations, Jerusalem, 1967.

Sugawara, K., Koyama, T., Terada, K.: Co-precipitation of iodide ions by some metallic hydrated oxides with special reference to iodide accumulation in bottom water layers and interstitial water of muds in some Japanese lakes. J. Earth Sci. Nagoya Univ. **16**, 52 (1958).

Sugawara, K., Terada, K.: Iodine distribution in the Western Pacific Ocean. J. Earth Sci., Nagoya Univ. **5**, 81 (1957).

Sugawara, K., Terada, K.: Oxidized iodine in sea water. Nature **182**, 250 (1958).

Swink, L. N., Carpenter, G. B.: The crystal structure of iodine monobromide, IBr. Acta Cryst. **B 24**, 429 (1968a).

Swink, L. N., Carpenter, G. B.: The crystal structure of iodine monobromide, IBr. Acta Cryst. **B 24**, 1702 (1968b).

Tensho, K., Yen, K.: Study on iodine and bromine in soil-plant systems in relation to the "Reclamation-Akagare" disease of lowland rice by means of radioisotope-techniques. Radioisotopes (Tokyo) **19**, 574 (1970).

Tower, H. L., Brewer, H. C.: Natural chilean nitrate of soda production and use in agriculture; fertilizer nitrogen—its chemistry and technology. Am. Chem. Soc. Monograph **161**, 315 (1964).

Trotter, J.: The crystal structure of arsenic triiodide. Z. Krist. **121**, 81 (1965).

Trotter, J., Zobel, T.: The crystal structure of SbI_3 and BiI_3. Z. Krist. **123**, 67 (1966).

Tsunogai, S.: Iodine in the deep waters of the ocean. Deep Sea Res., Oceanogr. Abstr. **18**, 913 (1971).

Tsunogai, S., Henmi, T.: Iodine in the surface water of the ocean. Nippon Kaiyo Gakkai-Shi **27**, 67 (1971).

Tsunogai, S., Sase, T.: Formation of iodide-iodine in the ocean. Deep Sea Res., Oceanogr. Abstr. **16**, 489 (1969).

Turekian, K. K., Wedepohl, K. H.: Distribution of the elements in some major units of the earth's crust. Geol. Soc. Am. Bull. **72**, 175 (1961).

Turner, G.: Extinct iodine-129 and trace elements in chondrites. J. Geophys. Res. **70**, 5433 (1965).

Underwood, E. J.: Trace Elements in Human and Animal Nutrition, 3rd. ed. New York-London: Academic Press 1971.

van Bolhuis, F., Koster, P. B., Migchelsen, T.: Refinement of the crystal structure of iodine at 100° K. Acta Cryst. **23**, 90 (1967).

Vinogradov, A. P.: The Elementary Chemical Composition of Marine Organisms. New Haven: Sears Foundation for Marine Research 1953.

Vinogradov, A. P.: The Geochemistry of Rare and Dispersed Chemical Elements in Soils, 2nd. ed. New York: Consultants Bureau 1959.

Vought, R. L., Brown, F. A., London, W. T.: Iodine in the environment. Arch. Environ. Health **20**, 516 (1970).

Walters, L. J., Winchester, J. W.: Neutron activation analysis of sediments for halogens using Szilard-Chalmers reactions. Anal. Chem. **43**, 1020 (1971).

Weast, R. C. (Ed.): Handbook of Chemistry and Physics, 52nd. ed. Cleveland, Ohio: The Chemical Rubber Co. Ltd. 1971.

WELLS, A. F.: Structural Inorganic Chemistry. Oxford: Clarendon Press 1962.

WHITE, D. E., HEM, J. D., WARING, G. A.: Chemical composition of sub-surface waters. In: FLEISCHER, M. (Ed.) Data of Geochemistry, 6th. ed. F. U.S. Geol. Surv. Profess. Paper 440-F (1963).

WILKE-DÖRFURT, E.: Über den Jodgehalt einiger Gesteine und seine Beziehungen zum chemischen Teil des Kropfproblems. Ann. Chem. **453**, 288 (1927).

WILKE-DÖRFURT, E., ROMERSPERGER, H.: Der Jodgehalt von Kohlen. Z. Anorg. Chem. **186**, 159 (1930).

WINCHESTER, J. W., DUCE, R. A.: Coherence of iodine and bromine in the atmosphere of Hawaii, northern Alaska, and Massachusetts. Tellus **18**, 287 (1966).

WINCHESTER, J. W., DUCE, R. A.: The global distribution of iodine, bromine and chlorine in marine aerosols. Naturwissenschaften **54**, 110 (1967).

YOSHIDA, M., MAKINO, I., YONEHARA, N., IWASAKI, I.: The fractionation of halogen compounds through the process of the volatilisation and the sublimation from volcanic rocks on heating. Bull. Chem. Soc. Japan **38**, 1436 (1965).

YOSHIDA, M., TAKAHASHI, K., YONEHARA, N., OZAWA, T., IWASAKI, I.: The fluorine, chlorine, bromine and iodine contents of volcanic rocks in Japan. Bull. Chem. Soc. Japan **44**, 1844 (1971).

ZYRIN, N. G., BYKOVA, L. N.: Iodine in some soils in the Moscow area. Vestn. Mosk. Univ., Ser. VI. Biol. **15**, 55 (1960); Chem. Abstr. **61**-3641 b (1964).

ZYRIN, N. G., IMADI, T. K.: Iodine in soils of the Russian Plain and Crimea. Agrokhimiya 100 (1967).

Xenon 54

and the other Noble Gases 10, 18, 36, 86

see Helium 2

Cesium 55

A	G. Cocco, L. Fanfani and P. F. Zanazzi	(Istituto di Mineralogia dell'Università di Perugia, Perugia, Italy)
B—C*, D—G, I—O	K. S. Heier and	(Mineralogisk-Geologisk Museum, Oslo, Norway)
	G. K. Billings	(SYN-AN INC. P. O. Box 1735 Socorro, New Mexico, U.S.A.)
***C—IV**	E. Steinnes, K. S. Heier and G. K. Billings	(Institut for Atomenergi, Kjeller, Norway)

55-A. Crystal Chemistry

The electronic structure of the cesium atom is $1s^2\ 2s^2\ 2p^6\ 3s^2\ 3p^6\ 3d^{10}\ 4s^2\ 4p^6\ 4d^{10}\ 5s^2\ 5p^6\ 6s^1$. As cesium is the heaviest stable alkali metal, it exhibits a high reactivity within the alkali-metal group. In all its compounds cesium occurs as a monovalent cation which has the stable electronic configuration of the noble gas xenon. The usually accepted ionic radius of Cs^+ is 1.65 Å (see Table 12-8 of Vol. I of this handbook). This value places cesium closest to rubidium and monovalent thallium but, because of the lack of Rb-minerals and of the different geochemical behavior of Tl, the larger concentrations of Cs occur in K-minerals. Isomorphism between these two elements, however, is inhibited by the relative difference of the ionic radii (24%) so that K-minerals, sometimes containing significant concentrations of Rb, generally accomodate less than 100 ppm of cesium (see section 55-D). This is partly the effect of a limited supply of Cs. Larger Cs-contents are observed in some beryl varieties — in morganite (upto 7.2%), in rhodizite (upto 7.1%) and in avogadrite (upto 7.0%). Pollucite, with a Cs-content ranging from 22% to 36%, represents the only independent Cs-mineral.

The present crystallochemical review is confined to the halogen- and oxygen-containing compounds. In Table 55-A-1 there are listed several structures of Cs-compounds from which reliable information concerning Cs coordination can be obtained.

Table 55-A-1

		Range, Å	Mean, Å	Reference
CsF	$Cs^{[6]}$—F	3.00	3.00	after Sysiö (1969)
$CsLiF_2$	$Cs^{[6+2]}$—6F	2.96—3.15	3.07	Burns and Busing (1965)
	—2F	3.50—3.53		
$CsCoF_3$	$Cs_I^{[12]}$—F	3.10—3.14	3.12	Babel (1969)
	$Cs_{II}^{[12]}$—F	3.11—3.30	3.17	
$CsNiF_3$	$Cs_I^{[12]}$—F	3.13—3.32	3.22	Babel (1969)
	$Cs_{II}^{[12]}$—F	3.23	3.23	
$CsBeF_3$	$Cs^{[8]}$—F	2.96—3.40	3.13	Steinfink and Brunton (1968)
$Cs[UF_6]$	$Cs^{[12]}$—F	3.10—3.15	3.12	Rosenzweig (1967)
$Cs_4[Mg_3F_{10}]$	$Cs_I^{[11]}$—F	3.05—3.47	3.22	Steinfink and Brunton (1969)
	$Cs_{II}^{[10]}$—F	2.94—3.19	3.09	
CsCl (low temp.)	$Cs^{[8]}$—Cl	3.57	3.57	after Sysiö (1969)
CsCl (high temp.)	$Cs^{[6]}$—Cl	3.54	3.54	after Poyhonen and Ruuskanen (1964)
$CsCl \cdot \frac{1}{3}H_3OHCl_2$	$Cs^{[9]}$—Cl	3.44—3.69	3.59	Schroeder and Ibers (1968)

Table 55-A-1 (continued)

		Range, Å	Mean, Å	Reference
$CsCoCl_3$	$Cs^{[12]}$—Cl	3.60—3.75	3.68	SOLING (1968)
$Cs_2[PuCl_6]$	$Cs^{[12]}$—Cl	3.70—3.72	3.71	ZACHARIASEN (SR 1948, 413)
$Cs_2[UCl_6]$	$Cs^{[12]}$—Cl	3.62—3.76	3.70	SIEGEL (SR 1956, 249)
$Cs_2[ThCl_6]$	$Cs^{[12]}$—Cl	3.63—3.82	3.74	SIEGEL (SR 1956, 249)
$Cs_2[BkCl_6]$	$Cs_I^{[12]}$—Cl	3.59—3.88	3.74	MORSS and FUGER (1969)
	$Cs_{II}^{[12]}$—Cl	3.53—3.90	3.73	
$CsNa[BkCl_6]$	$Cs^{[12]}$—Cl	3.82	3.82	MORSS and FUGER (1969)
$Cs_3[Cr_2Cl_9]$	$Cs_I^{[12]}$—Cl	3.61—3.79	3.66	WESSEL and IJDO (SR 1957, 277)
	$Cs_{II}^{[12]}$—Cl	3.59—3.61	3.60	
$Cs_3[Re_3Cl_{12}]$	$Cs_I^{[12]}$—Cl	3.44—3.87	3.67	BERTRAND *et al.* (1963)
	$Cs_{II}^{[11]}$—Cl	3.29—3.89	3.65	
	$Cs_{III}^{[10]}$—Cl	3.38—3.77	3.62	
CsBr	$Cs^{[8]}$—Br	3.72	3.72	after SYSIÖ (1969)
$CsBr \cdot \frac{1}{3}H_3OHBr_2$	$Cs^{[9]}$—Br	3.56—3.81	3.72	SCHROEDER and IBERS (1968)
Cs_2ZnBr_4	$Cs_I^{[11]}$—Br	3.72—4.27	4.02	MOROSIN and LINGAFELTER (SR 1959, 317)
	$Cs_{II}^{[9]}$—Br	3.58—4.02	3.72	
$Cs_2[TeBr_6]$	$Cs^{[12]}$—Br	3.89	3.89	DAS and BROWN (1966)
CsI	$Cs^{[8]}$—I	3.96	3.96	after SYSIÖ (1969)
$Cs_3[Bi_2I_9]$	$Cs_I^{[12]}$—I	4.21—4.24	4.22	LINDQVIST (1968)
	$Cs_{II}^{[12]}$—I	4.20—4.31	4.26	
$CsMnCl_3 \cdot 2H_2O$	$Cs^{[12]}$—8 Cl	3.48—3.67	3.60	JENSEN *et al.* (1962)
	—4 O	3.78—3.84	3.81	
$Cs_2MnCl_4 \cdot 2H_2O$	$Cs^{[10]}$—8 Cl	3.35—3.70	3.54	JENSEN (1964)
	—2 O	3.56—3.62	3.59	
$Cs_2RuCl_5 \cdot H_2O$	$Cs_I^{[11]}$—9 Cl	3.41—3.85	3.54	HOPKINS *et al.* (1966)
	—2 O	3.72	3.72	
	$Cs_{II}^{[7]}$—6 Cl	3.62—3.70	3.65	
	—1 O	4.15		
$Cs_2[(UO_2)Cl_4]$	$Cs^{[11]}$—8 Cl	3.49—3.73	3.60	HALL *et al.* (1966)
	—3 O	3.57—3.94	3.82	
$Cs_{0.9}[(UO_2)OCl_{0.9}]$	$Cs^{[8]}$—3 Cl	3.50	3.50	ALLPRESS *et al.* (1963)
	—5 O	2.92—3.22	3.10	
$Cs[PO_2F_2]$	$Cs^{[12]}$—6 F	3.26—3.62	3.52	TROTTER and WHITLOW (1967)
	—6 O	3.07—3.29	3.17	
$CsAl[SO_4]_2 \cdot 12H_2O$	$Cs^{[12]}$—O	3.37—3.45	3.41	CROMER *et al.* (1966)
${}^3_\infty Cs[B^{[4]}B_8^{[3]}O_{14}]$	$Cs^{[10]}$—O	3.14—3.40	3.30	KROGH-MOE and IHARA (1967)
$Cs[V_3^{[5+1]}O_8]$	$Cs^{[12]}$—O	2.99—3.68	3.30	EVANS and BLOCK (1966)
${}^2_\infty Cs_2[(U^{[7]}O_2)_2V_2^{[5]}O_8]$	$Cs^{[6+5]}$—6 O	3.01—3.28	3.16	APPLEMAN and EVANS (1965)
	—5 O	3.47—3.60	3.52	
$Cs_{0.7}Na_{0.3}[AlSi_2O_6] \cdot 0.3\ H_2O$ pollucite	$Cs^{[12]}$—O	3.40—3.57	3.48	NEWNHAM (1967)

I. Halogen Compounds

CsF has the "NaCl" structure type, $Cs^{[6]}$—F bond lengths being 3.004 Å (Sysiö, 1969). The radius ratio for CsCl = 0.93 allows 8-fold coordination but is so near to the ratio for 6-fold coordination that cesium chloride is dimorphous. At room temperature the cesium chloride structure is characterized by a body-centered lattice: each Cs atom is at the center of a cube of Cl atoms with $Cs^{[8]}$—Cl distances of 3.571 Å (Sysiö, 1969). At 445° C the CsCl structure changes into the NaCl arrangement with $Cs^{[6]}$—Cl bonds (at 469°) of 3.537 Å (calculated from Poyhonen and Ruuskanen, 1964). CsBr and CsI show only one modification corresponding to the arrangement in CsCl which is stable at room temperature; $Cs^{[8]}$-halogen distances are 3.720 Å and 3.956 Å respectively (Sysiö, 1969).

In $CsCl \cdot \frac{1}{3}H_3OHCl_2$ and in $CsBr \cdot \frac{1}{3}H_3OHBr_2$, cesium is 9-coordinated; average $Cs^{[9]}$—Cl and $Cs^{[9]}$—Br bond lengths are 3.59 and 3.72 Å respectively (Schroeder and Ibers, 1968). $CsHF_2$, together with the K- and Rb-bifluoride, exists in an α phase at room temperature and in a β phase at higher temperature. The transition temperature for the cesium term is 61° C (Kruh *et al.*, SR 1956, 215). In $CsLiF_2$, isostructural with $RbLiF_2$, 8 fluorine atoms in an asymmetric array surround Cs^+ ions. Six of them are at an average distance of 3.07 Å while the other ones are at a distance of 3.50 and 3.53 Å from the cation (Burns and Busing, 1965).

A large number of $CsMeX_3$ compounds — where Me represents a divalent cation and X an element of the halogen series — have a more or less distorted perowskite arrangement. Among these double halogenides, which often exhibit more than one modification, are $CsCaF_3$, $CsZnF_3$, $CsMgF_3$, $CsCoF_3$, $CsNiF_3$, $CsMnF_3$, $CsFeF_3$, $CsCdCl_3$, $CsHgCl_3$, $CsPbCl_3$, $CsCdBr_3$, $CsHgBr_3$ and $CsPbBr_3$. In $CsCoF_3$ (structure-type: $BaRuO_3$) and in $CsNiF_3$ (structure type: $BaNiO_3$), Cs is in 12-fold coordination with Cs—F distances upto 3.32 Å (Babel, 1969). The crystal structure of $CsCoCl_3$ is of the $CsNiCl_3$-type; Cs^+ ions are 12-coordinated with an average $Cs^{[12]}$—Cl distance of 3.68 Å (Soling, 1968). $Cs_2AgAuCl_6$ and $Cs_2Au_2Cl_6$, ordinarily cubic, become tetragonal after heating to 350° for several days and their structures are characterized by superlattices on the perowskite arrangement (Elliott and Pauling, SB 1938, 126). In $CsCuCl_3$ cesium and chlorine atoms form a hexagonal close-packing, the average $Cs^{[12]}$—Cl bond lengths being 3.63 Å (Schlueter *et al.*, 1966). The same arrangement has been ascribed to $CsNiCl_3$ and $CsVCl_3$.

There are three structural modifications of $CsBeF_3$ stable in ranges of increasing temperature (α-$CsBeF_3 > 360° > \beta$-$CsBeF_3 > 140° > \gamma$-$CsBeF_3$). A recent investigation on crystals probably belonging to the β-form shows an 8-fold coordination polyhedron around Cs with Cs—F distances in the range 2.96—3.40 Å (Steinfink and Brunton, 1968). $CsBF_4$ is isostructural with $RbBF_4$. The atomic arrangement is baryte-like and the coordination around Cs is 12-fold (Clark and Lynton, 1969). Cs_2ZnBr_4 has essentially the same structure as Cs_2ZnCl_4, Cs_2CuCl_4 and Cs_2CuBr_4. The coordination around the two different Cs^+ ions is 11-fold and 9-fold with Cs—Br distances from 3.581 to 4.266 Å (Morosin and Lingafelter, SR 1959, 317). $Cs_3[AlF_6]$ has a behavior closely similar to that of $Rb_3[AlF_6]$; this compound, which is tetragonal at room temperature, undergoes a transition in a cubic phase after heating (Holm, 1965).

$Cs[UF_6]$ is isostructural with $K[OsF_6]$. The Cs atom has 12 neighbors, 6 at 3.101 Å and 6 at 3.147 Å (ROSENZWEIG, 1967). $Cs[PuF_6]$, $Cs[NpF_6]$ and many other compounds with a general chemical formula of $Cs[Me^{V}F_6]$ show an arrangement closely related to $Cs[UF_6]$. The structures of many cubic cesium halogenide complexes, $Cs_2[Me^{IV}F_6]$, such as $Cs_2[CrF_6]$, $Cs_2[GeF_6]$ and $Cs_2[GeCl_6]$, can be referred to that of $K_2[PtCl_6]$. In $Cs_2[TeBr_6]$ which belongs to this group of compounds, $Cs^{[12]}$—Br bond lengths are 3.89 Å, in good agreement with the sum of the ionic radii corrected for a 12-coordination (DAS and BROWN, 1966). According to ZACHARIASEN (SR 1948, 413), $Cs_2[PuCl_6]$ is trigonal and cesium is bonded to twelve Cl atoms with a $Cs^{[12]}$—Cl distance of 3.71 Å. $Cs_2[UCl_6]$ and one form of $Cs_2[ThCl_6]$ are isostructural with it. The average $Cs^{[12]}$—Cl distances are 3.70 Å and 3.74 Å respectively (SIEGEL, SR 1956, 249). $Cs_2[BkCl_6]$ is not isomorphous with $Cs_2[PuCl_6]$. Its symmetry is hexagonal and its structure is typified by $Rb_2[MnF_6]$. Average $Cs_{I}^{[12]}$—Cl is 3.74 Å and $Cs_{II}^{[12]}$—Cl is 3.73 Å (MORSS and FUGER, 1969).

Another class of compounds, with the chemical formula $Cs_3[Me_2^{III}X_9]$, crystallizes in the hexagonal system. In $Cs_3[Bi_2I_9]$, coordination around the two different Cs atoms is 12-fold with distances from 4.205 to 4.308 Å (LINDQVIST, 1968). In $Cs_3[Cr_2Cl_9]$, Cs—Cl distances range from 3.59 to 3.79 Å (WESSEL and IJDO, SR 1957, 277).

II. Oxygen Compounds

Cs_2O crystallizes in the hexagonal system with a cubic packing of cations and the anti-$CdCl_2$ arrangement; the Cs—O bond length is 2.86 Å (TSAI *et al.*, SR 1956, 260).

$CsIO_3$ has a perowskite-type structure (NARAY-SZABO, SR 1947, 454) while $CsVO_3$ is orthorhombic (CALVO, SR 1958, 433) and $CsNO_3$ rhombohedral (FERRONI *et al.*, SR 1957, 352). The crystal structure of the modification of $CsIO_4$ stable at room temperature appears to be a relatively slight distortion of the $CaWO_4$ arrangement (BEINTEMA, SB 1937; 89); $CsReO_4$ and $CsTcO_4$ are isostructural with it. Cs_2SO_4 is dimorphic; the orthorhombic low-temperature form and the hexagonal high-temperature form are isotypic with the corresponding modifications of K and Rb sulfates (TABRIZI *et al.*, 1968).

Because of the large size of the cation, $CsAl[SO_4]_2 \cdot 12H_2O$ belongs to the β alum series. In this structure cesium can accomodate 12 oxygen neighbors of 12 water molecules at a mean distance of 3.41 Å (CROMER *et al.*, 1966). $CsCr[SO_4]_2 \cdot 12H_2O$ also has the β structure (LEDSHAM and STEEPLE, 1968).

In $Cs[PO_2F_2]$, isomorphous with K- and Rb-difluorophosphate, Cs^+ ions are surrounded by 6 oxygen and by 6 fluorine atoms with distances upto 3.62 Å; this compound has the $Ba[SO_4]$ structure (TROTTER and WHITLOW, 1967). In $Cs[B_9O_{14}]$, each cesium atom is coordinated by 10 oxygen atoms with distances ranging from 3.14 Å to 3.40 Å (KROGH-MOE and IHARA, 1967). In CsV_3O_8, the alkali-cation has a 12-fold coordination with average $Cs^{[12]}$—O bond length of 3.30 Å (EVANS and BLOCK, 1966). A structural investigation on $Cs_2[(UO_2)_2V_2O_8]$, the cesium analogue of anhydrous carnotite, has been performed by APPLEMAN and EVANS (1966); in this structure Cs^+ ions connect the sheets built up by U and V coordination polyhedra; there are 11 Cs—O bonds in the range 3.01—3.60 Å.

Few data concerning Cs-minerals are available. The crystal structure of pollucite, determined by NEWNHAM (1967) on a sample with the approximate chemical composition $Cs_{0.7}Na_{0.3}AlSi_2O_6 \cdot 0.3\ H_2O$, shows cesium coordinated to a distorted close-packed arrangement of twelve oxygen atoms, six at 3.40 Å and six at 3.57 Å. The atomic arrangement of rhodizite has been determined assuming the ideal formula of the mineral to be $CsBe_4B_{12}Al_4O_{28}$, although the sample analyzed by FRONDEL and ITO (1965) reveals large amounts of K and Rb and a sum of $Cs + Rb + K < 1 \cdot$ A $Cs^{[12]}$—O distance of 3.243 Å is reported. However uncertainties in the chemical formula allow only an approximate description of the atomic arrangement (TAXER and BUERGER, 1967).

Revised manuscript received: August 1970

References: Section 55-A

ALLPRESS, J. G., WADSLEY, A. D.: The crystal structure of caesium uranyl oxychloride $Cs_xUO_2OCl_x$ (x approximately 0.9). Acta Cryst. **17**, 41 (1964).

APPLEMAN, D. E., EVANS, H. T., JR.: The crystal structures of synthetic anhydrous carnotite, $K_2(UO_2)_2V_2O_8$, and its cesium analogue, $Cs_2(UO_2)_2V_2O_8$. Am. Mineralogist **50**, 825 (1965).

BABEL, D.: Die Kristallstrukturen der hexagonalen Fluorperowskite. Z. Anorg. Allgem. Chem. **369**, 117 (1969).

BERTRAND, J. A., COTTON, F. A., DOLLASE, W. A.: The crystal structure of cesium dodecachlorotrirhenate (III), a compound with a new type of metal atom cluster. Inorg. Chem. **2**, 1166 (1963).

BURNS, J. H., BUSING, W. R.: Crystal structures of rubidium lithium fluoride, $RbLiF_2$, and cesium lithium fluoride, $CsLiF_2$. Inorg. Chem. **4**, 1510 (1965).

CLARK, M. J. R., LYNTON, H.: Crystal structures of potassium, ammonium, rubidium and cesium tetrafluoborates. Canad. J. Chem. **47**, 2579 (1969).

CROMER, D. T., KAY, M. I., LARSON, A. C.: Refinement of the alum structures. I. X-ray and neutron diffraction study of $CsAl(SO_4)_2 \cdot 12H_2O$, a β alum. Acta Cryst. **21**, 383 (1966).

DAS, A. K., BROWN, I. D.: A refinement of the crystal structures of $(NH_4)_2TeBr_6$ and Cs_2TeBr_6. Canad. J. Chem. **44**, 939 (1966).

EVANS, H. T., BLOCK, S.: The crystal structures of potassium and cesium trivanadates. Inorg. Chem. **5**, 1808 (1966).

FRONDEL, C., ITO, J.: Composition of rhodizite. Tschermaks Mineral. Petrog. Mitt. **10**, 409 (1965).

HALL, D., RAE, A. D., WATERS, T. N.: The crystal structure of dicaesium tetrachlorodioxouranium (VI). Acta Cryst. **20**, 160 (1966).

HOLM, J. L.: Phase transitions and structure of the high-temperature phases of some compounds of the cryolite family. Acta Chem. Scand. **19**, 261 (1965).

HOPKINS, T. E., ZALKIN, A., TEMPLETON, D. H., ADAMSON, M. G.: The crystal structure of cesium aquopentachlororuthenate. Inorg. Chem. **5**, 1431 (1966).

JENSEN, S. J.: The crystal structure of $Cs_2MnCl_4 \cdot 2H_2O$ and $Rb_2MnCl_4 \cdot 2H_2O$. Acta Chem. Scand. **18**, 2085 (1964).

— ANDERSEN, P., RASMUSSEN, S. E.: The crystal structure of $CsMnCl_3 \cdot 2H_2O$. Acta Chem. Scand. **16**, 1890 (1962).

KROGH-MOE, J., IHARA, M.: The crystal structure of caesium enneaborate, $Cs_2O \cdot 9B_2O_3$. Acta Cryst. **23**, 427 (1967).

LEDSHAM, A. H. C., STEEPLE, H.: The crystal structures of sodium chromium alum and caesium chromium alum. Acta Cryst. B **24**, 1287 (1968).

LINDQVIST, O.: The crystal structure of caesium bismuth iodide, $Cs_3Bi_2I_9$. Acta Chem. Scand. **22**, 2943 (1968).

MORSS, L. R., FUGER, J.: Preparation and crystal structures of dicesium sodium berkelium hexachloride. Inorg. Chem. **8**, 1433 (1969).

NEWNHAM, R. E.: Crystal structure and optical properties of pollucite. Am. Mineralogist **52**, 1515 (1967).

POYHONEN, J., RUUSKANEN, A.: X-ray investigation of the transition of CsCl at 469°. Ann. Acad. Sci. Fennicae, Ser. A VI **146**, 12 (1964).

ROSENZWEIG, A.: The crystal structure of $CsUF_6$. Acta Cryst. **23**, 865 (1967).

SCHLUETER, A. W., JACOBSON, R. A., RUNDLE, R. E.: A redetermination of the crystal structure of $CsCuCl_3$. Inorg. Chem. **5**, 277 (1966).

Schroeder, W., Ibers, J. A.: The bihalide ions $ClHCl^-$ and $BrHBr^-$: Crystal structures of cesium chloride — $\frac{1}{3}$ — (hydronium bichloride) and cesium bromide — $\frac{1}{3}$ — (hydronium bibromide). Inorg. Chem. **7**, 594 (1968).

Soling, H.: The crystal structure and magnetic susceptibility of $CsCoCl_3$. Acta Chem. Scand. **22**, 2793 (1968).

Steinfink, H., Brunton, G. D.: The crystal structure of $CsBeF_3$. Acta Cryst. B **24**, 807 (1968).

— — The crystal structure of $Cs_4Mg_3F_{10}$. Inorg. Chem. **8**, 1665 (1969).

Sysiö, P. A.: On the additivity of crystal radii in alkali halides. Acta Cryst. B **25**, 2374 (1969).

Tabrizi, D., Gaultier, M., Pannetier, G.: Analyse radiocristallographique des formes "basse" (β) et "haute" (α) temperature de sulfate de césium. Bull. Soc. Chim. Franç. 935 (1968).

Taxer, K. J., Buerger, M. J.: The crystal structure of rhodizite. Z. Krist. **125**, 423 (1967).

Trotter, J., Whitlow, S. H.: The structures of caesium and rubidium difluorophosphates. J. Chem. Soc. (A) 1383 (1967).

Revised manuscript received: August 1970

55-B. Isotopes in nature

^{133}Cs is the only natural isotope of cesium.

Revised manuscript received: December 1969

55-C. Abundance in Cosmos, Meteorites and Tektites

I. Cosmic Abundance

The cosmic abundance of Cs as estimated by various authors is given in Table 55-C-1.

Table 55-C-1. *Cosmic atomic abundance of Cs (Silicon 1×10^6)*

	Goldschmidt (1937)	Brown (1949)	Urey (1952)	Aller (1961b)	Suess and Urey (1956)	Cameron (1959)	Clayton *et al.* (1961)
Cs	0.1	0.1	1.3	0.5	0.456	0.25	0.13

Even though the strongest line of Cs falls in the observable part of the solar spectrum, it has not been detected in the sun (Aller, 1961a).

II. Meteorites

Cesium contents in meteorites are listed in Table 55-C-2. The determinations by Gordon *et al.* are high when compared with later results and would appear to be affected by some systematic analytical error. The spectrographic determinations by Ahrens *et al.* (1960) are also suspect.

Smales *et al.* (1964) combined the data of their own and earlier determinations of Cs for 24 chondrite falls which gave a mean value of 0.073 ppm Cs (range 0.005 to 0.2 ppm) and showed that this Cs value is distinctly lower than the interpolated cosmic abundance of 0.4 ppm (Suess and Urey, 1956; Cameron, 1959), and somewhat lower than the value of 0.116 ppm calculated by Clayton and Fowler (1961).

III. Tektites

There is a lack of Cs determinations in tektites. Taylor (1962) gives an average of 2.5 ± 0.83 ppm Cs in 24 australites.

55-B. Isotopes in Nature

^{133}Cs is the only natural isotope of cesium.

Revised manuscript received: December 1969

55-C. Abundance in Cosmos, Meteorites, Tektites and Lunar Materials

I. Cosmic Abundance

The cosmic abundance of Cs as estimated by various authors is given in Table 55-C-1.

Table 55-C-1. *Cosmic atomic abundance of Cs (Silicon 1 × 10⁶)*

	GOLDSCHMIDT (1937)	BROWN (1949)	UREY (1952)	ALLER (1961 b)	SUESS and UREY (1956)	CAMERON (1959)	CLAYTON *et al.* (1961)
Cs	0.1	0.1	1.3	0.5	0.456	0.25	0.13

Even though the strongest line of Cs falls in the observable part of the solar spectrum, it has not been detected in the sun (ALLER, 1961 a).

II. Meteorites

Cesium contents in meteorites are listed in Table 55-C-2. The determinations by GORDON *et al.* are high when compared with later results and would appear to be affected by some systematic analytical error. The spectrographic determinations by AHRENS *et al.* (1960) are also suspect.

SMALES *et al.* (1964) combined the data of their own and earlier determinations of Cs for 24 chondrite falls which gave a mean value of 0.073 ppm Cs (range 0.005 to 0.2 ppm) and showed that this Cs value is distinctly lower than the interpolated cosmic abundance of 0.4 ppm (SUESS and UREY, 1956; CAMERON, 1959), and somewhat lower than the value of 0.116 ppm calculated by CLAYTON and FOWLER (1961).

III. Tektites

There is a lack of Cs determinations in tektites. TAYLOR (1962) gives an average of 2.5 ± 0.83 ppm Cs in 24 australites.

Table 55-C-2. *Cesium concentrations in meteorites*

Ordinary chondrites	ppm Cs	Analytical method	Reference
Allegan	0.060	N	5
	0.067	I	5
Atoka	0.0045	N	5
Beardsley	0.913	I	1
	0.193	I	3
Beaver Creek	0.038	N	5
	0.034	I	5
Bjurböle	0.11	I	2
Bluff	0.01	I	2
	0.14	S	4
Bluff A	0.0086	N	5
	0.0088	I	5
Bluff B	0.0037	N	5
	0.0038	I	5
Bremervörde	0.192	N	5
Chandakapur	0.20	I	2
Chateau-Renard	0.01	I	2
Crumlin	0.058	N	5
	0.062	I	5
Dhurmsala	0.10	S	4
	0.023	N	5
Eli Elwah	0.057	N	5
Elm Creek	0.032	N	5
Forest City	0.10	N	2
	0.106	I	3
Futtehpur	0.011	N	5
Gilgoin	0.140	N	5
	0.136	I	5
Hendersonville	0.0085	N	5
Hessel	0.015	N	5
Holbrook	0.489	I	1
	0.28	I	2
	0.146	I	3
Homestead	0.05 (5)	I	2
	0.07	N	2
Khor Temiki	0.060	N	5
Knyahinya	0.0084	N	5
	0.0080	I	5
Limerick	0.10	N	2
	0.11	S	4
	0.092	I	5

Table 55-C-2 (Continued)

Ordinary chondrites	ppm Cs	Analytical method	Reference
Long Island	0.01	N	2
	0.16	S	4
	0.0106	N	5
	0.0110	I	5
Mangwendi	0.10	S	4
	0.047	N	5
	0.050	I	5
Marion	0.0157	N	5
	0.0166	I	5
Merua	0.072	N	5
Modoc	0.08	N	2
	0.09	I	3
	0.11	S	4
Mt. Browne	0.0046	N	5
Ness County	0.04	N	2
Nininger 1349	0.913	I	3
Ochansk	0.12	I	2
Ochansk A	0.100	N	5
Ochansk B	0.092	N	5
Olivenza	0.074	N	5
Richardton	0.088	I	3
Achondrites:			
Johnstown	0.0076	N	5
Moore County	0.0052	I	3
Nuevo Laredo	0.018	I	3
Pasamonte	0.284	I	1
	0.011	I	3
white fraction	0.011	I	3
Sioux County	0.0118	I	3
Carbonaceous chondrites:			
Felix	0.045	N	5
Ivuna	0.181	N	5
	0.188	I	5
Mighei	0.119	N	5
	0.134	I	5
Mesosiderites:			
Estherville	0.020	N	5
Pallasites:			
"Bolivia"	0.036	N	5
Imilac	0.0027	N	5

References: 1. Gordon *et al.* (1957). 2. Webster *et al.* (1958). 3. Gast (1960a). 4. Ahrens *et al.* (1960). 5. Smales *et al.* (1964).

Revised manuscript received: December 1969

IV. Lunar Materials

Data on cesium in lunar fines and common lunar rock types are given in Table 55-C-3. Like K and Rb, Cs is depleted in lunar rocks. The highest concentrations are found in KREEP-type rocks and the lowest in anorthosites.

Cesium data for lunar samples are mainly obtained by neutron activation. Some data were obtained by spark source and isotope dilution mass spectrometry.

Table 55-C-3. *Cesium content of lunar rocks and fines* (in ppm)[a]

Rock type	$\bar{x}$	s	Range	n	Rb/Cs
Mare basalts					
Apollo 11 A	0.18	0.018	0.16–0.20	5	27–36
Apollo 11 B	0.036		0.028–0.049	4	24–30
Apollo 12	0.058	0.018	0.030–0.085	11	13–34
Apollo 15	0.039	0.011	0.028–0.062	7	12–27
Apollo 17	0.022		0.015–0.029	2	26–28
KREEP type rocks					
Apollo 12 KREEP	0.68			1	26
Apollo 12 sample 12013	1.9			1	20
Apollo 14 Breccias	1.07	0.22	0.80–1.4	5	13–25
Apollo 14 basalts	0.52		0.47–0.57	2	24–27
Apollo 16 KREEP	0.46		0.39–0.54	3	19–26
Apollo 17 noritic breccias	0.27		0.23–0.31	4	24–26
Highland rocks					
Apollo 15 anorthositic rocks,	0.023			1	7
Apollo 16 anorthositic rocks, >31% Al_2O_3	0.03		0.0012–0.07	5	10–20
Apollo 16, 25–31% Al_2O_3	0.12	0.09	0.04–0.28	10	15–25
Apollo 16, 21–24% Al_2O_3	0.33	0.15	0.13–0.53	5	10–24
Fines					
Apollo 11, sample 10084	0.117			1	25
Apollo 12	0.25		0.22–0.29	3	20–26
Apollo 12, high K	0.42		0.38–0.45	2	22–26
Apollo 14	0.66	0.077	0.57–0.79	7	21–25
Apollo 15	0.189	0.057	0.07–0.27	16	19–32
Apollo 16	0.106	0.024	0.059–0.14	16	11–29
Apollo 17	0.118	0.054	0.047–0.195	13	17–28
Luna 16	0.071	0.010	0.060–0.082	5	22–29
Luna 20	0.070		0.070	2	21–23

[a] $\bar{x}$ is derived by first averaging all reliable existing data on one sample; the different samples are then averaged (each sample is given equal weight). Standard deviation(s) is calculated from (n) which is the number of samples for which data exist.

Table 55-C-3 (continued)

References from which concentration data are obtained

Proceedings of the Apollo 11 Lunar Science Conference. Geochim. Cosmochim. Acta, Suppl. 1, **2** (1970).
Proceedings of the Second Lunar Science Conference. Geochim. Cosmochim. Acta, Suppl. 2, **1** and **2** (1971).
Proceedings of the Third Lunar Science Conference, Geochim. Cosmochim. Acta, Suppl. 3, **2** (1972).
Proceedings of the Fourth Lunar Science Conference, Geochim. Cosmochim. Acta, Suppl. 4, **2** (1973).
Luna 16 issue. Earth Planet. Sci. Letters **13**, 223 (1972).
Luna 20 issue. Geochim. Cosmochim. Acta **37** (4), 719 (1973).
Chamberlain, J. W., Watkins, C. (eds.): The Apollo 15 Lunar Samples. Houston: Lunar Science Institute 1972.
Watkins, C. (ed.): Lunar Science-III. Houston: Lunar Science Institute 1972.
Chamberlain, J. W., Watkins, C. (eds.): Lunar Science IV. Houston: Lunar Science Institute 1973.
Lunar Science V. Houston: Lunar Science Institute 1974.

Manuscript received: September 1974

55-D. Abundance in Rock Forming Minerals

Most of the Cs in the crust is contained in the micas and K-feldspars. No data exists on Cs concentrations in minerals from ultramafic rocks. K/Cs ratios decrease with differentiation and pegmatite formation. The ionic differences between K and Cs are sufficient to cause formations of independent Cs-minerals in some pegmatites.

K-feldspars from rocks other than pegmatites contain from less than 1 (in intermediate rocks) to some tens ppm (in fractionated granites) of Cs. Pegmatite K-feldspars may occasionally contain some thousand ppm Cs, but are most commonly in the range 10—100 ppm. Less than 1 ppm Cs may also be present in such feldspars.

IIYAMA (1968) studied experimentally the distribution of Cs (and Rb, Sr, Ba) between potassium feldspar and plagioclase of different compositions at 600° C and 1,000 bars. He found the partition coefficient [ppm Cs(Or)/ppm Cs (Plag)] to be about 4 for all pairs of feldspars.

The micas have the highest Cs contents of ordinary rock forming minerals. Biotite from granulite facies rocks contains less than 10 ppm Cs (HEIER, 1960) but higher concentrations are found in biotites from amphibolite facies gneisses and from granites. Muscovites typically contain more Cs than biotites, and muscovites from pegmatites may contain some thousand ppm Cs. HEIER and ADAMS (1964) reported a maximum of 3,400 ppm and a minimum of 140 ppm Cs in 11 analysed muscovites. The maximum Cs concentrations are found in Li-micas where more than 1 per cent Cs has been reported (FOSTER, 1960).

Cs has such a marked tendency to accumulate in the volatile phase that it forms its own mineral pollucite in some pegmatites. Pollucite contains 34.3 per cent Cs (and up to 3.4 per cent Rb). Determinations of Cs in pollucite are by AHRENS (1945, 1947) and NICKEL (1961).

A maximum of 3.9 per cent Cs has been reported from beryls (SOSEDKO, 1957).

Revised manuscript received: December 1969

55-E. Abundance in Common Igneous Rock Types and Terrestrial Abundance

I. Ultramafic Rocks

HORSTMAN (1957) indicated less than 1 ppm Cs in "ultramafic" rocks. TUREKIAN and WEDEPOHL suggested 0.X ppm and VINOGRADOV 0.1 ppm Cs in ultramafic rocks. HEIER and ADAMS (1964) suggested that the Cs content in ultramafic rocks could be less than 0.01 ppm. Due to the lack of direct Cs determinations in such materials, the K/Cs ratio in meteorites could serve as a first approximation of the Cs content. The average K/average Cs ratio in ordinary chondrites is 841/0.073, the average K content of ultramafic rocks about 20 ppm (Chapter 19) which could indicate an average Cs concentration of 0.0017 ppm.

GURNEY *et al.* (1966) found an average of 5.9 ppm Cs (K/Cs = 1.817) in eclogites from the Roberts Victor Mine, South Africa, and 6 ppm Cs (K/Cs = 5.100) in kimberlites. THEY find it impossible to explain the high Cs content without invoking some secondary enrichment process. In one case they found the secondary material containing the Cs to have an analcite-type structure. FORNASERI and PENTA (1960) showed that analcites from cavities in basalt contain on average 13 ppm Cs (see Section 55-F).

II. Volcanic Rocks

Recent determinations of Cs in basalts are by AHRENS *et al.* (1960), AHRENS and EDGE (1961), and GAST (1960a, b, 1965). ERLANK and HOFMEYER (1966) give Cs concentrations in Karroo dolerites, and HORSTMAN (1957) and TAYLOR and WHITE (1966) in andesites. Cs contents in rhyolites are given by JENK and GOLDICH (1956) and HORSTMAN (1957). The data are summarized in Table 55-E-1.

Added in proof: HART [Earth and Planetary Science Letters *6*, 295—303, (1969)] reported 0.024 ppm Cs and an average K/Cs ratio of 70,000 in unaltered submarine basalts. HART and NALWALK [Geochim. Cosmochim. Acta *34*, 145—155 (1970)] found an average Cs content of 0.1191 (range 0.038 to 0.2714) ppm and K/Cs ratio of 31,936 (range 4,200 to 81,000) in ten Puerto Rican trench basalts.

III. Plutonic Rocks

There has been little data on Cs in plutonic igneous rocks since the review article of HEIER and ADAMS (1964). The data are listed in Table 55-E-2. Information about Cs in gabbroic rocks is very limited, but may be estimated as approximately 0.1 ppm. The Cs concentration increases in the more differentiated rocks and is on average between 2 and 3 ppm in granodiorites and 3 to 6 ppm in granites. The high Cs content given by GERASIMOVSKII for some miascitic nepheline syenites refer to old analyses and may be in error, but Cs can reach high levels in these rocks.

Table 55-E-1. *Cesium in volcanic rocks*

	No. of samples	Analytical method	% K	ppm Cs	References
Olivine basalts	1	S	0.75	0.89	Ahrens *et al.* (1960)
Basalt	3	S	1.24	1.70	Ahrens *et al.* (1960)
Basalt ("diabase") (W-1)	1	S	0.53	1.08	Ahrens *et al.* (1960)
Basalt	1	S	1.30	1.53	Ahrens *et al.* (1960)
Olivine basalt, Mid-Atlantic Ridge	1	I	0.14	0.04	Gast (1960a)
Olivine basalt, Hawaii	1	I	0.36	0.10	Gast (1960a)
Basalt, New Zealand	1	S	0.33	0.31	Ewart *et al.* (1968)
Basalt, pigeonite bearing, Karroo	1	S	0.58	1.6	Erlank and Hofmeyer (1966)
Basalt, main dike, Karroo	1	S	0.51	1.0	Erlank and Hofmeyer (1966)
Basalt, vertical series, Karroo	1	S	0.60	1.2	Erlank and Hofmeyer (1966)
Basalt, chilled contact, Karroo	1	S	0.60	0.27	Erlank and Hofmeyer (1966)
Basalt, pigeonite bearing, Karroo	1	S	0.66	0.98	Erlank and Hofmeyer (1966)
Basalt, pigeonite bearing, Karroo	1	S	0.47	4.2	Erlank and Hofmeyer (1966)
Olivine basalt, Karroo	1	S	0.60	0.69	Erlank and Hofmeyer (1966)
Olivine basalt, Karroo	1	S	0.52	0.24	Erlank and Hofmeyer (1966)
Basalt, chilled contact, Tasmania	1	S	0.66	1.3	Erlank and Hofmeyer (1966)
Basalt (estimated average)			0.83	1.1	Turekian and Wedepohl (1961)
Basalt (estimated average)			1.26	1.7	Heier and Adams (1964)
Basalt (estimated average)			0.83	1.0	Taylor (1964)
Basalt (estimated average)			0.59	1.0	Taylor and White (1966)
Mafic andesite, New Zealand	4	S	0.58	1.2	Taylor and White (1966)
Andesite, New Zealand	2	S	0.95	1.5	Taylor and White (1966)
Andesite, New Zealand	6	S	1.23	2.0	Ewart *et al.* (1968)
Andesite, Japan	7	S	0.75	1.5	Taylor and White (1966)
Andesite (estimated average)			1.69	2.3	Heier and Adams (1964)
Andesite	1	S		4	Horstman (1957)
Dacite, New Zealand	1	S	1.99	1.1	Ewart *et al.* (1968)

Table 55-E-1 (Continued)

	No. of samples	Analytical method	% K	ppm Cs	References
Rhyolitic ignimbrites, New Zealand	13	S	3.20	4.0	EWART *et al.* (1968)
Rhyolite (estimated average)			3.80	5.1	HEIER and ADAMS (1964)
"Diabase"-granophyre sequence:					
1. Dillsburg, Pa.					
"diabase"	3	N	0.72	2.60	GOTTFRIED *et al.* (1968)
pegmatitic facies	1	N	0.84	0.89	GOTTFRIED *et al.* (1968)
transitional	1	N	1.43	2.7	GOTTFRIED *et al.* (1968)
granophyric basalt	1	N	1.15	1.8	GOTTFRIED *et al.* (1968)
granophyres	2	N	1.18	1.5	GOTTFRIED *et al.* (1968)
2. Great Lake intrusion, Tasmania					
chilled margin	1	N	1.29	3.0	GOTTFRIED *et al.* (1968)
lower zone basalts	7	N	0.58	0.63	GOTTFRIED *et al.* (1968)
central zone basalts	7	N	1.19	1.7	GOTTFRIED *et al.* (1968)
granophyres	2	N	2.10	3.5	GOTTFRIED *et al.* (1968)

Table 55-E-2. *Cesium in plutonic rocks*

	No. of samples	ppm Cs	Reference	Analytical method
Gabbro, Skaergaard chilled marginal	1	0.10	CABELL and SMALES (1957)	N
Gabbro	1	0.46	AHRENS *et al.* (1960)	S
Granites, Russia, pre-Hercynian	60	2.5	BEUS and FABRIKOVA (1961)	F
Granites, Hercynian and post-Hercynian	46	9.5	BEUS and FABRIKOVA (1961)	F
Granites, Russia (all samples)	—	5.5	BEUS and FABRIKOVA (1961)	F
Alkaligranites, Northern Nigeria	44	3.4	BUTLER and THOMPSON (1963)	N
Granodiorites and adamellites	20	4.8	BUTLER and THOMPSON (1963)	S
Gneissic granites	4	5.9	BUTLER and THOMPSON (1963)	S

Table 55-E-2 (Continued)

	No. of samples	ppm Cs	Reference	Analytical method
Leucogranites	8	12.3	Butler and Thompson (1963)	S
High calcium granites		2	Turekian and Wedepohl (1961)	
Low calcium granites		4	Turekian and Wedepohl (1961)	
Granodiorite		3	Heier and Adams (1964)	
Granite		6.8	Heier and Adams (1964)	
Granite		5	Taylor (1964)	
Syenite		0.6	Turekian and Wedepohl (1961)	
Nepheline syenite, Lovozero massif; phase 1		0.6	Gerasimovskii (1966)	
Urtites lujaurites, Lovozero massif; phase 2		1.3	Gerasimovskii (1966)	
Eudialyte lujaurites-tawites, Lovozero massif; phase 3		2.2	Gerasimovskii (1966)	
Average agpaitic nepheline syenite		1.5	Gerasimovskii (1966)	
Average miascitic nepheline syenite		5.6—38	Gerasimovskii (1966)	

IV. Terrestrial Abundance

For reasons pointed out in Chapter 19, the chondritic Cs-concentration is probably not a good guide to the Cs concentration in the earth (volatility). Because of scant knowledge of Cs concentrations in mafic and ultramafic rocks, and also in the high pressure metamorphic rocks believed to be representative of the lower crust, estimates of Cs distribution in the earth can be little more than a guess. The value of 0.0051 ppm for the earth is derived by assuming all Cs is above the lower mantle using the valves of 1 ppm Cs for the oceanic crust, 3 ppm for the continental crust, and 0.01 ppm for the upper mantle.

Revised manuscript received: December 1969

55-F. Behavior in Magmatogenic Processes

Cesium has a marked tendency to concentrate in the volatile phase during magmatic crystallization and may be strongly concentrated in some pegmatites and pegmatite minerals (see Chapter 55-D) where its concentration may be high enough for the formation of pollucite $(Cs, Na)(AlSi_2O_6) \cdot (H_2O)_{<1}$. Evidently the 25 per cent difference in ionic radius between K^+ and Cs^+ makes Cs altogether unstable in K minerals, and it is not concentrated in these minerals before the very late stages of crystallization. FORNASERI and PENTA (1960) showed that analcites from cavities of effusive rocks (basalt) contain on average 13 ppm Cs, but analcites that ave formed by replacement of primary leucite contained on average 615 ppm Cs. The primary leucite contained 169 ppm Cs.

The greater volatility of Cs as compared with Rb and K causes the enrichment of Cs in the sublimates of K-rich volcanic rocks. The lavas of Vesuvius contain, as a sublimate, the borofluoride avogadrite $(K, Cs)BF_4$ with from 0 to 19 per cent $CsBF_4$.

Two described pegmatites are particularly noted for their Cs concentrations: Varuträsk, Sweden (QUENSEL, 1956) and Bernic Lake, Canada (NICKEL, 1961).

Revised manuscript received: December 1969

55-G. Behavior during Weathering and Abundance in Soils

BERTRAND and BERTRAND (1949) found a range of 0.3 to 25.7 ppm Cs in soils of France and Italy. Largest quantities of Cs occurred in soils over igneous rocks and in alkaline soils. VINOGRADOV (1959) reported values (from other researchers) of approximately 5 ppm Cs in soils of the Russian Plain and 1 ppm in Japanese soils.

Cesium released by weathering or derived from atmospheric precipitation is rapidly and strongly absorbed by solid soil material (DAVIS, 1963).

Revised manuscript received: December 1969

55-H. Adsorption by Minerals

Although there is no single universal order of the replacing power of cations, it is often found to be Li$<$Na$<$K$<$Rb$<$Cs (WIKLANDER, 1964). This has been found on such diverse materials as NH_4-montmorillonite, humus, and beidellite. It should be stressed that exchange adsorption is a function of concentration in the solution. The very low concentration of Cs with respect to Na and K in natural solutions tends to decrease its exchange capability. A careful study of cesium adsorption on clay minerals indicated: a) the logarithm of the distribution coefficient for cesium adsorption, at very low cesium concentrations, is approximately a linear function of the logarithm of the competing ion's concentration, and b) cesium will adsorb rapidly when sodium is the dominant competing ion (WAHLBERG and FISHMAN, 1962). Consideration of ion exchange as a function of "equivalent anion" size (negative electrostatic field strength) (EISENMAN, 1962) leads to the conclusion that Cs should be preferentially adsorbed by most clays (JENNE and WAHLBERG, 1968).

Revised manuscript received: December 1969

55-I. Abundance in Natural Waters

I. Continental Waters

SREEKUMARAN *et al.* (1968) reported values from eight western U.S. rivers of 0.004 to 0.024 ppb Cs and an average value of 0.018 ppb. They found an average value for Lake Mead water of 0.057 ppb Cs. KHARKAR *et al.* (1968) investigated the stream supply of dissolved Cs to the oceans. They found a range of Cs in ten U.S. rivers of 0.011 to 0.028 ppb, a value for the Rhone river of France of 0.042 ppb and an overall average of 0.020 ppb. LIVINGSTONE (1963) reported six analyses of Japanese rivers ranging between 0.05 and 0.2 ppb Cs.

COLLINS (1963) developed a flame photometric method for determining Cs in oil field brines. From a few randomly-selected samples, he found a range up to 2 ppm (mg/liter) Cs. No brines were found that contained Cs without also containing Rb.

II. Sea Water

BOLTER *et al.* (1964) reported an average Cs^+ content in 28 samples of sea water of 0.30 ppb (at 19.0‰ chlorinity). Cesium did not appear to be in exchange equilibrium with deep-sea clays and seemed to have slightly higher values in Gulf of Mexico waters. RILEY and TONGUDAI (1966) reported an average Cs^+ value of 0.55 ± 0.06 ppb in sea water from the Irish Sea and the North Atlantic. SMALES and SALMON (1955) found 0.5 ± 0.07 ppb Cs^+ in North Atlantic water by activation. GOLDBERG (1965) suggests 0.5 ppb Cs^+ in average sea water. The residence time of Cs^+ is reasonably long [6.5×10^4 yrs., BOLTER *et al.* (1964); 4×10^4 yrs., GOLDBERG (1965)].

III. Thermal Waters

ELLIS and MAHON (1964) reported 17 values of Cs in New Zealand thermal waters ranging from 0.13 to 2.6 ppm Cs and values ranging from primarily < 0.1 to 0.55 ppm in warm springs.

GOLDING and SPEER (1961) reported values of Cs in New Zealand thermal waters ranging from 0.02 to 4.7 ppm. ELLIS and MAHON (1964) also performed hot-water leaching experiments on various volcanic rocks and greywackes of the area. They did not find detectable Cs (< 0.1 ppm) in the hot waters at temperatures up to 350° C and periods up to 480 hours. At temperatures of 600° C some Cs (up to 1.4 ppm) was released by the rock.

Revised manuscript received: December 1969

55-K. Abundance in Common Sediments and Sedimentary Rock Types

Chester (1965) discussed the distribution of alkalies in marine sediments. He concluded that the rare alkalies have complex distributions primarily related to clay mineral contents of the sediments and that insufficient data are available for Cs. Much of the information concerning Cs in sedimentary rocks has been reported by Horstman (1957). Horstman reported general values for clay minerals: kaolinite (14 ppm); mixed-layer minerals (17 ppm); K-bentonites (8 ppm); and illites (7 ppm). These relative values do not agree well with predictions based on alkali ion-exchange experiments with clays. Horstman reported average values of 5 ppm Cs for marine shales, 1 ppm for sandstones and limestones, and 6 ppm for modern sediments.

Welby (1958) found that 40% of Gulf of Mexico silicate sediments contained between 6 and 9 ppm Cs. The quartz-rich near-shore sediments generally contained less than 6 ppm. Globigerina oozes from the Gulf of Mexico averaged 8 ppm which compares with his value of 8 ppm for a composite of the Selma Chalk formation.

Spencer (1966) found an average of 14 ppm Cs in a Silurian graptolite band in Wales. He treated the data by factor analysis of 27 chemical elements and concluded the Cs distribution was best explained by a factor involving ion-exchange and clay minerals.

Sreekumaran *et al.* (1968) found an average of 1.1 ppm Cs in seven samples of suspended matter in western U.S. streams and an average of 2.6 ppm Cs in their bottom sediments. They reported Cs concentrations on a carbonate-free basis in deep sea sediments of 0.32, 0.91, and 1.35 ppm Cs. This compares with values of 0.34, 0.51, and 1.48 ppm reported from deep-sea sediments by Smales and Salmon (1955). Sreekumaran *et al.* (1968) reported one value for near-shore sands of 2.17 ppm.

Hirst (1962) reported the Cs content of 15 modern marine sands to range from 1 to 3.2 ppm and of modern marine clays to range from 7.9 to 13 ppm Cs. These values demonstrate the expected increase in Cs as a function of finer grain size, increased clay mineral content, and increased number of ion exchange sites. Jenne and Wahlberg (1968) have shown high adsorption rates for Cs on clay minerals by tracer studies. Hirst reported that varying rates of sedimentation seemed to have little affect on K/Cs ratios suggesting that most of the Cs was detrital.

Revised manuscript received: December 1969

55-L. Biogeochemistry

Bertrand and Bertrand (1949b) detected natural Cs in all of 13 species of invertebrates and 22 species of vertebrates. The invertebrates averaged 138 ppm of dry weight and the vertebrates averaged 32 ppm of dry weight. Pendleton and Hanson (1958) found lake animals to concentrate ^{137}Cs by factors of 600 to 11,000 over the ^{137}Cs concentrations in the lake water.

McKerrow *et al.* (1956) reported Cs to be concentrated relative to K or Rb in fossil trilobites and to be concentrated relative to the Cs content of adjacent sediments.

Bertrand and Bertrand (1949a) determined natural Cs in flowering species of plants and found a range of 3 to 89 ppm on a dry weight basis (average = 22 ppm). Smales and Salmon (1955) reported Cs values of 0.096—0.13 ppm of dry weight in seven seaweed species. Pendleton and Hanson (1958) found lake plants to concentrate ^{137}Cs by factors of 50 to 25,000 compared to lake water.

Revised manuscript received: December 1969

55-M-N. Abundance in Common Metamorphic Rocks and Behavior in Metamorphic Reactions

There are no studies particularly pertaining to the movement of Cs during regional metamorphic processes but Heier and Adams (1964) pointed out that Cs is probably strongly depleted in high grade metamorphic rocks (granulite facies). This would coincide with the disappearance of biotite as a stable phase. Heier (1960) also found the less abundant biotites in granulite facies rocks to contain less Cs than biotite in amphibolite facies rocks. Regional metamorphic processes acting on crustal material are considered to be effective in concentrating Cs in the upper levels of the continental crust.

Revised manuscript received: December 1969

55-O. Geochemical Behavior of Cs

Its large size makes cesium unsuitable for the lattices of minerals stable in the mantle and it is very strongly concentrated upwards in the earth's crust. Most of the cesium is contained in K-minerals, feldspars and micas, but under late stage hydrothermal pegmatite conditions its concentration can be so high that independent Cs minerals may form [pollucite; $(Cs, Na)(AlSi_2O_6)(H_2O)_{<1}$]. Avogadrite $(K, Cs)BF_4$, which is found as sublimate at Vesuvius may contain up to 19 per cent $CsBF_4$.

The low concentration of Cs in common rocks and minerals makes its analytical determination difficult with rapid methods without pre-enrichment. Thus the detailed geochemistry and abundance data of Cs are the least known among the alkali metals. Recent reviews of the geochemistry of Cs are by HORSTMAN (1957) and HEIER and ADAMS (1964).

Element 87, francium (Fr), occurs in nature only in minute quantities as a short-lived radio-isotope found by the branching decay of actinium. This element is not considered among the alkali elements in this handbook.

Revised manuscript received: December 1969

References: Sections 55-B to 55-L, 55-M/N, 55-O

AHRENS, L. H.: Quantitative spectrochemical examination of the minor constituents in pollucite (Norway, Maine). Am. Mineralogist **30**, 616—622 (1945).

— Analyses of the minor constituents in pollucite. Amer. Mineralogist **32**, 44—51 (1947).

—, and R. A. EDGE: The K/Cs ratio in some basic rocks. Geochim. Cosmochim. Acta **25**, 91—94 (1961).

— —, and S. R. TAYLOR: The uniformity of concentration of lithophile elements in chondrites — with particular reference to Cs. Geochim. Cosmochim. Acta **20**, 260—272 (1960).

ALLER, L. H.: Solar and stellar abundances of the elements. Phys. Chem. Earth **4**, 1—26 (1961a).

— The abundance of the elements. New York: Interscience Publishers 1961b.

BERGGREN, T.: Minerals from Varuträsk pegmatite. XV. Analyses of the mica minerals and their interpretation. Geol. Foren. Forh. **62**, 182—193 (1940).

BERTRAND, G., et D. BERTRAND: Sur la presence et la teneur en cesium des terres arables. Compt. Rend. **229**, 533—535 (1949).

— — Existence normale du cesium chez les vegetaux. Compt. Rend. **229**, 553—555 (1949a).

— — Existence normale du cesium chez les animaux. Compt. Rend. **229**, 609—610 (1949b).

BEUS, A. A., and E. A. FABRIKOVA: Distribution of cesium in the granites of the USSR. Geochemistry **1961**, 970—976 (1961).

BOLTER, E., K. K. TUREKIAN, and D. F. SCHUTZ: The distribution of rubidium, cesium and barium in the oceans. Geochim. Cosmochim. Acta **28**, 1459—1465 (1964).

BOROVIC-ROMANOVA, T. F., and A. F. SOSEDKO: Rubidium contents in beryls from pegmatite veins of the Kola peninsula. Dokl. Akad. Nauk SSSR **118**, 534—536 (1958).

BUTLER, J. R., and A. J. THOMPSON: Cesium in some alkali granites (younger granites) of Northern Nigeria. Geochim. Cosmochim. Acta **27**, 769—773 (1963).

CABELL, M. J., and A. A. SMALES: The determination of rubidium and caesium in rocks, minerals and meteorites by neutron-activation analysis. Analyst **82**, 390—405 (1957).

CAMERON, A. G. W.: A revised table of abundances of the elements. Astrophys. J. **129**, 676 (1959).

CHESTER, R.: Elemental geochemistry of marine sediments, In: Chemical Oceanography, vol. II (RILEY, J. P., and G. SKIRROW, eds.). Academic Press 1965.

CLAYTON, D. D., and W. A. FOWLER: Abundances of heavy nuclides. Ann. Phys. **16**, 50 (1961).

COLLINS, A. G.: Flame spectrophotometric determination of cesium and rubidium in oil field waters. Anal. Chem. **35**, 1258—1261 (1963).

DALY, R.: Igneous rocks and the depth of the earth. New York: McGraw Hill 1933.

DAVIS, J. J.: Cesium and its relationship to potassium in eclogy. Radioeclogy (SCHULTZ and KLEMENT (eds.). New York: Reinhold Pub. Corp. 1963.

EISENMAN, G.: On the elementary origin of equilibrium ionic specificity. In: (A. KLEINZELLER and A. KOTYK (eds.), Symposium on Membrane Transport and Metabolism New York: Academic Press 1962.

ELLIS, A. J., and W. A. J. MAHON: Natural hydrothermal systems and experimental hot-water/rock interactions. Geochim. Cosmochim. Acta **28**, 1323—1358 (1964).

ERLANK, A. J., and P. K. HOFMEYER: K/Rb and K/Cs ratios in Karroo dolerites from South Africa. J. Geophys. Res. **71**, 5439—5445 (1966).

EWART, A., S. R. TAYLOR, and A. C. CAPP: Trace and minor element geochemistry of the rhyolitic volcanic rocks, Central North Islands, New Zealand. Contr. Mineral. and Petrol. **18**, 76—104 (1968).

FORNASERI, M., and A. PENTA: Minor alkali elements in analcime and their behaviour in analcimization of leucite. Periodico Mineral. (Rome) **29**, 85—102 (1960).

FOSTER, N. D.: Interpretation of the composition of Li-micas. U.S. Geol. Surv. Profess. Papers **354-E**, 115—147 (1960).

GAST, P. W.: Alkali metals in stone meteorites. Geochim. Cosmochim. Acta **19**, 1—4 (1960a)

— Limitations on the composition of the upper mantle. J. Geophys. Res. **65**, 1287—1297 (1960b).

— Terrestrial ratio of potassium to rubidium and the composition of earth's mantle. Science **147**, 858—860 (1965).

GERASIMOVSKII, V. I.: Geochemical features of agpaitic nepheline syenites. Chemistry of the Earth's Crust, vol. 1 (A. P. VINOGRADOV, ed.). Israel Program Scientific translations (1966).

GOLDBERG, E. D.: Minor elements in sea water. In: Chemical Oceanography, vol. 1 (RILEY and SKIRROW, eds.) Academic Press 1965.

GOLDING, R. M., and M. G. SPEER: Alkali ion analysis of thermal waters in New Zealand. New Zealand Sci. **4**, 203—213 (1961).

GORDON, B. M., L. FRIEDMAN, and G. EDWARDS: Caesium in stony meteorites. Geochim. Cosmochim. Acta **12**, 170—171 (1957).

GOTTFRIED, D., P. GREENLAND, and E. Y. CAMPBELL: Variation of Nb—Ta, Zr—Hf, Th—U and K—Cs in two diabase-granophyre suites. Geochim. Cosmochim. Acta **32**, 925—947 (1969).

GURNEY, J. J., G. W. BERG, and L. H. AHRENS: Observations on caesium enrichment and the potassium/rubidium/caesium relationship in eclogites from the Roberts Victor mine, South Africa. Nature **210**, 1025—1027 (1966).

HEIER, K. S.: Petrology and geochemistry of high-grade metamorphic and igneous rocks on Langöy, Northern Norway. Norg. Geol. Undersokelse **207** (1960).

—, and J. A. S. ADAMS: The geochemistry of the alkali metals. Phys. Chem. Earth **5**, 235—380 (1964).

HIRST, D. M.: The geochemistry of modern sediments from the Gulf of Paria. II. The location and distribution of trace elements. Geochim. Cosmochim. Acta **26**, 1147—1188 (1962).

HORSTMAN, E. L.: The distribution of lithium, rubidium and caesium in igneous and sedimentary rocks. Geochim. Cosmochim. Acta **12**, 1—28 (1957).

IIYAMA, J. T.: Etude experimentale de la distribution d'elements en traces entre deux feldspaths. Feldspath potassique et plagioclase coexistants. I. Distribution de Rb, Cs, Sr et Ba a 600° C. Bull. Soc. Fr. Mineral. Cristallogr. **91**, 130—140 (1968).

JENK, W. F., and S. S. GOLDICH: Rhyolitic tuff flows in southern Peru. J. Geol. **64**, 156—172 (1956).

JENNE, E. A., and J. S. WAHLBERG: Role of certain stream-sediment components in radioion sorption. U.S. Geol. Surv. Profess. Papers **433-F** (1968).

JOHANSEN, A. A.: Descriptive petrography of the igneous rocks. Chicago 1949.

KHARKAR, D. P., K. K. TUREKIAN, and K. K. BERTINE: Stream supply of dissolved silver, molybdenum, antimony, selenium, chromium, cobalt, rubidium and cesium to the oceans. Geochim. Cosmochim. Acta **32**, 285—298 (1968).

KOLBE, P., and S. R. TAYLOR: Geochemical investigation of the granitic rocks of the Snowy Mountains area, New South Wales. J. Geol. Soc. Australia **13**, 1—25 (1966).

LIVINGSTONE, D. A.: Chemical composition of rivers and lakes. U.S. Geol. Survey Profess. Papers **440-G** (1963).

MCKERROW, W. S., S. R. TAYLOR, A. L. BLACKBURN, and L. K. AHRENS: Rare alkali elements in tribolites. Geol. Mag. **93**, 504—516 (1956).

NICKEL, E. H.: The mineralogy of the Bernic Lake pegmatite, Southeastern Manitoba. Dept. of Mines and Techn. Surv. Ottawa, Mines Branch. Tech. Bull. TB **20**, 1—38 (1961).

NOCKOLDS, S. R.: Average chemical composition of some igneous rocks. Bull. Geol. Soc. Am. **65**, No. 10 (1954).

PENDLETON, R. C., and W. C. HANSON: Absorption of ^{137}Cs by components of an aquatic community. In: Proc. 2nd Intern. Conf. Peaceful Uses Atomic Energy, United Nations, Geneva **18**, 419—422 (1958).

QUENSEL, P.: Paragenesis of the Varuträsk pegmatite including a review of its mineral assemblage. Arkiv Mineral Geol. **2**, No. 2, 9—125 (1956).

RILEY, J. P., and M. TONGUDAI: Caesium and rubidium in sea water. Chem. Geol. **1**, 291—294 (1966).

SMALES, A. A., T. C. HUGHES, D. MUPPER, C. A. J. MCINNES, and R. K. WEBSTER: The determination of rubidium and cesium in stony meteorites by neutron activation analysis and by mass spectrometry. Geochim. Cosmochim. Acta **28**, 209—233 (1964).

—, and L. SALMON: Determination by radioactivation of small amounts of rubidium and cesium in sea water and related materials of geochemical interest. Analyst **80**, 37—50 (1955).

SOSEDKO, T. A.: Structure changes and changes in the properties of beryls with increased alkali contents. Zap. Vses. Mineralog. Obshchestva **86**, 495—499 (1957).

SPENCER, D.: Factors affecting element distributions in a Silurian graptolite band. Chem. Geol. **1**, 221—249 (1966).

SREEKUMARAN, C., K. C. PILLAI, and T. R. FOLSOM: The concentrations of lithium, potassium, rubidium and cesium in some western American rivers and marine sediments. Geochim. Cosmochim. Acta **32**, 1229—1234 (1968).

SUESS, H. E., and H. C. UREY: Abundances of the elements. Rev. Mod. Phys. **28**, 53 (1956).

TAYLOR, S. R.: The chemical composition of australites. Geochim. Cosmochim. Acta **26**, 685—722 (1962).

— Abundance of chemical elements in the continental crust; a new table. Geochim. Cosmochim. Acta **28**, 1273—1285 (1964).

— Trace element abundances and the chondritic earth model. Geochim. Cosmochim. Acta **28**, 1989—1998 (1964).

—, and A. J. R. WHITE: Trace element abundances in andesites. Bull. Volcanol. **29**, 177—194 (1966).

TUREKIAN, K. K., and K. H. WEDEPOHL: Distribution of the elements in some major units of the earth's crust. Bull. Geol. Soc. Am. **72**, 175—192 (1961).

USSING, N. V.: Geology of the country around Julianehaab, Greenland. Meddel. Groenland **38** (1911).

VINOGRADOV, A. P.: The geochemistry of rare and dispersed chemical elements in soils. New York: Consultants Bureau 1959.

— Average contents of chemical elements in the principal type of igneous rocks of the earth's crust. Geochemistry **1962**, 641—664 (1962).

WAHLBERG, J. S., and M. J. FISHMAN: Adsorption of cesium on clay minerals. U.S. Geol. Soc. Bull. **1140-A**, 30 pp. (1962).

WEBSTER, R. K., J. W. MORGAN, and A. A. SMALES: Caesium in chondrites. Geochim. Cosmochim. Acta **15**, 150—152 (1958).

WELBY, C. W.: Occurrence of alkali metals in some Gulf of Mexico sediments. J. Sediment. Petrol. **28**, 431—452 (1958).

WIKLANDER, L.: Cation and anion exchange phenomena. In: Chemistry of the Soil (F. E. BEAR, ed.), p. 163—205. New York: Reinhold Publ. Corp. 1964.

Revised manuscript received: December 1969

Barium 56

A K. FISCHER (Lehrstuhl für Kristallographie, Universität des Saarlandes, Saarbrücken, Germany)

B—O H. PUCHELT (Mineralogisch-Petrographisches Institut der Universität, Tübingen, Germany)

56-A. Crystal Chemistry

I. General

In the crystal structures of barium minerals, Ba is surrounded by O, OH, H_2O or a halogen ion as nearest neighbors. Its valency is two and the bond type is predominantly ionic. Of the divalent positive ions, Ba^{2+} has the largest ionic radius (except Ra^{2+}). Isostructural replacement of, or by, other large cations such as Pb^{2+} and Sr^{2+} is frequently observed. Less common is replacement of K^+ and of the smaller ion Ca^{2+}. A few examples of crystal chemical relations which are not listed in the tables below are the following synthetic compounds: $BaFe_{12}O_{19}$, $BaAl_{12}O_{19}$, $SrFe_{12}O_{19}$, $SrAl_{12}O_{19}$ etc. in the magnetoplumbite series; pandaite with Ba partially replacing Ca in the pyrochlore group; isotypy of $K[NO_3]$ and witherite $Ba[CO_3]$; Ca—Ba-mimetisite; heinrichite and meta-uranocircite in the torbernite and meta-torbernite group.

The coordination of Ba has been reviewed by MANOHAR and RAMASESHAN (1964). The coordination number ranges from 6 to 12 for O, OH, H_2O. The existence of a large variety of coordination polyhedra is illustrated in Tables 51-A-1, 51-A-2, and 51-A-3 in which crystal structures of Ba minerals and a few selected synthetic compounds are listed. In general, oxygen atoms with distances larger than 3.3 Å are not considered as sharing one coordination polyhedron (exception: barite). This limitation appears to be justified as in the majority of the structures the distance of the "next-nearest" O neighbors is substantially larger than the range of distances attributed to the first-order coordination. In other cases, however, the distances show a broad and rather uniform range of variation (e.g. Ba_3 in taramellite) so that it is difficult to define the coordination number. The type of coordination polyhedron of Ba apparently has a minor influence on the energy balance of these structures. Nevertheless it can control the structure type as has been demonstrated for silicates by LIEBAU (1962, 1968).

The following averages for atomic distances were obtained for the different coordination numbers indicated:

Coordination number	Atomic distance, Å	
6	2.7_6	(BaO, $BaZnO_2$, benitoite)
7	2.7_8	(paracelsian, $Ba[GeO_3]$ high)
8	2.8_1	($BaNiO_2$, $Ba(OH)_2 \cdot 8H_2O$, $BaUO_4$, gillespite, taramellite)
9	2.8_5	($Ba[S_5O_6] \cdot 2H_2O$, $Ba[B(OH)_4]_2$, $Ba[B_4O_7]$, celsian, sanbornite)
10	2.8_7	(BaO_2, $BaTi_4O_9$, $Ba[B_4O_7]$, $Ba_3[PO_4]_2$)
12	2.9_3	(psilomelane, high $BaTiO_3$, nitrobarite, barite, norsethite, $Ba_3[PO_4]_2$, "hexagonal" $Ba[Al_2Si_2O_8]$, α-$BaO_2 \cdot 2H_2O_2$)

These data demonstrate a general increase of distance with increasing coordination number. This can be considered only as a first approximation because of the restricted selection of minerals and artificial compounds; (for some other compounds used for the computation of the averages, which are not listed in Tables 56-A-1, 56-A-2 and 56-A-3, see Manohar and Ramaseshan, 1964).

II. Crystal Chemistry of Some Minerals and Synthetic Compounds

a) Oxides, Halides

The structure of psilomelane $(Ba, H_2O)_2Mn_5O_{10}$ consists of MnO_6 octahedra and $(Ba, H_2O)O_{12}$ cubo-octahedra (only oxygen atoms with distances $\leqq 3.16$ Å are considered). Ba and H_2O occupy the same equipoint (partial ordering assumed), probably because of very similar distances of Ba—O and O—H_2O atoms and molecules (H-bridges?). The crystal structure of hollandite $BaMn_8O_{16}$ has the same coordination polyhedron for Ba (Byström and Byström, SR **13**, 1950, 186). In synthetic oxides and hydroxides almost all coordination numbers from 6 to 12 occur;

Table 56-A-1. *Coordination of Ba in some oxides and halides*

Mineral or synthetic compound	Coordinated atoms, distances to Ba in Å		Coordination polyhedron	References
$Ba^{[6]}O^{[6]}$	6O	2.76	octahedron	1
$Ba^{[6]}Zn^{[4]}O_2$	6O (2O	2.64—2.97 3.36)	octahedron, severely distorted	2
$Ba^{[8]}Ni^{[4]}O_2$	8O	2.80; 2.84	quadratic prism, severely distorted	3
$Ba^{[8]}(OH)_2 \cdot 8H_2O$	8OH	2.69—2.77	distorted quadratic antiprism	4
$Ba^{[8]}U^{[6]}O_4$	8O	2.71—2.99	irregular	5
$Ba^{[10]}O_2$	8O 2O	2.79 2.68	tetragonal prism with tetragonal pyramids on two base faces	6
$Ba^{[10]}Ti_4^{[6]}O_9$	10O	2.81—3.09	pentagonal prism	7
psilomelane, $(Ba, H_2O)_2^{[12]}Mn_5^{[6]}O^{10}$	8O 4H_2O	2.85—3.16 2.78; 2.88	distorted cubo-octahedron	8
$Ba^{[12]}Ti^{[6]}O_3$ (high-temp. phase)	12O	2.83	cubo-octahedron	9
$Ba^{[8]}F_2^{[4]}$	8F	2.68	cube	10
$Ba^{[9]}Cl_2 \cdot 2H_2O$	7Cl 2H_2O	3.14—3.38 2.76	distorted trigonal prism with 2Cl and 1H_2O on top of prismatic faces	11

References: 1. Gerlach (SB 1913—1928, 119). 2. v. Schnering, Hoppe and Zemann (SR **24**, 1960, 454). 3. Lander (SR **15**, 1951, 180); for structure, see Lander (1951). 4. Manohar and Ramaseshan (1964). 5. Samson and Sillén (SR **11**, 1947—48, 441). 6. Abrahams and Kalnajs (SR **18**, 1954, 364). 7. Templeton and Dauben (SR **24**, 1960, 326). 8. Wadsley (SR **17**, 1953, 429). 9. Goldschmidt (SB **1**, 1913—1928, 333). 10. Davey (SB **1**, 1913—1928, 187). 11. Jensen (SR **10**, 1945—1946, 95).

halides and hydrated halides are known with coordination numbers of 8 and 9. A selection of these compounds (especially oxides and hydroxides with coordination polyhedrons not described for minerals in this section) is included in Table 56-A-1.

b) Nitrates, Carbonates, Borates

The BaO_{12} icosahedron found in nitrobarite $Ba[NO_3]_2$ was also observed in $Ba[ClO_4]_2 \cdot 3H_2O$, α-$BaO_2 \cdot 2H_2O_2$, $Ba[SiF_6]$ and $Ba[GeF_6]$ (MANOHAR and RAMASESHAN, 1964). In the carbonate minerals, Ba replaces Ca in the aragonite series (witherite and solid solutions) with a coordination polyhedron which can be described as a distorted cube with one of its edges elongated for housing the 9th O atom above its centre; the same coordination polyhedron is also reported for sanbornite $Ba_2[Si_4O_{10}]$. Barytocalcite $BaCa[CO_3]_2$ has an independent structure also with a coordination number of 9 (two distances of 3.28 and 3.30 Å are not included). In norsethite, $BaMg[CO_3]_2$, Ba has a 6+6 coordination forming a distorted ditrigonal prism (LIPPMANN, 1968). No barium borate mineral structures are known at present. Of the synthetic borates, $Ba[B(OH)_4]_2$ is isostructural with the corresponding Sr borate (see subsection 38-A-II), whereas $Ba[B_4O_7]$ is not (coordination numbers 9 and 10).

c) Sulfates, Phosphates

Barite $Ba[SO_4]$ is isostructural with anglesite $Pb[SO_4]$ and celestite $Sr[SO_4]$. (Solid solutions are known as angleso-barite, baryto-celestite and calcio-barite.)

Table 56-A-2. *Coordination of Ba in some oxo-salts*

Mineral or synthetic compound	Coordinated atoms, distances to Ba in Å	Coordination polyhedron	References
Nitrobarite, $Ba^{[12]}[NO_3]_2$	6+6O 2.86; 2.95	distorted icosahedron	1
Witherite, $Ba^{[9]}[CO_3]$	9O ca. 2.8	see text	2
Barytocalcite, $Ba^{[9]}Ca^{[6]}[CO_3]_2$	9O 2.56—2.99 (2O 3.28; 3.30)		3
Norsethite, $Ba^{[12]}Mg^{[6]}[CO_3]_2$	6+6O 2.72; 3.18	distorted ditrigonal prism	4
$Ba^{[9]}[B(OH)_4]_2$	Ba_1: 9OH 2.73—2.90 Ba_2: 9OH 2.77—2.99	trigonal prism with centered faces	5
$Ba^{[9]}Ba^{[10]}[B_4^{[3]}B_4^{[4]}O_{14}]$	Ba_1: 9O 2.61—3.08 Ba_2: 10O 2.71—3.12	irregular	6
Barite, $Ba^{[2]}[SO_4]$	10O 2.76—3.08 (2O 3.30)	see text	7
$Ba^{[12]}Ba_2^{[10]}[PO_4]_2$	Ba_1: 6+6O 2.80; 3.23 Ba_2: 10O 2.71—2.83	distorted icosahedron see text	8

References: 1. VEGARD and BILBERG (SB **2**, 1928—1932, 386); cf. LUTZ (SR **24**, 1960, 421). 2. COLBY and LA COSTE (SB **3**, 1933—35, 407). 3. ALM (SR **24**, 1960, 425). 4. LIPPMANN (1968). 5. KRAVCHENKO (1965). BLOCK and PERLOFF (1965). 7. SAHL (1963); see also COLVILLE and STAUDHAMMER (1967). 8. ZACHARIASEN (SR **11**, 1947—48, 388).

The coordination polyhedron based on a coordination number of 12 consists of two nearly parallel rings of 5 and 6 oxygens on both sides of the Ba atom, with a 12th O on top of the 6-ring (for an alternative description, see MANOHAR and RAMASESHAN, 1964). In this coordination polyhedron, 2 O with distances of 3.30 Å are included. Omitting them from the 5-ring would lead to a rather one-sided coordination polyhedron (SAHL, 1963). $Ba_3[PO_4]_2$, isostructural with $Sr_3[PO_4]_2$, has two coordination polyhedra for Ba with coordination numbers of 12 and 10; the latter one consists of a distorted "hexagonal" pyramid (symmetry 3m) slightly above the Ba atom and a triangle below.

d) Silicates

$Ba_2^{[6]}[SiO_4]$ and $Sr_2^{[6]}[SiO_4]$ are isostructural with olivine (O'DANIEL and TSCHEISCHWILI, SR **9**, 1942—44, 261). Another silicate with isolated tetrahedra, $BaO \cdot SiO_2 \cdot 6H_2O$, has two different coordination polyhedra with coordination numbers of 8 and 10 which share faces with each other but have no common corner with the $[SiO_4]$ tetrahedron. The ten-fold coordination can be described as a pentagonal pyramid above a square. In benitoite $BaTi[Si_3O_9]$ (isomorphous to pabstite $BaSn[Si_3O_9]$), the coordination polyhedron is severely distorted. (The next 6 O atoms beyond the nearest ligands are not considered because of their distance of 3.43 Å.) Taramellite $Ba_2(Fe, Ti, Mg)_2[Si_4O_{12}](OH)_2$, with rings of $4[SiO_4]$ tetrahedra, has three different coordination polyhedra for Ba with a coordination number of 6, 6 and 7 (?). The last coordination polyhedron can also be described as having a coordination number of 9 (two additional O's at a distance of 3.12 Å) or with a coordination number of 11 (two additional oxygen atoms at a distance of 3.21 Å). $Ba[GeO_3]$, with a germanate chain of the "Zweierketten"-type, is one of the few examples for a coordination number of 7. (Higher condensed silicate chains with Ba as cations have been investigated by KATSCHER and LIEBAU (1965, 1966)). In the sheet silicates, small percentages of Ba have been found in some muscovites. The mineral anandite $(Ba, K)(Fe, Mg)_3[(Si, Al, Fe)_4O_{10}](O, OH)_2$ was recently described as a member of the trioctahedral mica family (PATTIARATCHI, SAARI and SAHAMA, 1967). Other Ba silicates with single sheet structures do not have the usual plane silicate sheet of the mica type and do not contain OH or H_2O (LIEBAU, 1968). Sanbornite, $Ba[Si_2O_5]$, with undulating sheets built of 6-membered rings has the same coordination polyhedron for a coordination number of 9 as described for witherite. In the structure of gillespite $BaFe[Si_4O_{10}]$ (with folded sheets of 4- and 8-membered rings of $[SiO_4]$ tetrahedra), the coordination polyhedron of Ba can be derived from a cube whose faces are each divided into 2 triangles to form a triangular dodecahedron of symmetry $\bar{4}$ (cf. also cuprorivaite: PABST, 1959; MAZZI and PABST, 1962). Ba replaces Sr or Ca in synthetic compounds of the melilite series ($Ba_2[Fe^{[4]}Si_2O_7]$, $Ba_2[Mn^{[4]}Si_2O_7]$, cf. WYCKOFF, 1968, p. 225—227)[1]. For the non-feldspar form of $Ba[Al_2Si_2O_8]$, containing double sheets, a hexagonal (or pseudo-hexagonal) prism has been described as the coordination polyhedron (on the orthorhombic distortion, cf. TAKÉUCHI, SR **22**, 1958, 501). The same coordination polyhedron has been found in cymrite $Ba[AlSi_3O_8(OH)]$. In the Ba-feldspar celsian $Ba[Al_2Si_2O_8]$, the coordination polyhedron is similar to that in witherite (if a tenth oxygen atom at a

[1] The classification of tetrahedral structures follows ZOLTAI's suggestion (1960).

Table 56-A-3. *Coordination of Ba in some silicates*

Mineral or synthetic compound	Coordinated atoms, distances to Ba in Å		Coordination polyhedron	Ref.
$BaO \cdot SiO_2 \cdot 6H_2O$	Ba_1: 8OH, H_2O	2.83—2.90	distorted quadratic antiprism	1
	Ba_2: 10OH, H_2O	2.83—2.90	see text	
Benitoite, $Ba^{[6]}Ti^{[6]}[Si_3O_9]$	6O	2.77	heavily distorted octahedron or trigonal prism	2
Taramellite, $Ba_2(Fe, Ti, Mg)_2^{[6]}(OH)_2[Si_4O_{12}]$	Ba_1: 6O	2.83; 2.90	distorted hexagon	3
	Ba_2: 6O	2.73; 2.75	distorted trigonal prism	
	Ba_3: 7O	2.71—3.00	see text	
	(2+2O	3.12;3.21)		
$^1_\infty Ba^{[7]}[Ge^{[4]}O_3]$ (high)	7O	2.62—2.94	irregular	4
Sanbornite, $^2_\infty Ba^{[9]}[Si_2O_5]$	7O 2O	2.74—2.94 3.14	see text	5
$^2_\infty$Gillespite, $Ba^{[8]}Fe^{[4p2]}[Si_4O_{10}]$	8O	2.73; 2.98	heavily distorted cube or trigonal dodecahedron	6
$Ba^{[12]}[Al_2Si_2O_8]$ ("hexagonal celsian")	12O	~2.89	hexagonal prism	7
Cymrite, $Ba^{[12]}[AlSi_3O_8(OH)]$	12O $(2 \times \frac{1}{2} H_2O$	3.05 3.39)	hexagonal prism	8
Celsian, $Ba^{[9]}[Al_2Si_2O_8]$	9O (1O	2.67—3.14 3.42)	see text	9
Paracelsian, $Ba^{[7]}[Al_2Si_2O_8]$	7O (2O	2.73—2.83 3.32;3.37)	see text	10
Barylite, $Ba^{[12?]}[Be_2^{[4]}Si_2O_7]$	11O (1O	2.82—3.12 3.34)	distorted cubo-octahedron (minus one?)	11
Harmotome, $Ba^{[10]}[Al_2Si_6O_{16}] \cdot 6H_2O$	8O, H_2O 2O	2.77—3.08 3.26	see text	12

References: 1. Höhne and Dornberger-Schiff (SR **26**, 1961, 513). 2. Fischer (1969). 3. Mazzi and Rossi (1965). 4. Hilmer (1962). 5. Douglass (SR **22**, 1958, 490). 6. Pabst (SR **9**, 1942—44, 249); cf. Mazzi and Pabst (1962), Pabst (SR **23**, 1959, 486). 7. Matsumoto (SR **15**, 1951, 304); cf. Takéuchi (SR **22**, 1958, 501). 8. Kashaev (1966). 9. Newnham and Megaw (SR **24**, 1960, 491). 10. Bakakin and Belov (SR **24**, 1960, 492); cf. Smith (SR **17**, 1953, 556). 11. Abrashew, Ilyukhin and Belov (1964). 12. Sadanaga, Marumo and Takéuchi (SR **26**, 1961, 532).

distance of 3.42 Å is not counted), with one surprisingly short distance of 2.67 Å (see Newnham and Megaw, SR **24**, 1960, 491). In the framework structure of paracelsian (with the same chemical composition as celsian), which has a structure closely related to danburite $Ca[B_2^{[4]}Si_2O_8]$, a similar coordination polyhedron with a coordination number of 9 has been described, including two distances larger than 3.3 Å. Omitting these latter two oxygen atoms, the coordination polyhedron consists of a distorted rectangle with Ba nearly in its plane plus two oxygen atoms forming a gable-roof on top of it and the seventh oxygen below its centre. The distorted cubo-octahedra around Ba in barylite $Ba[Be_2Si_2O_7]$ (with one distance larger than 3.3 Å), form chains by linked triangles. Two crystal structures of natural Ba-zeolites, edingtonite $Ba[Al_2Si_3O_{10}] \cdot 4H_2O$ (TAYLOR and JACKSON, SB **3**, 1933 to 1935, 529) and harmotome $Ba[Al_2Si_6O_{16}] \cdot 6H_2O$ are known. In harmotome, Ba is surrounded by ten O or H_2O in a coordination polyhedron similar to that of $BaO \cdot SiO_2 \cdot 6H_2O$. Two of the 6 distances from Ba to framework oxygen atoms are rather long (3.26 Å).

Revised manuscript received: September 1970

56-B. Isotopes in Nature

Natural Ba is a mixture of seven stable isotopes (Table 56-B-1). The atomic weight is 137.34 (Atomic Weight Conference of the IUPAC, 1961, Butterworth Scientific Publications, 1962), calculated on the basis that $^{12}C = 12.00000000$.

Table 56-B-1. *Stable Ba isotopes in nature*
(STROMINGER *et al.*, 1958; MATTAUCH *et al.*, 1965; EUGSTER *et al.*, 1969)

Stable isotopes	% in natural mixtures	Isotope masses $^{12}C = 12.00000000$
^{130}Ba	0.1056 ± 0.0002	129.9062
^{132}Ba	0.1012 ± 0.0002	131.9051
^{134}Ba	2.417 ± 0.003	133.9046
^{135}Ba	6.592 ± 0.002	134.9056
^{136}Ba	7.853 ± 0.004	135.9043
^{137}Ba	11.232 ± 0.004	136.9055
^{138}Ba	71.699 ± 0.007	137.9050

Measurements (I) made by UMEMOTO (1962) indicate an enrichment of the lighter Ba isotopes in the Pasamonte and Nuevo Laredo achondrites and Bruderheim chondrite with a maximum of 2.5% for ^{130}Ba in the Bruderheim achondrite. This potential accumulation decreases almost linearly to mass ^{138}Ba.

Recent investigations of Ba isotope distribution in meteorites (EUGSTER *et al.*, 1969) showed that differences in isotopic composition between meteoritic and terrestrial samples are always $< 0.1\%$ for all isotopes.

A very small amount of the short-lived ^{140}Ba occurs in nature as a fission product of ^{238}U (KURODA and EDWARDS, 1957; HEYDEGGER and KURODA, 1959).

Revised manuscript received: September 1971

56-C. Abundance in Cosmos, Meteorites, Tektites, and Lunar Samples

I. Cosmos

Ba has been detected spectroscopically in stars of all spectral classes except classes O and B. These classes correspond to such high temperatures that all resonance transitions of the highly ionized atoms fall into the ultraviolet range, which is inaccessible for terrestrial observation. Ba abundances are normally given as atomic ratios per 10^{12} H atoms or 10^6 Si atoms, or as ratios of Ba atoms/Fe atoms.

Atmospheres of main sequence stars always seem to contain Ba in similar concentrations. In Table 56-C-1, the sun is given as one example. Pronounced differences of Ba concentrations have been reported for two other groups of stars which do not belong to the main sequence. Ba II stars, which are considered to be carbon stars with higher temperatures than N stars (GORDON, 1968), exhibit an overabundance of s process elements, including Ba (BIDELMAN and KEENAN, 1951; BURBIDGE and BURBIDGE, 1957; WARNER, 1965). FUJITA and TSUJI (1965) found s process

Table 56-C-1. *Barium abundance in stars*

Type of star	Name	log Ba for log H = 12.00	Reference
High velocity stars:			
F 6 IV-V	γ Serpentis	2.09	KEGEL (1962)
Subdwarf	HD 140283	0.11	BASCHEK (1962)
Horizontal branch	HD 161817	0.94	KODAIRA (1964)
Subdwarf	HD 19445	0.35	ALLER, GREENSTEIN (1960)
Subdwarf	HD 122563	−2.65	WALLERSTEIN, PARKER, GREENSTEIN, HELFER, ALLER (1963)
Main sequence stars:			
G 2 V	sun	2.10	GOLDBERG *et al.* (1960)
Other stars			
G 8 III	ε Virginis	2.01	CAYREL, CAYREL (1966)
		$(\log Ba/\log Fe)_{star} - (\log Ba/\log Fe)_{sun}$	
Main sequence stars:			
G V	β Com	0.06 ± 0.20	HELFER *et al.* (1963)
G V	99 Her A	0.11 ± 0.20	HELFER *et al.* (1963)
G V	85 Peg A	-0.10 ± 0.20	HELFER *et al.* (1963)
Other stars (subgiant):			
G IV	ζ Her	0.02 ± 0.15	HELFER *et al.* (1963)

elements in N star Y CVn to be overabundant by factors of 10^2 to 10^3 when compared to main sequence stars. Strong enhancement of Ba lines in N stars has been detected by GORDON (1968) and UTSUMI (1967). A certain relationship of BaII stars to S type stars was observed (WARNER, 1965) whereas R stars show normal Ba abundances. CLAYTON (1964) has developed the idea that in a special type of s process nucleosynthesis, the elements heavier than Zr are favored and this process is assumed to work in BaII stars.

Population II stars which are characterized by their high velocity show higher atomic ratios, H/Ba, than main sequence stars. Nevertheless, the abundance ratios of elements heavier than carbon are similar to main sequence stars (UNSÖLD, 1967). BURBIDGE and BURBIDGE (1957) found that the subdwarfs HD 106223 and λ Bootis are depleted in Ba by factors of 20 to 30.

ALLER (1961) gives the logarithms of the ratio, number of Ba atoms in the sun/number of Ba atoms in the star, in the subdwarf for three typical subdwarfs: HD 140283, $\log (N_{sun}/N_{star}) = 2.59$; HD 19445, $\log (N_{sun}/N_{star}) = 2.15$; HD 219617, $\log (N_{sun}/N_{star}) = 1.40$. Spectral measurements for the CH stars HD 26 and HD 626 (WALLERSTEIN and GREENSTEIN, 1964) and comparison to the G 8III star ε Virginis showed them to be Ba deficient by a factor of 30 to 50. HELFER *et al.* (1963) gave the logarithms of atomic ratios $(N_{Ba}/N_{Fe})_{star} - (N_{Ba}/N_{Fe})_{sun}$ for four αG stars: ζ Herculis ($+0.02 \pm 0.15$), β Comae Berenicis ($+0.06 \pm 0.20$), 99 Herculis ($+0.11 \pm 0.20$), 85 Pegasi (-0.10 ± 0.20). Logarithms of atomic abundances of Ba in cosmos are published by ALLEN (1963) as 2.11 for H = 12. The weight ratio is 4.25. SUESS and UREY (1956) and CAMERON (1959) calculated the ratio of Ba atoms per 10^6 Si atoms as 3.66; in the sun the same ratio is 3.978 (ALLER, 1961).

II. Meteorites

The published data are summarized in Tables 56-C-2 and 3. Distribution patterns of Ba are plotted in Figures 56-C-1a through d. Generally meteorite "finds" show higher Ba values than "falls", indicating a probable terrestrial contamination of the "finds". Thus a log of the normal distribution of Ba can be observed only in chondrite falls, whereas the "finds" of the same class exhibit a random distribution.

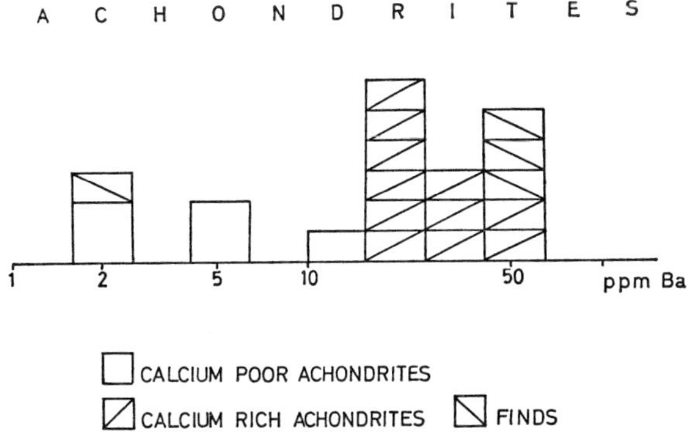

Fig. 56-C-1a. Distribution of Barium in achondrites

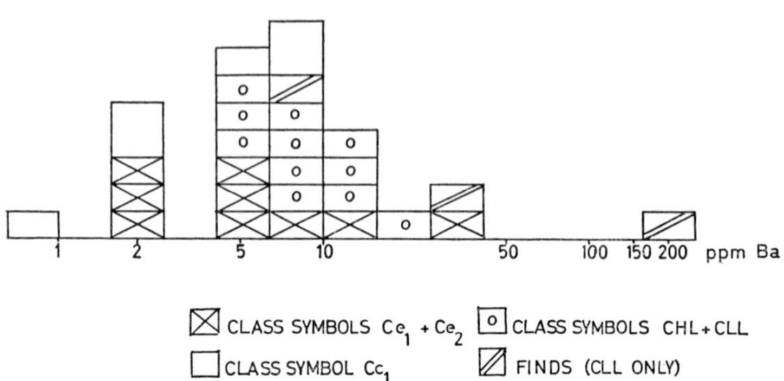

Fig. 56-C-1b. Distribution of Barium in carbonaceous and enstatite chondrites etc.

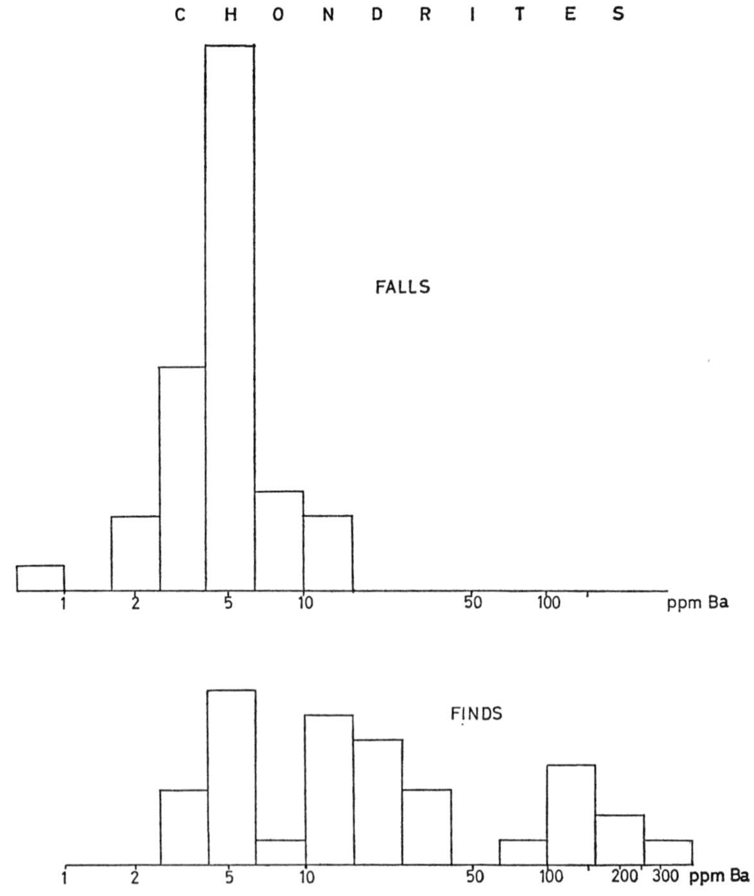

Fig. 56-C-1c. Distribution of Barium in ordinary chondrites, Class symbol CH

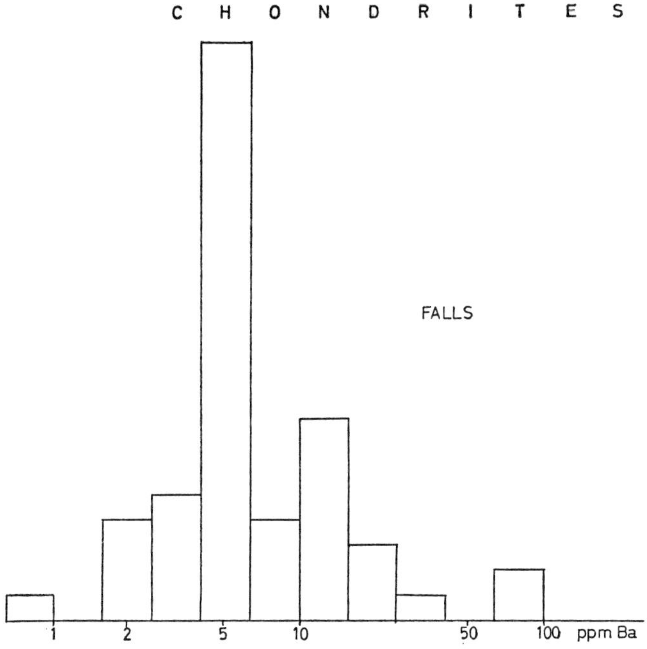

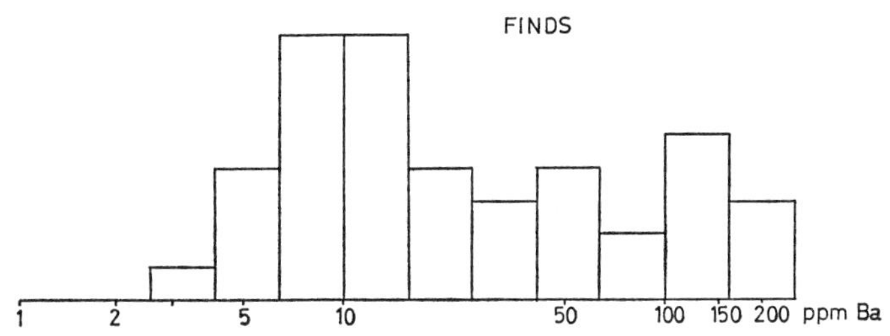

Fig. 56-C-1 d. Distribution of Barium in ordinary chondrites, Class symbol CL

Calcium rich achondrites are also high in Ba, whereas Ca-poor achondrites show low Ba values, close to chondrites. In calculations of Ba averages in meteorites, the finds have to be omitted. The few data available for nonsilicate meteorite phases (all <9 ppm) indicate that Ba is definitely a lithophilic element. Table 56-C-4 lists the mean concentrations of Ba in individual meteorite groups. Numbers of analyses used are given in brackets. Class CH and CL meteorites are by far the most frequent stony meteorites and by using their data, the Ba average for chondrites is calculated as 6.3 ppm.

Table 56-C-2. *Barium in stony meteorites*

Name of meteorite	Ba content ppm	Reference
Calcium poor achondrites		
Ae (falls):		
Cumberland Falls	14	10
Norton County	2	10
Ab (falls):		
Johnstown	2.5	10
	5	12
Shalka	4	10
Calcium rich achondrites		
Aor (falls):		
Bununu	18.5	11
Ab (falls):		
Juvinas	9—27	2
	26	1
	30.2	15
Moore County	17	1
	22	5
Pasamonte	26	1
	28.6—29.8	15
	38	5
	28.2	3
Sioux County	20	1
	25	5
	27.2	15
Stannern	62	1
	48	2
	53	15
Bereba	28.6	15
Jonzac	29—29.3	15
Angra dos Reis	21.5	15
Ap finds):		
Binda	2	1
Nuevo Laredo	40	1
	46	7
	44	13
	39.3	3
	39.3	15
Chondrites		
Ce_1 (falls):		
Abee	11.8	6
	1.8	13
	1.6	14
	2.41	3
	2.25	15
Indarch	6.4	6
	1.9	13
St. Marks	34.6	6
Ce_2 (falls):		
Hvittis	5.9	6
Khairpur	6.3	6
Pillistfer	5.5	6
CH (finds):		
Acme	120	10
Alamagordo	26	10
Aurora	20	10
Cavour	4	10
Colby (Kansas)	170	10
Coldwater	10	10
Cook	20.3	6
Coolidge	32	10
Covert	115	10
Dimboola	12.8	6
Estacado	6	12
Farley	290	10
Gladstone	17	10
Hugoton	200	10
Kaldoonera Hill	16.8	6
Kansas City	4	10
Kissij	5	8
Marsland	3	10
Modoc	3.6	7
	5	10
	3.8	13
Morland	5	10
Morven	10.1	6
Nardoo	12.9	6
Orlovka	18	10
Petropavlovka	4	10
Plainview	10	10
Ransom	28	10
	7	12
Seibert	125	10
Texline	13	10
Tulia	120	10
Wilmot	82	10
CH (falls)		
Alexandrowsky	10	10
Allegan	4.6	6
	4	10
Aviles	<1	2
Barbotan	1—3	2
Beardsley	5	10
	3	13
Beaver Creek	5	12
Cangas de Onis	5	12
Erxleben	1—3	2

Table 56-C-2. (Continued)

Name of meteorite	Ba content ppm	Reference	Name of meteorite	Ba content ppm	Reference
Forest City	3.37	3	Hayes Center	32	12
	3.7	7		30	10
	4	10	Herimitage Plains	52.4	6
	9	12	Kingfisher	5	10
	3.3	13	Kulnine	21.2	6
	3.30—3.37	15	Ladder Creek	94	10
Forest Vale	4.3	6	Lake Brown	8.4	6
Hessle	8	12	La Lande	210	10
Ichkala	4	10	Long Island	100	12
Kernouve	6	12		190	10
Kesen	4	10	Loongana (Forest Lake)	13.1	6
Lumpkin	6	12			
Monroe	5	12	McKinney	6	10
Mount Browne	7.9	6	Melrose	155	10
	4	10	Nardoo (2)	15.5	6
Nanjemoy	13	10	Ness County (1894)	20	10
Ochansk (1)	5	10	Otis	7	10
Ochansk (2)	2	10	Potter	155	10
Olmedilla de Alarcon	5	10	Rawlinna	15	10
Pantar	4	4	Roy	72	10
	3	4	Rush Creek	5	10
	5	10	Silverton (N.S. Wales)	10.3	6
	3.2	13			
	2.9	13	Tryon	190	10
Pultusk	3	10	Vincent	15.8	6
	7	12	Waconda	7	12
Richardton	3.2	7	Yalgoo	18.2	6
	4	10	*CL (falls):*		
Uberaba	4	10	Alfianello	3	10
Weston	6	10	Bjurböle	<1	2
Yatoor	6	10		8	12
Zhovtnevyi	11.3	6		5	10
	6	10		14.2	6
CL (finds):			Bruderheim	3.37; 3.4	3
Accalana	16.0	6		4.3	14
Adelie Land	11.3	6		3.37	15
Arriba	53	10	Chateau Renard	6	10
Asson	8	12	Chantonnay	1 to 3	2
Barratta	6.6	6	Colby (Wisconsin)	4.5	10
	7	12	Dhurmsala	6	10
Beenham	34	10	Elenovka	5	10
Berdyansk	7	10		10.5	6
Bluff	3	10	Farmington	9.1	6
Brisco	150	10	Holbrook	2.7—9	2
Cadell	5.8	6		5.8	3
Cocunda	10.0	6		12.2	6
Cook	40.6	6		4	7
Coolamon	46.5	6		26	10
De Nova	115	10		9	12
Goodland	10	10		3.6	13
Harrisonville	7	10			

Table 56-C-2. (Continued)

Name of meteorite	Ba content ppm	Reference	Name of meteorite	Ba content ppm	Reference
Homestead	11.0	12	Tané	5	10
	5.1	6	Tenham	15.8	6
Khohar	13.8	6	Tennasilm	10	12
Knyahinya	1—3	2	*CLL (falls):*		
	5	10	Ensisheim (light)	6.9	13
Krasnoi-Ugol	5	10	(dark)	7.1	13
Kuleschovka	3.5	10	Mangwendi	12.3	6
Kunashak	4	10	Olivenza	6	10
(light)	4.8	9	Savtschenkoje	4	10
(dark)	5.2	9	*CLL (finds):*		
L'Aigle	2.7—9	2	Kelly	165	10
Leedey	3.76; 3.85	3	Lake Labyrinth	9.9	6
	3.64—3.76	15	Shaw	26	10
Marion	3	10	*CHL:*		
Maziba	4	10	Felix	4	10
Mocs	21.7	6	Karoonda	11.9	6
	5	10	Lancé	21.8	6
	6	12		8.2	13
Narellan	7.6	6	Mokoia	9.5	6
New Concord	4	10	Warrenton	10.5	6
Ni Kolskoje	2	10	Cc_1:		
Olivenza	16.4	6	Orgueil	9.8	6
Parnallee	3.5	10		<1	12
Pavlograd	6.5	10		2.4	13
Perpeti	8.6	6	Hessle	8	12
Pervomaisky	5	10	Mighei	2.5	12
Saint Michel	13.3	6	Murray County	4	10
	4	10			
Saratov	22	10			
Stavropol	2.5	10			

References (methods in brackets): 1. Duke and Silver, 1967 (?); 2. von Engelhardt, 1936 (S); 3. Eugster *et al.*, 1969 (I); 4. Fredrikson and Keil, 1963 (S); 5. Gast, 1965 (I); 6. Greenland and Lovering, 1965 (S); 7. Hamaguchi *et al.*, 1957 (N/R); 8. Hey, 1966 (?); 9. Lavrukhina *et al.*, 1966 (N/R); 10. Moore and Brown, 1963 (S); 11. Philpotts *et al.*, 1967 (I); 12. Pinson *et al.*, 1953 (S); 13. Reed, 1963 (N/R); 14. Shima and Honda, 1967 (N/R); 15. Tera *et al.*, 1970 (I).

Table 56-C-3. *Barium in stony irons and irons*

Name of meteorite	Ba content ppm	Reference	Name of meteorite	Ba content ppm	Reference
M (fall):			*Om* (find):		
Estherville	5	1	Ni poor ataxite		
M (find):			Toluca (troilite)	<0.1	13
Pallasite olivine	7	12	El Taco	2.37	3
Og (find):			Octahedrite, Weekeroo Station	8.70	3
Canon Diablo (troilite)	0.4	13			
	<0.006	13			
	<0.3	13			

References: see Table 56-C-2.

Table 56-C-4. *Average Ba concentrations of stony meteorites*

		Falls	Finds
Ae	Enstatite achondrites	8 (2)	
Ab	Bronzite achondrites	3.8 (3)	
Ap	Pigeonite-plagioclase achondrites (eucrites)	30.0 (11)	34.2 (5)
Ce_1	Enstatite chondrites	8.6 (7)	
Ce_2	Enstatite chondrites	5.9 (3)	
CH	High iron (H)-group chondrites	4.9 (42)	43 (33)
CL	Low iron (L)-group chondrites	7.2 (53)	47.1 (42)

All calculations of Ba averages in meteorites suffer from one or both of the following uncertainties: inhomogeneity of the samples and difficulties in analytical methods. The first point was demonstrated by MOORE and BROWN (1963), when they analyzed different parts of the Holbrook chondrite, which was a fall. They got a range of 8 to 110 ppm Ba for the different parts of the specimen.

III. Tektites

Ba concentrations in tektites are reported to be in the range of 300 to 7,700 ppm (Table 56-C-5, Fig. 56-C-2). Older data which, are generally lower, are reviewed by GMELIN (1960). All the more recent papers show Ba values to be much higher in tektites than in any meteoritic material. Whereas most investigators found Ba con-

Table 56-C-5. *Barium in tektites and other natural glasses*

Locality	Ba concentration		No. of anal.	Method	Reference
	range	mean			
		Tektites			
Africa					
Ivory Coast	648—665	657	2	I	SCHNETZLER *et al.* (1967)
Asia					
Indomalaysia	300—320	310	2	S	PINSON *et al.* (1953)
Philippines	420		1	S	PINSON *et al.* (1953)
Indochina	900—2,000	1,300	6	S	VOROB'EV (1959)
Australia					
Australia	340—420	375	6	S	CHAO (1963)
	540—800	620	43	S	TAYLOR and SACHS (1964)
	540—800	630	24	S	TAYLOR (1962)
Europe					
Czechoslovakia	2,000—7,700	3,600	10	S	VOROB'EV (1960)
North America					
Texas	370—1,100	610	21	S	CHAO (1963)
	380—1,100	580	10	S	CUTTITTA *et al.* (1967)
Georgia	340—715	566	7	S	CUTTITTA *et al.* (1967)
Martha's Vineyard	390		1	S	CUTTITTA *et al.* (1967)

Table 56-C-5. (Continued)

Locality	Ba concentration		No. of anal.	Method	Reference
	range	mean			
		Natural glasses			
Tasmania					
Darwin Glass	290—360	340	8	S	TAYLOR and SOLOMON (1964)
Darwin Glass, dark	550		1	M	CHAPMAN *et al.* (1967)
Darwin Glass, light	300		1	M	CHAPMAN *et al.* (1967)
Australia					
Macedon, dark	400		1	M	CHAPMAN *et al.* (1967)
Macedon, light	200		1	M	CHAPMAN *et al.* (1967)
Henbury impact glass	600—700	650	2	S	TAYLOR and KOLBE (1964)
Africa					
Bosumtwi Crater	533—624	579	2	I	SCHNETZLER *et al.* (1967)

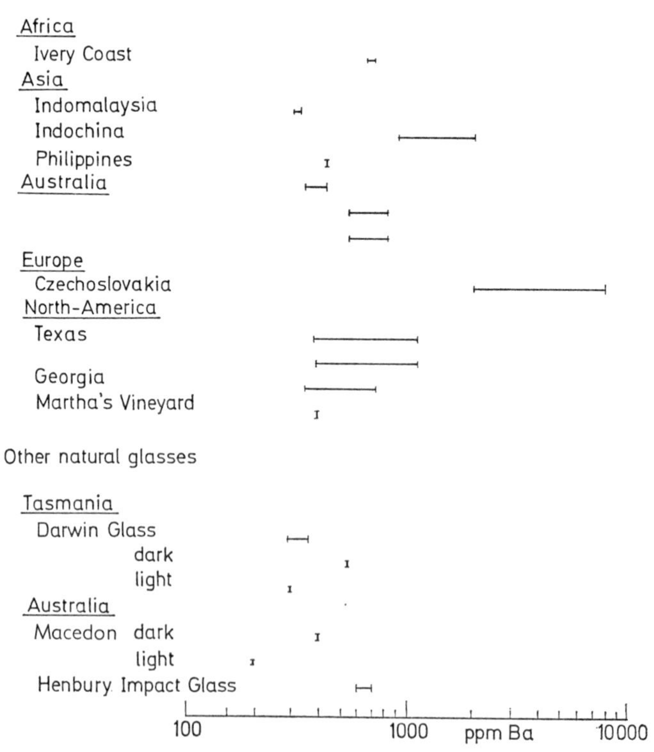

Fig. 56-C-2. Barium in tektites and other natural glasses

centrations in the range of 330 to 1,100 ppm, the values of VOROB'EV (1959, 1960) both for moldavites and indochinites are higher by a factor as great as ten, compared with tektite values from the same areas investigated by other scientists. PREUSS (1935) gives a mean of 450 ppm Ba for 36 tektites, 14 of which were moldavites. BOUSKA and POVONDRA (1964) published semiquantitative spectrographic data for moldavites showing that six out of seven samples had 100 to 1,000 ppm Ba, and the other had less than 100 ppm Ba. Thus, it cannot be excluded that VOROB'EV's values are too high due to a systematic error, especially since no simple parent material can be imagined which would provide so much Ba. TAYLOR (1965, 1966) could relate australites by their Ba concentration and the content of other elements to Henbury impact glass and Henbury greywacke as parent materials. Similar investigations of SCHNETZLER *et al.* (1967) showed the consistency of Ba and RE concentrations in Ivory Coast tektites and Bosumtwi Crater impact glasses and strong similarities to Bosumtwi phyllites. Calculations by TAYLOR (1962) made it obvious that no meteorite splash could affect the Ba concentrations of the resulting impact glasses and tektites.

IV. Lunar Samples

Considerable effort has been expended on analytical investigations of lunar samples from Apollo 11 and 12 missions. From data of the individual investigators, averages were calculated for the different types of material according to the classification established by the *Lunar Sample Preliminary Examination Team* (1969, 1970), (Table 56-C-6).

The overall unweighted average for Apollo 11 material calculated from the means of the particular batches (33) is 240 ppm Ba, s = 76. The range of data for

Table 56-C-6. *Barium in lunar rocks*

Sample type	Barium content (ppm)		No. of analyses
	arith. mean	stand. dev.	
Apollo 11 (Mare Tranquillitatis):			
Type A: fine grained vesicular rocks	251	130	37
Type B: medium grained vuggy igneous rocks	176	72	23
Type C: breccia	230	66	37
Type D: "lunar soil" (fines)	169	40	16
Apollo 12 (Ocean of Storms)			
Basaltic rocks	72	50	13
Breccia	420	210	3
Fines	586	189	7
Sample 12013	3,088	2,256	24

References for Apollo 11 samples (methods in brackets): ANNEL and HELZ (1970) (S); BROWN *et al.* (1970) (X); COMPSTON *et al.* (1970) (X); GAST *et al.* (1970) (I); GOLES *et al.* (1970) (N/R); HASKIN *et al.* (1970) (N/R); MAXWELL *et al.* (1970) (S); MORRISON *et al.* (1970) (N/R, M); MURTHY *et al.* (1970) (I); PHILPOTTS and SCHNETZLER (1970) (I); SMALES *et al.* (1970) (I); TAYLOR *et al.* (1970) (S); TERA *et al.* (1970) (I); WAKITA *et al.* (1970) (N/R); WÄNKE *et al.* (1970) (N/R).

Reference for Apollo 12 samples: DRAKE *et al.* (1970) (M); HUBBARD *et al.* (1970) (I); HUBBARD *et al.* (1971) (I); LSPET (1970) (S); MAXWELL and WIIK (1971) (S); SCHNETZLER *et al.* (1970) (I); WAKITA and SCHMITT (1970) (N/R).

total rock analyses covers 10 to 370 ppm Ba, but values of upto 55,900 ppm and 24,400 ppm have been found (M) in interstitial K rich (9.4 and 10.9% K_2O) phases from a 4 mm fragment from lunar soil (10085-LR-1) (ALBEE and CHODOS, 1970). These K rich phases consist predominantly of glass and extremely fine-grained crystalline material. Their chemical composition approaches but does not reach K feldspar composition. The K rich phase, which is present in about 6 vol.-% in sample 10085-LR-1, may also be responsible for Ba, Rb, and K concentrations in other samples, where a close relationship between these elements was observed. For several samples, a grouping to separate high K, Rb, Ba material from low K, Rb, Ba material has been tried. A few samples do not fit this separation. Still, the K/Ba ratios are rather constant in general. Inhomogeneities of Ba content in different chips of one sample have been reported, especially by WAKITA *et al.*, N/R, (1970). Different chips of their batch 10019 differ by as much as 210 ppm in their Ba content (130 and 340 ppm). Analyses (N/R) of size fractions of lunar soil showed a Ba enrichment in the finer fraction (WAKITA *et al.*, 1970).

Enrichment of Ba in moon rocks from Mare Tranquillitatis is 26 to 110 times (GAST *et al.*, 1970) compared to chondritic abundances. Within a single rock, concentration factors are similar for the "incompatable elements" Ba, U, Th, Zr, and REE except Eu.

Several individual minerals have been analyzed for Ba. The results are given in Table 56-C-7.

Apollo 12 samples from the Ocean of Storms show lower Ba content in the basaltic rocks (average 72 ppm), but higher values for breccia (average 420 ppm) and for "fines" (average 586 ppm). A most unusual rock is sample 12013 with 61% SiO_2, 2% K_2O and an average of 3088 ppm Ba. Microprobe investigations of individual points in alkali feldspars of this sample gave upto 8% BaO and 13.1% K_2O, i.e., compositions of celsian-orthoclase solid solution series (*Lunatic Asylum*, 1970). In rock 12013, Ba is enriched upto more than 2,000 times compared to chondrites (HUBBARD *et al.*, 1970).

Table 56-C-7. *Ba content of individual minerals from Apollo 11 and 12 samples*

Mineral	Lunar sample	Ba concentration ppm	Method	Reference
Apollo 11:				
Plagioclase	10085	<1; <1; 15; 1,500	M	ANDERSEN *et al.* (1970)
Plagioclase	10044—24	271	I	PHILPOTTS *et al.* (1970)
Plagioclase	10062—29	70.1	I	PHILPOTTS *et al.* (1970)
Clinopyroxene	10085	<1; 45	M	ANDERSEN *et al.* (1970)
Pyroxene	10044—24	93.7	I	PHILPOTTS *et al.* (1970)
Pyroxene	10062—29	30.9	I	PHILPOTTS *et al.* (1970)
Ilmenite		81.2	I	MURTHY *et al.* (1970)
Apollo 12:				
Plagioclase	12013	910	M	DRAKE *et al.* (1970)
Alk. feldspar	12013	10650	M	DRAKE *et al.* (1970)
Alk. feldspar	12013—10 (40)	895; 910; 8,680; 20,400	M	LUNATIC ASYLUM (1970)

Revised manuscript received: September 1971

56-D. Abundance in Rock-Forming Minerals (I) and Barium Minerals (II)

I. Rock-Forming Minerals

In igneous rocks of the earth's crust Ba usually does not form minerals of its own, but is distributed among a number of silicate structures, mainly feldspars and micas. The most important substitution is for potassium due to the nearly identical ion sizes, even with the somewhat more covalent character of the Ba—O bond. Substitution for Ca is observed in plagioclases, pyroxenes and amphiboles. Apatite and calcite are the most important rock forming non-silicates containing Ba.

a) Feldspars

K feldspars are the most important Ba carriers. Investigations by Roy (1965, 1967) and Gay and Roy (1968) with synthetic members have shown that a continuous series of solid solutions exist at high temperatures between K feldspars and celsian. In the subsolidus region, two gaps of miscibility seem to exist, one close to the microcline composition and the other between hyalophane and celsian (Fig. 56-D-1).

The system $BaO—Al_2O_3—SiO_2$ was recently investigated by Lin and Foster (1968, 1969).

In nature, K feldspars with BaO $>2\%$ are rare. They are mostly restricted to alpine type fissures (adularia) and deposits of manganese oxide.

Ba concentrations upto 9.5% in feldspars of the orthoclase-celsian series are reported from the alkalic rock complex, Magnet Cove (Erickson and Blade, 1963) and from phonolites in South West Germany (Weiskirchner, 1969). Ba distribution in K feldspars has been investigated by many authors. As a general pattern, it was found that in magmatic sequences the early crystallized K feldspars show the highest Ba concentrations, whereas microclines from pegmatites were low (S) in this element (Shimer, 1943; Bray, 1942). Metasomatic alteration of a granite body (Black Forest, Germany) was discussed as a cause of abundant K feldspar phenocrysts with a high and rather homogeneous Ba distribution by Emmermann (1968), (X).

Distribution of Ba abundances in K feldspars is shown for igneous rocks, pegmatites, and alpine fissures in Fig. 56-D-2 using the data of Table 56-D-1. The histograms clearly demonstrate that, in general, average pegmatitic K feldspars contain less Ba than those from granitic rocks.

Ba may enter the plagioclase structure depending on composition in the An-Ab series, temperature and pressure, of competing elements. Duchesne (1968) found a correlation between potassium and Ba concentrations. Many indications exist supporting the view that all the Ba in feldspars can be plotted in the four component system Ab, An, Or, Cn. Plots of Ba concentrations for different members

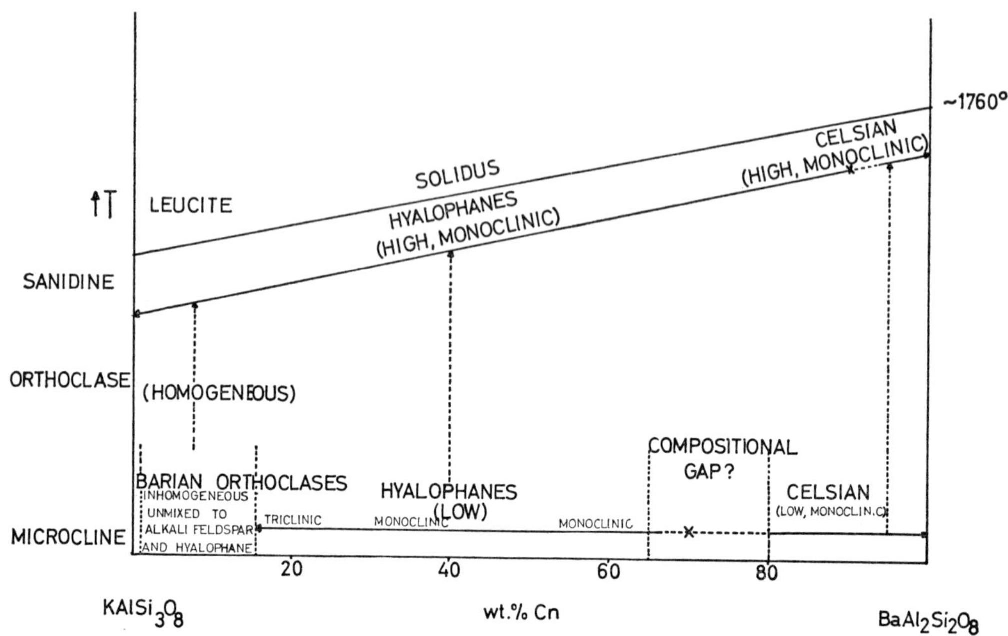

Fig. 56-D-1. Phases in the system $KAlSi_3O_8$—$BaAl_2Si_2O_8$ (GAY and ROY, 1968)

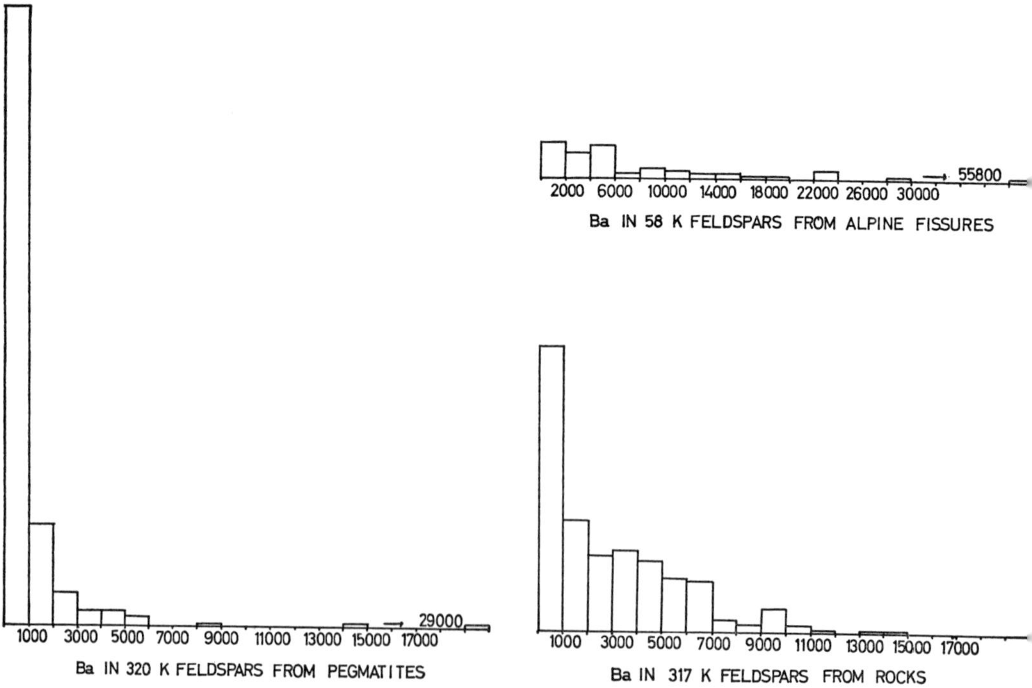

2000 6000 10000 14000 18000 22000 26000 30000 → 55800

Ba IN 58 K FELDSPARS FROM ALPINE FISSURES

1000 3000 5000 7000 9000 11000 13000 15000 17000 → 29000

Ba IN 320 K FELDSPARS FROM PEGMATITES

1000 3000 5000 7000 9000 11000 13000 15000 17000

Ba IN 317 K FELDSPARS FROM ROCKS

Fig. 56-D-2. Distribution of Ba concentrations in K feldspars (see Table 56-D-1 for references)

Table 56-D-1. *Barium concentrations in K feldspars*

Rock type	No. of analyses	Barium concentration range ppm	arith. mean	Method	Reference
Pegmatites	7	79— 1,130	270	S	Bray (1942)
	38	19— 450	100	X	Correia Neves (1964)
	6	3— 470	132	S	von Engelhardt (1936)
	5	200— 6,000	1,600	S	Erickson and Blade (1963)
	44	50— 3,400	710	S	Heier and Taylor (1959)
	17	20— 10,000	911	S	Higazy (1953)
	15	100— 1,700		S	Higazy (1949)
	5	435— 5,620	2,408	S	Hitchon (1960)
	9	190— 4,750	1,290	X	Markart and Preisinger (1964)
	11	150— 600	355	S	Oftedal (1961)
	40	150— 1,000	436	S	Oftedal (1962)
	44	< 10— 9,000	1,660	S	Oftedal (1958)
	16	171— 10,525	4,581	S	Pirani and Simboli (1963)
	64	76— 397	251	S	Shcherba *et al.* (1964)
	11	27— 3,550	495	S	Taylor *et al.* (1960)
	4	236—> 6,000		S	Townend (1966)
		< 10— 10,525	*863*		*average for 320 samples*
Alpine fissures	13	220— 55,800	7,730	N	Rybach and Nissen (1967)
	3	2,400— 3,270	2,690	X	Markart and Preisinger (1964)
	15	2,600— 29,000	14,050	S	Weibel (1957)
	33	630— 15,300	5,000	S	Weibel and Meyer (1957)
	1	1,970	1,970	S	Hewlett (1959)
		220— 55,800	*7,481*		*average for 65 samples*
Igneous rocks					
Granite	11	162— 2,340	1,150	S	Bray (1942)
Granodiorite	4	322— 1,440	810	S	Bray (1942)
Quartzmonzonite	3	1,080— 1,260	1,170	S	Bray (1942)
Trachyte	2	5,100— 8,600	6,850	S	Carmichael (1965)
Granite	10	108— 361	201	S	Emiliani and Vespignani (1964)
Granite[a]	150	800— 5,400	4,700	X	Emmermann (1968)
Granite-larvikite	5	90— 14,300	3,920	S	von Engelhardt (1936)
Alkalic rock	4	900— 4,000	2,475	S	Erickson and Blade (1963)
Monzonite, granite	25	130— 12,000		S	Heier (1960)
Granite, gneiss	41	1,000— 9,500	3,840	S	Heier and Taylor (1959)
Granite, granodiorite, gneiss	12	90— 5,480	2,765	S	Herz and Dutra (1966)
Trachyte	10	360— 13,600	3,370	S	Hewlett (1959)
Granite	3	420— 2,700	1,550	S	Hewlett (1959)
Charnockite	4	2,500— 5,000		S	Howie (1955)
Monzonitic accumulate[a]	16	1,070— 12,100	5,000	X	Jasmund and Seck (1964)
Granite, gneiss	19	364— 4,620	2,220	X	Markart and Preisinger (1964)

Table 56-D-1. (Continued)

Rock type	No. of analyses	Barium concentration range ppm	arith. mean	Method	Reference
Alkalic rock	4	630— 5,900	314	X	PERCHUCK and RYABCHIKOV (1968)
Granite	70	26— 9,480	2,124	X	RHODES (1969)
Granite	9	4,350— 4,770	4,560	X	RICHTER (1966)
Quartz, monzonitic	29	3— 3,000	990	S	ROGERS (1958)
Granite	4	835— 1,020	950	S	SCHARBERT (1966)
Granite-tonalite	5	2,000— 3,000	2,500	S	SEN *et al.* (1959)
Granite	91	45— 392	176	S	SHCHERBA *et al.* (1964)
Granodiorite	14		836	S	SHCHERBA *et al.* (1964)
Adamellite	3		460	S	SHCHERBA *et al.* (1964)
Syenite	4		162	S	SHCHERBA *et al.* (1964)
Monzonite	2		155	S	SHCHERBA *et al.* (1964)
Syenodiorite	6		595	S	SHCHERBA *et al.* (1964)
Different rocks	32	< 200— 18,000	3,559	M	SMITH and RIBBE (1966)
Different rocks[a]	23	< 200— 6,200	1,183	M	SMITH and RIBBE (1966)
Granite	12	277—>6,000		S	TOWNEND (1966)
Granite	7	766— 1,526	1,240	X	WHITE (1966)
		3— 18,000	*2,626*		*average for 598 samples*

[a] Sanidine.

of the plagioclase solid solution series show that the pure components contain less Ba than the intermediate plagioclases (Table 56-D-2, Fig. 56-D-3). This may be due to the fact that the pure end members are mostly very late or secondary crystals which formed from a Ba-poor liquid. Distribution of Ba between coexisting K feldspars and plagioclases is discussed as a potential geothermometer by HEIER (1960, 1961). BARTH (1961) showed that a straight line relation exists between log ratios of Ba in coexisting feldspars and log inverse absolute temperatures. He deduced that below 250° C, plagioclase is the preferred host mineral for Ba. IIYAMA (1968) studied Ba distribution between potassium feldspars and plagioclases experimentally by hydrothermal runs at 600° C and 1,000 bars. His results contradict the observations in natural rocks as he finds higher Ba incorporation in the plagioclases than in the coexisting K feldspars. RUDERT (1970) experimentally found an incorporation of Ba in albite at 930° C/1 kb, upto 30 weight % of celsian molecules; BRUNO and GAZZONI (1970) substituted large amounts of Ca in the modifications of $BaAl_2Si_2O_8$. In the hexagonal modification synthesized by solid state reaction at 1,200° C, Ca replaces Ba upto 37% (atomic fraction); in the hexagonal modification obtained by crystallization of a melt, the replacement is limited to 25%; the same value was found for the monoclinic modification by heating the mentioned modifications to 1,450° C.

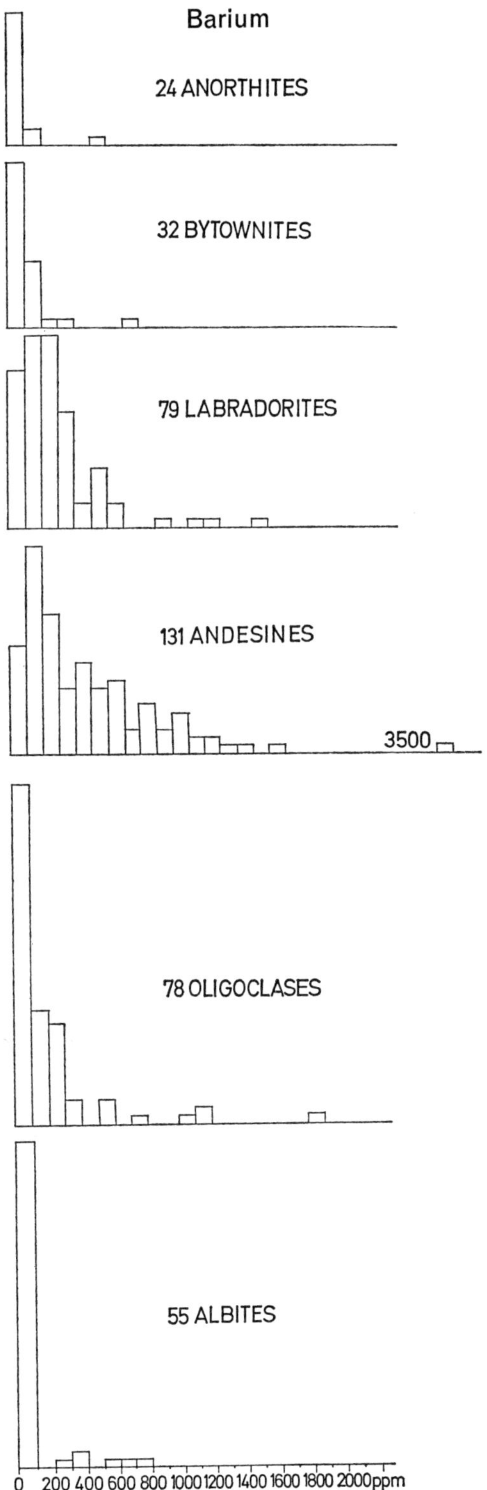

Fig. 56-D-3. Distribution of Ba concentrations in plagioclases (see Table 56-D-2 for references)

Table 56-D-2. *Barium concentrations in plagioclases from igneous and metamorphic rocks and rocks of hydrothermal origin*

Rock type	No. of analyses	Barium concentration range ppm	Barium concentration arith. mean	Method	Reference
Anorthites from:					
Basalt	1	60	60	S	Byers (1961)
Pumice	1	10	10	S	Coats (1952)
Anorthosite, gabbro and others	13	< 100— 500	<217	M	Corlett and Ribbe (1967)
Metam. rock (amphibolite facies)	3	<45	<45	S	Sen (1960)
Metam. rock (granulite facies)	2	<45— 80	<60	S	Sen (1960)
Dacite-andesite	3	<45— 80	<63	S	Sen (1960)
Norite	1	10	10	S	Sen *et al.* (1959)
Bytownites from:					
Anorthosite, norite and others	12	<100— 100	<100	M	Corlett and Ribbe (1967)
Gabbro-anorthosite	9	40— 320	130	S	Emmons (1952)
Basalt	2	22— 270	146	S	Muir *et al.* (1964)
Metam. rock (amphibolite facies)	7	<45— 54	<46	S	Sen (1960)
Gabbro	2	80+ 85	82	S	Sen *et al.* (1959)
Labradorites from:					
Anorthosite, gabbro and others	38	<100— 300	<195	M	Corlett and Ribbe (1967)
Basalt	9	200— 800	400	S	Cornwall and Rose (1957)
Gabbro-anorthosite	9	130— 580	347	S	Emmons (1952)
Anorthosite	2	85+ 160	122	S	Papezik (1965)
Metam. rock (amphibolite facies)	3	115— 150	160	S	Sen (1960)
Metam. rock (granulite facies)	4	270— 480	295	S	Sen (1960)
Norite	3	45— 130	75	S	Sen *et al.* (1959)
Gabbro	2	50— 80	65	S	Wagner and Mitchell (1951)
Teschenite-basalt, gabbro	9	80— 1,500	451	S	Wilkinson (1959)
Andesines from:					
Anorthosite	9	630— 3,500	1,250	M	Anderson (1966)
Granite, granodiorite	6	370— 1,000	310	S	Bray (1942)
Gneiss, anorthosite and others	19	<100— 600	<177	M	Corlett and Ribbe (1967)
Granodiorite-anorthosite	5	90— 330	224	S	Emmons (1952)
Monzonitic accumulates	12	440— 1,020	735	X	Jasmund and Seck (1964)
Gneiss	11	900—18,000	6,100	S	Oftedal (1958)
Pegmatite	12	10— 1,430	300	S	Oftedal (1958)
Anorthosite	10	150— 1,075	384	S	Papezik (1965)

Table 56-D-2. (Continued)

Rock type	No. of analyses	Barium concentration range ppm		arith. mean	Method	Reference
Metam. rock (amphibolite facies)	15	<45—	990	<224	S	SEN (1960)
Metam. rock (granulite facies)	12	45—	700	380	S	SEN (1960)
Dacite-andesite	4	450—	870	560	S	SEN (1960)
Granite	2	250+	320	285	S	SEN (1960)
Tonalite, granodiorite	6	80—	255	174	S	SEN *et al.* (1959)
Gabbro	6	200—	600	390	S	WAGER and MITCHELL (1951)
Teschenite, gabbro	2	350+	600	475	S	WILKINSON (1959)
Oligoclases from:						
Anorthosite	1	3,480		3,480	M	ANDERSON (1966)
Pegmatite, granite and others	43	<100—	1,100	<134	M	CORLETT and RIBBE (1967)
Basalt	2	700+	1,000	850	S	CORNWALL and ROSE (1957)
Granite	4	70—	380	132	S	EMMONS (1952)
Pegmatite	1	365		365	S	HITCHON (1960)
Granite	2	150—	268	209	S	PIRANI and SIMBOLI (1963)
Granite	4	180—	250	210	S	SCHARBERT (1966)
Metam. rock (amphibolite facies)	5	45—	540	205	S	SEN (1960)
Metam. rock (granulite facies)	7	<90—	1,170	<425	S	SEN (1960)
Rhyolite-rhyodacite	2	45—	1,800	922	S	SEN (1960)
Granite	5	63—	380	155	S	SEN (1960)
Granite, granodiorite	2	120+	235	175	S	SEN *et al.* (1959)
Albites from:						
Pegmatite, alpine fissure	36	<100—	650	<145	M	CORLETT and RIBBE (1967)
Basalt	1	700		700	S	CORNWALL and ROSE (1957)
Pegmatite	6	54—	80	69	X	CORREIA NEVES (1964)
Pegmatite	1	1		1	S	EMMONS (1952)
Pegmatite	1	2		2	S	VON ENGELHARDT (1936)
Pegmatite	1	5		5	S	HIGAZY (1953)
Metam. rock (greenschist facies)	1	<45		<45	S	SEN (1960)
Different rocks	2	<200		<200	M	SMITH and RIBBE (1966)
Pegmatite	5	12—	49	~24	S	TAYLOR *et al.* (1960)
Granophyre	1	30		30	S	WAGER and MITCHELL (1951)
Anorthoclases from:						
Basalt, trachyte	2	1,000+	1,100	1,050	S	WILKINSON (1962)
Phonolite	1	2,160		2,160	S	HEWLETT (1959)

b) Micas and Other Important Rock Forming Minerals

Micas are the next important Ba carriers in rocks. Ranges of Ba concentrations in these and other minerals are given in Table 56-D-3 and in Fig. 56-D-4. Ba is a concentrated in biotites as in muscovites. Only lepidolites have much lower Ba values. Generally micas from pegmatites show lower Ba values than those from the adjacent country rocks (Takubo and Tatekawa, 1954), but each pegmatite may have its own "Ba level". Extensive material on Ba contents, mainly from pegmatite muscovites, was published by Heinrich *et al.* (1956). These authors found Ba concentrations from 1 to 9,900 ppm in 162 samples, predominantly pegmatite muscovites. They calculated an average of 1,020 ppm Ba.

Petrov *et al.* (1965) found that the relative concentrations of Ba in biotites increase with metamorphic grade. In rocks free of K feldspars, biotite is often the main Ba carrier, sometimes together with pyroxenes and amphiboles. In skarns

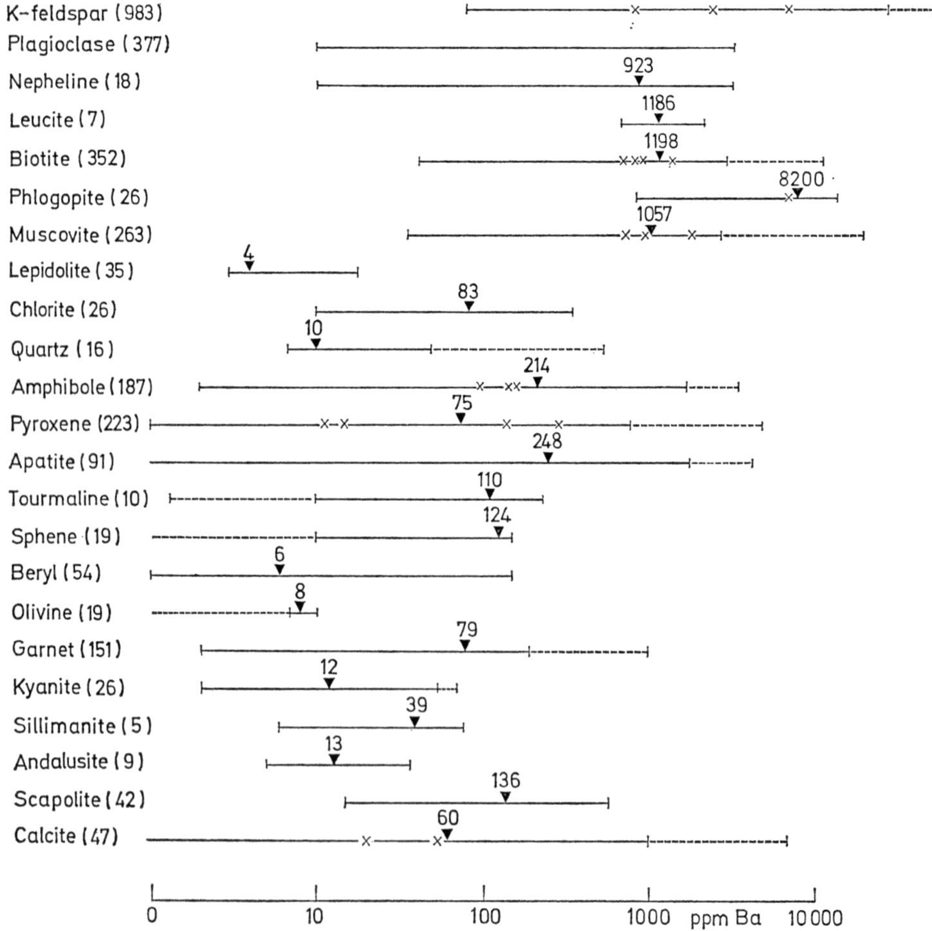

Fig. 56-D-4. Ba concentrations in rock forming minerals (see Table 56-D-4 for references). ▼ Indicates total averages. × Indicates group averages

Table 56-D-3. *Barium concentrations in rock forming minerals of igneous and metamorphic rocks*

Rock type	No. of analyses	Barium concentration range ppm	arith. mean	Method	Reference
Nepheline					
Alkalic rock	3	<45— 3,320	1,270	S	von Eckermann (1952)
Alkalic rock	5	100— 3,000	1,200	S	Erickson and Blade (1963)
Pegmatite	6	<10— 20	20	S	Oftedal (1962)
Alkalic rock	4	<100— 3,400	1,670	X	Perchuck and Ryabchikov (1968)
		<10— 3,400	*923*		*average for 18 samples*
Leucite					
Leucite-basanite	2	1,160— 1,250	1,200	S	von Engelhardt (1936)
Leucitophyre-leucite basanite	5	710— 2,250	1,180	S	Henderson (1965)
		710— 2,250	*1,186*		*average for 7 samples*
Biotite					
Granite, granodiorite	12	67— 720	294	S	Bray (1942)
Pegmatite	3	103— 470	241	S	Bray (1942)
Schist	18	600— 1,100	790	S	Butler (1967)
Metapelite	5	500— 2,000	1,500	S	Card (1964)
Rhyolite	8	1,430— 6,730	5,200	M	Carmichael (1967a)
Granitic rock	34	240— 5,000	1,405	S	Dodge *et al.* (1969)
Alkalic rock (carbonatite)	8	1,700— 6,360	4,065	S	von Eckermann (1952)
Granite	2	84+ 100	90	S	Emiliani and Vespignani (1964)
Paragneiss	25	300— 2,200	930	S	Engel and Engel (1960)
Alkalic rock	4	900—12,000	4,500	S	Erickson and Blade (1963)
Quartzdiorite-granite	13	400— 2,700	1,830	S	Haslam (1968)
Charnockite	2	900+ 2,300	1,600	S	Heier (1960)
Pegmatite	7	45— 2,320	717	S	Hitchon (1960)
Gneiss, schist, amphibolite	35	100— 1,200	750	S	Hunzicker (1966)
Gneiss, amphibolite	8	900— 3,180	~2,400	S	Kretz (1959)
Gneiss, schist	20	120— 2,450	810	S	Moxham (1965)
Pegmatite	3	200— 400	300	S	Oftedal (1962)
Quartzmonzonite, granodiorite	24	450— 8,500	1,212	X	Rimsaite (1964)
Gneiss, schist	17	900— 2,700	808	X	Rimsaite (1964)
Metaarkose	7	472— 1,086	780	S	Schwarcz (1966)
Granite, norite	10	500— 2,500	1,822	S	Sen *et al.* (1959)
Pegmatite, wallrock	48	<150— 2,000	873	S	Stern (1966)
Metamorphic rock (staurolite zone)	22	42— 1,410	800	S	Turekian and Phinney (1962)
Migmatite, granite	8	240— 1,335	880	X	White (1966)
Basalt	1	2,500	2,500	S	Wilkinson (1962)
Cordierite-biotite, gneiss	8	60— 530	290	S	Wynne-Edwards and Hay (1962)
		42— 8,500	*1,198*		*average for 352 samples*

Table 56-D-3. (Continued)

Rock type	No. of analyses	Barium concentration range ppm	arith. mean	Method	Reference
Phlogopite					
Wyomingite, orendite	12	3,670—10,100	7,350	M	CARMICHAEL (1967b)
Wolgidite, wyomingite, fitzroyite	9	3,140—14,300	12,000	M	CARMICHAEL (1967b)
Alkalic rock	2	1,000+ 4,000	2,500	S	ERICKSON and BLADE (1963)
Lamprophyre, basalt	3	2,700— 4,850	4,016	X	RIMSAITE (1964)
		1,000—14,300	*8,202*		*average for 26 samples*
Muscovite					
Granite	7	108— 900	490	S	BRAY (1942)
Pegmatite	5	23— 360	125	S	BRAY (1942)
Schist	22	1,400— 2,900	1,930	S	BUTLER (1967)
Granite	2	1,000+ 1,100	1,050	S	EMILIANI and VESPIGNANI (1964)
Pegmatite	168	2— 9,800	1,020	S	HEINRICH *et al.* (1953)
Pegmatite	1	36	36	S	HEINRICH (1967)
Pegmatite	13		160	S	HITCHON (1960)
Gneiss, schist	14	110— 1,200	850	S	HUNZICKER (1966)
Alpine fissure	2	230+ 300	265	S	SCHWANDER *et al.* (1968)
Metamorphic rock	1	21,400	21,400	W	SNETSINGER (1966)
Pegmatite	28	180— 1,650	797	S	STERN (1966)
		2—21,400	*1,057*		*average for 263 samples*
Lepidolite					
Pegmatite	26	2— 9	3	S	HEINRICH *et al.* (1953)
Pegmatite	9	3— 18	7	S	HEINRICH (1967)
		2— 18	*4*		*average for 35 samples*
Chlorite					
Basalt	11	20— 300	120	S	CORNWALL and ROSE (1957)
Granite	3	10— 50	25	S	EMILIANI and VESPIGNANI (1964)
Schist	11	17— 148 (2,100)	36	S	GRESENS (1967)
Garnet, chlorite, schist	1	358	358	S	MOHR (1956)
		10— 2,100	*83*		*average for 26 samples*
Quartz					
Granite	4	280— 550	416	S	BRAY (1942)
Pegmatite	11	30— 50	11	S	HITCHON (1960)
Basic intrusion	1	7	7	S	WAGER and MITCHELL (1951)
		7— 550	*112*		*average for 16 samples*
Amphibole					
Metamorphic rock (amphibolite)	3	<50— 200	100	S	CARD (1964)
Wolgidite, orendite	4	1,800— 5,400	3,700	M	CARMICHAEL (1967b)

Table 56-D-3. (Continued)

Rock type	No. of analyses	Barium concentration range ppm		arith. mean	Method	Reference
Schist, glaucophane	9	2—	120	20	S	Coleman and Papike (1968)
Schist, riebeckite	4	<4—	140	74	S	Coleman and Papike (1968)
Schist, actinolite	2	12+	410	211	S	Coleman and Papike (1968)
Granite	17	7—	50	30	S	Dodge *et al.* (1968)
Granodiorite, quartz-diorite	3	15—	95	60	S	Dodge *et al.* (1968)
Metamorphic rock (amphibolite)	24	24—	320	100	S	Engel and Engel (1962)
Diorite, porphyr	15	100—	300	150	S	Engel (1959)
Schist	13	44—	104	66	S	Gresens (1967)
Quartzdiorite	6	10—	36	20	S	Haslam (1968)
Granite	4	10—	36	20	S	Haslam (1968)
Rhyodacite	1	180		180	S	Haslam (1968)
Gneiss, schist	8	<10—	195 (1,200)	80	S	Hunzicker (1966)
Monzonitic accumulate	11	90—	1,780	755	X	Jasmund and Seck (1964)
Gneiss, amphibolite	16	<90—	300	160	S	Kretz (1959)
Amphibolite	9	<90—	360	230	S	Kretz (1960)
Gneiss, schist	20	10—	485	150	S	Moxham (1965)
Nepheline-syenite, pegmatite	2	10+	20	15	S	Oftedal (1962)
Ophiolite	1	27		27	S	Plas and Hügi (1961)
Tonalite, granodiorite	12	10—	80	40	S	Sen *et al.* (1959)
Essexite	2	25—	150	90	S	Simpson (1954)
Basalt	1	100		100	S	Wilkinson (1962)
		2—	*5,400*	*214*		*average for 187 samples*
Pyroxene						
Basic layered rock	14	12—	22	18	S	Atkins (1969)
Basalt	1	60		60	S	Byers (1961)
Basalt etc.	8	30—	800	195	S	Cornwall and Rose (1957)
Alkalic rock	9	<45—	540	160	S	von Eckermann (1952)
Alkalic rock	32	<10—	5,000	300	S	Erickson and Blade (1963)
Peridotite	2	10—	18	14	S	Green (1964)
Quartzdiorite	4	5—	27	16	S	Haslam (1968)
Granite	1	10		10	S	Ishioka (1967)
Metamorphic rock (skarn)	10	<90—	360	<145	S	Kretz (1960)
Charnockite	20	<5—	22	≦11	S	Leelanandam (1967)
Basalt, trachyte	2	10+	50	30	S	LeMaitre (1962)
Metamorphic rock (skarn)	38	<1—	44	12	S	Moxham (1960)
Basalt	8	5—	45	21	S	Muir *et al.* (1964)
Kimberlite	9	<10—	150	33	S	Nixon *et al.* (1963)

Table 56-D-3. (Continued)

Rock type	No. of analyses	Barium concentrations range ppm		arith. mean	Method	Reference
Pegmatite	1	10		10	S	OFTEDAL (1962)
Olivine basalt	1	1.5		1.5	N/R	ONUMA *et al.* (1968)
Metamorphic rock (skarn)	38	1—	65	15	S	SHAW *et al.* (1963)
Syenite, gabbro, essexite	5	<10—	30	<21	S	SIMPSON (1954)
Gabbro	9	<5—	60	16	S	WAGER and MITCHELL (1951)
Teschenite, basalt, gabbro	11	<5—	60	11	S	WILKINSON (1959)
		<1—	*5,000*	*75*		*average for 223 samples*

from the Greenville province, KRETZ (1960) reports that the Ba content decreases in the following sequence: biotite > amphibole > pyroxene. The structure of chain silicates, where Ba occupies Ca positions, seems to accept more Ba at higher temperature of formation; ENGEL and ENGEL (1962) found an increasing Ba content (9, 18, 86, 106 ppm) in hornblendes from metamorphic rocks formed at 400, 500, 525, and 625° C.

Quartz has a rather low Ba content, and it is not certain whether contamination by other minerals was avoided in all analyses published.

From minerals of metamorphic origin, garnets exhibit the highest Ba values. Kyanites, sillimanites, and andalusites are much lower and always contain less than 100 ppm Ba.

Calcite from alkalic rocks in the form of carbonatites is somewhat enriched in Ba. Crystals from hydrothermal veins generally have very low Ba concentrations. The very rare Ca-Ba-carbonates, alstonite, barytocalcite, and benstonite are of no importance in rock formation and do not form solid solution series with the end members.

Depending on type and environment of formation, zeolites contain different amounts of Ba in their structures. Upto 300 ppm Ba were found in analcime and natrolite (ERICKSON and BLADE, 1963; WILKINSON, 1959). In stilbite, thomsonite, chabasite, gmelinite, and gismandine, concentrations between 600 and 5,200 ppm Ba were reported (HOSS and ROY, 1960). Phillipsites, or constituents of the series phillipsite-harmotome, which occur in manganese nodules and in larger amounts in certain deep sea sediments, may be extremely enriched in Ba, upto 44,500 ppm (SHKABARA, 1950; HOSS and ROY, 1960).

c) Barium Partition between Mineral and Host Rock and Between Coexisting Minerals

During the first stages of differentiation of basaltic magmas, Ba is enriched in the liquid phase. With progressing crystallization Ba is incorporated, especially in K feldspars and micas, which extract this element from the melt. Pegmatitic stages of differentiation series are consequently often impoverished in Ba.

Table 56-D-4. *Barium distribution between minerals and total rock*

Pair: mineral	Pair: rock	No. of pairs	Barium concentration range: mineral ppm	Barium concentration range: rock ppm	Distribution coefficient Ba_{min}/Ba_{rock}	Method	Reference
Sanidine	Trachyte	7	140—6,900	120—1,800	1.17— 8.95	S	Berlin and Henderson (1969)
Plagioclase	Trachyte	5	560—3,500	770—1,100	0.72— 3.98	S	Berlin and Henderson (1969)
Anorthoclase	Phonolite	2	2,600—3,000	1,100—1,400	2.14— 2.37	S	Berlin and Henderson (1969)
Biotite	Dacite, rhyolite trachyte	8	1,400—6,700	400—1,100	1.6 — 15	M	Carmichael (1967a)
Garnet	Eclogite	11	150— 165	<10—1,140	0.13—>16.3	S	Hahn-Weinheimer and Lücke (1963)
Amphibole	Rhyodacite	1	180	580	0.31	S	Haslam (1968)
Biotite	Metaarkose	7	472—1,086	1,190—1,730	0.29— 0.65	S	Haslam (1968)
Pseudoleucite	Juvite, tinguaite	7	50—4,000	50—6,500	0.23— 1.00	X	Henderson (1965)
Leucite	Leucite-porphyre, leucite-basanite	5	710—2,250	1,950—2,250	0.36— 0.81	X	Henderson (1965)
Biotite	Amphibolite	9	575—1,150	270— 760	1.22— 4.60	S	Hunzicker (1966)
Biotite	Gneiss	19	290—1,150	250—1,050	0.40— 2.44	S	Hunzicker (1966)
Muscovite	Amphibolites	5	1,000—1,200	270— 760	1.33— 4.08	S	Hunzicker (1966)
Muscovite	Gneiss	9	390—1,100	310—1,050	0.85— 3.55	S	Hunzicker (1966)
Plagioclase	Granulite	12	110— 475	160—2,500	0.11— 3.37	S	Sen (1960)
Orthoclase	Granitic gneiss	5	3,330—5,380	706— 731	4.67— 7.38	X	White (1966)
Biotite	Granitic gneiss	2	847—1,110	706— 730	1.20— 1.52	X	White (1966)
Biotite	Biotite gneiss	8	60— 530	103— 380	0.41— 3.95	S	Wynne-Edwards *et al.* (1962)

Table 56-D-5. *Barium distribution between coexisting minerals*

Pairs: mineral I	Pairs: mineral II	No. of pairs	Rock type	Range mineral I ppm Ba	Range mineral II ppm Ba	Range of distrib. coeffic.	Method	Reference
Biotite	Muscovite	5	Gneiss, migmatite	1,250[a]	630[a]	1.98	X	1
		12	Schist, paragneiss	1,525[a]	1,970[a]	0.80	X	1
		18	Schist	600 — 1,100	1,400 — 2,900	0.31— 0.56	S	2
Biotite	Orthoclase	7	Migmatite, gneiss	240 — 1,335	1,176 — 6,718	0.20— 0.25	X	3
Biotite	Garnet	8	Paragneiss	350 — 2,200	35 — 1,100	2.00—15.0	S	4
		22	Metam. sequence (42)	580 — 1,410	11 — 38	21.0 —75.8	S	5
Biotite	Hornblende	5	Metamorphic rock	1,000 — 3,000	15 — 60	33 —68		6
		20	Gneiss, schist	120 — 2,450	10 — 485	2.50—60	S	7
Biotite (phlogopite)	Pyroxene	2	Phonolite, rhyodacite	2,480 + 4,670	21.3 + 55.0	116 +85	I	8
Biotite	Ilmenite	7	Metaarkose	472 — 1,086	7 — 132	0.01 — 0.28	S	9
Alk. feldsp.	Plagioclase	4	Trachyte	2,700 — 6,900	560 — 3,500	0.93 —10.8	S	10
Sanidine	Plagioclase	12	Monz. cumulates	900 —12,100	394 — 1,165	2.30 —15.9	X	11
Sanidine	Amphibole	11	Monz. cumulates	900 —12,100	90 — 1,770	1.50 —22.1	X	11
Orthoclase	Nepheline	3	Alkalic rock	990 — 2,300	100 — 3,400	0.38 —25.0	X	12
Plagiocl.	Amphibole	10	Monz. cumulates	395 — 1,170	415 — 1,780	0.51 — 1.69	X	11
Plagiocl.	Clinopyroxene	3	Oceanite, andesite	23 — 58.2	1.26— 10.2	4.75 —18.7	I	8
Pyroxene	Amphibole	5	Skarn	2.6— 26	12 —>500	<0.016— 2.00	S	13
Pyroxene	Scapolite	19	Skarn	1 — 31	14 — 210	0.013— 0.20	S	13
Clinopyrox.	Olivine	2 (6)	Ankoranite, alkali basalt	4.9— 175	1.9 — 5.1	2.59 —34.5	I	8

Reference: 1. Rimsaite (1964); 2. Butler (1967); 3. White (1966); 4. Engel and Engel (1960); 5. Turekian and Phinney (1962); 6. Hietanen (1971); 7. Moxham (1965); 8. Philpotts and Schnetzler (1970a); 9. Schwarcz (1966); 10. Berlin and Henderson (1969); 11. Jasmund and Seck (1964); 12. Perchuck and Ryabchikov (1968); 13. Shaw *et al.* (1963).

[a] average.

The degree of relative incorporation of an element into a specific mineral is given by the ratio:

$$D = \frac{\text{concentration of Ba in the phenocryst}}{\text{concentration of Ba in the melt}}.$$

This distribution coefficient can be calculated if equilibrium phenocrysts are compared with total ground mass, for instance, of volcanic rocks. If the amount of phenocrysts is low, the Ba content of the total rock can be used in the denominator of the above given equation. Distribution coefficients can also be calculated for metamorphic rocks, which constitute equilibrium mineral assemblages.

A number of these coefficients have been calculated from published data (Table 56-D-4). It must be mentioned that the distribution coefficients depend on several factors: pressure, temperature, the amount of Ba available, presence of competing elements, etc.

Berlin and Henderson (1969) determined Ba in sanidine and plagioclase crystals, and the embedding ground mass of trachytes and phonolites. They could reconstruct, by Ba and Sr determinations, the sequence of crystallization (sanidine or plagioclase first) and even detected in one case indications for secondary redistribution.

Philpotts and Schnetzler (1970a), analyzing (I) phenocrysts and a matrix of basic to intermediate volcanic rocks, calculated partition coefficients (Ba concentration in mineral/Ba concentration in matrix) for: plagioclases, D = 0.0537 to 0.589 (9 pairs); K feldspars, D = 6.12; clinopyroxenes, D = 0.0129 to 0.388 (10 pairs); orthopyroxenes, D = 0.121 and 0.141; micas (biotite-phlogopite), D = 1.09, 6.36 15.3; hornblendes, D = 0.0996, 0.417, 0.731; garnet, D = 0.0172; and olivines, D = 0.00864 and 0.0112. Only K feldspars and micas concentrate Ba relative to the matrix. They observed a strong coherence of Ba and K in almost all phenocrysts. The D-value of the K/Ba ratio is within the range of 0.5 to 2.0 in nearly all their examples. Since this K/Ba ratio is found to be extremely consistent in basic rock types, these authors recommend it as an "indexing criterion" for solar systems.

Ba distribution coefficients between coexisting minerals (Table 56-D-5) give some idea of the relative importance of minerals for the Ba content of a rock. It must be kept in mind that analyses of mineral phases isolated from a certain rock do not necessarily represent equilibrium conditions. Ba concentration in a definite mineral is affected by pressure, temperature, Ba availability, structure and position in the sequence of separation from the melt, and other parameters.

A useful index of fractionation is the ratio Ba/Rb. Ba has a tendency to be captured in early K minerals, whereas Rb is enriched in the residual melts due to its smaller charge and larger ion size. Taylor and Heier (1960) report a variation of the Ba/Rb ratio in feldspars from gneisses, granites and pegmatites from 54 to 0.04.

II. Barium Minerals

Because of their abundance, the most important Ba minerals and consequently the main Ba carriers in the earth's crust are: in igneous rocks, K feldspars, which contain a certain percentage of celsian molecules; and in sedimentary rocks and hydrothermal deposits, barite. Under special conditions, a large number of well defined Ba minerals can form. Sometimes they are reported from only one locality. In Table 56-D-6 the nonsilicates and silicates are listed. Criteria for the presentation in these lists have been: (1) reported by STRUNZ (1966), and/or (2) approved by the Commission on New Minerals of IMA.

Table 56-D-6. *Barium minerals*

Mineral	Formula
Oxides	
Billietite	$(BaO \cdot 6UO_3) \cdot 11H_2O$
Hollandite	$Ba_2\,Mn_8O_{16}$
Pandaite	$(Ba, Sr, Ca)(Nb, Ti, Ta)_2O_6 \cdot H_2O$
Priderite	$(K, Ba)_{1,3}(Ti, Fe)_8O_{16}$
Psilomelane	$(Ba, H_2O)Mn_5O_{10}$
Rijkeboerite	$Ba_{1-x}(Ta, Nb)_2O_5(H_2O)$
Todorokite	$(Mn, Mg, Ca, Ba, Na, K)_2Mn_5O_{12} \cdot 3H_2O$
Carbonates	
Alstonite (Ba-aragonite)	$BaCa[CO_3]_2$
Barytocalcite	$BaCa[CO_3]_2$
Benstonite	$(Ca, Mg, Mn)_7(Ba, Sr)_6[CO_3]_{13}$
Burbankite	$(Na, Ca, Sr, Ba, Ce)_6[CO_3]_5$
Carbocernaite	$(Ca, RE, Na, Sr, Ba)[CO_3]$
Ewaldite	$Ba(Ca, RE, Na, K, Sr, U, \square)[CO_3]_2$
Huanghoite	$BaCe[CO_3]_2F.$
Kordylite	$Ba(Ce, La, Nd)_2F_2/[CO_3]_3$
Mckelveyite	$Na_2Ba_4CaY_2[CO_3]_9 \cdot H_2O$
Norsethite	$BaMg[CO_3]_2$
Stenonite	$(Sr, Ba, Na)_2Al[CO_3]F$
Witherite	$BaCO_3$
Nitrate	
Nitrobarite	$Ba[NO_3]_2$
Sulfate[a]	
Barite	$BaSO_4$
Selenite	
Guilleminite	$Ba(UO_2)_3[SeO_3]_2(OH)_4 \cdot 3H_2O$
Phosphates, Arsenates, Vanadates	
Babefphite	$Be_5Ba_4[PO_4]_4O\,F_4 \cdot 0.35H_2O$
Bergenite[b]	$Ba(UO_2)_4[PO_4]_2(OH)_4 \cdot 8H_2O$
Dussertite	$BaFe^{3+}{}_3H[AsO_4]_2(OH)_6$
Ferrazite[c]	$(Pb, Ba)_3[PO_4]_2 \cdot 8H_2O$
Francevillite	$(Ba, Pb)(UO_2)_2[VO_4]_2 \cdot 8H_2O$
Gamagarite	$Ba_4(Fe, Mn)_2U_4O_{15}(OH)_2$
Gorceixite	$BaAl_3(OH)_5[PO_4]_2 \cdot H_2O$
Heinrichite	$Ba(UO_2)_2[AsO_4]_2 \cdot 10H_2O$
Metaankoleite	$(K, Ba)(UO_2)_2[PO_4]_2 \cdot 6H_2O$
Metaheinrichite	$Ba(UO_2)_2[AsO_4]_2 \cdot 8H_2O$
Metauranocircite I	$Ba(UO_2)_2[PO_4]_2 \cdot 8H_2O$
Metauranocircite II	$Ba(UO_2)_2[PO_4]_2 \cdot 6H_2O$
Strontiumapatite[b]	$(Sr, Ba)_6(Ca, RE, Mg, Na)_4[PO_4]_6(F, OH)_2$
Uranocircite I	$Ba(UO_2)_2[PO_4]_2 \cdot 12H_2O$
Uranocircite II	$Ba(UO_2)_2[PO_4]_2 \cdot 10H_2O$
Vesignietite	$BaCu_3[VO_4]_2(OH)_2$
Weilerite[d]	$BaAl_3H_{0-1}[AsO_4, SO_4]_2(OH)_{7-6}$

Table 56-D-6. (Continued)

Mineral	Formula
Nesosilicates	
Bariumuranophane	$BaH[UO_2/SiO_4]_2 \cdot 5H_2O$
Garrelsite	$(Ba, Ca)_4H_6Si_2B_6O_{20}$
Sorosilicates	
Bafertisite	$BaFe_2TiSi_2O_9$
Barylite	$BaBe_2Si_2O_7$
Hyalotektite	$(Pb, Ca, Ba)_4B[Si_6O_{17}](F, OH)$
Innelite	$Ba_2(Na, K, Mn, Ti)_2Ti(O, OH, F)_2[(S, Si)O_4/Si_2O_7]$
Labuntsovite	$(K, Ba, Na)(Ti, Nb)(Si, Al)_2(O, OH)_7 \cdot H_2O$
Nenandkevichite	$(Na, K, Ca, Ba)(Nb, Ti)[Si_2O_7] \cdot 2H_2O$
Shcherbakovite	$(K, Na, Ba)_3(Ti, Nb)_2[Si_2O_7]_2$
Yoshimuraite	$(Ba, Sr)_2(Mn, Fe, Mg)_2(Ti, Fe)(OH, Cl)_2[(S, P, Si)O_4/Si_2O_7]$
Ring silicates	
Armenite	$BaCa_2Al_6Si_8O_{28} \cdot 2H_2O$
Baotite	$Ba_4(Ti, Nb)_8Si_4O_{28}Cl$
Benitoite	$BaTiSi_3O_9$
Cappelenite	$(Ba, Ca, Ce, Na)_3(Y, Ce, La)_6[BO_3]_6[Si_3O_9]$
Muirite	$Ba_{10}Ca_2MnTiSi_{10}O_{30}(OH, Cl, F)_{10}$
Papstite	$Ba(Sn, Ti)Si_3O_9$
Taramellite	$Ba_2(Fe^{3+}, Ti, Fe^{2+})_2(OH)_2[Si_4O_{12}]$
Traskite	$Ba_9Fe_2Ti_2Si_{12}O_{36}(OH, Cl, F)_6 \cdot 6H_2O$
Verplanckite	$Ba_2(Mn, Fe, Ti)Si_2O_6(O, OH, Cl, F)_2 \cdot 3H_2O$
Chain silicates	
Batisite	$Na_2BaTi_2[Si_2O_7]_2$
Krauskopfite	$BaSi_2O_5 \cdot 3H_2O$
Walstromite	$BaCa_2Si_3O_9$
Sheet silicates	
Anandite	$(Ba, K)(Fe, Mg)_3(Si, Al, Fe)_4O_{10}(O, OH)_2$
Bariumphlogopite	$(K, Ba)Mg_3(F, OH)_2[AlSi_3O_{10}]$
Barium-vanadium-muscovite	$(K, Na, Ba)(Al, Ti, V, Mg)_2(OH)_2[AlSi_3O_{10}]$
Gillespite	$BaFe[Si_4O_{10}]$
Oellacherite	$(K, Ba)(Al, Mg)_2 (OH, F)_2[AlSi_3O_{10}]$
Sanbornite	$Ba_2[Si_4O_{10}]$
Tectosilicates (without zeolites)	
Banalsite	$BaNa_2[Al_2Si_2O_8]$
Celsian	$Ba[Al_2Si_2O_8]$
Paracelsian	$Ba[Al_2Si_2O_8]$
Bariumalbite	solid solution Ab-Or-Ce
Bariumplagioclase	solid solution Ab-An-Ce
Calciocelsian	solid solution Ce-An
Hyalophane	solid solution Or-Ce
Bariumsanidine	solid solution
Cymrite	$BaAlSi_3O_8(OH)$
Wenkite	$(Ba,Ca)_9[SO_4]_2Al_9Si_{12}O_{42}(OH)_5$

Table 56-D-6. (Continued)

Mineral	Formula
Zeolites	
Barium heulandite	$(Ca,Ba)[Al_2Si_7O_{18}]\cdot 6H_2O$
Brewsterite	$(Sr, Ba, Ca)_2\ Al_4Si_{12}O_{32}\cdot 10H_2O$
Edingtonite	$Ba\ Al_2Si_3O_{10}\cdot 3H_2O$
Harmotome	$Ba_2\ Al_4Si_{12}O_{32}\cdot 12H_2O$
Wellsite	solid solution phillipsite-harmotome
Unclassified silicates	
Fresnoite	$Ba_2TiSi_2O_8$
Joaquinite	$NaBa(Ti, Fe)_3Si_4O_{15}$
Leukosphenite	$BaNa_4(TiO)_2[Si_2O_5]_5$
Macdonaldite	$BaCa_4Si_{15}O_{35}\cdot 11H_2O$
Tienshanite	$Na_2BaMnTiB_2Si_6O_{20}$

Calciobarite (Ca), Baritocelestite, Celestobarite (Sr), Baritoanglesite, Anglesobarite, Hokutolite, Weisbachite (Pb), Radiobarite (Ra). Reviews on composition, occurence, and crystallographic and physical properties are given by DANA (1951), HINTZE (1930, 1938, 1960). X-ray evidence for the existence of a barite-celestite isomorphous series has been obtained by SABINE and YOUNG (1954).

BOSTRÖM *et al.* (1968) studied subsolidus phase relations and lattice constants in the system $BaSO_4$-$SrSO_4$-$PbSO_4$.

[a] Barite forms more or less continuously, solid solutions with Ca, Sr, Ra, and Pb sulfates. In keeping with respective compositions, different names are used for the members of the series:

[b] No decisive vote of the IMA new mineral commission.

[c] Doubtful mineral species.

[d] Not yet approved by IMA new mineral commission.

Revised manuscript received: September 1971

56-E. Abundance in Common Igneous Rock Types

A survey of Ba concentrations of rocks published by v. ENGELHARDT (1936) showed that Ba content in igneous rock series normally increases with increasing SiO_2 concentration. Recent data of Ba concentrations in common igneous rocks are summarized in Tables 56-E-1 and 2. The grouping follows the outlines of WEDEPOHL in volume 1 of this handbook. The listing contains the range of individual values, the arithmetic means and the standard deviation (s) of these means. Ba distribution in the main rock types is briefly discussed, as well as its bearing on genetic interpretations.

I. Ultramafic Rocks

The most important ultramafic rocks in the development of magmatic series are dunites and peridotites. From the available data, mostly spectrographic, an average of 8.8 ppm for dunite and of 25 ppm for peridotite was calculated. In their table of elemental distribution in the earth's crust, TUREKIAN and WEDEPOHL (1961) preferred a neutron activation value of 0.4 ppm Ba for dunite rather than a spectrographic determination average of 6 ppm.

All data for these rocks suffer from potential contamination (see discrepancy between finds and falls of chondrites) and from analytical difficulties occurring close to the detection limit of Ba.

Pyroxenites contain a little more Ba (average 23 ppm); biotite pyroxenites have concentrations upto 3,200 ppm Ba (HIGAZY, 1954), as do kimberlites, where phlogopite is the main Ba carrier. Very high values are also reported for carbonatites (average 3,520 ppm) which sometimes even contain barite.

II. Gabbroic and Basaltic Rocks

Gabbroic rocks of intrusive occurence (average Ba content 246 ppm) closely resemble continental tholeiitic basalts (average Ba content 246 ppm).

Basalts can be divided into three groups according to their trace element concentration (Fig. 56-E-1). The averages for Ba are as follows:

Oceanic tholeiitic basalts	14.5 ppm Ba
Tholeiitic basalts of continents and oceanic islands	246 ppm Ba
Alkali basalts	613 ppm Ba.

All trace element data for oceanic tholeiitic basalts indicate a general uniformity for widely separated parts of the oceans. MUIR *et al.* (1964), however, determined Ba from the rift zone of the Midatlantic Ridge, at 45° N, and found 55 to 220 ppm (S). The continental tholeiitic basalts contain more Si, K, Ba, Cs, Pb, and Rb than oceanic tholeiitic basalts (GAST, 1960).

Table 56-E-1. *Barium in intrusive rocks*

Rock type	No. of localities or No. of mean values used	No. of individual values	Range of individual values in ppm	Arith. mean of means grouped by locality in ppm	s[a]	References[b]
Alkali granite	3	19	22— 2,100	857		11, 12, 73
Granite	28	608	22— 3,000	732	453	6, 17, 21, 24, 26, 31, 32, 49, 50, 54, 55, 57, 62, 67, 72, 73, 75, 80, 83, 85, 87, 89, 93, 106, 121, 126
Granodiorite	7	53	400— 1,815	888		6, 73, 83, 87, 124, 131
Quartzdiorite	5	18	150— 1,250	811		87, 90, 131
Quartzmonzonite	3	18	233— 6,000	1,605		12, 87, 124
Alkali syenite	4	10	300— 2,070	1,067		21, 34, 75, 109
Syenite	9	30	230—18,000	2,753		24, 34, 44, 58, 63, 97, 100, 109, 111
Monzogabbro	1	4	7,500—14,000	10,360		100
Diorite	3	25	126— 1,150	714		75, 87, 90
Gabbro	10	148	5— 1,250	246	228	4, 21, 75, 87, 91, 100, 101, 109, 111, 127
Anorthosite	3	66	65— 475	171		75, 91
Norite	4	18	5— 310	78		75, 87, 99
Nepheline syenite	12	524	10— 6,300	1,427	1,206	27, 34, 35, 43, 44, 58, 59, 61, 75, 111, 125
Essexite	3	8	850— 2,800	1,598		64, 88, 111
Teschenite	1	3	650— 2,000	1,500		63
Ijolite, melteigite, jacupyrangite	5	66	88— 2,430	973		27, 35, 43, 44, 75
Dunite	9	18	0.3— 40	8.8		34, 39, 56, 75, 98, 100
Peridotite	6	9	1— 70	38		38, 39, 64, 75, 98
Pyroxenite	2	13	5.5— 67	23		75, 98
Kimberlite	4	35	20— 2,860	847		23, 29, 53, 68
Carbonatite	9	215	88—54,000	3,799		23, 27—29, 35, 41, 46, 64, 69, 105, 123

[a] s is the standard deviation of the mean values of the different authors, standard deviations are given only when 10 or more mean values were available.

[b] For list of references see Page 56-E-4.

Table 56-E-2. *Barium in volcanic rocks*

Rock type	No. of localities or No. of mean values used	No. of individual values	Range of individual values in ppm	Arith. mean of means grouped by locality in ppm	s[a]	References[b]
Alkali rhyolite	8	126	1— 700	118		25, 30, 36, 86, 102, 103, 109
Rhyolite	20	153	5— 3,650	1,127	632	8, 9, 18, 25, 26, 30, 74, 79, 88, 89, 92, 103, 108, 109, 112, 121
Rhyodacite	3	12	550— 1,800	1,210		18, 30, 89
Dacite	7	22	150— 1,250	629		87, 89, 118, 122
Quartz latite	1			550		112
Alkali trachyte	2	2	800+ 1,500	1,150		88, 128
Trachyte	10	30	20— 3,000	1,177	667	3, 36, 71, 88, 92, 107, 128
Latite minette	3	4	137— 2,500	1,379		18, 88, 112
Latite andesite	2	7	300— 2,250	841		88, 112
Andesite	23	185	80— 2,700	703	475	2, 3, 8, 39, 45, 51, 74, 79, 88, 89, 108, 110, 112, 115, 118, 120, 122, 128
Tholeiitic basalt	51	555	20— 1,160	246	197	3, 5, 8, 10, 13—16, 19, 20, 22, 37, 48, 51, 52, 60, 65, 74, 76, 77, 81, 87—89, 92, 94—96, 99, 113, 114, 117, 119, 127, 128
Alkali basalt	16	100	30— 1,350	613	223	1, 3, 33, 51, 66, 76, 78, 81, 88, 116, 117, 129, 130, 132
Oceanic tholeiite	41	41	3— 46	14.5	11.3	33, 42, 70, 82, 84
Phonolite	5	42	10— 2,000	999		27, 44, 71, 88, 104
Nepheline basanite, Nepheline tephrite, Leucite basanite, Leucite tephrite	8	95	250— 9,360	1,976		7, 40, 64, 88, 104, 107, 130, 133
Nephelinite	2	5	600— 5,900	3,444		64, 130
Ankaratrite	1	2	1,800+ 4,000	2,900		64
Leucitite	1	9	1,800— 7,000	3,510		64
Melilitite	1	2	2,200+10,000	6,100		64
Alnöite	2	22	450— 6,930	1,890		27—29, 47

[a] s is the standard deviation of the mean values of the different authors, standard deviations are given only when 10 or more mean values were available.

[b] For list of references see Page 56-E-4.

References (methods in brackets): 1. Baker, 1969 (X); 2. Baker, 1968 (S); 3. Baker *et al.*, 1964 (S); 4. Baragar, 1960 (S); 5. Bartel *et al.*, 1963 (S); 6. Bräuer, 1965; 7. Brown and Carmichael, 1969 (X); 8. Byers, jr., 1961 (S); 9. Carmichael and McDonald, 1961 (S); 10. Clarke, 1970; 11. Clifford *et al.*, 1962 (S); 12. Clifford *et al.*, 1969 (S); 13. Coats, 1952 (S); 14. Coats, 1953 (S); 15. Coats, 1959 (S); 16. Coats *et al.*, 1961 (S); 17. Cocco, 1953 (S); 18. Cornwall, 1962 (S); 19. Cornwall and Rose jr., 1957 (S); 20. Cox and Hornung, 1966 (S, X); 21. Cox *et al.*, 1965 (S); 22. Cox *et al.*, 1967 (S); 23. Dawson, 1962 (S); 24. Dietrich and Heier, 1967; 25. Dixon *et al.*, 1968 (S); 26. Dunham, 1968 (S); 27. v. Eckermann, 1948 (S); 28. v. Eckermann, 1966 (S); 29. v. Eckermann, 1967 (S); 30. El-Hinnawi, 1969 (S); 31. Emiliani and Vespignani, 1964 (S); 32. Emmermann, 1968 (X); 33. Engel *et al.*, 1965 (S); 34. v. Engelhardt, 1936 (S); 35. Erickson and Blade, 1963 (S); 36. Ewart *et al.*, 1968 (S); 37. Fairbairn *et al.*, 1953 (S); 38. Fisher and Engel, 1969 (S); 39. Flanagan, 1969 (S, N/R, X); 40. Fornaseri *et al.*, 1963 (W); 41. Garson, 1967; 42. Gast, 1965; 43. Gerasimovskii, 1966 (S); 44. Gerasimovskii and Belyaev, 1963 (S); 45. Giusca and Ionescu, 1965; 46. Gold, 1963; 47. Gold, 1967; 48. Greenland and Lovering, 1966 (S); 49. Grohmann and Schroll, 1966 (S); 50. Grout, 1935 (W); 51. Gunn, 1965 (X); 52. Gunn, 1966 (X); 53. Hahn-Weinheimer, 1959 (S); 54. Hahn-Weinheimer and Ackermann, 1967 (X); 55. Hall, 1967 (X); 56. Hamaguchi *et al.*, 1957 (N/R); 57. Heier, 1960 (S); 58. Heier, 1964 (S); 59. Heier, 1965 (S); 60. Heier *et al.*, 1966 (S); 61. Henderson, 1965 (S); 62. Herz and Dutra, 1960 (S); 63. Higazy, 1952 (S); 64. Higazy, 1954 (S); 65. Hotz, 1953 (S); 66. Huckenholz, 1969 (X); 67. Hügi and Swaine, 1963 (S); 68. Janse, 1962; 69. Johnson, 1961 (S); 70. Kay *et al.*, 1970 (I); 71. King and Sutherland, 1967; 72. Kolbe, 1964; 73. Kolbe and Taylor, 1966 (S); 74. Kuno *et al.*, 1957 (S); 75. Liebenberg, 1960 (S); 76. Lipman, 1969 (S); 77. MacDonald and Eaton, 1964 (S); 78. Le Maitre, 1962 (S); 79. Markhinin *et al.*, 1964 (S); 80. Marmo and Siivola, 1966 (S); 81. Mathias, 1957 (S); 82. Melson *et al.*, 1968 (S); 83. Moenke, 1960 (S); 84. Muir *et al.*, 1964 (S); 85. Mukherjee, 1968 (S); 86. Noble and Haffty, 1969 (S); 87. Nockolds and Allen, 1953 (S); 88. Nockolds and Allen, 1954 (S); 89. Nockolds and Allen, 1956 (S); 90. Okrusch and Richter, 1969 (X); 91. Papezik, 1965 (S); 92. Patterson, 1951 (S); 93. Patterson, 1953 (S); 94. Patterson and Swaine, 1955 (S); 95. Patterson *et al.*, 1955 (S); 96. Peck *et al.*, 1966 (S); 97. Pecora, 1962 (S); 98. Pinson *et al.*, 1953 (S); 99. Prinz, 1964 (S); 100. Read and Haq, 1963 (S); 101. Read *et al.*, 1965 (X); 102. Renfrew *et al.*, 1966 (S); 103. Renfrew *et al.*, 1968 (S); 104. Ridley, 1970 (S, X); 105. Russell *et al.*, 1954 (S); 106. Sahama, 1945 (S); 107. Savelli, 1967 (X); 108. Shelton, 1955; 109. Siedner, 1965 (S); 110. Siegers *et al.*, 1969 (X); 111. Simpson, 1954 (S); 112. Sinha and Tiwari, 1964 (S); 113. Sinha and Karkare, 1964a (S); 114. Sinha and Karkare, 1964b (S); 115. Smith, 1964 (S); 116. Smith and Carmichael, 1969 (X); 117. Snavely jr. *et al.*, 1968 (S); 118. Staritsin, 1964 (S); 119. Stark and Tracey, 1963 (S); 120. Taylor and White, 1966 (S); 121. Taylor *et al.*, 1968 (S); 122. Taylor *et al.*, 1969 (S); 123. Temple and Grogan, 1965 (X); 124. Townend, 1966 (S); 125. Vlasov *et al.*, 1966 (S); 126. Volborth, 1962 (W); 127. Wager and Mitchell, 1951 (S); 128. Wager and Mitchell, 1953 (S); 129. Wedepohl, 1954 (S); 130. Wedepohl, 1961 (S); 131. Weibel, 1960 (S); 132. Wilkinson, 1959 (S); 133. Wilkinson, 1968.

Differences in Ba concentration among tholeiitic basalts of particular regions were found in South Africa, where Rhodesian tholeiites show unusually high mean Ba values upto 1,020 ppm (Cox *et al.*, 1967), while the basalts of Basutoland and Swaziland are of normal tholeiitic geochemistry. Greenland and Lovering (1966) studied differentiation within a tholeiitic sill in Tasmania (Australia). Ba concentrations increase here, from the bottom to the 1735 ft-high top of the flow, from 160 to 500 ppm. Elements with similar enrichment trends are F and Ga, whereas nonparallel behavior was found for Ni, Co, Cr and Sc.

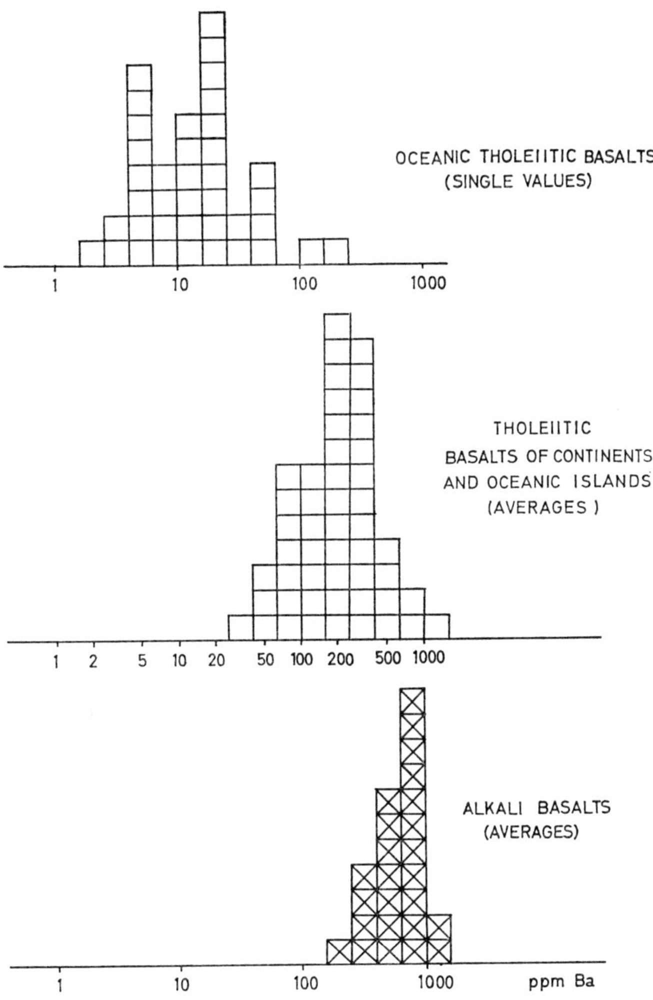

Fig. 56-E-1. Barium distribution in basaltic rocks

ENGEL *et al.* (1965) gave the ratio of the masses of alkali olivine basalts to tholeiitic basalts as 2:98. Consequently, an average of 253 ppm Ba is to be assumed for average continental basaltic rock. PRINZ (1967), in his summary of trace element data for basalts, found an arithmetic mean for all basalts of 303 ppm and a geometric mean of 220 ppm for 253 analyses. From the wide range of Ba concentrations within similar petrographic types and from similar Ba values for different petrographic types within any region, it is indicated that differences in the initial abundance of Ba in basaltic magmas exist (PRINZ, 1967). The generation of such differences is attributed to different degrees of partial melting, mantle inhomogeneity or wall rock reactions (JAMIESON and CLARKE, 1970).

Anorthosites and norites are both lower in Ba than tholeiitic basalts. The normally observed relationship between K and Ba is not found in Canadian anorthosites. In these rocks, which are rich in Ca, the Ba-Ca diadochy is superimposed on the more common Ba-K relation.

III. Granitic Rocks

a) High Ca Content

Granitic rocks with a high Ca content, as in granodiorites and quartz-diorites, are usually high in Ba. Averages are 888 ppm and 811 ppm, respectively; within particular areas the Ba content may vary considerably. An average for both rock types, calculated with the percentages of WEDEPOHL (1969), is 873 ppm.

b) Low Ca Content

Granitic rocks with low Ca content, which are represented by granites and quartzmonzonites, show extremely different Ba values (Fig. 56-E-2). This may be

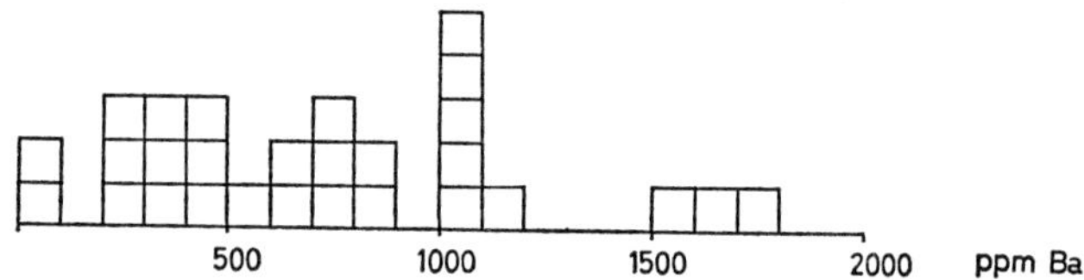

Fig. 56-E-2. Ba distribution in granites (28 local means, see Table 56-E-1)

partly due to origin and to differences in the respective processes of formation. Metasomatical alterations may also have changed the Ba content in some granites (cf. EMMERMANN, 1969). The granite average from 28 regional mean values is 732 ppm. Within certain granite bodies, the Ba concentration may exhibit a narrow spread, while the means are low or high. Gradational change of Ba content was also observed (EMMERMANN, 1969). For three quartzmonzonite areas, an average of 1,605 ppm was calculated.

The effusive equivalents of granitic rocks contain more Ba (average 1,127 ppm) than granites. The single quartz latite value, however, is much lower than the average quartzmonzonite content.

IV. Intermediate Rocks

Intermediate rocks, such as syenites and trachytes, are strongly enriched in Ba. While the syenite average is 2,753 ppm, the effusive equivalent averages only 1,177 ppm. TUREKIAN and WEDEPOHL (1961) gave a value of 1,600 ppm, which was calculated from data of v. ENGELHARDT (1936) and SAHAMA (1945).

V. Alkalic Rocks

Alkalic rocks are all considerably enriched in Ba. All averages are higher than 1,000 ppm, with nepheline-syenite showing 1,427 ppm, phonolite 1,000 ppm and nepheline-basanite 1,976 ppm Ba. In rocks of the Lovozero alkali massif (USSR), Ba is enriched upto 3,300 ppm (VLASOV *et al.*, 1966). The Kola alkali complex (USSR) contains upto 1,600 ppm Ba in its nepheline-syenites and upto 1,350 ppm Ba in the foyaites (GERASIMOVSKII and BELYAEV, 1963).

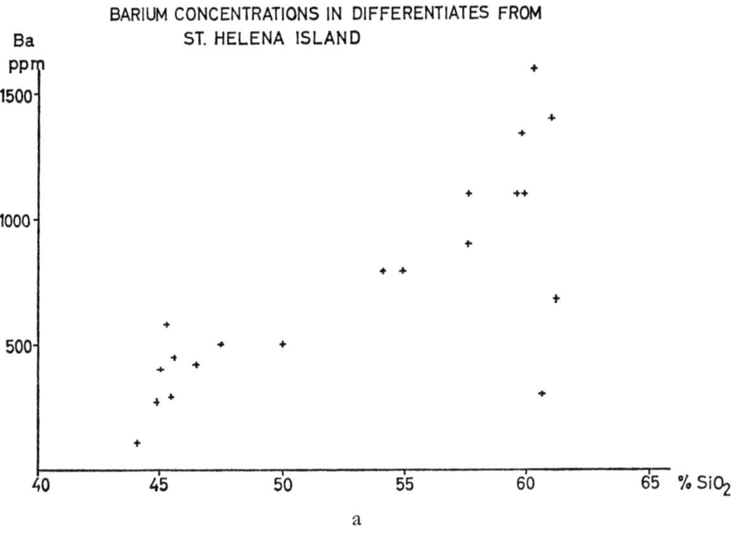

a

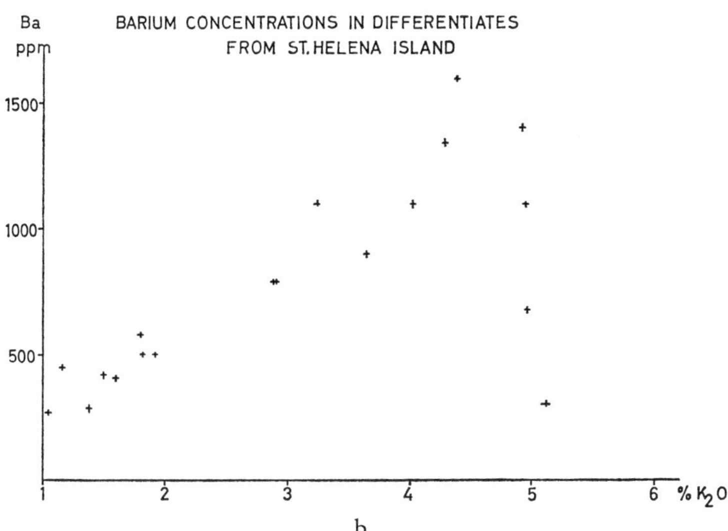

b

Fig. 56-E-3. Correlation of Ba concentration and SiO_2 (upper plot) and K_2O content in a differentiation series (BAKER, 1969)

Many investigators (v. ENGELHARDT, 1936; WAGER and MITCHELL, 1951; NOCKOLDS and ALLEN, 1953, 1954, 1956; WILKINSON, 1959; MARKHININ *et al.*, 1964; P. E. BAKER, 1968) found that Ba concentrations increase during progressing differentiation. One example of correlation between concentrations of Ba and SiO_2 and K_2O (as parameter of differentiation) is given for volcanic rocks of Saint Helena island, South Atlantic (Fig. 56-E-3).

Sometimes Sr shows the same trend, while a nonparallel behavior is observed for Ca, V, Cu, Co, Ni, and Sc. In several instances the relation between Ba and other elements is not as simple; plotting versus differentiation, solidification, mafic index or Larsen factor provides a better insight into differentiation behavior.

In series which proceed far in differentiation, Ba concentrations normally pass a maximum. Since Ba substitutes mostly for K, this indicates that a distribution coefficient $\left(\frac{Ba}{K}\right)_{crystal} \Big/ \left(\frac{Ba}{K}\right)_{melt}$, larger than unity is effective in acid magmas for at least one major mineral — mostly potash feldspar or mica.

This type of Ba distribution prevails in the example given in Fig. 56-E-3, in the Scottish Caledonian rocks, and the East Central Sierra Nevada rocks analyzed by NOCKOLDS and ALLEN (1953). As a consequence of this differentiation behavior, Ba is always very low in pegmatites of truly magmatic origin (cf. Table 56-D-1 and v. ENGELHARDT, 1936). A trend of K/Ba ratios to increase from 26 to 36 in a tholeiitic sequence was found by GUNN (1965).

Data for averages of Ba concentration in igneous rocks have been subject to changes in the last year due to the fact that more and better data have been published. Consequently, the mean calculated Ba content in igneous crustal rocks has changed, since better insight into the relative abundance of rock types was obtained. With the Ba concentrations of Table 56-E-1, one arrives at a Ba average of 728 ppm for the upper continental crust of the earth, using the data for the relative amounts of intrusive rocks by WEDEPOHL (1969, handbook; Tables 7—8). Examples for Ba distribution between minerals of certain rocks are given in Table 56-E-3.

Compilation and interpretation of Ba data from literature suffer somewhat from the difficulties in analytical determination. This is shown graphically by the Ba values for the standard rocks G1 and W1 (Fig. 56-E-4). The considerable spread of values — even with the same analytical method — cannot be explained by too coarse grain size of the reference samples (KLEEMANN, 1967), since similar observations were made with the much more finely ground new standard rocks of the USGS (FLANAGAN, 1969). As the individual authors use their own different values for internal reference, difficulties arise when results from different sources are compared. This must be considered when the data from these paragraphs are evaluated.

Table 56-E-3. *Barium distribution*

Rock type	Total rock Ba ppm	Quartz Ba ppm	Ortho-clase Ba ppm	Plagio-clase Ba ppm	Biotite Ba ppm
Hortonolite ferro gabbro	50			200	
Olivine norite	5			10	
Quartz biotite norite	300			110[a]	1,600
Granodiorite	1,000		3,000	90[a]	2,500
Granite	1,000		2,500	125[a]	2,000
Predazzo granite	70		108	65	100
Adamellite porphyrite	600			275	1,250
Adamellite	110		225		120
Sphen pyroxenite	500	100	900		
Teschenite	275			350	

[a] Quartz and plagioclase.

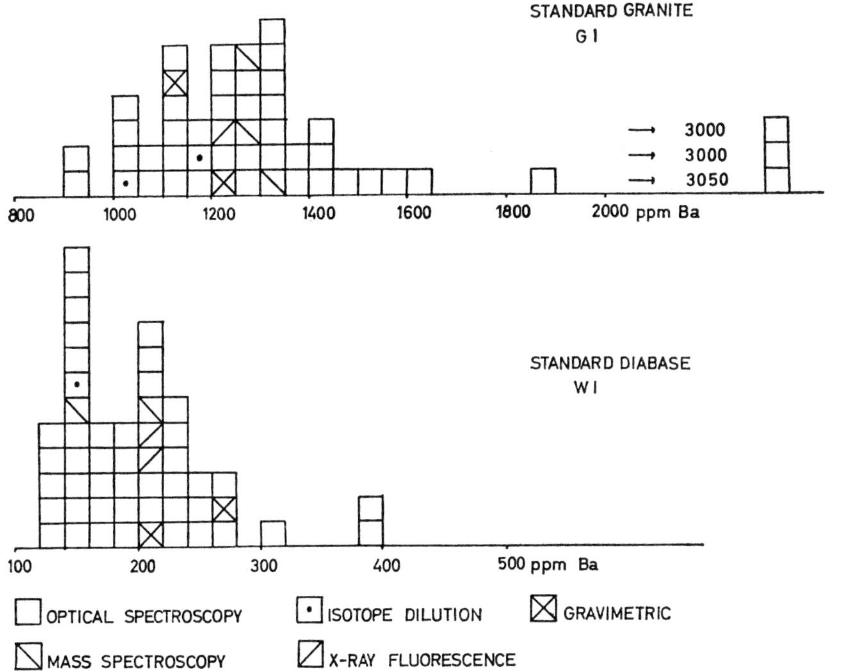

Fig. 56-E-4. Ba determinations on standard rocks (FAIRBAIRN *et al.*, 1951; STEVENS *et al.*, 1960; FLEISCHER and STEVENS, 1962; FLEISCHER, 1965, 1969; FLANAGAN, 1969)

in rocks

Muskovite Ba ppm	Amphibole	Pyroxene Ba ppm	Olivine Ba ppm	Method	References
		10	10	S	WAGER and MITCHELL (1951)
	15			S	SEN *et al.* (1959)
	5	5		S	SEN *et al.* (1959)
				S	SEN *et al.* (1959)
				S	SEN *et al.* (1959)
				S	EMILIANI and VESPIGNANI (1964)
	120	35			WILKINSON *et al.* (1964)
10				S	BUTLER (1953a)
		20		S	ERICKSON and BLADE (1963)
		10		S	WILKINSON (1959)

Revised manuscript received: September 1971

56-F. Behavior during Processes Connected with Magmatism

I. Pegmatites

Pegmatites of magmatic origin generally contain less Ba than their embedding wallrocks of magmatic or metamorphic origin. This feature is most apparent in the main Ba carriers, feldspar and mica. By analytical studies of alkali feldspars from the South Norwegian Precambrian basement complex, HEIER and TAYLOR (1959) found that the concentration of Ba in potash feldspars decreases with increasing differentiation. The K feldspars of large pegmatites in the surveyed area always contain less Ba than the host rock. This general trend was diagramatically shown by HEIER (1962) and can also be seen in Fig. 56-D-2, for which additional analyses were used. TAYLOR and HEIER (1960) stressed the importance of the Ba/Rb ratio in feldspars for judging the degree of fractionation. Since Rb is more discriminated against in the K feldspar structure than Ba, Rb is continuously enriched in the fluid during crystallization. Under the assumption of constant distribution coefficients for both elements, this leads to the highest Rb values in the last crystallization. Low Ba values in granite-pegmatite feldspars are also reported by v. ENGELHARDT (1936), OFTEDAL (1958), and TAKUBO and TATEKAWA (1954). Within a pegmatite body, OFTEDAL (1959) found the younger microcline to be poorer in Ba than the older one. Similar Ba impoverishment was found in nepheline-syenite pegmatites (v. ENGELHARDT, 1936; OFTEDAL, 1962) as compared to the lardalite and larvikite from which they are supposed to be derived. Biotite from granite pegmatite also is impoverished in Ba compared to the biotite from the mother granite (TAKUBO and TATEKAWA, 1954).

Ba behavior is different in the small pegmatite bodies which sometimes form by metamorphic processes. In the plagioclase gneiss area of Justøy, Norway, small pegmatitic veins occur with high Ba values. HITCHON (1960) investigated pegmatites of three metamorphic complexes in Scotland. While two complexes showed the usual Ba relation between pegmatite and country rock, pegmatites of the third complex (Laxfordian) were markedly enriched in Ba in all minerals (microcline-perthite 3,785 and 5,620 ppm; oligoclase 365 ppm; biotite 900 ppm, 760 ppm, 1,100 ppm and 2,320 ppm Ba). In these pegmatites, the Ba content of the individual minerals increases towards the core of the respective body. It is possible that in the case of Laxfordian pegmatites metasomatic processes caused a later Ba enrichment in the pegmatite minerals.

II. Metasomatism, Wallrock Alteration, Greisenization

Metasomatic changes of Ba content of rocks occur sometimes with emplacement of pegmatites. Evidence for a large scale metasomatic Ba addition in a rock body was found for the Albtal granite, Germany, by EMMERMANN (1968, 1969). This author investigated the distinct differences in the distribution patterns of Ba in the two occurring K feldspar generations and reached the conclusion that the potash feldspar megacrysts (mean Ba content 4,600 ppm) had grown during a postmagmatic stage from a metasomatic, Ba rich fluid, whereas the groundmass K feldspar (mean Ba content 1,600 ppm) represents the first generation.

Metasomatic alterations in granodiorite, dacite, and gabbro related to serpentinization of ultramafic rocks in the Western United States, generally resulted in a distinct Ba impoverishment. In a few cases, small intermediate zones with Ba enrichment were observed (COLEMAN, 1967).

LUR'YE (1963), analyzing spectrographically the silicic wallrocks of Zambarah zinc-lead ore deposit, Central Asia, found a pronounced decrease in the concentration of Ba and Sr towards the veins. The Ba content drops from 3,000 to 6,000 ppm, in fresh rock containing abundant K feldspar, to 300 ppm as a maximum in completely sericitized, feldspar free rocks close to the ore. He concludes that all barite and its Sr content originates from the feldspar decomposition in the wall rocks. Granitic wallrocks of hydrothermal veins in the Black Forest, Germany, were analysed for Ba distribution by DEGENS (1956). Extensive studies of wallrock alterations were carried out by TOOKER (1963) in the Front Range Mineral Belt, Colorado, USA. His spectrographic analyses showed that Ba, as with other large ions, normally tends to be removed veinward from all Precambrian and Tertiary metamorphic and igneous rocks he investigated. During alteration in the rock, pH drops along with the decrease of K, Ba etc.

Greisenization normally proceeds with the removal of Ba from the affected rock. In a mixture of samples from 24 greisen, v. ENGELHARDT (1936) found an abundance of 160 ppm Ba. This value is far below the content in the respective unaltered igneous silicic rocks. SOLOMON (1966) reports Ba removal from granites (430 ppm and 250 ppm) by greisenization (Ba in the altered rocks: 160 ppm and 105 ppm respectively) in the North Pennine ore field, Great Britain.

III. Ore Deposition

Probably, magmatic-hydrothermal fluids originally do not contain any significant amount of Ba, but obtain this element by leaching suitable wallrocks. The importance of this mechanism, mentioned in the previous paragraph, was already stressed by v. ENGELHARDT (1936). Another mechanism working in sedimentary environments is dissolution of barite from the sediment by bacterial sulfate reduction in suitable diagenetic environment (PUCHELT, 1967). As a third way of producing Ba containing fluids, certain metamorphic reactions may give off Ba because structures which had incorporated high Ba concentrations become unstable.

The most common Ba ore is barite. It is deposited by fluids with a high oxidation potential, where sulfur is present as sulfate. Suitable conditions of this kind occur close to the earth's surface or in subsurface areas where sulfate solutions mix with reduced Ba containing waters (cf. Subsection 56-I-IV). In solution, Ba migrates to the region of sulfate stability and thus is often bound to a narrow zone close to the earth's surface.

Modes of barite formation and possibilities of $BaSO_4$ transport in solution are discussed by PUCHELT (1967). Barite contains, in all cases, certain amounts of isomorphous Sr. Concentrations of this element in marine barites (upto 3.36%) are summarized by CHURCH (1970). STARKE (1964) analysed a large number of vein barites which contain upto 12% $SrSO_4$. The conditions necessary for witherite formation from ore forming fluids are not often fulfilled. Calculations for $BaCO_3$ formation, at 250°C and CO_2 fugacities of 0.1, 1.0 and 100 atm., were carried out by HOLLAND (1965) for varying sulfur and oxygen fugacities.

Products of hydrothermal activities are also the alpine fissure type minerals of which adularia are sometimes most apparently enriched in Ba (cf. Table 56-D-1).

IV. Volcanic Exhalations, Gas Transport

The possibility of Ba transport through hydrous gas phases was demonstrated by the experiments of STRÜBEL (1967). Results are discussed in chapter 56-H. NABOKO (1945) found traces of Ba in fumarole sublimates (mostly NaCl, KCl and NH_4Cl) of Klyucherskoy volcano, USSR. MINGUZZI (1948) determined traces of Ba in fumarole products of Vesuvius, Italy.

Revised manuscript received: September 1971

56-G. Behavior during Weathering and Alteration of Rocks

Experimental weathering of K feldspar in distilled water (PUCHELT, 1967) showed that Ba is preferentially released from this silicate structure into the solution. The weight ratio K_2O/BaO, being 18.9 in the mineral, is much lower in the weathering solution (8.1).

In the naturally occurring weathering series biotite — hydrobiotite — vermiculite, BOETTCHER (1966) observed, by spectrographic analysis, a decrease of the BaO content from 4,500 ppm to 300 ppm.

ROSENQUIST (1939) leached a granite powder (0.11% BaO) with distilled water. In the residue of the weathering solution BaO was enriched to 0.91%.

Extensive studies on different rocks and their weathering products were carried out by BUTLER (1953b, 1954) (Table 56-G-1). In three types, Ba is enriched in the silt and clay fraction of the weathered material, while in a fourth example even the weathering residue is leached with respect to Ba.

Table 56-G-1. *Barium distribution during rock weathering* (BUTLER, 1953b, 1954)

Rock type	Ba content (ppm)			Reference (Analytical method)
	fresh rock	from weathered rock		
		silt fraction	Ca saturated clay fraction	
Granite	110	180	500	BUTLER, 1953b (S)
Hornblende schist	10	190	500	BUTLER, 1953b (S)
Quartz free syenite	1,000	300	870	BUTLER, 1954 (S)
Hypersthene monzonite	1,000	45	15	BUTLER, 1954 (S)

Both Ba increase and Ba decrease have been observed in the weathering products in studies by a great number of investigators. Among the factors which influence Ba behavior in this process are: climate; type of clay minerals, which form during the decomposition; amount and kind of organic material present; and sulfur or sulfate content. Since barium sulfate is a compound of very low solubility, this last factor predominates in sequences originally rich in sulfides.

A survey of Ba content of soils was published by SWAINE (1955). Most soils contain 100 to 3,000 ppm Ba of which only trace amounts can normally be extracted by means of 1N NH_4 acetate. The highest value which SWAINE reports (3.3% Ba) comes from Tennessee, USA, from areas where barite had been mined. Older literature is listed in GMELIN (1960). Under special arid conditions of weathering in deserts, varnishes form, which always show Ba enrichment (ENGEL and SHARP, 1958) (S).

Revised manuscript received: September 1971

56-H. Solubilities of Compounds which Control Concentrations of Barium in Natural Waters (I), Adsorption Processes (II)

I. Solubilities

Only two Ba compounds exist which can control the Ba content of natural waters: $BaSO_4$ — barite, and $BaCO_3$ — witherite.

a) Solubility of $BaSO_4$

$BaSO_4$ is the least soluble and most abundant Ba mineral in the earth's crust. Its solubility in water upto 100° C has been repeatedly determined. Some of the available data are given in Fig. 56-H-1. STRÜBEL (1967) published data for $BaSO_4$ solubility in the hydrothermal range upto 600° C. At this temperature and 1,084 bars, he obtained a solubility of 9.61 ± 1.95 mg $BaSO_4$/1,000 g H_2O.

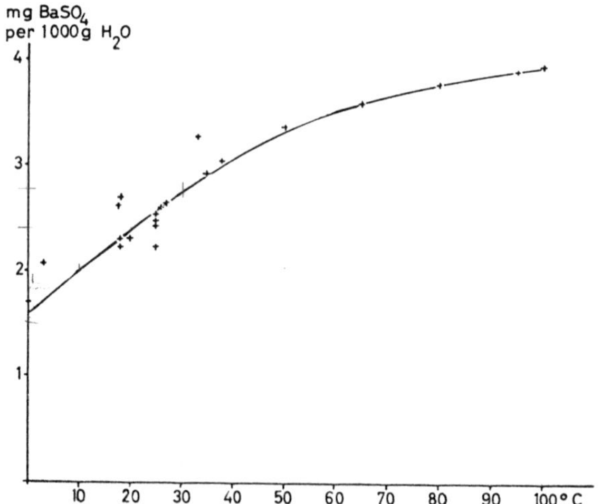

Fig. 56-H-1. Solubility of barium sulfate in distilled water (KOHLRAUSCH, 1908; MELCHER, 1910; NEUMANN, 1933; ROSSEINSKY, 1958; TEMPLETON, 1960; BURTON *et al.*, 1968)

Electrolytes considerably increase the $BaSO_4$ solubility. TEMPLETON (1960) has investigated sodium chloride influence upto 95° C with solutions upto 5 molal NaCl. PUCHELT (1967) radiochemically determined $BaSO_4$ solubility at 25 and 50° C in upto 6.08 molal NaCl solutions. Experiments upto 350° C were carried out by UCHAMEYSHVILI *et al.* (1966) with 0.25 N, 1.0 N and 2.0 N sodium chloride solutions. Investigations upto 600° C with upto 2 molal (or 11.69% ?) NaCl solutions were performed by STRÜBEL (1967). Solubility data upto the boiling point are plotted versus NaCl molality in Fig. 56-H-2. For 600° C, 1,990 bars and 2 N (?) NaCl solutions, STRÜBEL reports a solubility of 971 mg/kg H_2O. In the hydrothermal range, $BaSO_4$ solubility sensitively increases with pressure. STRÜBEL's investigations show that an area of retrograde $BaSO_4$ solubility exists between 350 and 450° C.

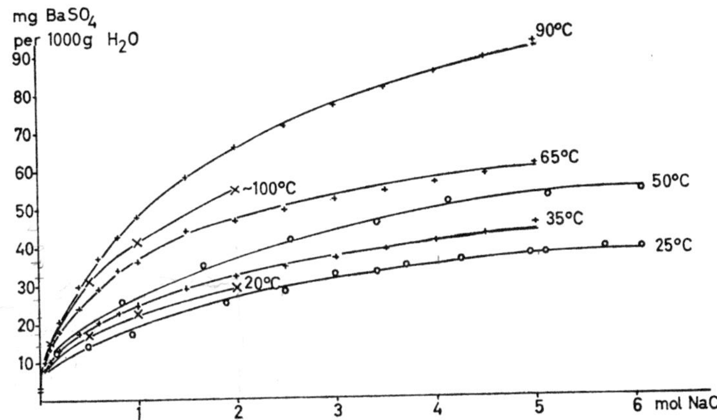

Fig. 56-H-2. Solubility of barium sulfate in NaCl solution. + TEMPLETON, 1960; × STRÜBEL, 1967; ○ PUCHELT, 1967

Influences of other electrolytes on $BaSO_4$ solubility in aqueous solutions were studied by: NEUMANN (1933), KCl, KNO_3, $MgCl_2$, $Mg(NO_3)_2$, $LaCl_3$, $La(NO_3)_3$; UCHAMEISHVILI *et al.* (1966), KCl, $MgCl_2$, $CaCl_2$; and PUCHELT (1967), KCl, $MgCl_2$, $CaCl_2$. COLLINS and ZELINSKI (1966) investigated the effect of synthetic brines containing NaCl, $MgCl_2$, $CaCl_2$, and $NaHCO_3^-$.

The results of PUCHELT are:

Solution	Maximum solubility found at		$BaSO_4$ solubility mg/1,000 g H_2O
	temperature	ionic strength	
KCl	25° C	5.0	40.8
KCl	50° C	5.0	54.0
$CaCl_2$	25° C	5.5	42.4
$CaCl_2$	50° C	5.5	59.8
$MgCl_2$	25° C	6.0	47.2
$MgCl_2$	50° C	6.0	71.6

UCHAMEISHVILI *et al.* (1966) found a strong increase in $BaSO_4$ solubility in $CaCl_2$ solutions with temperatures between 100 and 255° C. In $MgCl_2$ solutions, too, a barium sulfate solubility higher than that in NaCl and KCl solutions was observed by these investigators.

Barium sulfate solubility in sea water was calculated by CHOW and GOLDBERG (1960), and experimentally studied at one atmosphere pressure by PUCHELT (1967), who also made investigations regarding the kinetics of $BaSO_4$ precipitation in sea water. The solubility of 89 μg $BaSO_4$/l (at 20° C), found by PUCHELT agrees well with the value of 87.9 μg $BaSO_4$/l (at 25° C) of CHOW and GOLDBERG. Recently, BURTON *et al.* (1968) obtained a mean value of 81 μg/l. Close to saturation, complete equilibrium is reached slowly. Despite seeding, PUCHELT (1967) needed about 80 days. He also made experiments to study the influence of salinity of sea water on solubility and covered the range upto 87.5‰ salinity. HANOR (1969) and CHURCH (1970) calculated the effect of aqueous complexing and presence of Sr, Ca, and K on the solubility of $BaSO_4$—$SrSO_4$ mixed crystals in sea water.

Pressure increases the solubility of $BaSO_4$. In pure water the solubility product is, by a factor of 5.4, larger at 1 kilobar than at 1 atm. pressure. In a sodium chloride solution of 0.727 molality the same ratio is only 4.2.

b) Solubility of $BaCO_3$

Barium carbonate solubility depends largely on the CO_2 partial pressure of equilibrium atmosphere. At 25° C and 1 atm. CO_2 pressure, GARRELS *et al.* (1960) determined a dissociation constant of $10^{-8.64}$ for witherite. The solubility product of ^{14}C-labeled $BaCO_3$ in basic aqueous solutions at 25° C was found to be $4.0 \cdot 10^{-10} \pm 0.5 \cdot 10^{-10}$ by BACCANARI *et al.* (1968).

Increase in carbon dioxide pressure causes an increase in $BaCO_3$ solubility. This effect is much smaller at higher temperatures than at lower temperature (MALININ, 1963), (Fig. 56-H-3).

TOWNLEY *et al.* (1937) showed that LiCl, NaCl, and KCl at 25° C and 40° C, increased $BaCO_3$ solubility according to their respective concentration. In all concentrations investigated (upto 3 molal) LiCl produced the strongest increase in solubility. The effect of KCl in the same molality was the least.

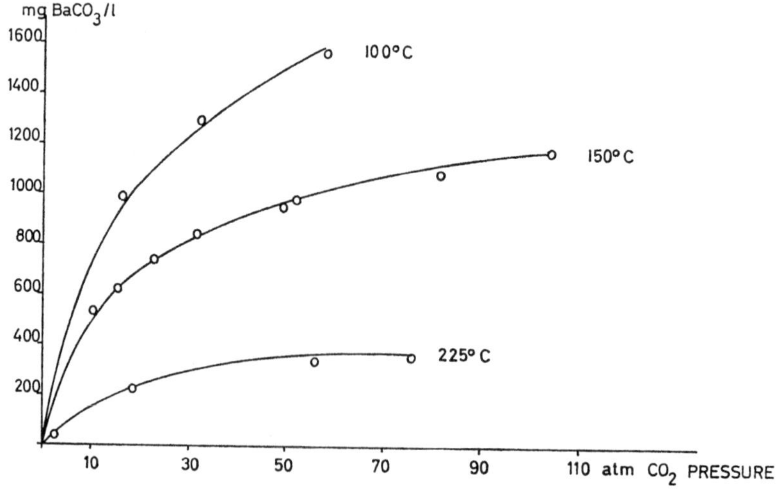

Fig. 56-H-3. Solubility of $BaCO_3$ in water with increasing CO_2 pressure (MALININ, 1963)

II. Adsorption Processes

Ba is adsorbed from solutions by clays, hydroxides, and organic matter. In addition to $BaSO_4$ solubility, these processes control the amount of Ba present in natural waters.

a) Clays

Ba adsorption on the sodium charged form of standard montmorillonite, illite, and kaolinite, at 25° C was studied by PUCHELT (1967) for pure and electrolyte containing solutions. Ba adsorption decreases with ionic strength of the exchange solution. In rivers, the ratio of Ba which is adsorbed by suspended matter depends on the type of suspension and the concentration of ions competing for adsorption sites.

Ba exchange on vermiculite and bentonite was investigated by LEVI and SCHIEWER (1965). Bentonite adsorbs Ba more strongly than NH_4^+, Mg^{++}, and Ca^{++} (KOMLEV *et al.*, 1965). CARLSON and OVERSTREET (1967) found a high adsorption of incompletely dissociated Ba hydroxide by bentonite at pH 6. The heat of exchange of Ba ions on bentonite with H^+, Na^+, and K^+ were calorimetrically determined by TADZIEV and MUKSINOV (1967). GANGULY and MUKHERJEE (1951) investigated Ba exchange on bentonite, kaolinite, illite, and mica.

b) Hydroxides

Ba adsorption by hydrous ferric oxide was investigated by DUVAL and KURBATOV (1952). PUCHELT (1967) studied, experimentally, Ba adsorption on γ MnO(OH) and found that NaCl concentrations upto 3.5% do not influence the amount of Ba adsorption. γ MnO(OH) can adsorb as much as 20% (by weight) of its Mn content of Ba. These results probably can explain the Ba content of deep sea manganese nodules. PUCHELT also observed that γ MnO(OH) adsorbs more than 85% of the Ba concentration which exists in equilibrium with a $BaSO_4$ precipitate. Adsorption of Ba ions on silica gels from acid solutions depends on time of exchange, specific surface, and pore size of the gel (KIRICHENKO *et al.*, 1965).

c) Organic Substances

BEL'KEVICH *et al.* (1966) equilibrated 0.05 to 0.2N earth alkali solutions with the H-form of peat. They found Ba to have the strongest tendency to substitute for H in peat. Adsorption of ^{137}Ba by coal humic acid was studied by MATSUMARA and ISHIYAMA (1966). PUCHELT (1967) observed that bacteria may extract Ba from solutions, but it is not yet clear, whether this happens by adsorption or incorporation.

Revised manuscript received: September 1971

56-I. Abundance in Natural Waters

I. Springs and Fresh Water Wells

Only very few spring and fresh waters are free of sulfate. Thus the solubility product of $BaSO_4$ is the limiting factor for the Ba concentration. As spring waters normally have only low amounts of dissolved solids and moderate temperatures, no considerable increase of $BaSO_4$ over the distilled water solubility is to be expected. These waters originate normally from rain water which had only a limited time for equilibration in sediments and soils. The Ba content is mainly controlled by the solution of Ba compounds (mostly barite), and exchange of Ba from silicate structures. Several analyses are published for water which served medical purposes. A survey of older data (methods: W) is given by DELKESKAMP (1900, 1902); some values published before 1949 are tabulated by GMELIN (1960). PUCHELT (1967) surveyed the more recent literature grouping the waters in hydrocarbonate, chloride, sulfide and sulfate types according to their prevailing anion.

Table 56-I-1. *Ba concentrations in European spring waters.* (PUCHELT, 1967)

Type of water	No. of springs	Ba range ppb	Arith. mean ppb	Standard deviation
Hydrocarbonate	16	4—22,900	1,757	5,672
Chloride	22	12— 9,500	1,340	2,681
Sulfide	3	150— 750		
Sulfate	9	1— 230		

In very few cases of spring water, higher values were observed than were expected from the $BaSO_4$ solubility product. Their existence is explained by supersaturation which sometimes occurs for a short time after adding sulfate to Ba solutions.

Drinking water from fresh water wells was analysed in the USA by DURFOR and BECKER (1964). For 10 wells from all parts of the country they found 4.6 to 34 ppb Ba (S).

Additional new data for ground and spring waters were published for: South Africa (KENT, 1949; KENT and RUSSELL, 1949), Bulgaria (PENCHEV *et al.*, 1958, 1960), Czechoslovakia (RUBESKA and MIKSOVSKY, 1963), Finland (WILSKA, 1952), Germany (FRICKE, 1968), Hungary (STRAUB, 1950), Japan (IKEDA, 1955a/b; ICHIKUNI, 1966; IWASAKI *et al.*, 1963), and the Soviet Union (BABINETS and RADKO, 1956; GRUSHKO and SHIPITSYN, 1948; KONTOROVICH *et al.*, 1963; OSTROUMOV and RUSSKIKH, 1965; SHINKARENKO, 1948; YUSUROVA, 1957). The highest value reported comes from a Japanese warm spring of sodium chloride type and is 62 ppm Ba.

II. Rivers and Lakes

Only North American rivers have been extensively analysed for Ba. The investigations of DURUM *et al.* (1960) and DURUM and HAFFTY (1961) cover long periods of time, climatic conditions and discharge for a number of rivers. In all cases they found, over a year, a considerable and complex variance of Ba content. One example is given in Table 56-I-2.

Table 56-I-2. *Variation of Ba concentration in Mississippi water near Baton Rouge, Lousiana, U.S.A.*

Date of sampling	Run-off m^3/sec	Dissolved solids ppm	Ba ppb
May 10, 1958	24.550	160	78
Oct. 14, 1958	7.400	223	127
March 13, 1959	18.600	184	72
May 18, 1959	11.800	255	84

DURUM *et al.* (1960) covered 30% of the total run-off of the North American rivers with their analyses, and DURUM and HAFFTY (1963) found a median Ba concentration of 45 ppb from the available data. Ba/Sr ratios (by weight) vary between 0.2 and 3.9 (PUCHELT, 1967), but give a geometric mean of 0.87 for North American rivers (DURUM and HAFFTY, 1963), which may be compared with $0.78 \cdot 10^{-3}$ in the oceans. A number of rivers and lakes being used as drinking water resources have been analysed for Ba by DURFOR and BECKER (1964). By spectrographic analyses, they determined a range of Ba content in rivers between 3.1 and 340 ppb (arithmetic mean: 75.7 ppb). Lakes and brackish water in North America and Europe ranged from 3 to 140 ppb Ba (DURFOR and BECKER, 1964; BROWN *et al.*, 1962; WILSKA, 1952; LANDERGREN and MANHEIM, 1963). Concentration ranges for several rivers are plotted in Fig. 56-I-1.

Local variations of Ba content due to rock composition of the drained area were found by MILLER (1961) in New Mexico and BROWN *et al.* (1962) in Alaska. They observed the highest Ba values from regions with sediments (sandstone, slates), less from granites, and obtained the lowest means from quarzite. The distribution of Ba along a river was studied by LEUTWEIN and WEISE (1962) for the Mulde in Germany. They observed values of 5 to 100 ppb Ba in true solution in the river itself but upto 730 ppb in certain adjoining creeks which drained mining areas. According to these authors, in the upper part of the river 70% and more of the Ba is transported in true solution as the ion. On flowing into the Elbe after 245 km, only 20% of the Ba is still in the ionic form, 80% having been adsorbed onto clays and organic matter. In regions with extreme sulfate concentrations, Ba content is low in accord with the $BaSO_4$ solubility product. TUREKIAN *et al.* (1967) describe the Ba variation of the Neuse river (North Carolina, USA). Ba concentration decreases in the upper part of the river, (16 to 5.7 ppb) in slate and granite areas, but increases regularly downstream to 22 ppb in slate, schist, sand and limestone

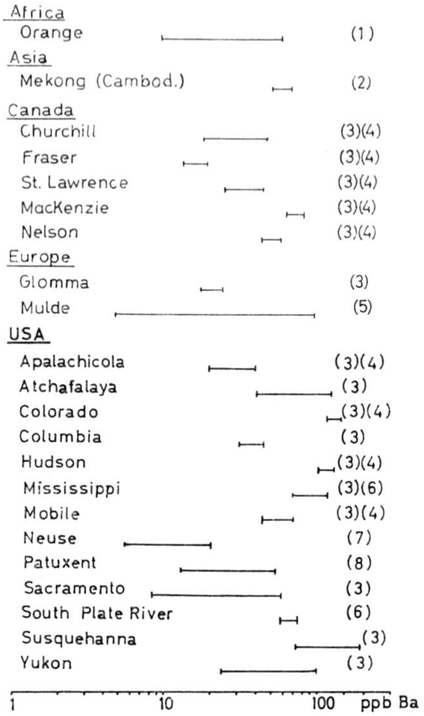

Africa
Orange (1)
Asia
Mekong (Cambod.) (2)
Canada
Churchill (3)(4)
Fraser (3)(4)
St. Lawrence (3)(4)
MacKenzie (3)(4)
Nelson (3)(4)
Europe
Glomma (3)
Mulde (5)
USA
Apalachicola (3)(4)
Atchafalaya (3)
Colorado (3)(4)
Columbia (3)
Hudson (3)(4)
Mississippi (3)(6)
Mobile (3)(4)
Neuse (7)
Patuxent (8)
Sacramento (3)
South Plate River (6)
Susquehanna (3)
Yukon (3)

1 10 100 ppb Ba

Fig. 56-I-1. Ba concentration ranges of rivers (all data obtained by spectrographic methods). 1. DEVILLIERS (1962); 2. DURUM and HAFFTY (1963); 3. DURUM *et al.* (1960); 4. DURUM and HAFFTY (1961); 5. LEUTWEIN and WEISE (1962); 6. DURFOR and BECKER (1964); 7. TUREKIAN *et al.* (1967); 8. HEIDEL and FRENIER (1965)

areas. Where the petrographic composition influences the drainage waters of a certain area distinctly, no general conclusion can be drawn from trace element data from large river basins regarding the origin of the particular trace element.

III. Oceans

From all oceans, Ba determinations are available in surface to bottom profiles. The concentration ranges upto 78 ppb Ba and the estimated mean is about 20 ppb (TUREKIAN and JOHNSON, 1966). Analytical data are compiled in Table 56-I-3.

In general, the Ba concentration seems to be lower in the Atlantic than in the Pacific. In most cases the surface layers are depleted in Ba. Special features have been observed for the distribution of Ba in the Pacific close to the equator; CHOW and GOLDBERG (1960) have found a steady increase of Ba concentration with depth in the respective profiles (Fig. 56-I-2a). They explained this by high biological activity in the surface layers of this region, incorporation or adsorption on organic matter, and a downward transport with the organic debris. They found Ba to resemble radium in distribution with depth. WOHLGEMUTH and BROECKER (1970) could also correlate Ba content with concentrations of other bio-important elements and found parallels with Ra distribution. These authors sampled from very deep

Table 56-I-3. *Ba concentrations in the oceans*

Locality	Maximum sampling depth (m)	No. of samples	Ba range ppb	Method	References
Pacific Ocean					
Central part, close to equator	4,752	18	10 —63	I	CHOW and GOLDBERG (1960)
	4,350	4	19 —33	N/R	TUREKIAN and JOHNSON (1966)
Philippine Sea	4,000	10	11 —33		TUREKIAN and JOHNSON (1966)
Antarctic	5,120	125	8 —56		TUREKIAN and JOHNSON (1966)
South East Pacific	1,500	8	9 —17		TUREKIAN and JOHNSON (1966)
South West Pacific	5,000	5	19 —78		TUREKIAN and JOHNSON (1966)
East Pacific	4,580	13	8.5 —31.2	I	WOLGEMUTH and BROECKER (1970)
East Pacific	4,000	13	6.1 —23.5	I	WOLGEMUTH (1970)
Indian Ocean					
Central part	surface	1	14	N/R	TUREKIAN and JOHNSON (1966)
West Indian Ocean	400	3	21 —46		TUREKIAN and JOHNSON (1966)
South Indian Ocean	4,500	3	10 —15	N/R	BOLTER *et al.* (1964)
Atlantic Ocean					
Long Island Sound	39	60	9 —32 (65)	N/R	TUREKIAN and JOHNSON (1966)
Caribbean, Gulf of Mexico	3,019	17	5 —23		TUREKIAN and JOHNSON (1966)
South Atlantic	5,000	3	15 —21		TUREKIAN and JOHNSON (1966)
North Atlantic	5,061	21	9 —31		TUREKIAN and JOHNSON (1966)
	4,100	3	12.9 —13.1	N/R	BOLTER *et al.* (1964)
	3,000	8	12 —18	I	CHOW and PATTERSON (1966)
	2,098	16	0.04—22.8	F	ANDERSEN and HUME (1968)
Equatorial	4,387	35	0.80—37.5		ANDERSEN and HUME (1968)
English Channel	surface	1	6.3	N/R	BOWEN (1956)
Caribbean	4,729	20	7 —23	S	SZABO and JOENSUU (1967)
Puerto Rico Trench and off Barbados	7,540	28	7.9 —19.1	I	WOLGEMUTH and BROECKER (1970)

ocean regions but did not find a marked Ba increase towards the sea floor. TUREKIAN and JOHNSON (1966) observed, in some places, a maximum of Ba concentration in depths of 600 to 1,200 m (Fig. 56-I-2b) which, however, coincides with the region of lowest Ba sulfate solubility (35 μg/l) in agreement with the interaction of tem-

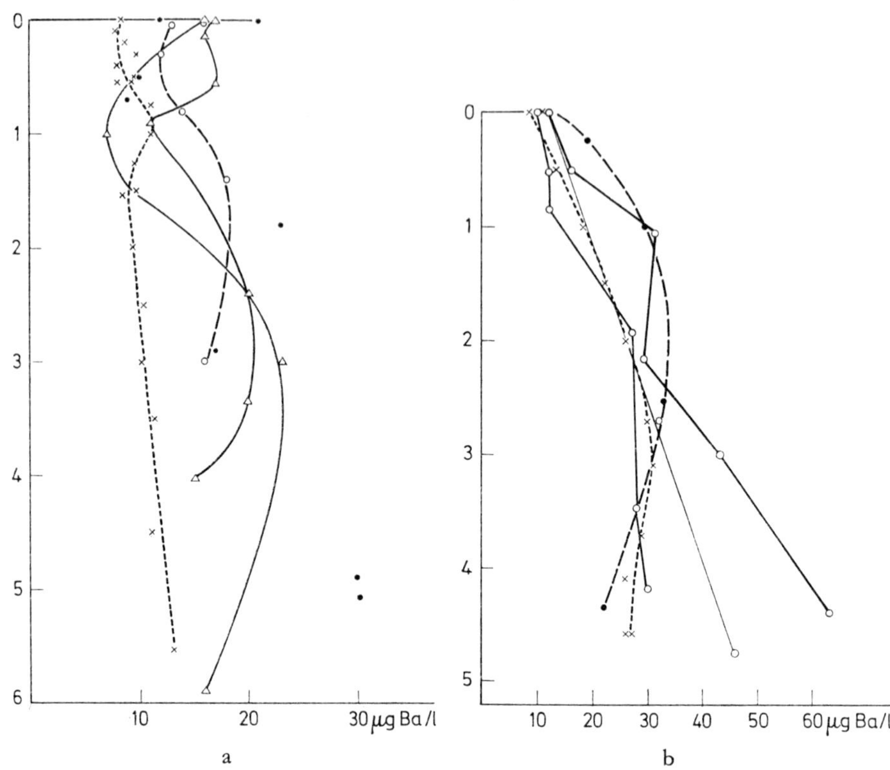

Fig. 56-I-2a and b. Barium distribution in the oceans. a Barium profiles in the Atlantic. × WOLGEMUTH and BROECKER, 1970; ○ CHOW and PATTERSON, 1966; • TUREKIAN and JOHNSON, 1966; △ SZABO and JOENSUU, 1967. b Barium profiles in the Pacific. × WOLGEMUTH and BROECKER, 1970; ○ CHOW and GOLDBERG, 1960; TUREKIAN and JOHNSON, 1966

perature decrease and pressure increase (CHOW and GOLDBERG, 1960). One possible explanation for these high values could be that microcrystals of barite have contaminated the samples. SZABO and JOENSUU (1967) found in profiles in the Caribbean the lowest concentration in the depths of about 1,000 m (Fig. 56-I-2a). In two areas, the equatorial Pacific and the Atlantic off southwest Africa, high values for Ba were found both in the sediments and in sea water, whereas in other places no correlation could be detected.

From the Ba supply of the streams, saturation of Ba sulfate should almost be reached in the oceans. From stream discharge ($3.6 \cdot 10^4$ km^3/yr) and Ba concentration of 45 ppb (DURUM and HAFFTY, 1963), ocean-mass of $1{,}372 \cdot 10^6$ km^3, and Ba content of 20 μg/l (TUREKIAN and JOHNSON, 1966), the residence time of Ba in the sea can be calculated to be $17 \cdot 10^3$ years. Using a sedimentation rate of 0.05 g SiO_2/yr for diatomaceous ooze (based on ^{32}Si), a Ba content of 6,000 ppm and a mean ocean depth of 5,000 m, TUREKIAN and JOHNSON computed a residence time of only 33 years above this particular sediment. Possibly this low value and the observed increase of Ba content with depth indicate an additional supply of Ba from some volcanic source on the ocean floor. With an average of 20 μg/l Ba, the total content of the oceans is $27.4 \cdot 10^9$ tons of Ba.

Table 56-I-4. *Barium concentration in formation waters*

Period	Country	Total No. of samples	No. of samples >1 ppm Ba	Mean Ba conc. of samples with Ba >1 ppm	Maximum Ba conc. ppm	Analytical method	Reference
Precambrian	U.S.A.	1	1	4	4	S	WHITE (1965)
Ordovician	U.S.A.	4	3	61	100	S	McGRAIN and THOMAS (1951)
Silurian	U.S.A.		1	320	320	S	McGRAIN and THOMAS (1951)
Devonian	Canada		2	2	2	S	WHITE (1965)
	Germany	2	1		1.1	W	FRESENIUS (see MICHEL 1963)
	U.S.A.	37	21	218	1,140	S	POTH (1962)
		13	7	205	700	W, S	PRICE *et al.* (1937), WHITE *et al.* (1963)
		69	12	828	2,000	S	HOSKINS (1947)
	U.S.S.R.	73	67		971	S	KOZIN (1964)
Carboniferous (Mississippian, (Pennsylvanian)	Belgium	1	1	347	347	S	CAMERMAN (1951)
	Germany	4	3	1,007	1,260	S	JAKOBSHAGEN and MÜNNICH (1964)
		249	46	1,006	2,860	W	MICHEL (1963)
		225	69	832	2,806	W, S	PUCHELT (1964), (1967)
		152	38	798	2,000	W, S	*Wasserwirtschaftsstelle* (in PUCHELT, 1967)
	Great Britain	38	22		5,100		ANDERSON (1945), GIBSON (1963)
	U.S.A.	152	112	447	5,530	W, S	PRICE *et al.* (1937)
		77	34	265	1,080	S	POTH (1962), HOSKINS (1947)
		72			600		COLLINS (1969)
		36	20	236	1,980	S	McGRAIN and THOMAS (1951)
	U.S.S.R.	111	≧36		190	S	KOZIN (1964)
Permian	Germany		4	11	21	S	HERRMANN (1961)
	U.S.A.		3	5.2	9.6	S	WHITE *et al.* (1963)
	U.S.S.R.	48	≧5		73	S	KOZIN (1964)
Jurassic	Germany	26	7	14.8	30	S, W	PUCHELT (1967)
Cretaceous	Sweden	8	2	1	1	S	ASSARSON (1948)
	U.S.A.		24	14	72		BUCKLEY *et al.* (1958), WHITE *et al.* (1963)
Tertiary	Japan	3	1	59	59	S	BAILEY *et al.* (1961)
	U.S.A.		15	37	150	S	BAILEY *et al.* (1961), WHITE (1965)
Quarternary	Poland	6	2	3.6	4		DOWGIALLO (1965)
	U.S.A.	2	2	16	28	S	WHITE *et al.* (1963)

IV. Formation Waters

Frequently, Ba was discovered in formation waters which had lost their initial sulfate content through bacterial activity during diagenesis. As these bacteria require a reducing environment and organic substances to live on, their areas of activity and thus high Ba concentrations in formation waters, are always connected with occurences of organic matter (oil, bitumen, coal, or gas). Ba has been found in such waters from beds of all geological ages from all over the world. The Ba concentration may reach 5,500 ppm, but no correlation of the Ba concentration with any other parameter of the solutions (dissolved solids, Sr, Ca, K concentration) could be found. Often the Ba/Sr weight ratio is larger than unity. An extensive survey of data is given by PUCHELT (1967). A more condensed compilation was prepared for this chapter (Table 56-I-4).

In the Soviet Union, Ba concentrations in formation waters have been successfully used for a correlation of stratigraphic horizons (KOZIN, 1964; NIKOLAEV *et al.*, 1960). Ba has been assayed in formation waters of a few areas, especially in connection with oil field investigations (AKHUNDOV and SAPPO, 1960; DODONOV *et al.*, 1949; KATCHENKO and FLEGONTOVA, 1955, 1956; KAVEEV and VASIL'EV, 1956; KOROLEV, 1938; KUKABAEV and SYDYKOV, 1962; SKROBOV and SMIRNOV, 1939; SUKHAREV, 1961; TIMASHEVA, 1963; VAROV and ROMM, 1942).

Mixing of Ba-containing formation waters with sulfate waters is often the reason for a scale formation in oil wells (GATES and CARAWAY, 1965; TEMPLETON, 1960) and mines (PATTEISKY, 1954; ANDERSON, 1945). PUCHELT (1967) presented evidence for the abundant formation of certain types of barite deposits from those waters.

V. Brines

Geothermal brines were tapped by deep wells near the Salton Sea, California, USA, an area characterized by rhyolites and Tertiary sediments. With a total of 319,000 ppm evaporation residue (180° C), 200 ppm Ba were reported for a 1963 sample (WHITE, 1965) whereas 1966 samples from two wells gave 235 and 250 ppm Ba (SKINNER *et al.*, 1967). A hot brine from the Atlantis II Deep in the Red Sea containing more than 300 g dissolved solids per liter had almost 1,100 ppb Ba (MILLER *et al.*, 1966).

Revised manuscript received: September 1971

56-K. Abundance in Common Sediments and Sedimentary Rock Types

A compilation of data on the Ba distribution in recent and fossil sediments has been published by PUCHELT (1967). An abbreviated review, but with the important new data included, is given below.

I. Recent Sediments

a) Deep Sea Sediments

Deep sea clays were often analysed for Ba in recent years. Generally, they are enriched in Ba compared to shales. WEDEPOHL (1960) found a definite difference between Atlantic clays (average: 750 ppm Ba) and Pacific clays (average: 4,000 ppm Ba). Since the rate of Ba deposition is about the same in both oceans, he assumed the difference to be caused by a lower rate of detrital accumulation in the Pacific. GOLDBERG and ARRHENIUS (1958) found Ba enrichment in sediments under equatorial waters and related this observation to the high biological productivity in the surface layers of the waters. Several planktonic organisms are known to accumulate Ba in their tests which carry the Ba bottomward after death. They may cause a Ba enrichment in layers close to the bottom, where they dissolve, and may generate local $BaSO_4$ precipitations (PUCHELT, 1967; BRONGERSMA-SANDERS, 1967; cf. Table 56-L-2). Barium is not deposited homogeneously on the sea floor. The zones of high biological activity as well as the ocean ridge systems usually have higher Ba concentrations (TUREKIAN, 1968) than normal deep sea sediments. The origin of Ba in deep sea sediments is a matter of discussion.

BOSTRÖM and PETERSON (1966) determined upto 3.1% Ba in cores from the flanks of the East Pacific Rise. From additional data for other elements and data for the heat flow, it can be concluded that volcanic activity adds several of the enriched elements. TUREKIAN (1968) concluded from data of Ba supply to the oceans by streams, from average Ba content of clays (shales), from the model of Ba enrichment by plankton and from some additional information, that a volcanic or hydrothermal Ba supply need not be assumed to explain the observed Ba data. Ba concentrations in deep sea matter do not often correlate with any of the major constituents. It has to be concluded that Ba adsorption on clays is of less importance. The main Ba carrier is most likely barite. Locally, manganese oxides and phillipsite will accumulate Ba. Calcium carbonate of organic origin is usually very low in Ba (<100 ppm, often only 10 to 30 ppm). In order to reach comparable data for clay sediments, TUREKIAN and TAUSCH (1964) have calculated their Ba determination of Atlantic cores on calcium carbonate free basis. Because the original data are not given, only these corrected values are included in Table 56-K-1. These authors found higher Ba values in the South Atlantic than in the North Atlantic. The areas

Table 56-K-1. *Ba concentrations in deep sea sediments*

Origin	No. of samples	Ba content ppm	Method	References
		a) Deep sea clays		
Atlantic				
Area north of equator	1	200	S	v. ENGELHARDT (1936)
South Atlantic	2	590	S	v. ENGELHARDT (1936)
Area north of 20° S	37	596[a]	S	ERICSON *et al.* (1961)
Area north of 10° N	62	1,100[a]	S	TUREKIAN and TAUSCH (1964)
Area south of 10° N	63	2,000[a]	S	TUREKIAN and TAUSCH (1964)
Area north of equator	5	910	S	EL WAKEEL and RILEY (1961a)
North American trench, south part	15	725	X	WEDEPOHL (1960)
Kap Verde Trench	3	470	X	WEDEPOHL (1960)
Mid Atlantic Ridge	1 (9)	4,400[a]	N/R	TUREKIAN (1968)
Average of Atlantic clays	189	1,260		
Pacific and Indian Ocean				
Area south of equator	8	1,150	S	GOLDBERG and ARRHENIUS (1958)
Area north of equator	5	5,700	S	GOLDBERG and ARRHENIUS (1958)
Baja Californian seamounts	3	5,800	S	GOLDBERG and ARRHENIUS (1958)
North Pacific	2	2,450	S	GOLDBERG and ARRHENIUS (1958)
Area north of equator	4	6,100	W ?	GRIM *et al.* (1949)
Trench NNW Sixty Mile Bank	2	2,700	W ?	GRIM *et al.* (1949)
Indian Ocean	18	610	S	KATCHENKO and FLEGONTOVA (1964)
Area north of equator	8	2,120	S	EL WAKEEL and RILEY (1961a)
Indian Ocean equator area	2	1,330	S	EL WAKEEL and RILEY (1961a)
Area north of equator	20	8,050	S	YOUNG (1954)
Area south of equator	1	300	X	WEDEPOHL (1960)
From all over the Pacific	9	6,700	X	WEDEPOHL (1960)
Average of Pacific clays	82	4,160		
		b) Deep sea carbonates[b]		
Atlantic and Mediterranean				
Caribbean Sea	2	210	S	ERICSON and WOLLIN (1956)
Mid Atlantic Ridge	1 (9)	689	N/R	TUREKIAN (1968)
Various places	4	190	S	TUREKIAN and WEDEPOHL (1961)
North Atlantic (with clay)	4	840	S	EL WAKEEL and RILEY (1961a)
South Atlantic off Africa	1	1,900	S	EL WAKEEL and RILEY (1961a)
Mediterranean	2	1,600	S	EL WAKEEL and RILEY (1961a)
Pacific and Indian Ocean				
Equatorial area, East Pacific (manganiferous)	2	5,000	S	GOLDBERG and ARRHENIUS (1958)

Table 56-K-1. (Continued)

Origin	No. of samples	Ba content ppm	Method	References
Equatorial area, Central Pacific	1	680	S	El Wakeel and Riley (1961a)
North equator area, Central Pacific	3	540	S	Young (1954)
		c) Deep sea siliceous muds		
Atlantic				
Atlantic off African coast	1	700	S	El Wakeel and Riley (1961a)
Equatorial area, Central Pacific	3	3,470	S	El Wakeel and Riley (1961a)
Central northern Pacific	6	10,400	S	Young (1954)
Central equatorial Pacific	3	8,100	S	Young (1954)

[a] Deep-sea clays calculated on $CaCO_3$ free basis.
[b] Raw analyses of carbonate rich cores but not necessarily indicative of the pure carbonate fraction.

closer to the continents are normally lower than the Central Ocean areas, but just off the coast of Africa (20 to 25° S), an area with values of $>$4,000 ppm Ba was detected.

Turekian (1968) has analyzed samples from different depths in a deep sea core for Ba. Concentrations range between 1,700 and 6,700 ppm Ba (calculated $CaCO_3$ and salt free); they indicate changes in the rate of Ba deposition within the last 30,000 years. Accumulation rates for Ba reported by Turekian (1968) vary from $<$90 μg/cm² per 1,000 years to 790 μg/cm² per 1,000 years within the special core and are about 1,000 μg/cm² per 1,000 years in the Antarctic. In Table 56-K-1, averages for Ba are calculated using the "reduced $CaCO_3$ free" data. This means that the actual data may be lower.

Deep sea carbonates. Foraminifera ooze high in carbonate contains low Ba concentrations (10 to 30 ppm, Table 56-L-2). The barium content in carbonate sediments is either due to $BaSO_4$ or to manganese oxides or clay. Since all these sources may be active at the same time and do not work coherently, a recalculation to "pure carbonates" is very difficult. Turekian and Tausch (1964) extrapolated deep-sea cores in the North Atlantic to 100% $CaCO_3$ and got 10 to 30 ppm Ba for the pure carbonate. The available data are given in the table. The value of 190 ppm Ba by Turekian and Wedepohl (1961) for deep sea carbonates derived from 4 globigerina oozes from Atlantic cores is as yet the best information.

Manganese nodules cover wide areas of the deep sea bottom of all oceans and contain upto 20,000 ppm Ba. A survey by Puchelt (1967) of published literature shows that the Ba means are 4,500 ppm, 5,200 ppm and 3,700 ppm Ba for nodules from the Pacific, Atlantic, and Indian Oceans, respectively. In manganese nodules, Ba is either adsorbed, incorporated in acid soluble compounds (zeolites), or occurs as barite (ARRHENIUS, 1963).

Siliceous sediments occurring in deep areas of the oceans where carbonates are no longer stable can locally contain more than 1% Ba.

b) Shallow Water Sediments

The barium content of near shore and shelf sediments is influenced by the amount and kind of detrital matter and the barium content of the rivers. A review of literature by PUCHELT (1967) shows that clay sediments of these areas are generally higher in Ba than sand and silt fractions. Clays from the Mississippi delta are especially high in Ba.

Three studies of reef carbonates (STEHLI and HOWER, 1961; SENAKOLIS, 1964; FRIEDMAN, 1968) demonstrated that reef debris, reef material and oolitic muds contain only limited amounts of Ba. STEHLI and HOWER (1961) found a range from 10 to 61 ppm in 59 samples and an average of 18.4 ppm Ba. FRIEDMAN (1968) obtained spectrographically, 18 to 62 ppm Ba in corals with encrusting coralline algae, and 15.5 to 68 ppm Ba in carbonate sands from reef aprons. Carbonate sands with admixed terrigenous debris showed 130 to 280 ppm Ba. In the analysed samples, Ba concentrations change parallel to the "insoluble residue". Similar observations are reported by SENAKOLIS (1964). Within the internal parts of the reef, average Ba content in clastic limestone was 3.1 ppm, peripheral parts of the reef had 436 ppm Ba.

II. Barium in Consolidated Sediments

a) Sandstones, Cherts, Graywackes

Pure quartz sandstones are very low in Ba, but since most sandstones contain considerable amounts of feldspars, these minerals are the most important Ba carriers besides micas, which occasionally occur. PETTIJOHN (1963) calculated the proportion of sandstone types to be 34% quartzite, 26% graywacke, 25% subgraywacke, and 15% arkose. This combination has an average K_2O content of 1.3%, to which a Ba content should be proportional. Comparison of the K/Ba ratios in low Ca granites and in sandstones and graywackes gives additional support to this proportionality. Single sandstones vary widely in Ba content, since even barite concentration (as cement) occurs locally (cf. PUCHELT, 1967). If barite is a sandstone constituent, normally the weight ratio Ba/Sr is greater than 10, for barite generally contains much less than 10% $SrSO_4$. The Ba content of sandstones and graywackes ranges from 5 to 900 ppm. An average composition calculated from the European, Russian, and American sandstone and graywacke is 316 ppm Ba. This value is subject to changes when the individual data can be properly weighted. Nevertheless it is a more realistic value than the X0 ppm guess of TUREKIAN and WEDEPOHL (1961). Cherts constitute a special group in silica sediments and always exhibit higher Ba means.

Table 56-K-2. *Barium in quartz sandstones, cherts, and graywackes*

Locality	No. of samples	Barium concentration			Method	References
		range ppm	group mean ppm	total mean ppm		
Quartz sandstone						
Germany	73	50—810		406		
			134		X	Bertsch (1964)
			313		X	Wedepohl (1961)
			770		S	Zurlo (1963) see Puchelt (1967)
Sunda Islands	9	5—900	150		S	v. Tongeren (1938)
U.S.A.	289			280		Shoemaker *et al.* (1958)
			1,800		S	Young (1954)
U.S.S.R.	399	230—820		249		
			340		S	Babina and Kotorovich (1966) see Puchelt (1967)
			223		S	Lebedev (1967) see Puchelt (1967)
			140		S	Litvin (1961)
			250		S	Litvin (1963)
			290		S	Sinkarenko (1948)
Chert						
C.S.S.R.	2	300— 500		400	S	Leutwein (1957)
Finland	46	360— 630		495	S	Sahama (1945)
Germany	311	30—1,000		440	S	Leutwein (1957)
					S	Prashnowsky (1957)
Indonesia	22	50—1,900		420		Audlee-Charles (1965)
Different places		350		350		Maxwell (1953) see Puchelt (1967)
Graywacke and arcose						
Africa				258	S	Danchin (1970) see Puchelt (1967)
Europe	51	189—670		370		
			290		S	v. Engelhardt (1936)
			360		S	Klein (1935)
			447		S	Kuenen (1941)
			389		S	Rivalenti and Sighinolfi (1969)
			270		X	Wedepohl (1961)
			335			Westermann (1961)
			500		S	Zurlo (1963) see Puchelt (1967)
North America	95	30—830		330		
					S	MacPherson (1958)
					S	Weber and Middleton (1961)
Indonesia } New Zealand }	12	30—480		252	S	McLaughlin (1955)
					S	v. Tongeren (1938)

b) Shales

Ba averages for shales reported in the literature vary from 250 to 800 ppm (PUCHELT, 1967; VINOGRADOV, 1956). The average of this paper is 546 ppm (s for the 25 means used: 212). Individual samples gave values from 10 to 5,000 ppm. From the data summarized in Table 56-K-3, it can be observed that especially low values have been found in shales of the Dnieper-Donets depression (Russia) and the west Siberian depression (LITVIN, 1961, 1963; TOLKACHEV, 1968). If these low

Table 56-K-3. *Barium in shales*

Locality	No. of samples[a]	Barium concentration range ppm	Barium concentration mean ppm	Method	References
Africa:	2	270— 450	360		JUNNER and JAMES (1947)
	25 (323)	394—1,004[b]	681	S	DANCHIN (1970)
Asia:					
Japan	1 (14)		540	X	WEDEPOHL (1960)
U.S.S.R.	8 (521)	150— 370[b]	270	S	BABINA and KOTOROVICH (1966)
	6 (920)	260— 520[b]	394		LEBEDEV (1963)
	19	30— 450	83		LITVIN (1961)
	6 (32)	140— 230	188		LITVIN (1963)
	1		360		SINKARENKO (1948)
	5 (185)	140— 220	182	S	TOLKACHEV (1968)
Europe:	1 (36)		800	X	WEDEPOHL (1960)
Finland	17 (105)	9—2,700	654	S	SAHAMA (1945)
Germany	3	480— 540	513	S	HEIDE and CHRIST (1953)
	66	50—5,000	750	S	LEUTWEIN (1951)
	3 (9)	700— 900	824	S	PRASHNOWSKY (1957)
	3	390— 730[b]	527	S	ZURLO (1963)
Great Britain	24	210—2,240	739	S	MOHR (1959)
	6	330—1,050	555	S	NICHOLLS and LORING (1962)
	27	325—1,280	723	S	SPENCER (1966)
Sweden			500	S	LANDERGREN and MANHEIM (1963)
	9	450—2,150	866		LARSSON (1932)
North America:	33	250—1,000	526	S	DEGENS *et al.* (1957)
	31	100— 750	393	S	FENNER and HAGNER (1967)
	41	10—1,020	470	S	MACPHERSON (1958)
	26	190— 610	359	S	MURRAY (1954)
	15		580		SHAW (1954, 1957)
	17		720		TOURTELOT (1957)
	3 (?)		750	S	YOUNG (1954)
Pacific Islands:	2	500—1,050	775	S	EL WAKEEL and RILEY (1961b)
	4	20—1,800	840	S	v. TONGEREN (1938)

[a] No. of individual samples used for bulk samples, or for means, are given in parantheses.

[b] Range of means, not of individual samples.

values for Russia are omitted from averaging, the mean is 628 ppm with a standard deviation of 157 ppm.

Ba does not seem to be an environmental indicator for shales. VINE (1966) and LEBEDEV (1967) found a Ba increase in shales from fresh water to marine environment, while MURRAY (1954) reports opposite observations from Indiana and Illinois, USA. In general, shales have higher Ba contents than graywackes or sandstones, but locally, siltstones or sandstones may be higher in Ba (ALANOV, 1963).

The mode of Ba binding in shales is complex. Several indications exist which point to a correlation of Ba with mica; parallelism with the amount of illite present has been found (FENNER and HAGNER, 1967); and $BaSO_4$ was shown to be another possible carrier. Black shales often contain more Ba than normal shales thus suggesting a connection of Ba with organic matter. While certain shales retained their Ba content of deposition, others gained or lost some. Redistribution in diagenetic processes is possible.

c) Carbonate Rocks

Literature on Ba in carbonate rocks is summarized by GRAF (1960) and PUCHELT (1967). The Ba content of carbonate rocks varies from 1 to 10,000 ppm (cf. PUCHELT, 1967). Using the literature cited in Table 56-K-4, an average Ba concentration for carbonate rocks of 90 ppm was calculated. This value is within the range reported by GRAF (1960) (150 ± 110 ppm). TUREKIAN and WEDEPOHL (1961) based their average limestone value (10 ppm) on the Ba content of modern molluscan shells. This value seems to be too low even for average carbonate, excluding detrital material. Means calculated by area as listed in PUCHELT (1967) are plotted in Fig. 56-K-1.

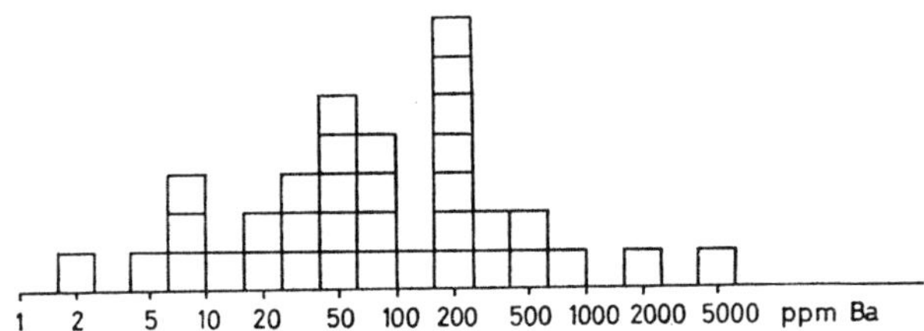

Fig. 56-K-1. Frequency distribution of Ba concentrations in carbonates (PUCHELT, 1967)

High average Ba values in relation to the overall mean are reported for 10 Ordovician dolomites from Missouri, USA (620 ppm, KELLER *et al.*, 1950), Cretaceous limestones, USA (900 and 1,800 ppm, YOUNG, 1954), 91 Pennsylvanian limestones from Illinois, USA (260 ppm, OSTROM, 1957) and 6 limestones from Africa (1,330 ppm, JUNNER and JAMES, 1947).

Ba in carbonate rocks originates mainly from 3 sources or processes:

1. detrital clay material,
2. redistribution during diagenetic processes, in which $BaSO_4$ can be precipitated,
3. incorporation of Ba in carbonate minerals.

In most cases processes 1 and 2 are quantitatively more important. Reference is made here to the difference between "pure" carbonates within recent reefs and carbonate sands from their peripheral parts (cf. chapter 56-K-I). In carbonate sands with detrital clay, FRIEDMAN (1968) observed an increase of Ba concentrations parallel to "insoluble residue". Fossil carbonate sediments exhibit similar features (VINOGRADOV *et al.*, 1952).

During evaporation of sea water, Ba is precipitated as barite which usually occurs disseminated in the calcium carbonate (PUCHELT, 1967). Diagenetic alterations may cause local $BaSO_4$ concentrations. Thus CAMERON (1966) found between 1 and 6,100 ppm Ba in a core from carbonate rocks within a distance of 80 feet. This diagenetic barite usually can be recognized under the microscope. Manganese-containing carbonate sediments are sometimes enriched in Ba (MOHR and ALLEN, 1965).

Table 56-K-4. *Barium in carbonate rocks*

Locality	No. of samples[a]	Barium concentration range ppm	Barium concentration mean ppm	Method	References
Asia:					
Indonesia	4	9—450	220	S	v. TONGEREN (1938)[b]
U.S.S.R.	8	30—240	65	S	GORLITSKY and KALYAEV (1962)
				S	LEBEDEV (1967)
					LITVIN (1963)
	59 (2,973)	1—250	36	S	RONOV (1956)
	198	12—690	47		SINKARENKO (1948)
					VINOGRADOV and RONOV (1956)
Europe:					
Germany	41 (131)	1—300	62	S	v. ENGELHARDT (1936)
				S	HEIDE and CHRIST (1953)
				S	PRASHNOWSKY (1957)
	125	15—403	92	X	BAJOR (1965)
Great Britain	183	<5—8,000	220	S	MUIR *et al.* (1956)
Roumania	15	10—300	100	S	IMREH and ECATERINE (1965)
Scandinavia	5	3—330	78	S	v. ENGELHARDT (1936)
				S	HENRIQUES (1964)
				S	SAHAMA (1945)
North America:					
U.S.A.	10	200—2,000	620	S	KELLER *et al.* (1950)[b]
	420	2—10,000	106		CANNON (1955)
				S	LAMAR and THOMPSON (1956)
					MOORE (1960)
				S	OSTROM (1957)
				S	RUNNELS and SCHLEICHER (1956)

[a] Number of individual samples used for bulk samples, or for means, are given in parantheses.
[b] Data not included in calculation of average.

Revised manuscript received: September 1972

56-L. Biogeochemistry

Barium is present in recent and fossil plants, in animals and fuels. Ba accumulation in plants and animals was found by several investigators, but there is no evidence that this element is physiologically necessary. Ba is moderately toxic for plants and slightly toxic for mammals. Reviews of barium biogeochemistry are published by PUCHELT (1967) and BOWEN (1966), who also makes reference to earlier compilations.

BOWEN (1966) summarizes the available literature for Ba content in dry weights as follows:

marine plants	30 ppm
land plants	14 ppm
marine animals	0.2—3 ppm (higher in hard tissues)
land animals	0.75 ppm

I. Plants

In ash of marine plants, Ba varies over a wide range (Table 56-L-1). Coccoliths, which form the main constituents of marine carbonates, contain 10 to 30 ppm Ba in ash (TUREKIAN and TAUSCH, 1964). Ash of phytoplankton species from the Black Sea (*Chaetoceros Curvistus* and *Rhizosolenia Calcar Avis*) is very high in Ba (4,000 and 20,000—30,000 ppm, respectively; VINOGRADOVA and KOVAL'SKIY, 1962). Since these diatoms are very abundant in surface waters in summer, they form a considerable Ba enrichment, which may contribute to the Ba content of pelagic sediments. The tests of rhizosolenia and chaetoceros are very delicate and under marine conditions not stable. Thus they are subject to dissolution and are not to be found in sediments, although they certainly serve as barium conveyors to the sea floor (BRONGERSMA-SANDERS, 1967).

BOWEN (1966) reports concentration factors (ppm Ba in fresh organisms/ppm Ba in sea water) for plankton and brown algae to be 120 and 260, respectively.

Algae from the coast of Great Britain show seasonal variance in their Ba content in ash (upto 900 ppm Ba; BLACK and MITCHELL, 1952).

For terrestrial plants, an extensive study exists for bryophytes (SHACKLETTE, 1965). Some of the species investigated concentrate Ba considerably. The highest enrichment factor for Ba in bryophyte ash versus soil is 2,000. Equisetum (horsetail) ash was analysed by CANNON *et al.* (1968) and BOROVIK-ROMANOVA (1939). Ba content in ash of this plant (70 to 4,500 ppm) resembles approximately the concentration in the respective substrata.

Phanerogames reabsorb distinct amounts of Ba from the soil. Fir and spruce have 500 to 6,200 ppm Ba in ash with the highest concentrations in twigs (LOTSPEICH and MARKWARD, 1963). Black walnut, hickory, and red-ash leaves contain 870 to 2,570 ppm. Relative Ba enrichment was reported for oaks, which gave upto 2.30% Ba in the ash of twigs (BLOSS and STEINER, 1960). The average for Ba in ash of legumes is 1,420 ppm (CANNON, 1964).

Table 56-L-1. *Barium in plants*

Plant	Ba in dry tissue ppm	Ba in ash ppm	References
Schizophyta			
Bacteria	—62,000		FOERSTER and FOSTER (1966)
Phycophyta			
Coccoliths		10 — 30	TUREKIAN and TAUSCH (1964)
Diatoms		20 —30,000	VINOGRADOVA and KOVAL'SKIY (1962)
Brown algae	0.4—120 ⌀ 31	270 — 900	PUCHELT (1967)[a]
Red algae	50	0.6— 5.6	PUCHELT (1967)[a]
Bryophyta	5—200 ⌀ 150	200 —50,000	PUCHELT (1967)[a] BOWEN (1966)[a]
Pteridophyta			
Equisitinae		30 — 4,500	CANNON *et al.* (1968)
Ferns	8		BOWEN (1966)[a]
Spermatophyta			
Conifers		10 — 100 500 — 6,200	PUCHELT (1967)[a] LOTSPEICH and MARKWARD (1963)
Angiosperms	14		BOWEN (1966)[a]
Deciduous trees		10 — 2,700 —23,000	PUCHELT (1967)[a] BLOSS and STEINER (1960) ROBINSON *et al.* (1950)
Legumes	average: 1,420		CANNON (1964)

[a] Compilations.

II. Animals

Organisms are only important for trace element geochemistry, if they occur in large amounts. Zooplankton (especially crustacees) from the Black Sea shows Ba accumulation upto 2,000 ppm in ash (VINOGRADOVA and KOVAL'SKIY, 1962). Since these animals constitute about 80% of the planktonic population in the survey area, they may also contribute to the Ba content of marine sediments.

Protozoan skeletons and shells consisting of $CaCO_3$ or SiO_2 contain upto 270 ppm Ba. They are the source for Ba in pelagic globigerina and radiolarian oozes. From investigations of ARRHENIUS (1963) it must be concluded that in Recent planktonic foraminifera, Ba is mostly bound to organic matter (upto 700 ppm in ash). In Table 56-L-2 the radiolaria Acantharia and the rhizopod Xenophyophora, which concentrate $BaSO_4$ in their skeletons, are listed. After death these skeletons are dissolved, but during this process they transport Ba towards the sediment. No fossil skeletons of these species are reported.

GOLDBERG and ARRHENIUS (1958) assume a certain Ba accumulation by digestion of benthonic organisms, since they found Ba enrichment in fecal pellets from the sea floor.

Table 56-L-2. *Barium in animals.* (Compilations by BOWEN, 1966; PUCHELT, 1967)

	Ba in dry tissue ppm	Ba in ash ppm	Ba in hard tissue ppm	Material in hard tissue
Protozoa[a]	10—270			$CaCO_3$
Foraminifera			180	$CaCO_3$
			500	SiO_2
Coelanterates	11—450		8.6—35[b]	$CaCO_3$
Corals				
Ctenophora		40—2,000		
Echinodermata	20— 50[c]		35	$CaCO_3$
Annelida (Vermes)		17— 50[d]		
Tentaculata				
Bryozoes		12—2,000[e]		
Brachiopodes				
Mollusca	3		<1—90[b]	$CaCO_3$
Lamellibranchiates	4— 75	4— 500	4—75[f]	
Gastropodes			4—50[g]	
Cephalopodes } Scaphopoda }	7.2+20			
Arthopoda		15— 800		
Mammalia	2.3		6.9	apatite

[a] Radiolaria Acantharia contains 5,400 ppm Ba (ARRHENIUS, 1963), $BaSO_4$ is the hard tissue in the rhizopod Xenophyophora (VINOGRADOV, 1953).

[b] Including data by FRIEDMAN (1969).

[c] Upto 5,000 ppm Ba were found in dry tissue of *Asterias Linkii* from the Barents Sea.

[d] High values (1,000 to 1,500 ppm) are reported from Black Sea plankton (VINOGRADOVA and KOVALSKIY, 1962).

[e] Including data by SCHOPF and MANHEIM (1967).

[f] With certain species of Anodonta, Pecten, Astarte and Tellina, higher Ba concentrations (upto 500 ppm) were found.

[g] Some samples of Helix species, Littorina species, and Neptuna species contain upto 500 ppm Ba in their shells.

Much support for Ba in molluscan shells has been published (cf. PUCHELT, 1967). LEUTWEIN (1963) and PILKEY (1963) found a distinct Ba enrichment in Recent mollusc shells from fresh and brackish water environments. Obviously, the structure of the shells (calcite or aragonite) is of less importance as a cause of Ba incorporation than Ba content in the water and the environment. A reconstruction of palaeoenvironments from Ba concentrations of unaltered fossil shells has been attempted by PROKOF'EV (1964), FRIEDMAN (1967) and others. TUREKIAN and ARMSTRONG (1960, 1961), investigating Recent and fossil shells, found Bato be much higher in molluscan shells of the Fox Hill Formation (Cretaceous), South Dakota, than in Recent species. According to their studies diagenetic alterations might considerably shift the initial trace element composition, even with only slightly altered mineralogy of the shell. Animals do not contain appreciable Ba concentrations. A few data from BOWEN (1966) are included in Table 56-L-2. Additional references of detailed investigations are compiled by GMELIN (1960).

Table 56-L-3. *Barium in fuels*

Locality	Stratigraphy	Ba content in ash ppm	Method	References
Brown coal				
Australia	Permian-Tertiary	upto 800[a]	S	SWAINE (1967) BROWN and SWAINE (1964)
Czechoslovakia	Tertiary	100— 1,000		HONEK and JIRELE (1965), see PUCHELT (1967)
Germany	Tertiary	700— 7,600	X	PIETZNER and WOLF (1964)
	Tertiary	200— 2,800	S	RÖSLER and LANGE (1965), see PUCHELT (1967)
U.S.A.	Tertiary	100—10,000	S	BREGER *et al.* (1955), see PUCHELT (1967)
			S	BREWER and RYERSON (1935), see PUCHELT (1967)
	Cretaceous	100— 1,000	S	DEUL and ANNEL (1956)
U.S.S.R.		140— 2,730		TKACHEV *et al.* (1965)
Hard coal				
Australia	Permian	1,000—10,000	S	CLARKE and SWAINE (1962)
Canada		20— 2,200	S	HAWLEY (1955)
Finland		360— 1,600	S	LOKKA (1943)
Germany	Carboniferous	100—27,000	S	v. ENGELHARDT (1936)
			S	THILO (1934), see PUCHELT (1967)
			S	LEUTWEIN and RÖSLER (1956), see PUCHELT (1967)
			S	RADMACHER (1965), see PUCHELT (1967)
	Permian	100— 500	S	LEUTWEIN and RÖSLER (1956), see PUCHELT (1967)
	Triassic, Jurassic	100	S	LEUTWEIN and RÖSLER (1956), see PUCHELT (1967)
Great Britain	Carboniferous	90— 710	S	NICHOLLS and LORING (1962), see PUCHELT (1967)
			S	GIBSON (1963)
New Zealand		250— 8,400	S	BROWN and TAYLOR (1960) COAL RES. COMPANY (1949)
Norway		average: 4,000	S	BUTLER (1953)
U.S.A.		270—22,000	S	HEADLEE and HUNTER (1953), see PUCHELT (1967)
Oils and bitumina				
Germany	Triassic—Cretaceous	—10,000	S	HEIDE (1938), see PUCHELT (1967)
U.S.A.	Cambrian—Tertiary	X—X 00,000	S	BELL (1960), see PUCHELT (1967)
			S	ERICKSON *et al.* (1954), see PUCHELT (1967)
			S	HYDEN (1961)
U.S.S.R.	Devonian—Tertiary	100— 3,000		KATCHENKOV (1951), see PUCHELT (1967)

[a] Ba concentration in dried coal.

III. Fuels (Including Coal)

Fuels of all geologic ages contain Ba in their ashes; often in amounts considerably above the earth's crust mean (Table 56-L-3).

From Ba data in Recent plants it can be deduced that at least part of the barium originates from living plants. During diagenetic alteration, humic acids may absorb additional Ba from the involved solutions. Extremely high Ba values of coal ashes (upto 4.76%), as reported from Great Britain (REYNOLDS, 1939), may be caused by a secondary $BaSO_4$ mineralization. No general trend of Ba concentration with maturity of coal could be observed. In one instance, LEUTWEIN (1966) paralleled Ba content with the amount of clays in a brown coal profile. ERSHOV (1958) carried out several electrodialysis experiments on coal samples and concluded that Ba—in addition to other elements—was present either as soluble minerals or in weakly absorbed form. It is not bound as strongly as Ge, which could not be extracted by this procedure.

It is to be assumed that Ba is incorporated in certain metallic-organic compounds in oil, but no investigations on the specific types have been published.

Revised manuscript received: September 1971

56-M. Abundance in Common Metamorphic Rock Types

Barium concentrations in metamorphic rocks exhibit a large variation within each type (Table 56-M-1). They vary as widely as do values for all igneous and sedimentary rocks. Consequently, no meaningful Ba averages can be calculated. The only exception seems to be eclogites which form under special pressure conditions. Comparison of Ba concentrations in typical metamorphic minerals such as sillimanite, staurolite, garnet etc., show that these structures have no appreciable tolerance for Ba.

Data on Ba distribution between coexisting minerals in metamorphic rocks are included in Table 56-D-5.

Table 56-M-1. *Barium in metamorphic rocks*

Rock type and locality	No. of samples	Barium concentration range ppm		mean ppm	Method	Reference
Gneiss, Randesund, Norway	8	<100—	1,060	<605	X	Ball (1966)
Slightly altered gneiss, Adirondacks, U.S.A.	10	220—	1,400	610	S	Engel and Engel (1958)
"Granitized" gneiss, Adirondacks, U.S.A.	18	125—	2,600	980	S	Engel and Engel (1958)
Gneiss, Montana, U.S.A.	7	1,800—	3,800	2,580	S	Foster (1962)
Gneiss, Langøy, Norway	5	823—	1,300	1,050	S	Heier (1960)
Gneiss, Lewisian, Inverness-shire, Scotland	4	300—	920	520	S	Lambert (1964)
Basic gneiss, Scotland	8	20—	300	110	S	O'Hara (1961)
Gneiss, S.W. Finland	40	<340—	654	≦430	S	Parras (1958)
Glaucophane schist, California, U.S.A.	11	7—	300	92	S	Coleman and Lee (1963)
Pelitic schist, Connemara, Eire	16	550—	1,850	1,300	S	Evans (1964)
Sillimanite schist, Montana, U.S.A.	2	1,300+	1,900	1,600	S	Foster (1962)
Schist, Moine, Inverness-shire, Scotland	5	800—	1,440	970	S	Lambert (1964)
Phyllite, Finland	174	90—	1,500	552	S	Lonka (1967)
Quartz-albite-biotite schist, New Zealand	8	450—	1,500	690	S	Taylor (1955)
Greenschist, New Zealand	6	10—	125	34	S	Taylor (1955)

Table 56-M-1. (Continued)

Rock type and locality	No. of samples	Barium concentration range ppm	mean ppm	Method	Reference
Hornfels, Connemara, Eire	11	670— 2,000	1,210	S	EVANS (1964)
Hornfels, Palaman district, India	7		160	S	GHOSE (1966)
Amphibolite, Randesund, Norway	17	260— 680	450	X	BALL (1966)
Amphibolite, Brazil	20	21— 270	67	S	BARROS-GOMES *et al.* (1964)
Amphibolite, Adirondacks, U.S.A.	16	42— 140	84	S	ENGEL and ENGEL (1962)
Sericitite and biotite-amphibolite, Adirondacks, U.S.A	11	140— 1,650	220	S	ENGEL and ENGEL (1962)
Amphibolite, Australian shield	61	<610—> 990	817	X	LAMBERT and HEIER (1968)
Amphibolite, S. W. Finland	20		251	S	PARRAS (1958)
Granulite, Australian shield	89	<420—>1,090	720	X	LAMBERT and HEIER (1968)
Metabasite, Saxonia, Germany	9	100— 125	110	S	MATHÉ (1969)
Charnockite, Finland	29		570	S	PARRAS (1958)
Paracharnockite, Finland	24		672	S	PARRAS (1958)
Eclogite, Naustdal, Norway	1		<5	S	BINNS (1967)
Eclogite, Nordfjord, W. Norway	6	<10— 30	≦20	S	BRYHNI *et al.* (1969)
Eclogite, California, U.S.A.	2	15+ 300	160	S	COLEMAN and LEE (1963)
Eclogite, different places	5	5.6—136	56	I	GRIFFIN and MURTHY (1968)
Eclogitic rock, Münchberg, Germany	18	<100— 355	≦190	S	HAHN-WEINHEIMER (1959)

Revised manuscript received: September 1971

56-N. Behavior in Metamorphic Reactions

Only very few investigations exist on Ba behavior under metamorphism. LONKA (1967), analyzing Precambrian phyllites of Finland, observed no differences in Ba concentration between phyllites of lower and higher degree of metamorphism. Trace element data including Ba concentrations have been used by TAYLOR (1955) to discuss the origin of New Zealand metamorphic rocks under the assumption of isochemical metamorphism.

In the Adirondacks, New York, ENGEL and ENGEL (1958) studied progressive metamorphism and granitization of the major paragneiss. They found Ba to decrease with increasing metamorphism, while the Ba content in biotites increased (580, 717, 888, 1,766 ppm). Granitized gneisses of this area generally showed much higher Ba values than normal gneisses. TUREKIAN and PHINNEY (1962) could not detect characteristic changes of Ba content in garnets and coexisting biotites in a metamorphic sequence from Nova Scotia.

A special feature of transport during metamorphism is skarn formation. HIGAZY (1952) observed an increase of Ba content from epidiorite, 90 ppm (S), to biotite-epidiorite, 270 ppm, to biotite skarn, 910 ppm. The subsequent alteration to lepidomelan-skarn (720 ppm) and chlorite-skarn (270 ppm) caused distinct decreases in Ba. NESTERENKO *et al.* (1958) found Ba to be depleted from biotite hornfels when this rock was altered to pyroxene-garnet-skarn.

Revised manuscript received: September 1971

56-O. Relations to Other Elements, Crustal Distribution, Economic Importance etc.

I. Inter-element Relationships

In igneous rocks, Ba generally substitutes for K in silicate structures. A certain correlation of Ba and Ca in igneous rocks with low K was demonstrated in sections 56-D and E. In the sedimentary cycle, Ba preferentially occurs as barite, in clays and in feldspar. Presence of barite is dependant on sulfate abundance, which in turn requires suitable redox conditions. In sediments, including evaporites, the correlation Ba-K is much less pronounced than in igneous rocks. The substitution Ba-Ca observed in few carbonate minerals is of less general importance.

II. Distribution in the Earth's Crust

Details discussed in the preceding sections are summarized in the following table:

Table 56-O-1. *Abundance of Ba in important masses of the earth's crust. (Means calculated on the basis of* WEDEPOHL'S, 1969, *data on the abundance of rock units)*

Igneous intrusive rocks (mean)	728 ppm
Gabbroic rocks	246 ppm
Granites	732 ppm
Granodiorites and quartzdiorites	873 ppm
Diorites	714 ppm
Consolidated sediments (mean)	538 ppm
Sandstones (including graywackes)	316 ppm
Shales	628 ppm[a]
Carbonate rocks	90 ppm
Sea water	0.020 ppm

[a] 546 ppm, if the Russian shales low in Ba are included.

A discrepancy exists between the Ba means for consolidated sediments and magmatic rocks. Part of the Ba missing in the fossil sediments is bound in pelagic clays (mean Ba content 2,000 to 3,000 ppm; cf. Table 56-K-1) which constitute at least 10% of the total sediment mass (WEDEPOHL, 1969). But part of the discrepancy may be due to the fact that the individual data used for the averages could not be weighted properly for the compilation.

III. Technical and Economic Importance

Barite and witherite are the only barium minerals of economic interest. Descriptions of deposits and their production are given by GMELIN (1960), BROBST (1970) and others. World production of barite has been almost constant since 1964 at about 4 million short tons per year (ORR, 1970; for detailed information see U.S. Geol. Surv. Bulletin 1321).

Barite is used for drilling muds in oil and gas geology (consuming about 75% of the world's production) and for radiation-shielding concrete. Chemically treated $BaSO_4$ of fine particle size is an important filling material in the rubber, paper, and fabric industries. It is one of the components of the white pigment "lithopone". Ba compounds are used in glass and enamel production as flows and for glasses with special optical properties. Barium chloride serves as a rat poison and insecticide; barium chlorate causes the green colour of pyrotechnics; barium titanate is a ferroelectric substance used in the electro-industry; certain barium compounds are the active substances of fluorescent screens.

Reviews on barium compounds and their industrial uses are given by STÖHR and FLASCH (1953), in GMELIN (1960), in RÖMPP (1966) and in KIRK and OTHMER (1958). A bibliography on barium chemistry was compiled by SCHWIND (1952).

Acknowledgements

I am very grateful to Dr. MICHAEL FLEISCHER, U.S.G.S., Washington, D.C., for making available to me his literature surveys and giving many valuable suggestions. Thanks are also due to Professor WEDEPOHL for his many suggestions for improvement.

Revised manuscript received: September 1971

References: Section 56-A to 56-O

ABRASHEV, K. K., ILYUKHIN, V. V., BELOV, N. V.: Crystal structure of barylite, $BaBe_2Si_2O_7$. Kristallografija **9**, 691 (1964), translated.

AKHUNDOV, A. R., SAPPO, P. V.: Distribution of some trace elements in formation water of the pay stratum in the Balakhany-Sabunchi-Ramaninsk oil pool. Azerb. Neft. Khoz. **39**, 9 (1960).

ALANOV, A.: Geochemistry of the Lower Cretaceous deposits in the Gaurdak Region. Izv. Akad. Nauk. Turkm. SSR, Ser. Fiz.-Tekhn. Khim. i Geol. Nauk 116 (1963).

ALBEE, A. L., CHODOS, A. A.: Microprobe investigations on Apollo 11 samples. Proceedings of the Apollo 11 Lunar Science Conference **1**, 135 (1970).

ALLEN, C. W.: Astrophysical Quantities. London: University of London, Athlone Press 1963.

ALLER, L. H.: The Abundance of the Elements. Interscience Monographs and Texts in Physics and Astronomy **7** (1961).

— GREENSTEIN, J. L.: The abundances of the elements in G-type subdwarfs. Astrophys. J., Suppl. **5**, 139 (1960).

ANDERSEN, C. A., HINTHORNE, J. R., FREDRIKSSON, K.: Ion microprobe analysis of lunar material from Apollo 11. Proceedings of the Apollo 11 Lunar Science Conference **1**, 159 (1970).

ANDERSEN, N. R., HUME, D. N.: Strontium and barium content of sea water. Advan. Chem. Ser. **73**, 296 (1968).

ANDERSON, A. T., Jr.: Mineralogy of the Labrieville anorthosite, Quebec. Am. Mineralogist **51**, 1671 (1966).

ANDERSON, W.: On the chloride waters of Great Britain. Geol. Mag. **82**, 267 (1945).

ANNELL, C. S., HELZ, A. W.: Emission spectrographic determination of trace elements in lunar samples from Apollo 11. Proceedings of the Apollo 11 Lunar Science Conference **2**, 991 (1970).

ARRHENIUS, G.: Pelagic sediments. In: The Sea. Ideas and Observations on Progress in the Study of the Seas, vol. III, ed. by M. N. HILL. New York-London: Interscience Publ. 1963.

ASSARSSON, G.: On the winning of salt from the brines in southern Sweden. Sveriges Geol. Undersökn., Ser. C **501** (1948).

ATKINS, F. B.: Pyroxenes of the Bushveld intrusion, South Africa. J. Petrol. **10**, 222 (1969).

BABINA, N. M., KOTOROVICH, A. E.: Alkali and alkaline-earth metals in sedimentary rocks of the Western Siberia Lowland. Geochem. International, 508 (1966).

BABINETS, A. E., RAD'KO, N. I.: Microelements in the mineral waters of the southern slopes of the Soviet Carpathians. Geol. Zh., Akad. Nauk Ukrain. RSR **16**, 21 (1956).

BACCANARI, D. P., BUCKMAN, B. A., YEVITZ, M. M., SWAIN, H. A., Jr.: The solubility of carbon-14-labelled barium carbonate in aqueous systems. Talanta **15**, 416 (1968).

BAILEY, E. H., SNAVELY, P. D., Jr., WHITE, D. E.: Chemical analyses of brines and crude oil, Cymric field, Kern County, California. U.S. Geol. Surv., Profess. Papers **424**-D, 306 (1961).

BAJOR, M.: Geochemistry of Tertiary lacustrine beds of the Steinheim basin in Steinheim on Albuch (Württemberg). Jahresh. Geol. Landesamtes Baden-Württemberg **7**, 355 (1965).

BAKER, I.: Petrology of the volcanic rocks of St. Helena Island, South Atlantic. Bull. Geol. Soc. Am. **80**, 1283 (1969).

BAKER, P. E.: Petrology of Mt. Misery Volcano, St. Kitts, West Indies. Lithos **1**, 124 (1968).

— GASS, I. G., HARRIS, P. G., LEMAITRE, R. W.: The volcanological report of the Royal Society expedition to Tristan da Cunha, 1962. Phil. Trans. Roy. Soc. London **256**, 439 (1964).

BALL, T. K.: The geochemistry of the Randesund Gneisses. Norsk. Geol. Tidskr. **46**, 379 (1966).

BARAGAR, R. A.: Petrology of basaltic rocks in part of the Labrador Trough. Bull. Geol. Soc. Am. **71**, 1589 (1960).

BARKER, F.: Sapphirine-bearing rock, Val Codera, Italy. Am. Mineralogist **49**, 146 (1964).

BARROS GOMES, C. DE, SANTINI, P., DUTRA, C. V.: Petrochemistry of a Precambrian amphibolite from the Jaraguá area, Sao Paulo, Brazil. J. Geol. **72**, 664 (1964).

BARTEL, A. J., FENNELLY, E. J., HUFFMAN, C., Jr., RADER, L. F., Jr.: Some new data on the arsenic content of basalt. U.S. Geol. Surv. Profess. Papers **475**-B, B20 (1963).

BARTH, T. F. W.: The feldspar lattices as solvents of foreign ions. Instituto Lucas Mallada, Cursillos y Conferencias **8** (1961).

BASCHEK, B.: Häufigkeitsbestimmung für Kohlenstoff aus CH-Banden im Unterzwerg HD 140283 und in der Sonne. Z. Astrophys. **56**, 207 (1962).

BEL'KEVICH, P. I., CHISTOVA, L. R., STROGONOVA, L. F.: Effect of temperature and form of ion and anion on the ion-exchange equilibrium in peat. Vestsi Akad. Nauk Belarusk. SSR, Ser. Khim. Nauk **4**, 29 (1966).

BERLIN, R., HENDERSON, C. M. B.: The distribution of Sr and Ba between the alkali feldspar, plagioclase and groundmass phases of porphyritic trachytes and phonolites. Geochim. Cosmochim. Acta **33**, 247 (1969).

BERTSCH, W.: Barium in Buntsandsteinen. Master's Thesis, University of Tübingen, 26 pp. 1964.

BIDELMAN, W. P., KEENAN, P. C.: The Ba II stars. Astrophys. J. **114**, 473 (1951).

BINNS, R. A.: Barroisite — bearing eclogite from Naustdal, Sogu og Fjordane, Norway. J. Petrol. **8**, 349 (1967).

BLACK, W. A. P., MITCHELL, R. L.: Trace elements in the common brown algae and in sea water. J. Marine Biol. Assoc. U. K. **30**, 575 (1952).

BLOCK, A., PERLOFF, S.: The crystal structure of barium tetraborate, $BaO \cdot 2B_2O_3$. Acta Cryst. **19**, 297—300 (1965).

BLOSS, F. D., STEINER, R. L.: Biogeochemical prospecting for manganese in north-east Tennessee. Bull. Geol. Soc. Am. **71**, 1053 (1960).

BOETTCHER, A. L.: Vermiculite, hydrobiotite, and biotite in the Rainy Creek igneous complex near Libby, Montana. Clay Minerals Bull. **6**, 283 (1966).

BOLTER, E., TUREKIAN, K. K., SCHUTZ, D. F.: The distribution of rubidium, cesium and barium in the oceans. Geochim. Cosmochim. Acta **28**, 1459 (1964).

BOROVIK-ROMANOVA, M. F.: Spectroscopic determination of barium in the ash of plants. Tr. Biogeokhim. Lab. Akad. Nauk SSSR **5**, 175 (1939).

BOSTRÖM, K., PETERSON, M. N. A.: Precipitates from hydrothermal exhalations on the east Pacific Rise. Econ. Geol. **61**, 1258 (1966).

— FRAZER, J., BLANKENBURG, J.: Subsolidus phase relations and lattice constants in the system $BaSO_4$—$SrSO_4$—$PbSO_4$. Arkiv Mineral. Geol. **4**, 477 (1968).

BOUSKA, V., POVONDRA, P.: Correlation of some physical and chemical properties of moldavites. Geochim. Cosmochim. Acta **28**, 783 (1964).

BOWEN, H. J. M.: Strontium and barium in sea water and in marine organisms. J. Marine Biol. Assoc. U.K. **35**, 451 (1956).

— Trace Elements in Biochemistry. London-New York: Academic Press 1966.

BRÄUER, H.: Spurenelementgehalte in Graniten Thüringens und Sachsens. Ph. D. Thesis, Bergakademie Freiberg, 1965.

BRAY, J. M.: Spectroscopic distribution of minor elements in igneous rocks from Jamestown, Colorado. Bull. Geol. Soc. Am. **53**, 765 (1942).

BROBST, D. A.: Barite: world production, reserves and future prospects. Bull. U.S. Geol. Surv. **1321**, 46 pp. (1970).

BRONGERSMA-SANDERS, M.: Barium in pelagic sediments and in diatoms. Proc. Koninkl. Ned. Akad. Wetenschap., Ser. B **70**, 93 (1967).

BROWN, F. H., CARMICHAEL, I. S. A.: Quaternary volcanoes of the Lake Rudolf region: 1. The basanite-tephrite series of the Korath Range. Lithos **2**, 239 (1969).

BROWN, G. M., EMELEUS, C. H., HOLLAND, J. G., PHILLIPS, R.: Mineralogical, chemical and petrological features of Apollo 11 rocks and their relationship to igneous processes. Proceedings of the Apollo 11 Lunar Science Conference **1**, 195 (1970).

BROWN, H. R., SWAINE, D. J.: Inorganic constituents of Australian coals. I. Nature and mode of occurrence. J. Inst. Fuel **37**, 422 (1964).

— TAYLOR, G. H.: Metamorphosed coal from the Theron Mountains (Falkland Islands Dependencies). Trans-Antarctic Expedition 1955—58, Sci. Rept. **12**, 1 (1960).

BROWN, J., GRANT, C. L., UGOLINI, F. C., TEDROW, J. C. F.: Mineral composition of some drainage waters from arctic Alaska. J. Geophys. Res. **67**, 2447 (1962).

BRUNO, E., GAZZONI, G.: On the system $Ba(Al_2Si_2O_8)$-$Ca(Al_2Si_2O_8)$ I. Ca-Ba substitution in polymorphic modifications of $BaAl_2Si_2O_8$. Contr. Mineral. and Petrol. **25**, 144 (1970).

BRYHNI, I., BOLLINGBERG, H. J., GRAFF, P. R.: Eclogites in quartzofeldspathic gneisses of Nordfjord, West Norway. Norsk Geol. Tidsskr. **49**, 193 (1969).

BUCKLEY, S. E., HOCOTT, C. R., TAGGART, M. S., Jr.: Distribution of dissolved hydrocarbons in subsurface waters. In: L. G. Weeks (ed.), Habitat of Oil. Tulsa: Am. Assoc. Petrol. Geol. 1958.

BURBIDGE, E. M., BURBIDGE, G. R.: Chemical composition of the BaII star HD 46407 and its bearing on element synthesis in stars. Astrophys. J. **126**, 357 (1957).

BURTON, J. D., MARSHALL, N. J., PHILLIPS, A. J.: Solubility of barium sulfate in sea water. Nature **217**, 834 (1968).

BUTLER, B. C. M.: Chemical study of minerals from the Moine Schists of the Ardnamurchan Area, Argyllshire, Scotland. J. Petrol. **8**, 233 (1967).

BUTLER, J. R.: Geochemical affinities of some coals from Svalbard (Spitzbergen). Kong. Ind.-, Handverk, Skips-fartdept., Norsk Polarinst., Skrifter **96**, 1 (1953a).

— The geochemistry and mineralogy of rock weathering. I. The Lizard area, Cornwall. Geochim. Cosmochim. Acta **4**, 157 (1953b).

— The geochemistry and mineralogy of rock weathering. II. The Nordmarka area, Oslo. Geochim. Cosmochim. Acta **6**, 268 (1954).

BYERS, F. M., Jr.: Petrology of three volcanic suites, Umnak and Bogoslof Islands, Aleutian Islands, Alaska. Bull. Geol. Soc. Am. **72**, 93 (1961).

CAMERMAN, C.: Composition d'une eau à forte salure du bassin houiller de Charleroi. Bull. Soc. Belge Géol. **60**, 361 (1951).

CAMERON, A. G. V.: A revised table of abundance of the elements. Astrophys. J. **129**, 676 (1959).

CAMERON, E. M.: Evaluation of sampling and analytical methods for the regional geochemical study of a subsurface carbonate formation. J. Sediment. Petrol. **36**, 755 (1966).

CANNON, H. L.: Geochemistry of rocks and related soils and vegetation in the Yellow Cat area, Grand County, Utah. U.S. Geol. Surv. Bull. **1176** (1964).

— SHACKLETTE, H. T., BASTRON, H.: Metal absorption by equisetum (Horsetail). U.S. Geol. Surv. Bull. **1278**-A (1968).

CARD K. D.: Metamorphism in the Agnew Lake area, Sudbury District, Ontario, Canada. Bull. Geol. Soc. Am. **75**, 1011 (1964).

CARLOSN, R. M., OVERSTREET, R.: The ion exchange behavior of the alkaline earth metals. Soil Sci. **103** 213 (1967).

CARMICHAEL, J. S. E.: Trachytes and their feldspars phenocrysts. Mineral. Mag. **34**, 107 (1965).

— The iron-titanium oxides of salic volcanic rocks and their associated ferromagnesian silicates. Contr. Mineral. and Petrol. **14**, 36 (1967a).

— The mineralogy and petrology of the volcanic rocks from the Leucite Hills, Wyoming. Contr. Mineral. and Petrol. **15**, 24 (1967b).

— McDONALD, A.: The geochemistry of some natural acid glasses from the North Atlantic Tertiary volcanic province. Geochim. Cosmochim. Acta **25**, 189 (1961).

CAYREL,R., CAYREL DE STROBEL, G.: Abundance determination from stellar spectra. Ann. Rev. Astronomy Astrophys. 4, 1 (1966).

CHAO, E. C. T.: The petrographic and chemical characteristics of tektites. In: J. O'KEEFE (ed.), Tektites. Chicago: University Chicago Press 1963.

Chapman, D. R., Keil, K., Annell, C.: Comparison of Macedon and Darwin glass. Geochim. Cosmochim. Acta **31**, 1595 (1967).

Chow, T. J., Goldberg, E. D.: On the marine geochemistry of barium. Geochim. Cosmochim. Acta **20**, 192 (1960).

— Patterson, C. C.: Concentration profiles of barium and lead in Atlantic waters off Bermuda. Earth Planet. Sci. Lett. **1**, 397 (1966).

Church, T. M.: Marine barite. Ph. D. Thesis, University of California, San Diego, 1970.

Clarke, D. B.: Tertiary basalts of Baffin Bay: possible primary magma from the mantle. Contr. Mineral. and Petrol. **25**, 203 (1970).

Clarke, M. C., Swaine, D. J.: Trace elements in coal. I. New South Wales coals. II. Origin, mode of occurrence, and economic importance. Fuel Research Tech. Comm. **45**, 115 (1962).

Clayton, D. D.: Implications of the solar-system abundances near atomic weight 90. J. Geophys. Res. **69**, 5081 (1964).

Clifford, T. N., Nicolaysen, L. O., Burger, A. J.: Petrology and age of the pre-Otavi basement granite at Franzfontein, Northern South West Africa. J. Petrol. **3**, 244 (1962).

— Rooke, J. M., Allsopp, H. L.: Petrochemistry and age of the Franzfontein granitic rocks of Northern South West Africa. Geochim. Cosmochim. Acta **33**, 973 (1969).

Coal Research Committee: D.S.I.R. Wellington N. Z., Coal Report No. **249** (1949).

Coats, R. R.: Magmatic differentiation in Tertiary and Quarternary volcanic rocks from Adak and Kanaga Islands, Aleutian Islands, Alaska. Bull. Geol. Soc. Am. **63**, 485 (1952).

— Geology of Buldir Island, Aleutian Islands, Alaska. Bull. U.S. Geol. Surv. **989**-A, 1 (1953).

— Geologic reconnaissance of Semisopochnoi Island, Western Aleutian Islands, Alaska. Bull. U.S. Geol. Surv. **1028**-O, 477 (1959).

— Nelson, W. H., Lewis, R. Q., Powers, H. A.: Geologic reconnaissance of Kiska Island, Aleutian Islands, Alaska. Bull. U.S. Geol. Surv. **1028**-R, 568 (1961).

Cocco, G.: Genesis of the Elba granite rocks: geochemistry of strontium and barium. Rend. Soc. Mineral. Ital. **9**, 48 (1953).

Coleman, R. G.: Low temperature reaction zones and alpine ultramafic rocks of California, Oregon and Washington. Bull. Geol. Soc. Am. **1247** (1967).

— Lee, D. E.: Glaucophane bearing metamorphic rock types of the Cazadero Area, California. J. Petrol. **4**, 260 (1963).

— Papike, J. J.: Alkali amphiboles from the blueschists of Cazadero, California. J. Petrol. **9**, 105 (1968).

Collins, A. G.: Chemistry of some Anadarko basin brines containing high concentrations of iodide. Chem. Geol. **4**, 169 (1969).

— Zelinski, W. P.: The solubilities of barium and strontium sulfates in oil field brines. Am. Chem. Soc. Div. Water Air Waste Chem., Preprints **6**, 7 (1966).

Colville, A., Staudhammer, K.: A refinement of the structure of barite. Am. Mineralogist **52**, 1877—1880 (1967).

Compston, W., Chappell, B. W., Arriens, P. A., Vernon, M. J.: The chemistry and age of Apollo 11 lunar material. Proceedings of the Apollo 11 Lunar Science Conference **2**, 1007 (1970).

Corlett, M., Ribbe, P. H.: Electron probe microanalysis of minor elements — plagioclase feldspars. Schweiz. Mineral. Petrog. Mitt. **47**, 317 (1967).

Cornwall, H. R.: Calderas and associated volcanic rocks near Beatty, Nye County, Nevada. Geol. Soc. Am. Petrologic Studies, Buddington volume, 357 (1962).

— Rose, H. J., Jr.: Minor elements in Keweenawan lavas, Michigan. Geochim. Cosmochim. Acta **12**, 209 (1957).

Correia Neves, J. M.: Genese des zonar gebauten Beryllpegmatits von Venturinha (Viseu, Portugal) in geochemischer Sicht. Beitr. Mineral. Petrog. **10**, 357 (1964).

Cox, K. G., Hornung, G.: The petrology of the Karroo basalts of Basutoland. Am. Mineralogist **51**, 1414 (1966).

— Johnson, R. L., Monkman, L. J., Stillman, C., Vail, J. R., Wood, D. N.: The geology of the Nuanetsi igneous province. Philos. Trans. Roy. Soc. London **257**, 71 (1965).

Cox, K. G., Macdonald, R., Hornung, G.: Geochemical and petrographic provinces in the Karroo basalts of Southern Africa. Am. Mineralogist **52**, 1451 (1967).

Cuttitta, F., Clarke, R. S., Jr., Carron, M. K., Annell, C. S.: Martha's Vineyard and selected Georgia tektites: new chemical data. J. Geophys. Res. **72**, 1343 (1967).

Dana, J. D., Dana, E. S.: In: C. Palache, H. Berman, Frondel, C.: The System of Mineralogy, 7th ed. New York: John Wiley & Sons 1951 and 1952.

Danchin, R. V.: Aspects of the geochemistry of some selected south African fine grained sediments. Ph. D. Thesis, University of Cape Town, 1970.

Dawson, J. B.: Basutoland kimberlites. Bull. Geol. Soc. Am. **73**, 545 (1962).

Degens, E. T.: Geochemical investigations into the country rock of the fluorite-barite-Co-Ni-Bi-Ag-Cl- ore veins of the middle Black Forest. Glückauf **92**, 842 (1956).

— Williams, E. G., Keith, M. L.: Environmental studies of Carboniferous sediments, part 1: Geochemical criteria for differentiating marine and fresh water shales. Bull. Am. Ass. Petrol. Geologists **41**, 2427 (1957).

Delkeskamp, R.: Die Schwerspathvorkommnisse in der Wetterau und Rheinhessen und ihre Entstehung. Notizbl. geol. Landesanst. Hessen, IV. F. **21**, 47 (1900).

— Die weite Verbreitung des Baryums in Gesteinen und Mineralquellen und die sich hieraus ergebenden Beweismittel für die Anwendbarkeit der Lateralsecretions- und Thermaltheorie auf die Genesis der Schwerspathgänge. Z. prakt. Geol. **10**, 117 (1902).

DeVilliers, P. R.: The chemical composition of the water of the Orange River at Vioolsdrif, Cape Province. Rep. Suid-Afrika, Dept. Mynwese Ann. Geol. Opname **1**, 197 (1962).

Dietrich, R. V., Heier, K. S.: Differentiation of quartz-bearing syenite (nordmarkite) and riebeckitic-arfvedsonite granite (ekerite) of the Oslo series. Geochim. Cosmochim. Acta **31**, 275 (1967).

Dixon, J. E., Cann, J. R., Renfrew, C.: Obsidian and the origins of trade. Sci. Am. **218**, 38 (1968).

Dodge, F. C. W., Papike, J. J., Mays, R. E.: Hornblendes from granitic rocks of the Central Sierra Nevada Batholith, California. J. Petrol. **9**, 378 (1968).

— Smith, V. C., Mays, R. E.: Biotites from granitic rocks of the Central Sierra Nevada Batholith, California. J. Petrol. **10**, 250 (1969).

Dodonov, Ya. Ya., Eferova, L. V., Kolosova, V. S.: Über die Salze der Erdalkalimetalle in den Wässern der Bohrungen von der Gaslagerstätte Saratov. Dokl. Akad. Nauk SSSR **65**, 887 (1949).

Dowgiallo, J.: The occurrence of brines within the Koobrzeg Unit, their genesis and relation to tectonics. Bull. Acad. Polon. Sci. Sér. Sci. Géol. Géograph. **13**, 305 (1965).

Drake, M. J., McCallum, I. S., McKay, G. A., Weill, D. F.: Mineralogy and petrology of Apollo 12 sample no. 12013: a progress report. Earth Planet. Sci. Lett. **9**, 103 (1970).

Duchesne, J. C.: Strontium vs. calcium and barium vs. potassium relations in plagioclases from southern Rogaland anorthosites. Ann. Soc. Geol. Belg. Bull. **90**, 643 (1968).

Duke, M. B., Silver, L. T.: Petrology of eucrites, howardites and mesosiderites. Geochim. Cosmochim. Acta **31**, 1637 (1967).

Dunham, A. C.: The felsites, granophyres, explosion breccias, and tuffisites of the north east margin of the Tertiary igneous complex of Rhum, Inverness-shire. Quart. J. Geol. Soc. London **123**, 327 (1968).

Durfor, G. N., Becker, E.: Public water supplies for the 100 largest cities in the United States 1962. U.S. Water Supply Papers **1812** (1964).

Durum, W. H., Haffty, J.: Occurrence of minor elements in water. U.S. Geol. Surv. Circular **445** (1961).

— — Implications of the minor element content of some major streams of the world. Geochim. Cosmochim. Acta **27**, 1 (1963).

— Heidel, S. G., Tison, L. J.: World wide runoff of dissolved solids. Int. Ass. Sci. Hydrology Gen. Assembly Helsinki 1960, Commission of Surface Waters, Publ. **51**, 618 (1960).

Duval, J. E., Kurbatov, M. H.: The adsorption of cobalt and barium ions by hydrous ferric oxide at equilibrium. J. Phys. Chem. **56**, 982 (1952).

Eckermann, H. von: The distribution of barium in the alkaline rocks and fenites of Alnö Island (Sweden). Intern. Geol. Congr. Rept. 18th Session Gt. Britain, Part II, 46 (1948).

— The distribution of barium and strontium in the rocks and minerals of the syenitic and alkaline rocks of Alnö island. Arkiv Mineral. Geol. **1**, 367 (1952).

— The strontium and barium contents of the Alnö carbonatites. Min. Soc. of India, IMA volume, 106 (1966).

— The strontium and barium contents of the Alnö sovites and the alkaline, carbonatitic, and ultrabasic rocks. Arkiv Mineral. Geol. **4**, 417 (1967).

El-Hinnawi, E. E.: Trace element distribution in Chilean ignimbrites. Contr. Mineral. and Petrol. **24**, 50 (1969).

Emiliani, F., Vespignani-Balzani, G. C.: Sr and Ba distribution in the Predazzo granite. Mineral. Petrogr. Acta **10**, 81 (1964).

Emmermann, R.: Differentiation und Metasomatose des Albtalgranits (Südschwarzwald). Neues Jahrb. Mineral. Abhandl. **109**, 94 (1968).

— Genetic relations between two generations of K feldspar in a granite pluton. Neues Jahrb. Mineral. Abhandl. **111**, 289 (1969).

Emmons, R. C.: Selected petrogenetic relationships of plagioclase. Chemical analyses. Geol. Soc. Am. Mem. **52**, 11 (1952).

Engel, A. E. J., Engel, C. G.: Progressive metamorphism and granitization of the major paragneiss, northwest Adirondack Mountains, New York. Bull. Geol. Soc. Am. **69**, 1369 (1958).

— — Progressive metamorphism and granitization of the major paragneiss, northwest Adirondack Mountains, New York. Bull. Geol. Soc. Am. **71**, 1 (1960).

— — Hornblendes formed during progressive metamorphism of amphibolites, northwest Adirondack Mountains, New York. Bull. Geol. Soc. Am. **73**, 1499 (1962).

— Havens, R. G.: Chemical characteristics of oceanic basalts and the upper mantle. Bull. Geol. Soc. Am. **76**, 719 (1965).

Engel, C. G.: Igneous rocks and constituent hornblendes of the Henry Mountains, Utah. Bull. Geol. Soc. Am. **70**, 951 (1959).

— Sharp, R. P.: Chemical data on desert varnish. Bull. Geol. Soc. Am. **69**, 487 (1958).

Engelhardt, W. von: Die Geochemie des Barium. Chem. Erde **10**, 187 (1936).

Erickson, R. L., Blade, L. V.: Geochemistry and petrology of the alkalic igneous complex at Magnet Cove, Arkansas. U.S. Geol. Surv., Profess. Papers. **425** (1963).

Ericson, D. B., Ewing, M., Wollin, G., Heezen, B. C.: Atlantic deep-sea sediment cores. Bull. Geol. Soc. Am. **72**, 193 (1961).

— Wollin, G.: Correlation of six cores from the equatorial Atlantic and the Caribbean. Deep-Sea Res. **3**, 104 (1956).

Ershov, V. M.: Relation of germanium to the organic matter in fossil coals. Geochemistry, 763 (1958).

Eugster, O., Tera, F., Wasserburg, G. J.: Isotopic analyses of barium in meteorites and terrestrial samples. J. Geophys. Res. **74**, 3897 (1969).

Evans, B. W.: Fractionation of elements in the pelitic hornfelses of the Cashel-Lough Wheelaun intrusion, Connemara, Eire. Geochim. Cosmochim. Acta **28**, 127 (1964).

Ewart, A., Taylor, S. R., Capp, A. C.: Geochemistry of the pantellerit of Mayor Island, New Zealand. Contr. Mineral. and Petrol. **17**, 116 (1968).

Fairbairn, H. W., Ahrens, L. H., Gorfinkle, L. G.: Minor element content of Ontario diabase. Geochim. Cosmochim. Acta **3**, 34 (1953).

— Schlecht, W. G., Stevens, R. E., Dennen, W. H., Ahrens, L. H., Chayes, F.: A cooperative investigation of precision and accuracy in chemical, spectrochemical and modal analysis of silicate rocks. Bull. U.S. Geol. Surv. **980** (1951).

Fenner, P., Hagner, A. F.: Correlation of variations in trace elements and mineralogy of the Esopus Formation, Kingston, New York. Geochim. Cosmochim. Acta **31**, 237 (1967).

Fischer, K.: Verfeinernug der Kristallstruktur von Benitoit $BaTi[Si_3O_9]$. Z. Krist. **129**, 222 (1969).

Fisher, R. L., Engel, C. G.: Ultramafic and basaltic rocks dredged from the near shore flank of the Tonga trench. Bull. Geol. Soc. Am. **80**, 1373 (1969).

FLANAGAN, F. J.: U.S. Geological Survey standards. II. First compilation of data for the new U.S.G.S. rocks. Geochim. Cosmochim. Acta **33**, 81 (1969).

FLEISCHER, M.: Summary of new data on rock samples G-1 and W-1, 1962—1965. Geochim. Cosmochim. Acta **29**, 1263 (1965).

— U.S. Geological Survey standards. — I. Additional data on rocks G-1 and W-1, 1965—1967. Geochim. Cosmochim. Acta **33**, 65 (1969).

— STEVENS, R. E.: Summary of new data on rock samples G-1 and W-1. Geochim. Cosmochim. Acta **26**, 525 (1962).

FOERSTER, H. F., FOSTER, J. W.: Endotrophic calcium, strontium, and barium spores of bacillus megaterium and bacillus cereus. J. Bacteriol. **91**, 1333 (1966).

FORNASERI, M., SCHERILLO, A., VENTRIGLIA, U.: La regione vulcanica dei Colli Albani. Consiglio Nazionale delle Ricerche Centro di Mineralogia e Petrografia Aziende Tipografiche Eredi Dott. G. Bardi (1963).

FOSTER, R. J.: Precambrian corundum bearing rocks, Madison Range, southwestern Montana. Bull. Geol. Soc. Am. **73**, 131 (1962).

FREDRIKSON, K., KEIL, K.: The light-dark structure in the Pantar and Kapoeta stone meteorites. Geochim. Cosmochim. Acta **27**, 717 (1963).

FRICKE, K.: Neue hydrogeologische Untersuchungen und Neubohrungen im Heilquellengebiet von Bad Hermannsborn, NRW. Das Gas- Wasserfach **109**, 251 (1968).

FRIEDMAN, G. M.: Obtaining paleoenvironmental information. U. S. Patent Office No. 3, 343, 917 (1967).

— Geology and geochemistry of reefs, carbonate sediments, and waters, Gulf of Aqaba (Elat), Red Sea. J. Sediment. Petrol. **38**, 895 (1968).

— Trace elements as possible environmental indicators in carbonate sediments. Soc. Econ. Paleontologists Mineralogists, Spec. Publ. **14**, 193 (1969).

FUJITA, Y., TSUJI, I.: Spectrophotometry of Y Canum Venaticorum. Publ. Dominion Astrophys. Obs. **12**, 339 (1965).

GANGULY, A. K., MUKHERJEE, S. K.: The cation exchange behaviour of heteroionic and homoionic clays of silicate minerals. J. Phys. and Colloid Chem. **55**, 1429 (1951).

GARRELS, R. M., THOMPSON, M. E., SIEVER, R.: Stability of some carbonates at 25° C and one atmosphere total pressure. Am. J. Sci. **258**, 402 (1960).

GARSON, M. S.: Carbonatites in Malawi, pp. 33—77. In: O. F. TUTTLE, GITTINS, J. (eds.), Carbonatites. New York: Interscience Publishers 1967.

GAST, P. W.: Limitations on the composition of the upper mantle. J. Geophys. Res. **65**, 1287 (1960).

— Terrestrial ratio of potassium to rubidium and the composition of the earth's mantle. Science **147**, 858 (1965).

— HUBBARD, N. J., WIESMANN, H.: Chemical composition and petrogenesis of basalts from Tranquillity Base. Proceedings of the Apollo 11 Lunar Science Conference **2**, 1143 (1970).

GATES, G. L., CARAWAY, W. H.: Oil-well scale formation in water flood operations using ocean brines, Wilmington, California. U.S. Bur. Mines, Rept. Invest. **6658** (1965).

GAY, P., ROY, N. N.: The mineralogy of the potassium-barium feldspar series. III. Subsolidus relationships. Mineral. Mag. **36**, 914 (1968).

GERASIMOVSKII, V. I.: Geochemical features of agpaitic nepheline-syenites, p. 104. In: VINOGRADOV (ed.), Chemistry of the Earth's Crust, I. Jerusalem: 1966.

— BELYAEV, Yu. I: Content of manganese, barium and strontium in alkaline rocks of the Kola Peninsula. Geochemistry, 1161 (1963).

GHOSE, N. C.: Behaviour of trace elements during thermal metamorphism and (oblique) or granitization of the metasediments and basic igneous rocks. Geol. Rundschau **55**, 608 (1966).

GIBSON, J.: Personal communication (1963).

GIUSCA, D., IONESCU, J.: Geochemistry of volcanic rocks in the Gutai Mountains. Probl. Geokhim., Akad. Nauk SSSR, Inst. Geokhim. i Analit. Kkim., 436 (1965).

GMELIN: Gmelins Handbuch der anorganischen Chemie, 8th edit. Barium Ergänzungsband. Weinheim: Verlag Chemie 1960.

GOLD, D. P.: Average chemical composition of carbonatites. Ec. Geol. **58**, 988 (1963).
— Alkaline ultrabasic rocks in the Montreal Area, Quebec, p. 288. In: WYLLIE (ed.), Ultramafic and Related Rocks. New York: John Wiley & Sons 1967.
GOLDBERG, E. D., ARRHENIUS, G. O. S.: Chemistry of Pacific pelagic sediments. Geochim. Cosmochim. Acta **13**, 153 (1958).
GOLDBERG, L., MÜLLER, E. A., ALLER, L. H.: The abundances of the elements in the solar atmosphere. Astrophys. J., Suppl. **5**, 1 (1960).
GOLES, G. G., RANDLE, K., OSAWA, M., SCHMITT, R. A., WAKITA, H., EHMANN, W. D., MORGAN, J. W.: Elemental abundances by instrumental activation analyses in chips from 27 lunar rocks. Proceedings of the Apollo 11 Lunar Science Conference, **2,** 1165 (1970).
GORDON, C. P.: The BaII and N stars as a temperature sequence. Astrophys. J. **153**, 915 (1968).
GRAF, D. L.: Geochemistry of carbonate sediments and sedimentary carbonate rocks. III. Minor element distribution. Illinois State Geol. Surv. Circ. **301** (1960).
GREEN, D. H.: The petrogenesis of the high-temperature peridotite intrusion in the Lizard area, Cornwall (England). J. Petrol. **5**, 134 (1964).
GREENLAND, L. P., LOVERING, J. F.: Minor and trace element abundances in the chondritic meteorites. Geochim. Cosmochim. Acta **29**, 821 (1965).
— — Fractionation of fluorine, chlorine, and other trace elements during differentiation of a tholeiitic magma. Geochim. Cosmochim. Acta **30**, 963 (1966).
GRESENS, R. L.: Tectonic-hydrothermal pegmatites. II. An example. Contr. Mineral. and Petrol. **16**, 1 (1967).
GRIFFIN, W. L., MURTHY, V. R.: Abundances of K, Rb, Sr, and Ba in ultramafic rocks and minerals. Earth Planet. Sci. Lett. **4**, 497 (1968).
GRIM, R. E., DIETZ, R. S., BRADLEY, W. F.: Clay mineral composition of some sediments from the Pacific Ocean off the Californian coast and the Gulf of California. Bull. Geol. Soc. Am. **60**, 1785 (1949).
GROHMANN, H., SCHROLL, E.: Seltene Elemente in Granitoiden der südlichen Böhmischen Masse. Tschermaks Mineral. Petrog. Mitt. **11**, 348 (1966).
GROSS, E.: Pabstite, the tin analogue of benitoite. Am. Mineralogist **50**, 1164 (1965).
GROUT, F. F.: The compositions of some African granitoid rocks. J. Geol. **43**, 281 (1935).
GRUSHKO, Ya. M., SHIPITSYN, S. A.: Toxic substances in the drinking waters of Irkutsk from spectral analyses. Gigiena i Sanit. **13**, 4 (1948).
GUNN, B. M.: K/Rb and K/Ba ratios in Antarctic and New Zealand tholeiites and alkali basalts. J. Geophys. Res. **70**, 6241 (1965).
— Modal and element variation in Antarctic tholeiites. Geochim. Cosmochim. Acta **30**, 881 (1966).
HAHN-WEINHEIMER, P.: Geochemische Untersuchungen an den ultrabasischen und basischen Gesteinen der Münchberger Gneismasse (Fichtelgebirge). Neues Jahrb. Mineral., Abhandl. **92**, 203 (1959).
— LUECKE, W.: Garnets from the eclogites of the Muenchberger gneiss massif (N. E. Bavaria). Can. Mineralogist **7**, 764 (1963).
— ACKERMANN, H.: Geochemical investigation of differentiated granite plutons of the Southern Black Forest — II. The zoning of the Malsburg granite pluton as indicated by the elements titanium, zirconium, phosphorus, strontium, barium, rubidium, potassium, and sodium. Geochim. Cosmochim. Acta **31**, 2197 (1967).
HALL, A.: The variation of some trace elements in the Rosses granite complex, Donegal. Geol. Mag. **104**, 99 (1967).
HAMAGUCHI, H., REED, G. W., TURKEWITSCH, A.: Uranium and barium in stone meteorites. Geochim. Cosmochim. Acta **12**, 337 (1957).
HANOR, J. S.: Barite saturation in sea water. Geochim. Cosmochim. Acta **33**, 894 (1969).
HASKIN, L. A., ALLEN, R. O., HELMKE, P. A., PASTER, T. P., ANDERSON, M. R., KOROTEV, R. L., ZWEIFEL, K. A.: Rare earths and other trace elements in Apollo 11 lunar samples. Proceedings of the Apollo 11 Lunar Science Conference **2**, 1213 (1970).
HASLAM, H. W.: The crystallization of intermediate and acid magmas at Ben Nevis, Scotland. J. Petrol. **9**, 84 (1968).

HAWLEY, J. E.: Germanium content of some Nova Scotian coals. Ec. Geol. **50**, 517 (1955).
HEIDE, F., CHRIST, W.: Zur Geochemie des Strontiums und Bariums. Chem. Erde **16**, 327 (1953).
HEIDEL, S. G., FRENIER, W. W.: Chemical quality of water and trace elements in the Patuxent River Basin. Maryland Geol. Surv. Rept. Invest. No. **1**, 1 (1965).
HEIER, K. S.: Petrology and geochemistry of high-grade metamorphic and igneous rocks on Langøy, Northern Norway. Norsk Geol. Undersökning **207** (1960).
— The amphibolite-granulite facies transition reflected in the mineralogy of potassium feldspars. Instituto Lucas Mallada, Cursillos y Conferencias fasc. **8** (1961).
— Trace elements in feldspars — a review. Norsk Geol. Tidsskr. **42**, 415 (1962).
— Geochemistry of the nepheline-syenite on Stjernøy, North Norway. Norsk Geol. Tidsskr. **44**, 205 (1964).
— A geochemical comparison of the Blue Mountain (Ontario, Canada), and Stjernøy (Finnmark, North Norway) nepheline syenites. Norsk Geol. Tidsskr. **45**, 41 (1965).
— TAYLOR, S. R.: Distribution of Ca, Sr and Ba in southern Norwegian pre-Cambrian alkali feldspars. Geochim. Cosmochim. Acta **17**, 286 (1959).
— CHAPPELL, B. W., ARRIENS, P. A., MORGANS, J. W.: The geochemistry of four Icelandic basalts. Norsk. Geol. Tidsskr. **46**, 427 (1966).
HEINRICH, E. W. M.: Micas of the Brown Derby pegmatites, Gunnison County, Colorado. Am. Mineralogist **52**, 1110 (1967).
— LEVINSON, A. A., LEVANDOWSKI, D. W., HEWITT, C. H.: Studies in the natural history of micas. Univ. Mich., Final Rept. Project M 978 (1953).
— — — — Geochemistry of muscovite and related micas. 20th Internat. Geological Congr. Mexico 1956.
HELFER, H. L., WALLERSTEIN, G., GREENSTEIN, J. L.: Metal abundance in the sub-giant Hercules and three other dG stars. Astrophys. J. **138**, 97 (1963).
HENDERSON, C. M. B.: Minor element chemistry of leucite and pseudoleucite. Mineral. Mag. **35**, 596 (1965).
HERRMANN, A. G.: Über das Vorkommen einiger Spurenelemente in Salzlösungen aus dem deutschen Zechstein. Kali Steinsalz **3**, 209 (1961).
HERZ, N., DUTRA, C. V.: Minor element abundance in a part of the Brazilian shield. Geochim. Cosmochim. Acta **21**, 81 (1960).
— — Geochemistry of some kyanites from Brazil. Am. Mineralogist **49**, 1290 (1964).
— — Trace elements in alkali feldspars, Quadrilátero ferrifero, Minas Gerais, Brazil. Am. Mineralogist **51**, 1593 (1966).
HEWLETT, C. G.: Optical properties of potassic feldspars. Bull. Geol. Soc. Am. **70**, 511 (1959).
HEY, M. H.: Catalogue of Meteorites, 3rd edit. British Museum 1966.
HEYDEGGER, H. R., KURODA, P. K.: Natural occurrence of the short-lived barium and strontium isotopes. J. Inorg. Nucl. Chem. **12**, 12 (1969).
HIETANEN, A.: Distribution of elements in biotite-hornblende pairs and in an orthopyroxene-clinopyroxene pair from zoned plutons, Northern Sierra Nevada, California. Contr. Mineral. and Petrol. **30**, 161 (1971).
HIGAZY, R. A.: Petrogenesis of perthite pegmatites in the Black Hills, South Dakota. J. Geol. **57**, 555 (1949).
— Behaviour of the trace elements in a front of metasomatic-metamorphism in the Dalradian of Co. Donegal. Geochim. Cosmochim. Acta **2**, 170 (1952).
— Observations on the distribution of trace elements in the perthite pegmatites of the Black Hills, South Dakota. Am. Mineralogist **38**, 172 (1953).
— Trace elements of volcanic ultrabasic potassic rocks of south-western Uganda and adjoining parts of the Belgian Congo. Bull. Geol. Soc. Am. **65**, 39 (1954).
HILMER, W.: Die Struktur der Hochtemperaturform des $BaGeO_3$. Acta Cryst. **15**, 1101 (1962).
HINTZE, C.: Handbuch der Mineralogie, Bd. I, 3. Abt., 1. Hälfte. Berlin: Walter de Gruyter 1930.
— Handbuch der Mineralogie, Bd. I, 3. Abt., 2. Hälfte. Berlin: Walter de Gruyter 1930.
— Handbuch der Mineralogie. Ergänzungsband I. Berlin: Walter de Gruyter 1938.
— Handbuch der Mineralogie. Ergänzungsband II. Berlin: Walter de Gruyter 1960.

HITCHON, B.: The geochemistry, mineralogy, and origin of pegmatites of three Scottish Precambrian metamorphic complexes. Intern. Geol. Congr. Rept. 21st session Copenhagen XVII, 36 (1960).

HOLLAND, H. D.: Some applications of thermochemical data to problems of ore deposits. II. Mineral assemblages and the composition of ore-forming fluids. Ec. Geol. **60**, 1101 (1965).

HOSKINS, H. A.: Analyses of West Virginia brines. State of West Virginia Geol. and Ec. Surv. Rept. Invest. **1** (1947).

HOSS, H., ROY, R.: On natural phillipsite, gismondite, harmotome, chabasite, gmelinite. Beitr. Mineral. Petrog. **7**, 389 (1960).

HOTZ, P. E.: Petrology of granophyre in diabase near Dillsburg, Pa. Bull. Geol. Soc. Am. **64**, 675 (1953).

HOWIE, R. A.: The geochemistry of the charnockite series of Madras, India. Trans. Roy. Soc. Edinburgh **62**, 725 (1953—55).

HUBBARD, N. J., GAST, P. W., WIESMANN, H.: Rare earth, alkaline and alkali metal and $^{87/86}Sr$ data for subsamples of lunar sample 12013. Earth Planet. Sci. Lett. **9**, 181 (1970).

— MEYER, C., Jr., GAST, P. W.: The composition and derivation of Apollo 12 soils. Earth Planet. Sci. Lett. **10**, 341 (1971).

HUCKENHOLZ, G.: Personal communication (1969).

HÜGI, TH., SWAINE, D. J.: The geochemistry of some Swiss granites. J. Proc. Roy. Soc. New S. Wales **96**, 65 (1963).

HUNZICKER, J. C.: Zur Geologie und Geochemie des Gebietes zwischen Valle Antigoro (Provincia di Novara) und Valle di Campo (Kt. Tessin). Schweiz. Mineral. Petrog. Mitt. **46**, 473 (1966).

HYDEN, H. J.: Uranium and other metals in crude oils. II. Distribution of uranium and other metals in crude oils. U.S. Geol. Surv. Bull. **1100**-B, 17 (1961).

ICHIKUNI, I. M.: Barium content of barite-forming hydrothermal waters as exemplified by Tamagawa Hot Spring waters. Bull. Chem. Soc. Japan **39**, 898 (1966).

IIJAMA, J. T.: Etude experimentale de la distribution d'élément en traces entre deux feldspaths. Feldspath potassique et plagioclase coexistants. I. Distribution de Rb, Cs, Sr et Ba à 600° C. Bull. Soc. Franc. Mineral. Crist. **91**, 130 (1968).

IKEDA, N.: Chemical studies on the hot springs of Nasu. VII. Spectrochemical determination of rarer element contents of Motoyu spring. J. Chem. Soc. Japan, Pure Chem. Sect. **76**, 711 (1955a).

— Chemical studies on the hot springs of Nasu. VIII. The spring water of the Motoyu. J. Chem. Soc. Japan, Pure Chem. Sect. **76**, 713 (1955b).

IMREH, J., ECATERINE, J.: Geochemical study of some Eocene limestones from the Transylvanian Basin, Romania. Bull. Serv. Carte Geol. Alsace Lorraine **18**, 277 (1965).

ISHIOKA, K.: A clinopyroxene granitic rock from Myogase, Japan. Geochem. J. **1**, 95 (1967).

IWASAKI, I., KATSURA, T., TARUTANI, T., OZAWA, T., YOSHIDA, M.: Geochemical studies of Tamagawa Hot Spring. Geochem. Tamagawa Hot Springs, 7 (1963).

JAKOBSHAGEN, V., MÜNNICH, K. O.: ^{14}C-Altersbestimmung und andere Isotopenuntersuchungen an Thermalsolen des Ruhrkarbons. Neues Jahrb. Geol. Palaeontol. Monatsh. **9**, 561 (1964).

JAMIESON, B. G., CLARKE, D. B.: Potassium and associated elements in tholeiitic basalts. J. Petrol. **11**, 183 (1970).

JANSE, A. J. A.: Monticellite-peridotite from Mt. Brukkaros, South West Africa. Univ. of Leeds Res. Inst. African Geol. 86th Ann. Rept., 21 (1962—1963).

JASMUND, K., SECK, H. A.: Geochemische Untersuchungen an Auswürflingen (Gleesiten) des Laacher-See-Gebietes. Beitr. Mineral. Petrog. **10**, 275 (1964).

JOHNSON, R. L.: The geology of the Dorowa and Shawa carbonatite complexes Southern Rhodesia. Trans. Geol. Soc. S. Africa **64**, 101 (1961).

JUNNER, N. R., JAMES, W. T.: Chemical analyses of gold coast rocks, ores, and minerals. Gold Coast Geol. Surv. Bull. **15**, 1 (1947).

KASHAEV, A.: Crystal structure of cymrite. Dokl. Akad. Nauk SSSR **169**, 201 (1966).

KATCHENKOV, S. M., FLEGONTOVA, E. I.: Trace elements in the Devonian (petroleum) deposits of the Volga-Ural region, as found by spectral analysis. Tr. Vses. Neft. Nauchn.-Issled. Geologorazved. Inst., Geol. Sbornik **83**, 466 (1955).

— — Distribution of chemical elements in sedimentary rocks, waters, and in the ashes of the crudes of the Gronznyi-Dagestan region. Tr. Vses. Neft. Nauchn.-Issled. Geologorazved. Inst., Geol. Sbornik **95**, 481 (1956).

— — Trace elements in bottom muds of the Indian Ocean. Tr. Vses. Neft. Nauchn.-Issled. Geologorazved. Inst. **227**, 202 (1964).

KATSCHER, H., LIEBAU, F.: Über die Kristallstruktur von $Ba_3Si_3O_8$, ein Silikat mit Dreifachketten. Naturwissenschaften **52**, 512 (1965).

— — Triple, quadruple and quintruple chains in barium silicates. Acta Cryst. **21**, Suppl., A 58 (1966).

KAVEEV, M. S., VASIL'EV, B. V.: Hydrogeology of petroleum formations in Devonian deposits of southeast Tatar. Tr. Soveshch. Probl. Neftegaz. Uralo-Volzhskoi Oblasti, Tr. Soveshchaniya 10—15 Maya, 337 (1954, 1956).

KAY, R., HUBBARD, N. J., GAST, P. W.: Chemical characteristics and origin of oceanic ridge volcanic rocks. J. Geophys. Res. **75**, 1585 (1970).

KEGEL, W. H.: Die Atmosphäre des F6 IV–V-Sternes γ Serpentis. Z. Astrophys. **55**, 221 (1962).

KELLER, W. D., KLEMME, A. W., PICKETT, E. E.: Detailed survey of the chemical composition of rock layers in an agricultural limestone quarry. Ec. Geol. **45**, 461 (1950).

KENT, L. E.: The thermal waters of the Union of South Africa and South West Africa. Trans. Proc. Geol. Soc. S. Africa **52**, 231 (1949).

— RUSSELL, H. D.: The warm spring on Buffelshoek, near Thabazimbi, Transvaal. Trans. Roy. Soc. S. Africa **32**, 161 (1949).

KING, B. C., SUTHERLAND, D.S.: The carbonatite complexes of Eastern Uganda, p. 73. In: TUTTLE, O. F., GITTINS, V. (eds.), Carbonatites. New York: Interscience Publishers 1967.

KIRICHENKO, L. F., CHERTOV, V. M., VYSOTSKII, Z. Z., STRAZHESKO, D. N.: Sorption of cations from acid solutions on silica gels obtained by the hydrothermal method. Dokl. Akad. Nauk SSSR **164**, 618 (1965).

KIRK, R. E., OTHMER, D. F.: Encyclopedia of Chemical Technology. New York: Interscience Publishers 1963.

KLEEMAN, A. W.: Sampling error in the chemical analysis of rocks. J. Geol. Soc. Australia **14**, 43 (1967).

KODAIRA, K.: Atmosphärenstruktur und chemische Zusammensetzung des Schnelläufers HD 161817. Z. Astrophys. **59**, 139 (1964).

KOHLRAUSCH, F.: Über gesättigte wäßrige Lösungen schwerlöslicher Salze II. Teil: Die gelösten Mengen mit ihrem Temperaturgang. Z. Physik. Chem. **64**, 129 (1908).

KOLBE, P.: Geochemistry of some African and Australian granitic rocks. Ph. D. Thesis, Australian National University 1964.

— TAYLOR, S. R.: Major and trace element relationships in granodiorites and granites from Australia and South Africa. Contr. Mineral. and Petrol. **12**, 202 (1966).

KOMLEV, O. I., TIKHA, N. I., ZHERDEVA, N. I.: Tricationic exchange on Transcarpathian bentonite. Visn. L'viv. Derzh. Univ., Ser. Khim. **8**, 54 (1965).

KONTOROVICH, A. E., SADIKOV, M. A., SHVARTSEV, S. L.: Abundances of certain elements in the surface and ground waters of the northwestern part of the Siberian platform. Dokl. Akad. Nauk SSSR **149**, 168 (1963).

KOROLEV, A.: The water content of the petroleum deposit of Kala and the chemical classification of this water. Azerb. Neft. Khoz. **15**, 35 (1938).

KOZIN, A. N.: Barium in formation waters of oil pools in the Volga area near Kuibyshev. Geokhimiya **9**, 937 (1964).

KRAVCHENKO, V. B.: Crystal structure of $BaB_2O_4 \cdot 4H_2O \equiv Ba[B(OH)_4]_2$. Zh. Strukt. Khim. **6**, 724 (1965), translated.

KRETZ, R.: Chemical study of garnet, biotite, and hornblende from gneisses of southwestern Quebec, with analysis on distribution of elements in coexisting minerals. J. Geol. **67**, 371 (1959).

— The distribution of certain elements among coexisting calcic pyroxenes, calcic amphiboles and biotites in skarns. Geochim. Cosmochim. Acta **20**, 161 (1960).

KUKABAEV, B., SYDYKOV, ZH.: Hydrochemistry of subsurface waters of the Permian-Triassic deposits in the southwestern part of the Ural and Emba interfluve. Izv. Akad. Nauk Kaz. SSR, Ser. Geol. **4**, 58 (1962).

KUNO, H., YAMASAKI, K., IIDA, CH., NAGASHIMA, K.: Differentiation of Hawaiian magmas. Japan. J. Geol. Geography, Trans. **28**, 179 (1957).

KURODA, P. K., EDWARDS, R. R.: Radiochemical measurement of the natural fission rate of uranium and the natural occurrence of ^{140}Ba. J. Inorg. Nucl. Chem. **3**, 345 (1957).

LAMBERT, I. B., HEIER, K. S.: Geochemical investigations of deep-seated rocks in the Australian shield. Lithos **1**, 30 (1968).

LAMBERT, R. ST.: The relationship of the Moine schists and Lewisian gneisses near Mallaigmore, Inverness-shire. Proc. Geologists' Assoc. **75**, 1 (1964).

LANDER, J.: The crystal structures of NiO · 3BaO, NiO · BaO, BaNiO and intermediate faces with composition near $Ba_2Ni_2O_5$; with a note on NiO. Acta Cryst. **4**, 148 (1951).

LANDERGEEN, ST., MANHEIM, F. T.: Über die Abhängigkeit der Verteilung von Schwermetallen von der Fazies. Fortschr. Geol. Rheinld. u. Westf. **10**, 173 (1963).

LARSSON, W.: Chemical analyses of Swedish rocks. Bull. Geol. Inst. Univ. Upsala **24**, 47 (1932).

LAVRUKHINA, A. K., KOLESOV, G. M., KALICHEVA, I. S., AKOL'ZINA, L. D.: Activation analysis for Ce, Eu, Sc, Ba, U, and P in the dark and light varieties of the Kunashak and Pervomayskiy Poselok chondrites. Geochem. Int., 217 (1966).

LEBEDEV, B. A.: Trace elements in Jurassic and Lower Cretaceous formations of the Caspian depression. Dokl. Akad. Nauk SSSR **173**, 192 (1967a).

— Comparison of trace element contents in marine and fresh water clays. Geochem. Int., 821 (1967b).

LEELANANDAM, C.: Chemical study of pyroxenes from the charnockitic rocks of Kondagalli (Andrah Pradesh), India, with emphasis on the distribution of elements in coexisting pyroxenes. Mineral. Mag. **36**, 153 (1967).

LEMAITRE, R. W.: Petrology of volcanic rocks, Gough Island, South Atlantic. Bull. Geol. Soc. Am. **73**, 1309 (1962).

LEUTWEIN, F.: Geochemische Untersuchungen an Alaun- und Kieselschiefern Thüringens. Arch. Lagerstättenforsch. **82**, 1 (1951).

— Spurenelemente in rezenten Cardien verschiedener Fundorte. Fortschr. Geol. Rheinld. u. Westf. **10**, 283 (1963).

— Geochemische Charakteristica mariner Einflüsse in Torfen und anderen quarternären Sedimenten. Geol. Rundschau **55**, 97 (1966).

— WEISE, L.: Hydrogeochemische Untersuchungen an erzgebirgischen Gruben- und Oberflächenwässern. Geochim. Cosmochim. Acta **26**, 1333 (1962).

LEVI, H. W., SCHIEWER, E.: Bariumaustausch an Vermikulit und Bentonit. Z. Anorg. Allgem. Chem. **337**, 105 (1965).

LIEBAU, F.: Die Systematik der Silikate. Naturwissenschaften **49**, 481 (1962).

— Ein Beitrag zur Kristallchemie der Schichtsilikate. Acta Cryst. B **24**, 690 (1968).

LIEBENBERG, C. J.: The trace elements of the rocks of the Bushveld igneous complex. Publikasies Univ. Pretoria, Nuwe Reeks **12** (1960).

LIN, H. C., FOSTER, W. R.: Studies in the system $BaO\text{-}Al_2O_3\text{-}SiO_2$. I. The polymorphism of celsian. Am. Mineralogist **53**, 134 (1968).

— — Studies in the system $BaO\text{-}Al_2O_3\text{-}SiO_2$. III. The binary system sanbornite-celsian Mineral. Mag. **37**, 459 (1969).

LIPMAN, P. W.: Alkalic and tholeiitic volcanism related to the Rio Grande depression, Southern Colorado and Northern New Mexico. Bull. Geol. Soc. Am. **80**, 1343 (1969).

LIPPMANN, F.: Die Kristallstruktur des Norsethit, $BaMg(CO_3)_2$. Tschermak's Min. Petr. Mitteil. **12**, 299 (1968).

LITVIN, I. I.: Minor elements in Lower Cretaceous rocks of the Dnieper-Donets depression. Dokl. Akad. Nauk SSSR **139**, 450 (1961).

LITVIN, S. V.: Trace elements in the Upper Carboniferous sedimentary rocks of the Donets Basin and in northeastern part of the Dnieper-Donets syncline. Dokl. Akad. Nauk SSSR **152**, 1453 (1963).

LONKA, A.: Trace elements in the Finnish Precambrian phyllites as indicators of salinity at the time of sedimentation. Bull. Comm. Géol. Finlande **228** (1967).

LOTSPEICH, F. B., MARKWARD, E. L.: Minor elements in bedrock soil, and vegetation at an outcrop of the Phosphoria Formation on Snowdrift Mountain, southeastern Idaho. U.S. Geol. Surv. Bull. **1181**-F, F1 (1963).

LSPET (Lunar Sample Preliminary Examination Team): Preliminary examination of lunar samples from Apollo 11. Science **165**, 1211 (1969).

— Preliminary examination of lunar samples from Apollo 12. Science **167**, 1325 (1970).

Lunatic Asylum: Mineralogic and isotopic investigations on lunar rock 12013. Earth Planet. Sci. Lett. **9**, 137 (1970).

LURYE, L. M.: Migration of barium and strontium during country rock metasomatism in the Zambarak ore field. Dokl. Akad. Nauk SSSR **149**, 1167 (Engl. translation 185) (1963).

MACDONALD, G. A., EATON, J. P.: Hawaiian volcanoes during 1955. Bull. U.S. Geol. Surv. **1171** (1964).

MACPHERSON, H. G.: A chemical and petrographic study of Pre-Cambrian sediments. Geochim. Cosmochim. Acta **14**, 73 (1958).

MALININ, S. D.: An experimental investigation of the solubility of calcite and witherite under hydrothermal conditions. Geochemistry, 650 (1963).

MANOHAR, H., RAMASESHAN, S.: Crystal coordination of the barium ion. Proc. Ind. Acad. Sci. **60**, 317 (1964).

— — The crystal structure of barium hydroxide octahydrate $Ba(OH)_2 \cdot 8H_2O$. Z. Kirst. **119**, 357 (1964).

MARKART, H., PREISINGER, A.: Zur Bestimmung der Feldspate in Gesteinen. Tschermaks Mineral. Petrog. Mitt., 3. F., 315 (1964/65).

MARKHININ, E. K., SAPOZHNIKOVA, A. M., STRATULA, D. S.: Barium in the volcanic rocks of Kamchatka and the Kurile islands. Geochemistry, 933 (1964).

MARMO, V., SIIVOLA, J.: The barium content of some granites of Finland. Bull. Comm. Géol. Finlande **222**, 169 (1966).

MATHÉ, G.: Die Metabasite des sächsischen Granulitgebirges. Freiberger Forschungsh. C **251** (1969).

MATHIAS, M.: The geochemistry of the Messum Igneous Complex, South-West Africa. Geochim. Cosmochim. Acta **12**, 29 (1957).

MATSUMURA, T., ISHIYAMA, T.: Adsorption of radioactive materials by coal humic acid. Ann. Rept. Radiat. Center Osaka Prefect **7**, 14 (1966).

MATTAUCH, J. H. E., THIELE, W., WAPSTRA, A. H.: 1964 atomic mass table. Nucl. Phys. **67**, 1 (1965).

MAXWELL, J., PECK, L. C., WIIK, H. B.: Chemical composition of Apollo 11 lunar samples 10017, 10020, 10072 and 10084. Proceedings of the Apollo 11 Lunar Science Conference **2**, 1369 (1970).

— WIIK, H. B.: Chemical composition of Apollo 12 lunar samples 12004, 12033, 12051, 12052 and 12065. Earth Planet. Sci. Lett. **10**, 285 (1971).

MAZZI, F., PABST, A.: Re-examination of cuprorivaite. Am. Mineralogist **47**, 409 (1962).

— ROSSI, G.: The crystal structure of taramellite $Ba_2(Fe, Ti, Mg)_2H_2[O_2(Si_4O_{12})]$. Z. Krist. **121**, 243 (1965).

MCGRAIN, P., THOMAS, G. R.: Preliminary report on the natural brines of Eastern Kentucky. Kentucky Geol. Surv. Rept. Invest., Ser. 9 **3**, 1 (1951).

MELCHER, A. C.: The solubility of silver chloride, barium sulphate and calcium sulphate at high temperatures. Am. Chem. Soc. **32**, 50 (1910).

MELSON, W. G., THOMSON, G., ANDEL, T. H. VAN: Volcanism and metamorphism in the Mid-Atlantic Ridge, 22°N latitude. J. Geophys. Res. **73**, 5925 (1968).

MICHEL, G.: Untersuchungen über die Tiefenlage der Grenze Süßwasser-Salzwasser im nördlichen Rheinland und anschließenden Teilen Westfalens, zugleich ein Beitrag zur Hydrogeologie und Chemie des tiefen Grundwassers. Forschungsber. Landes Nordrhein-Westfalen **1239** (1963).

— Zur Mineralisation des tiefen Grundwassers in Nordrhein-Westfalen. J. Hydrology **3**, 73 (1965).

MILLER, A. R., DENSMORE, C. D., DEGENS, E. T., HATHAWAY, J. C., MANHEIM, F. T., MCFARLIN, P. F., POCKLINGTON, R., JOKELA, A.: Hot brines and recent iron deposits in deeps of the Red Sea. Geochim. Cosmochim. Acta **30**, 341 (1966).

MILLER, J. P.: Solutes in small streams draining single rock types, Sangre de Cristo Range, New Mexico. U.S. Geol. Surv. Water Supply **1535**-F (1961).

MINGUIZZI, C.: Spectrography of some products of Vesuvius fumaroles. Rend. Soc. Mineral. Ital. **5**, 60 (1948).

MOENKE, H.: Trace elements in Variscan and pre-Variscan German granites. A spectrochemical analysis of granitic rocks of different age. Chemie Erde **20**, 227 (1960).

MOHR, P. A.: Trace element distribution in a garnet-chlorite-rock from Foel Ddu, near Harlech, Merionethshire. Mineral. Mag. **31**, 319 (1956).

— A geochemical study of the shales of the Lower Cambrian manganese shale group of the Harlech Dome, North Wales. Geochim. Cosmochim. Acta **17**, 186 (1959).

— ALLEN, R.: Further considerations on the deposition of the Middle Cambrian manganese carbonate beds of Wales and New Foundland. Geol. Mag. **102**, 328 (1965).

MOORE, C. B., BROWN, B.: Barium in stony meteorites. J. Geophys. Res. **68**, 4293 (1963).

MORRISON, G. H., GERARD, J. T., KASHUBA, A. T., GANGADHARAM, E. V., ROTHENBERG, A. M., POTTER, N. M., MILLER, G. B.: Elemental abundances of lunar soil and rocks. Proceedings of the Apollo 11 Lunar Science Conference **2**, 1383 (1970).

MOXHAM, R. L.: Minor element distribution in some metamorphic pyroxenes. Can. Mineralogist **6**, 522 (1960).

— Distribution of minor elements in coexisting hornblendes and biotites. Can. Mineralogist **8**, 204 (1965).

MUIR, A., HARDIE, H. G. M., MITCHELL, R. L., PHEMISTER, J.: Limestones of Scotland: chemical analyses and petrography. Mem. Geol. Surv. Gt. Brit. Mineral Resources Gt. Brit. **37**, 1 (1956).

MUIR, T. D., TILLEY, C. E., SCOON, J. H.: Basalts from the Northern Part of the rift zone of the Mid-Atlantic Ridge. J. Petrol. **5**, 409 (1964).

MUKHERJEE, B.: Genetic significance of trace elements in certain rocks of Singhlburn, India. Mineral. Mag. **36**, 661 (1968).

MURRAY, H. H.: Genesis of clay minerals in some Pennsylvanian shales of Indiana and Illinois. Proc. 2nd Nat. Conf. Clays and Clay Minerals 1953, 47 (1954).

MURTHY, V. R., EVENSEN, N. M., COSCIO, M. R.: Distribution of K, Rb, Sr and Ba and Rb-Sr isotopic relations in Apollo 11 lunar samples. Proceedings of the Apollo 11 Lunar Science Conference **2**, 1393 (1970).

NABOKO, S. I.: The sublimates of the Klyuchevskoy volcano. Bull. Acad. Sci. USSR, Geol. Ser. **1**, 50 (1945).

NESTERENKO, G. V., STUDENIKOVA, Z. V., SAVINOVA, E. N.: Rare and disseminated elements in skarns of Tyrny-Auz (Soviet Armenia). Geochemistry, 287 (1958).

NEUMANN, E. W.: Solubility relations of barium sulfate in aqueous solutions of strong electrolytes. J. Am. Chem. Soc. **55**, 879 (1933).

NICHOLLS, G. D., LORING, D. H.: The geochemistry of some British Carboniferous sediments. Geochim. Cosmochim. Acta **26**, 181 (1962).

NIKOLAEV, V. M., SOKIRKO, L. E., PESTOVA, N. M.: Spectral analysis for correlation of formation waters in Mesozoic deposits of the Eastern Caucasus Region. Tr. Groznensk. Neft. Nauchn.-Issled. Inst. **8**, 200 (1960).

NIXON, P. H., KNORRING, O. VON, ROOKE, J. M.: Kimberlites and associated inclusions of Basutoland: a mineralogical and geochemical study. Am. Mineralogist **48**, 1090 (1963).

NOBLE, D. C., HAFFTY, J.: Minor-element and revised major-element contents of some Mediterranean pantellerites and comendites. J. Petrol. **10**, 502 (1969).

NOCKOLDS, S. R., ALLEN, R.: The geochemistry of some igneous rock series. I. Calc-alkalic rocks. Geochim. Cosmochim. Acta **4**, 105 (1953).

— — The geochemistry of some igneous rock series. II. Alkalic rocks. Geochim. Cosmochim. Acta **5**, 245 (1954).

— — The geochemistry of some igneous rock series. III. Tholeiitic rocks. Geochim. Cosmochim. Acta **9**, 34 (1956).

OFTEDAL, I.: On the development of granite pegmatite in gneiss areas. Norsk Geol. Tidsskr. **38**, 231 (1958).

— Distribution of Ba and Sr in microcline in sections across a granite pegmatite band in gneiss. Norsk Geol. Tidsskr. **39**, 343 (1959).

— Remarks on the variable contents of Ba and Sr in microcline from a single pegmatite body. Norsk Geol. Tidsskr. **41**, 271 (1961).

— Contribution to the geochemistry of nephelinesyenitic pegmatite in the Langesundsfjord area. Norsk Geol. Tidsskr. **42**, 167 (1962).

O'HARA, M. J.: Zoned ultrabasic and basic gneiss masses in the early Lewisian metamorphic complex at Scourie, Scotland. J. Petrol. **2**, 248 (1961).

OKRUSCH, M., RICHTER, P.: Zur Geochemie der Dioritgruppe. Vergleichende Untersuchungen an Gesteinen des Bayerischen Waldes, des Spessarts und des Odenwaldes (Süd-Deutschland). Contr. Mineral. and Petrol. **21**, 75 (1969).

ONUMA, N., HIGUCHI, H., WAKITA, H., NAGASAWA, H.: Trace element partition between two pyroxenes and the host lava. Earth Planet. Sci. Lett. **5**, 47 (1968).

ORR, A. R. D.: Barytes. Mining Annual Review, London 108 (1970).

OSTROM, M. E.: Trace elements in Illinois Pennsylvanian limestones. Illinois State Geol. Surv. Circ. **243**, 1 (1957).

OSTROUMOV, V. M., RUSSKIKH, A. M.: Sarazon spring. Izv. Altai. Geogr. Obshchest. SSSR **6**, 45 (1965).

PABST, A.: Structures of some tetragonal sheet silicates. Acta Cryst. **12**, 733 (1959).

PAPEŽIK, V. S.: Geochemistry of some Canadian anorthosites. Geochim. Cosmochim. Acta **29**, 673 (1965).

PARRAS, K.: The charnockites in the light of a highly metamorphic rock complex in southwestern Finland. Bull. Comm. Géol. Finlande **181** (1958).

PATTEISKY, K.: Die thermalen Solen des Ruhrgebietes und ihre juvenilen Quellgase. Glückauf **90**, 1334, 1508 (1954).

PATTERSON, E. M.: A petrochemical study of the Tertiary lavas of north-east Ireland. Geochim. Cosmochim. Acta **2**, 283 (1951).

— Petrochemical data for some acid intrusive rocks from the Mourne Mountains and Slieve Gullion. Proc. Roy. Irish Acad., Sect B **55**, 171 (1953).

— MITCHELL, W. A., SWAINE, D. J.: Tertiary volcanic succession in the western part of the Antrim Plateau. Proc. Roy. Irish Acad., Sect. B **57**, 155 (1955).

— SWAINE, D. J.: A petrochemical study of Tertiary tholeiitic basalts: the middle lavas of the Antrim Plateau. Geochim. Cosmochim. Acta **8**, 173 (1955).

PATTIARATCHI, D. B., SAARI, E., SAHAMA, TH. G.: Anandite, a new barium iron silicate from Wilagedera, North Western Province, Ceylon. Mineral. Mag. **36**, 1 (1967).

PECK, D. L., WRIGHT, T. L., MOORE, J. G.: Crystallization of tholeiitic basalt in Alae Lava Lake, Hawaii. Bull. Volcanol. **29**, 629 (1966).

PECORA, W. T.: Carbonatite problem in the Bearpaw Mountains, Montana. Bull. Geol. Soc. Am., Buddington Volume 83 (1962).

PENCHEV, N. P., PENCHEVA, E. N., BONCHEV, P. R.: Spectrographic investigation of the trace elements in Bulgarian mineral waters. Compt. Rend. Acad. Bulgare Sci. **11**, 375 (1958).

— — — Spectrographic investigations of the microcomponents of Bulgarian mineral waters. II. Compt. Rend. Acad. Bulgare Sci. **13**, 55 (1960).

PERCHUCK, L. L., RYABCHIKOV, I. D.: Mineral equilibria in the system nepheline-alkali feldspar-plagioclase and their petrological significance. J. Petrol. **9**, 123 (1968).

PETROV, V. P., PREDOVSKII, A. A., SERGEEV, A. S., GALIBIN, V. A.: Some peculiarities in the distribution of trace elements in biotites from crystalline schists and gneisses in the Northern Ladoga Region. Vestn. Leningr. Univ. Ser. Geol. i Geogr. **4**, 5 (1965).

PETTIJOHN, F. J.: Data of Geochemistry, 6th ed. U.S. Geol. Surv., Profess. Papers **440-S** (1963).

PHILPOTTS, J. A., SCHNETZLER, C. C.: Phenocryst-matrix partition coefficients for K, Rb, Sr and Ba, with applications to anorthosite and basalt genesis. Geochim. Cosmochim. Acta **34**, 307 (1970a).

PHILPOTTS, J. A., SCHNETZLER, C. C.: Apollo 11 lunar samples: K, Rb, Sr, Ba and rare-earth concentrations in some rocks and separated phases. Proceedings of the Apollo 11 Lunar Science Conference 2, 1471 (1970b).

— — THOMAS, H. H.: Rare-earth and barium abundances in the Bununu howardite. Earth Planet. Sci. Lett. 2, 19 (1967).

PIETZNER, H., WOLF, M.: Geochemical study of brown coal ashes, and coal-petrographic study of brown coals from the Lower Rhine district. Fortschr. Geol. Rheinland Westfalen **12**, 517 (1964).

PILKEY, O. H.: Trace elements in Recent marine mollusk shells. Thesis, Ann Arbor, Michigan, 1963.

PINSON, W. H., AHRENS, L. H., FRANCK, M. L.: The abundance of Li, Sc, Sr, Ba and Zr in chondrites and some ultramafic rocks. Geochim. Cosmochim. Acta **4**, 251 (1953).

PIRANI, R., SIMBOLI, G.: Genesis of Sardinian granite of the Buddoso platea. Geochemistry and structure of the chief mineralogical components. I. K-feldspar. Mineral. Petrogr. Acta **9**, 179 (1963a).

— — Plagioclase and the use of the 2 coexisting feldspars in geothermal determinations. Mineral. Petrogr. Acta **9**, 211 (1963b).

PLAS, L. VAN DER, HÜGI, T.: A ferrian sodium-amphibole from Vals, Switzerland. Schweiz. Mineral. Petrog. Mitt. **41**, 371 (1961).

POTH, CH. N.: The occurrence of brine in Western Pennsylvania. Pennsylv. Bull. Geol. Surv., 4th Ser. **M 47** (1962).

PRASHNOWSKY, A. A.: Sedimentpetrographische und geochemische Untersuchungen im südlichen Rheinischen Schiefergebirge. Neues Jahrb. Geol. Palaeontol. Abhandl. **105**, 47 (1957).

PREUSS, E.: Spektralanalytische Untersuchung der Tektite. Chem. Erde **9**, 365 (1935).

PRICE, P. H., HARE, C. E., MCCUE, J. B., HOSKINS, H. A.: Salt brines of West Virginia. West Virg. Geol. Surv. Rept. **8** (1937).

PRINZ, M.: Geologic evolution of the Beartooth Mountains, Montana, and Wyoming. Part 5. Mafic Dike Swarms of the Southern Beartooth Mountains. Bull. Geol. Soc. Am. **75**, 1217 (1964).

— Geochemistry of basaltic rocks: trace elements. In: HESS, H. H., POLDERVAART, A. (ed.), Basalts. The Poldervaart Treatise on Rocks of Basaltic Composition, vol. I, New York: Interscience Publishers 1967.

PROKOF'EV, V. A.: Elemental chemical composition of shells of the Paleozoic brachiopods from spectral analysis data. Geokhimiya, 75 (1964).

PUCHELT, H.: Zur Geochemie des Grubenwassers im Ruhrgebiet. Z. Deutsch. Geol. Ges. **116**, 167 (1964).

— Zur Geochemie des Bariums im exogenen Zyklus. Sitzungsber. Heidelb. Akad. Wiss. Math.-nat. Kl. 4. Abh. (1967).

READ, H. H., HAQ, B. T.: The distribution of trace elements in the dunite-syenite differentiated series of the Insch Complex. Aberdeenshire. Proc. Geologists' Assoc. **74**, 203 (1963).

— SADASHIVAIAH, M. S., HAQ, B. T.: The hyperstene-gabbro of the Insch Complex, Aberdeenshire. Proc. Geologists' Assoc. **76**, 1 (1965).

REED, G. W.: Heavy elements in the Pantar meteorite. J. Geophys. Res. **68**, 3531 (1963).

RENFREW, C., DIXON, J. E., CANN, J. R.: Obsidian and early cultural contact in the Near East. Proc. Prehistoric Soc. **32**, 30 (1966).

— — — Further analysis of near eastern obsidians. Proc. Prehistoric Soc. **34**, 319 (1968).

REYNOLDS, F. M.: Bemerkungen über das Vorkommen von Ba in der Kohle. J. Soc. Chem. Ind. **58**, 64 (1939).

RHODES, J. M.: On the chemistry of potassium feldspars in granitic rocks. Chem. Geol. **4**, 373 (1969).

RICHTER, W.: Die Feldspate des Granites von Eisenkappel (Kärnten) und seines Randporphyres. Tschermaks Mineral. Petrog. Mitt., 3. F. **11**, 439 (1966).

RIDLEY, W. I.: The petrology of the Las Canadas volcanoes, Tenerife, Canary Islands. Contr. Mineral. and Petrol. **26**, 124 (1970).

RIMSAITE, J. H. Y.: On micas from magmatic and metamorphic rocks. Beitr. Mineral. Petrog. **10**, 152 (1964).

RIVALENTI, G., SIGHINOLFI, G. P.: Geochemical study of graywackes as a possible starting material of para-amphibolites. Contr. Mineral. and Petrol. **23**, 173 (1969).

ROBINSON, W. O., WHETSTONE, R. R., EDGINGTON, G.: Barium in soils and plants. U.S. Dept. Agr. Tech. Bull. **1013** (1950).

RÖMPP, H.: Chemie Lexikon, vol. I. Stuttgart: Francksche Verlagsbuchhandlung 1966.

ROGERS, J. J. W.: Textural and spectrochemical studies of the White Tank quartz monzonite, California. Bull. Geol. Soc. Am. **69**, 449 (1958).

ROSENQVIST, I. TH.: Note on leaching of granite with special reference to lead, radium and barium. Norsk Geol. Tidsskr. **19**, 110 (1939).

ROSSEINSKY, D. R.: The solubilities of sparingly soluble salts in water. Trans. Faraday Soc. **54**, 116 (1958).

ROY, N. N.: The mineralogy of the potassium-barium feldspar series. I. The determination of the optical properties of natural members. Mineral. Mag. **35**, 508 (1965).

— II. Studies on hydrothermally synthesized members. Mineral. Mag. **36**, 43 (1967).

RUBESKA, I., MIKSOVSKY, M.: Strontium and barium in plain and mineral waters of the Karlovy Vary and Teplice areas and their determination. Vestn. Ustredneko Ustavu Geol. **38**, 153 (1963).

RUDERT, V.: Phasenbeziehungen im System $NaAlSi_3O_8$-$BaAl_2Si_2O_8$-H_2O. Fortschr. Mineral. **48**, 86 (1970).

RUSSELL, H. D., HIEMSTRA, S. A., GROENEVELD, D.: The mineralogy and petrology of the carbonatite at Loolelop, eastern Transvaal. Trans. Proc. Geol. Soc. S. Africa **58**, 197 (1954).

RYBACH, L., NISSEN, H. U.: Zerstörungsfreie Simultanbestimmung von Na, K und Ba in Adular mittels Neutronenaktivierung. Schweiz. Mineral. Petrog. Mitt. **47**, 189 (1967).

SABINE, P. A., YOUNG, B. R.: Cell size and composition of the baryte-celestite isomorphous series. Acta Cryst. **7**, 630 (1954).

SAHAMA, TH., G.: Spurenelemente der Gesteine im südlichen Finnisch-Lappland. Bull. Comm. Géol. Finlande **135** (1945a).

— On the chemistry of east Fennoscandian Rapakivi granites. Bull. Comm. Géol. Finlande **136**, 15 (1945b).

SAHL, K.: Die Verfeinerung der Kristallstrukturen von $PbCl_2$ (Cotunnit), $BaCl_2$, $PbSO_4$ (Anglesit) und $BaSO_4$ (Baryt). Beitr. Mineral. Petrog. **9**, 111 (1963).

SAVELLI, C.: The problem of rock assimilation by Somma-Vesuvius magma. I. Composition of Somma and Vesuvius lavas. Contr. Mineral. and Petrol. **16**, 328 (1967).

SCHARBERT, S.: Mineralbestand und Genesis des Eisgarner Granits im niederösterreichischen Waldviertel. Tschermaks Mineral. Petrog. Mitt. 3. F. **11**, 388 (1966).

SCHNETZLER, C. C., PHILPOTTS, J. A., BOTTINO, M. L.: Li, K, Rb, Sr, Ba and rare-earth concentrations, and Rb-Sr age of lunar rock 12013. Earth Planet. Sci. Lett. **9**, 185 (1970).

— — THOMAS, H. H.: Rare earth and barium abundances in Ivory Coast tektites and rocks from the Bosumtwi Crater area, Ghana. Geochim. Cosmochim. Acta **31**, 1987 (1967).

SCHOPF, T. J. M., MANHEIM, F. T.: Chemical composition of Ectoprocta (Bryozoa). J. Paleontol. **41**, 1197 (1967).

SCHWANDER, H., HUNZICKER, J., STERN, W.: Zur Mineralchemie in Hellglimmern in den Tessiner Alpen. Schweiz. Mineral. Petrog. Mitt. **48**, 357 (1968).

SCHWARCZ, H. P.: Chemical and mineralogic variations in an arkosic quartzite during progressive regional metamorphism. Bull. Geol. Soc. Am. **77**, 509 (1966).

SCHWIND, S. B.: Barium. A bibliography of unclassified literature. U.S. Atomic Energy Comm. Nat. Sci. Foundation Wash. **TID-369** (1952).

SEN, N., NOCKOLDS, S. R., ALLEN, R.: Trace elements in minerals from rocks of the S. Californian batholith. Geochim. Cosmochim. Acta **16**, 58 (1959).

SEN, S. K.: Some aspects of the distribution of barium, strontium, iron, and titanium in plagioclase feldspars. J. Geol. **68**, 638 (1960).

SENAKOLIS, A. F.: Lithology and geochemistry of the Cambrian carbonate rocks in some sections of the Sayan-Altai territory. Materialy po Geol. i Polezn. Iskop. Zapadn. Sibiri 193 (1964).

SHACKLETTE, H. T.: Element content of bryophytes. Contr. Geochem. Prosp. Minerals, Geol. Surv. Bull. **1198-D** (1965).

SHAW, D. M.: Trace elements in pelitic rocks I, II. Bull. Geol. Soc. Am. **65**, 1151, 1167 (1954).

— Some aspects of the determination of barium in silicate rocks. Spectrochim. Acta **10**, 125 (1957).

— MOXHAM, R. L., FILBY, R. H., LAPKOWSKY, W. W.: The petrology and geochemistry of some Grenville skarns. II. Geochemistry. Can. Mineralogist **7**, 578 (1963).

SHCHERBA, G. N., GUKOVA, V. D., KUDRYASHOV, A. V., SENCHILO, N. P.: Alkali feldspars from feldspar-quartz veinlets in molybdenum-tungsten deposits, Kazakhstan. IV. Quartz and feldspar-quartz veins and veinlets. Geochem. Int. **1**, 141 (1964).

SHELTON, J. S.: Glendova volcanic rocks, Los Angeles Basin, Cal. Bull. Geol. Soc. Am. **66**, 45 (1955).

SHIMA, M., HONDA, M.: Distributions of alkali, alkaline earth and rare earth elements in component minerals of chondrites. Geochim. Cosmochim. Acta **31**, 1995 (1967).

SHIMER, J. A.: Spectrographic analysis of New England granites and pegmatites. Bull. Geol. Soc. Am. **54**, 1049 (1943).

SHINKARENKO, A. L.: The gas component and content of microelements in mineral springs of the Caucasian mineral waters. Tr. Lab. Gidrogeol. Probl. **3**, 253 (1948).

SHKABARA, M. N.: On the mineralogy of wellsite. Zentr. Mineralogie, 180 (1950).

SIEDNER, G.: Geochemical features of a strongly fractionated alkali igneous suite. Geochim. Cosmochim. Acta **29**, 113 (1965).

SIEGERS, A., PICHLER, H., ZEIL, W.: Trace element abundances in the "andesite" formation of northern Chile. Geochim. Cosmochim. Acta **33**, 882 (1969).

SIMPSON, E. S. W.: The Okonjeje igneous complex, southwest Africa. Trans. Proc. Geol. Soc. S. Africa **57**, 125 (1954).

SINHA, R. C., KARKARE, S. G.: Distribution and behavior of trace elements in some of the Deccan basalts. Geol. Soc. India Bull. **1**, 21 (1964a).

— — Geochemistry of Deccan basalts: a study of the behaviour of major and trace elements in the basaltic flows of India. Intern. Geol. Congress, Rep. 21, Session India, Part VII, 85 (1964b).

— TIWARI, B. D.: Geochemistry of the volcanic rocks of Pavagarh. Intern. Geol. Congr. Rep. 21, Session India, Part VII, 104 (1964).

SKINNER, B. J., WHITE, D. E., ROSE, H. J., MAYS, R. E.: Sulfides associated with the Salton Sea geothermal brine. Ec. Geol. **62**, 316 (1967).

SKROBOV, A. A., SMIRNOV, V. J.: Die natürlichen Mineralwässer des nördlichen Gebiets. Nördl. Geol. Verw. Archangelsk (1939).

SMALES, A. A., WEBB, M. S. W., WEBSTER, R. K., WILSON, J. D.: Elemental composition of lunar surface material. Proceedings of the Apollo 11 Lunar Science Conference **2**, 1575 (1970).

SMITH, A. L., CARMICHAEL, I. S. E.: Quaternary trachybasalts from southeastern California. Am. Mineralogist **54**, 909 (1969).

SMITH, G. J.: Geology and volcanic petrology of the Lava Mountains, San Bernardino County, California. U.S. Geol. Surv. Profess. Papers **457** (1964).

SMITH, J. V., RIBBE, P. H.: X-ray-emission microanalysis of rock-forming minerals. III. Alkali feldspars. J. Geol. **74**, 197 (1966).

SNAVELY, P. D., Jr., MACLEOD, N. S., WAGNER, H. C.: Tholeiitic and alkalic basalts of the Eocene Siletz river. Am. J. Sci. **266**, 454 (1968).

SNETSINGER, K. G.: Barium-vanadium muscovite and vanadium tourmaline from Mariposa County, California. Am. Mineralogist **51**, 1623 (1966).

SOLOMON, M.: Origin of barite in the North Pennine ore field. Inst. Mining Met., Trans. **75**, 230 (1966).

SPENCER, D.: Factors affecting element distribution in a Silurian graptolite band. Chem. Geol. **1**, 221 (1966).

STARITSIN, F. V.: Stratigraphic subdivision of Mesozoic extrusives of eastern Transbaikaliya by trace elements. Geochemistry 83 (1964).

STARK, J. T., TRACEY, J. T.: Petrology of volcanic rocks of Guam. U.S. Geol. Surv. Profess. Papers **403-C** (1963).

STARKE, R.: Die Strontiumgehalte der Baryte. Freiberger Forschungsh. C **150** (1964).

STEHLI, F. G., HOWER, J.: Mineralogy and early diagenesis of carbonate sediments. J. Sediment. Petrol. **31**, 358 (1961).

STERN, W. B.: Zur Mineralchemie von Glimmern aus Tessiner Pegmatiten. Schwei z Mineral. Petrog. Mitt. **46**, 137 (1966).

STEVENS, R. E., FLEISCHER, M., NILES, W. W., CHODOS, A. A., FILBY, H., LEININGER, R. K., FLANAGAN, F. J.: Second report on a cooperative investigation of the composition of two silicate rocks. Bull. U.S. Geol. Surv. **1113** (1960).

STÖHR, H., FLASCH, H.: Barium und Bariumverbindungen. In: Ullmanns Encyklopädie der technischen Chemie, vol. IV, München-Berlin: Urban & Schwarzenberg 1953.

STRAUB, J.: Chemical composition of mineral waters of Transylvania: their rare constituents and the biochemical importance. Magy. Allami Foldt Int. Evkonyve **39** (1950).

STROMINGER, D., HOLLANDER, J. M., SEABORG, G. T.: Table of isotopes. Rev. Mod. Phys. **30** (1958).

STRÜBEL, G.: Zur Kenntnis und genetischen Bedeutung des Systems $BaSO_4$-NaCl-H_2O. Neues Jahrb. Mineral. Monatsh., 223 (1967).

STRUNZ, H.: Mineralogische Tabellen, 4. Aufl. Leipzig: Akademische Verlagsgesellschaft Geest & Bortig 1966.

SUESS, H. E., UREY, H. C.: Abundances of the elements. Rev. Mod. Phys. **28**, 53 (1956).

SUKHAREV, G. H.: The composition of connate waters in Mesozoic deposits in the Caucasus region and their connection with the prospecting for oil and gas. Geol. Nefti i Gaza **5**, 17 (1961).

SWAINE, D. J.: The trace-element content of soils. Commonwealth Agr. Bur. Tech. Com. **48**, 16 (1955).

— Inorganic constituents in Australian coals. Mitt. Naturforsch. Ges. Bern (N. F.) **24**, 49 (1967).

SZABO, B. J., JOENSUU, O.: Emission spectrographic determination of barium in sea water using a cation-exchange concentration procedure. Environ. Sci. Technol. **1**, 499 (1967).

TADZHIEV, F. KH., MUKSINOV, T. KH.: Change in the thermodynamic functions of some ion-exchange reactions on Oglanly bentonite. Uzbeksk. Khim. Zh. **11**, 49 (1967).

TAKUBO, J., TATEKAWA, M.: Distribution of minor elements in pegmatites and granites from Oku-Tango District, Kyoto Prefecture, Japan, I. Distribution of barium and strontium. Kobutsugaku Zasshi **1**, 301 (1954).

TAYLOR, S. R.: The origin of some New Zealand metamorphic rocks as shown by their major and trace element composition. Geochim. Cosmochim. Acta **8**, 182 (1955).

— Consequences for tektite composition of an origin by meteoritic splash. Geochim. Cosmochim. Acta **26**, 915 (1962).

— Similarity in composition between Henbury impact glass and australites. Geochim. Cosmochim. Acta **29**, 599 (1965).

— The application of trace element data to problems of petrology. Phys. Chem. Earth **6**, 133 (1965).

— Australites, Henbury impact glass and subgreywacke: a comparison of the abundances of 51 elements. Geochim. Cosmochim. Acta **30**, 1121 (1966).

— CAPP, A. C., GRAHAM, A. L.: Trace element abundances in andesites. II. Saipan, Bougainville and Fiji. Contr. Mineral. and Petrol. **23**, 1 (1969).

— EWART, A., CAPP, A. C.: Leucogranites and rhyolites: trace element evidence for fractional crystallisation and partial melting. Lithos **1**, 179 (1968).

— HEIER, K. S.: The petrological significance of trace element variations in feldspars. Proc. XXI. Inter. Geol. Congr. Norden 47 (1960).

— JOHNSON, P. H., MARTIN, R., BENNETT, D., ALLEN, J., NANCE, W.: Preliminary chemical analyses of Apollo 11 lunar samples. Proceedings of the Apollo 11 Lunar Science Conference **2**, 1627 (1970).

— KOLBE, P.: Henbury impact glass: parent material and behaviour of volatile elements during melting. Nature **203**, 390 (1964).

TAYLOR, S. R., SACHS, M.: Geochemical evidence for the origin of australites. Geochim. Cosmochim. Acta **28**, 235 (1964).

— SOLOMON, M.: The geochemistry of Darwin glass. Geochim. Cosmochim. Acta **28**, 471 (1964).

— — SVERDRUP, T. L.: Contributions to the mineralogy of Norway. V. Trace element variations in three generations of feldspars from the Landsverk I pegmatite Evje, Southern Norway. Norsk Geol. Tidsskr. **40**, 133 (1960).

— WHITE, A. J. R.: Trace element abundances in andesites. Bull. Volcanol. **29**, 177 (1966).

TEMPLE, A. K., GROGAN, R. M.: Carbonatite and related alkalic rocks at Powder Horn, Colorado. Ec. Geology **60**, 672 (1965).

TEMPLETON, CH. C.: Solubility of barium sulfate in sodium chloride solutions from 25° to 95° C. J. Chem. Eng. Data **5**, 514 (1960).

TERA, F., EUGSTER, O., BURNETT, D. S., WASSERBURG, G. J.: Comparative study of Li, Na, K, Rb, Cs, Ca, Sr and Ba abundances in achondrites and in Apollo 11 lunar samples. Proceedings of the Apollo 11 Lunar Science Conference **2**, 1637 (1970).

TIMASHEVA, E. E.: Distribution of trace elements in formation waters of the Carpathian syncline and in southwestern boundaries of the Russian Platform. Tr. Ukr. Nauchn.-Issled. Geol. Razved. Inst. **3**, 344 (1963).

TKACHEV, Yu. A., SKIBA, N. S., BONDARENKO, C. F.: Distributions of strontium and barium in coals of Kirgizia. Litol., Geokhim. i Polezn. Iskop. Osad. Obrazov. Tyan-Shanya, Akad. Nauk Kirg. SSR, Inst. Geol., 24 (1965).

TOLKACHEV, M. V.: Alkaline-earth elements in sedimentary rocks of the west Siberian depression. Geokhimiya **4**, 489 (1968).

TOOKER, E. W.: Altered wallrocks in the central part of the Front Range mineral belt, Gilpin and clear Creek Counties, Colorado. U.S. Geol. Surv. Profess. Papers **439** (1963).

TOURTELOT, H. A.: Chemical composition of the Pierre shale and equivalent rocks of late Cretaceous age, Great Plains region. Bull. Geol. Soc. Am. **68**, 1806 (1957).

TOWNEND, R.: The geology of some granite plutons from Western Connemara, Co. Galway. Roy. Irish Acad. Proc. **65**, 157 (1966).

TOWNLEY, R. N., WHITNEY, W. B., FELSING, W. A.: The solubilities of barium and strontium carbonates in aqueous solutions of some alkali chloride. J. Am. Chem. Soc. **59**, 631 (1937).

TUREKIAN, K. K.: Deep-sea deposition of barium, cobalt, and silver. Geochim. Cosmochim. Acta **32**, 603 (1968).

— ARMSTRONG, R. L.: Magnesium, strontium, and barium concentrations and calcite-aragonite ratios of some Recent molluscan shells. J. Marine Res. **18**, 133 (1960).

— — Chemical and mineralogical composition of fossil molluscan shells from the Fox Hills formation, South Dakota. Bull. Geol. Soc. Am. **72**, 1817 (1961).

— HARRISS, R. C., JOHNSON, D. G.: The variations of Si, Cl, Na, Ca, Sr, Ba, Co, and Ag in the Neuse River, North Carolina. Limnol. Oceanog. **12**, 702 (1967).

— JOHNSON, D. J.: The barium distribution in seawater. Geochim. Cosmochim. Acta **30**, 1153 (1966).

— PHINNEY, W. C.: The distribution of Ni, Co, Cr, Cu, Ba, and Sr between biotite-garnet pairs in a metamorphic sequence. Am. Mineralogist **47**, 1434 (1962).

— TAUSCH, E. H.: Barium in deep-sea sediments of the Atlantic Ocean. Nature **201**, 696 (1964).

— WEDEPOHL, K. H.: Distribution of the elements in some major units of the earth's crust. Bull. Geol. Soc. Am. **72**, 175 (1961).

UCHAMEISHVILI, N. E., MALININ, S. D., KHITAROV, N. J.: Solubility of barite in concentrated solutions of chlorides of some metals at elevated temperatures in relation to the genesis of barite deposits. Geochemistry, 951 (1966).

UMEMOTO, S.: Isotopic composition of barium and cerium in stone meteorites. J. Geophys. Res. **67**, 375 (1962).

UNSÖLD, A.: Der neue Kosmos. Heidelberger Taschenbücher **16/17** (1967).

UTSUMI, K.: Spectral analysis of some carbon stars in the visual region. Publ. Astron. Soc. Japan **19**, 342 (1967).

Varov, A. A., Romm, I. F.: Occurrence of strontium and barium in oilfield waters of the Ural-Volga-Region. Dokl. Akad. Nauk. SSSR **35**, 114 (1942).

Vine, J. D.: Elemental distribution in some shelf and eugeosynclinal black shales. U.S. Geol. Surv. Bull. **1214-E** (1966).

Vinogradov, A. P.: The Elementary Chemical Composition of Marine Organisms. New Haven: Memoir Sears Foundation for Marine Research, No. II, Yale University 1953.

— Regularity of distribution of chemical elements in the earth crust. Geochemistry, 1 (1956).

— Ronov, A. B.: Composition of sedimentary rocks of the Russian platform in relation to the history of their tectonic movements. Geochemistry, 533 (1956).

— — Ratynsky, V. M.: Changes of chemical composition of carbonate rocks of the Russian Platform. Akad. Nauk SSSR Izv. Geol. Ser. **1**, 33 (1952).

Vinogradova, Z. A., Koval'skiy, V. V.: Elemental composition of the Black Sea plankton. Dokl. Akad. Nauk SSSR **147**, 1458 (1962).

Vlasov, K. A., Kuzmenko, M. Z., Es'kova, E. M.: The Lovozero Alkali Massif. Edinburgh-London: Oliver & Boyd Publishers 1966.

Volborth, A.: Rapakivi-type granites in the Precambrian complex of Gold Butte, Clark County, Nevada. Bull. Geol. Soc. Am. **73**, 813 (1962).

Vorob'ev, G. G.: Composition of tektites. I. Indochinites. Meteoritika **17**, 64 (1959).

— Composition of tektites. II. Moldavites. Meteoritika **18**, 35 (1960).

Wänke, H., Rieder, R., Baddenhausen, H., Spettel, B., Teschke, F., Quijano-Rico, M., Balacescu, A.: Major and trace elements in lunar material. Proceedings of the Apollo 11 Lunar Science Conference **2**, 1719 (1970).

Wager, L. R., Mitchell: R. L. The distribution of trace elements during strong fractionation of basic magma — a further study of the Skaergaard intrusion, East Greenland. Geochim. Cosmochim. Acta **1**, 129 (1951).

— — Trace elements in a suite of Hawaiian lavas. Geochim. Cosmochim. Acta **3**, 217 (1953).

Wakeel, S. K. El, Riley, J. P.: Chemical and mineralogical studies of deep-sea sediments. Geochim. Cosmochim. Acta **25**, 110 (1961 b).

— — Chemical and mineralogical studies of fossil red clays from Timor. Geochim. Cosmochim. Acta **24**, 260 (1961 a).

Wakita, H., Schmitt, R. A.: Elemental abundances in seven fragments from lunar rock 12013. Earth Planet. Sci. Lett. **9**, 169 (1970).

— — Rey, P.: Elemental abundances of major, minor and trace elements in Apollo 11 lunar rocks, soil and core samples. Proceedings of the Apollo 11 Lunar Science Conference **2**, 1685 (1970).

Wallerstein, G., Greenstein, J. L.: Chemical composition of two CH stars, HD 26 and HD 626. Astrophys. J. **139**, 1163 (1964).

— — Parker, R., Helfer, H. L., Aller, L. H.: Red giants with extreme metal deficiencies. Astrophys. J. **137**, 280 (1963).

Warner, B.: The barium stars. Monthly notices of the Roy. Astron. Soc. **129**, 263 (1965).

Wedepohl, K. H.: Der trachydoleritische Basalt (Olivin-Andesin-Basalt) des Backenberges bei Güntersen, westlich von Göttingen. Heidelberger Beitr. Mineral. Petrog. **4**, 217 (1954).

— Spurenanalytische Untersuchungen an Tiefseetonen aus dem Atlantik. Geochim. Cosmochim. Acta **18**, 200 (1960).

— Written communication August 1961.

— Composition and abundance of common igneous rocks, p. 227. In: K. H. Wedepohl, Handbook of Geochemistry, I. Berlin-Heidelberg-New York: Springer 1969.

Weibel, M.: Zum Chemismus alpiner Adulare (II). Schweiz. Mineral. Petrog. Mitt. **37**, 545 (1957).

— Chemismus und Mineralzusammensetzung von Gesteinen des nördlichen Bergeller Massivs. Schweiz. Mineral. Petrog. Mitt. **40**, 69 (1960).

— Meyer, F.: Zum Chemismus alpiner Adulare (I). Schweiz. Mineral. Petrog. Mitt. **37**, 153 (1957).

Weiskirchner, W.: Personal communication 1969.

White, A. J. R.: Genesis of migmatites from the Palmer region of south Australia. Chem. Geol. **1**, 165 (1966).

White, D. E.: Saline waters of sedimentary rocks. In: Fluids in subsurface environments. Am. Ass. Petr. Geol. Memoir **4**, 342 (1965).

— Hem, J. D., Waring, G. A.: Data of Geochemistry. 6th Edition. Chapter F.: Chemical composition of subsurface waters. Geol. Surv. Profess. Papers **440-F** (1963).

White, R. W.: Ultramafic inclusions in basaltic rocks from Hawaii. Contr. Mineral. and Petrol. **12**, 245 (1966).

Wilkinson, J. F. G.: The geochemistry of a differentiated teschenite sill near Gunnedah, New South Wales. Geochim. Cosmochim. Acta **16**, 123 (1959).

— Mineralogical, geochemical, and petrogenetic aspects of an analcite-basalt from the New England district of New South Wales. J. Petrol. **3**, 192 (1962).

— Analcimes from some potassic igneous rocks and aspects of analcime rich igneous assemblages. Contr. Mineral. and Petrol. **18**, 252 (1968).

— Vernon, R. H., Shaw, S. E.: The petrology of an adamellite-porphyrite from the New England Bathylith (New S. Wales). J. Petrol. **5**, 461 (1964).

Wilska, S.: Trace elements in Finnish ground and mine waters: a spectroanalytical study. Ann. Acad. Sci. Fennicae Ser. A II. Chem. **46**, 7 (1952).

Wolgemuth, K.: Barium analyses from the first geosecs test cruise. J. Geophys. Res. **75**, 7686 (1970).

— Broecker, W. S.: Barium in sea water. Earth Planet. Sci. Lett. **8**, 372 (1970).

Wyckoff, R.: Crystal Structures, 2nd ed., vol. 4. New York and London: Interscience Publishers 1968.

Wynne-Edwards, H. R., Hay, P. W.: Coexisting cordierite and garnet in regionally metamorphosed rocks from the Westport area, Ontario. Can. Mineralogist **7**, 453 (1962).

Young, E. J.: Studies of trace elements in sediments. Thesis, Mass. Inst. Techn. 1954.

Yusurova, S. M.: Geochemistry of the mineral water of thermal springs of Tadzhikistan (rare elements in thermal waters of Tadzhikistan). Dokl. Akad. Nauk Tadzh. SSR **21**, 19 (1957).

Zoltai, T.: Classification of silicates and other minerals with tetrahedral structures. Am. Mineralogist **45**, 960 (1960).

Zurlo, P.: Geochemische Untersuchungen im Unterkarbon von Doberlug und im Oberkarbon von Zwickau. Geologie **9**, 1112 (1963).